Monographs on
Theoretical and Applied Genetics 10

Monographs on Theoretical and Applied Genetics

Volume 1 **Meiotic Configurations**
A Source of Information for Estimating Genetic Parameters
By J. Sybenga

Volume 2 **Pollination Mechanisms, Reproduction and Plant Breeding**
By R. Frankel and E. Galun

Volume 3 **Incompatibility in Angiosperms**
By D. de Nettancourt

Volume 4 **Gene Interactions in Development**
By L. I. Korochkin

Volume 5 **The Molecular Theory of Radiation Biology**
By K. H. Chadwick and H. P. Leenhouts

Volume 6 **Heterosis**
Reappraisal of Theory and Practice
Editor: R. Frankel

Volume 7 **Induced Mutations in Plant Breeding**
By W. Gottschalk and G. Wolff

Volume 8 **Protoplast Fusion**
Genetic Engineering in Higher Plants
By Y. Y. Gleba and K. M. Sytnik

Volume 9 **Petunia**
Editor: K. C. Sink

Mohan L. H. Kaul

Male Sterility in Higher Plants

With 140 Figures and 205 Tables

Springer-Verlag
Berlin Heidelberg New York
London Paris Tokyo

Professor Dr. Mohan L. H. Kaul
Botany Department
University, Kurukshetra
Haryana – 132119, India

ISBN 3-540-17952-6 Springer-Verlag Berlin Heidelberg New York
ISBN 0-387-17952-6 Springer-Verlag New York Berlin Heidelberg

Library of Congress Cataloging-in-Publication Data. Kaul, Mohan L. H. Male sterility in higher plants. (Monographs on theoretical and applied genetics ; 10) Bibliography: p. Includes index. 1. Male sterility in plants. I. Title. II. Series. QK828.K38 1987 581.2'1662 87-14670

Typesetting: Overseas Typographers, Inc., Makati, Philippines
Offsetprinting: Druckhaus Beltz, Hemsbach/Bergstraße
Bookbinding: J. Schäffer GmbH & Co. KG., Grünstadt
2131/3130-543210

Preface

"...... Nature has something more in view than that its own proper males should fecundate each blossom."
Andrew Knight
Philosophical Transactions, 1799

Sterility implicating the male sex solely presents a paradoxical situation in which universality and uniqueness are harmoniously blended. It maintains a built-in outbreeding system but is not an isolating mechanism, as male steriles, the "self-emasculated" plants, outcross with their male fertile sibs normally. Both genes (nuclear and cytoplasmic) and environment, individually as well as conjointly, induce male sterility, the former being genetic and the latter nongenetic. Genetic male sterility is controlled either exclusively by nuclear genes (*ms*) or by the complementary action of nuclear (*fr*) and cytoplasmic (*c*) genes. The former is termed genic and the latter gene-cytoplasmic male sterility. Whereas genic male sterility exhibits Mendelian inheritance, gene-cytoplasmic male sterility is non-Mendelian, with specific transmissibility of the maternal cytoplasm type. Genetic male sterility is documented in 617 species and species crosses comprising 320 species, 162 genera and 43 families. Of these, genic male sterility occurs in 216 species and 17 species crosses and gene-cytoplasmic male sterility in 16 species and 271 species crosses. The Predominance of species exhibiting genic male sterility and of species crosses exhibiting gene-cytoplasmic male sterility is due to the fact that for the male sterility expression in the former, mutation of nuclear genes is required, but in the latter, mutations of both nuclear and cytoplasmic genes are necessary. Since simultaneous or successive occurrence of both these mutations in a cell are rare, hybridization bringing the mutated cytoplasm and nuclear genome together and leads to gene-cytoplasmic male sterility. Little is known, however about the molecular interaction between the nuclear and cytoplasmic genes. Much circumstantial evidence, especially restriction endonucleases analysis, isolated in-vitro mitochondrial translation products and drug and fungal toxin sensitivity implicate mitochondria as site of the cytoplasmic male sterility (*c*) gene. It is also evidenced by protoplast fusions permitting cytoplasmic transfer with or without mitochondria and chloroplast. Further confirmation of *c* gene location in mitochondrial genome is provided by the existence of intramitrochondrial DNA sequences some of which behave as plasmids.

The book is written to embody the principles, concepts and details about male sterility known in 617 species and species crosses. It is divided into two parts. Part I includes principles, generalizations and fundamental concepts, chemical induction and environmental influence, critical discussions and specific conclusions on the nature, type, inheritance pattern, biochemistry, molecular basis, major features, limitations, lacunae and breeding prospects of different male sterility types found in higher plants. Suppression of anther or viable pollen development and differentation infuses morphological, anatomical, physiological and biochemical alterations coupled with self-sterility. Details of all these facets are included. Descriptions of male sterility types, inheritance, gene action and control, chemical induction, breeding significance etc of individual plant species of various flowering families are presented in Part II. To facilitate location, species and their families are arranged in alphabetical order. For species, the traditional botanical name is used, followed by the less known new botanical names and by the common names and synonyms, placed in parenthesis. After a brief introduction of the genus or species, available details about various aspects of genic, gene-cytoplasmic and chemical male sterility are presented. In addition, the effect of chemicals and environment on male sterility induction and expression is described. The breeding value of male sterility is indicated for the economically useful species. Appropriately with each family, major findings like gene control, time and type of gene action, chemical influence etc, are tabulated to provide an insight into the known and unknown aspects of male sterility. The concluding chapter includes concepts, facts, postulations, conclusions and recommendations about the known cases of male sterility listed in Table 55.5. For each species, the page number for reference in the text is given. Thus this table serves as an index of the species and species crosses in which male sterility is known and recorded in this book. The figures and tables obtained from various authors are adapted after modifications. References indexed are cited at the end. Due to the huge number of papers published in an increasingly large number of periodicals and in a multitude of languages, it was not possible to consult and include all the papers on male sterility. Moreover, to keep the monograph within reasonable bounds, a high degree of literature selection, with vast background reading, was necessary. To those researchers whose contribution has been excluded or inadequately represented, I express my apologies in the two words "pure ignorance". The responsibility for errors of commission and omission is mine. These, brought to my notice in a constructive manner, will be appreciated and acknowledged gratefully.

This book is dedicated to the scientists whose hard toil unravelled the intricacies of male sterility and set free its use for increased productivity.

Kurukshetra, India
July 1987

MOHAN L. H. KAUL

Acknowledgements

To Professor Dr. H.F. Linskens, Botanisch Laboratorium, Nijmegen, the Netherlands, I am indebted for cooperation, encouragement, help and moral support especially during the difficult and frustrating period of past 4 years. His confidence in my work provided me with the motivation and stimulus to complete this book. The idea to work and write on male sterility owes its inception to my post-doctoral research (1972–74, 1978–1980) with Prof. W. Gottschalk, Bonn University, FRG, to whom I am grateful for inspiration and help, and to Prof. R. G. Palmer, Iowa State University Ames, USA, who offered me his personal collection of reprints and library facilities during 1981–1982. My heartfelt thanks to Ursula, Neelangini, Usha and Anjali, who were cheerful and unflappable in meeting the many demands on their time. My special thanks to scientists and researchers for promptly providing reprints and photographs of their publications used in this book. For transmission help, I thank Professor H. Schulte, Mr. J.B.H. Hoffmann and Dr. B. Brandi-Dohrn of DAAD. For printing, I thank the editorial staff of Springer-Verlag for their excellent cooperation and high technical assistance. To my family, I am grateful for their understanding during the period of neglect.

MOHAN L. H. KAUL

Contents

Part I General Account

Part I
General Account

CHAPTER 1
Introduction

Flowering represents the end result of many gene-controlled physiological processes, biochemical sequences and morphological events. The principle functions of a flower are: microspore and megaspore formation, pollen release and reception, guided male and female gametophyte development, gametic union and seed formation. All these constitute time sequence-programmed gene-controlled functions. Breakdown of any one of these functions occurs due to mutation of the genes controlling them (Kaul and Murthy 1985). One such event causes male sex abolition and leads to the evolution of new sex types, the male steriles. Their natural development, subsequent establishment, genetic features and breeding utility comprise the text of this monograph.

1.1 Discovery

Shortly after the detection of sex in higher plants by Camerarius in 1694, Kölreuter (1763) observed anther abortion both within species and species hybrids. The former was designated as male contabescens and the latter as hybrid contabescens. Such contabescent anthers were found by Gärtner (1844), Herbert (1847) and Darwin (1890) to occur regularly in certain plant species of Caryophyllaceae, Ericaceae and Liliaceae. The genetic potential of the male contabescens, which in sensu stricto is male sterility (mst), became evident due to its regular recurrence in the plants of many flowering families. Darwin (1893) attributed to this type of sterility a definite role in the evolution of dioecism, which represents an obligate cross-fertilizing adaptation in amphimictic species. Bateson et al. (1908) showed that mst is not simply a random sporadic event but a genetically conditioned system in sweet pea, in which the sterility is controlled by one recessive gene. Salaman (1910) detected mst in potato as conditioned by one dominant gene. Correns (1908) demonstrated maternal control of gynodioecy. This represents the first critical evidence for non-Mendelian inheritance in early genetic literature. Subsequent investigations revealed its control through gene-cytoplasm (g-c) interactions. Due to the various different modes controlling the whole phenomenon leading to mst in higher plants, investigations over it accumulated rapidly. This led to the discovery of sex types, sex reversals, sex suppressions and phasic, temporary, partial and complete mst in several higher plants.

1.2 Occurrence

The mst arises spontaneously due to the mutation of either nuclear or cytoplasmic genes or of both. Compared to female sterility (fst), mst is of widespread occurrence in higher plants (Tables 2.1, 2.2, 3.1 to 3.4). This is because the male sporophyte and gametophyte are more vulnerable to intrinsic and extrinsic forces than are the protected ovule and embryo sac. Furthermore, because of the ease of availability and the large amount of pollen in the anther, detection of mst is easier than that of fst. Pollen fertility can be easily and quickly tested as compared to ovular fertility. Usually the former needs simple staining by carmine, lacto-phenol or iodine, whereas the latter requires artificial crossing and developmental studies of seed formation in order to select the apomictic and other aberrant types of seed development. Moreover, the mst's have propagation potential in nature and breeding value in cultures. The fst's lack both these features.

1.3 Virtues

The occurrence of mst was initially considered as an intrinsic and undesirable anomaly that had no evolutionary potential and/or breeding utility. The selection value of mst plants was taken to be negligible, since most of them were eliminated through natural selection because of either no or little seed set, especially in inbreeders. However, as the appreciation of hybrid vigour grew, methods were urgently required that would eliminate, obviate or greatly alleviate the tedium of hand emasculation. This requirement prompted mst to become an asset and a breeder's tool for use in hybrid seed development. Utilisation of mst in sorghum was pointed out by Stephens (1937) and in onion by Jones and Emsweller (1937), who outlined schemes for its use in commercial hybrid seed and bulb production. Its potential value was soon realised and later utilised in hybrid seed production of many auto- and allo-gamous species. This utility enhanced its role in plant breeding and has been documented from time to time by Kley (1954, 1955), Gabelman (1956), Bianchi (1968), Frankel (1973a,b), Driscoll and Barlow (1976) and Frankel and Galun (1977). The breeding value of mst led to its being sought in higher plants, in the majority of which mst's were found to be commonly occurring in nature. Its general occurrence, potential and genetic significance were highlighted by Jain (1959), Chowdhury and Verghese (1968), Edwardson (1970), Gottschalk and Kaul (1974), Gottschalk and Klein (1976), Grün (1976), Frankel and Galun (1977), Pearson (1981), Van Damme (1983a) and Hanson and Conde (1985).

1.4 Initial Reviews

Early descriptions by Correns (1928), Loehwing (1938), Allen (1940), Lewis (1941) and Edwardson (1956) represent individual aspects, and that of Chowdhury and Verghese (1968) the generalised view points of mst in higher plants. Edwardson (1970) reviewed gc-mst, Duvick (1959) indicated its use in hybrid seed production and Laser and Lersten (1972) described microsporogenesis of gc-mst plants. Gottschalk and Kaul (1974) and Gottschalk and Klein (1976) analysed the cytogenetic causes of sterility in g-mst plants. Proposals for its occurrence and perpetuation, terminology and nature and mechanisms were forwarded by Kley (1955), Lacadena (1968) and Pearson (1981). Inferences from g-c interactions affecting mft of a few major crop plants were given by Hanson and Conde (1985). An overview of mst, its genetic control and breeding potential was given by Jain (1959), Frankel (1973a,b,c) and Frankel and Galun (1977). The present work provides a coherent picture of the various facets of mst. This has facilitated understanding of its control and the nature of its diversity, and has indicated that the earlier notion that mst is merely a by-product of the ever-sportive nature of the biological world is no longer valid. Instead, it represents a well-established genetic mechanism that broadens the gene pool and cytoplasmic base of higher plants in order to increase their genetic fitness and environmental adaptation. The latter widens their habitat niches and ecological amplitude.

1.5 Definition and Concept

The failure or inability of an individual to produce functional gametes or viable zygotes under a given set of environmental conditions is known as sterility. It concerns both the male and female sex and denotes a condition in which both pollen and ovules are nonfunctional. The term sterility covers all cases of infertility or barrenness that result from irregularities in the sexual reproductive systems. This infertility, a hindrance to sexual mating, is differentiated from incompatibility, in which there is inability of genetically identical viable gametes or nuclei to undergo karyogamy and fertilization. In incompatibility, the same gamete may successfully unite with the genotypically different partners, giving rise to viable zygotes. Therefore, incompatibility is a form of infertility caused by the failure of plants with normal pollen and ovules to set seeds and leads to selective restriction of mating competence by extrasexual genes. Thus, whereas in the incompatibility system, male and female gametes are normal and functional, in the sterility system, spores and gametes are either absent or, when present, are abnormal and nonfunctional. Specifically, sterility refers to the failure of formation or development of functional male or female organs, spores or gametes in an organism. When the sterility is due to the failure of plants to produce functional anthers, pollen or male gametes, it is termed male sterility. Of the two sterility types, male sterility is more prevalent and is a stabler genetic system than female sterility. This has infused male sterility (mst) with genetic significance and breeding value under both natural and cultural conditions.

Dorsey (1914) considered mst to be a condition resulting from defects leading to the nonformation of pollen or to the lack of functional power in it when formed. Thus, in a mst either the pollen is not formed or if formed, it is either nonviable or viable, but incapable of effective fertilization under normal conditions. In all such cases, female fertility is completely or nearly unimpaired. Thus, while the male sex is nonfunctional, the female sex is functional. The recognition of females as mst, occurring often in gynodioecious species, is justified under this definition. Similarly, inclusion of sex repressions or sex reversals leading to deficiency or complete extermination of maleness seems feasible. In all these cases, the net effect of mst in a genotype is the nonavailability of its own pollen or male gamete for fertilization. The sterility attains a variety of diverse forms and encompasses the following:

i. complete absence or high deficiency of male individuals in a dioecious strain,
ii. absence, atrophy, mis- or mal-formation of male organs in a normal bisexual plant,
iii. failure to develop normal microsporogenous tissue,
iv. anomalies in microsporogenesis yielding imperfect, inviable, deformed or aborted pollen,
v. inability of pollen to mature or its inability to germinate on compatible stigma,
vi. viable pollen formed but anthers nondehiscent,
vii. barriers other than incompatibility preventing pollen from reaching sigma or ovule.

1.6 Classification

1.6.1 Previous Systems

The events leading to mst are multifarious. This has prompted the classification of mst into various types. Gabelman (1956) differentiated mst into pollen, staminal and functional types. In the pollen steriles, the pollen is extremely scarce or totally absent. In the staminal steriles, stamens or male flowers are malformed and in the functional steriles, viable pollen liberation is prevented almost fully. Based on phenotypic identity and incidence, Jain (1959) differentiated mst into teratological, anatomical, developmental, environmental and unclassified types. Heslop-Harrison (1971) classified it into gametophyte- and sprophyte-mediated types. Depending on the developmental stage during which the anomally appeared, the author distinguished three types of mst control:

1. *Anomalies preceding meiosis.* This involves anther suppression, phylody, petaloidy and pistiloidy. These anomalies have sporophytic control.
2. *Aberrant male meiosis during gametophyte development.* This includes imperfect meiosis or meiotic failure after initiation or arrested or irregular gametophytic development or production of genomically unbalanced spores. These anomalies are either sporophytically or gametophytically mediated.
3. *Failure of anther dehiscence or anthesis.* In this case, fertile pollen is produced but is not released for pollination. This is because either anther dehiscence or flower opening is prevented. The anomaly is sporophytically conditioned.

Gottschalk (1976) perceived two ways by which mst is expressed in higher plants. In the "functional male sterility" he included mutants in which either stamens or archesporial tissue are not differentiated. In other functional mutants, viable male gametes are formed but fertilization is prevented either due to anther non-dehiscence or spatial separation of anthers and stigma. In this type he also included genes causing staminal transformation into carpels or other floral parts. In the second type of mst, the author included genes that influence the course of microsporogenesis or microgametogenesis without influencing megasporogenesis or megagametogenesis. Johns et al. (1981) referred the term "structural sterility" to those in which sterility is caused by abnormalities in the flowers or in the reproductive organs. For those in which the sterility is caused by microsporogenesis or microgametogenesis breakdown, the term "nonstructural sterility" was utilised by these authors. Both male and female sterility were included in this classification.

Edwardson (1956, 1970) distinguished gc-mst on the basis of origin from intra- and inter-specific and intergeneric crosses. Following a nearly similar criterion, Lacadena (1968) classified the sterility into autoplasmic, homoplasmic and alloplasmic types. Autoplasmic types included plants whose cells are constituted by nuclei and cytoplasm which, due to a long nucleus-cytoplasm co-adaptation, are specific to the genetic population to which the individuals belong. In this, species, varieties and strains are considered as distinct genetic populations. Homoplasmic types included mst plants that originated as a consequence of interaction between cytoplasm and chromosomes from two different genetic populations both belonging to the same species. In this, mst plants originating after intraspecific crosses were also included. Alloplasmic types later included mst plants that developed due to the interaction of the chromosomes and cytoplasm of different species or genera. The alloplasmy can be thus obtained by substitution backcrosses and may be of interspecific or intergeneric type, depending upon whether the species involved belong to the same or different genera.

1.6.2 Proposed System

None of these classifications is based either on phenotypic or on genotypic basis. This has now been done in Table 1.1. Whereas the phenotypic basis distinguishes mst on the expressional pattern, the genotypic basis relies on the genetic nature and inheritance pattern of a genotype, irrespective of its phenotype. The mst may be genetic or nongenetic and may arise either spontaneously or become induced by chemical treatments, environmental manipulations or physiological and biochemical alterations. The general pattern of classification of mst as devised at present is given in Tables 1.1 and 1.2, and details on phenotypic and genotypic basis are indicated below.

Table 1.1. General classification of male sterility

Male sterility type			
Genetic (spontaneous or induced)		Nongenetic (induced)	
1. Phenotypic	Sporogenous Structural Functional	1. Chemical	Male gametocides Other chemicals
2. Genotypic	Genic Cytoplasmic (?) Gene-cytoplasmic	2. Physiological	Temperature Photoperiod Enzyme balance Other biochemicals
		3. Ecological	Climatic Biotic Edaphic Physiographic

1.6.2.1 Phenotypic Basis

Irrespective of genetic constitution, certain phenotypic anomalies, especially in anther development form, and function, lead to mst. They are broadly categorised into the following three types:

i. *Structural Male Sterility.* This sterility is due to structural anomalies predominantly in the male sex organs and very rarely in the female sex organs.

Table 1.2. Male sterility types based on phenotype and gene action

Sterility type[a]	Phenotypic alteration	Effect on male sex organ	Action of the male sterility gene	Environmental Influence on male sterility gene action	Influence of other genes on male sterility gene action
1. Structural male sterility	a) No anthers, b) Mis-, mal- or transformed anthers c) Faulty androecial development and differentiation	Form and function impaired	During anther development and differentiation before the onset of microsporogenesis	Profound	Pronounced
2. Sporogeneous male sterility	d) Irregular sporocyte formation (no meiosis) e) Abnormal microsporogenesis	Function impaired	Just before, or during or after microsporogenesis	High	Medium
3. Functional male sterility	a) Viable pollen produced but not liberated b) Viable pollen fail to reach stigma due to defective exine or pollen tube or some mechanical barriers (e.g. lack of secretions)	Functionability inhibited	After microsporogenesis, before anther dehiscence	Low	Low

[a]Genotype-environment interaction very high.

The male reproductive organs are either completely absent or deformed, misformed or malformed so that no microsporogenous tissue is developed or microsporogenesis fails to occur completely.

ii. *Sporogenous Male Sterility.* In this type, the stamens develop but the sporogenous tissue is either mis- or mal-formed or develops normally, but microsporogenesis or gametogenesis is impaired so that either non-, mis- or mal-formations of pollen or its premature abortion occurs in otherwise normal homomorphic flowers. Hence the pollen is either completely absent or non-functional, or extremely scarce. The abortion of microsporogenous cells may take place before, during or after meiosis. Thus it predominantly includes premeiotic, meiotic and postmeiotic mst mutants.

iii. *Functional Male Sterility.* In this, viable pollen is formed but it is incapable of fertilization due to some barriers that render it unable to reach its own or foreign stigma. The main barriers are: anther indehiscence, faulty or no exine formation, inability of pollen to migrate to stigma or affect fertilization.

In a plant species, mst can have any of the three above-mentioned manifestations, most of which are distinct. However, in a few these may be diffuse. Such mst genotypes possess features intermediate between any of the two or three above-mentioned phenotypic types. For instance, in some barley, maize and tomato mutants, stamens are either reduced or deformed. In them, a few PMC's develop in which meiotic breakdown may occur either at the beginning or at the end of microsporogenesis. Thus they share features of structural and sporogenous mst types. Likewise, one g-mst mutant each in pea and tomato (this Vol.) has dwarf stamens and reduced anthers. Microsporogenous tissue is highly reduced in them. A few PMC's adjacent to tapetal cells undergo meiosis and develop a few pollen grains which are never liberated. Thus these two mutants exhibit some features common to structural, sporogenous and functional types of mst. Late-produced flowers in these mutants exhibit structural mst and no PMC's are ever developed in them. Moreover, the influence of temperature on the stage of microsporogenesis breakdown is high in both of them. Transferring them to the glasshouse with 30°±2°C temperature induced early PMC abortion and some mutants became sp-mst types. All such cases exhibiting intermediate features are classified primarily on the basis of their anomaly and the resultant effect. Moreover, the anomaly should be stable and exhibit maximum frequency. For instance, if the anomaly predominantly involves alterations in stamen form and function, the sterility should be classified as structural, even though in some flowers a few anthers have developing PMC's. If anther phenotype is normal or nearly normal and meiosis breaks down in most of them and continues in a few, the anomaly is considered a sporogenous type. Therefore, under the presently devised classification system, the above-mentioned barley, maize and tomato g-mst mutants are structural mst's. In many gc-mst's, especially in alloplasmic forms (e.g. wheat, tobacco, *Solanum*, *Streptocarpus*), the mst expression likewise passes through various phases of mst types. In them, the final anther phenotype will characterise the mst type, as it is predominant, stable and maintained constantly in subsequent generations.

1.6.2.2 Genotypic Basis

As mentioned above, some mst plants exhibit more than one phenotypic manifestation. Therefore, classification of mst types on a genetic basis becomes vital. Accordingly, depending on the nuclear and cytoplasmic genome constitution as evidenced by the inheritance pattern, mst's have primarily been classified into genic, cytoplasmic and gene-cytoplasmic types, and the sterility can have any of the phenotypic manifestations mentioned earlier.

i. *Genic Male Sterility* (g-mst). The sterility is controlled by nuclear genes whose action is not influenced by the cytoplasm type. Therefore, the inheritance pattern and expression of the sterility are entirely Mendelian. It exhibits minimum reciprocal differences and is conditioned by recessive genes in the majority of the cases (Tables 2.5, 2.6).
ii. *Cytoplasmic Male Sterility* (c-mst). This sterility, unlike the genic type, is controlled by a specific male sterility-inducing cytoplasm known as sterile (S) cytoplasm. Thus at least two cytoplasm types exist in a species exhibiting cytoplasmic mst, viz., normal fertile (N) and sterile (S). Irrespective of the nuclear genotype, the plants having S-cytoplasm are mst and those with N-cytoplasm are mft. Thus the influence of the nuclear genotype on this type of S-cytoplasm is negligible and the sterility is inherited maternally. Such plants, unlike genic male steriles, exhibit reciprocal differences in male sterility-fertility relationships in F_1 generation.
iii. *Gene-Cytoplasmic Male Sterility* (gc-mst). In this type, the mst results from the cooperative action of certain particular nuclear genes designated as *fr* genes and a specific sensitive cytoplasm type termed as S-cytoplasm. Thus plant species exhibiting gc-mst possess both *Fr* (mft) and *fr* (mst) nuclear genes as well as N- and S-cytoplasm types. The cytoplasm-specific *fr* genes are ineffective in N-cytoplasm in producing mst, but in S-cytoplasm the combination *fr* nuclear genes and S-cytoplasm leads to mst. The S-cytoplasm possesses a specific type of cytoplasmic genes termed *c* genes, which in association with *fr* genes cause mst. Combinations like $N_{fr\,fr}$ or S_{Frfr} or S_{Frfr} are mft. Since the *Fr* nuclear genes interact specifically with S-cytoplasm plants and revoke mft in S-cytoplasm, these genes represent the fertility-restoring *Fr* genes. The symbols used depend on the recessive (*fr*) or dominant (*Fr*) nature of these fertility-restoring *Fr* genes. The majority of these genes are dominant and I shall utilize this convention in the further text while using the gene symbols.

A diagrammatic representation showing the inheritance pattern of the three mst types is given in Fig. 1.1. Although both g-mst and gc-mst exhibit a different inheritance pattern, their nuclear genes exhibit analogy in having allelic and nonallelic diversity. Accordingly, a large number of such genes are known in the genomes of many cultivated plants. The number of nonallelic *ms* genes is much higher that that of the *fr* genes in a plant genome, e.g. *Hordeum, Pennisetum, Petunia, Solanum, Sorghum, Triticum, Zea* etc. Similar nonallelism is exhibited by

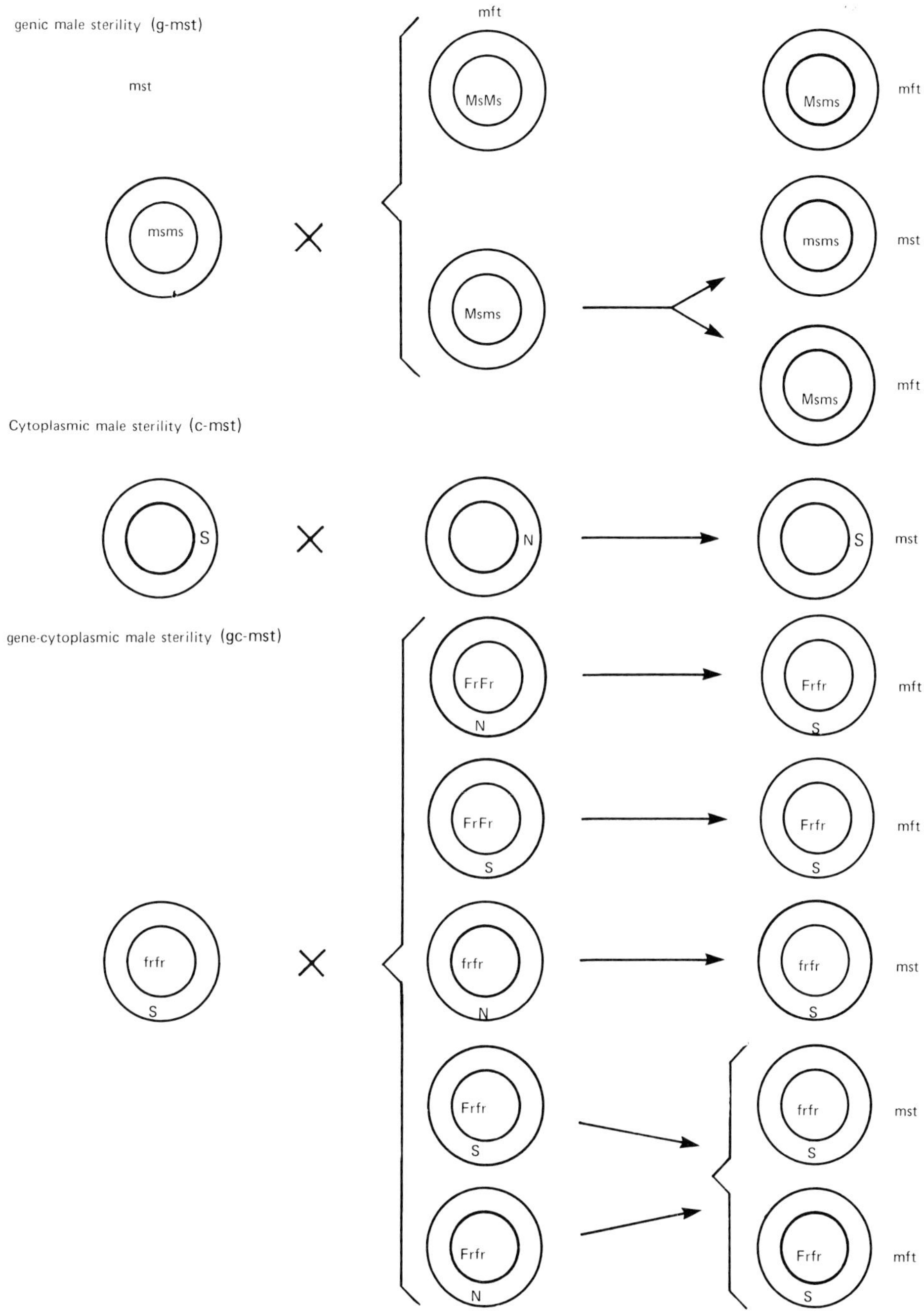

Fig. 1.1. Male-sterility inheritance pattern (Sears 1947). Letters in *inner* circles represent nuclear gene and in the *outer circles* the cytoplasm types. *N* normal fertile; *S* sterile cytoplasm type. Cytoplasm is assumed to be transmitted only by the female parent. mft : male fertile, mst : male sterile

c-genes and this leads to S-cytoplasm differentiation in crop plants like maize, wheat, sorghum, potato, pearl millet, etc. The *fr* and *ms* nuclear genes are distinct from each other and both may or may not be located on the same chromosome. In barley, petunia and maize, both these types of genes are located at different loci, segregate independently of each other and do not interfere with their individual expressions. This has been amply demonstrated by Evenor and Izhar (1984) and Evenor et al. (1984) in petunia, in which both *ms* and *fr* genes were placed in N- and S-cytoplasm backgrounds.

1.7. Genetic Types

Fundamentally genetic male sterility is of two types only: genic and gene-cytoplasmic, symbolically represented as g-mst and gc-mst types, respectively, throughout this text. The g-mst mutants appear spontaneously and regularly. Because of their predominant recessive gene control, they identify themselves more frequently in self-pollinating plants than in cross-pollinating ones. The mst becomes cytoplasmic when in addition to N-cytoplasm, one or more S-cytoplasms exist in a species and *Fr* genes are not detected in it, e.g. *Cirsium acaule, C. oleraceum* (Correns 1916) and *Satureja hortensis* (Wettestein 1924). On the other hand, the same sterility becomes gc-mst if *Fr* genes are detected. Many such and other mst's considered initially as cytoplasmic mst's (c-mst's) were later found to be gc-mst's as *Fr* genes were found in them, e.g. *Aquilegia, Epilobium, Linum, Streptocarpus*, sugarbeet, etc. In *Chenopodium quinoa*, mst is controlled by a recessive gene but the cytoplasmic determination once generated, overrides the nuclear control (Simmonds 1971) and the sterility is transmitted maternally. Since basically a *ms* gene generates *fr-c* interaction, loss of the *ms* gene from the population would thus leave the *gc*-mst system in *quinoa*. Rarely, this does occur, and the mst becomes gc-mst. Events like the loss of *Fr* genes or of the genotypes having these genes in a breeding population after initial wide crosses may explain the c-mst expression observed in *Cirsium acaule* and *C. oleraceum* by Correns (1916) and in *Satureja hortensis* by Wettestein (1924). Since mutation of N- to S-cytoplasm in the presence of *fr* genes leads to mst, detection of *Fr* genes in the parental or related species or genera is essential. This necessitates searching for the appropriate *Fr* genes. In a few species, *Fr* genes arise later than N-cytoplasm mutations to the S-type. In them, the mst persists even though the genome of the species is replaced partially or fully through substitution backcrosses. In such cases of recent N-cytoplasm mutations, the c-mst becomes an initial step towards gc-mst development and establishment. Being transitory in the evolution of gc-mst, c-mst may not be considered as a distinct mst type. Hence only two established mst types may be recognised, viz. genic male sterility (g-mst) and gene-cytoplasmic male sterility (gc-mst). Both these sterility types are designated as genetic male sterility types in this text, as the sterility expression is basically conditioned by genes. The term c-mst has also been retained, especially for those cases where distinct *Fr* genes still remain to be detected.

From the practical viewpoint, distinction between genic and gene-cytoplasmic male sterility appears valid and feasible because the two gene systems act independently. Moreover, interaction between *ms* genes and S-cytoplasm has not been demonstrated. Furthermore, the coexistence of *ms* and *fr* genes is known in maize and petunia (this Vol., Evenor et al. 1984). These genes segregate and act independently. Whereas Hermsen (1968) considered this noninteraction to be due to lack of variability among the cytoplasms studied, the noninteraction between *ms* genes and S-cytoplasm has been repeatedly demonstrated in barley, maize, sorghum, wheat and pearl millet, in which considerable cytoplasm diversity is known.

1.8 Symbols Proffered and Abbreviations Used

In genic male sterility (g-mst), the sterility genes are designated *ms* or *Ms*, depending upon their recessive or dominant nature. Thus g-mst genotypes have *msms* genes if the genes conditioning sterility are recessive, or *MsMs* or *Msms* if the genes are dominant. Since the majority of such genes are recessive, the mst genotypes have *msms* genes. Where more than one nonallelic *ms* genes are known in a species, e.g. barley, maize, pea, tomato etc., they are symbolised as ms_1, ms_2, ms_3. . . . This gives the total number of nonallelic *ms* genes known in a species. In gene-cytoplasmic male sterility (gc-mst), the nuclear genes restoring fertility, the fertility restorers, are symbolised as *Fr* or *fr* depending on their dominant or recessive nature. The majority of these genes are dominant. Where more than one such nonallelic genes are known in a species, they are designated as Fr_1, Fr_2, Fr_3. . . . The symbols *Fr* and *fr* adopted here are more appropriate than the conventionally used ones, *Rf* and *rf*, as the former represent the abbreviations for fertility-restoring genes. Likewise, the sterile cytoplasm is symbolised as S-cytoplasm and the fertile cytoplasm as N or F. Maize, sorghum, potato, pearl millet, barley and mustard exhibit considerable S-cytoplasm differentiation. In maize, different alphabets like C, S, and T have been used for each of the three sterile cytoplasm types. It is appropriate to designate different sterile cytoplasm types of a species by assigning to them the usual serial numbers, as is suggested for *ms* or *fr* genes. Thus the cytoplasms of a species having three S-cytoplasm types would be designated S_1, S_2 and S_3. This would introduce uniformity in symbolization and also indicate the number of S-cytoplasm types each species has. In this proposed symbolization, a distinction exists between *ms* and *fr* genes as the plants having *ms* genes have g-mst and those with *fr* genes have gc-mst. Thus these two symbols differentiate and characterise g-mst and gc-mst distinctly. These proposed symbols have been used in Part I, but the conventional symbols used by various authors are retained in Part II of this monograph. Furthermore, the abbreviation "mst" used in the present text refers to male sterile genotypes irrespective of their type and nature (genic or gene-cytoplasmic with recessive, dominant or polygenic gene control) and also to the phenomenon of male sterility. Since some words appear repeatedly in this monograph, their abbreviations as given in Table 1.3 are used to economise space.

Table 1.3. List of abbreviations used for some general terms and words

Abbreviation	Words and/or terms
A) *General*	
AI, AII	anaphase I, II of meiosis
Anon	anonymous
ACP	artificial cross-pollination
BC	backcross or backcrossing
c-gene	male sterility-inducing cytoplasmic gene
ch-mst	chemically induced male sterility
c-mst	cytoplasmic male sterile genotype, phenomenon or type
com	complementary genes
cv	cultivar or cultigen
D	dominant gene(s)
ds	double-stranded (RNA or DNA)
dup	duplicate gene(s)
EM	electron micrograph
emc	egg mother cell
cp	chloroplast
fft or fst	female fertile or sterile gene(s), genotype or phenomenon
fn-mst	functional male sterile gene(s), genotype or phenomenon
fr	nuclear gene inducing male sterility in presence of *c*-gene
Fr	dominant fertility-restoring nuclear gene(s) restoring fertility to S-cytoplasm or the phenomenon of fertility restoration
g-c or gc	gene-cytoplasm
gc-mst	gene-cytoplasmic male sterile genotype(s) or phenomenon
g-gene	nuclear male sterility gene
g-mst	genic male sterility gene(s), genotype or phenomenon
MI, MII	metaphase I, II of meiosis
meiocyte	a cell destined to undergo or undergoing meiosis
mft	male fertile gene(s), genotype or phenomenon
mt	mitochondria
mmc	microspore mother cell(s)
mms	meiotic male sterile(s)
ms	genic(nuclear) male sterile gene(s), genotype or phenomenon
mst	male sterility gene(s), genotype or phenomenon of male sterility
mt	mitochondria
NCP	natural cross-pollination
nk	not known
PI, PII	prophase I, II of meiosis
PMC	pollen mother cell
p-fst	partial female sterile or sterility or phenomenon
p-mft	partial male fertile or fertility or phenomenon
p-mst	partial male sterile or sterility or phenomenon
po	polygene(s)
ps	pollen sterile, sterility or phenomenon
r	recessive gene(s)
SEM	scanning electron micrograph
sp-mst	sporogenous male sterility gene(s), genotype or phenomenon
ss	single-stranded (DNA, RNA)
st-mst	structural male sterility gene(s), genotype or phenomenon
TI, TII	telophase I, II of meiosis
TS	transverse section
x	ploidy level with somatic chromosome number
B) *Mutagens*	
DES	diethyl sulphate
EB	ethidium bromide
EI	ethylene imine
EMS	ethyl methane sulphonate
NEU	nitrosoethyl urea
NMU	N-nitroso N-methyl urethane
SA	sodium azide

CHAPTER 2
Genic Male Sterility

2.1 Occurrence and Perpetuation

2.1.1 Spontaneous

Genic male sterility (g-mst) is of a wide occurrence in flowering plants and the majority of the steriles have arisen as spontaneous mutants (Table 2.1). Only a few have been induced after mutagen treatments (Table 2.3). In about 175 species in which mst has spontaneously arisen, the majority are dicots in species as well as in species crosses (Fig. 2.1). The ratio dicot to monocot is about 2.5 in induced mutants. Since the repeated appearance of g-mst in many plant species is a regular phenomenon, mutation of *mf* to *ms* genes is of high frequency. This is evidenced from the existence of vast numbers of *ms* genes especially in the cytogenetically intensively investigated species, viz. over 60 in maize, 55 in tomato, 48 in barley, 24 in garden pea, 12 in chillies and 10 in cotton. In addition to these known nonallelic genes, numerous other unidentified nonallelic and many more allelic *ms* genes will be found in these as well as in other plant species if they are intensively investigated. The *ms* genes have arisen in high frequency both in self- as well as in cross-pollinating species, and more recessive *ms* genes may be detected if cross-fertilized species are inbred or subdivided into smaller populations. For instance, selfing of cucurbits, *Lolium*, maize, pearl millet and rye leads to segregation of mst plants. Likewise, subdivision of many *Hebe* species produced progenies having mst segregants (Frankel 1940).

Since the majority of the mst mutations are recessive, elimination of *ms* genes is certain in selfers unless a chance of outcrossing occurs in them. The elimination depends to a great extent on the outcrossing rate. With a little outcrossing, elimination is gradual; but after its appearance, the chances of survival of mst's exist even without crossing because the expression is sporophytic in the majority of cases. Thus, whereas all the gametes of heterozygotes are functional, only female gametes of recessive homozygotes function. This is in contrast to dominant mst mutations or those affecting female fertility, since these are promptly removed from the populations. For this reason, in the majority of the spontaneously arisen mutants, mst is conditioned by monogenic recessive genes (Tables 2.4 to 2.6). Further, chances of survival and maintenance of *ms* genes, whether dominant or recessive increase as many mst mutants, especially in gynodioecious plants, have two times more female fertility than the hermaphrodites (Lewis 1941; Kley 1955) and the mst mutants combine higher seed fertility with heterozygote advantage. On an average, one monogenic recessive mst mutant occurs in about 20,000 plants of tomato (Rick 1945), in lima beans (Allard 1953), the frequency is 5×10^{-5}. The occurrence of a large number of nonallelic *ms* genes in many species indicates a

Table 2.1. Spontaneously arisen genic male sterile mutants

Species	References
Abelmoschus esculentus (1r, 6x = 120)	This Vol.
Ageratum conyzoides (2–3r, 4x = 40)	This Vol.
A.conyzoides (1r, 2x = 20)	This Vol.
A.houstonianum (1, 2r, 2x = 20)	Reimann-Philipp and Fuchs (1971), this Vol.
Allium ascalonicum (1r, 2x = 16)	Darlington and Haq (1955), this Vol.
A.fistulosum (1r, 2x = 16)	Nishi and Hiraoka (1958)
A.longicuspis (2x = 16)	Novâk (1972)
A.porrum (1–3r, 4x = 32)	Schweisguth (1970)
A.sativum (2x = 16)	Koul and Gohil (1970), Novâk (1972), Konviĉka et al. (1978), this Vol.
A.schoenoprasum (2x = 16)	Singh and Kobabe (1969), this Vol.
Alopecurus myosuroides (1-many r, 2x = 14)	Johnsson (1944)
A.pratensis (6x = 42)	This Vol.
Antirrhinum majus (1r, 2r, 2x = 16)	Baur (1924), Goldsmith (1968), Cachon (1980), this Vol.
Arachis hypogaea (1–3r, 4x = 40)	Hayes (1933), Patel et al. (1936), Hull (1937), Ashri (1968)
Argemone maxicana (1r, 4x = 28)	This Vol.
Avena byzantina (2r, 6x = 40)	Garber (1922)
[a]*A.byzantina* (6x = 42)	Ramage and Suneson (1958)
[a]*A.fatua* (6x = 42)	This Vol.
A.strigosa (1r, 2x = 14)	Sadanaga (1965)
Begonia semperflorens (1r, 2x = 36)	Reimann-Philipp and Seidel (1963)
Beta macrocarpa (1r, 2x = 18)	This Vol.
B.maritima (1r, 2x = 18)	Kinoshita (1971), Kinoshita and Takahashi (1972)
B.vulgaris (1r, 2x = 18)	Owen (1952), this Vol.
Brassica campestris (1r, 2x = 20)	Das and Pandey (1961), Chowdhry and Das (1966, 1968)
B.juncea (1r, 4x = 36)	Banga and Labana (1983)
B.napus (1r, 2r, 1D, 4x = 38)	Koch and Peters (1953), Heyn (1973), Mathias (1985)
B.oleracea var.*botrytis* (1r, 2x = 18)	Nieuwhof (1961), Borchers (1966), Chatterjee and Swarup (1972)·
B.oleracea var.*capitata* (1r, 2x = 18)	Nishi and Hiraoka (1958), Kototani and Yamada (1958), Rundfeldt (1961)
B.oleracea var.*gemmifera* (1r, 2x = 18)	Haigh (1954), Johnson and Faulkner (1956, 1957), Nieuwhof (1968a)
B.oleracea var.*italica* (1r, 2x = 18)	Cole (1959), Sampson (1966a), Dickson (1970)
Cajanus cajan (1r, 2x = 22)	Reddy et al. (1978), Wallis et al. (1980), Dundas et al. (1981, 1982)
[a]*Canna glauca* (2x = 18)	Honing (1928), Öfferijns (1938)
Cannabis sativa (1r, 2x = 20)	Breslavetz (1935), Pate and Joyner (1958)
Capsicum annuum (1r, 2x = 24)	Martin and Crawford (1951), Hirose (1965), Shifriss and Frankel (1969), Shifriss and Rylski (1972), Shifriss (1973), Murty and Lakshmi (1979), Hirose and Fujime (1981), Meshram and Narkhede (1982), Pochard (1970b)
Carica papaya (1r, 2r, 2x = 18)	Storey (1953)
Carthamus tinctorius (1r, 2x = 24)	Ebert (1964), Ebert and Knowles (1966)
[a]*Castanea crenata* (2x = 24)	Jaynes (1963), this Vol.
[a]*C.dentata* (2x = 24)	Jaynes (1963)
[a]*C.mollisima* (2x = 24)	This Vol.
[a]*C.pumila* (2x = 24)	This Vol.
[a]*C.sativa* (2x = 24)	This Vol.

Table 2.1 *(continued)*

Species	References
[a]*Centaurea americana* (1r)	This Vol.
[a]*C.imperialis* (1r)	This Vol.
[a]*C.scabiosa* (2x = 20)	Fröst (1963)
Centranthus ruber (2D, 1r, 2x = 14)	Sobrinho (1946), this Vol.
Chenopodium quinoa (1r, 4x = 32)	Simmonds (1971)
Cheiranthus cheiri (1r, 3r, 2x = 14)	Chittenden (1914), Sirks (1924), this Vol.
[a]*Chrysanthemum balsamita* (2x = 18)	This Vol.
C.carinatum (2x = 18)	This Vol.
[a]*C.parthenium* (2x = 18)	This Vol.
[a]*C.morifolium* (6x = 54)	Broertjes and Jong (1984)
[a]*Cirsium arvense* (2x = 34)	Delannay (1979)
[a]*C.montanum* (2x = 34)	Delannay (1979)
[a]*C.oleraceum* (2x = 34)	Delannay (1979)
[a]*C.palustre* (2x = 34)	Delannay (1979)
[a]*C.spinosissimum* (2x = 34)	Delannay (1979)
[a]*Citrus aurantifolia* (2x = 18)	Uphof (1931), Nakamura (1934), Krug and Bacchi (1943)
[a]*C.reticulata* (2x = 18)	This Vol.
[a]*C.sinensis* (2x = 18)	Osawa (1912), Oppenheim and Frankel (1929), Nakamura (1934)
[a]*C.unshiu* (2x = 18)	Osawa (1912), Nagai and Tanikawa (1928)
[a]*C.yatsushiro* (2x = 18)	Nakamura (1943)
[a]*Citrullus vulgaris* (2x = 22)	Watts (1962)
Coleus blumei (1D, 4x = 24)	Rife (1948), Ford (1950)
C.hybridus (1r, 8x = 48)	This Vol.
Corchorus capsularis (2x = 14)	Rakshit (1967), Mitra (1977)
Crotalaria juncea (1r, 2x = 16)	This Vol.
Cucumis melo (1r, 2x = 24)	Bohn and Whitaker (1949), Bohn and Principe (1964), Chauhan and Singh (1968)
C.sativus (1r, 2x = 14)	Shifriss (1950), Groff and Odland (1963), this Vol.
Cucurbita maxima (1r, 4x = 40)	Scott and Riner (1946), Rhodes (1959), Singh and Rhodes (1961), Francis and Bemis (1970), Kályágin (1974)
C.pepo (1r, 4x = 40)	Shifriss (1945), Eisa and Munger (1968)
[a]*Dactylis glomerata* (4x = 28)	Filion and Christie (1966), Nâsinee (1973), this Vol.
Daucus carota (1r, 2x = 18)	Michalik (1974)
D. carota (2x = 18)	Welch and Grimball (1947), McCollum (1966), Litvinova (1973)
D.carota (1D + 1r, 2x = 18)	Hansche and Gabelman (1963a)
Echinocloa colone (1r, 6x = 54)	Ayyangar and Srinivasa Rao (1941)
Eleusine indica (1r, 2x = 18)	This Vol.
Eruca sativa (2r, 2x = 22)	Verma (1984)
Fagopyrum esculentum (1r, 2x = 16)	This Vol.
[a]*Festuca arundinacea* (6x = 42)	Goral et al. (1981)
Fragaria sp. (2x = 14)	Longley (1926)
[a]*Geranium argenteum* (2x = 28)	This Vol.
[a]*G.grandiflorum* (2x = 28)	This Vol.
Glycine max. (1r; 4x = 40)	Stewart and Wentz (1926), Bernard and Jaycox (1969), Brim and Young (1971), Bernard and Cremeens (1975), Chaudhari and Davis (1977), Rubaihayo and Gumisiriza (1978), Albertson and Palmer (1979), Palmer et al. (1980), Delannay and Palmer (1982)

Table 2.1 *(continued)*

Species	References
G.max (pmst) (1r, 4x=40)	Singh (1972), Caviness et al. (1970), Caviness and Fagala (1973), Stelly and Palmer (1980a,b, 1982)
Gossypium barbadense (1r, 1D, 4x=52)	Yang (1982)
G.hirsutum (1r, 4x=52)	Justus and Leinweber (1960), Richmond and Kohel (1961), Justus et al. (1963), Seth et al. (1981), Tang et al. (1981)
G.hirsutum (2r, 4x=52)	Weaver (1968), Rhyne and Rhyne (1971), Kohel (1973), Murthi and Weaver (1974)
G.hirsutum (1D, 4x=52)	Allison and Fisher (1964), Weaver and Ashley (1971), Murthi and Weaver (1974), Bowman et al. (1978), Bowman and Weaver (1979)
[a]*Hebe parviflora* (4x=80)	Frankel (1940)
[a]*H.salicifolia* (2x=40)	Frankel (1940)
[a]*H.subalpina* (4x=80)	Frankel (1940)
[a]*H.townsonii* (2x=40)	Frankel (1940)
[a]*H.traversii* (6x=120)	Frankel (1940)
Helianthus annuus (1r, 2x=34)	Kuptsov (1935), Anonymous (1964), Pirev (1966), Putt and Heiser (1966), Vranceanu (1970), Kovacik and Skaloud (1973), Paun (1974), Nakashima and Hosokawa (1974a)
[a]*Hevea brasiliensis* (4x=36)	Ramaer (1935)
Hordeum vulgare (1r, 2x=14)	Singh and Kaul (1966), Roath and Hockett (1970), Sharma and Reinbergs (1972), Mian et al. (1974), Ahokas (1975a), this Vol.
[a]*Ipomoea* sp. (2x=30)	Correns (1926)
[a]*Iris douglasiana* (4x=40)	Uno (1982)
[a]*I.pallida* (2x=24)	Maleci and Maugini (1977)
[a]*I.persica* (2x=[illegible])	This Vol.
[a]*Jasminum* sp. (2x=26)	Raman (1955)
[a]*J.pubescens* (2x=26)	Bhargava (1986)
[a]*Kniphofia* sp. (2x=12)	Moffet (1932)
Lactuca sativa (1r+1D, 2x=18)	Ryder (1963)
L.sativa (1r, 2x=18)	Lindqvist (1960), Ryder (1967, 1971)
Lathyrus luteus (2x=14)	This Vol.
L.odoratus (1r, 2x=14)	Gregory (1905), Punnett (1923, 1925), Faberge (1937), Upcott (1937)
L.sativus (1r, 2x=14)	Srivastava and Somayajulu (1981)
Lavandula angustifolia (2–3r, 8x=48)	This Vol.
Lens culinaris (1r, 2x=14)	Sharma and Sharma (1978)
Limnanthes douglasii (1D, 2x=10)	Jain et al. (1978)
Lolium multiflorum (1r, 2x=14)	Nitzsche (1971)
[a]*L.perenne* (2x=14)	Hayward and Manthiratna (1972)
L.perenne (1r, 2x=14)	Fejer (1966)
[a]*Lotus filicaulis* (2x=12)	de Nettancourt and Grant (1963)
[a]*Lupinus mutabilis* (4x=48)	Pakendorf (1970)
L.regalis (4x=48)	This Vol.
Lycopersicon esculentum (1r, 2x=24)	Crane (1915), Lesley and Lesley (1939), Lewis (1942), Larsen and Paur (1948), Rick and Robinson (1951), Bishop (1954), Hafen and Stevenson (1955), Rick and Butler (1956), Mittal and Thomas (1961), Tronichkova (1962), Rick and Boynton (1967), Simonov (1973), Sawhney and Greyson (1973a), Tsikova and Abadzhieva (1974), Zamir et al. (1980), this Vol.

Table 2.1 *(continued)*

Species	References
L.esculentum (2r, 2x = 24)	Solovjeva (1970)
[a]*Manihot esculenta* (4x = 36)	Jos et al. (1967), Magoon et al. (1968)
Medicago sativa (2r, 3po, 4x = 32)	Childers (1952)
M.sativa (3r, 4x = 32)	Childers and McLennan (1960)
[a]*Melilotus alba* (2x = 16)	Castetter (1925), this Vol.
[a]*Mercurialis perennis* (8x = 64)	This Vol.
[a]*Mimulus luteus* (8x = 64)	Brozek (1926)
Nicotiana alata (1r, 2x = 18)	Krishnamurthy and Bhat (1959)
N.tabacum (2r, 4x = 48)	Raeber and Bolton (1955), Bhat and Krishnamoorthi (1956)
N.tabacum (1r, 4x = 48)	This Vol.
[a]*Oenothera alata* (2x = 14)	Pohl (1895), Gates (1907)
[a]*O.flava* (2x = 14)	Oehlkers (1972)
[a]*O.gigas* (4x = 28)	Gates (1911)
[a]*O.lamarckiana alata* (2x = 14)	Oehlkers and Harte (1943)
O.strigosa (1r, 2x = 14)	Oehlkers (1923, 1926), Harte and Bissinger (1952)
[a]*Origanum majorana* (4x = 32)	This Vol.
O.vulgare (2D, 4x = 32)	Lewis and Crowne (1952, 1956)
Oryza sativa (1r, 2x – 24)	Nagai (1926a,b), Ishikawa (1927), Miyazawa (1932), Takezaki (1932), Ramanujam (1935), Hara (1946), Jones (1952), Craigmiles et al. (1968), Hoff and Chandrapanya (1973), Shibuya (1973), Razzaque (1975), Trees and Rutger (1978), Mahadevappa and Coffman (1980), this Vol.
Papaver somniferum (2x = 22)	Singh and Khanna (1970)
Parthenium hysterophorus (1r, 2x = 18)	This Vol.
Paspalum spp. (1r, 2x = 20)	Burton (1943)
[a]*Pelargonium crispum* (2x = 22)	Tokumasu (1974, 1976)
[a]*P.denticulatum* (10x = 90	Tokumasu and Tamai (1964)
[a]*P.radula* (9x = 81)	Conagin (1964)
[a]*P.roseum* (8x = 72)	Tamai et al. (1958)
P.tomentosum (1–2r, 4x = 44)	This Vol.
Pennisetum typhoides (1r, 2x = 14)	Krishnaswamy and Ayyanger (1942), Rao and Koduru (1978a,b), Rao and Devi (1983)
P.typhoides (3r, 2x = 14)	Gill et al. (1973)
Petunia axillaris (1r, 2x = 14)	Levan (1942), this Vol.
P.hybrida (1r, 2x = 14)	Frankel (1962, 1971), Izhar (1973), this Vol.
[a]*Phaseolus atropurpureus* (2x = 22)	Pritchard and Hutton (1972)
P.atropurpureus (1r, 2x = 22)	This Vol.
P.lunatus (1r, 2x = 22)	Allard (1953)
P.vulgaris (1r, 2x = 22)	Ibrahim et al. (1974), Agbo and Wood (1977), Rheenen et al. (1979), Wyatt (1984a,b)
P.vulgaris (2x = 22)	Lamprecht (1945)
P.vulgaris (2D, 2x = 22)	Mutschler and Bliss (1980)
Phleum nodosum (1r, 2x = 14)	May and Kasha (1971, 1980)
Pisum sativum (1r, 2x = 14)	Gottschalk (1961), Klein (1969), this Vol.
[a]*Plantago coronopus* (2x = 10)	Paliwal and Hyde (1959), Gorenflot (1964), Ross (1970a,b)
P.insularis (1r, 2x = 8)	Murr and Stebbins (1971)
P.lanceolata (1D, 2x = 12)	Ross (1969)
[a]*Primula officinalis* (2x = 12)	Dahlgren (1932)
[a]*Prunus amygdalus* (2x = 16)	Smith (1927)
[a]*P.avium* (2x = 16)	This Vol.
[a]*P.mume* (2x = 16)	Nakanishi (1982)

Table 2.1 *(continued)*

Species	References
[a]*P.persica* (2x = 16)	Smith (1927)
[a]*Ranunculus acris* (4x = 28)	Whyte (1929)
[a]*Rapanea thwaitesii*	Vasanthy and Pocock (1981)
Raphanus sativus (1r, 2x = 18)	Tokumasu (1957), Nishi and Hiraoka (1958)
Ricinus communis (1r, 2x = 20)	Claasen and Hoffman (1950)
Rubus idaeus (1r, 2x = 14)	Crane and Lawrence (1931)
Rumex acetosa (1r, 2x = 14)	Löve (1943)
[a]*Saccharum officinarum* (8x = 80)	Janaki-Ammal (1941), Dutt et al. (1954)
[a]*Salvia nemorosa* (2x = 12)	Linnert (1958)
[a]*S.officinalis* (2x = 14)	Linnert (1955)
[a]*Secale cereale* (2x = 14)	Putt (1954)
S.cereale (1r, 2x = 14)	Adamĉhuk (1971, 1972), Kobyliansky and Katerova (1973b), Kolobaeva (1974), Garlicka and Madej (1975), Kubicki et al. (1981)
S.cereale (1-many r, 2x = 14)	Cebrat and Zadecka (1978)
Sesamum indicum (1r, 2x = 26)	Brar (1982), Osman and Yermanos (1982)
[a]*Silene acaulis* (2x = 24)	Bock (1976)
S.hookeri (1r, 6x = 72)	This Vol.
S.vulgaris (2r, 4x = 48)	Horovitz and Dulberger (1983)
[a]*Solanum integrifolium* (2x = 24)	Habib and Rao (1973)
S.melongena (1r, 2x = 24)	Jasmin (1954)
S.nigrum (1r, 2x = 24)	This Vol.
S.tuberosum (1–2r, 4x = 48)	Young (1923)
S.tuberosum (many r, 4x = 48)	Fukuda (1927)
S.tuberosum (1D, 4x = 48)	Salaman (1910)
[a]*S.tuberosum* (4x = 48)	Breeze (1921), Fukuda (1927), Heyn (1930), Bleier (1931), Ellison (1936), Williams (1954), Kishore et al. (1974)
S.verrucosum (many r, 2r, 2x = 24)	Abdalla (1970), Abdalla and Hermsen (1972)
Sorghum vulgare (1r, 4x = 20)	Ayyanger and Ponnaiya (1937), Stephens (1937), Karper et al. (1938)
S.vulgare (many r, 4x = 20)	Stephens et al. (1952)
Spinacea oleracea (1r, 2x = 12)	Kobabe (1967)
S.oleracea (2r, 2x = 12)	This Vol.
Streptocarpus hybrid (1D)	Zeven (1972a)
Tagetes erecta (1r, 2x = 12)	Bolz (1961), Towner (1961), this Vol.
[a]*Taraxacum balticum* (4x = 32)	Malecka (1971)
[a]*T.fulvum* (4x = 32)	Malecka (1971)
[a]*T.nordstedti* (6x = 48)	Gustafsson (1935)
[a]*T.pseudofulvum*	Gustafsson (1935)
[a]*T.silesiacum* (3x = 24)	Malecka (1971)
[a]*Thespesia populnea* (2x = 26)	This Vol.
Trifolium pratense (1r, 2x = 14)	Smith (1971), Taylor et al. (1978)
T. pratense (2r, 2x = 14)	Kazimerski and Tsikova (1981)
T.pratense (1D, 2x = 14)	Nijdam (1932)
T.repens (1r, 4x = 32)	Atwood (1944)
[a]*Tripsacum laxum* (4x = 72)	Dodds and Simmonds (1946)
Triticum aestivum (1r, 6x = 42)	Pugsley and Oram (1959), Lupton and Bingham (1966–1967), Krupnov (1968), Lemekh et al. (1971)
T.aestivum (3r, 6x = 42)	Jan (1974)
T.aestivum (po, 6x = 42)	Athwal et al. (1967)
T.aestivum (1D, 6x = 42)	Xiangmin and Herong (1981), Deng and Gao (1982)

Table 2.1 *(continued)*

Species	References
T.durum (1r, 4x = 28)	Bozzini and Scarascia-Mugnozza (1968)
[a]*T.monococum* (2x = 14)	Smith (1938b)
Vaccinium angustifolium (many r, 2x = 24)	Hall and Aalders (1961)
[a]*Valeriana officinalis* (2x = 14)	Shugaeva (1979)
Vicia faba (1r, 2x = 12)	Bond et al. (1964a)
Vigna angularis (1r, 2x = 22)	Nakashima et al. (1980)
V.sinensis (1r, 2x = 24)	Sen and Bhowal (1962), Rachie and Rawal (1976), this Vol.
Viola orphanidis (1r, 2x = 20)	Clausen (1930)
[a]*Vitis vinifera* (2x = 38)	Olmo (1943)
Zea mays (1r, 2x = 20)	Kempton (1920), Young (1927), Phipps (1928), Singleton and Jones (1930), Beadle (1931, 1932b), Emerson (1932), Mangelsdorf (1932), Fraser (1933), Eyster (1933), Rhoades and Rhoades (1939), Sprague (1939), Clark (1940), Burnham (1941), Nelson and Clary (1952), Jain (1956), Palmer (1971), Golubovskaya (1977), this Vol.
Zinnia elegans (3r, 2x = 24)	Duffy (1982), this Vol.
Z.linearis (1r, 2x = 24)	This Vol.
Z.pauciflora (1r, 2x = 24)	This Vol.

[a]Gene control not known, D = dominant, r = recessive, po = polygene control.

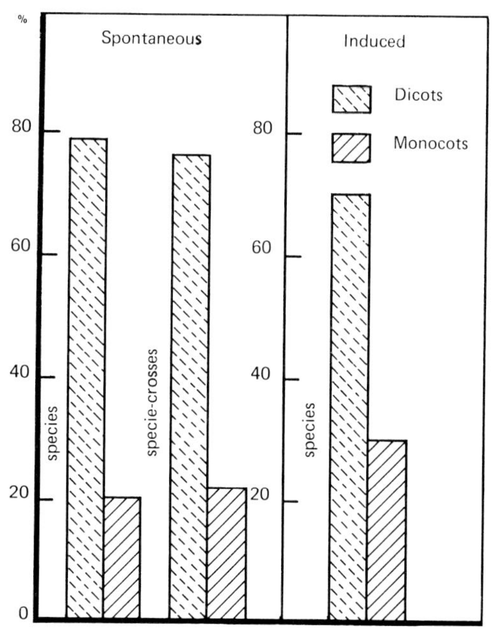

Fig. 2.1. Frequency distribution of spontaneously arisen and induced genic male sterile plants

high frequency of g-mst mutations and spontaneous recurrence of these mutations from time to time. Removal of such recessive *ms* mutant genes from a population is very slow. For instance, with over 50 nonallelic *ms* genes in tomato, the mutation rate towards mst equals 10^{-5}. In addition to heterozygote advantage, *ms* genes in barley and pea are adaptively favoured (Jain and Suneson 1966, this Vol.). This boosts the survival of the *ms* gene further and thus indicates the adaptive role that g-mst may have in regulating the gene recombination system in self-fertilizing species.

2.1.2 Species Crosses

Male sterility is confined not only to species, but appears in the offspring of many interspecific and some intergeneric crosses also (Table 2.2). Although it has been considered as g-mst by the authors, in most of them viz. *Epilobium hirsutum* × *E. roseum*, *E. montanum* × *E. parviflorum*, *E. roseum* × *E. parviflorum*, *E. roseum* × *E. luteum* and *Lolium multiflorum* × *L. perenne*, reciprocal crosses have not been done or studied. Therefore in these, cytoplasmic influence cannot be ruled out. Whereas in the interspecific crosses involving *Hebe*, *Hevea*, *Lactuca*, *Lolium* and *Vitis* the chromosome number of the two parents is the same, it differs in *Helianthus*, *Pelargonium* and *Viola* crosses (Table 2.2). Thus, in the latter species hybrids, the observed sterility could be due to meiotic anomalies caused by chromosomal imbalances and would exist in the reciprocal cross hybrids also. In the crosses where the chromosomal number of the parents is the same, genic causes of sterility cannot be eliminated. Therefore, unless reciprocal crosses also yield mst hybrids with unimpaired female fertility, all the examples given in Table 2.2 cannot be considered as confirmed cases of mst.

2.2 Mutagen-Induced

Both physical and chemical mutagens singly and/or in combination have produced mst mutants in about 35 species (Table 2.3). Of these, γ-rays have induced maximum mst mutations and colchicine and NEU the least (Fig. 2.2). However, this does not reflect the effectiveness or efficacy of these and other mutagens indicated in Fig. 2.2 as far as mst induction is concerned. Instead, the use of these mutagens has been limited and many mutagens have not been tested on the same species. If this is done, an effective and efficient mutagen for mst induction may be found. For instance, in barley (Yamashita and Ukai 1979) and rice (Ko and Yamagata 1980) ethylene imine induces more mst mutations than gamma rays. Thus, this chemical or any other mutagen may prove more potent for mst induction in a certain plant species and a certain mutagen may be species-specific. If so, mst induction may be directed and controlled. This needs investigation. In the majority of the species, the mst mutations induced are monogenic recessive; the EMS-induced mst mutations in alloplasmic *Triticum aestivum*, having *T. tauschii* cytoplasm, and in *Vicia faba* are, however, controlled by one dominant gene in each case.

Table 2.2. Genic male sterility in some species crosses

Crosses	References
a) Interspecific	
Epilobium hirsutum × *E.montanum* (2x = 36) (2x = 36)	Michaelis and Wertz (1935)
E.hirstum × *E.roseum* (2x = 36) (2x = 36)	Michaelis and Wertz (1935)
E.roseum × *E.hirsutum* (2x = 36) (2x = 36)	Michaelis and Wertz (1935)
E.montanum × *E.parviflorum* (2x = 36) (2x = 36)	Michaelis and Wertz (1935)
E.parviflorum × *E.montanum* (2x = 36) (2x = 36)	Oberreuter (1925)
E.parviflorum × *E.roseum* (2x = 36) (2x = 36)	Oberreuter (1925)
E.roseum × *E.parviflorum* (2x = 36) (2x = 36)	Oberreuter (1925)
E.luteum × *E.roseum* (2x = 36) (2x = 36)	Michaelis and Wertz (1935)
E.roseum × *E.luteum* (2x = 36) (2x = 36)	Michaelis and Wertz (1935)
Gossypium armourianum × *G.hirsutum* (2x = 26) (4x = 52)	Roux (1961)
Hebe leiophylla × *H.salicifolia* (4x = 80) (4x = 80)	Frankel (1940)
H.parviflora × *H.leiophylla* (4x = 80) (4x = 80)	This Vol.
Helianthus grossseserratus × *H.annuus* (1r, 2x = 32) (2x = 34)	Georgieva-Todorova (1974)
Hevea sprucaena × *H.brasiliensis* (4x = 36) (4x = 36)	Ramaer (1935)
[a]*Lactuca sativa* × *L.serriola* (2x = 18) (2x = 18)	Lindqvist (1960)
Lolium perenne × *L.multiflorum* (2x = 14) (2x = 14)	Ahloowalia (1981)
L.multiflorum × *L.perenne* (2x = 14) (2x = 14)	Ahloowalia (1981)
Lotus tenuis × *L.filicaulis* (2x = 12) (2x = 12)	de Nettancourt and Grant (1963)
Oenothera rubricalyx × *O.alata* (2x = 14) (2x = 14)	Gates and Thomas (1914)
O.suaveolens × *O.strigosa* (2x = 14) (2x = 14)	Oehlkers (1927)
Papaver somniferum × *P.setigerum* (2x = 22) (4x = 44)	Hrishi (1960)
Pelargonium denticulatum × *P.roseum* (10x = 90) (8x = 72)	Tokumasu (1974)
Viola arvensis × *V.tricolor* (2x = 34) (2x = 26)	Clausen (1926)
Vitis vinifera × *V.riparia* (2x = 38) (2x = 38)	Staudt (1973)
b) Intergeneric crosses	
[b]*Festuca pratensis* × *Lolium perenne* (2x = 14) (2x = 14)	Jenkin (1933)
L.multiflorum × *F.arundinacea* (2x = 14) (6x = 42)	Anderson (1966)
L.perenne × *F.arundinacea* (2x = 14) (6x = 42)	Anderson (1966)

Gene control of *mst* not known. [a]3r and [b]1–2r gene control.

Table 2.3. Induced genic male sterile mutants

Species	Mutagen	References
Abelmoschus esculentus (1r, 6x = 120)	γ-rays	Dutta and Nath (1974)
Antirrhinum majus L.(1r, 2x = 16)	X-, γ-rays	Baur (1924), Schick and Stubbe (1932), Goldsmith (1968), this Vol.
Arabidopsis thaliana (1r, 2x = 10)	Many	McKelvie (1963)
	X-rays	Müller (1968)
	EMS	Veen and Wirtz (1968)
Arachis hypogaea (4x = 40)	EB	Levi and Ashri (1975)
Brassica napus (1r, 2x = 38)	γ-rays	Takagi (1970)
Capsicum annuum (1r, 2x = 24)	X-and r-rays	Daskaloff (1968, 1971, 1971–72, 1973a,b, 1974)
	X-rays, EMS and r-rays	Pochard (1970a,b)
Carthamus tinctorius (1r, 2x = 24)	Colchicine	Heaton and Knowles (1982)
	γ-rays	Joshi et al. (1983)
Chionachne koenigii (1r, 2x = 20)	Colchicine	Rao (1975)
Chrysanthemum morifolium (6x = 54)	X-rays	Broertjes and Jong (1984)
Citrullus vulgaris (1r, 2x = 22)	X-rays	Watts (1962)
Corchorus capsularis (1r, 2x = 14)	X-rays	Rakshit (1967)
Cucumis sativus (1r, 2x = 14)	γ-rays	Whelan (1972)
Dactylis glomerata (4x = 28)	NEU, DES	Kremnina (1973)
Daucus carota (1r, 2x = 18)	NMU, DMS	Litvinova et al. (1980)
Glycine max (1r, 4x = 40)	γ-rays	Singh et al. (1974), Patil and Singh (1976), Jha and Singh (1978), Singh and Jha (1978)
Hordeum vulgare (1r, 2x = 14)	Atomic rays	Moh and Nilan (1953), Kamra and Nilan (1959)
	EI, γ-rays	Yamashita and Ukai (1979)
	EMS	Sharma and Reinbergs (1976)
	Acetone	Hockett and Eslick (1968)
	Radiation-induced	Tsuchiya (1969)
	γ-rays	Deshpande et al. (1981)
	γ-rays, EMS	Kumar and Singh (1983)
Lens culinaris (1r, 2x = 14)	γ-rays, NMU	Sharma and Sharma (1978)
Lycopersicon esculentum (1r, 2x = 24)	Radium	Lindström (1933)
	γ-rays	Yamakawa (1968), Tsikova and Abadzhieva (1974), Hernandez and Wollard (1978)
	32_p	Lesley and Lesley (1958)
	EI	This Vol.
Matthiola incana (2x = 14)	X-rays	Kappert (1956)
Nicotiana tabacum (1r, 4x = 48)	γ-rays	This Vol.
Oryza sativa (1r, 2x = 24)	EI	Fujimaki et al. (1977), Singh and Ikehashi (1979), Ko and Yamagata (1980)
	EMS	Nair and Gopalakrishnan (1973), Pavithran and Mohandas (1976)
	γ-rays	Fujimaki et al. (1977), Lu and Rutger (1984a) Mese et al. (1984)
	MNU	Kitada et al. (1983)
Papaver somniferum (2x = 22)	γ-rays	Singh and Khanna (1970), this Vol.
Phaseolus vulgaris (1r, 2x = 22)	EMS, γ-rays	Hussein and Disouki (1974), this Vol.

Table 2.3 *(continued)*

Species	Mutagen	References
Pennisetum typhoides (1r, 2x = 14)	X-rays	Krishnaswamy and Ayyangar (1942), Powell and Burton (1969)
Petunia hybrida (1r, 2x = 14)	X-rays	Welzel (1954)
	EMS, γ-rays	Cornu (1970), Cornu and Dommerguess (1974), Singh (1975)
Pisum arvense (1r, 2x = 14)	γ-rays, EMS	This Vol.
P. sativum (1r, 2x = 14)	X- and γ-rays	Gottschalk (1960, 1976), Klein and Milutinovic (1971), Gottschalk and Kaul (1974), this Vol.
Raphanus sativus (1r, 2x = 18)	EMS	This Vol.
Sorghum vulgare (1r, 4x = 20)	X-rays, colchicine	Barabas (1962)
	γ-rays	Andrews and Webster (1971)
Tagetes erecta (1r, 2x = 12)	EMS	This Vol.
Triticum aestivum (1r, 6x = 42)		
euplasmic	X-rays	Fossati and Ingold (1970), Kleijer and Fossati (1976)
alloplasmic	EMS	Franckowiak et al. (1976)
	NEU	Latypov (1974)
	γ-rays	Driscoll and Barlow (1976)
T. durum (1r, 4x = 28)	Neutron	Bozzini (1974)
T. monococcum (1r, 2x = 14)	X-rays	Smith (1938b)
Vicia faba (1r, 2x = 12)	EMS, EMS + EB	Picard et al. (1981), Duc (1982)
Vigna mungo (1r, 2x = 22)	DMS, γ-rays	Roy and Jana (1983)
V. sinensis (1r, 2x = 24)	EMS	This Vol.
Zea mays (1r, 2x = 20)	NEU, NMU	Mashnenkov and Golubovskaya (1980)

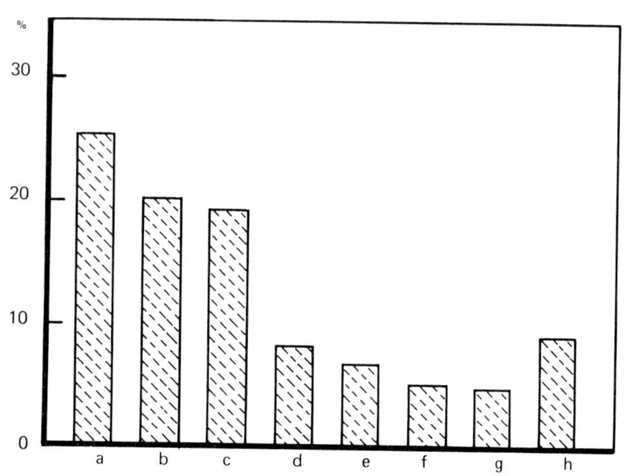

Fig. 2.2. Frequency distribution of male sterile mutants induced by γ-rays (*a*), X-rays (*b*), EMS (*c*), NMU (*d*), EI (*e*), colchicine (*f*), NEU (*g*) and others (*h*)

2.3 Chromosome Loss or Addition

In a self-fertilizing species, the *mft* gene(s) are always present on their normal complements. Loss of the chromosome carrying *mft* gene(s) results in mst. Due to the absence of many other vital traits present on the lost chromosome, such plants are weak, slow-growing and inviable. This is known in the deficient chromosome plants of diploid *Oenothera lamarckiana* (Pohl 1895), *Viola orphanidis* (Clausen 1930) and *Pisum sativum* (this Vol.). In contrast to this deficiency, the addition of a specific chromosome carrying a dominant mft gene suppressor induces mst in diploid *O. lamarckiana*.

2.4 Gene Control

Of the known 175 species of spontaneously arisen mst's, gene control of the sterility has been determined in only 58% of them (Fig. 2.3). In these, in about 64, 24, 10 and 2% of the species, the mst is controlled by single and many recessive genes, dominant genes and polygenes, respectively (Fig. 2.6). On the other hand, in the majority of the induced mutants (94%), the sterility is controlled by single recessive *ms* genes, in the remaining 6%, gene control is unknown (Fig. 2.3) (Tables 2.4, 2.5).

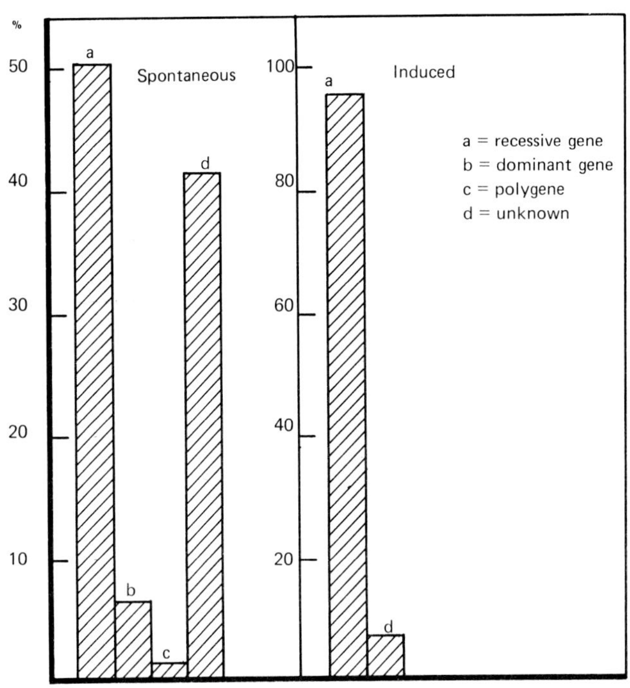

Fig. 2.3. Gene control in spontaneous and induced genic male-sterile plants

Table 2.4. Key to Tables 2.5 to 2.13

Symbol	Table	Gene control	Female fertility	Gene action	Specific gene control
a	2.5	+	+	+	Single recessive gene
b	2.6	+	+	–	Single recessive gene
c	2.7	+	+	±	Many recessive genes
d	2.8	+	+	±	Dominant genes
e	2.9	–	+	+	nk
f	2.10	–	+	–	nk
g	2.11	–	–	+	nk
h	2.12	–	–	–	nk
i	2.13	–	–	–	nk

(+ known, – unknown) (a-e confirmed, f-i unconfirmed cases of g-mst)

Table 2.5. Genic male sterility (confirmed[a]) controlled by single recessive (1r) genes

Species	References
Ageratum conyzoides (2x = 20)	This Vol.
A.houstonianum (2x = 20)	Reimann-Philipp and Fuchs (1971), this Vol.
Allium ascalonicum (2x = 16)	This Vol.
A.porrum (4x = 32)	Schweisguth (1970)
A.schoenoprasum (2x = 16)	This Vol.
[a]*Alopecurus myosuroides* (2x = 14)	Johnsson (1944)
Antirrhinum majus (2x = 16)	This Vol.
Arabidopsis thaliana (2x = 10)	Veen and Wirtz (1968)
Argemone mexicana (4x = 28)	This Vol.
Avena strigosa (2x = 14)	Sadanaga (1965)
Beta macrocarpa (2x = 18)	This Vol.
B.maritima (2x = 18)	Kinoshita (1971), Kinoshita and Takahashi (1972)
B.vulgaris (2x = 18)	Owen (1952), this Vol.
Brassica campestris (2x = 20)	Das and Pandey (1961), Chowdhry and Das (1966)
B.oleracea var.*capitata* (2x = 18)	Kototani and Yamada (1958), Nishi and Hiraoka (1958)
B.oleracea var.*gemmifera* (2x = 18)	Haigh (1954), Johnson and Faulkner (1956, 1957)
B.oleracea var.*italica* (2x = 18)	Cole (1959)
Cajanus cajan (2x = 22)	Reddy et al. (1978), Wallis et al. (1980), Dundas et al. (1981)
Capsicum annuum (2x = 24)	Hirose (1965), Shifriss and Frankel (1969), Pochard (1970b), Daskaloff (1971, 1971–1972, 1973a,b, 1974), Hirose and Fujime (1975, 1981), Meshram and Narkhede (1982)
Centaurea americana	This Vol.
C.imperialis	This Vol.
Centranthus ruber (2x = 14)	This Vol.
Cheiranthus cheiri (2x = 14)	This Vol.
Coleus hybridus (4x = 48)	This Vol.
Crotalaria juncea (2x = 16)	This Vol.
Cucumis melo (2x = 24)	Bohn and Whitaker (1949), Bohn and Principe (1964), Chauhan and Singh (1968)
C.sativus (2x = 14)	Whelan (1974)
Cucurbita maxima (4x = 40)	Rhodes (1959), Singh and Rhodes (1961), Francis and Bemis (1970) Kalyagin (1974)
C.pepo (4x = 40)	Eisa and Munger (1968)
Datura stramonium (2x = 24)	This Vol.
Eleusine indica (2x = 18)	This Vol.
Fagopyrum esculentum (2x = 16)	This Vol.
Glycine max. (4x = 40)	Brim and Young (1971)
Geranium himalayense	This Vol.
G.meeboldi	This Vol.

Table 2.5 *(continued)*

Species	References
Glycine max (4x = 40)	Brim and Young (1971), Singh et al. (1974), Bernard and Cremeens (1975), Patil and Singh (1976), Rubaihyo and Gumisiriza (1978), Albertson and Palmer (1979), Palmer et al. (1980), Stelly and Palmer (1980a,b, 1982), Delannay and Palmer (1982)
Helianthus annuus (2x = 34)	Kuptsov (1935), Pirev (1966), Putt and Heiser (1966), Nakashima and Hosokawa (1974a,b), Paun (1974)
Hordeum vulgare (2x = 14)	Kaul and Singh (1966), Singh and Kaul (1966), Roath and Hockett (1970), Sharma and Reinbergs (1972), Miah et al. (1974), Ahokas (1975a), this Vol.
Lathyrus luteus (2x = 14)	This Vol.
L.odoratus (2x = 14)	Bateson et al. (1903, 1908), Punnett (1923, 1925), Faberge (1937)
Lolium multiflorum (2x = 14)	Nitzsche (1971)
Lycopersicon esculentum (2x = 24)	Rick (1948), Rick and Butler (1956), Tsikova and Abadzhieva (1974), Zamir et al. (1980), this Vol.
Malus sylvestris (2x = 34)	Gagnieu (1951)
Melilotus alba (2x = 16)	This Vol.
Oenothera strigosa (2x = 14)	Oehlkers (1923, 1926), Harte and Bissinger (1952).
Origanum majorana (4x = 32)	This Vol.
Oryza sativa (2x = 24)	Kitada et al. (1983), Lu and Rutger (1984a,b)
Parthenium hysterophorus (2x = 18)	This Vol.
Pelargonium tomentosum (4x = 44)	This Vol.
Pennisetum typhoides (2x = 14)	Powell and Burton (1969), Rao and Koduru (1978a,b)
Petunia axillaris (2x = 14)	Levan (1942), this Vol.
Phaseolus atropurpureus (2x = 22)	This Vol.
P.vulgaris (2x = 22)	Wyatt (1984a,b)
Phleum nodosum (2x = 14)	May and Kasha (1971, 1980)
Pisum arvense (2x = 14)	This Vol.
P.sativum (2x = 14)	Klein (1969), Klein and Milutinovic (1971), Gottschalk and Kaul (1974), this Vol.
Primula sinensis (2x = 24)	This Vol.
Raphanus sativus (2x = 18)	Tokumasu (1957), Nishi and Hiraoka (1958)
Ricinus communis (2x = 20)	Claassen and Hoffman (1950)
Rumex acetosa (2x = 14)	Löve (1943)
Secale cereale (2x = 14)	Putt (1954), Adamĉhuk (1971, 1972), Kolobaeva (1974), Cebrat and Zadecka (1978), this Vol.
Silene hookeri (6x = 72)	This Vol.
Solanum nigrum (2x = 24)	This Vol.
S.tuberosum (4x = 48)	Young (1923)
Sorghum vulgare (4x = 20)	Ayyangar and Ponnaiya (1937), Stephens (1937)
Spinacea oleracea (2x = 12)	Kobabe (1967)
Triticum aestivum (6x = 42)	Lupton and Bingham (1966–67), Athwal et al. (1967), Krupnov (1968), Fossati and Ingold (1970), Lemekh et al. (1971), Jan (1974)
T.durum (4x = 28)	Bozzini and Scarascia-mugnozza (1968), Bozzini (1974)
T.monococcum (2x = 14)	Smith (1938a,b)
Vigna angularis (2x = 22)	Nakashima et al. (1980)
V.sinensis (2x = 24)	Sen and Bhowal (1962), Rachie and Rawal (1976)
Viola orphanidis (2x = 20)	Clausen (1930)
V.tricolor (2x = 26)	This Vol.
Zea mays (2x = 20)	Beadle (1931, 1932b), Emerson (1932), Mangelsdorf (1932), Rhoades and Rhoades (1939), Clark (1940), Nelson and Clary (1952), Weijer (1952), Golubovskaya (1977, 1979), Cheng et al. (1979), Albertsen and Phillips (1981), Coe et al. (1981) this Vol.

[a]Gene control, gene action and female fertility known.

Since most of the mutations are recessive and monogenic, the predominant single recessive gene control of induced as well as of spontaneously arisen mst mutants is justifiable. However, in certain plant species, the sterility is controlled by either recessive or dominant genes or by both, e.g. *Centranthus ruber, Daucus carota, Gossypium barbadense, G. hirsutum, Lactuca sativa, Phaseolus vulgaris, Solanum tuberosum, Trifolium pratense*. In *Medicago sativa* it is controlled by recessive and polygenes. *Triticum aestivum* represents a unique example in which the sterility is controlled by recessive, dominant and polygenes. However, the gene control appears to be specific so that the sterility in each mst genotype is conditioned by only one of the three gene types. In some species, the sterility is conditioned by one or more than one recessive or dominant genes or by polygenes (Tables 2.6 to 2.8). In those where the control is indicated to be by one to two genes or one to three or one to many genes (Table 2.7), the sterility is actually conditioned by two, three or many recessive genes, respectively. The reason for variable gene number reported by various authors is because the genotypes they studied had already a few *ms* genes in their genomes and they became mst when the remaining required number of genes were added to them. Thus control of g-mst is either monogenic, digenic, trigenic or multigenic and not mono- to di- or mono- to trigenic as visualised by some authors. In some populations of *Cheiranthus cheiri*, androecium is replaced by a cylindrical column bearing supernumerary pistils. Chittenden (1914) found monogenic control, Sirks (1924) obtained segregations indicating trigenic control for this mst. On the other hand, the microspore abortion in normally developed anthers is due to one recessive gene in this species (this Vol.). This indicates a variable gene control of mst in the species. In *Medicago sativa*, mst is conditioned by three genes, but it is inherited in a disomic manner (Childers and Mclennan 1960). In another strain, additive action of three genes leads to complete mst. In *Daucus carota* and *Lactuca sativa* mst results due to joint action of a dominant and a recessive *ms* gene (Table 2.8).

Table 2.6. Genic male sterility (confirmed[a]) conditioned by single recessive genes

Species	References
Antirrhinum majus (2x = 16)	Baur (1924), Schick and Stubbe (1932), Goldsmith (1968)
Arabidopsis thaliana (2x = 10)	McKelvie (1963), Müller (1968)
Arachis hypogea (4x = 40)	Hayes (1933), Patel et al. (1936), Hull (1937), Ashri (1968)
Begonia semperflorens (3x = 36)	Reimann-Philipp and Seidel (1963)
Brassica oleracea var. *botrytis* (2x = 18)	Nieuwhof (1961), Borchers (1966), Chatterjee and Swarup (1972)
B. oleracea var. *capitata* (2x = 18)	Rundfeldt (1960)
B. oleracea var. *gemmifera* (2x = 18)	Nieuwhof (1968a)
B. oleracea var. *italica* (2x = 18)	Sampson (1966a), Dickson (1970)
Cannabis sativa (2x = 20)	Pate and Joyner (1958)
Capsicum annuum (2x = 24)	Martin and Crawford (1951), Shifriss and Rylski (1972), Shifriss (1973)
Carthamus tinctorius (2x = 24)	Heaton and Knowles (1982)
Centrathus ruber (2x = 14)	This Vol.
Chenopodium quinoa (4x = 32)	Simmonds (1971)

Table 2.6 *(continued)*

Species	References
Chionachne koenigii (2x = 20)	Rao (1975)
Coleus blumei (1x, 4x = 24)	Ford (1950)
Corchorus capsularis (2x = 14)	Rakshit (1967), Mitra (1977)
Daucus carota (2x = 18)	Litvinova et al. (1980)
Dianthus chinensis (2x = 30)	This Vol.
Echinocloa colone (6x = 54)	Ayyangar and Srinivasa Rao (1941)
Glycine max (4x = 40)	Chaudhari and Davis (1977), Singh and Jha (1978)
Gossypium barbadense (4x = 52)	Yang (1982)
G.hirsutum (4x = 52)	Justus and Leinweber (1960), Justus et al. (1963), Seth et al. (1981), Tang et al. (1981), Yang (1982)
Helianthus annuus (2x = 34)	Kuptsov (1935), Anonymous (1964), Putt and Heiser (1966), Vranceanu (1970), Pogorletskii (1973), Kovačik and Skaloud (1973), Pogorletskii and Borlov (1973), Paun (1974)
Hordeum vulgare (2x = 12)	Kumar and Singh (1983)
Lactuca sativa (2x = 18)	Lindqvist (1960), Ryder (1967)
Lathyrus sativus (2x = 14)	Spirova (1978), Srivatsava and Somayajulu (1981)
Lycopersicon esculentum (2x = 24)	Crane (1915), Lindström (1933), Lesley and Lesley (1939), Lewis (1942), Roever (1948)
Malus sylvestris (2x = 34)	Heilborn (1932)
Oryza sativa (2x = 24)	Nagai (1926a,b), Ishikawa (1929), Miyazawa (1932), Takezaki (1932), Ramanujam (1935), Hara (1946), Jones (1952), Craigmiles et al. (1968), Hoff and Chandrapanya (1973), Shibuya (1973), Razzaque (1975), Trees and Rutgers (1978), Mahadevappa and Coffman (1980)
Papaver somniferum (2x = 22)	This Vol.
Paspalum sp. (2x = 30)	Burton (1943)
Pennisetum typhoides (2x = 14)	Krishnaswamy and Ayyangar (1941), Gill et al. (1973)
Phaseolus lunatus (2x = 22)	Allard (1953)
P.vulgaris (2x = 22)	Ibrahim et al. (1974), Hussein and Disouki (1974), Agbo and Wood (1977), Rheenen et al. (1979)
Pisum arvense (2x = 14)	Gritton (1980)
Plantago insularis (2x = 8)	Murr and Stebbins (1971)
Rubus idaeus (2x = 14)	Crane and Lawrence (1931)
Secale cereale (2x = 14)	Kobyliansky and Katerova (1973a), Garlicka and Madej (1975)
Sesamum indicum (2x = 26)	Osman and Yermanos (1982)
Sorghum vulgare (4x = 20)	Stephens (1937), Karper et al. (1938), Stephens et al. (1952), Kajjari and Chavan (1953), Barabâs (1962), Andrews and Webster (1971)
Trifolium pratense (2x = 14)	Smith (1971), Taylor et al. (1978)
Triticum aestivum (6x = 42)	Pugsley and Oram (1959), Latypov (1974), Driscoll and Barlow (1976), Kleijer and Fossati (1976), Sasakuma et al. (1978)
Vicia faba (2x = 12)	Bond et al. (1964a,b)
Zea mays (?x = 20)	Mashnenkov and Golubovskaya (1980)
Zinnia linearis (2x = 24)	This Vol.
Z.pauciflora (2x = 24)	This Vol.

[a]Gene control and female fertility known, gene action not known.

Table 2.7. Genic male sterility (confirmed[a]) conditioned by many recessive genes

Species	References
A. Gene control: 1-2r	
Carica papaya (2x = 18)	Storey (1953)
Lycopersicon esculentum (2x = 24)	Rick and Robinson (1951)
Pelargonium tomentosum (4x = 44)	This Vol.
Solanum tuberosum (4x = 48)	Young (1923)
B. Gene control: 2r	
Arachis hypogaea (2r, 4x = 40)	Patel et al. (1936)
Avena byzantina (6x = 42)	Garber (1922)
Brassica napus (2x = 38)	Heyn (1973)
Capsicum annuum (2x = 24)	Pochard (1970b), Hirose and Fujime (1975)
Carica papaya (2x = 18)	Storey (1953)
Crotalaria mucronata (2x = 16)	Edwardson (1967)
Eruca sativa (2x = 22)	Verma (1984)
Gossypium hirsutum (4x = 52)	Weaver (1968), Rhyne and Rhyne (1971), Kohel (1973), Murthi and Weaver (1974)
Lupinus regalis (4x = 48)	This Vol.
Lycopersicon esculentum (2x = 24)	Lesley and Lesley (1939), Solovjeva (1970)
Medicago sativa (4x = 32)	Childers (1952), this Vol.
Nicotiana tabacum (4x = 48)	Raeber and Bolton (1955), Bhat and Krishnamoorthi (1956[a])
Plantago lanceolata (2x = 12)	Ross (1969)
Solanum verrucosum (2x = 24)	Abdalla and Hermsen (1972)
Spinacea oleracea (2x = 12)	This Vol.
Trifolium pratense (2x = 14)	Kazimerski and Tsikova (1981)
Gene control: 1-3r	
Allium porrum (4x = 32)	Schweisguth (1970)
Arachis hypogaea (4x = 40)	Hayes (1933), Patel et al. (1936), Hull (1937), Ashri (1968)
C. Gene control: 2-3r	
Ageratum conyzoides (4x = 40)	This Vol.
Arachis hypogaea (4x = 40)	Ashri (1968)
Lavendula angustifolia (8x = 48)	This Vol.
Lycopersicon esculentum (2x = 24)	Lesley and Lesley (1939)
D. Gene control: 3r	
Cheiranthus cheiri (2x = 14)	Sirks (1924)
Lactuca sativa (2x = 18) × *L.serriola* (2x = 18)	Lindqvist (1960)
Medicago sativa (4x = 32)	Childers and McLennan (1960)
Pennisetum typhoides (2x = 14)	Gill et al. (1973)
Solanum verrucosum (2x = 24)	Abdalla (1970)
Triticum aestivum (6x = 42)	Jan (1974)
Zinnia elegans (2x = 24)	Duffy (1982), this Vol.
E. Gene control: 1-many r	
Alopecurus myosuroides (2x = 14)	Johnsson (1944)
Secale cereale (2x = 14)	Cebrat and Zadecka (1978)
F. Gene control: many r	
Solanum tuberosum (4x = 48)	Fukuda (1927)

Table 2.7 *(continued)*

Species	References
S. verrucosum (2x = 24)	Abdalla and Hermsen (1972)
Sorghum vulgare (4x = 20)	Stephens et al. (1952)
Vaccinium angustifolium (2x = 24)	Hall and Aalders (1961)
G. Gene control: polygenes	
Medicago sativa (4x = 32)	Childers and McLennan (1960)
Triticum aestivum (6x = 42)	Athwal et al. (1967)
H. Gene control: complex	
Secale cereale (2x = 14)	Putt (1954)

[a]Gene control and female fertility known, gene action either known or unknown.

Table 2.8 Genic male sterility (confirmed[a]) controlled by dominant genes

Species	References
Brassica napus (4x = 38, 1D)	Mathias (1985)
Centranthus ruber (2x = 14, 2D)	Sobrinho (1946)
Coleus blumei (4x = 24, 1D)	Rife (1940)
Daucus carota (2x = 18, 1D + 1r, 1D)	Thompson (1961), Hansche and Gabelman (1963a), Banga et al. (1964)
Gossypium barbadense (4x = 52, 1D)	Yang (1982)
G. hirsutum (4x = 52, 1D)	Allison and Fisher (1964), Weaver and Ashley (1971), Murthi and Weaver (1974), Bowman et al. (1978), Bowman and Weaver (1979)
Hordeum vulgare (2x = 14, 1D)	Deshpande et al. (1981)
Lactuca sativa (2x = 18, 1D + 1r, 1D)	Ryder (1963, 1971)
Limnanthes douglasii (2x = 10, 1D)	Jain et al. (1978)
Origanum vulgare (4x = 32, 2D)	Lewis and Crowe (1952, 1956)
Phaseolus vulgaris (2x = 22, 2D)	Mutschler and Bliss (1980)
Plantago lanceolata (2x = 12, 1D)	Ross (1969)
Solanum tuberosum (4x = 48, 1D)	Salaman (1910)
Streptocarpus hybrid (1D)	Zeven (1972a)
Trifolium pratense (2x = 14, 1D)	Nijdam (1932)
Triticum aestivum (6x = 42, 1D)	Sasakuma et al. (1978), Wang et al. (1980), Xaingmin and Herong (1981), Deng and Gao (1982)
Vicia faba (2x = 12, 1D)	Picard et al. (1981), Duc (1982)
Zea mays (2x = 20, 1D)	Sprague (1939), Schwartz (1951)

[a]Gene control and female fertility known, gene action known or unknown.

2.4.1 Epistatic and Inhibitory Genes

In *Centranthus ruber*, *Origanum vulgare* and *Phaseolus vulgaris*, the sterility is basically conditioned by a single dominant *Ms* gene but its expression is regulated by another dominant gene which dominates the *Ms* gene expression. Since the two genes are nonallelic, epistasis is involved. In *Origanum*, Lewis and Crowe (1956) proposed a digenic model in which gene *F* was supposed to cause anther abortion, but its action was suppressed by gene *H*. Sometimes many disrupted segregations occurred. These were interpreted as the result of a lethal genotype (*ffhh*) and the reduced viability of *HH* hermaphrodites. Similar double recessive lethals occur in *Phaseolus vulgaris*. However, in this, the action of a dominant *Ms* gene is inhibited by another dominant gene. On the other hand, a good fit to 13:3 ratio of mft:mst plants occurs in some *Lactuca sativa* breeding lines (Ryder 1963); the mst genotypes have *ms4 ms4 Ms5* genes. Segregations suggest inhibitory gene action.

2.4.2 Dominant Genes

Although the majority of mst mutations are recessive, a few are dominant and mostly monogenic (Table 2.8). The dominant *Ms* genes, like recessive *ms* genes, are mostly stage- and site-specific in their action on various microsporogenesis stages. In *Limnanthes*, the gene action is variable and microsporogenesis breakdown may occur at any stage. Influence of environment on *Ms* gene action is low and in a few is not known. In *Plantago lanceolata*, the influence seems to be high (Ross 1969). However, recent evidence indicates multigenic control of mst in this species. Tapetal behaviour in almost all of them is not known except in barley where it is persistent and in cotton in which it is defective. In *Daucus* and *Lactuca*, complementary action of a dominant and recessive gene transforms stamens into petals. This transformation is brought about by a single *Ms* gene in cotton.

2.4.3 Polyploidy

Whereas in about 130 diploids gene control of mst is ascertained, it is known only in about 45 polyploid species. In about 70% of diploid dicots and monocots, mst is controlled by single recessive genes (Fig. 2.5) and multigene control occurs in 23% of diploid dicots compared to 15% in diploid monocots (Figs. 2.5, 2.6). One to many recessive gene control is uniformly known in 2x and 4x monocot and dicot plant species but in all cases, dicots outnumber the monocots (Fig. 2.4). Dominant *Ms* genes are more prevalent in dicots in which polyploidy is known, e.g. *Centranthus ruber* (2x, 4x), *Coleus blumei* (4x), *Daucus carota* (2x), *Gossypium barbadense* (4x, 6x), *G. hirsutum* (4x, 6x), *Lactuca sativa* (2x, 4x), *Origanum vulgare* (4x), *Solanum tuberosum* (4x), *Trifolium pratense* (2x, 4x) *Triticum aestivum* (6x) etc., than in those lacking it. This is also evidenced by Fig. 2.5. However, all the inferences drawn above cannot be generalized, because compared to the high number of diploids in which mst is known, it has been detected only in a few polyploids. Moreover, determination of genetic control of sterility is easier in diploids than in polyploids because of the complex segregations and low seed fertility in the latter plant types.

2.4.4 Confirmed and Unconfirmed Male Steriles

Of the various known mst types (Table 1.1), genetic mst's have a high breeding value. This is because in addition to lack of development or functioning of male sex, these have the two basic characteristics viz. genetically conditioned mst, and unimpaired female fertility. However, both these features are investigated in only about 53% of the species (Tables 2.5 to 2.8, Fig. 2.7) and either one or both of these

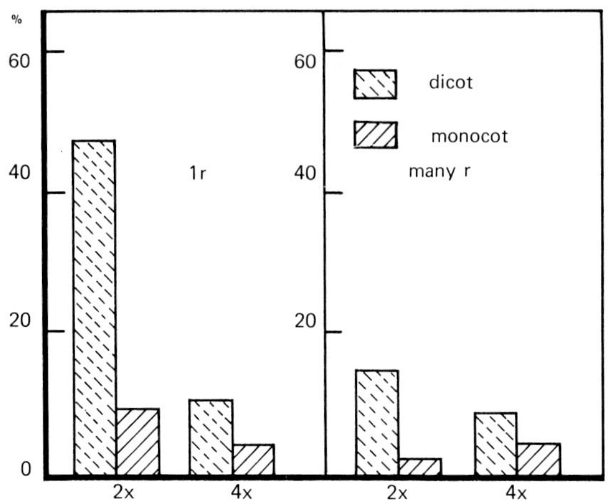

Fig. 2.4. Frequency distribution of recessive ms genes in diploid (2x) and tetraploid (4x) genic male steriles

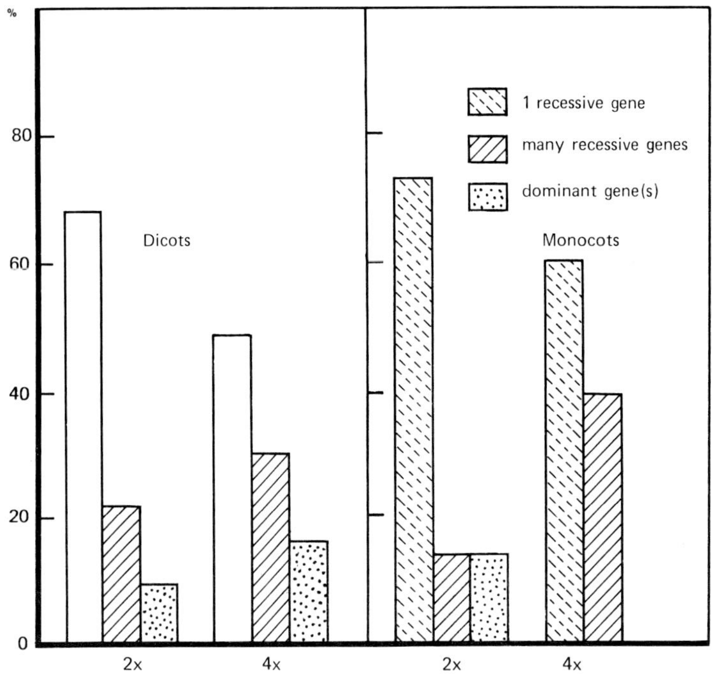

Fig. 2.5. Frequency of ms gene control in 2x and 4x angiosperms

remain undetermined in the remaining 47% of the species (Tables 2.9 to 2.12). Thus, while the former represent confirmed cases of g-mst, the latter are unconfirmed ones. In many of them, the time of action of *ms* genes is not known (Tables 2.7, 2.12). Based on this available information, Tables 2.5 to 2.13 have been prepared and a key is given in Table 2.4. Of the confirmed cases listed in Tables 2.5

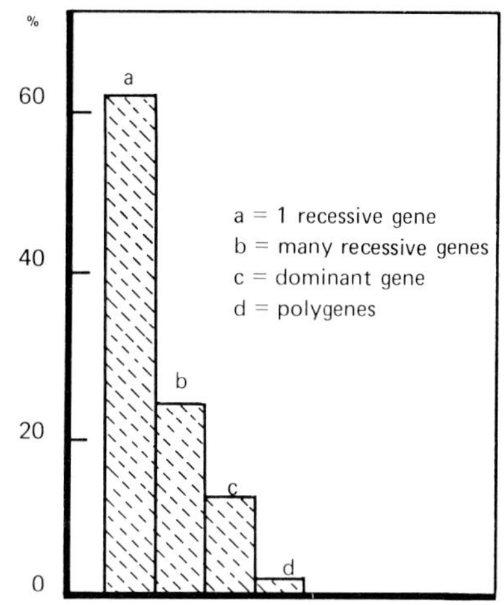

Fig. 2.6. Frequency distribution of gene control types in some genic male sterile plants

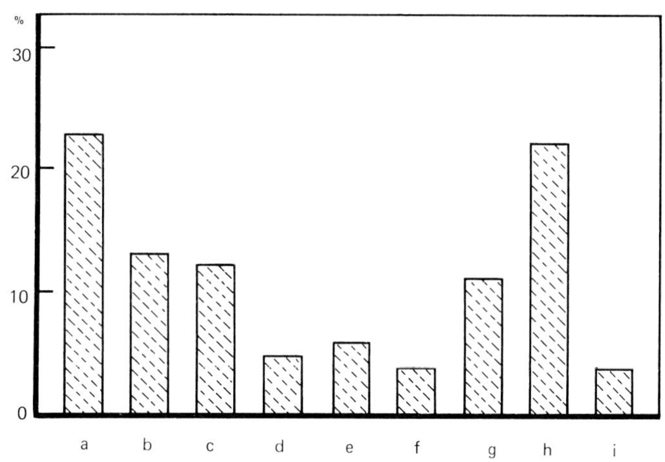

Fig. 2.7. Frequency of confirmed and unconfirmed cases of male sterility, for symbols *a-i* refer to Table 2.4

to 2.8, *ms* gene action is unknown in many of them and allelic nature of the different *ms* genes of a species is not determined in most of them. Their relative proportion is given in Fig. 2.7. The proportion of completely confirmed and investigated ones and of unconfirmed and uninvestigated ones is almost nearly equal (Fig. 2.7,a,h). The species listed in Tables 2.9 to 2.12 represent unconfirmed cases, as gene control of their anomaly is not known. In addition, the female fertility is not tested in the species listed in Tables 2.11 and 2.12. Therefore, unless the gene control and female fertility of the genotypes listed in Tables 2.9 to 2.12 are determined, these cannot be considered as g-mst mutants, as some or many of these could be gc-mst's, meiotic mutants, other genetic or nongenetic steriles. Species listed in Table 2.13, although considered to be mst's by various authors, are both male-female sterile mutants.

Table 2.9 Genic male sterility (unconfirmed[a])

Species	References
Allium sativum (2x = 16)	Novăk (1972), Koul and Gohil (1970), Konvička et al. (1978), this Vol.
Alopecurus pratensis (6x = 42)	This Vol.
Avena fatua (6x = 42)	This Vol.
Cajanus cajan (2x = 22)	Dundas et al. (1982)
Cannabis sativa (2x = 20)	Breslavetz (1935)
Crotalaria juncea (2x = 16)	Mitra (1976)
Cucumis sativus (2x = 14)	Shifriss (1950)
Dactylis glomerata (4x = 28)	Filion and Christie (1966), Kremnina (1973), Năsinec (1973), this Vol.
Daucus carota (2x = 18)	McCollum (1966), Litvinova (1973)
Geranium argentium (2x = 28)	This Vol.
G. grandiflorum	This Vol.
G. himalayense	This Vol.
Iris persica (2x = 26)	This Vol.
Melilotus alba (2x = 16)	Castetter (1925)
Nicotiana tabacum (4x = 48)	Raeber and Bolton (1955)
Papaver somniferum (2x = 22)	Singh and Khanna (1970)
Petunia hybrida (2x = 14)	This Vol.
Phaseolus atropurpureus (2x = 22)	Pritchard and Hutton (1972)
Plantago coronopus (2x = 10)	Paliwal and Hyde (1959), Gorenflot (1964), Ross (1970a)
Ranunculus acris (4x = 28)	Whyte (1929)
Saccharum officinarum (8x = 80)	Janaki-Ammal (1941)
Solanum tuberosum (4x = 48)	Breeze (1921), Fukuda (1927), Heyn (1930), Bleier (1931), Ellison (1936), Williams (1954), Schark (1957), Kishore et al. (1974)
Valeriana officinalis (2x = 14)	Shugaeva (1979)
Vigna mungo (2x = 22)	Roy and Jana (1983)

[a]Gene action and female fertility known, gene control unknown.

Table 2.10. Genic male sterility (unconfirmed[a])

Species	References
Cucurbita pepo (4x = 40)	Sinnott and Durham (1922), Shifriss (1945), this Vol.
Epilobium hirsutum (2x = 36)	Michaelis and Michaelis (1948)
E. hirsutum (2x = 36) × *E. roseum* (2x = 36)	Michaelis and Wertz (1935)
E. roseum (2x = 36) × *E. luteum* (2x = 36)	Michaelis and Wertz (1935)
Festuca arundinacea (6x = 42)	Goral et al. (1981)
Iris douglasiana (4x = 40)	Uno (1982)
I. pallida (2x = 24)	Simonet (1932)
Rhus integrifolia	Young (1972)
R. ovata	Young (1972)

[a]Female fertility normal, gene control and action unknown.

Table 2.11. Genic male sterility (unconfirmed[a])

Species	References
Allium ascalonicum (2x = 16)	Darlington and Haque (1955)
A.longicuspis (2x = 16)	Novâk (1972)
A.schoenoprasum (2x = 16)	Singh and Kobabe (1969)
Canna glauca (2x = 18)	Honing (1928), Öfferijns (1938)
Cirsium arvense (2x = 34)	Delannay (1979)
C.montanum (2x = 34)	Delannay (1979)
C.oleraceum (2x = 34)	Delannay (1979)
C.palustre (2x = 34)	Delannay (1979)
C.spinosissimum (2x = 34)	Delannay (1979)
Citrus aurantifolia (2x = 18)	Uphof (1931), Nakamura (1934), Krug and Bacchi (1943)
C.reticulata (2x = 18)	This Vol.
C.sinensis (2x = 18)	Oppenheim and Frankel (1929)
C.yatsushiro (2x = 18)	Nakamura (1943)
Centaurea scabiosa (2x = 20)	Fröst (1963)
Hebe leiophylla (2x = 20)	Frankel (1940)
H.parviflora (4x = 80)	Frankel (1940)
H.salicifolia (2x = 40)	Frankel (1940)
H. subalpina (4x = 80)	Frankel (1940)
H.townsonii (2x = 40)	Frankel (1940)
H.traversii (6x = 120)	Frankel (1940)
Hevea brasiliensis (4x = 36)	Ramaer (1935)
H.spruceana (4x = 36) × *H.brasiliensis* (4x = 36)	Ramaer (1935)
Kniphofia sp.	Moffet (1932)
Lathyrus odoratus (2x = 14)	Upcott (1937)
Lolium perenne (2x = 14)	Hayward and Manthiratna (1972)
Manihot esculenta (4x = 36)	Jos et al. (1967), Magoon et al. (1968)
Oenothera alata (2x = 14)	Pohl (1895), Gates (1907)
O.flava (2x = 14)	Oehlkers (1927)
O.gigas (4x = 28)	Gates (1911)
O.lamarckiana (2x = 14)	Oehlkers and Harte (1943)
Pelargonium crispum (2x = 22) var.*crispum minor* var.*lemon crispum*	Tokumasu (1974, 1976)
P.denticulatum (10x = 90)	Tokumasu and Tamai (1964)
P.radula (9x = 81)	Conagin (1964)
P.roseum (8x = 72)	Tamai et al. (1958)
Prunus mume (2x = 16)	Nakanishi (1982)
Saccharum officinarum (8x = 80)	Janaki-Ammal (1941), Purseglove (1972)
Salvia nemorosa (2x = 12)	Linnert (1955)
Taraxacum balticum (4x = 32)	Malecka (1971)
T.fulvum (4x = 32)	Gustafsson (1935)
T.nordstedti (6x = 48)	Gustafsson (1935)
T.pseudofulvum	Gustafsson (1935)
T.silesiacum (3x = 24)	Malecka (1971)
Thespesia populnea (2x = 26)	This Vol.
Tripsacum laxum (4x = 72)	Dodds and Simmonds (1946)
Vitis vinifera (2x = 38)	Olmo (1943)

[a]Gene action known, gene control and female fertility unknown.

Table 2.12. Genic male sterility (completely unconfirmed[a])

Species	References
Allium schoenoprasum (2x = 16)	Kobabe (1965)
Anacardium occidentale (4x = 48)	This Vol.
Brassica oleracea (2x = 18)	Rundfeldt (1960)
Capsella bursa-pastoris (4x = 32)	Shull (1927)
Castanea crenata (2x = 24)	Shull (1927)
C.dentata (2x = 24)	Jaynes (1963)
C.mollissima (2x = 24)	This Vol.
C.pumila (2x = 24)	This Vol.
C.sativa (2x = 24)	This Vol.
Chrysanthemum balsamita (2x = 18)	This Vol.
C.carinatum (2x = 18)	This Vol.
C.parthenium (2x = 18)	This Vol.
Cucumis sativus (2x = 14)	Barnes (1961), Groff and Odland (1963)
Dactylis glomerata (4x = 28)	Jenkin (1931b), Stapledon (1931)
Datura stramonium (2x = 24)	Avery et al. (1966)
Daucus carota (2x = 18)	Beiyernick (1885), Staes (1889), Warnstorf (1896), Knuth (1898), Rubashevskia (1931), Braak and Kho (1958), Thompson (1961), Konokov and Mokhov (1972)
Dianthus caryophyllus (2x = 30)	Reimann-Philipp (1976)
Eleusine coracana (4x = 36)	Ayyangar (1931)
Glycine max (4x = 40)	Stewart and Wentz (1926), Bernard and Jaycox (1969)
Gossypium herbaceum (4x = 52)	Kumar (1937)
Hevea brasiliensis (4x = 36)	Ramaer (1935)
Hibiscus esculentus (8x = 72)	Dutta and Nath (1974)
Iris douglasiana (4x = 40)	Uno (1982)
I.pallida (2x = 24)	Simonet (1932)
Jasminum pubescens (2x = 26)	This Vol.
Lolium perenne (2x = 14)	Jenkin (1931a,b), Hayward and Manthiratna (1972)
Malus sp. (2x = 34)	Gagnieu (1951)
Mangifera indica (4x = 40)	This Vol.
Medicago sativa (4x = 32)	Clarke and Fryer (1930), Engelbert (1932), Armstron and White (1935), Brink and Cooper (1936), Bolton and Fryer (1937), Julein (1944), Grün (1951), Sprague (1959), Pettem (1951)
Nicotiana tabacum (4x = 48)	Goodspeed (1930, 1954), Goodspeed and Avery (1939), Raeber and Bolton (1955), Bhat and Krishnamoorthy (1956)
Olea europea (2x = 46)	Villemur et al. (1984)
Parthenium argentatum (6x = 54)	Powers and Gardner (1945)
Petunia hybrida (2x = 14)	This Vol.
Phleum pratense (6x = 42)	Nordensköld (1945–1946)
Pistacia vera (2x = 30)	This Vol.
Plantago serraria (2x = 10)	Fröst (1959)
P.lagopus (3x = 12) *P.major* (2x = 12) *P.ovata* (2x = 8)	Knuth (1909), Chandler (1954), Atal (1958), Ross (1969, 1970a), Mittal and Issar (1970)
P.maritima (2x = 12)	Ross (1970a,b)
Polemonium caeruleum (2x = 18)	Ostenfelds (1923)
Prunus communis (2x = 16)	Crane (1915)
P.persica (2x = 16)	Knowlton (1924), Anonymous (1973b), Vitkovskii and Ruban (1974)

Table 2.12 *(continued)*

Species	References
Ranunculus acris (4x = 28)	Marsden-Jones and Turill (1925)
Raphanus sativus (2x = 18)	Ogura and Etoh (1975)
Rosaceae (many species)	(see Chap. 46)
Rubus sp. (2x = 14)	Lewis (1939)
Secale cereale (2x = 14)	Frimmel and Baranek (1935), Müntzing (1946), Putt (1954)
Sesamum indicum (2x = 26)	Kumar and Abraham (1941), Kumar and Rao (1945), Malaguti and Mazzani (1958), Dabral (1968), Dabral and Mandloi (1974)
Solanum integrifolium (2x = 24)	Habib and Rao (1973)
S.tuberosum (4x = 48)	East (1908), Dorsey (1919), Breeze (1921), Stout and Clark (1924), Stow (1926), Longley and Clark (1930), Krantz et al. (1939), Okuno (1952a)
Streptocarpus cyaneus	This Vol.
S.dunni (2x = 32)	This Vol.
S.kewensis (2x = 32)	This Vol.
S.luteus (2x = 32)	This Vol.
S.polyanthus (2x = 32)	This Vol.
S.saundersii (2x = 32)	This Vol.
S.watsonii (2x = 32)	This Vol.
Veronica alba	This Vol.
V.alpina (2x = 18)	This Vol.
V.candida	This Vol.
V.dubia	This Vol.
V.elegans	This Vol.
V.rosea	This Vol.
V.rubra	This Vol.
V.variabilis	This Vol.
Vitis vinifera (2x = 38)	Portele (1883), Rathay (1888–89), Olav (1930), Kaczmarek (1938), Olmo (1943), Neagum and Georgescu (1971), Negi et al. (1974), Singh et al. (1979)
Wistaria frutescens (2x = 16)	Roscoe (1927), this Vol.
W.floribunda (2x = 16)	Roscoe (1927), this Vol.
var.*alba* (2x = 16)	This Vol.
var.*rosea* (2x = 16)	This Vol.
W.macrostachya (2x = 16)	Roscoe (1927), this Vol.
W.sinensis (2x = 16)	Roscoe (1927), this Vol.
W.venusta (2x = 16)	Roscoe (1927), this Vol.
Zea mays (2x = 20)	Eyster (1921, 1933)

[a]Gene action, gene control and female fertility not known.

Table 2.13. Male-female sterile mutants[a]

Species	References
Antirrhinum majus (2x = 16)	Bergfeld (1956)
Avena byzantina (6x = 42)	Ramage and Suneson (1958)
Brassica oleracea (2x = 18) var.*capitata*	Nieuwhof (1961)
Cajanus cajan (2x = 22)	Deshmukh (1959)
Capsicum annuum (2x = 24)	Martin and Crawford (1951)
Carthamus tinctorius (2x = 24)	Claassen (1948, 1952), Carapetian and Knowles (1976)
Citrus sinensis (2x = 18)	Osawa (1912), Nakamura (1934)
C.unshiu (2x = 18)	Osawa (1912), Nagai and Tanikawa (1928)
Cucumis sativus (2x = 14)	Shifriss (1950)
Cucurbita maxima (4x = 40)	Hutchins (1944)
C.pepo (4x = 40)	Curtis and Scarchuk (1948), this Vol.
Dianthus caryophyllus (2x = 30)	Reimann-Philipp (1976)
Festuca pratensis (2x = 14) × *Lolium perenne* (2x = 14)	Jenkins (1933)
Lolium perenne (2x = 14)	Fejer (1958)
Lycopersicon esculentum (2x = 24)	Hafen and Stevenson (1956), Rick and Robinson (1951)

[a]Considered as male sterile by various authors.

2.5 Types

Based on anther development and phenotype, mst's have been classified into structural, sporogenous and functional types as detailed in Section 1.6.2. Examples for each type and specific anomalies in some of them are given here.

2.5.1 Structural

As mentioned earlier, the sterility is caused by alterations in differentiation and development of the androecium so that no viable pollen is produced. The majority of the species having this anomaly are diploids, few are tetraploids and only two are hexaploids (Table 2.14). The action of the *ms* gene is highly diverse, ranging from complete absence to the promotion of maldeveloped androecium. Thus, the stamens may be absent, aborted, deformed or transformed into other floral or foliar structures, the transformation being referred to as sepaloidy, petaloidy, staminoidy, carpeloidy and phyllody. These are given below.

2.5.1.1 Staminal Absence

In certain mutants of *Antirrhinum majus* (Schick and Stubbe 1932), *Gossypium hirsutum* (Allison and Fisher 1964), *Lycopersicon esculentum* (Hafen and Stevenson 1958; this Vol.), *Pisum sativum* (Gottschalk 1976; this Vol.), *Tagetes erecta* (Bolz 1961) stamens are absent. Whereas complete abortion of androecium

Table 2.14. Structural male sterile plants

Species	References
Abelmoschus esculentus (1r, 8x = 72)	This Vol.
[a]*Antirrhinum majus* (1r, 2x = 16)	Kuckuck and Schick (1930), Schick and Stubbe (1932), Sink (1966)
Begonia semperflorens (1r, 2x = 36)	Reimann-Philipp and Seidel (1963)
Brassic oleracea (1r, 2x = 20)	Nieuwhof (1961)
Carica papaya (2r, 2x = 18)	Storey (1953)
Castanea crenata (2x = 24)	Jaynes (1963)
C.dentata (2x = 24)	Jaynes (1963)
C.mollisima (2x = 24)	This Vol.
C.pumila (2x = 24)	This Vol.
C.sativa (2x = 24)	This Vol.
Cheiranthus cheiri (1r, 2x = 14)	Chittenden (1914), Meyer (1966c), this Vol.
C.cheiri (3r, 2x = 14)	Sirks (1924)
Chionachne koenigii (1r, 2x = 20)	Rao (1975)
Chrysanthemum balsamita (2x = 18)	This Vol.
C.carinatum (2x = 18)	This Vol.
C.parthenium (2x = 18)	This Vol.
Cucumis sativus (1r, 2x = 14)	This Vol.
Cucurbita maxima (1r, 4x = 40)	Scott and Riner (1946)
C.pepo (1r, 4x = 40)	Shifriss (1945), this Vol.
Daucus carota (2x = 18)	McCollum (1966)
Fragaria sp. (2x = 14)	Longley (1926)
Gossypium hirsutum (1D, 4x = 52)	Allison and Fisher (1964), Meyer (1967)
[a]*Hordeum vulgare* (1r, 2x = 14)	Suneson (1940), Moh and Nilan (1953), Kamra and Nilan (1959), Tsuchiya (1969)
Ipomoea sp. (1r, 2x = 30)	Correns (1926)
Jasminum sp. (2x = 26)	Raman (1955)
Lolium perenne (1r, 2x = 14)	Fejer (1966)
Lycopersicon esculentum (1r, 2r, 2x = 24)	Rick (1945), Rick and Robinson (1951), Bishop (1954), Hafen and Stevenson (1958), Dorossiev (1962), Bianchi and Soressi (1969), Simonov (1973), Sawhney and Greyson (1973b), Hernandez and Wollard (1978), this Vol.
L.esculentum (2x = 24)	Solovjeva (1970)
Manihot esculenta (4x = 32)	Jos et al. (1967), Magoon et al. (1968)
Nicotiana alata (1r, 2x = 18)	Krishnamurthy and Bhat (1959)
N.tabacum (2r, 4x = 48)	Raeber and Bolton (1955)
[a]*Papaver somniferum* (2x = 22)	Singh and Khanna (1970)
Paspalum sp. (1r, 2x = 20)	Burton (1943)
Pennisetum typhoides (1r, 2x = 14)	Goyal (1962), Manga (1977)
Petunia hybrida (1r, 2x = 14)	Sink (1973)
Pisum sativum (1r, 2x = 14)	Lamprecht (1933)
[a]*P.sativum* (2x = 14)	Gottschalk (1960, 1976), this Vol.
Primula officinalis (2x = 22)	Dahlgren (1932)
Prunus amygdalus (2x = 16)	Smith (1927)
P.persica (2x = 16)	Smith (1927)
Ricinus communis (1r, 2x = 20)	Claassen and Hoffman (1950)
Rubus idaeus (1r, 2x = 14)	Crane and Lawrence (1931)
Secale cereale (1r, 2x = 14)	Kubicki et al. (1981)
Silene sp. (1r, 2x = 24)	Correns (1928)
S.acaulis (2x = 24)	Bock (1976)
S.hookeri (1r, 6x = 72)	This Vol.

Table 2.14 *(continued)*

Species	References
Sorghum sudanensis (1r, 4x = 20)	Karper and Stephens (1936), Townsend (1960a,b), Jaisani and Drolson (1971, 1972)
[a]*S. vulgare* (1r, 4x = 20)	Barabâs (1962)
S. vulgare (1r, 4x = 20)	Karper and Stephens (1936), Ayyangar and Ponnaiya (1939), Ayyangar (1942)
Streptocarpus sp.	Oehlkers (1938)
Tagetes erecta (1r, 2x = 24)	Bolz (1961), Towner (1961), Horn (1974), this Vol.
Trifolium repens (4x = 32)	Kazimerski and Tsikov (1981)
Triticum aestivum (3r, 6x = 42)	Jan (1974), Jan and Qualset (1977b)
Zea mays (1r, 2x = 20)	Weatherwax (1925), Phipps (1928), Emerson et al. (1935)
Z. mays (1D, 2x = 20)	Sprague (1939)
Zinnia elegans (1r, 3r, 2x = 24)	Cachon (1980), Duffy (1982), this Vol.

[a]Mutagen-induced.

occurs in one *Cucurbita pepo* mutant (Shifriss 1945), in another the abortion is gradual (this Vol.). On the other hand, in the vestigeal glume maize mutant (Sprague 1939), outer glume is atrophied, stamens are developed, but the exposed anthers wither and fall off prematurely.

2.5.1.2 Faulty Stamen Development

This is known in a mutant of pea (Gottschalk and Kaul 1974) and tomato (Hafen and Stevenson 1958), in which rudimentary stamens are developed but no sporogenous tissue is formed. On the other hand, in a *Gossypium hirsutum* mutant, stamens are normally differentiated but their anthers remain rudimentary. In some they are absent. These anomalies are conditioned by a dominant gene, *Ms*4 (Allison and Fisher 1964). In some mutants of *Brassica oleracea* (Nieuwhof 1961), *Hordeum vulgare* (Suneson 1940) and *Lycopersicon esculentum* (Hafen and Stevenson 1958), anthers are poorly developed. In a pea and tomato mutant, stamens are developed but no archesporial tissue is formed (Gottschalk 1976; this Vol.; Sawhney and Greyson 1973a). In the tomato mutant, external ovules are borne on these aberrant stamens.

2.5.1.3 Staminal Transformation

Microsporophylls may be transformed fully or partially into foliar or floral organs other than stamens. With complete transformation microsporogenous tissue is not produced, but when transformation is partial, some pollen may be formed. Examples of the latter type are mostly excluded in this volume. Included are those in which stamens are transformed either into an intermediate to full petaloid or to carpeloid structures or into fertile carpels or normal-looking petals. In all cases the flowers are mst (Table 2.14). Carpeloidy of stamens occurs in *Antirrhinum majus*

(Bergfeld 1956), *Cheiranthus cheiri* (Chittenden 1914), *Gossypium hirsutum* (Meyer 1967), *Hordeum vulgare* (Moh and Nilan 1953; Tsuchiya 1969), *Lolium perenne* (Fejer 1966), *Nicotiana alata* (Krishnamoorthy and Bhat 1959), *Pennisetum typhoides* (Manga 1977), *Pisum sativum* (Gottschalk 1960), *Primula officinalis* (Dahlgren 1932), *Sorghum vulgare* (Barabâs 1962), and petaloidy in *Nicotiana tabacum* (Raeber and Bolton 1955; Krishnamoorthy and Bhat 1959). Petaloidy has been used in the production of double-flowered cultivars of ornamentals such as *Mathiola incana* and *Begonia semperflorens*. In some mst mutants, male sex suppression is associated with the absence of petals, e.g. antherless and apetalous mutants occur in thyme (Appl 1933), snapdragon (Sink 1966), zinnia (Goldsmith 1968), petunia (Sink 1973). In the *Silene* mutant polycarpellata (Correns 1928), both petals and anthers are transformed into carpels. In an atypical gynantherous mutant of *Cheiranthus cheiri*, the androecium is replaced by a cylindrical column bearing supernumerary pistils.

2.5.2 Sporogenous

Of the three sterility types, this mst type is the most prevalent in g-mst plants (Fig. 2.8, Table 2.15), thus indicating that most of the *ms* genes affect microsporogenesis or gametogenesis. In majority of such species, the *ms* genes act either during early or late meiosis, only a few act before premeiosis (Fig. 2.11). Thus, maximum *ms* genes act mostly during prophase I or during tetrad formation, a few being variable in action. The majority of the genes have arisen spontaneously, and a few have been obtained after mutagenic treatments (Tables 2.1, 2.3). The genes are most prevalent in diploids followed by tetraploids; in hexa- and octaploids, their frequency is very low. As the *ms* genes affect microsporogenesis and/or microgametogenesis, their details are given in Section 2.6. Except in *Epilobium* crosses and *Lolium perenne* × *L. multiflorum* cross, female fertility and mst in the reciprocal species cross hybrids listed in Table 2.15 are not known. Therefore, their inclusion is arbitrary.

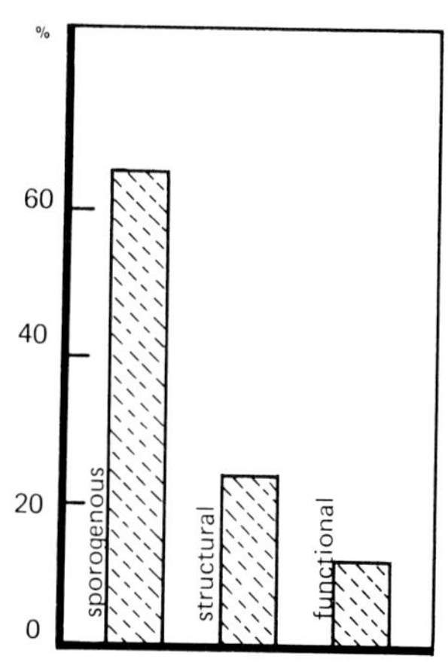

Fig. 2.8. Relative proportion of male sterility types in angiosperms

Table 2.15. Sporogenous male sterile plants

Species	References
Ageratum conyzoides (2–3r, 4x = 40)	This Vol.
A.conyzoides (1r, 2x = 20)	This Vol.
A.houstonianum (1r, 2x = 20)	Reimann-PhilippandFuchs(1971),thisVol.
Allium ascalonicum (1r, 2x = 16)	Darlington and Haque (1955), this Vol.
A.fistulosum (1r, 2x = 16)	Nishi and Hiraoka (1958)
A.longicuspis (2x = 16)	Novâk (1972)
A.porrum (1–3r, 4x = 32)	Schweisguth (1970)
A.sativum (2x = 16)	Novâk (1972), Koul and Gohil (1970), this Vol.
A.schoenoprasum (2x = 16)	Singh and Kobabe (1969), this Vol.
Alopecurus myosuroides (1-many r, 2x = 14)	Johnsson (1944)
A.pratensis (6x = 42)	This Vol.
Antirrhinum majus (1r, 2x = 16)	Baur (1924), Goldsmith (1968), this Vol.
[a]*Arabidopsis thaliana* (1r, 2x = 10)	Veen and Wirtz (1968)
Avena fatua (6x = 42)	This Vol.
Beta macrocarpa (1r, 2x = 18)	This Vol.
B.maritima (1r, 2x = 18)	Kinoshita (1971), Kinoshita and Takahashi (1972)
B.vulgaris (1r, 2x = 18)	Owen (1952), this Vol.
Brassica campestris (1r, 2x = 20)	Das and Pandey (1961), Chowdhry and Das (1966, 1968)
B.napus (1r, 2r, 4x = 38)	Takagi (1970)
B.oleracea var.*capitata* (1r, 2x = 18)	Nishi and Hiraoka (1958)
B.oleracea var.*gemmifera* (1r, 2x = 18)	Haigh (1954), Johnson and Faulkner (1956, 1957)
B.oleracea var.*italica* (1r, 2x = 18)	Cole (1959)
Cajanus cajan (1r, 2x = 22)	Reddy et al. (1978), Wallis et al. (1980), Dundas et al. (1981)
Canna glauca (2x = 18)	Honing (1928), Öfferijins (1938)
Cannabis sativa (1r, 2x = 20)	Pate and Joyner (1958)
Capsicum annuum (1r, 2x = 24)	Hirose (1965), Murty and Lakshmi (1979), Hirose and Fujime (1981), Meshram and Narkhede (1982)
Centaurea americana (1r)	This Vol.
C.imperialis	This Vol.
C.scabiosa (1r, 2x = 14)	Fröst (1963)
Citrus aùrantifolia (2x = 18)	Uphof (1931), Nakamura (1934), Krug and Bacchi (1943)
C.reticulata (2x = 18)	This Vol.
C.sinensis (2x = 18)	Osawa (1912), Nagai and Tanikawa (1928)
C.yatsushiro (2x = 18)	Nakamura (1943)
Coleus hybridus (1r, 4x = 48)	This Vol.
Cucumis melo (1r, 2x = 24)	Bohn and Whitaker (1949), Bohn and Principe (1964), Chauhan and Singh (1968)
C.sativus (1r, 2x = 14)	Shifriss (1950), Groff and Odland (1963), this Vol.
Cucurbita maxima (1r, 4x = 40)	Rhodes (1959), Singh and Rhodes (1961), Francis and Bemis (1970), Kalyagin (1974)
C.pepo (1r, 4x = 40)	Eisa and Munger (1968)

Table 2.15 *(continued)*

Species	References
Dactylis glomerata (4x = 28)	Filion and Christie (1966), Nâsinec (1973), this Vol.
Datura stramonium (1r, 2x = 24)	This Vol.
Daucus carota (1r, 2x = 18)	Litvinova (1973)
Dianthus chinensis (1r, 4x = 60)	This Vol.
Eleusine coracana (4x = 36)	Ayyangar (1931)
E.indica (1r, 2x = 18)	This Vol.
Fagopyrum esculentum (1r, 2x = 16)	This Vol.
Geranium argenteum (2x = 28)	This Vol.
G.himalayense (1r)	This Vol.
G.meeboldi (1r)	This Vol.
Glycine max (1r, 4x = 40)	Brim and Young (1971), Bernard and Cremeens (1975), Patil and Singh (1976), Jha and Singh (1978), Rubaihayo and Gumisiriza (1978), Albertsen and Palmer (1979), Palmer et al. (1980), Delannay and Palmer (1982), Stelly and Palmer (1982)
Gossypium hirsutum (1D, 4x = 52)	Richmond and Kohel (1961), Tang et al. (1981)
G.hirsutum (4x = 52)	Weaver (1968), Rhyne and Rhyne (1971), Kohel (1973), Murthi and Weaver (1974), Bowman et al. (1978)
Hebe parviflora (4x = 80)	Frankel (1940)
H.salicifolia (2x = 40)	Frankel (1940)
H.subalpina (4x = 80)	Frankel (1940)
H.townsonii (2x = 40)	Frankel (1940)
H.traversii (6x = 120)	Frankel (1940)
Helianthus annuus (1r, 2x = 34)	Pirev (1966), Putt and Heiser (1966), Paun (1974), Nakashima and Hosokawa (1974)
Hevea brasiliensis (4x = 36)	Ramaer (1935)
Hordeum vulgare (1r, 2x = 14)	Singh and Kaul (1966), Roath and Hockett (1970), Sharma and Reinbergs (1972), Mian et al. (1974), Ahokas (1975a), this Vol.
Iris persica (2x = 26)	This Vol.
Kniphofia sp. (2x = 12)	Möffet (1932)
Lathyrus luteus (1r, 2x = 14)	This Vol.
L.odoratus (1r, 2x = 14)	Gregory (1905), Punnett (1923, 1925), Faberge (1937)
Limnanthes douglasii (1D, 2x = 10)	Jain et al. (1978)
Lolium multiflorum (1r, 2x = 14)	Nitzsche (1971)
L.perenne (1r, 2x = 14)	Hayward and Manthiratna (1972)
Lupinus mutabilis (4x = 48)	Pakendorf (1970)
L.regalis (2r, 4x = 48)	This Vol.
Lycopersicon esculentum (1r, 2x = 24)	Rick and Robinson (1951), Rick and Butler (1956), Rick and Boynton (1967), Tsikova and Abadzhieva (1974), Zamir et al. (1980), this Vol.
Mannihot esculenta (4x = 32)	Jos et al. (1967), Magoon et al. (1968)
Medicago sativa (3r, 3po, 4x = 32)	Childers (1952)
Melilotus alba (2x = 16)	Castetter (1925)

Table 2.15 *(continued)*

Species	References
Mercurialis perennis (8x = 64)	This Vol.
Nicotiana tabacum (1r, 2r, 4x = 48)	Raeber and Bolton (1955), Bhat and Krishnamoorthi (1956), this Vol.
Oenothera alata (2x = 14)	Pohl(1895), Gates (1907)
O.flava (2x = 14)	Oehlkers (1927)
O.gigas (4x = 28)	Gates (1911)
O.lamarckiana (2x = 14)	Oehlkers and Harte (1943)
O.strigosa (2x = 14)	Oehlkers (1923, 1926), Harte and Bissinger (1952)
Origanum majorana (2D, 4x = 32)	This Vol.
Oryza sativa (1r, 2x = 24)	Kitada et al. (1983)
Papaver somniferum (1r, 2x = 22)	This Vol.
Parthenium hysterophorus (1r, 2x = 18)	This Vol.
Pelargonium crispum (2x = 22)	Tokumasu (1974, 1976)
P.denticulatum (10x = 90)	Tokumasu and Tamai (1964)
P.radula (9x = 81)	Conagin (1964)
P.roseum (8x = 72)	Tamai et al. (1958)
P.tomentosum (1–2r, 4x = 44)	This Vol.
Pennisetum typhoides (1r, 3r, 2x = 14)	Powell and Burton (1969), Rao and Koduru (1978a,b), Rao and Devi (1983)
Petunia axillaris (1r, 2x = 14)	Levan (1942), this Vol.
[a]*P.hybrida* (1r, 2x = 14)	Welzel (1954), Frankel (1962, 1971), Izhar (1973), Cornu and Dommergues (1974), Singh (1975), this Vol.
Phaseolus atropurpureus (1r, 2x = 22)	This Vol.
P.vulgaris (1r, 2D, 2x = 22)	Agbo and Wood (1977)
Phleum nodosum (1r, 2x = 14)	May and Kasha (1971, 1980)
[a]*Pisum arvense* (1r, 2x = 14)	This Vol.
[a]*P.sativum* (1r, 2x = 14)	Gottschalk (1961), Klein (1969), Gottschalk and Kaul (1974), this Vol.
Plantago coronopus (2x = 10)	Paliwal and Hyde (1959), Gorenflot (1964), Ross (1970a,b)
Prunus mume (2x = 16)	Nakanishi (1982)
Ranunculus acris (4x = 28)	Whyte (1929)
Rapanea thwaitessi (2x = 24)	Vasanthy and Pocock (1981)
Raphanus sativus (1r, 2x = 18)	Tokumasu (1957), Nishi and Hiraoka (1958)
Rumex acetosa (1r, 2x = 14)	Löve (1943)
Saccharum officinarum (8x = 80)	Janaki-Ammal (1941)
Salvia nemorosa (2x = 12)	Linnert (1955)
S.officinalis (2x = 14)	Linnert (1955)
Secale cereale (1-many r, 2x = 14)	Putt (1954), Adamčhuk (1971, 1972), Kolobaeva (1974), Cebrat and Zaděcka (1978)
Silene hookeri (1r, 6x = 72)	This Vol.
Solanum nigrum (1r, 2x = 24)	This Vol.
S.tuberosum (1-many r, 1D, 4x = 48)	Fukuda (1927), Heyn (1930), Bleier (1931), Ellison (1936), Williams (1954), Kishore et al. (1974)
S.verrucosum (2r, many r, 2x = 24)	Abdalla and Hermsen (1972)
Sorghum vulgare (1r, 4x = 20)	Ayyangar and Ponnaiya (1937)

Table 2.15 *(continued)*

Species	References
Spinacea oleracea (1r, 2x = 12)	Kobabe (1967)
Tagetes erecta (1r, 2x = 12)	This Vol.
Taraxacum balticum (4x = 32)	Malecka (1971)
T.fulvum (4x = 32)	Gustafsson (1935)
T.nordstedti (6x = 48)	Gustafsson (1935)
T.pseudofulvum	Gustafsson (1935)
T.silesiacum (3x = 24)	Malecka (1971)
Thespesia populnea (2x = 26)	This Vol.
Tripsacum laxum (4x = 72)	Dodds and Simmonds (1946)
Triticum aestivum (1r, 6x = 42)	Lupton and Bingham (1966–1967), Athwal et al. (1967), Krupnov (1968), Lemekh et al. (1971)
T.durum (1r, 4x = 28)	Bozzini and Scarascia-Mugnozza (1968)
T.monococcum (2x = 14)	Smith (1938b)
Vaccinium angustifolium (many r, 2x = 24)	Hall and Aalders (1961)
Valeriana officinalis (2x = 14)	Shugaeva (1979)
Vicia faba (1r, 2x = 12)	Bond et al. (1964a)
Vigna angularis (2x = 22)	Nakashima et al. (1980)
[a] *V.mungo* (2x = 22)	Roy and Jana (1983)
V.sinensis (1r, 2x = 24)	Sen and Bhowal (1962)
Viola orphanidis (1r, 2x = 20)	Clausen (1930)
Vitis vinifera (2x = 38)	Olmo (1943)
Zea mays (1r, 2x = 20)	Beadle (1932b), Albertsen and Phillips (1981), Cheng et al. (1979), this Vol.
Species hybrids	
Epilobium hirsutum (2x = 36) × *E.montanum* (2x = 36)	Michaelis and Wertz (1935)
E.hirsutum (2x = 36) × *E.roseum* (2x = 36)	Michaelis and Wertz (1935)
E.luteum (2x = 36) × *E.hirsutum* (2x = 36)	Michaelis and Wertz (1935)
E.roseum (2x = 36) × *E.hirsutum* (2x = 36)	Michaelis and Wertz (1935)
E.roseum (2x = 36) × *E.luteum* (2x = 36)	Michaelis and Wertz (1935)
E.montanum (2x = 36) × *E.parviflorum* (2x = 36)	Oberreuter (1925)
E.parviflorum (2x = 36) × *E.montanum* (2x = 36)	Oberreuter (1925)
E.parviflorum (2x = 36) × *E.roseum* (2x = 36)	Oberreuter (1925)
E.roseum (2x = 36) × *E.parviflorum* (2x = 36)	Oberreuter (1925)
Festuca pratensis (2x = 14) × *Lolium perenne* (2x = 14)	Jenkin (1933)
Hebe leiophylla (2x = 40) × *H.salicifolia* (2x = 40)	Frankel (1940)
Helianthus grosseserratus (2x = 32) × *H.annuus* (2x = 34)	Georgieva-Todorova (1974)
Lactuca sativa (2x = 18) × *L.serriola* (2x = 18)	Lindqvist (1960)
Lolium perenne (2x = 14) × *L.multiflorum* (2x = 14)	Ahloowalia (1981)
L.multiflorum (2x = 14) × *L.perenne* (2x = 14)	Ahloowalia (1981)
L.multiflorum (2x = 14) × *Festuca arundinacea* (6x = 42)	Anderson (1966)
L.perenne (2x = 14) × *F.arundinacea* (6x = 42)	Anderson (1966)

[a]Induced mst.

2.5.3 Functional

In the majority of the mst's, sterility arises spontaneously and is conditioned by single recessive *ms* genes (Table 2.16). Most of the species are diploid, a few tetraploid and only one is octaploid (viz. *Mimulus*). In all these, irrespective of ploidy level, anther development and differentiation of tapetal and sporogenous tissue are normal. However, the *ms* genes impair and inhibit anther dehiscence so that pollen is not released naturally. Artificial prodding of the anther thecae releases pollen which is viable and functional; but in soybean and a pea mutant, although viable pollen is produced, it remains glued together and is not released after shaking the pricked anther sacs. In some pea, tomato, carrot and soybean mutants, abnormal filaments are associated with anther nondehiscence. Likewise, in some tomato and pea mutants, deformed filaments create a spatial separation

Table 2.16. Functional male-sterile plants

Species	References
Brassica campestris (1r, 2x=20)	Chowdhry and Das (1968)
Carthamus tinctorius (1r, 2x=24)	Ebert (1964), Ebert and Knowles (1966)
Chrysanthemum morifolium (6x=54)	Broertjes and Jong (1984)
Daucus carota (2D, 2x=18)	Braak and Kho (1958), Hansche and Gabelmen (1963a)
Glycine max (2r, 4x=40)	Singh (1972), Johns and Palmer (1982)
[a]*G.max* (2r, 4x=40)	Singh and Jha (1978)
Hebe townsonii (2x=40)	Frankel (1940)
Hevea brassiliensis (4x=36)	Ramaer (1935)
Hordeum vulgare (2x=14)	Roath and Hockett (1971)
Lycopersicon esculentum (2r, 2x=24)	Currence (1944)
L.esculentum (1r, 2x=24)	Crane (1915), Rick (1947a), Larsen and Paur (1948), Roever (1948), Rick and Robinson (1951), Mittal and Thomas (1961), Tronichkova (1962), Clayberg et al. (1966)
Manihot esculenta (4x=32)	Jos and Bai (1981)
Mimulus luteus (1r, 8x=64)	Brozek (1926)
Oryza sativa (1r, 2x=24)	Hoff and Chandrapanya (1973), this Vol.
Phaseolus vulgaris (2x=22)	Wyatt (1984a,b)
[a]*Pisum sativum* (1r, 2x=14)	Gottschalk (1976), this Vol.
Sesamum indicum (1r, 2x=26)	Brar (1982)
Solanum melongena (1r, 2x=24)	Jasmin (1954)
Streptocarpus (cv. Constant Nymph)	Marston (1964–65), Zeven (1971a,b)
S.cooksonii	Story (1955), Hillard (1966), this Vol.
S.cyaneus	Story (1955), Hillard (1966), this Vol.
S.kentaniensis	Story (1955), Hillard (1966), this Vol.
S.primulifolius	Story (1955), Hillard (1966), this Vol.
[a]*S.hybrid* (cv. Constant Nymph)	Zeven (1972)
Trifolium repens (1r, 4x=32)	Atwood (1944)
Vigna sinensis (1r, 2x=24)	Rachie and Rawal (1976)
[a]*V.sinensis* (1r, 2x=24)	This Vol.
Vitis vinifera (2=28)	Dorsey (1914), Susa (1927), Olmo (1943)
Zea mays (1r, 2x=20)	Beadle (1932b), Sprague (1939)

[a]Mutagen-induced.

between anthers and stigma. In pea and soybean mutant (this Vol.; Johns and Palmer 1982) filaments fail to elongate, and in another pea mutant, filaments elongate normally but collapse before anthesis (this Vol.). In a tomato mutant, the mutant gene *ps* (Larsen and Paur 1948) causes stomium cells to remain tenacious, the abnormally hooked petals hold anthers excessively close to the carpel and inhibit corolla unfurling. In mutants of eggplant (Jasmin 1954), barley (Roath and Hockett 1971), rice (Hoff and Chandrapanya 1973), pea (this Vol.) and tomato (Roever 1948; this Vol.), stomium is absent. This prevents anther dehiscence.

In tomato, the mutant gene Cl_2 (Rick and Robinson 1951) not only prevents stomial opening, but the flowers remain cleistogamous and develop abnormal corolla. In another such tomato mutant, an unusually extended style completely separates stigma from anthers (Currence 1944). Some mutant plants of tomato (Crane 1915) and grapes (Dorsey 1914; Susa 1927; Olmo 1943) develop reflexed stamens while their stigmas extrude normally. This anatomical defect leads to anther nondehiscence and the viable pollen remains locked up inside the anthers. In mst mutants of white clover (Atwood 1944), maize (Sprague 1939), tomato (Clayberg et al. 1966), grapes (Olmo 1943), rubber (Ramaer 1935), the anthers are reflexed or hard and the filaments do not elongate. In a rubber plant, the entire whorl of stamens is transformed into a hard central column and no anther dehiscence ever occurred in this mutant plant (Ramaer 1935). The sterility of indirect form occurs in *Mimulus luteus* var *tigrinoides*, in which petaloid organs (paracolla) develop within anthers separating them from the style (Brozek 1926) and prevent selfing. Similar indirect fn-mst occurs in *Cirsium arvense* in which transformed stigma and absence of papillae on the stigma prevent pollen fixation and germination (Delannay 1977) even though a normal embryo sac is present. The anomaly is conditioned by a dominant gene and cannot be considered as a typical example of fn-mst since pollen is shed normally but selfing is prevented due to stigma anomaly.

2.5.4 Distributional Pattern

In majority of the angiospermic species mst has arisen spontaneously (Table 2.1) and in only a few it has been induced after mutagen treatments (Table 2.3). Of the various mst types, the sporogenous is the most prevalent and the least is functional (Fig. 2.8). This confirms my earlier contention that most of the *ms* genes affect microsporogenesis in higher plants. These as well as other *ms* genes are more prevalent in dicots than in monocots (Fig. 2.9). This could be due to (a) the relatively higher number of dicots than monocots existent in nature (b), the greater mutation rate to mst in dicots than monocots or (c), the higher number of dicots investigated for mst compared to monocots, or (d) some or all these three (a-c). However, the least frequency of fn-msts in both dicots and monocots (Fig. 2.9) may not entirely be due to low mutation frequency of genes but due to their difficult detection in flowering plants as their anther phenotype is normal. In obligate selfers, such fn-mst mutants are seed-sterile but some seed set occurs over them following outcrossing. Barring a low proportion of sp-mst's in species hybrids, most

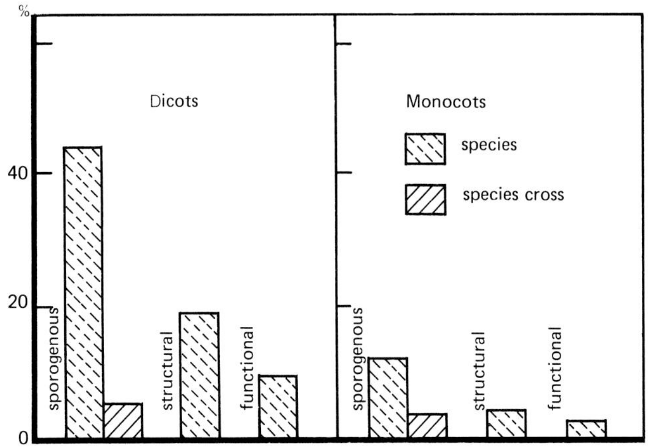

Fig. 2.9. Distribution of genic male sterility types in dicots and monocots

of which are unconfirmed, st- and fn-mst's are almost absent in hybrid plants (Fig. 2.9). This is because the majority of the g-mst's arise spontaneously due to the mutation of single dominant genes to the recessive state. Therefore, monogenic recessive control is most predominant in all three mst types (Fig. 2.5). Dominant gene control occurs in very few species (Table 2.8) and polygenes do not seem to control st- or fn-mst. Furthermore, 70–80% of the diploids comprise a mst type, the remaining being most tetraploids (Fig. 2.10); hexa- and octaploids are extremely rare. This extremely low number of mst's in polyploids is due to their multiple genome that buffers the mutation rate and suppresses *ms* mutant gene expression. Moreover, most of the polyploids, especially autoploids, are infertile. Hence, the mst's are quickly lost due to low female fertility.

2.6 Gene Action

Higher plants, including dioecious species, are potentially bisexual. Although flowers of monoecious plants also have bisexual potentialities, their sex expression depends on the suppression of one or the other type of sex. In the majority of such species, the potentiality of sex expression is genetically controlled, but the degree of its expression is environmentally influenced. Likewise, maintenance of bisexuality is gene-controlled and most of the genes have to be present in a dominant state. Mutation of one or more of these genes results in male or female sterility. Such genes are designated as sterility genes. Whereas some of them cause male-female sterility, others produce only male sterility (mst) and a few only female sterility (fst). In their mode of action most of the mst genes, like other mutant genes, are time-, stage- and site-specific. Due to their specificity they induce mst which assumes diverse expressions in higher plants, some of which are given below:

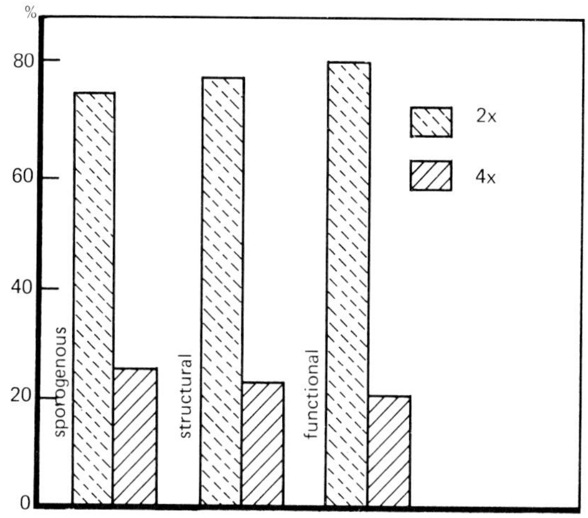

Fig. 2.10. Male sterility types in diploid and tetraploid plants

2.6.1 Male Sex Suppression

The earliest action of *ms* genes is during anther development and differentiation so that either no anthers are formed or they are mal- or mis-formed. The anther cells may develop supernumerary petals or stigmoid anther tips. Such antherless mutants occur in many plant species viz. maize, barley, pea, tomato, sorghum, tobacco etc. (Table 2.14). In all of them, the anomaly is caused by single recessive genes. The gene expression is modified slightly by environmental factors, but highly by exogenous application of growth regulators like GA in tomato (Phatak et al. 1966).

2.6.2 Defective Microsporangium Development

Instead of complete staminal suppression, the *ms* genes may cause abnormal or irregular staminal development or its transformations into other floral organs. The anther sacs are either vestigeal, rudimentary, warty, or lack microsporogenous tissue. These are included in structural mst (Sect. 2.5.1).

2.6.3 Irregular Sporocyte Formation

Normal anther development and its wall differentiation are prerequisites to sporocyte formation and sporogenous tissue differentiation. Both wall layers and sporogenous cells initiate from the archesporial cells. Some *ms* genes suppress or alter this initiation. This leads to the development of abnormal, nonfunctional sporogenous tissue as in two mutants of pea and tomato. In them the anomaly is conditioned by a single recessive mutant gene which causes disturbances in sporogenous cell differentiation so that no PMC's are differentiated (this Vol.). In a few tomato, sweet pea, and garden pea mutants, the *ms* gene causes abnormal archesporial differentiation so that PMC's are not developed (Rick 1948; Faberge 1937; Gottschalk and Kaul 1974).

2.6.4 Abnormal Microsporogenesis

In some plant species, sexual organ development is normal but the *ms* genes impair microsporogenous tissue differentiation and development. Normally, microsporogenous cells undergo a series of changes during their differentiation from the meristematic state to the formation of PMC's, tetrads and finally pollen grains. The changes, although continuous, have presently been differentiated into premeiosis, meiosis and postmeiosis stages. During these changes, the *ms* genes interrupt or disturb the developmental sequence either completely or partially, due to which either microsporogenesis is inhibited or disturbed so that abortive, inviable or nonfunctional pollen is produced.

2.6.4.1 Premeiosis

In a normally developed anther, sporogenous cells usually divide by asynchronous mitoses. When a certain number of PMC's is formed, the somatic mitoses stop and premeiotic interphase begins which is followed by meiosis in the PMC's. Many *ms* genes act before meiotic initiation (Table 2.17). The earliest action of *ms* genes occurs before part of the cells of the microsporangium are transformed into meiotic cells. Therefore, in such plants no archesporial tissue is differentiated. In a few plants, *ms* gene acts only after the archesporial tissue is differentiated. In a relatively large number of plants, PMC's are formed but they degenerate before meiosis. In these, the accumulation of RNA and proteins is inhibited in PMC's and the switchover to meiosis does not occur. In all the cases of premeiotic *ms* gene action, the tapetum may develop normally or abnormally.

2.6.4.2 Meiosis

Different *ms* genes act at almost all these stages (Table 2.18) but most of them act either at the beginning or at the end of meiosis (Fig. 2.11). Determination of the precise time of action of *ms* genes over each phase of prophase I is hampered due to tangled nature of the chromosomes. Many *ms* genes act during the pachytene and diplotene, only a few act during zygotene and diakinesis. In many others, the precise time of action over prophase I is not known. These *ms* genes either interrupt or inhibit pachytene pairing or suppress diplotene-chiasmata formation in addition to causing asynapsis, desynapsis, sticky chromosomal formations. The tapetum is developed either normally or abnormally. Relatively fewer *ms* genes act during MI (Table 2.18). They either impair or inhibit congression, orientation or distribution of bivalents by causing spindle anomalies, or act directly over the chromosome form or function. Very few *ms* genes act during AI, interphase II and

Table 2.17. *ms* gene action over premeiosis and tapetum

Species	Tapetum	References
Premeiotic stage		
No archesporial tissue formed		
Cheiranthus cheiri (1r, 2x = 14)	–	This Vol.
Gossypium hirsutum (1D, 4x = 52)	b	Allison and Fisher (1964)
Iris persica (2x = 26)	–	This Vol.
Lathyrus odoratus (1r, 2x = 14)	a	Gregory (1905), Faberge (1937)
Papaver somniferum (2x = 22)	–	Singh and Khanna (1970)
Pennisetum typhoides (1r, 2x = 14)	–	Rao and Koduru (1978a,b), Rao and Devi (1983)
Pisum sativum (1r, 2x = 14)	a	Gottschalk (1961), Klein and Milutinovic (1971), Gottschalk and Kaul (1974), this Vol.
Secale cereale (1-many r, 2x = 14)	b	Cebrat and Zadecka (1978)
Silene hookeri (1r, 6x = 72)	–	This Vol.
Solanum tuberosum (1r, 1D, 4x = 48)	–	Salaman (1910), Young (1923), Fukuda (1927)
Zea mays (1r, 2x = 20)	–	Beadle (1932b)
Archesporial tissue formed, no PMC's differentiated		
Cirsium arvense (2x = 34)	f	Delannay (1979)
C.montanum		Delannay (1979)
Epilobium parviflorum (2x = 36) × *E.montanum* (2x = 36)	a	Oberreuter (1925)
E.parviflorum (2x = 36) × *E.roseum* (2x = 36)		Oberreuter (1925)
Oryza sativa (1r, 2x = 24)	–	Lu and Rutger (1984)
Pisum sativum (1r, 2x = 14)	d	This Vol.
PMC's degenerate before meiosis		
Capsicum annuum (1r, 2x = 24)	–	Hirose (1965)
Cirsium oleraceum (2x = 34)	f	Delannay (1979)
[a]*Citrus aurantifolia* (2x = 18)	d	Uphoff (1931), Nakamura (1934), Krug and Bacchi (1943)
[a]*C.unshiu* (2x = 18, 36)	–	Osawa (1912), Nagai and Tanikawa (1928)
Cucurbita maxima (1r, 4x = 40)	d	Rhodes (1959), Singh and Rhodes (1961)
Fagopyrum esculentum (1r, 2x = 16)	e	This Vol.
[a]*Geranium argenteum* (2x = 28)	–	This Vol.
Hordeum vulgare (1r, 2x = 14)	–	This Vol.
Limnanthes douglasii (1D, 2x = 10)	–	Jain et al. (1978)
Lycopersicon esculentum (1r, 2x = 24)	a	Rick and Butler (1956), Zamir et al. (1980), this Vol.
[a]*Oenothera gigas* (4x = 28)	–	Gates (1911)
Pisum sativum (1r, 2x = 14)	–	Gottschalk (1961), Gottschalk and Kaul (1974), Klein and Milutinovic (1971), this Vol.
Rubus idaeus (1r, 2x = 14)		Crane and Thomas (1949)
[a]*Solanum tuberosum* (4x = 48)	–	Breeze (1921), Fukuda (1927), Heyn (1930)
Sorghum vulgare (1r, 4x = 20)	–	Ayyangar and Ponnaiya (1937), Stephens (1937)
Zea mays (1r, 2x = 20)	a	Beadle (1932b), Palmer (1971), Albertsen and Phillips (1981), this Vol.

[a]Tapetum normal (a), abnormal (b), persistent (c), disintegrates later (d), simultaneously (e) or before (f), sporogenous tissue breakdown; in others not known (–), female fertility not known (g) or reduced (h).

Table 2.18. *ms* gene action over meiosis tapetum[a]

Species	Tapetum	References
Meiotic stage		
During zygotene		
Crotalaria juncea (1r, 2x = 16)	—	This Vol.
Medicago sativa (3 po, 4x = 32)	b	Childers (1952)
Melilotus alba (2x = 16)	—	This Vol.
Oryza sativa (1r, 2x = 24)	—	Kitada et al. (1983)
Primula sinensis (1r, 2x = 24)	—	This Vol.
Zea mays (1r, 2x = 20)	a	Beadle (1932b)
During pachytene		
Cajanus cajan (1r, 2x = 22)	b,g	Dundas et al. (1982)
Glycine max (1r, pmst, 4x = 40)	—	Jha and Singh (1978), Stelly and Palmer (1980a,b, 1982)
Hebe parviflora (4x = 80)	—	Frankel (1940)
H.subalpina (4x = 80)	—	
Helianthus annuus (1r, 2x = 34)	a	Paun (1974)
Jasminum pubescens (2x = 26)	—	This Vol.
Lathyrus odoratus (1r, 2x = 14)	a,g	Upcott (1937)
Phleum nodosum (1r, 2x = 14)	d	May and Kasha (1971, 1980)
Rumex acetosa (1r, 2x = 14)	—	Löve (1943)
Salvia nemorosa (2x = 12)	a,g	Linnert (1955)
Viola orphanidis (1r, 2x = 20)	—	Clausen (1930)
Zea mays (1r, 2x = 20)	a	Beadle (1932b), Albertsen and Phillips (1981), this Vol.
During diplotene		
Canna glauca (2x = 18)	g	Honning (1928), [g]Öfferijns (1938)
Capsicum annuum (1r, 2x = 24)	h	Murty and Lakshmi (1979)
Lycopersicon esculentum (1r, 2x = 24)	a or c	Rick and Butler (1956)
Nicotiana tabacum (1r, 4x = 48)	a	This Vol.
During diakinesis		This Vol.
Lycopersicon esculentum (1r, 2x = 24)	a	This Vol.
Petunia hybrida (1r, 2x = 14)	b	Welzel (1954)
During prophase-I		
Alopecurus pratensis (6x = 42)	a	This Vol.
Cannabis sativa (2x = 20)	—	Breslavetz (1935)
Capsicum annuum (2x = 24)	—	Pochard (1970b)
Centaurea americana (1r)	a	This Vol.
C.imperialis (1r)		This Vol.
Coleus hybridus (1r, 8x = 48)	a	This Vol.
Cucurbita maxima (4x = 40)	—	Kalyagin (1974)
Hordeum vulgare (1r, 2x = 14)	a	Kaul and Singh (1966), this Vol.
	c	Deshpande et al. (1981)
Iris pallida (2x = 24)	—	Lowig (1928)
Lycopersicon esculentum (1r, 2x = 24)	a or c	Rick and Butler (1956)
	d	Rick and Butler (1956), this Vol.
Medicago sativa (3 po, 4x = 32)	b	Childers (1952)
Mercurialis perennis (8x = 64)	a	This Vol.
Pelargonium denticulatum (10x = 90)	—	Tokumasu and Tamai (1964)
Petunia hybrida (1r, 2x = 14)	b,c	Welzel (1954)
	—	Cornu and Dommerguess (1974), this Vol.

Table 2.18 *(continued)*

Species	Tapetum	References
Pisum sativum (1r, 2x = 14)	–	Gottschalk and Kaul (1974), this Vol.
Raphanus sativus (1r, 2x = 18)	b	Tokumasu (1957)
Secale cereale (1-many r, 2x = 14)	–	Putt (1954), Kolobaeva (1974)
	b	Cebrat and Zadecka (1978)
Solanum tuberosum (1r, 4x = 48)	–	Fukuda (1927), Heyns (1930), Bleier (1931), Ellison (1936), Williams (1954), Schark (1957)
Taraxacum balticum (4x = 32)	f	Gustafsson (1935)
T.silesiacum (3x = 24)	b,g	Malecka (1971)
Triticum aestivum (1r, 6x = 42)	–	Lemekh et al. (1971), Xiangmin and Herong (1981)
Vigna sinensis (1r, 2x = 24)	–	Sen and Bhowal (1962)
During metaphase-I		
Alopecurus myosuroides (1-many r, 2x = 14)	a	Johnsson (1944)
Hevea brasiliensis (4x = 36)	g	Ramaer (1935)
Hordeum vulgare (1r, 2x = 14)	–	Sharma and Reinbergs (1972)
Jasminum pubescens (2x = 26)		Bhargava (1986)
Kniphofia sp.	g	Möffet (1932)
Lathyrus odoratus (1r, 2x = 14)	–	Gregory (1905), Faberge (1937)
Pelargonium denticulatum (10x = 90)	a,g	Tokumasu and Tamai (1964)
Pisum arvense (1r, 2x = 14)	–	This Vol.
Saccharum officinarum (8x = 80)	–	Janaki-Ammal (1941)
Thespesia populnea (2x = 26)	–	This Vol.
Tripsacum laxum (4x = 72)	–	Dodds and Simnonds (1946)
Triticum monococcum (2x = 14)	–	Smith (1938b)
Valeriana officinalis (2x = 14)	e	Shugaeva (1979)
Zea mays (1r, 2x = 20)	a	Beadle (1932b), Nelson and Clary (1952, Golubovskaya (1977, 1979), Albertsen and Phillips (1981)
During interphase		
Hordeum vulgare (1r, 2x = 14)	–	Herd and Steer (1984)
Pisum sativum (1r, 2x = 14)	–	Klein (1969)
Salvia officinalis (2x = 14)	–	Linnert (1955)
Solanum tuberosum (1r, 4x = 48)	–	Arnason (1941)
During metaphase II		
Capsicum annuum (1r, 2x = 24)	–	Novák et al. (1971)
Cucumis melo (2x = 24)	–	Bohn and Principe (1964)
Lathyrus odoratus (1r, 2x = 14)	a	Gregory (1905), Punnett (1923, 1925), Faberge (1937)
Zea mays (1r, 2x = 20)	a	Clark (1940)
During anaphase-II		
Pisum sativum (1r, 2x = 14)	–	Klein (1969), Gottschalk and Klein (1976)
During tetrad formation (before microspore release)		
Ageratum conyzoides (1r, 2x = 40)	a	This Vol.
A.houstonianum (1r, 2x = 20)	–	Reimann-Philipp and Fuchs (1971), this Vol.

Table 2.18 *(continued)*

Species	Tapetum	References
Allium fistulosum (1r, 2x = 16)	b	Nishi and Hiraoka (1958)
Antirrhinum majus (1r, 2x = 16)	a	This Vol.
Brassica campestris (1r, 2x = 20)		
var.*sarson*	b,c	Chowdhry and Das (1966, 1968)
B.campestris (1r, 2x = 20)	—	Das and Pandey (1961)
B.oleracea (1r, 2x = 18)		
var.*capitata*	f	Nishi and Hiraoka (1958), Kototani and Yamada (1958)
var.*italica*	—	Cole (1959)
Cajanus cajan (1r, 2x = 22)	c	Reddy et al. (1978)
	b	Wallis et al. (1980), Dundas et al. (1981)
Capsicum annuum (1r, 2x = 24)	—	Shifriss and Frankel (1969), Hirose and Fujime (1975, 1981), Pochard (1970b)
Cirsium palustre (2x = 34)	f	Delannay (1979)
C.spinosissimum (2x = 34)		
Cucumis melo (1r, 2x = 24)	c	Bohn and Whitaker (1949), Chauhan and Singh (1968)
	—	Bohn and Principe (1964)
Cucurbita maxima (1r, 4x = 40)	b	Francis and Bemis (1970)
Epilobium montanum (2x = 36) × *E.parviflorum* (2x = 36)	a	Oberreuter (1925)
E.roseum (2x = 36) × *E.parviflorum* (2x = 36)		Oberreuter (1925)
E.parviflorum (2x = 36) × *E.montanum* (2x = 36)		Oberreuter (1925)
E.parviflorum (2x = 36) × *E.roseum* (2x = 36)		Oberreuter (1925)
Glycine max (1r, 4x = 40)	a	Brim and Young (1971), Rubaihayo and Gumisiriza (1978), Albertson and Palmer (1979)
	—	Singh et al. (1974), Patil and Singh (1974, 1976), Bernard and Cremeens (1975)
	d	Jha and Singh (1978), Delannay and Palmer (1982), Stelly and Palmer (1982)
Hebe salicifolia (4x = 80)	—	Frankel (1940)
Hordeum vulgare (1r, 2x = 14)	d,e	Singh and Kaul (1966)
	d,f	Roath and Hockett (1970)
Lathyrus luteus (1r, 2x = 14)	—	This Vol.
L.sativus (1r, 2x = 14)	—	Zelenskaya and Postova (1974)
Lupinus mutabilis (4x = 48)	c	[g]Pakendorf (1970)
Lycopersicon esculentum (1r, 2x = 24)	f	Rick and Butler (1956), Zamir et al. (1980), this Vol.
Manihot esculenta (4x = 36)	—	Jos et al. (1967), Magoon et al. (1968)
Melilotus alba (2x = 16)	g	Castetter (1925)
Oenothera gigas (4x = 28)	g	Gates (1911)
Origanum majorana (4x = 32)	a	This Vol.
Parthenium hysterophorus (1r, 2x = 18)	a	This Vol.
Petunia axillaris (1r, 2x = 14)	—	Levan (1942)
P.hybrida (1r, 2x = 14)	—	Frankel (1962, 1971)
Pisum arvense (1r, 2x = 14)	a	This Vol.
P.sativum (1r, 2x = 14)	—	Klein (1969)

Table 2.18 *(continued)*

Species	Tapetum	References
Tagetes erecta (1r, 2x = 12)	d	This Vol.
Taraxacum balticum (4x = 32)	f,g	Malecka (1971)
T.fulvum (4x = 32)	a,g	Gustafsson (1935)
T.nordstedti (6x = 48)		Gustagsson (1935)
T.pseudofulvum		Gustafsson (1935)
Triticum aestivum (1r, 6x = 42)	—	Lupton and Bingham (1966–1967), Deng and Gao (1982)
Vaccinium angustifolium (many r, 2x = 24)	a,e	Hall and Aalders (1961)
Wistaria sp. (2x = 16)	—	Roscoe (1927)
Zea mays (1r, 2x = 20)	a	Singleton and Jones (1930) [g]Beadle (1932b), Cheng et al. (1979), [g,h]Albertsen and Phillips (1981), this Vol.

[a]For symbols refer to Table 2.17.

AII, but the largest number of these act during tetrad formation before microspores are released from the quartets (Fig. 2.11). Microspore formation is a crucial meiotic stage marking the end of meiosis, and is most susceptible to *ms* gene action as a large number of *ms* genes act during this stage. The genes induce anomalies either in chromosomal behaviour or in quadripartition of PMC's. Some genes inhibit the release of microspores from the tetrad binding due to lack of enzymatic digestion of the callose envelope surrounding the microspores in the tetrads. This starves the young developing microspores and inhibits viable pollen production in some petunia, tomato and pea mutants (Frankel et al. 1969; Izhar and Frankel 1971; this Vol.).

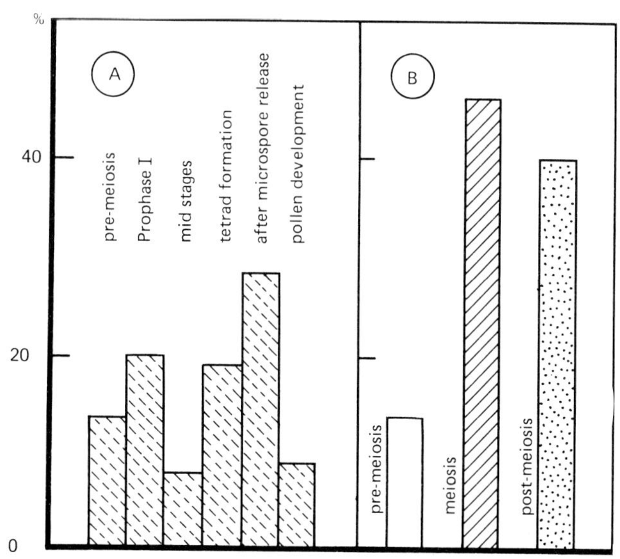

Fig. 2.11A,B. Distributional time of ms gene action during microsporogenesis

2.6.4.3 Postmeiosis

The majority of the *ms* genes act during postmeiosis mostly either immediately after microspores are released from quartets or during microgametophyte development and pollen maturity (Fig. 2.11, Table 2.19). This immediate degeneration of microspores after release either prior to, during or after mitosis, is coupled with abnormal tapetal form and function, e.g. premature or delayed degeneration, malfunction or anomalous development in some mst mutants (Table 2.19A). In *Vaccinium angustifolium*, the microsporogenesis breakdown and tapetal degeneration is simultaneous but in *Zea mays* tapetal development is normal. When *ms* gene action is delayed until pollen development and mitosis, tapetal behaviour, where known, is nearly normal except in barley (Table 2.19C).

Table 2.19. ms Gene action over postmeiosis and tapetum[a]

Species	Tapetum	References
Microsporogenesis stage		
After microspore liberation		
Ageratum conyzoides (1r, 2x = 20)	a	This Vol.
Allium ascalonicum (1r, 2x = 16)	a	This Vol.
A.fistulosum (1r, 2x = 16)	—	Virnich (1967)
A.longicuspis (2x = 16)	b,g	Novâk (1972)
A.porrum (1-3r, 4x = 32)	—	Schweisguth (1970)
Antirrhinum majus (1r, 2x = 16)	a	This Vol.
Arabidopsis thaliana (2x = 10)	—	Veen and Wirtz (1968)
Avena fatua (6x = 42)	h	This Vol.
A.strigosa (1r, 2x = 14)	h	Sadanaga (1965)
Beta macrocarpa (1r, 2x = 18)	c	This Vol.
B.maritima (1r, 2x = 18)	c	Kinoshita (1971), Kinoshita and Takahashi (1972)
B.vulgaris (1r, 2x = 18)	—	Owen (1952)
	c,d	This Vol.
Brassica juncea (1r, 4x = 36)	a	Banga and Labana (1983)
B.oleracea (1r, 2x = 18) var.*gemmifera*	—	Haigh (1954), Johnson and Faulkner (1956, 1957)
B.napus (1r, 2r, 4x = 38)	—	Takagi (1970)
Centaurea scabiosa (2x = 20)	—	Fröst (1963)
Centranthus ruber (2D, 1r, 2x = 14)	—	This Vol.
Cheiranthus cheiri (1r, 2x = 14)	—	This Vol.
Citrus sinensis (2x = 18)	g	Oppenheim and Frankel (1929)
Dactylis glomerata (4x = 28)	b	Filion and Christie (1966)
Datura stramonium (1r, 2x = 24)	e	This Vol.
Dianthus chinensis (1r, 4x = 60)	e	This Vol.
Eleusine coracana (4x = 36)	—	Ayyangar (1931)
E.indica (1r, 2x = 18)	d	This Vol.
Eruca sativa (2r, 2x = 22)	—	Verma (1984)
Glycine max (1r, 4x = 40)	f,d	Palmer et al. (1980)
	—	Patil and Singh (1976), Singh et al. (1974)
Gossypium hirsutum (2r, 4x = 52)	—	Weaver (1968)
Helianthus annuus (1r, 2x = 34)	—	Putt and Heiser (1966)
	c	Nakashima and Hosokawa (1974)
Hordeum vulgare (1r, 2x = 14)	—	Singh and Kaul (1966), Herd and Steer (1984)

Table 2.19 *(continued)*

Species	Tapetum	References
Hordeum vulgare (1r, 2x = 14)	d,e,f	Mian et al. (1974), Roath and Hockett (1970, 1971)
Limnanthes douglasii (1D, 2x = 10)	–	Jain et al. (1978)
Lupinus regalis (2r, 4x = 48)	a	This Vol.
Malus sylvestris (1r, 2x = 34)	–	Gagnieu (1951)
Manihot esculenta (4x = 36)	f	Jos et al. (1967), Magoon et al. (1968)
Medicago sativa (3r, 4x = 32)	b	Childers and Mclennan (1960)
Nicotiana rustica (2r, 4x = 48)	–	This Vol.
N.tabacum (2r, 4x = 48)	–	Bhat and Krishnamoorthi (1956)
Oenothera alata (2x = 14)	b	Pohl (1895), Gates (1907)
O.flava (2x = 14)		Oehlkers (1927)
O.lamarckiana (2x = 14)		Harte and Bissinger (1952)
O.strigosa (2x = 14)	b	Oehlkers (1923, 1926), Oehlkers and Harte (1943)
Oryza sativa (1r, 2x = 24)	–	Lu and Rutger (1984b), this Vol.
Papaver somniferum (1r, 2x = 22)	a	Singh and Khanna (1970), this Vol.
Parthenium hysterophorus (1r, 2x = 18)	a	This Vol.
Pelargonium radula (9x = 81)	b,g	Conagin (1964)
P.roseum (8x = 72)	g	Tamai et al. (1958)
P.tomentosum (4x = 44)	–	This Vol.
Pennisetum typhoides (1r, 2x = 14)	–	Powell and Burton (1969)
Petunia axillaris (1r, 2x = 14)	–	This Vol.
Phaseolus atropurpureus (2x = 22)	b	Pritchard and Hutton (1972), this Vol.
P.vulgaris (2D, 2x = 22)	–	Mutschler and Bliss (1980)
Pisum sativum (1r, 2x = 14)	–	Klein (1969), Myers (1984), this Vol.
Rannunculus acris (4x = 28)	–	Whyte (1929)
Rapanea thwaitesii (2x = 24)	h	Vasanthy and Pocock (1981)
Solanum nigrum (1r, 2x = 24)	c	This Vol.
S.tuberosum (4x = 48)	–	Kishore et al. (1974)
S.verrucosum (2r, many r, 2x = 24)	–	Abdalla and Hermsen (1972)
Spinacea oleracea (1r, 2x = 12)	a	Kobabe (1967)
Triticum aestivum (1r, 6x = 42)	–	Athwal et al. (1967), Krupnov (1968), Fossati and Ingold (1970)
Vaccinium angustifolium (many r, 2x = 24)	e	Hall and Aalders (1961)
Vicia faba (1r, 2x = 12)	–	Bond et al. (1964a,b)
Vigna angularis (1r, 2x = 22)	b,h	Nakashima et al. (1980)
V.mungo (1r, 2x = 22)	c	Roy and Jana (1983)
V.sinensis (1r, 2x = 24)	b	Rachie and Rawal (1976)
Viola tricolor (1r, 2x = 26)	e	This Vol.
Zea mays (1r, 2x = 20)	a	Eyster (1921), Beadle (1931, 1932), Emerson (1932), Mangelsdorf (1932), Rhoades and Rhoades (1939), Nelson and Clary (1952), Weijer (1952), Cheng et al. (1979), Albertsen and Phillips (1981), Coe et al. (1981), this Vol.
Around microspore mitosis		
Allium sativum (2x = 16)	b	Koul and Gohil (1970), Novâk (1972), Konvička et al. (1978), this Vol.
Capsicum annuum (1r, 2x = 24)	–	Daskaloff (1971, 1971-1972, 1973a,b, 1974)
Dactylis glomerata (4x = 28)	–	Našinec (1973), this Vol.
Gossypium hirsutum (1r, 4x = 52)	–	Richmond and Kohel (1961)
	d	Bowman and Weaver (1979), Tang et al. (1981)

Table 2.19 *(continued)*

Species	Tapetum	References
Gossypium hirsutum (1r, 4x = 52)	f	Murthi and Weaver (1974)
Lolium perenne (2x = 14)	c	Hayward and Manthiratna (1972)
Lycopersicon esculentum (1r, 2x = 24)	d,f	Rick and Butler (1956), Zamir et al. (1980), this Vol.
Pelargonium tomentosum (1-2r, 4x = 44)	–	This Vol.
Petunia hybrida (1r, 2x = 14)	–	Izhar (1973), this Vol.
Rapanea thwaitesii (2x = 24)	–	Vasanthy and Pocock (1981)
Raphanus sativus (1r, 2x = 18)	b	Nishi and Hiraoka (1958)
Secale cereale (1r, 2x = 14)	–	Adamĉhuk (1971, 1972)
Solanum tuberosum (4x = 48)	d	Cho and Kim (1984)
Triticum durum (1r, 4x = 28)	–	Bozzini and Scarascia – Mugnozza (1968), Bozzini (1974)
T.monococcum (1r, 2x = 14)	–	Smith (1938b)
During and after pollen mitosis		
Allium ascalonicum (1r, 2x = 16)	–	Darlington and Haque (1955)
Arabidopsis thaliana (2x = 10)	–	Veen and Wirtz (1968)
Canna glauca (2x = 18)	–	Honing (1928), Öfferijns (1938)
Capsicum annuum (1r, 2x = 24)	–	Meshram and Narkhede (1982)
Citrus yatsushiro (2x = 18)	g	Nakamura (1943)
Cucumis sativus (1r, 2x = 14)	a	Whelan (1972)
Cucurbita pepo (1r, 4x = 40)	b	Eisa and Munger (1968)
Gossypium hirsutum (1r, 4x = 52)	a	Weaver and Ashley (1971), Murthi and Weaver (1974)
Hebe leiophylla (4x = 80)	–	Bowman et al. (1978)
× *H.salicifolia* (4x = 80)	–	Frankel (1940)
H.parviflora (4x = 80) × *H.leiophylla* (4x = 80)	–	This Vol.
H.townsonii (2x = 40)	–	Frankel (1940)
H.traversii (4x = 80)	–	Frankel (1940)
Helianthus annuus (1r, 2x = 34)	–	Pirev (1966), Putt and Heiser (1966)
Hordeum vulgare (1r, 2x = 14)	b	Roath and Hockett (1970), Ahokas (1975a), this Vol.
Lolium multiflorum (1r, 2x = 14)	–	Nitzsche (1971)
Oryza sativa (1r, 2x = 24)	–	Lu and Rutger (1984)
Pisum sativum (1r, 2x = 14)	–	Klein (1969)
Plantago coronopus (2x = 10)	–	Paliwal and Hyde (1959)
Prunus mume (2x = 16)	M	Nakanishi (1982)
Solanum tuberosum (many r, 4x = 48)	–	Fukuda (1927)
Triticum aestivum (6x = 42)	–	Lemekh et al. (1971)
Vitis vinifera (2x = 38)	g	Olmo (1943)
Zea mays (1r, 2x = 20)	a	Rhoades (1950)

[a]For symbols refer to Table 2.17.

2.6.5 Variable *ms* Gene Action

Stage and site specificity is a characteristic feature shared by the majority of *ms* genes (Tables 2.17–2.19). This specificity, however, is not absolute, as *ms* gene action is influenced by the genotype, the environment, and their interactions. In pearl millet the cytoplasmic background alters the time of *ms* gene action (Rao and Devi 1983). These influences render *ms* gene action over microsporogenesis highly variable so that in a few genotypes the genes act anywhere between pre- to postmeiotic stages (Table 2.20). In this table, *Canna*, *Citrus* and *Oenothera* are included. The mst in these is unconfirmed as female fertility of the steriles is not known. In others, the *ms* genes are sensitive so that with suitable manipulations of genotype and environment, the gene action can be initiated during the early or very late stage of microsporogenesis. In three such mutants, one each of pepper, maize and tomato, the gene action can be delayed so much that a few anthers escape *ms* gene action completely and viable pollen is produced. However, the number of such malleable *ms* genes is extremely low.

Table 2.20. ms genes with variable action over microsporogenesis (from pre- to postmeiosis)

Species	References
Canna glauca[a] (2x = 18)	Honing (1928), Öfferijns (1938)
Capsicum annuum[b] (2r, 2x = 24)	Pochard (1970b)
Citrus aurantifolia[a] (2x = 18)	Uphof (1931), Nakamura (1934), Krug and Bacchi (1943)
Limnanthes douglasii[b] (1D, 2x = 10)	Jain et al. (1978)
Lycopersicon esculentum[b,c] (1r, 2x = 24)	Rick and Butler (1956), Zamir et al. (1980), this Vol.
Oenothera gigas[a] (4x = 28)	Gates (1911)
Papaver somniferum[b] (2x = 22)	Singh and Khanna (1970)
Pennisetum typhoides[b,d] (1r, 2x = 14)	Rao and Koduru (1978a,b), Rao and Devi (1983)
Petunia hybrida[b] (1r, 2x = 14)	Cornu and Dommerguess (1974), Singh (1975)
Triticum aestivum[b] (6x = 42)	Lemekh et al. (1971)
T.monococcum[b] (2x = 14)	Smith (1938b)
Zea mays[b,d] (1r, 2x = 20)	Beadle (1932b), Cheng et al. (1979), Albertsen and Phillips (1981)

[a]Female fertility not known, or [b]normal; [c]tapetal behaviour abnormal, or [d]normal, in others not known.

2.6.6 Stage-Specific *ms* Gene Action

Barring a few *ms* genes whose action over microsporogenesis is diffuse and variable, the action of the majority of *ms* genes over male organ development or function is stage-specific, as most of these affect certain specific stages of anther and pollen development (Figs. 2.12, 2.13). A relative proportion of breakdown stages during premeiosis, prophase I and postmeiosis is nearly similar in dicots and monocots but is considerably greater in dicots at the tetrad stage (Fig. 2.12). The maximum number of *ms* genes affect meiotic and postmeiotic stages and the minimum the anther dehiscence and pollen liberation stages. During microsporogenesis, the majority of these genes influence either the beginning or the end stages of meiosis (Fig. 2.11). The mature pollen development and meiotic stages between prophase I and tetrad formation are least influenced by *ms* genes. In meiosis, prophase I is a long-duration stage during which vital cytogenetical events take place. All these events are controlled by a large number of dominant genes, and mutation of any of these impairs the coordinated and balanced progression of

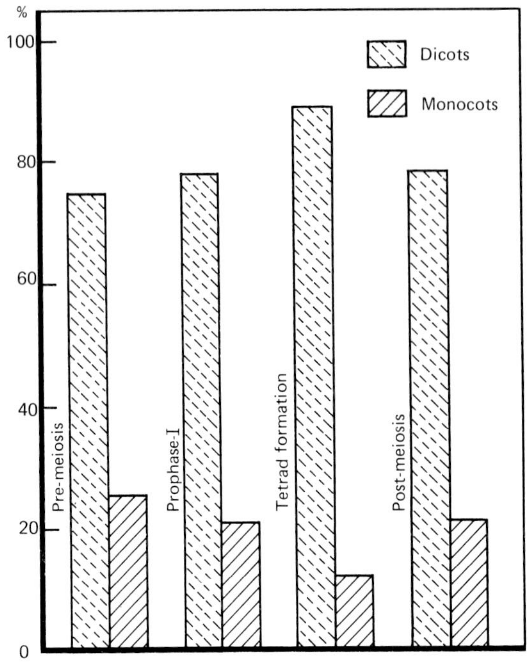

Fig. 2.12. Comparative ms gene action during microsporogenesis in dicots and monocots

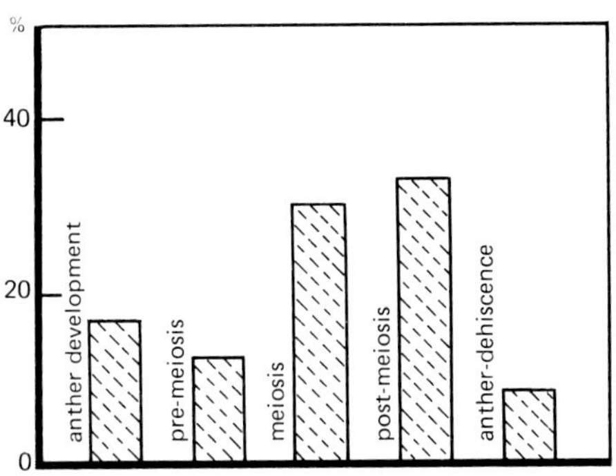

Fig. 2.13. Relative proportion of ms genes acting during male sex formation and function

meiotic course (Kaul and Murthy 1985). Accordingly, a high number of *as* and *ds* genes exist in plant genomes which impair synapsis and chiasma formation in higher plants (Gottschalk and Kaul 1980a,b; Kaul and Murthy 1985). Likewise, the number of *ms* genes influencing the prophase I stage is very high (Table 2.18). However, unlike *as* and *ds* genes, the *ms* genes do not affect synapsis, synaptinemal complex or chiasma building, instead they alter chromosome function and cytoplasmic cell contents. Furthermore, whereas *as* and *ds* genes do not influence microspore development, *ms* genes predominantly influence microspore formation and function but do not affect megasporogenesis or megaspore development and function. Thus *ms* genes are distinct from the meiotic mutant genes *as* and *ds* in exhibiting certain unique features and genc action confined specifically to the male sex. Their action over prophase I and after microspore release from tetrads is genetically significant because during prophase I genetic recombination takes place and genetic and metabolic autonomy of spores begins only after microspores are released from PMC's. However, like other meiotic mutant genes, many allelic and nonallelic *ms* genes affect the same meiotic phase in maize, tomato, barley and pea.

2.7 Tapetal Form and Function

Tapetum, the innermost wall layer of the microsporangium, is considered to have a nutritive function for the maturing heterotrophic spores (Echlin 1971) and thus plays a major role in microsporogenesis by providing the transporting enzymes, hormones and other food materials to the developing PMC's and microspores. Since growth and maturation of the spore depends on nutrients supplied by the tapetum, mst has been assigned to the anomalous tapetal form or function, viz. lack of tapetum, premature or delayed tapetal degeneration, developmental discrepancy between meiocyte or microspore and tapetal development, abnormal nuclear behaviour or malfunction of tapetum, hypertrophy or atrophy of tapetal cells, periplasmodial formation in glandular tapetum etc. This assignment is based on the predominant occurrence of mst in plants exhibiting anomalous tapetal form, behaviour or function. The major anomalies range from premature degeneration to persistence, abnormal behaviour and/or malfunctioning of the normally developed tapetum in mst plants. However, the tapetal development and differentiation and its form and breakdown can be both normal or abnormal in g-mst plants in species as well as in species crosses (Figs. 2.14, 2.15, 2.16). This is evidenced by the examples listed in Tables 2.21 to 2.24, in which confirmed cases of g-mst have been included. In many species, although tapetal development, differentiation and disintegration are normal, microsporogenesis breakdown occurs at various stages (Table 2.21). In them, the *ms* genes may act during any stage of microsporogenesis, but the modal value (40%) lies around tetrad formation (Fig. 2.17A). On the other hand, in those in which tapetum is abnormal, the *ms* genes in over 50% of cases act after microspores are released (Fig. 2.17B). Thus whereas in the majority (66%) of the cases with normal tapetum, the *ms* genes act during meiosis, they act after meiosis in 55% and during meiosis in 35% of the plants having an abnormal tapetum. This dominant postmeiotic *ms* gene action in plants has led to the common concept of a tapetal role in mst induction.

Persistence or premature breakdown of the tapetum has been considered as the main cause of pollen sterility in higher plants. This is evidenced by the examples listed in Table 2.23, in all of which tapetum is persistent and the microspores abort either just before or just after release from the quartets. Likewise, in the majority of the g-mst's in which tapetal form or function is abnormal, microspores degenerate either before or mostly after liberation from PMC's (Table 2.22). Since the tapetum has a nutritive role, its premature disintegration or constant persistence during microspore development deprives the microspores of food materials and starves them to death. But this is not uniformly known in all g-mst's. For

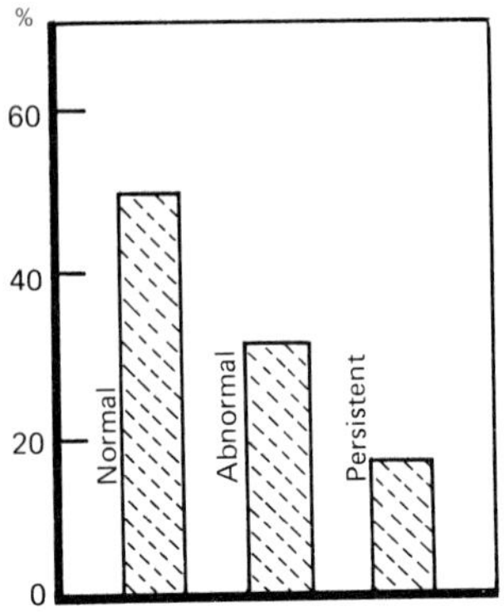

Fig. 2.14. Frequency distribution of tapetal types in genic male steriles

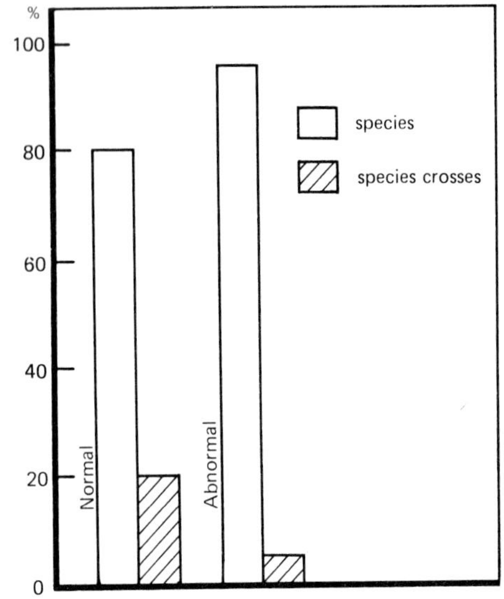

Fig. 2.15. Frequency distribution of tapetal types; in species and species crosses

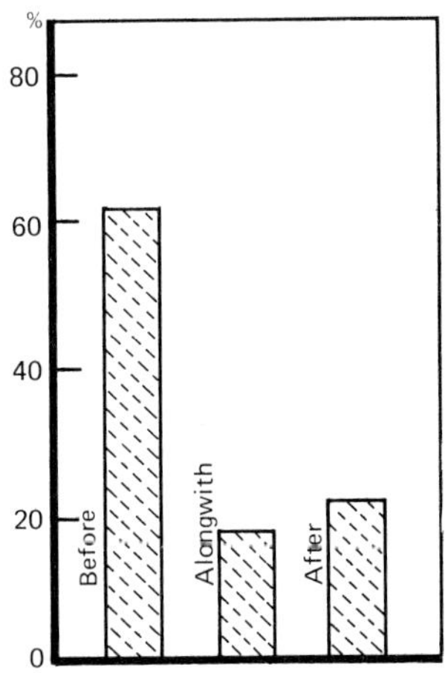

Fig. 2.16. Tapetal breakdown in relation to time of sporogenous tissue degeneration

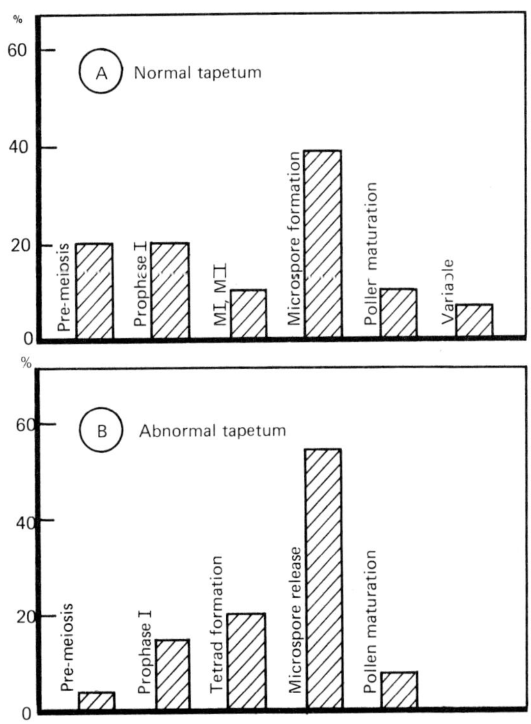

Fig. 2.17A,B. Microsporogenesis breakdown stages in genic male steriles with normal (A) and abnormal (B) tapetum

Table 2.21. Microsporogenesis breakdown in genic male steriles with normal tapetum

Species	References
1 Breakdown stage during:	
Pre-meiosis	
Citrus sinensis (2x = 18)	Nakamura (1934)
Epilobium parviflorum (2x = 36) × *E.montanum* (2x = 36)	Oberreuter (1925)
E.parviflorum (2x = 36) × *E.roseun* (2x = 36)	Oberreuter (1925)
Hordeum vulgare (1r, 2x = 14)	This Vol.
Hordeum vulgare (1r, 2x = 14)	This Vol.
Pisum sativum (1r, 2x = 14)	Gottschalk (1971), this Vol.
2 Pachytene	
Helianthus annuus (1r, 2x = 34)	Paun (1974)
Hordeum vulgare (1r, 2x = 14)	This Vol.
Nicotiana tabacum (1r, 4x = 48)	This Vol.
3 Prophase-I	
Centaurea americana (1r)	This Vol.
C.imperialis (1r)	This Vol.
Coleus hybridus (1r, 8x = 48)	This Vol.
4 Metaphase-I	
Pisum arvense (1r, 2x = 14)	This Vol.
5 Metaphase-II	
Lupinus regalis (2r)	This Vol.
Spinacea oleracea (1r, 2x = 12)	Kobabe (1967)
6 Tetrad formation	
[a]*Epilobium montanum* (2x = 36) × *E.parviflorum* (2x = 36)	Oberreuter (1925)
E.parviflorum (2x = 36) × *E.montanum* (2x = 36)	Oberreuter (1925)
E.parviflorum (2x = 36) × *E.roseum* (2x = 36)	Oberreuter (1925)
E.roseum (2x = 36) × *E.parviflorum* (2x = 36)	Oberreuter (1925)
Glycine max (1r, 4x = 40)	Brim and Young (1971), Albertsen and Palmer (1979), Rubaihayo and Gumisiriza (1978)
Vaccinium angustifolium (many r, 2x = 24)	Hall and Aalders (1961)
7 Microspore formation	
Lathyrus luteus (1r, 2x = 14)	This Vol.
Lycopersicon esculentum (1r, 2x = 24)	Rick (1948), this Vol.
Papaver somniferum (1r, 2x = 22)	This Vol.
Pelargonium crispum (2x = 22)	Tokumasu (1976)
Pisum arvense (1r, 2x = 14)	This Vol.
P.sativum (1r, 2x = 14)	This Vol.
Vaccinium angustifolium (many r, 2x = 24)	Hall and Aalders (1961)
8 Pollen maturation	
Gossypium hirsutum (1D, 4x = 52)	Weaver and Ashley (1971), Murthi and Weaver (1974)
9 Premeiosis, meiosis or postmeiosis	
Lycopersicon esculentum (1r, 2x = 24)	Rick (1948)
Zea mays (1r, 2x = 20)	Beadle (1932a), Albertsen and Phillips (1981), Cheng et al. (1979)
10 Anywhere during meiosis (Variable)	
Hordeum vulgare (1r, 2x = 14)	Sharma and Reinbergs (1972)
Pisum sativum (1r, 2x = 14)	This Vol.

Table 2.22. Microsporogenesis breakdown in genic male steriles with abnormal tapetal form and/or function

Species	References
Breakdown stage during:	
Premeiosis	
Gossypium hirsutum (1r, 1D, 4x = 52)	Bowman et al. (1978), Tang et al. (1981)
Prophase	
Medicago sativa (2r, 3po, 4x = 32)	Childers (1952)
Petunia hybrida (1r, many r, 2x = 14)	Welzel (1954), Cornu and Dommerguess (1974), Singh (1975)
Raphanus sativus (1r, 2x = 18)	Tokumasu (1957)
Secale cereale (1-many r, 2x = 14)	Cebrat and Zadecka (1978)
Tetrad formation	
Allium fistulosum (1r, 2x = 16)	Nishi and Hiraoka (1958)
A.schoenoprasum (1r, 2x = 16)	Tatliogu (1982)
Cajanus cajan (1r, 2x = 22)	Wallis et al. (1980), Dundas et al. (1981)
After microspore release	
Helianthus annuus (1r, 2x = 34)	Nakashima and Hosokawa (1974a,b)
Hordeum vulgare (1r, 2x = 12)	Mian et al. (1981)
Medicago sativa (2r, 3po, 4x = 32)	Childers and McLennan (1960)
Oenothera lamarckiana (2x = 14)	Oehlkers and Harte (1943), Harte and Bissinger (1952)
O.strigosa (2x = 14)	Oehlkers (1923, 1926), Harte and Bissinger (1952)
Pelargonium crispum (2x = 22)	Tokumasu (1974, 1976)
P.radula (9x = 81)	Conagin (1964)
Phaseolus atropurpureus (2x = 22)	Pritchard and Hutton (1972), this Vol.
Solanum tuberosum (4x = 48)	Cho and Kim (1984)
Vigna angularis (1r, 2x = 22)	Nakashima et al. (1980)

Table 2.23. Microsporogenesis breakdown in genic male steriles with persistent tapetum

Species	References
Breakdown stage during:	
1 Tetrads and microspore formation	
Allium schoenoprasum (2x = 16)	Singh and Kobabe (1969)
Brassica campestris (1r, 2x = 20)	Chowdhry and Das (1966, 1968)
Cajanus cajan (1r, 2x = 22)	Reddy et al. (1978)
Cucumis melo (1r, 2x = 14)	Bohn and Whitaker (1949), Chauhan and Singh (1968)
2 After microspore release	
Beta macrocarpa (1r, 2x = 18)	This Vol.
B.maritima (1r, 2x = 18)	Kinoshita (1971), Kinoshita and Takahashi (1972)
B.vulgaris (1r, 2x = 18)	This Vol.
Helianthus annuus (1r, 2x = 34)	Nakashima and Hosokawa (1974)
Solanum nigrum (1r, 2x = 24)	This Vol.
S.tuberosum (4x = 48)	Kishore et al. (1974)
3 Variable	
Hordeum vulgare (1r, 2x = 14)	Kaul and Singh (1966), Roath and Hockett (1970), Mian et al. (1974)
Pisum sativum (1r, 2x = 14)	This Vol.

Table 2.24. Tapetal breakdown in relation to sporogenous tissue breakdown in genic male steriles

Species	References	*ms* Gene action
Breakdown stage		
Before sporogenous tissue degeneration		
Cirsium arvense (2x = 34)	Delannay (1979)	Premeiotic
C.montanum (2x = 34)	Delannay (1979)	Premeiotic
C.oleraceum (2x = 34)	Delannay (1979)	Premeitoic
C.palustre (2x = 34)	Delannay (1979)	Premeiotic
C.spinosissimum (2x = 34)	Delannay (1979)	During tetrad formation
Glycine max (1r, 4x = 40)	Palmer et al. (1980)	After microspore liberation
Gossypium hirsutum (2r, 4x = 52)	Murthi and Weaver (1974)	Microspores abort during maturation
Hordeum vulgare (1r, 2x = 14)	Kaul and Singh (1966)	During prophase-I
	Mian et al. (1974), this Vol.	After microspore liberation
Lycopersicon esculentum (1r, 2x = 24)	Rick (1948)	Tetrad formation to microspore release
Raphanus sativus (1r, 2x = 18)	Nishi and Hiraoka (1958)	After microspore liberation
Taraxacum balticum (4x = 32)	Gustafsson (1935),	During prophase or during
T.silesiacum (3x = 24)	Malecka (1971)	tetrad formation
Along with sporogenous tissue degeneration		
Fagopyrum esculentum (2x = 16)	This Vol.	After PMC formation
Hordeum vulgare (1r, 2x = 14)	Singh and Kaul (1966), Roath and Hockett (1970)	During microspore formation, before release
Pennisetum typhoides (1r, 2x = 14)	Rao and Koduru (1978a,b), Rao and Devi (1983)	Before archesporial differentiation or during microspore formation
Vaccinium angustifolium (many r, 2x = 24)	Hall and Aalders (1961)	After tetrad formation or microspore release
After sporogenous tissue degeneration		
Cucurbita maxima (1r, 4x = 40)	Rhodes (1959), Singh and Rhodes (1961)	Before meiotic initiation
Glycine max (1r, 4x = 40)	Delannay and Palmer (1982)	Before cytokinesis
	Palmer et al. (1980)	After microspore liberation
Lycopersicon esculentum (1r, 2x = 24)	Rick (1948), this Vol.	Prophase-I, Prophase-I to tetrad formation
Phleum nodosum (1r, 2x = 14)	May and Kasha (1971, 1980)	During pachytene
Pisum sativum (1r, 2x = 14)	This Vol.	Prophase-I

instance, in a turnip and an alfalfa g-mst mutant, male meiosis breaks down during early zygotene and in late prophase I stages, respectively (Tokumasu 1957; Childers 1952). In both these mutants, tapetal development is normal. Compared to their respective mft's, however, tapetal breakdown in both of these is considerably delayed. On the other hand, in a barley and a pea mutant, microsporogenesis breakdown is highly variable but tapetal persistence is constant (Kaul and Singh 1966; this Vol.). In a soybean mutant, both male meiosis as well as tapetal development are highly abnormal (Stelly and Palmer 1982). Likewise, a number of g-mst plants exhibit sporogenous tissue degeneration either prior to or during or even after tapetal breakdown (Table 2.24). Thus in these, the tapetal degeneration is not correlated to any specific stage of meiotic breakdown. This indicates an independence of sporogenous tissue and tapetal breakdown in some g-mst plants. The contention is strengthened by the fact that in many of the g-mst's with normal tapetal development, differentiation and behaviour, microsporogenesis breakdown is mostly stage-specific and in only a few it is variable (Table 2.21). Disparities in the tapetal breakdown stage exist within some g-mst mutants of barley (Singh and Kaul 1966; Roath and Hockett 1970) and tomato (Rick 1948). In them, the tapetal behaviour is variable. It may break down prior to or later than that of their respective mft lines. In three mst barley mutants, tapetal alterations precede pollen abortions but in two others they follow. In all of them microspores degenerate after liberation from PMC's. Mian et al. (1974), after analysing the effect of several barley mst genes on DNA and histone turnover, concluded that *ms* gene effects can first become evident either at the microspore or at the tapetum but that subsequently both exhibit biochemical alterations.

2.7.1 Dual Role of *ms* Genes

These descriptions and inferences clearly indicate that the *ms* genes act either on nonsporogenous cells forming the anther wall, mainly the tapetum, or on microsporogenous cells or on both these tissues. Among the four layers forming the anther wall, the epidermis, the endothecium, the middle layer and the tapetum, the innermost of them, the tapetum, undergoes many cytohistological alterations in mst's and is therefore implicated as the site of *ms* gene action. In such cases, the tapetum develops or functions abnormally (Table 2.22, 2.23) and causes mst by impeding nutrition and starving the developing microsporogenous tissue to death. Where the *ms* gene acts directly on the microsporogeneous cells, the tapetal development and differentiation are normal but microsporogenesis breakdown occurs during various stages of its progression (Table 2.21). When the *ms* gene acts on both tapetal and microsporogenous tissues, they exhibit anomalous development or degenerate simultaneously or successively (Table 2.24). Thus mst can result by *ms* gene action on nonsporogenous anther tissue (mostly tapetum) or on microsporogenous cells or on both. In other words, the *ms* gene action is either indirect via anther wall or direct on microsporogenous cells, or dual on both these tissues. It is equally probable that in mst plants with normal tapetal form and development, the *ms* gene acts on tapetal function by causing biochemical disturbances. Some such disturbances, although known, are considered as the consequences but not causes of mst in higher plants (Gottschalk and Kaul 1974).

2.8 Environmental Influence

2.8.1 Sensitive Male Steriles

Sexuality in higher plants is a delicate and fragile system susceptible to the influence of genes (nuclear and cytoplasmic), chemicals, environment, and their interactions. Whereas sex alterations induced by chemicals and environment are excellently reviewed by Frankel and Galun (1977) mst induced by genes, chemicals and environment are reviewed in this present volume. Considering the large number of *ms* genes known in plants, the influence of environment is determined in only a few of them (Table 2.25). Of these in about 44%, the major factor influencing *ms* gene action and expression is temperature. In 12%, it is photoperiod and in the remaining 44% the precise environmental factor is not known (Figs. 2.18, 2.19). Sterility in the majority of the sensitive genotypes is conditioned by single recessive *ms* genes, but in wheat and *Vaccinium* the control is by three recessive genes or by polygenes. In most of them, the *ms* gene acts just before or just after the microspores are liberated (Table 2.25). Such *ms* genes are not, however, more sensitive to environmental influence than others, since they are also known in environmental nonsensitive mst genotypes (Table 2.26). Three mst mutants, one each in cabbage, pepper and tomato, are sensitive both to temperature and photoperiod combinations. Of these, the cabbage mutant is mst in summer and mft in winter (Rundfeldt 1960) and two mutants of pepper and tomato are mft in summer but mst in winter. Thus while mft in the cabbage mutant can be revoked by growing it in winter, in the pepper and tomato mutants, it can be induced by growing them in summer.

A barley mutant, ms_9, (Ahokas and Hockett 1977) exhibits sensitivity to photoperiod and physiography so that it is a complete mst in Finland but partially mst in the USA. Some mst genotypes are sensitive to temperature and light intensity, so that manipulation of these two factors revokes fertility in them. This is known in two mutants each of pepper and tomato (Martin and Crawford 1951; this Vol.). One mutant each of these is mst in the field and the other in the glasshouse. Transfer of these from glasshouse to field or vice versa revokes fertility or sterility in them. Sterility in each of the two tomato mutants is conditioned by two nonallelic recessive *ms* genes (this Vol.). Nearly similar environmental sensitivity is exhibited by a sesame mutant which likewise is mst in field and mft in glasshouse (Brar 1982). The cumulative influence of altered temperature, light intensity and photoperiod appear to revoke fertility in certain mst genotypes of cabbage, tomato, pepper and sesame. In a wheat (Jan 1974) and a rice mutant (this Vol.), early-formed tillers are fully mst but late-formed ones are partially mst. Likewise, two pea mst mutants produce 9–11 flowers at the end of flowering. Of these, the last formed 3–4 are partially mft (this Vol.). Whether temperature and/or light alterations or some physiological changes revoke fertility in these mutants is not known.

2.8.2 Nonsensitive Male Steriles

The response of mst genotypes or *ms* genes to different environmental conditions has been studied in only a few cases. In them, a few genotypes maintain sterility under altered environmental conditions (Table 2.26), indicating the nonsensitivity of these *ms* genes to the environmental alterations and the stability of their action under changed environments. In all these genotypes, mst is conditioned by single recessive genes, except in cotton, in which two genes condition the sterility. In most of them, the *ms* genes act during microspore formation and development and only in a few does the gene action occur during the other meiotic stages. In general, the ratio of sensitive to stable mst genotypes is nearly 2.3 (Fig. 2.18A). This indicates a high susceptibility of *ms* genes to environmental factors. The ratio may change

Table 2.25. Environmental sensitive genic male steriles

Species	References	Gene action
Temperature-influenced		
Brassica oleracea (1r, 2x = 18)		
var. *capitata*	Rundfeldt (1961)	After tetrad formation
var. *gemmifera*	Nieuwhof (1968b)	After tetrad formation
var. *italica*	Dickson (1970)	
Capsicum annuum (1r, 2x = 24)	Martin and Crawford (1951), Daskaloff (1971–1972, 1973a)	Between microspore formation and pollen maturation
	Peterson (1958)	nk
Daucus carota (1r, 2x = 18)	Michalik (1974), Chadha and Freese (1981)	nk
Glycine max (1r, 4x = 40)	Caviness et al. (1970), Caviness and Fagala (1973), Albertsen and Palmer (1979)	After T_2 stage
Gossypium hirsutum (1r, 4x = 52)	Justus et al. (1963)	nk
Hordeum vulgare (1r, 2x = 14)	Sharma and Reinbergs (1976)	nk
Lycopersicon esculentum (1r, 2x = 24)	Rick (1948), Rick and Boynton (1967), Sawhney (1983), this Vol.	From prophase to microspore formation
Petunia hybrida (1r, 2x = 14)	Cornu and Dommergues (1974), Singh (1975)	Early prophase-I to free microspore abortion
Phaseolus vulgaris (2x = 22)	Estrada et al. (1984)	After microspore liberation
Sesamum indicum (1r, 2x = 26)	Brar (1982)	nk
Photoperiod-influenced		
Brassica oleracea (1r, 2x = 18)		
var. *capitata*	Rundfeldt (1961)	After tetrad formation
Hordeum vulgare (1r, 2x = 14)	Ahokas and Hockett (1977)	Microspore liberation to maturation
Sorghum vulgare (1r, 4x = 20)	Barabâs (1962)	nk
Unknown environmental factor(s)-influenced		
Beta maritima (1r, 2x = 18)	Kinoshita (1971), Kinoshita and Takahashi (1972)	After microspore liberation
Carica papaya (1r, 2r, 2x = 18)	Storey (1953)	nk
Chionachne koenigii (1r, 2x = 20)	Rao (1975)	nk
Citrullus vulgaris (1r, 2x = 20)	Watts (1962)	nk
Citrus reticulata (2x = 18)	This Vol.	nk
Pelargonium crispum (2x = 22)	Tokumasu (1974, 1976)	After microspore liberation
Plantago coronopus (2x = 10)	Paliwal and Hyde (1959), Ross (1970)	After microspore liberation
P.lanceolata (2r, 2x = 12)	Ross (1969)	nk
Ricinus communis (1r, 2x = 20)	Claassen and Hoffman (1950)	nk
Triticum aestivum (1r, 6x = 42)	Pugsley and Oram (1959)	nk
(3r)	Athwal et al. (1967)	After microspore liberation
(3 com.)	Jan (1974)	After microspore liberation
Vaccinium angustifolium (many r, 2x = 24)	Hall and Aalders (1961)	Tertrad formation to microspore release

Table 2.26. Environmental nonsensitive genic male steriles

Species	References	Gene action
Chionachne koenigii (1r, 2x = 20)	Rao (1975)	nk
Cucumis sativus (1r, 2x = 14)	Whelan (1972)	During pollen mitosis
Cucurbita pepo (1r, 4x = 40)	Eisa and Munger (1968)	Before pollen mitosis
Gossypium hirsutum (2r, 4x = 52)	Weaver (1968)	After microspore release
Lycopersicon esculentum (1r, 2x = 24)	This Vol.	After microspore release
Nicotiana tabacum (1r, 4x = 48)	This Vol.	After pachytene
Phaseolus lunatus (1r, 2x = 22)	Allard (1953)	nk
Pisum sativum (1r, 2x = 14)	This Vol.	Prophase-I or after microspore release
Sesamum indicum (1r, 2x = 26)	Osman and Yermanos (1982)	nk
Silene hookeri (1r, 6x = 72)	This Vol.	Premeiotic
Triticum aestivum (1r, 6x = 42)	Lupton and Bingham (1966–1967)	After microspore release

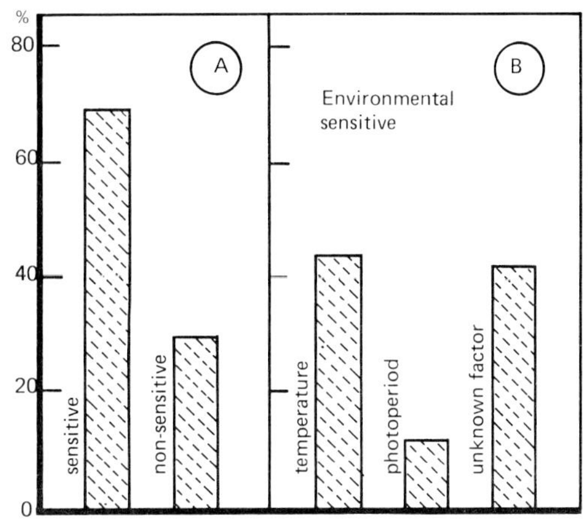

Fig. 2.18A,B. Environmental influence over genic male sterile genotypes: **A** relative proportions of sensitive and non-sensitive genotypes. **B** relative proportions of genotypes sensitive to environmental factors

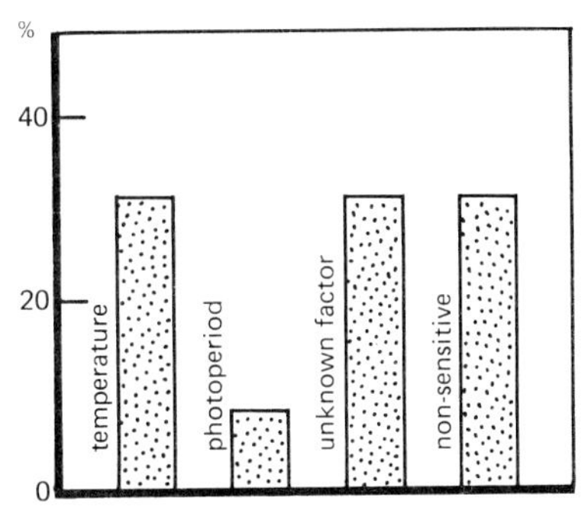

Fig. 2.19. Environmental sensitivity of genic male steriles

when the environmental sensitivity of more mst genotypes is tested. If so, it will deviate from the existing equal proportion of temperature-sensitive, unknown environmental factor-sensitive, and nonsensitive mst genotypes given in Fig. 2.18B.

2.9 Identification and Genetic Characterisation

2.9.1 Phenotypic Detection

The wide occurrence of spontaneously arisen monogenic recessive mst mutants (Table 2.1) indicates a high mutation rate of *Ms* genes to *ms*. The rate is still higher in inbreeders, as most of the mst mutants arise in them but are lost due to self-sterility. In habitual inbreeders, such mst plants are seedless, unlike in outbreeders in which seeds are set. Thus mst plants in inbreeders can be detected only after flowering and fruiting of mft plants is over. Detection at such a late stage usually prevents their controlled propagation unless new flowers are produced that could be used for ACP. However, some mst plants may segregate in subsequent selfed progenies of such populations, as the mft plants may be heterozygous for the *ms* genes. Although usually the search for mst plants is made by visual observations of anther and pollen, some exhibit phenotypic alterations. For instance, in barley, cotton, pea and tomato, some mst genotypes are bushy, late with luxuriant vegetative growth, and remain green even when mft plants have seeded and dried. In barley, at maturity mft plants turn yellow and lodge but some mst genotypes are bright, green and fresh. The ms_3 mutant is dwarf, stiff, short-leaved and has a pronounced head density. In pearl millet, a mst mutant is very densely haired and develops more productive tillers than mft plants. A pea mst mutant produces wide open flowers having malformed corolla. These alterations, although they provide phenotypic identification, are uncommon for most other mst plants. Instead, the majority of mst's do not exhibit any major phenotypic alterations that would assist in their easy field identification.

2.9.2 Linkage and Pleiotropism

Some mst mutants exhibit identifiable altered phenotypical and/or phenological alterations. This is because their *ms* genes are either closely linked to such genes whose phenotypic effect is easily identifiable (Table 2.27) or exhibit pleiotropic effects that influence many phenotypic traits (Table 2.28). These two features of the *ms* genes assist in field identification of mst genotypes. Most of the *ms* genes linked to other genes are not of negative selection value to the mst plants except in *Plantago*, in which the mst plants are albinos. Compared to linkages, the pleiotropic effect of *ms* genes introduces some negative features of breeding value in *Arachis, Carthamus, Citrullus, Corchorus, Nicotiana, Oryza* and *Vigna* (Table 2.28). In the remaining ones given in the table, the effects induced are not harmful. It is usually hard to distinguish pleiotropy from close linkage unless the segregating

Table 2.27. Male sterile genes exhibiting linkage with other genes

Species	References	*ms* Gene linked to gene(s) for
Brassica oleracea var. *italica* (1r, 2x = 18)	Sampson (1966a)	Bright green hypocotyl colour
Coleus blumei (1r, 4x = 24)	Ford (1950)	Deep leaf lobing
Corchorus capsularis (2x = 14)	Rakshit (1967)	Ribbon-shaped leaves
Gossypium hirsutum (1D, 4x = 52)	Kohel (1973)	Nectarlessness
Helianthus annuus (1r, 2x = 34)	Vranceanu and Stoenescu (1970b)	Anthocyanin distribution
	Pogorletskii (1973)	Spoon-shaped leaf
Hordeum vulgare (1r, 2x = 14)	Kasha and Walker (1960), Wiebe (1960)	Brachytic-dwarf, DDT resistance
	Falk and Kasha (1980)	Shrunken endosperm
Lycopersicon esculentum (1r, 2x = 24)	Currence (1944), Philouze (1974c)	Potato leaf shape, and green stem colour, absence of anthocyanin
	Hernandez and Wollard (1978)	Potato leaf shape
	Soressi and Salamini (1975)	Parthenocarpic fruits
	Tanksley et al. (1984)	Enzyme marker
Phaseolus vulgaris (1r, 2x = 22)	Durand (1981),	Woolly trait
	Lamprecht (1945)	Hypogeal cotyledons and exterior stigma
Plantago insularis (1r, 2x = 8)	Murr and Stebbins (1971)	Albino phenotype
Sorghum vulgare (1r, 4x = 20)	Stephens and Quinby (1945)	Awned and virescent traits

populations observed are too large and enough recombinations occur. This is known in a BC derivative of a cross *Lycopersicon esculentum* × *L. parviflorum* in which the cotyledons are wide and have round tips. The mature plants have wide and obtuse leaflets and split and recurved sterile stamen tips. Kesicki (1980) considers it a pleiotropic expression of *ms* gene, but F_2 and F_3 segregations reveal these two traits as conditioned by two recessive genes, the *ms* gene being closely linked to them. Since simultaneous occurrence of multiple mutations, although known, is extremely rare, pleiotropism is assumed in those cases where *ms* genes exhibit a constellation of other phenotypic alterations (Table 2.28). Furthermore, since such alterations are lost in restored or reverted mft plants, these have been considered as pleiotropic effects of *ms* genes rather than as linkages.

2.9.3 Anther Phenotype

This is an easily distinguishable and reliable phenotypic marker by which mst's are usually detected in field. Whereas in a few genotypes, the anthers are absent, aborted, deformed or malformed or transformed (st-mst, Table 2.14), in most of them (sp-mst, Table 2.15), the anthers are light-coloured, shrivelled, shrunken and

Table 2.28. The *ms* genes exhibiting pleiotropism

Species	References	Pleiotropic effect
Arachis hypogaea (1–3r, 4x = 40)	Ashri (1968)	Brachytic dwarfs, reduced stems, 5 instead of 4 leaves, diminitive pollenless anthers, reduced female fertility
Carthamus tinctorius (1r, 2x = 24)	Ebert (1964), Ebert and Knowles (1966)	Stem thin, anthers indehiscent, hulls thin and papery
Chionachne koenigii (1r, 2x = 20)	Rao (1975)	Plants shorter, profusely tillered, early flowering and long lived
Chrysanthemum morifolium (6x = 54)	Broertjes and Jong (1984)	Diminutive foliar and floral organs
Citrullus vulgaris (1r, 2x = 22)	Watts (1962)	Cleistogamous male flowers and glabrous plant form
Corchorus capsularis (1r, 2x = 14)	Mitra (1977)	Ribbon-like leaves, small flowers, short narrow petals having distinct notch at tip, stigma receptivity delayed
Glycine max (1r, 4x = 40)	Stelly and Palmer (1980b)	Reduced and highly abnormal corolla, p-mst temperature-sensitive
	Chen et al. (1985)	Polyembryony and polyploid progeny
Lactuca sativa (1r + 1D, 2x = 18)	Ryder (1967)	Leaf distortions high, flower size reduced, colour pale yellow, leaves narrow and twisted
L.sativa (1D, 2x = 18)	Ryder (1971)	Abnormal flowers with rolled and folded petals
L.sativa × *L.serriola* (2x = 18) (2x = 18)	Lindquist (1960)	Narrow, sharply cut leaves
Lens culinaris (1r, 2x = 14)	Sharma and Sharma (1978)	Plants dwarf, profusely branched, late with dark green leaves
Limnanthes douglasii (1D, 2x = 10)	Jain et al. (1978)	Short petals and white anthers
Lycopersicon esculentum (2x = 24) × *L.parviflorum* (BC_1) (1r, 2x = 24)	Kesicki (1980)	Wide and obtuse cotyledons and leaves
Nicotiana tabacum (1r, 4x = 48)	Bhat and Krishnamoorthy (1956)	Diminutive vegetative and reproductive parts
Oryza sativa (1r, 2x = 24)	Trees and Rutgers (1978)	Many floral anomalies and deviations in floral organ size and number
Phaseolus lunatus (1r, 2x = 22)	Allard (1953)	Seedling and plant morphology altered
Sesamum indicum (1r, 2x = 26)	Brar (1982)	Plant type and form considerably altered
Trifolium pratense (ir, 2x = 14)	Taylor et al. (1978)	Wrinkled petals
Vigna sinensis (ir, 2x = 24)	Sen and Bhowal (1962)	Luxuriant and brisk growth, reduced floral size, ill developed whitish minute dehiscent anthers.

Table 2.29. Anther phenotype in some genic male steriles

Species	Anther phenotype
Allium cepa (2x = 16)	Translucent, green, nondehiscent; brown and hard with normal filaments
Beta vulgaris (2x = 18)	White to yellow, shrunken, dehiscent or nondehiscent
Daucus carota (2x = 18)	Brown, shrivelled and pollenless; transformed into petals, rudimentary and degenerate, or stamens with unrolled filaments, dehiscent shrunken anthers
Glycine max (4x = 40)	Anthers normal to shrunken, reduced, flat and white
Hordeum vulgare (2x = 14)	Normal or highly reduced, dwarf, shrunken, lacking stomium
Lycopersicon esculentum (2x = 24)	Normal, anthers indehiscent; reduced, abnormal; rudimentary, deformed
Nicotiana tabacum (4x = 48)	Anthers rudimentary, deformed and indehiscent; dwarf with no pollen or archesporial tissue; normal, initially elongated and curved later; transformed into petals
Oryza sativa (2x = 24)	Filaments shortened, anther sacs compressed; anthers appendaged, thin and papery or minute, white, lacking sporogenous tissue; shrunken and indehiscent
Pennisetum typhoides (2x = 14)	Underdeveloped, indehiscent, shrivelled to normal anthers
Secale cereale (2x = 14)	Shrivelled to normal, fully to partially indehiscent
Solanum tuberosum (4x = 48)	Anthers dwarf, thin or fused to style, indehiscent, shrivelled or transformed
Zea mays (2x = 20)	Normal or shrunken, indehiscent, slender, stiff, needle-like or pebbly and refractive, indehiscent

nondehiscent (Table 2.29). In a few, anthers look normal but are indehiscent (fn-mst, Table 2.16). Thus considerable phenotypic diversity exists between the various mst genotypes of a species so that a few genotypes have a structural, others a functional and some others a sporogenous type of mst, e.g. barley, cucurbits, maize, pea, pearl millet, tobacco, tomato etc. (Tables 2.14–2.16). Since the diversity in anther phenotype occurs among the mst genotypes of a species but not within the plants of a genotype, genetic control of the anther phenotype is suggestive. The degree of anomalies in the anther phenotype is influenced by the genotype, the environment and their interactions.

2.9.4 Pollen Form and Function

One of the most reliable methods of mst plant identification is through pollen form, stainability and viability determinations. Barring the functional mst plants (Table 2.16) in which pollen is normal and viable but caged within anther sacs, and structural mst plants which lack pollen completely (Table 2.14), the sporogenous type of mst occurs in the majority of the plants (Table 2.15). In these, the pollen when formed is sterile, shrunken, deformed and has degenerated cytoplasmic and nuclear contents. Considerable diversity in this pollen phenotype exists among the sporogenous mst genotypes of many species, some of which are given in Table 2.30. Acetocarmine, acetoorcein, iodine-potassium iodide and cotton-blue stainings are used for pollen fertility determinations. These stains, although fairly useful for differentiation of mst and mft pollen, are not always reliable. For instance, pollen of some mst mutants of rice, beet, wheat, barley, pea, tomato, maize are phenotypically normal and stain well in the pollen stains, but they are nonfunctional. In them, the *ms* gene action is delayed until after pollen mitosis is completed and cytoplasm degeneration is also slow. In a mst rice line, 21% pollen is phenotypically normal and stains well with acetocarmine (Pan et al. 1982), but germination of this pollen is abnormal and the pollen fails to fertilize the ovary. In vitro germination tests are limited because of the difficulty in achieving germination in many taxa, particularly in three-celled systems. Moreover the medium which is optimal for fresh pollen may not be optimal for stored pollen. Also the stored pollen of many species fails to germinate in the optimal medium but germinates on the stigma.

In ms_1 soybean, large-sized, coenocytic pollen grains are formed (Albertsen and Palmer 1979). The coenocytes develop a pollen-type wall comprising tectum, columellae and endexine as in a normal pollen. Unlike it, however, they possess a variable number of randomly positioned colpi and furrows. Some of these produce pollen tube-like structures but the majority of them are nonfunctional. The mature pollen of two mst mutants of barley (Ahokas 1975a; this Vol.) has normal cytoplasmic contents but mildly formed exine and inapertulate pollen that does not germinate on barley stigma or on artificial media. Likewise, pollen of some mst genotypes is phenotypically normal and exhibits perfect stainability with acetocarmine and iodine solutions, but it is physiologically weak and develops an

Table 2.30. Pollen phenotype in some genic male steriles

Species	Pollen phenotype
Allium cepa (2x = 16)	Shrunken, shrivelled, deformed, uninucleate
Beta vulgaris (2x = 18)	Abortive and shrunken pollen, phenotypically normal, nonviable or viable
Glycine max (4x = 40)	Multi- or uninucleate, complete partially aborted, shrivelled and deformed
Hordeum vulgare (2x = 14)	Shrunken to normal, degenerated or less protoplasmic, inapertulate with poorly formed exine, stainable but nonviable
Oryza sativa (2x = 24)	Deformed with or without cytoplasm, phenotypically normal, stainable but nongerminative
Sorghum vulgare (2x = 14)	Multiploid sporocytes, deformed microspores lacking protein and starch, uninucleate with fuzzy, weakly reticulate exine having little starch inside
Triticum aestivum (6x = 42)	Microspores free or clumped with reduced cytoplasm contents, lacking starch, nuclei degenerating or crumpled, mono- or bi-nucleate
Vicia faba (2x = 12)	Intine absent, ectexine sculpture highly reduced, aperture configurations modified variously
Zea mays (2x = 20)	Highly shrivelled, yellow to white, aborted, adhering to walls, or clumped together

inefficient pollen tube that fails to fertilize the ovary (e.g. pea, rice, tomato; this Vol.). Since usually the enzymatic activity of mst pollen is much lower than that of mft pollen, evaluation of this activity would facilitate reliable mst pollen identification. Peroxidase or dehydrogenase activities have been used to differentiate mst from mft pollen. These enzyme tests are, however, not fool-proof. For instance, over 95% of stored jojoba pollen developed colour reaction in tetrazolium solution even from samples which did not exhibit any germination (Beasley and Yermanos 1976). Heslop-Harrison and Heslop-Harrison (1970) standardized a pollen viability test based on enzymatically induced fluorescence readily observable under a fluorescent microscope. All these methods involve loss of the pollen sample tested. A nondestructive method to test pollen viability developed by Dumas et al. (1983) depends on the use of nuclear magnetic resonance spectroscopy and is based on the changes in pollen water content as it becomes nonviable. In practice, for differentiation of mst pollen from mft pollen, acetocarmine, iodine or cotton blue staining is supplemented by tetrazolium testing. Some use various pollen germination tests, but even such tests carried out under artificial conditions do not indicate pollen sterility, since growth of pollen tubes in a few mst plants of rice and pea is insufficient to induce fertilization. Also, formation of well-developed pollen tubes may not indicate pollen fertility, as male gametes may be sterile at fertilization time. This was found by Stort (1984) in some orchid hybrids. Thus whenever phenotype and stainability of pollen are normal, the easy and reliable method to identify mst's would be to bag the young flower buds of suspected mst's, pollinate some of their flowers by their own and some by foreign pollen. Also the pollen of suspected mst plants should be used to pollinate normal mft's whose flower buds have been emasculated and bagged. Seed set would reliably identify mst's from mft's.

2.9.5 Gene Symbolisation and Location

Each mutant gene is symbolised by the major phenotypic alteration it produces. Accordingly the mutant genes causing g-mst are symbolised as *ms* genes. As many *ms* genes are known to occur in some plant species, they are numbered after being confirmed as nonallelic. Thus the first known *ms* gene of a species receives the numeral 1 and is designated as ms_1. The next one nonallelic to this gene becomes ms_2 and so the number increases. Since all the known mst genotypes are not available for allelism tests, the number lineage cannot always be maintained. Also many mst materials are lost and do not receive any gene symbol. Whenever the mst materials are available prior to allelism tests, the mst genotypes are assigned temporary letter symbols like a, b, c,. . . to z, aa, ab, ac,. . . to az until the allelism tests with the known *ms* genes prove them to be different from the basic available set. If they differ, the next numeral designation is given to them and the previously assigned temporary letter symbol is dropped. This has been done in maize, barley, tomato, cotton, soybean, etc. but not in many others in which *ms* genes occur. Location of *ms* genes on chromosomes is known in only a few plants and in them also only some *ms* genes are located (Table 2.31). For instance, in maize and barley,

Table 2.31. Distribution of *ms* genes over chromosomes

Species	*ms* Genes		Carrier chromosomes	Linkage groups
	Known	Located		
Hordeum vulgare	48	15	7	7
Lycopersicon esculentum	55	25	8	12
Pisum sativum	38	10	5	7
Zea mays	60	33	10	10

only half of the known *ms* genes are located on their chromosomes. The maximum number of *ms* genes (12 and 4) are located on chromosome 1 of maize and of barley, respectively. Chromosome 6 of maize has six *ms* genes; over other chromosomes the genes are randomly distributed. On the contrary, all located *ms* genes in tomato are present only on eight chromosomes, the remaining four either possess the unlocated ones or only a few *ms* genes. In pea, of the seven chromosomes, five carry eight *ms* genes, the remaining have not been located so far.

2.9.6 Allelic ms Genes

Male fertility in higher plants is controlled by a vast number of genes, most of which are present in the dominant state. Mutation of any of these to the recessive state leads to mst (Gottschalk and Kaul 1974). This is evidenced by the existence of a large number of recessive *ms* genes present in higher plants (Tables 2.6, 2.7). Many such *ms* genes existing in the genomes of barley, cotton, maize, pea, pepper, sorghum, soybean, tomato, etc. are nonallelic and have been numbered following the procedure given in section 2.9.5. However, in many others allelism tests have not been done. In addition to nonallelic mutations, certain allelic mutations occur at the *ms* locus. This is known for ms_1 in soybean, in which four independently arisen allelic mutations are known (Palmer 1976; Palmer et al. 1978). The mutants exhibit certain minor differences in different genetic backgrounds. In pea, three alleles of ms_5 and two each of ms_6 and ms_{10} were identified by Myers (1984). Likewise, two similar mutations at *ms* loci occur in winter squash and tomato (Francis and Bemis 1970; this Vol.). Since allelic mutations are difficult to differentiate, such mutations of *ms* genes may also be common in other plant species exhibiting g-mst, but they have escaped detection.

2.10 Breeding Value

For crossing, bisexual and self-pollinated plants have to be emasculated, which is tedious and time-consuming, requiring technical skill and labour, especially when the flower size is small and stamen number large. Male sterility obviates emasculation and facilitates ACP and NCP. It may either be complete so that no viable pollen is formed and no seed-set occurs in the mst flowers, provided they have been protected to exclude pollen from external sources or it may be partial, permitting small amounts of viable pollen and seed set formation. The expression is highly influenced by environment. Thus, with varying environmental conditions, high to very low pollen viability may be obtained. Unless a *ms* gene causes complete pollen sterility eliminating self-pollination fully and is stable in a wide range of environments, its utility in plant breeding is limited. Once a genotype with such a *ms* gene is detected, it is to be maintained and propagated.

2.10.1 Maintenance and Propagation

Due to the predominantly recessive and monogenic nature of *ms* genes, the maintenance of mst genotypes within a population poses certain problems. Their detection in a heterozygous state is not possible. Furthermore, a pure g-mst population cannot be produced, but *ms* genes can be carried along in a high frequency if seeds from mst plants only are harvested and used to produce the next generation. The seeds harvested may have been pollinated either by homozygous (*MsMs*) or heterozygous (*Msms*) mft plants. In the former case, the progeny will be heterozygous (*Msms*) and in the latter, it will segregate 50% *Msms* : 50% *msms*. The proportions of mst and mft plants that would be maintained through five generations after a cross, with only the mst plants being harvested, are given in Table 2.32. When an mst plant (*msms*) is pollinated by a homozygous mft plant (*MsMs*), all F_1 progeny plants are heterozygous (*Msms*) and mft. The F_2 generation will segregate 25% *MsMs* : 50% *Msms* : 25% *msms*. The proportions of mst and mft plants in succeeding generations can be predicted from the proportion of pollen grains with *Ms* genes and those with *ms* genes. For instance, in F_2 about 66.6% of pollen cells will be *Ms* and 33.3% *ms*. Random mating of male gametes in these proportions with mst eggs will produce 66.6% heterozygous (*Msms*) and 33%

Table 2.32. Male sterile and fertile plants (%) obtained after mst × mft cross (only mst's harvested). (After Poehlman 1979)

Generation	Parents		
	♀ *msms*	×	♂ *MsMs*
F_1	*MsMs* (self-pollinated)		
F_2	25% *msms*	×	25% *MsMs* 50% *MsMs*
F_3	33% *msms*	×	66% *Msms*
F_4	50% *msms*	×	50% *Msms*
F_5	50% *msms*	×	50% *Msms*

homozygous (*msms*) F_3 populations. Since all F_3 mft plants will be heterozygous, 50% of the pollen produced will have the dominant gene (*Ms*) and 50% the recessive gene (*ms*). This proportion would be maintained in subsequent generations as well. Thus, in practice, a g-mst line is maintained through pollinations by a line with an identical genotype (isoline) except for having the dominant *Ms* gene.

2.10.2 Gene Transfer

The majority of *ms* genes have arisen spontaneously and only a few after induced mutations (Tables 2.2, 2.2). Genotypes of most of the economically useful plants in which *ms* genes arise are not agronomically useful. This necessitates *ms* gene transfer from source to desirable genotypes. This is done through backcrossing (BC), whose steps depend upon the control of mst by dominant or recessive genes.

2.10.2.1 Dominant *Ms* Genes

When the mst is controlled by a dominant gene (*Ms*), the sterility can be easily transferred from its original source variety B to an elite variety A through a usual backcrossing procedure depicted in Fig. 2.20. Here the average proportion of the B genome is represented by dot numbers, which, initially 64, are reduced by one half with each backcross. Thus, with each cross or backcross to the recurrent elite parent A, the average proportion of the germplasm from the donor parent B is reduced to one half. Assuming that the F_1 from the original cross had one half of its germplasm from the donor parent, the proportion in BC_1 will be one quarter. With the n number of crosses and BC's to the recurrent parent, the proportion of germplasm from the donor parent B is $(\frac{1}{2})^n$, after five BC's, the proportion would be $(\frac{1}{2})^6 = 1/64$. In Fig. 2.20, this is represented by one dot out of 64 from the *Ms* gene donor variety B. The homozygosity level is given by the same formula as that used in the selfing series viz. $\frac{2^b-1}{2^b}$, where b is the number of backcrosses. Thus, after 5 BC's, the number of homozygous loci in the population will be 31/32 or 96.9%. The main distinction from the selfing is that in backcrossing, the homozygous loci are those of recurrent parent and are not different combinations of loci from both the parents. The proportion of homozygous genotypes in the population is given by the formula $\left(\frac{2^b-1}{2^b}\right)^n$, where b is the number of BC's and n the number of heterozygous loci in the F_1 of the original cross. With ten heterozygous loci, 72.8% of the BC_5 plants would be homozygous for the ten loci of the recurrent parent. In other words, almost all the homozygous renotypes would be those of the elite recurrent parent. With one more backcross, the number of genotypes, like that of the recurrent parent, would be 85.4%. The number of BC's mostly varies from three to ten. Where ten are used, almost complete identity with the recurrent parent is achieved. In general, five to six BC's suffice and the material may later be handled by pedigree or bulk methods.

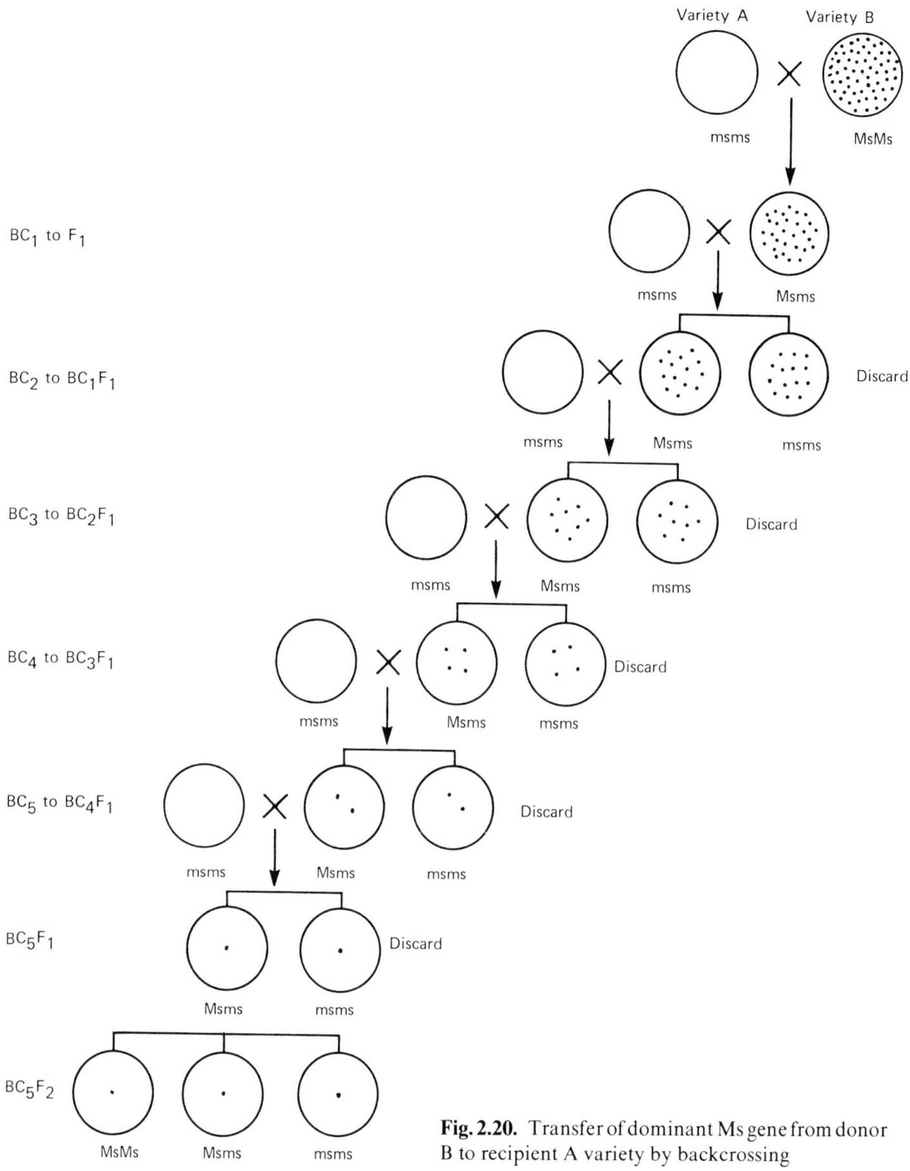

Fig. 2.20. Transfer of dominant Ms gene from donor B to recipient A variety by backcrossing

2.10.2.2 Recessive *ms* Genes

The g-mst in majority of the cases is recessive monogenic. With such genes, progeny from each BC segregates into *MsMs* and *Msms* genotypes. Since the heterozygote which contains the *ms* gene cannot be identified, the progeny has to be selfed to find mst plants before backcrossing to the recurrent parent. For this, either BC generations may be grown to F_2 or backcrossed two or three times to yield BC_2 or BC_3

plants homozygous for the *ms* gene under transfer. Another way is to backcross both the homozygous (*MsMs*) and heterozygous (*Msms*) plants to the recurrent parent and at the same time to self each plant to detect mst plants (Fig. 2.21). The backcross progeny from the homozygous plants are discarded. Details regarding development and utilization of *ms* genes are given in Section 2.10

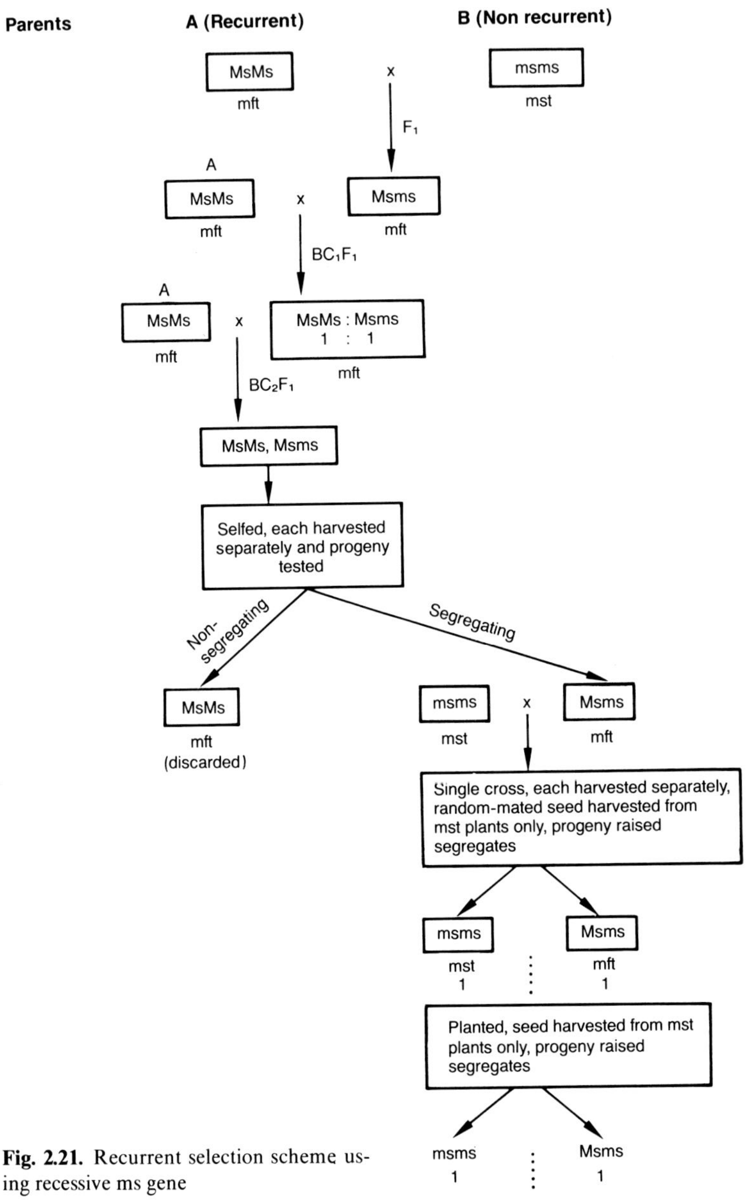

Fig. 2.21. Recurrent selection scheme using recessive ms gene

2.10.2.3 Polygenes

In rare cases, the mst is conditioned by polygenes (Table 2.1). In them each BC generation has to be grown through to F_3 lines before the next BC is made. Since in such cases the heritability of the mst traits is low and several genes are involved, BC F_2 and F_3 populations have to be very large and sometimes even F_4 populations have to be raised for mst plant isolations.

2.10.2.4 Limitations in Transfer

The predominant occurrence of recessive *ms* genes is a major obstacle in their transfer to elite varieties since the generation time involved is much longer than that required for the transfer of the dominant *Ms* gene. Linkage and/or pleiotropy may prove advantageous or troublesome and may facilitate or complicate backcross *ms* gene transfer. Where the *ms* gene either exhibits pleiotropic spectrum comprising phenotypically visible and desirable traits, or where it is linked to a distinct phenotypic marker gene or to some other useful genes, selection of mst genotypes is easy and less complicated. The occurrence of such a pleiotropy or linkage is, however, known in only a few cases (Tables 2.27, 2.28). The complication increases when the *ms* gene is linked to an undesirable gene or a block of genes because selection for the *ms* gene enforces undesirable gene selection, the intensity of the selection for the undesirable gene being directly proportional to the tightness of the linkage. Where the effect of the undesirable gene is conspicuous, a selfing series following a cross is effective in breaking the linkage, because an opportunity exists for breakage of the linkage in both male and female gametes. In backcrosses this cannot occur in the recurrent parent. However, where selection for breaking the linkage is not possible either because the effect of undesirable gene or genes are not expressed under the conditions of selfing and backcrossing or their heritability is low, backcrossing is more effective in breaking linkage than selfing. This is, however, limited where the linked traits are located near centromere and in certain other areas of the chromosomes where relatively little or no crossing-over may occur for the characteristics from either the recurrent parent or from the nonrecurrent parent. In such cases, the chromosome segments associated closely with the gene brought in from the nonrecurrent parent conditioning desirable or undesirable characteristics may persist unaltered in small segregating BC populations through many generations. Success in BC is thus related to the ease and freedom with which recombinations take place.

Since backcrossing involves several generations and is thus time-consuming, the additional labour and delay involved in the procedure is unjustifiable and impracticable unless the *ms* gene recipient variety is superior, stable and serves as a parent in many crosses for several years. In barley, over 200 spring and winter varieties have been converted to mst's and are available from USDA. These lines are maintained by pollinating with a line having identical genotype (isolines) except in having the respective dominant *Ms* gene. A major problem in using the gene for population improvement is the difficulty of transferring this gene to other cytoplasms, since plants carrying it have to be used as female. One way is to identify either the partial steriles or the environmental conditions in which *Ms* gene action is suppressed. Pollen from these plants can be used to fertilise hand-emasculated lines, cultivars or hybrids having a different cytoplasm. If the gametes carrying *Ms* gene are competitive, the progeny of the crosses will segregate 1:1 mst:mft. It is equally possible that the *Ms* gene does not express in different cytoplasms equally well, but this is rare.

2.11 Development and Utilization

The common objective of most plant breeding programs is population improvement. For this, various breeding procedures are applied but their utilization is restricted by (a) limited diversity of germplasm, (b) low genetic variability and recombination potential, (c) absence of intercrossing among hybrid progenies. Recurrent selection combats these disadvantages and has been adopted for population improvement in inbreeders like barley, sorghum, soybean and rice by using an *ms* gene to compose a random crossing population. Since g-mst is mostly conditioned by recessive and in some cases by dominant genes, the recurrent selection techniques used for them are described separately below.

2.11.1 Recessive *ms* Genes

To develop a population, a desirable mft parent and an acceptable mst parent are used. Two main considerations underlie the population development viz. (1) percentage of acceptable genes from the mst parent other than the *ms* gene, a percentage of less than 50% requires backcrossing to mft parents, (2) number of recombination generations required before selection begins. Although the recombination is easily accomplished after the *ms* gene has been incorporated into the population, each generation of recombinants takes additional time. Procedure for population development using a monogenically conditioned recessive *ms* gene is given in Table 2.33 and its use in recurrent selection in Fig. 2.21. It is based on assumptions that (a) 12.5% genetic contribution from the mst parent to the final population is acceptable, (b) the desired genetic proportion of the mst parent is achieved in the least possible time, and (c) two generations of recombination are required after the population has the desired proportion of genes from the mst parent. If the cytoplasm desired in the population is that of the recurrent parents, they should be used as females during one crossing generation. The backcrossing procedure given in Table 2.33 necessitates hand emasculation of mft plants. An alternative procedure outlined in Table 2.34 obviates many hand emasculations, facilitates crossing, but takes more generation time. In both these methods, the same parent is utilized in each BC generation. An alternative is to vary the matings with the mft parents in each generation of crossing, e.g. F_1 could be (*ms ms* × Parent A) and BC_1 could be (*ms ms* × Parent A) × Parent B. Each male parent should be used in the same number of crosses to maintain a similar genetic contribution from each in the final population.

Table 2.33. Population development using one ms gene. (W. Fehr, personal communication)

Generation	Steps
1	The mft recurrent parents (*MsMs*) are manually crossed to mst parents (*msms*), F_1 being heterozygous (*Msms*) has 50% genes from the mst parent
2	To recover recurrent parent cytoplasm, F_1 plants (male) are backcrossed to their recurrent parent. The BC_1F_1 progeny has MsMs : Msms 1:1. Average genetic contribution of the ms parent in BC_1F_1 generation is 25%
3	The BC_1F_1 plants (1 MsMs : 1 MsMs) are manually crossed to the recurrent parent making a number of crosses in order to be certain that the *ms* gene is present in BC_2 generation. Average genetic contribution of the *ms* parent to BC_2F_1 generation is 12.5%, the level desired in the final population
4	The BC_2F_1 plants (*MsMs* and *Msms*) are self-pollinated and each one is harvested separately
5	Each BC_2F_1 plant is progeny-tested. Progeny derived from homozygous mft plant (*MsMs*) is discarded but that from heterozygous mft plants (*Msms*) segregates for mst and is used to make single crosses among backcross-derived parents. These are made by crossing mft's as male parent onto mst's as female parent. Each single cross is harvested separately and similar quantities of each bulked to form a single population
6	Random mating of population is done by planting it in isolation and permitting open pollination to take place. The population comprises ⅓ mst (*msms*) and ⅔ mft (*Msms*) plants. Seed is harvested from each mst plant individually and a similar quantity of seed from each is bulked for the next season. The seeds of the population are 50% *Msms* and 50% *msms*
7	The next generation of recombination is conducted by planting seed of the population in isolation allowing open pollination to occur, harvesting the seed from each mst plant individually and bulking a similar quantity of seed from each plant. The seed harvested from the mst plants is 50% *Msms* and 50% *msms*
8	Each additional recombination generation is raised like generation 7, the segregation ratio being 1 Msms : 1 msms. The seed harvested from mst genotypes will have the same proportion of mst and mft genotype

Table 2.34. Population development by modified method using one ms gene. (W. Fehr, personal communication)

Generation	Steps
1.	Each mft parent (*MsMs*) is mated to mst plants (*msms*)
2.	F_1 plants obtained and selfed
3.	F_2 population is raised to segregate ¼ mst progeny. Desirable mst plants are manually pollinated by pollen of the recurrent parent to yield BC_1F_1 generation
4.	The BC_1F_1 plants (*Msms*) are selfed
5.	F_2 population is raised. BC generations are raised by repeating procedure of generations 2 and 3. To recover recurrent parent cytoplasm each must be used as female in one of the crossing generations. After the last BC is completed, the population is intermated in the manner described for generations 4 to 7 in Table 2.33

For population improvement using g-mst, many methods of recurrent selection can be used, viz. selection among individual plants, e.g. sorghum (Doggett 1972), soybean (Burton and Brim 1981), half-sib families, within half-sib families, e.g. soybean (Burton and Brim 1981) and selfed progeny, e.g. sorghum (Doggett 1972) etc. Some of these selection methods could be combined to fit the available resources and demands of the situation. Interpopulation improvement can be made by either reciprocal half- or full-sib selection. With g-mst, direct plant-to-plant crosses for reciprocal full-sib selection are not possible. Whereas mft can be crossed to mst, mst cannot be selfed. However, reciprocal full-sib selection can be made by using selfed progeny of heterozygous mft plants. Although several schemes for using g-mst in hybrid seed production exist, none has been commercially utilized so far. Using mst-facilitated recurrent selection, many crops have been developed and used to a limited extent. For instance, Suneson (1945) used a g-mst mutant discovered earlier (Suneson 1940) in hybrid barley breeding. Gardner (1972) developed many sorghum populations by using g-mst mutants at Nebraska. Using ms_3 gene, Doggett (1972) developed sorghum populations in Uganda. Sorghum population KP_2BR was developed by crossing 11 mft parents with greenbug resistance to a g-mst antherless population KP_1BR. Burton and Brim (1981) used ms_1 gene to develop soybean populations. In barley, mst-facilitated recurrent selection populations have been established (Ramage 1977) for a number of useful traits like earliness, shattering resistance, large awns, short straw, drought tolerance, powdery mildew resistance and winter-summer adaptability. In Japan, *ms* genes were used for some initial trials of recurrent selection and backcrossing in rice (Fujimaki et al. 1977; Fujimaki 1978). Later, Fujimaki (1979) proposed the use of *ms* gene in recurrent selection and its applications in rice improvement when undesirable linkages exist between agronomically important traits or when useful alleles are to be brought together from various genetic sources (Figs. 27.5 to 27.7). Using *ms* genes, Ikehashi and Fujimaki (1980) proposed a model of recurrent selection for rice improvement involving a breeding program that begins with composing a foundation population to which selection is applied. The population which is constructed to segregate for mst is to be derived from various genetic sources such as biparental crosses, double-crosses, poly-crosses or diallel-crosses and grown in isolation to advance outcrosses between mst segregants and mft plants.

The g-mst is more useful for recurrent selection schemes than g-mst as it is easy to manipulate a single *ms* gene but more difficult to bring together a specific *fr-c* gene combination and *Fr* gene into a specific genetic background. Despite the theoretical merits of recurrent selection, little work exists on selfers because of the difficulty in obtaining adequate crossed seeds. Use of g-mst has been proposed to overcome this problem, but such schemes have associated drawbacks in testing progeny that are segregating for mft and mst plants. The number of years per cycle of selection may also be greater for populations maintained with mst than for populations developed and maintained by hand pollination. The use of g-mst in recurrent selection for soybean yield is not practicable because hand pollination provides adequate crossed seeds (Fehr and Ortiz 1975). Therefore, g-mst will be useful in recurrent selection programs where a limited hybrid seed is to be produced.

2.11.2 Dominant *Ms* Genes

The majority of the *ms* genes being recessive, several breeding procedures utilising them in recurrent selection programs have been proposed (Gilmore 1964; Dogget 1968; Brim and Stuber 1973; Ramage 1977). Two major features favouring the use of *Ms* over *ms* genes are (a) selfing generation is not required to obtain segregants that breed true for fertility and (b) progenies of mst plants always segregate 1:1 for the *Ms* allele. Sorrells and Fritz (1982) proposed six procedures (A-F, Fig. 2.22) using dominant *Ms* gene in recurrent selection (Fig. 2.22) and backcrossing (Fig. 2.23). In Fig. 2.22, method A is the simplest. It consists of phenotypic recurrent (mass) selection of mst plants in a population followed by composing the seed from selected plants to reconstitute the population. A dense population planted in a single block would minimise assortative mating. Further minimisation could be achieved by combining mass selection with half-sib or S_1 selection. In method B, progenies of mass-selected mst plants are planted in separate rows or hills. Selections can be done both among and in their half-sib families. Seeds of mst plants from the latter are bulked to recompose the population. To promote random mating, method C is used for half-sib selection every year. For this, mst plants from selected half-sib families are planted directly back to half-sib rows. Random mating can be improved by alternating mst rows with the pollinator. Procedure D

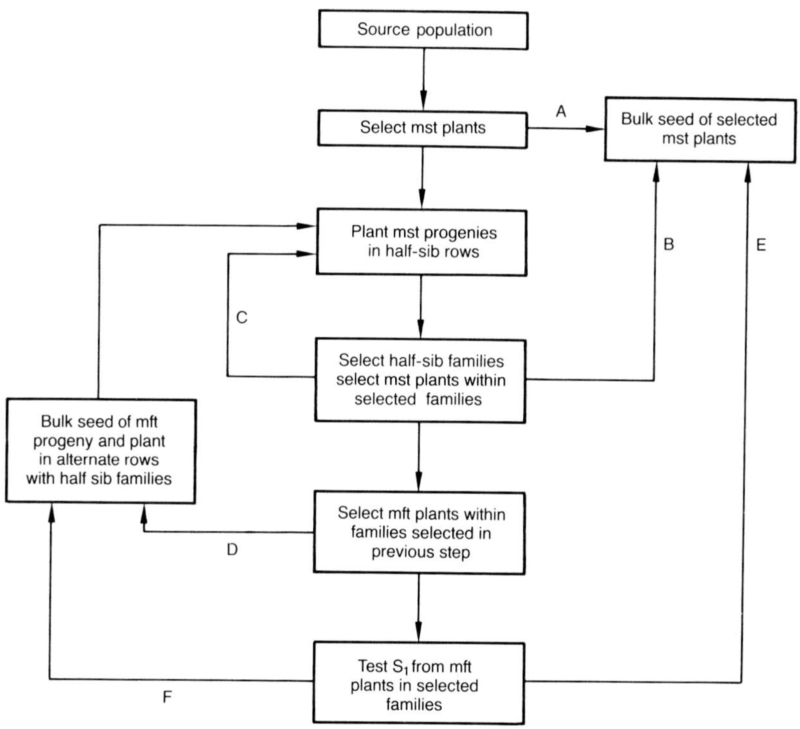

Fig. 2.22. Recurrent selection schemes using a dominant Ms gene

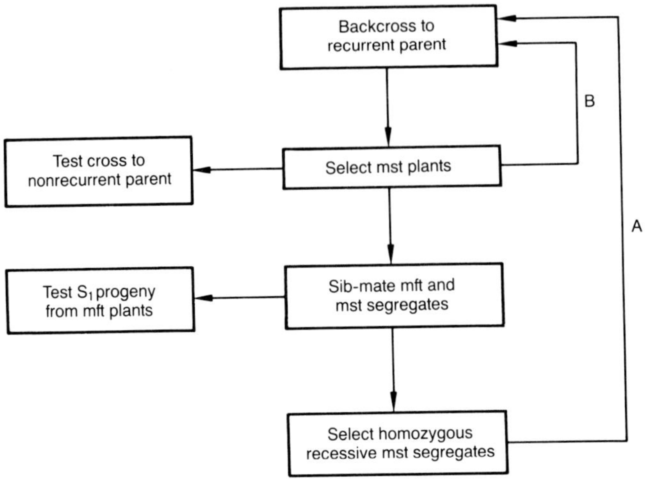

Fig. 2.23. Backcrossing scheme using a dominant ms allele

involves selection of mft and mst plants in half-sib families. Seeds of mft selections are bulked and planted in alternate pollinator rows and mst plants are planted in half-sib rows. Method E is more labour-intensive but enables testing S_1s from selected half-sib families. If selected half-sib progenies are planted directly back to half-sib rows, the cycle can be shortened by one generation in method F. Assuming one generation per year, a cycle of selection can be completed in 1 year by method A, C and D, in 2 years by B and F and in 3 years by E.

The *Ms* allele is rarely present in recurrent or nonrecurrent parents. Therefore, the first cross is done to the mst source. Time saving is possible if the recurrent parent is crossed to both mst source and nonrecurrent parent simultaneously. The mst F_1 plants from the cross mst source × recurrent parent and crossed to F_1 plants are obtained from the cross recurrent × nonrecurrent parents. To maintain *Ms* gene in the population (necessary, if the recurrent parent lacks *Ms* gene) and to obtain homozygous recessive segregants for the desired trait, mst and mft plants must be sibbed in the BC generation, as indicated in method A (Fig. 2.23). In method B, after BC_1, a large number of crosses are required to ensure that the recessive allele is not lost before sibbing or test-crossing progeny. To minimize the number of crosses, this method may be alternated with method A. Test crosses to a nonrecurrent parent would indicate which mst plants carry the recessive allele.

Compared to recessive *ms* traits, dominant *Ms* traits are easily managed in BC schemes, as the plants exhibiting the desired trait before anthesis can be directly backcrossed. For one dominant *Ms* gene expressed after anthesis, seven crosses ensure that at least one BC progeny possesses the *Ms* allele. Changing cytoplasm is not a problem in backcrossing where the endproduct is a fertile line or cultivar, since fertile progeny from the last BC may be used as male and the recurrent parent hand-emasculated. Dominant *Ms* genes occur in wheat, maize, cotton, potato, carrot, beans, etc. (Table 2.8). Because of the shorter generation time required for dominant genes, the feasibility of using recurrent selection and the BC method is higher in dominant than in recessive gene-controlled mst. However, recurrent selection supplements and is a component of other plant breeding methods.

2.12 Maintenance and Perpetuation

It is hard to breed mst plants, as they cannot be selfed unless fertility is restored by changed environment or by some chemical manipulation. Monogenic recessive g-mst occurs in the majority of plants. For breeding, one line (B) is required for maintenance and another (C) for fertility restoration (Fig. 2.24). The B-line is mft, nonfertility-restoring and, with the exception of the gene *Ms*, is mostly isogenic to the A-line. It is maintained by self- or sib- pollination. In seed production plots, roguing of mft plants is done manually. This not only involves labour but reduces plant number in seed fields. In practice, the maintenance of the A-line is ensured by planting it in rows adjacent to the B-line in an isolated field. The A-line is naturally or hand-pollinated by the B-line. B- and C-lines are maintained by bagging heads and harvesting self-pollinated or sibbed seeds from isolated fields, each line being grown in separate well-isolated fields. For producing single-cross hybrids, the A-line is planted in a second isolated field and pollinated by the C-line. Where reproductive parts serve as the agricultural product, recessive *ms* genes should not be present in the pollen parent C-line while using the mst plants for hybrid seed production.

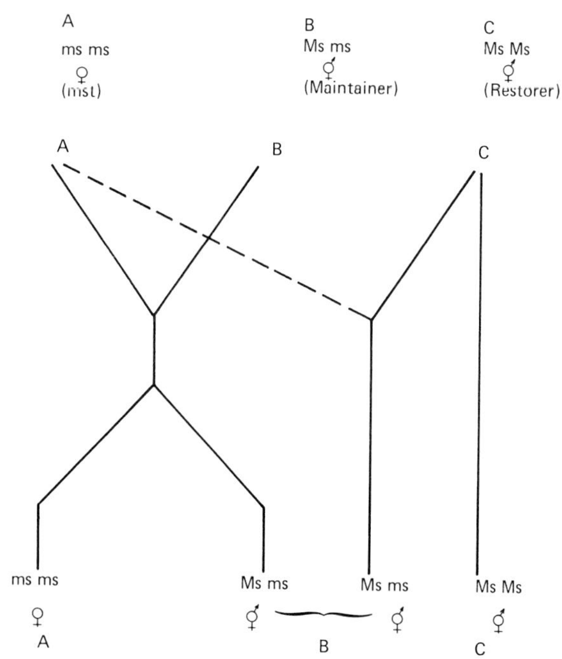

Fig. 2.24. Environmental sensitivity of gene-cytoplasmic male steriles

2.13 Contrivances to Usage

With almost all the available breeding procedures, maintenance of 100% g-mst stands is not possible. Whereas the maintainers segregate mst:mft in a 1:1 ratio, the restorers produce only mft plants (Fig. 2.24). In addition, the majority of the plants in which g-mst is known are selfers. This necessitates artificial cross-pollination (ACP). These two features restrict the use of g-mst in hybrid seed production to those plants in which gc-mst is not known or satisfactory, or *Fr* genes have not been detected. To obviate these restrictions, Frankel and Galun (1977) suggested certain genetic and field management procedures. These and others are described here briefly.

2.13.1 Artificial Maintenance

Usually selfing of mst mutants does not produce any seed. However, in a few mst mutants, mft is restored under certain environmental conditions (Table 2.25) or by subjecting them to certain chemical treatments (viz. GA_3 treatment in barley, tomato and pea) so that selfed seed is produced. Its plant progeny yields a nonsegregating pure stand of mst plants. Such conditions, if defined accurately, could be applied for the increased seed of the pure A-line. These optimal environmental conditions must, however, be determined for various mst mutants of different plants and are hard to define and provide in large-scale plantation areas. Moreover, various mst mutants of a plant exhibit a differential, sometimes diametrically opposite response to the same environmental factor. For instance, one mutant each of pepper and tomato is mft in the glasshouse but mst in the field and their other two mutants exhibit reverse behaviour (Martin and Crawford 1951; this Vol.). Thus, alterations in environmental conditions may restore mft in some but not in other mst mutants. Even in the former, the restoration is partial so that some mst and mft plants are intermingled in mst seed stands. To overcome this difficulty, it may be useful to have recessive marker genes in the A-line which are absent in the C-line. By this method, the contamination degree of a hybrid seed lot can be monitored, and in some instances the selfed progeny of the A-line can be eliminated. Genes linked to *ms* genes are known in a few species (Table 2.27) and in a few others *ms* genes exhibit pleiotropy (Table 2.28). Some linkages are not tight, and recombinations between the marker and the *ms* genes occur. These contaminate the mst lines in seed fields. In general, due to variable phenotype and unpredictable environmental conditions, environmentally sensitive mst mutants cannot reliably be used for commercial hybrid seed production.

2.13.2 Structural Male Steriles

These represent ideal female partners for hybrid seed production as in most of them either the anthers are absent or sporogenous tissue is degenerated (Table 2.14). Only a few of these, however, have been used (viz. snapdragon, chrysanthemums, barley, tomato, pea, almond, castor bean, rye, wheat) as in others, female fertility is impaired and fruits are deformed in shape, size or form. However, a mutant each of pea, tomato, snapdragon, and barley produces normal fruits and the hybrids outyield the parents. Except in snapdragons, in other plants, hand pollination is required for pollen transfer. Since this sterility type is not prevalent among mst plants, its use in commercial hybrid breeding is limited.

2.13.3 Clonal Propagation

A pure stand of mst plants may be established by vegetative propagation especially in the plants exhibiting versatile reproduction, e.g. many ornamentals, stem and tuber crops etc. This method is economical and can be used in *Allium, Canna, Chrysanthemum, Citrus, Dactylis, Geranium, Iris, Oryza, Saccharum, Vitis* etc. In them g-mst is known and clonal propagation is possible. The mst plants of rice, cotton, mustard and many ornamentals can be propagated vegetatively using their tillers, or other vegetative propagules for hybrid production, but the majority of the economically useful plants, especially cereals and legumes, do not reproduce vegetatively. Tillers of rice and wheat act as vegetative propagules but the tillering is limited and restricted to the growing season. Therefore, the asexual propagation of mst plants is a restricted method reserved for a few clonally propagated species.

2.13.4 Heterozygote Fertiles

In mst mutants in which mft is restored under certain environmental or other conditions, the stability of mst is not dependable under field conditions. Hence it is desirable to use stable mst mutants and remove heterozygous mft plants from the seed production fields. Double-density planting can be done to ensure a full stand in the field. The mft plants have to be removed before anthesis. In the absence of genetic markers, linkages or pleiotropic expressions associated with *ms* genes, reliable identification of mft plants needs careful survey and scrutiny and in many cases anther examinations and pollen fertility determinations are essential. This involves manual labour and technical skill. However, in species which are basically self-pollinating, where no natural vector for pollen transfer exists and where hand pollination is predicted, we have to wait for the first fruit to set upon selfing and then remove all fruit-bearing plants. Phenotypic markers (deviant plant forms, floral structure alterations etc.), genetic linkages or pleiotropic expressions facilitate mst plant detection in the fields in some species (Tables 2.27, 2.28). For this purpose, seedling markers are of special value, since roguing at the seedling stage in the nursery is cheaper than removal of adult plants in the field. This has been done successfully in tomato (Clayberg 1966; Soressi 1968; Philouze 1974a,b), cabbage (Sampson 1970) and cucurbits (Whelan 1974; Watts 1962). Certain physiological markers such as resistance to phytocides (DDT resistance in mst barley; Wiebe 1960) can be used to eliminate mft plants leaving mst plants in the seed production fields. Height differentials resulting from the linkage of stature with *ms* genes (e.g. barley; Wiebe 1960) can be used for roguing and cutting of taller heterozygous phenotypes at pre-anthesis stage, leaving a pure stand of shorter homozygous mst plants. To ensure a pure stand of the mst parent in maize hybrid seed production fields, Singleton and Jones (1930) proposed utilizing the linkage between endosperm colour and *ms* genes. However, this linkage was not close and tight, and hence could not be used for mft roguing. Using a close linkage, Galinat (1975) proposed a method for hybrid seed production of bicolor sweet corn. The method appears to resolve the problem of recombinants when linkage is fairly close between *ms* gene and endosperm colour and the scheme could be tried in crops having *ms* genes linked to a conspicuous endosperm coloration. The method uses a homozygous seed parent and a heterozygous maintainer (in the coupling phase) for white endosperm (y) and *ms* gene so that the A-line is *y-ms/y-ms* and B-line is *Y-ms/y-ms* (Y = non-white/coloured endosperm gene). The male parent of the hybrid is homozygous for fertility and yellow endosperm in maize, so that the C-line is *y-Ms/y-Ms*. The white seed from the maintenance field is separated by an electronic eye seed sorter to furnish the A-line mixed with a small proportion of recombinants. The separation of yellow seed from the hybrid seed production field removes all the selfed seeds (white-coloured) of the recombinants.

2.13.5 Extra Chromosomes

The use of balanced tertiary trisomics or plants carrying an extra homologous chromosome derived from a related species provides possibilities to obtain pure g-mst stands for use in seed production. Since chromosome segments on both sides of the breakpoint of the extra chromosome or the extra homologous chromosome are not transmitted through pollen, the fertility and marker genes located on them can assist in identification of the mst genotypes. In the desired tertiary trisomics stock, the interchange chromosome carries the normal *Ms* gene and two normal chromosomes carry *ms* gene. The only functional pollen has *n* chromosomes and the *ms* allele. When crossed with *msms*, it produces only mst progeny. The basic principle of this method was reported by Ramage and Tuleen (1964) and later schemes for utilizing balanced tertiary trisomics for hybrid seed production of diploid crops were outlined by Ramage (1965) and were used for barley hybrid seed production. These schemes use height, phytocide susceptibility, seed size and shape differentials to selectively combine harvest or separate diploids, trisomics and F_1 hybrids. Driscoll (1972) proposed the use of g-mst and a corresponding gene *mf* on a homologous alien chromosome.

2.13.6 Duplications

A duplication may owe its origin to an insertion, translocation or may arise following unequal exchange of end segments of the same arm of two homologs occurring after treatment of sporophytic tissue with mutagenic agents. These duplications are unpredictable and therefore do not have breeding value. Duplications that can be predicted to recur may be effectively used in breeding programs. Interchanges with one very short interchange segment may generate haploviable deficiency duplications. They are useful for placing and locating genes on chromosomes and in conjunction with *ms* genes can be utilized for producing all mst progeny for subsequent use in hybrid seed production. This was elegantly done by Patterson (1973) in producing hybrid maize, and the method was subjected to a patent right held by the University of Illinois foundation. Stocks needed to use the same method have been established using the polymitotic gene that also causes mst (Phillips 1975).

Another method of producing "all mst" progeny entails the utilisation of double duplications produced by type-2b intercrosses between interchange stocks involving the same two chromosomes (Phillips and Springer 1972). Tests of the feasibility of this "double duplication" method of producing "all mst" progeny indicate that if the Ms_1 locus is in the duplication, then crossovers within the duplication between either of the breakpoints and that locus will transfer the Ms_1 gene to the normal chromosome and produce undesired mft plants. The same is true if the Ms_1 locus is not in the duplication but closely linked to one breakpoint. One possible solution for this is to use a very short duplication that includes that locus, but a very long duplication from the other chromosome to preclude functioning of pollen carrying the duplication. An important feature was the use of yellow versus white (Y/y) endosperm marker, linked closely to the *ms* gene, which could be scored on self-pollinated ears and on yy, *ms, ms* ears crossed with putative double duplication heterozygotes. This excluded the necessity of growing the progeny in the field for classifying mst and mft plants. A reliable seedling marker linked to the *ms* gene could be used in the test if no closely linked endosperm markers are available.

2.13.7 Pollination Control

For profitable hybrid seed production, high natural cross-pollination of mst plants is most desirable. The predominant self-fertilizing nature of most of these species in which g-mst for hybrid seed production is of potential importance, and the unattractive nature of mst flowers to pollen-collecting insects limits the use of g-mst in hybrid seed production. Consequently, the pollen transfer from the C- to the A-line by insects or wind does not often result in sufficient seed set on the mst plants. This happens even in flowers in which nectar is the main attractant, as the insect visits to the mst flowers are infrequent. NCP could be improved variously viz. by adjustments of the ratio of pollen donors to seed parents and by spacing and location of pollen donors in relation to seed parents (planting design, row-to-row and plant-to-plant distances), by selection of a pollen donor attractive to vectors with high pollen-shedding capacity, with floral morphology favouring insect visits and wind pollen dispersal, by selection of suitable locations, season and proper nicking and handling of A- and C-lines. Even after many cultural and agronomic manipulations, hand pollination is supplemented to achieve high seed set on the mst lines of self-fertilizing species (e.g. pea, tomato, barley, rice, wheat). In tomato, Lapushner and Frankel (1967) found this economically useful. Thus, even when hand pollination supplementations are necessary in high-value autogamous plants like tomato, pepper, brinjal, pea and ornamentals, g-mst lines become a valuable asset for economically profitable hybrid seed production.

2.13.8 Planting Design

For improving F_1 seed yield, planting design is an important factor. For instance, in comparison to separate row design, a mixed design in which each mst plant is surrounded by mft plants increases F_1 seed yield. A random mixed design that consists of sowing mixed seeds of the mst A-line with the pollinator C-line in different ratios may be effective if recessive gene markers on mst plants, seeds or fruits are available. For this, the extent of natural cross-pollination should be high in the mst plants, otherwise hand pollination must be done. The occurrence of high NCP and gene markers in mst plants is rare and limits the efficacy of planting designs.

2.14 Use in Breeding

Despite many constraints, especially in the selfer, g-mst has been utilized for hybrid seed production on a limited scale mostly in plants where gc-mst was either not detected or available or an efficient *Fr* gene source had not been isolated. A few examples are cited below.

2.14.1 Cereals

Initially, g-mst was utilized for hybrid seed production in rice, wheat, maize and barley, but this was later replaced and/or supplemented by gc-mst. In rice g-mst lines have been used in recurrent selection and backcross breeding for rice improvement (Fujimaki 1979; Ikehashi and Fujimaki 1980). The selection has helped to dissipate undesirable linkages and increase desirable gene recombinations. Rice being a selfer, random crossing for recombinations in each recurrent cycle necessitates hand pollination. Some mst plants have slightly opened florets. This improves their natural crossability. Singh and Ikehashi (1981) obtained 15–33% seed set on four g-mst rice lines. In maize, initially g-mst was used for hybrid production but was later replaced by gc-mst. Maintenance of g-mst lines in the pure state has been achieved by the use of endogenous growth regulators and chromosomal genic systems. The major systems used are tertiary trisomics, pollen abortion genes, deficient chromosome segment, duplication deficiency etc. Due to positive high parent heterosis, some Russian and English monogenic recessive g-mst mutants have been used for developing hybrid wheat. Driscoll (1972) proposed a chromosomal system for producing hybrid bread wheat. Of the numerous known *ms* genes in barley some have been used directly and others through various genetic systems like balanced tertiary trisomic systems, *ms* gene linkages, pleiotropic expressions or phenotypic markers, auxins, chemicals etc. In sorghum, using g-mst mutants, single-, double- and triple-cross hybrids have been produced. Although these exhibit grain and forage yield heterosis, hand pollination and mft plant removal increase labour cost and reduce the economic utility of the g-mst lines.

2.14.2 Legumes

Unlike gc-mst, g-mst is common in most of the economically useful legumes like *Arachis, Cajanus, Crotalaria, Glycine, Medicago, Lathyrus, Phaseolus, Pisum, Trifolium, Vicia* and *Vigna.* It has been utilized for hybrid seed production and gaining monogenic heterosis. In *Pisum,* some mst plants have partially opened flowers that enhance their natural outcrossing. In *Cajanus,* anther colour acts as a marker phenotype and over 80% seed set occurs on the mst plants. In *Vigna,* the *ms* gene is used in recurrent selection and backcross breeding. In *Trifolium, ms* and *S* alleles are used conjointly for hybrid seed production. *Vicia faba* exhibits high positive grain yield and seed protein heterosis. For self-pollination, it needs petal movement brought about by bees, which also cause 30% cross-pollination. In mst flowers, the NCP percent increases and the hybrid flowers are less dependent upon bees for petal movement. However, for satisfactory seed set, hand pollination of the mst plants is necessary.

2.14.3 Oil Seed Crops

In mustard, rape, sunflower, sesame, cotton and castor, g-mst has been used in hybrid production. In sunflower, the MS_1 gene has been used for commercial hybrid seed production because of its linkage with a dominant gene controlling anthocyanin production (Vranceanu and Stoenescu 1970b) and high natural cross-pollination of msts. Hybrids usually outyield parents by over 50% in seed yield and 30% in oil content, but roguing of mft seedlings involves labour. Also g-mst becomes altered due to modifiers, and linkage is partial in some lines. Therefore, use of g-mst is limited. Because of the instability and erratic behaviour of many p-mst plants, they cannot be used in commercial hybrid sunflower seed production. Borchers (1971) used the temperature sensitivity of ms_6 and Dickson (1975) developed cytosteriles in *Brassica oleracea* for use in hybrid seed production. In *Gossypium hirsutum,* Huang et al. (1981) obtained 12–34% increase in lint yield of F_1 plants using g-mst with a local Chinese cotton line. A similar increase in seed oil content occurs in hybrid cotton seed. Suggestions of Weaver (1968) to use ms_5 ms_6 genotypes for hybrid cotton production are not appealing, as one out of 16 F_2 plants would be mst. Through environmental manipulations, some mst plants of *Sesamum indicum* can be selfed as well as crossed. This facilitates their maintenance, multiplication and utilisation in hybrid seed production. Although many mst plants are stable and exhibit high monogenic heterosis in grain yield, hand pollination is necessary in sesame for high economic yield.

2.14.4 Vegetable Crops

In the majority of vegetable crops like tomato, carrot, onion, pea, chillies, cabbage, cauliflower, broccoli, melons, pumpkins, cucumbers, gourds, squashes, turnip, lettuce, etc., g-mst mutants have been detected and utilized for F_1 production. In *Capsicum annuum,* g-mst has an edge over gc-mst because of relatively better stability to various environmental conditions and simple inheritance. Insect visits are higher by over 50% in mst than in mft plants. The mst lines of *Daucus carota* derived from wild carrot are characterized by distinct petal colour, reduced petal size and petaloid anthers. Honey bee visits over mst flowers are, however, reduced, due to their low nectar and aroma production. However, some open-pollinated carrot mst lines exhibit high outcrossing. F_1 hybrid tomato cultivars exhibit a high degree of positive heterosis over open-pollinated tomatoes in many characters especially in economic yield, disease resistance, net productivity and adaptability. Israel was the first country to develop hybrid tomatoes commercially (Lapushner and Frankel 1967, 1979). This was followed by France, Germany, Bulgaria and the USA. Penetrance of most of the *ms* genes is complete and therefore no danger of contamination by selfing exists. Elimination of mft heterozygotes has to be done and in only a few cases has linkage permitting identification of mst and mft plants in field been found. The "short anther" mst plants produce parthenocarpic fruits, but they are smaller with deep red colour and low acid content (Falavigna et al. 1978). Such fruits are valuable for processing. Fruit production of some mst tomato mutants is of poor quality. This has been improved by GA application, which also restores anther development and pollen fertility in some stamenless tomato mutants (Phathak et al. 1966; Sawhney and Greyson 1973b, Schmidt and Schmidt 1975), but fruits developed in these treated plants were deformed. Since bees are the vectors for pollen transfer, insecticide applications reduce NCP and in all cases hand pollination is necessary. A functional mst tomato mutant (Mittal and Thomas 1961) produced fruits without green shoulders and its fruit colour upon ripening was uniformly bright. Moreover, it exhibits a good combining ability and therefore shows a promise in tomato breeding. G-mst has been profitably used in F_1 fruit production in cucurbits as fruit growth and size exhibit high positive heterosis. Furthermore, bee and insect pollinations are high in them, so that hand pollination is unnecessary.

2.14.5 Ornamental Composites

Hybrid seed is obtained by rubbing the mst and mft heads of *Ageratum, Carthamus, Chrysanthemum, Tagetes, Zinnia* etc. Some F_1 seed is also produced in field plots by alternating mst and mft rows and harvesting the seed produced on mst plants. Since in many cases natural cross-pollination is low, hand pollinations by rubbing mft and mst flower heads are preferred. Colchicine-induced sunflower mst mutants are stable and have higher breeding value than others, many of which are generally unstable and sensitive to environment.

CHAPTER 3

Gene-Cytoplasmic Male Sterility

3.1 Introduction

Typically, mst represents a condition in which female sex is normal and male sex is nonfunctional. Two types of nuclear genes, *fr* and *ms*, control this sterility. The action of *ms* genes is cytoplasm-insensitive, whereas *fr* gene action is cytoplasm-dependent and -sensitive as these genes act only in a certain cytoplasm type, the S-cytoplasm. This cytoplasm has specific *c*-genes which interact with nuclear *fr* genes to produce mst phenotype so that only S_{frfr} individuals are mst. Since both nuclear (*fr*) and cytoplasmic (*c*) genes are required to produce mst, the sterility is due to g-c interaction and is termed gene-cytoplasmic male sterility (gc-mst). Plants having this sterility have thus two cytoplasm types or plasmatypes viz. N or F (normal or fertile) and S (sterile). N-cytoplasm may have a fertility-restoring gene *C*. The nuclear *fr* genes are ineffective in this cytoplasm in causing mst. Likewise, S cytoplasm is ineffective in the presence of *Fr* nuclear genes or their isoalleles. Thus, fertility to S_{frfr} genotypes can be restored by either replacement to S- by N-cytoplasm or of *fr* by *Fr* genes. This effect of the *Fr* gene is similar to the action of a suppressor allele in that the S-cytoplasm of restored lines is not permanently freed of mst and if the restoring alleles are removed, the S-cytoplasm expresses the original mst condition. In other words, the cytoplasmic condition is not reversed by the restorer alleles but it is merely suppressed. Furthermore, since normal-restoring nonfertility lines do not become cytoplasmically mst, no nuclear gene responsible for initiating the condition seems to exist. The combination S_{frfr} producing mst is highly sensitive and delicate, as mutation of S- to N-cytoplasm or any of the *fr* genes to *Fr* restores mft in the organism. On the other hand, the combination N_{FrFr} is an ideal one for stable fertility in an organism because for mst, the mutations of all the *Fr* genes to *fr* as well as of N-cytoplasm to S-type are needed. Such simultaneous mutations, although rare, do occur. In maize, pearl millet and sorghum, mutation rates of *Fr* and *c* genes are fairly high. Compared to *c*-genes, *fr*-gene mutation rate is higher in these as well as in many other crops.

Kley (1955) derived formulae giving relationships between frequency of natural occurrence of g-mst and gc-mst in self- and cross-fertilized species, their relative fertility and mutation rate. Accordingly, the expected percentages of mst plants in gene-cytoplasmic and in monogenic recessive and dominant are p/1-q, p/2-q and 2p/2-q%, respectively, where p is the probability of its occurrence and q is the relative fertility expressed as the ratio between the average seed set per plant by mst and mft plants. Thus, the gc-mst type is expected to be more frequent in allogamous than in autogamous species. Based on the origin of nucleus and cytoplasm, Lacadena (1968) classified gc-mst into autoplasmic and alloplasmic types. Autoplasmic types would include gc-mst plants that arise spontaneously in various plants due to cytoplasmic and nuclear mutations (Table 3.1). The mst arising following intraspecific crosses (Table 3.2) would fall under homoplasmic types. In the alloplasmic types, mst plants resulting after interspecific and intergeneric crosses (Table 3.3, 3.4) would be included. Thus, the examples given in Table 3.6 include interspecific and intergeneric alloplasmic types.

Table 3.1. Spontaneously arisen gene-cytoplasmic male sterility

Species	References
Allium cepa (1r, 2x = 16)	Monosmith (1926), Jones and Emsweller (1937)
A.fistulosum (2x = 16)	Virnich (1967)
A.schoenoprasum (1r, 2x = 16)	Tatliogu (1982)
Aquilegia vulgaris (2x = 14)	Kappert (1944)
Beta macrocarpa (2r, 2x = 18)	Oldemeyer (1957)
B.maritima (2r, 2x = 18)	Coe and Stewart (1977)
B.vulgaris (2r, 2x = 18)	Owen (1942, 1945, 1949, 1950)
Brassica campestris (2r, 2x = 24)	Ohkawa and Shiga (1981)
B.juncea (4x = 36)	Rawat and Anand (1979)
B.napus (4x = 38)	Yamaguchi and Kanno (1963)
Capsicum annuum (1–2r, 2x = 24)	Peterson (1958), Shifriss and Frankel (1971), Novâk et al. (1971)
Castanea hybrida (2x = 18)	Jaynes (1963)
C.mollissima (2x = 24)	Jaynes (1963)
C.sativa (2x = 24)	Jaynes (1963)
Centaurea scabiosa (2x = 20)	Fröst (1963)
Chenopodium quinoa (1r, 4x = 32)	Simmonds (1971)
Cirsium acaule (2x = 34)	Correns (1916)
C.eriophorum (2x = 34)	Correns (1916, 1937)
C.oleraceum (2x = 34)	Correns (1916)
C.palustre (2x = 34)	Correns (1916, 1937)
Crotalaria mucronata (2r, 2x = 16)	Kempana and Sastry (1958)
C.pallida (2x = 16)	Gupta (1975)
Dactylis glomerata (1D, 4x = 28)	Myers (1946)
Daucus carota (1–3D, 1D + 1r, 2x = 18)	Welch and Grimball (1947), Gabelman (1956), Banga (1959), Thompson (1961), Morelock (1974), Timin and Dobrutskaya (1981)
Epilobium hirsutum (2x = 36)	Michaelis (1949, 1954)
Gilia capitata (2x = 18)	Grant (1956)
Godetia whitheyi (1r, 2x = 14)	Hiorth (1948a)
Gossypium hirsutum (2r, 4x = 52)	Thombre and Mehetre (1979)
Helianthus annuus (po, 2x = 34)	Gundaev (1966)
Hordeum vulgare (2x = 14)	Favret and Ryan (1966)
Humulus lupulus (2x = 20)	Cech and Pozdena (1962)
Lolium multiflorum (2x = 14)	Nitzsche (1971)
Medicago sativa (2-many r, po, 4x = 32)	Davis (1967), Davis and Greenblatt (1967), Bradner and Childers (1968), Pederson and Stucker (1969), Staszewski (1970, 1976)
Mercurialis annua (1r, 2x = 16, 32)	Durand and Louis (1974), Louis and Durand (1978)
Origanum vulgare (1r, 4x = 32)	Kheyr-Pour (1980, 1981)
Oryza sativa (2x = 24)	Shuhua (1982)

Table 3.1 *(continued)*

Species	References
Pelargonium crispum (2x = 22)	Tokumasu (1974)
Pennisetum typhoides (1–3r, po, 2x = 14)	Burton (1958), Menon (1958, 1959), Singh and Sharma (1963), Athwal (1965b), Prakash (1967), Yadav (1974), Akénova and Chheda (1981)
Plantago ovata (1D, 2x = 8)	Mittal and Issar (1970)
Raphanus sativus (1r, 2x = 18)	Ogura (1968)
Ricinus communis (1r, 2x = 20)	Parkey (1957)
Secale cereale (1–3r, 2x = 14)	Putt (1954), Madej (1975), Garlicka and Madej (1975), Geiger and Schnell (1970), Scoles and Evans (1979c), Ahokas (1980c)
Sorghum vulgare (4x = 20), 1r	Maunder and Pickett (1959), Kidd (1961), Hadley and Singh (1961)
2r	Stephens and Holland (1954), Damon (1961), Erichsen and Ross (1963a), Pi and Wu (1963), Miller and Pickett (1964), Quinby (1980)
3r	Craigmiles (1961, 1962), Alam and Sandal (1967b)
4r	Tripathi et al. (1980)
5r	Webster and Singh (1964)
nk	Narkhede et al. (1968)
Tabernaemontana coronaria (2x = 22)	Dubey (1970)
Trifolium pratense (1r + 1D, 2x = 14)	Macewicz (1976a,b)
Vicia faba (1r, 2x = 12)	Bond et al. (1966a), Berthelem (1970)

Table 3.2. Gene-cytoplasmic male sterility in intraspecific hybrids

Species	References
Aegilops caudata (2x = 14)	Kihara (1968)
Allium cepa (2x = 16)	Peterson and Foskett (1953)
Aquilegia vulgaris (2x = 14)	Kappert (1944)
Beta vulgaris (2r, 2x = 18)	Owen (1945)
Brassica napus (1D, 4x = 38)	Kanno et al. (1963a), Yamaguchi and Kanno (1963)
Capsicum annuum (1r, 2r, 2x = 24)	Peterson (1958)
Crotalaria mucronata (2r, 2x = 16)	Edwardson (1967)
Epilobium hirsutum (1r, 2x = 36)	Michaelis (1939, 1948, 1954)
Gilia capitata (2x = 18)	Grant (1956)
Godetia whitneyi (2r, 2x = 14)	Hiorth (1948a,b)
Helianthus annuus (2x = 34)	Duvick (1966), Heiser (1982)
Hordeum vulgare ssp. *spontaneum* (1r, 2x = 14)	Ahokas (1979a, 1982b)
Linum usitatissimum (2x = 30)	Bateson and Gairdner (1921), Chittenden and Pellew (1927), Burnham et al. (1981)
Lolium perenne (2x = 14)	Jenkin (1931a)
Oryza sativa (2x = 24)	
(*indica* × *japonica*, 3r)	Hsu (1945), Shinjyo and Omura (1966), Kinoshita et al. (1980)
(*burmese* × *japonica*)	Selim (1931), Watanabe et al. (1968), Erickson (1969)

Table 3.2 *(continued)*

Species	References
(*spontanea* × *japonica*, po)	Katsuo and Mizushima (1958), Katsuo (1959)
(*phillipines* × *japonese*, 1r)	Kitamura (1962), Chu et al. (1972), Tamaru et al. (1980)
Petunia hybrida (2r, 2x = 14)	Frankel (1956), Edwardson and Corbett (1961), Edwardson and Warmke (1967)
Ricinus communis (1D, 2x = 20)	Parkey (1957)
Satureja hortensis (1r, 9x = 45)	Correns (1904, 1908)
Solanum tuberosum (Many D, 4x = 48)	De la Puente and Peloquin (1968), Grün (1970a,b, 1973)
Sorghum vulgare (4x = 20)	Alam and Sandal (1967a,b)
po	Stephens and Holland (1954)
1r	Maunder and Pickett (1959), Brooks and Brooks (1967)
2–3r	Craigmiles (1962), Narkhede et al. (1968)
Triticum aestivum (6x = 42)	Savchenko and Lastovych (1965)
Vicia faba (2x = 12)	Sirks (1931a, 1932),
(minor, major)	Bond et al. (1966a)
Zea mays (2x = 20)	Rhoades (1933)
1–2r	Duvick (1956), Jones et al. (1957a), Duvick et al. (1961)
1r	Buchert (1961), Beckett (1971)

Table 3.3. Gene-cytoplasmic male sterility in some interspecific crosses

Species crosses	References
Aegilops caudata (2x = 14) × *Ae. persica* (4x = 28)	Kihara and Kondo (1943)
Aegilops caudata (2x = 14) × *Ae.typica*	Kihara and Kondo (1943)
Aegilops caudata (2x = 14) × *Ae.umbellulata* (2x = 14)	Kihara and Kondo (1943)
Ae.longissima (2x = 14) × *Ae.aucheri* (2x = 14)	Kihara (1951)
[*Ae.triuncialis* (4x = 28) × *Ae.caudata* (2x = 14)] × *Ae.umbellulata* (2x = 14)	Kihara and Kondo (1943)
Allium cepa (2x = 16) × *A.ascalonicum* (2x = 16)	Little et al. (1944)
Allium cepa (2x = 16) × *A.drobovii* (2x = 16)	Saini and Davis (1969)
Allium cepa (2x = 16) × *A.galanthum* (2x = 16)	Saini and Davis (1969)
Allium cepa (2x = 16) × *A.pskemense* (2x = 16)	Saini and Davis (1969)
A.roylei (2x = 16) × *A.drobovii* (2x = 16)	Saini and Davis (1969)
Aquilegia flabellata (2x = 14) × *A.truncata* (2x = 14)	Skalinska (1931), Kappert (1944)
A.vulgaris (2x = 14, 28) × *A.chrysantha* (2x = 14)	Skalinska (1928a,b)
Begonia schmidtiana (4x = 32) × *B.acuminata*	Villerts (1942)
Begonia schmidtiana (4x = 32) × *B.hirtellà* (4x = 36)	Villerts (1942)
Begonia schmidtiana (4x = 32) × *B.subvillosa*	Villerts (1942)
Beta vulgaris (2x = 18) × *B.maritima* (2x = 18)	Dalke et al. (1983)
Brassica campestris (2x = 20) × *B.oleracea* (2x = 18)	Nishi and Hiraoka (1958)
Brassica campestris (2x = 20) × *B. napus* (4x = 38)	Heyn (1978)
B.napus (ms, 4x = 38) × *B.carinata* (4x = 34)	Shiga (1980)
B.napus (ms, 4x = 38) × *B.juncea* (4x = 36)	Shiga (1980)
B.nigra (2x = 16) × *B.oleracea* (2x = 18)	Mezushima and Katsuo (1958), Pearson (1972)
Bryonia alba (2x = 20) × *B.dioica* (2x = 20)	Heilbronn and Basarman (1942)
Capsicum annuum (2x = 24) × *C.pendlum* (2x = 24)	Rusenova-Kondareva (1968)

Table 3.3 *(continued)*

Species crosses	References
Capsicum annuum (2x = 24) × *C.pubescens* (2x = 24)	Rusenova-Kondareva (1968)
C.baccatum (2x = 24) × *C.annuum* (2x = 24)	Shifriss, personal communication, this Vol.
C.peruvianum (2x = 24) × *C.annuum* (2x = 24)	This Vol.
Cirsium oleraceum (2x = 34) × *C.canum* (2x = 34)	Correns (1916, 1928, 1937)
Cirsium oleraceum (2x = 34) × *C.palustre* (2x = 34)	Correns (1916, 1928, 1937)
Citrus limon (2x = 18) × *C.paradisi* (2x = 18)	Fröst (1948)
Citrus limon (2x = 18) × *C.sinensis* (2x = 18)	Fröst (1948)
C.unshiu (2x = 18) × *C.sinensis* (2x = 18)	Iwamasa (1966)
Cucurbita moschata (2x = 24) × *C.foetidissima* (4x = 40)	Bemis (1970)
Cucurbita moschata (2x = 24) × *C.palmata* (4x = 40)	Bemis (1973)
Desmodium sandwicense (2x = 22) × *D.intortum* (2x = 22)	McWhirter (1969)
Epilobium hirsutum (2x = 36) × *E.luteum* (2x = 36)	Michaelis (1929, 1932, 1933)
Epilobium hirsutum (2x = 36) × *E.montanum* (2x = 36)	Michaelis and Wertz (1935)
Epilobium hirsutum (2x = 36) × *E.parviflorum* (2x = 36)	Michaelis (1943)
Epilobium hirsutum (2x = 36) × *E.roseum* (2x = 36)	Michaelis (1943)
E.luteum (2x = 36) × *E.hirsutum* (2x = 36)	Michaelis (1943)
E.luteum (2x = 36) × *E. roseum* (2x = 36)	Michaelis (1943)
E.montanum (2x = 36) × *E.hirsutum* (2x = 36)	Michaelis (1943)
E.montanum (2x = 36) × *E.roseum* (2x = 36)	Oberreuter (1925)
E.montanum (2x = 36) × *E.parviflorum* (2x = 36)	Michaelis (1943)
Fragaria vesca (2x = 14) × *F.nilgerrensis* (2x = 14)	Nürnberg-Krüger (1958)
Geranium albiflorum × *G.pratense* (2x = 28)	Sansome (1936)
G.endressi (2x = 28) × *G.striatum* (2x = 28)	Sansome (1936)
Gilia exilis bizonata (2x = 18) × *G.tenuiflora tenuiflora* (2x = 18)	Grant (1956)
Gossypium anomalum (2x = 26) × *G.arboreum* (2x = 26)	Rhyne (1965), Meyer (1968), Chen and Meyer (1979)
Gossypium anomalum (2x = 26) × *G.hirsutum* (4x = 52)	Meyer and Meyer (1965), Rhyne (1965)
[*Gossypium anomalum* (2x = 26) × *G.thurberi* (2x = 26)] × *G.hirsutum* (4x = 52)	Meyer and Meyer (1965)
G.arboreum, (2x = 26) × *G.thurberi* (2x = 26)	Meyer and Meyer (1965), Richmond and Kohel (1961)
G.harknessii (2x = 26) × *G.hirsutum* (4x = 52)	Meyer (1975), Murthi and Weaver (1974)
G.hirsutum (4x = 52) × *G.barbadense* (4x = 52)	Zainyshev (1974)
Helianthus giganteus (2x = 34) × *H.annuus* (2x = 34)	Whelan (1981)
H.lenticularis (2x = 34) × *H.annuus* (2x = 34)	Anashchenko (1981)
H.maximiliani (2x = 34) × *H.annuus* (2x = 34)	Whelan and Dorrell (1981)
H.petiolaris (2x = 34) × *H.annuus* (2x = 34)	Leclercq (1969), Whelan (1979)
[*Hordeum jubatum* (2x = 14) × *H.compressum* (2x = 14)] × [*H.bulbosum* (2x = 14) × *H.vulgare* (2x = 14)]	Schooler (1967)
Hypericum acutum (2x = 16) × *H.montanum* (2x = 16)	Noack (1932)
Kohleria amabilis (2x = 26) × *K.bogotensis* (2x = 26)	Clayberg (1974)
K.tubiflora (2x = 26) × *K.bogotensis* (2x = 26)	Clayberg (1974)
K.tubiflora (2x = 26) × *K.digitaliflora* (2x = 26)	Clayberg (1974)
Linum floccosum (2x = 30) × *L.usitatissimum* (2x = 30)	Gajewski (1937)
Lolium perenne (2x = 14) × *L.multiflorum* (2x = 14)	Ahloowalia (1981)
Lycopersicon esculentum (2x = 24) × *L.hirsutum* (2x = 24)	Siegemund (1979a,b)
Lycopersicon esculentum (2x = 24) × *L.parviflorum* (2x = 24)	Kesicki (1980)
L.peruvianum (2x = 24) × *L.hirsutum* (2x = 24)	Vulkova-Achkova (1980)
Nicotiana amplexicaulis (2x = 36) × *N.tabacum* (4x = 48)	Berbeč, personal communication
N.benthamiana (2x = 38) × *N.tabacum* (4x = 48)	Ramavarma et al. (1978)

Table 3.3 *(continued)*

Species crosses	References
N.bigelovii (4x = 48) × *N.tabacum* (4x = 48)	Burk (1960), Schweppenhauser and Maan (1968)
N.debneyi (4x = 48) × *N.tabacum* (4x = 48)	Clayton (1950), Sand (1968), Hicks et al. (1977), Tsikov and Nikova (1980), Gerstel et al. (1980)
N.exigua (2x = 32) × *N.tabacum* (4x = 48)	Berbeč, personal communication
N.glauca (2x = 24) × *N.tabacum* (4x = 48)	Berbeč (1966)
N.goodspeedi (4x = 40) × *N.tabacum* (4x = 48)	Tsikov et al. (1974)
N.gossei (4x = 36) × *N.tabacum* (4x = 48)	Burns and Gerstel (1981)
N.knightiana (2x = 24) × *N.tabacum* (4x = 48)	Berbeč, personal communication
N.langsdorffii (2x = 18) × *N.sanderae* (2x = 18)	East (1919, 1929, 1932), Smith (1962)
N.megalosiphon (4x = 40) × *N.tabacum* (4x = 48)	Clayton (1950), Chaplin (1959), Burk (1960)
N.plumbaginifolia (2x = 20) × *N.tabacum* (4x = 48)	Chaplin (1964), Cameron and Moav (1957)
N.raimondii (2x = 24) × *N.tabacum* (4x = 48)	Berbeč, personal communication
N.repanda (4x = 48) × *N.tabacum* (4x = 48)	Burk (1967), Burns et al. (1978), Gerstel et al. (1978)
N.rustica (4x = 48) × *N.tabacum* (4x = 48)	Hart (1965)
N.suaveolens (4x = 32) × *N.tabacum* (4x = 48)	Izard and Hitier (1955a,b), Schweppenhauser and Maan (1968)
N.tabacum (4x = 48) × *N.bigelovii* (4x = 48)	Chaplin (1959), Burk (1960)
N.tabacum (4x = 48) × *N.glutinosa* (2x = 24)	Burk (1960)
N.tabacum (4x = 48) × *N.plumbaginifolia* (2x = 20)	Burk (1960)
N.tomentosiformis (2x = 24) × *N.otophora* (2x = 24)	Goodspeed (1954), Gerstel and Burns (1974)
N.undulata (2x = 24) × *N.tabacum* (4x = 48)	Chaplin (1964)
Oenothera biennis (2x = 14) × *O.argillicola* (2x = 14)	Stubbe (1964)
O.parviflora (2x = 14) × *O.biennis* (2x = 14)	Stubbe (1964)
O.parviflora (2x = 14) × *O.hookeri* (2x = 14)	Stubbe (1964)
Oryza glaberrima (2x = 24) × *O.sativa* (2x = 24)	Erickson (1969), Carnahan et al. (1972)
O.perennis (2x = 24) × *O.sativa* (2x = 24)	Katsuo and Mizushima (1958)
O.rufipogon (2x = 24) × *O.sativa* (2x = 24)	Cheng, unpubl.
O.sativa (2x = 24) × *O.glaberrima* (2x = 24)	Yabuno (1977), Sano (1983)
Paspalum urvillei (4x = 40) × *P.malacophyllum* (4x = 40)	Burton (1943)
Pelargonium denticulatum (10x = 90) × *P.roseum* (8x = 72)	Tokumasu (1974)
Petunia axillaris (2x = 14) × *P.hybrida* (2x = 14)	Everett and Gabelman (after Duvick 1959), Bianchi (1963), Hamilton (1965), Edwardson and Warmke (1967), Frankel et al. (1969), Izhar and Frankel (1971), Izhar (1975), Izhar et al. (1984), this Vol.
P.integrifolia (2x = 14) × *P.hybrida* (2x = 14)	Izhar et al. (1948), this Vol.
P.parodii (2x = 14) × *P.hybrida* (2x = 14)	Izhar et al. (1948), this Vol.
Phaseolus vulgaris (2x = 22) × *P.multiflorus* (2x = 22)	Rudorf (1954)
Saccharum spontaneum (8x = 80) × *S.barberi* (10x = 120)	Raghavan (1951, 1953, 1964)
Saccharum spontaneum (8x = 80) × *S.narenga* (4x = 40)	Raghavan (1951, 1953, 1964)
Saccharum spontaneum (8x = 80) × *S.officinarum* (8x = 80)	Raghavan (1951, 1953, 1964)
Saccharum spontaneum (8x = 80) × *S.sclerostachya* (4x = 48)	Raghavan (1951, 1953, 1964)
Secale cereale (2x = 14) × *S.kuprijanovii* (2x = 14)	Lapinski (1972)
Secale cereale (2x = 14) × *S.montanum* (2x = 14)	Lapinski (1972)

Table 3.3 *(continued)*

Species crosses	References
Solanum crosses	(refer to Table 49.10)
Sorghum vulgare (4x = 20) × *S.halepense* (4x = 20)	Wilhite (1960a), McClure (1962)
Sorghum vulgare (4x = 20) × *S.almum* (8x = 40)	McClure (1962)
Streptocarpus soleanthus (2x = 32) × *S.rexii* (2x = 32)	Oehlkers (1964)
S.comptonii (2x = 32) × *S.gardeni* (2x = 32)	Oehlkers (1961, 1964)
S.comptonii (2x = 32) × *S.grandis* (2x = 32)	Oehlkers (1961, 1964)
S.comptonii (2x = 32) × *S.rexii* (2x = 32)	Oehlkers (1961, 1964)
S.comptonii (2x = 32) × *S.soleanthus* (2x = 32)	Oehlkers (1961, 1964)
S.wendlandi (2x = 32) × *S.baudertii* (2x = 32)	Oehlkers (1938, 1940, 1941)
S.wendlandi (2x = 32) × *S.grandis* (2x = 32)	Oehlkers (1938, 1940, 1941)
S.wendlandi (2x = 32) × *S.meyeri* (2x = 32)	Oehlkers (1938, 1940, 1941)
S.wendlandi (2x = 32) × *S.michelmorei* (2x = 32)	Oehlkers (1961, 1964)
S.wendlandi (2x = 32) × *S.montigena* (2x = 32)	Oehlkers (1961, 1964)
S.wendlandi (2x = 32) × *S.rexii* (2x = 32)	Oehlkers (1961, 1964)
S.wendlandi (2x = 32) × *S.soleanthus* (2x = 32)	Oehlkers (1961, 1964)
S.wendlandi (2x = 32) × *S.vandeleuri* (2x = 32)	Oehlkers (1961, 1964)
Tragopogon pratensis (2x = 12) × *T.porrifolius* (2x = 12)	Ownbey and McCollum (1953)
Triticum araraticum (4x = 28) × *T.aestivum* (6x = 42)	
T.boeoticum (2x = 14) × *T.aestivum* (6x = 42)	
T.boeoticum (2x = 14) × *T.spelta* (6x = 42)	
T.boeoticum (2x = 14) × *T.turgidum* (4x = 28)	
T.compactum (6x = 42) × *T.aestivum* (6x = 42)	(refer to Tsunewaki 1980b,
T.monococcum (2x = 14) × *T.aestivum* (6x = 42)	Panayotov 1980, Wilson and
T.monococcum (2x = 14) × *T.spelta* (6x = 42)	Driscoll (1983)
T.monococcum (2x = 14) × *T.turgidum* (4x = 28)	
T.timonovum × *T.aestivum* (6x = 42)	
T.timopheevi (4x = 28) × *T.aestivum* (6x = 42)	
T.zhukovskyi (6x = 42) × *T.aestivum* (6x = 42)	
Viola tricolor (2x = 26) × *V.arvensis* (2x = 34)	Clausen (1926)
Vitis vinifera (2x = 38) × *V.riparia* (2x = 38)	Staudt (1973)
Zea mexicana (2x = 20) × *Z.mays* (2x = 20)	Duvick (1965)

Table 3.4. Gene-cytoplasmic male sterility in intergeneric crosses

	Female parent		Male parent
1	*Aegilops aucheri* (2x = 14)	×	*Triticum aestivum* (6x = 42)
2	*Ae.bicornis* (2x = 14)	×	*T.aestivum* (6x = 42)
	Ae.bicornis (2x = 14)	×	*T.durum* (4x = 28)
3	*Ae.biuncialis* (4x = 28)	×	*T.aestivum* (6x = 42)
	Ae.biuncialis (4x = 28)	×	*T.spelta* (6x = 42)
	Ae.biuncialis (4x = 28)	×	*T.sphaerococcum* (6x = 42)
4	*Ae.caudata* (2x = 14)	×	*T.aestivum* (6x = 42)
	Ae.caudata (2x = 14)	×	*T.durum* (4x = 28)
	Ae.caudata (2x = 14)	×	*T.spelta* (6x = 42)
	Ae.caudata (2x = 14)	×	*T.vulgare* (6x = 42)
5	*Ae.columnaris* (4x = 28)	×	*T.aestivum* (6x = 42)
	Ae.columnaris (4x = 28)	×	*T.macca* (6x = 42)
	Ae.columnaris (4x = 28)	×	*T.spelta* (6x = 42)

Table 3.4 *(continued)*

	Female parent		Male parent
	Ae.columnaris (4x = 28)	×	*T.sphaerococcum* (6x = 42)
6	*Ae.comosa* (2x = 14)	×	*T.aestivum* (6x = 42)
7	*Ae.heldreichii* (2x = 14)	×	*T.aestivum* (6x = 42)
	Ae.heldreichii (2x = 14)	×	*T.spelta* (6x = 42)
8	*Ae.kotschyi* (4x = 28)	×	*T.aestivum* (6x = 42)
	Ae.kotschyi (4x = 28)	×	*T.macca* (6x = 42)
	Ae.kotschyi (4x = 28)	×	*T.spelta* (6x = 42)
	Ae.kotschyi (4x = 28)	×	*T.timopheevi* (4x = 28)
9	*Ae.macrochaeta* (2x = 28)	×	*T.aestivum* (6x = 42)
10	*Ae.mutica* (2x = 14)	×	*T.aestivum* (6x = 42)
	Ae.mutica (2x = 14)	×	*T.spelta* (6x = 42)
11	*Ae.ovata* (4x = 28)	×	*T.aestivum* (6x = 42)
	Ae.ovata (4x = 28)	×	*T.durum* (4x = 28)
	Ae.ovata (4x = 28)	×	*T.macca* (6x = 42)
	Ae.ovata (4x = 28)	×	*T.spelta* (6x = 42)
	Ae.ovata (4x = 28)	×	*T.sphaerococcum* (6x = 42)
12	*Ae.recta* (6x = 42)	×	*T.aestivum* (6x = 42)
13	*Ae.sharonensis* (2x = 14)	×	*T.aestivum* (6x = 42)
	Ae.sharonensis (2x = 14)	×	*T.spelta* (6x = 42)
14	*Ae.speltoides* (2x = 14)	×	*T.aestivum* (6x = 42)
15	*Ae.squarrosa* (2x = 14)	×	*T.spelta* (6x = 42)
16	*Ae.triaristata* (6x = 42)	×	*T.aestivum* (6x = 42)
	Ae.triaristata (6x = 42)	×	*T.macca* (6x = 42)
	Ae.triaristata (6x = 42)	×	*T.spelta* (6x = 42)
	Ae.triaristata (6x = 42)	×	*T.sphaerococcum* (6x = 42)
17	*Ae.triuncialis* (4x = 28)	×	*T.aestivum* (6x = 42)
	Ae.triuncialis (4x = 28)	×	*T.macca* (6x = 42)
	Ae.triuncialis (4x = 28)	×	*T.spelta* (6x = 42)
18	*Ae.umbellulata* (2x = 14)	×	*T.aestivum* (6x = 42)
	Ae.umbellulata (2x = 14)	×	*T.spelta* (6x = 42)
19	*Ae.uniaristata* (2x = 14)	×	*T.aestivum* (6x = 42)
	Ae.uniaristata (2x = 14)	×	*T.durum* (4x = 28)
	Ae.uniaristata (2x = 14)	×	*T.spelta* (6x = 42)
20	*Ae.variabilis* (4x = 28)	×	*T.aestivum* (6x = 42)
	Ae.variabilis (4x = 28)	×	*T.macca* (6x = 42)
	Ae.variabilis (4x = 28)	×	*T.spelta* (6x = 42)
	Ae.variabilis (4x = 28)	×	*T.timopheevi* (4x = 28)
21	*Ae.ventricosa* (4x = 28)	×	*T.aestivum* (6x = 42)
	Ae.ventricosa (4x = 28)	×	*T.timopheevi* (4x = 28)
22	*Agroelymus turneri* (4x = 28)	×	*Agropyron distachymum* (4x = 28)
	Agroelymus turneri (4x = 28)	×	*A.repens* (4x = 28)
	Agroelymus turneri (4x = 28)	×	*Elymus canadensis* (4x = 28)
23	*Diplotaxis muralis* (2x = 22)	×	*Brassica campestris* (2x = 20)
24	*Euchlaena mexicana* (2x = 20)	×	*Zea mays* (2x = 20)
25	*Lolium multiflorum* (2x = 14) × *L.perenne* (2x = 14)	×	*Festuca arundianacea* (6x = 42)
26	*L.perenne* (2x = 14) × *L.multiflorum* (2x = 14)	×	*Festuca pratensis* (2x = 14)
27	*Lycopersicon cheesemannii* (2x = 24)	×	*Solanum pennelli* (2x = 24)
28	*L.esculentum* (2x = 24)	×	*S.pennelli* (2x = 24)
29	*L.hirsutum* (2x = 24)	×	*S.pennelli* (2x = 24)
30	*L.minutum* (2x = 24)	×	*S.pennelli* (2x = 24)

Table 3.4 *(continued)*

	Female parent		Male parent
31	*L.minutum* (2x=24) × *L.hirsutum* (2x=24)	×	*S.pennelli* (2x=24)
32	*L.peruvianum* (2x=24)	×	*S.pennelli* (2x=24)
33	*Raphanus sativus* (2x=18)	×	*Brassica campestris* (2x=20)
34	*R.sativus* (2x=18)	×	*B.napus* (2x=38)
35	*Saccharum officinarum* (8x=80)	×	*Narenga prophyrocoma* (6x=30)
36	*Triticum aestivum* (6x=42)	×	*Secale cereale* (2x=14)
37	*T.durum* (4x=28)	×	*S.cereale* (2x=14)

References 1–21 refer to Panaytov (1980), Tsunewaki (1980b), Wilson and Driscoll (1983); 22 Nielsen (1961a); 23 Hinata and Kanno (1976); 24 Mazoti and Velasquez (1962); 25 Anderson (1966); 26 Wit (1974); 27–28 Andersen (1963a,b); 29 Andersen (1963b); Vulkova-Achkova (1980); 30–31 Andersen (1964); 32 Vulkova-Achkova (1980); 33 Heyn (1978); 34 Bannerot et al. (1974); 35 Janaki-Ammal (1942); 36 Lein (1948); 37 Vega and Lacadena (1979).

3.2 Occurrence

3.2.1 Spontaneous and Species Crosses

Requirement of conjoint occurrence of nuclear *fr* genes and *c*-cytoplasmic genes for the expression of gc-mst reduces the frequency of spontaneous occurrence of gc-mst to ¼ g-mst genotypes. This is supported by the ratio of the total number of spontaneously arisen g-mst and gc-mst plants (Tables 2.1 175:44, and Table 3.1 4:1). Since gc-mst requires a specific nuclear (*fr* gene) and cytoplasm (*c* gene) combination, about 75% of gc-mst plants known have arisen after interspecific and intergeneric hybridizations (Fig. 3.1c,d,g,h). The sterility arisen spontaneously or after intra- or inter-specific hybridizations is more prevalent in dicots than in monocots (Fig. 3.1). On the other hand, most of the monocots gc-mst's have arisen after intergeneric hybridizations. In both dicots and monocots, the majority of the diploid gc-mst arise spontaneously; the proportion of tetraploids is low (Fig. 3.2). In *Dactylis glomerata* and *Plantago ovata*, a single dominant gene causes mst and in *Trifolium pratense*, the association of a dominant and a recessive gene is required for the development of stable mst. In carrots, either two to three duplicate dominant *Fr* genes cause mst or complementary action of one recessive and one dominant gene results in mst. In S-cytoplasm *Mercurialis annua*, three dominant genes I, R_1 and R_2 interact with each other differently (Louis and Durand 1978). Gene I induces mst, whereas R_1 restores mft partially and R_2 complements R_1 to revoke full mft; gene R_2 is inactive in itself. In some alfalfa, pearl millet and sunflower genotypes, mst is conditioned by polygenes with the result that mst expression increases with each generation of selfing that increases the number of polygenes. The influence of environment is profound in these. In contrast to rampant *fr* gene mutations, *c*-gene mutations are rare. These are however, frequent in maize, sorghum and solanum in which three or more S-cytoplasm types exist.

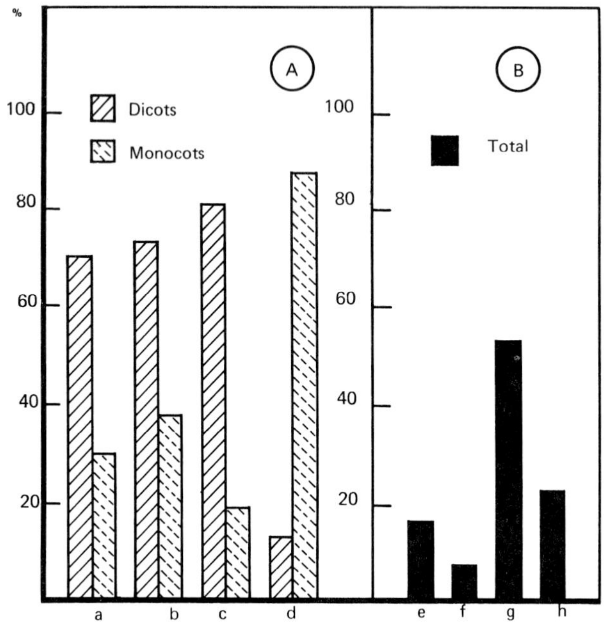

Fig. 3.1A,B. Relative proportion of gene-cytoplasmic male steriles arisen spontaneously (*a,e*) or after intraspecific (*b,f*) interspecific (*c,g*) and intergeneric (*d,h*) crosses

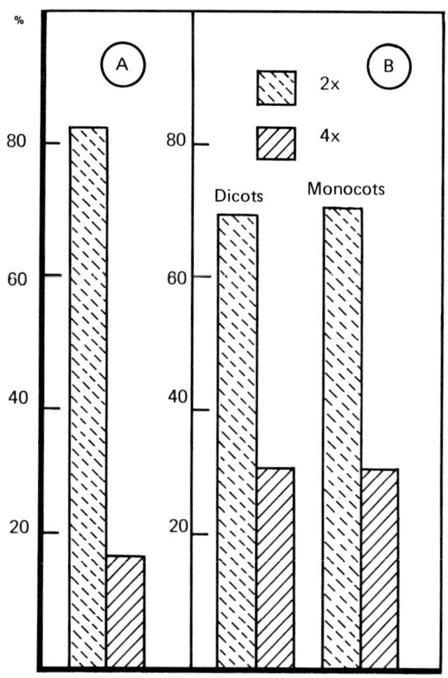

Fig. 3.2A,B. Frequency of spontaneously arisen diploid and tetraploid male steriles; general **(A)**, dicots and monocots **(B)**

3.2.2 Induced

Gc-mst induction is arduous as simultaneous or successive mutations at the nuclear as well as at the cytoplasmic level are required for its expression. Such mutations are rare and can neither be easily induced nor directed. Instead it seems easier to induce cytoplasmic mutations in plants having *fr* genes and N-cytoplasm. This has been achieved by using chemical or physical mutagens in beet, pearl millet and grass pea (Table 3.5). No mst plants were obtained after treating grass pea genotypes with 0.1 to 0.5% EMS and 1 to 5 kR γ-rays, singly and in combination (this Vol.). It seems that the cytoplasmic mutations, like nuclear gene mutations, are genotype-specific. This needs confirmation by extensive investigation.

Table 3.5. Mutagen-induced gene-cytoplasmic male sterility

Species	References	Mutagen
Beta vulgaris (2r, 2x = 18)	Kinoshita (1971, 1980), Mikami et al. (1978)	γ-rays
Lathyrus sativus (2x = 14)	Chekalin (1972), Chekalin et al. (1973)	NEU + DAB, γ-rays
Pennisetum typhoides (1r, 2x = 14)	Burton and Hanna (1976)	EB, mitomycin, streptomycin
Sorghum vulgare (1r, 4x = 20)	Malinovsky et al. (1973)	Colchicine

3.2.3 Alloplasmic

Alloplasmy refers to the phenomenon wherein cells of biological organisms have the cytoplasm of the one and the nucleus of the other species. It is produced by placing a nuclear genome into an alien cytoplasm. This is achieved by substitution BC's in which an interspecific or intergeneric hybrid is repeated backcrossed as female to the male parent whose nuclear genome is to be placed in the cytoplasm of the female parent. Five to six such substitution BC's replace the nuclear genome of the female parent by that of the male parent so that the female parent contributes cytoplasm and the male parent the nuclear genome in the alloplasmic hybrid. This genome transfer and development of the alloplasmic line is depicted in Fig. 3.3. One of the major effects of genome-cytoplasm interaction in some alloplasmic lines is the development of mst. This is evidenced by the examples given in Table 3.6. In this, the examples of interspecific and intergeneric alloplasmic mst genotypes have been grouped together. Since the resultant phenotype is mainly governed by the nuclear genome, the table is arranged alphabetically on the nuclear genome basis which is followed by donor species cytoplasm. In most of these, F_1 plants arising after crossing of cytoplasm and genome-donor species are meagre to partial mst. However, the proportion and degree of mst increases with each BC generation of the hybrid plants by genome donor recurrent male parent. After five to six BC generations, the mst is almost complete and stable in them.

More species genome + cytoplasm combinations produce alloplasmic mst than genera + genera combinations (Fig. 3.4). The maximum number of species

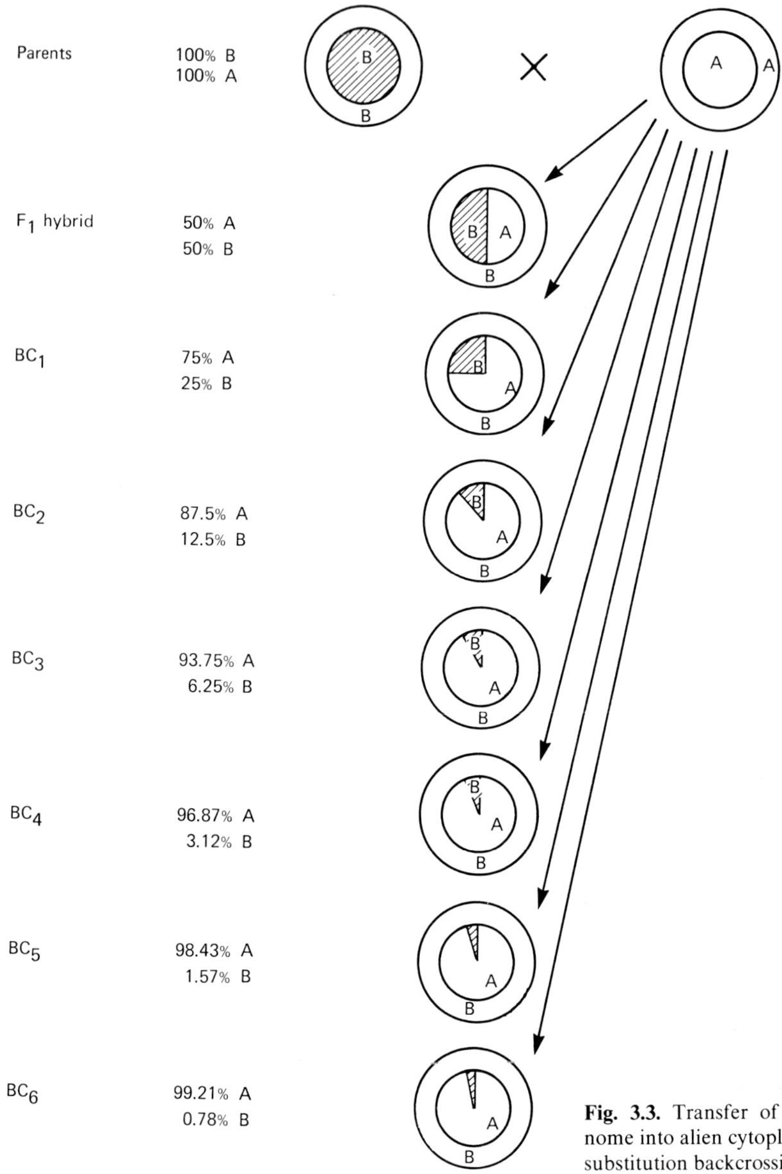

Fig. 3.3. Transfer of nuclear genome into alien cytoplasm through substitution backcrossing

Table 3.6. Alloplasmic male sterility in higher plants

Genome	Cytoplasm
Allium ascalonicum (1r, 2x = 16)	*A.cepa* (2x = 16)
A.drobovii (1r)	*A.cepa* (2x = 16)
A.galanthum (1r, 2x = 16)	*A.cepa* (2x = 16)
A.pskemense (1r, 4x = 32)	*A.cepa* (2x = 16)
Beta macrocarpa (2r, 2x = 18)	*B.vulgaris* (2x = 18)
B.maritima (2r, 2x = 18)	*B.vulgaris* (2x = 18)
Brassica campestris (2r, 2x = 20)	*Raphanus sativus* (2x = 18)
B.juncea (1r, 4x = 36)	*Brassica napus* (4x = 38)
B.juncea (1r, 4x = 36)	*B.carinata* (4x = 38)
B.juncea (1r, 4x = 36)	*B.tournefortii* (2x = 20)
B.juncea (1r, 4x = 36)	*Diplotaxis muralis* (6x = 42)
B.napus (4x = 38)	*R.sativus* (2x = 18)
B.oleracea (2x = 18)	*R.sativus* (2x = 18)
Capsicum annuum (2x = 24)	*C.peruvianum*
C.pendulum (2x = 24)	*C.annuum* (2x = 24)
C.pubescens (2x = 24)	*C.annuum* (2x = 24)
Gossypium anomalum (2x = 26)	*G.hirsutum* (4x = 52)
G.arboreum (2x = 26)	*G.anomalum* (2x = 26)
G.barbadense (4x = 52)	*G.hirsutum* (4x = 52)
G.hirsutum (4x = 52)	*G.harknessii* (2x = 26)
Helianthus annuus (2x = 34)	*H.giganteus* (2x = 34)
H.annuus (2x = 34)	*H.maximiliani* (2x = 34)
H.annuus (2x = 34)	*H.petiolaris* (2x = 34)
Lycopersicon hirsutum (2x = 24)	*L.esculentum* (2x = 24)
L.hirsutum (2x = 24)	*L.peruvianum* (2x = 24)
L.parviflorum (2x = 24)	*L.esculentum* (2x = 24)
Nicotiana glutinosa (2x = 24)	*N.tabacum* (4x = 48)
N.otophora (2x = 24)	*N.tomentosiformis* (2x = 24)
N.plumbaginifolia (2x = 20)	*N.tabacum* (4x = 48)
N.sanderea (2x = 18)	*N.langesdorffii* (2x = 18)
N.tabacum (4x = 48)	*N.amplexicaulis* (2x = 36)
N.tabacum (4x = 48)	*N.benthamiana* (2x = 38)
N.tabacum (4x = 48)	*N.bigelovii* (4x = 48)
N.tabacum (4x = 48)	*N.debneyi* (4x = 48)
N.tabacum (4x = 48)	*N.exigua* (2x = 32)
N.tabacum (4x = 48)	*N.glauca* (2x = 24)
N.tabacum (4x = 48)	*N.glutinosa* (2x = 24)
N.tabacum (4x = 48)	*N.goodspeedii* (4x = 40)
N.tabacum (4x = 48)	*N.gossei* (4x = 36)
N.tabacum (4x = 48)	*N.knightiana* (2x = 24)
N.tabacum (4x = 48)	*N.megalosiphon* (4x = 40)
N.tabacum (4x = 48)	*N.plumbaginifolia* (2x = 20)
N.tabacum (4x = 48)	*N.raimondii* (2x = 24)
N.tabacum (4x = 48)	*N.repanda* (4x = 48)
N.tabacum (4x = 48)	*N.rustica* (4x = 48)
N.tabacum (4x = 48)	*N.suaveolens* (4x = 32)
N.tabacum (4x = 48)	*N.undulata* (2x = 24)
Oryza glaberrima (2x = 24)	*O.sativa* (2x = 24)
O.rufipogon (2x = 24)	*O.sativa* (2x = 24)
O.sativa (2x = 24)	*O.glaberrima* (2x = 24)
O.sativa (2x = 24)	*O.rufipogon* (2x = 24)

Table 3.6 *(continued)*

Genome	Cytoplasm
Petunia hybrida (2x = 14)	*P.axillaris* (2x = 14)
P.hybrida (2x = 14)	*P.integrifolia* (2x = 14)
P.hybrida (2x = 14)	*P.parodii* (2x = 14)
Secale cereale (2x = 14)	*S.kuprijanovii* (2x = 14)
S.cereale (2x = 14)	*S.montanum* (2x = 14)
Solanum acaule (4x = 48)	*S.phureja* (2x = 24)
S.acaule (4x = 48)	*S.tuberosum* (4x = 48)
S.ajanhuiri (2x = 24)	*S.tuberosum* (4x = 48)
S.andigenum (4x = 48)	*S.tuberosum* (4x = 48)
S.berthaultii (2x = 24)	*S.demissum* (6x = 72)
S.bulbocastanum (2x = 24)	*S.acaule* (4x = 48)
S.bulbocastanum (2x = 24)	*S.verrucosum* (2x = 24)
S.caniarense (2x = 24)	*S.verrucosum* (2x = 24)
S.chacoense (2x = 24)	*S.antipoviczii* (4x = 48)
S.chacoense (2x = 24)	*S.demissum* (6x = 72)
S.chacoense (2x = 24)	*S.phureja* (2x = 24)
S.chacoense (2x = 24)	*S.polytrichon* (4x = 48)
S.chacoense (2x = 24)	*S.rybinii* (2x = 24)
S.chacoense (2x = 24)	*S.stenotomum* (2x = 24)
S.chacoense (2x = 24)	*S.verrucosum* (2x = 24)
S.chomatophilum (2x = 24)	*S.verrucosum* (2x = 24)
S.curtilobum (2x = 24)	*S.tuberosum* (4x = 48)
S.famatinae (2x = 24)	*S.verrucosum* (2x = 24)
S.gibberulosum (2x = 24)	*S.demissum* (6x = 72)
S.gibberulosum (2x = 24)	*S.verrucosum* (2x = 24)
S.gigantophyllum (2x = 24)	*S.demissum* (6x = 72)
S.gigantophyllum (2x = 24)	*S.verrucosum* (2x = 24)
S.goniocalyx (2x = 24)	*S.demissum* (6x = 72)
S.goniocalyx (2x = 24)	*S.polytrichon* (4x = 48)
S.goniocalyx (2x = 24)	*S.tuberosum* (4x = 48)
S.goniocalyx (2x = 24)	*S.verrucosum* (2x = 24)
S.infundibuliforme (2x = 24)	*S.verrucosum* (2x = 24)
S.kurtzianum (2x = 24)	*S.chacoense* (2x = 24)
S.kurtzianum (2x = 24)	*S.verrucosum* (2x = 24)
S.leptophyes (2x = 24)	*S.verrucosum* (2x = 24)
S.macolae (2x = 24)	*S.verrucosum* (2x = 24)
S.megistacrolobum (2x = 24)	*S.verrucosum* (2x = 24)
S.multidissectum (2x = 24)	*S.verrucosum* (2x = 24)
S.pennelli (2x = 24)	*Lycopersicon cheesemannii* (2x = 24)
S.pennelli (2x = 24)	*L.esculentum* (2x = 24)
S.pennelli (2x = 24)	*L.hirsutum* (2x = 24)
S.phureja (2x = 24)	*S.antipoviczii* (4x = 48)
S.phureja (2x = 24)	*S.demissum* (6x = 72)
S.phureja (2x = 24)	*S.polytrichon* (4x = 48)
S.phureja (2x = 24)	*S.tuberosum* (4x = 48)
S.phureja (2x = 24)	*S.verrucosum* (2x = 24)
S.pinnatisectum (2x = 24)	*S.etuberosum* (2x = 24)
S.raphanifolium (2x = 24)	*S.verrucosum* (2x = 24)
S.saltense (2x = 24)	*S.demissum* (6x = 72)
S.simplicifolium (2x = 24)	*S.demissum* (6x = 72)
S.simplicifolium (2x = 24)	*S.verrucosum* (2x = 24)
S.soukupii (2x = 24)	*S.demissum* (6x = 72)

Table 3.6 *(continued)*

Genome	Cytoplasm
S.stenotomum (2x = 24)	*S.chacoense* (2x = 24)
S.stenotomum (2x = 24)	*S.tuberosum* (4x = 48)
S.stenotomum (2x = 24)	*S.verrucosum* (2x = 24)
S.toralapanum (2x = 24)	*S.verrucosum* (2x = 24)
S.tuberosum (4x = 48)	*S.chacoense* (2x = 24)
S.tuberosum (4x = 48)	*S.demissum* (6x = 72)
S.tuberosum (4x = 48)	*S.phureja* (2x = 24)
S.tuberosum (4x = 48)	*S.stoloniferum* (4x = 48)
S.tuberosum (4x = 48)	*S.tuberosum* (4x = 48)
S.tuberosum (4x = 48)	*S.verrucosum* (2x = 24)
S.vernie (2x = 24)	*S.verrucosum* (2x = 24)
S.yabari (2x = 24)	*S.verrucosum* (2x = 24)
Triticum aestivum (6x = 42)	*Aegilops aucheri* (2x = 14)
T.aestivum (6x = 42)	*Ae.bicornis* (2x = 14)
T.aestivum (6x = 42)	*Ae.biuncialis* (4x = 28)
T.aestivum (6x = 42)	*Ae.caudata* (2x = 14)
T.aestivum (6x = 42)	*Ae.columnaris* (4x = 28)
T.aestivum (6x = 42)	*Ae.comosa* (2x = 14)
T.aestivum (6x = 42)	*Ae.heldreichii* (2x = 14)
T.aestivum (6x = 42)	*Ae.kotschyi* (4x = 28)
T.aestivum (6x = 42)	*Ae.macrochaeta* (2x = 28)
T.aestivum (6x = 42)	*Ae.mutica* (2x = 14)
T.aestivum (6x = 42)	*Ae.ovata* (4x = 28)
T.aestivum (6x = 42)	*Ae.recta* (6x = 42)
T.aestivum (6x = 42)	*Ae.sharonensis* (2x = 14)
T.aestivum (6x = 42)	*Ae.speltoides* (2x = 14)
T.aestivum (6x = 42)	*Ae.triaristata* (6x = 42)
T.aestivum (6x = 42)	*Ae.triuncialis* (4x = 28)
T.aestivum (6x = 42)	*Ae.umbellulata* (2x = 14)
T.aestivum (6x = 42)	*Ae.uniaristata* (2x = 14)
T.aestivum (6x = 42)	*Ae.variabilis* (4x = 28)
T.aestivum (6x = 42)	*Ae.ventricosa* (4x = 28)
T.durum (4x = 28)	*Ae.bicornis* (2x = 14)
T.durum (4x = 28)	*Ae.caudata* (2x = 14)
T.durum (4x = 28)	*Ae.ovata* (4x = 28)
T.durum (4x = 28)	*Ae.uniaristata* (2x = 14)
T.macha (6x = 42)	*Ae.columnaris* (4x = 28)
T.macha (6x = 42)	*Ae.kotschyi* (4x = 28)
T.macha (6x = 42)	*Ae.ovata* (4x = 28)
T.macha (6x = 42)	*Ae.triaristata* (6x = 42)
T.macha (6x = 42)	*Ae.triuncialis* (4x = 28)
T.macha (6x = 42)	*Ae.variabilis* (4x = 28)
T.spelta (6x = 42)	*Ae.biuncialis* (4x = 28)
T.spelta (6x = 42)	*Ae.boeoticum* (2x = 14)
T.spelta (6x = 42)	*Ae.caudata* (2x = 14)
T.spelta (6x = 42)	*Ae.columnaris* (4x = 28)
T.spelta (6x = 42)	*Ae.heldreichi* (2x = 14)
T.spelta (6x = 42)	*Ae.kotschyi* (4x = 28)
T.spelta (6x = 42)	*Ae.mutica* (2x = 14)
T.spelta (6x = 42)	*Ae.ovata* (4x = 28)
T.spelta (6x = 42)	*Ae.sharonensis* (2x = 14)
T.spelta (6x = 42)	*Ae.squarrosa* (2x = 14)

Table 3.6 *(continued)*

Genome	Cytoplasm
T.spelta (6x=42)	*Ae.triaristata* (6x=42)
T.spelta (6x=42)	*Ae.triuncialis* (4x=28)
T.spelta (6x=42)	*Ae.umbellulata* (2x=14)
T.spelta (6x=42)	*Ae.uniaristata* (2x=14)
T.spelta (6x=42)	*Ae.variabilis* (4x=28)
T.sphaerococcum (6x=42)	*Ae.biuncialis* (4x=28)
T.sphaerococcum (6x=42)	*Ae.columnaris* (4x=28)
T.sphaerococcum (6x=42)	*Ae.ovata* (4x=28)
T.sphaerococcum (6x=42)	*Ae.triaristata* (6x=42)
T.timopheevi (4x=28)	*Ae.kotschyi* (4x=28)
T.timopheevi (4x=28)	*Ae.variabilis* (4x=28)
T.timopheevi (4x=28)	*Ae.ventricosa* (4x=28)
T.turgidum (4x=28)	*Agropyron intermedium* (6x=42)
T.vulgare (6x=42)	*A.caudata* (2x=14)
T.aestivum (6x=42)	*T.araraticum* (4x=28)
T.aestivum (6x=42)	*T.boeticum* (2x=14)
T.aestivum (6x=42)	*T.compactum* (6x=42)
T.aestivum (6x=42)	*T.monococcum* (2x=14)
T.aestivum (6x=42)	*T.timopheevi* (4x=28)
T.aestivum (6x=42)	*T.zhukovskyi* (6x=42)

genome + cytoplasm combinations produce mst in *Solanum* followed by *Nicotiana* and *Triticum*; in the remaining the proportion is low and is grouped together in Fig. 3.4B. In many such alloplasmic mst lines, mft can be restored after pollinating them by cytoplasm donor pollen parents (Fig. 3.5) which possess *Fr* genes that maintain mft in S-cytoplasm plants. Even one such *Fr* gene, because of its dominant nature, is sufficient to maintain fertility. Until this *Fr* gene is brought into the genome of alloplasmic mst lines, the fertility is not restored in them. This explains the delayed fertility restorations in which many crossing cycles to cytoplasm donors are required to restore mft in some alloplasmic mst lines. Furthermore, in many plants, especially polyploids like wheat, mft is regulated by a balance of *Fr* and modifier genes some of which are epistatic and others inhibitory in nature. The influence of environment over expression of both *Fr* and *fr* genes is high (Tables 3.25, 3.26). In addition to mst, alloplasmy usually produces other effects also. For instance in wheat, rice, tobacco and mustard, certain alloplasmic combinations produce haploid and twin seedlings, polyembryonic and apomictic seeds, reduced plant vigour, delayed heading, partially open florets, complete to partial anther supression and/or transformation etc. A characteristic feature of alloplasmic mst is the variability in anther, sporogenous tissue or pollen degeneration stages. This variability is more pronounced in allogamous than in autogamous species. For instance, in the allogamous mst rye (Vega and Lacadena 1979) pistilloidy, malformed anthers, anther loculi fusions, plasmodial PMC's, haplochromosomes, aborted microspores or pollen, indehiscent and normal to rudimentary anthers occur. This differential expression is due to different genetic backgrounds, since rye is an allogamous plant.

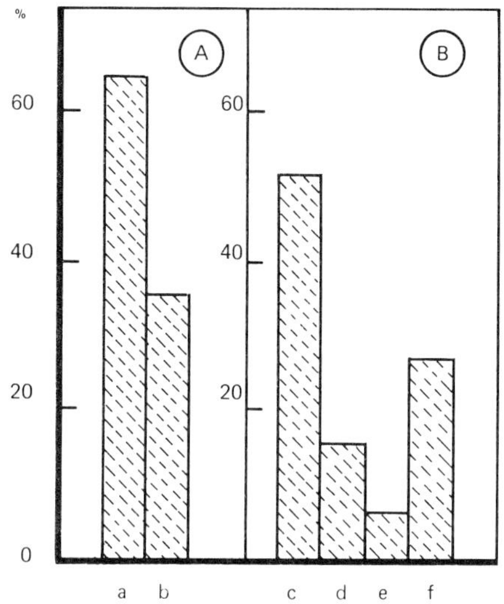

Fig. 3.4

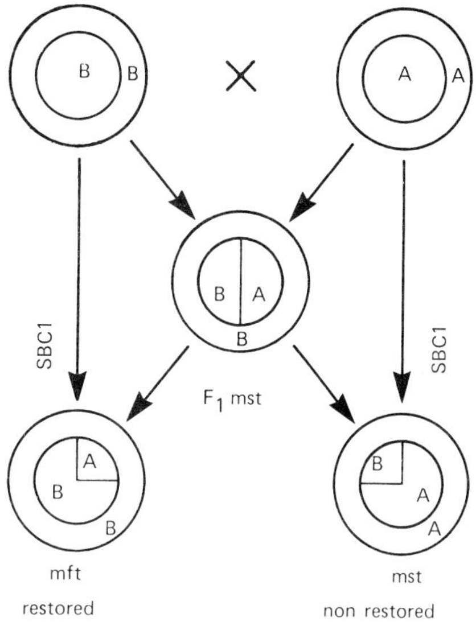

Fig. 3.5

Fig. 3.4A,B. Relative frequency of alloplasmic male sterility in species + species (*a*), genera + genera (*b*) and in *Solanum* (*c*), *Nicotiana* (*d*), *Triticum* (*e*) and others (*f*)

Fig. 3.5. mst Maintenance and stabilisation and mft restoration through substitution backcrossing (SBC) method

3.3 Types

To recapitulate, gc-mst is caused by the interaction of nuclear *fr* genes with cytoplasmic *c* genes. The cytoplasm having such mst causing *c* genes is known as the S-cytoplasm. Although the presence of S-cytoplasm is a basic necessity for gc-mst expression, neither the nature of this cytoplasm nor the precise location of the *c* genes is known. Instead, the presence of *fr* genes in the nucleus and their action in the S-cytoplasm is well known. Therefore, in the following text, the *fr* gene action described implies the automatic existence of *c* gene carrier S-cytoplasm. Accordingly, the gc-mst assumes the phenotypic expression like that of g-mst and is classified into the structural, sporogenous and functional types described below.

3.3.1 Structural

Like *ms* genes, *fr* genes suppress male sex expression to varying degrees during its developmental phases in many plants (Table 3.7). The stage of suppression depends upon time and mode of gene action. The earliest action is during staminal primordia differentiation or just later so that the plants are stamenless or antherless or produce rudimentary, vestigeal, warty, minute staminodes instead of normal stamens (Table 3.8a). The genes may act during staminal differentiation after primordia are differentiated. This causes failure of orderly stamen differentiation (Table 3.8b). In alloplasmic (*N. tabacum* × *N. glutinosa*) having *N. bigelovi* cytoplasm and in *Brassica oleracea* or *B. nigra* having *Raphanus sativus* cytoplasm, abnormal differentiation of epidermis and primary parietal layers and sometimes even of the primary sporogenous layers occurs so that anthers are misdeveloped and sterile. Some *fr* genes lead to aberrant anther differentiation ranging from stigma like extensions of the connective tissue in otherwise normal anthers to stigmoid mst anthers and petaloid mst anthers to the development of carpels (Table 3.8c,d). In cotton and tobacco mst plants, petaloidy and carpeloidy are common types of staminal transformation. These vary both in number of stamens per flower and degree of transformation. Some stamens are transformed into intermediate carpeloid structures but are rarely transformed into a fuctional fertile carpel. They may bear external ovules (Table 3.8e) or develop abnormally thickened nonfunctional staminal tube (Meyer 1966c). In the alloplasmic *Brassica* having radish S-cytoplasm, staminal atrophy is associated with reduced floral organ size. In some plants, the *fr* gene action is very variable. For instance, in *Trifolium*, either stamens are completely absent or are developed abnormally (Macewicz 1976a,b). The anomaly is conditioned by the complementary action of one dominant and one recessive gene. In *Begonia* (Villerts 1942), the stamens are either absent or highly reduced or are transformed into minute leaflike appendages. In some mst plants, the stamens lacking pollen grains are developed but they abscise prematurely, e.g. *Capsicum, Epilobium, Lolium* and *Solanum*. In *Gilia* (Grant 1956), 50% of the plants exhibit staminal abortion and others pollen abortion.

3.3.2 Sporogenous

The majority of the gc-mst plants, like those of g-mst, exhibit this mst type (Figs. 3.6, 3.7). This type, although prevalent in plants arising after intra- and interspecific crosses (Table 3.9), may be equally rampant in intergeneric and alloplasmic hybrids but the time of *fr* gene action in most of them has not been determined, hence their placement cannot be finalized. The majority of the genotypes exhibiting sporogenous gc-mst are dicots and the sterility abounds mostly in interspecific dicot hybrids (Fig. 3.8). In intergeneric hybrids, although monocots exceed dicots, the number of known cases is too low. Among the gc-mst plants arisen spontaneously or after intraspecific hybridizations, the ratio of occurrence of sporogenous mst in dicots and monocots is 3.5 : 1. Most of these are diploids and only a few are tetraploids (Fig. 3.9). Barring wheat, which is a hexaploid, the majority of the sp-mst plants arising spontaneously or after intra- or interspecific crosses are diploids and only a few are tetraploids. In many others, the chromosome number of the parents that produced the mst hybrids is not known (Tables 3.1–3.3). The mst plants arising after intergeneric hybridizations have 2x, 4x or 6x parents. Except for the cross *Aegilops ovata* × *Triticum durum*, in others the chromosome numbers of the two parents differ widely (Table 3.4). Both *Saccharum* and *Sclerostachya* are complex polyploids. In the gc-mst plants arising spontaneously or after intraspecific crosses, majority of the *fr* genes act mostly during

Table 3.7. Structural gene-cytoplasmic male sterile plants

Genotypes	References
a) **Species**	
Daucus carota (1D, 2x = 18)	Thompson (1961), Timin and Dobrutskaya (1981)
Ricinus communis (2x = 20)	Parkey (1957)
Trifolium pratense (1r + 1D, 2x = 14)	Macewicz (1976a,b), this Vol.
b) **Species crosses**	
Aquilegia vulgaris (2x = 14) × *A.chrysantha* (2x = 14)	Kappert (1944)
Begonia schmidtiana (2x = 32) × *B.acuminata*	Villerts (1942)
Begonia schmidtiana × *B.hirtella* (4x = 36)	Villerts (1942)
Begonia schmidtiana × *B.subvillosa*	Villerts (1942)
Brassica campestris (2x = 20) × *B.napus* (4x = 38)	Heyn (1978)
B.nigra (2x = 16) × *B.oleracea* (2x = 18)	Mezushima and Katsuo (1958), Pearson (1972)
Capsicum annuum (2x = 24) × *C.pendlum* (2x = 24)	Rusenova-Kondareva (1968)
Capsicum annuum (2x = 24) × *C.pubescens* (2x = 24)	Rusenova-Kondareva (1968)
Geranium albiflorum (2x = 28) × *G.pratense* (2x = 28)	Sansome (1936)
G.endressi (2x = 28) × *G.striatum* (2x = 28)	Sansome (1936)
Gilia exilis-bizonata (2x = 18) × *G.tenuiflora-tenuiflora* (2x = 18)	Grant (1956)
Gossypium anomalum (2x = 26) × *G.arboreum* (2x = 26)	Rhyne (1965), Meyer (1968), Chen and Meyer (1979)
Nicotiana amplexicaulis (2x = 36) × *N.tabacum* (4x = 48)	Berbeč, personal communication
N.benthamiana (2x = 38) × *N.tabacum* (4x = 48)	Ramavarma et al. (1978)
N.bigelovi (4x = 48) × *N.tabacum* (4x = 48)	Burk (1960), Schweppenhauser and Maan (1968)
N.debneyi (4x = 48) × *N.tabacum* (4x = 48)	Clayton (1950), Sand (1968), Tsikov and Tsikova (1971), Hicks et al. (1977), Gerstel et al. (1980)
N.exigua (2x = 32) × *N.tabacum* (4x = 48)	Berbeč, personal communication
N.glauca (2x = 24) × *N.tabacum* (4x = 48)	Berbeč (1966)
N.goodspeedi (4x = 40) × *N.tabacum* (4x = 48)	Tsikov et al. (1977)
N.gossei (4x = 36) × *N.tabacum* (4x = 48)	Burns and Gerstel (1981)
N.knightiana (2x = 24) × *N.tabacum* (4x = 48)	Berbeč, personal communication
N.megalosiphon (4x = 40) × *N.tabacum* (4x = 48)	Clayton (1950), Chaplin (1959)
N.plumbaginifolia (2x = 20) × *N.tabacum* (4x = 48)	Cameron and Moav (1957), Chaplin (1964)
N.raimondii (2x = 24) × *N.tabacum* (4x = 48)	Berbeč, personal communication
N.repanda (4x = 48) × *N.tabacum* (4x = 48)	Burk (1967), Burns et al. (1978), Gerstel et al. (1978)
N.suaveolens (4x = 32) × *N.tabacum* (4x = 48)	Izard and Hitier (1955b), Schweppenhauser and Maan (1968)
N.undulata (2x = 24) × *N.tabacum* (4x = 48)	Chaplin (1964)
Solanum antipoviczii (4x = 48) × *S.chacoense* (2x = 24)	Koopmans (1952)
S.phureja (2x = 24) × *S.chacoense* (2x = 24)	Grün et al. (1962)
S.rybinii (2x = 24) × *S.chacoense* (2x = 24)	Koopmans (1951)
c) **Genera crosses**	
Aegilops caudata (2x = 14) × *Triticum durum* (4x = 28)	Kihara (1951)
Ae.caudata (2x = 14) × *T.vulgare* (6x = 42)	Kihara and Tsunewaki (1961)
[*Lolium multiflorum* (2x = 14) × *L.perenne* (2x = 14)] × *Festuca arundinacea* (6x = 42)	Anderson (1966)
Raphanus sativus (2x = 18) × *Brassica campestris* (2x = 20)	Heyn (1978)
R.sativus (2x = 18) × *B.napus* (4x = 38)	Bannerot et al. (1974)
Triticum aestivum (6x = 42) × *Secale cereale* (2x = 14)	Lein (1948), Sasaki (1956), Vega and Lacadena (1979)

Table 3.8. Major stamen alterations in structural gene-cytoplasmic male steriles

	Phenotypic alterations	Genera
a)	Stamens absent, aborted or diminutive	*Begonia, Brassica*[a], *Festuca, Geranium, Gilia, Trifolium, Triticum*[a]
b)	Faulty stamen differentiation	*Festuca, Gossypium*[a], *Lycopersicon, Nicotiana*[a], *Triticum*[a]
c)	Anthers abnormal (stigmoid-pistiloid)	*Brassica*[a], *Festuca, Gossypium*[a], *Nicotiana*[a], *Raphanus, Triticum*[a]
d)	Petaloidy	*Brassica*[a], *Gossypium*[a], *Nicotiana*[a], *Triticum*
e)	External ovules abnormal	*Brassica*[a], *Gossypium*[a], *Nicotiana*[a]

[a]Alloplasmic, others hybrids.

microspore formation and after their release (Fig. 3.17). But in those arising after interspecific crosses, the *fr* genes act mostly during prophase-I and microspore formation and development (Fig. 3.18). In many, *fr* gene action is delayed until pollen maturation stages. On the other hand, *fr* gene action in mst plants arising after intergeneric hybridizations is confined to premeiosis and to the stage after microspores are liberated (Fig. 3.19). Thus *fr* gene action during microsporogenesis depends on the origin of mst's.

3.3.3 Functional

Compared to structural and sporogenous mst types, the proportion of fn-mst's is low (Fig. 3.6, Table 3.10), and the tetraploids exceed diploids (Fig. 3.9). In most of these, all the pollen formed is viable but the anthers are indehiscent. However, in certain rice, potato and rye genotypes, some sterile pollen grains occur interspersed

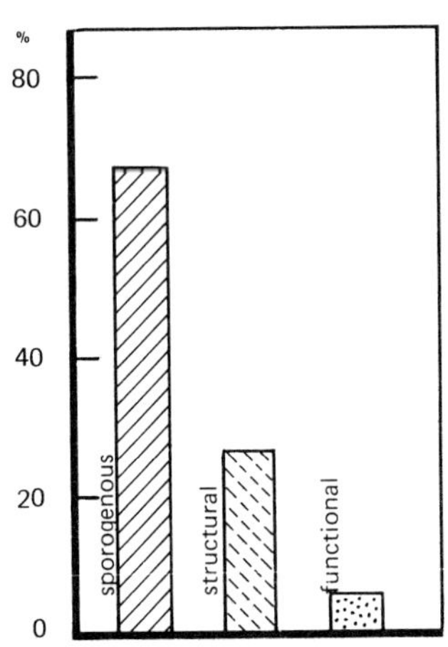

Fig. 3.6

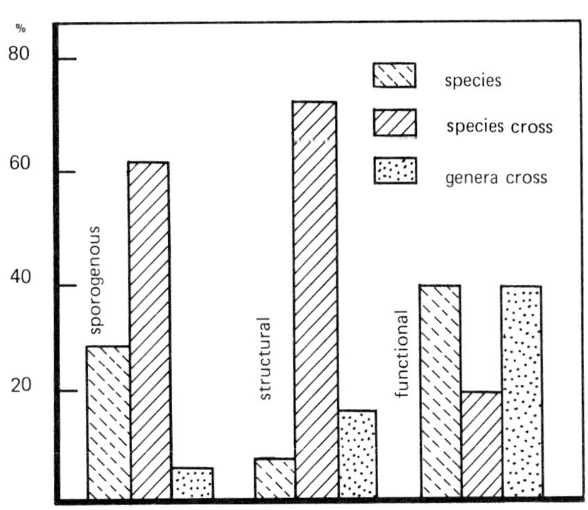

Fig. 3.7

Fig. 3.6. Frequency of gene-cytoplasmic male sterility types in angiosperms

Fig. 3.7. Frequency distribution of male sterility types in angiosperms

Table 3.9. Sporogenous gene-cytoplasmic male sterile plants

Species	References
a) **Intraspecific crosses**	
Allium cepa (1r, 2x = 16)	Monosmith (1926), Jones and Emsweller (1937), Tatebe (1952), Yen (1959), Virnich (1967), Patil et al. (1973), Dyki (1973a)
Beta vulgaris (2r, 2x = 18)	Artschwager (1947), Nagao and Kinoshita (1962), Kinoshita (1971)
Capsicum annuum (1r, 2x = 24)	Pochard (1970b), Novâk et al. (1971)
Crotalaria mucronata (2r, 2x = 16)	Edwardson (1967)
Cucumis melo (2x = 24)	Chauhan and Singh (1968)
Dactylis glomerata, (1D, 4x = 28)	Filion and Christie (1966)
Daucus carota (1D, 2x = 18)	Thompson (1960), Zenkteler (1962), Timin and Dobrutskaya (1981)
Gossypium hirsutum (2r, 4x = 52)	Thombre and Mehetre (1979)
Hordeum vulgare (1r, 2x = 14)	Schooler (1967), Ahokas (1979a, 1982b)
Lathyrus sativus (2x = 14)	Zelenskaya and Pestova (1974)
Lolium multiflorum (2x = 14)	Nitzsche (1971)
Lupinus mutabilis (4x = 48)	Pakendorf (1970)
Lycopersicon sp. (2x = 24)	Nishi and Hiraoka (1958)
Medicago sativa (many r, 4x = 32)	Bradner and Childers (1968), Pederson and Stucker (1969), Staszewski (1970, 1976)
Oryza sativa (2x = 24)	Chu et al. (1972), Anon (1978a), Tamaru et al. (1980), Pan et al. (1982), Shuhua (1982)
O. s.indica (2x = 24) × *O. s.japonica* (2x = 24)	Kinoshita et al. (1980)
O.spontanea (2x = 24) × *O. s.japonica* (2x = 24)	Katsuo (1954), Katsuo and Mizushima (1958)
Pelargonium crispum (2x = 22)	Tokumasu (1974)
Pennisetum typhoides (1–3r, 2x = 14)	Burton (1958), Singh and Sharma (1963)
Petunia hybrida (2x = 14)	Hamilton (1965), Edwardson and Warmke (1967), Frankel et al. (1969), Izhar (1977)
Plantago ovata (1D, 2x = 8)	Mittal and Issar (1970)
Raphanus sativus (1r, 2x = 18)	Nishi and Hiraoka (1958), Ogura (1968)
Secale cereale (1–2r, 2x = 14)	Geiger and Schnell (1970), Garlicka and Madej (1975), Scoles and Evans (1979), Artemova (1980)
Sorghum vulgare (1, many r, 4x = 20)	Singh and Hadley (1961), Erichsen and Ross (1963a), Alam and Sandal (1967a), Narkhede et al. (1968), Raj (1968)
Tabarnaemontana coronaria (2x = 22)	Dubey (1970)
Trifolium pratense (1r + 1D, 2x = 14)	Macewicz (1976a,b)
Triticum aestivum (6x = 42)	DeVries and Ie (1970)
Zea mays (1–2r, 2x = 20)	Lee (1976), Warmke and Lee (1977), Lee et al. (1979, 1980), Colhoun and Steer (1981)
b) **Interspecific crosses**	
Allium cepa (2x = 16) × *A.drobovii* (2x = 16)	Saini and Davis (1969)
A.cepa (2x = 16) × *A.galanthum* (2x = 16)	Saini and Davis (1969)
A.cepa (2x = 16) × *A.pskemense* (2x = 16)	Saini and Davis (1969)
A.roylei (2x = 16) × *A.drobovii* (2x = 16)	Saini and Davis (1969)
A.cepa (2x = 16 × *A.ascalonicum* (2x = 16)	Little et al. (1944)
A.cepa (2x = 16) × *A.fistulosum* (2x = 16)	Virnich (1967)

Table 3.9 *(continued)*

Species	References
Aquilegia flabellata (2x = 14) × *A.truncata* (2x = 14)	Skalinska (1931)
A.vulgaris (2x = 14) × *A.chrysantha* (2x = 14)	Skalinska (1928b)
Beta vulgaris (2x = 18) × *B.maritima* (2x = 18)	Dalke et al. (1983)
Brassica campestris (2x = 20) × *B.oleracea* (2x = 18)	Nishi and Hiraoka (1958)
Citrus limon (2x = 18) × *C.paradisi* (2x = 18)	Fröst (1948)
C.limon (2x = 18) × *C.sinensis* (2x = 18)	Fröst (1948)
C.unshiu (2x = 18) × *C.sinensis* (2x = 18)	Iwamasa (1966)
Desmodium intortum (2x = 22) × *D.sandwicense* (2x = 22)	McWhirter (1969)
Epilobium hirsutum (2x = 36) × *E.luteum* (2x = 36)	Michaelis (1929, 1932, 1933, 1954)
E.hirsutum (2x = 36) × *E.montanum* (2x = 36)	Michaelis and Wertz (1935)
E.h.jena (2x = 36) × *E.parviflorum* (2x = 36)	Michaelis (1943)
E.montanum (2x = 36) × *E.roseum* (2x = 36)	Oberreuter (1925)
Fragaria vesca (2x = 14) × *F.nilgerrensis* (2x = 14)	Nürnberg-Krüger (1956, 1958)
Gossypium harknessi (2x = 26) × *G.hirsutum* (4x = 52)	Murthi and Weaver (1974), Meyer (1975)
Helianthus petiolaris (2x = 34) × *H.annuus* (2x = 34)	Leclercq (1969), Paun (1974), Whelan (1979)
Linum floccosum (2x = 30) × *L.usitatissimum* (2x = 30)	Dubey and Singh (1965)
Lolium perenne (2x = 14) × *L.multiflorum* (2x = 14)	Ahloowalia (1981)
Nicotiana bigelovii (4x = 48) × *N.tabacum* (4x = 48)	Burk (1960), Chaplin (1964)
N.langsdorfi (2x = 18) × *N.sanderae* (2x = 18)	East (1919, 1929)
N.tabacum (4x = 48) × *N.glutinosa* (2x = 24)	Tsikov et al. (1977)
N.tomentosiformis (2x = 24) × *N.otophora* (2x = 24)	Gerstel and Burns (1974)
Solanum acaule (4x = 48) × *S.bulbocastanum* (2x = 24)	Hermsen and Ramanna (1969)
S.chacoense (2x = 24) × *S.kurtzianum* (2x = 24)	Grün et al. (1962), Grün and Aubertin (1966)
S.chacoense (2x = 24) × *S.phureja* (2x = 24)	Grün et al. (1962), Grün and Aubertin (1966)
S.chacoense (2x = 24) × *S.stenotomum* (2x = 24)	Grün et al. (1962), Grün and Aubertin (1966)
S.chacoense (2x = 24) × *S.tuberosum* (4x = 48)	Ramanna and Hermsen (1974)
S.demissum (6x = 72) × *S.tuberosum* (4x = 48)	Dionne (1961)
S.demissum (6x = 72) × *S.berthaulti* (2x = 24)	Magoon et al. (1958c)
S.demissum (6x = 72) × *S.gibberulosum* (2x = 24)	Magoon et al. (1958c)
S.demissum (6x = 72) × *S.gigantophyllum* (2x = 24)	Magoon et al. (1958c)
S.demissum (6x = 72) × *S.goniocalyx* (2x = 24)	Magoon et al. (1958c)
S.demissum (6x = 72) × *S.phureja* (2x = 24)	Magoon et al. (1958c)
S.demissum (6x = 72) × *S.saltense* (2x = 24)	Magoon et al. (1958c)
S.demissum (6x = 72) × *S.etuberosum* (2x = 24)	Hermsen and Taylor (1979)
S.demissum (6x = 72) × *S.pinnatisectum* (2x = 24)	Ramanna and Hermsen (1979, 1981)
S.phureja (2x = 24) × *S.acaule* (4x = 48)	Koopmans (1952)
S.polytrichon (4x = 48) × *S.phureja* (2x = 24)	Abdalla and Ramanna (1971)
S.tuberosum (4x = 48) × *S.acaule* (4x = 48)	Lamm (1953)
S.verrucosum (2x = 24) × *S.bulbocastanum* (2x = 24)	Abdalla and Hermsen (1972)
S.verrucosum (2x = 24) × *S.gibberulosum* (2x = 24)	Magoon et al. (1958b)
S.verrucosum (2x = 24) × *S.yabari* (2x = 24)	Magoon et al. (1958b)
S.verrucosum (2x = 24) × *S.phureja* (2x = 24)	Buck (1960)
S.verrucosum (2x = 24) × *S.tuberosum* (4x = 48)	Buck (1960)
S.polytrichon (4x = 48) × *S.chacoense* (2x = 24)	Toxopeus (1960, 1964)
S.polytrichon (4x = 48) × *S.goniocalyx* (2x = 24)	Toxopeus (1960, 1964)
Triticum compactum (6x = 42) × *T.aestivum* (6x = 42)	Xiangmin and Herong (1981)
T.timonovum × *T.aestivum* (6x = 42)	Fedorova and Nettevich (1969)
T.timopheevi (4x = 28) × *T.aestivum* (6x = 42)	Joppa et al. (1966), Jain et al. (1981), Xiangmin and Herong (1981)

Table 3.9 *(continued)*

Species	References
T.zhukovskyi (6x = 42) × *T.aestivum* (6x = 42)	Fedorova and Nettevich (1969)
Vitis vinifera (2x = 38) × *V.riparia* (2x = 38)	Staudt (1973)
c) Intergeneric crosses	
Aegilops caudata (2x = 14) × *Triticum durum* (4x = 28)	Hervás (1974)
Ae.ovata (4x = 28) × *T.aestivum* (6x = 42)	Chauhan and Singh (1966)
Ae.ovata (4x = 28) × *T.durum* (4x = 28)	Fukasawa (1953, 1956b)
Diplotaxis muralis (2x = 22) × *Brassica campestris* (2x = 20)	Hinata and Kanno (1976)
Raphanus sativus (2x = 18) × *Brassican apus* (4x = 38)	Kanno et al. (1963a), Yamaguchi and Kanno (1963), Thompson (1972), Shiga and Baba (1971, 1972), Bartkowiak-Broda et al. (1979)
Saccharum spontaneum (8x = 80) × *Sclerostachya fusca* (4x = 48)	This Vol.

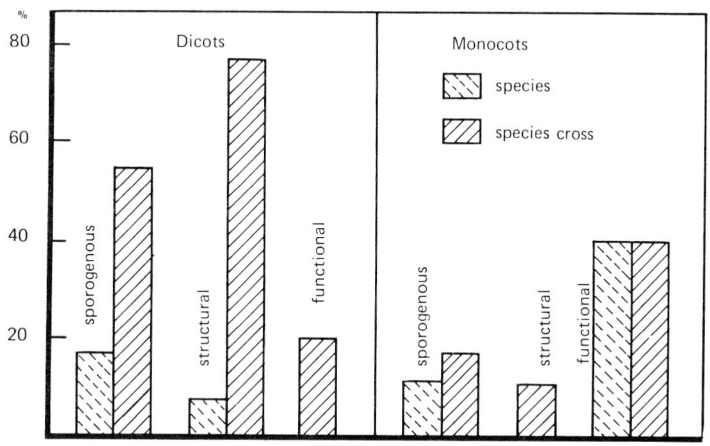

Fig. 3.8. Distribution of male sterility types in dicots and monocots

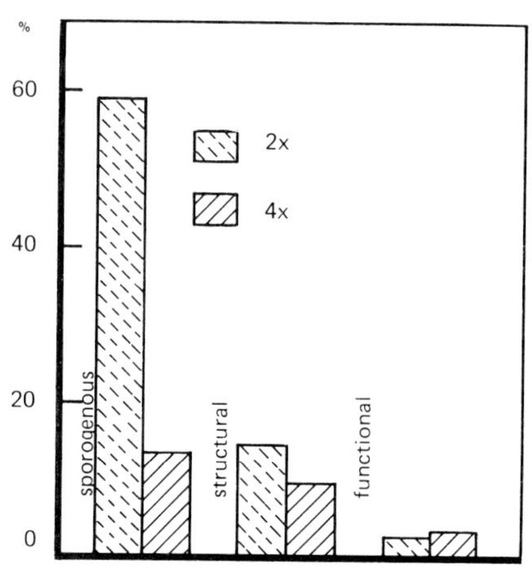

Fig. 3.9. Sterility types in relation to ploidy level in gene-cytoplasmic male steriles

Table 3.10. Functional gene-cytoplasmic male sterile plants

Species	References
Oryza sativa (1r, 2x = 24)	Kitamura (1962), Erickson (1969), this Vol.
Solanum tuberosum (4x = 48) × *S. andigenum* (4x = 48)	De la Puente and Peloquin (1968), Grun (1973)
Sorghum vulgare (5r, 4x = 20)	Webster and Singh (1964)
Triticum aestivum (6x = 42) × *Secale cereale* (2x = 14)	Lein (1948)
Triticum durum (4x = 28) × *Secale cereale* (2x = 14)	Vega and Lacadena (1979)

with the fertile pollen. In other mst genotypes of these species, only fertile pollen is present in the indehiscent anthers. The disparity could be due to cytoplasm differences existing in these alloplasmic genotypes. Likewise, some differences in anther dehiscence per cent occur among different genotypes of a few alloplasmic mst rye, rice and wheat. In rye (Vega and Lacadena 1979) fn-mst occurs in conjunction with other anomalies like pistilloidy, anther malformation, locular fusion, plasmodial PMC's etc. Two naturally occurring disorders, "crazy-top" in *Gossypium hirsutum* and "hollow boll" in *G. barbadense*, cause anther nondehiscence (Feaster and Turncotte 1959). These anomalies are nongenetic and are induced by environmental stress conditions. In an fn-mst rice genotype, the *fr* gene influences endothecial structure. In it, the endothecium is strongly thickened and the anther wall exhibits a specific pattern of cell orientation different from the dehiscent anthers (Cheng and Huang 1980). Moreover, the epidermal cells are highly thickened. This prevents the intersporangial stripe from separating from the underlying parenchyma along each side of the connective tissue during anthesis. The outer surface of intermicrosporangial anthers is smooth, but it is ridged in the dehiscent anthers. This probably impedes pore development and results in anther indehiscence.

3.3.4 Frequency

About 74% of the gc-mst's have arisen either after interspecific or intergeneric hybridizations and only 26% either spontaneously or after intraspecific crosses (Fig. 3.1B). Of these, whereas 70–80% of gc-mst's arisen spontaneously or after intraspecific or intergeneric hybridizations are dicots, 87% gc-mst monocots arise after intergeneric hybridizations (Fig. 3.1A). Thus, while monocot gc-mst's owe their origin mostly to intergeneric hybridizations, dicot gc-mst's mostly arise spontaneously or after intra- and interspecific hybridizations. Comparing diploids and tetraploids, sporogenous mst types are most prevalent in diploids and comprise about 72% of the total known gc-mst's (2x + 4x). Of the 23% structural mst's, 14% are diploids and 9% tetraploids (Fig. 3.9). The functional mst's constitute only about 5%, distributed almost equally in diploids and tetraploids. Of the 72% gc-mst's having sp-mst, recessive *fr* genes condition sterility in 70% of them, in the remaining 30% mostly dominant genes control the sterility (Fig. 3.12). Comparing

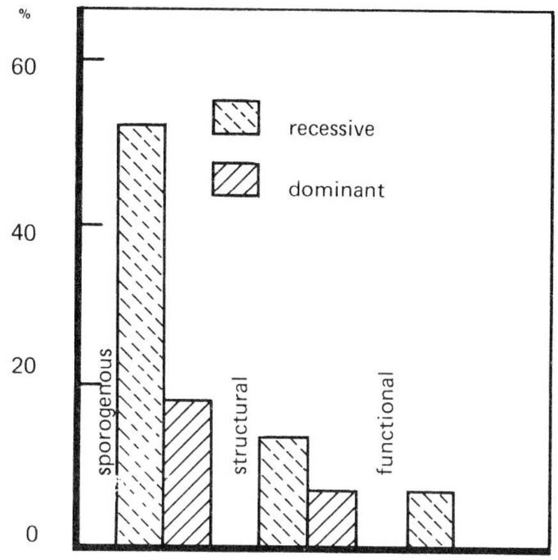

Fig. 3.10. Frequency of recessive and dominant genes in gene-cytoplasmic male steriles

the three sterility types, the frequency of species having recessive gene control is significantly higher in diploid sporogenous and structural mst's than in their respective tetraploids; it is about 52% in diploid sp-mst's (Fig. 3.10).

3.4 Gene Control

Precise determination of genetic control of gc-mst's is hampered due to the complicated interaction of *fr* genes with modifiers and environment. Thus, from about 320 recorded cases of gc-mst arising spontaneously (49) or after intra- (24) or inter- (171) specific crosses or intergeneric (74) hybridizations (Tables 3.1–3.4), genetic control of the sterility is known only in about 110 cases, majority of which have arisen after intra- or inter-specific hybridizations (Tables 3.11–3.14). Of these, about 75% are dicots and 25% monocots (Fig. 3.11). In dicots, the mst is predominantly conditioned by dominant and many recessive genes (Fig. 3.12) but in monocots, it is almost exclusively conditioned by recessive genes, the dominant gene control being almost unknown. Furthermore, whereas the frequency of dicot and monocot gc-mst with single gene control is almost equal, it is significantly higher in dicots in which many recessive, dominant or polygenes control mst (Fig. 3.13). In general, in about 65% gc-mst angiosperms, the mst is controlled by one or more recessive genes, in about 30% by dominant genes and only in 5% by polygenes (Fig. 3.14). Recessive gene control (1 to many) is more common in gc-mst's arising spontaneously or after intraspecific crosses (Fig. 3.15). Whereas single recessive gene control is most common in gc-mst's arising after interspecific crosses, dominant gene control is predominant in the mst's arising after intergeneric crosses. In general, in the species crosses which include interspecific and intergeneric crosses, recessive and dominant gene control frequency is nearly similarly distributed (Fig. 3.16). This could be because the majority of the gc-mst's with known gene control are dicots (Fig. 3.11) and have arisen after species crosses

Table 3.11. Gene-cytoplasmic male sterility controlled by single recessive gene

Genotype	References
a) Intraspecific hybrids	
Allium cepa (2x = 16)	Jones and Emsweller (1937), Tatebe (1952), Kobabe (1958), Dyki (1973a), Schweisguth (1973), Meer (1978)
A.schoenoprasum (2x = 16)	Tatliogu (1982)
Capsicum annuum (2x = 24)	Peterson (1958)
Chenopodium quinoa (4x = 32)	Simmonds (1971)
Godetia whitneyi (2x = 14)	Hiorth (1948a)
Hordeum vulgare (2x = 14)	Ahokas (1979a)
Linum usitatissimum (2x = 30)	Burnham et al. (1981)
Medicago sativa (4x = 32)	Bradner and Childers (1968), Barnes et al. (1972)
Mercurialis annua (1r, 2x = 16)	Louis and Durand (1978)
M.perennis (1r, 8x = 64)	This Vol.
Oryza sativa (2x = 24)	Kitamura (1962)
Pennisetum typhoides (2x = 14)	Burton (1958), Athwal (1965b)
Ricinus communis (2x = 20)	Parkey (1957)
Satureja hortensis (9x = 45)	Correns (1904, 1908), East (1934)
Secale cereale (2x = 14)	Madej (1975), Garlicka and Madej (1975)
Sorghum vulgare (4x = 20)	Maunder and Pickett (1959), Hadley and Singh (1961), Kidd (1961), Duvick (1966)
Vicia faba (2x = 12)	Bond et al. (1966a)
Zea mays (2x = 20)	Buchert (1961), Beckett (1971)
b) Interspecific hybrids	
Allium cepa (2x = 16) × *A.ascalonicum* (2x = 16)	Little et al. (1944)
Allium cepa (2x = 16) × *A.drobovii* (2x = 16)	Saini and Davis (1969)
Allium cepa (2x = 16) × *A.galanthum* (2x = 16)	Saini and Davis (1969)
Geranium endressi (2x = 28) × *G.striatum* (2x = 28)	Sansome (1936)
Gossypium anomalum (2x = 26) × *G.hirsutum* (4x = 52)	Meyer and Meyer (1965)
[*G.anomalum* (2x = 26) × *G.thurberi* (2x = 26)] × *G.hirsutum* (4x = 52)	Meyer and Meyer (1965)
[*G.arboreum* (2x = 26) × *G.thurberi* (2x = 26)] × *G.hirsutum* (4x = 52)	Meyer and Meyer (1965)
Kohleria amabilis (2x = 26) × *K.bogotensis* (2x = 26)	Clayberg (1974)
Linum floccosum (2x = 30) × *L.usitatissimum* (2x = 30)	Gajewski (1937)
Oryza sativa (2x = 24) × *O.glabberima* (2x = 24)	Yabuno (1977)
Secale cereale (2x = 14) × *S.kuprijanovii* (2x = 14)	Lapinski (1972)
S.cereale (2x = 14) × *S.montanum* (2x = 14)	Lapinski (1972)
Solanum antipovicizii (4x = 48) × *S.chacoense* (2x = 24)	Koopmans (1952)
S.phureja (2x = 24) × *S.chacoense* (2x = 24)	Grun et al. (1962)
S.rybinii (2x = 24) × *S.chacoense* (2x = 24)	Koopmans (1951)
Triticum timopheevi (4x = 28) × *T.aestivum* (6x = 42)	Tahir and Tsunewaki (1969)
c) Intergeneric hybrids	
Lolium perene (2x = 14) × *L.multiflorum* (2x = 14)	Wit (1974)
L.perenne (2x = 14) × *Festuca pratensis* (2x = 14)	Wit (1974)

Table 3.12. Gene-cytoplasmic male sterility controlled by more than one recessive gene

Genotype	References
a) Intraspecific hybrids	
Allium cepa (2r, 2x = 16)	Schweisguth (1973)
Beta macrocarpa (2x = 18)	Oldemeyer (1957)
B.maritima (2x = 18)	Coe and Stewart (1977)
B.vulgaris (2r, 2x = 18)	Owen (1942), Artschwager (1947), Nagao and Kinoshita (1962), Bliss and Gabelman (1965), Rohrbach (1965a), Kinoshita (1971), Kinoshita et al. (1972), Chauhan and Kinoshita (1980a)
B.vulgaris (4r, 4x = 36)	Kinoshita (1971), Kinoshita et al. (1972)
B.vulgaris (semi-sterile, 2r, 2x = 18)	Kinoshita (1971)
Brassica campestris (2r, 2x = 20)	Ohkawa and Shiga (1981)
Capsicum annuum (2r, 2x = 24)	Novăk et al. (1971)
Daucus carota (many r, 2x = 18)	Morelock (1974)
Desmodium sandwicense (2r, 2x = 22)	McWhirter (1969)
Godetia whitneyi (1-many r, 2x = 14)	Hiorth (1948 a,b)
Gossypium hirsutum (2, 4x = 52)	Thombre and Mehetre (1979)
Lathyrus sativus (2r, 2x = 14)	Chekalin (1972, 1975)
Linum usitatissimum (1-many r, 2x = 30)	Burnham et al. (1981)
Medicago sativa (4x = 32)	Davis and Greenblatt (1967)
M.sativa (2-many r, 4x = 32)	Pederson and Stucker (1969), Staszewski (1970, 1976)
Oryza sativa (2r, 2x = 24)	Hsu (1945), Shinjyo and Omura (1966), Kinoshita et al. (1980)
Pennisetum typhoides (2r, 2x = 14)	Menon (1958, 1959), Yadav (1974)
(3r)	Akénova and Chheda (1981)
Secale cereale (2r, 2x = 14)	Geiger and Schnell (1970), Garlicka and Madej (1975), Scoles and Evans (1979c)
Sorghum vulgare (2–5r, 4x = 20)	Stephens and Holland (1954), Damon (1961), Craigmiles (1961, 1962), Pi and Wu (1963), Erichsen and Ross (1963a), Miller and Pickett (1964), Alam and Sandal (1967a), Quinby (1980), Tripathi et al. (1980)
Zea mays (2r, 2x = 20)	Duvick (1956), Jones et al. (1957a), Duvick et al. (1961)
b) Interspecific hybrids	
Beta vulgaris (2x = 18) × *B.maritima* (2x = 18)	Dalke et al. (1983)
Brassica juncea (4x = 36) × *B.napus* (4x = 38)	Brar et al. (1980)
Gossypium anomalum (2x = 26) × *G.hirsutum* (4x = 52)	Meyer and Meyer (1965)
[*G.anomalum* (2x = 26) × *G.thurberi* (2x = 26)]	Meyer and Meyer (1965)
× *G.hirsutum* (4x = 52)	Meyer and Meyer (1965)
[*G.arboreum* (2x = 26) × *G.thurberi* (2x = 26)] × *G.hirsutum* (4x = 52)	Richmond and Kohel (1961), Meyer and Meyer (1965)
Linum floccosum (2x = 30) × *L.usitatissimum* (2x = 30)	Gajewski (1937)
Lycopersicon esculentum (2x = 24) × *L.parviflorum* (2x = 24)	Kesicki (1980)
Nicotiana langsdorfi (2x = 18) × *N.sanderea* (2x = 18)	East (1919, 1929)
Petunia axillaris (2x = 14) × *P.hybrida* (2x = 14)	Edwardson and Warmke (1967)

Table 3.12 *(continued)*

Genotype	References
Triticum timopheevi (4x = 28) × *T.aestivum* (6x = 42)	Livers (1964a), Maan and Lucken (1967), Yen et al. (1969)
c) Intergeneric hybrids	
[*Lolium perenne* (2x = 14) × *L.multiflorum* (2x = 14)] × *Festuca pratensis (2xc14)*	Wit (1974)
Raphanus sativus (2x = 18) × *Brassica campestris* (2x = 20)	Heyn (1978)
R.sativus (2x = 18) × *B.napus* (4x = 38)	Bannerot et al. (1974)
R.sativus (2x = 18) × *B.oleracea* (2x = 18)	Bannerot et al. (1974), Mc Collum (1979)
Saccharum spontaneum (8x = 80) × *Sclerostachya fusca* (4x = 48)	This Vol.

Table 3.13. Gene-cytoplasmic male sterility controlled by dominant *Fr* genes

Species	References
a) Intraspecific hybrids	
Daucus carota (1D, 2x = 18)	Gabelman (1956), Banga (1959)
2–3D	Thompson (1961)
1D + 1r	Timin and Dobrutskaya (1981)
3D	Morelock (1974)
Plantago ovata (1D, 2x = 8)	Mittal and Issar (1970)
Trifolium pratense (1r + 1D, 2x = 14)	Macewicz (1976a,b)
b) Interspecific hybrids	
Fragaria vesca (2x = 14) × *F.nilgerrensis* (2x = 14)	Nürnberg-Krüger (1958)
Kohleria tubiflora (2x = 26) × *K.digitaliflora* (2x = 26)	Clayberg (1974)
Kohleria tubiflora (2x = 26) × *K.bogotensis* (2x = 26)	Clayberg (1974)
Solanum chacoense (2x = 24) × *S.kurtzianum* (2x = 24)	Grün et al. (1962)
Solanum chacoense (2x = 24) × *S.phureja* (2x = 24)	Grün and Aubertin (1966)
Solanum chacoense (2x = 24) × *S.stenotomum* (2x = 24)	
S.polytrichon (4x = 48) × *S.phureja* (2x = 24)	Abdalla and Ramanna (1971)
S.verrucosum (2x = 24) × *S.bulbocastanum* (2x = 24)	Grün et al. (1962), Abdalla (1970), Abdalla and Hermsen (1972)
S.verrucosum (2x = 24) × *S.canasense* (2x = 24)	Grün et al. (1962), Abdalla (1970), Abdalla and Hermsen (1972)
S.verrucosum (2x = 24) × *S.chacoense* (2x = 24)	Grün et al. (1962), Abdalla (1970), Abdalla and Hermsen (1972)
S.verrucosum (2x = 24) × *S.chomatophilum* (2x = 24)	Grün et al. (1962), Abdalla (1970), Abdalla and Hermsen (1972)
S.verrucosum (2x = 24) × *S.gigantophyllum* (2x = 24)	Grün et al. (1962), Abdalla (1970), Abdalla and Hermsen (1972)
S.verrucosum (2x = 24) × *S.goniocalyx* (2x = 24)	Grün et al. (1962), Abdalla (1970), Abdalla and Hermsen (1972)
S.verrucosum (2x = 24) × *S.infundibuliforme* (2x = 24)	Grün et al. (1962), Abdalla (1970), Abdalla and Hermsen (1972)
S.verrucosum (2x = 24) × *S.kurtzianum* (2x = 24)	Grün et al. (1962), Abdalla (1970), Abdalla and Hermsen (1972)

Table 3.13 *(continued)*

Species	References
S.verrucosum (2x = 24) × *S.leptophyes* (2x = 24)	Grün et al. (1962), Abdalla (1970), Abdalla and Hermsen (1972)
S.verrucosum (2x = 24) × *S.megistacrolobum* (2x = 24)	Grün et al. (1962), Abdalla (1970), Abdalla and Hermsen (1972)
S.verrucosum (2x = 24) × *S.multidissectum* (2x = 24)	Grün et al. (1962), Abdalla (1970), Abdalla and Hermsen (1972)
S.verrucosum (2x = 24) × *S.phureja* (2x = 24)	Grün et al. (1962), Abdalla (1970), Abdalla and Hermsen (1972)
S.verrucosum (2x = 24) × *S.raphanofolium* (2x = 24)	Grün et al. (1962), Abdalla (1970), Abdalla and Hermsen (1972)
S.verrucosum (2x = 24) × *S.simplicifolium* (2x = 24)	Grün et al. (1962), Abdalla (1970), Abdalla and Hermsen (1972)
S.verrucosum (2x = 24) × *S.stenotomum* (2x = 24)	Grün et al. (1962), Abdalla (1970), Abdalla and Hermsen (1972)
S.verrucosum (2x = 24) × *S.tarijense* (2x = 24)	Grün et al. (1962), Abdalla (1970), Abdalla and Hermsen (1972)
S.verrucosum (2x = 24) × *S.toralapanum* (2x = 24)	Grün et al. (1962), Abdalla (1970), Abdalla and Hermsen (1972)
S.verrucosum (2x = 24) × *S.tuberosum* (4x = 48)	Grün et al. (1962), Abdalla (1970), Abdalla and Hermsen (1972)
S.verrucosum (2x = 24) × *S.vernei* (2x = 24)	Grün et al. (1962), Abdalla (1970), Abdalla and Hermsen (1972)
Triticum timopheevi (4x = 28) × *T.aestivum* (6x = 42)	Goujon and Ingold (1967)

Table 3.14. Gene-cytoplasmic male sterility controlled by polygenes

Species	References
Intraspecific crosses	
Helianthus annuus (2x = 34)	Gundaev (1966)
Linum usitatissimum (2x = 30)	Burnham et al. (1981)
Medicago sativa (4x = 32)	Bradner and Childers (1968)
Origanum vulgare (4x = 32)	Kheyr-Pour (1980, 1981)
Pennisetum typhoides (2x = 14)	Prakash (1967)
Trifolium pratense (2x = 14)	This Vol.

(Figs. 3.1, 3.4). Polygenic control present in the mst's arising spontaneously or after intraspecific crosses is absent in intergeneric hybrids. A wide polymorphism due to differential *fr-c* genes number and type is known in *Brassica, Gossypium, Nicotiana, Origanum, Solanum, Streptocarpus* and *Triticum*. This polymorphism leads to diverse mst expressions within and sometimes between segregating families. In *Solanum verrucosum* such differential polymorphism produces normal to highly abnormal types of pollen (Abdalla and Hermsen 1972). In tobacco, anther malformations of diverse type (Fig. 49.7) result due to aberrant anther differentiation and development (Frankel and Galun 1977).

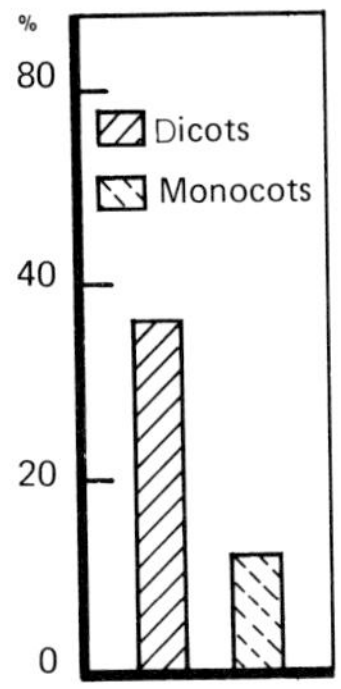

Fig. 3.11

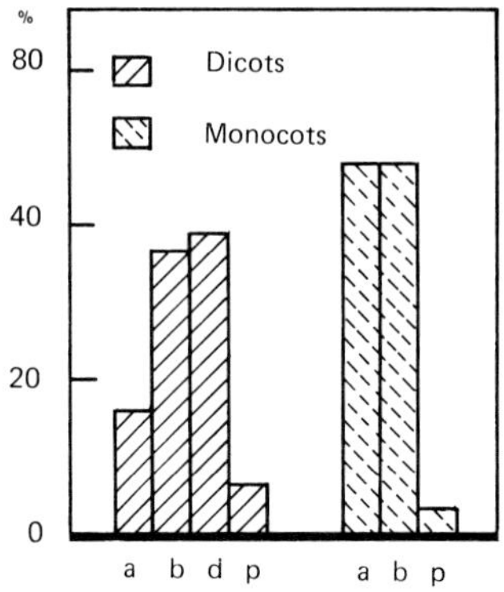

Fig. 3.12

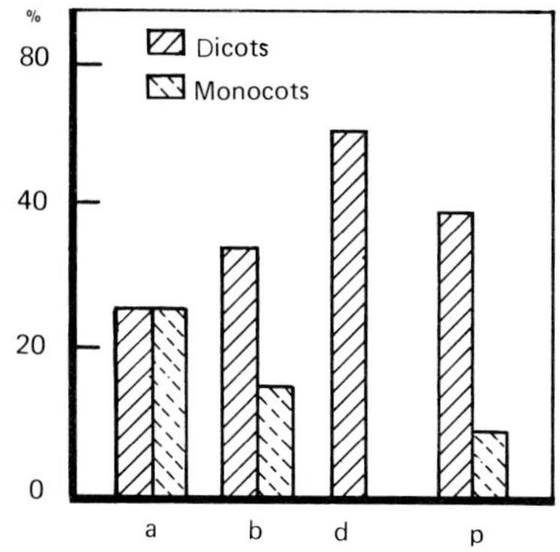

Fig. 3.13

Fig. 3.11. Relative proportion of dicots and monocots with known gene control in gene-cytoplasmic male steriles

Fig. 3.12. Proportion of dicots and monocots having 1 recessive (*a*), many recessive (*b*), dominant (*d*) and polygene (*p*) in gene-cytoplasmic male steriles

Fig. 3.13. Relative proportion of gene control types in dicot and monocot gene-cytoplasmic male steriles

◀ **Fig. 3.14.** Frequency of gene control types in gene-cytoplasmic male steriles

Fig. 3.15. Frequency of gene control types in species (▭), species crosses (▨) and genera crosses (▦) in gene-cytoplasmic male steriles ▼

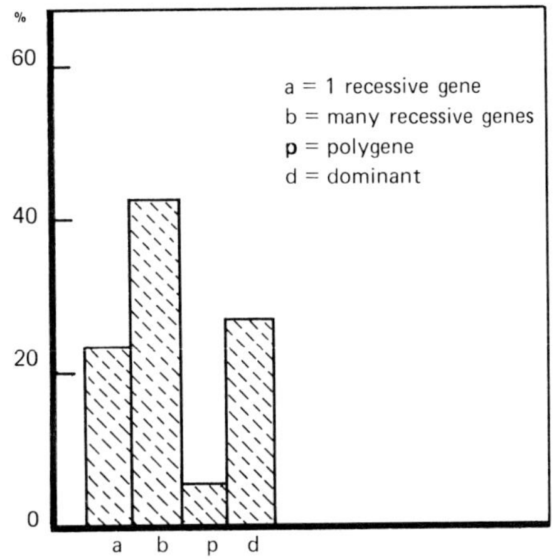

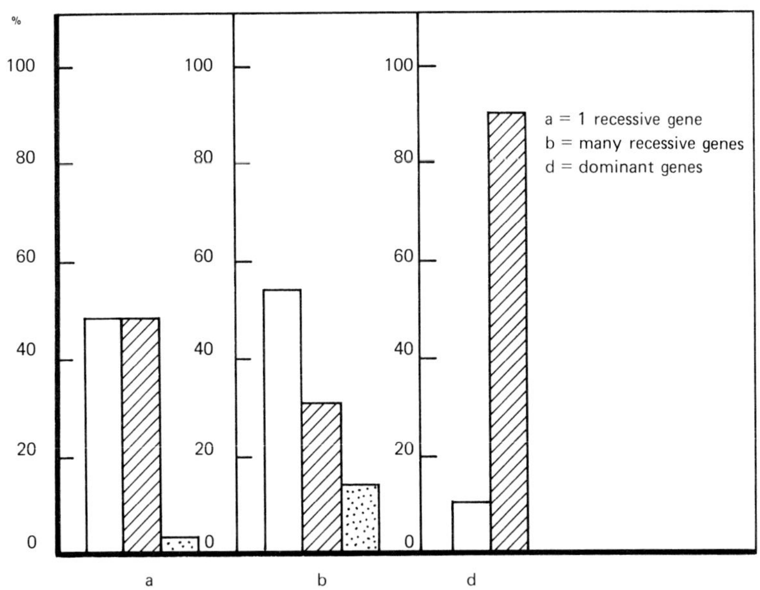

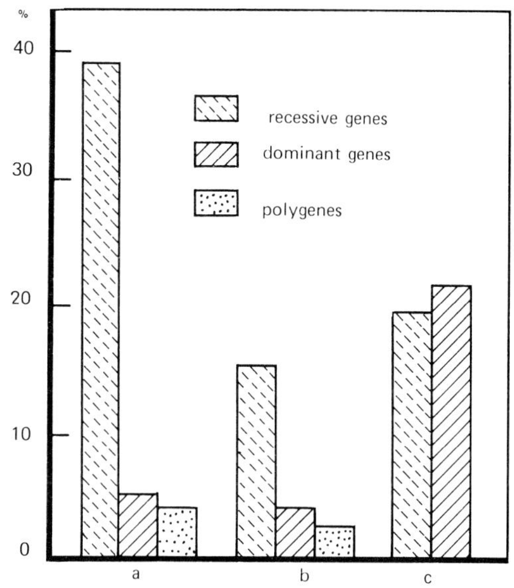

Fig. 3.16. Relative frequency of gene-cytoplasmic male steriles arisen spontaneously (*a*), or following intraspecific (*b*) or interspecific (*c*) hybridisation in relation to gene control

3.5 Gene Action

Like *ms* genes, *fr* genes also suppress male sexuality in diverse ways and likewise lead to structural, sporogenous and functional types of mst mainly because of their stage specificity and differential time of action. The *fr* gene male sex suppression assumes various modalities and dimensions. These include suppression of staminal development, formation of aborted or diminitive stamens, faulty stamen differentiation, abnormal anther development, anther transformations and malfunctioning of anther tissue (Table 3.8). All these features lead to structural (st) mst (Table 3.7). Certain *fr* genes permit normal anther and pollen development but inhibit pollen release. The viable pollen produced is thus not available to the stigma and the plants become functionally (fn) male sterile (Table 3.10). In contrast to the few *fr* genes that cause st- and fn-mst, the majority of *fr* genes affect three major microsporogenesis stages (Fig. 3.17B) and lead to sporogenous (sp) mst. The mode of action of these genes is described in the following.

3.5.1 Spontaneous and Intraspecific Hybrids

In most of the species, the *fr* genes act either during tetrad formation or during microspore development before or after release (Fig. 3.17A), only in a few they act during premeiosis or prophase-I (Table 3.15). In them, meiosis may be arrested or continues abnormally. The final breakdown occurs either after tetrad formation or during microspore development. This breakdown stage varies between species and even within mst genotypes of a species (Table 3.18). For instance, in one of the two gc-mst strains of red clover, cytoplasm degeneration follows nuclear degeneration and in the other, the reverse occurs. Since in both the genotypes, the S-cytoplasm

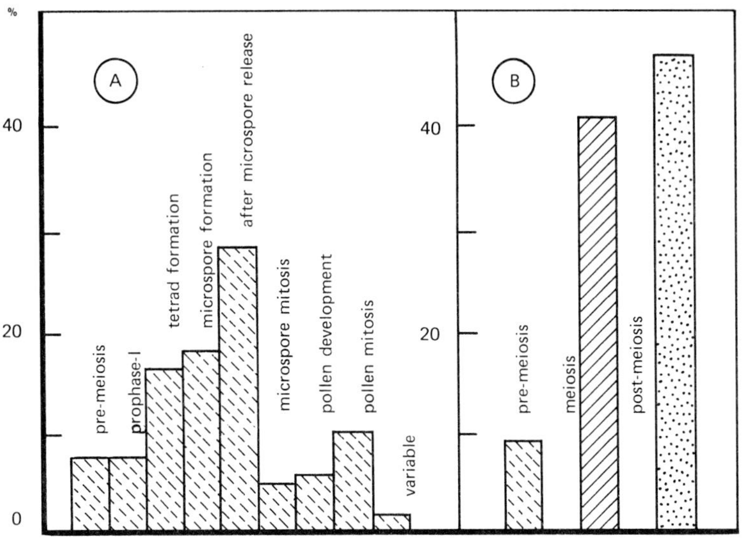

Fig. 3.17A,B. Microsporogenesis breakdown stages (specific, **A** and general, **B**) in gene-cytoplasmic male steriles arisen spontaneously or after intraspecific hybridisations

Table 3.15. *fr* Gene action over microsporogenesis in some spontaneously arisen and intraspecific hybrids

Breakdown stage during:	Tape-tum[a]	References
Premeiosis		
Capsicum annuum (1r, 2x = 24)	–	Pochard (1970b)
Daucus carota (2x = 18)		
(2–3D)	–	Thompson (1961)
(1D + 1r)	–	Timin and Dobrutskaya (1981)
Gossypium hirsutum (2r, 4x = 52)	d	Thombre and Mehetre (1979)
Oryza sativa (1r, 2x = 24)	–	This Vol.
Trifolium pratense (1r + 1D, 2x = 14)	–	Macewicz (1976a,b)
Meiosis		
Prophase-I		
Allium cepa (1r, 2x = 16)	b	Dyki (1973a)
Daucus carota (2x = 18)	b	Zenkteler (1962)
Mercurialis perennis (1r, 8x = 64)		This Vol.
Oryza sativa (2x = 24)	f	Pan et al. (1982), Yang and Li (1984)
Sorghum vulgare (2x, 4x = 20)	–	Damon (1961), Erichsen and Ross (1963a)
Tetrad formation		
Allium cepa (1r, 2x = 16)	–	Patil et al. (1973)
A. fistulosum (1r, 2x = 16)	b,c	Nishi and Hiraoka (1958) Virnich (1967)
Beta vulgaris (2r, 2x = 18)	–	Owen (1942), Artschwager (1947), Hosokawa et al. (1954)
	b	Nagao and Kinoshita (1962), Bliss and Gabelman (1965), Röhrbach (1965), Chauhan and Kinoshita (1980a), this Vol.
Beta vulgaris (4r, 4x = 36)	b	Kinoshita (1971), Kinoshita et al. (1972)

Table 3.15 *(continued)*

Breakdown stage during:	Tape-tum[a]	References
Capsicum annuum (2r, 2x = 24)	b	Peterson (1958), Pochard (1970b), Horner and Rogers (1974)
	c	Novâk et al. (1971)
	—	Hirose and Fujime (1975)
Crotalaria mucronata (2x = 16)	b,c	Edwardson (1967)
Lolium multiflorum (2x = 14)	c	Nitzsche (1971)
Lycopersicon sp. (2x = 24)	c	Nishi and Hiraoka (1958)
Oryza sativa (1r, 2x = 24)	c	This Vol.
Pennisetum typhoides (1r, 2x = 24)	c	Burton (1958)
Triticum aestivum (6x = 42)	—	This Vol.
Zea mays (2x = 20)	—	Herich (1965), Spasojevic (1966), Orel (1967)
	b,c	Colhoun and Steer (1981)
Microspore formation		
Allium schoenoprasum (1r, 2x = 16)	c,d	Tatliogu (1982)
Beta vulgaris (semi-sterile, 2r, 2x = 18)	d	Kinoshita (1971)
B.vulgaris (4r, 4x = 36)	c	Kinoshita (1971)
Capsicum annuum (1r, 2x = 24)	b,f	Peterson (1958), Novâk and Betlach (1970), Horner and Rogers (1974)
Crotalaria mucronata (2x = 16)	b	Edwardson (1967)
Lathyrus sativus (2x = 14)	—	Zelenskaya and Pestova (1974)
Lolium multiflorum (2x = 14)	c	Nitzsche (1971)
L.perenne (2x = 14)	—	Hayward and Manthiratna (1972)
Lycopersicon sp. (2x = 24)	c	Nishi and Hiraoka (1958)
Medicago sativa (1–4 po, 4x = 32)	e	Bradner and Childers (1968)
(2r, 4x = 32)	f	Staszewski (1970, 1976)
Secale cereale (2r, 2x = 14)	b,e	Scoles and Evans (1979)
	—	Artemova (1980)
Sorghum vulgare (3r, 4x = 20)	c	Alam and Sandal (1967b)
		Laser and Lersten (1972)
Zea mays (2x = 20)	b,c	Lee (1976), Warmke and Lee (1977)
Variable (anywhere in meiosis)		
Sorghum vulgare (4x = 20)	b	Raj (1968)
Zea mays (2x = 20)	a	This Vol.
Postmeiosis		
After microspore liberation		
Allium cepa (1r, 2x = 16)	b	Monosmith (1926), Jones and Ensweller (1937), Kobabe (1958)
Allium cepa (1r, 2x = 16)	d	Virnich (1967)
Beta vulgaris (2r, 2x = 18)	b	Owen (1942), Artschwager (1947), Nagao and Kinoshita (1962)
Beta vulgaris (4r, 4x = 36)	b	Kinoshita (1971), Kinoshita et al. (1972), Chauhan and Kinoshita (1980a), Mikami et al. (1978), Mikami (1982), this Vol.
Capsicum annuum (1r, 2r, 2x = 24)	b,f	Novâk and Betlach (1970), Novâk et al. (1971)
Cucumis melo (2x = 24)	b	Chauhan and Singh (1968)
Dactylis glomerata (4x = 38)	f	Artschwager (1947), Filion and Christie (1966)
Daucus carota (2–3D, 2x = 18)	—	Thompson (1960)
Hordeum vulgare (1r, 2x = 14)	c	Schooler (1967), Ahokas (1979a, 1982b)

Table 3.15 *(continued)*

Breakdown stage during:	Tapetum[a]	References
Medicago sativa (1–4 po, 4x = 32)	e	Bradner and Childers (1968)
(few r genes)	f	Pederson and Stucker (1969)
Oryza sativa (2x = 24)	—	Chu et al. (1972), Shuhua (1982)
	c	This Vol.
Pennisetum typhoides (2x = 14)	c	Singh and Sharma (1963)
Phaseolus vulgaris (1r, 2x = 22)	—	Bassett and Shuhua (1982)
Plantago ovata (1D, 2x = 8)	—	Mittal and Issar (1970)
Raphanus sativus (1r, 2x = 18)	—	Ogura (1968)
	c	Nishi and Hiraoka (1958)
Secale cereale (2r, 2x = 14)	c	Geiger and Schnell (1970)
	d	Garlicka and Madej (1975)
Tabernaemontana coronaria (2x = 22)	c	Dubey (1970)
Vitis vinifera (2x = 38)	—	Staudt (1973)
Zea mays (2x = 20)	b,c	Rhoades (1933), Gabelman (1949), Chang (1954), Jones et al. (1957a), Lee (1976), Warmke and Lee (1977)
Microspore mitosis		
Beta vulgaris (4r, 2x = 18)	b	Kinoshita (1971), Kinoshita et al. (1972)
Hordeum vulgare (2x = 14)	c	Schooler (1967)
Oryza sativa (2x = 24)	—	Anon (1978a)
Vicia faba (1r, 2x = 12)	—	This Vol.
Pollen development		
Desmodium sandwicense (2r, 2x = 22)	—	McWhirter (1969)
Epilobium hirsutum (2x = 36) × *E.parviflorum* (2x = 36)	—	Michaelis (1943)
Pelargonium crispum (2x = 22)	b	Tokumasu (1974)
Oryza sativa (1r, 2x = 24)	c	This Vol.
Sorghum vulgare (1r, 4x = 20)	c	Singh and Hadley (1961)
Zea mays (2x = 20)	—	Rogers and Edwardson (1952)
	a	Colhoun and Steer (1981)
Or after pollen mitosis		
Allium cepa (1r, 2x = 16)	a	Tatebe (1952), Yen (1959)
Oryza sativa (2x = 24)	a	Tamaru et al. (1980)
Sorghum vulgare (4x = 20)	d	Narkhede et al. (1968)
Triticum aestivum (6x = 42)	a	De Vries and Ie (1970)
Zea mays (2x = 20)	—	Khoo and Stinson (1957)
	b,c	Lee et al. (1979)
	—	Lee et al. (1980)

[a]Tapetum normal (a), abnormal (b), persistent (c), disintegrates later (d), simultaneously (e), or before (f) sporogenous tissue breaks down, in others not known (—).

is similar, the anomaly is caused by two iso- or nonallelic *fr* genes. In a pepper gc-mst genotype, *fr* genes disturb PMC development and differentiation and only a few PMC's develop. In them, meiosis continues abnormally until tetrad formation and the microspores degenerate later. In carrot, *fr* genes inhibit normal anther development so that either petaloids stamens or stamens with reduced anthers lacking either sporogenous tissue or having abnormal PMC-like cells are developed. Similar *fr* gene action is known in some gc-mst genotypes of cotton and red clover. In a gc-mst carrot line (Zenkteler 1962) translocation heterozygosity accompanies *fr* gene action in early prophase-I. Meiosis continues abnormally in the PMC's. On the other hand, in a cotton gc-mst genotype (Thombre and Mehetre 1979), no PMC's develop in early-formed flowers. In the late-formed ones, some develop but degenerate before meiosis. Thus, in these instances, *fr* genes act prior to the onset of meiosis.

In a few gc-mst lines of carrot, onion, petunia, rice and sorghum, *fr* genes reduce either chromosome synapsis or chiasma building so that many unpaired chromosomes occur randomly at MI. This leads to anomalous meiosis and the final breakdown occurs either during tetrad or microspore formation. However, none of these *fr* genes causes complete asynapsis, but desynapsis to a varying degree is induced. In most of these, female meiosis is influenced to a considerably lesser extent than the male meiosis, so that whereas mst is complete, female fertility is slightly impaired. In this respect, the *fr* genes do not act like *as* or *ds*-genes (Gottschalk and Kaul 1980a,b). Thus, whereas *as* and *ds* genes impair synapsis and chiasma formation respectively in PMC's and EMC's to almost equal level and affect both male and female sex nearly equally (Gottschalk and Kaul 1980a,b), *fr* genes suppress male sexuality but do not significantly impair female fertility. Another feature of *fr* gene action is that most of these act during tetrad formation or microspore development before or after release (Fig. 3.17A). In many of these the tapetum develops but functions abnormally (Table 3.15). Based on the contention that anomalous tapetum development or function impedes nutrient supply to the developing microspores and starves them to death, the microsporogenesis breakdown has been correlated to the tapetal anomalies.

Some *fr* genes may suppress cytokinesis or inhibit microspore development or their release from PMC's. The latter occurs mostly either through repression of callase activity so that the microspores are not released at all (e.g., *Beta, Capsicum, Lycopersicon*) or due to faulty timing of enzymatic digestion of callose around the PMC's (e.g. *Beta, Capsicum, Medicago, Petunia, Secale*). Izhar and Frankel (1971) evidenced that *fr* gene-induced deviations in the timing of enzymatic digestion of callose caused mst in one g-mst and two gc-mst lines of *Petunia hybrida*. After establishing that in vitro activity of callose is optimum at about pH 5.0 and inductive above pH 6.5, they measured in vivo pH during different microsporogenesis stages. It was found that in mft anthers, pH was 5.0 until the late tetrad stage after which it was about 6.0. This was followed by callose dissolution. On the other hand, whereas in one gc-mst line (RM), pH was uniformly low all through microsporogenesis and PMC callose digested early leading to PMC abortion, in the other gc-mst line (PR), callose activity became delayed until the late tetrad stage and abortion ensued during the early microspore stage.

The immediate phase after release of microspores from tetrad binding appears to be critical as most of the *fr* genes act during this stage (Fig 3.17A). The genes either arrest microspore wall formation, nuclear division and separation or microgametophyte development (Table 3.15). After release from the quartet callose capsule, the primary microgametophyte normally enters into a rapid stage of growth and develops vacuolation and exine. Degeneration centered around this stage in microgametogenesis is prevalent in many of the gc-mst genotypes. In some, the development either stops immediately after release from PMC's or when microspores appear vacuolate and are mononucleate (amitotic and degenerate).

The amitosis is usually followed by gradual collapse of mononucleate microspore walls (e.g. sugarbeet, alfalfa, rice, rye). Cessation of further development associated with lack of food accumulation occurs in some gc-mst lines of barley, onion, pepper, radish, rice, rye and sorghum. The tapetal nondegeneration appears to be correlated physiologically with the lack of buildup of food reserves of the microgamete. Many *fr* genes act quite late in microsporogenesis after the microspores are liberated and before they undergo pollen transformation (Table 3.15). Some of these genes act still later especially during pollen development, mostly either during or after pollen mitosis and exine wall development. Thus, in two rice and wheat gc-mst lines, microsporogenesis progresses normally until pollen development. The *fr* genes impair exine development which is followed by nuclear degeneration. In a few gc-mst lines of rice and petunia, *fr* genes either inhibit or slow down pollen mitosis so that no functional pollen is available at anthesis. Late *fr* gene action results in the production of sticky pollen grains in some gc-mst genotypes of petunia, rice and sorghum. Release of pollen without exine formation occurs in a wheat and a rice genotype. In these, the *fr* genes suppress primexine development but mitosis occurs normally.

3.5.2 Interspecific and Intergeneric Hybrids

Of the known 320 cases of gc-mst, 245 (75%) arise after interspecific or intergeneric crosses (Tables 3.3, 3.4). However *fr* gene action over microsporogenesis is known in only 40% of such crosses (Tables 3.16, 3.17). In the majority of the interspecific gc-mst genotypes, *fr* genes mostly act in prophase-I and during tetrad and microspore formation, a few act during premeiosis or pollen development (Fig. 3.18). Of the vast number of intergeneric and alloplasmic hybrids exhibiting gc-mst (Tables 3.4, 3.6), *fr* gene action is known in only a few genotypes (Table 3.17). In the majority of these, the genes act during premeiosis, after tetrad formation and microspore release from PMC's (Fig. 3.19, 3.20). In intergeneric hybrids, the number of cases exhibiting *fr* gene action during premeiosis is higher than those in which it is during pollen development. The reverse is, however, true for interspecific hybrids (Fig. 3.18). Hence, the time of *fr* gene action depends on the origin of gc-mst's. Whereas in those arising spontaneously and after intraspecific hybridizations, the genes act mostly during microspore formation and after its release (Fig. 3.17), in those arising after interspecific hybridizations, the *fr* genes act in prophase-I or during microspore formation and development (Fig. 3.18). In gc-mst's arising after intergeneric hybridizations, the gene action is confined to pre- and postmeiotic stages (Fig. 3.19), but in gc-mst's in general, the *fr* gene action is specific and is mostly confined to microspore formation and its developmental stages; in a few the genes act during prophase-I (Fig. 3.20A). Only in 2% of the cases is the *fr* gene action variable. The general breakdown stages (Fig. 3.20B) exhibit the same trend as in the mst's arising spontaneously and after intraspecific hybridizations (Fig. 3.17). In both these, the *fr* genes act mostly during meiotic and postmeiotic stages. A characteristic feature of some gc-mst genotypes arising after interspecific crosses involving rye and wheat (Table 3.16) is the development of sticky pollen or grains lacking exine ornamentation. These pollen grains stain well in carmine but are nonfunctional on compatible stigma.

Table 3.16. *fr* gene action in some interspecific hybrids

Breakdown stage during:	Tapetum[a]	References
Premeiosis		
Citrus limon (2x = 18) × *C.sinensis* (2x = 18)	d	Fröst (1948)
C.limon (2x = 18) × *C.paradisi* (2x = 18)	–	Fröst (1948)
C.unshiu (2x = 18) × *C.sinensis* (2x = 18)	d	Iwamasa (1966)
Gossypium harknessii (2x = 26) × *G.hirsutum* (4x = 52)	d	Murthi and Weaver (1974), Meyer (1975)
Helianthus petiolaris (2x = 34) × *H.annuus* (2x = 34)	d	Leclercq (1969), Paun (1974)
Nicotiana langesdorffii (2x = 18) × *N.sanderae* (2x = 18)	–	East (1919, 1929)
N.tomentosiformis (2x = 24) × *N.otophora* (2x = 24)	–	Gerstel and Burns (1974)
Petunia axillaris (2x = 14) × *P.hybrida* (2x = 14)	–	Frankel et al. (1969)
	d	Bino (1985a,b)
Meiosis		
Prophase-I		
Lolium perenne (2x = 14) × *L.multiflorum* (2x = 14)	–	Ahloowalia (1981)
Nicotiana bigelovii (4x = 48) × *N.tabacum* (4x = 48)	–	Burk (1960), Chaplin (1964)
N.megalosiphon (4x = 40) × *N.tabacum* (4x = 48)		
Petunia axillaris (2x = 14) × *P.hybrida* (2x = 14)	–	Izhar (1977)
	e	Bino (1985a,b)
Solanum chacoense (2x = 24) × *S.kurtzianum* (2x = 24)	–	Grün and Aubertin (1966)
Solanum chacoense (2x = 24) × *S.phureja* (2x = 24)	–	Grün and Aubertin (1966)
Solanum chacoense (2x = 24) × *S.stenotomum* (2x = 24)	–	Grün and Aubertin (1966)
Solanum chacoense (2x = 24) × *S.tuberosum* (4x = 48)	–	Ramanna and Hermsen (1974)
S.demissum (6x = 72) × *S.berthaulti* (2x = 24)	–	Magoon et al. (1958c)
S.demissum (6x = 72) × *S.gibberulosum* (2x = 24)	–	Magoon et al. (1958c)
S.demissum (6x = 72) × *S.gigantophyllum* (2x = 24)	–	Magoon et al. (1958c)
S.demissum (6x = 72) × *S.goniocalyx* (2x = 24)	–	Magoon et al. (1958c)
S.demissum (6x = 72) × *S.phureja* (2x = 24)	–	Magoon et al. (1958c)
S.demissum (6x = 72) × *S.pinnatisectum* (2x = 24)	–	Ramanna and Hermsen (1979, 1981)
S.demissum (6x = 72) × *S.saltense* (2x = 24)		
S.demissum (6x = 72) × *S.etuberosum* (2x = 24)	–	Hermsen and Taylor (1979)
S.demissum (6x = 72) × *S.tuberosum* (4x = 48)	–	Schnell (1948), Dionne (1961)
S.polytrichon (4x = 48) × *S.phureja* (2x = 24)	–	Abdalla and Ramanna (1971)
S.tuberosum (4x = 48) × *S.acaule* (4x = 48)	–	Lamm (1953)
Triticum compactum (6x = 42) × *T.aestivum* (6x = 42)	–	Xiangmin and Herong (1981)
Anaphase-I		
Solanum chacoense (2x = 24) × *S.tuberosum* (4x = 48)	–	Ramanna and Hermsen (1974)
Tetrad to microspore formation		
Aquilegia flabellata (2x = 14) × *A.truncata* (2x = 14)	c	Skalinska (1931)
Brassica campestris (ms, 2x = 20) × *B.oleracea* (2x = 18)	c	Nishi and Hiraoka (1958)
Linum floccosum (2x = 30) × *L.usitatissimum* (2x = 30)	c	Dubey and Singh (1965)
Lycopersicon esculentum (2x = 24) × *L.hirsutum* (2x = 24)	–	Siegemund (1979b)
Petunia axillaris (2x = 14) × *P.hybrida* (2x = 14)	–	Edwardson and Warmke (1967)
Solanum verrucosum (2x = 24) × *S.canasense* (2x = 24)	–	Grün et al. (1962)
Solanum verrucosum (2x = 24) × *S.chacoense* (2x = 24)	–	Abdalla (1970), Abdalla and Hermsen (1972)
Solanum verrucosum (2x = 24) × *S.chomatophilum* (2x = 24)	–	Abdalla (1970), Abdalla and Hermsen (1972)
Solanum verrucosum (2x = 24) × *S.gigantophyllum* (2x = 24)	–	Abdalla (1970), Abdalla and Hermsen (1972)
Solanum verrucosum (2x = 24) × *S.goniocalyx* (2x = 24)	–	Abdalla (1970), Abdalla and Hermsen (1972)
Solanum verrucosum (2x = 24) × *S.infundibuliforme* (2x = 24)	–	Abdalla (1970), Abdalla and Hermsen (1972)

Table 3.16 *(continued)*

Breakdown stage during:	Tapetum[a]	References
Solanum verrucosum (2x = 24) × *S.kurtzianum* (2x = 24)	–	Abdalla (1970), Abdalla and Hermsen (1972)
Solanum verrucosum (2x = 24) × *S.leptophyes* (2x = 24)	–	Abdalla (1970), Abdalla and Hermsen (1972)
Solanum verrucosum (2x = 24) × *S.megistacrolobum* (2x = 24)	–	Abdalla (1970), Abdalla and Hermsen (1972)
Solanum verrucosum (2x = 24) × *S.multidissectum* (2x = 24)	–	Abdalla (1970), Abdalla and Hermsen (1972)
Solanum verrucosum (2x = 24) × *S.phureja* (2x = 24)	–	Abdalla (1970), Abdalla and Hermsen (1972)
Solanum verrucosum (2x = 24) × *S.raphanifolium* (2x = 24)	–	Abdalla (1970), Abdalla and Hermsen (1972)
Solanum verrucosum (2x = 24) × *S.simplicifolium* (2x = 24)	–	Abdalla (1970), Abdalla and Hermsen (1972)
Solanum verrucosum (2x = 24) × *S.stenotomum* (2x = 24)	–	Abdalla (1970), Abdalla and Hermsen (1972)
Solanum verrucosum (2x = 24) × *S.tarijense* (2x = 24)	–	Abdalla (1970), Abdalla and Hermsen (1972)
Solanum verrucosum (2x = 24) × *S.toralapanum* (2x = 24)	–	Abdalla (1970), Abdalla and Hermsen (1972)
Solanum verrucosum (2x = 24) × *S.tuberosum* (4x = 48)	–	Abdalla (1970), Abdalla and Hermsen (1972)
Solanum verrucosum (2x = 24) × *S.vernie* (2x = 24)	–	Abdalla (1970), Abdalla and Hermsen (1972)
Postmeiosis		
After microspore release		
Allium cepa (2x = 16) × *A.ascalonicum* (2x = 16)	a	Little et al. (1944)
Allium cepa (2x = 16) × *A.drobovii* (2x = 16)	a	Saini and Davis (1969)
Allium cepa (2x = 16) × *A.galanthum* (2x = 16)	–	Saini and Davis (1969)
Allium cepa (2x = 16) × *A.pskemense* (2x = 16)	–	Saini and Davis (1969)
A.roylei (2x = 16) × *A.drobovii* (2x = 16)	a	Saini and Davis (1969)
Aquilegia flabellata (2x = 14) × *A.truncata* (2x = 14)	c	Skalinska (1931)
Beta vulgaris (2x = 18) × *B.maritima* (2x = 18)	b,d	Dalke et al. (1983)
Epilobium montanum (2x = 36) × *E.roseum* (2x = 36)	a	Oberreuter (1925)
Fragaria vesca (2x = 14) × *F.nilgerrensis* (2x = 14)	–	Nürnberg-Krüger (1956, 1958)
Helianthus petiolaris (2x = 34) × *H.annuus* (2x = 34)	d	Whelan (1979)
Linum floccosum (2x = 30) × *L.usitatissimum* (2x = 30)	c	Dubey and Singh (1965)
Nicotiana tabacum (4x = 48) × *N.glutinosa* (2x = 24)	c	Tsikov et al. (1977)
Petunia axillaris (2x = 14) × *P.hybrida* (2x = 14)	–	Hamilton (1965)
Solanum verrucosum (2x = 24) × *S.bulbocastanum* (2x = 24)	–	Abdalla and Hermsen (1972)
Solanum verrucosum (2x = 24) × *S.gibberulosum* (2x = 24)	–	Magoon et al. (1958b)
Solanum verrucosum (2x = 24) × *S.yabari* (2x = 24)		
Triticum timopheevi (4x = 28) × *T.aestivum* (6x = 42)	c	Joppa et al. (1966), Jain et al. (1981)
Triticum timopheevi (4x = 28) × *T.aestivum* (6x = 42)	–	Xiangmin and Herong (1981)
Vitis vinifera (2x = 38) × *V.riparia* (2x = 38)	–	Staudt (1973)
Pollen development		
Brassica napus (4x = 38) × *B.juncea* (4x = 36)	–	Banga, personal communication
Epilobium hirsutum (2x = 36) × *E.luteum* (2x = 36)	–	Michaelis (1929, 1932, 1933, 1954)
Petunia axillaris (2x = 14) × *P.hybrida* (2x = 14)	–	Izhar (1977)
Solanum acaule (4x = 48) × *S.bulbocastanum* (2x = 24)	–	Hermsen (1966), Hermsen and Ramanna (1969)

Table 3.16 *(continued)*

Breakdown stage during:	Tapetum[a]	References
S.phureja (2x = 24) × *S.acaule* (4x = 48)	–	Koopmans (1952)
S.polytrichon (4x = 48) × *S.chacoense* (2x = 24)	–	Toxopeus (1960, 1964)
S.polytrichon (4x = 48) × *S.goniocalyx* (2x = 24)	–	Toxopeus (1960, 1964)
S.verrucosum (2x = 24) × *S.macolae* (2x = 24)	–	Abdalla and Hermsen (1972)
S.verrucosum (2x = 24) × *S.phureja* (2x = 24)	–	Buck (1960)
S.verrucosum (2x = 24) × *S.tuberosum* (4x = 48)	–	Buck (1960)
During or after pollen mitosis		
Aquilegia vulgaris (2x = 14) × *A.chrysantha* (2x = 14)	–	Skalinska (1928a)
Oryza sativa (2x = 24)		
(*indica* × *japonica*)	–	Kinoshita et al. (1980)
(*spontanea* × *japonica*)	–	Katsuo (1954), Katsuo and Mizushima (1958)
Petunia axillaris (2x = 14) × *P.hybrida* (2x = 14)	–	Izhar (1977)
Triticum timonovum × *T.aestivum* (6x = 42)	–	Fedorova and Nettevich (1969)
T.timopheevi (4x = 28) × *T.aestivum* (6x = 42)	–	Fedorova and Nettevich (1969)
T.zhukovskyi (6x = 42) × *T.aestivum* (6x = 42)	–	Fedorova and Nettevich (1969)

[a]For tapetal symbols see Table 3.15.

Table 3.17. *fr* gene action in some intergeneric hybrids

Breakdown stage during:	Tapetum[a]	References
Premeiosis		
Aegilops ovata (4x = 28) × *Triticum aestivum* (6x = 42)	b	Chauhan and Singh (1966)
Raphanus sativus (2x = 18) × *Brassica napus* (4x = 38)	–	Kanno et al. (1963a,c), Yamaguchi and Kano (1963), Shiga and Baba (1971, 1973)
Meiosis		
Prophase-I		
Saccharum spontaneum (8x = 80) × *Sclerostachya fusca* (4x = 48)	a	This Vol.
Tetrad formation		
Aegilops ovata (4x = 28) × *Triticum durum* (4x = 28)	–	Fukasawa (1953, 1956b)
Ae.ovata (4x = 28) × *T.aestivum* (6x = 42)	b	Chauhan and Singh (1966)
Raphanus sativus (2x = 18) × *Brassica campestris* (2x = 20)	c	Nishi and Hiraoka (1958)
R.sativus (2x = 18) × *B.oleracea* (2x = 18)	c	Nishi and Hiraoka (1958)
Microspore release		
Aegilops caudata (2x = 14) × *Triticum durum* (4x = 28)	–	Hervás (1974)
Diplotaxis muralis (2x = 22) × *Brassica campestris* (2x = 20)	–	Hinata and Kanno (1976)
Raphanus sativus (2x = 18) × *Brassica napus* (4x = 38)	–	Thompson (1972), Bartkowiak-Broda et al. (1979)
Triticum durum (4x = 28) × *Secale cereale* (2x = 14)	c	Vega and Lacadena (1979)
Pollen development		
Triticum durum (4x = 28) × *Secale cereale* (2x = 14)	c	Vega and Lacadena (1979)
Variable (anywhere in meiosis)		
Raphanus sativus (2x = 18) × *Brassica napus* (4x = 38)	–	Kanno et al. (1963a,c), Yamaguchi and Kanno (1963)

[a]For tapetal symbols see Table 3.15.

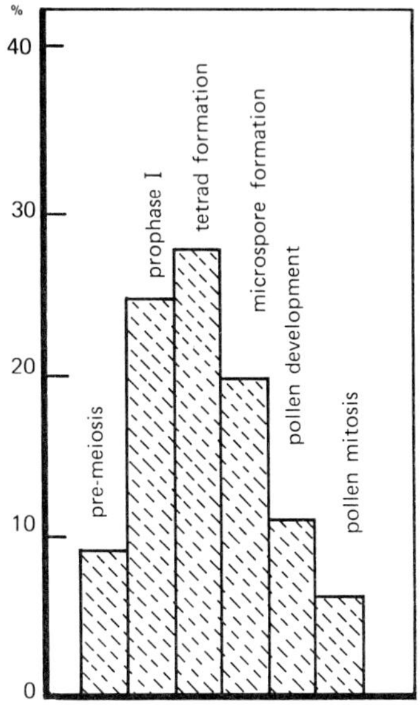

Fig. 3.18. Microsporogenesis breakdown stages in gene cytoplasmic male steriles arising after interspecific hybridisation

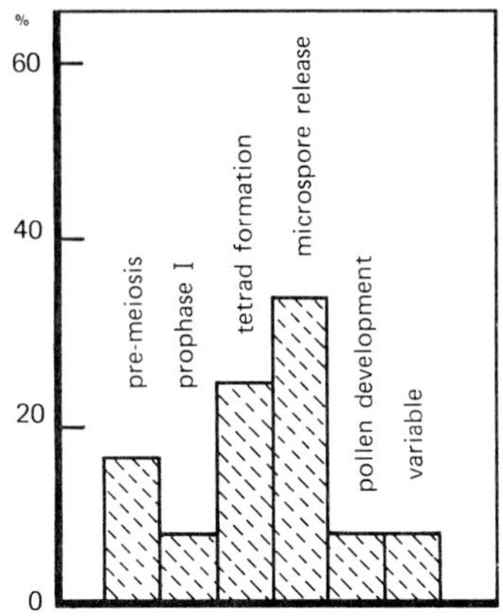

Fig. 3.19. Microsporogenesis breakdown stages in gene-cytoplasmic male steriles arising after intergeneric hybridisation

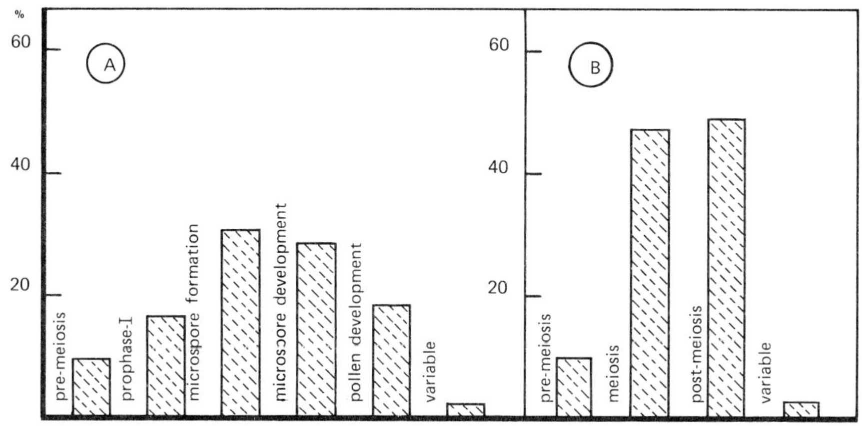

Fig. 3.20A,B. Microsporogenesis breakdown stages in gene-cytoplasmic male steriles

3.5.3 Differential *fr* Gene Action

The precise action of *fr* genes on microsporogenesis in a majority of gc-mst plants is not determined. In a few investigated ones, the time of *fr* gene action within a species is differential and divergent (Table 3.18). The genes may act either prior to or during premeiosis, meiosis and postmeiosis. Extreme difference in the timing of *fr* gene action occurs in gc-mst onion, carrot, mustard, rice, petunia, sorghum,

Table 3.18. Differential *fr* gene action within spontaneously arisen (intraspecific) and interspecific hybrids

Species hybrids	References	*fr* Gene action
Intraspecific		
Allium cepa (1r, 2x = 16)	Dyki (1973a)	During prophase-I
	Patil et al. (1973)	During tetrad formation
	Monosmith (1926)	After microspore release
	Tatebe (1952), Yen (1959)	During or after pollen mitosis
Beta vulgaris (2r, 2x = 18)	Artschwager (1947), Hosokowa et al. (1954), Rohrbach (1965a), Chauhan and Kinoshita (1980a)	During tetrad formation
	Owen (1942), Artschwager (1947), Nagao and Kinoshita (1962), Kinoshita et al. (1972), Mikami et al. (1978)	During microspore formation, after release
	Kinoshita (1971), Kinoshita et al. (1972)	Around microspore mitosis
Capsicum annuum (2x = 24)	Peterson (1958), Horner and Rogers (1974), Hirose and Fujime (1975)	During tetrad formation
	Novâk et al. (1971), Novâk and Betlach (1970)	After microspore release
Daucus carota (2x = 18)		
1D + 1r	Timin and Dobrutskaya (1981)	During premeiosis
2-3D	Thompson (1961)	
	Zenkteler (1962)	During meiosis
	Thompson (1960)	After microspore release
Oryza sativa (2x = 24)	Pan et al. (1982)	Prophase-I
	Chu et al. (1972)	After microspore release
	Anon (1978a)	Around microspore mitosis
	Tamaru et al. (1980)	During or after pollen mitosis
Pennisetum typhoides (1r, 2x = 14)	Burton (1958)	During anaphase-I
	Singh and Sharma (1963)	After microspore release
Petunia hybrida (2x = 14)		
1r	Frankel et al. (1969)	During premeiosis
	Izhar (1977)	During prophase-I
2r	Edwardson and Warmke (1967)	During tetrad formation
	Hamilton (1965)	After microspore release
	Izhar (1977)	During pollen development
	Izhar (1977)	During or after pollen mitosis
Sorghum vulgare (2r, 4x = 20)	Damon (1961), Erichsen and Ross (1963a)	During prophase-I

Table 3.18 *(continued)*

Species hybrids	References	*fr* Gene action
Sorghum vulgare (2r, 4x = 20)	Laser and Lersten (1972), Alam and Sandal (1967b)	During microspore formation
Sorghum vulgare (2r, 4x = 20)	Raj (1968)	Variable (anywhere in meiosis)
Sorghum vulgare (2r, 4x = 20)	Singh and Hadley (1961)	During pollen development
Sorghum vulgare (2r, 4x = 20)	Narkhede et al. (1968)	During or after pollen mitosis
Zea mays (2x = 20)	Herich (1965), Spasojevic (1966), Orel (1967), Warmke and Lee (1977), Colhoun and Steer (1981)	During tetrad formation
	Rogers and Edwardson (1952) Colhoun and Steer (1981)	During pollen development
	Khoo and Stinson (1967) Lee et al. (1979, 1980)	During or after pollen mitosis
Interspecific		
Helianthus petiolaris (2x = 34) × *H. annuus* (2x = 34)	Leclercq (1969), Paun (1974)	During premeiosis
	Whelan (1979)	After microspore release
Solanum verrucosum (2x = 24) × *S. tuberosum* (4x = 48)	Grün et al. (1962), Abdalla (1970)	During tetrad to microspore formation
	Abdalla and Hermsen (1972)	During pollen development

sunflower and wheat. In most of them, the *fr* genes act either at the beginning (prophase-I) or at the end (during pre- or postmicrospore release stage) of meiosis. The time of *fr* gene action is less diverse in beet, potato, pepper, maize and rye. In them, the genes act after meiotic division is nearly completed. The gene action is confined to microspore and pollen developmental stages. In pearl millet, the genes act either at AI or immediately after the microspores are released from the quartets. In all the cases where *fr* genes act during prophase-I, male meiosis proceeds abnormally and the microsporogenesis mostly breaks down during microspore formation or its developmental stages.

Diversity in the time of *fr* gene action is due either to the existence of many nonallelic *fr* genes within a species genome or to the presence of many S-cytoplasm types in the cytoplasm donor species. Existence of nonallelic *fr* genes and of different S-cytoplasm types is known in onion, rice, petunia, *Solanum*, sorghum, wheat, maize etc., but the action of different *fr* genes on one S-cytoplasm type and of different S-cytoplasms on the same *fr* gene has not been precisely investigated. In maize, three sterile cytoplasms exist and each cytoplasm is sensitive to specific *fr* genes, so that a specific combination of *fr* gene and cytoplasm type results in pollen abortion. Alteration of cytoplasm type or *fr* genes revokes fertility. This is true for almost all the gc-mst systems.

Both expressivity and penetrance of *fr* genes and their time of action during microsporogenesis are influenced by environmental factors, the major ones being temperature and photoperiod (Table 3.25). Action of *fr* genes that act during prophase-I in rice and sorghum (Table 3.18) can be delayed until TII if the mst plants are grown in the glass house having 7°–12°C higher temperature than the field. In *Crotalaria, Cucumis* and *Medicago, fr* gene action timing is altered by changing either the genotypic or the S-cytoplasmic background. In them the *fr* genes usually act either before or after the microspore release from the PMC's.

3.5.4 Linkage and Pleiotropy

It is hard to differentiate linkage effects from pleiotropic expressions unless large populations segregating in uniform environment are available. Where mst is associated with a constellation of altered phenotypic traits, pleiotropy has been assumed or else linkage is inferred. Therefore, the presently given two instances of these two phenomena are not fully confirmed cases. Compared to the high number of known gc-mst's, only a few *fr* genes exhibit linkage with other genes or pleiotropy (Table 3.19). In *Beta, Brassica, Lolium* and *Lycopersicon, fr* genes are considered to be linked to the traits indicated in Table 3.19, as these traits are free from cytoplasmic effects. Pleiotropy in *Raphanus* is the result of *fr-c* gene interaction as the genotypes of *Brassica napus* and *B. oleracea* in which the S-radish cytoplasm is substituted exhibit the pleiotropic expression which involves development of abnormal flowers, minute floral buds and crooked style. This expression is constant, but is influenced by genome-cytoplasm interactions and environmental factors. Alterations in nuclear genome suppress the extent and degree of this pleiotropic expression.

Table 3.19. *fr* genes exhibiting linkage or pleiotropism

Species	References	Linkage or pleiotropism (trait)
Beta vulgaris (2x = 18)	Stein and Gabelman (1959)	Red pigmentation of anthers
	Hogaboam (1957)	Monogerm
Brassica napus (4x = 38)	Heyn (1978)	White flower colour
Lolium perenne (2x = 14) × *L.multiflorum* (2x = 14)	Ahloowalia (1981)	Desynapsis
Lycopersicon esculentum (2x = 24) × *L.parviflorum* (2x = 24)	Kesicki (1980)	Wide obtuse cotyledons and leaflets, split and recurved stem tips
Pleiotropy		
Allium cepa (2x = 16)	Meer (1978)	Perianth reduced and swollen chloranthy of filaments
Raphanus sativus (2x = 18)	Ogura (1968)	Abnormal flowering, minute floral buds and crooked style

3.5.5 Induced Polyploidy

Induction of polyploidy and retention of mst by polyploidy are known in *Beta* and *Solanum*. Nearly 19% 4x mst plants segregate from the cross *Solanum acaule* (6x) × *S. bulbocastanum* (2x) (Hermsen 1966). Male meiosis is normal in them but microspores degenerate after liberation (Hermsen and Ramanna 1969). Their colchicine-induced 8x plants are also mst although 30% of the pollen that they produce is stainable but nonfunctional. In these polyploids, the *fr* gene action is considerably delayed until the late pollen developmental stage. In *B. vulgaris*, 4x mst plants were obtained by Savitsky (1954) and Kinoshita (1971) after colchicine treatment of 2x mst plants. These yield a few aneu- and anisoploid plants, indicating instability in the induced tetraploids. However, these 4x steriles can be maintained by using appropriate maintainers. 2x mft beet plants when diplodized produce highly fst and fully mst 4x plants; the sterility is due to abnormal meiosis caused by tetraploidy. The mst in induced polyploids obtained from 2x mst plants is due to the carryover of the sterilizing effects of *fr*- and *c*-gene combinations.

3.6 Tapetum

During the course of microsporogenesis, food material, growth substances, water and other essential supplies have to pass through the tapetal cells. Therefore, the tapetum assumes a vital nutritive role especially during and after microsporogenesis. Much indirect and circumstantial evidence implicates abnormal tapetal behaviour as causing mst in higher plants (Chap. 2.7). This is evidenced by some instances in which microsporogenous tissue breakdown is correlated with the occurrence of tapetal anomalies or its persistence. Compared to the large number of documented instances of gc-mst's (about 320, Tables 3.1 to 3.4), tapetal development, the course of microsporogenesis and its breakdown stages are known only in 12% of gc-mst plants. Of these, in 16% the tapetum is normal (Table 3.22), in 35% it is abnormal (Table 3.20) and in 49% it is persistent (Table 3.21), but the microsporogenous tissue breaks down in all these cases (Fig. 3.21). Tapetal persistence occurs in the majority of the gc-mst's, whether they arise after intra- or interspecific crosses (Fig. 3.22).

3.6.1 Abnormal

The earliest action of *fr* genes in plants with abnormal tapetal form and function is known in petunia and wheat, in which the breakdown occurs either during premeiosis or just after meiotic initiation (Table 3.20). In carrots and onion, the *fr* gene action is slightly delayed so that microsporogenesis breaks down in prophase-I. In the majority of gc-mst's with abnormal tapetum, the breakdown occurs after tetrad formation and during microspore development (Table 3.20). In the alloplasmic wheat having *Aegilops* cytoplasm (Chauhan and Singh 1966), the *fr*

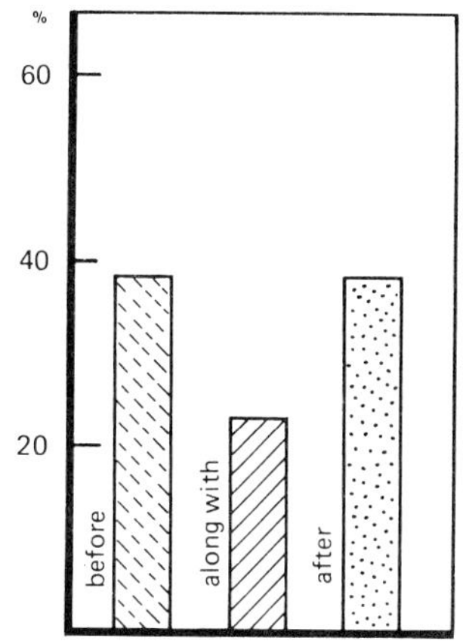

Fig. 3.21. Tapetal breakdown in relation to sporogeneous tissue degeneration

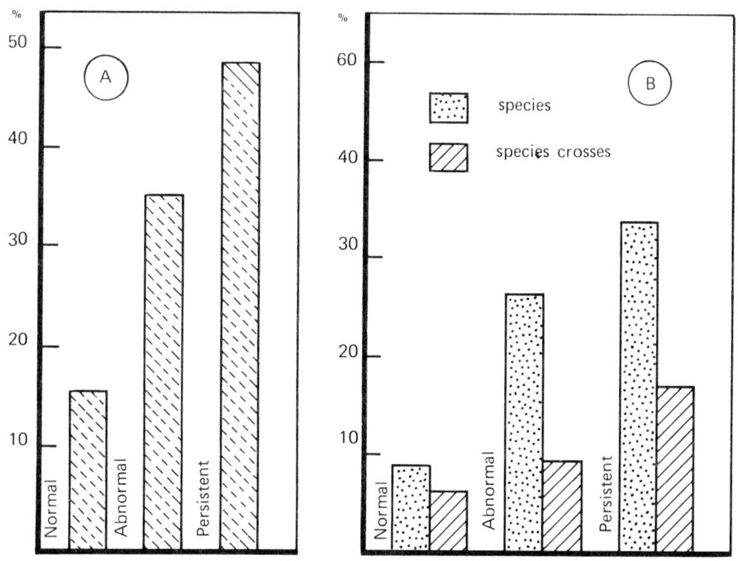

Fig. 3.22A,B. Frequency distribution of tapetal types in gene-cytoplasmic male steriles, total **(A)**, species and species crosses **(B)**

genes may act either in premeiosis or during tetrad formation. Whereas meiosis continues until microspore release in the meiocytes lying at the periphery of the anther sacs, it breaks down either in prophase-I or during the tetrad stage in some plants. In certain carrot lines, the *fr* genes act during prophase-I, but meiosis continues until microspores are formed. In *Sorghum*, the gene(s) may act anywhere during meiosis, but degeneration occurs mostly after microspores are developed. In some gc-mst strains of onion, beet, carrot, alfalfa and pepper, the tapetum swells up considerably and appresses the developing microspores to death.

Table 3.20. Microsporogenesis breakdown in gene cytoplasmic male steriles with abnormal tapetum

Breakdown stage during:	References
Pre- to early meiosis	
Aegilops ovata (4x = 38) × *Triticum aestivum* (6x = 42)	Chauhan and Singh (1966)
Petunia axillaris (2x = 14) × *P. hybrida* (2x = 14)	Binŏ (1984)
Prophase-I	
Allium cepa (2x = 16)	Dyki (1973a)
Daucus carota (2x = 18)	Thompson (1960), Zenkteler (1962)
Tetrad formation	
Aegilops ovata (4x = 28) × *Triticum aestivum* (6x = 42)	Chauhan and Singh (1966)
Allium fistulosum (2x = 16)	Nishi and Hiraoka (1958)
Beta vulgaris (2x = 18)	Artschwager (1947), Nagao and Kinoshita (1962), Chauhan and Kinoshita (1980a), this Vol.
Capsicum annuum (2x = 24)	Horner and Rogers (1974)
Crotalaria mucronata (2x = 16)	Edwardson (1967)
Secale cereale (2x = 14)	Scoles and Evans (1979)
Zea mays (2x = 20)	Lee (1976), Warmke and Lee (1977)
Microspore release	
Allium cepa (2x = 16)	Monosmith (1926), Jones and Emsweller (1937), Kobabe (1958)
A. cepa (2x = 16) × *A. pskemense* (2x = 16)	Saini and Davis (1969)
Beta maritima (2x = 18)	Kinoshita (1971), Kinoshita and Takahashi (1972)
B. vulgaris (2x = 18) × *B. maritima* (2x = 18)	Dalke et al. (1983)
Brassica napus (4x = 38) × *B. juncea* (4x = 36)	Banga et al. (1984)
B. carinata (4x = 34) × *B. juncea* (4x = 36)	Banga et al. (1984)
Capsicum annuum (2x = 24)	Novâk and Beltach (1970)
Cucumis melo (2x = 24)	Chauhan and Singh (1968)
Raphanus sativus (2x = 18)	Ogura (1968)
Sorghum vulgare (4x = 20)	Brooks et al. (1966)
Triticum timopheevi (4x = 38) × *T. aestivum* (6x = 42)	Jain et al. (1981)
Zea mays (2x = 20)	Chang (1954), Lee (1976), Warmke and Lee (1977)
Microspore development	
Aegilotrichum	Fukasawa (1956b)
Beta vulgaris (4x = 36)	Mikami et al. (1978), Mikami (1982)
Pelargonium crispum (2x = 22)	Tokumasu (1974)
Pollen mitosis	
Allium cepa (2x = 16)	Tatebe (1952), Yen (1959)
Zea mays (2x = 20)	Lee et al. (1979)
Anywhere in meiosis	
Sorghum vulgare (4x = 20)	Raj (1968)

Tapetal anomalies of subtle nature occur in the three sterile cytoplasms of mst maize. In the mst anthers of some T-cytoplasm plants, shortly after the initiation of male meiosis, tapetal mitochondria breakdown (Warmke and Lee 1977). This involves mitochondrial swelling, loss of its cristae and matrix density. In other T-anthers, tapetal mitochondria remain intact (Colhoun and Steer 1981). In C-anthers, the tapetum either persists or vacuolates, then enlarges and gradually disintegrates (Lee et al. 1979). In S-anthers, tapetal development is normal, but at late stages an unusual loss of rough endoplasmic reticulum and cisternae occurs in the tapetal cells (Colhoun and Steer 1981).

3.6.2 Persistent

The reason for assigning tapetal malfunction as the prime cause of mst in higher plants is because in the majority of the known instances of mst, the tapetum is either abnormal or persistent (Tables 3.20, 3.21). In almost all of them, the microspores mostly abort either before or just after release from PMC's or during their subsequent development (Fig. 3.23B). This is more pronounced in gc-mst plants having persistent tapetum (Table 3.21). In sunflower, *fr* genes may act as early as before meiotic initiation or as late as after microspore release. In the former case, meiosis progresses irregularly and microspores degenerate after liberation from PMC's. In the latter case, although meiosis is normal, microspores likewise abort after liberation from the quartets. In beet, onion and rye, tapetal persistence is associated with abnormal endothecium development.

3.6.3 Normal

In a few gc-mst plants, although the tapetum is apparently normal, the microspores mostly degenerate during tetrad formation or immediately after release (Fig. 3.23A). In the alloplasmic durum wheat having *Aegilops* cytoplasm, tapetal degeneration is normal but tetrads degenerate abruptly. Compared to the

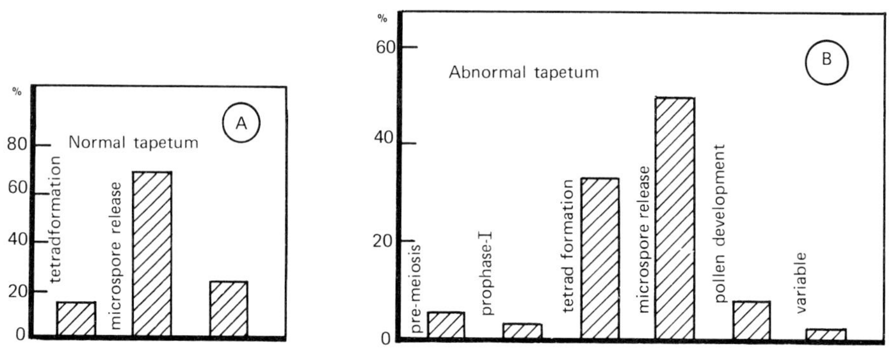

Fig. 3.23A,B. Microsporogenesis breakdown stages in gene-cytoplasmic male steriles with normal **(A)** and abnormal **(B)** tapetum

Table 3.21. Microsporogenesis breakdown in some gene-cytoplasmic male sterile plants with persistent tapetum

Species/crosses	References
Breakdown stage during:	
Meiosis	
Helianthus petiolaris (2x = 34) × *H.annuus* (2x = 34)	Paun (1974)
Tetrad formation	
Brassica campestris (2x = 20) × *B.oleracea* (2x = 18)	Nishi and Hiraoka (1958)
Lupinus mutabilis (4x = 48)	Pakendorf (1970)
Tetrad to microspore formation	
Allium cepa (2x = 16)	This Vol.
A.fistulosum (2x = 16)	Virnich (1967)
A.schoenoprasum (2x = 16)	Tatliogu (1982)
Aquilegia flabellata (2x = 14) × *A.truncata* (2x = 14)	Skalinska (1931)
Lolium multiflorum (2x = 14)	Nitzsche (1971)
Lycopersicon sp. (2x = 24)	Nishi and Hiraoka (1958)
Raphanus sativus (2x = 18)	Nishi and Hiraoka (1958)
Microscpore release	
Linum floccosum (2x = 30) × *L.usitatissimum* (2x = 30)	Dubey and Singh (1965)
Sorghum vulgare (4x = 20)	Alam and Sandal (1967a), Laser and Lersten (1972).
Zea mays (2x = 20)	Lee (1976), Warmke and Lee (1977)
After Microspore release	
Allium cepa (2x = 16)	Virnich (1967)
Beta vulgaris (2x = 18)	Kinoshita (1971)
Helianthus petiolaris (2x = 34) × *H.annuus* (2x = 34)	Whelan (1979)
Hordeum vulgare (2x = 14)	Ahokas (1979a, 1982b)
Linum floccosum (2x = 30) × *L.usitatissimum* (2x = 30)	Dubey and Singh (1965)
Lolium perenne (2x = 14)	Hayward and Manthiratna (1972)
Nicotiana tabacum (4x = 48) × *N.glutinosa* (2x = 24)	Tsikov et al. (1977)
Oryza sativa (2x = 24)	This Vol.
Pennisetum typhoides (2x = 14)	Singh and Sharma (1963), Prakash (1967)
Secale cereale (2x = 14)	Geiger and Schnell (1970), Garlicka and Madej (1975), Laser, unpubl.
Tabernaemontana coronaria (2x = 22)	Dubey (1970)
Triticum aestivum (6x = 42)	This Vol.
T.durum (4x = 28) × *Secale cereale* (2x = 14)	Vega and Lacadena (1979)
T.timopheevi (4x = 28) × *T.aestivum* (6x = 42)	Jain et al. (1981)
Microspore mitosis	
Triticum timopheevi (4x = 28) × *T.aestivum* (6x = 42)	Joppa et al. (1966)
Pollen development and mitosis	
Hordeum vulgare (2x = 14)	Schooler (1967)
Sorghum vulgare (4x = 20)	Singh and Hadley (1961), Narkhede et al. (1968)
Zea may (2x = 20)	Lee et al. (1979)

number of g-mst's with normal tapetum, the number of gc-mst's with such tapetum is low, but the examples demonstrate that visible tapetal anomalies may not be a necessary prerequisite for mst. This is further strengthened by the following.

3.6.4 Relative Breakdown

Studies on the relative breakdown of tapetum and microsporogenous tissue reveal that tapetum in some gc-mst plants may breakdown either before or along with or later than the microsporogenous tissue (Table 3.23). In the plants in which the tapetum breaks down before sporogenous tissue degenerates, the *fr* genes act mostly during microspore formation and maturation; but the *fr* gene action is much more variable in gc-mst plants in which tapetal breakdown follows sporogenous tissue degeneration. In them the microsporogenesis breakdown occurs during early to late stages of meiosis. The earliest is known in *Citrus* in which degeneration of either archesporial tissue or of young immature PMC's occurs immediately after their formation. A similar premeiotic degeneration occurs in cotton. Breakdown of microsporogenesis during late stages of meiosis occurs in a pepper line and alloplasmic bread wheat. In them, the microspores degenerate during their formation and after liberation, respectively. In both these, the tapetum breaks down immediately after sporogenous tissue has degenerated.

3.6.5 Form and Function

Alterations in tapetal form and function are considered as the primary causes of mst (Edwardson 1970; Laser and Lersten 1972), but there are mst's in which both tapetal form and function are normal (Table 3.22). In such genotypes, the majority of the genes (64%) act during pollen development (Fig. 3.23A). In the plants with abnormal and persistent tapetum, *fr* genes in 70% of the cases act during

Table 3.22. Microsporogenesis breakdown in gene-cytoplasmic male sterile plants with normal tapetum

Breakdown stage during:	References
Tetrad formation	
Aegilops ovata (4x = 28) × *Triticum durum* (4x = 28)	Fukasawa (1956b)
Microspore release	
Allium cepa (2x = 16) × *A. ascalonicum* (2x = 16)	Little et al. (1944)
A. cepa (2x = 16) × *A. drobovii* (2x = 16)	Saini and Davis (1969)
A. cepa (2x = 16) × *A. galanthum* (2x = 16)	Saini and Davis (1969)
A. roylei (2x = 16) × *A. drobovii* (2x = 16)	Saini and Davis (1969)
Epilobium montanum (2x = 36) × *E. roseum* (2x = 36)	Oberreuter (1925)
Helianthus annuus (2x = 34)	This Vol.
Sorghum vulgare (4x = 20)	This Vol.
Pollen development and mitosis	
Triticum aestivum (6x = 42)	DeVries and Ie (1970)
Zea mays (2x = 20)	Lee et al. (1980)

Table 3.23. Tapetal disintegration in relation to sporogenous tissue degeneration in gene-cytoplasmic male steriles

Tapetal breakdown/species	Reference	*fr* gene action
a) Before sporogenous tissue degeneration		
Beta vulgaris (4x = 36)	Kinoshita (1971), Kinoshita et al. (1972)	During microspore mutaration
Capsicum annuum (2x = 24)	Novâk and Betlach (1970), Novâk et al. (1971)	Microspores abort after release
	Horner and Rogers (1974)	Microspores abort before release while in tetrads
Dactylis glomerata (4x = 28)	Filion and Christie (1966)	Microspores degenerate after release
Medicago sativa (4x = 32)	Staszewski (1976)	During microspore formation, before release
Raphanus sativus (2x = 18)	Nishi and Hiraoka (1958), Ogura (1968)	After microspore release
b) Along with sporogenous tissue degeneration		
Gossypium harknessii (2x = 26)	Murthi and Weaver (1974)	PMC degeneration before meiosis
Medicago sativa (4x = 32)	Bradner and Childers (1968), Pederson and Stucker (1969)	After microspore release
Petunia axillaris (2x = 14) × *P.hybrida* (2x = 14)	Bino (1985a,b)	During leptotene
Secale cereale (2x = 14)	Scoles and Evans (1979)	After microspore formation before release
c) After sporogenous tissue degeneration		
Capsicum annuum (2x = 24)	Betlach and Dubovsky (1971)	Microspore degeneration after T II
Gossypi um hirsutum (4x = 52)	Thombre and Mehetre (1979)	Premeiotic degeneration
Citrus limon (2x = 18) × *C.paradisi* (2x = 18)	Fröst (1948)	Archesporial degeneration
C.limon (2x = 18) × *C.sinensis* (2x = 18)	Fröst (1948)	Archesporial degeneration
C.unshiu (2x = 18) × *C.sinensis* (2x = 18)	Iwamasa (1966)	Early PMC degeneration
Lycopersicon esculentum (2x = 24)	This Vol.	Microspore degeneration after release
Triticum timopheevi (4x = 28) × *T.aestivum* (6x = 42)	Jain et al. (1981), This Vol.	After microspore release

tetrad formation and just after the microspores are released from the quartets (Fig. 3.23B). For other stages, the frequency is low. Hence in gc-mst's the *fr* genes predominantly act during microspore release and developmental stages. This *fr* gene action is independent of tapetal form and function in many of them.

3.6.6 Developmental Behaviour

Tapetal development and behaviour have been investigated in only a few gc-mst species. In the few well-studied species, its development is variable in different mst genotypes. For instance, in American mst onions, the tapetum enlarges drastically, appressing the microspores (Monosmith 1926); in the Japanese type, it breaks down abruptly, forming plasmodial masses (Tatebe 1952) and in some European mst onions, it remains intact (this Vol.). In an Indian mst onion, the tapetal cells swell up considerably immediately after microspore release, almost fill the whole anther and degenerate abnormally or remain intact until microspores abort. The endothecium in some mst onions is also abnormally developed. As in mst onion, tapetal behaviour is variable in many other mst plants. For instance, in mst pepper, tapetal disintegration may initiate during premeiosis (Novâk 1971) or may be postponed until microspores degenerate. In other mst peppers, the tapetum remains intact. In mst carrot, the tapetum may become plasmodial (Zenkteler 1962) or enlarge and crush developing microspores to death (Artschwager 1947). In mst sunflower, the tapetum either disintegrates along with tetrads (Horner 1977) or enlarges and becomes periplasmodial and appresses the microspores to death. In one mst line, the tapetum is normal (this Vol.). In alfalfa, the tapetum may disintegrate when the PMC's are at the MI stage (Staszewski 1976) or swell and degenerate during microspore development stages (Pederson and Stucker 1969). In both cases, tapetal and sporogenous tissue degeneration is almost simultaneous. In rice, the tapetum enlarges abnormally, the major part of it is persistent and the rest bursts out into locules discharging its contents (Cheng et al. 1981a,b, 1982). In other mst rice lines, the tapetum is persistent, whereas in two lines, it disintegrates during prophase-I (this Vol.). In mst Sudan grass, the tapetum is either persistent (Alam and Sandal 1967b) or develops and functions abnormally (Brooks et al. 1966).

In mst rye, microspore and tapetal degeneration is either simultaneous (Scoles and Evans 1979C) or is delayed until microspores abort completely (Garlicka and Madej 1975). However, in one line, the tapetum disintegrates during prophase-I, but final breakdown occurs during the tetrad stage. In alloplasmic mst bread wheat having *T. timopheevi* cytoplasm, tapetal behaviour is variable (Jain et al. 1981). The tapetum may either breakdown prematurely during early meiosis or enlarge and become plasmodial during microspore development or remain persistent and breakdown after microspores abort. In alloplasmic bread wheat having *Aegilops ovata* cytoplasm, the tapetum may disintegrate during premeiosis or become periplasmodial during microspore development or remain persistent (Chauhan and Singh 1966). In alloplasmic mst durum wheat having *Ae. ovata* cytoplasm, the tapetum may either degenerate prematurely or become vacuolated, enlarged and periplasmodial (Fukasawa 1956a). Subtle and distinct differences exist in tapetal development and its subsequent breakdown in maize having sterile cytoplasms, T, C and S. In T-anthers, mitochondrial breakdown accompanied by loss of cristae and matrix density occurs during microspore development (Warmke and Lee 1977). In C-anthers, orbicule development is impaired or completely inhibited (Lee et al. 1980; Colhoun and Steer 1981). In S-anthers, unusual loss of rough endoplasmic reticulum and some other cellular abnormalities occur in tapetum (Colhoun and Steer 1981).

3.6.7 Causes of Variability

Both *ms* and *fr* genes act either on sporogenous or on nonsporogenous tissue or on both the tissues of the anther. Since gc-mst results from the interaction of *fr* genes and sterile cytoplasm that possesses *c*-genes, the difference in *c*- or *fr*-genes may alter the pattern of gene action and lead to the observed variability in microsporogenesis course and tapetal development and its breakdown. This is evidenced by a few examples given in Table 3.24. For instance, in onion, where sterile cytoplasm differences are known, differences in microsporogenesis breakdown and tapetal behaviour exist between the two cytoplasm types. Similar differences exist in alloplasmic wheats having two different cytoplasms. Likewise, certain fundamental differences occur between the three sterile maize cytoplasms in the pattern of microsporogenesis breakdown (Table 27.19). Not only *c*-gene differences but *fr*-gene differences alter the microsporogenesis breakdown pat-

Table 3.24. Tapetal behaviour and microsporogenesis breakdown in some male steriles with different sterile cytoplasm types

Species (genome)	Cytoplasm	Tapetal behaviour	Microsporogenesis breakdown stage
Allium cepa (2x = 16)	S	Enlargement and abnormal degeneration, malfunction	During microspore or pollen development
	T	During prophase-I delayed degeneration, persistent	Before microspore release or during prophase I
Triticum aestivum (6x = 42)	*T. timopheevi*	Persistent or plasmodial	After microspore release
	Aegilops ovata	Periplasmodial or premature degeneration	Premeiotic or postmeiotic
T. durum (2x = 28)	*Aegilops ovata*	Periplasmodial or premature degeneration	During microspore formation
Zea mays (2x = 20)	T	Abnormal, prolonged persistence	During microspore development
	C	Abnormal prolonged persistence	During pollen maturation
	S	Normal or persistent	During pollen maturation or slightly earlier

tern and tapetal behaviour in gc-mst plants (Tables 3.20–3.23). The environmental influence over the gene-cytoplasm (*fr-c*) interaction is equally profound (Tables 3.25, 3.27). This further enhances the dimension of variability.

3.7 Environmental Influence

Like other mutant genes, *ms* and *fr* genes are also susceptible to environmental factors (Tables 2.25, 2.26, 3.25, 3.27). Whether *c*-genes are equally susceptible or not cannot be precisely determined, but since *fr-c* gene interaction results in gc-mst,

Table 3.25. Environment-sensitive gene-cytoplasmic male steriles

Environmental factor/ Species/Species Cross	References	*fr* gene action
a) Temperature		
Allium cepa (1r, 2x = 16)	Clarke and Pollard (1949), Barham and Munger (1950), Kobabe (1958), Lichter and Mundler (1961), Virnich (1967), Jiřik and Novâk (1969), Jiřik (1973), Meer and Bennekom (1969, 1978)	Either before or after microspore release
A.schoenoprasum (1r, 2x = 16)	Tatliogu (1982)	Before microspore release
Beta vulgaris (2r, 2x = 18)	Cleij (1967), Theurer (1970)	
Capsicum annuum (1r, 2x = 24)	Peterson (1958), Novâk et al. (1971), Shifriss and Guri (1979)	After microspore release
Daucus carota (1–3D, 2x = 18)	Gabelman (1956), Banga (1959), Thompson (1961), Zenkteler (1962), Timin and Dobrutskaya (1981)	After microspore liberation or during development
Medicago sativa (4x = 32)		
(po)	Davis and Greenblatt (1967)	
(1–4 po)	Suginobu (1979)	After microspore release
Pennisetum typhoides (2x = 14)	Prakash (1967)	After microspore release
(1r)	Reddy and Reddi (1970)	After microspore release
Secale cereale (1r, 2r, 2x = 14)	Garlicka and Madej (1975) Scoles and Evans (1979b)	After microspore release
Sorghum vulgare (many r, 4x = 20)	Kidd (1961), Downes and Marshall (1971), Kongtian and Hongyi (1981), Zhang and Fu (1982)	After microspore release or during microspore development or prior to pollen mitosis.
Vicia faba (1r, 2x = 12)	Nouy and Le Guen (1977), Duc (1978)	During pollen development

Table 3.25 *(continued)*

Environmental factor/ Species/Species Cross	References	*fr* gene action
Species crosses		
Gossypium anomalum (2x = 26) × *G.hirsutum* (1-many, 4x = 52)	Meyer and Meyer (1965)	nk
(*G.anomalum* (2x = 26) × *G.thurberi*) (1-many, 2x = 26) × *G.hirsutum* (1-many, 4x = 52)	Meyer and Meyer (1965) Meyer (1966b)	nk
Oryza sativa (2x = 24)	Cheng and Huang (1978, 1979)	After microspore release
O.glaberrima (2x = 24)		
Petunia axillaris (2x = 14) × *P.hybrida* (2x = 14)	Marrewijk (1969), Izhar (1977) this Vol.	During microspore formation
Raphanus sativus (2x = 18) × *Brassica napus* (2x = 38)	Shiga (1976), Shiga and Baba (1971), Shiga et al. (1976)	After microspore release
R.sativus (2x = 18) × *B.oleracea* (2x = 18)	Bannerot et al. (1974) Heyn (1978), McCollum (1979)	After miscrospore release
b) Light intensity		
Vicia faba (1r, 2x = 12)	Nouy and Le Guen (1977), Duc (1978)	Around microspore mitosis
c) Unknown		
Beta vulgaris (2r, 2x = 18)	Owen (1942), Nagao and Kinoshita (1962), Rohrbach (1965a), Kinoshita (1971)	During tetrad formation or after microspore release
Helianthus annuus (po, 2x = 34)	Gundaev (1966)	nk
Lathyrus sativus (2x = 14)	Zelenskaya and Pestova (1974)	After tetrad formation
Medicago sativa (2r, 4x = 32)	Pederson and Stucker (1969)	After microspore release
	Staszewski (1970, 1976)	After tetrad formation
Pelargonium crispum (2x = 22)	Tokumasu (1974)	During microspore development
Pennisetum typhoides (1r, 2x = 14)	Vittal Rao (1969)	nk
Sorghum vulgare (1r, 4x = 20)	Maunder and Pickett (1959), Hadley and Singh (1961), Kidd (1961), Duvick (1966)	During pollen mitosis
(2r, 4x = 20)	Pi and Wu (1963), Erichsen and Ross (1963a), Miller and Pickett (1964)	Before or during pollen mitosis.
(3r, 4x = 20)	Craigmiles (1961), Alam and Sandal (1967)	During pollen mitosis
Species crosses		
Linum floccosum (2x = 30) × *L.usitatissimum* (2x = 30)	Gajewski (1937)	nk
Lolium perenne (2x = 14) × *L.multiflorum* (2x = 14)	Ahloowalia (1981)	During pachytene
Petunia axillaris (2x = 14) × *P.hybrida* (2x = 14)	Edwardson and Warmke (1967)	After tetrad formation
Secale cereale (2x = 14) × *S.montanum* (2x = 14)	Lapinski (1972)	nk
Vitis vinifera (2x = 38) × *V.riparia* (2x = 38)	Staudt (1973)	After microspore release

this circumstantial evidence indicates cooperational susceptibility of *fr* and *c*-genes, although this has not been unequivocally ascertained. Since in most of the gc-mst's that are influenced by environmental factors, the S-cytoplasm is common and the nuclear genotypic background differs, the susceptibility of *fr* genes to environment is inferred. Although in many gc-mst's the precise environmental factors influencing *fr* gene action are not known (Table 3.25), the potent environmental factors are climatic factors like temperature, photoperiod and light intensity and edaphic factors like soil micronutrients and pH. Of these, temperature and photoperiod are the major ones (Fig. 3.24B). Compared to temperature influence, photoperiodic influence on *fr* gene action is known in fewer gc-mst's (Table 3.25, 3.27) even though photoperiod induces mst in mft's by influencing many stages of microsporogenesis variously (Table 3.26).

3.7.1 Temperature

Temperature is a potent factor that influences the mst expression the most (Table 3.25, Fig. 3.24B). In gc-mst *Beta, Capsicum, Daucus, Pennisetum, Secale, Sorghum* and *Vicia*, increase in mst degree occurs following low temperature treatments during premeiotic stages. High temperature treatments given to the plants during these stages lead to reductions in mst degree. In grain sorghum, Kongtian and Hongyi (1981) and Zhang and Fu (1982) restored a high degree of mft in mst plants by subjecting them to 40°C at the premeiotic stage. In sugarbeet, Lichter (1978) obtained mft plants from mst plants following 36°C temperature treatment and subsequent meristem culture. Exposure of germinated seeds of mst sugarbeet to 55°C yielded four mft plants (Cleij 1967). In mst carrots, low winter temperature maintains mst and higher temperatures reduce its expression considerably. Thus

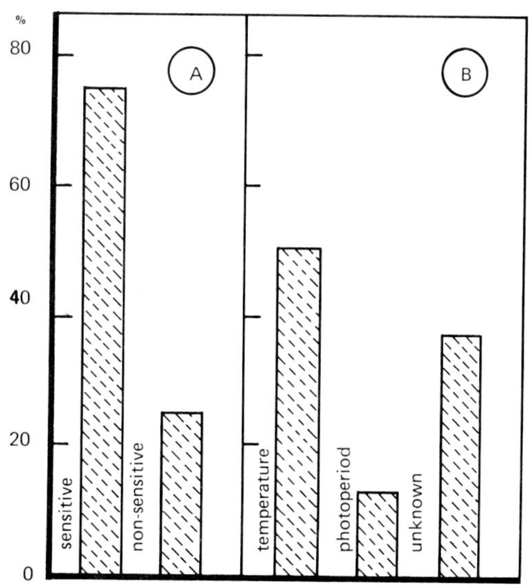

Fig. 3.24A,B. Environmental sensitivity of gene-cytoplasmic male steriles

Table 3.26. Influence of photoperiod on microscporogenesis

Microsporogenesis stage affected	*Cosmos bipinnatus* (A)	*Dichanthium aristatum* (B)	*Oryza sativa* (C)	*Zea mays* (D)
Early anther development and differentiation	+	nk	+	+
Premeiosis	+	+	+	+
Meiotic onset	+	+	+	nk
PMC formation	+	+	+	+
Prophase-I absent	+	+	–	+
Early prophase-I	+	+	+	+
Mid prophase-I	+	–	+	–
End prophase-I	–	–	–	+
Tetrads formed	+	nk	–	nk
Microspore development (pre-release stage)	–	+	–	–
Microspore release	+	+	+	nk
Microspore development after release	+	nk	+	nk
Abnormal male meiosis	+	+	+	+
Late prophase-I to telophase-II	+	+	+	+
Late prophase-I to microspore liberation	+	nk	+	nk

A and C = this Vol., B and D from Knox and Heslop-Harrison (1966), Moss and Heslop-Harrison (1968), + = yes, – = no, nk = not known.

transferring some mst carrot genotypes from field to glasshouse leads to the development of mft flowers, so that the late-formed flowers are fully mft. Return to the field revokes mst in them. The response of all the gc-mst genotypes of a species to low temperature is not uniform, as some mst populations are resistant to temperature changes and others sensitive. This is known in carrot, in which some mst lines are highly sensitive and others stable to temperature alterations. In *Capsicum*, one genotype is fully mst at low temperature and mft at high temperature, and the other exhibits a reverse response (Martin and Crawford 1951). Still more variable and diverse response to temperature is shown by various p-mst genotypes of some species. The differential response is considered to be mainly due to the gene differences, but the influence of genotype-environment interaction over the observed variability is highly significant. In onion, carrot, lucerne, beet and broad bean, considerable variability to the same temperature regime is exhibited by mst and p-mst plants segregating from the same cross. Some populations of these p-mst genotypes reveal considerable variability in mst degree within the same environment, so that pollen sterility differs from stand to stand, harvest to harvest, and year to year. Whether these are due to genetic or microclimatic differences is not clearly known. The differential response of mst genotypes to temperature is not only intra- but inter-genotypic. The intergenotypic differences in *fr* gene response are due to the differences in genetic background in which the *fr* gene is placed and to the differential genotype-environment interactions.

Low temperature treatments increase mst expression in alloplasmic *Brassica oleracea*, *B. campestris*, *Oryza sativa* and *Petunia hybrida*, but reduce it in *B. napus*, *Gossypium hirsutum* and *Zea mays*. In this latter group of species, high temperature treatments enhance the mst expression. Completely reverse response to two temperature regimes (15°C, 30°C) is exhibited by different gc-mst genotypes of rice, petunia and maize. This is due to the presence of different *fr*- and/or *c*-genes known to exist in these genotypes. Polygenes conditioning mst are likewise highly amenable to environment. In rye, in which mst results due to cumulative action of polygenes, the influence of temperature over mst expression is profound.

The precise stage of temperature sensitivity of PMC's is not clearly known in gc-mst plants. In mst wheat, sorghum and rye, low temperature induces mst. The premeiotic period is the stage most sensitive to temperature. This stage being the DNA synthesis phase, damages DNA and its replication and repair and results in mst. In rice, low temperature sensitivity is maximum during the microspore release stage (Satake and Hayase 1970) and to a lesser degree during early and pre-leptotene stages (Satake and Hayase 1974). These microsporogenesis stages are highly sensitive to *ms* and *fr* gene action, as these genes act either during the initiation of meiosis or just after its completion in g-mst and gc-mst rices (Table 27.6). In almost all the gc-mst plants in which mst is influenced by temperature or by some other environmental factor, *fr* genes act mostly after meiosis is completed and microspores are released from the quartets (Table 3.25). Thus, the *fr* genes acting at this microsporogenesis stage may be especially sensitive to environmental factors. However, in a few sensitive gc-mst strains of *Allium cepa* and *Petunia hybrida*, the *fr* genes may act during or just before microspores are released. Certain temperature regimes suppress or delay the action of *fr* genes in some gc-mst plants of onion, alfalfa, pearl millet and rye plants so that some viable pollen is produced. This causes mst expressional breakdown and fertility restoration in them.

3.7.2 Photoperiod

3.7.2.1 Pollen Sterility

In most of the higher plants, flowering is the result of an induction produced by photo- and thermoperiods. Photoperiod, an experimental means easy to utilize, has been used to control the development of many plants. In many grasses of Andropogoneae and Maydeae, pollen fertility is readily modified by environmental factors and particularly by the light regimes. This has been investigated in *Zea mays*, *Dichanthium aristatum* (Heslop-Harrison 1957; Knox and Heslop-Harrison 1966; Napp-Zinn 1967; Moss and Heslop-Harrison 1968), *Cosmos bipinnatus* and *Oryza sativa* var. Basmati (this Vol.). In maize, day length, particularly in combination with high temperature, affects the level of node at which flowering begins, the number of branches and incidence of femaleness and hermaphroditism in the normally wholly male tassel. Although maize flowers in a wide range of day lengths, its development is hastened by short days and it is regarded as a short-day plant. However, continued exposure short days during early differentiation of staminate terminal inflorescence induces pollen sterility (Heslop-Harrison 1961). Thus, compared to the long-day-grown controls, plants of the maize cultivar Golden Bantam grown in an 8-h day photoperiod exhibited acceleration of tassel and silk emergence, reduced leaf number on the principal axis, fewer tassel branches, a smaller total number of alicoles in the tassel, greatly reduced pollen fertility and tendency to sex reversal in the tassel (Moss and Heslop-Harrison 1968). In extreme cases, no fertile anthers were formed and the plants became mst. Since all the effects were negated or reduced by night interruptions of low light intensity levels, they are, therefore, direct or indirect effects of the photoperiodic reaction. In *D. aristatum*, with the exposure to 40 short days at an age of 135 days, the first emergent inflorescences were partially mst (Knox and Heslop-Harrison 1966). Under long-day conditions the later-formed inflorescences were mft, but with continuous short-day treatments from an age of 135 days, anther sterility was almost total. *Cosmos* and rice are short-day plants. Subjecting them to long photoperiods of 14 h after floral induction by short-day cycles led to pollen sterility that increased considerably in the later-formed florets (this Vol.).

In *Silene pendula*, as in many other Caryophyllaceae members, mst is induced by certain photoperiodic treatments. When the inductive long-day treatment was given prematurely before plants formed basal rosettes, a high proportion of mst flowers developed (Heslop-Harrison 1963). Some photoperiodic treatment given to the rosetted plants produced fertile flowers in both the sexes. However, the anther-sterilizing effect of the premature long-day photoperiod is quantitatively variable within plants and flowers, and induces a variable proportion of mst flowers having a variable degree of mst. α-Naphthalene acetic acid (NAA) likewise suppresses androecial development in this species. Like the photoperiodic treatment, a quantitative variation in the incidence of mst occurs in the treated plants. Both these treatments arrest anther development during the filament extension period. Subjecting wheat plants at a stage when stamen primordia were visible on the first floret of the most advanced spikelet of the main shoot to 10-h photoperiodic treatment led to the transformation of stamens into ovaries (Fisher 1972). Ovules developed on the anther lobes and the plants became mst.

3.7.2.2 Microsporogenesis

Photoperiodically induced pollen sterility is due to departures from the normal course of development at most stages on ontogeny and even when the breakdown begins during the same ontogenetic period more than one abnormal pathway may occur in the same inflorescence (Table 3.26). The disturbance in microsporogenesis may occur at any stage of meiosis except during MI to AII. The most sensitive stage to photoperiod appears to be premeiosis and PMC formation. Both these stages were consistently affected in all the four plant species given in Table 3.26. In such flowers, meiosis stops and PMC's degenerate. In majority of the flowers, microsporogenesis was arrested at various stages, in a few it continued abnormally but no viable pollen was produced. Both in rice and *Cosmos*, in the florets in which male meiosis had already initiated prior to the photoperiodic treatment, male meiosis proceeded until TII or to the microspore liberation stage; but in the florets where meiosis was to begin, the treatment led to abnormal meiosis and to its arrest at tetrad formation or during microspore development. In the later-formed flowers, meiosis stopped around prophase-I. In the last-developed few flowers, no PMC's were formed or they degenerated soon after formation.

3.7.2.3 Tapetal Development

In *Dichanthium annulatum*, short-day photoperiods induce mst (Knox and Heslop-Harrison 1966). In the mst's, tapetal anomalies preceded meiocyte or microspore abortion. This led the authors to conclude that induced sterility is a consequence of tapetal malfunction. Malfunction was also found by Moss and Heslop-Harrison (1968) in some anthers of the short-day-treated maize plants. In many flowers, however, PMC degeneration occurred without any tapetal anomalies. Furthermore, both in maize and *D. annulatum*, the tapetum degenerated along with the meiocytes at prophase-I or with the microspores that aborted immediately after formation or after release from the quartets. In *Cosmos bipinnatus* and *Oryza sativa*, tapetal anomalies accompanied, preceded or followed the onset of disturbances in the sporogenous tissue (this Vol.). This was followed by degeneration of tapetum and microspores. On the other hand, in some anthers, meiocytes degenerated at prophase-I but the tapetum appeared normal. In contrast, in many anthers tapetal anomalies appeared at prophase-I but meiosis continued until microspore liberation, the microspores degenerating thereafter. In rice, in the first-formed few flowers that appeared immediately after photoperiodic treatment, meiosis proceeded normally, but prior to release of microspores, the tapetum exhibited anomalous behaviour. It contracted towards the centre of the anther locules and aborted. The released miscrospores aborted immediately thereafter. In the later-formed flowers, the tapetum became abnormal during the end of prophase-I. It formed multinucleate coenocytes and crushed the meiocytes completely. Thus, the photoperiod appears to influence both tapetum and sporogenous tissue; the time of action is variable and probably depends upon the developmental stage at which the floral tissue is subjected to photoperiodic treatments. This explains the findings given in Table 3.26.

3.7.2.4 Seed Set

In rice and *Cosmos*, long-day photoperiodic treatments reduced seed set considerably. On hand-pollinated spikes, the seed set ranged between 31 and 34% (this Vol.). It was around 67–70% in the florets in which male meiosis had initiated prior to the photoperiodic treatment but it gradually dropped in the later-formed florets and no seeds were set in the last-formed florets in which male meiosis broke down during prophase-I or earlier. Although megasporogenesis was not investigated, it appears that the photoperiodic treatment reduced female fertility in *Cosmos* and rice plants grown continuously in long days.

3.7.2.5 Sensitive Genes

Certain short-day photoperiodic treatments in maize and *Dichanthium* and long-day treatments in rice and cosmos induce mst (Knox and Heslop-Harrison 1966; Moss and Heslop-Harrison 1968; this Vol.). The influence of such "effective" photoperiodic treatments over the sterility expression and fertility reversion of gc-mst rice and maize has not been determined. The treatments may annul, negate, inflate or drastically alter mst expression in these and other mst genotypes. The sensitivity of *fr* genes to photoperiod is known in only a few alloplasmic gc-mst plants, in all of which the sterility is conditioned by the interaction of two *fr* genes and the S-cytoplasm (Table 3.27). In the alloplasmic *Lycopersicon hirsutum* having tomato S-cytoplasm and in *Brassica* having radish S-cytoplasm, *fr* gene acts only after microspores are released. In these, photoperiodic variations delay the *fr* gene action so that some pollen escapes the gene action and is released as viable. This results in selfed seed set.

Table 3.27. Photoperiod-sensitive gene-cytoplasmic male steriles[a]

Genotypes	References
Lycopersicon esculentum (2x = 24) × *L.hirsutum* (2x = 24)	This Vol.
Raphanus sativus (2x = 18) × *Brassica campestris* (2x = 20)	Heyn (1978)
R.sativus (2x = 18) × *B.napus* (4x = 38)	Shiga (1976), Shiga et al. (1976)
R.sativus (2x = 18) × *B.oleracea* (2x = 18)	Bannerot et al. (1974)

[a]Gene control = 2r, gene action = after microspore release.

3.7.3 Stable Genes

Environmental factors profoundly influence the action of many *fr* genes (Table 3.25) so that the mst expression of the genotypes is related to the environmental conditions in which they are grown. Except in a few cases, the precise determination of the effect of specific environmental factors on *fr* gene expression has not been investigated (Table 3.25). Whereas many of these genes are susceptible to temperature, photoperiod, light intensity or to some other undetermined environmental factors (Fig. 3.24), a few seem to exhibit stability to some of these environmental components (Table 3.28). This has not been experimentally verified but inferred from the observations that these mst genotypes maintain their stability when grown in different seasons or regions. Since these areas differ in climatic, edaphic and physiographic factors, the *fr* genes are presumed to be stable to the alterations of any or all of these factors. Most of the species having stable *fr* genes are diploids and only two (*Triticum aestivum* and *Solanum demissum*) are hexaploids. In most of these, the mst is conditioned by dominant genes, in the others, gene control is not known. Whether the dominant nature of *fr* genes has any relationship to the environmental nonsensitivity is not known. The *fr* genes seem

Table 3.28. Environmental nonsensitive gene-cytoplasmic male steriles

Crosses and species	Reference	*fr* gene action
Aegilops caudata (2x = 14) × *Triticum durum* (4x = 28)	Hervás (1974)	After microspore release
Allium schoenoprasum (1r, 2x = 16)	Tatliogu (1985)	Before microspore release
Daucus carota (3D, 2x = 18)	Timin and Dobrutskaya (1981)	Premeiotic
Helianthus petiolaris (2x = 34) × *H.annuus* (2x = 34)	Leclercq (1969), Paun (1974)	Pre- or postmeiotic
Solanum chacoense (2x = 24) × *S.kurtzianum* (2D, 2x = 24)	Grün et al. (1962)	
S.chacoense (2x = 24) × *S.phureja* (2D, 2x = 24)	Grün and Aubertin (1966)	During early prophase-I
S.chacoense (2x = 24) × *S.stenotomum* (2D, 2x = 24)		
S.demissum (6x = 72) × *S.chacoense* (2x = 24)	Dionne (1961)	nk
S.demissum (6x = 72) × *S.phureja* (2x = 24)	Dionne (1961)	
S.demissum (6x = 72) × *S.simplicifolium* (2x = 24)	Dionne (1961)	
S.demissum (6x = 72) × *S.soukupi*	Dionne (1961)	
S.demissum (6x = 72) × *S.tuberosum* (2D, 4x = 48)	Schnell (1948)	Early prophase-I
Triticum timopheevi (4x = 28) × *T.aestivum* (1D, 6x = 42)	Xiangmin and Herong (1981)	After microspore release

to act anywhere during microsporogenesis, but the gene action is mostly confined to the initiation or termination the stages of meiosis.

The stability is not only *fr-c*-gene-specific but depends on the presence or absence of other genes. This is evidenced by the fact that a few other lines of many of the gc-mst genotypes listed in Table 3.25 exhibit environmental susceptibility. For instance, in carrot, of the two gc-mst lines studied by Timin and Dobrutskaya (1981), one is stable to temperature alterations but the other is highly susceptible. The majority of the gc-mst lines of sunflower are stable to environmental factors, but one exhibited susceptibility to photoperiod and the other to temperature. Likewise, of the nine lines tested, the mst expression of two lines of alloplasmic mst *Solanum chacoense* having *S. phureja* cytoplasm is susceptible to high temperatures. Of these, one has a mutated *fr*-gene and the other different S-cytoplasm type. Thus the stability of *fr* genes is not absolute but relative and depends on the genotypic and cytoplasmic background of a species. From about 50 genotypes tested for environmental sensitivity, about 12 (25%) are nonsensitive; the rest are sensitive to environmental factors (Fig. 3.24A). About 50% of the sensitive genotypes are susceptible to temperature, 12% to photoperiod and in the remaining, the precise environmental factor influencing mst expression is not known (Fig. 3.24B).

3.8 Sterile Cytoplasm Features

3.8.1 Inheritance Pattern

Nuclear genes are mainly responsible for heredity. Their interaction with cytoplasmic genes results in differentiation because cells in the different tissues of an organism have the same nuclear genome but the cytoplasmic contents vary according to the necessities of the tissues and functions. The genome of eukaryotic organisms is localized not only in the nucleus but also in the cytoplasm. Therefore, genes in a multigenomic eukaryotic cell are distributed into nuclear genome and extranuclear genome. For the latter, Wettstein (1926, 1928) coined the term "plasmon", which represents the totality of heredity transmitted through the cytoplasm, in other words the genetic information present in the cytoplasm. The information is localized in "plasmagenes" – the extranuclear nucleic acid particles in the cell transmitted by cytoplasm. The term "cytoplasmic-factor" has been commonly used as synonymous with plasmon. Hence, nuclear and cytoplasmic inheritance represent the two principal components of the cell's hereditary information. Whereas nuclear inheritance is carried by chromosomes, the cytoplasmic inheritance is mainly carried by (a) cell organelle-like plastids, mitochondria, kinetosomes, centrioles etc., (b) intracellular symbionts, (c) plasmids etc.

In contrast to nuclear inheritance, cytoplasmic inheritance is non-Mendelian and maternally transmitted. A certain confusion exists between maternal effects and cytoplasmic inheritance. Both "predetermination" and "dauermodifications" are maternal effects and do not fall under cytoplasmic inheritance. Occasionally a gene product of the female parent occurs in the unfertilized egg cytoplasm. The product influences the character of the developing zygote and a trait is thus already predetermined in the egg before fertilisation. This phenomenon is known as predetermination. On the other hand, certain nongenetic cytoplasmic changes induce some phenotypic alterations. These persist for several generations and gradually diminish and finally disappear. Such temporary nongenetic phenotypic alterations are known as dauermodifications. They are distinct from cytoplasmic inheritance (Kaul 1967b), which according to Caspari (1948) exists "only when a character proves to be more or less constant in a series of backcrosses to the paternal strain, or when genic transmission is excluded by substitution of all chromosomes in a species, and, dauermodifications are unlikely as no decrease in the action of cytoplasm appears in successive generations." Cytoplasmic inheritance is maternal but its phenotypic effects represent the net results of the interaction of nuclear and cytoplasmic genes. It has been comprehensively reviewed by Grun (1976), Gilham (1978), Srivastava (1981) and Hanson and Conde (1985).

3.8.2 Establishment Basis

3.8.2.1 Reciprocal F_1 Differences

Barring the sex chromosome, wherever they are distinct, the chromosomes of reciprocal hybrids are usually identical but their cytoplasms differ. The phenotypic differences between reciprocal F_1 are considered to be due to cytoplasm and are caused by gene-cytoplasm interactions. Thus, reciprocal cross differences, whenever found, were usually employed to infer cytoplasmic diffrences. However, under certain conditions this interpretation is not warranted, especially when the cytoplasmic inheritance is confused with dauermodifications, predeterminations, maternal effects etc. For establishment of cytoplasmic inheritance, the following methods are used.

3.8.2.2 Reciprocal Advanced Generation Differences

To eliminate these effects, usually F_2 and BC progenies are produced from reciprocal F_1 combinations and compared for their phenotypic effects. This was initially done by Oehlkers (1938, 1964) in *Streptocarpus*, by Michaelis (1954) in *Epilobium* and later by many others in *Solanum, Triticum, Brassica, Oryza* etc. These investigations clearly differentiated cytoplasmic effects from maternal effects, predeterminations and dauermodifications.

3.8.2.3 Genome Transfer

In this method the genome of a species is replaced by substitution BC. Each BC of a given female by a male parent halves the genome of the female (nonrecurrent) parent. Thus, after eight generations of recurrent crossing of a given female line by the same male line (BC8), 255/256 of the progeny genome becomes that of the male. In other words, the male parent's genome is almost fully substituted into the cytoplasm of the female parent. Comparison of the male parent having its own cytoplasm with those of the substituted male parent, which has the female parent's cytoplasm, indicates g-c interactions and reveals the cytoplasmic influence on phenotype. This is a frequently used test, and many alloplasmic mst lines have been developed, especially in *Allium, Gossypium, Nicotiana, Pennisetum, Solanum, Sorghum, Streptocarpus, Triticum, Zea mays* etc. (Table 3.6). In many cases no reciprocal crosses were done or they failed, yielding either nonviable embryos, zygotes, seeds or seedlings (*Epilobium, Nicotiana, Solanum, Streptocarpus* etc.). This deficiency leads to dubious interpretation. Also, usually the genome of one species is substituted into the cytoplasm of another through BC_4 or so, but even after such four recurrent backcrosses (BC_4) nearly 3% of the original female parent's genes remain in the genome of the substituted backcross line. Thus, the genome of the recurrent male parent placed in the cytoplasm of the female line has 3% deficit genes which could account for the differences between the recurrent parent and substitution line. Therefore, a large number of BC's are needed before genome-cytoplasm interaction can reliably be inferred. This, however, necessitates certain assumptions such as: (a) cytoplasmic genes of the female parent remain stable through the BC substitution process, (b) no interaction of male cytoplasmic genes with the male nucleus is operative, (c) no selective retention of specific genes of the female parent in the substituted progeny occurs, and (d) selective eliminations of certain major or minor genes causing phenotypic alterations, interspecific gene and chromosomal interactions etc. do not occur. These are elaborately discussed by Grün (1976).

3.8.2.4 Nuclear Transplantation

To avoid the problems arising due to gradual genome transfer through substitution BC's, transplantation of intact nucleus into artificially enucleated cells would be ideal. However, the recipient cytoplasm contains not only self-replicating cytoplasmic genes (*c*-genes) of the original parent, but also left-over products of the nuclear genes of the first nuclear parent. Comandon and de Fonbrune (1939) developed a method by which the nucleus of an amoeba was pushed into the cytoplasm of the enucleated one. This method has been used to study g-c interactions, antibiotic resistance and phenotypic variability of *A. proteus* and *A. discoides* (Danielli 1963).

3.8.3 *c*-Genes

Whereas the plasmagenes in cooperation with nuclear genes determine the phenotype, a specific type of plasmagenes, designated by the present author as *c*-genes, in association with specific cytoplasm-sensitive nuclear genes designated as *fr* genes, result in mst. Such *c*-genes transform a normal fertile (N) cytoplasm to male sterile (S) type. Genes which control or affect the functioning of a specific cytoplasm so as to produce male sterility or fertility, vigour or its loss, or alterations in phenotype were termed "cytoplasm-differentiating genes" by Maan and Lucken (1972). These genes belong to the general plasma-gene group and the *fr* genes to the cytoplasm-differentiating genes. In general, intraspecific cytoplasmic differences are controlled by one to two *c*-genes, and interspecific and intergeneric by two or more *c*-gene differences. These genes, in association with appropriate sensitive *fr* genes, lead to mst along with some other secondary anomalies, such as the production of haploids, twin seedlings, parthenogenetic embryos (e.g. *Triticum*, *Solanum* etc.). For reproductive perpetuation, individuals with S-cytoplasm should have nonsensitive fertility-restoring *Fr*-genes so that they are fertile and reproductively fit. Occasional mutation of these *fr*-genes to sensitive *fr*-genes would produce mst phenotypes which may become eliminated immediately or gradually depending on the breeding system and the magnitude of the phenotypic change. Therefore, a fertile panmictic population is expected to have (1) N-cytoplasm and nonsensitive *Fr* or sensitive *fr* genes or (2) S-cytoplasm and nonsensitive *Fr*-genes. In plants with S-cytoplasm, a close linkage between nonsensitive *Fr*-genes would ensure enbloc inheritance of all the *fr*-genes. This would contribute to the stability of such g-c mechanism, enhance population fitness through heterozygote advantage and ensure survival.

Simultaneous or successive mutations of N-cytoplasm to S-cytoplasm and of nonsensitive *Fr*- to sensitive *fr*-genes are rare but known to occur in *Solanum*. Such mutations would persist only if they are followed by specific compensating *fr*-gene and/or *c*-gene mutations. Certain cytoplasm nonspecific homoeoalleles may partially compensate for the effects of *c*-gene mutations until a complementary *fr* gene mutation revoking fertility has a chance to appear and spread in the population. A mutation causing less pollen fertility reduction has greater chances of survival because a complementary mutation may not be immediately required for the survival of the affected individual. Allelism tests or linkage relationships cannot be done among the cytoplasmic genes from different cytoplasmic sources in higher plants because these genes exhibit predominantly maternal inheritance. Therefore, different cytoplasm types are differentiated on the basis of their differential interactions and effects with specific nuclear genes or genomes. One of the major effects of such a genome-cytoplasm interaction is the development of mst. This is vivid in many interspecific crosses followed by successive substitution BC of the genome of one species with the alien cytoplasm of the other. In the resulting genome-cytoplasm combination (gc-combination) the gc effects on pollen fertility become apparent after sufficient BC's have eliminated almost all the genes of the cytoplasm donor species. This has led to the origin of alloplasmic mst wheat, tobacco, maize, *Solanum*, etc. (Table 3.6).

Cytoplasmic differences in higher plants are essentially maternal, as cytoplasmic *c*-genes are exclusively maternally transmitted and seldom through the male gametes. Therefore, phenotypic differences in F_1 hybrids from reciprocal crosses and differential segregation in F_2 or backcross progenies from reciprocal crosses are inferred to indicate cytoplasmic differences between the parents. Among closely related species, cytoplasmic differences become apparent from the different phenotypes of hybrid progenies of direct and reciprocal crosses because hybrid genotypes have differential sensitivity to the parental cytoplasms. However, certain hybrid-gene combinations will not produce noticeable differential phenotypic effects if they share certain nuclear genes which control cytoplasmic effects of both the parental cytoplasms. In such instances, the parents are considered to possess the same or similar cytoplasms until another nuclear genome produces differential sensitivity to one or both of these cytoplasms. This is known in *Aegilops*, in which *Triticum durum* differentiates its species into two major cytoplasmic groups but *T. aestivum* does not (Maan 1978a,b).

3.8.4 S-Cytoplasm Differentiation

Like genic variations caused by mutations, cytoplasm variations are also common in certain plants. This was recognised early by two German geneticists, Michaelis (1954) and Oehlkers (1964), in *Epilobium* and *Streptocarpus*, respectively. Many races and strains of *E. hirsutum* are specific in having cytoplasm types unique to each race or strain. The *c*-genes also exhibit specific variation caused by mutations. This has resulted in S-cytoplasm variability and in subsequent differentiation in some plants (Table 3.29). Based on differential fertility restoration reactions, 160 main sources of gc-mst maize can be classified into three major cytoplasm types viz. T, C and S. The mst plants having these cytoplasms differ from each other in anther and pollen phenotype, pollen abortion pattern, *Helminthosporium* and drug susceptibility, mt-DNA structure, fertility reversion incidence and fertility restoration control. These sterile cytoplasms maintain their individuality and are characterised variously (Table 27.19). Considerable genetic diversity exists in the *Aegilops* and *Triticum* cytoplasms (Tsunewaki 1980b). Based on cytoplasmic differences and similarities among cytoplasm donor species and on the presence or absence of plasmon-sensitive nuclear genes among the genome of donor species,

Table 3.29. S-cytoplasm differentiation in some plants

Species	S-cytoplasm types (number)	References
Aegilops	Many	[a]Maan (1979), Tsunewaki (1980b)
Allium	2	Jones and Emsweller (1937), Berninger (1965)
Beta	2	Owen (1942), Oldemeyer (1957)
	4	Kinoshita (1980), Mikami (1982)
Epilobium	Many	[a]Michaelis (1954)
Hordeum	2	Ahokas (1978a, 1982b)
Medicago	2–3	Davis and Greenblatt (1967), Staszewski (1970)
Nicotiana	Many	[a]Gerstel (1980)
Origanum	Many	Kheyr-Pour (1980, 1981)
Pennisetum	4	[a]Rachie and Majmudar (1980)
Plantago	2	Van Damme and Van Delden (1982)
Solanum	Many	[a]Grün (1976, 1979)
Sorghum	5–6	[a]Schertz and Pring (1982)
Streptocarpus	Many	[a]Oehlkers (1964)
Triticum	Many	[a]Tsunewaki et al. (1980b)
Vicia	2	Bond et al. (1966a), Berthelem (1970)
Zea	3	[a]Duvick and Noble (1978)

[a]Reviewed.

cytoplasmic descent in some *Aegilops* and *Triticum* species has been traced (Fig. 27.19). The study indicates that interspecific cytoplasm differentiation is maximum in diploid species, less in tetraploids and least in hexaploids. The genus *Solanum* is unusual in that a large number of contrasting *c*-genes have been identified in the species belonging to the section *Tuberarium*. Eight different *c*-genes known within the group of plants are thought to be involved in *S. tuberosum* evolution (Grün 1979). These *c*-genes can be recognised by their responses to different nuclear *fr* genes and each *fr-c* gene combination has a specific morphological expression either on anthers or on spores (Table 49.14). Two *fr* genes are recessive and six are dominant. Similar cytoplasmic diversity is also known in many other species of *Solanum*, especially in *S. chacoense, S. demissum, S. phureja, S. verrucosum* (Chap. 49.6.10). Thus, like *Fr* genes, *c*-genes also mutate but their expression is visualised only when either sensitive *fr* genes are already present in the nuclear genome or when *Fr*-genes also mutate to *fr* genes sensitive to *c*-genes.

3.8.5 S-Cytoplasm Effects

The main effect of *c-fr* gene interaction is mst, which is expressed in many forms even within a single genus like *Epilobium, Nicotiana, Solanum, Streptocarpus* or *Triticum*. Depending on the g-c combination and the surrounding genomes, a block can appear at any stage during stamen development, differentiation or maturation process and this leads to structural, sporogenous and functional types of mst. Some of its manifestations are: (a) absence of normal anthers, (b) reduction in anther size, (c) varying degree of petaloidy, (d) abnormal meiosis followed by formation of empty, shrivelled microspores, (e) normal meiosis followed by abnormal microspore development, (f) normal meiosis but failure of opening of terminal anther pores so that pollen is not shed. In addition, the *c-fr*-gene interaction leads to many floral anomalies in some species of the above-mentioned genera. The interaction produces aneuploids and haploids in alloplasmic wheat (Chap. 27.20.1.2.1). In maize, sterile T-cytoplasm exhibits fungal and chemical sensitivity (Chap. 27.21.2.1.2.4–5).

3.8.6 N-Cytoplasm Mutations

Genetic material in multigenomic cells is mainly seated in the nuclear genome and a fraction of it is located in plasmon, especially in the cytoplasmic organelle mitochondria and chloroplast. Evidence favours the location of organelle DNA, which is capable of coding for certain components. Hence mutant organelle-producing cytoplasmic mutations should result after mutagenic treatments. For this, N_{frfr} genotypes are utilized. The recovered mst mutants are tested for cytoplasmic mutations by crossing to the initial untreated parental line. If the progeny is mst, cytoplasm mutation from N- to S-type is inferred. This was found by Lysikov et al. (1969), who obtained cytoplasmic mst mutants in maize after γ-irradiations and ultrasonic wave treatments to N-cytoplasm maize. Similar cytoplasmic mutants arose following γ-ray and/or neutron irradiations (Stoilov 1970). A US patent (No. 3,594,152) was issued to a researcher from the Soviet Union for a technique of producing gc-mst with streptomycin treatment. This has neither been used, nor have attempts to produce such cytoplasmic mutants been successful. Moreover, it is not known if the cytoplasmic or nuclear genes mutated to produce the gc-mst plants.

γ-Irradiations in *Beta vulgaris* led to four S-cytoplasmic mutations (Mikami 1982) and this enlarged the S-cytoplasm base in sugarbeet. In tobacco, Berbeĉ (1974) isolated a cytoplasmic mutant that had large F-1-P characteristics of alloplasmic tobacco (*N. tabacum*) *N. glauca*. Of the 12 mutations obtained by Franckowiak et al. (1976) in the alloplasmic breadwheat variety Chris (*T. tauschii*) *T. aestivum*, 11 are nuclear and only one is a cytoplasmic mutation. In contrast to nuclear mutations, no single mutagen appears to induce S-cytoplasm-specific mutations. Instead, different mutagens induce these mutations in various crop plants (Table 3.30). EB, which successfully induced cytoplasmic mutations in *B. vulgaris* and *Pennisetum typhoides*, failed to induce mutations in *Sorghum vulgare*, *Vicia faba* and *Zea mays*. Colchicine induced cytoplasmic mutations in red clover and sorghum but failed to induce such mutations in sugarbeet and rice (this Vol.). Instead, it induced polyploidy. Likewise, EMS that induces mutations in sugarbeet, rice and barley does not induce cytoplasmic mutations in sorghum, onion and tobacco. However, this does not rule out the possibility of finding cytoplasm specific potent mutagens. These could be EB or EMS or both, as they induce N-cytoplasm mutations to the S-type and should be extensively tried using varying concentrations and timing. Like nuclear mutagenesis, we may achieve directed cytoplasmic mutagenesis by these chemicals. The existing S-cytoplasm variability and diversity is too low for developing "multiplasms" for which more different S-cytoplasms are required. If they are created by using

Table 3.30. Induced mutations of N- to S-cytoplasm

Species	Mutagen	References
Arachis hypogaea (4x = 40)	EB	Levi and Ashri (1975)
Beta vulgaris (2x = 18)	γ-rays, EMS, EB, acriflavine, streptomycin	Kinoshita (1971, 1980, 1983)
Oryza sativa (2x = 24)	EMS	Mallick (1980)
Pennisetum typhoides (2x = 14)	EB, mitomycin, streptomycin	Burton and Hanna (1976)
Hordeum vulgare (2x = 14)	EMS, X-rays	Favret and Ryan (1964)
Sorghum vulgare (4x = 20)	Colchicine	Erichsen and Ross (1963a), Malinovsky et al. (1973)
Trifolium pratense (2x = 14)	Colchicine	Zvyagina (1973), Scheglov and Zvyagina (1975)
Triticum aestivum (6x = 42)	Acridine	Davydenko (1981)
	EMS	Franckowiak et al. (1976)
Zea mays (2x = 20)	γ-rays, ultrasonic waves	Lysikov et al. (1969)
	Neutrons	Stoilov (1970)

physical and chemical mutagens, and then compared critically, factors that are common to all S-cytoplasms may be discovered and the basis of gc-mst may be determined. Moreover, recovery of S-cytoplasm plants from N-cytoplasm plants by using mutagens suggests that by this treatment the time and effort required to create and have mst mutants in production fields would be much less than that required for five or more BC generations used currently.

Of the cytoplasmic mutagens the phenanthridine dye, EB, is of particular interest because mitochondria are the prime targets of this drug (Naas 1974). The chemical intercalates between the bases of double-stranded DNA and exhibits differential binding affinities for covalently closed circular and nicked circular or linear duplex DNA. The physical structure of covalently closed mt-DNA is drastically altered upon EB treatment. Strand breakages and some rejoining of strands are manifested in nicked and linear DNA and in some closed circular DNA. Such DNA alterations in organelle genomes may account for N-cytoplasm mutations to S-type in higher plants.

3.8.7 S-Cytoplasm Mutations

Such mutations of S- to N-cytoplasm type are known in broad beans, maize, pearl millet and sunflower (Table 3.31). The mst in Tift 23A pearl millet is highly stable, but sometimes whereas the primary tillers are mst, the secondary tillers are mft. Clements (1975) found similar spontaneous mutations in four mst stocks of this millet. *Vicia faba* S-447 exhibits two characteristic fertility reversion features viz. some mst plants produce small amounts of viable pollen in almost all the flowers

Table 3.31. Mutations from S- to N-cytoplasm

Species	Mutagen	Cytoplasm types	References
Helianthus annuus (2x = 34)	Sp[a], EMS	–	This Vol.
Pennisetum typhoides (2x = 14)	Sp	Tift 23 A	Vittal Rao (1969), Burton (1972, 1977b), Clements (1975)
Triticum aestivum (6x = 42)	EMS	*T. timopheevii*	Cerny (1984)
Vicia faba (2x = 12)	Sp, EMS	S-447	Bond et al. (1966a), Berthelem and LeGuen (1975), Thiellement (1977a), Duc (1978)
Zea mays (2x = 20)	EMS, X-rays	T and S	Singh and Laughnan (1972)

[a]Sp = Spontaneous.

of the plants and some others bear one or more fully fertile tiller, while the remaining are completely mst (Bond et al. 1966b). Another feature is the permanent restoration of mst plants, for once the mst plants are crossed by a restorer (*Fr Fr*), F_1 and F_2 progenies are fully fertile even though 25% F_2 plants carry (*fr fr*) genes. In such a plant, mutation of S-cytoplasm to N-type has apparently occurred.

Of the three known sterile cytoplasm types in maize, T and S cytoplasms exhibit mutations from S- to N-cytoplasm. After culturing young T-cytoplasm mst embryos with HmT pathotoxin added in culture, Gengenbach et al. (1977) regenerated HmT-resistant mft plants. These represent mutants in which T-cytoplasm mutated to the N-type. Brettell et al. (1980) obtained such mft mutants with or without the addition of HmT pathotoxin to the cultures. The former represent spontaneously arisen and the latter induced mft mutants. Induced mft mutants were obtained by Cassini et al. (1977) and Cornu et al. (1980) after treating T-cytoplasm mst maize seeds with EMS and X-rays. Spontaneous reversion of S-cytoplasm mst plants to mft occurs by mutations at either nuclear or cytoplasmic level. Whereas some genetic backgrounds are highly favourable to these reversions, others are relatively stable (Laughnan and Gabay-Laughnan 1982). Genetic background also influences the proportion of cytoplasmic versus nuclear mutations. Strain M825 exhibits a high frequency of cytoplasmic and a low frequency of nuclear mutations. In the strain WB_4, the frequency of cytoplasmic mutations is low but that of nuclear mutations is high and strain M825/Oh07 has about an equal frequency of both.

3.8.8 Nature of the S-Cytoplasm

After Correns (1909) discovered maternal inheritance of a spontaneous variegated mutation in *Mirabilis jalapa* and Baur (1909) described non-Mendelian biparental transmission of plastid mutation in *Pelargonium zonale*, mitochondria, plastids, kinetosomes and centrioles were considered as the seats of cytoplasmic genes. Mitochondria are implicated as the carrier of *c*-genes in maize, sorghum, *Brassica* etc. In S-cytoplasm mst maize, plasmid like mt-DNA's occur (Pring et al. 1977). These DNA's have molecular weights of about 4.06×10^6 and 3.45×10^6 and are designated S-1 and S-2, respectively. Each of these molecules has a terminal inverted repeat in which identical DNA sequences lie in reverse polarity at either end of the molecule. Such inverted repeats are common in mechanisms by which transposable elements are inserted into bacterial chromosomes. The spontaneous cytoplasm reversion is associated with what appears to be a transpositional event in the mt genome involving S-1 and S-2 plasmid like molecules (Levings et al. 1980). This is because when S-cytoplasm mutates to the mft type, some, perhaps all, S-1 and S-2 plasmid sequences are transposed and integrated into the main high molecular weight mt-DNA. Whether a corresponding transposition and integration of S-1 and S-2 mt-DNA sequences into nuclear chromosomes is involved in the nuclear mutation is not known. However, it is suggestive that the new sequences in the mt genome arose by incorporation of plasmid like DNA's and that these DNA's might have arisen by excision from mt genome, an event that could be linked with the origin of S-cytoplasm of maize. Thus the episomal element when free in cytoplasm may lead to mst in S-cytoplasm maize, but when integrated either in the mt- or the nuclear genome, the element induces mft. This unique property of the plasmid like DNA suggests its use as a vehicle for inserting DNA sequences into a maize genome having a specific nuclear background. Clements (1975) proposed a hypothesis in which the cytoplasm mutates from S to N at a rate depending on the nuclear background. The above interpretations do not seem to be valid in mst *Vicia faba* S-447, in which complete, partial and feeble mst tillers or flowers are produced on the same plant. Thiellement (1977a) proposed the co-existence of N- and S-cytoplasms in different proportions in the same plant. Whereas with the stable proportion of N- and S-cytoplasms the phenotype is stable, altered proportions modify the phenotype, the modifications occurring chiefly by unequal mitotic segregations. Many segregation patterns appear erratic and/or nonrandom. For these Thiellement (1979b) proposed a "grand-mother" effect to explain the instability of phenotypically unstable mst plants.

3.8.9 Broadening the S-Cytoplasm Base

i. *Maize.* Until 1970, T-cytoplasm was used for the production of nearly all hybrid maize in the US. However, after the 1970 epidemic of southern leaf blight (SLB) caused by race T of *Helminthosporium maydis* and yellow leaf blight (YLB) caused by *Phyllosticta maydis*, that severely damaged the maize crop in most of the US corn belt, the use of T-cytoplasm to sterilize hybrid females was virtually abandoned. This genetic vulnerability cautioned the plant breeders to introduce genetic variability and diversity in *c*-genes and led to the search for new sterile cytoplasm types. In maize, two new types, S and C, were soon added (Duvick and Noble 1978). These three sterile cytoplasm types differ amongst each other in many features (Table 27.18). The majority of the 160 sterile cytoplasm sources belong to one of these three types, but a few are not grouped in any of them and may represent an additional source of sterile cytoplasm in maize. This enhanced cytoplasmic diversity in maize led to the proposal of a technique producing hybrids using a blend of several kinds of sterile cytoplasms – the multiplasm – for hybrid seed production.

ii. *Sorghum.* In Sorghum, gc-mst was first detected in milo S-cytoplasm with kafir nuclear *fr*-genes. The milo source found by Stephens and Holland (1954) forms the component of virtually all mst sorghum parents used for hybrid seed production. Thus all sorghum hybrids had a common type of S-cytoplasm. Lack of S-cytoplasm diversity produced genetic uniformity. A strong need arose to identify and utilize alternate S-cytoplasm sources to safeguard against possible hazards like corn leaf blight or ergot of pearl millet. Other S-cytoplasm sources were reported by Rao (1961), Webster and Singh (1964), Hussaini and Rao (1964), Ross and Hackerott (1972), Nagur and Menon (1974), Schertz (1977), Schertz and Ritchey (1978). Definitive genetic testing has just been completed. Four S-cytoplasm types under the series A_1–A_4 are known in sorghum (Table 27.9). These exhibit differences in fertility restoration, anther aminogram, soluble protein and esterase isoenzyme patterns (Tripathi et al. 1981, 1982b).

iii. *Wheat.* S-cytoplasm diversity of hybrid wheat is limited, as almost all hybrid wheat breeding programs are based on *T. timopheevi* S-cytoplasm. The wide use of this cytoplasm in hybrid wheat production is due to its apparently neutral effect on agronomic and quality characters. Most other *Triticum* and *Aegilops* species have deleterious effects on various traits. If susceptibility to a particular disease, insect or environment should become associated with *T. timopheevi* cytoplasm, hybrid production would be jeopardized. Therefore, cytoplasmic diversity is needed to protect against these production hazards. Additional sources of S-cytoplasm for developing gc-mst bread wheat are known from *Ae. kotschyi* and *Ae. variabilis* (Mukai and Tsunewaki 1979) and *Ae. speltoides* (Ghiasi and Lucken 1982b).

3.8.10 S-Cytoplasm Specific Bodies

Two types of sterile cytoplasm, designated S-447 and S-350, are known in *Vicia faba.* Of these, S-447 cytoplasm exhibits four features, viz. instability, somatic reversion, permanent restoration and presence of cytoplasmic spherical bodies (CSB's). The CSB's are unique to S-447 cytoplasm mst plants but are absent in the revertants, restored sterile plants, fertile maintainers and restorers, as also in S-350 mst plants. The bodies are nearly 70 nm in diameter and are composed of a circular unit membrane enclosing an electron-dense central zone, and contain a high molecular weight double-stranded (ds) RNA (Edwardson and Corbett 1961; Edwardson et al. 1976; Scalla et al. 1981; Grill and Garger 1981). The CSB's are associated with mst and their ds-RNA appears to have characteristics similar to a virus. Transfer of the ds-RNA and sterility from mst to mft *V. faba* plants through a dodder bridge (Grill and Garger 1981) indicates the viral nature of mst in S-447 broad beans and associates CSB's with the mst trait. Initially, the number of CSB's is low but it increases during the microspore formation stage. When the microspores abort, the CSB's are destroyed by an autophagic process. This indicates a strong correlation between the ds-RNA contained in CSB's and mst. Similar ds-RNA occurs in five cytoplasms of LBN maize (Sisco et al. 1984), but there are certain dissimilarities between the mst trait or ds-RNA's in maize and faba beans. Fertility restoration caused by *Fr* nuclear genes is reversible in maize but irreversible in faba beans. Moreover, the CSB's and their ds-RNA's disappear in faba beans restored to fertility, but the LBN ds-RNA's of maize are not reduced or eliminated by such *Fr*-genes. Similar ds-RNA is known in Ru cytoplasm of maize not associated with mst (Weissinger et al. 1982). Thus, the LBN ds-RNA's in maize may not be associated with mst and may represent the replicative form of ss-RNA's, as is also known in many plant viruses.

3.8.11 Aneuploids

The mst's, asynaptics and desynaptics comprise rich sources of aneuploids (Gottschalk and Kaul 1974, 1980a,b). Many g-mst mutants of soybean, tomato, pea and wheat produce aneuploids, but some of these are viable only in tomato. In others, seedlings die after the four- to six-leaved stages. From among the progeny of ms_1 soybean plants, Kenworthy et al. (1973) and Beversdorf and Bingham (1977) detected aneuploids ranging from haploids to hexaploids. Other soybean mst mutants also produce aneuploids, but these do not survive to flowering stages. Likewise, the haploids produced by ms_1 soybean (Kenworthy et al. 1973, Beversdorf and Bingham 1977) and two g-mst mutants of pea (this Vol.) do not survive to maturity. They may survive after diploidization by colchicine and produce isogenic lines. Progeny of alloplasmic *Solanum* and *Triticum* gc-mst plants produce of few viable aneuploid seedlings. Their frequency is high in bread wheat having *Aegilops biuncialis, Ae. columnaris* or *Ae. variabilis* cytoplasms (Kobayashi and Tsunewaki 1980). Although both diploid and autotetraploid mst sugarbeet produce aneuploid progeny (Savitsky 1954; Kinoshita 1971), the frequency is low in tetraploids. This is due to reduced seed fertility of the tetraploids.

3.8.12 Haploids

Kihara and Tsunewaki (1962b) obtained 30% haploid seedlings from the S_1 progeny of the alloplasmic wheat *T. aestivum* Salmon having *Ae. caudata* cytoplasm. Later, Kobayashi and Tsunewaki (1980) found haploid plants in the progeny of the alloplasmic wheat *T. aestivum* or *T. macha* plants having *Ae. biuncialis, Ae. caudata, Ae. columnaris, Ae. kotschyi, Ae. triuncialis, Ae. variabilis* cytoplasms. These haploids develop parthenogenetically from the unfertilized egg cells. Their production is conditioned by a specific gene, *h*, whose interaction with the cytoplasm may not always produce mst. In certain alloplasmic combinations of *T. aestivum* Salmon, the genes conditioning mst and haploid production seem closely placed on the same chromosome. Association of a specific satellited chromosome in haploids indicates the haploid gene *h* to be located on this chromosome. Mukai (1981) associated translocation with haploid production. Probably translocations suppressed crossing-over between *fr* gene and haploid producing gene *h*.

3.8.13 Disease Susceptibility

Association of disease susceptibility with T-sterile cytoplasm is known in maize, in which over a span of 20 years (1950 to 1970) the hybrid seed maize industry changed from mechanical detasseling to genetic emasculation by using gc-mst and back again to mechanical detasseling. This was because of the infection of T-cytoplasm mst plants by southern leaf blight (SLB) caused by the fungus *Helminthosporium maydis* race T (HmT) and by yellow leaf blight (YLB) caused by the fungus *Phyllosticta maydis*. A pathotoxin present in the crude filterate of cultures of *H. maydis* race-T and also in the extracts from leaves injected with fungus, when injected into non-diseased leaves, causes a sensitive reaction in plants carrying T-cytoplasm but does not significantly affect plants with other cytoplasm types like N, C or S (Hooker et al. 1970; Lin and Hooker 1972; Gracen et al. 1972). The *H. maydis* race-O also produces a toxin but shows no specificity for the T-cytoplasm. Thus the T-cytoplasm exhibits a specific sensitivity to HmT fungus and its toxin. Sensitivity to the fungus or its toxin is maternally inherited and the gene(s) appears to reside within the mitochondria of T-cytoplasm plants because the toxin specifically affects the functions of isolated mitochondria of T-cytoplasm plants.

Like T-cytoplasm maize, the Tift 23A pearl millet is susceptible to *Sclerospora graminicola*. Due to this, India's pearl millet yield dropped from 8 to 3.6 million metric tons (Jain and Pokhriyal 1975) because of the common mst line Tift 23A used in the production of all hybrids popular up to that time. The S-cytoplasm-associated disease susceptibility is maternally transmitted and its gene(s) is closely linked to the *c*-gene, as many induced mst mutants resistant to the disease were obtained by Murthy (1974), Raut et al. (1973), Gill et al. (1977) and Andrews and Anand Kumar (1982). This disease susceptibility and genetic vulnerability of gc-mst lines has led to the use of multiplasms and the isolation of many sterile cytoplasm sources especially in maize, pearl millet and sorghum.

3.8.14 Methomyl Sensitivity

When maize seedlings with the cytoplasm N, C, S and T and their maintainers are treated with 300 ppm methomyl, only the inbreds and maintainers having T-cytoplasm severely affected (Ohmasa 1982). Thus, like HmT sensitivity, T-cytoplasm of maize exhibits specific sensitivity to the chemical methomyl. Reasons for the specificity are not known.

3.8.15 Somatic Hybridization

Sexual reproduction imposes a limitation to free recombination of genetic information located in the cytoplasmic organelles like the chloroplast, mitochondria, kinetosome, centrioles, etc. as the recombination is restricted to the nuclear genome only. Protoplast fusion technology permits symmetric combinations of nuclear, chloroplast and mitochondrial genomes between and among different species and genera. Although protoplasts have been isolated from primary tissues of most plant organs such as roots, leaves, petals, fruits, coleoptiles and storage organs, as well as from microspore tetrads and legume root nodules, leaf mesophylls are preferentially used because of high yields and apparent uniformity. The fusion of leaf protoplasts of two plants having different cytoplasm types allows genetic analysis of cytoplasmic genes. The protoplast fusion technique permits achieving a heteroplasmic state of the extra nuclear genetic elements in addition to and independent of the nuclei. This allows a direct study of g-c interactions and gives information about the transmission of cytoplasmic genes and their location. Through protoplast fusions, transfer of mst or mft from one line to another has been possible without going through a series of generation-consuming backcrossing and genome substitution procedures. Furthermore, through such fusions, somatic hybrids and cybrids at nuclear and cytoplasmic level have been obtained in *Nicotiana* and *Petunia.* Of these, the cybrids are more useful than the hybrids, as they have one genome and a mixture of two cytoplasms. Using gc-mst *Nicotiana tabacum,* Belliard et al. (1978) and Gleba et al. (1978) obtained novel tobacco cybrid plants having *N. tabacum* nuclear genome and cytoplasms of this and other species. Zelcer et al. (1978) found that X-irradiation of a protoplast parent caused preferential transfer of cytoplasm alone into the fusion product, resulting in cybrid rather than hybrid formation. Aviv et al. (1980) produced cybrids by fusing protoplast of *N. sylvestris* and X-irradiated gc-mst *N. tabacum* having *N. debneyi* cytoplasm. Aviv and Galun (1980) restored mft by fusing protoplasts from mst *N. sylvestris* plants with X-irradiated protoplasts derived from mft *N. tabacum* plants. Uchimiya (1982) isolated a new mst hybrid having two different cytoplasms and a mixture of nuclear genomes of *N. tabacum* and *N. glutinosa.* In *Petunia,* using leaf mesophyll protoplast fusion, Izhar and Power (1979) transferred mst from gc-mst *P. hybrida* to *P. axillaris.* Later 4x somatic hybrids between *P. hybrida* and *P. parodii* were raised (Izhar 1984). Most of these were mft but some were mst, while a few had both mst and mft flowers. These mixture plant types segregated mst and mft flowers and the trait was inherited maternally.

After artificial inoculation by *Helminthosporium maydis,* Gengenbach and Green (1975) selected a few toxin-resistant cell lines from T-cytoplasm *Zea mays* cultures which are highly susceptible to the toxin produced by the fungus. Some of the toxin-resistant plants were mft and others were mst. From maize cultures, Brettell et al. (1980) obtained one mft toxin-sensitive plant without any toxin inoculation. This indicates a breakage of extreme close association of T-cytoplasm, toxin susceptibility and mst in maize. The above examples indicate the potential of novel phenotype production through cell cultures and protoplast fusion. In addition to plant types with deviated vegetative and/or floral morphology (in mst's and mft's), interspecific mst transfer has been achieved through protoplast fusion in tobacco and petunia (Izhar and Power 1979; Uchimiya 1982). The technique has demonstrated the dominance-recessive relationships of N- over S-cytoplasms, cytoplasmic segregations of heteroplasmons in cybrids following first and second meiotic cycles (petunia, tobacco), independent segregation of chloroplast type and mst-mft traits (petunia), DNA recombination in mitochondria and chloroplast (maize, petunia, tobacco), association of *c*-gene with mt-DNA (maize, sorghum, *Brassica*) and cytoplasmic inheritance of sensitivity to tentoxin (tobacco).

3.9 Fertility Restoration

The gc-mst is conditioned by interaction of *fr* nuclear genes and *c*-cytoplasmic genes present in the S-cytoplasm. Alterations in either the *fr* or *c*-gene revoke fertility. Thus fertility in a gc-mst is restored either by introducing a fertility-restorer gene, designated usually as *Fr*, or by replacing S- by N-type cytoplasm. Introduction of the nuclear gene is easier and quicker than replacement of the S-cytoplasm. Therefore, fertility is restored to the S-cytoplasm by introducing one or more *Fr* genes. This has divulged the genetic control of fertility restoration in some plants (Table 3.32), the nature of which is described below.

Table 3.32. Genetic control of fertility restoration in some gene-cytoplasmic male steriles

Species	Fertility restoration	Remarks (mst expression)
Allium cepa (2x = 16)	1-2 dominant genes	Partial to complete, 2 cytoplasm types temperature influence very high
Beta vulgaris (2x = 18)	6 nonallelic, independent dominant genes	Partial to complete, modified by genes and environmental factors
Brassica campestris (2x = 20)	2 dominant genes	Temperature- and photoperiod-sensitive
B.napus (4x = 38)	2-4, or, 2-complementary dominant genes	Temperature, plant development stage and genotype-sensitive
B.oleracea (2x = 18)	2 dominant genes	Partial, temperature and genotype-sensitive
Capsicum annuum (2x = 24)	1-2 dominant genes independent or complementary	Influenced by modifiers and temperature
Daucus carota (2x = 18)	2 complementary dominant genes, 2-3 duplicate dominant genes, epistasis, inhibitory gene action	Influenced by temperature and photoperiod
Gossypium hirsutum (4x = 52)	1 dominant gene or 3 genes with at least one dominant allele at 2 loci	Temperature, photoperiod and modifier genes influence considerable
Helianthus annuus (2x = 34)	1-2 complementary dominant genes 3 complementary genes or 2 dominant genes with cumulative action	Stable to environment, slight influence by temperature in one genotype
Hordeum vulgare (2x = 14)	1 dominant gene	Stable to environment, 2 genotypes influenced by physiographic factors and or temperature and photoperiod

Table 3.32 *(continued)*

Species	Fertility restoration	Remarks (mst expression)
Medicago sativa (4x = 32)	2 complementary or additive genes, polygenes	Both environmental factors and major genes influence considerably
Nicotiana tabacum (4x = 48)	1 or many genes	Environmental influence high, *fr* genes located near sat chromosomes
Oryza sativa (2x = 24)	1-2 dominant genes or polygenes	Both gametophytic and sporophytic type of mst, influenced considerably by temperature and humidity
Pennisetum typhoides (2x = 14)	3 dominant genes	Highly influenced by environment and modifiers
Petunia hybrida (2x = 14)	1-2 dominant genes	Highly temperature-sensitive
Secale cereale (2x = 14)	1 or many dominant genes	Temperature and genotype sensitivity very high
Solanum tuberosum (4x = 48)	1-3 dominant genes	Environmental influence high
Sorghum vulgare (4x = 20)	1-4 dominant genes, duplicate dominant genes, epistasis	Highly influenced by modifier genes, genotype and environment
Triticum aestivum (6x = 42)	1-3 dominant genes, 2 complementary, 2 dominant + 1recessive	Influenced by genotype, environment and genome-cytoplasm interaction
Vicia faba (2x = 12)	1-2 dominant genes	Influenced by modifiers and environment, restorer genes recessive in some and dominant in other genotypes, epistasis and reversal of dominance, restoration permanent in S-447 cytoplasm
Zea mays (2x = 20)	1 or 2 dominant genes	Cytoplasm diversity and specificity very high, restoration gametophytic and sporophytic

3.9.1 Genetic Basis

Use of gc-mst in hybrid seed production necessitates F_1 fertility restoration through incorporation of *Fr* genes. Usually the sources of *Fr* genes have been detected either in the cytoplasm donor parental species or in other allied species. Since information about the control of fertility restoration is vital for the development of restorer lines and hybrid seeds, it is known in many species. In most of these, the genetic control of fertility restoration is conditioned by one or more dominant genes which interact with many other major or minor genes (Table 3.32). This interaction is positive in some and negative in other lines of the same species and involves the genetic phenomena of complementation, inhibition, epistasis, cumulation, modifications, etc. This renders restoration control highly subtle, delicate and fragile. Moreover, the action of *Fr* genes is influenced by the environment, especially by temperature and photoperiod. In rice, the restoration is controlled by one or two dominant genes or by polygenes. The action and expressional degree of the polygenes is highly influenced by the environment and by the number of polygenes present in the restored plants. A multiple gene system for full mft restoration in rice as postulated by Maekawa et al. (1980) is operative only under certain environmental conditions. In broad beans, reversal of dominance of *Fr* gene control occurs in some crosses, in others epistasis and partial dominance is apparent.

Restoration of mft in wheat is controlled and coordinated by a complex gene system comprising recessive, dominant, duplicate-dominant, inhibitory or complementary genes (Table 3.32). The restoration is influenced by cytoplasm type and source, modifying genes, environment and genotype-environment interaction. For the alloplasmic mst wheat (*T. timopheevi*) *T. aestivum*, dominant *Fr* genes have been detected on at least seven wheat chromosomes, but not more than three chromosomes carry the *Fr* genes in any one restorer line and up to seven chromosomes are known to carry modifier genes in the same cultivar. Of the 21 wheat chromosomes, two have not been implicated in restoration of mft, the others carry either *Fr* or modifier genes influencing restoration. In maize, three distinct sterile cytoplasm (T, C, S) types exist. Restoration is sporophytic for T- and C- and gametophytic for S-cytoplasm. Two dominant *Fr* genes, Fr_1 and Fr_2, restore fertility to T-cytoplasm mst plants, and one each, Fr_3 and Fr_4, to S- and C-cytoplasms, respectively. The existence of many more *Fr* alleles in the maize genome is documented by Kheyr-Pour et al. (1981). In some cytoplasms, a combination of different *Fr* genes is required for restoration. Some of these are stable, others highly sensitive to environmental and physiological alterations and in a few mst breaks down 7–10 days after silking. In rye, one to many major genes restore fertility. The relative expressivity of three dominant restorer genes is in the order $Fr_1 > Fr_2 > Fr_3$ (Scoles and Evans 1979b). Partial fertility occurs when certain genotypes carry two of the three *Fr* genes, the extent and degree of expression is genotype-environment-dependent. In *Petunia*, in addition to a major Fr_1 dominant gene, a recessive gene (fr_2) and a single or complex set of genes (fr_3) condition restoration (Edwardson and Warmke 1967). In sugarbeet, restoration is conditioned by six nonallelic dominant genes (Table 18.3). The complementary action of two *Fr* genes *X* and *Z* condition fertility restoration in a peculiar way. Whereas gene *X* restores full fertility, gene *Z* restores partially and X is epistatic to Z. The N_{XXZZ} plants restore full fertility and N_{XxZz} genotypes restore 25% full and 50% partial fertility, in the remaining plants fertility is not restored. A dominant gene, *Sh*, enhances the pollen-producing ability of S_{xx} mst plants, so that they appear mft (Hogaboam 1957). Cleij (1967) restored fertility by subjecting germinating seeds of mst sugarbeet to 55°C for 10–60 min. By heat treatment and subsequent meristem culture, Lichter (1978) obtained fertile plants from mst sugarbeet. In alfalfa, restoration is controlled either by three dominant genes acting independently or additively (exhibiting dosage or compensation effects) or by one to four polygenes, whose expression is influenced by modifiers and environment. In onion, in one S-cytoplasm, fertility is restored by one dominant (*Fr*) gene, in the other (T-cytoplasm) by two genes. In general, in the majority of the species, fertility restoration is of sporophytic type and all the pollen grains of plants heterozygous for restorer gene(s) are functional. However, in the S-cytoplasm type of maize, in an Indian rice variety Chinsurah, in an onion and a sugarbeet line, the restoration is of gametophytic type. In these, only those pollen grains are functional that carry the dominant form of restorer gene (*Fr*).

3.9.2 Chromosomal Control

The *Fr* genes are located in the nuclear genome. Since most of the S-cytoplasm donor species have these genes, the retention or reintroduction of *Fr* genes carrying specific chromosome(s) of the cytoplasm donor species restore mft in the mst plants. This is known in tobacco, petunia and wheat. In tobacco, some *Fr* genes are located on the satellite arms of a highly specialized and thoroughly arranged chromosome fragment of *Nicotiana debneyi*, *N. repanda* and *N. undulata* genome (Gerstel et al. 1978; Burns and Gerstel 1981). These genes are close to NOR but not necessarily linked to this region. The *N. undulata* chromosomal fragment restores mst partially, but the anther phenotype is restored fully. *N. debneyi* fragment restores anther morphology but not mft. The *N. repanda* chromosomal fragment restores both anther phenotype and mft fully. This restoration in degrees indicates the presence of at least two *Fr* genes, one restoring the anther phenotype and the other pollen fertility in *Nicotiana* genome. Besides restoring pollen fertility, the chromosomal element derived from the genome of *N. repanda* has an active centromere that ensures its regular transmission in somatic and meiotic divisions. In addition, it has satellites and an active nucleolar organiser at both the ends. Whether present in single or double doses, the chromosomal fragment inhibits the nucleolar organising ability of *N. tabacum* chromosomes. This nucleolar dominance or amphiplasty is almost complete in PMC's but partial in root tip cells.

3.9.3 Environmental Influence

The influence of temperature and/or photoperiod on the penetrance and expressivity of *ms* and *mf* genes (*Ms, ms, Fr, fr*) is well known. Whereas the data on the effect of temperature on restoration are often incomplete and contradictory (Scoles and Evans 1979a), the precise impact of photoperiod on *Fr* gene expression is not known. However, the total effects of these two environmental components on restoration is high in almost all the plants of economic value in which the gc-mst system is known (Table 3.25). Marrewijk (1969) concluded that in every species there is an optimum temperature for fertility restoration which is highly dependent upon genotypic and environmental factors. In fact, each species appears to have an optimum temperature for fertility restoration and precise relationships between temperature and restoration degree can be found in each species exhibiting mst. In wheat (Johnson and Patterson 1973), maize (Duvick 1965) and petunia (Marrewijk

1969), restoration is more complete under cool conditions than at the normal growing conditions of these crops. Rye, a cool-climate crop, exhibits poor restoration at low and full restoration at high temperatures. Duvick (1965) found temperature as the major environmental factor that affected pollen production of restored maize lines. Temperature-dependent partial restoration is known in onion, cotton and wheat (Wilson 1968a,b; Meer and Bennekom 1969; Marshall et al. 1974). Thus temperature is a major factor influencing fertility restoration of gc-mst plants.

3.9.4 Heat Treatment and Meristem Culture

Cleij (1967) subjected germinating seeds of gc-mst sugarbeet to temperatures up to 55°C for 10 to 60 min. Four plants produced stainable pollen and were regarded as mft. However, the results of selfing, intermating and outcrossing these transformed plants were inconsistent and did not prove S-cytoplasm mutation to the N-type. On the other hand, from the diplodised haploid sugarbeets derived from gc-mst plants (Yuce 1973), Lichter (1978) obtained mft plants by heat treatment (+36°C) and subsequent meristem culture. Such plants were stable for two to three generations but segregated mst plants subsequently.

3.9.5 Mutation in Cultures

Cell cultures of mst plants have not been grown as extensively as those of mft plants. In some cell cultures of mst plants of *Datura*, *Nicotiana*, *Petunia* and *Zea*, mft plants arise in very low frequency due to spontaneous mutations of the S-cytoplasm to N-type. From T-cytoplasm HMT toxin-susceptible mst maize cell lines, Gengenbach and Green (1975) selected toxin-resistant cell lines after HMT toxin inoculation in the cell cultures. Some of these were mft and others mst. Brettell et al. (1980) obtained one toxin-sensitive mft plant from the T-cytoplasm cell cultures without toxin inoculations.

3.9.6 Protoplast Fusion

Fusing irradiated or unirradiated protoplasts of mft and mst lines has led to the isolation of mft hybrids, a few of which are stable over generations and environments in *Nicotiana* and *Petunia*. Aviv and Galun (1980) fused X-irradiated protoplast of *N. tabacum* var. Xanthi with an alloplasmic mst line (*N. suaveolens*) *N. sylvestris*. Of the calli developed, 25% produced recombinants and cybrids (*N. sylvestris*) *N. tabacum*, half of these were fully mft and the rest p-mft. This technique led to the restoration of alloplasmic *N. sylvestris* plants. Similar mft plants were obtained by fusing *N. tabacum* protoplast with the alloplasmic mst (*N. debneyi*) *N. tabacum* (Belliard et al. 1977, 1978). The mft tobacco plants had mostly *N. tabacum* and a little of *N. debneyi* cytoplasm. Thus through protoplast fusion, mft was restored in tobacco. Uchimiya (1982) likewise fused the protoplast of mft *N. glutinosa* with the alloplasmic mst (*N. debneyi*) *N. tabacum*. From the calli grown plants, mft, intermediate and mst phenotypes were obtained. In all these cases, X-irradiations selectively eliminated nuclei, inhibited the unfused protoplasts from dividing and enhanced cytoplasm transfers.

3.9.7 Gametophytic and Sporophytic Sterility

In the majority of gc-mst plants, the mst expression in all the pollen grains is determined by the interaction of the *fr* nuclear genes and *c*-cytoplasmic genes of the plant. In other words, the genotypic expression of the sporophyte and not of the gametophyte decides the mst or mft expression. This is known as the sporophytic type of mst. However, in *Oryza sativa* (Shinjyo and Omura 1966; Shinjyo 1969), *Sorghum vulgare* (Brooks et al. 1966; Brooks and Brooks 1967) and S-cytoplasm *Zea mays* (Buchert 1961; Duvick 1965), the mft/mst determination is conditioned by the genotype of the individual pollen grains and not by the sporophyte. This is known as gametophytic determination. In the sporophytic type, the pollen phenotype depends on the genotype of the sporophyte, but in the gametophytic type, the pollen phenotype is determined by its own genotype and not by that of the sporophyte. Whereas only one type of pollen grain is produced in the sporophytic type of mst, in the gametophytic type both fertile and sterile pollen grains may be present. For instance, in S-cytoplasm plants with *FrFr* genes, pollen is fertile, in *frfr* plants the pollen is aborted. *Frfr* plants produce half-fertile and half-aborted pollen grains, the fertile pollen has the gene *Fr* and the sterile the *fr* gene. Such anthers of the heterozygous plants have a pebbly appearance due to the random distribution of aborted-shrivelled and fertile-plump pollen grains within the same anther.

3.10 Breeding Value

Hybrid production is the primary aim of gc-mst lines as they act as females, being genetically emasculated. Before use in commercial breeding these lines are to be developed, maintained and improved. This involves many steps and evokes various problems described below.

3.10.1 Development of Breeding Lines

i. *Male Sterile A-Line.* A gc-mst system producing mst plants banks upon a balanced and appropriate combination of nuclear *fr* genes and *c*-cytoplasmic genes. This combination forms the source for developing other promising gc-mst lines. Commercial varieties or desirable lines that have undergone inbreeding (especially if selfers) and selection for several generations are crossed initially to the gc-mst source. Thereafter, the desirable commercial line to be converted is used as the recurrent parent in the BC procedure and the final progeny after five to six BC cycles is genetically similar to the recurrent parent except that it is mst (Fig. 3.25).

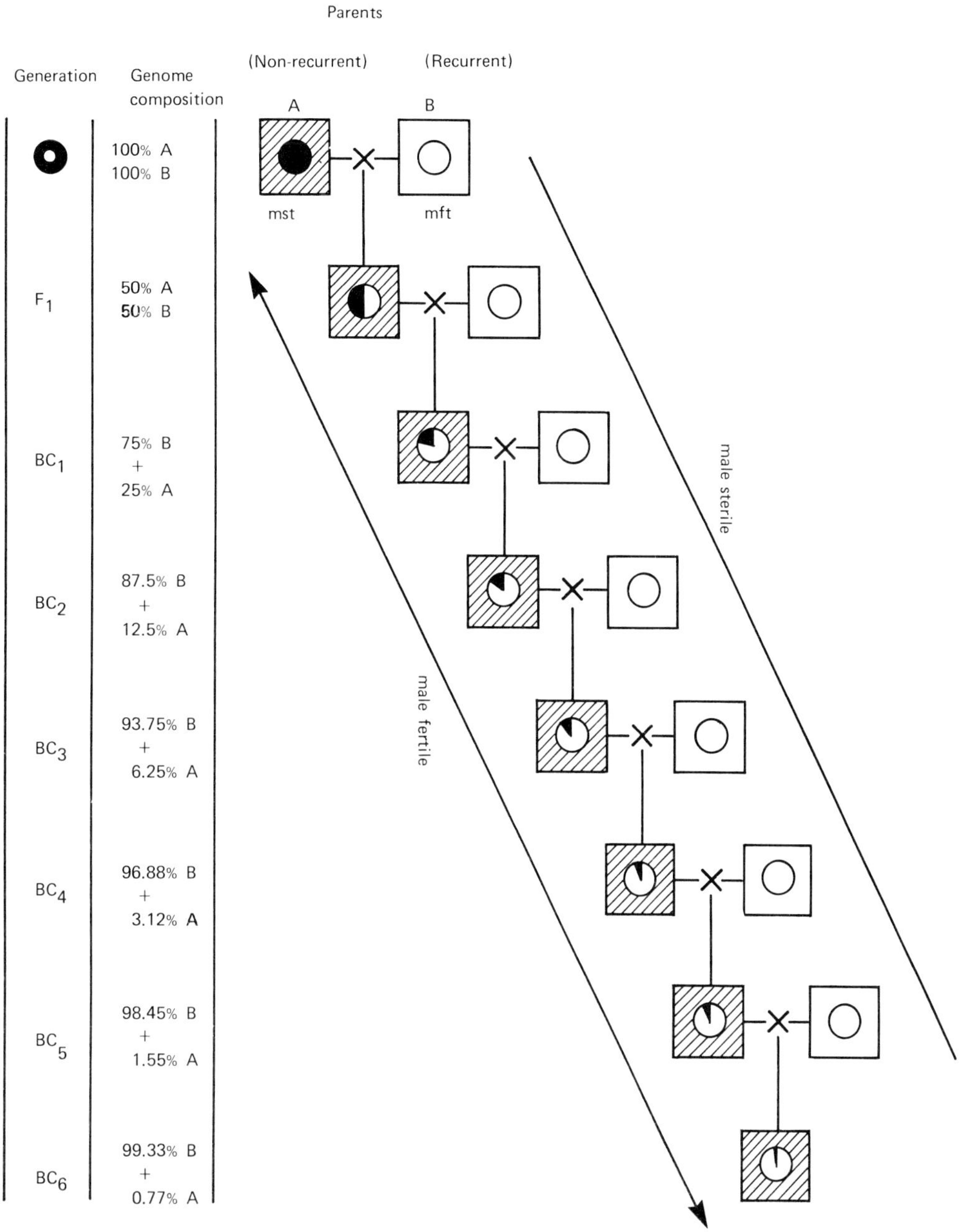

Fig. 3.25. Backcross method of developing a gene-cytoplasmic male sterile line. The final progeny will be identical with the recurrent parent B in most characteristics except being mst instead of mft

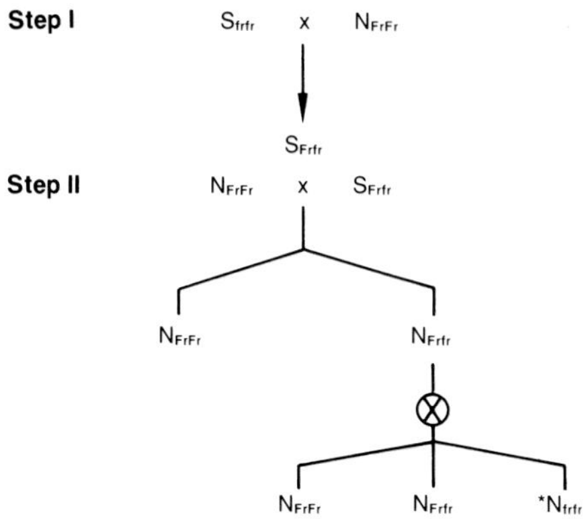

Fig. 3.26. Development of a maintainer (*B) line

The conversion of mft to mst inbred line can be accomplished in a relatively short time by growing hybrid plants in different regions and/or in a glasshouse under controlled conditions to obtain two to three generations a year.

ii. *Maintainer B-Line.* This is identical to the A-line except that it has N-cytoplasm and is hence mft. It can be maintained by bagging heads and harvesting self- or sib-pollinated seeds set over plants grown in isolated fields. The procedure for developing this line is given in Fig. 3.26 and the steps involved therein are as follows:

- Cross mst (S_{frfr}) plants with mft (N_{FrFr}), F_1 will be S_{FrFr} (mft).
- Hybridize N_{Frfr} plants as female by pollen of S_{Frfr} plants. Progeny will be mft and phenotypically identical but genetically of N_{FrFr} and N_{Frfr} types which can be distinguished by either selfing or progeny testing after crossing to S_{frfr} plants.
- Only N_{frfr} plants after crossing to S_{frfr} plants will yield all mst progeny. This N_{frfr} line becomes the B-line of the desired cultivar and its A-line can be developed by the BC procedure mentioned earlier (Fig. 3.25).

iii. *Restorer c-Line.* This is developed by crossing plants having dominant restorer genes (*Fr*) to the gc-mst line and subsequent selection of mft plants in later generations. This procedure, unlike BC procedure using lines with normal cytoplasm, does not require test crossing to a gc-mst plant after each generation of selection to determine if the selected plants carry the restorer (*Fr*) genes. Lines with *Fr* genes cannot be converted to mst lines. Restorer lines having S-cytoplasm can be developed by the method given in Fig. 3.27.

3.10.2 Maintenance and Restoration

Interaction of *fr* genes with S-cytoplasm leading to mst phenotype and its reversion by *Fr* genes permits utilization of a breeding system most favourable for hybrid seed production. Pure breeding mst A-line can be maintained by pollinating it by maintainer B-line, and fertility can be restored in it by crossing it to the restorer line C (Fig. 3.28).

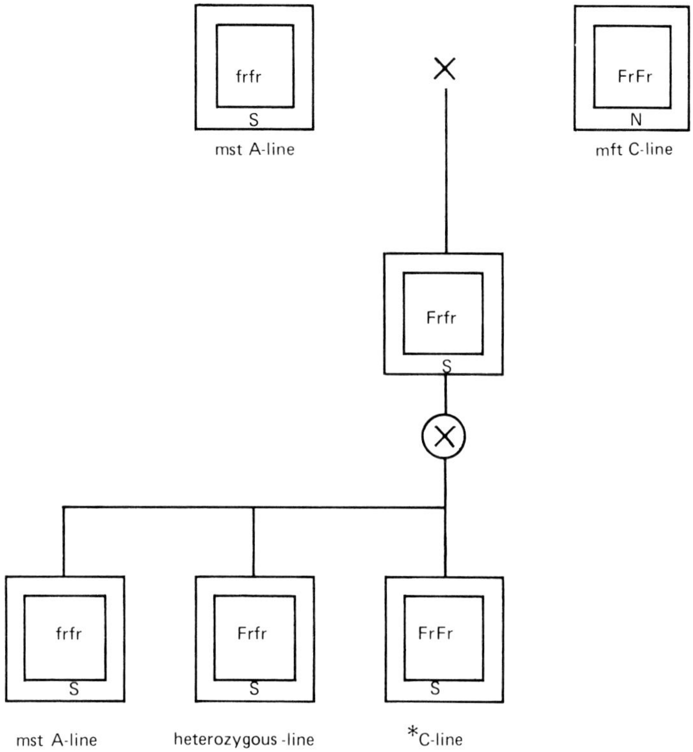

Fig. 3.27. Development of an isoplasmic restorer *C-line

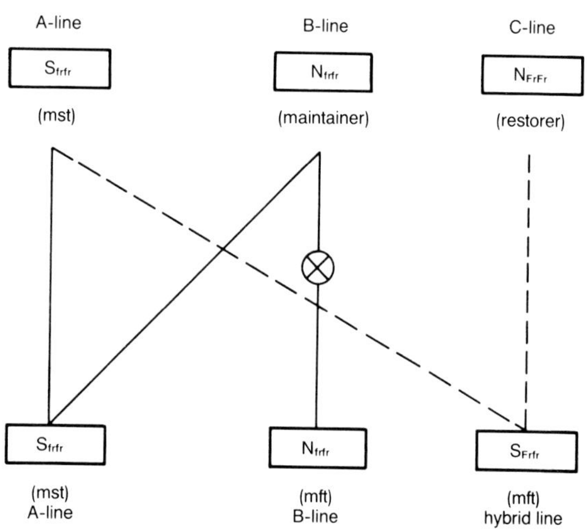

Fig. 3.28. Method for mst maintenance and mft restoration

3.10.2.1 N-Cytoplasm Incorporation

i. *Detection.* From F_1 or F_2 phenotypes, Rutger and Jensen (1967) and Hermsen (1965, 1968) proposed methods (Table 3.33) to detect an N-cytoplasm type of unknown parent by crossing them to S_{frfr} or S_{Frfr} genotypes. F_2 segregations of the cross involving the unknown line as male parent and S_{frfr} mst plants as female parent indicate genotypes of the male parent (Table 3.33b). When only mft plants occur, the unknown parent has N-cytoplasm, but when 1 : 1 mst : mft plants occur, the unknown parent has S-cytoplasm. F_1 progeny phenotype will reveal the genotype of the unknown parent, when S_{frfr} is used as female and the unknown parent as male (Table 3.33a). In this method F_1 progeny is needed and hand emasculation is obviated. Some reciprocal crossings also reveal genotypes of the unknown parent (Table 3.33c).

Table 3.33. Method for detecting N-cytoplasm by using S-cytoplasm

Parental genotypes	F_1 population	F_2 segregation mft	mst
a) $S_{fr.fr} \times N_{fr.fr}$	$S_{fr.fr}$	0	100
$S_{fr.fr} \times N_{Fr.Fr}$	$S_{Fr.fr}$	66	33
$S_{fr.fr} \times S_{Fr.Fr}$	$S_{Fr.fr}$	66	33
b) $N_{fr.fr} \times S_{Fr.fr}$	$N_{Fr.fr} + N_{Fr.fr}$	100	0
$N_{Fr.Fr} \times S_{Fr.fr}$	$N_{Fr.Fr} + N_{Fr.fr}$	100	0
$S_{Fr.fr} \times S_{Fr.fr}$	$S_{Fr.Fr} + S_{Fr.fr}$	50	50
c) $S_{Fr.Fr} \times N_{fr.fr}$	$S_{Fr.fr}$	66	33
$N_{fr.fr} \times S_{Fr.Fr}$	$N_{Fr.fr}$	100	0
$S_{Fr.Fr} \times N_{Fr.Fr}$	$S_{Fr.Fr}$	100	0
$N_{Fr.Fr} \times S_{Fr.Fr}$	$N_{Fr.Fr}$	100	0
$N_{Fr.Fr} \times N_{fr.fr}$	$N_{Fr.fr}$	100	0
$N_{fr.fr} \times N_{Fr.Fr}$	$N_{Fr.fr}$	100	0

ii. *Introduction.* Replacement of S- by N-cytoplasm is done by backcrossing (Hermsen 1965) by using N-cytoplasm line as female to the heterozygous S_{Frfr}-line, until the N_{Frfr} genotype is recovered. After n generations, the idiotype of BC_n is N_{Frfr}. Selfing BC_n will result in a population consisting of 25% N_{FrFr} + 50% N_{Frfr} + 25% N_{frfr}, the last one represents the desired phenotype in which N replaces S-cytoplasm. Test crossing the selfed BC_n-population on S_{frfr} as female detects N_{frfr} plants on the basis of following results:

- $S_{frfr} \times N_{FrFr}$ — 0% mst
- $S_{frfr} \times N_{Frfr}$ — 50% mst
- $S_{frfr} \times N_{frfr}$ — 100% mst

N_{frfr} plants can be used to increase S_{frfr} plants at will by using the former as pollinators to the latter genotypes.

3.10.3 Breeding Requirements

3.10.3.1 Line Selection

For a breeding program, suitable parental lines for hybrids have to be selected. One parent in most hybridisation and selection programs is an elite strain with high combining ability. Since this ability can be recognised in yield trials of hybrids, recognition of superior male or female parents follows the selection of lines, the crossing to produce hybrid seed and the yield testing of hybrids. Usually high yield testing is started in F_5 and testing of female parents has to await male sterilization of the strain. To advance the program as rapidly as possible, the male sterilization process is sometimes begun by F_4 generation of a group of B-lines (female parents). The presence of many C-lines (male parents) is basic to a breeding program. To develop improved lines, B×B and C×C line crossings are done to reduce the possibility of the emergence of strains combining characteristics of both. Most crosses to develop inbred lines are made using a female parent with N-cytoplasm so that any strain that emerges from the breeding program will possess N (fertility-inducing) cytoplasm. B-lines must have N-cytoplasm, but a male parent (C-line) may have S- or N-cytoplasm since it does not contribute cytoplasm to the hybrid. After initial selection and standardization of B- and C-lines, they are finally selected after crossing them to A-lines. Whereas the F_1 of A×B should be mst, that of A×C is mft. Some varieties when crossed to the A-line produce partial mst F_1 plants. Such plants cannot usually be completely male sterilized and it is best to discard them as soon as they are recognised. The main goal of the breeding process is to obtain B-lines that produce perfect mst and C-lines that restore mft completely to the F_1 hybrid. It is best to discard B- and C-lines that do not conform to this high standard.

3.10.3.1.1 A-, B- and C-Lines

With regard to *fr* genes, A- and B-lines are isogenic, but A is mst due to the S-cytoplasm and B is mft due to the N-cytoplasm. After a strain has been identified as a suitable B-line and is deemed to have potential as a female parent, it is male sterilized. By transferring the genome of B into A by paired progeny selection in a BC process (Fig. 3.25), B-lines are male sterilized to produce A-lines. Usually the process of male sterilization of a B-line is begun before the line is homozygous, and selection for desirable agronomic traits proceeds as the line is being male sterilized. Paired progeny breeding is necessary as mst is expressed only with hybrid progeny, but selection can occur only in the recurrent (B-line) parent. The original cross of the B-lines to be male sterilized is made on some suitable A-line. Progeny of each B-line selection and the progeny of its cross onto A-line are grown in adjacent (paired) rows. After flowering is well advanced and mst of the plants in A-line rows (progeny from crossing) can be evaluated, one or more heads in the row with the best mst intensity are crossed with single plants in the paired B-line row. The identity of each pair involved in the cross is recorded. Seeds from the new crosses and from their B-line parents are then planted in A- and B-paired progenies in the next generation. Subsequent crosses are made only between A- and B-line plants whose A-line rows are mst and B-line rows are mft and agronomically acceptable. This process is continued until within a family the B-line rows are similar to one another and the A-line rows are similar and mst. The completeness of mst in the A-line can easily be determined by observing seed set on bagged heads that have not been pollinated. When the B-line is crossed to the A-line for the first time, it is not always possible to recognise the B-lines that cannot be successfully male sterilized; but progeny of BC_1 involving a partial mst line will be partial mst. Such lines cannot usually be successfully male sterilized and are therefore discarded. Whereas A-lines can be maintained by pollinating them with pollen from their respective B-lines, B-lines can be maintained simply by selfing. After a line is established as a female parent of a commercial hybrid, both A- and B-line seeds are produced in a parent crossing field.

If in a breeding program C-lines are crossed only to C-lines, most resulting lines will restore fertility completely when crossed to A-lines. It is often necessary to cross C- and B-lines, as some traits desired in a C-line are present only in a B-line. In such a case, a suitable C-line can sometimes be selected from a partial restoring line that is still heterozygous for *Fr* genes. The process is somewhat similar to paired progeny selection. Plants of the incipient C-line are test crossed to the A-line and selections are made in the next generation only in rows whose parents restored full fertility in the test crosses. If sound restorers are not recognised immediately, probably none can be obtained later.

3.10.3.2 Line Testing

Well-planned breeding programs produce many new potential male parents per year. This is known in crop plants like wheat, rice, maize, sorghum, pearl millet etc. However, all the hybrids that result from crossing new males to the few good combiner females (mst) cannot be immediately tested at many locations. Fortunately, heterosis can be recognised in yield trials at locations other than those where the hybrids will be grown. Therefore, after first screening is done at the breeding location, parents whose hybrids survive the first screening test should be crossed to other likely parents and the resulting hybrid should be more widely tested at distant locations. Hybrids with superior performance are usually put into production for sale in one area, but are often so widely adapted that they may ultimately be grown over a large area within the country and perhaps outside also. Usually, male parents with high combining ability produce high-yielding superior hybrids when crossed to most of the female parents (mst lines). Likewise, a female parent (mst) with high combining ability results in high-yielding hybrids with most of the pollen parents. For this reason, it is possible and practicable to screen male parents by using one female parent and to screen female parents by using one male parent.

3.10.3.3 Seed Production

Seeds of newly produced hybrids have to be produced in quantities large enough to plant tests for 3–4 years. For hybrid seed production, two parents are grown in paired rows, the A-line in the first row and the C-line in the second. In cereals like wheat, rice and barley, the plants are thinned to allow about 5 g or more seed to develop on each panicle of the female line. Panicles of the A-line plants are bagged prior to flowering and pollinated when each panicle has flowered completely. If the bags are dated when put over panicles, those that have completed flowering can be recognised easily a few days later when pollination is required. Single-, double- or triple-cross hybrids (Fig. 3.29) or multiple-cross hybrids obtained by crossing many single-cross hybrids may be produced using the available mst lines as female parents and desirable or commercial varieties as the C-line male parents. In certain crosses involving sugarbeet, mustard, clover, etc. mst lines may be used in conjunction with self-incompatible lines to obtain double-cross hybrids (Fig. 3.30).

3.10.4 Male Sterile Propagation

3.10.4.1 Environmental Manipulation

For use in commercial hybrid seed production, a population with 100% mst would be ideal. This is possible if a method is devised to perpetuate a homozygous mst population. Normally mst plants cannot be selfed, but those fertile under one set of environmental conditions and sterile in others could be used for selfing and hybrid seed production, respectively. Some genotypes of cotton, sorghum, carrot, maize and barley are fertile in field and sterile in glasshouse or vice versa. By increasing lines homozygous for the recessive *fr* gene in an environment similar to greenhouse or field, pure-breeding mst lines could be established in them. These grown under another set of conditions would be mst and could be used as seed parents in crossing blocks. This was done by Justus and Leinweber (1960) in a genotype of American Upland Cotton which was mft in greenhouse and mst in field. A carrot line (Hansche and Gabelman 1963b) was completely mst in California but p-mst in Wisconsin. The authors suggested the lines be maintained in Wisconsin and used as female in California to produce hybrid seed. Using different environments to maintain near-female lines was suggested in a monoecious castor bean by Shifriss (1960). Strains with a dominant gene for femaleness plus a recessive gene for femaleness plus a recessive gene for occasional interspresed male flowers whose expressivity is environment-dependent can be maintained by self- or sib-pollinations in environments where the latter gene is expressed. These lines could be used as mst seed parents in other environments where the gene for interspersed male flowers is suppressed. Shifriss (1960) also selected a castor bean inbred homozygous for the *fr* gene (gene *f*; Claassen and Hoffman 1950). It produced occasional male flowers but was entirely female in some environments and not in others.

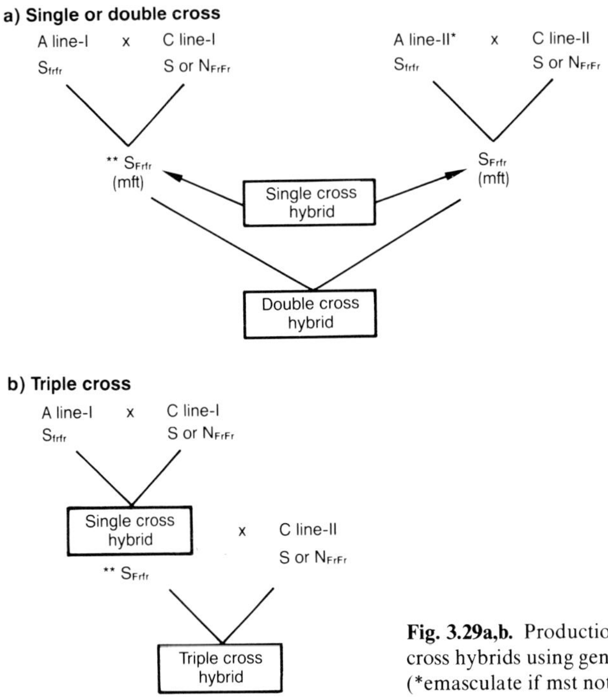

Fig. 3.29a,b. Production of single-, double- and triple-cross hybrids using gene-cytoplasmic male steriles line. (*emasculate if mst not available, **emasculate)

3.10.4.2 Dioecism

This provides a population which regularly segregates 1:1 males and females but it is necessary to save seed only from mst (as is done in g-mst or gc-mst) because no monoecious plants exist in a strictly dioecious population. Roguing male plants is necessary if a line is used as seed parent for F_1 seed production. In dioecious plants, sex-determining genes located on sex chromosomes are affected by modifiers which may or may not be located on sex chromosomes (Westergaard 1958). These modifiers alter the sex expression so that a "genetic male" can produce a small proportion of perfect or female flowers. By selfing such an altered female, a pure-breeding "almost mst" line can be established. In many instances (castor, papaya, *Cannabis*) the environment influences the mst expression of such sub-gynoecious plants so that they can be maintained in one environment and used as female to produce hybrid seed in another. Alternatively, if a small amount of selfed seed in the hybrid seed is of no consequence, the subgynoecious plants could be used as females without a special environment.

3.10.4.3 Chemical Treatments

Most of the male gametocides are toxic and reduce female fertility. Since in some cucurbits male flowers develop after gibberellic acid (GA_3) treatment of genetically female plants, such treated plants could be used for hybrid seed production. This

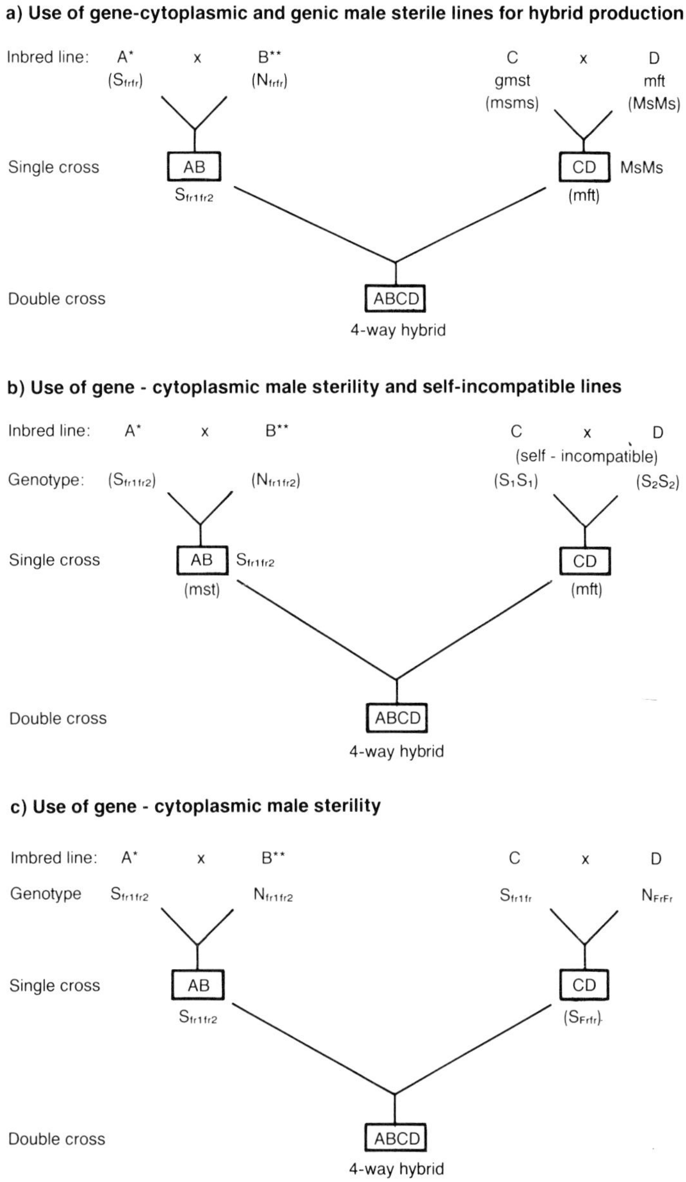

Fig. 3.30a-c. Production of double-cross hybrids using genic and gene-cytoplasmic male sterility and S-alleles (*gc-mst, **O-bollinator)

is possible in cucumbers in which a dominant gene *Acr* induces complete mst or nearly complete femaleness by accelerating the conversion rate from male to female type flowers. Lines homozygous for this gene are entirely female, but they can be reproduced because GA_3 treatment produces some male flowers (Shifriss 1961b). Such lines could also be used for F_1 hybrid seed production. Since cucumber hybrid is valued for fruit whose development requires seed setting, the male parent used in hybrid seed production should have a genetic background which suppresses mst in hybrid plants sufficiently to allow production of male flowers to pollinate the female flowers. Otherwise, pollinators will have to be interplanted in the commercial fields as is done with triploid watermelon. Certain heterozygous lines (*Acr, acr*) are monoecious (Shifriss et al. 1964), the gene *Acr* being incompletely dominant in them.

3.10.4.4 Haploids

Randall and Rick (1945) described a method to produce mst lines in dioecious *Asparagus*. Double haploid XX (female) plants would be crossed to homozygous YY (male) lines developed by selfing XY (male) plants with some female modifiers. All the hybrid plants would be male (XY). A method for maintaining the females (XX) is not known.

3.10.4.5 Functional Male Steriles

In these, normal pollen is produced but it is not shed, mainly due to failure of anther dehiscence. They can be selfed by pricking or prodding their anthers, but the necessity for hand labour to maintain the lines places a restriction on their utility. Therefore, they have not been utilized on a commercial basis.

3.10.5 Use in Breeding (Examples)

In large-scale hybrid production gc-mst has been used all over the world. From the early figures available, it is known that production in the USA based on this system is about 90% of the cultivated area for hybrid sorghum, about 60% for hybrid sugarbeet and up to 85% for hybrid field maize (Duvick 1966). Hybrid cultivars of vegetable crops produced after using gc-mst comprises 94% of sweet corn, 27% onion and 7% of carrot acreage in the USA (Gabelman 1974). In the other developed and fast developing countries, the statistics are similar. The main reason for such a large-scale use of gc-mst lies in the simplicity of the procedures providing pure breeding mst A-lines and fertility restoration wherever required in the hybrids. The mst-inducing S-cytoplasms of autoplasmic origin in allogamous plants and of alloplasmic origin in autogamous plants have been found almost universally. Experience reveals that once S-cytoplasm has been found in a crop, it has been followed by the detection of many more such sources or different types (e.g. maize, sorghum, pearl millet, barley, mustard, sunflower, potato etc.). Thus, it appears that S-cytoplasms, *fr* and *Fr* genes can be found in all the crops after making various genome cytoplasm substitutions. This is evidenced by the examples given in Table 3.6. A brief account of the use of gc-mst in some plants of economic importance is given in the following.

3.10.5.1 Cereals

In the grass family the occurrence of S-cytoplasms and *fr* genes is widely known in wheat, rice, maize, barley, sorghum, pearl millet, rye etc. In these, all the necessary components required for hybrid seed production are available viz. gc-mst maintainer, restorer sources and heterotic expression. Except for maize, pearl millet and rye, all the other above-mentioned cereals are self-pollinated, and this represents a major handicap in their use in commercial hybrid seed production.

Wheat. Many alloplasmic wheats are mst. Placing the bread wheat genome into the cytoplasms of *Triticum timopheevi, Aegilops caudata, Ae. kotschyi, Ae. ovata, Ae. speltoides, Ae. variabilis* results in mst. This has been profitably used in hybrid seed production on a limited scale. Apart from a few restoration difficulties, low NCP and pollen transfer limit hybrid seed production. It is further limited in some g-c combinations due to undesirable features like seed setting variability, fertility variations, premature seed sprouting etc.

Rice. Japonica rices with N-cytoplasm and indica varieties with S-cytoplasm, *fr*- and *Fr*-genes have been detected. Using *Oryza glaberrima, O. perennis, O. rufipogon* and *O. sativa* as cytoplasm donors, gc-mst has been obtained in cultivated rices and NCP up to 40% in these mst rices is known. Seed set over mst's is being improved by adjusting mft and mst lines and other crop managements in China and Japan. Ratooning and clonal propagation of mst lines for their use in recurrent selection and BC breeding has been done in Japan. Since hand pollination is necessary for hybrid seed production in rice, the labour-intensive process renders production uneconomic on a commercial basis. Therefore, the development of mst lines with open florets is desirable for obtaining high NCP.

Maize. Prior to 1950, hybrid seed maize was produced by the conventional procedure of detasseling the seed-producing rows which would then be pollinated by the pollen-producing rows. During the 1950's the utilisation of gc-mst (T-cytoplasm) and *Fr*-genes replaced detasseling. Due to the susceptibility of the T-cytoplasm to southern corn leaf blight, two new cytoplasmic sources (C and S) for use in gc-mst were isolated. This increased the cytoplasm base and improved the scope of the gc-mst system in hybrid maize production that is now used profitably and extensively throughout maize-growing regions. Thus we have a large number of hybrid maize lines based on mst stocks.

Sorghum. With the discovery of the gc-mst *Fr*-gene system, hybrid sorghum having milo-cytoplasm and kafir-nuclear genes was developed on a commercial basis. Over 97% of grain hybrid seed production in the USA is owed to this origin. A few more S-cytoplasm sources were identified, transferred to other sorghum species and hybrids developed in Asia and Africa. In some of them, dwarf and early genes have been introduced. Both *ms* and *fr* genes are utilized in population development and in seed and forage improvement.

Pearl millet. After the first discovery of gc-mst in pearl millet (Tift 23A) by Burton (1958), several S-cytoplasm sources have been identified and utilised profitably in commercial hybrid production. Tift 23A has excellent agronomic characteristics and combining ability and is widely used as a parent of hybrid pearl millet in India and the USA.

Barley and Rye. A large number of nonallelic *ms* genes are known in barley. A few of these have been utilized in hybrid seed production. Three S-cytoplasm sources have recently been detected. Their commercial utility awaits further results. In rye, likewise, gc-mst-*Fr* systems are well known, but compared to their high potentiality their use is still limited and needs broadening.

3.10.5.2 Vegetable Crops

Potato. This is both a food and vegetable crop. The gc-mst has been used successfully not only in hybrid potato tuber production, but also for the introduction of resistance to diseases and pests, in genome transfer and in development of lines from unreduced gametes. The transfer of the intact genome into alien cytoplasm by the protoplast fusion technique has great potential, as it is time-saving, efficient and produces stable mst plants. This has been done in potato. From mst dihaploids, tetraploids are obtained and used in F_1 tuber production. Although hand pollinations are necessary for hybrid seed production on mst lines, the hybridity advantage can be perpetuated and propagated through tubers and clonal reproduction.

Tomato. Like barley, the tomato genome possesses numerous nonallelic *ms* genes, but no *fr* genes are known from its genome. Thus, the gc-mst known in tomato and its related species is hard to use commercially. Instead of *Lycopersicon esculentum* genome, the genome of *Solanum pennelli* is more sensitive of the S-cytoplasm found in tomato and its relatives. It is desirable to transfer the *L. esculentum* genome into the *S. pennelli* cytoplasm and transfer *fr* genes of *S. pennelli* into the tomato genome. This alloplasmic type can be used to develop hybrids based on gc-mst system using *S. pennelli* as seed parent. Unilateral incompatibility may, however, prevent this gene transfer.

Onion. Ten years after the discovery of gc-mst in a downy mildew-resistant onion by Jones and Clarke (1943), its commercial utilisation was started first in the USA and then spread all over the world. The main reason for its rapid spread was the over 100% bulb yield increase in hybrids, efficiency of *Fr* genes, stability of *fr* alleles, and above all because mst lines could be crossed in isolation plots with the aid of insect pollinators to any mft line without hand emasculation and roguing. Its utilization in Europe led to the identification of another S-cytoplasm source, the T-source. This increased the S-cytoplasm base and has enhanced the utility of gc-mst in commercial onion bulb production. Thus, the population structure of onion varieties was changed from a heterogenous to heterozygous and to homogeneous but still heterozygous, with a maximum amount of hybrid vigour. A maximum use could be made of both additive and nonadditive types of gene action in hybrid vigour.

Carrot. The scope of gc-mst-*fr* system in carrots is limited mainly due to instability, incomplete restoration and low NCP in mst lines which have poor nectar and aroma production reducing bee visitation and pollen transfer. However, a line found by Timin and Dobrutskaya (1981) appears promising but needs extensive testing.

Radish. The gc-mst systems found in Japanese and French radish varieties have been extensively used in hybrid production, as the hybrids are homogeneous and uniform and the plant roots are large, robust, solid, long-lasting and tasty. Due to low bee visitation, NCP in mst plants is low, leading to diminished seed production. The S-cytoplasm has been transferred to mustard, rapeseed and cabbage for use in their hybrid development.

Cabbage. Alloplasmic mst cabbage lines having radish S-cytoplasm are being developed into agronomically useful fast-breeding lines and used in hybrid production. Partial restoration, instability of mst expression and reduced chlorophyll development inhibit mst utilization on a commercial basis. Utilisation of a single locus, multi-allelic sporophytic incompatibility system in conjuction with the gc-mst system may pave the way for commercial hybrid production if agronomically superior normal mst lines are developed.

Pepper. Due to high positive heterosis in fruit size and yield, gc-mst systems have been used in common peppers to a great extent. Most pepper varieties are highly variable in almost all the characters of economic importance, and the stability of the S-cytoplasms is relatively low. This, coupled with very low NCP, has restricted the use of gc-mst in commercial fruit production.

3.10.5.3 Oil-Seed Crops

Sunflower. The gc-mst-*Fr* system being well established in sunflower, hybrid sunflowers are being increasingly used for seed production, as they yield 15–20% more than open-pollinated varieties. In addition, most hybrids are disease-resistant and exhibit high uniformity in height and maturity, which facilitates their mechanical harvesting.

Mustard and Rape. The S-cytoplasm of Japanese mst carrot was initially transferred to *Brassica napus* (rape) and *B. oleracea* and subsequently to *B. campestris* from *B. napus.* Plants of many *B. napus* mst lines have smaller flower and poorly developed nectaries, resulting in low insect pollination in them. However, a few lines have normal-sized petals and nectaries with fragrance. These have high breeding value. Similar mst lines developed in *B. campestris* and *B. juncea* (mustard) are being used for hybrid seed production in combination with S-alleles. High female fertility, absence of chlorosis, normal nectary function and multiple disease resistance, have been introduced in mst Chinese cabbage through recurrent backcrossing and selection (Williams and Heyn 1981).

Flax. Gc-mst lines available in flax have not been used commercially for hybrid production, as flowers of most of the agronomically useful mst lines remain closed and prevent bee visitation. The open corolla mst lines developed by Thompson and Miller (1978) have better breeding value, provided their corollas remain open all through flowering season.

Castor Bean. The gc-mst system needs to be exploited in castor hybrid bean production as it forms a rich source of seed oil. For this a stable and reliable C-line has yet to be developed.

3.10.5.4 Ornamental Plants

The mst is known in *Ageratum, Antirrhinum, Aguilegia, Begonia, Carthamus, Geranium, Gilia, Godetia, Iris, Pelargonium, Petunia, Streptocarpus, Tagetes, Vaccinium, Valeriana* and *Zinnia.* However, it has not been widely used in hybrid production except in *Geranium* and *Streptocarpus*, even though heterosis for flower and seed size is high. In composites, hybrid flowers are attractive and showy. Hence, use of mst in hybrid production of ornamentals is highly desirable. The main reasons for nonutilisation are the sparse availability of *fr* genes, instability of *fr* and *ms* gene expression and low NCP. Transfer of *Petunia hybrida* genome into cytoplasms of some *Petunia* species led to a few mst progenies. These hybrids were further backcrossed to several named varieties of *Petunia,* using the named varieties always as pollinator. In some BC progenies, mst plants segregated and from these, gc-mst lines have now been developed through continual BC's. Whereas some of these bear showy, attractive and delicate flowers that have a high market value, in a few floral development was suppressed, impeded or blasted. The main advantage of mst *Petunia* hybrids is that they tend to stay in bloom indefinitely and do not seed themselves out of flower as do some mft varieties.

3.10.5.5 Other Economic Plants

Cotton. Gc-mst was developed in cultivated cotton, *Gossypium hirsutum,* when its genome and perhaps some genes from *G. thurberi* were transferred to *G. anomalum* or *G. arboreum* cytoplasm. Likewise, mst plants occurred when the *G. hirsutum* genome was placed in the cytoplasm of a synthesized amphidiploid *G. arboreum* var. *neglectum.* In all these, the mst expression is sensitive to genotype (g), environment (e) and g × e interactions. However, the hybrids are more productive, luxuriant and vigorous, having fine staple, long lint length and strong and higher uniformity ratio, but NCP is low and breeding of mst cottons attractive to bees or bees attractive to cotton is required.

Sugarbeet. Many hybrid varieties using gc-mst sources have been developed and mst is used for commercial hybrid sugarbeet production throughout the USA and Europe. Single-, double- and triple-cross hybrids and triploids have been produced after using mst germplasm. New S-cytoplasm sources have been detected in the USA, the USSR, Europe and Japan and are being used profitably for hybrid beet production. Discovery of the monogerm trait has facilitated the mst hybrid production and mechanical harvesting.

Sugarcane. Transferring *Saccharum officinarum* genome into *S. spontaneum* cytoplasm leads to mst progeny that exhibits vigorous vegetative and low reproductive growth. Since the market for its seeds is nil and reproduction is by vegetative cuttings, use of gc-mst is desirable in this species, as its stem yields sugar. The introduction of agronomic superiority, disease and pest resistance is needed in mst A- or C-lines from which hybrids can be produced and propagated vegetatively.

Tobacco. Extensive use of alloplasmic mst has been made in hybrid tobacco production. The hybrids exhibit vegetative luxuriance, quick growth and high nicotine content. Specific parental combinations have yielded high positive heterosis and many lines of Burley tobacco using various mst lines have been developed and commercially utilized. Low NCP is a major obstacle to its commercial utilisation.

Citrus. In a large number of wild and cultivated lemons, oranges, citrons and limes, mst exists, but it has not been utilized in hybrid production because most of these are highly heterozygous due to outbreeding, apomixis and predominant vegetative propagation.

Alfalfa. In this forage crop, a sterility system like S-alleles mst and *fr*-genes and S-cytoplasms favouring cross pollination exists. These have been exploited for hybrid production using a variety of mating systems. Due to diminished bee visitation over mst plants, NCP is low and consequently hand pollinations must also be done. This makes F_1 seed uneconomical to randomly pollinated synthetic varieties. Being grown for forage, fertility restoration is unnecessary. Therefore, the search for *Fr* genes is not necessary.

Broad Bean. This protein-rich seed crop exhibits positive superior parent heterosis in grain yield, winter hardiness, seed protein content and yield stability. Therefore, the use of mst in hybrid production is highly desirable. Both gc-mst and g-mst are known in it, but they have not been utilised due to certain limitations, viz.: (a) very low pollen transport and pollination requiring petal movement, (b) permanent fertility restoration occurs following *Fr* gene introduction even after its substitution by *fr* genes later, (c) unstable mst expression, and (d) environmental influence preventing use of sterility systems in commercial hybrid breeding.

3.10.6 Breeding Problems and Remedies

Even though gc-mst has been utilized for hybrid production, its use is restricted by certain problems, most of which are akin to g-mst. Some of the special ones infrequent in g-mst but encountered in gc-mst plants are mentioned below.

i. *Low Natural Cross Pollination.* Extremely low NCP and inefficient ACP are the major obstacles for gc-mst exploitation in small flowered selfers like rice, wheat and barley. The gc-mst flowers usually do not expose stigma to receive pollen, and flax flowers remain closed. Such cleistogamous flowers do not even serve as pollen donors and genetic manipulation of anther extrusion becomes important in them.

In entomophilous plants, effective transfer of pollen from the C- to the A-line is hampered due to nonvisitation of insects to the A-line from the C-line. To overcome the limitations in cross pollination, judicious selection of phenotypes with appropriate flower morphology, planting design, choice of season and region for growth and manipulation of pollen vectors etc. should be helpful to optimise NCP. Concurrently, stigma receptivity of female and pollen shedding of male parent are of utmost importance and may require cultural, chemical or genetical manipulation.

ii. *Inbreeding Depression.* This is a major problem in allogamous plants and leads to reduced seed production and enhancement in floral and vegetative abnormalities. It can be overcome by using hybrids obtained after double and three-way crosses, as is a general practice in hybrid maize production. For this, various combinations of S-cytoplasms and *Fr* genes may be used in such multiple cross hybrids (Fig. 3.29). Alternatively, when hybrid fertility is not required, nonrestored double-cross hybrid seed may be mixed with seeds of the same cross combination, using the emasculation procedure.

iii. *Polyploid Breeding.* When hybrid fertility is of no importance, ploidy combinations are utilised. Triploid sugarbeet or begonias are conveniently produced by a 4x mst A line and 2x C-line. Usually, female fertility of 4x plants is better than their male fertility and haploid pollen is far more functional than diploid pollen.

iv. *Pleiotropy.* The polyploid nature (wheat, potato, sugarcane, etc.) and linkages complicate fertility restorations. In addition, many undesirable pleiotropic effects of the plasma type cause inferior agronomic hybrid performance especially in wheat, rye and potato. In maize, the yield advantage of gc-mst hybrids is in some instances partly due to the absence of pollen (Duvick 1965), as restoration of pollen production removes this advantage. Mixing seed of an identical hybrid combination made with detasseling and with mst enhances hybrid yield.

v. *Search for Naturally Occurring mst Plants.* Since to develop A- and B-lines in a desirable cultivar and then to maintain and perpetuate them is very time-consuming, it is feasible to search for naturally occurring S_{frfr} plants in developed varieties. Such genotypes have been found in some crops and have greatly facilitated the development of other mst lines and production of hybrid seeds on a commercial basis. Also no undersirable genes are introduced by crossing them with other cultivars.

CHAPTER 4

Chemical Male Sterility

4.1 Introduction

Utilisation of mst permits hybrid seed production and commercial exploitation of heterosis in crops where emasculation on a large scale is not possible. The occurrence of commercially utilisable g-mst or gc-mst is, however, uncommon in economically useful plants. Accordingly, induction of mst using mutagens has been tried and a few mst's have been obtained (Tables 2.3, 3.5), but such mutations are rare, recessive and associated with negative and agronomically inferior features. However, when a mst trait exists in a genotype, its transfer into an agronomically useful variety is possible through backcrossing. This transfer, however, without disturbance of the recipient variety genotype, is tedious, time-consuming and sometimes untenable due to hybrid sterility, linkage or other crossing barriers. Furthermore, whenever mst is naturally available or developed through gene or genome transfers, detection and roguing of mst's in g-mst and restoration of mst's in gc-mst are the major problems limiting the use of mst in commercial hybrid production. In addition, low NCP over mst's necessitates hand pollination supplementations. The instability of mst expression produces selfed seeds among hybrid seeds and reduces the market value of the commercial product. These difficulties have led to mst induction using certain chemicals (Tables 4.1, 4.2), which kill the male gametes, spores or organs and render the treated plant mst. These chemicals are known as male-gametocides or chemical hybridizing agents and the selective abolition of male sex is termed chemical male sterility (ch-mst). In the following, effectiveness, efficacy and efficiency of some chemicals tried as male gametocides are given. Their effects on plant development, growth, morphology and female fertility are indicated with a view to evaluate their breeding utility and economic feasibility in commercial hybrid production.

4.2 Chemical Gametocides

4.2.1 Dalapon (Dow Chemical Co., USA)

This is an effective rapidly absorbed postemergent herbicide which induces high to complete mst in *Abelmoschus, Coriandrum, Gossypium, Foeniculum, Hibiscus, Linum, Ranunculus, Triticum, Vicia* and *Vigna;* in others the sterility induced is partial (Table 4.3). In all of them, except in *Abelmoschus,* female fertility is highly reduced. It causes leaf necrosis, interferes with the mitotic activity of dividing

Table 4.1. Gametocides used in inducing male sterility

No.	Name/abbreviation	Chemical name
1	6-BAP	6-Benzylaminopurine
2	CCC	Chlorocholine chloride (2-chloroethyl) trimethylammonium chloride
3	CCDP	3-Carboxy-1-(p-chlocophenyl) -4, 6, dimethyl-2 pyridine
4	Colchicine	An alkaloid
5	Coumarin	A bicyclic organic compound
6	Dalapon	Sodium 2, 2-dichloroporpionate
7	DPX	3-(p-Chlorophenyl)-6-methoxy-3-triazine-2, 4-(1H, 3H) dione
8	2, 4-D	2, 4-Dichlorophenoxy-acetic acid
9	DMA	Dimethylarsenic acid
10	Ethrel	2-Chloroethyl phosphonic acid
11	EB	Ethidium bromide
12	Estrone	2,7-Diamino-10-ethyl-9-phenylphenantridium bromide
13	Embutox-4	γ-(2, 4-Dichlorophenoxybutyric acid)
14	GA	Gibberellic acid
15	IAA	Indole-3-acetic acid
16	IBA	Indole butyric acid
17	MH	Meleic hydrazide (1, 2-dihydropyridazine-3, 6-dione)
18	Mendok	Sodium 2, 3-dichloro-2-methyl-propionate
19	Morphactin	Fluorine-9-carboxylic acid derivative
20	NAA	Naphthalene acetic acid
21	Phosphon	Tributyl-2, 4-dichlorobenzylphosphonium chloride
22	RH	Sodium p-(chlorophenyl)-1-2-dihydro-4, 6-dimethyl-2-oxonicotinate (and its derivatives)
23	SADH	Succinic acid, 2, 3 dimethyl hydrazide
24	Sodium nucleate	Sodium salt of nucleic acid
25	TIBA	2, 3, 5-Tri-iodobenzoic acid
26	ZMA	Zinc methyl arsenate

Table 4.2. Chemical male sterility induction in some plants

Species	Chemical number[b]
Abelmoschus esculentus	17, 18, 14
Allium cepa	6, 14, 17, 21, 24
Antirrhinum majus	18
Avena sativa	7, 14, 17
Beta patula	18
B.vulgaris	7, 9, 10, 12, 18
B.webbiana	18
Brassica napus	26
B.oleracea	14
Cajanus cajan	17, 18
Cannabis sativa	1, 10
[a]*Callistephus sinensis*	17

Table 4.2 *(continued)*

Species	Chemical number[b]
Capsicum annuum	17, 18
Carthamus tinctorius	14
Citrullus vulgaris	18, 25
[a]*Commelina virginica*	17
[a]*Coriandrum sativum*	6, 8, 17, 18
Crotalaria juncea	17, 18
Cucumis melo	18, 20, 23
C.sativus	10, 23
Cucurbita maxima	17
C.pepo	17, 20
[a]*Datura alba*	17
[a]*Foeniculum vulgare*	6, 8, 17, 18
Glycine max	18
Gossypium hirsutum	6, 17, 18
Helianthus annuus	18, 25
[a]*Hibiscus cannabinus*	6, 17, 18
Hordeum vulgare	3, 10, 14, 22
Lactuca sativa	10
Linum usitatissimum	6, 18
[a]*Luffa cylindrica*	18
Lycopersicon esculentum	17, 18, 19, 25
Medicago sativa	18
[a]*Momordica charantia*	18
Nicotiana rustica	17, 18
N.tabacum	17, 18
[a]*Nigella damascena*	19
Oryza sativa	7, 10, 22
Pennisetum typhoides	6, 7, 9, 11, 18
Petunia alba	18
Phaseolus aureus	18
P.mungo	17
Pisum sativum	18
Ranunculus muricatus	6, 17
R.sceleratus	17
Ricinus communis	14
[a]*Rubus idaeus*	17
Sesamum indicum	6, 17, 18
Silene pendula	15
Solanum melongena	8, 10, 18
S.nigrum	18, 25
Sorghum vulgare	4, 6, 17, 18
Trifolium pratense	18
T.repens	18
[a]*Trigonella foenum-gracum*	5, 17, 18
Triticale	7
Triticum aestivum	6, 7, 10, 12, 17, 22
Vicia faba	5, 6, 16, 17, 20
[a]*Vigna mungo*	6, 17, 18
V.sinensis	17, 18
Zea mays	7, 14, 17, 18

[a]Only chemical male sterility known, in others genic and/or gene-cytoplasmic male sterility also known.
[b]Refer to Table 4.1 for chemical numbers.

Table 4.3. Dalapon-induced male sterility

Species	References	Male sterility induction	Female fertility reduction	Side effects
Abelmoschus esculentus	Dubey and Singh (1968)	Complete	Low	nk
Capsicum annuum	Chauhan (1980)	Highly variable	Variable	nk
Coriandrum sativum	Dubey and Singh (1968)	Complete	High	nk
Crotalaria juncea	Kaul and Singh (1967b)	Partial	High	Vegetative and reproductive growth reduced
Foeniculum vulgare	Dubey and Singh (1968)	Complete	High	nk
Gossypium hirsutum	Scott (1961), Meyer et al. (1958)	High	nk	nk
	Dubey and Singh (1968)	Complete	Very high	nk
Hibiscus cannabinus	Dubey and Singh (1968)	Complete	Low to high	nk
Linum usitatissimum	Rustagi and Mohan Ram (1971)	High	High	High toxicity, reduced vegetative and reproductive growth
Pennisetum typhoides	Miller (1961)	Partial	Very high	Highly phytotoxic
Ranunculus muricatus *R.sceleratus*	Chauhan et al. (1976)	Complete	Very high	Reduced vegetative and reproductive growth
Sesamum indicum	Chauhan and Singh (1971)	Partial	High	Highly toxic
Sorghum vulgare	Miller (1961)	Partial	High	nk
Triticum aestivum	Kaul and Singh (1967b)	Complete	High	Reduced vegetative and reproductive growth
Triticum aestivum	Porter and Weise (1961)	Partial	High	Highly toxic
Vicia faba	Kaul and Singh (1976)	High	High	nk
Vigna sinensis	Kaul and Singh (1967b)	High	nk	nk

meristem, and induces toxicity in most of these species, reducing their vegetative and reproductive growth and vigour. Therefore, this chemical is not promising for use as a male gametocide in commercial breeding. In *Crotalaria juncea*, 0.25% and 0.50% dalapon treatments cause exudation of pollen cytoplasm, lower concentrations in *C. juncea*, *Phaseolus aureus* and *P. mungo* caused heterostyly (Kaul and Singh 1967b). The mst induced is unstable and of a short duration.

4.2.2 DPX-3778 (E.I. du Pont de Nemours & Co., Wilmington, Delaware, Ohio, USA)

This prevents pollen shedding by inhibiting anther dehiscence. Microsporogenesis in treated plants has not been investigated, but in most of the plants in which it acts as male gametocide (Table 4.4), pollen formation is normal and thus male meiosis is not influenced by the chemical. Likewise the chemical does not reduce female fertility but delays floral development, emergence, opening and maturity (Laible 1974; Johnson and Brown 1976; Theurer 1979). The delayed anthesis necessitates synchronisation of pollination with female receptivity. In wheat the first-formed spikes emerging after treatment are affected slightly to partially and hence selfing occurs. In *Triticale*, in which 9 kg ha^{-1} of this chemical is effective (Johnson and Brown 1976), considerable anther dehiscence occurs in the glassine bags. In rice,

Table 4.4. DPX-induced male sterility[a]

Species	References	Male sterility induction	Female fertility reduction	Side effects
Avena sativa	Johnson and Brown (1976)	Partial	Partial	Reduces shoot height, spikelet number and grain yield
Beta vulgaris	Theurer (1979)	Partial	nk	Causes flower closure leaf rolling, chlorophyll destruction, necrosis and inhibited plant growth
Oryza sativa	Long et al. (1973)	Partial	nk	nk
Pennisetum typhoides	Hanna (1977)	Partial to complete	nk	nk
Triticum aestivum	Johnson and Brown (1976, 1978)	Nearly complete	Slight	Delays anthesis, reduces plant height and is later toxic
Triticale	Johnson and Brown (1976)	Nearly complete	High	nk
Zea mays	Long et al. (1973), Laible (1974)	Complete to partial	Nil	Delays anthesis

[a]High genotype specificity in dose and duration.

pearl millet and sugarbeet (Long et al. 1973; Hanna 1977; Theurer 1979), foliar application of the chemical likewise prevents anther dehiscence in some varieties but not in others. The flowers of sugarbeet do not open and leaf rolling, chlorophyll destruction and plant necrosis occur in the treated plants. In oats, the chemical is totally ineffective (Johnson and Brown 1976) and causes reduction in shoot height, spikelet number and seed yield. Like other male gametocides, it also exhibits genotype specificity within a species. Results with this chemical are variable and vary with the stage of plant development. In subsequent years, ovular fertility reduction and phytotoxicity in the treated plants persists (Johnson and Brown 1978). The effects of this chemical on animals and humans are not known, as the chemical is translocated and stored in the seeds of treated plants. All these factors limit its use as a male gametocide for commercial hybrid seed production.

4.2.3 2,4-D

This is an excellent selective post-emergence synthetic herbicide used for the control of herbaceous annual dicot weeds in crop fields, grasslands, lawns etc. Its 0.5% aqueous solution given as foliar spray to run off before and at bud initiation led to complete pollen sterility in *Coriandrum* and *Foeniculum* and to partial sterility in *Nicotiana* (Dubey and Singh 1968). In these, the treatments reduced ovular fertility considerably (Table 4.5). However, for producing a limited amount of hybrid seed in *Coriandrum* and *Foeniculum*, the chemical could have been useful as a male gametocide, as only 20–30% of ovular fertility is reduced by it. However, fruit production is much lower. In *N. tabacum*, induced mst is about 60% but the female fertility reduction is 35%. However, when induced mst is high, female sterility is still higher (Dubey and Singh 1968). Thus the chemical is not useful as a male gametocide in these plants.

Table 4.5. 2, 4-D-induced male sterility[a]

Species	References	Female sterility reduction	Side effects
Coriandrum sativum	Dubey and Singh (1968)	Low	nk
Foeniculum vulgare	Dubey and Singh (1968)	Low	nk
Nicotiana rustica	Jos and Singh (1967b)	Complete	Highly toxic, reduced vegetative and reproductive growth
Nicotiana tabacum	Dubey and Singh (1968)	Low	nk
	Jos and Singh (1967b)	Very high	Highly toxic
Solanum melongena	Jyotishi and Hussain (1968), Helal and Zaiki (1981)	Low	nk
	Diki and Anikeenko (1975)	Very high	nk

[a]Male sterility induction almost complete in all cases except in *N.tabacum* in which it is partial.

4.2.4 Ethrel. (Ethephon, CEPA, Amchem Products Inc., Ambler, Pennsylvania, USA)

This is an ethylene-generating synthetic compound that acts as a plant growth regulator having a role in abscission, apical dominance, dormancy breaking and sprouting, fruit thinning, and as a harvest aid in many fruits. In cucumber and squashes, it enhances femaleness by increasing female floral number. The gametocidal concept of ethrel was given by McMurray and Miller (1969) and Robinson et al. (1969), who obtained an increase in pistillate flower number in monoecious cucumber after ethrel treatments. Its potential as a rapid and flexible system to produce hybrid wheat was indicated by Rowell and Miller (1971), who found that application of 2000 to 3000 ppm ethrel at early, mid and late boot stages in wheat produced up to 100% mst. However, all these treatments cause plant dwarfing and female fertility reductions to varying degrees not only in wheat but also in other plants (Table 4.6). Effectiveness of ethrel as a male gametocide depends upon genotype concentration and application time (Bennett and Hughes 1972; Hughes et al. 1974; Vandam 1974). These specificities make its use highly sensitive. In wheat, ethrel application before male meiosis initiation leads to mst without significant impairment of female fertility (Hughes et al. 1974). Treatment at other times diminishes female fertility considerably. Furthermore, each variety, especially of wheat, rice and barley, is sensitive to a specific ethrel dose and duration. Due to these variables, inconsistent and inefficient male sterilizing effects with ethrel on wheat were obtained by Bonali et al. (1972), Vandam (1974), Fairy and Stoskopf (1975) and in barley and pea by Kaul (this Vol.). In these, plant growth and female fertility were reduced by the most effective male sterilizing ethrel treatment. Ethrel-treated bread wheat plants exhibit reduced female fertility, delayed vegetative growth, height reductions and imperfect heading (Dotlacil and Apltauerova 1978). In rice, ethrel induced partial male and female sterility and the higher doses that enhanced mst caused stunted growth, severe leaf burning, suppression of panicle development and its emergence, defoliation, mortality of tender tillers and yellowing of the whole plant, leading to its death (Perez et al. 1973; Parmer et al. 1979; this Vol.). In *Cannabis sativa*, ethrel treatment causes drooping and epinasty of leaves and leads to the development of female, intersexual and abnormal female flowers (Mohan Ram and Jaiswal 1970). High toxicity accompanies high chemical dosage treatments. At 960 ppm treatment, 70% female flowers on male plants are developed, but complete mst is not obtained. A similar suppressive effect on male fertility was found by Heslop-Harrison and Heslop-Harrison (1958) in long-day plants of *Silene pendula*.

Ethrel decomposes spontaneously in aqueous solutions and yields ethylene which accomplishes hormonal interactive functions by diffusing rapidly through plant tissues. How it induces mst is not known. Its known regulatory activities are, however, so varied that it defies classification (Bidwell 1979). Indirect but circumstantial evidence indicates that ethylene acts through the regulation of differential gene action. In carrot, ethylene treatment causes two fold increase of the average polyribosome size and leads to the appearance of new mRNA's (Christoffersen and Laties 1982). In cucumbers and muskmelon, ethylene

Table 4.6. Ethrel-induced male sterility

Species	References	Male sterility induction	Female fertility reduction	Side effects
Beta vulgaris	Hecker et al. (1972), Hecker and Smith (1975)	Partial	High	Phytotoxic
Cannabis sativa[a]	Mohan Ram and Jaiswal (1970)	Partial	Nil	Phytotoxic, causing drooping and epinasty of leaves
Cucumis sativus[a]	Minina and Tylkina (1942), Rudich et al. (1969, 1970)	High to nearly complete	Nil	Reduced plant growth and vigour and deformed fruits
	Iwahori et al. (1970)	High to nearly complete	Nil	nk
	Augustine et al. (1973)	High to nearly complete	Nil	nk
Cucurbita maxima[a]	Robinson et al. (1970)	Nearly complete	Nil	nk
C.pepo	Rudich et al. (1969, 1970)	Nearly complete	Nil	Reduced plant growth and development
	Borghi and Pironi (1976)	Complete	Nil	—
Hordeum vulgare	Stoskopf and Law (1972), Law and Stoskopf (1973)	Low	High	Many morphological ill effects
	Kumar et al. (1976), Verma and Kumar (1978)	Nearly complete	Very low	Nil
Lactuca sativa	Han and Lee (1972)	High	Low	nk
Oryza sativa	Perez et al. (1973), Cheng and Huang (1978), Parmar et al. (1979), Chan and Cheah (1981), this Vol.	Partial to high	High	Severe phytotoxicity and plant death with high dosage
	Wang and Que (1981)	High	High	nk
Solanum melongena	Helal and Zaiki (1981)	Complete	Low	nk
Triticum aestivum	Rowell and Miller (1971), Hughes et al. (1974)	Complete	Low	Dwarfing, poor spike emergence
	Bennett and Hughes (1972), Hughes et al. (1974)	Complete	Low	Dwarfing, poor spike emergence
	Stoskopf and Law (1972), Bonali et al. (1972), Vandam (1974), Fairey and Stoskopf (1975)	High	High	Dwarfing, increased tillering and high floret sterility
T.aestivum	Jan and Rowell (1981)	Low	nk	nk
	Dotlaĉil and Apltauerova (1978)	Complete	Low	Reduced and slow vegetative growth, imperfect heading

[a]Female floret number increased, gametocidal effects variable.

regulates femaleness and its production is much lower in androecious and monoecious phenotypes compared to hermaphrodite and gynoecious types (Rudich et al. 1976). Thus, high ethylene evolution is associated with femaleness through suppression of staminate buds. Removal of ethylene by hypobaric ventilation or enrichment of the atmosphere with CO_2 (a competitive inhibitor of ethylene) promotes maleness in gynoecious muskmelon (Byers et al. 1972). Exogenous ethrel application (Rudich et al. 1969) or enrichment with ethylene (Minina and Tylikina 1942) induce femaleness in monoecious cucumbers. Short-day conditions enhance female expression and ethylene evolution (Rudich et al. 1972).

The use of ethrel as a male gametocide is limited. In cucurbits, it suppresses but does not eliminate male floral development completely. In cereals, its concentration and application time are highly critical and specific to each cultivar. The mst induction is incomplete and the dosage that induces high mst in some wheat and rice cultivars inhibits spike emergence and impairs female fertility. GA_3 additions may circumvent this effect, but economically its use is not feasible as the chemical is expensive and induces undesirable morphological anomalies like shoot elongation and lodging. Moreover, imperfect heading and its ineffectiveness in late-formed tillers reduces the value of ethrel as a gametocide in wheat and rice. Ethylene applications induced high male and female sterility in rice (Wang and Que 1981). In *C. pepo*, its treatment increases early yield in summer cultivars but not the total yield (Sams and Krueger 1977). The fruit shape deteriorates with the treatment. Its applications cause epinasty, senescence and abscission in rice, oat, sorghum, pea, gram and beans. Therefore, in these crops, ethrel cannot be used as a male gametocide for commercial hybrid production.

4.2.5 Gibberellic Acid

Gibberellins, the endogenous regulators of growth, differentiation and development and almost ubiquitous in the plant kingdom, causing stimulation of cell elongation and division, induce partial to complete mst in some plants (Table 4.7). In plants other than maize, castor and pepper the female fertility, wherever known, is impaired. The excessive internode elongation caused by GA treatments weakens the treated plants, which lodge considerably. In maize, its treatment suppresses staminal formation and induces female floret development at places where male florets normally develop. However, the mst accompanies stem weakness within the tassel and the stems either break off or lodge to the ground. This considerably limits its use in breeding. Use of GA_3 as a male gametocide becomes more limited especially in maize because of the critical timing and concentration (foliar spray of 5 mg GA_3 1–3 days prior to meiotic initiation). Genetic diversity and environmental variability prevent large populations from uniformly reaching a suitable physiological state for GA_3 treatment. When GA_3 is applied to an mft population,, the earlier maturing plants shed normal pollen and the late ones collapse due to weakened stems. Its use is further limited by various practical difficulties in accurately predicting the exact stage of reproductive (especially male) development. Usually floral dissections, anther size and phenotypic examination of anthers may suffice to give an idea of microsporogenous developmental stage in plants, but

Table 4.7. GA-induced male sterility

Species	References	Male sterility induction	Female fertility reduction	Side effects
Allium cepa	Meer and Bennekom (1973), Meer (1978)	High	High	nk
Avena sativa	Frohberg and Frey (1970)	High	To half	Excessive internode elongation and reduced spikelet size
Brassica oleracea	Meer and Vandam (1979)	Variable	nk	nk
Capsicum annuum	Sawhney (1981)	Nearly complete	Not much affected	Staminal feminization, extraovular formation, fruit deformation and distortion
Carthamus tinctorius	Yermanos and Knowles (1960)	Complete	Complete	Internode elongation and chlorosis
Helianthus annuus	Schuster (1961)	Nearly complete	Very high	nk
	Seetharam and Kumari (1975)	Nearly complete	High	Shoot elongation, early flowering
	Frank and Koves (1977), Vear (1981)	Complete	Very high	—
	Piquemal (1970), Anashchenko (1972a,b)	Complete	Low	nk
Hordeum vulgare	James and Lund (1960)	Complete	nk	nk
	Paleg and Aspinali (1958)	Only in tips	nk	nk
Lactuca sativa	Eenink and Vereijken (1978a)	Complete	nk	nk
Ricinus communis	Shifriss (1961b)	Partial	Nil	Increased internode length, reduced flower number, delayed flowering
Zea mays	Nickerson (1959), Hansen et al. (1976)	Complete	Nil	Increased height, thin, weak, lodging shoots
	Wittwer and Bukovac (1957), Nelson and Rossman (1958), Cherry et al. (1960)	Partial	Nil	nk

examination of anther smears has to be made for accurate predictions in most of the GA_3-treated plants. In maize, the tassel near the meiosis period is hidden deep within the whorl and develops very fast during and after meiosis. Although methods of predicting physiological maturity using vegetative development or some other secondary features, such as leaf number, flower form etc., are known in some crop plants like wheat, barley and rice, in many others they are still not determined. Even the available methods are not reliable because of environmental influence and differential genotype-environment interaction.

Ricinus communis is a monoecious, potentially everflowering perennial. Its main stem and lateral shoots terminate in racemes that develop both staminate and pistillate florets. Suppression of male flower development and regulation of sex by GA_3 treatment was documented by Shifriss (1961b), who found that repeated GA_3 treatments cause some racemes to produce only female flowers. However, the treatment led to a substantial reduction in flower number and seed yield. Spraying GA_3 three times in concentrations of 1500, 1500 and 500 ppm at two-, three- and four-leaf stages, respectively, led to 100% pistillate flower development in the first three sequential racemes but not in the later-formed ones. Also, the treatment caused an increase in internode elongation, decrease in leaf size, suppression of shoot development at some lateral leaf axils and delay in flowering. In sunflower, GA_3 applications either retard anther development and exsertion (Schuster 1961) or suppress anther development considerably (Seetharam and Kumari 1975). Female fertility of the treated plants is highly reduced and seed set drops to 20% of the control (Schuster 1961). Thus in both the oil-yielding plants, sunflower and castor, GA_3 is not an effective gametocide for use in commercial hybrid seed production.

Gibberellins move quickly from the application site to other plant parts, but the site of GA_3-controlled sex expression is not exactly known. Impediment of nutrient flow to developing anthers occurs in maize and pea plants after GA_3 applications. This may lead to mst in them. However, once meiosis is completed, GA_3 applications are ineffective in preventing pollen and anther development. In maize and wheat, the chemical induces alterations in peduncle, lemma and palea. This indicates that the biochemical alterations are not limited to anthers only but to other plant organs as well. The stage preceding male meiosis is sensitive to GA_3 treatments in barley, maize, wheat and rice. In barley (Batch and Morgan 1974) and wheat (Fisher 1972), staminal development is inhibited by reducing the photoperiod given during this critical premeiotic stage. In these, the role of GA is not implicated directly as stimulation of GA synthesis occurs by light exposure.

Whereas in certain plants GA suppresses maleness (Table 4.7), in mst mutants of barley (Kasembe 1967), maize (Nickerson 1960) and tomato (Sawhney and Greyson 1973b) it induces maleness without impairing the female fertility. In many monoecious and gynoecious cucumbers, it enhances maleness (Rudich and Halevy 1974) and induces the development of female flowers in a gynoecious line of cucumis (Shifriss 1961b). Similarly, whereas in a single-gene tomato mutant, Stamenless-2 (sl_2sl_2), stamens are developed after GA_3 applications, its treatment to normal tomato plants led to induction of polycarpellery ovaries with no apparent effect on stamen development (Sawhney and Greyson 1973b). Conflicting results after GA treatments are known in *Cannabis* (Herich 1960; Heslop-Harrison and Heslop-Harrison 1961), *Luffa* (Bose and Nitsch 1970; Krishnamoorthy 1972) and

Morus (this Vol.). Moreover, the effectiveness of different GA types in inducing mst varies considerably with genotypes, environmental factors (soil fertility, temperature, photoperiod), plant developmental stage and GA type. Therefore, no specific GA concentration or its application mode can be recommended that can male sterilize a crop variety effectively and efficiently without any side effects.

4.2.6 Maleic Hydrazide

A heterocyclic base is a growth-retardant and mitotic poison that has been tried as a male gametocide in some plants (Table 4.8). In most of them, it induces female sterility as well. In addition, growth and vigour are highly reduced as the chemical is toxic. In general, it causes killing or temporary inhibition of terminal buds, narrowing of leaves, chlorosis, water absorption interference (Moore 1950). Its gametocidal effect is genotype-, environment- and plant development stage-dependent (McIlrath 1957). This explains its nonfunctionability in some maize genotypes, as found by Warren and Dimock (1954). Keeping in view its toxic effects like induction of plant damage, female fertility reductions, floral abscission and variability in effectiveness, its use as a male gametocide for commercial hybrid seed production is not warranted.

4.2.7 Mendok (FW-450, Rohm and Haas, Philadelphia, USA)

High to complete mst is caused by this chemical in most of the species listed in Table 4.9, but in *Beta* and *Nicotiana*, mst induced is partial, and in *Trifolium* the induction is variable. In many species, the effect of mendok on female fertility is not known. In others where it is known, the fertility is highly impaired. In *Antirrhinum majus nanum* L., it did not considerably impair female fertility (Kho and De Bruyn 1962), but in another snapdragon variety it caused p-mst and high female sterility (Sink and Walker 1964). In *Nicotiana rustica*, the chemical induced complete mst in apical and middle inflorescence regions but not in the basal regions (Jos and Singh 1967a). In wheat, it damaged the plant fully without inducing significant mst (Chopra et al. 1960; Porter and Weise 1961; Kaul and Singh 1967d). Whereas in *Phaseolus aureus*, *P. mungo* and *Crotalaria juncea*, heterostyly and anther dehiscence occurred following this chemical treatment (Kaul and Singh 1967b), in *Trigonella* and *Medicago* in situ pollen germination occurred but the pollen was not shed (Kaul and Singh 1967b; Miller and Hittle 1963). The germination is either due to the stimulatory effect of the chemical on the pollen growth within the anther cavity or to the prevention of normal dehiscence which makes mature pollen germinate in situ. In cotton, within 15 min of spraying, a measurable amount of the chemical is absorbed and it moves more readily into younger leaves and flower buds than into mature leaves (Mc Rae and Usdin 1958). In anthers, it is selectively accumulated (Scott 1960a, 1961) and combines with co-enzyme A and competes for site in pantothenate synthesis. Additions of D-ribose or other precursors of co-enzyme A reverse the effects of this gametocide. In many plants, the chemical is toxic, and reduces vegetative and reproductive growth, female fertility and fruit set. In *Cajanus*, *Glycine* and *Linum*, floral bud abscission occurs on chemical treatment. Therefore its use as male gametocide is not warranted.

Table 4.8. Maleic hydrazide-induced male sterility

Species	References	Male sterility induction	Female fertility reduction	Side effects
Abelmoschus esculentus	Dubey and Singh (1968)	Complete	Low	nk
Allium cepa	Chopra et al. (1960[a])	Complete	nk	nk
	Kaul and Singh (1967c,d)	Complete	Much high	nk
	Meer and Bennekom (1973)	Complete	Complete	toxic
Avena sativa	Frohberg and Frey (1970)	Complete	High	Reduction in anther, stigma and top growth, toxicity high
Cajanus cajan	Kaul and Singh (1967b)	Complete	Very high	High floral abscission
	Kaul and Singh (1967a)	–	–	nk
Callistephus sinensis	Moore (1950)	Complete	High	nk
Capsicum annuum	Chauhan and Singh (1972), Chauhan (1980)	Variable to complete	High	–
Commelina virginica	Moore (1950)	Partial	High	Floral bud abortion
Coriandrum sativum	Dubey and Singh (1968)	Complete	nk	nk
Crotalaria juncea	Kaul and Singh (1967a,b)	Complete	Complete	Highly phytotoxic
Cucurbita maxima	Chauhan (1976)	Complete	nil	nk
C.pepo	Wittwer and Hillyer (1954)	Suppresses staminate buds	nil	Female flower number reduced
Datura alba	Chauhan and Singh (1972)	Complete	nk	nk
Foeniculum vulgare	Dubey and Singh (1968)	Complete	Very high	nk
Gossypium hirsutum	Singh (1964), Meyer et al. (1958), this Vol.	Partial to complete	High	Phytotoxicity
Hibiscus cannabinus	Dubey and Singh (1968)	Complete	Very high	nk
Lycopersicon esculentum	Rehm (1952)	Complete	Nil	nk
	Jain (1956), Chopra et al. (1960)	Complete	Very high	nk
Nicotiana rustica	Jos and Singh (1976b)	High	High	High toxicity
N.tabacum	Jos and Singh (1967a)	High	High	Considerable floral abscission, stunted plant growth
Phaseolus mungo	Kaul and Singh (1967b)	High	nk	nk
Ranunculus muricatus	Chauhan et al. (1976)	Complete	Very high	Highly reduced vegetative and reproductive growth

Table 4.8. *(continued)*

Species	References	Male sterility induction	Female fertility reduction	Side effects
R.sceleratus	Chauhan et al. (1976)	Complete	Very high	
Rubus idaeus	David (1950)	Partial	High	Delayed flowering
Sesamum indicum	Chauhan and Singh (1971)	Partial	High	Highly toxic
Sorghum vulgare	Mcllrath (1957)	Nearly complete	Highly reduced	Reduced plant growth
Trigonella foenum-graecum	Kaul and Singh (1967b)	High	High	Reduced plant growth and yield
Triticum aestivum	Hoaglund et al. (1953)	Complete	Very high	Reduced height and top growth
	Chopra et al. (1960)	Complete	Slight[a]	nk
	Porter and Weise (1961)	Complete	Very high	Severe plant damage, delayed maturity
	Kaul and Singh (1967d)	High	High	nk
Vicia faba	Kaul and Singh (1967b)	Low	High	
	Awasthi and Dubey (1983)	High	High	Highly toxic, reduced vegetative and reproductive growth
Vigna mungo	Kaul and Singh (1967b)	High	nk	nk
Vigna sinensis	Kaul and Singh (1967b)	High	nk	nk
Zea mays	Denisen and Haber (1950)	Complete	Nil	nk
	Eskew and Willard (1950)	Partial	nk	nk
	Moore (1950), Naylor and Davis (1950)	Complete	Nil	nk

[a]Actual seed set not known.

Table 4.9. Mendok (FW-450) – induced male sterility

Species	References	Male sterility induction	Female fertility reduction	Side effects
Abelmoschus esculentus	Dubey and Singh (1968)	Complete	Low	nk
Antirrhinum majus nanum	Kho and DeBruyn (1962)	Complete	Nil	Highly phytotoxic
A.majus	Sink and Walker (1964)	Partial	Very high	nk
Beta patula[a]	Wit (1960)	High to partial	nk	nk
B.vulgaris[a]	Butterfass (1960), Ruebenbauer and Schultis (1960), Wit (1960), Salmon (1963), Bronckers and Amand (1966), Kinoshita (1971)	Partial	Nil to very high	Phytotoxic and high yield reduction
B.webbiana[a]	Wit (1960)	Partial	nk	nk
Cajanus cajan	Kaul and Singh (1967b)	Complete	Very high	Considerable floral bud abscission
Capsicum annuum	Hirose (1965), Markus (1968)	Considerable	Very high	nk
Citrullus vulgaris	Hensz and Mohr (1959)	High	Low	nk
Coriandrum sativum	Dubey and Singh (1968)	Complete	nk	nk
Crotalaria juncea	Kaul and Singh (1967b)	Complete	High	Reduced vegetative and floral growth, high phytotoxicity
Cucumis melo	Pundir and Singh (1965)	Complete	nk	nk
Foeniculum vulgare	Dubey and Singh (1968)	Complete	High	nk
Glycine max	Starnes and Hadley (1962)	Nearly complete	nk	Highly toxic, inhibited vegetative and reproductive growth, floral bud abscission
Gossypium hirsutum	Eaton (1957)	High, unstable	Nil	nk
	Roux and Chirinian (1959)	High	nk	nk
	Pate and Duncan (1960)	Complete	Complete	nk
	Scott (1961)	High	nk	nk
	Richmond (1962)	High	High	nk
	Dubey and Singh (1968)	Complete	Very high	nk
Helianthus annuus	Schuster (1961)	Very low	nk	nk
Hibiscus cannabinus	Dubey and Singh (1968)	Complete	Very high	nk
Linum usitatissimum	Dubey and Singh (1963), Rustagi and Mohan Ram (1971)	Nearly complete	Very high	Highly toxic, chlorosis, burning of foliar and floral buds and flower closure

Table 4.9. *(continued)*

Species	References	Male sterility induction	Female fertility reduction	Side effects
Luffa cylindrica	Chauhan (1980)	Complete	nk	nk
Lycopersicon esculentum	Moore (1959, 1964)	Complete	Medium	nk
	Walkof (1959)	Complete	nk	nk
	Aung and Byrne (1977)	High	Medium	nk
Medicago sativa	Pedersen (1959), Srivastava (1961), Miller and Hittle (1963)	High	High	High phytotoxicity, leaf burning
Momordica charantia	Chauhan (1980)	Complete	nk	nk
Nicotiana rustica	Jos and Singh (1967b)	Variable and differential	nk	nk
N.tabacum	Jos and Singh (1967a)	High	Low	nk
Pennisetum typhoides	Sharma (1979)	High	Nil	nk
Petunia[a] sp.	Sink and Gunesch (1966)	Considerable	Very high	Stunted growth, shoot fasciation, corolla tube splitting, pistil malformation
Phaseolus aureus	Kaul and Singh (1967)	Complete	nk	nk
Pisum sativum	This Vol.	High	High	Phytotoxicity
Sesamum indicum	Kumar and Singh (1963), Chauhan and Singh (1971)	Complete	nk	nk
Solanum melongena	Nasrallah and Hopp (1963)	Nearly complete	nk	nk
S.nigrum	Dulieu (1963)	Partial	High	Altered floral morphology
Sorghum vulgare	Sayers (1959), Miller (1961), this Vol.	Partial	Complete	nk
Trifolium pratense	Wit (1960), Summing (1959)	High, variable	High	Reduced plant growth
T.repense	Wit (1960)	Complete	nk	nk
Trigonella foenum-graceum	Kaul and Singh (1967b)	Complete	High	Reduced growth and development
Veronica alba, V.candida, V.rubra	This Vol.	Partial	High	Phytotoxic
Vicia faba	Kaul and Singh (1967b)	High	High	nk
Vigna mungo	Kaul and Singh (1967b)	High	nk	nk
V.sinensis	Kaul and Singh (1967b)	High	nk	nk
Zea mays	Cameron and Eaton (1959)	Low-high	High	nk

[a]Results highly variable.

4.2.8 RH-531, -532, -2956 (Rohm and Haas, Philadelphia, USA)

These are synthetic pyridines which act as chemosterilants and were developed as early-generation wheat gametocides. They induce mst in wheat and barley (Table 4.10). In addition, they cause spike compactness, poor flower opening and reduce female fertility and grain yield. In wheat, RH-531 treatment did not favour outcrossing and 98% of the open-pollinated seeds were selfs. Additionally, many undesirable cytogenetic effects are caused by these chemicals in wheat (Rajendra and Bates 1981). Therefore, these chemicals do not have immediate prospects for use as male gametocides in cereals.

Table 4.10. RH-induced male sterility

Species	References	Male sterility induction	Female fertility reduction	Side effects
Hordeum vulgare	Wang and Lund (1975)	Nearly complete	Low-medium	nk
	Hockett and Feltner (1978)	Partial	Partial	Decreased grain yield, kernel size, shoot height, spike compact
Oryza sativa	Perez et al. (1973)	High	High	Strong phytotoxic effects
Triticum aestivum	Jan et al. (1974, 1976), Jan and Rowell (1981), Rajendra and Bates (1981)	Nearly complete	Low	Spike compact, poor flower opening, undesirable cytogenetic effects, hazardous genetically

4.2.9 SADH

This is a growth retardant that reduces the auxin level, either by enhancing auxin destruction (Halevy 1963) or by interfering with its biosynthesis (Reed et al. 1965; Williams and Stahly 1970), and inhibits ethylene production (Abeles and Rubinstein 1964; Looney 1968). It enhances female flower development in muskmelon when sprayed alone or in combination with ethrel (Table 4.11) and suppresses male floral production temporarily for 2–3 weeks completely in some and partially in other muskmelon and cucumber varieties. In some, female flower

4.11. SADH-induced male sterility[b]

Species	References	Male sterility induction	Side effects
Cucumis melo	Halevy and Rudich (1967)	High	Plants dwarfed, tendrils suppressed
C.melo[a]	Rudich et al. (1969)	Nearly complete	nk
C.sativus[a]	Rodriquez and Lambeth (1972)	Nearly complete	Reduced plant growth

[a]SADH + Ethrel. [b]Female fertility reduction nil.

number is also reduced. Due to these anomalies, its use as male gametocide is limited. However, SADH increases tolerance to stress conditions (Halevy 1967), delays senescence, and thus increases longevity (Halevy and Wittwer 1966). Therefore, if used in combination with ethrel or some other plant hormones, apart from its male gametocidal properties, it may improve the utility value of seed parent plants by enhancing their tolerance to stress and increase the longevity and keeping quality of their fruits.

4.2.10 Other Chemicals

Apart from the above-mentioned often tried chemicals, the effectiveness of many others used sporadically as male gametocide has been tried (Table 4.12). A few of these warrant description because of their initial and limited partial success.

i. *6-BAP* (6-Benzylaminopurine). This is a synthetic cytokinin and a plant growth regulator. Addition of 5 mg/1,6-BAP to the nutrient medium in which *Cannabis sativa* seedlings were cultured led to the complete suppression of male florets (Chailakhyan and Khryanin 1980). No staminate flower suppression occurs when the chemical is foliar-sprayed prior to floral initiation. Within a week of treatment leaf curling and hardening occur.

ii. *Coumarin.* This is an antiauxin, and a metabolic inhibitor that inhibits longitudinally directed processes but stimulates radially directed ones. In *Trigonella* 0.25, 0.50 or 0.75% of this chemical induces complete mst (Kaul and Singh 1967b). This is accompanied by reductions in shoot height, female fertility, grain yield and considerable delay in flowering and fruiting. Almost similar results were obtained by Awasthi and Dubey (1983) in *Vicia faba.*

iii. *Dimethylarsenic Acid* (Cacodylic acid). 200 ppm aqueous foliar spray causes pollen death in sugarbeet (Hecker *et al.* 1972). The chemical is highly phytotoxic and damages ovular fertility considerably. Nearly similar effects are known in pearl millet (Powell and Taylorson 1967), but the pollen abortion induced is low.

iv. *Ethidium bromide.* Known to cause cytoplasmic mutations (Table 3.30), this chemical causes partial to complete mst in pearl millet (Burton and Hanna 1976). In rice and pea this chemical induces neither cytoplasmic mutations nor mst (this Vol.).

v. *Estrone.* This is a female sex hormone and has a limited and low effect on wheat pollen (Fendal 1967). Iv sugarbeet, 30 ppm application through the leaf mid vein at pre- and postmeiotic stages induces nearly 50% pollen sterility (Hecker et al. 1972). However, the chemical is phytotoxic and causes considerable ovular damage.

vi. *Morphactin.* Morphactins, a group of synthetic regulators manufactured largely in the laboratory of the German chemist Dr. G. Schneider, affect many plant physiological and morphogenetic processes in a variety of ways (Schneider 1970). The effects on floral organs, especially the sexual parts, are diverse and varied. In general, morphactins inhibit transport of IAA and the plants exhibit depressed growth activity and delayed phenological events. Using 500 ppm

Table 4.12. Induced chemical male sterility

Chemical	Species	References	Male sterility induction	Female fertility reduction	Side effects
6-Benzylaminopurine	*Cannabis sativa*	Chailakhyan and Khryanin (1980)	Complete	Nil	nk
3-Carboxy-1-(p-chloro-phenyl)-4, 6-dimethyl-2-pyridone (CCDP)	*Hordeum vulgare*	Yih et al. (1971)	Nearly complete	nk	nk
Colchicine	*Sorghum vulgare*	Erichsen and Ross (1963a,b)	High	High	nk
Coumarin	*Cajanus cajan*	Kaul and Singh (1967b)	Complete	Very high	High floral abscission
	Gossypium hirsutum	Dubey and Singh (1968)	Complete	Very high	nk
	Trigonella foenum-graceum	Kaul and Singh (1967b)	Complete	High	Reduced vegetative and reproductive growth, delayed flowering and fruiting
	Vicia faba	Awasthi and Dubey (1983)	High	Very high	High toxicity
Dimethyl arsenic acid	*Beta vulgaris*	Hecker et al. (1972)	Low	High	Highly phytotoxic
	Pennisetum typhoides	Powell and Taylorson (1967)	Very low	High	Highly phytotoxic
Embutox-4	*Abelmoschus esculentus*	Dubey and Singh (1968)	Complete	Low	nk
Estrone	*Beta vulgaris*	Hecker et al. (1972)	Partial	High	Phytotoxic
	Triticum aestivum	Fendal (1967)	Low	nk	nk
Ethidium bromide	*Pennisetum typhoides*	Burton and Hanna (1967)	Partial to complete	nk	nk
IAA	*Silene pendula*	Heslop-Harrison and Heslop-Harrison (1958)	High	Low	nk
Indole butyric acid	*Vicia faba*	Awasthi and Dubey (1983)	High	Very high	Highly toxic
Morphactin	*Lycopersicon esculentum*	Jain and Mukherjee (1980)	High	High	nk
	Nigella damascena	Rani (1976)	Complete	nk	floral and fruit anomalies

Table 4.12. *(continued)*

Chemical	Species	References	Male sterility induction	Female fertility reduction	Side effects
Napthalene acetic acid (NAA)	*Cucumis melo*	Laibach and Kribben (1951)	Partial	Nil	nk
	Cucurbita pepo	Wittwer and Hillyer (1954), Hillyer (1956)	Partial	Nil	nk
	Vicia faba	Awasthi and Dubey (1983)	High	Very high	Highly toxic
3-(p-Chlorophenyl)-6-methoxy-s-triazine-2, 4-(1H, 3H) dione triethanol amine salt	*Oryza sativa*	Zhangxing and Chunnong (1980), this Vol.	Nearly complete	Very high	Highly toxic
Phosphon-5	*Allium cepa*	Cohan and Weigle (1966)	Partial to complete	High	nk nk
Sodium nucleate	*Allium cepa*	Chopra et al. (1960)	Complete	nk	nk
Tri-iodobenzoic acid (TIBA)	*Citrullus vulgaris*	Rehm (1952)	High	Very high	nk
	Helianthus annuus	Kiermayer (1959)	Complete	nk	nk
	Lycopersicon esculentum	Rehm (1952)	Complete	Not much	nk
	Solanum nigrum	Rehm (1952)	Complete	High-complete	nk
Zinc methyl arsenate	*Brassica napus*	Guan et al. (1981)	High	Very high	High bud abortions, plant death and toxicity

EMD-7311 W morphactin, Rani (1976) obtained mst in *Nigella damascena*, but this was accompanied by morphological anomalies and impaired female fertility. Chloroflurenol methyl ester (CME 74050), which belongs to a group of synthetic bioregulators, has a high morphogenetic potency and male gametocidal specificity in tomato, in which Jain and Mukherjee (1980) induced mst. Ovular fertility was considerably reduced and parthenocarpic fruits appeared on the treated plants.

vii. *Phosphon-5.* This is a growth retardant which acts primarily by inhibiting GA synthesis. In onion, complete mst in the greenhouse and p-mst in the field was induced by 750 and 1000 ppm spray (Cohan and Weigle 1966). The treatments damaged female fertility and seed production was drastically reduced.

viii. *Sodium nucleate.* The gametocidal properties have not been investigated except in onion, in which 3–4% aqueous solution injection led to complete pollen abortion (Chopra et al. 1960). Female fertility of the treated plants was not determined due to crop damage by hailstorm.

ix. *Tri-iodobenzoic acid.* This naturally synthesised auxin enhances floral production in some plants but induces mst in nightshade, sunflower and tomato (Table 4.12). Whereas in tomato, female fertility is not highly impaired, in nightshade the impairment is considerable. In sunflower, the effects are not known.

4.3 Induced Sterility

Fairly high to complete ch-mst has been induced in some plants in which genetic mst is not known (Table 4.13). The sterility accompanies a low to high degree of female sterility. Although in most of these plants, other effects on the treated plants are not known, these chemicals are temporarily effective, mostly phytotoxic, and lead to abnormal fruit and seed development. However, in the absence of genetic mst in them, the ch-mst may be used for hybrid seed production to a limited extent.

4.4 Male Gametocides

4.4.1 Effects

The majority of the chemicals which act as male gametocides are toxic to plants, reduce female fertility and induce mst for a short duration only. In addition, the use of male gametocides is beset with many limitations which have discouraged investigations on the mechanism of gametocidal action. Chauhan and Kinoshita (1980b) investigated the histological and biochemical changes that occur following applications of dalapon, MH and mendok. In *Datura alba*, potent doses induce complete pollen abortion and inhibit vascular differentiation of anthers partially or fully. This impairment completely blocks or appreciably reduces nutrient transmission to the tapetum and across it to the sporogenous tissue. Since PMC's and the developing microspores depend upon tapetum for their nutrition, they

Table 4.13. Induced chemical male sterility (complete) in plants not exhibiting genetic male sterility

Species	Reference	Chemical	Female fertility reduction
Callistephus sinensis	Moore (1950)	MH	High
[a]*Commelina virginica*	Moore (1950)	MH	High
Coriandrum sativum	Dubey and Singh (1968)	Dalapon	High
		2, 4-D	Low
		MH	nk
		Mendok	nk
Datura alba	Chauhan and Singh (1972)	MH	nk
Foeniculum vulgare	Dubey and Singh (1968)	Dalapon	High
		2, 4-D	Low
		MH	High
		Mendok	High
Hibiscus cannabinus	Dubey and Singh (1968)	Dalapon	Low to high
		MH	High
		Mendok	Low to high
Luffa cylindrica	Chauhan (1980)	Mendok	nk
Momordica charantia	Chauhan (1980)	Mendok	nk
[b]*Nigella damascena*	Rani (1976)	Morphactin	nk
Rubus idaeus	David (1950)	MH	High
[c]*Trigonella foenumgraceum*	Kaul and Singh (1967b)	Coumarin	High
		MH	High
		Mendok	High

[a]High floral bud abortion, [b]High floral anomalies, [c]Reduced and delayed vegetative and reproductive growth.

degenerate eventually because of the lack of nutrients and the consequent starvation. Furthermore, deficiency of carbohydrates, proteins and DNA occurs in the ch-mst anthers. In hypertrophied tapetal cells, poor acid phosphatase activity occurs. This indicates a low metabolic status of the ch-mst anther tissue. Impediment of nutrient flow to developing anthers occurs following GA applications in maize and pea. In maize, the chemical induces alterations in peduncle, lemma and palea (Hansen et al. 1976). Apart from inducing mst, these gametocides cause high ovular sterility, reduce vegetative and reproductive growth and vigour and induce floral abscission in many plants (Tables 4.3, 4.8, 4.9). It appears that the chemicals likewise impair biochemical activities of these plant organs, especially of ovules. However, since mst induction has been the prime purpose of gametocidal treatments, anther tissue has remained the main target of investigations. The study of the effect of male gametocides on other plant organs, especially ovule development, is worthwhile.

4.4.2 Utility Value

Most male gametocides induce partial to complete mst. The doses that induce complete mst mostly impair female fertility also. This female fertility impairment reduces the breeding utility of these chemical gametocides due to low seed production. However, a few gametocides induce nearly complete mst but reduce female fertility slightly (Table 4.14). Of these, ethrel in a few cucurbits and

Table 4.14 Some effective male gametocides

Chemical	References	Species	Side Effects
2, 4-D	Dubey and Singh (1968),	*Coriandrum sativum*	nk
	Dubey and Singh (1968),	*Foeniculum vulgare*	nk
	Jyotishi and Hussain (1968), Helal and Zaiki (1981)	*Solanum melongena*	nk
DPX-3778	Johnson and Brown (1976, 1978)	*Triticum aestivum*	Genotype-specific sensitivity, delays anthesis
	Long et al. (1973)	*Zea mays*	Reduces shoot height and is later toxic
	Laible (1974)		Genotype-specific sensitivity, delays anthesis
Ethrel	Minina (1938), Iwahori et al. (1970), Augustine et al. (1973)	[a]*Cucumis sativus*	nk
	Rudich et al. (1969, 1970)		Reduces plant growth, vigour and development
	Robinson et al. (1970)	[a]*Cucurbita maxima*	nk
	Rudich et al. (1969, 1970)	*C.pepo*	Reduces plant growth and development
	Kumar et al. (1976), Verma and Kumar (1978)	*Hordeum vulgare*	nil
	Helal and Zaiki (1981)	*Solanum melongena*	nk
	Rowell and Miller (1971), Bennett and Hughes (1972), Dotlacil and Apltauerova (1978)	*Triticum aestivum*	Causes dwarfing, high floret sterility, poor spike emergence, imperfect heading, increases tillering
GA_3	Sawhney (1981)	*Capsicum annuum*	Increases shoot height, forms extra ovules
	Nickerson (1959), Hansen et al. (1976)	*Zea mays*	nk
IAA	Heslop-Harrison and Heslop-Harrison (1958)	*Silene pendula*	nk
MH	Chauhan (1976)	*Cucurbita maxima*	nk
	Rehm (1952)	*Lycopersicon esculentum*	nk
	Denisen and Haber (1950), Moore (1950), Naylor and Davis (1950)	*Zea mays*	nk
Mendok	Kho and DeBruyn (1962)	*Antirrhinum majus nanum*	nk
	Hensz and Mohr (1959)	*Citrullus vulgaris*	nk
	Sharma (1979)	*Pennisetum typhoides*	nk
SADH + Ethrel	Rudich et al. (1969)	*Cucumis melo*	nk
	Rodriguez and Lambeth (1972)	*C.sativus*	Reduces plant growth
TIBA	Rehm (1952)	*Lycopersicon esculentum*	nk

bread wheat varieties is effective in inducing high mst accompanied by low female fertility impairment. Whereas in cucurbits, ethrel treatment hardens fruit texture and modifies fruit form, in wheat, dwarfing, poor spike emergence, imperfect heading and high floret sterility are caused by the chemical treatment.

Effects of 2,4-D, apart from the induction of mst in *Coriandrum* and *Foeniculum* plants, are not known. Since the chemical accumulates in the seeds, the hybrid becomes unfit for human consumption. In one snapdragon variety, mendok induces mst, in the other, it is ineffective. Although its side effects on snapdragon plants are not known, it causes severe phytotoxicity and leaf burning in *Medicago* and in many other plants (Table 4.9). Maleic hydrazide induces complete mst in maize and tomato, but the chemical is a well-known growth retardant and is phytotoxic in many plants, causing reduced plant growth, development and premature floral bud abscission (Table 4.8). Therefore, its use in commercial breeding is highly doubtful and not feasible in view of its known side effects. Due to similar side effects, the use of mendok is also not warranted. Certain specific GA_3 concentrations induce complete mst in maize without impairing female fertility, but the chemical causes internode elongation, making the stems so weak that they either lodge completely or break off. Moreover, in field-grown corn plants, the mst induced is not complete, as the earlier maturing plants shed normal pollen while the late ones collapse due to weakened stems (Hansen et al. 1976). Morphactins exhibit a high degree of selectivity and are not persistent in soil. Furthermore, they are too toxic and cause a number of growth anomalies. In tomato, chloroflurenol induces complete mst (Jain and Mukherjee 1980) but reduces seed fertility, deforms fruit form and greatly hardens fruit texture. Therefore this chemical cannot be used for tomato hybrid seed production. Moreover, the induced sterility lasts for 2–3 weeks. Chronic spraying may maintain mst, but fruit deformation and fruit texture deterioration multiply.

4.4.3 Limitations

A large number of chemicals can act as gametocides (Table 4.1) but none of them can be considered as an ideal male gametocide for a species, mainly due to the following reasons, which limit their use in mst induction.

4.4.3.1 Dose and Duration Specificity

For ch-mst induction, a specific male gametocidal dose is required. For every plant species, an optimal dose and duration of the chemical treatment exists at which maximum pollen sterility can be induced without significant female fertility reduction and detrimental side effects. Whereas treatments higher than the optimal dose induce male-female sterility and phytotoxicity, treatments below the dose (suboptimal) induce partial pollen sterility of almost no breeding value. Barring a few, in the majority of plant species in which male gametocides have been tested, optimal dose and duration of the treatment have not been determined. The treatments applied are based at random on a hit-or-miss system. Therefore in most of them either ch-mst is partial, or when high, it is mostly accompanied by reduced female fertility and detrimental side effects.

4.4.3.2 Genotype Specificity

Each variety or cultivar responds differently to a chemical dose, the response being highly genotype-specific. For instance, whereas mendok induces ch-mst unaccompanied by female fertility reductions in certain cotton varieties, in others both male and female sterility reductions are very high after such treatments. In four varieties, the same chemical dose induces p-mst of a low order and in two others, the dose is totally ineffective (this Vol.). Almost similar results were obtained by the present author in pea, rice and tomato with ethrel, dalapon and MH treatments (this Vol.). However, none of these chemicals in any of the varieties tested induces complete mst unaccompanied by female fertility reductions. *Sorghum* cultivars respond almost similarly (McIlrath 1957). Likewise in one snapdragon variety, mendok induces complete mst (Kho and DeBruyn 1962), while in another variety, the sterility induced is partial (Sink and Walker 1964).

4.4.3.3 Tissue Specificity

The gametocidal treatment is not only genotype- but tissue-specific. Seed treatments of coumarin, MH and mendok induce mostly seedling injury, plant mortality and male-female sterility in many cereals, legumes and umbellifers, but their foliar sprays in some of these results in pollen sterility. GA_3 seed treatments enhance a proportion of gynoecious plants in *Cannabis sativa* (Herich 1960), but its foliar application increases male flower development on the gynoecious plants (Mohan Ram and Jaiswal 1970).

4.4.3.4 Developmental Stage

The effectiveness of a chemical gametocide depends on the developmental stage of the plant organ receiving chemical treatment. Thus due to differential floral development, mendok-induced mst in *Antirrhinum* is localised to the central florets of the spike and not found in the peripheral ones. In *Beta*, early- and late-formed flowers of mendok-treated plants are p-mst, the rest are completely mst (Wit 1958). Similarly, in *Nicotiana rustica*, mendok treatment induces complete mst in the apical and middle regions of the spike, but not in the basal regions (Jos and Singh 1967b). In most of the plant species, since all the flowers borne by a plant are not synchronous in development, the gametocidal dose, even if optimal, induces p-mst. This reduces the effectiveness of even potent male gametocides given in optimal concentrations. Five mg GA_3 applied 1–3 days prior to male meiotic initiation causes mst in two maize varieties (Hansen et al. 1976). This critical timing and concentration limits its use in large populations because genetical and environmental diversity prevents the populations from uniformly reaching a suitable physiological state for GA_3 treatments. When the chemical is applied to a normal population growing in the field, the earlier maturing plants shed normal pollen and the late ones collapse due to weakened stems. In wheat and maize, the gametocidal effect is efficient when given during or just prior to premeiotic DNA synthesis. This is not synchronous in all the tillers and spikelets, and produces inconsistencies in mst as were obtained by Dotlacil and Apltauerova (1978) for ethrel in wheat and by Hansen et al. (1976) for GA_3 in maize. Whereas early-formed tillers are mst and later ones are p-mst, the latter are fertile in four ethrel-treated bread wheat cultivars.

4.4.3.5 Physiological State and Environmental Factors

The developmental stage specificity to effectiveness and the efficiency of a gametocide over a genotype is dependent upon the metabolic status of the plant tissue that becomes exposed to the chemical treatment. The net resultant effect of the treatment is modified by the physiological state of the whole plant and the prevalent environmental conditions. Both these factors are highly intricate, dynamic and difficult to define and control. They introduce so many variations that it is not possible to delimit a specific concentration and mode of application of a gametocide to achieve optimal results in a species. This explains the failure of repeatibility of many effective male gametocidal treatments identified by various investigators for certain plants. Due to strong ion antagonism, fertilizer, fungicide and insecticide applications as well as soil mineral content, the effectiveness of a gametocide is highly altered. The effect of environmental factors on gametocidal effectiveness is known in cotton in which Eaton (1957) obtained high pollen sterility after mendok treatments. Whereas some plants did not shed pollen, others shed it in the afternoon. The ethrel sensitivity of wheat is greater in the greenhouse than in the field (Rowell and Miller 1971). In onion, phosphon-5 induces complete mst in the greenhouse, but in the field its effect is highly variable (Cohan and Weigle 1966). Furthermore, even within a crop plantation some cole plants are completely mst and others p-mst after GA application (Meer and Van-dam 1979).

4.4.3.6 Structural Differences

Even small structural differences of gametocidal compounds yield significant differences in their action on the same variety. For instance, sodium salt of dichloroacetic acid (mendok) and the sodium salt of 2,2-dichloropropionic acid (dalapon) do not differ greatly structurally, but they differ completely in their gametocidal action in many plants. In general, in cereals and legumes, mendok is more effective than dalapon. In snapdragon and pea, mendok induces complete mst but dalapon is completely ineffective (Kho and de Bruyn 1962; this Vol.).

4.4.3.7 Chronic Spray

Pollen sterility induced by a chemical treatment is usually of short duration. To prolong it, the chemical treatment has to be given at regular intervals of time. This regular exposure, although maintaining mst, enhances female sterility and other phytotoxic effects. Such regular chemical applications lead to highly impaired female sterility, premature seed/fruit drop, abnormal or deformed seed and fruit development (e.g. morphactin in tomato and brinjals, ethephon in cucurbits and pea). The detrimental effects increase in intensity with advancing plant age.

4.4.3.8 Fertility Impairment

Reduction in female fertility through ovular damage is the major obstacle limiting gametocidal usage for chemical emasculation. The chemical dosage that induces complete to high mst impairs female fertility to a considerable extent (Tables 4.3 to 4.14). In addition, in cucurbits, composites, brinjals and tomato, premature floral abscission occurs, so that floral number is considerably decreased, resulting in diminished seed and fruit yield. This effectively reduces the breeding value of ch-mst.

4.4.3.9 Detrimental Side Effects

In addition to high ovular damage, almost all male gametocides induce some morphological and reproductive variations, most of which are of negative selection value and render the genotype inferior for use in breeding. The common effects are decrease of vegetative and reproductive growth and vigour leading to dwarfing and delay in phenological events like flowering, fruiting and seeding. Due to reduced vigour, the treated plants are more susceptible to insect infestations and fungal infections than the untreated ones. Apart from the general effects, certain other specific anomalies are caused by some effective gametocides. For instance, ethrel in wheat induces extreme dwarfing, poor spike emergence, high flower sterility. Mendok causes floral bud abscission, floral closure, chlorosis, burning of floral buds and, in extreme cases, plant death. GA treatments make stems weak within the tassel, and in maize they either break off or lodge to the ground.

4.4.3.10 Phytotoxicity and Health Hazards

Chemicals like dalapon, MH, mendok, phosphon, RH-531, 532 and RH-2956 are phytotoxic. They inhibit growth, delay maturity and induce foliar epinasty, leaf burning, premature leaf abscission etc. so that plants die early before maturity. Apart from phytotoxicity, some of the chemicals like DDT are health hazards and are known to cause certain serious irreparable damage to human genetic systems.

Since such chemicals are accumulated in the treated plants and their seed, they are dangerous for human consumption. Serious genetic hazards of RH-531 and -532 have already been indicated by Rajendra and Bates (1981).

4.4.4 Breeding Prospects

Induction of mst by chemicals without impairment in female fertility, plant growth and vigour does not seem to be possible. The type of chemical, its dose, duration and mode of treatment and the plant development stage at which the gametocide is to be tried need standardization, at least for the most promising cultivars of economically important plants in which hybrid production is needed. For high efficiency and efficacy, the gametocides should be applied either prior to or at the beginning of flowering. Since male organs are more sensitive to gametocides than female organs, the chemical dose should be kept at a minimum, but the number of applications may be increased. By this method a desirable mst level may be achieved without affecting female fertility and vegetative or reproductive growth. Gametocides hazardous to human health and toxic to plants should be eliminated. The requirements for an effective male gametocide are exacting. Its potential use and versatility also depend on genotypes × chemical, environment × chemical, and genotype × environment × chemical interactions. Tailoring application rates and timing specifically for the probable cultivar or environment will assure maximum seed production of certain exceptional hybrid combinations. Large interactions, however, could mean variable and less predictable performance and cause problems in hybrid seed production.

An easy, cheap and reliable method for mst induction would allow more economical production of hybrid seeds. Even if ch-mst induction is not 100%, and even if female fertility is slightly reduced, it could enable a breeder to produce a sufficient quantity of hybrid seed to test the value of hybrid populations on a limited scale. The most promising combination could be stabilized later by introducing a plasmic factor for mst in one or more lines. Induced ch-mst lines can be used as a substitute for emasculation when making interspecific and intervarietal crosses, and also recurrent backcrosses, because in these only a limited amount of seeds are required.

CHAPTER 5

Biochemistry

The absence of viable pollen or of normal anther differentiation and development and microsporogenesis or faulty anther dehiscence are the main characteristic features of mst plants. Such features introduce alterations in anther biochemistry, physiology and function. Thus, compared to mft anthers, reduced carbohydrate and protein content and a lower metabolic activity due to either absence or deficiency of some enzymes have often been found in mst anthers. Based on the observed differences in nucleic acid content, isoenzyme pattern, protein content and quality, and amino acid composition (aminogram), various causes leading to mst have been proposed from time to time. Whether these biochemical differences are causes or consequences of mst gene action is not known. In most of the cases, such differences appear as consequences rather than causes. These are discussed individually in the following.

5.1 Amino Acids

5.1.1 Total Amount

Amino acids, the initial products of nitrogen assimilation, are the building blocks of proteins. Characteristic differences in some bound and free amino acids have been detected between mst and mft anthers. For instance, compared to mft anthers, in mst anthers, either decrease or complete absence of proline occurs in wheat (Fukasawa 1954; Khoo and Stinson 1957; Kihara 1966a; Erickson 1967; Savchenko et al. 1968; Rai and Stoskopf 1974), sugarbeet (Hosokawa et al. 1963; Rohrbach 1965a), maize (Fukasawa 1954; Khoo and Stinson 1957), sunflower (Kovačik and Kryzanek 1969), rice (Ozaki and Tai 1961; this Vol.), lily and poppy (Britikov et al. 1964), tomato, sorghum and sunflower (Silvolap 1968) and pea and pepper (this Vol.). Likewise, a reduction in the amounts of isoleucine, leucine phenylalanine and valine occurs in the mst anthers of barley and sorghum (Whited 1967; Kern and Atkins 1972). An increase in asparagine amount occurs in mst anthers of wheat (Fukasawa 1954; Khoo and Stinson 1957; Rai and Stoskopf 1974), maize (Fukasawa 1954; Khoo and Stinson 1957), petunia (Izhar and Frankel 1973b), sunflower, tomato and sorghum (Silvolap 1968, Kovaĉik and Kryzanek 1969), pea, rice and chillies (this Vol.). A similar accumulation of other amino acids, either due to their increased synthesis or nonutilization, occurs in mst anthers of some plants, e.g. alanine in maize and rice (Khoo and Stinson 1957; Jones et al. 1957b; this Vol.) and sugarbeet (Hosokawa et al. 1963), glycine in sorghum, pea and tomato (Brooks 1962; this Vol.), arginine and aspartic acid in cotton (Sarvella and Stojanovic 1968), aspartic acid and glutamic acid in wheat (Rai and Stoskopf 1974), asparagine and glutamine in petunia (Izhar and Frankel 1973b) etc. Both asparagine and glutamine are important vehicles of nitrogen transport in higher plants, and their accumulation has significance in pollen development and physiology.

5.1.2 Aminogram Variability

Inconsistency in amino acid contents between mst and mft anthers of different mst genotypes occurs in sorghum, pea, cotton, rice, barley, sunflower and broad beans. For instance, in mst anthers glycine content at the pre-pollen stage may be high in some (Brooks 1962) or low in other mst genotypes of sorghum and rice (Alam 1967; Kern and Atkins 1972; Tripathi et al. 1981). Likewise, the threonine content may be very high (Tripathi et al. 1981) or too low (Alam 1967) in various sorghum and broad bean mst anthers. During meiotic initiation, the tyrosine content is high in mst sorghum (Kern and Atkins 1972) but during the latter stages of meiosis and gametogenesis the accumulation is rapid in mft anthers. This increases the tyrosine content in mft anthers. However, a reverse situation was detected by Tripathi et al. (1981). Similar variations occur in leucine and isoleucine in rice and methionine in pea (this Vol.). Likewise, alanine accumulation occurs in T-cytoplasm maize, but is absent in C- and S-type anthers. The variations in anther aminogram of different mst genotypes may be due to their *fr* or *ms* gene or cytoplasmic differences or due to differential g-c interactions which are known to occur in sorghum, wheat, solanum and cotton. In *Gossypium*, the total amount of acid-soluble amino acids in the mst buds of the plants having *G. anomalum* cytoplasm and *G. hirustum* genome is nearly the same as in the normal mft buds having *G. hirsutum* nucleus and cytoplasm (Sarvella and Stojanovic 1968). But the amino acid content is significantly low in buds of the plants containing *G. hirsutum* in *G. arboreum* cytoplasm. Thus, the amount of acid-soluble amino acids is reduced when *G. hirsutum* nucleus is placed in *G. arboreum* cytoplasm, but not when placed in *G. anomalum* cytoplasm. This differential g-c interaction at the biochemical level is not manifested at the sexual level, since mst is caused in both cases when *G. hirsutum* nucleus is placed in *G. anomalum* or *G. arboreum* cytoplasm. The differences in aminogram of anthers of different mst genotypes of a species may not be due to genes or g-c interactions only but also to environmental influence. In *sorghum*, contents of leucine, isoleucine and phenyl-alanine are lower in anthers of field-grown mst plants compared to those of greenhouse-grown mst plants (Kern and Atkins 1972).

Differences in aminogram pattern of mst and mft anthers have led to the contention that amino acids may play a direct role in the induction or control of mst in higher plants. Since no constant and specific differences occur between mst and mft anthers, the direct role of amino acids in mst cannot be visualized. Furthermore, the anther aminogram pattern is dependent upon genotype, cytoplasm type, environmental conditions and stages of meiosis and microspore development. Therefore it is certain that the differences in concentration in some amino acids between mst and mft anthers are associated with the deviated reactions or failure of reactions that occur in mst anthers during the course of microsporogenesis. Duvick (1965) suggested that accumulation of alanine and asparagine in anthers of sterile lines of maize may indicate that the abortive microspores are unable to utilize these amino acids. This may hold true also for mst wheat, sorghum, sunflower and tomato plants, whose anthers have a uniformly higher content of alanine. Both alanine and asparagine are mobile and highly reactive, and participate in reductive amination and other reactions vital to the nitrogen metabolism of plants. The g-mst pea and tomato anthers are deficient in glycine and proline. Injection of these amino acids into immature mst anthers does not revoke mft in them (this Vol.). The mst maize anthers have an increased asparagine content, but its injection into mft anthers does not induce mst in them. In *Zea mays*, Sarvella et al. (1967) did not find any consistent differences among mst, mft and restored lines in individual or total acid hydrolysable (bound) amino acids. The above examples, however, indicate that the accumulation or deficiencies of certain amino acids are not the causes but the consequences of mst in higher plants. The causal factors that lead to the changed aminogram in mst anthers are not clearly known. Since amino acid differences between mst and mft anthers do exist but are not consistently uniform in mst plants of different species, these deviations may be caused by the *ms* or *fr* gene action, whose expression and penetrance varies both within and between genotypes and environments.

5.1.3 Proline Content

In several amino acids, the proline content is either absent or uniformly low in the mst anthers of wheat, rice, maize, sunflower, tomato, petunia, pea, sugarbeet and pepper. This led Duvick (1965) to suggest that the absence or low proline content of mst anthers could be a direct cause of pollen abortion, if proline is essential for the formation of certain storage proteins or of enzymes necessary for the development of pollen. Free proline of pollen has high physiological activity and is quickly incorporated into protein molecules (Britikov et al. 1964). It is an important component of many biochemical substances produced during the course of sexual differentiation of flowers. Mostly proline is incorporated into peptide linkage and serves as a precursor to help adaptation to biological and environmental stress and strain. It has multifarious physiological functions and protects the plant from environmental stress, besides serving as a source of nitrogen in plant metabolism (Britikov et al. 1970; Dashek and Erickson 1981). Sarvella et al. (1967) injected proline in gc-mst maize. This improved fertile pollen production in one line but not in many other mst lines. Injection of several other amino acids also increased pollen production of p-mst maize lines, but not in the complete mst lines. Compared to the mft sorghum anthers, the proline content is either low (Silvolap 1968; Kern and Atkins 1972) or high (Tripathi et al. 1981) in mst anthers. Injecting proline in mst anthers does not restore pollen fertility in proline-deficient mst anthers of sorghum, pea and tomato. (this Vol.). All these observations indicate that mst is not caused by proline deficiency. Instead, this deficiency is a consequence of mst.

5.2 Nucleic Acid Starvation

Significantly low amounts of nucleic acids occur in mst anthers of sorghum (Brooks 1965; Alam and Sandal 1969), onion (Saini and Davis 1969), barley (Mian et al. 1974) sugarbeet (Chauhan and Kinoshita 1980a), sunflower (Kovačik and Kryzanek 1969) and Indian mustard (Banga et al. 1984). The complete absence of RNA and DNA synthesis occurs in the mst anther locules of gc-mst onion after the release of microspores from microsporocytes (Saini and Davis 1969). These microspores do not develop further and abort prematurely. This complete absence of nucleic acids at the microspore release stage is because of the low level of synthesis, probably due to a malfunction of the enzyme systems, precursors or nucleotides required for the nucleic acid synthesis (Saini and Davis 1969). In the mst onion anthers, initial nucleic acid content is nearly equal to that of the mft anthers (this Vol.), but at the microspore stage, the nucleic acids are not replenished in the mst anthers due to tapetal malfunction. Transfer of DNA from tapetum to microspore mother cells was documented by Cooper (1952) in *Lilium henryi* and *L. regale.* Linskens (1955, 1956) observed the escape of DNA in the form of nucleotides from degenerating tapetal cells and nuclei. This DNA is taken up by the PMC's for synthetic process and a part of it is converted into RNA. In vitro pollen development of excised anthers was studied by Linskens (1956), who concluded that the beneficial effect of coconut milk was mainly due to the nucleic acid content. The premature death of anthers in *in vitro* was due to lack of DNA. Thus, a low level of RNA and DNA in anther locules could be a limiting factor in the development of microspores into mature pollen grains. In two pea and one rice mst mutants, the total nucleic acid content in the mst anthers at the onset of meiosis was lower by 54.63% of the mft anthers. Meiosis did not proceed beyond mid-prophase-I and the meiocytes degenerated in late prophase-I (this Vol.). The tapetum appeared normal in them. Whether the initial nucleic acid content present at the onset of meiosis in anthers of all the mst plants in which microspores abort (like gc-mst onion) is nearly equal to that of mft anthers or not is not known. This needs to be determined so that the *ms* and *fr* gene actions and nucleic acid starvation could be correlated or linked.

It appears that in the mst plants in which meiosis continues normally, nucleic acid starvation during late stages of meiosis caused by the tapetal malfunction usually leads to microspore abortion. On the other hand, in those mst plants in which the nucleic acid contents are low at the beginning of meiosis, nucleic acid starvation causes meiotic arrest and meiocyte degeneration even before meiotic completion. These contentions need testing by using two specific groups of mst mutants of a species, viz., one in which male meiotic course is normal but microspores or pollen abort, and the other in which PMC's degenerate before meiotic completion, preferably before homotypic division. These two groups of genic mst mutants are available in barley, maize, pea and tomato, and can be utilized for the study.

5.3 Carbohydrate-Protein Metabolism

Once amino acids are altered, different or defective proteins are formed in mst anthers. A decreased or disturbed carbohydrate and protein metabolism occurs in the mst anthers of beets (Chauhan and Kinoshita 1980a), pepper (Markova and Daskaloff 1974, 1976), sunflower (Pirev 1966), wheat (Fukasawa 1957b; Erickson 1967; Savchenko et al. 1968), Indian mustard (Banga et al. 1984) and rice (this Vol.). During the pollen maturation stage, mst anthers in wheat are deficient in sugars, especially sucrose, whose addition to the culture medium in which the mst wheat plants were grown improved microsporogenesis but did not improve pollen fertility. Two major zones present in the polypeptides of mft and restored anthers are absent in mst anther polypeptides (Ahokas 1980a). Unlike in mft anthers, in mst anthers, the proteins become hydrolysed to short-chain peptides, indicating the existence of abnormal proteolysis in mst barley anthers. In pepper, an electrophoretic study of anther protein of mst and mft lines by Markova and Daskaloff (1974, 1976) revealed differences in protein bands and in their stainability and mobility. In mft anthers, 17 bands existed, but mst anthers had only 15. A similar reduction of protein bands occurs in sorghum mst anthers (Alam and Sandal 1969; Kongtian and Hongyi 1981). This indicates disturbed and deviated protein metabolism in the mst anthers. Both carbohydrate and protein deficiencies occur in mst anthers of maize, barley, pea, rice, tomato, sorghum, petunia, beet, pepper and sunflower. These cause reductions in sugars and certain amino acids in mst anthers. Since about 85% of the weight of amino acid molecules represent non-nitrogenous components, chiefly carbon, hydrogen and oxygen, amino acid synthesis cannot occur without an adequate supply of carbon and nitrogen sources. The supply is retarded in these plants either due to tapetal malfunction and/or impaired anther vascular bundle development.

5.4 Cytokinin Content

These purine derivatives influence many physiological processes including sexuality in higher plants. In *Vitis vinifera* (*sylvestris* form) and *Mercurialis annua*, cytokinin applications inhibit maleness and induce female flower development on genetically male plants (Negi and Olmo 1972; Kahlem 1976). The cytokinin contents of mst pea and tomato are much lower than their mft sibs. Both the biological activity and the total quantity of bound cytokinin are low in gc-mst barley having S-1 cytoplasm and high and similar in three mft lines and in the restored genotypes (Ahokas 1982d). The precise modus operandi of cytokinins in sex conversion is unknown. Cytokinins stimulate protein and nucleic acid synthesis and also act as gene derepressors. Interestingly, they are found in t-RNA preparations from mycoplasma to man but not in r-RNA. Probably they regulate the activity of t-RNA and their presence may be necessary for codon-anticodon interaction between m- and t-RNA's, as they have been located in the anticodon loop of certain t-RNA's. The *ms* and *fr* genes probably repress the gene responsible for cytokinin synthesis and in the absence of sufficient cytokinin content, the sex organ development or function is impeded. It is equally possible that cytokinins suppress *ms* and *fr* gene action through their association with t-RNA or by working as gene derepressors. Chloroplast and mitochondria are target sites for cytokinin action and both these organelles are implicated in mst.

5.5 Enzyme Activity

5.5.1 Callase

A pre-requisite for the initiation and completion of synchronous meiotic division is the regular callose deposition around PMC's during early prophase-I (Eschrich 1961; Heslop-Harrison 1964; Frankel and Galun 1977). For proper and autonomous transformation of the four products of meiosis into normal young microspores, callose deposition around tetrads is equally essential. Before this, the callose envelope around each meiocyte has to be degraded to effect microspore release from the tetrads. The degradation is brought about by the enzyme callase (1,3-B-glucanase), the degradation products being d-glucose and its polymers of different lengths. These products are used as the base materials in pollen-exine formation and pollen nutrition (Eschrich 1961; Waterkeyn 1964; Izhar and Frankel 1971). Callose production, accumulation and deposition, and formation and timing activity of enzyme callase are all under the control of different nonallelic independently functioning major genes, whose activity, like that of many other physiological genes, is time- and phase-specific. Non-production or low accumulation of callose in some barley, pea and tomato mst mutants leads to PMC degeneration either in prophase or during microspore development; the anomaly in each case is conditioned by a single recessive gene (this Vol.). Likewise, delayed or precocious callase enzyme activity disturbs microsporogenesis and leads to mst in *Petunia* (Frankel et al. 1969; Izhar and Frankel 1971), *Lolium* (Hayward and Manthiratna 1972), *Allium* (Novâk 1972; Nanda and Gupta 1974), *Sorghum* (Overman and Warmke 1972) *Phaseolus* (Pritchard and Hutton 1972), *Pisum* (Gottschalk and Kaul 1974) and *Capsicum* (Horner and Rogers 1974). The phase-specific activity of this enzyme varies in different mst types, but the activity shows a relationship with *fr* gene action in *Petunia.* In mft lines, callase activity is at a peak at the late tetrad stage, but in the gc-mst type 1 *Petunia,* meiotic breakdown occurs at early prophase and the activity of this enzyme rises considerably just prior to the breakdown stage (Izhar and Frankel 1971). In another gc-mst type no callase activity occurs in anthers and the microspores are not released from the tetrads. Significant differences between mst and mft sorghum anthers occur in callose formation and callase activity (Warmke and Overman 1972). In mst onion anthers in which the tapetum hypertrophies, tetrads lack the callose wall and shrivel up considerably (Nanda and Gupta 1974). In these, a higher callase activity of hypertrophied tapetum appears to lead to precocious callose degradation.

The phase-specific callase activity is pH-dependent. This dependence, coupled with lack of glucanase activity in the meiocytes at all stages of meiosis and its presence in the tapetum and anther locules at the time of normal degradation of the callose matrix of the microspore quartet (Eschrich 1961; Stieglitz and Stern 1973), leads to the conjuncture that faulty activation of the enzyme in the anther locule may be a primary cause for microspore abortion in a *Petunia* and a *Pisum* mst mutant. The rationale of exact timing of the callase activity, external to microsporocytes, becomes evident in view of the normal dynamics of callose and primexine synthesis which is at least in part internal to the microsporocytes (Frankel and Galun 1977). The polysaccharide reserves of starch in plastids disappear during microsporogenesis in microspores when callose and cellulosic primexine synthesis occurs (Dupis 1974). Recurrent transitory formation of callose walls around the normal generative cell of the microgametophyte (Gorska-Brylass 1967; Heslop-Harrison 1968; Mepham and Lane 1970), after the orderly dissolution of callose matrix of the tetrad capsule, accentuates the significance of the dynamics of synthesizing and degrading enzymes during microgametogenesis. The role of the callose wall as providing a molecular filter and thereby establishing a certain degree of transitory autonomy for meiocyte quartets of microspores and perhaps also for the generative cells of the microgametophyte has been proposed (Heslop-Harrison 1964) and questioned (Mascarenhas 1975). Irrespective of the exactly known role of the callose wall, any untimely or abnormal formation or dissolution is associated with mst in *Petunia* (Frankel et al. 1969; Izhar and Frankel 1971), *Lolium* (Hayward and Manthiratna 1972), *Allium* (Novâk 1972; Nanda and Gupta 1974), *Sorghum* (Overman and Warmke 1972), *Phaseolus* (Pritchard and Hutton 1972), *Pisum* (Gottschalk and Kaul 1974; this Vol.), *Capsicum* (Horner and Rogers 1974) and Indian mustard (Banga et al. 1984).

5.5.2 Cytochrome Oxidase

This copper or iron compound acts as hydrogen carrier in the biological oxidation-reduction system. The gene positively identified as being encoded in plant mt-DNA is that for subunit II of cytochrome oxidase (Fox and Leaver 1981). In mst wheat anthers, cytochrome oxidase content is higher than in mft anthers (Erickson 1967), but in T-cytoplasm mst anthers isozymatic bands for the enzyme are highly reduced (Watson et al. 1977). Likewise in mst sorghum (Alam and Sandal 1969), rice (Anonymous 1977; this Vol.) and maize (Watson et al. 1977) anthers, a deficient or abnormal cytochrome oxidase system occurs. This causes defective and inefficient oxidative phosphorylation in these anthers. Introduction of the *Fr* gene restores fertility and the normal enzyme activity in mft rice anthers.

5.5.3 Dehydrogenase

Anthers of mst wheat (*Ae. ovata*), *T. durum* and rice exhibit a lower reducing activity of this enzyme at the pollen degeneration stage, as also at the early meiotic stage, during which no signs of microspore degeneration are cytologically visible (Fukasawa 1957b; this Vol.). These enzyme differences are confined only to the anthers and not to the seed, germinating embryos and seedlings of mst and mft plants. This indicates a definite link of enzyme activity with mst expression.

5.5.4 Glucose-6-Phosphate Dehydrogenase

Reduced carbohydrate and protein metabolism in mst anthers of some plants may be due to the lower energy effectiveness of the oxidative processes in them. This is borne out by the fact that Loseva et al. (1974) found a lower intensity of oxidative phosphorylation in mitochondria of mst maize anthers and an increase in ATP-ase activity. This led to a lower ATP content in the mst maize anthers. In mst pepper anthers, the activity of glucose-6-phosphate dehydrogenase (G-6-PDH), an enzyme of the pentose phosphase pathway, is much lower than in mft anthers (Markova and Daskaloff 1976). This indicates a low synthetic capacity and energy effectiveness of mst pepper anthers. A direct association of G-6-PDH with staminal development and differentiation is known in some *Nicotiana* mst forms in which the activity of this enzyme is greatly reduced in mst's in which staminal organogenesis is impaired. This is known in mst tobacco lines having *N. tabacum* genome and *N. bigelovii* or *N. debneyi* cytoplasm (Markova and Tsikov 1979).

5.5.5 Phosphorylase

In addition to the reversible transformation of starch to glucose-1-phosphate brought about by phosphorylase, the latter plays an indispensible role in carbohydrate and protein metabolism. Decreased phosphorylase activity occurs in sugarbeet and rice mst anthers (Chauhan and Kinoshita 1980a; this Vol.) and reflects the reduced starch and protein metabolism in the mst anthers.

5.5.6 Phosphatase

High enzyme acid phosphatase activity occurs in the areas of high metabolic activity, rapid cellular differentiation, protoplasmic synthesis and vascularization. Compared to mft anthers, the activity of this enzyme is low in the mst anthers of *Allium cepa, Capsicum annuum, Cucumis melo, Cucurbita maxima, Datura alba, Solanum melongena* and *Triticum aestivum* (Chauhan 1976, 1979a) *Hordeum vulgare* (Kaul and Singh 1966, 1968), *Lycopersicon esculentum, Oryza sativa* and *Pisum sativum* (this Vol.).

5.5.7 Other Enzymes

Compared to mft anthers, a reduction in the isoenzyme bands of polyphenoloxidase and esterase isoenzymes occurs in the mst anthers of peppers (Markova and Daskaloff 1974). Similar reductions in isoenzyme composition of peroxidase occurs in mst anthers of *Nicotiana tabacum* having *N. bigelovii* or *N. debneyi* cytoplasm (Markova and Tsikov 1979). In barley, lipoxygenase (LOX) activity in leaves of mst plants is much lower than in those of normal and restored plants (Ahokas 1982a). The activity is higher in developing anthers, but decreases considerably during degeneration of the sterile anthers.

5.6 Enzyme Differences

In mst anthers, the low activity of enzymes, especially of those connected with the pentose pathway, has a biological significance because the pathway supplies the reduced forms of nicotinamide adenosine dinucleotide phosphatase (NADP) which are required in the biosynthesis of macroergic compounds like energy-rich ATP. The pathway also supplies pentose phosphatases that can be used as base materials for the nucleotide synthesis required for nucleic acid production. A strong decrease of Gl-6-PDH enzyme activity in anthers of mst peppers (Markova and Daskaloff 1976) indicates that these anthers have a highly reduced synthetic capacity and a low energy budget. The decreased and slow synthesis of ATP may lead to alterations or termination of some biochemical reactions, especially protein biosynthesis in which macroergic compounds like ATP participate. Reduction in the isoenzyme bands of polyphenol oxidase in anthers of mst peppers (Markova and Daskaloff 1974) indicates the cell's lowered energy exchange status in them. A similar reduced esterase isoenzyme synthesis in mst anthers (Markova and Daskaloff 1974) indicates disturbed protein biosynthesis in these anthers. However, the direct role of this enzyme in protein biosynthesis is not clearly established as yet.

Reduction in the activities of the metabolic enzymes like phosphorylase, phosphatase, esterase (Markova and Daskaloff 1974; Chauhan and Kinoshita 1980a,b,c; Chauhan 1980) in mst anthers of various plants further confirms the low energy budget of mst anthers. Loseva et al. (1974) detected a much lower intensity of oxidative phosphorylation in maize mitochondria of mst anthers. Enhanced ATP-ase activity leads to low ATP content in the mst anthers. Thus mst is associated with lowered energy effectiveness of the oxidative processes. An abnormal and deficient cytochrome oxidase system coupled with a lack of free histones in mst sorghum anthers led Alam and Sandal (1969) to suggest that in mst anthers the histones may be complexed with DNA and repress the regulator genes that support DNA-dependent RNA synthesis. This repression could block the synthesis of certain specific proteins and lead to pollen abortion.

5.7 Micronutrient Deficiency

Micronutrient deficiency delays flowering, fruiting, and maturity in many higher plants. Boron, copper, and molybdenum deficiencies induce pollen sterility in many cereals and other higher plants (Löhnis 1937, 1940; Kibalenko and Sidorshina 1971; Graham 1975; Agarwala et al. 1979; Dell 1981). Crop plants subjected to a micronutrient deficiency exhibit reduction in anther size and pollen abortion. The pollen grains are poorly developed, small, shrivelled, highly vacuolated and lack proper exine and cytoplasmic contents. With continued deficiency, seed and fruit development are impaired. Therefore, this type of induced mst is not of practical value in heterosis breeding. Specific effects of boron, copper and molybdenum deficiencies in relation to mst have been investigated in some plants and are described in the following.

5.7.1 Boron

Although not absolutely essential for animals and fungi, boron plays a key role in partitioning metabolism between the glycolytic and pentose shunt pathways. The in vitro borate application inhibits 6-phosphogluconate dehydrogenase activity in a tomato mst mutant (this Vol.). In garden pea roots, the absence of boron reduces histone quantity in chromatin and in isolated mitochondria of pea its deficiency leads to uncoupling of oxidative phosphorylation, thereby depleting the cell's energy budget (this Vol.). This impairs the meristematic activity of developing root and shoot. In wheat, oat, rye and barley, pollen structure, development and viability are highly affected by boron deficiency (Löhnis 1937, 1940). Pollen size and viability are drastically decreased in sugarbeet and pea plants grown in boron-deficient soils (Kibalenko and Sidorshina 1971; this Vol.). In rye and pea, fertile pollen has a higher boron content than the sterile pollen. This indicates a higher boron requirement and/or accumulation in the fertile pollen. However, continued boron deficiency impairs ovular fertility and impedes normal seed and fruit development in rye, pea and barley. Therefore, this deficiency cannot assume a direct role in mst.

5.7.2 Copper

Copper acts as a component of phenolase, laccase and ascorbic acid oxidase, and its role as part of these enzymes makes it vital to plant metabolic functions. In higher plants, copper deficiency affects the reproductive phase more than the vegetative phase. It delays flowering, fruiting and maturity (Riceman and Anderson 1943; Graves and Sutcliffe 1974; Graham 1976; Hill et al. 1978), and produces many virulent symptoms in cereals (Graham and Nambiar 1981). Copper deficiency induces mst in wheat (Graham 1975). The deficient plants develop minute anthers having scanty, dented, small, starchless abortive pollen. The lignified cell wall thickenings are reduced or absent in their endothecium (Dell 1981). The critical period of deficiency is the early boot stage (Graham 1976), that approximates to male meiosis initiation. Its continued deficiency impairs seed and fruit development. This led to 30% seed set obtained by Graham (1975) after hand pollinating the Cu-deficient plants. In wheat, oat, barley, sweet corn, sunflower, petunia and subclover, normal reproductive development, including microsporogenesis, is impaired due to Cu deficiency (Dell 1981). In wheat and oat, with increased Cu deficiency a progression from anthers with sterile pollen and reduced lignification, anthers without tetrads, staminodes, or no flower formation occurred. In subclover, sunflower and petunia, either staminodes developed or microsporogenesis was disturbed. Thus wheat, oat and barley are highly sensitive to copper and are copper-inefficient plants. Rye and triticale are nonsensitive to the deficiency and are thus copper-efficient plants. The efficiency is due to increased copper absorption and its quick translocation to reproductive structures. This is genetically conditioned and the gene has been located on a specific rye chromosome (Graham 1978). In oats, the efficiency is conditioned by one to two genes and is transferable from efficient to inefficient plants.

5.7.3 Molybdenum

Molybdenum is a constituent of the nitrate reductase enzyme and has therefore a role in nitrogen metabolism. After growing maize plants in refined sand cultures having severe or mild molybdenum deficiency along with control that had sufficient molybdenum, Agarwala et al. (1979) found reduction in tassel size, male flowers and anthers in the molybdenum-deficient plants. In these, anthesis was either suppressed or considerably delayed, the anthers had scanty, shrivelled, vacuolated and aborted pollen. Most of the anthers failed to dehisce due to lack of fibrous thickening in the endothecial layer. Cells of the anther wall were markedly reduced in size and lacked cytoplasmic inclusions. Female fertility of molybdenum-deficient plants is not known. In the molybdenum-deficient plants, decreased activity of starch phosphorylase in mature pollen and of invertase and acid phosphatase occurred throughout the pollen developmental stages (Agarwala et al. 1979), but the catalase and peroxidase enzyme activities increased considerably. Stanley and Linskens (1974) found low pollen viability to be associated with low catalase and peroxidase activity, but in the molybdenum-deficient inviable pollen, the activity of these enzymes is high. However, the activity of certain enzymes may not always reflect its pollen viability or nonviability, but the overall metabolic status of pollen grains (Petrochenko 1964; Goss 1968).

5.8 Chlorophyll Content

Many mst genotypes of barley, maize, pea, rice, tomato and wheat have higher leaf chlorophyll content than their mft counterparts. In wheat (Savchenko et al. 1974), barley (Ahokas 1980a) and pea (this Vol.), mst's have likewise a higher chlorophyll b/a ratio, but the reverse occurs in Indian mustard (Banga et al. 1984).

5.9 Biochemical Causes and Consequences

Major biochemical changes that occur in sporogenous and tapetal cells prior to, during and after mst gene action have not been investigated systematically. The limited available studies comparing some biochemical components and enzyme levels of mst and mft anthers in some plants reveal reductions in (a) carbohydrate and protein levels, (b) total nucleic acid contents and (c) enzyme activities of dehydrogenases, peroxidases, esterases, phosphatases and phosphorylases. Many of these biochemical alterations are primarily associated with the reduced carbohydrate and protein levels of mst anthers and diminished nutrient supply to the sporogenous tissue. Most of the anthers investigated for the biochemical components possess a subnormal vascular supply that diminishes the nutrient translocation to anther sacs. Consequently, the tapetal cells have insufficient food materials. The anthers of some mst plants have an apparently normal-looking vascular supply, but the delayed tapetal degeneration starves the developing spore cells and leads to pollen abortion. Anomalies in anther form, vascular supply and tapetal breakdown are primarily due to *ms* or *fr* gene action. The alterations in biochemical substances are consequences rather than causes of mst. In the g-mst anthers of barley, pepper, pea and wheat, decreased carbohydrate, sugar and protein levels occur. This is coupled with reduced dehydrogenase activity in mst anthers. The addition of sucrose or glucose to culture media in which g-mst wheat, barley and pea were grown improved microsporogenesis, but did not improve pollen fertility. Different alterations from the normal free amino acid pool and phase-specific changes in the amount of some of the free amino acids are known in the anthers of mst plants. These changes in the amino acid pool within the mst anther locules of cotton, maize, sorghum and wheat are influenced by environmental factors, gene-cytoplasm combinations, microsporogenesis stages, anther pH, and above all by the *ms* or *fr* gene causing mst. Since the balance of free amino acids in anthers is related both to the synthesis and breakdown of amino acids and to their translocation into and out of anther sacs, alterations in the amino acid profile in mst anthers may reflect either anabolic or catabolic processes. Thus, abnormally high asparagine accumulation in the mst anthers of wheat, maize, sorghum, sunflower, tomato, petunia etc. may be associated with the breakdown of sporogenous tissue or with some block in the metabolic pathway between asparagine and ornithine in the TCA cycle. The mst anthers of these plants, as also of rice and pea, possess decreased amount of proline in comparison to their respective mft anthers. Injection of proline in the mst anthers of tomato, rice and pea does not revoke pollen fertility. These results indicate that alterations in amino acid profile represent the consequences and not the causes of male sterility.

Reduced nucleic acid contents of mst anthers, especially during the spore cell abortion stage, may be due to malfunction of the enzyme systems, precursors or nucleotides required for nucleic acid synthesis. In mst onion anthers, nucleic acid contents are nearly similar until the T-II stage, after which they drop rapidly. Injection of nucleic acids reduces pollen sterility but does not eliminate it completely. Anthers of the majority of mst genotypes exhibit disturbed protein-carbohydrate metabolism. Sugar, nucleic acid and proline contents are reduced and the activities of peroxidase, phosphorylase and acid phosphatase are either low or negligible. But artificial additions of these biochemical components and enzymes does not eliminate mst in pea, pepper, tomato and rice (this Vol.). This confirms that the biochemical alterations are not the causes but consequences of *mst* gene action.

CHAPTER 6

Graft Transfer and Viral Nature

6.1 Transfer Across Grafts

Initially, mst was considered as a diseased condition caused by unknown agents, probably a virus. Since viruses are graft-transferrable, grafting of mst stocks over scions and of normal stocks over mst stocks was assayed, to transfer the mst across grafts. This was done by Frankel (1956), who grafted fertile maintainer scion on mst *Petunia hybrida* rootstocks. Scion mft remained unaltered but mst scion progeny was recovered when scion flowers were selfed or crossed to another maintainer line. A similar transmission was obtained by Edwardson and Corbett (1961). Similarly, however, the mst did not appear in the year of grafting. It increased in the succeeding generations. Graft transfer of this type of mst was also demonstrated in *Datura* (Blakeslee 1921), sugarbeet (Curtis 1967), tobacco (Corbett and Edwardson 1964) and alfalfa (Thompson and Axtell 1978). The mst transfer in alfalfa was similar to *Petunia* as scion autonomy was maintained in every case, while some sterile scion F_1 progeny was produced in the progeny of three graft combinations. In sugarbeet, scion fertility was altered (some flowers were sterile) and sterile scion progeny was also produced. Reciprocal grafts (sterile scion on maintainer stock) did not induce scion flower mst alterations. Transmission of mst across grafts through dodder *(Cuscuta subinclusa)* plants was achieved by Grill and Garger (1981). However, later investigations reveal that the transferred sterility is not stable. A similar unstable expression of mst occurred in pepper scions (Daskaloff 1968). In *Zea mays,* Borowsky (1960) transplanted normal maize embryos into endosperm of mst maize and recovered some mst plants in F_1 progenies. Using the technique and utilising both T and S cytoplasms, Edwardson (1970), however, could not obtain mst even up to F_2. Shumway and Bauman (1966) used mesocotyl graft technique to unite mst T-cytoplasm maize inbreds with mft plants. No mst progeny plants ever appeared.

In *Desmodium,* onion, soybean, sunflower, tobacco, tomato and wheat, transmission of mst across grafts did not occur. In pepper, Daskaloff (1968) found either stable or unstable expression of mst after grafting, but Ohta (1961b) failed to obtain sterility transmission across grafts in pepper. Failure to transmit mst asexually via grafts was attributed by Edwardson and Corbett (1961) to: (a) faulty inoculation techniques, (b) failure to observe sufficient generations, (c) presence of genotypes in scions or recipient materials which do not permit sterility expression or which inhibit movement of the cytoplasmic factor, (d) environmental conditions which influence movement and/or rate of multiplication of the cytoplasmic factor, and (e) environmental conditions which may influence expression of the sterile phenotype.

6.1.1 Specific Features

Transfer of mst across grafts and its expression in scion progeny was successful in a few cases (Table 6.1) and unsuccessful in others (Table 6.2). The success was considered to be due to certain specific features of stock and scion, some of which are unique to each plant. For instance, in all the cases of successful graft transmission, appropriate compatible cytogenetic background of the stock and scion was considered as essential. In addition, the transfer and expression of mst across the grafts depends upon the genotype of the scion. Only scions with certain specific genotypes are receptive to this transfer. The sterility occurs only among the sexual progeny obtained from the scions. However, this restriction does not apply to sugarbeet, as Curtis (1967) found alterations to mst in mft scions grafted on mst stocks. The graft-induced mst plants, when pollinated by other scions grafted onto

Table 6.1. Successful graft transfer of male sterility

Species	References
Beta vulgaris	Curtis (1967)
Capsicum annuum	Daskaloff (1968)
Datura alba	Blakeslee (1921)
Humulus lupulus	Cech and Pozděna (1962)
Medicago sativa	Thompson and Axtell (1978)
Nicotiana tabacum	Corbett and Edwardson (1964)
Petunia hybrida	Frankel (1956, 1962, 1971), Edwardson and Corbett (1961), Bianchi (1963), Hope (1979)
Vicia faba	Scalla et al. (1981)
Zea mays	Borowsky (1960)

Table 6.2. Unsuccessful graft transfer of male sterility

Species	References
Allium cepa	Meer and Bennekom (1970)
Beta vulgaris	Owen (1949), Cleij (1967), Theurer et al. (1968), Marrewijk (1970)
Capsicum annuum	Ohta (1961b), Merrewijk (1970)
Crotalaria mucronata	Edwardson (1967)
Desmodium sandwicense	McWhirter (1969)
Epilobium hirsutum	Michaelis (1940)
Glycine max	Caviness et al. (1970), Stelly and Palmer (1980b), this Vol.
Lycopersicon esculentum	Andersen (1965)
Nicotiana tabacum	Burk (1960), Sand (1960), Marrewijk (1970)
Petunia hybrida	Marrewijk (1968)
Triticum aestivum	Lacadena (1968), Zeven (1967), Marrewijk (1970), this Vol.
Pisum sativum	This Vol.
Vicia faba	Bond et al. (1966a,b)
Zea mays	Rhoades (1933), Shumway and Bauman (1966), Edwardson (1970)

mst plants, produced mst offspring. Furthermore, the mst across graft union was transmitted in both directions viz. from mst scions to o-type (nonrestorer) stocks (side shoots) and from sterile stocks to o-type scions. The capacity for mst induction is usually retained indefinitely by the grafted scion (e.g. *Petunia*) for years in some cases. To achieve induction, both normal parents, the female and the male, have to remain grafted onto gc-mst stocks. If either or both scions are detached, the inductive capacity becomes lost. However, in alfalfa (Thompson and Axtell 1978) only the scion used as female was grafted, while ungrafted fertile plants served as the male parents. Likewise, when RM_5 *Petunia* line was grafted on gc-mst stock (Frankel 1962, 1971), the selfed progeny segregated 3 mft : 1 mst, regardless of whether it was obtained from permanently grafted or grafted and then detached RM_5 scions. The Mendelian segregation pattern of the mst was maintained through subsequent generations, i.e. integration into a chromosome had taken place and was expressed as a recessive allele. The possible alternative that the mst appeared due to spontaneous gene mutation rather than due to grafting was regarded as possible by Frankel (1971).

6.2 Viral Transmission

Viral infection as a possible cause of mst in higher plants was proposed by Atanasoff (1964a), who suggested that infections could account for all cytoplasmically inherited traits, even though their presence has not been demonstrated. Successful mst transmission across grafts in some plants (Table 6.1) favoured this view, but attempts to transmit gc-mst asexually by rubbing leaves, inoculation of normal plants with juices expressed from mst plants, growing mst and mft plants in adjacent plots, as well as mft-mst plant grafts have failed in maize (Rhoades 1933), tobacco (Burk 1960; Sand 1960), pepper (Ohta 1961b), *Epilobium* (Michaelis 1964), wheat (Zeven 1967; Lacadena 1968), beet (Cleij 1967) and onion (Meer and Bennekom 1970). Likewise, heat treatments to inactivate viruses and to "cure" plants of gc-mst (Shumway and Bauman 1966; Zeven 1967) have failed. However, many viral diseases are difficult to transfer sexually, as is known from Blakeslee's (1921) demonstration that grafting was the only successful means of transmitting the Quercina trait (abnormal morphology + mst) in *Datura*. In all the cases in which mst is transmitted across the grafts (Table 6.1), the sterility is of gc-mst type. This led to the contention of a viral presence in S-cytoplasms of these plants and its transfer across grafts. But in all the plants listed in Table 6.2 in which viruses are known to cause pollen sterility, g-mst is known in almost all and gc-mst in a few of them. Thus, no relationship exists between viral infection or its induction of mst or the type of mst induced or developed following or preceding viral infection. In fact, in the majority of the plants listed in Table 6.3, both male and female fertility is impaired by the viral infection, the extent of female fertility reduction is much lower than male fertility reduction, because unlike ovules, anthers are exposed and unprotected against viral and other infections. The *ms* or *fr* genes in plants listed in Table 6.3 cause almost complete mst, but female fertility reductions are very low. Thus, while these genes induce true mst, viral infection

Table 6.3. Plants in which male sterility is caused by *ms* or *fr-c* genes or by viruses

Genotype	Viral	Genic	Gene-cytoplasmic
Allium ampeloprasum	+	+	+
A.cepa	+	–	+
A.sativum	+	+	–
A.schoenoprasum	+	+	–
Aquilegia vulgaris	+	+	–
Beta vulgaris	+	+	+
Brassica oleracea var. *capitata*	+	+	+
Daucus carota	+	+	+
Glycine max	+	+	–
Helianthus annuus	+	+	+
Nicotiana tabacum	+	+	+
Oryza sativa	+	+	+
Phaseolus lunatus	+	+	–
Pisum sativum	+	+	–
Prunus avium	+	+	–
Trifolium pratense	+	+	+
Triticum aestivum	+	+	+
Vigna sinensis	+	+	–

(+ = known, – = not known)

causes severe male and less female sterility. Since in most of the cases, pollen fertility reductions following viral infections were studied and ovular fertility impairment was not estimated, a viral basis of mst was postulated. The postulation was strengthened by graft transfer in petunia, sugarbeet and alfalfa. This led Ohta (1961b) to suggest that gc-mst is either of viral and infective type or of plasmon and noninfective type.

In a few cases some cytoplasmic agents may indicate virus or virus like entities, causing mst. However, it cannot be excluded that different reactions of the cytoplasm may be caused also by microorganisms like mycoplasma or ricketsia. If the concept of virus or virus like nature of gc-mst is valid, it should be possible to make mst plants mft by freeing them from the virus. This was possibly achieved by Lichter (1978) in sugarbeet through heat treatments and meristem cultures. In this case, however, mutation of the c-gene from S- to N-type by heat treatments cannot be ruled out as similar experiments with other mst sugarbeet genotypes did not yield mst plants. Some of the data on *Petunia* grafting experiments may be reconciled with the existence of either virus like or episomal elements propagated in the host cells of the mst lines, migrating through graft unions into sensitive mft graft symbionts and occasionally becoming integrated into the genome. This is comparable to the integration of *fr* elements either into the genome or the cytoplasmic organelle of S-cytoplasm maize. The behaviour of these elements in *Petunia* is, however, unusual and almost unique because they are to be transmitted through the male and female parent before mst induction actually occurs. The loss of induction capacity by scions removed from their mst grafts needs further explanation.

6.2.1 Latent Virus

Although viral infections do not explain all the cytoplasmically inherited traits, many normal-appearing plants may carry latent virus. This was demonstrated by Atanasoff (1925) in some potato varieties artificially inoculated with Y-virus. They exhibited slight or no symptoms even though the varieties carried the virus. Johnson (1925) and Schultz and Folsom (1925) transmitted virus diseases from apparently healthy potato to susceptible varieties. The potato variety King Edward is a classic example of a masked carrier of viruses (Salaman and Le Pelley 1930; Salaman 1939). Thus all these instances of latent virus favour the viral nature of mst in *Vicia* and *Petunia*.

6.2.2 Infectious Heredity

Despite conflicting evidence, mst transmission across grafts has been attributed to the presence and genetic information of latent viruses or other infectious agents in gc-mst plants (Edwardson and Corbett 1961; Atanasoff 1964a,b, 1971). The idea of "infectious heredity", as known in tulips, is supported by the fact that graft and seed transmission of some viruses causes mst of a varying degree, e.g. barley stripe mosaic virus (McKinney and Greeley 1965), tobacco ring spot virus (Valleau 1932; Desjardins et al. 1954) and broad bean wilt virus (Crowley 1957; Ohta 1975). In peppers, interference in *fr* gene action occurs by viral infection (Ohta 1975).

6.2.3 Silent Partner

The alloplasmic gc-mst in tobacco is not transmitted through graft unions, but the plant acts as a "silent partner" in this transmission process. This was demonstrated by Corbett and Edwardson (1964), who sandwich-grafted a portion of *Nicotiana tabacum* stem between mst *Petunia* stock and mft *Petunia* scion. Some plants grown from seeds of the fertile scion were mst.

6.2.4 Viral Basis

Due to the possibility of transfer of mst across grafts in some plants (Table 6.1), virus or virus like entities have been considered to cause mst (Atanasoff 1964a, 1971; Siegemund 1976; Grill et al. 1983). Transmission across grafts in *Petunia* (Frankel 1956, 1962, 1964; Edwardson and Corbett 1961; Bianchi 1963) closely resembles the graft transmission of many known viruses. For instance, the occurrence of fertile, semi-fertile and sterile flowers in some plants suggests the infection patterns exhibited by plants infected with virus, where symptomless tissues and tissues exhibiting mild or severe symptoms may occur in the same plant. The mst expression, like virus invasion, is influenced by cytoplasmic and nuclear genes. These analogies between mst and virus symptoms, although not decisive, led

Atanasoff (1964b) to suggest a viral nature of mst in *Petunia.* In fact, Atanasoff (1964a) proposed that all gc-mst cases should be reexamined for viral infection, but Michaelis (1964) argued against this extreme proposition and experimentally evidenced against viral participation in many mst plants. Brewbaker (1964) suspected that cytoplasmic inclusions were virus particles that survive only in plants with a virus-susceptible genotype and that the virus lived symbiotically in diploid somatic cells, killing the haploid pollen. Although initially Edwardson and Corbett (1961) also suggested viral transmission of mst, later Edwardson (1968) negated his theory, based on certain circumstantial evidence. Tobacco ring spot virus is transmitted by petunia seeds; it causes both male and female sterility (Henderson 1931). Evidence indicates the nonviral nature of mst in the following plants in which mst was initially considered as due to viruses.

i. *Datura.* Blakeslee (1921) reported viral attack to cause pollen sterility in *Datura.* The virus, in addition, induces morphological anomalies like narrow leaves, deep slitted petal lobes, shrivelled anthers lacking functional pollen, abnormal orientation of stigmatic surfaces, suppression of capsule spines, seedling abnormalities and cotyledons adhering to seed coat. These anomalies are graft-transmitted and are caused by viral invasion and not by induced pollen sterility. Moreover, female fertility and seed production are highly reduced after the viral infestations. Thus, a virus causes not only many morphological alterations but both male and female fertility reductions in *Datura.*

ii. *Broad Bean.* Due to the association of RNA with mst, Edwardson et al. (1976) suggested a viral nature of the sterility in *Vicia faba.* Grill and Garger (1981) isolated this RNA and transferred mst through dodder grafts, but could not transfer it through direct graftings. Since transmission of virus and viroids by dodder is known (Weathers 1965), Scalla et al. (1981) believe in the viral nature of mst as specific cytoplasmic spherical bodies (CSB's) present in S-447 mst plants are absent in mft plants. These bodies are considered as the sites or possibly the products of sterility factors and reminiscent of viruses. Such bodies are, however, absent in S-350 plants of broad bean, which are also mst. Moreover, similar CSB's exhibiting maternal inheritance but not associated with mst occur in *Tropaeolum* (Ie 1972).

iii. *Maize.* In T-cytoplasm corn plants, Edwardson (1962) found dense inclusions probably lodging RNA, but these are absent in mft as well as in other mst plants having either C- or S-cytoplasm. Therefore, these cytoplasmic bodies are not exclusively associated with mst in all three sterile maize cytoplasms.

iv. *Sugarbeet.* Lichter (1978) obtained mft plants by heat treatment (36°C and above) and subsequent meristem culture of mst plants. The mft conversion was believed to be due to thinning or elimination of viral RNA-containing particles which caused mst. Kassanis (1965) opines that since a virus cannot infect plants at a temperature above 36°C, it became eliminated from the infected plants kept at high temperatures. Viral infections like curly top, curly leaf, leaf and crinkle virus induce both male and female sterility in sugarbeet, but mst induction is extremely high due to heavy viral concentration in anther and developing pollen grains. The mst induction by yellow virus in beets is already known (Larsen 1977, 1981).

v. *Soybean.* The viral infection in p-mst plants found by Cavinees et al. (1970) and Stelly and Palmer (1980b) is not associated with the mst, as grafting fails to transmit the sterility. Moreover, the infection reduces female fertility as well. In a soybean cultivar, p-mst and p-fst were induced by tobacco ring spot virus infection. The virus particles were present in the intine of pollen wall, cytoplasm of generative cell, integuments, nucellus, embryo sac wall and megagametophyte cells (Stelly and Palmer 1980b). In another strain, a viral mosaic invasion drastically reduced shoot height and male-female fertility (this Vol.). The virus is aphid-transmitted.

vi. *Pea and Bean.* In *Pisum sativum, Phaseolus lunatus* and *Vigna sinensis*, mosaic viruses reduce both male and female fertility considerably and thus do not exclusively induce mst.

vii. *Sunflower.* Atanasoff (1971) proposed a viral cause of mst but could not verify it. The viral infection reduces both male and female fertility in this plant.

6.3 Fungal and Insect Infestations

Chauhan et al. (1981) reported the occurrence of pollen sterility following infections of *Coriandrum sativum* by the fungus *Protomyces macrosporous*, of *Capsicum annuum* by cucumber mosaic virus, of *Raphanus sativus* by the aphid *Lipaphis erysimi* and of *Solanum melongena* by the nematode *Meloidogyne incognita.* The plants inoculated with their respective parasites at different stages of growth exhibited pollen sterility of variable degree, and those inoculated at the pre-floral bud initiation stage were completely pollen-sterile. Anthers of such plants exhibited anomalies in the development and behaviour of endothecium, tapetum and vascular strand. Deficiency of total carbohydrates, proteins, histones, DNA and acid phosphatase activity occurred in the anthers of infected plants. Since vascular differentiation in the anthers of infected plants was poor, the transfer of nutrients to developing pollen grains became inhibited and led to the above-mentioned deficiencies and pollen sterility. The infestation caused reductions in seed yield and fruit set. Thus, female fertility was also reduced following infection. Chauhan et al. (1981) did not study megasporogenesis or female fertility reductions in the infected plants. Both pollen and ovular sterility is induced in *Pisum sativum* following *Peronospora pisii* infection. This leads to significant reductions in seed and pod yield (Kaul 1977). Similar pollen as well as ovular sterility is induced in *Coriandrum sativum* following its artificial inoculation of *Protomyces macrosporous* chlamydospores. Thus, fungal and insect infections induce both pollen and ovular sterility, probably through depletion of nutrition and/or through malfunctioning of vascular tissue.

CHAPTER 7
Molecular Basis

7.1 *c*-Gene Location

Green plants have independent genetic and protein-synthesizing systems in the nucleus and the cytoplasm. In the latter, cytoplasmic organelle mitochondria and chloroplast, the major energy conversion sites with a vital role in the vigour and productivity determinations of a plant are considered as the carriers of cytoplasmic genes including *c*-genes, which are inherited and transmitted maternally. This is based on the premise that both these organelles are independently functioning and able to synthesise some proteins, independent of nucleus, due to the presence of specifically unique DNA, RNA, ribosomes and enzymes. The availability of site-specific restriction endonucleases provides a sensitive assay of sequence variability in chloroplast (cp) and mitochondria (mt) DNA, which is of relatively lower complexity compared to nuclear DNA. Differences between fragmentation patterns provide an index of genetic divergence, since the number and size of fragments is a function of the occurrence of a specific nucleotide sequence in DNA. Genes and DNA sequences have been characterised for nucleus, chloroplast and viruses of eukaryotes and for chromosomes, plasmids and viruses. Detailed restriction maps of cp-DNA of several higher plant species are now available (Leaver et al. 1982) and even location and sequence of chloroplast-specific r-RNA's, t-RNA's and several of the major protein-coding genes are known (Bedbrook and Koldnner 1979). In contrast, too little is known about the precise organisational pattern, informational content and functional properties of higher plant mt-DNA. Intramolecular heterogeneity, intraspecific variation, the unknown coding capacity of mt-DNA's, difficulties of transmission genetic studies and the lack of characterised mt-mutants have limited the available information about mt-genome of higher plants compared to that of yeast and mammals.

7.2 Mitochondrial Site

7.2.1 Genome Composition

Higher plant mitochondrial (mt)-genomes exhibit certain unique features like: (a) extraordinarily large size having six to nine fold variation within a single family, (b) independence of genome size to that of nuclei and chloroplasts, (c) nonproportionality of mt volume to its DNA content (unlike nuclear volume), (d) nonalteration from seed to seedling, and (e) genome comprising different sizes of circular mt-DNA molecules, none of which approach 200 MDa, which is the size of the smallest plant mt genome. Thus, mt genome is fractioned and its genes are independent, mobile and migratory. This last feature makes it a convenient site for the location of many unlinked independently functioning genes including cytoplasmic mst-genes designated presently *c*-genes. These genes are known from many plants including maize, sorghum, wheat and broad bean mst cytoplasms. In them they are located in mt-DNA genome (Forde and Leaver 1979; Pring et al. 1979a,b). This is based on evidence obtained especially from maize in which extensive investigation has been done.

Fragment Analysis. Rhoades (1950) first suggested that mutated mt might induce gc-mst in maize. Later it was found that cleaving mt-DNA from N, T, C and S mst cytoplasms of maize by restriction endonucleases and fractioning the fragments by agarose gel electrophoresis produces patterns specific to each cytoplasm type, irrespective of the endonuclease employed. Whereas most DNA fragments are common to all four maize cytoplasms, each cytoplasm is characterised by bands specific to that cytoplasm (Pring and Levings 1978; Levings et al. 1979b). Restriction endonuclease pattern differences occur between mt-DNA of the mst and mft lines of *Beta vulgaris* (Powling 1981, 1982), *Brassica napus* (Vedel et al. 1982), *Petunia hybrida* (Kool et al. 1983), *Sorghum vulgare* (Pring et al. 1979b, 1982b), *Vicia faba* (Boutry and Briquet 1981). However, no such differences occur in their cp-DNA's. In addition to the major mt-DNA, Boutry and Briquet (1982) detected five small DNA molecules of 1000–2000 base pairs in mst cytoplasms of *Vicia faba*. These are possibly similar to the low molecular weight circular and linear DNA found in mst maize and sugarbeet mt-DNA. The mt-DNA of mst S-350 cytoplasm of broadbean has one additional band. The mt-DNA patterns of S-447 and S-350 mst cytoplasms are identical to each other, but distinct by two to four fragments obtained for mft cytoplasms.

7.2.2 Genome Translation

To correlate maize mt-DNA variation with phenotypic expression, translation products of mt from mft and mst cytoplasms having the same nuclear background were compared by Forde et al. (1978, 1980). In addition to 18-20 polypeptides synthesised by mitochondria of maize cytoplasms, T-cytoplasm mt (T-mt) synthesise an additional unique 13,000 M_r polypeptide not synthesised by N, C and S-mt's. A 21,000 M_r polypeptide synthesised by these cytoplasm mt's is not detectable in T-mt. This deletion or mutation in the structural gene for the larger polypeptide may have led to the synthesis of an abbreviated 13,000 M_r from polypeptide in T-cytoplasm (Leaver and Gray 1982). Peptide mapping of the two labelled polypeptides has, however, failed to confirm this. C-cytoplasm mitochondria synthesise an additional soluble 17,500 M_r polypeptide, which replaces a membrane bound 15,000 M_r polypeptide synthesised by N-, T- and S-mt's. S-mt synthesizes eight characteristic minor polypeptides (in mw range 42,000–85,000) not synthesized by the mitochondria of other cytoplasms. Distinct mt-DNA differences exist between mst and mft *Sorghum*, *Petunia* and *Vicia faba* lines, and these are maintained in polypeptides synthesised by the isolated mitochondria of these lines (Boutry and Briquet 1981; Conde et al. 1982; Dixon and Leaver 1982; Pring et al. 1982b; Kool et al. 1983). Comparison of mt translation products from mft (Kafir) sorghum and a wide range of mst sorghum lines differentiates mst lines into three groups. The mt from the common milo mst source synthesise additional high molecular weight polypeptides (M_r 54,000–82,000) and contain two plasmid like DNA molecules (M_r 3.55×10^6 and 3.8×10^6) not found in other cytoplasms. 9E mst cytoplasm indicates mst to be associated with the synthesis of an altered, higher molecular weight form of subunit 1 of cytochrome oxidase. In maize, the introduction of genes that restore fertility to T-cytoplasm inhibits 13,000 M_r polypeptide synthesis in isolated mitochondria. However, synthesis of the missing 21,000 M_r polypeptide, a polypeptide synthesised by N-, C- and S-cytoplasms, does not occur in any of the restored lines investigated by Forde and Leaver (1980). On the other hand, the *Fr* genes that restore fertility to C- and S-cytoplasm mst lines have no effect on synthesis of the variant polypeptides by their mitochondria. Since the investigations were carried out on mitochondria from etiolated shoots, it is possible that the *Fr* genes which restore fertility to these cytoplasms exert their effect on mt-translation late during anther or pollen development.

7.2.3 Structural and Biochemical Alterations

i. *Structural.* Direct evidence implicating mitochondria with *fr* gene action was provided by Warmke and Lee (1977), who found mitochondrial degeneration in the tapetum and middle layers of anthers as the first sign of abnormality that preceded microspore degeneration in T-cytoplasm maize plants, although in some T-anthers it apparently did not happen. In tapetal cells of S-anthers, unusual loss of rough endoplasmic reticulum from the cells apart from cisternae associated with the organelle occurred (Colhoun and Steer 1981). Of the three sterile maize cytoplasms, T-cytoplasm is sensitive to *Helminthosporium maydis* race T. Toxin produced by this fungus that has recently been purified (Kono et al. 1980) specifically affects mitochondria and not chloroplast, and elicits a wide range of effects, most of which affect the permeability of the inner mt-membrane. This leads to uncoupling of oxidative phosphorylation and linkage of NAD^+ (Levings and Pring 1976; Gregory et al. 1977) and indicates that mitochondrial oligomycin-sensitive ATP-ase complex may act as a binding site for the toxin. Since only mst maize plants with T-cytoplasm are susceptible to *H. maydis* fungus and it is not possible to separate these two features, they appear as expressions of a single genetic defect. This is evidenced by the recovery of toxin-resistant mft plants (Cassini et al. 1977) derived by mutagenesis of the seeds obtained on T-cytoplasm mst plants. Restriction endonuclease analysis of the mt-DNA of HmT toxin-resistant mft plants regenerated from mst toxin-sensitive T-cytoplasm tissue cultures (Gengenbach et al. 1977; Brettel et al. 1980) reveals mt-DNA organisation of these mft plants as distinct from T- and N-cytoplasm maize plants. The altered mt-DNA appears to have arisen by recombination. In tobacco, physical recombination was evidenced by the occurrence of novel mt-DNA restriction patterns in hybrid progeny of two *Nicotiana tabacum* varieties (Belliard et al. 1979). Thus the mft toxin-resistant lines do not originate from selection of N-mt genomes co-existing previously within T-mt genome in T-cytoplasm line (Gengenbach et al. 1981). Furthermore, mt of mft plants obtained after regeneration from cultures do not synthesise the 13,000 M_r polypeptide distinctive of parental T-cytoplasm mt.

ii. *Biochemical.* Constant and specific differences in cytochrome oxidase content and isozymatic bands occur between mst and mft anthers of maize, rice, sorghum and wheat, but no such differences occur in succinic dehydrogenase. Since cytochrome oxidase is a partially mt-coded and succinic dehydrogenase a nuclear genome-coded enzyme, significant differences in cytochrome oxidase on a developmental level basis evidence mt-DNA involvement in gc-mst in these cereals. In mst sorghum (Alam and Sandal 1969), rice (Anonymous 1977; this Vol.) and maize (Ohmasa et al. 1976) anthers, a deficient or abnormal cytochrome oxidase system leads to defective and inefficient oxidative phosphorylation. Comparing mst and mft wheat anthers, Srivastava (1981) found significant differences not only in cytochrome oxidase but also in other biochemical functions of mitochondria like oxidative phosphorylation and ATP-ase activity. Moreover, specific structural differences occur in mt-DNA's of fertile and sterile anthers of maize, sorghum, petunia, brassica and broad bean. These differences are reflected in their translational products as well. Hence, constant and specific biochemical differences exist between the mitochondria of fertile and sterile anthers. These have been extensively investigated in maize, in which each cytoplasm type has a unique mt-DNA. Part of this investigation is described below and the details are included in Chapter 27.21.

T-cytoplasm mitochondria typically synthesize an additional characteristic 13,000 M_r polypeptide (designated T-polypeptide) which is not synthesized by other sterile or fertile maize cytoplasms. Three items of circumstantial evidence link this polypeptide with functional defects that arise in T-cytoplasm plants and cause mst:

a) polypeptides synthesized by N and T cytoplasm plants are virtually identical except that one 21,000 M_r polypeptide (N-polypeptide) detected in N-mt is missing in T-mt and is replaced by a T-polypeptide. The T-polypeptide is characteristic of T-mt in 11 nuclear backgrounds and is also synthesized by mt from 3 other mst sources classified as T-type (Forde et al. 1980),
b) *Fr* genes restoring fertility to T-cytoplasm inhibit T-polypeptide synthesis,
c) analysis of toxin-resistant mft plants obtained from T-mst tissue cultures with or without addition of HmT-toxin (Gengenbach et al. 1977; Brettell et al. 1980) reveal them to possess a distinct mt-DNA genome. Similar toxin-resistant maize plants were derived by seed mutagenesis of T-mst plants using EMS (Cassini et al. 1977). These plants did not synthesize 13,000 M_r T-polypeptide characteristic of T-mt but synthesized 21,000 M_r N-polypeptides typical for N-mt. Thus the selection of mutants resistant to HmT-toxin led simultaneously to both mft and loss of the variant T-polypeptide.

HmT toxin specifically affects inner mt-membrane of T-cytoplasm mft maize plants. In yeast mitochondria, all the membrane-bound translation products are components of the inner membrane (Schatz and Masson 1974). Since the polypeptide is also likewise membrane-bound protein, the variant is a *c*-gene product responsible for altering mt-DNA genome structure, its translation products and its response to HmT-toxin. Flavell (1974) suggested that T-mt to has an altered structure, making it susceptible both to HmT toxin and to an unidentified substance (X) produced during pollen development. Forde and Leaver (1979) proposed T-polypeptide as responsible for this altered structure (Fig. 27.26). Not the absence of 21,000 M_r N-polypeptide but the presence of a specific amount of variant 13,000 M_r T-polypeptide is essential for complete mst expression. This is indicated by the observation of Forde and Leaver (1980), who found a very low rate of polypeptide synthesis in restored T-cytoplasm plants having *Frfr* nuclear genes. N-polypeptide was not synthesized in mt of such plants. Such mt's are partially toxin-resistant (Barratt and Flavell 1975). This is probably due to the low rate of T-polypeptide synthesis. To the *c*-gene translation product T-polypeptide, Forde and Leaver (1979) assigned some possible roles like : (a) it provides a binding site for HmT toxin and compound X, (b) alters membrane conformations, creating or exposing a binding site on another membrane compound, or (c) weakens protein-lipid or protein-protein interaction in the membrane and thus makes it more susceptible to disruption by these substances. Since N-mt can be made susceptible to HmT toxin by treatments that remove their outer membranes (Watrud et al. 1975b), alternatives (b) and (c) seem feasible as they suggest conformational rather than compositional alterations to cause toxin susceptibility. Forde and Leaver (1979) gave a model for involvement of the polypeptide in mst and toxin sensitivity in T-cytoplasm maize plants (Fig. 27.26).

7.3 Plasmid Role

7.3.1 Behaviour

Plasmids are extrachromosomal genetic elements (Lederberg 1952) composed of circular closed DNA molecules, and occur in bacteria. These basically represent any genetic elements which, in addition to the normal genome, are propagated either independently in the cytoplasm or following insertion as an integral part of the host chromosome. The term "episome", introduced by Jacob et al. (1960), comprises the genetic material such as the F (fertility or sex) factor which can exist either integrated into the bacterial chromosome or separately and autonomously in the cytoplasm. Thus, while episomes are integrable, plasmids are not. In current usage, as also in maize and sorghum mst, this precision of meaning has become blurred and the terms episomes and plasmids are frequently used synonymously. This is evidenced from the following account.

i. *Maize.* Unlike T- and C-mst cytoplasms, S-cytoplasm of maize exhibits instability in mst expression. This instability is expressed as fertile revertants in nuclear backgrounds, not known to carry *Fr* genes. The reversion occurs spontaneously and in high frequency in some lines (Laughnan and Gabay 1973, 1975). In an inbred line, M_{825}, the authors detected two types of change viz. (a) cytoplasmic mutation to fertility (up to 10% frequency), and (b) low frequency nuclear genetic change in lines exposed to S- or Vg-cytoplasms. The nuclear revertants behaved as if they had acquired a *Fr* gene and, like the Rf_3 allele, exhibited a gametic mode of restoration. Their fertility was not due to rf_3 gene mutation to Rf_3, as each of the ten best-characterised nuclear revertants mapped at a unique site in different chromosomes and not on the long arm of chromosome 2, over which the Rf_3 gene is located. On this basis, the authors proposed that a mft element with the characteristic of a bacterial episome was involved in fertility restoration and reversion. The episome, they visualised, originated in the cytoplasm. When fixed in the cytoplasm, it induced cytoplasmic reversion to fertility, and when translocated to and incorporated into maize nuclear chromosomes, it caused nuclear reversions to fertility. Occurrence of these genetic elements was confirmed by Pring et al. (1977), who in addition to high molecular weight mt-DNA found S-cytoplasm to have equimolar amounts of two low molecular weight plasmid like DNA's (amounting to 5–15% of total mt-DNA). These were absent in T- or C-cytoplasms, and were designated S_1 and S_2, being 6.4 and 5.4 kb, respectively. They have terminal inverted repeats of ca.200 bp and carry a 1.4 kb homologous sequence at one end of the molecule (Kim et al. 1982). Sequences homologous to S_1 and S_2 DNA's have been identified in Wf_9 mt-DNA (Lonsdale et al. 1981), while homologous sequences are apparently deleted in T- and C-mst cytoplasms (Thompson et al. 1980; Lonsdale et al. 1983). In the S-cytoplasm, spontaneous revertants to mft exhibit the stable inheritance of the trait (Laughnan and Gabay 1975; Laughnan et al. 1981); transposition of S_1 and S_2 sequences in mt-DNA of these revertants is associated with these events. Levings et al. (1980) suggested this S-plasmid like DNA as equivalent to the mft elements (described earlier by Laughnan and Gabay 1975) and cytoplasmic reversion to mft as due to plasmid DNA integration into mt-genomic DNA. The transpositional events associated with cytoplasmic reversion are facilitated by the terminal inverted repeat sequences present in the S-plasmid. Such inverted repeats are characteristic of the transposable elements involved in insertional events in bacterial systems.

Since the DNA sequence of S-plasmid shares homology with N-maize mt-DNA (Thompson et al. 1980), it appears that S_1 and S_2 DNA's have arisen by excision from mt-DNA of mft maize cytoplasm and undergone alterations in form and function to become highly specialised in fertility control. Their structural specificity is evidenced by the fact that RNA isolated from S-cytoplasm hybridizes to S_1 and S_2 elements and to mt-genomic DNA. In addition, eight high molecular weight mt polypeptides characteristically synthesised by parental mst S-lines are not synthesised by the cytoplasmic revertants. Thus, a positive correlation exists between presence of the "free-living" plasmid like DNA's, synthesis of additional polypeptides, and the S-type of gc-mst in maize. Direct involvement of S_1 and S_2 episomes in fertility reversion in S-cytoplasm maize is complicated by the following evidence:

a) Fertile cytoplasms of Latin American races of maize have plasmids similar to S_1 and S_2 (Weissinger et al. 1982).
b) mst-S HMW mt-DNA contain virtually the entire S_2 episomal sequence and about 50% of the S_1 episomal sequence. Thus the new restriction bands in S-mft

revertant lines (Levings et al. 1980) may have originated from an S_1- and S_2-related sequence already present in HMW S-mt-DNA.

c) T-cytoplasm fertile revertants lack S_1 and S_2 sequences, implying that these sequences are not indispensable for normal pollen development in the absence of *Fr* genes and that mst in T- and C-cytoplasms is not due to the loss of these sequences (Kemble et al. 1982).

ii. *Sorghum.* As in S-cytoplasm maize, plasmid like linear DNA's were detected by Pring et al. (1982a) and Leaver et al. (1982) in mt-DNA of two mst sources M35-1 and IS 1112C (race durra) obtained from India. Designated as N-1 and N-2, they exhibit considerable homology with each other and some homology to S_1 and S_2 maize plasmid like DNA's. This is suggestive of an etiological relationship of the molecules to inheritance and expression of gc-mst in sorghum, perhaps in a manner analogous to the postulated role of these elements in maize.

7.3.2 Nature

Of the three known maize sterile cytoplasm types, only the S-cytoplasm possesses S_1 and S_2 episomal elements. The physical structure of S_1 and S_2 DNA's as described by Kim et al. (1982), their apparent integration upon cytoplasmic reversion to fertility (Levings et al. 1980), and their sequence homology with N-mt-DNA (Thompson et al. 1980) suggest that they arose as an IS-directed deletion mutation in S-mt-DNA from the normal mft mt-DNA. Such an IS-directed deletion in maize is consistent with, but by no means proves, Kleckner's (1977) speculation that "elements with very short terminal repeats might represent a further stage in evolution of an element in which one of the repeated units has been partially deleted to leave only a residual terminal segment still homologous to the opposite end". Each of the plasmid like DNA's from S-mt-DNA exhibits a short-inverted repeat at one terminus and what may be an IS-like element at the other. In addition to S-cytoplasm maize, two sorghum mst-cytoplasms possess plasmid like DNA's, the N-1 and N-2 elements (Pring et al. 1982b). However, similar plasmid like DNA's occur in a fertile primitive South American maize race (Weissinger et al. 1982). Likewise, a plethora of small circular or linear DNA's occur in maize mt-DNA's (Kemble and Bedbrook 1980). This occurrence indicates that these DNA molecules may be only fortuitously isolated in mst cytoplasms. However, the apparent transposon like behaviour in spontaneous heritable fertile revertants of S-cytoplasm maize (Levings et al. 1980) suggests the intimate role of these molecules in gc-mst expression. Mitochondrial DNA's from other mst maize sources share a variable extent of homology to S_1 and S_2 molecules (Spruill et al. 1980, Thompson et al. 1980, Koncz et al. 1981) and the T-, C- and S-mst sources exhibit deletion and rearrangement of sequences homologous to S_1 and S_2 (Lonsdale et al. 1983). The mt-DNA of mft maize cytoplasm possesses a near-complete copy of S_1 and S_2 sequences (Lonsdale et al. 1981). The homology between maize and sorghum plasmid like DNA's suggests that these mt-DNA's are related to a common function or trait (Pring et al. 1982a) especially to mst-mft expression, to which their relationship is strong and circumstantial. Such elements indicate that gc-mst in other species may also be related to the small, discrete, low molecular weight mobile genetic elements, the *c*-genes, which are primarily lodged in the cytoplasmic organelle, but can be transferred, transposed and translocated to nuclear chromosomes.

Maize and sorghum are not the sole examples of mt-DNA's that include the mobile genetic elements. Similar circular mt-DNA's of 1.3 to 1.5 kbp are also known in sugarbeet (Powling 1981). In all fertile cytoplasms, in addition to either 1.45- or 1.5-kbp DNA, 1.3 and 1.4 kbp DNA's occur. Since mt-DNA from S-cytoplasms has only 1.5 kbp DNA, the missing DNA characterises this cytoplasm. On the other hand, in other maize and sorghum mst cytoplasms, additional t-DNA is known. It appears that the genetic elements originate by excision and rearrangement and then undergo amplification that permits their easy detection. The mt-DNA of mft and mst sugarbeet also exhibits possible multimers of a small repeating unit, forming a "ladder" of species (Powling 1981). Similar observations were made by Dale (1981) in *Phaseolus* and N-cytoplasm maize tissue cultures.

7.4 Chloroplast Site

With the exception of S-maize cytoplasm digested by Hind III, the restriction fragment analysis of maize and sorghum chloroplast (cp) DNA's fail to demonstrate differences between N- and T-cytoplasm (Pring et al. 1979a,b). In tobacco, cp-DNA is not directly associated with mst expression. For instance, in a gc-mst *Nicotiana tabacum*, Chen et al. (1975) found a fraction-1 protein pattern of two small subunit polypeptides identical to those of normal *N. tabacum* and three subunit polypeptides identical to those of an Australian species that had contributed the cytoplasm. From seven gc-mst tobacco cultivars, Chen et al. (1976) showed that the mst factor lies in the chloroplast. The most convincing evidence for this is that in a spontaneously arisen mst tobacco cultivar, the mst mutation was accompanied by a mutation in cp-DNA genes coding for fraction-1 protein (Berbeč 1974). Of the six isonuclear alloplasmic mst tobacco lines studied by Frankel et al. (1979), four had the same cp-DNA fragmentation pattern as their maternal species, but in two it differed from either parental species due to nucleotide changes. Thus, genetic analysis of the large subunit of fraction 1 protein (Chen et al. 1977) and restriction endonuclease fragment analysis (Frankel et al. 1979) evidence involvement of cp-DNA in mst tobacco. However, using mst somatic hybrid plants, Belliard et al. (1979) opined that mt-DNA could play a role in gc-mst in *Nicotiana*. Aviv et al. (1980) and Aviv and Galun (1980) found no relationship between cp-DNA and mst in hybrids in which cp segregation and transmission was independent of mst and was fully random. In these mt-DNA differences between mst and mft lines were discrete. Chloroplast as carrier of mst in barley was documented by Ahokas (1978b), who found a higher carotenoid content, lower xanthophyll-B carotene ratio and deviated chlorophyll a, b ratio in restored plants compared to the nonrestored ones. The cp DNA restriction pattern differences between mst and mft *Sorghum vulgare* (Pring et al. 1982a,b) and *Brassica napus* (Vedel et al. 1982; Rémy and Ambard-Bretteville 1983) have been evidenced to implicate chloroplast as the *c*-gene carrier in these species.

CHAPTER 8

Origin and Causes

8.1 Origin

The fertility of an organism is controlled by three major gene groups (*st, mm,* and *mst*), the majority of which are in the dominant state. The mutation of any of these to the recessive state leads to anomalies in sex organ development, differentiation and sporo- or gameto-genesis. Of these, the most common are the sterility (*st*) genes which cause absence, abortion, deformation, malformation or transformation of sporophylls or inhibit pollination or fertilisation, leading to total or partial male-female sterility. In another group, the meiotic mutant (*mm*) genes do not influence sex organ differentiation or development, but affect male-female meiosis almost equally and impair gametic fertility partially to completely. In the third group, the male sterility (*mst*) genes suppress male sexuality, rendering the organism female fertile-male sterile. The *mst* genes impair androecial or PMC's differentiation and development, microsporogenesis and/or anther dehiscence. This produces an array of mst's classified on a phenotypic basis into structural, sporogenous and functional types (Table 1.1). Control of all these anomalies is either exclusively by nuclear or by the interaction of nuclear and cytoplasmic genes (Fig. 1.1). Both these mst types arise through mutations at either the genic or cytoplasmic or at both these levels.

8.1.1 Genic Male Sterility

Impairment of fertility by any of the three gene groups, viz *st, mm, mst,* is caused by mutations, the majority of which are recessive (Kaul and Murthy 1985). The g-mst is predominantly conditioned by monogenic recessive genes, in some by two or three recessive genes and only in a few by dominant genes (Fig. 2.6). Thus, g-mst has originated in nature by spontaneous mutations (Table 2.1). This is evidenced by the induction of a large number of mst mutants, in almost all of which the sterility control is by single recessive genes (Table 2.3).

8.1.2 Gene-Cytoplasmic Male Sterility

It results from *fr-c*-gene interaction and the sterility is under a coordinated control of both nuclear and cytoplasmic genes. Although the precise interaction of the two gene systems is not known, both nucleus and cytoplasm share genetic information for this type of sterility. This relationship provokes speculations that gc-mst arises when the genetic information necessary for mft is absent from the genomes of both cytoplasmic organelle and nucleus. Harvey et al. (1972) and Levings and Pring (1977) opine that the genes responsible for mft were originally located on organelle DNA, but in some, through an evolutionary process, these become transposed to a nuclear site. When such "mft minus" cytoplasms are introduced into nuclear backgrounds that lacked the mft (*Fr*) genes, mst results. Thus, when a redundancy of genetic information exists between the organelle and nuclear genomes, a portion of this information at one site (organelle) could be lost without any adverse effect as long as the same information is retained by the other (nucleus). Although the reciprocal situation is equally possible, the nucleus seems to have established a dominant role in this information coding system. Since parallel evolution is not necessarily equal, the distribution of mft genes between organelle and nuclear genomes need not be equivalent among all segments of the population (Leving and Pring 1979a,b). This may explain the existence of several gc-mst types within a species, that require specific *fr-c*-gene combination for mst development and certain *Fr* gene(s) for fertility restoration. Hybridization among population members could produce the nuclear-organelle combinations having *c*- and *fr*-genes. Such combinations would be mst. This hypothesis gains support from the Hind III and Sma 1 restriction endonuclease digestions of maize in which N mt-DNA possesses higher molecular weight than T mt-DNA. In other words, N-mt genome of maize has a larger gene compliment than T-mt genome. This suggests deletion rather than gene mutations as the cause of mst; but how *Fr* genes substitute for the deleted segment is not known. Furthermore, this hypothesis is not consistent with the protein differences detected by Forde and Leaver (1979) between mitochondria of N, mst T and T Fr (restored) lines. These authors gave an alternative hypothesis for T-cytoplasm origin and its *Fr* genes. In maize they proposed that 13,000 M_r polypeptide synthesis in T-cytoplasm is regulated by the *Fr* gene whose absence causes overproduction of polypeptide that alters the mt-membrane, making it susceptible to HmT toxin and impairs pollen development, through compound X production. Whereas the Fr_1 gene represses the polypeptide synthesis (Forde and Leaver 1980), Fr_2 and other modifier genes regulate compound "X" synthesis and reduce mt-membrane permeability to "compound X" or perhaps correct for the effect of 13,000 M_r polypeptide by some alterations in inner mt membrane properties. For this, Forde and Leaver (1979) presented a model (Fig. 27.26).

8.1.3 Fr-Genes

The gc-mst includes types that either lack *C*-cytogenes responsible for mft or have them in mutated *c* form. The defects can be corrected by genes (*Fr*) located in the nucleus. This relationship indicates that the same gene system may function in different genomes and that complementation occurs between nuclear and cytoplasmic genetic systems; both these systems appear to have evolved in harmony with each other. Not much is known about *Fr* genes, except that they restore fertility to S-cytoplasm plants. It is equally unclear how they substitute for *c* genes. But it is certain that they have evolved in the nucleus and have created an alternative pathway to circumvent cytoplasmic lesions. Alternatively, they may be *C*-genes which have transposed to a nuclear site. This is suggestive in maize S-cytoplasm, in which an episomal event has been proposed by Laughnan and Gabay (1975) to explain the origin of the newly arisen *fr*-genes, but the S-situation is unique and is not applicable to T and C mst maize.

8.2 Proposed Causes

Basically mst is conditioned by nuclear genes which may or may not be sensitive to cytoplasm. The nuclear *ms* genes function independently and are cytoplasm-nonsensitive, but the *fr* nuclear genes act only in cooperation with the S-cytoplasm that is assumed to have *C*-cytoplasmic genes. Thus, *fr*-genes, unlike *ms*-genes, are not independent of cytoplasm type. Accordingly, g-mst is an independent nuclear gene phenomenon but gc-mst depends on the interaction of specific nuclear *(fr)* and cytoplasmic *(c)* genes. The two mst-inducing nuclear genes, *fr* and *ms*, do not interact with each other, but act independently on androecial differentiation and development, tapetal structure and behaviour, microsporogenesis and/or on pollen form and function. Laser and Lersten (1972) proposed various factors which cause mst in higher plants. These and others are given in the following.

8.2.1 Androecial Maldevelopment

In some mst plants, misformed, malformed or aborted anthers are developed. Such structures lack anther sacs as well as microsporogenous tissue and consequently no pollen is ever produced by them, e.g. tomato, pea, barley, maize, cotton etc. (Tables 2.14, 3.7). This represents the earliest action of mst genes that act during stamen differentiation and development. In some plants, delay in the gene action results in normal androecial development, but the anthers produced are reduced and lack microsporogenous tissue so that no PMC's are formed, e.g. barley, pea, tomato and maize. Plants in which androecial misdifferentiation or abnormal development leading to nonproduction of microsporogenous tissue occurs are designated structural male sterile plants.

8.2.2 Feeble Vasculature

Investigations of anther morphology and anatomy of some mst plants have revealed certain structural changes in their stamen filaments, primarily in their vascular bundle. This has been considered as the reason for microspore abortion in such plants. For instance, in tomato, Andersen (1963a,b) observed that pollen abortion was associated with reduction in anther size and filament length. Likewise, comparison of normal and gc-mst sugarbeet by Rohrbach (1965b) indicates that while the filaments of normal stamens possess elongated cells with a large amount of cytoplasm, cells in the filaments of mst stamens are short, broad and with scanty cytoplasm. The author hypothesized that pollen abortion resulted from nutritional imbalances caused by malfunction of the vascular tissue of the stamens. On the other hand, Kinoshita (1971) suggested that stamen abnormality was not the cause but the effect of the mst gene. In mst rye anthers, cell layers around the vascular bundle of the connective are considerably thickened. That these thickenings impaired nutrient flow from vascular bundle to pollen chambers and led to microspore starvation was opined by Cebrat and Zadecka (1978); but in some other mst rye genotypes, although no such thickenings occurred, microspore abortion is complete. Likewise in *Triticum aestivum* (Joppa et al. 1966), *T. timopheevi* (Rao and Jain 1975), and *Sorghum* (Alam and Sandal 1967a), stamens of mst plants have poorly differentiated vascular bundles, unlike those of mft plants. These authors concluded that decreased starch accumulation in the tapetum and lack of starch storage in the maturing microspore might be due to reduced solute transport in stamens of mst plants. Light microscope, EM and SEM studies over stamen vasculature of sorghum have, however, failed to reveal any structural differences between mft and mst plants during all the developmental stages (Laser 1972). This eliminates anatomical differences as the cause of microspore abortion in sorghum.

8.2.3 Abnormal Tapetal Form and Function

The tapetum, the innermost wall layer of microsporangium, is the tissue in closest contact with the developing pollen grains. Tapetal cells and pollen grains develop from the same archesporial tissue. The mature pollen grains are haploid cells which have arisen from diploid PMC's and the tapetal cells are diploid, frequently polyploid cells arisen from diploid cells. The tapetum, after its development, plays an essential nutritive role in pollen grain formation, as the food materials entering the sporogenous cells must either pass through it or be metabolised by it. Although tapetum does contribute food material to the developing microspores, nothing precise is known as to what is contributed and how this contribution is affected. However, in addition to transmitting nutrients directly into the loculus, tapetal cells seem to be associated with the synthesis of certain components required during pollen development. For instance, the tapetum is probably responsible for the production of a β-1,3-gluconase (callase) which liberates the young microspores from thick callose walls of the meiotic tetrads (Eschrich 1961; Stieglitz and Stern 1973). During the final stages of pollen maturation, numerous derivatives of tapetum like lipids (Heslop-Harrison et al. 1973; Howlett et al. 1973) and also sporopollenin precursors (Dickinson 1973) are deposited on or within the pollen exine. Even incompatibility substances of certain cruciferae are also tapetal products (Heslop-Harrison et al. 1974). The tapetum, therefore, assumes an important role in pollen development, its final form and subsequent function. A biochemical dialogue between tapetal cells and developing pollen cells exists. When the normal communication between these two tissues is disturbed in any way, abnormalities in development occur giving rise to mst (Mascarenhas 1975).

A major cause of mst in higher plants is associated with the disharmony between sporogenous cells and the nearest closest tissue surrounding it, the tapetum. Whether it is the tapetum or other tissues of the sporophyte (Vasil 1967; Laser and Lersten 1972; Mascarenhas 1975; Frankel and Galun 1977), the general tendency is to assume that the primary cause of the breakdown is due to the tapetum. This is mainly based on the observations that tapetal anomalies, malfunctions and persistence are associated with pollen sterility in many mst plants (Tables 2.22 to 2.24, 3.20, 3.21). Moreover, abnormal tapetal behaviour induced by environmental factors (light, heat, drought, soil minerals, edaphic stress etc.), chemicals (male gametocides) etc. is invariably followed by a failure in pollen development. However, in many plants, although tapetal development and behaviour are normal, mst occurs (Tables 2.21, 3.22). Also in many others, tapetal and PMC or microspore abortion and tapetal breakdown may be almost simultaneous or successive (Tables 2.24, 3.23). In certain mst mutants of barley, maize, pea and tomato, meiotic breakdown occurs during early prophase-I. The abnormality is apparent in the sporogenous tissue without evidence of accompanying or preceding changes in the tapetum. This indicates that either the particular tapetal defect involved is not such as to be detectable by inspection or whatever influence disrupts normal development, it affects the PMC's but not the tapetal cells; even though it is chemical in nature, its route to the PMC's would be of necessity through the tapetum. Since extensive reorganisation occurs in the meiocyte during mid and late prophase, a higher sensitivity to disturbing influences during this period is not unexpected. It is equally possible that the primary event, the *mst* gene action, originates in the sporogenous tissue and triggers malfunctions or anomalies in tapetal form. In other words, the mst gene in some plants acts on sporogenous cells via sporophytic tissue. In fact, microsporogenesis, which is under the control of a large number of independently nonallelic major genes, and pollen development, which is a highly programmed and coordinated process (Kaul and Murthy 1985), cannot be entirely dependent on the malfunction of the tapetum alone. The influences exerted by the vegetative parts and organs of a plant or by the flower cannot be overlooked (Vasil 1967; Gottschalk and Kaul 1974).

8.2.4 Impaired Endothecial Development

Until microspore differentiation, endothecial cells usually remain small and thin-walled. After tapetal degeneration, active radial elongation of endothecial cells occurs and fibrous thickenings appear on their walls. This is followed by anther dehiscence and pollen liberation. Thus, pollen development is a programmed process that involves tapetal breakdown, endothecial development and anther dehiscence (De Fossard 1969). In g-mst, gc-mst and ch-mst plants of *Allium cepa, Beta vulgaris, Capsicum annuum, Cucumis melo, Cucurbita maxima, Datura alba, Ranunculus muricatus, Sesamum indicum, Triticum aestivum,* endothecial development is either feeble or fully inhibited. In the former case, no fibrous thickenings develop on the radial walls of endothecium (Chauhan 1980). In all such mst plants, the tapetum is persistent. On the other hand, in the mft anthers, endothecial development is normal, and radial thickenings develop following tapetal disintegration. This indicates that endothecial development is correlated to tapetal form and function, and that fibrous bands develop in the endothecium only after tapetal breakdown occurs. This correlation could be either due to inhibition or lack of some chemical stimulus or substance that impairs the programmed development of exine formation, tapetal breakdown, endothecium development and anther dehiscence. Thus, tapetal malfunction may not only be associated with pollen abortion but also with inhibition of endothecial development that renders anthers indehiscent. This control of the tapetum on endothecium development, known in some g-mst and gc-mst plants (Chauhan 1977, 1979a,b), was confirmed by Chauhan (1979a,b) in some ch-mst plants. Some chemical treatments induce p-mst and delay tapetal degeneration, but as soon as the tapetum degenerates, endothecial development proceeds normally in them. Thus, tapetal malformation is correlated to inhibit endothecial development. However, the role of endothecial development in causing mst could be indirect, because in almost all such known mst plants, tapetal malfunction is directly associated with impaired endothecial development.

8.2.5 Faulty Callose Deposition and Degradation

Callose production and accumulation prior to meiosis and its deposition around PMC's before prophase-I are prerequisites for the onset of a synchronous and normal meiotic division in higher plants. Likewise, callose deposition around each tetrad cell is essential for proper and autonomous transformation of four meiotic products into normal young microspores. On the other hand, callose degeneration is essential for microspore release from the tetrad capsule. This is affected by the enzyme callase. Both callose production, accumulation and deposition, and callase synthesis and timing activity are under the control of different nonallelic independently functioning major nuclear genes whose activity is time-, phase- and site-specific. Some mst genes affect the action of the genes responsible for callose synthesis or callase activity, and cause mst in certain barley, maize, pea and tomato mutants. On the other hand, a premature dissolution of callose occurs due to hypertrophy of the tapetum or noncoordinated time sequence of callase activity in petunia (Izhar and Frankel 1971), sorghum (Warmke and Overman 1972), onion (Nanda and Gupta 1974), maize (Lee et al. 1979), wheat and rice (this Vol.). When the tapetum degenerates too early, callose is not digested in mst pepper (Horner and Rogers 1974) and a rice mutant (this Vol.), in which microspores remain caged in the PMC quartet and degenerate subsequently. In late-developed rice tillers, tapetal degeneration is delayed. In these, callose is degraded by the callase enzyme, and microspores are released but degenerate soon after liberation. Thus, the mst in rice is not due to faulty callose dissolution or callase activity, but to mst gene action over the sporogenous tissue. The same may be true for other mst mutants exhibiting anomalies in callose deposition and dissolution or in callase activity.

8.2.6 Microsporogenesis Breakdown

The majority of the mst genes act directly over microsporogenesis anywhere during male meiosis and lead to either male meiotic arrest or to its abnormal progression. Most of the genes influence either prophase-I or microspore formation and development (Figs. 2.11, 3.20). However, some mst genes permit microsporocyte development but inhibit diploid microsporocytes from entering into meiosis. Such sporocytes simply develop vacuoles and disintegrate, e.g. winter squash (Singh and Rhodes 1961), tomato ms_3 (Rick 1948) and pea (Gottschalk and Kaul 1974) or undergo synchronous mitosis instead of meiosis, e.g. raspberry (Crane and Thomas 1949) and maize (Palmer 1971). Some other mst genes may permit microsporogenesis initiation, but may act anywhere later. The earliest action of such genes is known in some mutants of alfalfa (Childers 1952), maize (Beadle 1932b) and rice (Kitada et al. 1983), in which the chromosomes degenerate prior to synapsis. Some mst genes either completely inhibit homologous chromosomal synapsis, thereby causing asynapsis during male meiosis, e.g. maize (Beadle 1931), rubber (Ramaer 1935), *Godetia* (Hokansson 1943), ms_{10} tomato (Rick and Butler 1956), or permit initial synapsis but inhibit chiasma formation and lead to desynapsis, e.g. *Canna* (Offerijns 1938), *Alopecurus* (Johnsson 1944), alfalfa (Childers 1952), pea (Gottschalk and Kaul 1974), tomato (this Vol.). Some mst genes induce chromosomal stickiness in *Alopecurus myosuroides* (Johnsson 1944), *Tripsacum laxum* (Dodds and Simonds 1946), ms_{18} tomato (Rick and Butler 1956), cowpea (Sen and Bhowal 1962), and faulty chromosomal contraction in sweet pea (Faberge 1937), maize (Beadle 1932b); some other such genes arrest meiosis between diakinesis and MI and a "long-chromosome" phenotype is produced. Some genes affect either the MI spindle, leading to chromosome overcrowding or clumping at the spindle *(Tripsacum;* Dodds and Simmonds 1946) or to its absence (*Kniphofia;* Moffet 1932), (*Artemisia;* Kaul 1964, 1965), or to divergent spindle orientation (potato; Ellison 1936), or to irregular MII spindle orientation (muskmelon; Bohn and Principle 1964; pepper; Novâk et al. 1971). Some mst genes inhibit cytokinesis partially or fully during TI (ms_8 and ms_9 maize; Beadle 1932b) or during TII (*Kniphofia;* Moffet 1932; ms_4, ms_{10} maize; Beadle 1932b; pea; Gottschalk and Kaul 1974) so that the quadripartition is either complete or incomplete. Regular tetrads are never formed in such mst mutants.

8.2.7 Weird Postmeiotic Development

The majority of the mst genes act during postmeiotic stages, especially during or prior to haploid phase development (Tables 2.19, 3.15–3.17). In some mst mutants, the gene action is delayed until mature pollen is formed (e.g. *Oenothera;* Oehlkers 1927; Harte and Bissinger 1952; maize; Beadle 1932b; *Allium schoenoprasum;* Levan 1935; tomato; Rick 1948; this Vol.; *Petunia hybrida;* Welzel 1954; pea; Gottschalk and Kaul 1974; this Vol.). In certain mst tomato (Rick 1948; this Vol.), barley (Mian et al. 1974) and pea (Gottschalk and Kaul 1974), pollen is released without subsequent primexine formation. This pollen never functions. Mature pollen becomes sticky in *Brassica* (Cole 1959), *Lupinus* (Pakendorf 1970), pea and tomato (this Vol.) and is not released even though anthers dehisce. The plants are thus completely mst.

8.2.8 Anther Nondehiscence

Some mst genes do not impair anther or pollen development but inhibit anther dehiscence, so that the functional pollen produced by them is encaged in anther locules and aborts subsequently. This functional sterility is more prevalent in g-mst than in gc-mst plants (Tables 2.16, 3.10) and represents the most delayed action of mst genes, as viable pollen is produced but anther dehiscence inhibited, so that functional pollen is encaged in anther locules and aborts subsequently.

8.2.9 Mitochondrial Genome Mutations

Both mitochondria and chloroplast possess their own DNA which is distinct from and inherited independently of that in that nucleus. Chloroplast DNA (cp-DNA) of higher plants is traditionally considered as highly conserved. This is because it exists in multiple copies per organelle (Herrmann 1970; Kirk 1971). On the other hand, mitochondrial DNA (mt-DNA) genome of higher plants is astonishingly large and exhibits considerable size variation and genetic diversity within a single family. Even in closely related species, wide variations in complexity are evident. Extensive evidence suggests gc-mst in maize and sorghum as due to mutations in mt-DNA. The main evidence is based on the mt-DNA structure, its restriction endonuclease fragment analysis, translation products, disease susceptibility and chemical effects. These features distinguish three sterile cytoplasm types T, C and S in maize, and two types in sorghum. In S-cytoplasm reversions to fertility, both cytoplasmic and nuclear mutations occur frequently (Laughnan and Gabey 1973, 1978). The S-cytoplasm is unique in having two linear double standard DNA molecules (Pring et al. 1977) fertility elements with the characteristics of an episome. The episomal nature is supported by their transposon like structure (Kim et al. 1982) and by the disappearance of S_1 and S_2 episomes in the fertile revertants of S-cytoplasm plants coupled with the appearance of new restriction fragments hybridizing to S_1 and S_2. This suggests episomal integration as linked to the reversion phenomenon (Levings et al. 1980). Much evidence argues against the involvement of S_1 and S_2 episomes in mst (Lonsdale et al. 1983).

Alterations in mt-DNA restriction profiles between T, C and S cytoplasms of maize occur in the majority of cases at similar locations. This indicates that if mt-DNA mutations leading to mst were responsible for all the observed alterations in the mst mt-DNA genomes, then these mutations have occurred three times in evolution and the mechanism of this mutation was similar in all three cytoplasms, or alternatively that it occurred once in evolution, and the S-, C- and T-lines are recent evolutionary variants of maize. In many plants which exhibit gc-mst, differences in mt-DNA genome organisation are associated with the synthesis of additional characteristic polypeptides by isolated mitochondria and phenotypic expression of mst. Final proof for the involvement of the mt-genome in gc-mst will require the functional identification of the polypeptides involved and the sequencing of the mt-genome which encodes them.

8.2.10 Biochemical Disturbances

Compared to mft anthers, in mst anthers of wheat, rice, maize, sorghum, sunflower, sugarbeet, tomato, pea, lily, poppy, etc., the proline is either completely absent or significantly reduced, but the asparagine content is increased in the mst anthers of wheat, maize, sorghum, sunflower, tomato and pea. Whereas the increase could be due to accumulation or nonutilisation of asparagine, the decrease in proline may be due to either nonsynthesis or its active utilisation. Similar differences between mst and mft anthers in various other amino acids, as well as in enzymes like cytochrome oxidase, peroxidase, dehydrogenase and phosphatase, led many authors to conclude that these biochemical differences are the causes of mst. Such differences are neither constant nor uniform, but vary from stage to stage during anther development and microsporogenesis course, and exhibit differences between mst anthers of the same flower and from genotype to genotype of the same species. The influence of the environment on these chemical products is also high, but it is not correlative to any specific genotype or to its developmental stage. Differences in the level of many chemical substances, their accumulation or disappearance during the development between mst and mft anthers indicate normal versus disturbed biochemical communication between microsporocyte and their surrounding tissues. These differences represent the consequences and not the causes of mst. This is evidenced by the fact that the addition of proline to tomato, pea and rice mst anthers does not restore pollen fertility in them (this Vol.). The mst genes act autonomously either directly or indirectly through a distinct trigger, and regulation of their timing further conceals the source of the disturbed metabolism in the mst anthers.

8.2.11 Viral Infection

That cytoplasmic inheritance and infection genes are a "biological impossibility" was opined by Atanasoff (1971), who concluded that viral infections are the sole cause of mst at least in *Helianthus* and *Triticum.* Besides reviewing investigations indicating viral causes of pollen sterility, grafting experiments strengthened his views. He postulated that viral infection symptoms can often be overlooked by the novice, and that mst is not always characterised by gross morphological defects in the anthers and pollen. He suspected that faulty techniques and a general lack of understanding of viral infection have resulted in most of the negative results in grafting experiments. But even his own results do not give convincing proof in support of the viral nature of mst. Michaelis (1964) argued against extreme propositions of Atanasoff (1971), and evidenced experimentally against the viral basis of mst in many plants. Circumstantial and convincing evidence indicates the nonviral nature of mst in *Beta vulgaris, Datura alba, Glycine max, Helianthus annuus, Phaseolus lunatus, Pisum sativum, Vicia faba* and *Vigna sinensis.* In fact, the viral infection in these impairs both pollen and ovular fertility. A similar effect is caused by insect and fungal infestations. Moreover, no virus like inclusions, except probably in one mst-type broad bean (Grill and Garger 1981) are known to occur in mst anthers. Edwardson (1962) did show electron micrographs of root tip cells of T-maize with inclusions in "dense cytoplasmic areas" and similar inclusions were seen in tapetal cells of mst and maintainer lines. He did not believe that these inclusions could be viruses because "it seems improbable that only mst root tips would be virus-infected". All these findings indicate that viral infection cannot be considered as a cause of mst in higher plants.

8.3 Conclusions

All the above-mentioned proposed causes actually represent manifestations of mst and not the causes. Of these, in addition to misshapen androecial differentiation and development, the most common manifestation of mst is microsporogenesis breakdown either during prophase-I or microspore formation and development. Others are feeble vasculature development, impaired endothecial formation, abnormal tapetal form or function and faulty callose deposition or degradation. In maize and sorghum, mt-DNA mutations are well characterised. Viral infection usually induces both male and female sterility. Biochemical differences between mst and mft anthers represent the consequences of mst gene action rather than the cause of mst in higher plants. Therefore, the major cause of mst is the action of specific nuclear genes *(ms, fr)*, which act either with or without the cooperation of cytoplasmic *(c)* gene(s).

CHAPTER 9

Major Features

The mst genotypes are unique in exhibiting certain features, some of which can be generalised, while others are specific to these sex types and are not shared by others. Individual details for g-mst's and gc-mst's are given in Chapters 2 and 3, and the ones common to both that characterise typical mst's are indicated in this chapter. To economise space and avoid repetitions, the descriptions given are brief.

9.1 General Attributes

9.1.1 Origin

Mostly mst's arise either spontaneously or after species crosses. Unlike g-mst's that mostly arise spontaneously, the gc-mst's predominantly arise after species crosses (Fig. 9.1B) involving interspecific and intergeneric hybridizations. Since gc-mst results from the interaction of *fr* nuclear genes and *c* cytoplasmic genes, assemblage of *fr-c* genes in offspring is possible through hybridisations of plants having either *fr* or *c* genes. Hence most of the gc-mst's arise after species hybridisations.

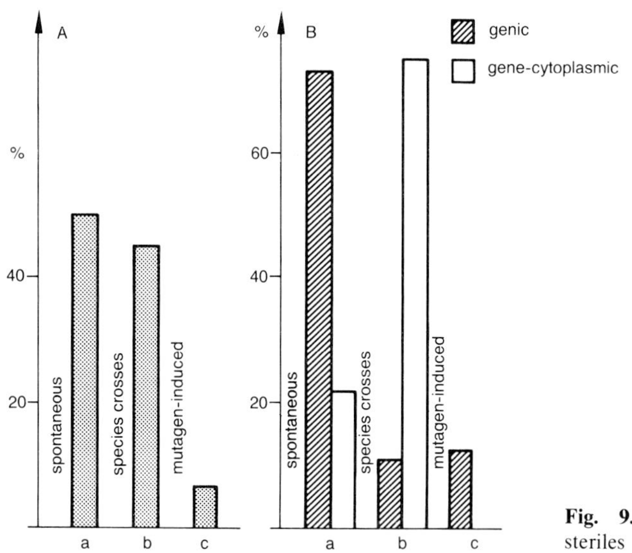

Fig. 9.1A,B. Frequency of male steriles

Frequency of g-mst's and gc-mst's arisen after induced mutations is 16 and 2%, respectively. The reduced frequency of the latter may be due to the lower rate of *C* gene mutations compared to that of *Ms* genes.

9.1.2 Types

In both g-mst's and gc-mst's, the sporogenous type is the most predominant (67%), followed by structural (25%) and functional (8%) types (Fig. 9.2). The high frequency of sporogenous mst's indicates mst gene action predominantly over microsporogenesis rather than on stamen development and differentiation or on anther dehiscence. In fact, inhibition of anther dehiscence by mst genes is reliably known in only a few mst genotypes (Tables 2.16, 3.10).

9.1.3 Gene Control

Single recessive gene control of mst is most predominant, occurring in about 64% of the mst's in which genetic control of mst is clearly known. Dominant gene control occurs in about 20% of such mst's (Fig. 9.3a). Multiple recessive gene control known in g-mst's is completely unknown in gc-mst's. Whereas polygene control present in gc-mst's is rare in g-mst's, dominant gene control occurs in a significantly higher number of gc-mst's than in g-mst's. Control of mst predominantly by single recessive genes indicates their origin by spontaneous mutations of *mft* genes. Thus, like other mutations, the majority of these are recessive and detrimental to the male sex and population fitness in hermaphrodites. Whether *Fr* and *c* gene mutations have been successive or simultaneous is not known. However, the former appears possible, as simultaneous mutations of nuclear and cytoplasmic genes leading to mst are rare. Since the majority of gc-mst's arise after species crosses, independent mutations of *Fr* and *c* genes are envisaged.

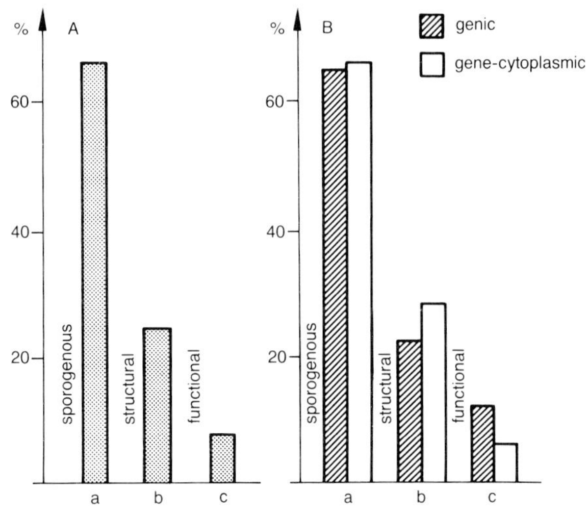

Fig. 9.2A,B. Male sterility types

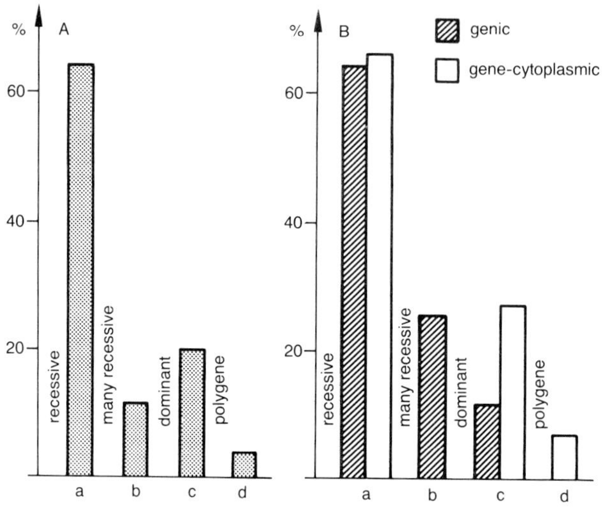

Fig. 9.3A,B. Gene control types in male steriles

9.1.4 Gene Action

The *mst* genes act before, during or after microsporogenesis and lead to the development of structural, sporogenous and functional mst's. Since the majority of these genes act over microsporogenesis, sporogenous types are the most frequent. In these, although breakdown occurs all through various microsporogenesis stages, in most of these it occurs during premeiosis, prophase-I, microspore formation and release, and during pollen development (Fig. 9.4). In 50% of the cases, breakdown

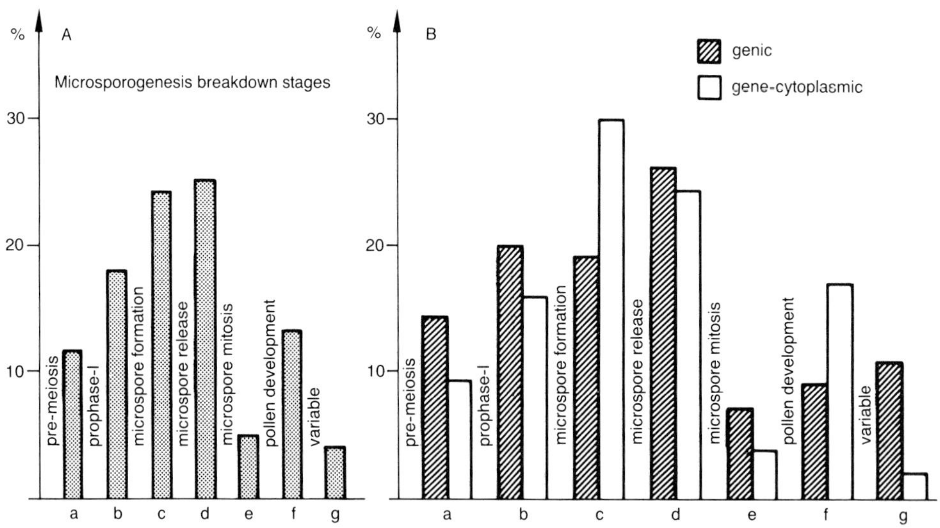

Fig. 9.4A,B. Action of male sterile genes on microsporogenesis

occurs during the last two stages. In g-mst's, the frequency of mst's in which microsporogenesis breakdown occurs during premeiosis and prophase-I is higher than those of gc-mst's. This comparative frequency is reversed during microspore formation and pollen developmental stages in these msts (Fig. 9.4b). In 60–70% of the mst's, the genes act mostly during and after microspore formation.

9.1.5 Tapetal Behaviour

In general, only 33% of mst's have normal tapetum, the remaining have either abnormal functioning (33.5%) or persistent (33.5%) tapetum. Whereas 50% of the g-mst's have normal tapetum, only 16% gc-mst's have such a tapetum. Persistent tapetum occurs in about (50%) g-mst's but only in 18% gc-mst's (Fig. 9.5). Frequencies of mst's with normal and persistent tapetal types in g-mst's is almost totally reversed in gc-mst's. However, the frequency of abnormal types is similar in both types, irrespective of tapetal form and function. In about 75% of mst's, microsporogenesis breakdown occurs during microspore development and pollen maturation stages irrespective of tapetal type (Fig. 9.6). This confirms my earlier contention that tapetal behaviour is independent of microsporogenesis breakdown and *ms* and *fr* gene action in mst's.

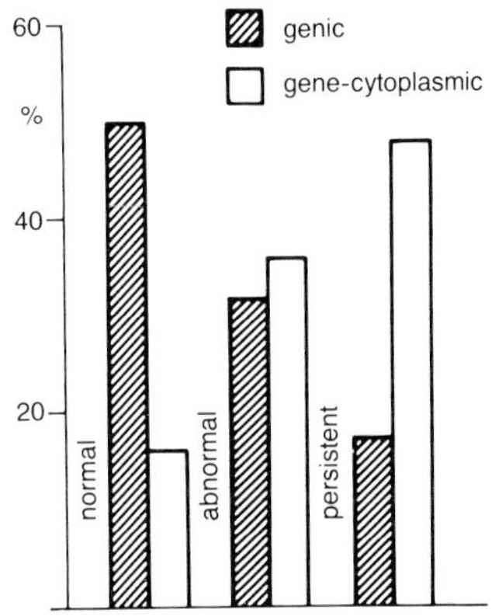

Fig. 9.5. Comparison of tapetal types in genic and gene-cytoplasmic male steriles

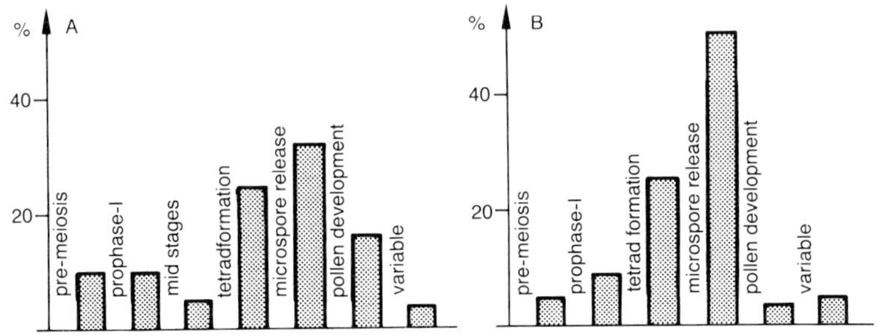

Fig. 9.6A,B. Microsporogenesis breakdown in plants with normal **(A)** and abnormal **(B)** tapetum

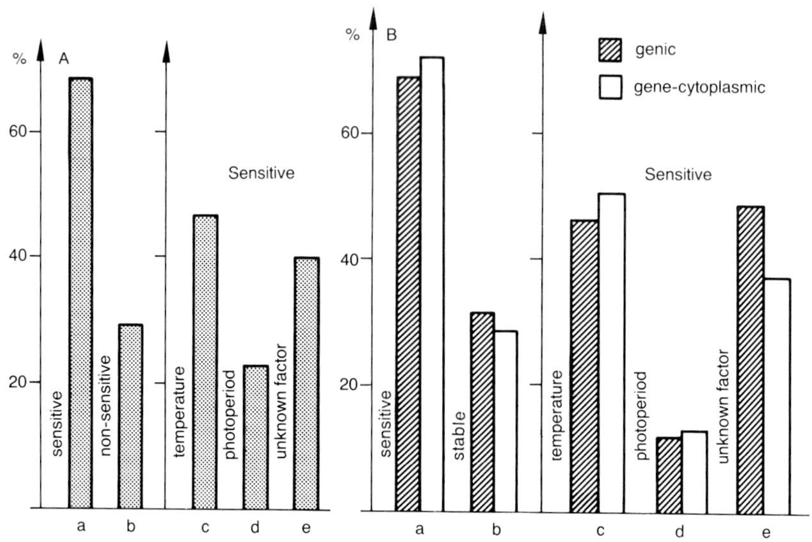

Fig. 9.7A,B. Influence of environment on male steriles

9.1.6 Environmental Sensitivity

Of the few mst's in which environmental sensitivity is tested, the mst expression is sensitive to environment in about 70%. Of these, about 50% are influenced by temperature, 13% by photoperiod and in the remaining the influencing factor is not known (Fig. 9.7). Since the differences between g-mst's and gc-mst's in the frequency of sensitivity to various environmental factors are insignificant, both exhibit almost equal sensitivity to the various environmental components.

9.1.7 Phenotypic Instability

In many alloplasmic hybrids (Table 3.6), the mst is highly sensitive to the nuclear genome, environment or physiological state. This results in instability by causing mst expressional breakdown in them, e.g. wheat, rice, maize, sorghum, pearl millet, sunflower, mustard, broad bean, sugarbeet, onion, tobacco, etc. In some alloplasmic mst wheats, *T. durum* having *Ae. umbellulata* or *Secale cereale* cytoplasm and *T. aestivum* having *T. timopheevi* cytoplasm, mft is restored partially in spring in the glasshouse but is lost in summer. In the later-formed branches of some mst *Brassica campestris, B. oleracea* and *B. napus* having *Raphanus sativus* cytoplasm, fertility is restored in the whole or part of the branch. Phenotypic instability of a high order occurs in *Vicia faba*. In this, reversion to mft is common, irreversible and hence permanent. The reversion covers whole or part of the inflorescence and renders the plant heterogenous. Similar sectorial reversions lead to phenotypic instability in S-cytoplasm maize. These are due to genetic changes at either the cytoplasmic or nuclear level. An episomal element capable of being fixed either in the cytoplasm or in the nucleus has been implicated in this reversion (Laughnan and Gabay-Laughnan 1983).

9.2 Special Attributes

9.2.1 Unimpaired Female Fertility

Male sex inhibition or suppression is the prime action of mst genes. This is associated with certain other traits given above that characterise the mst genotypes. Typical *mst* genes do not influence female fertility. Slight reductions in fertility are due to insufficient pollen received through NCP or due to limitations imposed by hand pollination. In fact, the genotypes with reduced male-female fertility are not mst mutants, but meiotic mutants (Kaul and Murthy 1985).

9.2.2 Increased Growth

Many mst's (pea, tomato, rice, cotton) exhibit slow and delayed reproductive growth, but are robust and luxuriant in vegetative growth. They remain green even when mft's have dried and turned yellow. Some mst's develop flowers on the late-formed branches or tillers. Barring a few in which female fertility is reduced, in most of the mst's reproductive fitness is enhanced due to increased flower number, higher seed and fruit set, better seed germination or greater vigour of offspring. This is well known in some gynodioecious species.

9.2.3 Altered Anther and Pollen Phenotype

The inability of plants to produce or release functional pollen leads to mst. Excepting fr-mst's, other mst's either lack normal anthers or have reduced, shrivelled, light-coloured anthers that dry up quickly. Unlike in mft flowers, anthers of msts are not exserted. Usually they are pollenless, but those that develop pollen have it in shrunken, deformed and degenerated form.

9.2.4 Coexistence of *ms* and *fr* Genes

The nuclear *fr* and cytoplasmic *c* genes are complementary and thus interdependent for gc-mst expression. The nuclear mst genes, *ms* and *fr*, are independent, nonallelic and do not exhibit interactions like complementations, supplementation, inhibition, epistasis etc. when they coexist in the same genome. A similar relationship exists also for the nuclear *ms* and cytoplasmic *c* genes in maize, sorghum, cotton and petunia. In maize, such independence of *ms*1, *fr*1 and T-cytoplasm *c* genes action was detected by Jones (1956). Similar independence of gene action occurs in *ms*8 and *fr*1 genes in maize (this Vol.). Male meiosis breaks down during leptotene in T *fr*1 *fr*1 *ms*8 *ms*8 plants. Early breakdown is due to the *ms* gene and late breakdown to *fr* gene action. Likewise in *Petunia*, microsporogenesis breakdown occurs during early meiosis in S *fr fr msms* plants, which is typical of the *fr* gene and at the tetrad stage in S *Frfr msms* plants, which is characteristic for *ms* gene. In *Petunia*, Evenor (1984) concluded that both *ms* and *fr* genes can coexist in one or two doses in the same genome, the *ms* genes can coexist in one or two doses with *fr* genes in somatic cells or female gametes; for this coexistence, the presence of *Fr* genes is not obligatory.

9.2.5 Stage-Specific Gene Action

Five types of genes, viz. sterility *(st)*, asynaptic *(as)*, desynaptic *(ds)*, female sterility *(fst)* and male sterility *(mst)*, impair meiosis and lead to gametic sterility. While the first three cause both male and female gamete sterility, *fst* and *mst* genes impair female and male fertility, respectively. Of all these, mst genes are most prevalent in plants (Gottschalk and Kaul 1974, 1980a,b; Frankel and Galun 1977; Kaul and Murthy 1985) and specifically impair male sex acting mostly during prophase-I and after microspore liberation (Fig. 9.4). They do not influence megasporogenesis. Their frequent occurrence and recurrence is due to their high mutation rate, as the majority of the mutant genes are nonallelic. Some of these nonallelic genes influence the same meiotic stage, e.g. barley, maize, pea, tomato etc.

9.2.6 Nuclear and Cytoplasmic Gene Diversity

The mst genes are characterised by a high mutation rate. This is evidenced by the presence of over 60, 55, 48 and 38 nonallelic (Table 2.31) *ms* genes in maize, tomato, barley and pea, respectively. Likewise, both *fr* and *c* genes exhibit a considerable mutation rate. This is known in potato, maize, wheat, pearl millet, sorghum and onion, in which many nuclear *fr* genes and cytoplasm types exist (Table 3.29).

9.2.7 Linkage and Pleiotropy

Some *ms* and *fr* genes are linked to other genes (Table 2.27, 3.19) while a few exhibit pleiotropy (Tables 2.28, 3.19). Special traits like increased vegetative and reproductive growth, altered anther and pollen phenotype, deviated tapetal form and function and many biochemical alterations unique to mst's represent either pleiotropic expressions of *mst* genes or their tight linkages with other genes. The possibility of linkages is ruled out, as the existence of so many closely linked genes is highly improbable. Additionally, they remain inseparable even in large segregating populations. Phenological, physiological and biochemical differences between sex phenotypes or between those having different S-plasmon types occur in *Allium, Beta, Capsicum, Helianthus, Lycopersicon, Pennisetum, Pisum, Solanum, Sorghum, Triticum* etc. These types support the existence of pleiotropy, as they are limited mainly to their flowering parts, especially to anther tissues.

9.2.8 Altered Tapetal Form and Function

For the maturing heterotrophic spore, the tapetum has a nutritive function as all food materials, growth substances, water and other vital nutrients must pass through the tapetal cells. In addition, the tapetum is associated with the synthesising callase enzyme, which liberates young microspores from the thick callose walls of meiotic tetrads. During the final stages of pollen maturation, lipids, proteins and sporopollenin precursors and some amino acids are transferred from the tapetum to the pollen. This transport is facilitated by channels and apertures in the maturing exine wall (Rowley 1976). Although in the majority of mst's details of tapetal development and its functioning are not known, in the few that are known, *mst* genes impair either tapetal development or its function or both. In some, the tapetal development is imperfect or abnormal (Tables 2.22, 3.20). This causes biochemical disturbances leading to inefficient tapetal functioning. In others, premature or delayed tapetal breakdown occurs during microsporogenesis (Tables 2.23, 3.21) due to which nutrient supply to the developing spores is denied, which starves them to death. In a few, tapetal development is not significantly altered but its function is impaired through biochemical disturbances. In such cases the tapetum breaks down normally (Tables 2.21, 3.22) but probably functions abnormally.

9.2.9 Biochemical Differences

In the majority of plant species, biochemical differences between mst and mft anthers are not investigated. In the few known ones (cereals and legumes) the mst anthers are characterised by lower metabolic activity, reduced carbohydrate, protein and proline contents and enhanced amount of asparagine. Also the mst and mft anthers differ in many specific enzymes like cytochrome oxidase, esterase, dehydrogenase and phosphorylase. In some, callase activity is precocious or delayed, e.g. *Capsicum, Lycopersicon, Oryza, Petunia, Phaseolus, Pisum* and *Zea*. Since such differences are constantly found, these biochemical alterations are considered as specific to these mst genotypes.

9.2.10 Environmental Influence

Of the five known types of sterility genes, mst genes are highly sensitive to environment, especially to temperature and photoperiod (Tables 2.25, 3.25, 3.27, Fig. 9.5). Of these, the p-mst's are most sensitive. This sensitiveness renders them mft in one and mst in another environment. Thus, through environmental manipulations, it is possible to produce hybrid or selfed seed over these genotypes. In addition to p-mst's, some g-mst mutants of pea, tomato, barley, maize, mustard, pepper etc. are highly sensitive to the environment, so that these mutants are mst in the glasshouse and mft in the field or vice versa. Such extreme environmental sensitivity is a common feature of p-mst genotypes. This type of sensitivity is unique to mst's and is therefore considered as their special feature.

9.2.11 Genotypic and Phenotypic Diversity

Of all the steriles, mst genotypes exhibit stupendous diversity in gene action, gene control, tapetal behaviour, origin and sterility type (Tables 9.1, 9.2). The mst genes act anywhere during anther differentiation and development, microsporogenesis or gametogenesis, irrespective of gene control of the sterility. The control ranges from monogenic to multigenic recessive, dominant or polygenic types. Likewise, the tapetal structure and behaviour are divergent. The tapetum is normal in tomato, maize, soybean, pepper and pea, and normal or persistent in barley, potato and sunflower (Table 9.2). In maize, onion, sunflower and barley, it may be either normal, abnormal or persistent (Tables 9.1, 9.2). The mst's arise spontaneously or after intra- and interspecific or induced mutagenesis. They exhibit sporogenous, structural and functional mst, the majority being of the sporogenous type. This enormous phenotypic and genotypic diversity among mst's within and between species makes them unique among sterile genotypes that have high genetic value and breeding potential.

Table 9.1. Some features of gene-cytoplasmic male steriles with diverse gene action

Species	Gene control	Tapetal behaviour	Origin and type[a]
Gene action			
a) Premeiotic to postmeiotic			
Aegilops caudata × *Triticum durum*	nk	Normal, abnormal	2,5
Ae.ovata × *T.aestivum*			
Ae.ovata × *T.durum*			
Brassica napus	1–2r	Normal, abnormal	2,5,7
Capsicum annuum	1–2r	Abnormal, disintegrates in early meiosis	1,5,7
Helianthus petiolaris × *H.annuus*	nk	Disintegrates normally or precociously	2,5
Petunia axillaris × *P.hybrida*	2r	nk	2,5,7
Solanum crosses	1r–5r, 1–4D, po	(Refer to Tables 47.13, 47.16)	1 to 7
Zea mays	1–2r	Normal, persistent or abnormal	1,5,7
b) Meiotic to postmeiotic			
Allium cepa	1r	Normal, abnormal, persistent or malfunction	1,5,7
Beta vulgaris	2r	Abnormal	1,2,5,7
Lycopersicon sp.	1r	Persistent	2,5,7
Medicago sativa	1–many r, po	Simultaneous disintegration	1,5,7
Oryza sativa	1r	Normal	1,2,5,6
Pennisetum typhoides	1–3r	Persistent	1,4,5,7
Secale cereale	1–2r	Persistent	1,5,7
Sorghum vulgare	1–5r	Persistent	1,2,5 to 7

[a]Arisen spontaneously (1) or after intraspecific or species crosses (2) or mutagen-induced (3), structural (4), sporogenous (5) and functional (6) male sterility types, *ms-* or *fr*-genes sensitive to environment (7).

Table 9.2. Some features of genic male steriles with diverse gene action

Species	Gene control	Tapetal behaviour	Origin and type[a]
Gene action			
a) Premiotic to postmeiotic			
Capsicum annuum	1–2r	Normal	1,3
Cucurbita maxima, C.pepo	1r	Abnormal form and function	1,4
Gossypium hirsutum	1r, 2r, 1D	Normal to defective	1,3,4,5
Hordeum vulgare	1r, 1D	Normal, abnormal, persistent	1,3 to 6
Lycopersicon esculentum	1r, 2r	Normal	1,3 to 6
Pennisetum typhoides	1r, 3 com	nk	1,3 to 5
Petunia hybrida	1-many r	Malfunction	1,4 to 7
Pisum sativum	1r	Normal	1,3 to 6
Secale cereale	1-many r	Abnormal	1,4,5
Solanum tuberosum	1-many r, 1D	Normal, persistent	1,5,7
Sorghum vulgare	1-many r	nk	1,3 to 5,7
Triticum aestivum	1–3r, po, 1D	nk	1,3 to 6
T.monococcum	1r	nk	1,3,5
Zea mays	1r, 1D	Normal	1,3 to 5
b) Meiotic to postmeiotic			
Brassica campestris	1r to 2r	Intact or degenerates prematurely	1,3 to 7
Glycine max	1r	Normal	1,3 to 7
Helianthus annuus	1r to 5r	Normal, abnormal, persistent	1,4,5
Medicago sativa	2r, 2 dup. 3 po.	Abnormal	1,5
Nicotiana tabacum	1r, 2r	Normal to abnormal	1,3 to 6
Vigna sinesis	1r	Hypertrophied	1,4 to 6
c) Normal to loss, abortion or transformation of anthers			
Daucus carota	1r, many r 1r+1D	nk	1,3 to 5
Gossypium barbadense	1r, 1D	nk	1,4,5
Pisum sativum	1–2r	Normal	1,3 to 6

[a]For numerals refer to Table 9.1.

CHAPTER 10

Gynodioecy

Nearly 85% of angiosperms are hermaphrodites and selfing occurs in most of them. This inbreeding enhances genetic homogeneity and narrows down the genetic base, limiting their adaptability and evolutionary potential. The limitation in them is negated by outbreeding, which is enforced by the development of forms having either a mixture of male and female flowers (monoecy) or of hermaphrodite and female (mst) flowers (gynomonoecy) or of hermaphrodites with male (fst) flowers (andromonoecy). The conjoint occurrence of hermaphrodites and female (mst) plants in natural populations was detected by Darwin (1877), who designated the phenomenon as gynodioecy, and the species having such populations as gynodioecious species. In some, hermaphrodites occur in conjunction with female sterile male plants. Such species are known as androdioecious species. Whereas gynodioecy is prevalent, androdioecy is extremely rare. Since mst's are common in gynodioecious plants, their constant maintenance is possible through outbreeding, as they must be pollinated by hermaphrodites. Thus gynodioecy constitutes an outbreeding mechanism whose breeding system is comparable to that of dioecy and self-incompatibility. Unlike them, however, it involves interbreeding of mst's and hermaphrodites. The hermaphrodites contribute pollen grains which fertilize their own ovules as well as those of the mst plants. Consequently they contribute their genes mostly via pollen and to a lesser extent through ovules. On a similar basis, female fertile plants of androdioecious populations transmit their genes mainly via ovules and to a lesser extent through pollen. Thus, the plants of dimorphic androdioecious and gynodioecious populations differ from those of monomorphic (hermaphrodite and monoecious) populations, as the former forms possess an outbreeding advantage which the latter lack. However, because of unidirectional pollen transfer, androdioecy cannot be maintained by an outbreeding advantage, whereas gynodioecy can. This difference may be one of the principal reasons why gynodioecy is much more common than androdioecy. Furthermore, the inability of cytoplasmically determined female sterile mutants to be maintained in an androdioecious system provides another reason to account for high frequency of gynodioecy (up to 3.5%) than that of androdioecy, which is extremely rare. In about 120,000 species, only ten are androdioecious (Yampolski and Yampolski 1922) and even these might be the result of year-to-year variation in sex expression (Bertin 1982).

10.1 Genetic Control

Compared to the large list of gynodioecious species (Table 10.1), the genetic control is known only in a few of them (Table 10.2). In many, precise gene control is still not established. For instance, initial studies indicated mst in gynodioecious species like *Cirsium oleraceum, C. acaule* (Correns 1928), *Plantago lanceolata, Satureja hortensis* (Correns 1908) as cytoplasmically transmitted, but later investigations over these and many others proved the control by interaction of dominant *(Fr)* or recessive nuclear *(fr)* genes with S-cytoplasm (Table 10.2). In *Fuchsia,* both recessive and dominant gene control of mst is known (Arroyo and Raven 1975). In

Table 10.1. Partial list of gynodioecious plant species

Family	Species
Amaranthaceae	*Chamissoa albida, C.nodiflora, C.pyramidalis, Charpentiera bracteata, C.ovata*
Anacardiaceae	*Rhus integrifolia, R.ovata*
Boraginaceae	*Anchusa officinalis, Echium vulgare, Myosotis laxa, M.palustris, M.scorpioides*
Campanulaceae	*Campanula bononiensis, C.carpatica, C.siberica, C.stevenii*
Caryophyllaceae	*Agrostemma githago, Arenaria biflora, A.serpyllifolia, Cerastium alpinum, C.arvense, C.brachypetalum, C.cerastoides, C.fontanum, C.glomeratum, C.latifolium, C.pumilum, C.semidecandrum, Holosteum umbellatum Lychnis flos-cuculi, L.viscaria, Minuartia sedoides, M.verna, Moehringia muscosa, Myosoton aquaticum, Sagina saginoides, Scleranthus annuus, S.perennis, Silene chlorantha, S.maritima, S.multiflora, S.nutans, S.viscosa, Spergula arvensis, S.media, S.morisonii, S.pentandra, Spergularia marina, S.rubra, Stellaria alsine, S.graminaea, S.holostea, S.media, S.nemorum, S.palustris*
Chenopodiaceae	*Chenopodium botrys, C.foliosium, C.rubrum, C.quiano,*
Compositae	*Anthemis tinctoria, Aster alpinus, A.fruticosus, A.umbellatus, Bellis sylvestris, Bidens triplinervia, Calendula officinalis, Carduus defloratus, Carthamus caeruleus, Centaurea jacea, C.montana, C.nigra, C.scabiosa, Cirsium acaule, C.canum, C.ciliatum,C.crinitum, C.dissectum, C.eriophorum, C.erisithales, C.esculentum, C.ferox, C.gracile, C.helenioides, C.incanum, C.lanceolatum, C.monspessulanum, C.montanum, C.niveum, C.oleraceum, C.palustre, C.pannonicum, C.rivulare, C.spinosissimum, Matricaria matricarioides, Serratula tinctoria*
Convolvulaceae	*Breweria cordata, B.elegans, B.evolvuloides, B.latifolia, Calystegia acetosaefolia, C.barbara, C.obtusa C.sepium, C.soldanella, Convolvulus arvensis, C.flavus, C.microphyllus, C.soldanella, C.villosa*
Crueiferae	*Arabis pumila, Cardamine amara, C.pratensis, Hirschfeldia incana*
Cyperaceae	*Eriophorum angustifolium*
Dipscaceae	*Cephalaria leucantha, Knautia arvensis, K.dipsacifolia, K.drymeia, K.fleischmanii, K.integrifolia, K.longifolia, K.magnifica, K.purpurea, Pterocephalus papposus, Scabiosa argentea, S.arvensis, S.atropurpurea, S.canescens, S.columbaria, S.graminifolia, S.lucida, S.ochroleuca, S.succisa*
Epacridaceae	*Cyathodes acerosa, C.colensoi, Leucopogon fasciculatus, L.melaleucoides, L.suaveolens*
Ericaceae	*Epigaea repens, Gaultheria antipoda, G.procumbens, G.rupestris, Pernettya mucronata*
Euphorbiaceae	*Euphorbia peplus, E.serrata, Fontainea* sp., *F.pancheri, F.picrosperma*

Table 10.1. *(continued)*

Family	Species
Gentianaceae	*Gentiana aspera, G.bellidifolia, Swertia perennis*
Geraniaceae	*Erodium cicutarium, Geranium columbinum, G.mulle, G.palustre, G.phaeum, G.pratense, G.pyrenaicum, G.robertianum, G.sanguineum, G.sylvaticum*
Gesneriaceae	*Cyrtandra bicolor, C.decurrens, C.humilis, C.pilosa*
Gramineae	*Arundo richardii, Bouteloua chondrosioides, Cortaderia araucana, C.archboldii, C.atacamensis, C.pilosa, C.richardii, C.selloana, C.splendens, C.toetoe, Danthonia archboldii, Poa hubbardiana, P.indigesta, P.pallens, P.resinulosa, P.sellovii, P.seoparia*
Grossulariaceae	*Ribes alpinum, R.petraeum, R.uva-crispa*
Hydrophyllaceae	*Nemophila menziesii*
Iridaceae	*Iris douglasiana, I.germanica, I.graminea, I.pumila*
Labiatae	*Acinos alpinus, Ajuga genevensis, A.reptans, Ballota nigra, Calamintha grandiflora, C.nepeta, Clinopodium vulgare, Colebrookia oppositifolia, C.ternifolia, Dracocephalum austriacum, D.moldavicum, D.ruyschiana, Galeopsis ladanum, G.tetrahit, Gleochoma hederacea, Hissopus officinalis, Horminum pyrenaicum, Lavandula angustifolia, Lycopus europaeus, Melissa clinopodium, M.officinalis, Mentha aquatica, M.arvensis, M.cunninghamii, M.hirsuta, M.longifolia, M.puelegium, M.suaveolens, Nepeta cateria, N.nuda, Origanum majorana, O.vulgare, Prunella grandiflora, P.laciniata, P.vulgaris, Salvia austriaca, S.dumetorum, S.glutinosa, S.nemorosa, S.offiicinalis, S.pratensis, S.sclarea, S.sylvestris, S.verticillata, Satureja hortensis, Scutellaria galericulata, S.hastifolia, Stachys germanica, Teucrium scorodonia, Thymus pannonicus, T.praecox, T.pulegioides, T.serpyllum, T.vulgaris*
Limnanthaceae	*Limnanthes douglasii*
Onograceae	*Epilobium dodonaei, E.hirsutum, Fuchsia colensoi, F.excorticata, F.microphylla, F.perscandens, F.ravenii, F.thymifolia, Oenothera octovalvis*
Piperaceae	*Lactoris ansonia, L.fernandeziana, Pepromia blanda, P.floridana*
Plantaginaceae	*Plantago amplexicaulis, P.coronopus, P.lanceolata, P.major, P.maritima, P.media, P.ovata*
Plumbaginaceae	*Armeria juncea, A.maritima, Limonium sinuatum, L.latifolium, L.vulgare*
Polemoniaceae	*Gilia aggregata, G.capitata, Phlox bifida, P.diffusa, P.ovata, P.pilosa, Polemonium caeruleaum*
Polygonaceae	*Fagopyrum esculentum, F.tataricum, Polygonum amphibium, P.historia, P.viviparum*
Ranunculaceae	*Aquilegia vulgaris, Hepatica nobilis, Ranunculus acris, R.bulbosus, R.hybridus, R.limosella, R.polyanthemos, R.repens*
Rosaceae	*Potentilla anserina, P.aurea, P.reptans, P.supina*
Rubiaceae	*Gouldia axillaris, G.coriacea, G.macrocarpa, G.terminalis, Ixora acuminata, I.chinensis, I.coccinea, Richardia scabra Sherardia arvensis*
Sapotaceae	*Planchonella firma, P.punctata, P.sericea, Pontederia alba, P.cordata, P.dubia, P.ovata, P.villosa*
Saxifragaceae	*Broussaisia* sp., *Carpodetus serratus, Philadelphus incanus, P.mexicanus, Ribes alpinum, R.rubrum, Saxifraga granulata, S.marginata, S.rosacea, S.sancta*
Scrophulariaceae	*Digitalis grandiflora, D.lutea, Hebe leiophylla, H.parviflora, H.salicifolia, H.subalpina, H.townsonii, H.traversii, Pedicularis kaufmanii, Verbascum densiflorum, V.lychnitis, V.phlomoides*
Thymeleaceae	*Daphne mezereum, Pimelea prostrata, P.traversi*
Umbelliferae	*Daucus carota, Falcaria vulgaris, Gingidium decipiens, G.trifoliata, Lignocarpa diversifolia, Seseli hippomarathrum, S.libanotis*
Valerianaceae	*Centrathus ruber, Valeriana montana, V.officinalis, V.tripteris*
Violaceae	*Viola arvensis, V.hirta, V.orphanidis*

Table 10.2. Gene control of male sterility in some gynodioecious species

Species	References
a) Dominant gene control	
Centranthes ruber	Sobrinho (1946)
Cortaderia richardii, C.selloana, Fuchsia excorticata, F.procumbens	Lloyd (1974)
Limnanthes douglasii var.*rosea*	Baker (1966), Jain et al. (1978)
Origanum vulgare	Lewis and Crowe (1956)
Plantago coronopus, P.lanceolata, P.ovata	Ross (1969, 1970a), Mittal and Issar (1970)
Salvia pratensis	Linnert (1958)
Thymus vulgaris	Allen (1940), Valdeyron et al. (1977)
b) Recessive gene control	
Campanula carpatica	Pellew (1917)
Centaurea scabiosa	Fröst (1963)
Chenopodium quinao	Simmonds (1971)
Cirsium oleraceum, C.palustre	Correns (1916)
Geranium sylvaticum var.*alpestre*	Ganders (1978)
Hirschfeldia incana	Horovitz and Galil (1972)
Nemophila menziesii	Ganders (1978)
Polemonium coeruleum	Ostenfelds (1923)
Satureja hortensis	Correns (1908)
Silene dichotoma	Correns (1906)
S.maritima, S.vulgaris	Marsden-Jones and Turrill (1925)

Centranthus ruber (Sobrinho 1946), *Cortaderia richardii* (Connor 1973), *Origanum vulgare* (Lewis and Crowe 1956), *Plantago lanceolata* (Ross 1969), *P. ovata* (Mittal and Issar 1970), control of mst is digenic, F_2 yielding 13 : 3 for hermaphrodite : mst plants. In them, mst is conditioned by a dominant gene (*Fr*) whose expression is inhibited by another dominant gene *(I)* so that only *Fr i* genotypes are mst. Digenic complementary gene control of mst in *Fuchsia procumbens* (this Vol.) needs confirmation, as many segregations suggest monogenic dominant gene control. S-cytoplasm and *fr*-nuclear gene diversity occur in *Thymus vulgaris* (Valdeyron et al. 1977), *Origanum vulgare* (Kheyr-Pour 1975, 1981), *Nemophila menziesii* (Ganders 1978) and *Plantago lanceolata* (Van Damme 1983a,b). In *Origanum vulgare*, gene *F* causes anther suppression, but its action is inhibited by another dominant gene, *H* (Lewis and Crowe 1956). Thus only the combinations *FFhh* and *Ffhh* yield mst's, others, like *FFHH, FFHh, FfHH, FfHh, ffHH, ffhH*, produce hermaphrodites. The *ffhh* combination is inviable. A near lethality of *ffhh* genotypes has not been found in other species in which this gene system has been proposed for mst. Also various genotypes under this model have to possess strong viability differences which is without any sound genetical or physiological basis. The control of mst in *O. vulgare* is not as simple as envisaged. Some mst populations in many gynodioecious species produce hermaphrodite progeny whose offspring may be mst. Also many intermediates, like hermaphrodites bearing some mst flowers, appear in *Cirsium, Origanum* and *Plantago*. The influence of the environment on mst suppression is also high.

In *Plantago lanceolata*, the mst is governed by g-c interactions in two forms and by only nuclear genes in the third form. Variation in the nuclear and cytoplasmic genes occurs in all possibilities in most of the populations (Van Damme 1983a,b).

These variations lead to morphological differences between mst's possessing different plasmon types. In all the three mst forms that have been distinguished by the authors, more than one nuclear gene, even up to five in one form, appears to be involved in the mst control. The nuclear inheritance is highly complex, but gynodioecy is stable and special, as successful establishment of mst has occurred on probably three different occasions, leading to the formation of three mst populations.

10.2 Occurrence and Frequency

Seventy to ninety percent of angiosperms are hermaphrodite except in some islands where they are around 55% only. In the sexually dimorphic types, gynodioecy occurs in about 2.5 to 3%. Delannay (1978b), however, recorded 7.5% for Belgium. Its distribution in some angiospermic families is scattered (Table 10.1) and occurs both in insect- and wind-pollinated species. Although several examples of gynodioecy exist in self-incompatible species, the system is generally associated with the ability for self-pollination (Charlesworth and Ganders 1979). The frequency of mst's in natural populations of gynodioecious plants varies from species to species and from locality to locality within a species. In general, it varies between 0 and 50%, but is below 35% on an average (Fröst 1963; Ross 1969; Horovitz and Galil 1972; Webb and Lloyd 1980; Webb 1981). For some Labiatae species, high frequencies are known viz. up to 62% for *Origanum vulgare* (Kheyr-Pour 1980) and up to 93% for *Thymus vulgaris* (Dommée et al. 1978). Even within populations of a species, the percentage of mst's varies considerably, e.g. 14–39%, 33–51% and 49–51% in *Pimelea prostrata, P. oreophila* and *P. sericoevillosa*, respectively (Burrows 1960). In 12 gynodioecious New Zealand populations of *Fuchsia*, mst's range from 4–40% (Godley 1955). At three locations, the ratio of hermaphrodites to mst's in *Pimelea prostrata* is 1.5, 2 and 3 (Burrows 1960). A similar variation occurs along line transects. Both location and seasonal differences in frequencies of the two phenotypes occur in *Ranunculus limosella* (Melville 1964) and *R. acris.* Whereas in some localities the mst's are predominant, in others they are rare. In *Centaurea nigra, Cirsium canum, Gentiana bellidifolia, Geranium pratense, Mentha longifolia* and *Nepeta nuda,* the ratio of hermaphrodites to mst's varies from 1 to 3. Whereas the frequency of mst's in natural populations of gynodioecious *Cortaderia richardii* is about 38%, it is over 50% in the introduced *C. selloana* (Connor 1965a,b) which is widely naturalized now in New Zealand. In *C. splendens* and *C. toetoe,* 51 and 43% of mst's occur, respectively, in New Zealand (Connor 1971). In New Guinea, 50% mst's occur in *C. archboldii* (Connor and Penny 1960). In India, their frequency is about 12%.

10.3 Sex Expression

The existence of two sex phenotypes, hermaphrodites and mst's, is a constant feature of gynodioecious plants. This sex dimorphism is genetically conditioned and is stable in many species, *Fuchsia colensoi, F. excorticata* and *F. perscandens* (Godley 1979). In a few, environmental factors like temperature, photoperiod and soil fertility alter sex expression considerably. For instance, in *Pimelea traversii*, in some localities, some mst flowers appear in early-formed flowers of hermaphrodites (Burrows 1960) which are also interspersed by intersex flowers. In other locations, sex expression is normal. The earlier-formed few flowers of mst *Arabis pumila, Dianthus armeria, Polygonum bisorta, Silene nutans* and *Stellaria alsine* plants are hermaphrodite in some populations and mst in others. A precise environmental or physiological factor influencing sex expression is not known.

10.4 Floral Size Differences

Reduction in stamen or anther size is a nearly constant feature of mst's. In addition, the flowers of *Fuchsia colensoi, F. excorticata, F. perscandens* (Godley 1955), *Pimelea* (Burrows 1960), *Cerastium fontanum, Silene multiflora, Spergula pentandra* (this Vol.) are smaller than those of hermaphrodites. The gynaecia of mst's are the only parts of the flower which are not significantly reduced in comparison to those of hermaphrodites (Kesseli and Jain 1984). In *Pimelea*, the mst flowers have much larger and prominent stigma than hermaphrodites. Ovaries of hermaphrodites of *Cortaderia richardii* are 0.6–0.8 times those of mst's (Connor and Penny 1960). In *C. selloana*, ovary size is nearly similar, but spikelet number per spike exceeds in mst's (Connor 1965a). In *Limnanthes douglasii*, mst's and their progeny have greater biomass and more flowers per plant than hermaphrodites (Kesseli and Jain 1984), but the results vary with environment.

10.5 Maintenance

Two to fifty percent of mst plants occur in stable gynodioecious populations of various species. This equilibrium between mst's and hermaphrodites is stable and well maintained in nature. For this a continuous and rigorous selection of mst's is needed in these populations. Outbreeding is the only source of maintenance of mst's, the pollen source being the hermaphrodites. In a population with a nearly equal number of mst's and hermaphrodites, each hermaphrodite contributes three times as many genes as does each mst. This is because the hermaphrodite contributes the genes through its pollen and egg nuclei and also through the pollen that fertilizes the mst plant. However, in a population with a low proportion of mst's, viz. 1–4% or so, the hermaphrodite contributes nearly twice as many genes as does one mst plant. This is because of the nearly negligible population of the mst's and the low probability of receiving the pollen from hermaphrodites by these mst plants. This asymmetry in reproductive roles is absent in dioecy and in other mating-type polymorphisms like self-incompatibility (Charlesworth and Charlesworth 1979; Heuch 1979) or heterodichogamy, a polymorphism of protogynous and protandrous plants (Gleeson 1982). Hence these systems can exist and perpetuate autonomously, unlike in gynodioecy in which mst's cannot leave offspring produced via pollen. Thus, compared to hermaphrodites, the mst's are at a disadvantage and should normally become lost from the populations unless they possess some compensating mechanisms or special features. Some of these mechanisms present in them are given below.

10.5.1 Enhanced Seed Production

Many mst's increase their seed set over hermaphrodites by having more spikelets per floret (*Cortaderia selloana;* Connor 1965a) or an increased number of ovules that develop into seeds (*Carpodetus serratus;* Shore 1978). The ratio of fruit set between mst's and hermaphrodites is 2, 10, 20 in *Pimelea traversii, P. prostrata* and *P. sericoevillosa* and *P. oreophila,* respectively, (Burrows 1960), in *Leucopogon melaleucoides* it is 1.44 (Carlquist 1966), in *Rhus integrifolium* and *R. ovata* it is 2.2–4.2 (Young 1972). In two *Thymus* species and in *Satureja hortensis,* the seed weight of mst's is significantly higher than that of hermaphrodites. Assuming mono- or di-genic inheritance, the increased seed set is insufficient to maintain the 35–44% mst's in *Pimelea traversii* found in natural conditions. That gynodioecy functions as a population strategy for increasing seed production in mst's from the resources becoming available after the inhibition of pollen production was hypothesized by Horovitz and Beiles (1980). This maximum seed hypothesis postulated earlier for dioecious species was severely criticised by Lloyd (1974), as it implies group selection. Since conditions for such selections to occur are stringent (Hull 1980), they may not apply to gynodioecious species. Under the assumption of nuclear inheritance, three times seed over mst's relative to hermaphrodites would maintain 25% mst's and a relative seed set of 10 and 100 would maintain 44.4 and 49.5% mst's respectively (Ross and Shaw 1971), but this assumption is unlikely to hold many gynodioecious species in nature. Therefore a supplementary mechanism for mst maintenance exists.

10.5.2 Heterozygote Advantage

In addition to differential seed fertility, the heterozygote advantage associated either with sex control loci or in the form of generalized heterosis has been regarded by Ross and Weir (1975) and Lloyd (1975a,b) to contribute to the maintenance of mst's in gynodioecious populations. The mst operates as a via media for hybridity maintenance and consequently heterosis in a population. Reciprocally, the maintenance of mst's through heterosis depends on its mode of inheritance. Based on the genic type of mst, the heterozygote advantage cannot account for the over 30% mst's in gynodioecious populations. However, partial or complete lethality of some homozygotes (Lewis and Crowe 1956; Linnert 1958; Jain 1968), as also the high mutation rate, may add to the differential seed set and heterozygote advantage for maintaining a high proportion of mst's in naturally occurring populations.

10.5.3 Inbreeding Depression

This, caused by selfing, has been considered to place hermaphrodites in a disadvantageous position compared to mst's in competitive situations. In *Limnanthes douglasii*, the mst's exhibit inbreeding depression lower than hermaphrodites (Kesseli and Jain 1984), but the selfing value should exceed the minimum of 0.5. Rarely, this is exceeded in *Thymus vulgaris*, one of the best-studied gynodioecious species (Valdeyron et al. 1977; Dommée 1981). The selfing rate in *Cirsium palustre* is 17% (Correns 1916), in *Origanum vulgare* 37% (Elena-Rosello et al. 1976) and in *Cortaderia richardii* it is 75% (Connor 1973). These values indicate that inbreeding depression, although important, is not the only mechanism involved in mst maintenance. In *Thymus vulgaris*, the maintenance is partly achieved by inbreeding depression and partly by pleiotropic effects (Assouad et al. 1978). Likewise, in *Fuchsia colensoi*, *F. excorticata* and *F. perscandens*, seeds from selfed hermaphrodites have lower germination and the progeny has less vigorous development than those of mst's or of outcrossed hermaphrodites.

10.5.4 Self-Incompatibility

The mst and self-incompatibility are independent genetic systems that prevent selfing. A combination of both these, although rare, is known in *Ameria maritima*, *Hirschfeldia incana*, *Limonium vulgare*, *Plantago lanceolata* (Baker 1966; Horovitz and Galil 1972; Ross 1967). Each system maintains heterozygosity and regulates sib- and outcrossing differentially, so that crosses possible with one system are prevented by the other. In *P. lanceolata*, a deficit of hermaphrodites occurs (Ross 1969). A linkage between mst and incompatibility genes may contribute to this deficit. If so, then this combination could add to the maintenance of mst's in gynodioecious populations.

10.5.5 Pleiotropic Effect

Most of the hermaphrodite species are self-compatible. In them, evidence for pleiotropic effects of *mst* genes are hard to obtain because the sexes also differ in pollination conditions and inbreeding amount (Charlesworth 1981). In self-incompatible species, fitness differences between sexes may be interpreted as pleiotropic effects, but the combination of self-incompatibility and gynodioecy is rare. In *Origanum vulgare*, the viability effects of mst's cause overdominance that helps to maintain mst (Jain 1968). Overdominance may also result from pleiotropic effects of sex chromosomes in subdioecious species when YY homozygotes are inviable (Ross 1980). Evidence of *mst* genes exhibiting pleiotropy do exist (Tables 2.28, 3.19), but now important pleiotropic effects are in gynodioecious species must be found out.

10.5.6 Ecological Factors

Environmental-dependent fitness is known in *Thymus vulgaris*. In it, the reproductive superiority of mst's relative to hermaphrodites enhances with increasing plant diversity (Bonnemaison et al. 1979) and the amount of inbreeding depression is much higher when offspring of selfed and crossed hermaphrodites are grown together than when they are grown apart (Perrot et al. 1982). Furthermore, while the mst's favour dense, relatively often disturbed vegetations, the hermaphrodites prefer rocky sites with low population density (Dommée et al. 1978). This differential distribution is due to unfavourable pollination conditions for mst's at low densities and high outbreeding advantages in the varying environments. In *Silene vulgaris,* mst's attain the highest frequency in meadowlands and the lowest in woodlands (Brockmann and Bocquet 1978). This is due to selection operating via differences in flowering propensity.

10.5.7 Resource Allocation

Hermaphrodites have to distribute their reproductive resources between the two sex functions and there is every likelihood that in the "sex mutants", the resources freed by loss of male or female function are transferrable to the sex function that remains. However, how much exactly the hermaphrodites distribute their resources between the two sexes is not clearly known. By analogy with the effect of inbreeding on sex ratio in dioecious species, Hamilton (1967) inferred that, compared to outbreeders, self-fertilizing hermaphrodites are likely to allocate lower amounts of resources to male function, but this may not alter their fitness value because a mst mutant in a hermaphrodite plant would not receive any such increment in fitness, as the flowers must be retained to subserve the remaining female function (Maynard-Smith 1978). Likewise, female sterility is also not advantageous in such a situation. Whereas theoretical studies of Charnov et al. (1976) and Charlesworth and Charlesworth (1981) suggested that the resources saved by mst's by producing sterile instead of fertile anthers can be invested in reproduction via ovules through the compensating mechanism of Darwin (1877), Lloyd (1982) pointed out that many mst mutations act late on anther development, so that savings on paternal costs are likely to be small. Furthermore, the time lapse between the moment of saving and of investment in maternal structures may be considerable, implying that there is no guarantee that the investment will really occur (Van Damme 1983a,b).

10.6 Conclusions

It appears that the equilibrium of mst's and hermaphrodites in gynodioecious populations is maintained through a balanced gene system appropriated by a combination of advantages offered by increased reproductive output, inbreeding depression, heterozygote advantage, self-incompatibility and ecological factors etc. Since in the majority of cases, the gynodioecy is governed by g-c interactions, a proportion of the mst's and hermaphrodites will achieve equilibrium when the potential of S- and N-cytoplasms is nearly equal. In many, the mst control is of bi- to multigenic type. This inheritance type, in association with a specific cytoplasm, constitutes an effective, perhaps mutually reinforcing and self-balancing mechanism, for the maintenance of a high proportion of mst's in gynodioecious populations. Thus, gynodioecious plants represent genetically atypical sample of mst mutations in heterotic species. For maintenance of mst's, many theoretical models supposing one nuclear gene (Lewis 1941; Jain 1961; Valdeyron et al. 1973; Lloyd 1974, 1976; Ross and Weir 1975; Charlesworth and Charlesworth 1978; Ross 1980; Gregorious et al. 1982), two or more nuclear genes (Ross and Shaw 1971; Ho and Ross 1973; Ross 1978), cytoplasmic (Lewis 1941; Lloyd 1974, 1975a,b) or nuclear-cytoplasmic inheritance of mst's (Charlesworth and Ganders 1979; Charlesworth 1981; Delannay et al. 1981) and based on inbreeding depression, overdominance, multiplicative or phenotypic effects have been given. In most of these, fitness values represent the relative number of successful gametes contributed to the next generation by each genotype. Most of these models are based on nuclear inheritance and only a few on nuclear cytoplasmic inheritance, although the latter is more common in natural gynodioecious species. Two different models, each with the assumption of inbreeding depression in hermaphrodites and a relatively higher ovule production in mst's, were studied by Charlesworth and Ganders (1979). Neither model permits joint equilibria, as either cytoplasmic genes becomes fixed and result in nuclear mst or nuclear mft *Ms* gene becomes fixed and result in neutral polymorphism for the cytoplasmic gene, and consequently no mst results. Joint equilibria that assume pleiotropism, variable pollination and an effect due to sex phenotype on ovule production favouring mst's were obtained in the models of Charlesworth (1981) and Delannay et al. (1981). Both the models are complex. In them, no simple relationship exists between relative seed production of mst's compared with hermaphrodites and mst frequency at equilibrium. Delannay et al. (1981) assumed complete selfing of hermaphrodites, full fertilization of mst's and the nuclear *ms* genes have pleiotropic effects on fitness. The last assumption implies heterogeneity among hermaphrodites. However, all these features are not present in naturally occurring gynodioecious populations. Since genetics and fitness values of mst's in most of the gynodioecious species are not clearly known, all the models presented so far are based only on theoretical assumptions.

CHAPTER 11

Utility, Limitations and Lacunae

11.1 Utility

11.1.1 Hybrid Production

There are no holidays in hybrid seed production, because for a single F_1 combination, emasculation and pollination have to be done repeatedly and carefully, the former being more tedious, time-consuming and harder than the latter. Furthermore, emasculation is especially difficult in plants with small and fragile flowers. This tedium of hand emasculation is greatly obviated by using mst lines. Even with most careful emasculation, over 5–15% selfing occurs, in contrast to mst lines, where it is 1–5%. Furthermore, unlike in emasculation, the mst plants are undamaged, suffer less from various infections, can be fertilized more easily and produce higher yields. Moreover, pollen sterility relieves plants from the burden of pollen production and from the stress on plants during flowering. This results in a higher yield in mst's compared to in those emasculated artificially especially in maize, rice, barley and pea. In maize, high plant populations, drought and hot weather are the types of stress which give advantage to mst lines (Duvick 1965). Since mst lines are effective females, in outbreeders the seed set on these lines is hybrid seed, being the product of NCP. On the other hand, in inbreeders pollination through pollen transfer from mft to mst parents has to be done. This pollen transfer is a formidable obstacle to large-scale hybrid seed production and is now being solved through use of mechanical and other devices. This has led to the use of mst in almost all the economically useful plants. With increasing labour cost, cheap and reliable methods of optimising NCP over inbreeding mst plants are being developed. The mst has special use in hybridising plants that have perfect flowers with few seeds per flower and where seed price cannot cover the expenses involved in hand emasculation. In some mst lines of pepper, barley, tomato and pea, the flowers open out fully, and sufficient hybrid seed is produced by them. In purple cauliflower, a mutant gene causes considerable stigma extrusion. The NCP on such mst plants is very high. Mass scale cultivation of mst rices in China has revolutionised hybrid rice yields.

In ornamentals where mostly g-mst is available, use of the mst gene in hybrid breeding has been limited. The main reason is the high cost of hiring technical skills, low seed output and poor sale. Moreover, ornamental plant seeds have low viability and less longevity. Therefore, plentiful hybrid seeds of high germination capacity, longevity and viability and high market value are needed by commercial growers. This has been achieved in *Ageratum*, *Antirrhinum* and *Zinnia*. Use of the *ms* gene in horticulture has a high value in plants that can be propagated vegetatively. In such cases, synthetic varieties can be developed using polycross techniques.

11.1.2 Recurrent Selection Breeding

This selection is an effective breeding technique to dissipate undesirable linkage and increase desirable gene combinations in a population. Due to difficulties in emasculation and low NCP, initially it was extensively used in cross-pollinated crops. Later, its use increased in self-pollinated plants. The main disadvantageous features of conventional breeding systems for self-pollinated crop improvement are limited gene pool size, restricted recombination and genetic variability and absence of intercrossing among hybrid progenies. All these disadvantages are obviated by use of mst genotypes in a recurrent selection breeding program. However, low NCP still remains the main obstacle. To facilitate NCP in a recurrent selection program, efficient pollen transfer is necessary. Voluntary or involuntary selection in a population for mechanisms that enhance pollen transfer efficiency may substantially change population characteristics in a positive or negative direction. In recurrent selection, g-mst is more useful than gc-mst, as it is easier to manipulate a single *ms* gene than to incorporate the specific *fr-c*-gene combination. In the latter, appropriate *Fr* genes are also to be sought. Therefore, g-mst becomes a suitable alternative to gc-mst where *Fr* genes are difficult to manage or find. Following Gilmore's (1964) suggestion to use this technique for population improvement, it has been used effectively in sorghum (Dogget 1972), soybean (Brim and Stuber 1973) and rice (Fujimaki 1979).

11.1.3 Development of Composites

Since a substantial amount of heterozygosity can be retained by using mst lines, composite crosses have now been established in cereals and some edible legumes. In maize, rice, barley, sorghum, pearl millet, sugarbeet, cotton and sunflower, many bi- to multiparental crossings and synthetic crosses have been performed and varieties developed. By combining highly selected genotypes into single- or multiple-cross hybrids, genetic gain is maximised, since both additive and dominance genetic variances are utilised. Development of random mating populations in selfers like barley by using mst was first demonstrated by Suneson (1940, 1945) and subsequently followed in many other plants. The mst barley was successfully used by Suneson and Houston (1942) for mass inoculation with a pathogen culture transmissible through floral infections.

11.1.4 Increased Reproductive Fitness

The majority of the mst mutants arise spontaneously as recessive mutants. In self-pollinating species, most of these are lost after origin due to inbreeding. Despite this bottleneck, mst mutants perpetuate in nature in numerous plant species. This is due to the recurrence of mutations and their increased reproductive fitness. Although total seed set is low mainly due to reduced NCP or impaired female fertility in some mst's, in many of them the reproductive potential is better than in hermaphrodites, due to greater flower number, higher seed and fruit set, better seed germination and greater vigour of offspring. This is well known in gynodioecious species of *Carpodetus, Crotalaria, Pimelea, Thymus*, etc. For their maintenance, such mst's should have more than twice the seed fertility of their corresponding hermaphrodites. Thus if sufficient pollination is ensured, mst's would yield higher than artificially produced hybrids, as the former would combine heterozygote advantage and increased reproductive fitness.

11.1.5 Combining Ability Testing

When an elite inbred line or variety has to be tested for its combining ability, crosses with a mst tester and subsequent evaluation of the F_1 hybrid provide information on the combining ability. In sunflower, several breeders in the USA have converted a large number of inbred lines to gc-mst, and then tested them for combining ability, using various restorer lines of population as testers.

11.1.6 Variability Control

Through the introduction of outcrossing, mst enhances gene recombination and genetic variability, especially in self-pollinating plants. By incorporating *ms* or *fr-c* gene combinations in self-pollinated plants, an increase in genetic recombination and variance can be achieved if the species is naturally wind- or insect-pollinated. In p-mst's, both selfing and crossing can be achieved by environmental manipulations, as their *mst* genes are highly sensitive to environment. Genetic variability can be increased by growing the p-mst in an environment that induces complete mst and collecting seeds borne by it after NCP or ACP. Homozygous mst's with reduced genetic variability can be produced by selfing p-mst's. Phenotypic comparisons of the progeny raised from hybrid and selfed seeds will reveal heterotic effects. Genetic variability can also be controlled by selfing or crossing fn-mst's, the former by puncturing anthers, and the latter by ACP or NCP. The availability of suitable seed markers like the *y* gene in bicolor sweet corn (Galinat 1975) or the seedling marker *w* gene in soybean would differentiate hybrid from selfed seed.

11.1.7 Pollen Movement and Pollination

The mst's provide a suitable material to determine mode and amount of pollen movement from mft to mst plants, and also indicate that extent of NCP in self-pollinating plants. Using different planting patterns, pollinating agents, cropping pattern, variable row to row and plant to plant distances, maximum NCP has been determined and utilized in soybean, tomato, pea, carrot, cabbage etc. for hybrid seed production. Through the amount of seed produced on the mst mutants, they have been utilized to determine pollinating effectiveness and efficiency of various insect types. This has indicated that not only insect type or its activity, but environmental factors, especially photoperiod, influence NCP rate and extent of pollen movement from mft to mst plants. In tomato, in addition, both location and site alter pollen movement and NCP rate. In soybean, in which bees are the main pollinators, use of mst mutants reveals pollen transfer to be influenced by weather conditions during anthesis, seasonal temperature conditions and water supply during flowering. Use of insecticide diminishes insect pollinations and reduces NCP over mst mutants. Because of weather vagaries and frequent insecticide use, insect pollination becomes unreliable and has to be supplemented by hand pollinations for hybrid seed production in insect pollinated species.

11.1.8 Planting Design Determination

This is an important factor for increasing F_1 seed yield. The mst mutants have effectively been used in autogamous species to evolve optimum planting design to maximise hybrid seed yield. In many of them, mixed planting design in which each mst plant is surrounded by mft's, increase in F_1 seed yield is more than that in the separate row design. Maximum yield gain is also achieved with a 1:1 mst:mft ratio, but this is uneconomical. In many plants, random mixed design is applied by sowing mixed seeds of mst A-line with the pollinator C-line in different ratios. In such cases fruit or seed markers are needed to identify F_1 seed.

11.1.9 Graft Transmission

For use in breeding, mst is to be transferred to desirable parents. This transfer is time-consuming, requiring technical skill and compatible genetic background. Asexual transfer of mst using the grafting technique is a possible approach as it is simple, rapid, cuts generation time, requires little space, less labour and can sterilize different genotypes simultaneously, thereby maintaining cytogenetic diversity. Thus it could be commercially used. However, such graft-transmissible mst is known only in a few cases, viz. alfalfa (Thompson and Axtell 1978), beet (Curtis 1967), broad bean (Grill and Garger 1981) *Datura* (Blakeslee 1921), hops (Cech and Pozdena 1962), petunia (Corbett and Edwardson 1964) and sunflower (Leclerco 1971). In broad bean, the sterility is transmitted through dodder plants, conversion occurring in 4–6 weeks. The graft transmission of mst is not universal, as it is highly stock-scion genotype-specific. Even in the known cases, the transmission has not been repeated by others. Hence, this transmission mode does not give any added advantage to these mst genotypes over others in which mst is not transmitted across grafts.

11.1.10 Apoximis and Parthenocarpy Estimation

These are detected by estimating seed and fruit set after bud baggings. Using mst lines, Burton (1948) determined the nature and type of apomixis in *Paspalum notatum* and Persson and Rapport (1958) refined testing procedures for the effectiveness of a chemical treatment on parthenocarpic-induced fruit set in tomato.

11.1.11 Polyploidy Breeding

Due to genome increase, complex segregations accompany fertility reduction in autoploids. Thus polyploidy imposes serious restrictions on inbreeding of both mst and mft plants. Also it complicates fertility restoration of polyploid mst's like wheat, cotton etc. where seed or fruit is the desired product. However, it finds use where hybrid fertility is not valued. For instance, triploid hybrids of sugarbeet, watermelon and begonia are easily produced by interplanting 4x mst A-lines and 2x mft C-lines. In these, female fertility of tetraploids is far better than male fertility and haploid pollen is far more functional than diploid pollen of 4x mft plants.

11.1.12 Genetic Control of Microsporogenesis

The majority of *mst* genes are recessive, induce sporogenous mst and influence male meiosis during prophase-I and after microspore release (Figs. 9.2 to 9.4). Since these genes are most prevalent in higher plants and exhibit a high mutation rate, it appears that male meiosis is controlled by a large number of genes, the majority of which are dominant. Of these, most of the meiotic genes control prophase-I and gametogenesis, or alternatively such genes exhibit a high mutability rate. Since the action of *mst* genes is not known in the majority of the higher plants, such investigations may reveal the genetic control of male meiosis more precisely than it is known at present. Moreover, *mst* genes reveal differential gene control of male and female meiosis in higher plants.

11.2 Concluding Remarks

For using mst to produce F_1 hybrids, no single adaptation is completely successful in eliminating hand labour required at some stage in the seed increase or seed production process. Marker genes render roguing easier, but unless linkage with mst is absolute, recombinations between marker and *mft* genes occur and mft's become a continual source of contamination. The p-mst, in theory, allows development of homozygous "mst" lines (being mst in one and mft in another environment). However, genes for this are always sensitive to slight environmental changes; the required environmental conditions for a given level of sterility are usually hard to define and still harder to control. Unless crosses are made in climate-controlled chambers, F_1 hybrids developed via the p-mst method are likely to contain many selfed plants. The use of systems in which mst is absolute and stable, with roguing of fertile segregates, is reliable but unprofitable unless the price and number of seeds per plant are high. In all cases, hybrid vigour and crop uniformity must be great enough to warrant the commercial usage of F_1 hybrid seed.

11.3 Limitations

Despite the numerous uses enumerated above, certain factors limit the use of mst in plant breeding. The specific ones for g-mst's and gc-mst's have been described earlier (Chaps. 2.10, 3.10.6). Some of the general ones are given below.

11.3.1 Phenotypic Identification

For practical mass-scale breeding, easy identification of mst's in the field is most vital, especially in g-mst's in which removal or roguing of fertile heterozygous segregates is required in seed production plots. Even though in the majority of mst's, anthers are usually shrunken, light-coloured and dry, microscopic examination of flower, pollen and stainability, viability and germinability tests of the pollen must be done for reliable detection of mst's. Such determinations are time-consuming and require efficient technical skill, especially when the flower size is small and the number great. Therefore, it is uneconomical. In self-fertilizing plants, mst's become conspicuous in the field by their slow growth, late maturity and unfruitfulness, but then usually it is too late to make controlled pollinations, as the pollen in mft's is already shed. Furthermore, only few mst mutants can be identified by using a hand lens or by prodding the anthers.

11.3.2 Maintenance and Perpetuation

Mostly, g-mst is recessive and monogenic. A pure population of g-mst's cannot be produced, but *ms* gene can be carried along if seeds from mst plants only are harvested and used to plant the next generation. Whereas the B-line maintainer is heterozygous for *ms* gene and segregates mst and mft in a 1:1 ratio, the C-line should be free from such *ms* genes, especially where the reproductive part serves as the agricultural product. Since the frequency of *mst* genes is low, virtually all such F_1 seeds are fertile. For maintenance of gc-mst line A, maintainer line B is needed, it has nuclear genes similar to line A but N-cytoplasm. Methods to develop such lines involve many BC cycles and generations (Chaps. 2.10.1, 3.10.1).

11.3.3 Transfer Difficulty

Usually *mst* genes arise in undeveloped or agronomically inferior genotypes of cultivated species. They are transferred to elite and developed lines through the various breeding procedures indicated earlier (Chaps. 2.10.2, 3.10.1). This transfer poses many problems, especially if the recipient genotypes are noncompatible or if the transfer accompanies many undesirable gene transfers due to pleiotropy or tight linkages. In many, the transfer necessitates indirect or bridging crosses, embryo cultures etc., e.g. *Brassica*, *Solanum* and *Triticum*.

11.3.4 Expressional Instability

This represents a major bottleneck in using mst in breeding, as most of the *mst* genes are sensitive to environment, a factor difficult to control or predict. The influence is more on p-mst's and fn-mst's. Usually the mst's of many crop plants develop fertile pollen in their late-formed tillers, so that the seed harvested on mst's is a mixture of hybrid and selfed seeds.

11.3.5 Reduced Female Fertility

Although typically *mst* genes suppress only male fertility, female fertility of mst's is also impaired. This leads to reduced seed set, which decreases further due to low or insufficient pollination over mst's.

11.3.6 Insufficient Pollination

The major cause of reduced seed set in mst's is the insufficient pollen their stigmas receive, as even hand pollination does not fully substitute for NCP. In insect-pollinated flowers, even if mst flowers have sufficient nectar, the NCP in them is low (Breuils and Pochard 1975). In entomophilous carrots, anthers of mft's flouresce (Erickson and Peterson 1979). This provides a visual cue for pollen-collecting honeybees, which is absent in mst flowers. Hence bees do not visit mst flowers, over which the seed set is consequently low. Likewise, bee visitation is poor on mst mustard as nectaries are rudimentary. Thus either mst flowers attractive to pollinating insects or insects not avoiding mst flower visitation have to be developed. Likewise, mst's with flowers suited to open pollination have to be developed, especially in cereals and legumes. The lack of such genotypes has slowed hybrid development and dampened acceptance by industry.

11.3.7 Inbreeding Depression

Use of mst, especially in allogamous species, accompanies this depression. It results in reduced plant growth and vigour and low seed production. In maize, rye and sunflower, it accompanies partial female sterility. In maize, the depression is obviated by production of double-cross hybrids. To eliminate depression, double, triple- and multiple-cross hybrids are being developed in sunflower. However, the development of such synthetics is not easy, especially in cereals and legumes whose flowers are self-pollinated, minute and numerous, where both emasculation and pollination are arduous.

11.4 Lacunae

Numerous instances of mst are known, but in the majority of these, especially in g-mst's, one or more of the following investigations, although necessary, have not been done. Determination of some of these, especially of the first three, are absolutely essential as they decide the breeding value of mst genotypes.

11.4.1 Pollen Sterility and Female Fertility

Nonfunctioning of male sex characterises mst's. In them, either male sex organs are absent, deformed, malformed (structural mst) or the sex organs are developed but the pollen grains are absent or abortive (sporogenous mst) or both stamens and pollen are normally developed but anther sacs are indehiscent (functional mst). Coupled with male sex impairment, female fertility of the mst's should be normal. This has not been determined in numerous known cases of mst's (Tables 2.11, 2.12, 3.11), some of which could be meiotic or sterility mutants. Hence, it is absolutely essential that once mst is doubted, female fertility should be tested by observing seed set after ACP, using the pollen of sibs. A genotype that lacks mft but has normal female fertility is a true mst genotype. Such mst's, if stable, have high breeding value, unlike those with reduced or impaired female fertility.

11.4.2 Gene Control, Action and Terminology

The majority of mst's arise spontaneously (Table 2.1). In most of them gene control is not known (Tables 2.9-2.12). In them the mst could be environmental, physiological, chemical or due to insect, fungal or viral infection. These factors, in addition to inducing mst, reduce female or seed fertility. Moreover, they are temporary and fortuitous and are therefore of no breeding value, as only genetically controlled mst's can be utilised in a breeding program. Hence, precise gene control determination is absolutely essential in order to assess the genetic value of a mst genotype. Furthermore, it is desirable to determine the time of *mst* gene action, as microgametogenesis breakdown during pollen maturation stages can be averted by altering temperature conditions or by chemical sprays like GA, e.g. tomato, pea or barley. Thus homozygous g-mst genotypes can be obtained.

Once genetically conditioned mst is found, it is desirable to know the allelic nature of the new genes so that gene symbols are assigned to it. This has been done in only a few cases like barley, chillies, cotton, maize, pearl millet, sorghum, soybean, tomato etc., and is especially required in those genotypes where many genes are known to induce mst viz., *Beta, Brassica, Cucurbita, Daucus, Helianthus, Oryza.* The procedure for assigning gene number and symbol outlined by Hockett and Reid (1981) for barley could be followed for others also.

11.4.3 Type Differentiation

Once mst is confirmed as genetically conditioned, it is necessary to determine its type, g-mst or gc-mst, for both of these are inherited, maintained and propagated differently. Furthermore, precise gene control determination is essential, as the fewer the number of genes conditioning mst, the easier is its transfer, maintenance, propagation and handling. All these features add to its breeding value and the genetic utility of mst's. For breeding use, gc-mst is preferred over g-mst, due to easier maintenance and propagation. Where the former is absent, the latter is used. However, in such cases the search for gc-mst's should be continued.

11.4.4 Biochemical Differences

Compared to the vast number of known mst genotypes, biochemical differences between mst and mft genotypes have been investigated only in a few cases (Chap. 5). In almost all of them, anther differences have been studied using mst and mft anthers selected at arbitrary developmental stages. Differences in enzymes, amino acids and sugars have been found to exist between such anthers. It is worthwhile to analyse anthers of mst and mft sibs biochemically before, during and after *mst* gene action. The plants should be grown in one field to eliminate environmental influence. This would reveal intrinsic biochemical alterations induced by *mst* genes, some of which may represent nor the consequences but the causes of mst. Manipulation of such biochemicals could probably alter the population phenotype and add to its breeding value, as both homozygous and heterozygous stocks of a desirable genotype would be raised and utilised.

11.4.5 Environmental Influence

Control of mst and self-fertility is the cherished goal of breeders, as these would facilitate mst population maintenance, perpetuation and use in commercial breeding. Since *mst* genes are sensitive to environment, especially to temperature and photoperiod, the environmental manipulations would facilitate fertility-sterility control of *mst* genotypes, enhancing their breeding value. Therefore, it is essential to determine environmental sensitivity of *mst* genes for their subsequent use in breeding.

11.4.6 Population Improvement

i. *Pollination Rate.* Despite the availability of numerous mst genotypes, both in open and in self-pollinated plants, their breeding value is limited due to low NCP and limited pollen transfer. These impose serious restrictions in their breeding utility, especially in self-pollinating cereals and legumes, which abound in mst genotypes. Hence, it is vital to develop a mst population with open flowers and extruded stigma to receive pollen. In insect-pollinated species like carrot, mst flowers produce less nectar and aroma, their anthers lack fluorescence and the necessary cue to attract bees. Hence bee visitations over these are infrequent and seed production is consequently low. Successful use of any cross-pollinated entomophilous hybrid crop system should involve selection for similar floral characteristics early in the breeding program to ensure maximum pollen transfer by insects from mft's to mst's.

ii. *Synchronous Flowering.* Mostly, mst's exhibit slower growth and flowering than mft's. This hampers NCP and ACP over mst flowers, as flowering and pollen shed in mft's is already over by the time mst's flower. Therefore, development of faster-growing mst's is highly desirable to achieve synchronous flowering and anthesis of mst's and mft's. Alternatively, mst's may be sown earlier than mft's.

iii. *Phenotypic Markers.* Identification of mst's in segregating populations is arduous, as it usually necessitates large-scale anther examinations. This requires skilled labour. As mentioned above, the late flowering of mst's delays identification and hampers their pollination. Hence, it is desirable to have phenotypic markers for early mst identification. Seedling markers, as in lima bean, would be useful, especially in g-mst's, where roguing of mft's is necessary.

Part II
Flowering Families

CHAPTER 12
Anacardiaceae

Spontaneously arisen partial to complete mst occurs in *Anacardium occidentale, Mangifera indica, Pistacia vera, Rhus integrifolia, R. ovata* (Table 12.1). Genetic control of sterility is not known in any of them.

12.1 *Anacardium occidentale* L. (cashew nut)

Its flowers usually have 8–10 stamens, some of which have aborted microspores. In one stand, two plants had aborted pollen in all the anthers (this Vol.). Male meiosis in these proceeded irregularly after the diplotene (Table 12.1) and shrunken, nonviable pollen resulted.

Table 12.1. mst Gene action in Anacardiaceae

Species	References	Gene action
Anacardium occidentale	This Vol.	After diplotene
Mangifera indica	This Vol.	After microspore release
Pistacia vera	This Vol.	After PMC formation
Rhus integrifolia }	Young (1972)	nk
R.ovata }		

12.2 *Mangifera indica* L. (mango)

Some mango varieties have one to five stamens, but mostly only one of these produces some fertile pollen. In five plants, aborted pollen was found in all the stamens (this Vol.). In these, male meiosis proceeded normally until microspores were liberated (Table 12.1), after which they degenerated abruptly.

12.3 *Pistacia vera* L. (pistachio)

Staminate flowers of a plant dropped off prematurely, rendering the plants mst (this Vol.). No meiosis occurred in the anthers, as their PMC's degenerated just after formation (Table 12.1).

12.4 *Rhus* L. (sumac)

In the gynodioecious species *R. integrifolia* (lemonade berry) and *R. ovata* (sugar bush), 11–26 and 6–20% mst plants respectively were detected by Young (1972). These had two flower types. Some had extremely small stamens and petals and others had normal-sized stamens exerted beyond petals as in the hermaphrodite flowers, but these stamens lacked pollen completely. Even though hermaphrodites greatly outnumbered mst's, only a small proportion of these set seeds. On the other hand, seed set in mst plants was twice that of the hermaphrodites. In both the *Rhus* species, progression from hermaphroditism to mst occurs (Young 1972).

CHAPTER 13

Begoniaceae

13.1 *Begonia semperflorens* Link & Otto

Among the easiest to grow and most freely flowering garden plants is the Brazilian *Begonia* in which both g-mst and gc-mst are known (Table 13.1). Various dalapon and mendok treatments induce pollen sterility but it is accompanied by very high ovular damage and reduction in seed set.

13.1.1 Genic Male Sterility

In this attractive floriferous ornamental plant, a monogenic recessive *ms* gene induces mishapen anther development (Reimann-Philipp and Seidel 1963) or transformation of anthers into pistil-like nonfunctioning organs (Table 13.1). The mutant known as Cinderella was utilized in hybrid production by Reimann-Philipp (1976) by using a hair-drier to blow pollen from mft to mst plants. Since only 50% of the plants of the maternal parent, being mst, are suited for hybrid seed production and in most of the countries F_1 hybrid seed production and F_1 hybrids with gracilis growth are preferred, polyploidy was induced in 2x mst's, as 3x and 4x plants have this growth habit. Since 4x mft's are poor fertile pollen producers, 4x mst's were pollinated by 2x mft pollen to produce 3x plants (Reimann-Philipp 1983). These triploids are becoming increasingly popular and have a high market acceptability and export demand. As bee visitation over mst's is low, pollen transfer is done either manually or by air blowers. In begonias, relatively few mother plants are required, as each plant yields about 0.5 g seed (40,000 kernels) after hand pollination. Thus vegetative propagation of an elite mst clone and its pollination would produce sufficient seed with high economic return.

Table 13.1. Male sterility in Begonia

Species	Reference	Remarks
Genic male sterility		
Begonia semperflorens[a]	Reimann-Philipp and Seidel (1963)	Structural male sterility
Gene cytoplasmic male sterility		
B.schmidtiana × *B.acuminata*	Villerts (1942)	Abscission of male flowers at various developmental stages
B.schmidtiana × *B.hirtella*		
B.schmidtiana × *B.subvillosa*		

[a]Gene control by 1r.

13.1.2 Gene-Cytoplasmic Male Sterility

Matous (1969) obtained some mst plants in certain interspecific crosses involving *B. semperflorens.* The author considered the sterility to be due to g-c interaction and indicated a possibility of using this type of mst in breeding. However, the details are not known. All the male flower buds abscised at an early developmental stage in F_1 generation of *B. schmidtiana* × *B. subvillosa* (Villerts 1942). The male flower buds dropped, but some at a late developmental stage, so that in some of them young pollen was formed. However, none of it was functional. Pollen was formed in the offspring of the cross *B. schmidtiana* × *B. hirtella* (Table 13.1). No such male floral fall occurred in the offspring of the reciprocal cross. A similar drop-off of male flowers occurred in F_1 plants of the cross *B. schmidtiana* × *B. acuminata.* In the hybrid plants, the flower fall is delayed until the flowers open partly. Anthers of the abscised flowers are green and contain no pollen. In BC_1 of the cross (*B. schmidtiana* × *B. hirtella)* × *B. schmidtiana,* 1:1 segregation occurred for plants abscising all male buds and for plants shedding buds at first but later retaining them fully. Villerts (1942) interpreted these results as due to interaction between *B. schmidtiana* cytoplasm and a gene or genes contributed by the pollinating parents.

CHAPTER 14

Cannaceae

The family Cannaceae possesses only one genus, *Canna*, which has over 55 species and a large number of hybrid cultivars. Chromosomal anomalies, especially translocations, inversions, eu- and aneuploidy, occur in many of its cultivars. These anomalies cause both male and female gametic sterility, and not only mst, as envisaged by many.

14.1 *Canna glauca* L.

Pure yellow F_3 plants obtained after crossing the varieties Java × Montevideo of *C. glauca* had anthers in which all the PMC's degenerated during the diplotene stage, in one or both the loculi (Öfferijns 1938). In a few others no such degeneration occurred and pollen grains were formed, apparently after a normal course of microsporogenesis. Neither the stainability nor the functionability of this pollen was determined. Hence, it is not clear if some viable pollen was really produced. In some plants, action of the *ms* gene is delayed and this enables microsporogenesis to proceed in certain anthers. Honing (1928) found stamens having one locule full of pollen and the other completely devoid of it. It is not known whether the pollen sterility is genic or not. If it is genic, then the *ms* gene(s) prevents transition from diplotene to metaphase I and leads to the abortion. However, once the timing of gene action is delayed in a loculus and the chromosomes reach MI, meiosis proceeds to produce nonviable pollen grains. Since neither female fertility nor genetic control of the anomaly is known, it cannot be regarded as true mst.

CHAPTER 15
Cannabinaceae

15.1 *Cannabis sativa* L. (hemp)

15.1.1 Genic Male Sterility

In this dioecious species, known popularly in the west as marijuana and as bhang, charas, hashish, ganja in the Orient, the sexuality is highly fragile and susceptible to environmental conditions. In a Russian collection comprising different races and grafts, male meiosis was abnormal in the majority of the anthers (Breslavetz 1935). This led to the formation of tetrads having either two large and two small nuclei or two large or one very large nucleus instead of the usual four nuclei found in the normal plants. In a few, many nuclei were formed in PMC's. Long before meiotic initiation, abnormalities begin in them and are vivid in early prophase stage (Table 15.1). Sometimes, PMC's become elongated and divide into two cells even before meiotic initiation. In some, PMC's grow abnormally in size and have three nuclei, one of them being gigantic in size. All other nuclei undergo abnormal meiosis. Three nuclear plates, incipient spindles, poor chromosome orientation, chromosomes lying away from the incipient spindles, nonbivalency and a large number of univalents (28–30) characterise MI. Anaphasic distribution and subsequent stages are extremely irregular. Pollen grains are empty and variable in size, shape and configuration. These abnormalities are retained by anthers of the male flowers developed after grafting such male plants on female plants. Nearly similar abnormalities in hemp were reported earlier also by Mac Phee (1925) and Hirata (1929). Furthermore, in many monoecious and dioecious hemp varieties, mst flowers are smaller and paler in colour than mft's. Their anthers are rugose and have fewer pores than those of the fertiles. Two plants in a progeny from the fourth generation of a selfed line had smaller flowers and lacked pollen completely (Pate and Joyner 1958). The sterility is controlled by one recessive gene.

15.1.2 Chemical Male Sterility

In hemp, plant form and function can be modified by introducing variations in day-length, temperature, soil nutrients, chemicals and moisture and also by exogenous application of growth hormones, regulators and inhibitors. Chailakhyan and Khryanin (1980) found that phytohormones not only influence sex expression but also growth, development and differentiation of this plant. Thus whereas 12.5 mg l^{-1} applications of GA_3 in the nutrient medium in which *C. sativa* seedlings were grown induced almost 100% maleness in the adult plants, the application of 5 mg l^{-1} of 6-benzylamino purine (6-BAP) led to the development of

Table 15.1. Genic male sterility in *Cannabis sativa*

Gene control	Gene action	References
nk	During prophase	Breslavetz (1935)
lr	nk	Pate and Joyner (1958)

Table 15.2. Chemical male sterility in *Cannabis sativa*

Chemicals	References	Male sterility induction	Female fertility reduction
6-BAP	Chailakhyan and Khryanin 1980	Complete	Nil
Ethrel	Mohan Ram and Jaiswal 1970	Partial	Nil

cent per cent females on this plant (Table 15.2). On the other hand, foliar applications of 960 ppm ethrel produced 70% female flowers on male plants (Mohan Ram and Jaiswal 1970). This treatment caused drooping and epinasty of leaves and the development of female, intersexual and abnormal male flowers on male plants. The female plants remained unaffected by the treatment.

15.2 *Humulus lupulus* L. (European hop)

In this common hop of hop-yards, pollen sterility occurs following severe mosaic and leaf blotch viral infections. Female fertility is also impaired resulting in numerous empty achenes produced over mst plants (this Vol.). Graft-transmissible gc-mst is known in infected hop plants (Cech and Pozdena 1962), but this has not been confirmed.

CHAPTER 16

Caricaceae

16.1 *Carica papaya* L. (papaya)

Of the known 45 species, mst is documented only in this horticulturally most useful species that is well known for its edible melonlike fruit and the proteolytic enzyme papain. Sexuality in this plant is determined by a single gene with three alleles, to which Hofmyer (1938) assigned the symbols M_1 (staminate), M_2 (hermaphrodite) and m (pistillate). Storey (1953) found all homozygous combinations of dominant genes like M_1 M_1, M_1 M_2 and M_2 M_2 to be lethal to the zygote. Therefore, while mft's ($M_1 M_1$) and hermaphrodites ($M_2 m_2$) are sex heterozygotes, mst's are homozygotes ($m_1 m_2$). Recent evidence indicates that both sex determination and mst are not as simply controlled as envisaged above. Instead, modifier genes, environment, nutrition, biotic stress etc. influence the sexuality of this plant considerably. A gene *sa,* when homozygous, induces a complete suppression of androecium (Storey 1953). Apparently the gene does not appear to be completely recessive to its normal allele, but influences androecium suppression in male and hermaphrodite plants only under certain environmental conditions like low temperature and high humidity.

CHAPTER 17

Caryophyllaceae

Spontaneous occurrence of g-mst is known in *Dianthus caesius*, *D. caryophyllus*, *D. chinensis*, *Silene acaulis* and *S. hookeri* (Table 17.1). In *S. pendula*, mst is induced by long-day photoperiodic and auxin treatments.

17.1 *Dianthus*

17.1.1 *D. caesius*

Plants exhibiting carpeloidy or petaloidy were found by Reimann-Philipp (1976). They lacked fertile anthers and had a high tendency to double. The anomaly is controlled by one recessive *ms* gene. This mst is of structural type and the *ms* gene acts during anther differentiation (Table 17.1)

17.1.2 *D. caryophyllus* L. (carnation)

Spontaneously arisen double-flowered types with anther feminizations or transformation into petals were found by Reimann-Philipp (1976). The mst is of structural type and the *ms* gene acts during anther differentiation (Table 17.1). In some flowers, partial transformation of stigmas and styles into petal-like structures occurs, and reduces their female fertility. Reimann-Phillip (1976) used such double flower types for hybrid flower production. These are highly attractive and useful. The mst homozygous double-flowered parent was propagated by growing under poor nutritional conditions that reduced doubling and improved female fertility. Crossing in mst's is done by rubbing their heads with the pollen parent. Genetic control of the anomaly is not clearly known, but segregations indicate a monogenic recessive control.

Table 17.1. Genic male sterility in Caryophyllaceae

Species	References	Gene control	Gene action
Dianthus caesius	Reimann-Philipp (1976)	1r	During anther differentiation
D.caryophyllus			
D.chinensis	This Vol.	1r	After microspore release
Silene acaulis	Bock (1976)	nk	nk
S.hookeri	This Vol.	1r	Before microsporogenous tissue differentiation

17.1.3 *D. chinensis* L. (*D. sinensis* Hort, rainbow pink)

A monogenic recessive spontaneous mutant having rose-lilac petals but lacking purplish eye arose in a stand of the species (this Vol.). It had reduced anthers and aborted pollen. Male meiosis was nearly normal until microspore formation. Simultaneous pollen and tapetal degeneration followed microspore release (Table 17.1).

17.2 *Silene* L. (catch fly, campion)

17.2.1 *S.acaulis* (L) Jacq. (moss campion, cushion pick)

Three types of plant were found by Bock (1976). These were plants having (a) abortive pistil but normal male sex organs, (b) strongly protandrous hermaphrodite flowers and (c) mst female fertile plants. In this trimorphic species, even among some mft plants, pollen lethality occurs. In them a few or several stamens possess indehiscent anthers having empty micropollen in isolated grains or in tetrads. The mst plants are of three types, viz., (a) anthers develop but remain indehiscent and degenerate rapidly, they either have no pollen or a few micropollen grains develop sporadically, (b) stamens poorly developed bearing rudimentary indehiscent anther lobes which are devoid of pollen, (c) stamens chlorophyllous, minute, poorly developed and differentiated, lacking anther sacs and pollen grains. The average frequency of mst occurs in the populations of subsp. *arctica,* in which micropollen is present even in mft anthers. The mst is genetically conditioned, as individual plants exhibit a specific type of anomaly which finally culminates in pollen abortion. According to the relative degree of development of androecium and gynoecium, more than three types of plants are present in the species. In them, the mst is of diverse types and appears to be conditioned by different alleles, the general control being multigenic. This needs confirmation.

17.2.2 *S. hookeri* Nutt

One purple-flowered plant with poorly developed stamens having minute indehiscent anthers lacking pollen was found (this Vol.). No meiocytes were differentiated in the anthers. Female fertility was normal. The mst is conditioned by one recessive gene, ms_1. Whereas anthers were rudimentary in the field-grown plants, they developed in glasshouse grown plants, but no PMC's were differentiated in them.

17.2.3 *S. vulgaris* (Moench) Garcke (bladder campion)

In a tetraploid wild population (2n = 48) of Israel, hermaphrodites, mst's and intermediates were detected by Horovitz and Dulberger (1983). In them, three sexual phenotypes occur. These are, (i) stable hermaphrodites with all bisexual flowers, (ii) stable mst's with all mst flowers and (iii) polygamous plants with bisexual, mst and/or partially mst flowers. Although precise genetic control of the sterility is unknown (Correns 1904, 1906; Marsden-Jones and Turill 1957), Horovitz and Dulberger (1983) postulated that the basic sex type is controlled at a single tetrasomic locus *AA/aa*. Quadruplex and triplex *AAAA* and *AAAa* individuals are hermaphrodite, simplex and nulliplex *Aaaa, aaaa* are mst and duplexes *AaAa* intermediate. In the intermediates, gene *A* is incompletely dominant over *a* and mst expression is confined to single flowers or single anthers, giving an appearance of randomness. Cytoplasmic influence over mst expression cannot be ruled out.

CHAPTER 18

Chenopodiaceae

Although mst occurs in beet, quinao and spinach (Tables 18.6, 18.7, 18.8), extensive investigations and utilization of mst are known only in sugarbeet, mainly because of its high economic utility and the early detection of *mst* genes.

18.1 *Beta* L. (beet)

Beetroot, which formed a common dietary ingredient in Egypt during the building of pyramids, is a major source of sugar now. The mst is known in three diploid species of the section *vulgaris,* viz. *B. macrocarpa, B. maritima* and *B. vulgaris.* In all these, sterility is conditioned by recessive genes which act with or without the association of S-cytoplasm (Tables 18.7, 18.8). In some, sterility has been induced by chemicals (Table 18.6).

18.1.1 *B. macrocarpa* Guss

18.1.1.1 Genic Male Sterility

Some mst plants segregated from the progeny of a European collection, (this Vol.). Their anthers are reduced and have abortive microspores. Meiosis proceeds normally until microspore formation, after which the mature microspores abort (Table 18.7). In mst anthers, the tapetum remains intact, whereas in mft anthers it disintegrates after microspore formation. The anomaly is conditioned by one recessive gene, whose penetrance and expressivity is almost complete. Female fertility of the mst plants, as evidenced from the seed set after hand pollination, is normal.

18.1.1.2 Gene-Cytoplasmic Male Sterility

Some mst plants segregated from progeny of the varietal cross GW 304 × GW 359, and in accession PI 206411 of an annual wild beet from Turkey (Oldemeyer 1957). Sterility in both these, although nearly similar, may not necessarily be conditioned by the same *ms* gene. In the mst plants obtained from the varietal cross, the sterility is gc-mst, the inheritance pattern being similar to that of *B. vulgaris.* However, one to two major modifier genes influence the restoration capacity of *Fr* genes or N-cytoplasm. Abortion occurs between the tetrad and pollen developmental stages (Table 18.8).

18.1.2 *B. maritima* L. (sea beet)

18.1.2.1 Genic Male Sterility

In a strain of this annual beet, Kinoshita and Nagao (1969) and Kinoshita (1971) found an mst mutant (BM-2) in which sterility was conditioned by a recessive *ms* gene whose penetrance was highly variable. The mst expression is considerably influenced by the environment. This produced an array of phenotypes ranging from complete to partial mst plants (Kinoshita and Takahashi 1972). In the former, indehiscent, white shrunken anthers that lacked viable pollen were developed. In the latter, dehiscent and nondehiscent anthers containing a variable proportion of stainable and nonstainable pollen were produced. The functionability of the stainable pollen is not known. Furthermore, flowers of the p-mst plants produced towards inflorescence apices sometimes contained normal and abnormal anthers within the same flower. The *ms* gene is nonallelic to the ms genes a_1 and a_2 present in *B. vulgaris* genome. In the mft anthers, the tapetum degenerates during pollen formation but in mst anthers, tapetal cells persist, enlarge, became multinucleate, vacuolar and plasmodial during microspore liberation (Fig. 18.1). The plasmodia bulge prominently into anther cavity (Kinoshita and Nagao 1969; Kinoshita and Takahashi 1972). The microspores stop further development and abort gradually. Endothecial development is highly impeded. Finally, microspores and tapetal plasmodium degenerate rapidly.

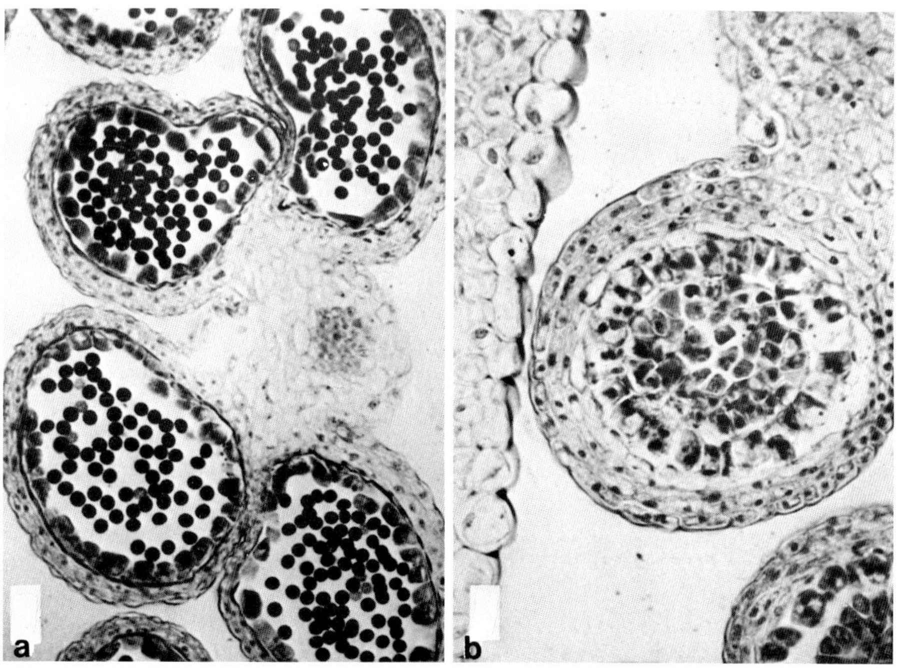

Fig. 18.1a,b. *Beta maritima* T.S. of fertile anther **(a)** showing stained viable pollen and of sterile anther **(b)** showing degenerating premeiotic PMC's and tapetum. (Kinoshita and Nagao 1969)

18.1.2.2 Gene-Cytoplasmic Male Sterility

Coe and Stewart (1977) found two monogerm mst plants in an accession PI 211874 collected at Warnbury Bay, near Plymouth, England. The sterility resulted due to the interaction of the S-cytoplasm with two *fr* genes (*xx*, *zz*). The inheritance pattern and fertility restoration are nearly similar to *B. vulgaris*. Maintainer O-line mft plants (N *xxzz*) for the mst plants occur in an ecotype of the species collected from Kalundberg, Denmark. Genes conditioning monogermness in mst *B. maritima* plants have the same locus as the monogerm gene in *B. vulgaris* but the precise allelic nature of the two genes is not known. From gc-mst *B. vulgaris*, Dalke (1983) transferred mst to *B. maritima* and found some *fr* genes present in its genome. In the mst plants, male meiosis proceeded normally, but microspores degenerated immediately after liberation (Table 18.8). This was followed by tapetal disintegration. Anther dehiscence was prevented due to absence of the characteristic thickenings in endothecial walls.

18.1.3 *B. vulgaris* L. (sugarbeet)

18.1.3.1 Genic Male Sterility

In some American and English sugarbeet varieties, Owen (1952) detected two *ms* genes, a_1 and a_2. Instead of lemon yellow-coloured anthers of mft plants, the mst plant anthers remain green, turn brown, shrivel and dry up immediately after anthesis (Fig. 18.2a,b). In these, meiosis proceeds normally but microspores degenerated after liberation from quartets. This degeneration is slightly delayed in the plants having a_2 genes and takes place just before the first microspore mitosis. In one Russian sugarbeet mst line, pollen abortion is conditioned by a recessive mutant gene, *a* (Peretyatko 1975). Its time of action and allelic nature to a_1 and a_2 genes is not known. Two more g-mst plants from Indian beet collections were detected (this Vol.). In each of them sterility is conditioned by one recessive gene. In their action over microsporogenesis and anther phenotype, these *ms* genes are identical to the gene a_1 found by Owen (1952), but allelic tests could not be done due to nonavailability of a_1 gene source. In the mst anthers, tapetal cells enlarge rapidly but remain intact during microspore abortion (Table 18.7), after which they degenerate gradually. The *ms* genes a_1 and a_2 found in the *B. vulgaris* genome (Owen 1952) and that in *B. maritima* (Kinoshita 1971) are nonallelic and should be designated ms_1, ms_2 and ms_3 respectively. The allelic nature of the ms genes of *B. macrocarpa* and *B. vulgaris* (this Vol.; Peretyatko 1975) has not been determined. This determination is highly desirable for appropriate gene symbol assignments.

18.1.3.2 Gene-Cytoplasmic Male Sterility

This was documented by Owen (1942), who found 2% plants of the first curlytop resistant variety US-1 as mst. Flowers of the mst plants had white shrunken anthers that progressively changed to dark brown or black colour. In contrast, the mft plants have lemon yellow plump anthers. The degree of mst is conditioned by the

Fig. 18.2a,b. *Beta vulgaris.* Male fertile **(a)** and male sterile **(b)** flower. (Kinoshita 1971)

interaction of two recessive nuclear genes *x* and *z* (≡ fr_1 and fr_2) with the S-cytoplasm (Owen 1942). Whereas S_{xxzz} plants produced abortive pollen, those with S_{Xxzz} or S_{xxZz} were considered semi-sterile by Owen (1942) even though their stainable pollen is mostly nonfunctional (Table 18.1). Although S_{xxZz} plants are partially mft, they are indistinguishable from normal plants on the anther pheno-

Table 18.1. Genotypic constitution and pollen phenotype of mst and mft beet plants

Genotype	Pollen phenotype
Male sterile	
S_{xxzz}	Complete mst, pollen abortive
S_{Xxzz}	Semi-sterile on pollen stainability basis but pollen nonviable
S_{xxZz}	Partially fertile with some viable pollen
S_{XxZz}	Low proportion of sterile pollen, sometimes indistinguishable from normal plant
Male fertile	
N_{xxzz}, N_{Xxzz}, N_{xxZz}, N_{XxZz}	Viable pollen, anthers normal

type basis. A similar difficulty in distinction occurs between S_{Xxzz} and S_{xxZz} plants (Owen 1948, 1950). This was because of the minor effect of *Z* gene under certain environmental conditions. Owen (1948) designated the genotypes N_{xxzz}, $N_{X\text{-}ZZ}$ and $N_{xxZ\text{--}}$ types 0, I, II pollen parents, respectively, to assure the use of only type 0 plants in hybrid beet production. The 0-type was also described as zero-restorer or maintainer, as it maintains sterility but does not restore it. The proportion of 0-type plants is highly variable. The highest frequency (50%) occurs in some US commercial beet varieties (Owen 1950). In others, it is low and variable viz. 4.3 to 15.5% (Peterson 1952), 0–20% (Knapp 1955), 1–30% (Imanshi 1968; Kinoshita 1971). Progeny analysis of crosses involving several different mst plants to the same set of pollinators indicates that, although the mode of g-c interaction leading to various mst expressions varies from one parental combination to another (Oldemeyer 1957), the basic control of mst remains as outlined in Table 18.2. This is also true for most of the European sugarbeet and fodder beet (Knapp 1955; Nagao and Kinoshita 1962; Kloen 1964).

18.1.3.2.1 Fertility Restoration

In sugarbeet, genes restoring partial fertility are common but those with complete restoration are relatively rare. Owen (1945) found that only one curlytop-resistant line possessed almost complete restoration capacity. When mst plants were pollinated by normal hermaphrodite beets, mst, p-mst and fertile plants segregated.

Table 18.2. Segregation ratios expected from S_{xxzz} × N-cytoplasm pollinators

Plant	Pollinator genotype	Progeny segregations (%)		
		Complete male sterile	Partial male sterile	Male fertile (normal)
0	N_{xxzz}	100	0	0
I	N_{Xxzz} or N_{xxZz}	50	50	0
II	N_{XxZz}	25	50	25
III	N_{XxZZ} or N_{XXZz}	–	50	50
IV	N_{XXzz} or N_{xxZZ}	–	100	–
V	N_{XXZZ}	–	–	100

Plants with cent per cent pollen fertility never occurred. It was assumed that the pollinators possessed S- instead of N-cytoplasm. The mst plants, when crossed to a subline of US 201–20, produced progeny plants with high pollen fertility. That complete fertility restoration is conditioned by the interaction of at least two complementary genes which act in association with modifiers and the S-cytoplasm was postulated by Theurer and Ryser (1969). But the *Fr* gene action is considerably influenced by modifiers and the environment. These lead to blurred segregations that do not fit any known genetic ratio.

Bliss and Gabelman (1965) confirmed Owen's basic contention of genetic control of fertility restoration, and like Rohrbach (1965c) concluded that complete restoration of S-cytoplasm plants is affected by a dominant gene *X*. Another gene *Z* conditions partial restoration. The difference between p-mst and mst plants is due to the segregation of gene *Z* in S-cytoplasm plants. Although genes *X* and *Z* are independent and nonallelic, *xxZ* is hypostatic to *X*. However, their results do not explain many segregations involving mst and p-mst crossed progenies. Owen (1942, 1945) and Nagao and Kinoshita (1962) postulated that p-mst plants are unstable and conditioned by heterozygous genotypes. However, they could not detect any mft plants that upon selfing could produce 3 mft : 1 mst segregating progenies. Their assumption that the genotypes S_{Xxzz}, S_{xxZz}, S_{xxZz}, S_{XxZz} condition p-mst suggests that cross pollination of mst S_{xxzz} by any mft plants either would produce progenies entirely free of mft offspring or any mft plants that did appear would be the result of phenotypic instabilities.

Extensive investigations of Kinoshita (1971) indicate that fertility restoration of gc-mst plants is mediated by a digenic system *X* and *Z* that complement each other's action, irrespective of the dose of dominant genes and their heterozygosity level. Thus nine genotypes would exist that could be the pollinators of mst beet. Kinoshita (1971) classified them into six types, depending upon their mode of segregation expected in progeny of their crosses with S_{xxzz} mst type (Table 18.2). In general, fertility restoration in beet is conditioned by at least six nonallelic dominant genes, listed in Table 18.3.

18.1.3.2.2 Sh gene

Hogaboam (1957) found that this dominant gene enhance the pollen-producing ability of p-mst S_{Xx} sugarbeet plants. The *Sh* allele has no effect upon S_{xxzz} plants, only S_{Xx} *shsh* plants appeared p-mft and phenotypically; but the inherit-

Table 18.3. Fertility restorer genes in sugar beet

Original designation	References	Present designation	Fertility restoration
X, x	Owen (1945)	Fr_1	Complete
Z, z	Owen (1942, 1945)	Fr_2	Partial
Rf_1	Theurer and Ryser (1969)	Fr_3	Nearly complete
Rf_2	Theurer (1970)	Fr_4	Nearly complete
Rf	Roundy and Theurer (1974)	Fr_5	Weak

ance of p-mst phenotypes conditioned by gene *xZ* (Bliss and Gabelman 1965) is not consistent with the findings of Hogaboam (1957), who assumed that a heterozygous genotype, S_{Xx} *shsh* produced p-mst plants. However, if it is assumed that the allele *X* is the same in both cases, then there is no evidence to suggest that the enhancer gene *Sh* is similar to or the same as the dominant gene *Z* which conditions p-mft in beets. Rohrbach (1965a) and Bliss and Gabelman (1965) have indicated that the proposed modifier gene *Sh* does not explain the segregation of fertiles and steriles in their material. Segregations led Rohrbach (1965a,b) to assume pollen stainability inheritance in beets as more complicated than that explained by the conventional dihybrid scheme proposed by Owen (1945). Bolz (1968) modified Owen's (1945) hypothesis on the plea that pollen-sterile phenotypes are produced due to the interaction of the S-cytoplasm and a genotype lacking either one or both dominant alleles *X* and *Z*, and p-mst occurs in the presence of one of the dominant alleles present either in the homozygous or the heterozygous state. He extended this modified hypothesis to tetraploid beets and described nine genotypes which were homozygous recessive for one or both gene loci and developed mst in the S-cytoplasm. There is increasing evidence which indicates that a greater number of *fr* genes interact with the S-cytoplasm and cause a different degrees of mst in sugarbeet (Knapp 1969).

18.1.3.2.3 Heat Treatment

Cleij (1967) induced mft by subjecting germinating seeds of gc-mst sugarbeet to 55°C for 10–60 min. The author hypothesized that this conversion was due to cytoplasmic mutation from S to N. All four mft plants obtained produced stainable pollen. The results of selfing, interpollination and crossing of these plants to gc-mst plants are inconsistent and do not prove S-cytoplasm conversion to the N-type. Likewise, by heat treatments and subsequent meristem culture, Lichter (1978) obtained mft plants from gc-mst sugarbeet. Thinning or elimination of RNA containing cytoplasmic particles was considered to cause mft. The viral nature of mst may be ruled out, as generally viruses are eliminated in plants maintained at or above 36°C (Kassanis 1965).

18.1.3.2.4 Fertility Reversion

Stein et al. (1959), after crossing gc-mst sugarbeet by mft table beet pollen, found that some segregants of the cross started as typical mst's, but their last-developed flowers produced 3–19% fertile pollen. The authors assume the reversion to be due to environmental factors and some internal conditions related to the relationship between vegetative and reproductive development. They suggested reversion as due to the accumulation of a "fertility substance" or exhaustion of a "sterility substance". Hornsey (1973) opined that sterility(s) genes in many sugarbeet genotypes might be present in a concentration below the threshold at which microsporogenesis is affected. The lines with a concentration of *S*-alleles close to the reversion threshold are more unstable under the influence of environment than those with higher concentrations.

Davies and Powling (1985) attempted to interconvert mst and mft forms by exposing their shoot cultures multiplied in vitro at 25°C for 6 weeks to 10 ppm streptomycin, for 1 week to 10 ppm ethidium bromide, for 2 weeks to 12 ppm mitomycin and for 4 weeks at 15°C to 10, 30 or 100 ppm acridine orange, tetracycline or tylosin. After these treatments, the shoots were transferred to a standard medium to recover. Later, after rooting, they were transplanted in fields to flower. No interconversions occurred by any of the treatments. Likewise, heat treatments of meristems followed by shoot tip isolation and its cultures to adult plants produced no changes in sterile or fertile plants.

.18.1.3.2.5 *Variable Expression*

Nielson and Nemazi (1967) and Theurer and Ryser (1969) opined that although basically two major genes in association with the S-cytoplasm cause gc-mst in sugarbeet, certain modifier genes conditioned the fertility of N_{xxzz} plants, as these plants produce p-mst progenies when crossed to certain biennial S_{xxzz} plants. Theurer (1970) found that all white anther mst sugarbeet were not uniformly of S_{xxzz} type, and even in these, modifier genes influenced sterility expression considerably. Thus both mst and mft expressions are influenced by modifier genes, whose number leads to variable mst and mft expressions in sugarbeet.

18.1.3.2.6 *Genetic Control*

As indicated earlier, Owen (1945) attributed gc-mst to the complementary action of two recessive homozygous gene pairs, *xx*, *zz*, that interact with the S-cytoplasm. However, this does not explain most of the exceptional types recovered from the progenies using complete and partial mst plants as female parents. In the S-cytoplasm, *X* or *Z* genes produce intermediate p-mst plants and preliminary tests do not distinguish N_{xxzz} and N_{xxZz} genotypes. This behaviour was explained by Owen (1950) by assuming that gene *X* exerted a major influence and gene *Z* a minor control over fertility restoration. Another gene *Sh*, in dominant state, enhances the pollen-producing ability of *Xx* plants (Hogaboam 1957) and alters the phenotypic ratios expected from Owen's (1945) complementary gene hypothesis. Therefore, the existence of minor modifier genes in mst expression is evident. Genetic models of Hogaboam (1957), Bliss and Gabelman (1965), Kovalenks (1966), Kinoshita (1971), Mikami et al. (1974) postulate the existence of two major genes, *X* and *Z*, and two cytoplasms, N and S, for the genetic control of sterility-fertility expression. Gene action is influenced by modifiers and environment, especially when one of these major genes is in recessive or heterozygous state. Penetrance and expressivity of gene *X* is complete, while that of *Z* is partial and environment-sensitive.

18.1.3.2.7 *Environmental Influence*

S_{xxzz} genotypes are stable mst's with reference to pollen sterility. In other S-cytoplasm genotypes, mst frequency and expression are variable. For instance, p-mst types may change over to complete mft or to mst types during mid- and late-flowering periods. This variation is exhibited by the same plant in different seasons (Kinoshita 1971). The degree of pollen sterility varies even within a flower and between different anthers or between lobes of a single anther. A similar tendency occurs between anthers, flowers and branches of a single plant. In them, pollen sterility is considerably influenced by genotype, environment and their interaction. Interesting variations were found by Cortessi (1967), some of which are:

a) plants arising from the root cuttings exhibit a higher amount of pollen sterility than those arising from seeds;
b) plants grown under 12-h illumination did not produce viable pollen but produced it under 16- or 24-h illumination;
c) interaction between illumination and temperature influenced pollen production considerably. Different families appear to respond differently to these environmental factors.

While Owen (1942) and Cortessi (1967) realised the environmental influence over mst, Bandlow (1964b) found that only p-mst's are influenced by environmental factors. Rohrbach (1965c) studied the influence of several climatic areas over mst in Germany. While half of the mst's were influenced by environmental conditions, one quarter of the complete mst's changed their expression. In this material, increasing daylength appears to have a positive influence on fertility. In others, the influence of temperature was predominant.

i. *Low Temperature.* This induces mst in mft plants having N-cytoplasm and *xxzz* genes (Kinoshita 1971). Sterility is considerably higher in plants grown in vermiculite to which no fertilizer is added than in those with normal fertilizer. Low temperature given during premeiosis causes reduction of PMC's entering meiosis, imbalanced meiosis, tapetal anomalies and microspore abortion (Kinoshita 1971). Tapetal anomalies include cell enlargement and plasmodial formations.

ii. *High Temperature.* Low light intensity and high temperature during floral initiation and flowering period induce almost complete mst. This is accompanied by 7–12% female fertility reduction. Exposure of germinated seeds of gc-mst annual beets to 55°C induced mft in four plants (Cleij 1967). The author hypothesized this as due to the conversion of the S-cytoplasm to the N-type. Although their pollens are stainable, selfing and crossing results do not support S-cytoplasm conversion to the N-cytoplasm type. Furthermore, when the seeds from these treated plants are grown under normal conditions, the majority of the progeny plants are mst. Therefore, the alteration is nongenetic, as cytoplasmic mutations cannot be of such high frequency. Lichter (1978) obtained mft plants after high temperature treatment (+36°C) and subsequent meristem cultures. Most of the mst plants died at this temperature, some recovered at 25°C and produced new flower-bearing branches. In such plants, a few had more or less fertile pollen but later-formed flowers were pollen-sterile. Seeds obtained from these plants after cross-pollination by mft pollen produced mst progeny. The third generation was completely mst. Thus reversion to mst occurred in the heat-transformed mft sugarbeet. Since this transformation from mst to mft was temporary, it could be a dauermodification (Kaul 1967b). Lichter (1978) considered the sterility to be due to a viruslike RNA particle that became thinned and eliminated after meristem culture. Actually only less than half of the parents produced mft plants, others retained sterility and three types of response were obtained, viz. (a) mother plants remaining mst, (b) mother plants becoming mft, meristem progeny mft (further progenies not tested), (c) mother plants becoming mft first, but reverting to mst later.

18.1.3.2.8 Anther Phenotype

The anther phenotype of S-cytoplasm beets exhibits diverse variability induced by some extrinsic and intrinsic factors. Hogaboam (1957) and Oldemeyer (1957) grouped such anther types into eight to ten classes. Owen (1945), Knap (1955), Bandlow (1964a), Rohrbach (1965a) and Kinoshita (1971) classified them into four classes viz. Complete mst, p-mst a and b and normal types. Under the latter type of classification, complete mst plants have white shrunken anthers devoid of pollen, the semi-sterile type-a plant possesses dehiscent and nondehiscent anthers in varying proportions and produces a variable amount of viable and nonviable pollen, semi-sterile type-b plants have nondehiscent reddish yellow anthers, in most of which pollen is shrunken and abortive, and the normal-anther type plants have *Fr* genes, due to which they produce a high proportion of viable pollen.

18.1.3.2.9 Microsporogenesis and Tapetal Development

Prior to meiosis, differentiation of the sporogenous tissue and anther wall layers into epidermis, endothecium, two to three middle layers and tapetum is normal in 2x and 4x mft, mst and p-mst plants (Kinoshita 1971). In 2x and 4x mst sugarbeet, microspores degenerate either immediately after tetrad formation or during microspore development (Table 18.4, 18.8, Fig. 18.3). Whereas male meiosis is normal in 2x mst plants, multivalency and consequent irregular A-I disjunctions are frequent in 4x plants. In mft anthers, tapetal cells lose contact with each other and degenerate during microspore development. The endothecial layer develops a banded appearance, lip cells become prominent, partition between two microsporangia of an anther lobe is dissolved and anther dehiscence follows. In complete and partial mst plants, tapetal behaviour is different and described as follows:

i. *Complete mst Plants (2x).* Male meiosis proceeds normally until tetrad formation, after which the microspores degenerate either during tetrad formation or immediately after liberation (Artschwager 1947; Nagao and Kinoshita 1962; Kinoshita 1971; Chauhan and Kinoshita 1980a,b, this Vol.) and the tapetum becomes either cellular or plasmodial (Fig. 18.3b). Both the types may be present in a flower cluster, but not within an individual flower. Tapetal degeneration in anthers with cellular tapetum is akin to that in mft anthers. In those with plasmodial type, the plasmodia enlarge, invade the anther locule and appress the developing microspores to death. The microsporangia collapse and the anther cavity lodges a strand of blackened matter and disintegrating microspores. In an mst material studied by Kinoshita (1971), the tapetum is persistent and disintegrates after the microspores abort.

Table 18.4. Microsporogensis and tapetal breakdown in mst sugarbeets

Genotype	References	Gene action	Tapetal behaviour
g-mst	Kinoshita (1971)	During microspore maturation	Delayed degeneration
gc-mst			
2x	Artschwager (1947), Nagao and Kinoshita (1962), Kinoshita (1971), Chauhan and Kinoshita (1980a,b) this Vol.	During tetrad formation or just later	Cellular or plasmodial or persistent
4x	Kinoshita (1971), Kinoshita et al. (1972)	During microspore maturation, after liberation	Abnormal, plasmodial or amoeboid
p-mst			
2x	Kinoshita (1971)	During microspore	Delayed degeneration
4x	Kinoshita (1971), Kinoshita et al. (1972)	maturation, after liberation	Delayed degeneration
Induced			
γ-ray	Mikami et al. (1978), Mikami (1982)	During microspore maturation, after liberation	Plasmodial formations
Low temperature	Kinoshita (1971)	During early meiosis	Abnormal

ii. *Complete mst Plants (4x).* Tapetal anomalies occur from meiotic initiation. The tapetal cells are irregular and low in protoplasmic content (Kinoshita 1971; Kinoshita et al. 1972). Plasmodial or amoeboid formations occur during the tetrad stage and periplasmodial masses protrude into the anther cavity. The microspores develop thin exine and minute germpores, but their cell contents degenerate either prior to or just after release from quatrets (Fig. 18.3c). In many, the partition between the two anther loculi breaks and microspores and periplasmodium disintegrate. At anthesis the anthers have a blackened mass with degenerated microspores.

iii. *Partial mst Plants (2x).* Tapetal development is normal but its degeneration is delayed until after microspores are released. Microspores develop slowly and exine formation is initiated in some of them, whereas others degenerate. The disintegrating tapetum adheres to the anther wall, producing a dark stained band. Kinoshita (1971) believes that microspore abortion is due to delayed tapetal disintegration and disturbed transfer of the nutrients from the tapetum to the developing microspores.

iv. *Partial mst Plants (4x).* Until microspore liberation from the tetrads, anther and tapetal development is normal. After liberation, the majority of the microspores abort and tapetal degeneration is considerably delayed. However, some develop into normal-looking pollen grains. Endothecium and lip cell development

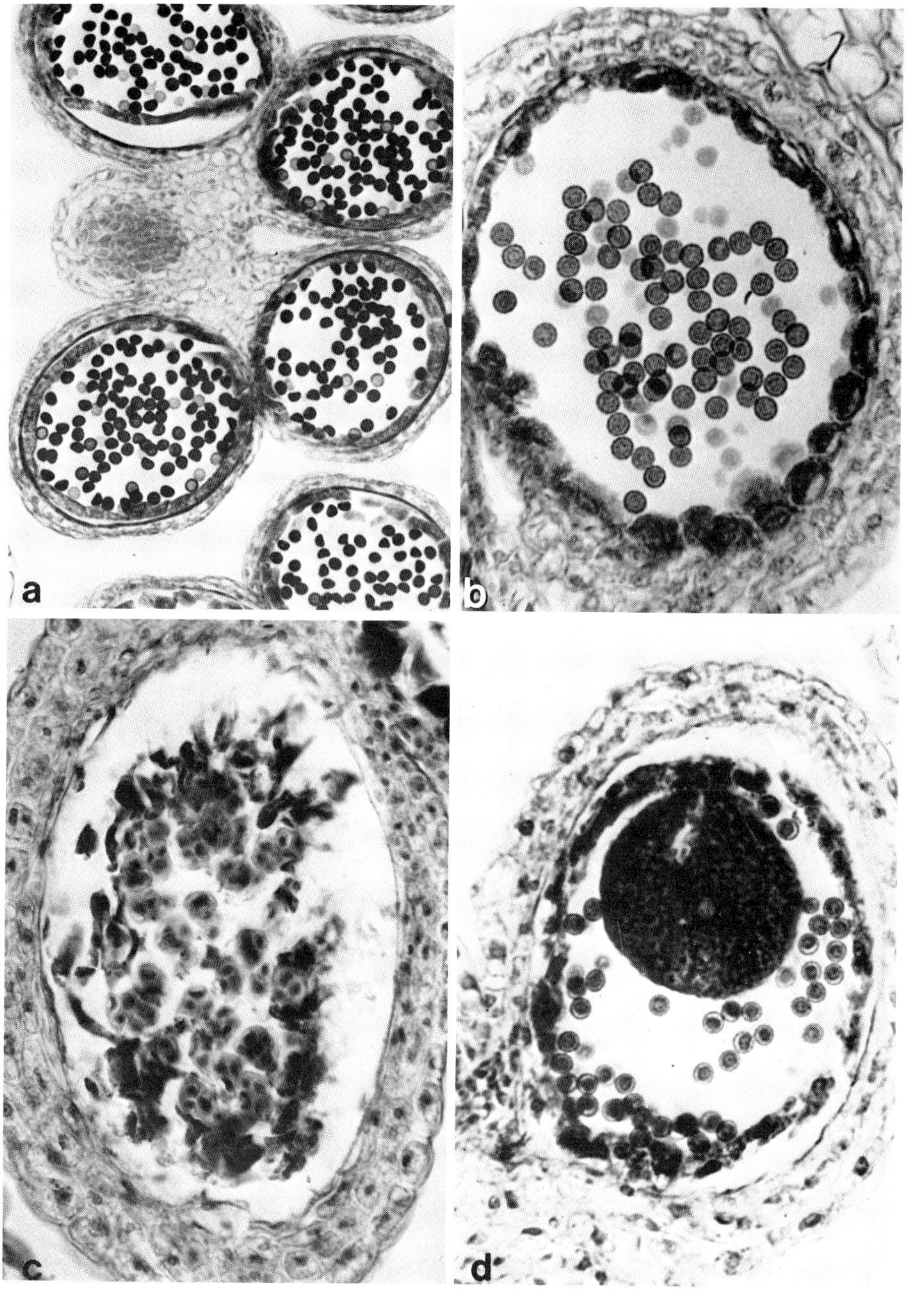

Fig. 18.3a-d. *Beta vulgaris.* T.S. of fertile and sterile anthers. **a** Stained viable pollen and degenerating tapetum of 2x male fertile anther. **b** Degenerating microspores and tapetum of male sterile anther. **c** PMC and tapetal degeneration in 4x male sterile anther. **d** Degenerating microspores and balloon-type tapetal hypertrophy in male sterile anther. (Kinoshita 1971; Mikami 1982)

is partially inhibited. This causes the development of feeble fibrous thickenings, and only a few anthers dehisce (Kinoshita 1971; Kinoshita et al. 1972). In a few plants, pollen and anther development is nearly normal and 40–50% of stainable pollen is viable.

In general, in anthers of 2x and 4x mst plants, tapetal cells exhibit erratic alignment, abnormal growth and become persistent, cellular or periplasmodial. The tapetum may also exhibit balloon-type (Fig. 18.3d) or hill-type (Mikami 1982) abnormalities. Periplasmodial cells encroach on anther locules and crush the microspores, whereas persistent cells deplete microspores of nutrition. Both the tapetum and microspores degenerate almost simultaneously. In 2x and 4x mst anthers, vascular development and differentiation are impaired. In p-mst anthers, the development of endothecium and lip cells is inhibited. All the anomalies lead to the production of inviable, abortive, clumped pollen (Fig. 18.4a,b) instead of fertile, viable, free pollen produced in mft anthers (Fig. 18.4c,d).

18.1.3.2.10 Linkage

i. *Monogerm Fruit.* A recessive gene, *m*, in the homozygous condition produces plants with single germ (monogerm) fruits and seeds. Hogaboam (1957) obtained 37.5% recombination between *Fr* gene *X* and gene *m*, but Kinoshita's (1971) values range between 21.7% and 24.5%. The disparity could be because Hogaboam (1957) considered white anther colour as dominant over yellow and the recombination value computed in the coupling phase was actually in the repulsion phase.

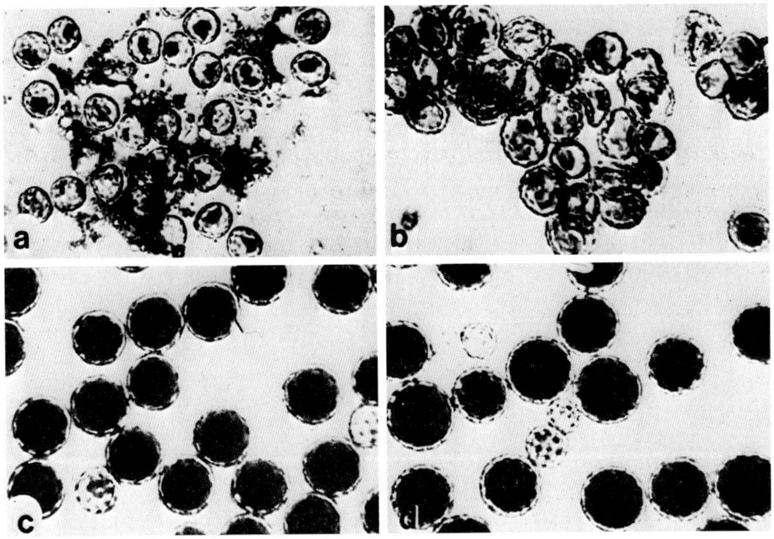

Fig. 18.4a-d. *Beta vulgaris.* Pollen of 2x and 4x male sterile **(a,b)** and male fertile **(c,d)** plants. (Kinoshita 1971)

ii. *Red Pigmentation.* No linkage of ms *a* and a_2 genes with any of the eight known genes was detected by Theurer (1968). Instead, a close linkage between red anther pigmentation and mst was obtained by Stein and Gabelman (1959). One to thirty percent of mst plants had a few red spotted anthers. The authors assumed that red pigmentation induced pollen sterility in anthers that are genetically pollen-fertile, as some anthers of p-mst plants have a red spot. Completely red anthers are entirely pollen-sterile.

18.1.3.2.11 Induced Polyploidy

Savitsky (1954) obtained autotetraploids after treating young seedlings of monogerm self-fertile and mst beets with colchicine. All the tetraploids were completely mst and developed minute, transparent, empty pollenless anthers. Their female fertility was slightly impaired and some seeds were set when pollinated by mft 4x self-incompatible or self-compatible pollen. Reduced seed set on 4x mst plants was not only due to low female fertility, but also to low pollen viability of 4x mft pollinator plants. Kinoshita (1971) obtained 86% and 51% eutetraploids respectively in mft and mst strains of sugarbeet. The frequency of eutetraploids in six mst tetraploid progenies ranged between 44–84%. The production of eutetraploids does not seem to be inhibited in mst plants, as in some 4x families a high proportion of such plants was produced. Moreover, as in 4x mst plants, aneuploidy and anisoploidy occurred also in normal mft 4x plants (Rommel 1965; Mitsuishi 1965; Takahashi and Kinoshita 1965). The aneuploid progeny is produced due to the production and fusion of gametes having variable chromosome numbers. The variability was because of the formation of a varying number of quadrivalents and their irregular disjunction at A-I. Although a proportion of meiotic anomalies varied considerably, all the plants produced after fusion of such gametes were uniformly mst. Thus mst is due to *fr-c* gene interaction and not due to pairing anomalies in 4x plants.

Since tetraploids exhibit tetrasomic inheritance, the 4x mst plants will be $S_{xxxx,zzzz}$ and 4x mft plants $S_{XXxx,ZZZZ}$ or $S_{XXXX,ZZzz}$, F_1 plants obtained from mst x mft would be $S_{XXzz,ZZzz}$ or $S_{XXxx,Zzzz}$ (normal mft) and $S_{XXxxzzzz}$ (p-mst). This was found by Kinoshita (1971) and Kinoshita et al. (1972), who concluded that digenic hypothesis of mft restoration at the diploid level holds true at the tetraploid level also. However, on the basis of chromatid assortment at the 4x level, Savitsky (1963) assumed a monofactorial inheritance. Bolz (1968) obtained monogenic control of sterility and Rohrbach (1965a,b) found it more complicated both at the diploid and at the tetraploid level.

18.1.3.2.12 Induced S-Cytoplasm Mutations

After 50 kR gamma irradiations of seeds of N-cytoplasm sugarbeet strains H-19 and H-2002, complete and p-mst M_1 plants were obtained by Kinoshita (1976, 1980). In seven lines, sterility was maternally maintained until the M_4 generation.

Table 18.5. Induced S-cytoplasm mutants and their *Fr* genes

Male sterile line	Cytoplasm type	*fr* Gene
γ-27	Si-1	Rf_3 (= Fr_3)
γ-20, γ-60	Si-2	Rf_4, Rf_5 (= Fr_4, Fr_5)
γ-130, γ-165	Si-3	Rf_6, Rf_7 (≡ Fr_6, Fr_7)
γ-114	Si-4	Rf_8, Rf_9 (≡ Fr_8, Fr_9)

These mst lines appear to possess four different S-cytoplasm types (Table 18.5). For Si-2 cytoplasm, restoration gene *rf*-4 (≡fr_4) is epistatic to *Rf*-5 (≡Fr_5) and for Si-3, *Rf*-6 (≡Fr_6) is epistatic to *rf*-7 (≡fr_7) (Kinoshita 1976, 1980; Mikami 1982). Partial and complete mst plants were obtained by Kinoshita (1980, 1983) after 100 mM EMS treatment, and only p-mst plants after acroflavin, ethidium bromide and streptomycin treatments to sugarbeet strains H-19 and H-2002. Barring a few that exhibited trisomy, triploidy or tetraploidy, other cytoplasmic mutants were uniformly diploids. In almost all the complete mst plants male meiosis is normal, but tapetal behaviour is abnormal (Mikami et al. 1978; Mikami 1982). After microspore liberation, the tapetum becomes plasmodial and bulges out, forming balloon- or hill-type structures. Usually both these types exist in the different anther loci of the same plant.

18.1.3.2.13 Major S-Cytoplasm Sources

In addition to the S-cytoplasm found by Owen (1942, 1945) in *B. vulgaris* variety US 1, two S-cytoplasm sources were detected in *B. macrocarpa* by Oldemeyer (1957) and Savitsky (1958), the S-cytoplasms of the latter two species appearing to differ from each other. Consistent differences occur in the proportion of mst plants produced by pollinating mst *B. vulgaris* or *B. macrocarpa* plants by the same pollinators. This indicates S-cytoplasm differences in the two species (Oldemeyer 1957). Such differences are due to N-cytoplasm mutations, whose frequency appears to be high, as Kinoshita (1976, 1980) obtained many induced mutants in beets, whose cytoplasms differ from each other on the basis of restoration by different *Fr* genes (Table 18.5).

18.1.3.2.14 Biochemical Differences

i. *Carbohydrates and Lipids.* With meiotic progression, the starch content decreases consistently and is almost absent at pollen development in mft anthers. On the other hand, in mst anthers, the starch content, although low, remains almost constant (Hosokawa et al. 1963). A marked deficiency in carbohydrates, DNA, histones and total proteins occurs in these anthers (Chauhan and Kinoshita 1980b). All along the outer tapetal cell membrane lipidlike droplets are present in mft anthers, but are absent in mst anthers (Nakashima 1975). Furthermore the pH of mst anthers is consistently lower than that of the mft anthers (Hosokawa et al. 1963).

ii. *Amino Acids.* Proline is abundant in mft but absent in mst anthers; the reverse occurs for alanine (Hosokawa et al. 1963; Rohrbach 1965b). Aspartic acid, initially present in both the anther types, decreases rapidly in mft anthers but increases in mst anthers (Rohrbach 1965b).

iii. *Enzyme Differences.* Peroxidase activity in cotyledons, hypocotyl, radicle and mature leaves of S-cotyplasm mst plants is higher than that of N-cytoplasm plants (Georgescu et al. 1979). This difference is so consistent that the authors suggested its use as a criterion for separation of N- and S-cytoplasm plants. Likewise during anther development stages, phosphorylase activity in mst anthers is much lower than in those of mft anthers (Tsuda et al. 1978).

iv. *Mitochondrial Form and Function.* During the tetrad stage, mitochondria of tapetal cells in mst anthers are abnormal and have degenerated cristae (Nakashima 1975). At the microspore stage, Ubisch bodies present at the inner surface of tapetal cells of mft anthers are absent in mst anthers. Considerable differences in mitochondrial respiration as judged by oxidative phosphorylation occur between mst and mft anthers (Cuelemans 1973). The altered mitochondrial form may account for such differences. Compared to mst lines, mft lines have higher phosphorylating potential, but this is compensated for by enhanced nonphosporylating respiration in mst anthers.

v. *Mitochondrial DNA Differences.* Using restriction endonuclease digestion of *cp* and *mt*-DNA, Powling (1981, 1982) and Powling and Ellis (1983) concluded that the mst trait in sugarbeet having US 1 S-cytoplasm is encoded by mt-DNA and not by cp-DNA. Sugarbeet mitochondria contain small, supercoiled DNA molecules in addition to high molecular weight (HMW) mt-DNA (Powling 1981). These are absent in cp-DNA (Fig. 18.5). On size basis, four different minicircles a,b,c,d are distinguishable. Whereas the mitochondria of all mft genotypes have either a and c or a, c and d or b, c and d, depending upon the genotype, mst mitochondria have the a minicircles only. The function of these minicircles is unknown. Since minicircle c is present in all the mft lines investigated by Powling and Ellis (1983), and absent in the mst line, it might encode information necessary for pollen production, although this is perhaps unlikely, due to its too small size. Circumstantial but compelling evidence of mitochondrial involvement in mst is provided by Mikami et al. (1984a), who obtained distinct mt-DNA restriction fragment patterns, invariant in different nuclear backgrounds in N and S cytoplasms. Although most of the DNA fragments are common to these two cytoplasms, each cytoplasm has a specific band unique to it. In contrast, the restriction fragment analysis of cp-DNA of the two cytoplasms exhibits few differences and less heterogeneity compared to mt-DNA. This suggests that mitochondrial genes are responsible for mst expression. Densitometric tracings and Southern hybridization experiments of Mikami et al. (1984b) reveal the absence of one band and its replacement by two fragments in S-cp-DNA.

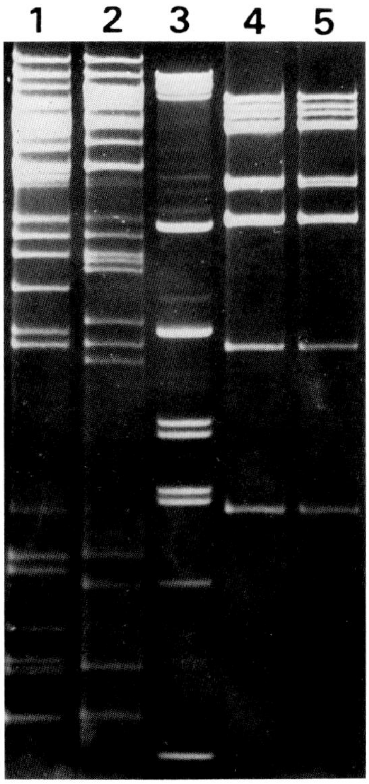

Fig. 18.5. *Beta vulgaris*. Agarose gel electrophoretograms of mitochondrial DNA of male fertile (*1*) and male sterile (*2*), bacteriophage λ DNA as marker fragment (*3*), chloroplast DNA of male fertile (*4*) and of male sterile (*5*) sugar beet. (Powling and Ellis 1983)

18.1.3.2.15 Grafting

No mst transmission across grafts in first- and second-year inflorescences was detected by Owen (1942), but sexual progeny of these grafted plants was not tested by him. Such progenies, obtained after grafting, aphid and mechanical inoculations, indicate nontransmission of mst in fertiles (Cleij 1967). Curtis (1967), however, reported transmission of mst across graft unions in both directions, viz. from gc-mst scions to 0-type (nonrestorer) stocks (side shoots), and from sterile stocks to 0-type scions. Complete to partial pollen abortion occurred in the grafts and first sexual generations in maintainers, grafted maintainer stocks on mst scions and its reciprocals. Thus many 0-type scions developed mst flowers, and in most progenies mst occurred even though the sterility could not be detected on the scions. Although this report represents the first one indicating an alteration in fertility in the graft generation, it could not be confirmed by Theurer et al. (1968), who found no such transmission in grafted mst and maintainer sugarbeet either in the graft generation or in offspring of selfed or outcrossed fertile flowers from grafted maintainers. Lichter (1978) used materials of Curtis, but could not transfer sterility across the grafts.

18.1.3.2.16 Viral Infection

Yellow leaf virus causes mst in Maribo sugarbeet variety (Larsen 1977, 1981). The infection destroys anthers fully or partially, so that anthers either degenerate completely or develop, but produce partially or completely aborted pollen, most of the anthers being indehiscent. Female fertility of the infected plants is nearly normal and the sterility is not seed-transmitted. The mst degree enhances with increasing host homozygosity, so that it is high in most inbreds and least in outbred heterozygotes (Larsen 1981).

18.1.3.3 Chemical Male Sterility

i. *Dimethyl Arsenic Acid.* Application of 200 ppm through leaf midvein as aqueous foliar spray caused some pollen sterility but was highly phytotoxic (Hecker et al. 1972) and damaged ovular fertility considerably (Table 18.6).

Table 18.6. Chemical male sterility in *Beta*

Species/Chemicals	References	Male sterility induction	Female fertility reduction	Side effects
Beta patula[a] Mendok	Wit (1960)	High-partial	nk	nk
Beta vulgaris DPX-3778	Theurer (1979)	Partial	nk	Phytotoxic, chlorophyll destruction, necrosis and inhibited plant growth
Dimethyl arsenic acid	Hecker et al. (1972)	Low	High	Highly phytotoxic
Ethrel	Hecker et al. (1972), Hecker and Smith (1975)	Partial	High	Phytotoxic
Estrone	Hecker et al. (1972)	Partial	High	Phytotoxic
Mendok[a]	Butterfass (1960), Reubenbauer and Schultis (1960), Wit (1960), Salmon (1963), Kinoshita (1971), Bronckers and Amand (1966)	Partial	Nearly nil to very high	Phytotoxic and high yield reductions
Beta webbiana[a] Mendok	Wit (1960)	Partial	nk	nk

[a]Results highly variable.

ii. *DPX.* Application 1250–2500 ppm aqueous DPX solutions at different stages of stalk development did not induce pollen sterility but reduced anther dehiscence (Theurer 1979). The treatments caused stunted plant growth, leaf rolling and yellowing, chlorophyll destruction, necrosis and cleistogamy. Higher concentration caused premature plant death.

iii. *Estrone.* Application of 30 ppm through leaf midvein at pre- and post-meiotic stages led to nearly 50% pollen sterility (Heckar et al. 1972); however, the treatment reduced ovular fertility and caused phytotoxicity.

iv. *Ethrel.* Using granular ethrel in various concentrations induces pollen sterility (Heckar and Smith 1975) but increases plant mortality and ovular damage. The phytotoxicity induced is also high. Nearly similar results were obtained by using 100 or 200 ppm ethrel applied weekly as spray or intravascularly (Hecker et al. 1972).

v. *Mendok.* In *B. patula, B. vulgaris* and *B. webbiana* 0.3, 0.8 and 0.6%, respectively, of the chemical treatments applied prior to flowering led to high pollen sterility, anther indehiscence and low ovular sterility (Wit 1960), but replications exhibited wide variations. Therefore, it is desirable to have a large sample. The induced sterility lasted for 4–5 weeks only and the earliest and the last-formed flowers were p-mst. Higher chemical concentrations considerably increased ovular damage and phytotoxicity. Butterfass (1960) and Salmon (1963) found two applications of 0.2% or 0.3% mendok to be effective male gametocides in sugarbeet, but Ruebenbauer and Schultis (1960) recommended higher concentrations, ranging from 0.5 to 1.5%, in this species. On the other hand, in both diploid and tetraploid beet, Kinoshita (1971) obtained 70% pollen sterility by 0.3% sprays at bud formation and preflowering stages, but leaf injuries, phytotoxicity and ovular abortions were very high. Similar results were obtained by Dudley (1960), Furste (1964), Bronckers and Amand (1966), Skoyen (1969) and Hecker et al. (1972).

18.1.3.4. Breeding Value

In sugarbeet, the discovery of mst by Owen (1945, 1952) revolutionized commercial hybrid beet production, as it provided economically the soundest means of developing superior varieties. Prior to this, hand-crossing was used for producing a limited amount of hybrid seed. Thousands of flowers, all of which are minute and bisexual, are produced by a sugarbeet plant. Their five anthers per flower must be removed in the young bud stage. Thus emasculation is not only tedious and time-consuming but requires a large number of workers for producing a small quantity of seed. Such a procedure is extremely laborious and economically impracticable. Sugarbeet, basically a self-incompatible, cross-pollinated crop, exhibits a versatile genetic and breeding system that includes outbreeding, vegetative reproduction and obligate selfing through the introduction of a self-fertility gene. Although the crop is biennial, bolting and early seed production can be induced by photothermal induction. This provides a broad option to breeders for adopting an appropriate breeding technology. Furthermore, both g-mst and gc-mst are known in sugarbeet. Using these, genetically diverse 2x, 3x, 4x sugarbeet have been developed by mass selection, family-line breeding, synthetic production and polyploidy breeding.

18.1.3.4.1 Monogerm Seed

The sugarbeet seed ball contains one to several seeds. When ordinary seed is planted, the seedling plants emerge in clumps that need to be thinned in order to reduce the stand to single plants. In order to employ mechanical methods of thinning, blocking, topping, and harvesting, which reduce the labour costs in growing beet, it is desirable to have a high percentage of single plants. Savitsky (1952a,b) detected five monogerm (single germ) mutant plants in a 4-acre seed field of Michigan Hybrid 18. The trait is determined by a recessive gene, m, and causes the seed ball to contain only one viable seed instead of several. A quick and easy way to utilize this trait for commercial production was to put it into a gc-mst line and then to use it, or a single cross between it and another sterilizable monogerm mst line, as seed parent for crossing with a satisfactory open-pollinated variety. Since only monogerm seed will be produced on the monogerm seed-bearing plant, the pollinator may be either mono- or multigerm. A large percentage of hybrid sugarbeet seed in the USA and Europe in the early 1960's was of this type.

18.1.3.4.2 Hybrid Production

Following the discovery of the gc-mst system in beets by Owen (1945), exploitation of positive heterosis in beet yield was vigorously followed in the USA and Europe. Initially, three types of hybrids, produced after single, double or three way crosses, were utilised for commercial production, but now only the latter two types are being used because of low seed set in the former.

i. *Single Cross Hybrid* ($A \times B$). These hybrids, designated as AB, are produced by crossing inbred A, a monogerm gc-mst (S_{xxzz}) type, to another mft inbred B type O (N_{xxzz}) plants. Inbred A is maintained by crossing to its fertile counterpart, viz. type O ($N_{xxzz} \equiv N_{fr1fr2}$). Inbred B should exhibit a high combining ability with A, its *Fr* genes (X and Z) are immaterial when seeds of the AB hybrid are not to be harvested. This cross becomes economically futile unless inbred A is vigorous and a high seed producer and inbred B an efficient pollen producer. Alternatively, inbred A may be pollinated by a vigorous open-pollinated variety. In any case, inbred A must be a monogerm trait. Usually seed set on inbred lines is low, therefore, single cross is not being extensively utilized in commercial sugarbeet production.

ii. *Three-Way Cross* [$(A \times B \times C)$]. To enhance seed set on inbred lines, a three-way cross is usually used for commercial sugarbeet production. For this, hybrid AB, obtained by the above-mentioned method, is pollinated by inbred C that must be a high pollen producer and good combiner in order to produce high-yielding desirable progeny. The status of X and Z genes in the C-line is of no consequence, since seed is not harvested from the commercial sugarbeet crop. However, both the inbreds A and B must be monogerm for AB to produce monogerm seeds. Inbred C or the open-pollinated pollinator may be mono- or multigerm.

iii. *Double-Cross* $[(A \times B) \times (C \times D)]$. Using self-incompatibility(s) alleles, g-mst and gc-mst, double-cross sugarbeet hybrids are produced commercially by the following three methods:

a) *Use of Genic and Gene-Cytoplasmic Male Steriles.* This method, proposed by Owen (1954), involves the production of single-cross mst hybrid AB by crossing A (S_{xxzz}) with B (N_{xxzz}) as given above. Inbred C is a line segregating for g-mst and is rogued to plants having ms genes *aa*. Inbred D is normal-crossed to normal AA. Thus D-line single cross CD will have *Aa* genes and is made to pollinate AB to produce the double-cross seed (Fig. 18.6a; Table 18.7).

Table 18.7. Genic male sterility in Chenopodiaceae

Species	References	Gene control	Gene action	Remarks
Beta macrocarpa	This Vol.	1r	After microspore release	Tapetum intact
B. maritima	Kinoshita (1971) Kinoshita and Takahashi (1972)	1r	After microspore release	Highly influenced by environment, tapetal cells intact, multinucleate and plasmodial
B. vulgaris	Owen (1952)	1r	Immediately after microspore release	Tapetum persistent
	This Vol.	1r	Immediately after microspore release	Tapetal degeneration delayed
Chenopodium quinoa	Simmonds (1971)	1r	nk	Cytoplasmic determination once generated overrides nuclear control of mst
Spinacia oleracea	Kobabe (1967)	1r	After microspore release	
	This Vol.	2r	nk	

b) *Use of Gene-Cytoplasmic Male Sterility and S-Alleles.* (Table 18.8) This method was outlined by Oldemeyer and Smith (1965) and involves production of mst hybrid AB by using a gc-mst female and O-pollinator line. Hybrid CD is produced by interplanting another two self-incompatible inbreds C and D. Seeds set on these lines is hybrid seed CD. Its plant progeny is used to pollinate mst plants AB over which double-cross seeds are produced (Fig. 18.6b). A little selfing occurs in self-incompatible lines and naturally the seed harvested comprises of single-cross CD with mixture of inbred C and D. Thus finally the hybrid seed will be a mixture of the double-cross AB × CD and three-way cross AB × C and AB × D.

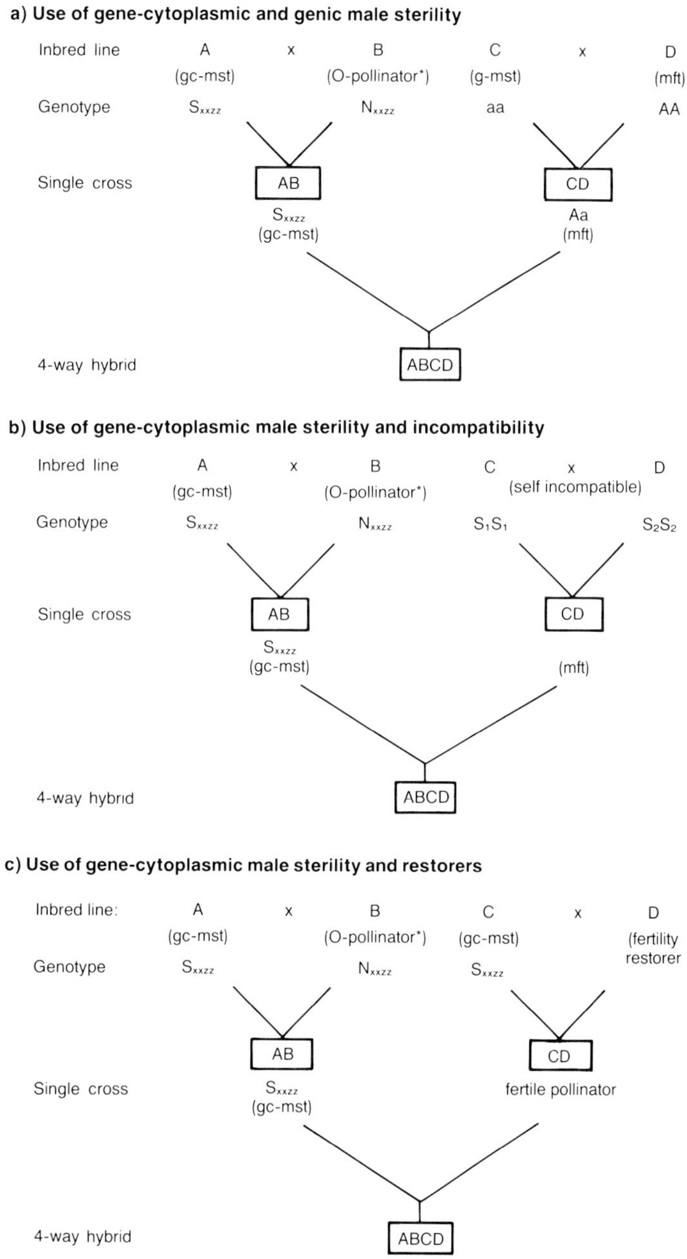

Fig. 18.6a-c. Production of single and double cross hybrids; *, maintainer

Table 18.8. Gene cytoplasmic male sterility in sugar beet

Species	References	Gene control	Gene action	Remarks
Beta macrocarpa	Oldemeyer (1957)	2r	After tetrad formation	Two dominant genes *X* and *Z* restore fertility. *X* is epistatic to *Z*
B.maritima	Coe and Stewart (1977)	2r	Immediately after microspore release	*X* and *Z* complementary, tapetal disintegration follows microspore degeneration
B.vulgaris (2x)	Owen (1942) Artschwager (1947) Nagao and Kinoshita (1962) Kloen (1964) Kinoshita (1971) Kinoshita et al. (1972), Chauhan and Kinoshita (1980a), this Vol.	2r	During tetrad development or immediately after microspore release	*fr* gene action highly variable, plasmodial tapetum Sh gene enhances pollen producing ability of semisterile plants
B.vulgaris (4x)	Kinoshita (1971), Kinoshita et al. (1972), Mikami et al. (1978)	4r	During microspore maturation	Plasmodial irregularities
B.vulgaris (pmst, 2x)	Kinoshita (1971)	2r	During microspore maturation	Delayed tapetal degeneration, *fr* gene linked with red pigmentation of anthers
B.vulgaris (pmst, 4x)	Kinoshita (1971)	4r	During microspore maturation	Delayed tapetal degeneration
B.vulgaris × *B.maritima*	Dalke et al. (1983)	2r	After microspore liberation	Tapetum degenerates after microspore degeneration, thickenings in endothocial walls absent

c) *Use of Gene-Cytoplasmic and Restorer Gene System.* The third procedure, outlined by Duvick (1959) and elaborated by Theurer and Ryser (1969), utilizes *fr* gene instead of *ms* gene. The single-cross mst AB is produced as indicated earlier. Inbred C is likewise a gc-mst line, but inbred D possesses a dominant *Fr* gene. The lines are mated as shown in Fig. 18.6c.

d) *Comparison of the Three Mating Systems.* In order to provide a pure mst population for D parent pollination in the first system (Fig. 18.6a), mft plants have to be rogued from inbred C in the early bud stage of flowering. For this a red seedling colour marker gene could be introduced into mst plants so as to identify them in the pure state. Roguing requires hand labour and is expensive. It also necessitates hand removal of 50% or more of the plant population because of mst:mft segregation. Therefore, only a few four-way hybrids have been produced by this method for commercial use. In the second system, use of self-incompatible inbred lines is made, but development and management of these is arduous and hazardous since in many such lines erratic and unpredictable incompatibility breakdown produces selfed instead of hybrid seed. To overcome incompatibility, paired crosses are utilized. These may be used as a form in inbreeding through sibpollinations. The third system (Fig. 18.6c), utilizing *fr* genes, provides an efficient method of producing double-cross seeds provided *fr* genes prove stable over a wide range of environment and genotypic backgrounds. Comparing performance of double-cross hybrids produced by methods (a) and (c), Theurer et al. (1969) found that the four-way hybrids produced after using method (a) had higher a sugar percentage and lower impurity index than that produced by method (c). This superiority occurs in some combination hybrids but not in others.

Depending upon the availability of inbred lines, one of these systems may be used for hybrid beet production. However, in each system, A and B inbreds have to be monogerm so that the final seed produced is monogerm. The commercial sugarbeet hybrid seed is produced either in strips of mst monogerm and mft pollinator (the germ number of pollinator is of no consequence) or as a mixture of monogerm mst and multigerm mft pollinator. In this case, multigerm pollinator seed is removed by thickness grading during processing of commercial seed. The combining ability of the inbreds has to be such as to produce high-yielding double-cross plants. The fertility of the double-cross ABCD hybrid plants is immaterial, as they are not harvested for seeds.

18.1.3.4.3 Polyploidy Breeding

Increased cell size and reduced seed production are the characteristics generally associated with polyploidy. For this reason, plants grown for vegetative organs are better adapted to polyploid breeding than plants grown for production of seeds. Use of polyploidy as a method of breeding sugarbeet found its greatest acceptance in Europe and was also taken up in the USA and Japan. Commercially, two types of polyploid are produced, anisoploid varieties are a mixture of approximately 10–20% tetraploid, 60–80% triploid and 10–20% diploid plants, specially formulated by growing for seed mixture of 10–20% diploid plants with 80–90% tetraploid plants. Triploids are produced by crossing diploids and tetraploids, generally using tetraploid as the seed parent and diploid as the pollinator. This combines the advantages of polyploidy and hybrid vigour. Using gc-mst tetraploid as seed parent has the advantage of less area devoted to pollinator in seed production, less chance for contamination from other diploid beets, and greater seedling vigour in hybrid plants because seeds from tetraploids are larger than seeds from diploid plants. Owen (1948) proposed the production of triploids using 2x and 4x gc-mst beets. Pollination of tetraploid mst plants by the pollen of diploid mft plants produces 3x hybrids agronomically superior to those produced in the reciprocal crosses. In crossings in which diploid mst plants are used without involvement of type-D plants, partial mst's appear and a mixture of triploid and selfed diploids segregates. Moreover, some 2x mst lines possess 2–14% unreduced egg cells (Ellerton and Hendricksen 1959; Bandlow 1964a) from which tetraploid seeds would result when fertilized by 2x pollen. Growth of haploid pollen tubes is much faster than diploid tubes produced by 4x plants. Therefore, haploid pollen is more efficient in fertilisation than the diploid pollen of 4x sugarbeet. Crossing 4x mst plants to diploid 2x mft plants produces a very low proportion of selfs and does not reduce the efficiency of triploid production. Triploid hybrids from such crosses are highly superior to parental tetraploid and diploid varieties. They possess equivalent or greater productivity in gross sugar yield to that of hybrids which can be produced from the crossings of the related lines at the diploid level (Kinoshita 1971). However, some triploids produced by using 2x mst × 4x mft plants are superior to their corresponding diploid control variety (Bandlow 1964b). Accordingly, pollinating a diploid mst variety to its tetraploid pollinator has led to the release of the improved variety Trirave or Triplex (Ellerton and Hendriksen 1959). In their gross sugar production, triploid hybrids of sugarbeet are superior to parental diploid and tetraploid varieties (Matsumara 1953; Knapp 1956; Helmerick et al. 1965; Kinoshita and Nagao 1969; Kinoshita 1971).

18.1.3.4.4 Uses in Breeding

18.1.3.4.4.1 Achievements

The major advantage of using mst is to combine increased yield (15–20%) with enhanced sugar content (a combination very hard to breed into open-pollinated varieties) plus the ease of combining disease-resistant genes from various sources. These have outweighed the difficulties existing in using mst lines. In addition, the hybrid sugarbeet exhibit more uniform seed germination, size and root shape. These features, being vital to the modern highly mechanised farming methods, have led to the production of hybrids that constitute about 80% of the US sugarbeet production; the percentage is rising fast with time. Initially, top and varietal crosses were utilized, but now double and multiple crosses are preferred. Hybrid beets have also been produced in Europe, after a Dutch company started their commercialization. The Dutch company not only utilised gc-mst, but took advantage of the polyploidy vigour by using a tetraploid pollinator on a diploid female so that the hybrid is a triploid. Such gc-mst triploids were also produced and utilized in Germany, England, France, Poland and in many other European countries. Many seed companies have hybrid seeds for sale. Recently, Vavilov et al. (1979) intercrossed many diploid and tetraploid gc-mst sugarbeet. Some of the cross combinations had the highest sugar and dry matter content. The hybrid between a p-mst and a diploid sugarbeet exhibited the highest root and dry matter yield, which remained unaffected by the ratio of the genotypes in the crossing blocks.

18.1.3.4.4.2 Problems

i. *Narrow S-Cytoplasm and fr Gene Base.* Although found in several open-pollinated varieties of sugarbeet *B. vulgaris*, no substantial evidence exists that the S-cytoplasm sources are contrastingly different. Over 90% of hybrid sugarbeet produced currently owes its S-cytoplasm to the US 1 variety. Thus the sugarbeet industry relies on an extremely narrow cytoplasm base, which is most hazardous, as shown by the fungal susceptibility of T-cytoplasm corn. Furthermore, the uniformity and narrowness of sugarbeet germplasm may easily be vulnerable to insidious pests and stress (Anonymous 1972). With the discovery of monogerm seed and gc-mst system in *B. maritima* (Coe and Stewart 1977), the possibilities of increasing the S-cytoplasm base and genotypic variability are strong. Some varieties of *B. maritima* possess many useful *fr* genes, and offer a means of diversifying sugarbeet germplasm.

ii. *Variable Expression of gc-mst Lines.* The mst expression of the progeny of gc-mst lines differs appreciably not only with the pollinators but also with environmental fluctuations. This is elucidated below.

iii. *Pollinator Selection.* Whereas certain pollinators produce 100% mst progeny when crossed to a gc-mst line, other pollinators with the same line produce a progeny that segregates into mst, p-mst and mft plants in varying proportions. A few pollinators produce only mft plants. Since the breeding behaviour of a pollinator has a great bearing on its use, a vital step in producing hybrid sugarbeet is to index the plants for the production of pollinator lines. Of course, this is a common procedure to be followed for all the crops in which mst is used in hybrid breeding; but once the maintainer pollinators are indexed, their efficiency in producing mst hybrids is not stable in sugarbeet.

iv. *Environmental Influence.*

a) *Temperature.* The impact of environmental factors, especially temperature, on mst expression is profound in mst sugarbeet. Low temperature prior to flowering enhances the mst degree, but high temperature reduces it. Monogerm beet are more sensitive to temperature variations than multigerm lines. Most of the mst sugarbeet is monogerm. Furthermore, some mst lines produce little viable pollen when bagged, whereas their unbagged sister plants remain mst. Bagging usually enhances temperature within a flower and this leads to the fertility of the bagged flowers.

b) *Ageing.* In many mst sugarbeet lines, anthers of the late-formed flowers on lateral branch apices of the seed stock produce some fertile pollen, whereas anthers of the earlier-formed flowers are devoid of pollen. Some p-mst plants having 15–20% fertile pollen produce flowers having 70–80% stainable pollen after 4–5 weeks.

c) *Intrinsic Factors.* Variability in single-flowering branches occurs in some mst sugarbeet lines. Whereas flowers borne at the proximal end are mst, a few borne at the terminal end produce some viable pollen. In a few mst plants, empty white anthers and yellow anthers filled with pollen are borne within the same flower. This type of variation appears to be due to some intrinsic physiological factors and is not genetical. Fertility of the stainable plump pollen is low and most of it is physiologically weak.

18.2 *Chenopodium quinoa* Wild (quinao)

Chenopodiums are summer and autumn blooming annuals, biennials, herbaceous perennials and subshrubs and comprise over 100 species. The mst is known only in *C. quinoa*, an amphidiploid economic plant of high Andes. From a seed sample collected from Peru and Bolivia, mst plants having abortive anthers and normal female fertility were detected by Simmonds (1971). The sterility is controlled by a recessive *ms* gene, which breeds as though it carries an erratically expressed or transmitted cytoplasmic factor for mst as well. The system of sterility is peculiar in that the cytoplasmic determination, once generated, overrides the nuclear control and the recessive mst becomes a maternally transmitted gc-mst. The influence of environment on mst expression is equally profound. Graft transmission of sterility is not possible. Evidence indicates that the sterility is due to cytoplasmic mutation from N- to S-type induced by a nuclear gene. Since basically the recessive gene *ms* generates a *c*-gene, loss of *ms* genes from the population would thus leave simply a rather inefficient cytoplasmic system in quinoa.

18.3 *Spinacia oleracea* L. (Spinach)

Of the four species native to West Asia, a monoecious local strain of *S. oleracea* segregated mst plants in low frequency. The sterility appears to be conditioned by duplicate recessive genes, as 15:1 mft:mst plant F_2 segregations were obtained (this Vol.). In the German cultivar, Wisemona, mst plants having light green shrivelled anthers and poorly developed filaments were detected by Kobabe (1967). In them, pollen degenerated after the tetrad stage and the tapetum broke down, forming a mass coalesced with the abortive microspores. The sterility is conditioned by a recessive gene not linked with the factor for sex expression or monoecism. Fertility-restoring genes were detected by Kobabe (1972).

Dioecism in *S. oleracea* is controlled primarily by a pair of sex chromosomes, X and Y (Bemis and Wilson 1953). The homozygote XX is a female plant and the heterozygote XY is a male. Under this system, heterozygotes produce pollen and homozygotes ovules. Spinach is therefore obligatorily cross-pollinated for morphological reasons, although the morphology is genetically determined. The existence of mst in monoecious varieties and the segregation of sexes in dioecious varieties make it simple to produce hybrid spinach seeds. Two varieties are interplanted and the males are rogued from the variety chosen as the seed parent. All seed produced on the "female" variety will be hybrid. This was practised in the USA and Europe. The method is laborious, but usable if the return for labour is profitable. Sneep (1958) outlined a novel way of developing purely female strains of dioecious spinach. In this, since the genotype of plants to be used as parents of the female (seed parent) strain cannot be determined phenotypically, all plants have to be identified by individual plant test crossing in each generation of maintenance. This mitigates the economic usefulness of the scheme.

CHAPTER 19

Compositae

Although over 800 genera constitute this family, mst is known only in *Ageratum, Carthamus, Centaurea, Chrysanthemum, Cirsium, Helianthus, Lactuca, Parthenium, Tagetes, Taraxacum* and *Zinnia* (Tables 19.1 to 19.3).

19.1 *Ageratum* L.

19.1.1 *A. conyzoides* L. (goat weed)

Kaul (1966) detected two cytotypes (2x, 4x) and two photoperiodic ecotypes, one being short-day and the other day-neutral in this species. In the short-day tetraploid ecotype, mst plants occurred in which male meiosis proceeded normally up to tetrad formation but the microspores degenerated within tetrads, (Table 19.1). Female fertility was nearly normal and F_3 segregations indicate sterility to be conditioned by two to three recessive *ms* genes. In the diploid it is conditioned by one recessive *ms* gene (this Vol.).

19.1.2 *A. houstonianum* Mill

In this ornamental garden species, mst's arise spontaneously (Reimann-Philipp and Fuchs 1971 and Reimann-Philipp 1976). In these, stamens are developed but no pollen is formed. At anthesis, whereas the mft flowers are refractive, being densely covered with pollen, the mst flowers lack pollen completely (Fig. 19.1). In these mst's, the male meiosis proceeds normally up to tetrad formation but tetrads degenerate before microspore differentiation. Female fertility is normal. The sterility is conditioned by one recessive gene, ms_1. Using mst plants, hybrids with increased plant and head size were produced by Reimann-Philipp (1976). To maintain mst, the hybrids are pollinated by the heterozygous *Msms* mft line. This necessitates elimination of 50% mft plants. F_1 hybrid propagation can also be done vegetatively, which is an archaic method. Recently in Germany, vegetative propagation has been done by in vitro culture (Reiman-Philipp 1983, personal communication).

Table 19.1. Genic male sterility in Compositae

Species	References	Gene control	Gene action	Remarks
Ageratum conyzoides	This Vol.	1, 2-3r	After tetrad formation	Female fertility normal
A.houstonianum	Reimann-Philipp and Fuchs (1971), this Vol.	1r	Before microspore formation	Female fertility nearly normal
Carthamus tinctorius	Heaton and Knowles (1982)	1r	nk	Female fertility highly reduced, seed size increased
	Ebert (1964), Ebert and Knowles (1966)	1r	Anther dehiscence delayed or impaired	Plants thin-hulled and weak-stemmed
Centaurea americana	This Vol.	1r	During early prophase I	
C.imperialis	This Vol.	1r	During early prophase I	
C.scabiosa	Fröst (1963)	nk	After microspore release	
Chrysanthemum balsamita, C.carinatum, C.parthenium	This Vol.	nk	nk	Anthers transformed into petal-like organs
C.morifolium	Broertjes and Jong (1984)	nk	Prior to anther dehiscence	*ms* gene reduces vegetative or floral organ size
Cirsium arvense, C.montanum, C.oleraceum	Delannay (1979)	nk	During meiosis	
C.palustre, C.spinosissimum	Delannay (1979)	nk	After tetrad formation	Tapetum degenerates prior to sporogeneous tissue degeneration
Helianthus annuus (complete)	Kuptsov (1935), Anonymous (1964)	1r	nk	
	Putt and Heiser (1966)	1r, 2r	After microspore release	
		1r	During pollen formation	
	Leclercq (1966), Pirev (1966) this Vol.	nk	During pollen formation	
	Vranceanu (1970)	1r	nk	
	Paun (1974)	1r	During pachytene	5 ms genes nonallelic, tapetal behaviour normal
	Nakashima and Hosokawa (1974a)	1r	After microspore release	Tapetum persistent
H.grossesarratus × *H.annuus* (partial)	Georgieva-Todorova (1974)	1r	During premeiosis or after microspore release	
	Putt and Heiser (1966)	1r	nk	2 Nonallelic *ms* genes
	Kovačik and Skaloud (1973)	1r	nk	

Lactuca sativa	Ryder (1963)	1D + 1r	nk	mst partial
		1r	nk	Female sterility partial, leaf distortions high, pleiotropy
L.sativa × *L.serriola*	Lindqvist (1960)	3r	During early prophase I	Narrow, deeply cut lobed leaves, mst sensitive to environment
Parthenium argentatum	Powers and Gardner (1945)	nk	After tetrad formation	
P.hysterophorus	This Vol.	1r	After microspore release	
Tagetes erecta	Bolz (1961), Goldsmith (1968), Horn (1974)	1r	nk	X-ray-induced
	Towner (1961)	1r	During stamen development	
	This Vol.	1r	After microspore liberation	EMS-induced
		1r	nk	Early and dwarf
Taraxacum balticum	Malecka (1971)	nk	During or after tetrad formation	Tapetum degenerates prior to sporogenous tissue
	Gustafsson (1935)	nk	During prophase-I	Tapetum degenerates prior to sporogenous tissue
T.fulvum, *T.pseudofulvum*,	Gustafsson (1935)	nk	During tetrad stage	Callase enzyme not secreted
T.nordstedti	This Vol.			Tapetum degenerates prior to sporogenous tissue
T.silesiacum	Malecka (1971)	nk	During premeiosis	Apetalous florets
Zinnia elegans	Cachon (1980),	1r	During staminal	
	Duffy (1982), this Vol.	3r	development	
Z.linearis	This Vol.	1r	nk	
Z.pauciflora	This Vol.	1r	nk	

Table 19.2. Gene-cytoplasmic male sterility in Composite[a]

Species/crosses	References	Gene action	Remarks
Cirsium acaule, C.eriophorum, C.oleraceum, C.palustre, C.velenovsky	Correns (1916, 1937)	nk	No *Fr* genes detected
Helianthus annuus	Gundaev (1966), Heiser (1982)	nk	Alloplasmy, cytoplasm specificity
H.giganteus × *H.annuus*	Whelan (1981)	After microspore liberation	
H.maximilliani × *H.annuus*	Whelan and Dorell (1980)	During staminal development	mst high but not complete
H.lenticularis × *H.annuus*	Anaschenko (1981)	nk	
H.petiolaris × *H.annuus*	Leclercq (1969), Paun (1974)	Before meiotic initiation, after microspore liberation	Anthers reduced highly, tapetum disintegrates after microspore liberation
	Whelan (1979)	After microspore liberation	

[a]Gene control not known except in *H.annuus*, in which it is polygenic.

Table 19.3. Chemical male sterility in Compositae

Species/chemicals	References	Male sterility induction[a]	Female fertility reduction	Side effects
Carthamus tinctorius				
GA	Yermanos and Knowles (1960)	a	Complete	Internode elongation and chlorosis
Helianthus annuus				
Ethrel	Campos (1974)	a	nk	nk
GA	Schuster (1961)	a	Very high	nk
	Seetharam and Kumari (1975), Vear (1981)	a	High	Plants tall, flower earlier
	Frank and Koves (1977)	a	Very high	Plants tall, flower earlier
	Piquemal (1970), Anaschenko (1972a,b)	a	Low	Seed weight reduced
Mendok	Schuster (1961)	c	nk	nk
TIBA	Kiermayer (1959)	a	nk	nk
Lactuca sativa				
Ethrel	Han and Lee (1972)	b	Low	nk
GA	Eenink and Vereijken (1978b)	a	nk	nk

[a]a = complete, b = high, c = low.

Fig. 19.1a,b. *Ageratum houstonianum*. Capituli of male fertile **(a)** and male sterile **(b)** plants. (Reimann-Philipp 1976)

19.2 *Carthamus tinctorius* L. (safflower, false saffron)

19.2.1 Genic Male Sterility

Spontaneously, few sterile plants occur in safflower fields. The two single gene-controlled steriles found by Claassen (1948, 1952) are both male-female steriles. Similar steriles were found by Carapetian and Knowles (1976). The sterility occurs due to complementary interaction of recessive gene *a* with *b* or *c* recessive alleles.

19.2.2 Functional Male Sterility

One single recessive gene-conditioned mst mutant was detected by Ebert (1964) and Ebert and Knowles (1966). In this, gene *th* reduces pericarp thickness, making the fruits thin-hulled. Such plants are weak-stemmed and exhibit delayed anther dehiscence. At anthesis, whereas the anther flaps of normal safflowers open abruptly, bringing the pollen effectively into contact with the style, in mutant plants, anther flaps collapse and prevent pollen release (Ebert and Knowles 1966, 1968). The delay in anther dehiscence is influenced by some environmental factors, so that mutant plants are not completely mst. This leads to some selfing (Urie and Zimmer 1970) in hybrid seed production plots.

19.2.3 Induced Male Sterility

From the diploid C_2 progeny of colchicine-treated safflower seeds obtained from Afghanistan, mst's lacking viable pollen segregated (Heaton and Knowles 1982). The sterility is conditioned by one recessive *ms* gene. Female fertility is reduced, but seeds produced on mst's are larger than those on mft's. Therefore, total seed yields of mst and mft plants do not differ significantly from each other. As in *Sorghum vulgare* (Erichsen and Ross 1963a,b), the mst plants are colchicine-induced mutants. A monogenic recessive mst mutant was detected by Joshi et al. (1983) in M_2 generation of a γ-ray treated population. Female fertility, as evidenced by seed set after natural and artificial pollination, was normal in the mutant.

19.2.4 Chemical Male Sterility

Pollen sterility is induced following GA treatments (Yermanos and Knowles 1960), but the treated plants are also female sterile and exhibit considerable internode elongation and chlorosis.

19.2.5 Breeding Value

Because of the high stability of mst gene expression and the insignificant yield loss, colchicine-induced mst produced by Heaton and Knowles (1982) has a high breeding value. The mst trait would prove useful in establishing random mating populations for facilitating hybrid seed production. For these, two lines incorporating the ms gene are available (Heaton and Knowles 1980). The *th* mutant (Ebert 1964) cannot be used in breeding due to its instability in delayed dehiscence. This, due to some environmental factors, renders the mutant useless in commercial hybrid seed production (Urie and Zimmer 1970).

19.3 *Centaurea* L.

19.3.1 *C. americana* Nutt. (basket flower) and *C. imperialis* Hort. (sweet sultan)

From the exotic seed stocks, mst's segregated in the greenhouse grown plants. In them, PMC's degenerated during early prophase I (this Vol.). Whereas in *C. americanum* stages up to pachytene were discernible in some PMC's, in *C. imperialis* no stage past leptotene could be detected. The anomaly in each is conditioned by a single recessive *ms* gene; their allelic nature is not known. The genes do not influence female fertility.

19.3.2 *C. scabiosa* L. (bachelor's button)

Progeny raised from the seeds obtained from Sweden, Denmark and England segregated some mst's (Fröst 1963). In these, male meiosis proceeded normally up to tetrad formation, after which the cytoplasm of the microsporocytes degenerated. This led to the formation of empty pollen grains. Hybridization between the steriles and fertiles segregated complete msts and p-mst progenies as classified on the basis of lacto-phenol stainability of the pollen grains. Genetic control of the anomaly is not known. While a varying number of accessory chromosomes in the offspring of steriles crossed by pollen from normal plants were found, no marked relationship between sterility and the presence of a number of accessory chromosomes could be found.

19.4 *Chrysanthemum* L. (costmary, nippon daisy)

Doubling is a common feature of this species. In 1–2% of the plants of *C. balsamita, C. carinatum, C. indicum* and *C. parthenium*, anthers become transformed into petal-like structures and in *C. alpinum, C. balsamita, C. carinatum, C. corymbosum, C. indicum, C. parthenium* both stamens and carpels exhibit such transformations. Genetic control of the anomaly is not known. The former type of plants are mst and can be used for F_1 seed production. Such F_1 hybrids in *C. parthenium* and *C. indicum* have large-sized flowers, but their stalks break off at full maturity due to strong winds. Therefore, staking these plants becomes necessary. In *C. morifolium* 3–12% mst mutants segregated after 1750 rad. In X-ray seed treatments to four cultivars Clinstar, No. 71, Copper and Yellow Rewilo (Broertjes and Jong 1984), although pollen formation was apparently normal, the pollen was not released.

Hence mst is of a functional type. Whereas some mst mutants are identical to untreated controls, others are reduced in height, or differ in flower form, size or colour. In some mst's, both vegetative and floral organ sizes are reduced (this Vol.). This represents a pleiotropic expression of the *ms* gene(s), as no segregation of mst and this phenotypic marker trait (diminitive foliar and floral organ) occurs.

19.5 *Cirsium* (Tourn.) Adans

This has three types of reproductive system viz. hermaphroditism, gynodioecy and dioecy. Gynodioecy exists in nearly 30% of its European species (Delannay 1978a), and is described in Chapter 10.

19.5.1 Genic Male Sterility

The mst is known in *C. arvense, C. montanum, C. oleraceum, C. palustre* and *C. spinosissimum* (Delannay 1979). The earliest action of the ms gene occurs in *C. arvense* and *C. montanum,* as degeneration occurs immediately after the differentiation of archesporial tissue and no PMC's develop. In *C. oleraceum,* ms gene action is delayed until PMC's are formed. Thereafter the PMC's degenerate before meiosis. Further delay in ms gene action occurs in *C. spinosissimum.* In this, meiosis proceeds until tetrad formation, after which both tapetum and tetrads degenerate. *C. palustre* resembles *C. spinosissimum* in this respect, but abortion in it occurs during tetrad formation. In all these species, except in a strain of *C. spinosissimum,* tapetal breakdown occurs prior to sporogenous tissue degeneration. This led Delannay (1979) to conclude that precocious tapetal degeneration is the cause of mst in *Cirsium.*

19.5.2 Gene-Cytoplasmic Male Sterility

Correns (1916, 1928, 1937) found that in *C. oleraceum* monoecious plants produced only monoecious offspring, and female plants (mst) produced only female offspring although they are always fertilized by pollen from monoecious plants. Upon crossing *C. oleraceum* by *C. canum* pollen, a species containing only monoecious plants produced only mst's in six progeny generations. Even when (*C. oleraceum* × *C. canum*) × *C. canum* mst's were crossed by *C. palustre* (a gynodioecious species), only mst offspring were produced. On crossing these mst's by *C. palustre* pollen, only mst progeny was produced. Correns (1937) concluded that the cytoplasm of female *C. oleraceum* plants inhibited genes responsible for the development of male floral parts in hybrids having some nuclear genes from *C. canum* and *C. palustre.* Segregations obtained by Correns (1916) are suggestive of a cytoplasmic inheritance of mst in *C. acaule* and *C. oleraceum* and of a g-c inheritance in *C. eriophorum* subsp. *velenovsky* and *C. palustre.* The species need to be reinvestigated, as it appears that whereas no genes have been found in *C. acaule* and *C. oleraceum*, they are known in the remaining species. Therefore, *fr* genes for these two species can also be found if tried.

19.6 *Helianthus annuus* L. (sunflower)

Of the four major oil-yielding crops, (rape seed, groundnut, soybean and sunflower), cultivated sunflower is gaining importance because of its high-quality oil, edible value as seed for man, birds and other small animals, and as forage for large domesticated animals. Both genetic (g-mst, gc-mst) and chemical mst are known in it (Tables 19.1 to 19.3).

19.6.1 Genic Male Sterility

19.6.1.1 Complete Male Sterility

In 1929, a sunflower plant from Banaras distinguished itself from others by the failure of its stamens to emerge from the flowers (Kuptsov 1935). Its stamens were under-developed and no pollen was formed in them. Cross-pollinations produced a high seed set, indicating normal female fertility. The sterility was conditioned by a recessive gene. From some Russian sunflower varieties, 16 mst plants were detected in France and sterility in one was controlled by a recessive *ms* gene (Anonymous 1964). After these initial documentations of mst, many such reports followed from Europe and the USA (Leclercq 1966; Putt and Heiser 1966; Vranceanu 1967). Vranceanu (1970) reported the isolation of over 30 sources of mst in Romania. The mst in the majority of these is conditioned by one recessive gene. Allelism tests of ten of these revealed the existence of five nonallelic genes, designated ms_1 to ms_5. The gene ms_1 is closely linked with a dominant gene controlling anthocyanin production (Vranceanu and Stoenescu 1970a) and has been used in breeding.

Two types of g-mst in cultivated sunflower were found by Putt and Heiser (1966). In one (type I) found at Bloomington, Indiana (USA), the anthers of an ornamental red-rayed sunflower variety were empty and split apart before exertion. In these pollen grains were either absent or they were empty, uneven-sized, and clumped. Early male meiotic events are normal but the precise stage of microspore degeneration is not known. Apparently, microspores rapidly aborted immediately after their liberation from the microspore tetrads. Whereas three crosses indicate mst to be conditioned by a single recessive gene, two others indicate the existence of two duplicate genes. The second type of g-mst (type II) was found by Putt and Heiser (1966) at Morden, Manitoba (Canada). In this, small-sized, nonstaining pollen grains were formed. They did not clump, unlike the type I mst plants of Bloomington. Male meiosis proceeded normally but microspore protoplasm degeneration was gradual, so that microspore abortion is completed just prior to anthesis. The sterility is conditioned by one recessive gene. The *ms* genes conditioning the two sterility types are nonallelic. The uniform-sized nonstaining pollen in the Morden stock and the lack of pollen or presence of irregularly sized, clumped, nonstaining pollen in the Bloomington stock suggest that the two *ms* genes differ from each other in their action on microsporogenesis. The existence of recessive *ms* genes in the genomes of various sunflower cultivars has been amply documented (Pirev 1966; Pogorletskii 1973; Pogorletskii and Barlov 1973; Kovačik and Skaloud 1973; Nakashima and Hosokawa 1974a; Paun 1974; Spirova 1978). In some, *ms* gene is linked to cotyledon colour. In the cultivated sunflower variety Karlik-68 (Dwarf-68) two recessive genes, msi_1 and msi_2, (Kovačik and Skaloud 1973) induce complete and stable mst.

19.6.1.2 Partial Male Sterility

This occurs in two lines, CM90 and CM30, both originating from the sunflower variety Mennonite (Putt and Heiser 1966). Pollen sterility ranged considerably. Although one had a mean of 75% normal pollen, the other had 32% only, and some of the other progeny plants were completely mst. Within- and between-family means differed significantly from each other. Since selection towards higher pollen sterility was possible, polygenic (with additive effect) control of the sterility cannot be ruled out. In one sunflower variety, Kavačik and Skaloud (1973) found a single recessive gene control of p-mst.

19.6.1.3 Male Sterility in Interspecific Hybrids

From the progeny of the cross between wild perennial sunflower *H. grosseserratus* Martens and cultivated annual sunflower *H. annuus*, mst plants were obtained in Bulgaria by Georgieva-Todorova (1974). The mst is conditioned by one recessive gene whose expression is considerably influenced by the environment. In one line, male meiosis proceeded normally up to the tetrad stage. Microspores degenerated after liberation from PMC's. In another line, the microsporogenous tissue degenerated prior to meiotic initiation.

19.6.1.4 Microsporogenesis

In most of the g-mst sunflowers, *ms* genes act after tetrad formation, causing abortion of developing or developed microspores. In two genotypes, the *ms* gene acts before or during early meiosis. The earliest action occurs in a plant of *H. grosseserratus* × *H. annuus* (Georgieva-Todorova 1974). In this, the *ms* gene acts before meiotic initiation and the whole sporogenous tissue degenerates. In the g-mst materials AS-110 and AS-116, studied by Paun (1974), which had ms_1 gene, the gene acts during pachytene, leading to asynapsis or desynapsis of one to two bivalents, so that two to four univalents are present during diplotene and diakinesis. Subsequent meiotic stages are abnormal and exhibit asynchronous and unequal chromosomal separation, univalent elimination, chromosome clumping and agglutinations, chromatin bridges and fragments etc. Unequal-sized tetrads with micronuclei are ultimately formed. At anthesis, 4–6% of the pollen stains with acetocarmine. This pollen is smaller than fertile pollen and lacks the spiny exine characteristic of *Helianthus* spp. Paun (1974) questioned the viability of this pollen and doubted that it could compete with normal pollen. Since acetocarmine staining is not a reliable criterion of pollen viability and selfing, such plants did not produce any seed, it is certain that the stained pollen is nonfunctional and thus mst is complete. The two mst materials studied by Georgieva-Todorova (1974) and Paun (1974) are interesting because whereas the *ms* gene of the former causes sporogenous tissue degeneration and consequently no meiosis occurs, in the latter the *ms* gene permits meiosis to continue, but the whole process is highly abnormal and erratic. In the material studied by Pirev (1966), the *ms* gene acts on the developing pollen so that male meiosis is normal, but pollen development halts abruptly at the uni- or bi-nucleate stage and the protoplasmic content of these developing pollen grains degenerates gradually. As anthesis approaches, exine and protoplasmic contents degenerate. In some genotypes, *ms* gene action is delayed further, so that pollen grains abort just prior to dehiscence or floral anthesis. Such grains stain feebly but are not functional. This accounts for the reduced pollen-staining ability of mst sunflower plants studied by Leclercq (1966). In a wet year, the mst plants developed pollen without spines on the exine. The pollen was inviable and contained oily granules. Under dry conditions, microspores degenerated early and no pollen was ever produced. Thus *ms* gene action is delayed under humid environmental conditions in this mst sunflower material.

19.6.1.5 Tapetal Behaviour

This is unknown in many mst sunflowers. In line P-21, tapetal cells remain intact, enlarge considerably and become vacuolated (Nakashima and Hosokawa 1974a). Meiosis proceeds normally, but microspores degenerate after release from PMC's. Nontapetal disintegration starves microspores to death. In another mst material, the tapetum degenerates along with microspores.

19.6.1.6 Gene Frequency

The g-mst in the majority of cases is conditioned by single recessive genes. The frequency of these genes is high in sunflower cultivars grown in Romania and Russia. With few exceptions, the genetic identity of *ms* genes has not been determined in most of the cases. Of the ten mst lines tested in Romania, five genes, ms_1 to ms_5, are nonallelic (Vranceanu and Stoenescu 1970b). The sterility in two mst materials studied by Putt and Heiser (1966) is conditioned by two nonallelic *ms* genes; their action over male meiosis is uniformly similar. The majority of these genes act during microspore development, two act before or during early prophase, and three act during pollen maturation.

19.6.1.7 Linkage

ms_1 is closely linked with the dominant gene *T* controlling anthocyanin production (Vranceanu and Stoenescu 1970b). A similar linkage of ms_1 and ms_2 genes with green hypocotyl colour, occurs in two out of four mst materials (Pogorletskii 1973). The *ms* gene of Karlik-68 variety is linked to a spoonlike leaf marker gene (*sp*). These two linkages facilitate selection of mst and mft plants at the seedling stage.

19.6.2 Gene-Cytoplasmic Male Sterility

Durand (1962) mentioned the discovery of gc-mst in sunflower by GDR researchers. Earlier in 1958, the GDR geneticist Stubbe (cited by Gundaev 1966, 1971) had already found sterility in some lines of sunflower. Duvick (1966) also indicated the occurrence of gc-mst in sunflower plants grown in the GDR. Apparently the sterility had appeared spontaneously in these sunflowers. A similar spontaneous occurrence of gc-mst was later reported from Russia by Gundaev (1966) and Volf (1966), from Romania by Vulpe (1968) and from Bulgaria by Stoyanova and Ivanov (1975). These reports indicated that most of the crosses of gc-mst plants with normal mft lines produced progenies with a highly variable percentage of mst plants. For instance, Gundaev (1966) found that the progenies of crosses involving the gc-mst source VNIIMK 8931 had as low as 10% or as high as 90% of complete mst plants. Almost all the progenies had p-mst's also. Through selection and test crosses, increase in mst plants from 23% to 75–90% was achieved in Russia (Panchenko 1966) and in Bulgaria (Atanasoff 1964a). Later, Gundaev (1966, 1971) developed gc-mst lines that produced 92 to 96% mst progeny. These were subsequently utilized for hybrid seed production. Following alloplasmy, hybrids having *H. annuus* nucleus and cytoplasm of other species yield mst plants.

i. *H. petiolaris* × *H. annuus*. From France, Leclercq (1969) obtained mst's by transferring the genome of *H. annuus* (2n = 34) into *H. petiolaris* (2n = 34) by backcrossing and selection. The degree of mst increased with each additional BC. BC_3 and BC_4 progeny was almost completely mst. The majority of the mst's have short anthers that do not project out of the corolla. They are fused at the bases, but their tips are free. In some mst lines, the anthers are of normal length and project out of the corolla, but all the anthers lack pollen. Female fertility of mst plants is normal. The mst is highly stable and results from the interaction of the *H. annuus* genome with the *H. petiolaris* cytoplasm. This represents the main cytoplasmic source used almost exclusively in breeding programs all over the world. It is designated cms_1 by Whelan (1980).

Another mst source, designated as cms_2 by Whelan (1980), was found in BC progeny from a cross between *H. petiolaris* collected from Canada and the open-pollinated cultivated sunflower Saturn. The mst plants appeared first in BC_1 and had indehiscent anthers with aborted pollen (Whelan 1979). In BC_2 almost 94% plants were mst and BC_3 and BC_4 progeny was completely mst. Thus interspecific substitution of the nucleus of *H. annuus* into the cytoplasm of *H.*

petiolaris led to this gc-mst. The female fertility of mst plants was normal. Male meiosis proceeded normally until microspore formation, after which abortion occurred. Fertility restoration in the mst plants was not possible with seven *Fr* sources and two collections of wild *H. annuus*. This led Whelan (1980) to postulate that this gc-mst was different from the one obtained by Leclercq (1969), even though both these originated from *H. petiolaris* cytoplasm.

ii. *H. giganteus* × *H. annuus*. Whelan (1981) found mst plants in progeny of the first interspecific hybrids between the diploid perennial species *H. giganteus* and the annual *H. annuus*. The hybridization involved a three-way cross, viz. *H. giganteus*, wild *H. annuus* and cultivated *H. annuus* Saturn, in which the nucleus had only partially been substituted by backcrossing. This partial substitution of the nucleus of cultivated sunflower *H. annuus* into the cytoplasm of *H. giganteus* was achieved by hybridizing *H. giganteus* first with wild *H. annuus*, and then backcrossing four times, using the open-pollinated sunflower Saturn as the recurrent pollen parent. The mst plants having indehiscent anthers and nonviable pollen appeared after two crosses with the Saturn pollen parent. In vivo pollen germination tests of 20 mst plants revealed complete pollen sterility in 15 plants. Seed set on them after cross-pollination was normal, indicating unimpaired female fertility. Using an inbred tester line lacking *fr* genes as pollen parent in crosses to an existing source of S-cytoplasm and four mst plants of the above-mentioned gc-mst source did not restore pollen fertility in the progeny. In similar crosses, using three *Fr* gene sources as pollen parents to the mst plants, pollen fertility was restored in some hybrid plants but not in others. However, one source restored pollen fertility in all the F_1 plants. This suggests *H. giganteus* as a S-cytoplasm source. This was designated as cms_3 by Whelan (1981).

iii. *H. maximiliani* × *H. annuus*. Whelan (1978a) obtained interspecific hybrids between diploid perennial species *H. maximiliani* × annual wild *H. annuus*. The subsequent four BC cycles by the cultivated *H. annuus* Saturn produce mst plants (Whelan and Dorrell 1980) whose anthers were highly modified and almost vestigeal in some of the plants (Fig. 19.2a). The sterility was not complete, but further backcrossing by Saturn pollen could lead to a complete mst progeny. This needs to be determined.

iv. *H. lenticularis* × *H. annuus*. Some pollen steriles segregated in the progeny of this cross (Anashchenko 1981). Frequency of mst's increased in subsequent BC's having *H. annuus* as the male recurrent parent.

v. *H. annuus* × *H. annuus*. Another mst source, designated as cms_4, was obtained by Heiser (1982), from a single mst plant of *H. annuus* subsp. *lenticularis*. Pollinating this mst plant by *H. annuus* cv. Commander, led to a nearly complete stabilization of mst in BC_3 generation. HA-89 and HA-265, which are respectively the maintainer and restorer for Leclercq's mst plants, restored pollen fertility to the mst plants. *Fr* genes for these mst plants occur in many sunflower varieties in addition to the stock from which the mst plants originated. Heiser (1982) registered this gc-mst source as Indiana-1.

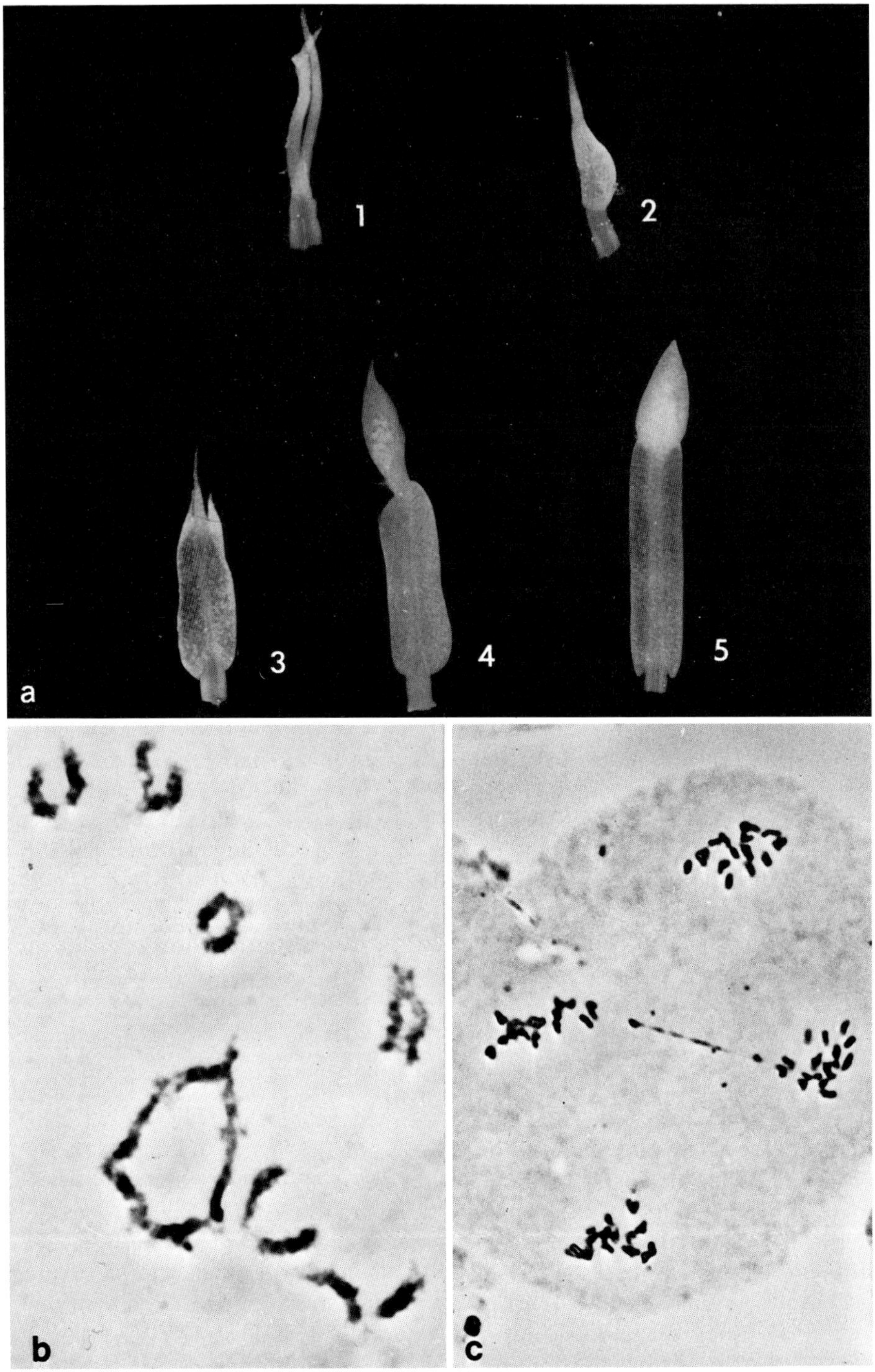
1
2
3
4
5
a
b
c

19.6.2.1 Microsporogenesis

Two types of meiotic behaviour are exhibited by the four mst lines that originated from the mst materials developed by Leclercq (1969) through interspecific hybridizations between *H. petiolaris* and *H. annuus* (Paun 1974). In two of the four lines, anthers are highly necrotic and sporogenous tissue degenerates before meiotic initiation. In the other two lines few, PMC's are differentiated. In them meiosis proceeds normally up to the tetrad stage, after which the microspores degenerate. Whelan (1978a,b) also found normal meiosis up to tetrad formation, but microspore degeneration in several gc-mst lines. In the mst cultivar Krasnodarets, developed by backcrossing to a gc-mst stock of Leclercq (1969), an AI bridge and a fragment indicated that the presence of paracentric inversion is common. However, this anomaly is not the cause of complete mst, as is evident from the account to follow. The F_1 plants of *H. petiolaris* × *H. annuus* exhibit multivalency, indicating 10 out of 34 chromosomes as heterozygous for chromosomal interchanges (Whelan 1979). A pair of chromosomes appear to have undergone paracentric inversion. Among many anomalies in F_1 plants, chromosomal rings (Fig. 19.2b) and anaphase II with a chromosomal bridge (Fig. 19.2c) are significant. Continued backcrossing by *H. annuus* pollen leads to a rapid progressive elimination of the meiotic aberrations present in F_1 plants. In BC_3 or BC_4 plants, no meiotic aberrations occur, but all the plants are completely mst. Using light and electron microscope, Horner (1977) compared behaviour of anther and sporogenous tissue in a normal fertile B-line and a gc-mst A-line, and divided microsporogenesis into 11 stages, stages 1–4 ranging from the first sporogenous tissue to the initiation of tetrads, and, stages 5–11 from late tetrad to the mature pollen stage. Up to stage 4, both A- and B-lines developed similarly. Whereas in the mst A-line, tetrads and tapetum disintegrated, in fertile B- and normal lines viable pollen were produced. By end of stage 5, enlarged tapetal cells had degenerated and the middle layer cells of the anther wall were highly enlarged, flattening and crushing the microspore tetrads. Both the tetrads and tapetum became grossly disorganized and complete pollen abortion resulted. Compared to normal fertile pollen, the sterile pollen stains feebly with iodine or carmine (Fig. 19.3a), is deformed and has a greater number of spines per unit area (Fig. 19.3b).

19.6.2.2 Fertility Restoration

Shortly after the discovery of gc-mst in sunflower by Leclercq (1969), fertility-restoring genes were identified by several researchers (Anonymous 1970; Enns et al. 1970; Kinman 1970; Leclercq 1971; Vranceanu and Stoenescu 1971b,c; Velkow

Fig. 19.2a-c. Sunflower. **a** Anther types (*1–5*) of male sterile plants obtained from backcross progeny of *H. maximilliani* × *H. annuus*. The anthers range from highly modified (*1–4*) to normal type (*5*). (Whelan and Dorell 1980). **b,c** Chromosomal ring at diakinesis **(b)** and chromosomal bridge at AII **(c)** in male sterile *H. petiolaris* × *H. annuus* sunflower meiocytes. (Whelan 1979)

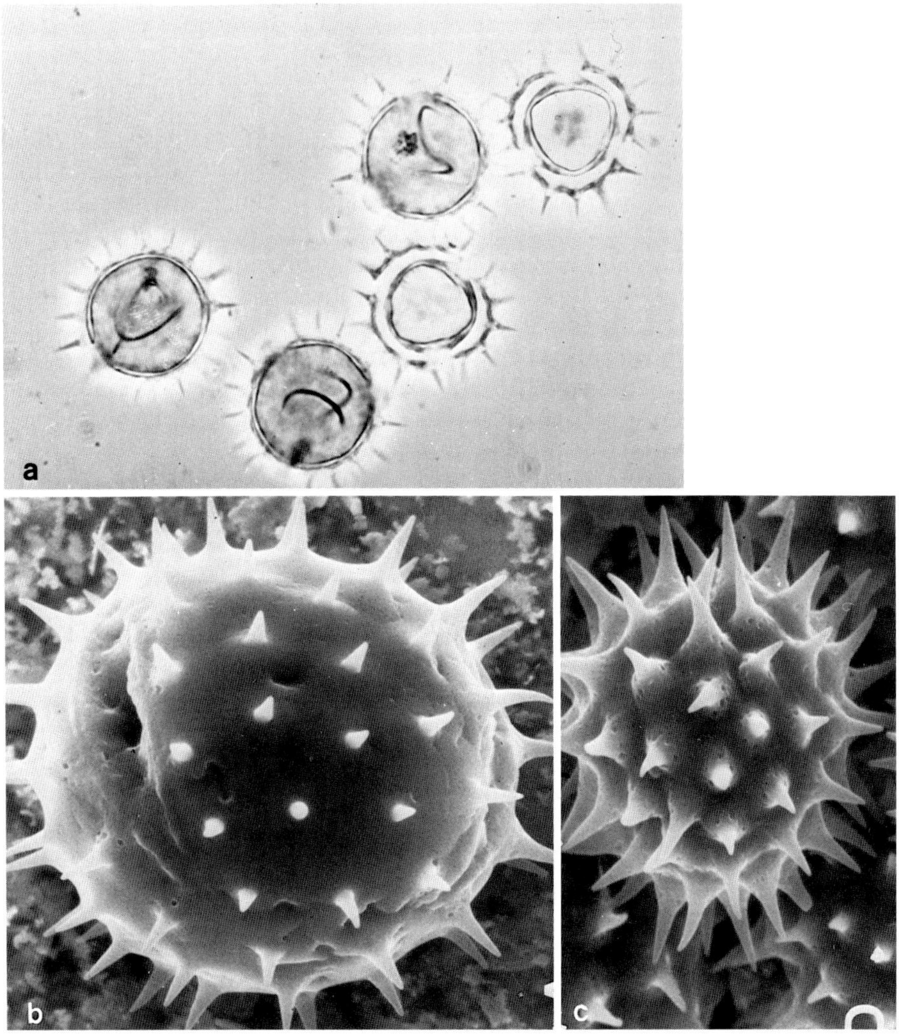

Fig. 19.3a-c. Sunflower. **a** Fertile (*dark*) and sterile (*light*) stained sunflower pollen. (Whelan and Dorrell 1980). **b,c** Scanning electron micrograph of a fertile pollen of *H. annuus* (**b**) and of sterile pollen of *H. petiolaris* × *H. annuus* plant. (Whelan 1978b)

and Stonyanova 1974). Whereas in the two crosses made by Fick and Zimmer (1974a) restoration was conditioned by a dominant gene, in the third cross, action of two independent complementary dominant genes was evident. Although a similar digenic control (Fr_1, Fr_2) of fertility restorer was found by Reddy and Thammirayu (1977), this needs confirmation using heterozygous tester lines. Whereas in some investigations pollen fertility restoration appears to be conditioned by a single dominant gene (Enns et al. 1970; Kinman 1970; Leclercq 1971; Vranceanu and Stoenescu 1971c), in others at least two independent, complementary dominant genes (Fick and Zimmer 1974a; Dominquez-Gimenez and Fick 1975; Reddy and Thammirayu 1977; Whelan 1980), or even up to four nonallelic genes (Dominquez-Gimenez and Fick 1975; Vranceanu and Stoenescu 1978; Whelan 1980) appear to control fertility restoration. Inheritance of fertility restoration of cms_2, cms_3 and cms_4 is not fully known. Further backcrossing is required to complete the substitution of nucleus of cultivated sunflower *H. annuus* with these cytoplasmic sources. Analysis from 16 sources of restorer genes by Vranceanu and Stoenescu (1978) revealed restoration to be conditioned in eight cases by one dominant gene, in three cases by two complementary genes, in four cases by three complementary genes and in one case by the cumulative action of the nonallelic dominant genes. The presence of minor genes was inferred to influence partial fertility restoration. Fick (1978) considers the fertility restoration as highly complex and as influenced considerably by environmental factors.

19.6.2.3 Restorer Line Development

This requires incorporation of the dominant restorer gene *Fr* in the male parent. This is usually done through backcrossing, using the established inbred lines as the recurrent parent. This procedure is, however, cumbersome because in each generation, the selected plants have to be crossed to a gc-mst line in order to find if the selected plants possess restorer genes. From these test crosses, the first generation plants will be mft if the selected plants possess the necessary *Fr* genes. Fick (1978) devised a short and easier procedure (Fig. 19.4) in which the restorer lines can be developed by crossing the line or population possessing *Fr* genes to a gc-mst plant. From the resultant mft hybrid, lines homozygous for the *Fr* genes and with other desirable traits may be isolated by continuous self-pollination without test crossing. This procedure led Fick and Zimmer (1974a) to develop lines having diffuse and extended branching, resistance to downy mildew and *Fr* genes. The former traits allow pollen production over an extended period of time. This is useful for producing hybrid seed with female lines that flower up to three weeks later than the fertility restorer lines. The majority of these sources now being used in breeding programs owe their origin to the wild species of sunflower, including the widely used T66006-2 source discovered by Kinman (1970). In the cultivated sunflower, such genes restoring complete fertility seem to be rare, although Vranceanu and Stoenescu (1971a,c) and Fick and Zimmer (1974a) have documented their frequent existence. Kinman's (1970) *Fr* genes are highly stable over a wide range of environments, but the search for new and renewable sources of S-cytoplasm and *Fr* genes has to be continued in order to reduce the genetic vulnerability to diseases and pests. Fick et al. (1974) and Dominguez-Gimenez and Fick (1975) documented a widespread occurrence of *Fr* genes from wild species. In many collections, restoration appeared to be controlled by genes other than the original Rf_1 (Fr_1) gene found by Kinman (1970). Thus possibilities exist for reducing the chances of genetic vulnerability associated with the use of restricted germplasm having a narrow genetic base.

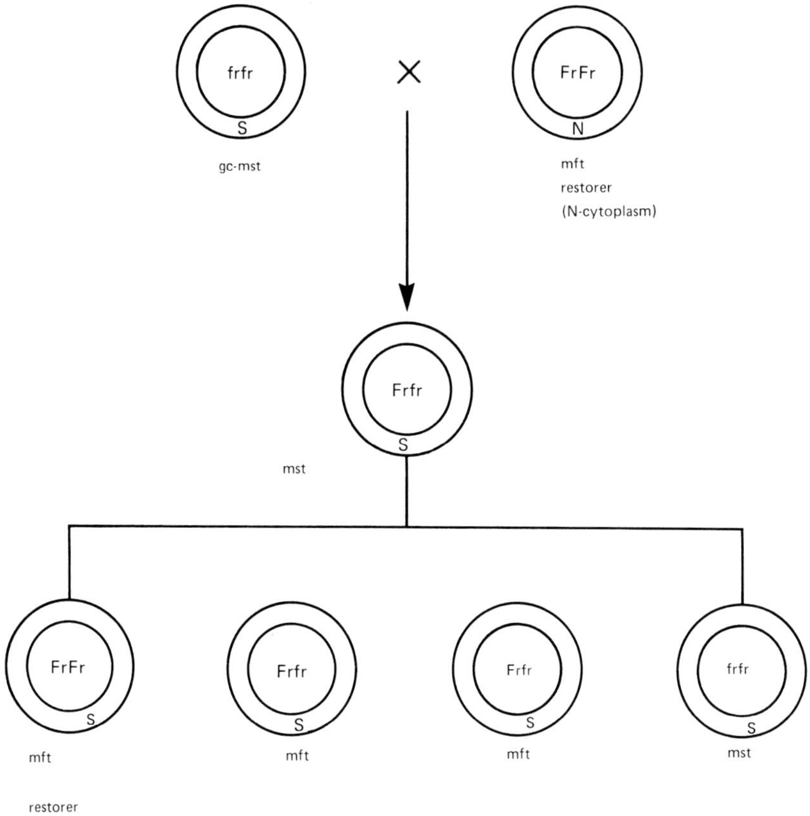

Fig. 19.4. Development of restorer C-line in S-cytoplasm (Fick 1978)

19.6.2.4 Biochemical Differences

In mft pollen, protein content enhances regularly all through its development. But this increase is slow and erratic in mst pollen (Pirev 1966). It has too low starch content compared to that of mft pollen. This starch difference enabled Pirev (1966) to differentiate mst from mft plants 4–5 days before flowering. The mst pollen has a lower sucrose content but the concentration of other saccharides is similar to mft pollen (Kovacik and Kryzanek 1969). In mst pollen, the amount of proline is lower and that of asparagine higher than in the mft pollen. Also phospholipid and nucleic acid contents are lower in the mst pollen, compared to those in mft pollen.

19.6.2.5 Viral Nature

Atanasoff (1971) proposed a viral cause as the basis of mst, but subsequent investigations failed to verify the viral nature.

19.6.3 Chemical Male Sterility

i. *Ethrel.* Campos (1974) obtained complete mst plants after ethrel treatments but female fertility reductions were equally high.

ii. *Gibberellic Acid.* After treating young flower heads, when they are just visible with 20 ppm GA, complete mst was obtained by Piquemal (1970) and Anashchenko (1972a,b). The seeds produced on the treated plants were lighter by 15–25% but exhibited good germinative capacity. Treatments of 25–100 ppm induced high mst but reduced seed production considerably (Vear 1981). After double treatment with 0.01% GA, to the sunflower variety Saratov dwarf, Beglaryan (1970) obtained a morphological mutant in the M_4 generation. It exhibited increased nucleoli per cell, chromosome bridges, laggards, polyploid cells and was both male and female sterile. In the Russian sunflower variety Armaverts, a high degree of pollen sterility was induced by spraying 90–100 ppm GA on the 3rd day after flower bud initiation; female fertility was not influenced by the treatment (Seetharam and Kumari 1975). For large-scale male sterilization, these authors recommended two sprays of 100 ppm GA at the rate of 0.5–0.75 ml per plant on 30th and 40th day after sowing. The treated plants became slightly taller and flowered 3–4 days earlier. Anther exertion was suppressed by GA treatment. At anthesis, anthers were shrunken with little or no viable pollen. With the recommended treatment, pollen sterility was 60%, anther suppression was complete in 72% and partial in 24% plants. Since such suppressed anthers remained indehiscent, the mst induced was high. Female fertility of the treated plants has not been determined, but is reduced following such treatments. Spraying 0.5 mg GA per plant prior to floral bud development (i.e. when plants are about 35 days old) retarded anther development and exertion and instead of protandry, protogyny occurred (Schuster 1961). Female fertility of the treated plants was highly reduced, as seed set was about 20% of the untreated ones. Likewise, with 0.033% aqueous GA_3 treatment 37 days after sowing, mst was complete but was accompanied by high female fertility reductions, 70% decrease of head size and increase of empty seeds (Frank and Koves 1977). In the 0.01% GA-treated plants, the cells of the inner tapetum were enlarged and the proportion of meiocytes was highly reduced (Konstantinova 1980). Cytokinesis was prevented and led to the production of coenocytes.

iii. *Mendok.* 7.5 mg of FW-450 retarded anther exsertion to some degree but the mst induced was meagre (Schuster 1961).

iv. *TIBA.* Spraying 5×10^{-2}% at 4-day interval caused dwarfing and reduction of stamen number (Kiermayer 1959). No pollen was produced. Female fertility of the treated plants is not known.

19.6.4 Breeding Value

Hybrid sunflowers utilising gc-mst in conjunction with the *Fr* gene system were produced and commercialised soon after the discovery of the gc-mst system. Since then, the use of hybrids has increased considerably. Earlier, more than half the sunflowers planted in Canada in 1978 were imported hybrids, although no hybrid varieties were licensed there (Dedio 1979). Besides yielding about 30–40% more in seed yield and 30% in oil content than the open-pollinated varieties, most hybrids are resistant to several diseases. *Verticillium* wilt, downy mildew and rusts are conditioned by dominant genes and can be easily incorporated into the hybrids. The hybrids are also fairly uniform in height, head size, flowering and maturity, which facilitates harvesting and enhances yield gain. In Romania, using mst, hybrids superior to varietal populations, Peredovik and Romson Record have been developed. Many of the single-cross hybrids developed, such as HS-52, HS-53 and HS-55, exceed the standard varieties by 24–29% in seed yield and 30–40% in oil yield per hectare (Vranceanu 1970), and cover over 50% of the area under sunflower cultivation. Their oil content in dry seeds ranges between 50.5 and 52.5%.

19.6.4.1 Gene-Cytoplasmic Male Sterility

The first sunflower hybrids utilised for commercial planting were developed in Canada by using a self-incompatibility system (Unrau and White 1944). However, the female parent used was not highly self-compatible or was a low pollen producer. This was interplanted with an abundant pollen producer and a part of the seed recovered was of hybrid origin. In commercial practice, the percentage of hybrid seed averaged from 39 to 45% with some samples being as low as 19% (Putt 1962). Currently natural hybridisations and commercial hybrid seed production in sunflower are achieved by using mst in conjunction with self-incompatibility. The gc-mst lines with some degree of self-incompatibility are interplanted with an abundant pollen-producer line having a low degree of self-incompatibility. The amount of hybrid seed recovered due to cross-pollination depends on the ratio of male to female plants, and degree of isolation and weather conditions that favour bee visitations. Usually in hybrid seed production fields, rows of gc-mst lines are interplanted with rows of a fertility-restorer line in ratios from 3:1 to 6:1. Where the flowering time between the parents differs, planting dates are adjusted accordingly in order to coincide with flowering and anthesis. When the difference in flowering is not great, planting the earlier-blooming parent deeper to delay emergence and blooming can be done. Conversely, fertiliser applications may be made to hasten flowering of the late parent.

To obtain a single-cross hybrid, three lines are required. Since the female line is gc-mst, a mft maintainer is needed before it can be multiplied. The line is increased by planting alternate strips of gc-mst and maintainer lines and harvesting only the mst heads. The maintainer line is increased by selfing with bags placed over individual heads. The third line is a fertility restorer and is increased in isolation through sib-crossing. No major problems have emerged on using the gc-mst and fertility-restorer system for hybrid seed production. The S-cytoplasm has no adverse effects on the agronomical or seed-oil traits when incorporated into inbred lines (Velkov and Stoyanova 1974; Fick 1978). The former authors found bee visitation necessary for good seed set in production fields similar in gc-mst and mft lines; but in certain mst lines seed set was reduced (Furgala et al. 1976). They had low nectar content and were less attractive to bees than the mft types. Since bee attraction depends upon nectar production by flowers, it appears that the nectar content of mst lines used by Velkov and Stoyanova (1974) was higher than that of Furgala et al. (1976). These nectar production differences could be either genotypic or environmental, or due to both.

Use of gc-mst in conjunction with self-incompatibility has helped to produce cent per cent hybrid sunflower seed, making it comparable to hybrid maize. This system has been adopted all over the world into the breeding programs of private and government agencies. In 1977, over 800,000 ha of this crop was grown in the USA (Kromer and Gazelle 1977). The first hybrids produced using gc-mst and Fr lines were made available for commercial production in 1972 in USA, and by 1976, it was estimated that these hybrids accounted for over 80% of the sunflower production in USA (Fick and Zimmer 1976). The best hybrids yielded over 20% more than the open-pollinated cultivar Peredovik, and most of them were resistant to the three major diseases of rust. In addition to developing new gc-mst lines, new restorer lines for use in hybrid seed production were developed by Fick and Zimmer (1974d); lines RHA-271, 273 and 274 are the first developed downy mildew-resistant fertility-restorer lines which possess recessive branching that allows pollen production over an extended period of time. This is useful for producing hybrid seed with gc-mst lines that flower up to 3 weeks later than the *Fr* lines.

19.6.4.2 Genic Male Sterility

i. *Partial Male Sterility*. Putt and Heiser (1966) advocated its use in hybrid seed production, as these plants hybridize well in natural crossing plots, selection for a low or high proportion of normal pollen is possible and F_1 hybrids between p-mst and normal lines yield normal pollen plants. However, the use of p-mst is limited because of the inherent difficulties associated with such plants, especially selfing and environmental sensitivity. Despite demerits, the oil seed hybrid Valley which was produced for commercial production using the p-mst line CM90RR as a female parent and the cultivar Peredovik as male parent, has an excellent seed yield (Jensen et al. 1970). Outcrossing percentage of this hybrid varies but seed lots with 60 to 80% hybrids are common.

ii. *Complete Male Sterility*. Use of complete g-mst for hybrid seed production necessitates maintenance of *ms* gene in a heterozygous condition. This is accomplished by sib-pollinations of mst (*ms,ms*) plants with heterozygous *mst* (*Ms,ms*) plants within the female parent. The resultant progeny segregates 1:1 for *mst* and *mft* plants. When such lines are used in hybrid seed production, the fertile plants have to be removed prior to flowering in order to have 100% hybridization with the male parental line. Early removal of mft plants can be done if a phenotypic marker gene is linked to the *ms* gene. For this, gene ms_1 has been exploited commercially because it is closely linked with a gene controlling anthocyanin production (Vranceanu and Stoenescu 1970b). In the coupling phase, this linkage enables the grower to eliminate purple seedlings which are mft; the mst's lack the purple coloration and are green. Theoretically, this system should eliminate all mft plants,

Table 20.1. Reciprocal cross differences in *Brassica*

Cross	Parents	F_1 Progeny
1a) (mst)	*B.napus* × *B.juncea*	mst
1b)	*B.juncea* × *B.napus*	mft + mst[a]
2a) (mst)	*B.napus* × *B.carinata*	mst
2b)	*B.carinata* × *B.napus*	mft[a] + mst[b]

[a,b]Some branches and some plants mst, respectively.

20.2.3.2.8 DNA Structure and Translation

The restriction pattern of cp- and mt-DNA isolated from well-expanded leaves of mft and gc-mst *B. napus* (having radish cytoplasm) were studied by Vedel et al. (1982). Some differences between the cp- and mt-DNA restriction pattern occurred in mst and mft forms distinguished by different locations of cp-RNA genes on Sma 1 cp-DNA fragments. In vitro RNA translation and protein synthesis of mitochondria isolated from 6-day-old etiolated seedlings was more efficient in mft than that in mst forms. Furthermore, electrophoresis of polypeptides synthesised by mitochondria obtained from mst and mft cotyledons and hypocotyls yielded distinct peptide differences not common to both the lines, as indicated by arrows in Fig. 20.5. Using two-dimensional gel separation, Remy and Ambard-Bretteville (1983) analysed the composition of leaf chloroplast peptides of normal and gc-mst *B. napus* lines with radish S-cytoplasm. Although the stromal compartments of the two lines were very similar, as also was the homology between protein maps of

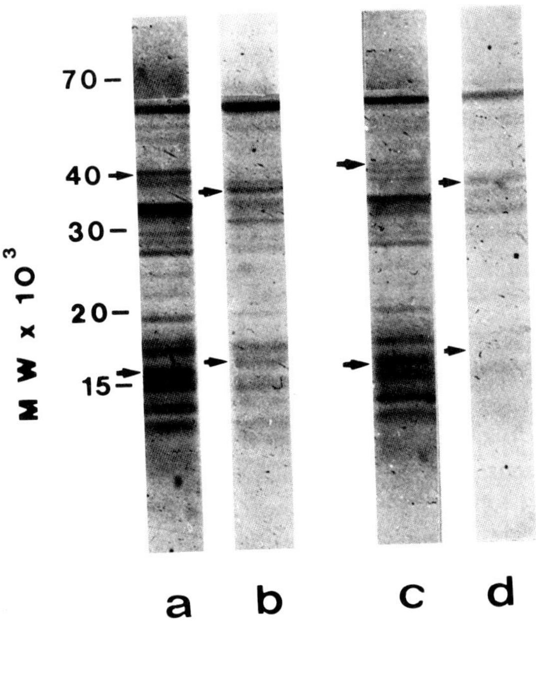

Fig. 20.5. *Brassica napus.* Electrophoretogram of polypeptides synthesised by mitochondria of male fertile *(a,c)* and male sterile *(b,d)* cotyledons and hypocotyls, respectively. (Vedel et al. 1982)

thyllakoids, the two lines could be distinguished by the spots corresponding to the subunits of the coupling factor CF_1 from the ATPase complex. The coupling factor (ATPase) subunits could thus serve as chloroplast markers in this 2D gel technique.

20.2.3.2.9 Chlorophyll Deficiency

In some crucifers, transfer of nucleus into alien cytoplasm may or may not cause mst but chlorophyll deficiency is common in the alloplasmic lines, especially in gc-mst *B. campestris, B. napus* and *B. oleracea* having S-cytoplasm of *R. sativus*. Iwasa (1963a,b) placed *B. pekinensis* (2n = 20) nucleus in *B. carinata* cytoplasm. All the completely substituted plants (2n = 20) had chlorophyll deficiency to a varying degree. This was absent in aneuploids (2n = 21 to 37) obtained during the genome transfer process. Likewise, when *Raphanus sativus* genome (2n = 18) was transferred into *B. pekinensis* (2n = 20) cytoplasm by repeated backcrossing of *Brassico raphanus* (2n = 37) to *R. sativus* as pollen parent, chlorophyll deficiency appeared in all the aneuploid plants having 18–26 somatic chromosomes (Kato and Tokumasu 1980). In both the above alloplasmic lines, mft was not impaired. Almost all the alloplasmic mst lines, aneuploids and stabilized lines of *B. campestris, B. napus* and *B.oleracea* having *Raphanus sativus* cytoplasm exhibit a severe to mild degree of chlorosis. In some, the chlorosis degree decreases with age. A few similar alloplasmic lines of *B. oleracea* var. *italica* are free from such chlorosis. Using chlorophyll-deficient alloplasmic (*B. carinata*) *B. pekinensis* as female in crosses with *B. chinensis, B. japonica, B. narinosa, B. pekinensis* and *B. rapa* (all having aa genomes). Iwasa (1963c) obtained F_1 hybrids, all of which are chlorophyll-deficient. Similarly chlorophyll-deficient F_1 progeny occurs after crossing chlorophyll-deficient (*B. japonica*) *R. sativus* as female parent with various cultivars of *R. sativus* and *R.raphanistrum* (Kato and Tokumasu 1980).

Chlorophyll-deficient F_1 progeny also develops from the crosses involving alloplasmic mst *B. campestris, B. nigra* and *B. oleracea* having *R. sativus* S-cytoplasm. Transmission of chlorophyll deficiency is maternal in the alloplasmic *Brassica* lines and appears to result from g-c interactions. Chlorophyll development is conditioned by many genes, as chlorosis degree could be gradually reduced and nearly eliminated by substituting the nuclear genes of the alloplasmic lines by the genes of closely related cultivars. By this method, Williams and Heyn (1981) developed mst alloplasmic Chinese cabbage lines that exhibit a low degree of chlorosis. Similar alloplasmic mst *B. napus* lines were developed by Ohkawa and Shiga (1981). The addition of one specific *B. carinata* chromosome to alloplasmic (*B. carinata*) *B. pekinensis* eliminated chlorophyll deficiency in it (Iwasa 1963c). However, over four chromosomes of *B. japonica* are required to eliminate chlorophyll defects in alloplasmic (*Raphanus sativus*) *B. japonica* (Kato and Tokumasu 1980). Thus, genetic control of chlorophyll deficiency may be mono- to multigenic; the genes may be present on the same or on different chromosomes of a complement. Whether these genes are closely linked to mst or represent a pleiotropic effect of g-c interaction is not known.

20.2.3.3 Chemical Male Sterility

A spray of 0.03% male gametocide No. 1 (zinc-methyl arsenate) at the floral initiation stage produced 60-80% partial to total sterile plants, plants with abortive floral buds, and numerous dead plants (Guan et al. 1981). In the sterile plants, PMC adhesion and microspore shrivelling was rampant. This chemical is not a male gametocide, as it kills the plants and causes high floral, male and female gamete abortion.

20.2.3.4 Breeding Value

Like mst *B. campestris*, the mst *B. napus* has the potential for heterosis breeding. However, the mst plants having *R. sativus* cytoplasm have small flowers and poorly developed nectaries, which is unfavourable for insect pollination (Heyn 1978). A few mst plants have nearly normal-sized petals and fragrant nectaries. On them, bee visitation is normal and seed set high. Under cool conditions, slight chlorophyll deficiency occurs in mst plants. This can be obviated by selection of plants having normal chlorophyll content in the restorer lines. Thus gc-mst lines of *B. napus* can effectively be used in F_1 hybrid seed production. Mesquida and Renard (1981) studied the effects of honeybee pollination on phenology and yield of mst and mft plants of *B. napus* var. *oleifera*, and obtained some favourable results. The mst material developed by Rousselle (1982) has a chlorophyll content half that of mft plants. This could not be improved effectively in successive BC's and selections. This has limited the use of alloplasmic gc-mst in rape hybrid production.

20.2.4 *B. oleracea* L. (cabbage)

This is a polymorphic cross-breeding species and includes important vegetables like broccoli, Brussels sprouts, cabbage, cauliflower, kale and kohlrabi. Although a distinct type of S-cytoplasm is unknown in this species, only a few genes causing g-mst and gc-mst are fully established (Tables 20.2, 20.3) in some of its varieties.

20.2.4.1 *B. oleracea* var. *botrytis* (cauliflower)

The presence of mst plants in cauliflower has been documented by Rundfeldt (1960) and Nieuwhof (1961). A mst mutant in an inbred line of purple cauliflower variety Early purple head was found by Borchers (1966). It was characterised by having anthers that dehisced normally, but the pollen grains remained fused with each other, forming a stiff pollen strip rather than the loose amorphous pollen powder produced by the normal plants (Borchers 1966). This pollen clumping is the main criterion to identify these mst's from mft's. Most of the pollen grains of the steriles are small, nonstaining, abortive and exhibit size variability. Normal-appearing pollen from the mst's germinates poorly. The sterility controlled by a recessive gene m_4 which is stable under field and greenhouse conditions. Furthermore, it remains unaffected by any of the several chemical treatments like IAA,

Table 20.2. Genic male sterility in Cruciferae

Species	References	Gene control	Gene action	Remarks
Arabidopsis thaliana	Veen and Wirtz (1968)	1r	During microspore formation to pollen development	EMS-induced
	McKelvie (1963)	1r	nk	Mutagen-induced
	Müller (1968)	1r	nk	X-ray-induced
Brassica campestris	Das and Pandey (1961)	1r	After tetrad formation, before microspore release	Petals reduced
	Chowdhury and Das (1966, 1968)	1r	After tetrad formation, before microspore release	Tapetum intact
		1r	During anther dehiscence	Functional mst
B.juncea	Banga and Labana (1983)	1r	After microspore liberation	
B.napus	Koch and Peters (1953)	1r	nk	
	Takagi (1970)	1r	After microspore release	γ-ray-induced, anther development abnormal, exhibits pleiotropy
	Heyn (1973)	2r	nk	
B.oleracea var.*botrytis*	Nieuwhof (1961)	1r	nk	Petals and stamens minute
	Borchers (1966)	1r	nk	mst Unaffected by IAA, GA, NAA
	Chatterjee and Swarup (1972)	1r	nk	*ms* gene combined with stigma protrusion gene favour cross pollination
B.oleracea var.*capitata*	Nishi and Hiraoka (1958)	1r	After tetrad formation	Premature degeneration of tapetum
	Rundfeldt (1961)	1r	nk	Highly sensitive to environment
B.oleracea var.*gemmifera*	Haigh (1954), Johnson and Faulkner (1956, 1957)	1r	After release from PMC's	Stable to high temperature
	Nieuwhof (1968a)	1r	nk	High temperature enhances mst degree

Table 20.2. *(continued)*

Species	References	Gene control	Gene action	Remarks
B.oleracea var.*italica*	Cole (1959)	1r	Before microspore release	
	Sampson (1966a)	1r	nk	Stable, linked to green hypocotyl colour gene
	Dickson (1970)	1r	nk	Complete mst at low and partial mft at high temperature
Cheiranthus cheiri	This Vol.	1r	Before sporogenous tissue differentiation	Tapetum degenerates prematurely
Daucus carota	Hansche and Gabelman (1963a,b)	1D + 1r	nk	Digenic control
	Michaelik (1974)	1r	nk	Partial mst, temperature-susceptible
	Welsch and Grimball (1947)	nk	nk	
	McCollum (1966)	nk	nk	Structural, petaloidy
	Litvinova (1973)	nk	During anaphase-I	Partial mst
	Litvinova et al. (1980)	1r	nk	Induced
Eruca sativa	Verma (1984)	2r	After microspore liberation	Inbreeding of a self-incompatible line
Raphanus sativus	Tokumasu (1957)	1r	During late prophase I	Differentiation and development suppressed
	Nishi and Hiraoka (1958)	1r	After microspore release, during microspore development	
	This Vol.	1r	During microspore formation	EMS-induced

Table 20.3. Gene-cytoplasmic male sterility in Cruciferae

Species/cross	Reference	Gene control	Gene action	Remarks
A) Species				
Brassica campestris var. *napifera*	Okhawa and Shiga (1981)	2r	nk	
B. juncea	Rawat and Anand (1979)	2r	nk	Petaloid or rudimentary anthers
B. napus	Kanno et al. (1963), Yamaguchi and Kanno (1963)	nk	During sporogenous tissue	*Fr* gene linked to white flower colour
	Thompson (1972)	2r	nk	Thin petals, pollenless anthers
	Shiga and Baba (1971, 1973)	nk	nk	Rugose petals, abnormal stamens
	Bartkowiak-Broda et al. (1979)	nk	After microspore liberation	*fr* gene action sensitive to temperature, photoperiod and plant development stage
Daucus caroia	Gabelman (1956), Banga (1959)	1D	nk	High environmental influence on mst expression
	Thompson (1967)	2–3D	During anther development and differentiation	2-3 duplicate dominant gene control fertility
	Morelock (1974)	3D	During anther development	Anther differentiation and or development inhibited
		many r	nk	nk
	Timin and Dobrutskaya (1981)	1D + 1r	During microspore formation	
		3D	During anther development and differentiation	Complementary action of three dominant genes causes complete mst
	Zenktler (1962)			
Raphanus sativus	Ogura (1968)	nk	During prophase I	Tapetum abnormal
	Nishi and Hiraoka (1958)	2r	After microspore liberation	Pleiotropy of *ms* genes
		nk	During tetrad to early microspore formation	
B) Interspecific crosses				
Brassica campestris × *B. oleracea*	Nishi and Hiraoka (1958)	nk	During tetrad formation	Tapetum persistent
B. juncea × *B. napus*	Brar et al. (1980)	2r	nk	
B. nigra × *B. oleracea*	Pearson (1972)	nk	nk	Petaloid and vestigeal anthers developed

Table 20.3 *(continued)*

Species/cross	Reference	Gene control	Gene action	Remarks
C) Intergeneric crosses[a]				
Diplotaxis muralis × *Brassica compestris*	Hinata and Kanno (1976)	nk	During microspore and pollen development	
Raphanus sativus × *B.campestris*	Heyn (1978)	2r	nk	Reduced floral organs, staminal atrophy high
R.sativus × *B.napus*	Bannerot et al. (1974)	2r	After microspore liberation	
R.sativus × *B.oleracea*	Bannerot et al. (1974), McCollum (1979)	2r	After microspore release	

[a]Temperature and photoperiod sensitivity high over *fr* gene action.

NAA, GA etc. which in some other steriles restore fertility. Nieuwhof (1961) detected mst's having slowly unfolding and poorly developed nondehiscent anthers. The sterility is controlled by a recessive gene that is nonallelic to the *ms* gene of Brussels sprouts. Another *ms* gene in Indian cauliflower was found by Chatterjee and Swarup (1972). The addition of a recessive gene to the mst mutant caused stigma protrusion in bud condition and led to high natural outcrossing and facilitated hybrid cauliflower production.

20.2.4.2 *B. oleracea* var. *capitata* (cabbage)

In a commercial variety of cabbage, Nishi and Hiraoka (1958) found some mst's in which male meiosis proceeded normally, but tetrads aborted soon after formation. The tapetum degenerated prematurely before tetrad development. A strain of cabbage found by Rundfeldt (1960) was mst in summer but mft in the greenhouse during winter. Genetic control of this anomaly is not known. Several mst's arose spontaneously in some Bulgarian and Japanese cabbage varieties and some hybrid lines of cabbage (Goshev 1974). Rundfeldt (1960) detected similar mst plants in some German cabbage lines. Likewise, two mst plants having short stamens and shrivelled anthers were found by Nieuwhof (1961) in a white cabbage variety grown in the Netherlands. The steriles were maintained by cutting, as pollinating them by fertile cabbage and cauliflower did not produce seeds, indicating fst in them. The presence of cellular outgrowths within their embryos led to female sterility. Neither genetic control nor cause of sterility is known in this material.

20.2.4.3 *B. oleracea* var. *gemmifera* (Brussels sprouts)

In a noninbred and some inbred lines of three varieties, several mst's segregated (Haigh 1954; Johnson and Faulkner 1956, 1957). Their flowers were pale and small. Male meiosis proceeded normally up to tetrad formation, after which the young pollen degenerated (Johnson 1958). In 1% of the anthers nearly 10% pollen tetrads produced pollen grains but these were nonfunctional. Such anthers developed to their normal length although they were emaciated and indehiscent. The sterility is due to the action of a recessive gene ms_2. The gene expression is uninfluenced by temperature variations (Johnson 1958). In the Brussels sprouts variety Roodnerf collected from the Netherlands, complete to partial mst occurs (Nieuwhof 1961). The sterility is controlled by recessive genes, but its inheritance pattern is not fully known. These, like ms_6ms_6 green sprouting broccoli plants, exhibit differential behaviour at lower and higher temperatures. While at 10°C more pollen is produced in both mst and p-mst plants, at 17°C, although staminal development is normal, pollen production is very low (Nieuwhof 1968b). It is intermediate at 14°C. Hence high temperature reduces pollen formation. Whereas at 14° and 17°C, seed set after hand pollination is similar in all the clones, at 10°C the seed set of mst's is lower than that of the mft's. Thus female fertility is not significantly influenced by such temperature treatments.

20.2.4.4 *B. oleracea* var. *italica* (green sprouting broccoli)

Some mst's, growing slowly and having brown shrivelled nondehiscent anthers, were detected by Moore (see Anstey and Moore 1954). Later, Cole (1957) found similar mst's in which the sterility is conditioned by one recessive gene. In them male meiosis proceeded normally up to tetrad formation, after which the four young microspores, instead of separating out, remain fused with each other within the microsporocyte and degenerated gradually (Cole 1959). Another gene, *ms*-1, when isolated (Sampson 1966a) exhibited linkage with a gene C_1 responsible for causing absence of anthocyanin and imparting a bright green colour to the hypocotyl. Unlike the mst's of Cole (1957), these steriles have fleshy pale green normal-sized anthers. Whereas a quadrivalent in addition to seven bivalents occurs regularly in the PMC's, a few higher chromosomal associations also occur sporadically. This quadrivalent formation is due to a homozygous chromosomal duplication (Sampson 1966b) rather than to heterozygous translocation. However, this presumed duplication does not cause mst. The ms_1 gene is stable in different environments and does not impair female sterility. These features, along with linkage, make it useful in identification of mst's immediately after seedling emergence and for use in breeding programs.

20.2.4.5 Temperature Sensitivity

From the line PI-189028, vestigeal anther mst plants growing at 24°C day and 17°C night were detected by Dickson (1970). At 17°C day and 12°C night, some abnormal pollen developed and with further temperature drop anthers developed normally, so that at 12°C limited viable pollen developed. At constant 10°C or at 15°C day and 7°C night, the plants are highly mft. This conversion from complete mst to mft takes about 30 days at 10°-11°C but over 2 months for partial fertility restoration at 12°C. The anomaly is conditioned by a recessive gene ms_6 and the control mechanism appears to operate early in the bud initials. The gene is nonallelic to other *ms* genes detected by Johnson (1958), Cole (1959), Nieuwhof (1961) and Borchers (1966). Furthermore, its temperature sensitivity is distinct and different from that found by Nieuwhof (1961) and Rundfeldt (1960) as the latter author isolated a cabbage strain which was mst in the field in summer but mft in the greenhouse in winter. Homozygous ms_6ms_6 lines could be maintained by selfing at low temperatures and used as female at high temperatures. Borchers (1971) utilized this temperature sensitivity for hybrid broccoli seed production and grew them in field and greenhouse under fluctuating temperatures of 17.8°C. Seed set on mst's was high after crossing them with mft's.

20.2.4.6 Gene-Cytoplasmic Male Sterility

Progeny of the cross black mustard *B. nigra* × common cabbage *B. oleracea* var. *capitata* is mostly total sterile but a few are mst (Mezushima and Katsuo 1958). Pearson (1972) transferred *B. oleracea* genome into *B. nigra* cytoplasm and developed fairly stable mst lines. Nishi and Hiraoka (1958) found mst plants in BC progeny of (*B. campestris* × *B. oleracea*) × *B. oleracea*. In these mst's, male meiosis is normal up to tetrad formation. The tapetum is persistent and tetrads degenerate gradually even before liberation. The F_1 plants of black mustard *B. nigra* (L) ♀ × broccoli *B. oleracea* L. var. *italica* ♂ were treated with colchicine by Pearson (1972) to produce amphidiploid (4n). By repeated BC's with broccoli pollen, thus placing *B. oleracea* genome in *B. nigra* cytoplasm, some diploid plants appeared. These progenies after BC_3 are identical with broccoli in appearance but differ in their content of sulphur oils (isothiocyanates) and therefore have a different flavour when eaten raw or cooked. From outcrosses of these early generation plants with cabbage, two mst types segregated in *B. nigra* cytoplasm (Pearson 1972). One of these exhibits a "petaloid" type of sterility in which stamens are transformed into petal- or carpel-like structures and flowers lack nectaries (Fig. 20.6b,c). The anomaly is conditioned by a recessive gene, *p*. Its maintainers, having *B. oleracea* cytoplasm and petaloid gene, were isolated from cabbage and narrow stem kale plants. Since the mst's lack nectaries which disfavour insect pollination, are self-incompatible and have deformed pistils, they do not have any breeding value. In another mst type, the vestigeal anther (Pearson 1972), anthers are rudimentary but nectaries develop and pistil is functional (Fig. 20.6c,f). In it seed set is normal and thus it has a higher breeding value than the petaloid type. The sterility is conditioned by a recessive gene, *vs*, which is hypostatic to gene *p*. Dickson (1975) improved Pearson's mst lines and incorporated the gene *p* into normal broccoli cytoplasm. This led to the development of stable maintainers. These were released in 1975 and had nearly fully restored female fertility and moderately sized functional nectaries. Likewise, mst *B. oleracea* var. *capitata* was developed by Engle (1974). The sterility is stable at low temperature. High temperatures and long day length revoke fertility in them.

20.2.4.6.1 Male Sterility Transfer from Radish to Cabbage

Bannerot et al. (1974) and Boulidard (1975) transferred mst from radish to *B. oleracea* var. *capitata* and var. *gemmifera* (Fig. 20.7). These crossings produced amphidiploids and amphitriploid *Raphano-brassicas*. Fertility of these derived plants was restored by chromosome doubling with colchicine. By repeated BC's up to four cycles of the fertile (6x) raphano-brassicas to cabbage and utilizing embryo culture technique to raise immature embryos (that normally aborted prematurely in vivo), from BC pollinations of a moderately seed fertile BC_3 plant ($2n = 3x = 27$), 35 plants were obtained. Of these, four were of *B. oleracea* type. In them, radish chromosomes were eliminated, seed fertility was partially restored, morphological characteristics and diploid chromosome number of *B. oleracea* were regained and complete mst was maintained. Efforts of McCollum (1979) to transfer mst from diploid radish to diploid cabbage via synthesis of amphidiploid *Raphano-brassica* were unsuccessful through BC_6 to cabbage. The reason was because in each BC generation, from BC_1 ($2n = 3x = 27$) to BC_6, radish chromosomes persisted in greater than expected numbers. No BC plant had less than two radish chromosomes and all the plants with less than four radish chromosomes were seed-sterile.

Fig. 20.6a-f. *Brassica oleracea.* Flowers **(a-c)** and stamens **(d-f)** of male fertile **(a-d)** and of petaloid **(b,e)** and vestigeal anther **(c,f)** male sterile alloplasmic *Brassica oleracea* having *B. nigra* cytoplasm. (Pearson 1972)

However, under nearly similar circumstances, Bannerot et al. (1974) obtained *B. oleracea* plants with 2n = 18 chromosomes by culturing BC_4 embryos from a 2n = 27 BC_3 plant pollinated by 2n = 18 *B. oleracea* pollen.

The alloplasmic mst *B. oleracea* having *R. sativus* cytoplasm exhibits a high degree of instability as long as the complete genome *B. oleracea* is not placed in *R.*

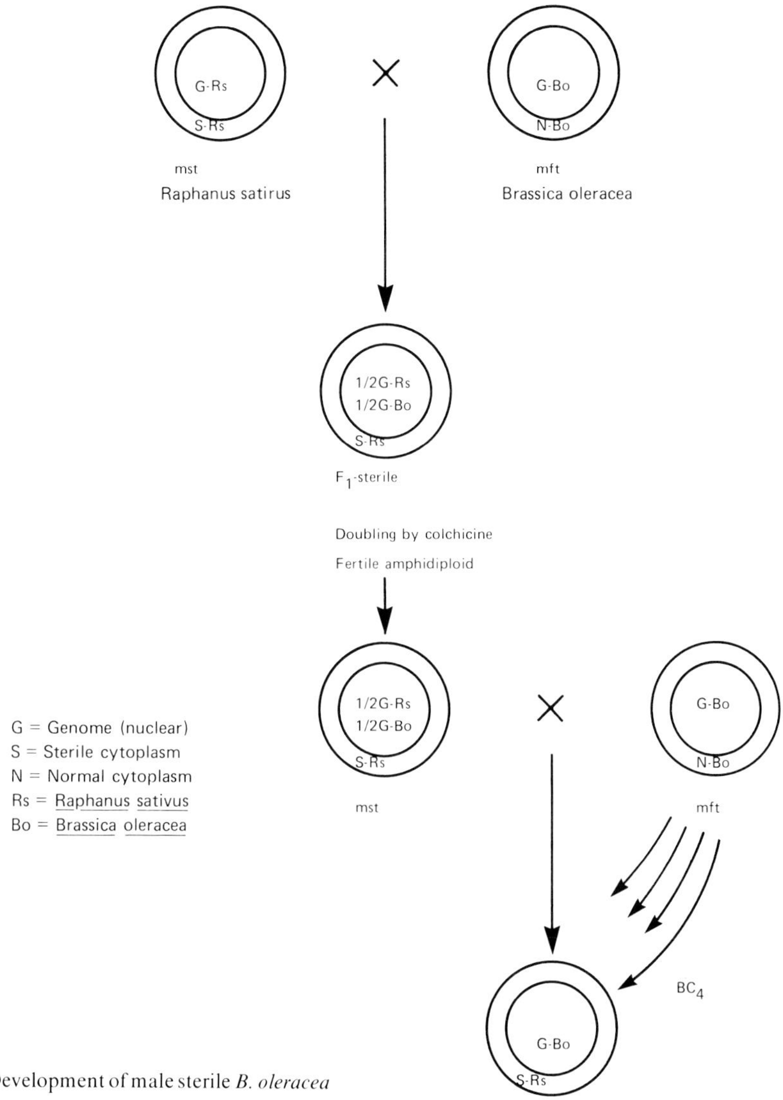

Fig. 20.7. Development of male sterile *B. oleracea*

sativus cytoplasm and until all *R. sativus* chromosomes are eliminated from (*R. sativus*) *B. oleracea* alloplasmic combination. However, the sterility is associated with chlorosis, especially of younger tissues, that enhances at temperatures below 12°C (Bannerot et al. 1977). Using the Early Scarlet Globe variety *R. sativus* as S-cytoplasm donor and *B. oleracea* var. *capitata* and var. *italica* as the male recurrent parent, McCollum (1981) produced mst *B. oleracea* plants. In first four BC's, radish chromosomes and high female sterility persisted. Using colchiploid 4x broccoli pollen, a BC_5 plant having $2n = 3x + 1 = 28$ chromosomes, improved seed set, and no radish trait was obtained. When this plant was pollinated with diploid broccoli pollen, 17 mst plants segregated, three of which had $2n = 18$, six, $2n = 19$ and the remaining $2n = 20$–25 chromosomes. Some of these had an increased level of seed setting. Like McCollum (1981), Bannerot et al. (1974) also obtained alloplasmic 2n = 18 (*R. sativus*) *B. oleracea* plants. However, whereas the former author obtained them directly from the plants, the latter obtained these through in vitro culture of immature embryos from BC pollination of a moderately seed-fertile BC_3 plant (2n = 3x = 27) that normally aborted prematurely in vivo. From 35 plants, four were *B. oleracea*-type plants having 18 or 19 chromosomes. On the other hand, from the seeds maturing naturally on the same plant, only one *B. oleracea*-type plant with 22 chromosomes was obtained by them in a total of 44 plants. Failure to obtain such alloplasmic mst parental seeds in vivo was due to the negative selection value of embryos containing few or no radish chromosomes and due to unfavourable conditions provided by the *Raphano-brassica*-type maternal tissue to such embryos. Since McCollum (1981) obtained alloplasmic mst seeds in vivo after using diploid pollen from induced tetraploid broccoli plants, the author opined that increasing the ploidy level of the embryo and endosperm relative to that of the surrounding triploid maternal tissues of mixed *R. sativus* and *B. oleracea* chromosome constitution aided the survival of a pure *B. oleracea* type embryo as much as did the embryo culture.

20.2.4.6.2 Fertility Restoration

From test crosses using *B. oleracea* mst stocks, Heyn (1976) obtained only mst offspring and concluded that the euplasmic forms of *B. oleracea* are functionally maintainers for mst in the alloplasmic (*R. sativus*) *B. oleracea.* Fertility restoration is conditioned by two dominant restorer genes designated Rf_1 and Rf_2 by Heyn (1976). Both fertility restoration and mst expression exhibit some degree of instability due to genetic and environmental factors. Using the pollen from one cauliflower and many broccoli varieties in crossing with alloplasmic *B. oleracea* mst plants, having the cytoplasm of American (Early Scarlet Globe) or Japanese carrots, mft plants segregated only in the progeny of American type and not in Japanese type alloplasmic brassica (McCollum 1981). The restoration was partial, stamens of such plants were less developed and transitions between vestigeal or petaloid to subnormal stamens occurred in them. A few anthers were fully developed and their pollen fertility ranged from 35 to 97%. Usually one to three anthers of a flower were dehiscent and shed pollen, the remaining were underdeveloped or absent.

20.2.4.6.3 Phenotypic Stability

The mst expression and phenotypic stability of the alloplasmic (*R. sativus*) *B. oleracea* developed by Bannerot et al. (1974) are high when the complete genome of *B. oleracea* is placed in *R. sativus* cytoplasm and when all the radish chromosomes are fully eliminated. On the other hand, a similar mst line, having the American carrot cytoplasm developed by McCollum (1981), is unstable in mst expression and produces a high frequency of fertile pollen. Similarly, high susceptibility to temperature and photoperiod is exhibited by the gc-mst line derived from the commercial cabbage *B. oleracea* var. *capitata* (Engle 1974). This line is mst at low temperature but mft at higher temperature and under long days. Due to environmental susceptibility, development of maintainers for this mst line is arduous.

20.2.4.7 Chemical Male Sterility

Meer and vanDam (1979) obtained mst *B. oleracea* var. *gemmifera* and var. *capitata* (Brussels sprouts and cabbage) after repeated spraying of 10 and 50 mg GA 4/7 per litre isopropyl alcohol during 2 weeks prior to bolting (Table 20.4). However, the sterility recovered was variable and incomplete. Only a few plants retained complete sterility up to 24 days after spraying. The effect of the chemicals on female fertility is not known. Since the sterility induced is low and variable, it does not seem to have immediate practical utility. In *B. oleracea* var. *capitata* (cabbage), maleic hydrazide failed to induce mst (Moore 1950) and was toxic.

Table 20.4. Chemical male sterility in *Brassica*

Species/(Chemical)	References	Male sterility induction	Female fertility reduction	Side effects
B.juncea (Ethrel)	Banga and Labana (1984)	High	High	nk
B.napus Zinc methyl arsenate)	Guan et al. (1981)	High	Very high	Severe plant toxicity and death, high floral abortion
B.oleracea var. *capitata*, var.*gemmifera* (Gibberellie acid)	Meer and Van Dam (1979)	Variable	nk	nk

20.2.4.8 Breeding Value

20.2.4.8.1 Gene-Cytoplasmic Male Sterility

Positive, mid and high parent F_1 heterosis in cabbage yield and total production per unit area of land is known in *B. oleracea*. Therefore, mst utilization for commercial hybrid production is highly profitable in this crop. The alloplasmic mst (*R. sativus*) *B. oleracea* developed by Bannerot et al. (1974) from the donor S-cytoplasm of Japanese radish (Ogura 1968) is stable and is not restored by *B. oleracea* pollen parents. Also crosses to the mst lines developed from Bannerot's material (Pearson 1977) produce stable mst progeny in the greenhouse (McCollum 1981). A similar mst alloplasmic (*R. sativus*) *B. oleracea* (ESG) line developed by McCollum (1981) is unstable, and high fertility is restored in it by cauliflower and many broccoli varieties. Additionally, the nectary development in this line is variable and it has multiple pistils and mis-shapen ovaries. Nearly similar flower types occur in mst alloplasmic lines having *B. nigra* cytoplasm (Pearson 1972). Seedlings of the ESG line exhibit a mild degree of chlorosis that decreases with age. Likewise, chlorosis occurs in the lines developed by Bannerot et al. (1974) and may be associated with g-c interaction involving *Brassica* genome and radish cytoplasm. Thus it is evident that alloplasmic mst lines (*R. sativus*) *B. oleracea* developed by Bannerot et al. (1974) are promising for use in hybrid production. Its mst expression is stable and its maintainers are available. Low temperature chlorosis, although a problem, can be selected against. Furthermore, some genetic backgrounds are better than others. In fact alloplasmic *B. oleracea* var. *italica* is free from such chlorosis. Thus it has high breeding value. By increasing CO_2 concentration, self-incompatibility can be broken in many cabbage lines. Maximum pollen tube penetration occurs at 4.5% CO_2 concentration (Nakanishi et al. 1969). Thus mst lines can be maintained by selfing if the incompatibility barrier is removed and mft restored. This can be done by CO_2 treatment of the mst plants that are maintained at higher temperatures under long light hours; high temperature and long days restore male fertility in mst plants. Thus the mst lines can be maintained by selfing and utilized for producing hybrid seeds.

20.2.4.8.2 Genic Male Sterility

Using mst stocks of Pearson, cytosteriles were developed by Dickson (1975). By rigorous selection, their female fertility was considerably improved in petaloid mst plants and three mst lines of broccoli having high female fertility and their maintainers were released for use in commercial hybrid production. The main advantage of this *ms* gene is that it permits selfing of the female and it is possible to maintain an mst line without recourse to the maintainer crossings in the greenhouse in winter. Thus while selfed progeny can be obtained in the greenhouse, hybrid seeds can be obtained in the field, as ms_6 ms_6 plants are mst at higher temperatures in field. This gene has been transferred to a promising broccoli type and is not linked to any deleterious characters (Dickson 1970). Furthermore, 20–30% of ms_6 plants are completely or partially self-compatible, and this enables selection of homozygous ms_6 self-compatible plants. Utilizing broccoli plants homozygous for ms_6 gene, a high amount of hybrid seed was obtained by Borchers (1971) in both greenhouse and field at relatively higher temperatures. However, at lower temperatures, partial breakdown of mst occurred and many selfs were produced which were used for maintenance of the mst line. The main problem in utilizing ms_6 ms_6 genotypes is the plant-to-plant variability obtained in pollen production as measured by anther dehiscence, especially under field conditions in India and elsewhere due to temperature fluctuations during the flowering period.

20.3 *Cheiranthus cheiri* L. (wallflower)

A clone of this species has rudimentary stamens in which no sporogenous tissue differentiates (this Vol.). The anomaly is conditioned by a recessive gene. Meyer (1966c) found many clones with pistilloid stamens and considers the anomaly to be due to recessive *ms* genes that mutate spontaneously and frequently. In another clone, meiosis proceeds normally but microspores abort after liberation from tetrads (this Vol). The sterility is monogenic recessive.

20.4 *Daucus carota* L. (carrot)

20.4.1 Genic Male Sterility

20.4.1.1 Spontaneous

This is one of the few species in which early documentation of mst exists. This is evidenced from the reports of Beiyernick (1885), Staes (1889) and Warenstorf (1896), who observed spontaneously arisen mst plants in cultivated carrot fields. These plants had greenish or reddish inflorescences in which anther degeneration was partial to complete. Likewise, the occurrence of physiologically female plants developing either leafy stamens or stamens with unrolled filaments and indehiscent anthers in carrot is documented in Knuth's (1898) book. In the cultivated carrot variety, Amsterdam Forcing, many mst plants occur frequently (Braak and Kho 1958). Similarly, a small proportion of wild carrot plants are mst. In these, filaments remain unrolled and anthers are indehiscent. In some plants, early staminal abortion occurs and their umbels have female and sexless flowers only. In general, carrots have a tendency to develop mst flowers which are few in the main umbel and many in the umbel of higher order. This female flower-producing tendency is genetically conditioned and environmentally influenced. On the other hand, some carrot lines have a genetically conditioned male-determining tendency. These flowers have either a rudimentary ovary or lack it completely. Although the mst detected by Braak and Kho (1958) does not influence the ratio existing between flowers with and without ovary in the normal mft plants, the percentage of sexless flowers in mst plants is higher than that of male flowers in normal plants. Furthermore, hermaphroditism was found in a wild carrot population by Rubashevskia (1931) in which the hermaphrodite flowers were larger in size and were lodged at the periphery of the umbellets, while staminate flowers occupied a more central position.

Genetically conditioned mst in a plant of commercial stock of carrot variety Tendersweet was reported by Welch and Grimball (1947). Its anthers were shrivelled, brown and nonexserted and female fertility was normal. Similar mst plants segregated in other breeding lines of carrot. While two such plants produced no pollen, in the other two, nonviable pollen was produced but seeds were set after open pollinations, indicating unimpaired female fertility. Later studies by Thompson (1961) have revealed the presence of many *Fr* genes and S-cytoplasm in this carrot variety. Michaelik (1974) detected mst plants from the progeny of a Polish carrot variety, Selecta. The plants have brown anthers with degenerated pollen and tapetal cells. In growth chambers, mft was partially restored in them, but sterility was complete in the field. From these, a line segregated 1:3 mst:mft plants in the field and nearly 50% p-mst's in growth chambers. Pollen fertility based on aceto-carmine staining ranged from 10–50% on first-order umbels and 20–70% on second-order umbels of these partial mst's in growth chambers. When the plants were transferred to field, pollen sterility was complete. The *ms* gene action is highly influenced by temperature; at constant high temperatures penetrance and expressivity of the *ms* gene is reduced by about 50%.

Many complete and a few partial mst carrot plants were found in Russian carrot varieties by Litvinova (1973). The sterility was due to meiotic anomalies occurring mainly at AI, MII, and TII, coupled with an arrest of microspore development after the uninucleate stage. This was followed by microspore degeneration. While in the fully sterile plants, developmental arrest occurred in all the microspores, in the partial types, some microspores developed normally, producing trinucleate stainable pollen but their functionability was not tested. Since carmine stainability does not always reliably indicate pollen viability, it is not certain if these trinucleate pollens were really functional. The same could be true for the mst material studied by Michaelik (1974). In two other Russian carrot varieties, Nantes-4 and Moskovskaya-aimnyaxa A-515, many mst plants segregated; they flowered and matured earlier than their mft counterparts (Kononkov and Mokhov 1972). These two genotypes differ from each other in the flowering rhythm and stigma receptivity. Sterility in each of them is conditioned by a recessive gene; allelic tests of the two genes have not been done. Female fertility of these steriles is likewise not known.

From a US plant introduction source a few mst plants segregated (Thompson 1961). Because of their low frequency, the sterility appears to be conditioned by recessive *ms* genes, but the exact gene number is not known. Plants of some of these lines exhibited petaloidy and no normal stamens were ever developed by these plants. Occurrence of petaloid stamens, staminodes and sterile stamens in a wild carrot population received from Sweden were detected by McCollum (1966). The progeny of these abnormal plants exhibited a varying degree of petaloidy which was genetically conditioned (Fig. 20.8). In a few plants, fleshy white organs resembling narrow elongated anthers on shortened filaments occurred attached singly or in two's and three's to the inner surface of petaloid stamens. They appeared as a whorl of rudimentary stamens within and opposite the whorl of five petaloid ones but had no pollen. Thus the plants were completely mst. Precise genetic control of the sterility is not known. In 17 segregating carrot progenies, mst plants were detected by Chadha and Freese(1981). The mst plants exhibited three different types of petaloidy viz., pure spoon-type and anther-type petaloidy. Progenies with high percentage of pure petaloid types remained sterile under higher temperatures in the glasshouse, whereas those with higher amounts of spoon and anther types developed more fertile umbels in the glasshouse than in the field. Although the precise genetic control of the sterility is not known, the gene(s) conditioning pure petaloidy are stable with a high degree of penetrance.

Of the 22 carrot lines segregating for mst, Hansche and Gabelman (1963a,b) found 20 lines segregating for p-mst; in the remaining two lines, a digenic control for complete mst was assumed. For instance, in line 1056, the sterility was controlled by the segregation of a dominant gene MS_4 and a recessive gene ms_5. Thus any plant of genotype ms_5ms_5 and carrying at least one dominant allels at the ms_4 locus is mst. Plants having $Ms_5 - ms_4ms_4$ would be mst. In them the collapse

Fig. 20.8a-i. Florets and floral parts of normal and male sterile carrots exhibiting petaloid stamens, **a-h** male fertile flower **(a)**, unlobed petaloid **(b)**, lobed petaloid **(c)** and partial petaloid **(d)** type flowers; petaloid lobed or unlobed anthers **(e,f,g)**, rudimentary anthers **(h)** and five normal petals and four petaloid stamens + one with rudimentary anther sacs **(i)** in male sterile carrot plants. (McCollum 1966)

and desiccation of stamens occur just before floral anthesis. No meiotic investigations have been made in them. In the remaining 20 carrot lines, heritable partial sterility that is phenotypically unstable, especially under different environmental backgrounds, was found. From the segregation studies of two generations of selfed and sib-crossed progenies, Hansche and Gabelman (1963b) assumed that their material was homozygous for the major sterility-inducing genes (Ms_4 or ms_5 or both) and that the segregation of partially fertile plants resulted from the segregating modifier genes. The phenotypic instability of these steriles was presumably due to the differential penetrance of the modifier gene(s) present in these lines along with the MS_4 and/or ms_5 genes. Furthermore, segregation of p-mst plants probably resulted from the segregating modifier genes. With the flowering season progression, mst frequency enhanced, indicating the role of physiological and biochemical and/or environmental factors on mst expression.

20.4.1.2 Induced

After treating carrot seeds with 0.012% N-nitroso N-methyl urea and 0.025% dimethyl sulphate, Litvinova et al. (1980) obtained many mst mutants which were used in breeding for heterosis. Sterility in each of them was conditioned by a single recessive *ms* gene whose allelic nature is not known.

20.4.2 Gene-Cytoplasmic Male Sterility

Jones (1950) and Lamprecht (1951) considered mst in carrots to be cytoplasmic, but neither any reason nor any reference or source from which such information is derived was given by them. Gabelman (1956) suggested the existence of some nuclear genes, (at least one dominant) interacting with the S-cytoplasm to cause mst in the carrot material obtained from the US plant introduction division. Likewise, Banga (1959) envisaged sterility to be conditioned by the interaction of a dominant nuclear gene with the S-cytoplasm. However, no clear data were presented by these authors. Thompson (1961) found mst plants in the carrot variety Tendersweet obtained from J.E. Welch and in a wild carrot variety selected originally by H.M. Munger. In these and their derivatives petaloidy of anthers occurred. The anthers were frequently replaced by a complete extra set of petals (Thompson 1961). The sterility was conditioned by interaction of the S-cytoplasm with at least two and probably three duplicate dominant nuclear genes designated MS_1, MS_2 and MS_3 ($\equiv Fr_1, Fr_2, Fr_3$). The combination of S-cytoplasm with one of these dominant *Fr* genes leads to mst. Thus while the combination S_{FrFr} and S_{Frfr} produced mst, other combinations like S_{frfr} or N_{FrFr}, N_{Frfr}, N_{frfr} led to mft. Hence the dominant *Fr* genes can act only in the presence of S-cytoplasm and are ineffective in N-cytoplasm. The existence of either one or more dominant epistatic genes conditioning mst in this material was also visualised by Thompson (1961). These epistatic genes appear to interact with *Fr* genes and render plants mft, even though they carry the dominant mst *Fr* genes or have a sterile cytoplasm. However, it is not known whether the nuclear gene MS_4 controlling mst in the cultivated carrot, as found by Hansche and Gabelman (1963a), is allelic to any of the three dominant genes postulated by Thompson (1961) in the cultivar Tendersweet. The digenic model of Hansche and Gabelman (1963a), although less complex than that of the four- or five-gene system proposed by Thompson (1961), is free of partial mst segregants which did occur and may have eliminated certain genes similar to those postulated by Thompson (1961). Nevertheless, the postulation of Welch and Grimball (1947) about the control of mst by a dominant *ms* gene obtains support from Gabelman's (1956) results. The other recessive *ms* genes seem to have been introduced in the line 1056 through its paternal grandparent PI line 169486, which was known to be segregating for mst.

Banga et al. (1964), while agreeing to the assumption that mst in carrots results from an interaction of duplicate nuclear genes (one recessive and one dominant) with the S-cytoplasm, believe that Hansche and Gabelman's material did not contain restorer genes. They interpret that segregation in their lines indicates the existence of two complementary dominant restorer genes *D* and *E*, whose presence in the homozygous or heterozygous condition restores pollen fertility; fertility being not restored when either of these genes is homozygous recessive. Although this hypothesis is based on their studies with structural mst in which stamens are degenerated, they believe that it accommodates Thompson's as well as Hansche and Gabelman's segregation data. However, no attention has been given to the two widely divergent phenotypic manifestations of mst embraced by these data. Timin and Dobrutskaya (1981) selected mst plants from two Russian carrot lines. In one of these, anthers were brown, shrivelled and pollenless. The sterility was conditioned by the complementary action of a dominant *Fr* and a recessive gene, designated by the authors as Ms_1 and ms_2, respectively. In other mst material, the stamens were transformed into petals. This anomaly was conditioned by the complementary action of three dominant genes designated Ms_3, Ms_4 and Ms_5. All these genes induce mst in the S-cytoplasm and not in the N-cytoplasm. Morelock (1974) documented that the petaloid type of mst is determined by at least two dominant *Fr* genes in combination with an epistatic gene, while a number of recessive *fr* genes determine the brown anther sterility type. The sterility expression of each type depends to some extent on environmental conditions (Eisa and Wallace 1969; Michaelik 1978).

20.4.2.1 Meiotic Breakdown

Male meiosis in most of the carrot mst material has not in general been studied. In a few cases where it is investigated, the action of the *mst* gene is not uniform or specific. Thus the pattern of meiotic breakdown differs from genotype to genotype and line to line. For instance, while in the mst derived from the wild carrot, pollen abortion occurs following microspore formation, in those originating from the varieties Tendersweet and Imperator and in the ones obtained by Gabelman (1956) from plant introduction division, no meiotic traces past prophase-I have been detected by Thompson (1961). Whereas in the sterile wild carrots breakdown occurs after meiotic completion, in others it occurs even before its initiation. In some, anomalies at AI, MII and TII are common and lead to inviable microspores. In other material investigated by Zenktler (1962), translocations associated with chromosome bridges and fragments are rampant at AI and AII. Further stages of meiosis are abnormal and lead to premature microspore abortion.

20.4.2.2 Tapetal Development

The causes of mst and the exact time and mode of action of *mst* genes in carrots have not been investigated in the majority of the mst materials. While in a few microsporogenesis is studied, only in two cases is tapetal behaviour known. In the material studied by Zenktler (1962), meiosis is abnormal and pollen abortion is complete. Tapetal development is also abnormal, the cells enlarge and form plasmodia. In other mst material the tapetum persists and starves the developing microspores to death.

20.4.2.3 Divergent Gene Control

Despite several reports about the existence of mst in carrots, genetic control of the sterility is not known in most of the cases. However, in the few where it is known, it is highly diverse, like its other mst features. While in some, the sterility is conditioned by one recessive gene, in others it is controlled either by two or more recessive genes or by a dominant gene or by complementation of a dominant and a recessive gene or by a combination of two or three dominant genes, working probably in association with epistatic and modifier genes. In some, the association of the S-cytoplasm is distinctly evident, while in others it is feeble. In one material, a single dominant gene in association with the S-cytoplasm induces mst, in another the sterility is conditioned by the interaction of the S-cytoplasm with at least two and probably three duplicate dominant nuclear genes, one of which is epistatic. The presence of epistatic genes working with these dominant genes has also been postulated in two carrot mst lines. In other material, the sterility results due to an interaction of one dominant and two recessive genes with the S-cytoplasm.

20.4.2.4 Phenotypic Variability

In carrot, mutations to mst are of high frequency and frequent records exist about the occurrence of g-mst and gc-mst plants arising spontaneously or through induced mutations or after intra- and interspecific crossing. In the mst's, phenotypic expression of sterility is quite diverse. While in some androecia are completely lacking, in others they are either rudimentary and degenerate, or appear as staminodes, or become transformed into petal-like structures. In some, stamens are normally developed but their filaments remain unrolled and anthers indehiscent. In others, archesporial tissue or PMC's degenerate, while in a few meiosis is abnormal. Where it is normal, microspores degenerate. In one mst mutant, viable microspores are formed, but they are appressed so hard by abnormal tapetal enlargements that they die (This Vol.). In another, viable pollen becomes caged in the anther loculus as the anthers remain indehiscent.

20.4.2.5 Environmental Influence

Although mst of diverse types exists in carrot, its expression is highly influenced by the environment. In most of the mst's, gene action is fully manifested at lower temperatures and inhibited partially or rarely completely at higher temperatures. In a mst line, partial fertility is revoked when the plants are transferred from field to growth chambers. Flowers developing subsequent to transfer to growth chambers are semi-male fertile. This is also evident in a gc-mst line studied by Timin and Dobrutskaya (1981). In a field-grown p-mst carrot line, the frequency of mst flowers increases with the progression of flowering season. Since no major environmental changes occurred during this period, the role of some physiological and biochemical factors in this mst expression alteration cannot be ruled out.

20.4.3 Chemical Male Sterility

Moore (1950) treated *Daucus carota* plants with 600 and 2400 ppm maleic hydrazide (MH). Both these treatments failed to induce an appreciable degree of mst. The chemical is toxic to the plants.

20.4.4 Breeding Value

Most of the modern horticultural carrot varieties lack uniformity when environmental conditions deviate from the optimum. By studying the combining ability of paired inbred lines, one in each combination possessing the mst trait, theoretically it is possible to obtain highly uniform carrot varieties that are superior to the available ones in appearance, productivity, quality and nutritional value. In perfect flowered plants like carrot, it is impractical to employ inbred lines in this way without the utilisation of mst genotypes. Sterility in lines derived from wild mst carrot is characterised by distinctive petal colour, reduced petal size and transformation of androecia into petaloid structures. These phenotypic traits identify mst from mft plants and could add to the utility of such mst genotypes in hybrid seed production. Thompson (1960, 1961) indicated the usefulness of petaloidy in heterosis breeding. But Erickson et al. (1979) found that the honeybee pollinator *Apis mellifera* visited the fertile carrots more frequently than the mst's. This bee visitation preference appears to be due to floral differences between fertile and sterile plants. Flowers of mst plants bloom later and exhibit delayed and reduced nectar and aroma production compared to mft's (Erickson and Peterson 1979). Anthers of mft plants fluoresce and this presumably provides a visual cue for pollen-collecting honeybees. This cue is absent from the sterile plants and could account for the reduced bee visitation frequency on these plants. Although Erickson and Peterson (1979) found many phenotypic and genotypic differences in the quantities of nectar taken directly from the mst and mft carrot flowers, no such differences in percentage of sucrose, fructose and glucose were found between the two genotypes by Erickson et al. (1979). However, differences in sugar content and nectar type may exist between mst and mft flowers, as the sample size analysed by Erickson et al. (1979) is small. Moreover, environmental factors influence sugar production differently in various carrot genotypes.

Complex problems ranging from nonrandom pollinator for aging activity and selective bee pollination, to genetic incompatibility between seed and pollen parents lead to poor hybrid seed production in mst carrots. Furthermore, many mst and mft lines which are desirable to intercross bloom at different times and are differentially attractive to pollinators even when they bloom simultaneously. This has been amply demonstrated by Erickson and Peterson (1979). Floral variations among maternal and paternal lines inhibit row-crossing by honeybees. On the other hand, open-pollinated cultivars of carrot are attractive to many pollinating insects. Thus, adequate cross-pollination of these is easily achieved in commercial production (McGregor 1976). High temperature during preflowering suppresses mst gene action and enhances fertility reversions in many gc-mst carrot lines. Thus in a gc-mst carrot material having brown anthers, cultivation under increased temperatures from the beginning of stem formation up to flowering enhanced mft reversion considerably (Timin and Dobrutskaya 1981). In the other gc-mst material having petaloid anthers this reversion does not occur. This material therefore has a higher breeding value than the brown anther mst types.

20.5 *Eruca sativa* Mill (rocket salad, taramira)

Inbreeding of two sibs of a selfed family of this cross-pollinated species produced 14 mst's whose sterility is controlled probably by two recessive nonallelic *ms* genes (Verma 1984). The mst's are healthy with profuse flowering but their petals are distinctly small, anthers narrow, thin, more or less needle-shaped and greenish-white in colour. Male meiosis is normal up to the tetrad stage but microspores degenerate after liberation.

20.6 *Raphanus sativus* L. (radish)

20.6.1 Genic Male Sterility

Believed to have originated in the Near East or in the Mediterranean region but unknown in the wild state, sporadically arising spontaneous mst mutants arise frequently in cultivated fields of N. India. In two cases, the sterility is controlled by single recessive genes (this Vol.). In the Japanese radish variety Tokinashi, Tokumasu (1951) found spontaneously arising mst mutants. The sterility is conditioned by a recessive *ms* gene which also supresses tapetal development completely. Thus no tapetal differentiation occurs even though other anther tissues develop and differentiate normally (Tokumasu 1957). Meiosis initiates normally but the majority of PMC's degenerate during late prophase I and the remaining few at MI, or very rarely at the tetrad stage. Meiocyte degeneration is so intense and abrupt that they lose their individuality completely and the anther cavity becomes filled up with a single chromatin body formed due to aggregation of the degenerated meiocytes. Later, this chromatin mass becomes pushed down due to pressure of the parietal tissues and forms a narrow striplike configuration. Towards anthesis, the chromatin body shrivels up and diminishes progressively in size. The anthers never dehisce and the plants are completely mst. The action of the *ms* gene is directly over sporogenous tissue development and differentiation. In another spontaneously arisen mst mutant of a Japanese radish variety, the tapetum develops normally but behaves abnormally during microspore formation. The cells enlarge rapidly and disproportionately and each cell lodges a large expanded vacuole (Nishi and Hiraoka 1958). During these tapetal abnormalities microspores are released from the tetrads and develop a thin rudimentary exine. They degenerate shortly after separation. In this mutant, the *ms* gene acts on the tapetum, making it function abnormally, so that nutrient supply to the developing microspores is impeded. In a third mst mutant of another Japanese radish variety Kuroba-riso, the exact time of action of *ms* gene is not known. The sterility is highly stable both at diploid and at higher polyploid levels (Ogura and Etoh 1975). In all the three spontaneously arisen mutants, sterility is conditioned by single recessive genes whose allelic nature remains to be determined. Equally important is to find the female fertility status of these mutants, and the environmental influence on mst expression. The influence of temperature on mst expression is high in carrots.

20.6.2 Gene-Cytoplasmic Male Sterility

In an unidentified Japanese radish variety growing as an escape, phenotypically identifiable mst plants were detected by Ogura (1968). They have small buds, crooked style in closed flower bud or projecting out of the closed bud and an abnormal flowering pattern. Their female fertility is normal. Meiosis proceeds normally up to early microspore formation, but microspores degenerate soon after their formation. This degeneration is preceded by the collapse of the tapetal tissue, which probably leads to microspore degeneration. Crossing these mst plants with their fertile sibs yielded mst and mft plants in F_1 progeny. While pollen from fertile F_1 plants applied to commercial radish varieties produced fertile progeny, reciprocal crosses yielded mst offspring and BC with pollen from commercial varieties also produced mst plants. Thus the sterility is gc-mst and is conditioned by interaction of S-cytoplasm with three *fr* genes. The environmental influence over the mst expression is not known. While normal anthers contain proline, sterile anthers lacked it and have much less sugars than the fertile anthers (Ogura 1968). A similar mst type was detected by Nishi and Hiraoka (1958). In these plants the tapetum is abnormal and microspores abort soon after release from quartets.

20.6.2.1 Male Sterility Through Genome Transfer

Through repeated BC's, successful introduction of cabbage *B. oleracea* nucleus into the S-cytoplasm of the Japanese radish line of Ogura (1968) was done by Bannerot et al. (1977). After BC_4, 18 cabbage chromosomes were recovered after which immature embryos were cultured in vitro and alloplasmic plants developed. All *B. oleracea* genomes introduced into this raddish cytoplasm exhibited a severe yellowing of young leaves, when the plantlets were grown at temperatures below 12°C. The most severe yellowing was with cauliflower nucleus and least with Brussel sprouts and savoy hybrids.

20.6.3 Breeding Use

All Japanese radish varieties used by Ogura (1968) possess homozygous recessive *fr* genes, but lack restorers. In France, Bonnet (1975, 1977) found some sterility maintainers and *Fr* genes in a few European radish varieties. A highly stable mst line was developed after crossing and later backcrossing with a European variety.

The line was stable over nine generations. Using the gc-mst source, several F_1 hybrids have been developed since 1972 and all of them manifest considerable hybrid vigour. Their roots are homogeneous, solid and tasty. One F_1 radish hybrid was registered in an official catalogue of species and varieties in 1977. Unfortunately, seed yield is poor despite the good fertility of the female line and its outbreeding nature. Bonnet (1977) suggested three-way crosses for F_1 seed production, using mst's as female parents. This has given promising results.

CHAPTER 21

Cucurbitaceae (gourd family)

A native of warm temperate regions, this family has about 110 genera which include many popular vegetables, salads and fruits, but mst is known only in *Bryonia*, *Citrullus*, *Cucumis*, *Cucurbita* (Tables 21.1 to 21.3). The sex expression is highly influenced by chemicals, genes and environment (Fig. 21.2).

21.1 *Bryonia* L. (bryony)

This comprises four fast-growing herbaceous perennial deciduous vine species that interbreed freely and exhibit considerable g-c interactions, especially in sex type, expression and sterility. However, gc-mst inheritance for stable mst segregating in a distinct cytoplasm is not known. Fertile females occur in the progeny of *B. dioica* × *B. alba*, the reciprocal cross has mst and fst in almost equal proportions. The sterility is due to sex suppression by *B. alba* cytoplasm (Heilbronn and Basaramann 1942). Fertile males and females are recovered when either of the two species is used to pollinate the hybrid plants, although 50% offspring of BC_1 (*B. dioica* × *B. alba*) × *B. alba*, should have been monoecious. It appears that *B. dioica* cytoplasm suppresses the bisexuality-determining genes of *B. alba* genome, since F_1 progeny of the cross *B. dioica* × *B. multiflora* is monoecious, while progeny of the reciprocal cross is dioecious. However, Heilbronn (1953, 1954) assumed that *B. dioica* cytoplasm possesses bisexual potency that overbalances relatively weaker genes inducing the dioecious condition of the *B. multiflora* genome, and that *B. multiflora* cytoplasm, with a weak monoecious potency, interacts with the *B. dioica* genome and produces mst and fst plants. This assumption is contradicted, since a high percentage of nonviable seeds occurred in BC progeny plants of (*B. dioica* × *B. alba*) × *B. alba* males. It appears that probably a lethal interaction occurs between *B. dioica* cytoplasm and the nuclei containing entirely or predominantly *B. alba* chromosomes. Heilbronn and Basarmann (1942) assumed that the bisexual potency of *B. dioica* cytoplasm is suppressed by *B. dioica* nuclear genes controlling the dioecious condition, but bisexuality is expressed in the presence of less potent *B. multiflora* genes. Westergaard (1958) considers Heilbronn's assumption as limited, being based on scanty data. Also amphidiploids from the cross *B. dioica* × *B. alba* and triploids obtained by crossing these amphidiploids with *B. dioica* pollen (Brabec 1962) produced dioecious plants in both the cytoplasms. Brabec (1962) opined that either the sterilizing capacity of *B. alba* cytoplasm assumed by Heilbronn and Basarmann (1942) is operative only in the Turkish *Bryonia* species used by these authors, or that the capacity is masked by genome doubling. Interaction of fst and mst genes with each other in certain cytoplasms is evident in this species.

Table 21.1. Genic male sterility in Cucurbitaceae[a]

Species	References	Gene action	Remarks
Citrullus vulgaris	Watts (1962)	nk	X-Ray-induced, male flowers closed, anthers indehiscent
Cucumis melo	Bohn and Whitaker (1949), Chauhan and Singh (1968)	After microspore differentiation	Tapetum persistent, delayed DNA transfer from tapetum to microspores
	Bohn and Principe (1964)	During or prior to tetrad formation	
C.sativus	Shifriss (1950)	nk	♂ Flowers closed, ♀ fertility low
	Barnes (1961)	nk	Staminate bud abortion
	Groff and Odland (1963)	nk	♂ Flowers closed
	This Vol.	During floral development	
	Whelan (1972)	After pollen mitosis	γ-Ray-induced, staminate flower abortion
Cucurbita maxima	Rhodes (1959), Singh and Rhodes (1961)	Before meiotic initiation	Tapetal disintegration delayed
	Scott and Riner (1946)	nk	Staminal abortion in buds
	Francis and Bemis (1970)	Before microspore release	Tapetal development and behaviour highly abnormal
	Kalyágin (1974)	During early meiosis	
C.pepo	Eisa and Munger (1968)	Before pollen mitosis	High tapetal hypertrophy
	Shifriss (1945)	Before meiosis	Staminal abortion in buds

[a]Gene control of sterility by single recessive *ms* genes.

Table 21.2. Gene-cytoplasmic male sterility in Cucurbitaceae[a]

Species cross	References	Gene action	Remarks
Bryonia alba × *B.dioica*	Heilbronn and Basarmann (1942)	nk	mst and fst Plants segregate
Cucurbita moschata × *C.palmata*	Bemis (1973)	After microspore release	
C.moschata × *C.foetidissima*	Bemis (1970)	Before meiotic initiation	

[a]Gene control not known.

21.2 *Citrullus vulgaris* Schrad (*C. lanatus*, watermelon)

21.2.1 Genic Male Sterility

From the progeny of X-irradiated seeds of Sugarbaby watermelon variety, a completely glabrous mst mutant was found by Watts (1962). The majority of its flowers did not open, a few opened but their anthers were indehiscent and had nonviable shrunken pollen. The glabrous phenotypic expression coupled with mst is controlled by a recessive *ms* gene.

21.2.2 Chemical Male Sterility

TIBA and MH. Rehm (1952) induced complete mst following chemical spraying 1 week before anthesis in varieties Hawksbury and Cape Mountain. Other treatments, like 10 ppm every week, 50 ppm repeated once after 4 weeks, 250 ppm once, led to pollen inviability. Female fertility was also highly impaired. MH sprays inhibited vegetative growth but did not induce mst.

Mendok. Its application prevented and/or delayed the opening of staminate buds but did not impair male-female fertility (Hensz and Mohr 1959). However, pollen released after dehiscence in these treated flowers did not reach female flowers because of their delayed opening. Chemical treatment of 0.5% drastically reduced male flower number. Thus whereas the proportion of male:female flowers in untreated controls was 675:45, it was 10:31 in the treated ones. Sprays of 0.25 and 0.5% eliminated open staminate flowers.

21.2.3 Environmental Influence on Sexuality

Like other cucurbits, sexuality in watermelon is greatly influenced by environmental factors. Using two cultivars, Sugarbaby and Malali, Rudich and Peles (1976) demonstrated under phytotron conditions that day length has less effect on sex expression than temperature. Higher day temperature (32°C) reduces female floral number, while lower temperature (27°C) increases it. Under field conditions, additional edaphic and climatic factors affect the sexuality of watermelons, so that mst can be induced either through delayed flower opening or premature bud abortion.

Table 21.3. Chemical male sterility in some cucurbits

Species/chemical	References	Male sterility induction	Female fertility reduction[a]	Side effects
Citrullus vulgaris				
Mendok	Hensz and Mohr (1959)	High	b	nk
TIBA	Rehm (1952)	High	a	nk
Cucumis melo				
Mendok	Lippert and Hall (1961)	Partial	a	Flowers abnormal
	Pundir and Singh (1965)	Complete	d	nk
SADH	Halevy and Rudich (1967)	Partial	c	Plants dwarfed, tendril development suppressed
SADH + Ethrel	Rudich et al. (1969, 1970)	Nearly complete	a	nk
SADH + GA	Rudich et al. (1972)	Low	c	Reduced shoot development
C.sativus				
Ethrel	Minina and Tylkina (1942), Rudich et al. (1969, 1970), Robinson et al. (1969, 1970)	High to nearly complete	c	Reduced plant growth and deformed fruits
NAA	Iwahori et al. (1970) Augustine et al. (1973)	High to complete	c	nk
SADH + Ethrel	Laibach and Kribben (1951)	Partial	c	nk
	Rodriquez and Lambeth (1972)	Nearly complete	c	Plant growth highly reduced
Cucurbita maxima				
Ethrel	Robinson et al. (1970)	Nearly complete	b,c	nk
MH	Chauhan (1976)	Complete	b,c	nk
C.pepo				
Ethrel	Rudich et al (1969, 1970), Robinson et al (1970)	Nearly Complete	c	Plant growth reduced
	Borghi and Pironi (1976)	Complete	b	Female floret number reduced
MH	Wittwer and Hillyer (1954)	Staminate bud development suppressed	c	Female flower number reduced
NAA	Wittwer and Hillyer (1954), Hillyer (1956)	Partial	c	nk

[a] a = high, b = low, c = nil, d = not known.

21.3 Cucumis

21.3.1 *C. melo* L. (muskmelon)

21.3.1.1 Genic Male Sterility

This species is genetically variable, comprising many biotypes including melons, muskmelons and cantaloupes, all of which have mst genotypes. Some mst plants segregated in the progeny of a naturally pollinated selection with commercially desirable characteristics and resistance to powdery mildew. The sterility is controlled by a recessive gene ms_1 (Bohn and Whitaker 1949). Anthers of mst's are small, indehiscent and lodge empty microspores. Meiosis proceeds normally until tetrad formation, thereafter further development ceases and the contents of the young microspores disintegrate. The tetrads remain cemented together and the microspores contain a little cytoplasm and degenerated nucleus. Tapetal cell invasion into anther sacs causes microspore disorganisation. While in fertile plants the tapetum degenerates as microspore development advances, in mst's, tapetal cells persist, become hypertrophied and encroach upon the anther sac cavity (Chauhan and Singh 1968). This tapetal persistence starves the developing microspores, that subsequently die for want of nutrition and become crushed by the mechanical pressure exerted by the hypertrophied tapetal cells. Chauhan and Singh (1968) found that while acid phosphatase enzyme activity increases with advancing development in mft tapetal cells, the enzyme was absent in the tapetal cells of mst plants. This lack of enzyme indicates a low metabolic activity in the tapetum of mst plants. Furthermore, transfer of DNA from the tapetum to the developing microspores is considerably delayed in mst anthers. This could be another cause for mst.

A second gene, ms_2 was found by Bohn and Principe (1964) in three plants of another powdery mildew-resistant muskmelon line. This gene is nonallelic to the gene found earlier by Bohn and Whitaker (1949). Anthers of this line are small, indehiscent and lack pollen. Meiosis is normal up to TII and in majority of the PMC's, 4-nucleate tetrad formation initiates. In a few, meiosis is arrested earlier and triads with one large and two small nuclei are formed. Cell growth in almost all the PMC's becomes awry and wall formation proceeds irregularly, producing cells that vary in size and spatial arrangements. The nuclear membrane is not reorganised and chromatin fragments disperse in cytoplasm. Cell content disorganisation occurs immediately after microspore separation initiates. The disintegration of the whole PMC protoplast is rapid, as it is complete in flowers that are slightly large than those at meiosis.

21.3.1.2 Chemical Male Sterility

i. *Mendok.* Pundir and Singh (1965) obtained fn-mst after two sprays of 0.3% mendok and sporogenous sterility after two sprays of 0.4% mendok treatment. The duration of anther indehiscence, pollen sterility, the status of female fertility and the effects of the treatments are not known. Lippert and Hall (1961) reported reduction in floral number, corolla growth and fruit set following 0.2%–0.7% chemical treatment.

ii. *SADH.* Three sprays (0.1 to 0.5%) at the three-leaved stage reduced male flower number by 50% during the first 3 weeks of flowering, and doubled the hermaphrodite flower number in muskmelon (Halevy and Rudich 1967). The treated plants dwarfed and their tendrils diminished. The treatments induced incomplete mst. Female fertility of treated plants is not known.

iii. *SADH + ETHREL.* Ethrel alone causes strong reduction in plant growth. Application of successive foliar sprays of 5000 ppm SADH and ethrel at second true leaf stage inhibits male flower formation for 3 weeks of flowering in two of the four muskmelon cultivars (Rudich et al. 1970). The treatment reduces female flower number considerably. Also hermaphrodite flowers are produced on the main axis where normally only male flowers are formed (Rudich et al. 1969). Pollen and ovular fertility of these flowers is not known.

iv. *SADH + GA.* Rudich et al. (1972) studied the effect of SADH and GA (GA_3, GA_{4-7}), singly and in combinations on growth and sex expression of an andromonoecious muskmelon cultivar Ananas-PMR, by spraying the chemicals to runoff at cotyledonary, 2nd and 4th leaf stage. The treatments were given by applying 10, 50 and 250 ppm GA concentrations with a constant SADH background (1500 ppm) and by applications of 750, 1500 and 3000 ppm SADH concentrations with a uniform GA background (50 ppm). SADH treatments effectively reduced internode length and increased the female tendency; the effect of GA_{4+7} treatments was the opposite; GA_3 was less effective in this respect. In combinational treatments, the two chemicals were mutually antagonistic. However, complete mst with unimpaired female fertility was not obtained in any of the single or combinational treatments. Furthermore, whereas gibberellin treatment enhances maleness in many muskmelon varieties, in a gynoecious variety such treatments had no effect (Peterson and Andher 1960).

21.3.1.3 Environmental Influence on Sexuality

Some lines of muskmelon are monoecious, but the majority of the commercial varieties are andromonoecious, bearing female and male flowers on the same plants. The environmental influence, especially of soil moisture and nutrients, photoperiod and light intensity and day-night temperature fluctuations alters the sex ratio considerably. The response of muskmelon is similar to cucumber (see Sect. 21.7).

21.3.1.4 Breeding Value

Most commercial varieties of muskmelon are andromonoecious, producing staminate (male) and perfect (hermaphrodite) flowers. The flowers are visited regularly by bees and rarely set fruits without manipulation by these insects. Outcrossing in different varieties varies considerably and some selective fertilisation and self-incompatibility also exists in certain muskmelon varieties. Thus muskmelon is well adapted to manipulation in producing F_1 hybrid seed if the mst line can be used as one of the parents. The two mst mutants in muskmelon found by Bohn and Whitaker (1949) and by Bohn and Principe (1964) have a potential in plant breeding and F_1 hybrid production. In a BC program, they may be used as recurrent parents to transfer desirable genes from various sources into these lines without resorting to laborious hand pollinations. Selected hybrids grown in isolation blocks with mst plants would produce fruits from self-pollinated flowers, while the mst plants would produce fruits from cross-pollinated flowers. However, separate isolation block would be necessary to maintain the mst line. Using *ms* genes, a F_1 hybrid program has been formulated by Bhattacharya et al. (1970), who found superiority in fruit weight, sugar content and fruit maturity in F_1 hybrids. Likewise Foster (1968c) reported a program of hybrid fruit and seeds in musk-melon using the mst genotypes, but this has not been widely adapted.

21.3.2 *C. sativus* L. (cucumber)

21.3.2.1 Genic Male Sterility

21.3.2.1.1 Spontaneous

From the heterogenous stocks of two horticultural cucumber varieties, Black Diamond and A and C, plants having cleistogamous male flowers occurred at a frequency of 1×10^{-5} (Shifriss 1950). The pollen stainability of these flowers ranged from 30–90%. Due to cleistogamy, pollen was not available to the female flowers. Thus such plants were effectively mst. Pistillate flowers of these plants were normal, but gave highly variable fruit set after hand pollination by fertile pollen. Thus their female fertility is also impaired. The anomaly is conditioned by the recessive gene, ms_1. It may not be a true case of mst, but of meiotic mutations, which diminish the gametic fertility of both the sexes. Another recessive gene designated as cl by Groff and Odland (1963) causes cleistogamy and pollen sterility. Female fertility of this mutant plant is not known. This determination is essential, as similar plants found by Shifriss (1950) exhibited impaired female fertility. Therefore it is not certain if the mutant is truly mst. In breeding line 307, Barnes (1961) detected complete staminate bud abortion. Genetic control of this anomaly is also not known, but the author opined involvement of a dominant and a recessive gene and/or S-cytoplasm. In a F_3 line, plants exhibiting staminate bud abortion segregated. The anomaly is conditioned by a recessive gene ms_3 (this Vol.). Female fertility, as evidenced by fruit set on the mst plants, is normal. In late-grown mst plants, minute dwarf male flowers bearing rudimentary anthers develop, but no archesporial tissue differentiates in them.

21.3.2.1.2 Induced

Some mst plants in two sister lines originating from inbred progeny of a single 16 kr γ-rayed seed of the cucumber cv. Burpless Hybrid were found by Whelan (1972). In these, staminate blossoms abort, but female fertility is unimpaired. In a few plants, male flowers develop and mature to the anthesis stage but their anthers are rudimentary, lacking pollen. In such flowers, pollen aborts after mitosis. Tapetal development is normal but the cells collapse, which is followed by complete anther disintegration. The sterility is conditioned by one recessive gene, ms_2. In one mst population, a translocation ring of four chromosomes occurs. The mst is neither linked with this interchange heterozygosity, nor with other traits like nonbitter cotyledons, glabrous or yellow cotyledons and crinkled leaf (Whelan 1974).

21.3.2.2 Chemical Male Sterility

i. *CCC.* An initial increase in female flower number occurred following 500, 1000 and 2000 ppm CCC application (Ghosh and Bose 1970). Later, many male flowers developed on the treated plants. In contrast, similar treatments inhibited female flower development in *Citrullus vulgaris* var. *fistulosus* and *Luffa cylindrica*.

ii. *Ethrel.* Minina (1938) obtained considerable increase of female flowers over males by the klin method that involves sealing the greenhouse containing cucumber plants at the three to four leaf stage and introducing flue gas having up to 10% CO. After 12-h treatment, the female flower number increases over that of the males. This method was used by the author to increase fruit production. Minina and Tylkina (1942) found that ethylene enrichment produced femaleness in monoecious cucumbers. The sex-modifying effect of flue gas was not due to CO, but to its ethylene component (Heslop Harrison 1959b). This was demonstrated by McMurray and Miller (1968), who obtained only female flowers on the first 19 nodes in monoecious cucumber after 240 ppm or four applications of 120 ppm ethrel spray. Foliar sprays of 250 or 500 ppm ethrel at second and fourth true-leaf stage led to the development of nearly complete female plants in two monoecious cucumber varieties (Rudich et al. 1969, 1970). The development of male flowers was inhibited and delayed by over 2 weeks. Female flowers appeared early and profusely in plants treated with 250 ppm ethrel solution. The treatment decreased staminate flower formation, increased female or hermaphrodite flower production and led to the formation of female flowers at lower nodes. A nearly similar ethrel effect was obtained by Robinson et al. (1969, 1970), who found the complete absence of male flowers on the first ten nodes of the main shoot in the monoecious cucumber variety Wisconsin and enhancement of female or hermaphrodite flowers in the dioecious cucumber variety Lemon following 250-ppm ethrel applications. However, all ethrel treatments caused vigour reduction and fruit shape deformation. In a monoecious cucumber variety Improved Long Green, 50 mg l^{-1} ethrel sprayed at second and third leaf stage induced almost complete mst through inhibition of male flower development without affecting total female flower number up to 20 nodes (Iwahori et al. 1970). Thus in controls, 74 male flowers were produced, but in the treated ones only two developed. The effect of the chemical on fruit size, shape and production is not known. Likewise, Augustine et al. (1973) obtained pistillate flowers after ethrel spray on an androecious (all male) muskmelon line. Thus the chemical caused a sex expression reversion of androecious to monoecious phenotype that could permit seed increase through self- and sib-pollinations. In this variety, optimum ethrel concentration for maximum pistillate flower production under greenhouse conditions is 50 ppm. The stage of application is highly critical, as twice as many pistillate flowers developed after 50 ppm application at the three- or four-leaf stage than at the two-leaf stage.

iii. *Gibberellins.* The action of gibberellins on cucumber sexuality is dependent upon genotype, plant development stage, application site and chemical concentration. Gibberellin applications promote staminate flower formation in gynoecious, gynomonoecious and monoecious cucumbers (Peterson and Andher 1960; Wittwer and Bukovac 1962; Fujieda 1963; Saito and Ito 1963; Robinson et al. 1969). The promotion is strongest in gynoecious, weaker in gynomonoecious and weakest in monoecious plants (Bhattacharya and Tokumasu 1970). Heterozygous female cucumbers which normally produce female flowers in the upper nodes of the main shoot and in the side branches could be converted by repeated GA_3 treatments into plants having the flowering pattern of a normal monoecious plant with well-developed male flowers (Galun 1959). Four applications at 2-day intervals of 250 ppm in one and 500 ppm GA in the other gynoecious variety of cucumber suppressed male flower formation. In two other varieties, male flowers increased dramatically after such GA applications. Thus, whereas GA applications enhance maleness in many cucumbers, in some femaleness is enhanced. This differential response may be due to genotypic differences, application mode, site etc.

iv. *NAA.* Application of 0.1% chemical in a lanolin paste to the cut surface of petioles of the first two leaves produced a high female flower number in the first seven leaf axils of cucumber (Laibach and Kribben 1951). This indicates the female-promoting tendency of this auxin; however, complete staminate flower suppression did not occur.

v. *SADH + Ethrel.* Foliar spray of SADH (2000 ppm) in combination with ethrel (100 ppm) sprayed to run off three times at 7-day intervals beginning at the second leaf stage led to almost complete suppression of staminate flowers and a dramatic enhancement of pistillate flowers in a monoecious *C. sativus* cultivar (Rodriquez and Lambeth 1972). Ethrel treatment (100 ppm) was the next most effective treatment. SADH treatment singly and in combination with ethrel reduces plant size by half and concentrates maturity, facilitating single mechanical harvesting.

21.4 *Cucurbita* L. (pumpkin, squash, yellow gourd)

Of the four cultivated species, *C. maxima, C. moschata, C. mixta* and *C. pepo*, g-mst is known in *C. maxima* and *C. pepo* which includes many varieties of pumpkins, gourds and winter turban and narrow squashes (Table 21.1).

21.4.1 *Cucurbita maxima* Duchesne (winter squash)

21.4.1.1 Genic Male Sterility

In the F_3 family of a cross between two inbred monoecious lines, a sterile plant occurred with blossoms having small, stiff petals that were ruffled at the tips (Hutchins 1944). Stamens of this plant were small with poorly developed anthers that lacked pollen. Although female flowers appeared normal, the plant was also female sterile. Natural crosses and hand pollination produced parthenocarpic fruits lacking embryos. The sterility is controlled by a single recessive gene. Being male-female sterile, the mutant cannot be considered as mst. Rhodes (1959) and Singh and Rhodes (1961), apparently utilising these plants for developing an interspecific gene pool, detected mst in the progeny of some crosses. In these, PMC's do not undergo meiosis, but instead the sporogenous cells enlarge, become vacuolated and their nuclei become hypertrophied. These cells disintegrate immediately thereafter; this is followed by tapetal degeneration. In some anthers, PMC's develop a pollen-like wall and then degenerate. In this mutant, the *ms* gene acts before meiotic initiation; the anomaly is conditioned by one recessive gene.

In another mst mutant, staminate buds appeared phenotypically normal but at anthesis the androecia were shrivelled and necrotic (Francis and Bemis 1970). Male meiosis proceeds normally up to the tetrad stage, but microspores degenerate during maturation. The sterility is monogenic recessively controlled. Whereas tapetal cells in normal plants remain one cell-layer thick, in mst's they undergo additional cell division, increase to double size and expand into the anther locule. The abnormal tapetal behaviour has been considered by Francis and Bemis (1970) as the prime cause of microspore destruction, although both microspore and tapetum degeneration are almost simultaneous. The mst mutant of Singh and Rhodes (1961) and the mutant of Francis and Bemis (1970) when intercrossed (each in the heterozygous state) segregate in 1:1 mst:mft, indicating two mutant genes as alleles at a common locus. Since the segregating populations were not examined meiotically, the allelic dominance through a study of *ms* gene action could not be determined.

In seven different lines of *C. maxima*, all of which had Golden Hubbard and Tapley variety parentage, mst plants segregated (Scott and Riner 1946). In them the androecium aborted in the bud stage before staminate flower opening and no viable pollen was formed. The sterility is controlled by a recessive *ms* gene. Three mst plants were detected in the Spanish *C. maxima* var. Spanskaya by Kalyágin (1974). In these, male meiosis is highly abnormal and pollen grains are sterile. Sterility in each is conditioned by a recessive gene. The allelic nature of these genes is not known. Some local varieties of *C. maxima* developed abnormal androecia lacking pollen grains, or lost the stamens completely when grown in the greenhouse, but were mft when grown in the field. (Shifriss 1945). In them mst appears to be induced by the high temperature prevalent in the greenhouse.

CHAPTER 22
Ericaceae

22.1 *Vaccinium angustifolium* Ait. (blue berry, bill berry)

As a source of excellent wild berries, this genus consists of 30 species common in the Himalayas and N. America, but mst plants were detected in only three clones out of 24 studied by Hall and Aalders (1961). All these have dwarf stamens and shrunken pollenless anthers. Hand pollination yielded 80–90% seed set, indicating normal female fertility. The mst is controlled by multiple genes whose allelic nature is not known. Cytological studies reveal differences between the three clones as far as *ms* gene action is concerned (Table 22.1). The gene action is influenced by environment. In clone Q_2, microspores and tapetal cells degenerate more or less simultaneously and no pollen is produced at all. Obviously the *ms* gene operates at the end of microsporogenesis. On the other hand, in clone N_{68}, degeneration sets in immediately when the microspore tetrads are just formed and a dark-stained mass is produced. Both the middle wall layers and tapetal cells also dissolve out in it. In the TH_8 clone, PMC's development is normal up to tetrad formation, after which microspore degeneration occurs very briskly and not a single viable pollen is produced when the flowers open out. The other anther tissues, including the tapetum, remain intact.

Table 22.1. ms gene action in *Vaccinium*

Species	References	Gene control	Gene action
V.angustifolium (3 clones)	Hall and Aalders (1961)	Many r	After microspore release After tetrad formation During tetrad formation

CHAPTER 23
Euphorbiaceae

From about 300 genera included in this family, genetic mst occurs in *Hevea, Manihot, Mercurialis* and *Ricinus* (Table 23.1). In all these, the influence of environment over mst expression is high.

Table 23.1 Genetic male sterility in Euphorbiaceae

Species	References	Gene action	Remarks
Genic			
[a]*Hevea brasiliensis*	Ramaer (1935)	During MI or microspore formation	Tapetum normal
[a]*H. sprucaena* × *H. brasiliensis*			
[a]*Manihot esculenta*	Jos et al. (1967),	After tetrad formation	Tapetum normal
	Magoon et al. (1968)	After microspore release	Early tapetal degeneration
	Jos and Bai (1981)	During anther dehiscence	Persistent tapetum
Mercurialis perennis	This Vol.	During early prophase I	
Ricinus communis	Claassen and Hoffman (1950)	During male flower initiation	Environmental effect high
Gene-cytoplasmic			
Mercurialis annua	Louis and Durand (1978)	During male flower development	two complementary *Fr* genes present
Ricinus communis	Parkey (1957)	During male flower initiation	Male flowers absent

[a]mst type not confirmed and gene control not known; in others 1r gene control.

23.1 *Hevea brasiliensis* Mell (rubber tree)

Of the 11 known species of this genus, spontaneous occurrence of mst plants is known in *H. brasiliensis*, the source of nearly all commercial natural rubber. Ramaer (1935) found three mst clones of this species in which male meiosis proceeded normally until MI, during which chromosomal organisation, congression and orientation of bivalents to the spindle became highly disturbed. The subsequent meiotic course was abnormal and led to nonviable microspore formation. This was followed by complete anther degeneration. Nothing is known about the genetic control of the anomaly or about the status of female fertility. Similar meiotic disturbances occurred in the hybrid plants of the cross *H. sprucaena* × *H. brasiliensis*, but unlike in the three above-mentioned clones, microspore abortion was not followed by anther disintegration. Instead, anthers of the hybrid plants looked phenotypically normal, although they had aborted pollen (Ramaer 1935). Genetic control of this anomaly and the status of female fertility of the hybrid plants are not known. Also, the pollen fertility of hybrid plants of the recriprocal crosses is not known. Due to the lack of this information, these clones and hybrid plants exhibiting abnormal meiosis cannot be considered as typical mst plants.

23.2 *Manihot esculenta* Crantz. (*M. utilissima,* cassava, tapioca, yucca)

Normally, cassava plants are monoecious, the female flowers being at the base and the males at the top of the inflorescence. In six mst cassava lines, while the majority of the male flower buds abscised prematurely, others had small narrow stunted and completely empty anthers (Magoon et al. 1968). Their female flowers were normal and fertile. Thus these lines are true mst's. Male meiosis proceeds normally up to microspore formation. Thereafter, the microspores fail to disjoin from the tetrads and degenerate subsequently. The tapetum also disintegrates subsequently. In other mst material, microspores separate normally from the tetrads, but the tapetum degenerates much more abruptly. The mature pollen grains gradually abort, probably due to nutritional disparity because of the premature tapetal breakdown. Genetic control of the sterility in both the mst materials is not known. Involvement of two different *mst* genes appears possible because of the difference in gene action over microsporogenesis.

In a seedling population derived from a tetraploid clone, a functional mst plant arose (Jos and Bai 1981). Its anthers were indehiscent although it had 73% stainable pollen. During microspore development and pollen grain formation, its tapetum remained intact, unlike in mft plants. Furthermore, the endothecium did not develop thick radial walls and fibrous thickenings. This inhibited anther dehiscence and pollen release and led to functional mst.

23.3 *Mercurialis* (herb mercury)

This N. American and temperate Asian and European genus possesses eight species, two of which, *M. annua* and *M. perennis*, exhibit mst (Table 23.1).

23.3.1 *M. annua* L. (mercury weed)

In a population of this dioecious weed, Louis and Durand (1978) found a spontaneously arisen mst mutant whose anthers were reduced and lodged shrunken abortive pollen (Fig. 23.1b,d) in contrast to normal mft's that have male flowers in rounded clusters forming an interrupted spike and producing well-filled fertile pollen (Fig. 23.1a,c). The spontaneously arisen mst mutant was propagated

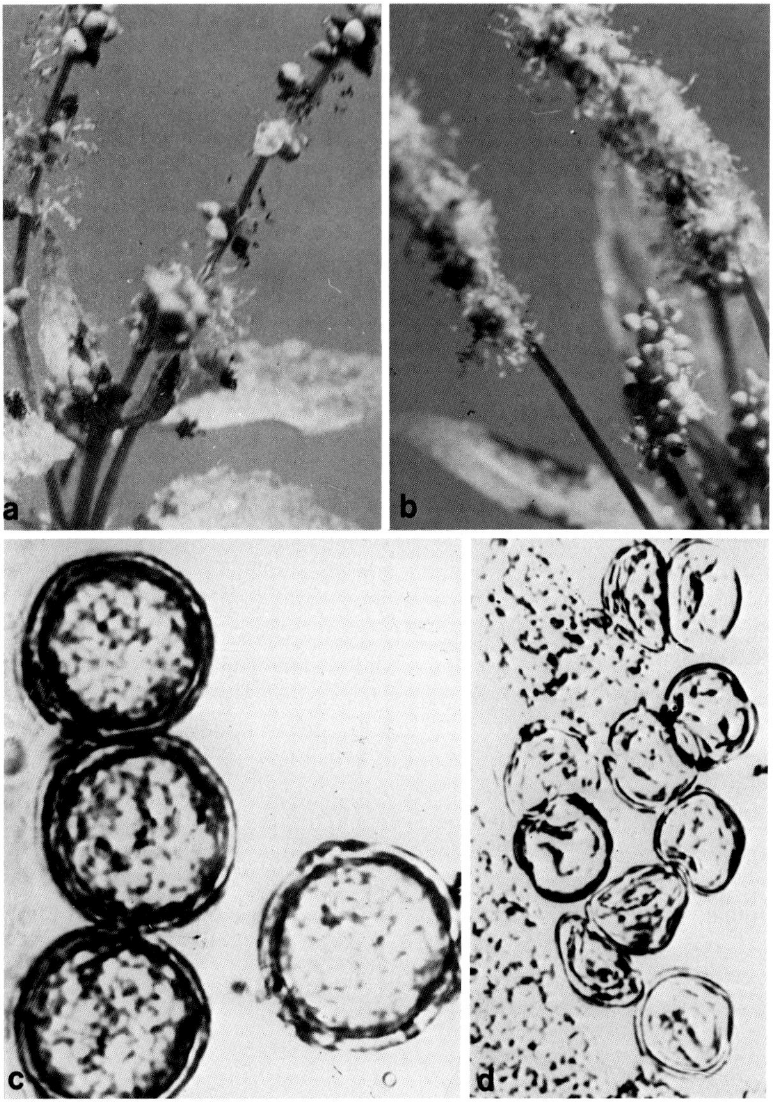

Fig. 23.1a-d. *Mercurialis annua*. Male fertile **(a)** and male sterile **(b)** plants having fertile **(c)** and sterile **(d)** pollen grains. (Louis and Durand 1978, Louis 1983)

vegetatively. By exogenous hormone treatments, this mst mutant was feminized and then used as a female by Louis and Durand (1978). Apart from the usual crossings, such conversions permitted self-fertilisations, reciprocal crosses between mst's and crosses with genetically mst individuals. When the mutant was feminised and fertilised by wild fertile males, the F_1 progeny included fully or partially restored mst's. This indicates that the sterility affects only the male organs. Using these constructed semi-sterile strains and the fully restored ones, it was found that mft restoration depends on the interaction of the S-cytoplasm with a system of three dominant nuclear genes, *I*, R_1 and R_2. In the S-cytoplasm, gene *I* induces sterility, R_1 confers partial fertility restoration and R_2 complements R_1 to induce full fertility restoration, but is inactive in itself. On the basis of this genetic interpretation, the inheritance pattern and genetic behaviour of the original total sterile strain and of the constructed, restored and semi-sterile strains used by the authors is satisfactorily explained. Furthermore, the sex segregations in crosses between mft stains selected for their various degrees of sensitivity towards feminizing hormones made it possible to explain sex determination and male plurality by a system of three complementary genes. Crosses between these strains indicate that the cytoplasm, in association with the R_1 gene, influences the sex distribution of this species. Comparative data between exogenous cytokinin levels, the phenocopies obtained by feminizing hormones and various crosses made by Louis and Durand (1978) reveal that all these strains constitute a series of male developmental mutants starting from strong males and continuing through to weak males, semi-steriles and mst's, and then finally to pure females in this dioecious species. An analysis of various crosses made by the authors demonstrated the existence of an unambiguous relation between cytokinin sensitivity of male and their fertility-restoring capacity. Genetic studies confirm that the mst is closely related to cytokinin sensitivity and probably to the hormonal level of a genotype. Since *M. annua* strains comprise a series of sex developmental mutants (Kahlem et al. 1975), these sterile strains may indicate the sequence of events that lead to mst. Louis and Durand (1978) found cytoplasmic genes also involved in mst control. If one of the two *Fr* genes is present in the dominant state and is carried by a female sporophyte, the progeny contains only males. Thus it appears that there is an interaction between female embryos and maternal female tissues. Probably female embryos could be lethal when they are surrounded by nucellus or endosperm tissues carrying both R_1 and *c* genes.

23.3.1.1 Biochemical Control of Sex Type

By exogenous cytokinin applications, pollen abortion is induced in mft mercury weed plants. The restoration power of the males is closely linked to their insensitivity to these hormones. Minimum restoration characterised by the reappearance of totally sterile males occurs in the progeny when pollen of most cytokinin-sensitive males is used, while maximum restoration occurs with less sensitive ones. These findings indicate cytokinin involvement in the mst expression of *M. annua*. Compared to each other, whereas cytokinin levels are higher in mst's, auxin levels are higher in fst's (Dauphin-Guerin et al. 1980). The relative amount of growth regulators determines the sex type in *M. annua*, and sex conversion in it is easily achieved through exogenous hormonal applications. While cytokinins induce the development of female flowers on male plants, auxin application produces male flowers on genetic females (Kahlem et al. 1975). These transformations are achieved through concomitant synthesis of certain specific macromolecules. A group of anodic isoperoxidases constitutes an early marker of stamen differentiation (Kahlem 1976) and a group of specific isoesterases, two subspecies of chloroplast tyrosin t-RNA's and one leucyl-t-RNA synthetase are specific for carpel differentiation (Bazin et al. 1975).

The synthesis of these sex markers appears to be hormone-level-dependent because free base zeatin is present only in the whole plant and in the differentiating apical tissues of females, and is absent in the corresponding tissues of the males, which instead have its nucleotide (Dauphin-Guerin et al. 1980. Similarly, only the females of *M. ambigua* (2n = 48) accumulate zeatin, while it is absent in the males (Dauphin *et al.* 1979). This high zeatin concentration and low level of its nuceotide are related to the absence of two dominant complementary genes determining maleness in *M annua* (Louis and Durand 1978). When one of the two genes is not dominant, the zeatin quantity decreases. Since the riboside, isopentaenyl-adenosine is present in plants of both the sexes, and as it is basic to cytokinin metabolism (Hall 1973), two higher levels of enzyme activity, a hydrolase in females and a phosphokinase in males, would account for the accumulation of zeatin in females and its nucleotide in males. Although this differential accumulation of nucleotide or free base reveals a quantitative difference in common metabolites, it is equally possible that males contain a base which is similar, but not identical to zeatin. However, the higher binding constant for zeatin sites in females probably saturated with endogenous zeatin (Chung et al. 1979), coupled with the data on endogenous metabolism (Dauphin-Guerin et al. 1980), indicate that the active form of cytokinin involved in female sex differentiation of *M. annua* is zeatin. Furthermore, one leucyl-t-RNA and two tyrosyl-t-RNA species are absent in male but present in female flowers (Kahlem 1976). In addition, female flowers have isoesterases and one antigen, but these are absent in males which specifically possess three isoperoxidases (Kahlem 1973; Durand et al. 1974). Since these specific male isorperoxidases are characteristic of the process of microsporogenesis and tapetal differentiation, and their synthesis starts at an early stage of staminogenesis, Kahlem (1976) regards these as the early markers of male flower differentiation and staminogenesis.

23.3.2 *M. perennis* L. (dog's mercury)

From a seed stock obtained from the FRG, two mst's segregated in which male blossom aborted prematurely (this Vol.). Their PMC's degenerated during early prophase I and subsequently the male flowers dropped off from the plants. The anomaly is controlled by one recessive *ms* gene.

23.4 *Ricinus communis* L. (castor bean, castor oil plant)

23.4.1 Genic Male Sterility

This monospecific genus of the spurge family, a native of Africa and naturalised throughout the warm parts of the world, is of great commercial and horticultural value. Its normal inflorescence is a monoecious raceme bearing pistillate flowers in the upper 30–50% and staminate flowers in the lower 50–70% part. Although usually a definite demarcation line exists between the staminate and pistillate flowers, many variations from the normal raceme were found by Claassen and Hoffman (1950). One of the extreme variations was the development of pistillate flowers and complete absence of staminate flowers upon inflorescences. This is governed by a recessive gene whose expression is influenced by the environment and by the modifying genes. Proposals to utilise such plants for hybrid seed production were made by the authors.

23.4.2 Gene-Cytoplasmic Male Sterility

In one line, mst is conditioned by the interaction of the S-cytoplasm with fr_1fr_1 genes (Parkey 1957). The dominant *Fr* gene designated *P* by the author is the fertility-restorer gene. This gc-mst has potential in hybrid production, as a *Fr* gene for the sterility restoration and maintenance exists. However, the restoration is sensitive to modifiers and environment. As an oil-yielder, the castor plant is gaining in importance, and mst for hybrid production could profitably be used and exploited on a commercial basis.

23.4.3 Chemical Male Sterility

Some GA treatments induce the suppression of male flower development and/or their abortion (Shifriss 1961b) but these treatments reduce female flower number, increase internode length and delay flowering and fruiting considerably. Many fruits formed are of abnormal shape.

23.4.4 Breeding Value

Two methods for using mst plants in commercial hybrid production were proposed by Claassen and Hoffman (1950). One method utilizes lines that breed true for 90–100% mst flowers, the other utilizes 100% mst plants in lines segregating heterozygous monoecious: dioecious plants. Use of the latter method would require roguing of heterozygous monoecious plants in crossing blocks before flowering begins. However, keeping in view high positive and superior parent heterosis, use of *mst* genes in hybrid production is highly desirable in castor oil plants.

CHAPTER 24
Fagaceae

24.1 *Castanea* Tourn (Chestnut)

This develops small unisexual flowers with both the sexes on the same tree or bush. Spontaneously arisen mst occurs in Japanese chestnut (*C. crenata*), American chestnut (*C. dentata*) Chinese chestnut (*C. mollissima*) Chinquapin (*C. pumila*) and *C. sativa* (this Vol.). In these, the anthers are rudimentary and lack PMC's. Genetic control of the sterility is not known, but the action of *ms* or *fr* genes is evident (Table 24.1). Jaynes (1963) reported failure of staminate primordia to develop anthers in some *C. mollissima*, *C. sativa* and many chestnut hybrid trees. Their female fertility was normal. Although the exact cause and control of sterility are not known, the sterility in the species is probably genetic or physiological, as only a few trees (one to three) out of a huge stand exhibited sterility, others growing side by side being perfectly normal. Omura and Akihama (1980) found mst plants in clones and cultivars of *Castanea crenata*, *C. mollissima*, *C. sativa*, *C. sequinii* and their interspecific hybrids. The sterility was manifested as: (i) antherless florets (ii) premature male floret fall and (iii) nonviable pollen. Some F_1 plants of *C. crenata* × *C. sativa* are also mst, as their florets develop staminodes.

Table 24.1. Male sterility in *Castanea*[a]

Species	References	Gene action	Additional effects
C.crenata *C.dendata* *C.mollissima* *C.pumila* *C.sativa* *C.seguinii*	Jaynes (1963) Omura and Akihama (1980) This Vol.	Before stamen differentiation or during premeiosis	Either male flower suppression, or sporogenous mst induction

[a]Gene control not known.

CHAPTER 25

Geraniaceae

.Five genera having 750 species comprise this cosmopolitan family. Of these, mst is known in *Geranium* and *Pelargonium* (Table 25.1) only.

Table 25.1. Genetic male sterility in Geraniaceae

Species	References	Gene action
Genic male sterility		
Geranium argenteum	This Vol.	During PMC development
G.himalayense[a]	This Vol.	After microspore liberation
G.meeboldii[a]		
Pelargonium crispum	Tokumasu (1974, 1976)	After microspore release
P.denticulatum	Tokumasu and Tamai (1964)	During prophase I
P.radula	Conagin (1964)	After microspore release
P.tomentosum[b]	This Vol.	After microspore release
P.hortorum	Harney and Kung (1967)	After microspore release
Gene cytoplasmic male sterility[a]		
Geranium albiflorum × *G.pratense*	Sansome (1936)	nk
G.endressi × *G.striatum*		
Pelargonium denticulatum × *P.roseum*	Tokumasu 1974	nk

Gene control by 1r[a] or 1–2r[b], in others not known.

25.1 *Geranium* L.

25.1.1 Genic Male Sterility

Sterile pollen grains occur in some plants of *G. himalayense* (*G. grandiflorum*) and *G. meeboldii* (this Vol.). Their female fertility is normal. Male meiosis proceeds normally until microspore liberation, after which microspores degenerate. This is followed by tapetal degeneration. The sterility in each case is conditioned by one recessive gene. In the dwarf tufted plants of a strain of *G. argenteum* introduced from Italy, the anthers were deformed and PMC's degenerated soon after their formation (this Vol.). Tapetum developed normally, but disintegrated after PMC degeneration. Female fertility could not be determined, as no mft plants were available as pollen donors.

25.1.2 Gene-Cytoplasmic Male Sterility

In the F_1 progeny of a cross between *G. endressi* × *G. striatum*, a single mst plant appeared. From BC's to *G. striatum* parent, Sansome (1936) postulated that a homozygous recessive gene of *G. striatum* controls mst in combination with *G. endressi* cytoplasm. Analysis of mst in many other interspecific hybrids of *Geranium* indicates a g-c interaction as the cause of sterility, in which two *fr* genes in combination with the S-cytoplasm lead to mst. Whereas *G. endressi* used by Sansome had S-cytoplasm and was heterozygous for *fr* genes, *G. striatum* had N-cytoplasm and was homozygous for recessive *fr* genes.

25.2 *Pelargomium* L.'Her.

This is a horticulturally important genus comprising about 285 species, many having a multitude of hybrids, cultivars, cultigens and horticultural types. Hybridizations among cultivars, varieties and species are frequently high. This makes species delimitation arduous mst is common in *P. crispum*, *P. denticulatum*, *P. radula*, *P. roseum* and *P. tomentosum* and in some of their interspecific and intervarietal hybrids.

25.2.1 *P. crispum* L.'Her.Ex.Ait. *(Geranium crispum*, lemon-scented geranium)

Its certain clones were fully mst, while those of Lemon Crispum were mft (Tokumasu 1974). Their F_1 plants had partially fertile and sterile anthers in varying proportions. While sterile anthers have only abortive pollen, fertile anthers contain both carmine stainable normal and small sized (Fig. 25.1) and nonstainable small-sized pollen in varying proportions (Tokumasu 1974, 1976). The functionability of carmine-stained fertile-looking pollen is not known. Although the precise nature of sterility is unknown, it is obviously gc-mst, as after using Prince Rupert as female parent, mst offspring appeared regularly, unlike in two reciprocal crossings. The environmental influence over the sterility is profound, as an array of divergent pollen phenotypes occur. Moreover, the sterility percent varies from anther to anther within the same flower and from flower to flower within the same plant.

As mentioned earlier, the majority of plants of cultivars Crispum Minor and Lemon Crispum produce functional pollen, but some are mst, developing sterile indehiscent anthers in varying proportions. Even fertile-looking anthers developed by some of these steriles contain abortive pollen in varying proportions. Among the anthers developed in the same flower, microspore breakdown and tapetal anomalies occur in some of them, whereas in others, pollen grains develop and are liberated normally. Among anthers borne on the same flower, in some, persistence of the partition between anther loculi occurs. In some, development of endothecia and lip cells is retarded and microspores abort. Penetrance and

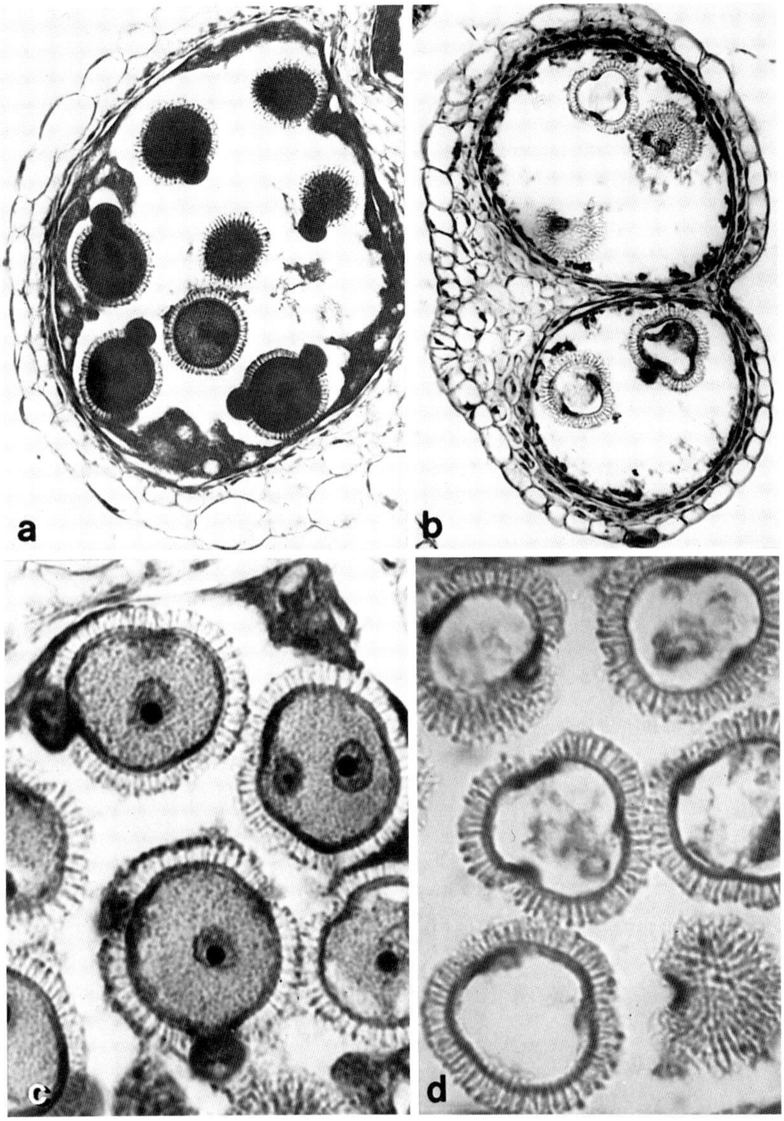

Fig. 25.1a-d. *Pelargonium crispum.* T.S. of mature anthers **(a,c)** and whole mount of pollen **(b,d)** of male fertile **(a,b)** and male sterile **(c,d)** plants. (Tokumasu 1976)

expressivity of the *mst* gene is incomplete, and its expression is sensitive to the environment and physiological condition of the plant. In the mst's, the nutrient supply to the developing microspores becomes retarded. Where retardation is complete, microspore abortion is total, and where it is meagre, abortion is partial. This causes a difference in microspore sterility among different anthers of the same flower.

25.2.1.1 Microsporogenesis

Meiosis proceeds normally in the PMC's of mst and mft cultivars and the microspores undergo an almost similar process of initial development until their cell walls are formed. Thereafter, unlike those of the fertile microspores, the cytoplasm of the steriles becomes congregated into a masslike structure, and further development stops in them. The majority of the microspores have only one nucleus, but a few have two, indicating the occurrence of mitotic division in the latter. This is followed by protoplasmic degeneration, leading to empty pollen grain formation (Tokumasu 1974, 1976).

25.2.1.2 Tapetal Behaviour

The tapetum of mft Lemon Crispum and Prince Rupert is basically of amoeboid type (Tokumasu 1976). In the latter, the cells form plasmodial masses which fuse to produce the periplasmodium, and finally degenerate. In mst anthers, endothecium and lip cells develop meagrely and are absent in many fibrous thickenings of their cell walls. Unlike in the fertiles, in the steriles, the partition between two neighbouring loculi does not break and both the endothecia of the same anther lobe remain connected with each other at the site of the stomium. Owing to reduced tension of the endothecium, each loculus becomes cramped and looks sickle-shaped. In such loculi only empty and shrunken pollen grains are formed. The endothecia never turn inside out, so that the anthers do not dehisce at the flowering time. These anthers finally dry up.

25.2.2 *P. denticulatum* Jacq.

In mst plants of this pine-scented species, male meiosis proceeds abnormally (Tokumasu and Tamai 1964) and nonviable shrunken pollen develops. The status of female fertility and genetic control of sterility are not known.

25.2.3 *P. radula* L.'Her. (P. multifidum Salisb.)

Conagin (1964) found some mst's in this aromatic pungent-scented species. In these, male meiosis proceeds normally, but during microspore maturation the tapetum loses its cell walls, forms plasmodial masses, enlarges rapidly and buries the microspores deep into the periplasmodial mass. The periplasmodium crowds the microspores together like pollinia, preventing them from dispersing from the anthers.

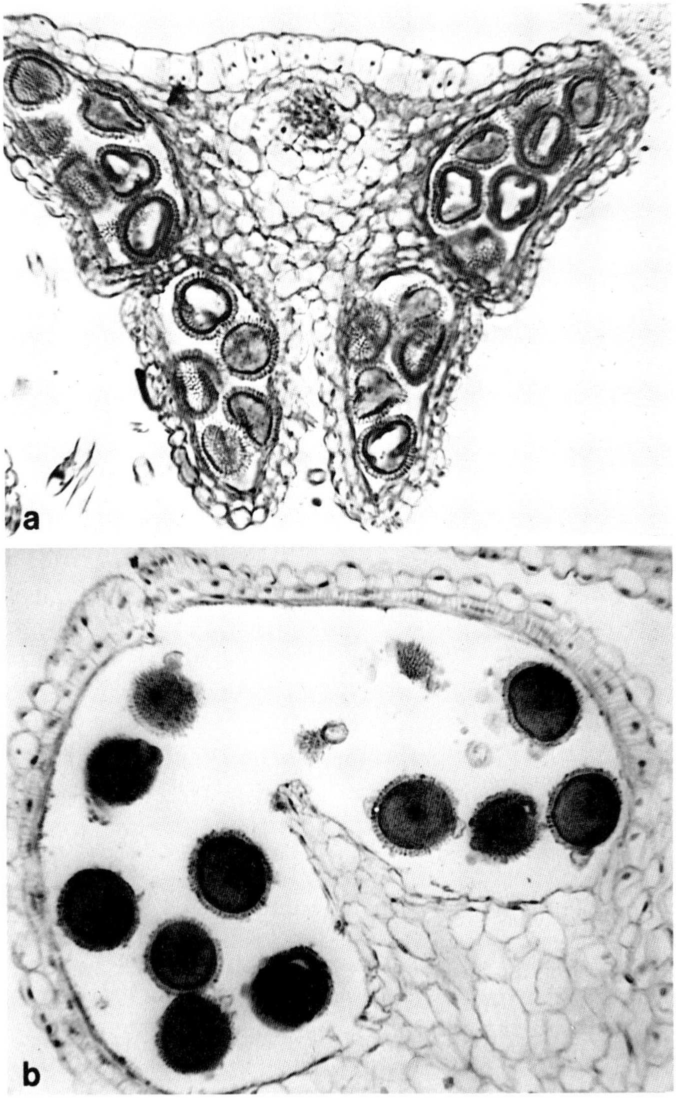

Fig. 25.2a,b. *Pelargonium roseum*. T.S. of anthers of 2x male sterile **(a)** and of 4x male fertile **(b)** plants. (Tokumasu 1970)

25.2.4 *P. roseum* R. Br. (Willd)

These plant, with 2n = 77, develop nondehiscent anthers, lodging abortive pollen (Fig. 25.2a). By treating leaf buds with 0.3% colchicine, Tamai et al. (1958) developed fertile colchiploids (2n = 154). The flowers of the colchiploids develop two kinds of anther, viz., large rose-coloured dehiscent and tiny yellow-coloured indehiscent anthers. Whereas the former have fertile functional and only a few sterile pollen grains (Fig. 25.2b), the latter have only abortive pollen (Tokumasu 1976). The meiotic course in the noncolchiploids ($2n = 77$) is not known, but microspores are formed in them (Tokumasu 1970). During exine development, their cytoplasm degenerates and the microspores abort almost simultaneously. In colchiploids, pollen formation is normal. Tapetal development is normal and similar in plants with $2n = 77$ and their colchiploids. The tapetal cells increase in size during the tetrad stage and form the periplasmodium during microspore development. Whereas this periplasmodium disappears abruptly and completely in noncolchiploids ($2n = 77$), in the colchiploids its disappearance is gradual. Neither female fertility nor male meiosis of the noncolchiploids is known. Therefore it is not certain if such plants are female fertile. An odd chromosome number is indicative of male-female gametic sterility, the fertility being restored by chromosome doubling.

25.2.5 *P. tomentosum* Jacq.

Out of 43 plants, two plants of var. *variegatum* ($2n = 44$) had broadly edged leaves with a creamy white margin with patches all along the lamina and nondehiscent anthers (this Vol.). Their microspores degenerate after liberation from the tetrads mostly prior to pollen mitosis. About 12-19% of the pollen grains stained with aceto-carmine, but no seeds were set upon selfing. Thus this pollen is nonfunctional. Hand pollination gave normal seed set, indicating unimpaired female fertility. The sterility is conditioned by one to two *ms* genes, but the exact gene number could not be determined due to tetrasomic segregations and environmental influence.

25.2.6 Sterility in *Pelargonium* Hybrids

i. *Intervarietal Hybrids.* A form of p-mst occurs in some plants of the semi-double commercial cultivar Jacqueline of *Pelargonium hortorum,* Bailey, the so-called commercial geranium (Harney and Kung 1967). The sterile plants have brown and shrunken anthers that lodge empty and shrunken microspores. Unlike in fertiles, in steriles, the tapetum remains intact and persistent. The sterility does not appear to be gc-mst, as crosses of p-mst Jacqueline to mft Jacqueline and other cultivars yielded mft progeny. However, possibly the male parents used carried *Fr* genes in a heterozygous state. Nugent and Snyder (1967) found a series of mst types in *P. hortorum* and believed it to be gc-mst.

ii. *Interspecific Hybrids.* From a natural cross of *P. denticulatum* × *P. roseum,* p-mst hybrids were detected by Tokumasu (1974). Although male meiosis in them was highly abnormal, some stainable pollen developed. Its functionability, however, is not known. The status of female fertility, genetic control of sterility and whether the sterility is maintained in the reciprocal cross hybrids or not are also not known. F_1 progeny of *P. denticulatum* ($2n = 90$) × *P. roseum* ($2n = 72$) is male-female sterile (this Vol.).

iii. *General.* The rampant occurrence of pollen sterility in *Pelargonium* led Tokumasu (1974) to remark that the trait is genetically fixed in a few species of this genus. Some of the pollen produced by these species, cultivars, varieties and their hybrids is stainable with carmine but it is nonfunctional. This also occurs in *Artemisia maritima* (Kaul 1965). If all the stainable pollen produced by mst *Pelargonium* is nonfunctional, much of the intra-, and interplant and floral variability in pollen stainability found by Tokumasu (1974) would vanish. On the other hand, if the stainable pollen proves functional, the p-mst in this genus is rewarding for plant breeders, as it provides them with an array of diverse pollen sterile lines with high chromosomal and genetic variability of breeding potential. The status of female fertility and the genetic control of sterility are not known in many mst *Pelargonium.* The sterility appears genic, as it is constantly maintained by the plants and their clones. Moreover, in almost all the mst genotypes, microspores abort after liberation from the tetrads, and wherever the tapetal behaviour has been studied, the *ms* gene acts via the tapetum by altering its form or function. *Pelargonium* outcrosses frequently and propagates vegetatively. The majority of the present-day pelargoniums are therefore complex hybrids. Thus, whether the recorded mst is true or is both male-female sterility has to be determined by systematic direct and reciprocal crossings. Moreover, the influence of the cytoplasm has to be determined by substitution crossings.

CHAPTER 26
Gesneriaceae

26.1 *Kohleria* Regel

With its tremendous unexploited economic potential as an ornamental pot plant in the florist's trade, because of its showy flowers and ease of vegetative propagation, obtaining hybridity through mst is useful in this genus. On the basis of F_1 and BC generation segregations, Clayberg (1974) proposed three genetic systems based on the assumption that *K. amabilis* and *K. tubiflora* possess S-cytoplasm and that the mst is conditioned by a recessive gene, *ms*, in *K. amabilis* × *K. bogotensis*, by a dominant gene, *Ms-2*, in *K-tubiflora* × *K. digitaliflora* and by two complementary dominant genes, *Ms-3* and *Ms-4*, in *K. tubiflora* × *K. bogotensis* (Table 26.1). In the compactly growing large-flowered ornamental *K. amabilis*, *fr* genes were found by Clayberg (1974) in the *K. bogotensis* genome. Their detection offers an opportunity to breed novel types of attractive compact flowering ornamentals that can be maintained sexually and asexually.

Table 26.1. Male sterility in Gesneriaceae

Species/cross	References
Kohleria amabilis × *K.bogotensis*[a]	Clayberg (1974)
K.tubiflora × *K.bogotensis*[a]	Clayberg (1974)
K.tubiflora × *K.digitaliflora*[b]	Clayberg (1974)
Streptocarpus cooksonii	Story (1955)
S.cyaneus	Story (1955)
S.kentaniensis	Story (1955)
S.primifolius	Story (1955)
S.comptonii × *S.grandis*	Oehlkers (1938, 1940 1941)
S.comptonii × *S.rexii*	Oehlkers (1938, 1940, 1941)
S.comptonii × *S.soleanthus*	Oehlkers (1938, 1940, 1941)
S.wendlandii × *S.baudertii*	Oehlkers (1938, 1940, 1941)
S.wendlandii × *S.grandis*	Oehlkers (1938, 1940, 1941)
S.wendlandii × *S.meyeri*	Oehlkers (1938, 1940, 1941)
S.wendlandii × *S.michelmorei*	Oehlkers (1938, 1940, 1941)
S.wendlandii × *S.montigena*	Oehlkers (1938, 1940, 1941)
S.wendlandii × *S.rexii*	Oehlkers (1938, 1940, 1941)
S.wendlandii × *S.soleanthus*	Oehlkers (1938, 1940, 1941)
S.wendlandii × *S.vandeleuri*	Oehlkers (1938, 1940, 1941)
S.soleanthus × *S.rexii*	Oehlkers (1964)
S.sp. × *S.soleanthus*	Oehlkers (1964)
S.sp. × *S.johannis*[b]	Zeven (1972a)

Gene control [a] = 1r, [b] = 1D, [c] = 2D, in others not known.

26.2 *Streptocarpus* Lindl. (Cape Brimrose)

26.2.1 Genic Male Sterility

Indehiscent anthers having nonviable, shrunken pollen or anthers filled with fertile-looking but nonfunctional pollen occur frequently in *S.cyaneus, S. dunnii, S. kewensis, S. luteus, S. polyanthus, S. saundersii, S. watsonii* (this Vol.). Genetic control of mst is not known, but the mst plants breed true and arise as spontaneous mutants.

26.2.2 Functional Male Sterility

In some plants of *S. cooksonii, S. cyaneus, S ketaniensis* and *S. primulifolius*, viable pollen is produced but anthers are indehiscent (Story 1955, Hillard 1966; Zeven 1972a). This trait appears constantly in them but its genetic control is not known. Plants of the hybrid variety Constant Nymph obtained after crossing Merton Blue ♀ × *S. johannis* ♂ (Marston 1964–1965) and its mutants like blue, cobalt, mason white and purple flowered Nymphs possess viable pollen but indehiscent anthers (Zeven 1971a,b, 1972a). Their sterility is conditioned by a dominant gene, *Da*, whose expresssion is highly influenced by the environment.

26.2.3 Gene Cytoplasmic Male Sterility

In *Streptocarpus*, whereas certain parental combinations produce mst progeny, their reciprocals, instead of being normal, are fst (Oehlkers 1964). Thus gc interaction is sex-dominated. By using *S. wendlandii* or *S. comptonii* as female parents in crosses to *S. grandis, S. rexii* or *S. soleanthus* (as males), mst plants appear in F_1 (Oehlkers 1938, 1940, 1941). Androecia of these hybrid plants are variously modified, deformed, underdeveloped or stigmoid. The frequency and intensity of sterility are enhanced with increased BC generations, obtained by using the same pollinator genome donors. In the reciprocals, progeny is mft, and p-fst. The fst frequency, as evidenced by the sterile ovule number, increases with increasing BC generations. On the basis of cytoplasmic homology, *S. comptonii* and *S. wendlandii* belong to the wendlandii-plasm group and *S. grandis, S. rexii* and *S. soleanthus* to rexii-plasm group (Oehlkers 1964). This author believes that the former plasm group has a "male-predominance" and the latter a "female-predominance" tendency. In such sex tendencies, *S-polyanthus* appears intermediate between these two groups.

26.2.4 Reciprocal Crosses into Advanced Generations

F_1 progeny of *S. wendlandii* ♀ × *S. rexii* ♂ is mst having reduced abortive anthers (Oehlkers 1938). Plants of reciprocal progeny are mst but partly fst, while BC (*S. rexii* × *S. wendlandii*) × *S. wendlandii* segregates 1 normal: 1 mft which are partly fst F_1 plants, BC of the reciprocal, (*S. wendlandii* × *S. rexii*) × *S. wendlandii*, segregate 1 normal: 1 mst plant. Genomically, these two BC progenies are identical in having 75% *S. wendlandii* genes and 25% *S. rexii* genes. The only difference between them is that while in the one, these genes are in *S. rexii* cytoplasm, in the other they are in *S. wendlandii* cytoplasm. Since in the former fst's occur and in the latter mst, the difference in sterility type is due to g-c interactions. This is confirmed by subsequent segregations. This complete mst progeny appears up to BC_4 (*S. wendlandii* × *S. rexii*) × *S. rexii*. The plants have anthers converted into style-like structures (Oehlkers 1938). The anomaly persists in subsequent BC's. Thus the mst results from the interaction of *S. rexii* genome with *S. wendlandii* cytoplasm.

26.2.5. Cytoplasmic Sex Determinants

To explain the above-mentioned g-c interactions, Oehlkers (1938,1952) proposed cytoplasmic effects as the major determining factors. For instance, hybrids between *S. rexii* × *S. wendlandii* were p-fst. When they were backcrossed to *S. wendlandii*, equal numbers of p-fst and fst forms segregated. In the reciprocals, F_1 hybrids were mst, but in being backcrossed to *S.rexii*, two types of mst form segregated, viz. one in which anthers changed to functional ovules (superfemales) and one in which the stamens were replaced by staminodes. The degree of sex expression in the hybrids differed according to the age of the plant and the temperature at which it was grown (Oehlkers 1956). The author interpreted the results on the assumption that the nucleus and cytoplasm share control of sex organ development. *S. rexii*, for example, was assumed to possess a female-promoting nucleus and a male-determining cytoplasm, while *S. wendlandii* had the reverse distribution of determinants. Thus intercrossing these two species led to the combination of male or female determinants in the nucleus and cytoplasm of the hybrid, leading either to superfemales or exaggerated males. Although these results indicate that the cytoplasm in which the hybrid nucleus functions has an effect on the gene action involved in sex organ development, the hypothesis invoking cytoplasmic male and female determinants cannot be accepted without further evidence (see also Goldschmidt 1955).

26.2.6 Partial Male Sterility

This occurred in F_1 progenies of the direct as well as of the reciprocal crosses among *S. cyaneus*, *S. soleanthus* and three races of *S. rexii* (Oehlkers and Ebell 1958). In all these, F_1 hybrid plants possessed some nonfunctional pollen grains, but when *S. cyaneus* or one of the *S. rexii* races was used as the female, the proportion of abortive pollen grains in the F_1 plants was considerably greater than when *S. soleanthus* was used as the female parent (Table 26.2). These differences, being common, consistent and uniform, indicate differential g-c- interaction, since the nucleus in the direct and reciprocal crosses is the same and only the cytoplasm changes. However, the interaction is not uniformly expressed in succeeding BC's. For instance, the amount of pollen sterility increases in further BC's of hybrid *S. soleanthus* × *S. rexii* base, but in the reciprocals, such an increase occurs only in BC_1; in subsequent backcrosses, the pollen sterility decreases (Table 26.1) Similarly, pollen sterility occurs in both direct and reciprocal crosses between *S. cyaneus* × *S. soleanthus*. In the direct cross, sterility is enhanced with BC's, but in the reciprocals, sterility decreases after BC_2. Apart from these variations, a high degree of variability in pollen sterility percentage occurs within families and within plants of a single family. In certain segregating families, pollen sterility differs even between different flowers of a single plant. Segregations reveal that the pollen sterility is conditioned by many major genes, which work in association with minor and modifier genes. Certain segregations of pollen sterility in some crosses indicate polygenic control, while others reveal dosage effects of major genes. The influence of environment over each g-c hybrid combination is profound. The segregation patterns and the specificity of each cytoplasm in its reaction to a genome indicate the existence of cytoplasmic diversity and differentiation in *Streptocarpus* species. This is substantiated by Oehlker's (1964) classification, which is based (except No. 1) on the expression of sex-determining genes in *S. wendlandii* cytoplasm. The grouping is as follows:

1. Species possessing N-cytoplasm, viz. *S. haygarthii.*
2. Normal F_1 progeny produced by crossing *S. wendlandii* as female to *S. baudertii, S. gracilis, S. johannis, S. meyeri* and *S. polyanthus* as males. Offspring of BC progeny obtained with pollen of these species exhibit various degrees of mst.
3. The mst F_1 progeny produced by crossing *S. wendlandii* as female to *S. dunnii, S. galpinii, S. grandis, S. michelmori, S. pole-evansii, S. rexii* (strain K 1275) and *S. soleanthus* as males; mst in BC is more pronounced than F_1.
4. *S. wendlandii* and *S. compactonii* when used as female and crossed with species given in (2) and (3) produce progeny with abnormal anthers.

Table 26.2. Mean pollen sterility in some *Streptocapus* hybrids and backrosses

Species cross	Generations				
	F_1	BC_1	BC_2	BC_3	BC_4
S.cyaneus × *S.soleanthus*	77	88	93	98	93
S.soleanthus × *S.cyaneus*	70	70	90	78	65
S.rexii Basel × *S.soleanthus*	62	88	80	20	23
S.soleanthus × *S.rexii* Basel[a]	54	82	84	97	100
S.rexii albus × *S.soleanthus*	56	93	70	41	50
S.soleanthus × *S.rexii* albus[a]	51	71	41	69	83
S.rexii type × *S.soleanthus*	34	91	74	42	55
S.soleanthus × *S.rexii* type[a]	28	80	85	74	99

Modified from Oehlkers (1964), first parent female, second male and recurrent male;
[a]gc-mst, in others hybrid sterility.

26.2.7 Cytoplasmic Differentiation

From the above it is evident that g-c interaction is of wide occurrence in *Streptocarpus*, and that it influences male as well as female fertility. Coupled with genome differentiation and diversity, cytoplasmic differentiation is of a high order in this genus. The functional mst in Constant Nymph, a hybrid variety derived from the cross Merton Blue × *S. johannis*, may represent gc-mst rather than the g-mst as visualized by Zeven (1972a). Since no BC of this hybrid or reciprocal crossings among the parents were done, it is desirable to test for g-c interactions in this hybrid as well as in *S. cooksonii, S. cyaneus, S. ketaniensis* and *S. primulifolius*, that also have functional mst.

21.4.1.2 Gene-Cytoplasmic Male Sterility

By crossing autotetraploids of *C. moschata* with *C. palmata*, an amphidiploid progeny developed (Bemis 1973). The sterility was genic and not chromosomal. Meiosis in these steriles was normal but microspores aborted immediately after their liberation from quartets. These amphidiploids, when pollinated by diploid *C. moschata*, resulted in an auto-allotriploid containing two *C. moschata* genomes and a single *C. palmata* genome. This polyploid was mst as it had no anthers. Another mst amphidiploid plant was developed by Bemis (1970) after crossing the diploid *C. moschata* with the wild semi-xerophytic perennial gourd *C. foetidissima*. Fertility was restored in it after pollinating with a different *C. moschata* strain that probably carried restorer genes; the parental *C. moschata* strain lacked these as it could not restore fertility. In the mst's, androecia failed to develop fully and never dehisced.

21.4.1.3 Chemical Male Sterility

Ethrel and MH. Complete to partial suppression of male flower development for varying periods of time (Robinson et al. 1970; Chauhan 1976) or shifting of male floret development to upper parts of shoots occur following ethrel and MH treatments (Table 21.3). The MH treatments reduce female fertility and cause phytotoxicity (this Vol.).

21.4.1.4 Environmental Influence on Sexuality

Most of the squashes are monoecious. Their sex ratio is highly influenced by various ecological factors like soil moisture, nitrogen content, day length and atmospheric temperature etc. Usually, low temperatures, high soil nutrients or short-day conditions promote female expression and the reverse conditions cause the male expression.

21.4.2 *C. pepo* L. (gourd, pumpkin)

21.4.2.1 Genic Male Sterility

In 1922, Sinnott and Durham (Shifriss 1945) found a number of inbreds of *C. pepo* that did not set seeds on selfing by seeded profusely when crossed. Their mst was genic. Shifriss (1945) found mst plants segregating from three inbred lines, each originating from a different commercial pumpkin variety. In them, androecium abortion occurred at the bud stage before staminate flowers opened (Fig. 21.1). In each of them, the sterility was controlled by a single recessive *ms* gene. The allelic nature of the three *ms* genes was not determined. From an Egyptian variety, Eskandarahy, mst plants segregated (Eisa and Munger 1968). Their anthers were

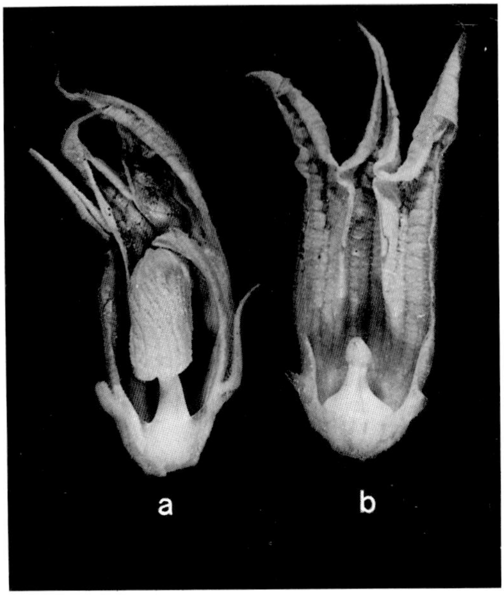

Fig. 21.1a,b. *Cucurbita pepo.* Male fertile **(a)** and male sterile **(b)** flowers. (Shifriss 1984, personal communication)

shrivelled, brown and pollenless. The sterility is controlled by a recessive gene ms_2. Its penetrance and expressivity are complete and consistent under a wide range of environmental conditions. Male meiosis proceeds normally up to microspore development, after which microspores abort to produce shrivelled and mis-shapen pollen grains, coupled with tapetal cell hypertrophy. Whereas a good fit for monogenic recessive ratio for ms_2 gene in F_2 of some crosses was obtained by Whitwood (1975), in others the ratio was inconsistent, especially when the variety Table King Bush Acorn was used as a parent. Allelism tests for the two *ms* genes mentioned above have not been done. The mutant detected by Curtis and Scarchuk (1948) is both male and female sterile, the sterility being conditioned by a single recessive gene (this Vol.). Shifriss (1945) found that certain varieties of *C. pepo* develop abnormal staminate flowers, having either highly reduced or no androecia in the greenhouse, but that they were normal when grown in the field. The sterility appears to be induced by the high temperature of the greenhouse. These varieties resemble those of *C. maxima* found by the author. In a local strain of *C. pepo,* two inbreds had scanty nonviable pollen while 13 had plump anthers lodging fertile pollen grains. Their female fertility was normal. Details of the genetic nature of the sterility are not known (this Vol.).

21.4.2.2 Chemical Male Sterility

i. *Ethrel.* Application of 250, 500 or 1000 ppm of ethrel to two monoecious squash varieties of *C. pepo* at first and third leaf stage led to the development of only female flowers during the first 2 weeks of flowering (Rudich et al. 1969, 1970; Robinson et al. 1970) (Table 21.4). After 3 weeks, male flowers reappeared on the treated plants.

Table 21.4. Induction of staminal suppression by chemicals in some cucurbits[a]

Species/chemicals	References	Male flower suppression (temporary)	Female flower increase[a]	Additional effects
Citrullus vulgaris				
Mendok	Hensz and Mohr (1959)	High	Nil	nk
TIBA	Rehm (1952)	Complete	Nil	nk
Cucumis melo				
Mendok	Laibach and Kribben (1951), Pundir and Singh (1965)	Partial	Nil	Floral anomalies induced
	Lippert and Hall (1961)	Complete	Nil	Floral number and fruit production reduced
SADH	Halevy and Rudich (1967)	50%	High	Plants dwarfed and tendril development suppressed
SADH + Ethrel	Rudich et al. (1969)	Nearly complete	Moderate	nk
SADH + GA	Rudich et al. (1972)	Low	Low	Shoot length reduced
Cucumis sativus				
Ethrel	Minina (1938)	Very high	Considerable	nk
	Rudich et al. (1969, 1970)	Nearly complete	Considerable	Plant growth reduced, fruits deformed
	Iwahori et al. (1970)	Nearly complete	Nil	nk
	Augustine et al. (1973)	Almost complete	Nil	nk
NAA	Laibach and Kribben (1951)	Partial	High	nk
SADH + Ethrel	Rodriquez and Lambeth (1972)	Nearly complete	High	Plant growth reduced
Cucurbita maxima				
Ethrel	Robinson et al. (1970)	Nearly complete	High	nk
MH	Chauhan (1976)	Complete	Nil	nk
Cucurbita pepo				
Ethrel	Rudich et al. (1969, 1970)	Nearly complete	High	Plant growth and development highly reduced
MH	Wittwer and Hillyer (1954), Chauhan (1976)	Nearly complete	Nil (decrease)	Growth highly reduced and high toxicity induced
NAA	Wittwer and Hillyer (1954)	Complete	Nil (decreased)	Highly toxic

[a]In many hermaphrodite flowers increase in number and frequency.

In addition to suppressing male floral development, the chemical treatments reduced plant growth and development, the least effect was by 250 ppm treatment, which thus represents the best treatment. In general, the treatment induced decrease in staminate flower number, increase in pistillate or hermaphrodite flower number and formation of female flowers at lower nodes. In contrast, Borghi and Pironi (1976) reported the best male-sterilizing effects by spraying 400 ppm ethrel first at the fourth to fifth true leaf stage and later after 7–10 days. This treatment kills anthers completely.

ii. *MH*. Wittwer and Hillyer (1954) induced mst in *C. pepo* by MH treatments, but pistillate flower number decreased considerably and their fertility was not impaired significantly.

iii. *NAA*. In *C. pepo* var. Table Queen, application of 900 ppm NAA, when the first leaf had expanded fully in the greenhouse, decreased the staminate pistillate flower ratio from 147:1 to 0.4:1, with occasional plants producing no staminate flowers (Wittwer and Hillyer 1954; Hillyer 1956). In field-grown plants, 100 ppm NAA first applied on seedlings with one to two true leaves only, with the treatment repeated even 10 days later at the fourth and fifth leaf stage, delayed the appearance of staminate flowers and the treated plants produced exclusively pistillate flowers for 8 days. Thus a remarkable suppression of staminate flower buds occurred following the chemical treatment.

21.4.2.3 Breeding Value

The g-mst in pumpkin has a high breeding value and can reduce the cost of hybrid seed production considerably (Shifriss 1945). Maintenance and increase of mst's is possible by propagating BC populations raised by staking mst plants and harvesting their seeds. Insects act as satisfactory pollinators for these. For commercial hybrid production, male and female parents are planted in alternate rows in ratios of 1:3 or 1:4. The female parent is represented by a BC population and mft's are removed during the roguing operation. The operation is easy because (a) only one flower per plant is needed to examine fertility, (b) staminate flowers are large and hence easy to detect, (c) mst's are easily distinguishable as their pollen sacs are black and shrivelled in contrast to the plump and yellow sacs of mft's, (d) a few staminate flowers of mst's may open, but they open about 2 weeks prior to pistillate flowers and hence sufficient time is available for roguing before open-sibbing is possible. Moreover, in most of the cases pollen is nonviable. Thus mst has a high breeding value in pumpkin hybrid production, and should be commercially utilised.

21.4.3 Chemical and Environmental Influence on Sex Expression in Cucurbitaceae

Cucurbitaceae comprise about 102 genera with 800 species, which are distributed in tropical, subtropical and temperate regions of the world. True hermaphrodites are rare and nearly all the species possess either only diclinous (male, female) or both male and female flowers. Whereas the first group has dioecious and monoecious species, the second group consists mainly of andromonoecious plants, gynomonoecious plants being rare. Basically each sex type is genetically conditioned, but the sex expression and ratio is highly influenced by chemical and ecological factors. The genetic control of sex expression of the commonly cultivated cucurbits has been described well by Frankel and Galun (1977). In the following, the influence of chemicals and environment on sex expression is briefly given.

i. *Chemical Influence.* Of the major gametocides and chemicals tested to alter sexuality in plants (Table 21.3), gibberellins and ethrel appear to influence the sex expression of cucurbits the most (Table 21.4). Gibberellins mostly suppress female flower development and induce male floral formation in the leaf axils otherwise occupied by female flowers. The effectiveness of various GA types differs considerably and their net effect is dependent upon the genotype and the environment in which the plant is grown. In some melon varieties, its effectiveness is unclear. The effect of ethrel is the reverse of that of GA, as it promotes a high degree of femaleness. TIBA, SADH and α-NAA also have similar, but relatively low female-promoting effects and delay male flower development considerably (Table 21.4). SADH, a growth retardant, suppresses endogenous gibberellin synthesis and promotes femaleness. This effect becomes negated by GA additions. Some combinations of SADH + ethrel completely suppress male floral development in cucumber (but not in muskmelon) and dramatically enhance female floral number per branch and total production per plant. This chemical effect has a high breeding value as a large number of fruits can be obtained if the treated plant is used as the female parent.

ii. *Environmental Influence.* Floral induction in cultivated cucurbits is usually not regulated by specific thermoperiods, as flowering occurs under almost all the climatic conditions that allow vegetative growth. Usually high temperatures and long photoperiods promote maleness and the reverse conditions femaleness. In many andromonoecious cucumber and melon varieties, high temperature and long days delay the appearance of hermaphrodite flowers, but male flowers appear normally. The pollen fertility of late-developed flowers is highly reduced. Too short daily illuminations suppress femaleness and lead to abortive pollen formation in male flowers.

21.4.3.1 Sex Expression Shifts

Frankel and Galun (1977) concluded that it appears possible to shift sex expression by chemical treatments in either direction in cucumbers, i.e. genetic female plants can be induced by GA to produce abundant male flowers, and genetic male plants can be induced by ethrel treatments to produce female flowers. However, as is evident from the preceding account, sex expression of cucurbits can be altered not only by these two chemicals but also by others like NAA, TIBA, SADH, FW-450, MH etc., as well as by many ecological factors (Fig. 21.2). However, none of the chemicals or environmental factors appears to be fool-proof for induction of a specific sex change in a species. The net effective change is the result of genotype-treatment interaction, and therefore each treatment may be specific for each

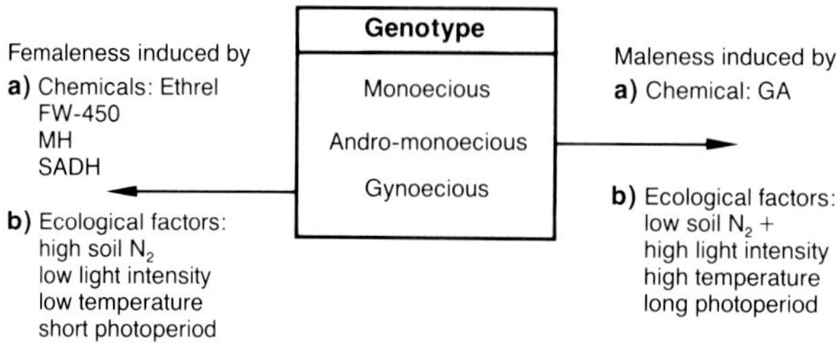

Fig. 21.2. Sex-expression shifts brought about by extrinsic factors in cultivated Cucurbits

genotype under a given set of environmental conditions in achieving optimal results. From the available findings, certain generalisations are made for the sex expression of cultivated cucurbits and are depicted in Fig. 21.2. To determine subtle genetic, biochemical and physiological differences, androecious and gynoecious forms of a species should be compared and subjected to chemical and environmental manipulations for sex-shift studies.

21.4.3.2 Sex Expression Alterations in Cucumbers

Cucumber has been under cultivation for over 4000 years and exhibits a tremendous diversity in sex forms and expression. Because of this, amongst the higher plants, sexuality in cucumbers has been most intensively studied and systematically detailed by Frankel and Galun (1977). Basic sex types and their alteration leading towards mst as induced by ecological factors and chemical treatments are indicated in the following account.

i. *Basic Sexual Types.* The flower buds, which differentiate in the leaf axils of the main shoot, are bisexual in their early developmental stages. Later they develop into either male (staminate) or female (pistillate) flowers. Usually male flowers differentiate at the lower nodes, followed by female flowers at the upper nodes. The number of nodes to the first female flower and the total number of female flowers form fairly reliable indices of sex expression. Cucumbers are mostly monoecious, but some gynoecious or andromonoecious forms also exist, gynoecism strains normally producing only pistillate flowers. Other cucumber and muskmelon strains produce staminate or pistillate, and in addition, hermaphrodite (perfect) flowers in various combinations. Andromonoecious strains produce staminate flowers first and hermaphrodite flowers later. The majority of the cucumber varieties are monoecious, bearing hypogynous male and epigynous female flowers. Considerable variability exists within the male-female flower distribution and ratio. In monoecious plants mostly there are either one to two female flowers or several male flowers in all the leaf axils barring a few basal nodes of the main shoot. In andromonoecious cucumbers both male (monoclinous) and hermaphrodite (diclinous) flowers are produced. In them the replacement of female flowers by hermaphrodites occurs. All grades, between hypogynous through perigynous to epigynous conditions, occur in them (Kubicki 1969a,g).

ii. *Environmental Influence.* Sex expression in cucumber plants is conditioned genetically but is modified by several environmental and hormonal factors. Whereas in general high soil nitrogen, low light intensity, short days, and low temperature favour femaleness, reverse conditions enhance maleness. This leads to a high degree of observed variability in sex expression and ratio under the various ecological conditions in which the cucumbers are cultivated almost all over the globe. Furthermore, whereas in many cucumber cultivars high temperature and long-day conditions promote a tendency towards more male flower production, some cucumber cultivars exhibit sensitivity to day length only. In these, short days enhance femaleness and in a few others no alteration occurs under such conditions. Another cultivar displayed a tendency to femaleness under long-day conditions (Fukushima et al. 1968). Similarly, the response to low temperatures is also equivocal. Low temperatures promote femaleness in some cultivars, while others remain unaffected. The specific response of various varieties to blight and temperature conditions appears to be typical for each variety and seems to be genetically controlled. Thus manipulations of light and temperature conditions may allow the development of complete mst plants with high female floral number and fertility for use in hybrid seed and fruit production. If the response to these ecological factors is genetically conditioned, the reliability of sex conversion in a cucumber genotype and its value in breeding increase considerably.

iii. *Chemical Influence.* Apart from ecological factors that influence sex expression in cucumbers, many chemicals like ethrel, NAA, ethrel + SADH enhance female flower production considerably and delay male flower development by 2–3 weeks (Table 21.4). Ethrel also induces sex reversion (Augustine et al. 1973). However, complete mst without impaired female floral number and fertility has not been obtained by any of the chemicals, except in a monoecious cultivar (see Rodriquez and Lambeth 1972). On the other hand, GA enhances maleness and plant growth in monoecious and gynoecious strains when applied singly, or in combination with ethrel, MH or SADH. Exogenous application of auxins and inhibitors of gibberellin biosynthesis promotes monoecious strains to form pistillate flowers and increase the femaleness. Strains with genetically strong female sex expression possess more auxin and less gibberellinlike substances than strains with male sex expression. However, certain differences between the species do exist. For example, gibberellin application does not result in male flower production in gynoecious muskmelon (Peterson and Andher 1960). In general, some chemicals and auxins alter the sex expression and ratio of cucumber considerably, and the sex expression appears to be controlled by an endogenous auxin-gibberellin balance.

iv. *Ethylene Effect.* In cucumbers, ethrel, an ethylene-releasing compound, enhances female floral number, lowers the number of nodes to the first female flower and shortens the days to anthesis. Furthermore, it reduces the number of male flowers, increases the node number at which the first male flowers develop and delays anthesis of the first few male flowers formed. Byers et al. (1972) correlated the sex expression in cucumber (*Cucumis sativus*) and muskmelon (*C. melo*) with endogenous ethylene production. This was based on the fact that plants of gynoecious (all female) sex types of the two species produced more ethylene than the monoecious (male-female) plants. Also *C. melo* plants of the gynoecious sex type normally produce only female flowers when grown at reduced atmospheric pressure combined with ventilation. This decrease in internal gases in the plant tissues causes change of sex expression towards maleness and leads to the production of perfect flowers in place of pistillate flowers. The addition of ethylene restored (partially or fully, depending upon the concentration) the original gynoecious sex expression, i.e. production of only pistillate flowers. Enrichment of the atmosphere at either normal or reduced pressure with CO_2, a competitive inhibitor of ethylene action, led to the production of hermaphrodite flowers. Foliar application of benzothiadiazole, a postulated inhibitor of ethylene action, results in the formation of perfect flowers on the gynoecious plants of *C. melo* and of male flowers on gynoecious *C. sativus*. These findings led Byers et al. (1972) to propose ethylene as an endogenous regulator of sex expression in *C. sativa* and *C. melo*. Thus this chemical has a major use in creating or terminating mst in cucumbers.

CHAPTER 27

Gramineae (grass family)

In this family of 400 genera, mst is known in *Aegilops, Agroelymus, Alopecurus, Avena, Chionachne, Dactylis, Echinocloa, Eleusine, Festuca, Hordeum, Lolium, Oryza, Pennisetum, Phleum, Saccharum, Secale, Sorghum, Tragopogon, Tripsacum, Triticum* and *Zea* (Tables 27.21 to 27.23).

27.1 *Aegilops* L.

Almost all the species of *Aegilops* and *Triticum* can be crossed as female to polyploid wheats and repeated BC's of their hybrids with the common wheat pollen have led to mst progenies in many cases (Tables 27.10, 27.11). Their description is given in *Triticum* (Sect. 27.20). Two *Aegilops* interspecific crosses are described below.

27.1.1 *Ae. caudata* × *Ae. umbellulata*

Amphidiploids of this cross are p-mst but the sterility increases considerably when these are pollinated by *Ae. persica, Ae. typica* and *A. umbellulata* pollen (Kihara and Kondo 1943). In reciprocal crosses, pollen sterility diminishes but does not disappear completely.

27.1.2 *Ae. longissima* × *Ae. aucheri*

In many intergeneric hybrids between *Aegilops* and *Triticum* mst results due to interaction of *Aegilops* cytoplasm and wheat genome. However, only in certain interspecific *Aegilops* crosses, progenies segregating mst plants occur. For instance, complete pollen sterility occurs in F_1 progeny of cross *Ae. longissima* × *Ae. aucheri*, progeny of the reciprocal cross is normal (Kihara 1951). Thus the mst is due to the interaction of *Ae. longissima* cytoplasm and *Ae. aucheri* genes. The sterility is maintained by pollinating the mst hybrid by *Ae. aucheri* pollen but is drastically reduced when the mst hybrid is pollinated by *Ae. longissima.*

27.2 *Agroelymus turneri*

This is considered to be a putative hybrid of *Agropyron dasytachum* × *Elymus innovatus* (Nielsen 1961a). Some mst plants segregate from the progenies derived from the outcrosses of *Agroelymus turneri* with *Agropyron dasytachum, A. repens* and *Elymus canadensis,* (Nielsen 1961b). In most of the PMC's of the mst plants, meiocyte degeneration occurs at prophase-I. It is coupled with accumulation of some extraneous chromatic and achromatic materials. In the remaining PMC's, although meiosis proceeds nearly regularly, pollen fertility is low. In these cells, microspores abort immediately after liberation from tetrads. Although Nielsen (1961b) considered mst as gc-mst type, female fertility of some of the steriles is also very low. Seed set on these after open pollinations varies from 23 to 72%. Hence all the sterile plants cannot be regarded as true mst's. Some of them, especially those with low seed set, are meiotic mutants (Gottschalk and Kaul 1980a,b). Moreover, reciprocal crossings to determine cytoplasmic effects have not been done to confirm the gc-mst type.

27.3 *Alopecurus myosuroides* Huds

Among the 25 known diploid and polyploid species, nearly 8.8% of plants of some inbred lines of *A. myosuroides,* an allogamous diploid grass (2n = 14), exhibit retarded male meiosis leading to pollen degeneration (Johnsson 1944). In their PMC's first division metaphasic arrest and a strong clumping of the chromosomes leading to the appearance of structureless lumps of chromatin occurs. While in PMC's of some plants meiosis stops, in others it continues sluggishly and is characterised by the presence of a few persistent nucleoli. Where tetrads are formed, advanced degeneration of the cell contents occurs, but the chromosomes maintain their distinctness until cells are finally destroyed. In those in which meiosis is arrested at M_1, monads with 14 chromosomes are formed. In them, every individual chromosome becomes surrounded by a more refractive layer of plasma, constituting thereby a mini-micronucleus, in some of which many other chromosomes may enter to form a micronucleus. One amongst these developed a conspicuous nucleolus (Johnsson 1944). In it the chromosomes never despiralized but the centromeres divided. In some of them, meiosis proceeds until dyad formation, after which degeneration occurs immediately. Differences in meiotic breakdown between and within the plants of the inbred lines are many and variable, but in all the cases, sterility is complete due to pollen degeneration. Not only the meiotic course but the spindle development and cytokinesis in these inbreds is greatly disturbed. Furthermore, with a marked delay in the onset of meiosis, the sporad stage is not reached and the division either proceeds or stops during the first or second meiotic stage. The mst appears to be controlled by recessive genes, as the frequency of mst's increases considerably following inbreeding. Certain population segregations indicate the sterility to be controlled by at least one recessive gene. A few other inbred lines of *A. myosuroides* segregate plants in which male meiosis proceeds normally, but their young pollen grains abort soon after formation (Johnsson 1944). Genetic control of pollen lethality is not known. It could either be haplontic, caused by specific genes lethal to gametes or by a specific disturbed relation between pollen and somatic tissues. These plants are also either fully or partially female sterile, so they cannot be regarded as true mst's, as considered by Johnsson (1944). Three female fertile mst plants of *A. pratensis* var. *aureo-variegatus* were found by the present author in a collection of 48 plants. In these, male meiosis terminated at the end of prophase-I and PMC's degenerated gradually. The tapetum developed normally and collapsed just prior to anthesis much later than PMC degeneration. Genetic control of the sterility is not known.

27.4 *Avena* L. (Oat)

The g-mst is known in *A. byzantina*, *A. fatua* and *A. strigosa* but not in the commonly cultivated hexaploid oat *A. sativa* (Table 27.21).

27.4.1 *A. byzantina* Koch. (red oat)

From the progeny of a nullisomic plant, a p-mst plant segregated (Ramage and Suneson 1958). Pollen produced by its normal-looking anthers are shrunken, with impaired viability. The bivalent formation is normal, but microspores abort after liberation. While selfing produces 10% seeds, hand pollination yields only 25% seeds. Since in controls such artificial pollinations produced about 60–70% seed, this plant cannot be regarded as true mst as it has low female fertility. Garber (1922) detected weak p-mst's exhibiting 8.5 to 18.5% pollen sterility. Crossing these to the oat variety White Russian yielded one F_1 plant that had 3.5% pollen abortion. F_2 plants from this F_1 plant had 1–14.5% pollen sterility. The author proposed a digenic recessive control for this sterility. Elliot (1925) concluded that the sterility is genic and is highly influenced by the environment.

27.4.2 *A. fatua* L. (wild oat)

In hexaploid oat stands, a mst mutant having dark green glumes and fully smooth lemmas appeared (This Vol.). Male meiosis up to tetrad formation is normal, but microspores degenerate after release from PMC's, so that no viable pollen is produced by this plant. Its female fertility, as evidenced by seed set after hand pollination, is reduced by 25–30%. Genetic control of the sterility is not known. From X-irradiated seed progeny, Sadanaga (1965) detected a mst mutant in which sterility was controlled by one recessive gene.

27.4.3 *A. sativa* L. (edible oat)

Making oat crosses is time-consuming, tedious and frequently results in a low percentage of seed set. The seed set after hand pollination is influenced by the environment, the skill of the crosser and the morphological traits of the female strains. These difficulties in obtaining crosses have limited the use of breeding methods like recurrent selection and BC for oat improvement. Neither g-mst nor gc-mst has been found in this hexaploid oat. Utilizing two isogenic lines of oats, Frohberg and Frey (1970) tried to induce mst using hot water (45°–48°C) and 0.25 to 2% of MH, GA, FW-676 (Calcium 2,3-dichloroisobutyrate), G-315 Magnesium 2,3 dichloroisobutyrate (Table 27.22). None of the treatments was effective and economically feasible for mass scale oat emasculation. Accordingly, a few more treatments described below, were tried.

i. *Hot Water Treatment.* 46°C hot water treatment given for 3 min at the boot stage led to a high degree of pollen sterility (Frohberg and Frey 1970), but hand pollination was still needed, as when the treated panicles were bagged with male panicles, no combination of temperature and immersion interval applied at either early or late boot stages of growth produced high seed set and a high percentage of hybrid seed. However, a distinction between hybrid and selfed seeds could be possible, 46°C hot water treatment for 3 min at the boot stage may be used for limited emasculation without impairing the female fertility significantly. However, this treatment is not better than the conventional emasculation method.

ii. *DPX-3778.* Using 15 *A. sativa* genotypes, Johnson and Brown (1976) could not obtain complete mst after spraying 18 kg ha^{-1} of the chemical. Even higher doses up to 27 kg ha^{-1}, when applied at early heading or in a split application at the early boot plus early heading stages, did not induce full mst in Noble oats. Such high doses decreased shoot height, spikelet number (due to spikelet abortion) and seed yield. In one set of experiments, phytotoxicity occurred.

iii. *Gibberellic Acid.* Four weekly applications of 0.05% GA given at the four to five leaf stage induced pollen abortion, but seed set, even after hand pollination, did not exceed 50%. The treatments reduced spikelet size and caused excessive internode elongation leading to lodging.

iv. *Maleic Hydrazide.* Knowles (1953) obtained mst by spraying 450 g acre^{-1} MH on *A. sativa* and *A. fatua* plants, but Frohberg and Frey (1970) found this chemical spray highly toxic, as it either inhibited the growth or killed the plants prematurely. Low concentrations of MH induced complete pollen and high ovular sterility accompanied by reduction in anther size, and stigma branching and top growth.

v. *Other Chemicals.* G-15 and FW-676 are ineffective as male gametocides and are highly toxic. Likewise, ethrel was ineffective in two spring oat varieties. It reduced height and yield and caused severe floret blasting (Brown and Early 1973). Since bagging, hot water or chemically treated pollen sterile panicles with male-fertile panicles produced less than 5% hybrid seeds, hand emasculation and pollination for oat hybrid breeding is recommended.

Conclusions. Utilisation of *ms* genes found in diploid *A. strigosa* or hexaploid *A. byzantina* and *A. fatua* oats is not likely because of crossability barriers and seed fertility problems. Transfer of genes from diploid to hexaploid oats is also arduous. In general, in all such crossings the *ms* gene reduces female fertility considerably. Lafever and Patterson (1964) suggested some possible uses of a mst nullisomic as the female parent to produce fertile F_1 monosomic, but by using this method they obtained only 11.7% NCP. Thus this method cannot be used in commercial oat breeding.

27.4.4 *A. strigosa* Schreber (bristle or small oat)

In this plant, anthers were shrivelled, nondehiscent and produced 5–10% stainable pollen (this Vol.). While the plants were completely sterile and set seed in the greenhouse, a few viable seeds were set in open fields due to NCP. Female fertility of these steriles is normal, as hand pollination yielded fertile F_1 seeds which segregated mft:mst in 3:1 proportion in F_2.

27.5 *Chionachne koenigii* Spreng Thw

In this Asiatic relative of maize, Rao (1975) induced mst by treating two to three leaved seedlings with colchicine. Whereas 0.1% colchicine treatment produced plants that had normal microsporogenesis but abortive pollen, 0.2% treatment yielded plants which lacked stamens completely. The mst in each case was conditioned by one recessive gene *ms* and *sta* (Rao 1975). Expressivity of the *sta* gene is complete and stable, but that of the *ms* gene is variable and unstable. Crosses between stamenless and p-mst mutants indicate that the *sta* is epistatic to gene *ms*. The tetraploid sectors of the mutants when crossed with 4x normals give similar genetic interactions. In all cases, the mst mutants are dwarf, profusely tillered, early flowering and live longer than the mft sibs.

27.6 *Dactylis glomerata* L. (orchard grass, cocksfoot grass)

Of the five known species, mst is documented only in this species, which is cultivated mainly for pasture, forage, hay and as an ornamental. The mst was first reported from Wales by Stapledon (1931) and Jenkin (1931a,b). These sterile plants were phenotypically normal but their anthers were characteristically extruded, light yellow-greenish, shrunken and nondehiscent. The status of the female fertility is not known. From a population of 6000, Filion and Christie (1966) found 12 plants with reduced, dark green, nondehiscent anthers and invisible pollen at anthesis. Of these, a clone OD-1 had up to six quadrivalents per PMC besides bivalents, univalents and laggards. This anomaly may be normal, as the species is an autopolyploid and may therefore have no connection with mst. In the remaining mst's, meiosis proceeded until microspore formation but immediately after microspores were released, they degenerated rapidly. Unlike in fertile plants, in steriles, the tapetum degenerates rapidly during early meiosis and no gradual elongation and periclinal division of epidermal and endothecial cells occur.

Striations in the endothecium necessary for anther dehiscence are absent in the mst anthers. The mst appears due to abnormal form and function of the tapetum which provides insufficient nutrition to the developing sporogenous tissue. Female fertility of the mst's is normal and they have been successfully utilised in hybrid production. Increase in forage yield of mst's is reported by Christie (1973).

Using N-nitroso-N-ethylurea and diethyl sulphate, mst's were isolated by Kremnina (1973). While most of such steriles of variety VIK 61 had pollenless anthers, hybrids of VIK 61 × Asta developed functionally fertile anthers, indicating the recessive nature of mst. The female fertility was normal. Năsinee (1973) found cocksfoot clones with reduced fertility and hybridising them with each other and subsequent self-pollination of the selected F_1 hybrids yielded plants with indehiscent anthers having scanty pollen of very low fertility. Intercrossing the scanty pollen plants produced a few complete mst's in which microspores aborted after liberation from the tetrads before pollen mitosis; their female fertility is normal (this Vol.). This breeding behaviour indicates the additive nature of *ms* genes and polygenic control of the sterility. This needs to be verified, as the species is hexaploid.

Myers (1946) found a mst type that depends for its expression upon the interaction of S-cytoplasms with a dominant *Fr* gene. Crossing mft's with mst's produced F_1 progeny that was either all sterile or sterile and fertile in varying proportion. These results are explainable if we assume that the sterile plants are produced when the female parent (mst) is quadriplex, triplex or duplex for the sterility gene, the fertile plants when the parent is nulliplex, and both sterile and fertile plants when the parent is simplex, the mst genes being effective in the S-cytoplasm. The segregations are explained by assuming that the p-mst plants carry S-cytoplasm and nuclear *fr* genes in varying numbers, whereas the male parent was heterozygous for such genes and had N-cytoplasm. The few exceptional ratios obtained by Myers (1946) could be due to the effect of modifying genes.

27.7 *Echinocloa colone* (barnyard millet)

In a spontaneously arisen mutant of this millet, a plant having nonexserting anthers with nonviable pollen but normal female fertility was found by Ayyangar and Srinivasa Rao (1941). The sterility is controlled by one recessive *ms* gene. The *ms* gene action is not known (Table 27.21).

27.8 *Eleusine*

27.8.1 E. coracana Gaertn

In a 4x (2n = 36) clone, Ayyangar (1931) detected pollen sterile plants whose female sterility was nearly normal. These plants exhibit normal male meiosis but mature microspore degeneration followed by tapetal breakdown occurred uniformly in them. Genetic control of the anomaly is not known.

27.8.2 *E. indica* (L) Gaertn (goose grass)

From a diploid (2n = 18) population, a mutant arose in which microspores degenerate after liberation from the quartets (this Vol.). The anomaly is controlled by one recessive *ms* gene. Whereas female fertility of mst's is normal, it is reduced in 4x colchicine-induced tetraploids which are also pollen sterile. Reduction in female fertility is due to autotetraploidy as multivalency reduced viable gamete formation.

27.9 *Festuca arundinacea* Schreber (tall fescue)

Six mst plants were identified in a population of the var. Brudzynska by Goral et al. (1981). These plants have pale green indehiscent anthers which are mostly empty, but some have sterile pollen grains. In three of these clones, pollen sterility is incomplete and is lower at 25°C than at 16°C. Sterility is complete in three other clones at both these temperatures. Male steriles are obtained in F_1 progenies when these clones are pollinated by the mft plants of the same variety. This indicates genetic control of the sterility, it being probably of gc-mst type. The two mst's, one sensitive to temperature and other nonsensitive, represent two distinct genotypes, but the precise nature of the mst is not known.

27.10 *Hordeum vulgare* L. (barley)

A self-pollinated diploid feed and food grain plant and cytogenetically intensively explored, this has over 50 *ms* genes in its genome, of which 33 are nonallelic. The *ms* genes have been backcrossed into a large number of spring and winter cultivars of two- and six-rowed types. About 176 spring barley mst's (70 six-rowed, 105 two-rowed and 1 deficiens type) comprise 130 separate mutational events. Forty six BC stocks and 147 winter barley (139 six-rowed, 8 two-rowed BC stocks) are available from Drs. Hockett and Wesenberg, respectively (Hockett and Reid 1981). In all g-mst cases, genes conditioning sterility are monogenic recessive. In addition, gc-mst is also known in barley. Formerly two-rowed *H. distichon* and six-rowed *H. vulgare* cultivated barleys were classified separately. These have been now grouped into a single species, *H. vulgare*, because they possess the same chromosome number (2n = 2x = 14), intercross freely to produce fertile hybrids and differ mostly by single genes.

27.10.1 Genic Male Sterility

Suneson (1940) first documented the existence of mst in field-grown barley in 1936. One recessive gene ms_1 controlled the sterility. Its female fertility was normal. Phenotypically the mutant was recognisable, as its glumes remained open and anthers at maturity were shrunken and pollenless. Austenson (1948) detected another *ms* gene (ms_2) linked closely to some phenotypic traits. In a brachytic dwarf mutant of barley, Kasha and Walker (1960) detected another ms_3 gene. These initial discoveries led to the subsequent isolation of about 50 *ms* genes, especially by Kasha and Walker (1960), Singh and Kaul 1966), Kaul and Singh (1966, 1968), Hockett and Eslick (1968, 1969, 1970a,b), Hockett et al. (1968), Roath and Hockett (1971), Hockett (1972a,b), Sharma and Reinbergs (1976), Deshpande et al. (1981), this Vol. From a composite stock of barley obtained from USDA, Kaul and Singh (1966,1968) Singh and Kaul (1966) isolated nine g-mst's designated MSB_1 to MSB_9. One more mst mutant was found by them from an Indian barley variety NP17. Allelism tests amongst these lines and with other available ones were not done. From two EMS-treated barley cultivars, Sharma and Reinbergs (1976) detected three mst mutants. Sterility in each of them is controlled by one recessive gene; the three genes being nonallelic. One of these exhibited low temperature sensitivity. Kumar and Singh (1983) obtained five mst mutants after treating six-rowed barley var. Jyoti with 20 and 25 kR γ-rays or 0.2% EMS. Sterility in each is conditioned by one recessive gene. The steriles bear more effective tillers and are taller than mft's.

The induction of mst by atomic bomb radiation was documented by Moh and Nilan (1953). Female fertility of this mutant is not known. Similar mst mutants were obtained by Yamashita and Ukai (1979) after treating barley seeds with ionising radiation and chemical mutagens. The chemical mutagen ethylene imine induced more mutations than γ-rays, the former being highly effective and efficient in inducing *ms* gene mutations. Deshpande et al. (1981) isolated two dwarf mst mutants in M_1 that had large flag leaves, long ears and matured late. They exhibited abnormal meiosis and their sporogenous cells degenerated prematurely but the tapetum persisted. In almost all these mst mutants, female fertility as evidenced by seed set after hand pollination is nearly normal.

27.10.1.1 Gene Symbolisation

Hockett and Reid (1981) compiled a list of g-mst barley mutants with known or unknown gene symbols. The *ms* gene symbolisation is done by Dr. E.A. Hockett, Montana State University, Bozeman, USA, the coordinator for the international collection of genetic male steriles (Tsuchiya 1972). After a mst stock is received by Hockett, it is assigned a number and permanent symbol. This symbol assignment is cleared and confirmed by Dr. Tsuchiya, (Colorado State University, Fort Collins, USA), who is also the overall coordinator of barley mutant gene symbol assignments. The gene symbols comprise three letters that designate the character, a number that represents a particular locus and a letter or letters that represent a particular allele at that particular locus. An unassigned letter symbol msg is given to a new mst mutant when it is added to the collection. After allelism tests are completed with the known mst stocks (ms_1 through ms_{33}), the new mutant is assigned the locus number to which it is allelic. This system has advantages, as the investigator can report on a new mutant mst genotype immediately and has not to wait for allelism tests necessary to assign a locus number; allelism tests require considerable efforts, labour and time. The letter part of the new symbol is maintained throughout for each mst genotype. Almost all the 33 nonallelic ms

Table 27.1. Chromosomal location of *ms* genes in barley

Chromosome	*ms* Genes located
1	ms_{10}, ms_{14}, ms_{22}, ms_{23}
2	ms_2, ms_3
3	ms_5
4	ms_{24}, ms_{25}
5	ms_1, ms_4, ms_{20}
6	ms_6
7	ms_{16}, ms_{19}

genes identified in the barley genome are from spring barley stocks and many more available ones await identification. Of these, 15 have been located on the seven barley chromosomes (Table 27.1). Künzel (1983) found ms_6 and ms_{24} to be located on short arms and close to centromere regions of their respective chromosomes.

27.10.1.2 Plant Phenotype

Barring a few whose spikes are compact and florets thin and appressed to the rachis (Fig. 27.1), most g-mst barley plants do not markedly differ phenotypically from the mft plants, but can be identified in the field at seeding time. Whereas mft plants lodge at maturity due to seed formation and their leaves turn yellow, plants of most of the mst mutants remain erect, seedless and their foliage is green and fresh. However, ms_9, ms_{15} and ms_{17} produce some seeds by selfing. Most mst plants possess flattened, shrunken and indehiscent anthers and the anther sacs of many mst genotypes are pale white at maturity. In ms_3 and ms_7, the anthers are reduced to rudiments. Sethi and Bhateria (1977) found a morphologically distinct mst mutant of barley. This was characterised by the transformation of filaments into flattened anther-like structures and the mutant was designated entire anther mutant by the authors. The sterility is controlled by one recessive gene, *ena*.

27.10.1.3 Anther and Pollen Phenotype

Anther and pollen development in mst genotypes varies from a nearly normal pattern (ms_6, ms_8 and ms_{16}) to formation of highly reduced stamens, ms_3 and ms_7 (Roath and Hockett 1970). Other genotypes exhibit abnormal and variable male sex organ development. Roath and Hockett (1971) categorised 18 gmst mutants into following four groups:

a) Anthers slightly shrunken, a little viable pollen produced, set a few seeds on selfing (ms_9).
b) Nearly normal-appearing anthers and pollen stomium present, some pollen of field-grown plants stained with TTC, pollen nonfunctional as no seed occurs following selfing (ms_6, ms_8, ms_{16}).

Fig. 27.1a-c. *Hordeum vulgare.* Spikes of male fertile **(a)**, partial male sterile **(b)** and complete male sterile **(c)** barley; lateral florets removed and awns clipped. (Qualset and Schaller 1968)

c) Anthers very small and shrunken, lacking stomium, filaments not elongated, pollen absent or shrunken and nonstaining (ms_3, ms_7).
d) Anthers subnormal and shrunken to highly reduced, no stomium, filaments not elongated, pollen shrunken or completely absent (ms_1, ms_2, ms_4 ms_5, ms_{10}, ms_{11}, ms_{13}, ms_{14}, ms_{18}, ms_{19}).

In most of the mst mutants, initial anther development is normal but as anthers mature, anomalies in anther form, organisation and contents appear. The anomalies involve abnormal tapetal form or function, irregular sporogenous tissue development of differentiation, incipient endothecial and vascular bundle development etc. The degree of anomalies during anther and pollen development varies not only between florets but also within anthers of a floret. Details of anther development in mst barley are given by Roath and Hockett (1970) and Mian et al. (1974).

27.10.1.4 Inapertulate Pollen

From the progeny of a cross between cultivated and wild barley, a monogenically recessive gene-controlled mst mutant arose spontaneoulsy. Male meiosis and initial pollen development was normal but the mature pollen was inapertulate (Ahokas 1975a), hence nonfunctional. Using scanning electron microscopy, the author found the mst mutant *msg 6 cf.* to have inapertulate pollen, in contrast to the results of Roath and Hockett (1971), who indicated a normal pollen development in the ms_6 mutant using light microscopy. The presence of inapertulate pollen was reaffirmed by transmission electron microscopy (Ahokas 1975b). Another inapertulate pollen was found in a mst mutant plant of a locally grown barley variety (this Vol.). Its anthers are normal, progression of male meiosis is slow, pollen grains have poorly formed exine and no germ pores. No seeds were set on selfing, but hand crossings yielded nearly normal seed set. The mst is controlled by one recessive gene.

27.10.1.5 Microsporogenesis

In almost all the known g-mst barleys, mst is conditioned by single recessive genes. The majority of these genes act upon the microsporogenesis after tetrad formation (Table 27.2). The earliest action of the *ms* genes occurs in a 0.2% EMS-induced mutant in which PMC's degenerate before meiotic onset (this Vol.). It is followed by tapetal degeneration. The anomaly is conditioned by a recessive *ms* gene, female fertility of this mutant being normal. In MSB_3, degeneration of PMC's in the uppermost three to four spikelets occurs soon after formation (Singh and Kaul 1966). In the basal region of the spikes, PMC's are fully formed and arranged on the periphery of the pollen sacs. Disintegration of the nuclei occurs around the pachytene. Thus *ms* gene action is earliest in the upper spikelets but is delayed in the lower ones. In two mst lines, *msg . . . ho* and *msg . . . fz*, of the barley variety Julia, the *ms* genes are allelic and cause defective cytokinesis at the end of meiosis I (Herd and Steer 1984). In another mst mutant, *msg . . . fg*, of this variety, *ms* gene action is nearly similar but nuclear restitution is incomplete. In *mst . . . gf*, failure of periclinal division of the inner secondary-parietal layer occurs so that the tapetal

Table 27.2. ms gene action over microsporogenesis in barley

Meiotic stage	Genotypes	References
Premeiotic PMC degeneration	ms, ms B3[a]	This Vol., Singh and Kaul (1966)
PMC degeneration during prophase-I	MsB3[b]	Singh and Kaul (1966)
Defective cytokinesis after meiosis I	*msg*...ho,fz,fg	Herd and Steer (1984)
Eccentric cell plate formation following I and II meiotic division	*ms*	Sharma and Reinbergs (1972)
During early microspore formation	*ms*7, *ms*B6[b]	Singh and Kaul (1966), Roath and Hockett (1970)
After microspore liberation	*ms*5, *ms*9, *ms*14, *ms*18, MSB1, MSB4	Singh and Kaul (1966), Mian et al. (1974)
During microspore maturation	*ms*8, *ms*10, msg...fp	Roath and Hockett (1970), Herd and Steer (1984)
During pollen grain maturation (pollen inapertulate)	*ms*6, *ms*g6cf, *ms*	Ahokas (1975a), Roath and Hockett (1970), this Vol.

[a]Apical spikelets.
[b]Basal spikelets.

layer is not formed at all. Male meiosis in *msg . . . fp* is normal until microspores are liberated, but normal sporopollenin production is inhibited in this mutant. In the mst mutant obtained by Sharma and Reinbergs (1972), the *ms* gene either suppresses cell plate formation or induces a high frequency of erratic cell plate formation in PMC's following meiosis I or II. This leads to the formation of bi-to multi-nucleate cells which develop into nonviable pollen grains. In ms_7, meiosis proceeds normally up to the quartet stage. As microspore formation is initiated, breakdown occurs. Microspores became shrivelled and periplasmodial. In a few PMC's microspore development continues, but no functional pollen grains are formed (Roath and Hockett 1970). Up to microspore liberation, male meiosis is normal in ms_5, ms_9, ms_{14}, ms_{18} (Mian et al. 1974), MSB_1 and MSB_4 (Singh and Kaul 1966), but microspores degenerated soon after their formation before karyokinesis takes place in them. Whereas in ms_8 and ms_{10}, the *ms* genes act during the microspore maturation stage, the ms_6 gene acts later during or after pollen grain maturation. If somehow the action of the ms_6 gene is delayed, pollen fertility can be restored. However, in the majority of barley mst mutants, male meiosis continues up to the quartet stage, but *ms* genes act after this stage. In some mst genotypes even when most of the microspores have aborted, a few PMC's continue microspore development but no functional pollen grains are formed.

27.10.1.6 Tapetal Breakdown

In mft barley plants, tapetal cells become binucleate and degenerate when PMC's are in interphase II. When the microspores are released from PMC's, the tapetal tissue is reduced to a thin layer of cells, so that at anther dehiscence and during pollen liberation only traces of tapetal tissue are left out. Tapetal behaviour differs in various mutants (Table 27.3). In ms_5, ms_7, ms_{10} and ms_{14}, the tapetum persists and disintegrates after microspore degeneration (Roath and Hockett 1970; Mian et al. 1974). In ms_9, the tapetum collapsed suddenly after the free microspore stage. This breakdown is early in ms_{18} when meiotic division is still going on (Mian et al. 1974). In the mst mutant studies by Sharma and Reinbergs (1972), tapetal breakdown occurs at the normal time, but anomalous cytokinesis leads to nonviable gamete formation. Within spike differences in tapetal and microsporogenesis, breakdown occur in MSB_3 (Singh and Kaul 1966; Kaul and Singh 1966, 1968). In the uppermost three to four spikelets of the spikes, the tapetum degenerates before meiosis begins, a little later PMC's also degenerate without undergoing meiosis. Anthers developed in the lower part of the same spike possess well-developed tapetum and PMC's. The PMC's are arranged on the periphery of the pollen sacs. Meiosis is initiated in them but microsporogenesis disrupts due to nuclear disintegration of PMC's. Tapetal tissue breakdown occurs later when PMC's have completely degenerated. Thus whereas in MSB_3, in the three to four uppermost spikelets, the tapetum degenerates much earlier to PMC degeneration, in the lower spikelets the reverse occurs. In MSB_6, PMC's degeneration is concurrent with tapetal degeneration. In MSB_1 and MSB_4, tapetal degeneration follows microspore degeneration. The endothecial cells of these two mst lines do not develop fibrous thickenings and this inhibits anther dehiscence. A strong vacuolation, reduced cytoplasm and lack of enzyme acid phosphatase activity in the tapetal cells of these mst lines indicate the low metabolic activity in the tapetal cells. In ms_6, ms_7, ms_8 and *ms*, gene action over the tapetum is early. In ms_6 and ms_8 a slight separation of tapetal cells occurs during premeiosis, and in ms_7 tapetal cells swell until TII (Roath and Hockett 1970). In ms_8, tapetal degeneration is slow and a considerable part of the tapetal tissue remains intact even during microspore maturation and preanthesis. Microspores became periplasmodial after the quartet stage in ms_7 but the tapetum persists, enlarges and becomes highly vacuolated. Thus whereas in some mst genotypes (serial 1–3, Table 27.3), the *ms* gene does not seem to act via the tapetum, in others (serial 4–6, Table 27.3) in acts via the tapetum.

Table 27.3. Tapetal behaviour in relation to microsporogenesis breakdown in genic male sterile barley

Tapetal behaviour	Genotypes	References
Normal tapetal degeneration (as in mft)	ms	Sharma and Reinbergs (1974)
Tapetal degeneration prior to meiocyte degeneration	ms18[a] MSB-3	Mian et al. (1974) Kaul and Singh (1966)
Concurrent tapetal and meiocyte degeneration	MSB-6, ms	Kaul and Singh (1966), This Vol. Roath and Hockett (1970)
Tapetal persistence and degeneration later than microsporogeneous tissue degeneration	ms5, ms7, ms10, ms14 MSB-1 msB-3[b], msB-4	Roath and Hockett (1970), Mian et al. (1974) Kaul and Singh (1966)
Sudden collapse at free microspore stage	ms9	Mian et al. (1974)
Anomalous tapetal development	ms6, ms7, ms8	Roath and Hockett (1970)

[a]Apical spikelets.
[b]Basal spikelets.

27.10.1.7 Linkage

Kasha and Walker (1960) found the ms_3 gene linked with the Brachytic dwarf trait, a short-statured mutant with short, stiff leaves and dense awns (Powers 1936; Swenson 1940). The gene ms_3 enhances the mutant expression of Brachytic dwarf as its florets exhibit the tendency to fasciate strongly. Its female fertility is normal. The existence of a close linkage between ms_{16} and DDT-resistant gene was detected (*ddt*) by Wiebe (1960). The two genes are seven crossover units apart. The gene ms_6 is located near the centromere of barley chromosome 6. Close to this is another gene *sex 1a*, causing the endosperm to become shrunken.

27.10.1.8 ms Gene Action

In almost all the g-mst's, female fertility is nearly unimpaired. This is evidenced from a near-normal seed set obtained after normal cross plus hand pollination in most of them. Some of the pollen produced by a mst genotype is stainable. While nonstainability confirms pollen sterility, stainability usually by TTC, carmine or iodine is not a sound indication of their viability in barley and onion. Stainable pollen produced by ms_1, ms_2 and ms_7 genotypes is not functional (Roath and Hockett 1971). The *ms* gene action is specific, limited to certain meiotic stages in barley. Whereas in some genotypes the gene action is directly over sporogeneous cells or spores, in others it is mediated via the tapetum (Table 27.3). The gene action thus becomes diverse. This is supported by the studies of Mian et al. (1974) in five mst genotypes. In ms_5, the gene disrupts the mechanism crucial for the prompt release of young microspores from the callose, while in ms_9 it causes DNA and histone deficiency, leading to deformation and developmental failure of the PMC's. In ms_{10}, the gene affects the permeability of tapetal and sporogenous cell walls and prevents the normal flow of nutrients, including DNA and histone precursors, from tapetal cells to the developing microspores. In ms_{14}, the *ms* gene blocks transport of nutrients from the tapetum to the microspores by inducing thickening in tapetal cell walls. The gene ms_{18} acts during the premeiotic stages and causes a deficiency of DNA and histones in the tapetum. This leads to meiotic failure in the meiocytes of the mst plants. The impact of ms_5 and ms_9 gene action over DNA and histone turnover occurs first in the sporogenous tissue and then in the tapetum (Mian et al. 1974). This indicates that ms_5 and ms_9 gene action are initiated within the microspores when they are still invested in the callose wall. In ms_{10} and ms_{14}, a reverse sequence occurs and thus the *ms* gene action is initiated in the tapetal tissue and the microspores are affected later. In ms_{18}, *ms* gene action is mediated via tapetal tissue even before meiosis is completed. In this mutant, tapetal breakdown occurs early when meiosis is still progressing. However, meiosis continues but the microspores abort. The earliest action of the *ms* gene over anther development occurs in ms_3 in which anthers are rudimentary and microsporogenous tissue differentiation is inhibited.

27.10.1.8.1 Environmental Influence

Like many other *ms* genes, these genes in barley are sensitive to temperature, photoperiod, physiography or location. Thus when six ms_9 lines were grown at Elimaki, Finland (61°N Lat., light period up to 24h) and at Bozeman, USA (46°N Lat., light period up to 16–17 h), ms_9 was completely mst at Elimaki but partially mft at Bozeman (Ahokas and Hockett 1977). Even at two US locations, its mst level differed. At Bozeman, 10% selfed seeds are produced by it but at Tucson the seed set is 24%. Ahokas and Hockett (1977) suggested fertility reversion as due to photoperiodic influence over the *ms* gene action. Variations in female fertility also occur at these two locations in some mst mutants (Hockett and Ahokas 1979). Of the three *ms* genes detected by Sharma and Reinbergs (1976), one is temperature-sensitive, as all three lines are mst at 30°C day/25°C night, but at 15°C one line became mft. The environmental influence on ms_9 and ms_{10} genotypes is high. Early-formed tillers of ms_{15} bear mst florets, but the late-formed ones are partially fertile at Bozeman, USA.

27.10.1.8.2 Total Suppression

Chemical Suppression. A foliar spray of 100–300 ppm of GA to two plants of a spontaneous mst mutant of the Maris Baldric variety restored partial fertility in them (Kasembe 1967). On selfing, the treated plants developed some seeds that produced complete mst progeny, indicating a temporary mft restoration by GA treatments.

Genome-Cytoplasm Suppression. A recessive *ms* gene in homozygous state induces mst in the Betzes barley variety (Pfeifer 1972). Crossing this mst Betzes barley with fertile barley yielded mft F_1 which segregated in F_2, producing 25% mst plants. When this *ms* gene was transferred into the genome of the barley variety Penrad, no sterility resulted. Since no reciprocal crosses could be made, cytoplasmic differences cannot be investigated. Thus either genome or cytoplasm or both suppress the expression of this *ms* gene in barley.

27.10.1.9 Biochemical Differences

Many g-mst mutants of barley are known but the biochemical activity of the anthers of only a few has been compared to their corresponding mft anthers. Either low or complete absence of enzyme acid phosphatase occurs in the anthers of MSB_1 and MSB_4 plants (Kaul and Singh 1966, 1968). In ms_5, ms_9, ms_{14} and ms_{18}, microspore nuclei undergo loss of DNA and histones, but in ms_{10} an initial increase of these two macromolecules is followed by a dramatic drop (Mian et al. 1974). In mature anthers of the mst mutant that has inapertulate pollen, the present author found a highly reduced activity of acid phosphatase (18% of mft plants) and lack of enzyme esterases. The formation of aperture in a pollen grain has been associated with specific chemical characteristics, including these two enzymes (Ahokas 1975b, 1976a). Cytochemical evidence indicates that the enzyme esterase is capable of hydrolysing sporopollenin of pollen grains.

27.10.2 Gene-Cytoplasmic Male Sterility

The first report of gc-mst in barley is by Favret and Ryan (1966), who found some p-mst's in a mutated barley population. One of these segregated mst and mft plants; the progeny of mst plants obtained after crossing to mft plant was mst. Restorer for this mst has not been found. Later Schooler (1967) detected mst's in F_1 progeny of two amphidiploid barley parents. The female parent was a 6x amphidiploid plant derived from the cross *Hordeum jubatum* (2n = 28) × *H. compressum* (2n = 14). The male parent was a 4x derivative of the cross *H. bulbosum* (2n = 14) × *H. vulgare* (2n = 28). Thus whereas the female parent was derived by pollinating tetraploid by diploid barley, both the parents of the pollinator single cross were autotetraploids. Using the F_1 as the female parents (*Hordeum jubatum* 2n = 28 × *H. compressum* 2n = 28) with (*H. bulbosum* 2n = 28 × *H. vulgare* 2n = 28), both the parents of pollinator single cross were autotetraploids. Using F_1

as the female parents and crossing it to the diploid hybrid obtained from a cross *H. vulgare* (2n = 14) var. Traill × *H. bulbosum* (2n = 14) mst progeny resulted in F_1 and in BC_1 progenies. When *H. vulgare* var. Dickson and Larker were used as pollen parent in crosses to mst BC_2 plants, the progeny was mst. In similar crossings, but using *H. vulgare* var. Traill as the pollen parent, complete to partial mst progeny developed. Thus Traill acted as maintainer of this sterility; the source of S-cytoplasm is *H. jubatum*. Satisfactory restorer genes for this gc-mst were not found (Foster and Scholler 1971).

Ahokas (1978a) found gc-mst in a naturally occurring variant in an Israeli strain of the wild progenitor of the cultivated barley *H. vulgare* ssp. *spontaneum*, (C. Koch) Thelling. This strain possessed S-cytoplasm and a dominant *Fr* gene designated by the author *Rfm la* (≡Fr_1). Cultivated barley varieties are mainly maintainers of sterility, 10–15% are partial restorers in Finland (Ahokas 1979a). Using 191 accessions of the wild progenitor of barley from Israel (Ahokas 1981) and pooling and results obtained from similar earlier studies in which 113 such accessions were tested (Ahokas 1979b, 1980b,c), the author found 56.7% accessions to be maintainers, 21.1% semi-partial restorers (0.1–5% fertility restoration), 16.4 partial restorers (5.1–60% fertility restoration) and 5.8% complete restorers with which over 90% seed set occurs after hand pollination. All the restorer accessions have N-cytoplasm and a single dominant *Fr* gene.

Another source of gc-mst (S2 type) was found by Ahokas (1982b) from *H. vulgare* ssp. *spontaneum* selection 79B S14–3 that originated from the southern coastal plains of Israel. The plant was normal fertile at Finland but on crossing to the Addora cultivar of barley, the F_1 progeny was p-mst. Evidently the selection possessed a recessive or a semi-dominant *Fr* gene. The dominant restorer gene *Rfm* la (Fr_1) for another gc-mst (Sl type) restores fertility to this S-cytoplasm as well. Female fertility of this gc-mst type 2 is normal and its sensitivity to partially restoring genotypes is much higher than that of gc-mst type-1 plants.

27.10.2.1 Cytoplasmic Differentiation

Two cytoplasmic types designated presently S_1 and S_2 were found by Ahokas (1978a,b, 1982) from an Israeli strain of *H. vulgare*, ssp. *spontaneum*. These exhibit different partial fertility restoration characteristics; the type S_2 exhibiting significantly higher fertility restoration than type S_1. When estimated with partially restored F_1 combinations, S_2 cytoplasm is nearly 50 times more sensitive to *fr* genes present in the *spontaneum* than S_1 cytoplasm (Ahokas 1982b,c).

27.10.2.2 Anther and Tapetal Anomalies

In the mst plants detected by Scholler (1967), male meiosis is normal but microspores abort immediately after their liberation. Tapetal and outer anther layers are thick and do not disintegrate during early microspore development but degenerate after microspores abort. Some mst plants possess only three instead of four anther lobes. Thus both anther and tapetal anomalies occur in this mst genotype. These anomalies prevent normal pollen development and maturation. The TEM micrograph of the anther locule of msm 1 (Ahokas 1978b) depicts microspores with degenerated cytoplasm and thick exine, well-developed intact tapetum possessing several plastids, many of which have a starch grain, highly thickened and aggregated sporopollenin deposits and homogenously thick locule wall (Fig. 27.2a). At anthesis, the mature anthers have degenerated epidermis, endothecium and tapetum (Fig. 27.2b), along with aborted pollen.

27.10.2.3 Environmental Influence

Ahokas and Hockett (1981) compared the mst expression and female fertility of ten gc-mst derivative plants having S_1 type of cytoplasm by growing them along with their corresponding restorers under two latitudes (46°N at Bozeman USA and 61°N at Elimaki, Findland) that had two photoperiods (16–17 h at Bozeman and up to 24 h at Elimaki). Of the ten derivatives tested, nine exhibited a near-stable expression at both the locations. Whereas Addora-like derivatives, heterozygous for the *Fr* gene where mft, those with partial (50%) ssp. *spontaneum* genotype (and heterozygous for *Fr* gene) exhibited a lower mft at Elimaki. At Bozeman such plants were fully mft. Since female fertility and mst expression of S_1-type cytoplasm was stable under two photoperiods and latitudes, a high stability of this gc-mst type is inferred. It enhances the value of this sterility for use in hybrid barley production.

27.10.2.4 Biochemical Differences

i. *Seed Protein Content.* Introducing S_1 type gc-mst in high lysine barley varieties decreases lysine and increases seed protein content in the derivatives (Ahokas 1979b). With such introductions in normal lysine barley varieties, no decrease in lysine content but a slight increase of total protein content occurs. Ahokas (1980a) explained this increase in seed protein content as due to a more complete proteolysis and mobilisation of amino acids from the vegetative organs to the grain-filling stage. A higher protein content and a faster proteolysis in kernels may underlie the faster germination observed by Ahokas (1980a), as envisaged through a rapid root elongation of mst seeds.

ii. *Polypeptide Pattern.* Ahokas (1980a) compared the polypeptide pattern of the restored (mft) and nonrestored gc-mst (S_1-type) barley anthers. In mst anthers, the polypeptide pattern was juvenile at the mid-anther development stage and two major zones were either weak or absent. When the anther development stopped, the proteins became hydrolysed to short-chain peptides, unlike in the restored plants (that were heterozygous for the restorer gene) and normal mft plants. Thus mst barley anthers have abnormal proteolysis occurring during the senescence of vegetative organs like leaves. Both senescence and proteolysis control are associated with cytokinins.

iii. *Soluble Amino-Nitrogen Content.* This is significantly lower in mst anthers than in mft anthers during 4–5 and 1–2 days prior to anthesis (Ahokas 1982c).

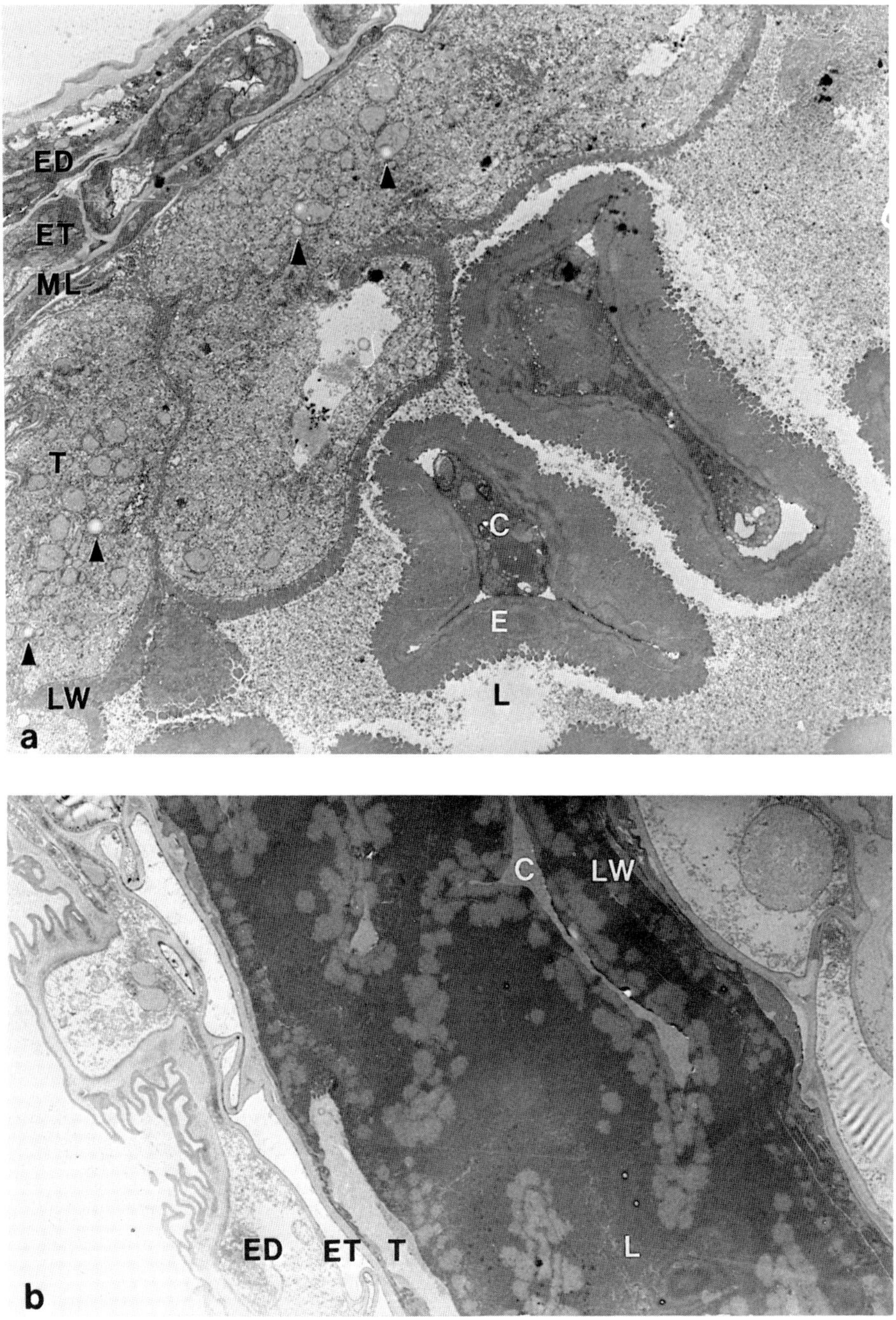

Fig. 27.2a,b. *Hordeum vulgare.* **a** TEM of male sterile anther locule showing degenerated microspore cytoplasm (*C*), thick exine (*E*), tapetum (*T*), locule (*L*) wall (*LW*); *arrows* depict starch grains. **b** TEM of male sterile anther locule showing degenerated epidermis (*ED*), endothecium (*ET*) and tapetum (*T*). (Ahokas 1978b)

During anther maturity this content is considerably enhanced in mft anthers. The depletion of the amino N-pool in mst anthers is normalised in restored plants. Reduction of the sink capacity and/or differential N-distribution in mst plants decreases the capacity of these anthers to synthesise the late-formed proteins which are either absent or present in low amounts in mst barley anthers.

iv. *Lipoxygenase Activity*. Either mitochondria or plastids are the carriers of *c*-gene(s). Since lipoxygenase (Lox) activity is located either entirely in plastids (Douillard and Bergeron 1979a,b; Douillard 1980) or partly in mitochondria (Siedow and Giruin 1979), their association with mst in barley is suspected. This led Ahokas (1982a) to evaluate Lox activity in three almost isogenic lines of barley viz. normal, mst and restored lines. Evaluated at the anthesis stage, the activity was lowest in the leaves of mst plants. In them, Lox activity in the developing anthers was higher than in the leaf tissue, but it decreased considerably during the degeneration stage of mst anthers.

v. *Cytokinin Content*. The involvement of cytokinins in the normal functioning of plants is well known. Compared to mft barley plants, mst plants possess a reduced amount of cytokinins. In the fully restored plants having S-cytoplasm, the cytokinin content increases or nearly equals that of mft plants having N-cytoplasm. Thus the *Fr* gene appears to be associated with an increase either in total cytokinin content or in one or more of the cytokinin types. This relationship is confirmed by certain physiological features of mst plants that are closely associated with deficient cytokinin levels. These features are: (a) increase in chlorophyll a/b ratio, xanthophyll/β-carotene ratio, susceptibility to induced leaf senescence and rate of germinal root elongation, (b) decrease in carotenoid content and nitrogen pool in anthers and (c) anther-protein instability (Ahokas 1980a, 1982a). Whereas both biological activity and total quantity of cytokinin are very low in gc-mst barley plants having S_1 cytoplasm, these are high and similar in three mft lines (Ahokas 1982d). The maintainer lines having Rf_1 gene have eight to nine times more bound cytokinin content, but the content is less in partially restored lines in which a limited cytokinin amount may be translocated and utilised by the anthers of the lowest spikes that becomes mft. Thus in partially restored plants, fertile anthers appear in the lowest florets of the spike. In fully restored plants, the cytokinin content is sufficient and freely available to most of the florets whose anthers become mft. Whether cytokinin alone conditions mst in barley or vice versa can be tested through artificial cytokinin enrichment of mst barley plants. In fact cytokinin under the influence of the *Fr* gene may suppress the *fr* gene action through direct association with t-RNA or by acting as gene repressor. Both chloroplast and mitochondria are target sites for cytokinin action. In barley, plastids (Ahokas 1976b) and mitochondria are transmitted maternally. Either one or both of these organelles are putative carriers of the *c*-gene responsible for maternally transmitted mst. Cytokinins are directly associated with the two cytoplasmic organelles, mitochondria and chloroplast, and are thus implicated in gc-mst systems.

vi. *Senescence*. Mature stem leaf pieces of mst barley senesce with 20 mM NaCl solution but such pieces of mft restored and normal barley plants do not senesce in NaCl or in distilled water (Ahokas 1980a). This NaCl-induced leaf senescence is prevented by exogenous kinetin application which reduces Na^+ uptake (Ilan 1971). The retardation of senescence by kinetin application is through inhibition of proteolysis, a phenomenon associated with senescence (Wittenbach 1978). Since compared to mft plants mst plants possess low amounts of cytokinin (Ahokas 1982c), senescence of mst stem leaf pieces in NaCl is due to the enhanced Na^+ ion uptake and the consequent toxicity.

27.10.2.5 c-Gene Location

Ahokas (1978b) regards plastids as the most probable organelle responsible for the transmission of gc-mst in barley. Mature leaves of the restored plants are characterised by a higher carotenoid content, lower xanthophyll/β-carotene ratio and a different chlorophyll a/b ratio than those of the nonrestored ones. Such changes in chloroplastidic pigments are due to cytokinins which are known to lower the chlorophyll a/b ratio (Kaul and Sabharwal 1971; Farineau and Roussaux 1975), increase carotenoid content (Farineau and Roussaux 1975; Mukherji and Biswas 1979) and lower xanthophyll/carotene ratio (Mukherji and Biswas 1979). One of the target sites of cytokinins is plastid (Buschmann and Sironval 1978) and the chloroplastidic fraction possesses some proteinase activity (Wittenbach 1978). These findings favour plastid transmission of gc-mst in barley. In other words, plastids may carry the *c*-genes in barley.

27.10.3 Photoperiodic Influence

Short-day photoperiod exposures cause young microspores to clump (Batch and Morgan 1974). Male meiosis is abnormal and polyads with micronuclei are formed. Female fertility of the plants is normal because hand pollination yields high seed set.

27.10.4 Chemical Male Sterility

i. *CCDP.* Yih et al. (1971) induced complete mst by applying this chemical after floral initiation. The status of female fertility of the treated plants is not known.

ii. *Ethrel.* Using about 750 ppm of this chemical, Stoskopf and Law (1972) induced pollen sterility in 36% of heads of the treated plants. Female fertility was also reduced as 53–72% seeds were set on the treated plants crossed by *H. bulbosum* pollen (Law and Stoskopf 1973). Apart from the height reductions, many morphological ill effects such as partial ear emergence from the boot leaf occurred in the plants treated at low concentrations such as 750 to 1500 ppm. Whereas many Canadian and European barley varieties appear to be highly sensitive to ethrel (Stoskopf and Law 1972), Indian barleys are more tolerant. Thus higher doses like 3000–4000 ppm, induce a high level of mst without impairing female fertility appreciably (Kumar et al. 1976; Verma and Kumar 1978). Optimal results with ethrel were obtained when awns were just visible through the boot. If the chemical works similarly in other barley varieties, its use may be economical and useful for hybrid barley seed production.

iii. *GA*. Treating after vernalizing seeds of the winter barley variety Wong with 500 ppm potassium gibberrelate solution for 20 min and spraying the resultant plants with this chemical every third day from 10–70 days after planting produced mst plants that had nonviable pollen in top third of the spike (James and Lund 1960). In some gibberrellin-treated plants, failure of the spikelet tips to develop led to short and seedless spikes (Paleg and Aspinall 1958). Female fertility of gibberrellin-treated plants is not known.

iv. *RH-531*. When a foliar spray of the chemical was applied at the rate of 6 kg ha^{-1}, 26 days and 1.5 to 6 kg ha^{-1}, 21 days before heading or at about the six-leaf stage (shortly before meiotic initiation), complete mst occurred in one winter and two spring cultivars of greenhouse-grown barley (Wang and Lund 1975). Female fertility was reduced to 19% in one and to 40% in the other variety. Hockett (1971) induced mst by applying the chemical after meiosis and 4 days prior to anthesis. Using 4 kg ha^{-1} spray of the chemical induced partial male and female sterility in three barley varieties. Female sterility increased when the plants were treated at the six-leaf stage. (Hockett and Feltner 1978). With the increased chemical dosage, the mst degree was enhanced, but this was accompanied by decreased grain yield, shoot height, flowers per spike, lodging, kernel size and spike length. The chemical treatment rendered the tapetum persistent, which caused carbohydrate starvation of the microspores and led to their subsequent degeneration. No meiotic anomaly occurred following the chemical treatment (Wang and Lund 1975). Using similar chemical concentrations, response to this chemical was different in the barley varieties tested by Wang and Lund (1975) and by Hockett and Feltner (1978). Whereas the former authors obtained complete mst accompanied by a low degree of fst, the latter authors obtained partial male and female sterility in the field-grown plants. With increasing chemical concentration, both male as well as female sterility increased. In fact, in barley, the greenhouse environment is more conducive to sterility than the field environment. Sterility occurred in untreated barley plants grown in the greenhouse (Hockett and Feltner 1978). Moreover, the response of barley varieties to the chemical differs. Keeping in view the partial male and female sterility induced by this chemical and its genetic toxicity, its commercial use as a male gametocide is not feasible.

27.10.5 Breeding Utility

Barley, a diploid-selfer, has low NCP ($<$ 0.2%; Stevenson 1928; Robertson and Deming 1931). Natural crossing is greater in awnless, awnleted and naked types than in the awned ones. Six-rowed types have more NCP than two-rowed types. Cultivar and seasonal differences are profound; within-cultivar variance exceeds seasonal variance. Although its breeding system narrows its genetic variability, the rate of spontaneous and induced mutations is high in barley. This has led to the isolation of numerous mst and other mutant genes. Due to the availability of many *ms* genes and translocations, many methods to maintain and use the g-mst lines in hybrid breeding have been proposed from time to time. Some of the main ones are:

a) DDT resistance and ms_{16} linkage (Wiebe 1967),
b) *ms* genes plus restorer cytoplasm (Hermsen 1965),
c) balanced tertiary trisomics in combination with g-mst (Ramage 1965),
d) balanced g-mst lethals, heterozygous translocations and dominant pre-flowering selective genes (Eslick and Hockett 1967),
e) fertility restoration of g-mst by chemical spray (Kasembe 1967),
f) fixing *ms* genes through apomixis (Wiebe 1967).

27.10.5.1 Linked ms Genes

To increase the mst proportion in the progeny or to allow removal of fertile plants, several procedures have been outlined (Wiebe 1960; Eslick 1970; Ramage 1975; Foster 1979). Jarvi and Eslick (1975) proposed linking a xenia-expressing shrunken endosperm gene to the *ms* gene and Falk et al. (1980) proposed the use of the shrunken endosperm marker, *sex* 1 gene closely linked to ms_{6}, for barley breeding. Initially the systems proposed by Wiebe (1960), Eslick (1970) and Foster (1979) were based upon the linkage of a ms gene with a dominant gene, which would allow the selection of mst plants prior to flowering. Wiebe (1960) made a proposal for hybrid barley production using ms_{16} linked to a gene-conditioning reaction to insecticide DDT. Dominant (Ddt) and recessive (ddt) genes cause susceptibility and resistance to DDT, respectively. The gene combination *ms*-ddt produces *ms*-DDT-resistant plants, so that in a segregating population used as a seed parent, spraying with DDT will kill plants heterozygous or homozygous for the Ms-DDT linkage, leaving only mst plants which are homozygous for *ms-ddt* genes. Seed harvested from the DDT-rogued rows would be 100% hybrid. With the appropriate combinations of parents, this technique can be used to produce single, triple- and double-cross hybrids. In the first two hybrids, 100% of the plants will shed pollen, in the double-cross hybrids 50% of the plants will shed pollen. The main obstacles to the use of this plan are: (a) such a linkage is known only in one *ms* gene, (b) seed set on mst barley is low, (c) any linkage less than complete will leak fertile resistant plants, which necessitates continual hand roguing of seed stocks and crossing fields. Inversions and translocations could be used to tighten linkage between *ms* and *ddt* genes. The DDT spray impairs female fertility and the consumption of hybrid seed would be harmful to the biological system. Therefore the utility of this linkage is limited in practice. On the other hand, the close linkage between shrunken endosperm and ms_{6} (Falk and Kasha 1980) permits a presowing selection of seed homozygous for mst (Falk and Kasha 1981; Falk et al. 1981). The mst seedlings can be planted in blocks to produce F_1 hybrid seed, as proposed by Falk and Kasha (1980). The mst plants can also be used as female in the traditional crosses or in synthesising recurrent selection populations where numerous male parents may be desired. Since crossover frequency is less than 1%, any contamination from shrunken endosperm fertile (crossover) plants will produce shrunken endosperm seed, whereas the outcrossed seed will be plump. The Orange lemma marker gene will show that the germplasm is in a pure state or it can be a case of selfing among the crossed seeds. In order to maintain and/or increase the genetic stock, the shrunken endosperm seeds should be removed and the plump seeds planted. The uniculm plants will have to be rogued to leave only the desired heterozygous plants. These will yield about 25% shrunken endosperm seed which can be used as source for mst plants and 75% of plump seeds that can be utilized for maintaining the germplasm. Shrunken endosperm, although it increases protein and lysine, is an undesirable trait for commercial seed production, hence the use of this linkage is limited.

27.10.5.2 Chemical Restoration

Fertility restoration by GA treatment (Kasembe 1967) in a mst barley provides a method for the maintainance of mst genotype, but GA treatments do not restore fertility in all the mst barley genotypes. Moreover, the treatments cause considerable stem elongations so that the mature culms either break off or lodge to the ground. Some of these treatments reduce female fertility and seed yield.

27.10.5.3 Apomictic Fixation

Fixing of *ms* genes through apomixis, as suggested by Wiebe (1967), is not tenable, as in many cases instead of apomictic lines, aneuploids, and haploids were produced.

27.10.5.4 Male Sterility Suppression

i. *Cytoplasm Source.* Since chemical suppression of *ms* gene action is not feasible for commercial hybrid seed production, Hermsen (1965) proposed a procedure by which *ms* genes could be placed in N-cytoplasm. This line N-*msms* would act as maintainer for S*msms*. This proposal is based on the supposition that g-mst plants possess not only *ms* genes but also S-cytoplasm. However, the N-cytoplasm envisaged by Hermsen (1965), capable of fertilizing and maintaining g-mst barley lines, could not be found.

ii. *Genome-Cytoplasmic Combination.* Pfeifer (1972) opined that Betzes barley variety has S-cytoplasm and with a recessive *ms* gene in homozygous state, mst results in it. The source of S-cytoplasm is from Betzes barley and the same cytoplasm is present in most of the spring barleys. According to this author, the barley variety Penrad has a recessive form of the restorer gene and N-cytoplasm. Crossing the two barley varieties Betzes × Penrad yielded mft F_1 progeny and 3:1 mft:mst F_2 progeny. When the *ms* gene is transferred to the Penrad variety, no mst results even when this gene is in a homozygous state. However, it is not clear if the genome alone or genome cytoplasm interaction suppresses the action of *ms* genes in the Penrad barley variety, as reciprocal crossings using Penrad var. as female and Betzes as male were not possible. Pfeifer (1972) was apparently able to develop maintainers and produce 100% mst plants. This indicates a probable g-c interaction leading to mst rather than genome-cytoplasmic suppression of *ms* gene action. Further studies are needed to clear this mechanism of *ms* gene action. On the basis of the g-c interaction concept leading to mst in Betzes barley, Pfeifer (1972) proposed a simple method for hybrid barley seed production. In this method, fertile barley plants having the *msms* gene in the Penrad variety could be used to wind-pollinate isogenic lines which are mst (i.e. Betzes mst lines). Seeds obtained from such crossings would produce 100% mst plants in the next generation. These could be pollinated by any normal barley variety to produce a self-fertile F_1 hybrid. A major advantage of this system is that it obviates the necessity of introducing restorer genes into male parental lines of potential hybrid varieties. Most of the *ms* genes in barley tested so far are of genic type and do not exhibit such genomic or cytoplasmic suppression. Therefore, Pfeiffer's method, although useful, may not be feasible with many other *ms* genes of barley.

27.10.5.5 Balanced Trisomics

Ramage (1965) produced balanced tertiary trisomics (BTT) carrying an *Ms*-allele on extra chromosome and an *ms*-allele on the two normal chromosomes for hybrid seed production in barley. As the extra chromosome carrying *Ms* is rarely transmitted through the male, nearly 70% of the selfed progeny of such a trisomic would consist of the desired mst diploids. In addition, the cross *ms-ms* line × trisomic would result in nearly 100% mst. However, practical use of this cross requires abundant pollen production by the trisomic. In the BTT system outlined by Wiebe and Ramage (1971), the dominant gene for mft (*Ms*) and for a green seedling (*A*) are placed on the interchange chromosome at the breakpoint and the corresponding recessive alleles for mst (*ms*) and albino seedlings (*a*) on separate normal chromosomes. Barley being diploid, the extra interchange chromosome is not transmitted through pollen, because those receiving this extra chromosome at meiosis exhibit delayed maturation so that they have no chance in pollination. Instead, eggs with an extra chromosome are functional. In the end, the extra chromosome is eliminated in one of the final breeding steps and thus is absent in the commercial hybrids. This extra interchange chromosome with its specific genetic information can be viewed as an interim program device to schedule certain required genetic events in the overall plan for producing a female parent of the hybrid.

27.10.5.6 Maintenance and Multiplication

Mst barleys are usually maintained by hand pollinating them with the pollen of normal mft plants. Foster (1972) proposed an alternative automatic roguing system for maintenance and multiplication of mst genotypes. In this system, heterozygous *Msms* stocks are allowed to self and segregate. The pollen produced by these fertiles will be 2*Ms*:1*ms* type and the seed set on the mst's would be two thirds fertile and one third mst, all the mft plants being heterozygous. When these fertile plants are made to pollinate the mst segregants, seed set on the mst plants in this and in every subsequent interpollination will segregate one heterozygous fertile to one mst. Maintenance and multiplication of the stock can be carried out repeatedly and extensively and in every generation 50% mst plants will segregate invariably. Thus this system obviates the necessity of repeated progeny testing and provides seed stocks that will repeatedly segregate 50% mst as opposed to 25% in the selfed progeny of *Msms* plants.

27.10.5.7 Problems in Breeding

i. *Pollination Limitations.* In all cases of hybrid seed production, hand pollination is necessary, as NCP in mst barley is extremely low. This limits hybrid barley production without additional manipulations of pollen shower, female receptivity and nicking time (Hockett and Eslick 1970). In the "entire anther mutant" found by Sethi and Bhateria (1977), lemma and palea are widened apart. This permits a higher rate of outcrossing in this mutant, but the hybrid seed set is low. Since awnless, awnleted and naked spikes of barley exhibit higher outcrossing, Suneson et al. (1970) suggested the use of a gene pool of the naked seeded barleys for use in hybrid breeding, because in addition to higher foreign pollen receptivity they have efficient potential as pollinators for producing hybrids and for subsequent mechanical seed separation based on specific gravity, seed colour or seed size. Using six bulk populations, each containing a recessive *ms* gene, and selecting for the seeds produced on mst plants, Campbell and Kasha (1981) obtained a highly variable seed set that fluctuated from year to year from 10.5–51.6%. No constant trend either upwards or downwards occurred over a 6-year period of testing. Thus NCP is neither sufficient nor reliable to produce hybrid barley seeds. It needs supplementation by hand pollination.

ii. *Outbreeding and Population Fitness.* As mentioned earlier, the main obstacle to the use of g-mst in barley is its extremely low, rather negligible amount of outcrossing. Jain and Suneson (1966) detected higher cross-pollination rates in some mst barley genotypes. This enhanced between- and within-family components of genetic variability, but it was not accompanied by any greater rate of improvement in population fitness. Instead, the mean fitness under a higher outbreeding system was lower than that achieved with regular low rates of outcrossing. Therefore, the predominant selfing system is advantageous to the fitness value of barley populations and the use of *ms* genes lies only in the exploitation of F_1 heterosis.

27.11 *Lolium*

Of the eight species of this inexpensive temporary lawn grass genus, the annual *L. multiflorum* and the perennial *L. perenne*, ryegrass constitute important forage species of temperate and mediterranean regions of Europe and N. America. Both these species exhibit a very close taxonomic similarity and are basically self-incompatible and diploid (2n = 14). In them spontaneously arisen or artificially induced tetraploids are available. The g-mst is known in both of them at the 2x level, but a spontaneously arisen mst occurs in 4x *L. perenne*. In some interspecific and intergeneric hybrids involving *Festuca* gc-mst occurs (Table 27.22).

27.11.1 *L. multiflorum* Lam. (Italian rye-grass)

In this multiflowered diploid annual or biennial grass, pollen sterile plants with or without one to two accessory chromosomes were found (this Vol.). Seed set in these plants is about 15–20% less than in pollen fertile plants, but whether seed set is due to outcrossing or apomixis is not known. Neither genetic control of the sterility nor *ms* gene action is known. In the mst plant detected by Nitzsche (1971), male meiosis proceeds normally, but pollen grains degenerate immediately after their formation. At anthesis anther lobes are either empty or have degenerated pollen. Nitzsche (1971) regards the sterility as cytoplasmic but could not develop the A line even up to BC_5 generations. Population segregation indicates a monogenic recessive control of sterility in this spontaneously arisen mst plant.

27.11.2 *L. perenne* L. (perennial ray or rye grass)

27.11.2.1 Genic Male Sterility

In this short-lived perennial rye grass, spontaneous occurrence of a few mst plants is documented by Jenkin (1931a,b) and Hayward and Manthiratna (1972) from the progeny of long term-stored aged seeds. Genetic control of sterility is not known, but since the occurrence of spontaneously arisen single gene recessive mutants is known in the progeny raised from aged seeds, the sterility appears to be due to recessive gene mutations. Female fertility of these spontaneously arisen mutants is not known. Hence it is not clear if the mutants are only male or both male-female sterile. Fejer (1958) documented the existence of a mst *L. perenne* plant whose sterility was conditioned by recessive genes. However, this plant is not only male but also female sterile as no seed set was obtained by the author after a series of hand pollinations. In another mst mutant, all three stamens became transformed into carpeloid structures. The sterility is conditioned by a single recessive gene (Fejer 1966).

27.11.2.2 Microsporogenesis and Tapetal Behaviour

In all the mst mutants abortive pollen grains occur in the mature anthers. The course of microsporogenesis is not known in any of them except in the one investigated by Nitzsche (1971) and Hayward and Manthiratna (1972). In this, male meiosis proceeds normally until microspore formation, after which microspores degenerate quickly. Tapetal breakdown was inhibited, unlike in mft anthers. The authors associate microspore degeneration to cellular persistence of the tapetum. This persistence impeded the movement of water and reserve food to the developing microspores. Thus *ms* gene acts on the tapetum.

27.11.2.3 Interspecific Hybrids

Segregation of partial to complete mst plants was reported by Ahloowalia (1981) from the F_2 progeny of the cross desynaptic and trisomic *L. perenne* × normal *L. multiflorum*. Some of these plants had indehiscent anthers but set seeds after open pollination. Such partial sterile plants occurred in BC_1 progeny when the desynaptic *L. perenne* was used as male parent for developing hybrids of the cross mentioned above. Similar plants were found by the author in F_2 progeny of the cross normal *L. multiflorum* × desynaptic *L. perenne*. The anomaly is not due to the *ms* gene but to the *ds* gene. This is because desynaptic-plants produce some viable and functional pollen and ovules and desynapsis in most of the cases is conditioned by single recessive genes (Gottschalk and Kaul 1980b). Thus occurrence of little viable seed set and high seedling mortality, as reported by Ahloowalia (1981), is due to unbalanced chromosome numbers caused by desynapsis and not due to mst. Spontaneously arisen and artificially produced interspecific hybrids between these two species have been found to be fertile by numerous investigators (Terrell 1966; p.143–144 for review). Furthermore, in the majority of the interspecific hybrids, mst results from the interaction of a single or many genes with alien cytoplasm and reciprocal differences are also apparent; but no such reciprocal progeny differences were found (Ahloowalia 1981). Therefore, the anomaly appears to be due to *ds* rather than to *ms* genes. However, the possibility of the existence of a very close linkage of *ds* and *ms* genes cannot be eliminated, as the linkage is known in the grass *Phleum nodosum*. Because of the recessive nature of *ds* genes and the weak selection potential of the gametes having the *ds* gene in the homozygous state, very few F_2 plants exhibit this anomaly. Since Ahloowalia (1981) investigated a small number of plants cytologically, desynaptics have been eliminated.

27.11.2.4 Intergeneric Hybrids

i. *Lolium* ssp. × *Festuca arundinacea*. All the natural and artificial F_1 hybrid plants obtained after crossing *Lolium perenne* × *L. multiflorum*, *L. multiflorum* × *L. perenne* var. Grassland Manawa (Corkill 1964) and (*L. multiflorum* × *L. perenne*) × *L. perenne* var. Grassland Ariki (Barckay 1963) as females to *Festuca arundinacea* as male are mst (Anderson 1966). The mst F_1 progeny was raised by Anderson (1966) after culturing the embryos (18 days after fertilisation) on agar medium and treating the seedlings with colchicine. A broad range of variability in mst expression was exhibited by F_1 mst plants. In 40% of the plants, initially phenotypically normal anthers appeared, but they gradually shrivelled at anthesis and had no viable pollen. In the remaining 60% of the plants, anther development from the beginning was highly abnormal. About 20% of such plants had flowers with one, two or all three anthers poorly develop or completely abortive. In the other 40% of the plants, stamens were transformed into pistil-like organs. Such plants were designated pistiloids. Not only a considerable variation in sterility existed between the progenies of various interspecific crosses, but variability is high within progeny plants. Also in plants that had no pistiloidy, the proportion of abnormal or abortive anthers in 5–10% of the flowers was much lower than in those with pistiloidy. In plants exhibiting pistiloidy, transformation of one to three anthers was common. When *L. multiflorum* was the female parent, floral anomalies were less than when *L. perenne* was the female parent. This lends support to the suggestions of Buckner et al. (1961), indicating a closer affinity of *L. multiflorum* to *F. arundinacea* than to *L. perenne*. The causes of the anomaly are not known. Neither successful reciprocal crosses nor details about female fertility of sterile plants is known. Therefore it is not certain if the hybrid plants are male or male-female sterile. Genetic control of the anomaly is also not known.

ii. *Lolium perenne* × *Festuca pratensis*. Distinct and stable mst (female fertile) plants were obtained by Wit (1974) from the diploid derivatives of a tetraploid intergeneric hybrid that originated from a cross made between the diploid F_4 hybrid (*L. perenne* × *L. multiflorum*) female and an autotetraploid plant of *Festuca pratensis* as the male parent. The author considered the sterility to be conditioned by interaction of one to two *fr* genes and S-cytoplasm. Further progeny testing of this intergeneric cross indicates sterility control by at least three fr genes (this Vol.). In these steriles, microspores degenerate soon after their release from the quartets.

iii. *Festuca pratensis* × *Lolium perenne*. In all the above-mentioned intergeneric crosses, mst plants segregate when *Lolium* is female and *Festuca* the male parent, but segregation of mst, p-mst and mft plants occurred in the fifth and subsequent BC generations of the cross involving *F. pratensis* as female and *L. perenne* as male parent (Connolly and Wright-Turner 1984). Based on anther and pollen phenotype, the segregants were classified into four types, two *Fr* genes appear to be involved in fertility restoration of the mst plants. Frequency and purity of maintainer genotypes is low and requires further breeding and selection. Thus the breeding value of this mst type is low.

Of the numerous investigators who crossed *L perenne* with *F. pratensis* (Terrell 1966 for references), Jenkin (1933) obtained seven mst F_1 plants after pollinating 1513 florets of one genotype of *F. pratensis* by the pollen of *L. perenne*. The plants had nondehiscent anthers and a very low percentage of stainable pollen. Female fertility of these plants is not known. In 1955, the author made similar crosses and found very low female fertility in such F_1 plants. Thus these F_1 plants cannot be considered as mst. A great variation in anther development and morphological features occurs in the mst's. While nearly half of the plants possess much reduced, thin, papery, waxy yellow pollenless anthers, the other half develops pale yellow to normal-looking anthers that have either mostly empty pollen or pollen filled with degenerated protoplasmic material. In about 10% of the latter type of pollen, cytoplasmic degeneration is meagre but the pollen is nonfunctional. Although the female fertility of these plants was between 44–68%, this was improved through selection of highly female fertile mst plants, indicating the role of polygenes in female fertility control of these mst's.

27.11.2.5 Breeding Value

The mst found by Wit (1974) is stable, and it is possible to select populations in which the sterility is readily identified and maintained by a single recessive gene system that permits an easy identification of the maintainer genotype in F_1. Effective maintainer lines can be isolated directly from an adapted material and a fully fledged restorer system is not needed. Furthermore, the mst can be combined with high seed fertility, since selection made in this direction by Wit (1974) proved to be fruitful. The author documents that Dijkstra successfully obtained 850 F_1 hybrids by pollinating a mst *Lolium* clone with the tall fescue pollen. Thus potential of gc-mst in this outbreeding forage grass is high and should be exploited. This is also stressed by Wit (1974) and Gaue (1977,1981).

27.12 *Oryza sativa* L. (cultivated rice)

In this and its wild relatives, both spontaneously arisen and induced mst is known (Table 27.4, 27.21, 27.24). The gc-mst occurs in many intervarietal and interspecific hybrids and their progenies (Table 27.5 and 27.22). Unlike in wild rice, temperature variations and certain chemicals induced high pollen sterility in cultivated rice (Table 27.7).

Table 27.4. Induced genic male sterility in rice

Rice variety	Mutagen used/conc	Effect and extent of mst	References
Nihonmasari	γ-rays (20–30 kR), EI (0.2% for 2 h)	No pollen grains in anthers, single pair of recessive genes	Fujamaki (1977)
Japanese rice varieties Etsunan N. 77 Kishihikari Toyonishiki	EI (0.5%) for 2.5 h	0.30 to 38% of mst plants in MII, 8 mutants, 6 with nonallelic ms genes	Ko and Yamagata (1980)
IR-36	EI (0.4% for 1 h or 0.2% for 3 h)	4 mst mutants, 2 having same ms genes, in 2 others not known. Meiosis normal, 0–4% stained pollen	Singh and Ikehashi (1979)
PTB-10	EMS (2%)	Monogenic	Nair and Gopal-akrishnan (1973),
		Trigenic	Pavithran and Mohandas (1976)
Basmati-370	EMS (1%)	Monogenic recessive, mst complete	This Vol.

Table 27.5. Gene-cytoplasmic male sterility in cultivated rice. (After Rutger and Shinyo 1980)

Genome donor Species	Variety	Cytoplasm donor Species	Variety	Restorer gene	References
O.glaberrima	WO 440	*O.sativa*	Akebono	Fr_1	Yabuno (1977), Sano (1983)
O.sativa	Calrose	*O.glaberrima*	–	–	Erickson (1969), Carnahan et al. (1972)
O.sativa	Fujisaka-5	*O.perennis*	Chinese strain Indian strain	–	Katsuo and Mizushima (1958)
O.sativa	Many	*O.perennis*	wild rice		Anonymous (1978a)
O.sativa	Taichung-65	*O.sativa*	Chinsurah Boro II	Fr_1	Shinjyo and Omura (1966)
O. sativa	Fujisaka-5	*O.sativa*	Burmese Lead rice	Fr_1, Fr_2	Watanabe et al. (1968), Shinjyo and Watanabe (1977)
O.sativa	Calrose	*O.sativa*	Bir-co		Erickson (1969), Carnahan et al. (1972)
O.sativa	Pankhari-203	*O.sativa*	Taichung native	*Fr*–	Athwal and Virmani (1972)
O.sativa	Norin 9-go	*O.sativa*	Assam-III	–	
O.sativa	Norin 9-go	*O.sativa*	Bhutmuri 36	–	Kinoshita et al. (1980)
O.sativa	Taichung-65	*O.rufipogon*	Bhutmuri 36	*fr*	Cheng unpublished
O.sativa	Reimei	*O.rufipogon*	Bhutmuri 36		Cheng unpublished

27.12.1 Genic Male Sterility

27.12.1.1 Spontaneous

This is recorded by Nagai (1926a,b), Ishikawa (1927), Miyazawa (1932), Takezaki (1932), Ramanujam (1935), Jones (1952), Hara (1946), Craigmiles et al. (1968), Hoff and Chandrapanya (1973), Shibuya (1973), Razzaque (1975), Trees and Rutgers (1978), Mahadevappa and Coffman (1980), Kaul (this Vol.). Whether *ms* genes causing mst in these rices are nonallelic or not is not known, but in almost all of them, the mst is conditioned by one recessive gene and female sterility is unimpaired. Some mst plants exhibit weak parthenogenesis and develop embryoless and endospermless caryopsis. The *ms* genes in addition influence certain floral traits described below.

i. *Floral Anomalies.* Four spontaneously arisen single gene recessive mst mutants, one in each of the four American rice varieties, Calady, Caloro, Earlyrose and CS-M3, obtained by Trees and Rutgers (1978), exhibit certain specific floral anomalies absent in their respective mft counterparts. In Earlyrose, mst flowers have five to seven stamens which possess shortened filaments and much compressed anthers. Their styles are elongated and have anthers-like appendages protruding from the ovary. In mst Caloro, palae are either missing or highly reduced, sterile glumes are too many and anthers are abnormally shaped, often less than six in number. This genotype closely resembles the "paleaceous-sterile" mutant detected by Nagai (1926b) except that it has 10% stainable pollen and perfect female fertility. The CS-M3 mst mutant has highly elongated ovary sacs at anthesis. In this respect it resembles the "awned sterile mutant" found by Nagai (1926a), except in having 5% stainable pollen. The Calady mst flowers have minute white anthers which have almost no fertile pollen. The floral anomalies appear to be a pleiotropic expression of the *ms* genes, as these are not separated in several segregating populations. In many cereals and legumes pollen stainability does not always infer pollen viability. This is also valid for many rices (Katsuo 1959; Sasaharo and Katsuo 1965) as also in four mst rices in which no selfed seed set occurs (Trees and Rutgers 1978); their stainable pollen is nonfunctional. Similar mst mutants detected by Nagai (1926a,b), Miyazawa (1932), Takezaki (1932), Hara (1946), Jones (1952), Craigmiles et al. (1968), Razzaque (1975), Mahadevappa and Coffman (1980) have some stainable pollen, but it is nonfunctional.

ii. *Anther Phenotype.* This is diverse in mst rice. In many examples the anthers are phenotypically normal (Ishikawa 1927, 1929; Miyazawa 1932) but no viable pollen is produced. In some, anthers are shrunken and indehiscent (Ramanujam and Parthasarathy 1935; Chao et al. 1960; Nair and Gopalakrishnen 1973) and pollen grains are nonviable, the mst's having two anther types segregated in F_2–F_3 populations (Hoff and Chandrapanya 1973). In one type, the plants have rudimentary anthers that lack pollen and in the other anthers are phenotypically normal but indehiscent. Fifty percent of their pollen is stainable by carmine, but its functionability is not known. Since no seed set occurred after selfing, either the pollen was nonviable or the anthers were completely indehiscent. A similar situation occurs in Caloro and CS-M3 mst plants. They possess stainable pollen (Trees and Rutgers 1978), but produce no seeds on selfing. The same is valid for many induced mst's, some of which have completely to partially stained pollen (Lu and Rutger 1984a), but none is functional. Therefore, carmine stainability does not reliably indicate pollen viability in rice; mercury-bromophenol blue staining is more reliable.

iii. *Gene Control.* In the majority of the spontaneously arisen mst mutants, mst is conditioned by single recessive genes whose allelic nature is not known. In some, *ms* gene action is influenced by modifier genes and environment. In a hybrid population, mst appeared after intervarietal hybridizations involving the parents Bluebonnet 50 × Gulfrose followed by six generations of selfing (Razzaque 1975). The F_1 plants were uniformly semi-sterile and their pollen fertility varied between 13 and 98%. The dominance of the *Ms* gene is partial, as the gene exhibits a variable degree of penetrance and expressivity in F_1 plants.

27.12.1.2 Induced

Not only is the frequency of spontaneously arisen mst mutants high, as evidenced from frequent documentation, but the rate of occurrence of recessively inherited single gene-induced mutations especially by ethyleneimine is very high in rice (Table 27.4). This is also evident from the recovery of 0.3–0.38% of mst plants from M2 populations of three Japanese rice varieties (Ko and Yamagata 1980). The high rate of occurrence of mst plants appears to be due to the presence of a large number of nonallelic *Ms* genes in the rice genome. Thus in eight induced mst mutants, six *ms* genes are nonallelic (Ko and Yamagata 1980). However, of four ethyleneimine-induced mst mutants of IR36, two have allelic *ms* genes, in two others allelism tests were not conducted (Singh and Ikehashi 1979). In all the induced mst mutants, whereas mst is nearly complete, female fertility is normal. Either no pollen or nonviable pollen is produced by these steriles, and natural outcrossing or artificial hand pollination yields viable seeds. Some (up to 4%) stainable pollen is produced by IR 36 mst mutants (Singh and Ikehashi 1979) but is not functional, as no seed set occurs after selfing. Fujimaki et al. (1977) obtained mst mutants following ethyleneimine and γ-ray treatments to the Japanese rice cultivar Nihonmasari. Anthers of the mst mutants are shrivelled and nondehiscent (Fig. 27.3b) and have aborted microspores. Kitada et al. (1983) isolated a mst mutant after seed treatments with NMU. In it, male meiosis breaks down during the zygotene. The anomaly is controlled by one recessive gene. Female fertility of the mutant is normal. As in spontaneously arisen mst mutants, mst in almost all the induced mutants is controlled by single recessive genes; their allelic nature, as in many spontaneously arisen one, is not known. Pavithran and Mohandas (1976) provide data that indicate trigenic control of mst mutant-induced after 2% EMS treatments by Nair and Gopalakrishnan (1973). Subsequent investigations of Pavithran (1981) reveal control of mst by a dominant gene or by a major gene interacting with one or two complementary genes, their recessives causing mst. In 11 mst mutants, single recessive genes conditioning nine of them were isolated by

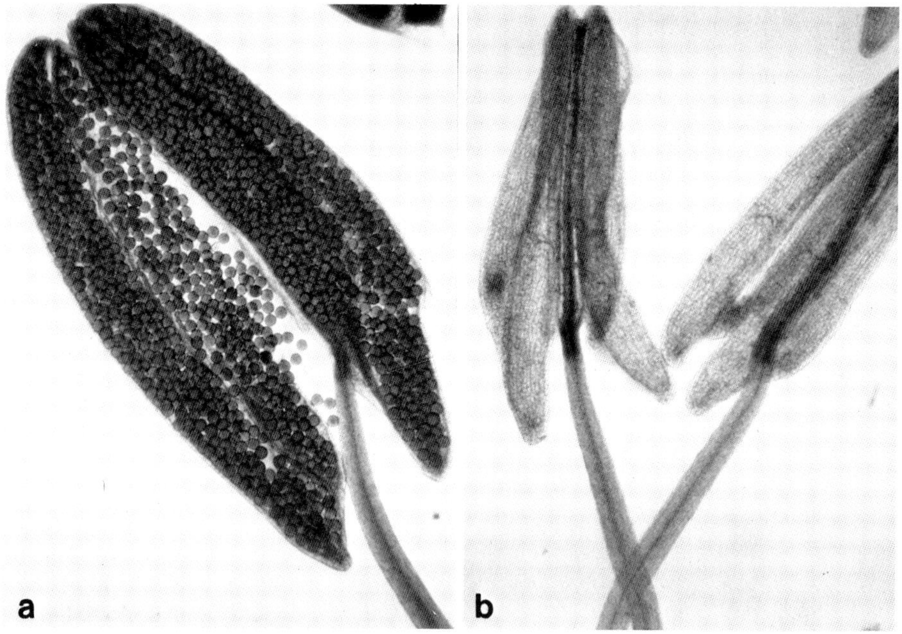

Fig. 27.3a,b. *Oryza sativa.* Dehiscent fertile **(a)** and nondehiscent sterile **(b)** rice anthers. (Fujimaki et al. 1977)

Lu and Rutger (1984a) after 30 kR γ-ray seed treatment to rice cultivar M-101. The allelic nature of the ms genes is not known. Of these mutants, one is semi-dwarf japonica type with an appreciable outcrossing rate (Mese et al. 1984). In all the mst mutants, anther width is uniformly reduced. Based on anther and pollen phenotypes, these lines become classified into four types: fully pollen stained, partially pollen stained, aborted pollen types and nonpollen types. The mutants exhibit diverse anomalies during male meiosis (Lu and Rutger 1984b). In the nonpollen types, meiosis never initiated, and microsporocytic pseudoplasmodia with variable nuclei developed in large numbers. In partially stained and aborted pollen types, meiosis broke down after initiation and in fully stained types, mature pollen aborted gradually. Thus, in sequence, the *ms* gene action is premeiotic, meiotic, postmeiotic and gametophytic in nonpollen, aborted, partially pollen stained and fully pollen stained mst rice types, respectively. Hence, the pollen phenotype depends on the time of *ms* gene action in these rices.

27.12.2 Gene-Cytoplasmic Male Sterility

Like other major cereal crops (wheat, maize, sorghum, pearl millet and barley), rice also exhibits mst due to gene-cytoplasmic interactions. The sterility is manifested variously in some intervarietal and interspecific crosses having certain specific parental combinations. (Table 27.5).

27.12.2.1 Intervarietal Crosses

i. *O. sativa indica and japonica*. Clear indications of the existence of gc-mst in rice are given by the occurrence of mst plants in the progeny of some intervarietal hybrids derived from the crosses involving *O. sativa indica* ♀ and *O. sativa japonica* ♂ parents. Hsu (1945) interpreted the mst to be conditioned by three pairs of genes. Since the progeny of the reciprocal crosses was fertile, the involvement of g-c interaction in causing mst in the progeny plants appears to be certain. Using similar parents, Sampath and Mohanty (1954) obtained F_1 partial sterile plants in both direct and reciprocal crosses. In crosses with *O. sativa indica* as the female parent, pollen sterility was significantly increased. This indicates that the pollen development is impeded when *O. sativa japonica* genes are placed in *O. sativa indica* cytoplasm. Actually the partial sterility of F_1 plants is due to chromosomal inversions.

A clear instance of mst caused by g-c interaction in rice was reported by Shinjyo and Omura (1966). They found F_1 progeny of the cross of an Indian rice *O. sativa indica* var. *chinsurah Boro II* × *O. sativa japonica* to be partially sterile, the plants have nearly 50% nonviable pollen. Reciprocals of such a cross were fully normal. The inviable pollen was smaller and stained less deeply than the viable pollen. Crossing the semi-sterile F_1 plants by *O. sativa japonica* pollen yielded offspring segregating 1:1 for completely and partially sterile plants. In F_2, both partially sterile and fully fertile plants segregate in nearly equal proportions. Shinjyo (1969) opined that the *japonica* parent contained *FrFr* genes and S-cytoplasm and the *indica* parent *frfr* genes but N-cytoplasm. Thus the plants with S-cytoplasm and the genes *FrFr, Frfr, frfr* were mft, p-mst and completely mst respectively, but those with N-cytoplasm were mft regardless of the constitution of the *Fr*/*fr* genes. The F_1 plants of (S) *frfr* ♀ × (N) *Frfr* ♂ segregated into complete and partial mst in 1:1 ratio, but the selfed progeny of (N) *Frfr* was mft. Therefore the control of genes is of the gametophytic type.

Two mst strains, K-11 and K-13, were developed by Kinoshita et al. (1980) through successive BC's of two *indica-japonica* crosses with the recurrent genome donor Japanese rice parent Norin-9–90, the cytoplasm donor being the Indian rice variety Assam-III and the Burmese variety Bhutmuri-36. The mst was stable and persisted up to BC_{10} after successive BC's to the recurrent parent Norin-9–90. In these substitution lines, male meiosis proceeded normally in mst's, and pollen grains developed normally up to second pollen mitosis. Further pollen development was, however, inhibited (Fig. 27.4b) and the sperm nuclei tended to degenerate gradually during the period when accumulation of pollen reserves occurs. The chromatin of sperm nuclei degenerated fully and the nuclei became elliptical 1 day prior to flowering (Tamaru et al. 1980). The occurrence of degeneration at such an advanced stage is rare in mst rices and is in contrast to the observations of Chu et al. (1972), who found disruption at an early stage. Using *O. sativa indica* var. Bir-co and var. Pai-co as the female parent in crosses with the Californian rice varieties Carlrose, Caloro, and Colusa, the F_1 plants were almost completely mst, the progeny of the reciprocal crosses yielded about 50% seeds on selfing (Erickson 1969). With further BC's that decreased Bir-co or Pai-co genome in their respective cytoplasms, the extent of mst increased, and in BC_3 the plants

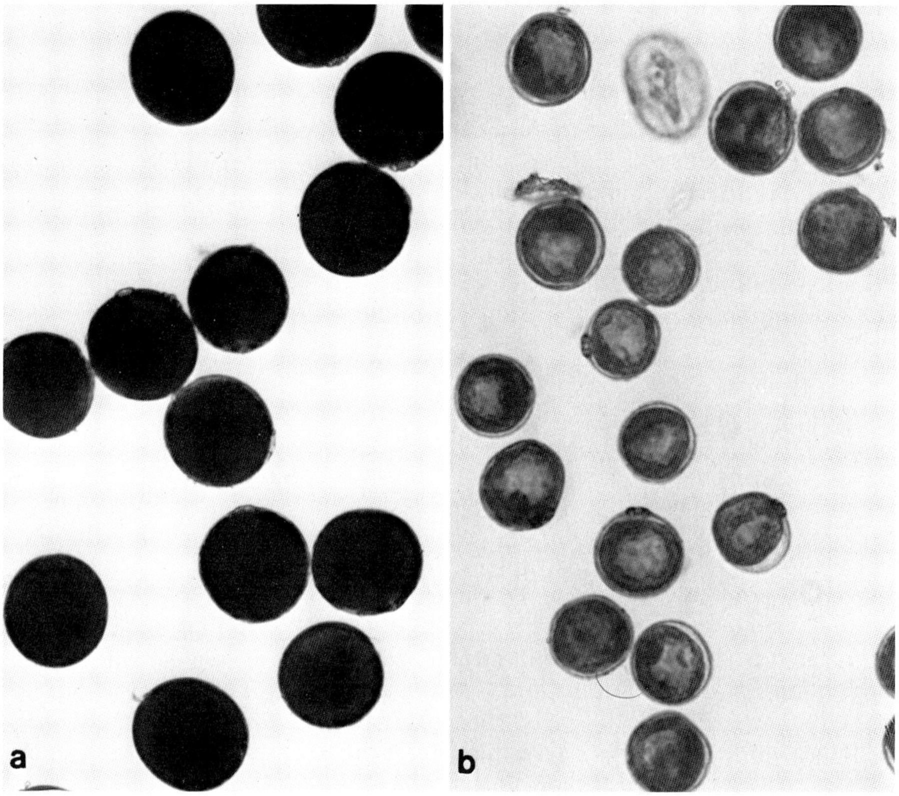

Fig. 27.4a,b. *Oryza sativa.* Pollen grains prior to anthesis from fertile **(a)** and sterile **(b)** rice plants. (Tamaru et al. 1980)

were completely mst. The g-c interaction leading to mst was manifested through anther indehiscence in two substitution lines with Pai-co cytoplasm. Pollen stainability in these two lines was about 85%, but selfed seed fertility was 0.2% because of anther indehiscence. In general, the mst plants could be classified into completely pollen fertile anther-indehiscent type, partially pollen fertile anther-indehiscent type and pollen sterile types. All three phenotypes are mst.

ii. *O. sativa burmese* × *O. sativa japonica.* The *fr* gene(s) of Japanese rice *O. sativa* cultivar Fujisaka-5 are effective in the S-cytoplasm of Burmese *O. sativa* rice variety Lead Rice. When F_1 of the cross Lead Rice ♀ × Fujisaka ♂ was backcrossed by Fujisaka pollen, the progeny was mst. (Watanabe et al. 1968) and offspring of the reciprocal cross were mft. Thus the sterility resulted due to interaction of Lead Rice S-cytoplasm and Fujisaka *fr* genes. Therefore S-cytoplasm sensitive to Fujisake *fr* genes is present in the Lead Rice variety.

iii. *O. sativa spontanea* × *O. sativa japonica.* Complete mst's were detected by Katsuo and Mizushima (1958) in BC_1 progeny of cross (Chinese rice *O. sativa spontanea* ♀ × *O. sativa* cultivar Fujisaka 5) × Fujisaka 5. Progeny of reciprocal cross combination giving *O. sativa spontanea* plants having *O. sativa* Fujisaka-5 cytoplasm was normal. Pollen grains of mst and mft plants did not differ phenotypically or in carmine stainability, but pollen of mst plants did not germinate on rice stigma (Katsuo 1959) or in culture media (Sasahara and Katsuo 1965), unlike those of the mft plants. Backcrossing the mst hybrid plants containing *O. sativa spontanea* cytoplasm with Fujisaka pollen for two generations yielded progenies that had over 90% mst plants (Katsuo and Mizushima 1958). On the other hand, backcrossing on hybrids containing Fujisaka-5 cytoplasm by *spontanea* pollen yielded fully mft progeny plants. Thus cytoplasmic differences exist between Fujisaka-5 and *spontanea*, the former having N- and the latter S-type cytoplasm. The S-cytoplasm of *spontanea* in association with *fr* genes of Fujisaka-5 induces mst.

27.12.2.2 Interspecific Crosses

Cultivated rice comprises basically of two species, *O. sativa* L. and *O. glaberrima* steud; the former is cultivated throughout almost all the rice-growing areas, but the latter is grown mostly in tropical West Africa. Although they have the same chromosome number and regular chromosomal synapsis in F_1 plants, hybrid sterility separates them. However, mst F_1, BC_1, and BC_2 plants were obtained by Carnahan et al. (1972) after crossing three strains of *O. glaberrima* as female to three Californian *O. sativa* varieties Caloro, Colusa and Earlyrose. The mst plants did not set seeds but yielded 0–5% seeds when backcrossed four times to the above-mentioned three rice varieties. Thus the resultant mst is due to the interaction of *O. glaberrima* cytoplasm and *fr* genes of the three *O. sativa* varieties. Genes of some *O. sativa japonica* varieties do not induce mst in *O. glaberrima* cytoplasm. This is evidenced by Yabuno (1977), who obtained highly sterile F_1 plants from the crosses involving two *O. glaberrima* varieties as female and three *O. sativa japonica* varieties as male. Pollen fertility of the hybrid plants ranged from 0.3 to 0.5%, and increased gradually with succeeding BC's to *japonica* males. Eventually, completely fertile plants with the characteristics of parental *japonica* variety appeared. Thus no mst results when *japonica* rice genome is placed in *O. glaberrima* cytoplasm. However, mst occurred in the reciprocal cross progenies of such five substitution lines having the cytoplasm of three *japonica* varieties and the genome of *O. glaberrima*. Four lines produced mst plants with abortive microspores and indehiscent anthers in BC generations. In the remaining one, a simultaneous segregation occurred for plants with abortive microspores and indehiscent anthers and those with fertile pollen and indehiscent anthers (fn-mst). From these findings it is apparent that, whereas mst appears due to interaction of *fr* genes of three American *O. sativa* rice varieties with *O. glaberrima* S-cytoplasm, a reciprocal combination produces mst when *japonica* lines are used as female to *O. glaberrima* males. Thus mst results when *O. glaberrima fr* genes are placed in *O. sativa japonica* S-cytoplasm. Progeny of the reciprocal crosses is mft. Likewise, progeny of *O. sativa* × *O. glaberrima* is mst but female fertile. The *O.sativa* (Taichung 65) and *O. glaberrima* (WO25) parents were considered to be Sa_3/Sa_3 and S_3/S_3, respectively, by Sano (1983). The S_3 locus has no deleterious effect on the development of microspores in homozygous condition, but induced pollen abortion in plants carrying it in the heterozygous state (S_3/Sa_3). No adverse effects of S_3 on megaspores or somatic cells occur. Thus S_3 differs in gene action from S_1 and S_2 alleles reported previously. The Sa_3 locus is tightly linked with the lazy habit causing gene (*la*) on chr. 9. Plants homozygous for S_3 are photoperiod-sensitive, having a heading date 10 days later than S_3/Sa_3 or Sa_3/Sa_3 plants. Thus at least three genes exist; two are effective in both male and female spores (S_1,S_2), one is effective only in microspores (Sa_3).

By placing the genome of some *O. sativa japonica* rice lines in *O. glaberrima* and/or *O. rufipogon* cytoplasm, stable mst plants were developed by Cheng and Huang (1979). The manifestation of sterility varied with each gene-cytoplasm combination. In two lines having Pai-co cytoplasm, and in one having *O. rufipogon* cytoplasm, anther dehiscence was poor or negligible. In a line with *O. sativa indica* var. Pai-co and in two lines with *O. sativa indica* var. Pai-co and *O. rufipogon* cytoplasm, both anther and microspore development was highly impeded. On the other hand, in the lines having *O. glaberrima* cytoplasm, anther development was normal, but the pollen produced was abortive-shrunken and its shedding was poor over the stigma. No pollen germinated on the stigma when dropped artificially.

27.12.2.3 Fertility Restoration

In genome A rices, an unequivocal existence of mst was documented for the first time by Katsuo and Mizushima (1958) in the Japanese rice plants *O. sativa* cultivar Fujisaka-5 having Chinese rice *O. sativa* f. *spontanea* cytoplasm. Later, many other g-c combinations leading to mst were documented (Table 27.5). Despite these documentations, only in a few sources have *fr* genes been identified. Fertility restoration in them is of the following types:

i. *Gametophytic Restoration.* Shinjyo and Omura (1966) and Shinjyo (1969) demonstrated that the Indian rice variety *O. sativa indica* Chinsurah Boro II possesses S-cytoplasm and restorer genes. For this sterility system the genome of *japonica* variety Taichung 65 and Ponlai acts as the maintainer. By successive BC's mst Taichung 65 plants were developed, the s-cytoplasm was designated cms-boro and its fertility-restoring gene Rf_1 ($\equiv Fr_1$). Plants with S-cytoplasm and $fr_1 fr_1$ and $Fr_1 fr_1$ genes are completely and partially mst respectively. Those with $Fr_1 Fr_1$ genes are fully fertile. In all of them no reduction in ovular fertility occurs, plants having N-boro cytoplasm being mft irrespective of the *fr* genes. Since F_1 plants having S-cytoplasm and being heterozygous for the Fr_1 gene exhibited 50% pollen sterility and those with homozygous $fr_1 fr_1$ genes are completely mst, the fertility restoration is of the gametophytic type.

ii. *Sporophytic Restoration.* This occurs in certain plants having the *O. glaberrima* strain Wo 401 genes and *O. sativa* var. Akebono S-cytoplasm (Yabuno 1977). In these, the *fr* genes in homozygous recessive state (*frfr*) cause microspore abortion and anther indehiscence. When the genes are present in the heterozygous state (*Fr fr*), fertile pollen is produced, but anthers are indehiscent, rendering the plants functional male sterile (fn mst). When the genes are present in the dominant state

(*Fr Fr*), the plants are completely mft and have dehiscent anthers and viable pollen even though they have S-cytoplasm. Whereas these plant types breed true on selfing, those with fertile pollen and indehiscent anthers segregate mft's and fn-mst's in a 1:1 ratio. This indicates a sporophytic type of *fr* gene action. The *fr* gene exhibits full penetrance and restores fertility in the mst plants having *O. glaberrima* genes and Norin 19 and Omuchi cytoplasms.

iii. *Polygenic Control.* Genes other than fr_1 and fr_2 appear to be responsible for pollen fertility restoration in mst material having boro-cytoplasm (Kinoshita et al. 1980). Since the restoration is weak to strong and shows an additive trend, it appears to be under polygenic control. This is also evident from F_3 pollen fertility segregation data of the authors, who used 1–127 MS (Taichung 65 go) and three pollen fertility restorers H-103, H-406 and 1–130 (fr_1) as parental material. Independent and additive action of restorer genes, *Rf*a, *Rf*b, and *Rf*d, is visualised by them. Their specific combination produces complementary effects and leads to complete fertility restoration. The polygenic control of fertility restoration is not fully established, as the action of modifiers cannot be completely eliminated. Moreover, environmental influence, especially of temperature, is profound on pollen fertility. This confounds genetic segregations and determination of exact gene control becomes arduous.

27.12.2.4 Purple Colour and *fr* Genes

Existence of a positive correlation between purple colour distribution and *fr* genes in rice was postulated by Peitian et al. (1981a), who obtained 44,54 and 83% fertility restoration in non-, little and most-purple varieties of rice respectively. It appears that wild rice (*O. sativa* var. *spontanea*) has maximum purple and *fr* genes and with the progression of evolution both these genes seem to have been lost. These authors have proposed distribution rules for *fr* and *Fr* genes in a cladogram of rice (Peitian et al. 1981b).

27.12.2.5 S-Cytoplasm Distribution

On the basis of documentations that rice varieties Bir-co, IR8 and Sakotira 55 carry S-cytoplasms (Swaminathan et al. 1972; Carnahan et al. 1972; Athwal and Virmani 1972), Parmar et al. (1981) used them in reciprocal and BC combinations with some improved Indian rice varieties. Whereas the presence of the S-cytoplasm is not clearly distinct in Sakotira and the *fr* gene action is highly complex in it, the presence of the S-cytoplasm and *Fr* genes are fairly evident in IR8 and Bir-co. This contrasts with the conclusion reported in the IRRI Annual Report (1969) that the behaviour of the IR8 cytoplasm with respect to mst manifestation is complex and difficult to interpret. However, it appears that while g-c interaction is direct in certain cross combinations and discrete in others it is influenced by modifier and inhibitory genes. Thus the extent and expression of mst in rice depends upon cross combination and genetic constitution of the parental lines.

Utilizing the difference in mean pollen sterility obtained after using a variety as female and as male in various cross combinations as an index, Parmar et al. (1981) determined the relative paternal and maternal influence in inducing mst of different rice varieties. They found rice selections or varieties like Cl 5309, Early Prolific, IR8, Jaya, Tadukan, more effective as maternal parents in inducing mst, and CO-13, DGWG, IR-22 Pusa 2–21 and T(N)1 to have the tendency to induce more pollen sterility when used as male rather than female parents. Rice varieties Bir-co, IR8, Jaya, Jeeragasamba and Tadukan appear to be usable sources of S-cytoplasm, and some of them seem to carry *fr* genes as well.

From these findings it appears that many *O. sativa* varieties possess S-cytoplasm along with *fr* genes. In order to know the frequency of the S-cytoplasm, Shinjyo (1972) crossed 150 Japanese low-land rice varieties with two testers viz., C-tester that had N-cytoplasm and fr_1 fr_1 genes and an N-tester that had S-cytoplasm and fr_1 fr_1 genes. He found that 87.3% of the varieties tested possessed N-cytoplasm and the remaining 12.7% had *fr* genes of weak restoration capacity. Crossing C-testers with 146 *O. sativa* varieties revealed that only four varieties had S-cytoplasm. Of 213 Japanese varieties tested, 27 have *fr* genes (Ishimine and Shinjyo 1978). The S-cytoplasm is present not only in some Asiatic rice varieties but also in the wild perennial rice *O. perennis*. Of 130 different maternal progenies of this rice, 62 have S-cytoplasm and the remaining 68 N-cytoplasm (Rutger and Shinjyo 1980). Whereas in Asian and American *O. sativa* rice strains frequency of S-cytoplasm is about 65% and 5%, no S-cytoplasm is known in African and Oceanian rice strains. S-cytoplasm differentiation occurs both within *O. sativa* and between it and in *O. perennis*. This is evidenced by the occurrence of different *fr* genes both within and between the two rice species. Within *O. sativa*, the S-cytoplasm appears to be highly differentiated and is at least of three different types. More investigations are needed to characterise these three types and discover many more S-types that seem to exist in the cultivated rices.

27.12.2.6 Induced N-Cytoplasm Mutation

From N-cytoplasm IR8, a mst mutant considered to have S-cytoplasm by Mallick (1980) was obtained after seed treatment with 1% EMS for 5 h. Pollinating the mutant plants by mft IR8 produced a partial mst progeny.

27.12.2.7 Functional Male Sterility

By selecting for fertility in BC generations of the Philippine rice variety Tadukan × Japanese variety Norin-8 with Norin 8 as the recurrent parent, Kitamura (1962) obtained a new strain of rice designated Ta strain. Both pollen and embryo sacs of this strain are normal and functional. While all F_1's of Japanese rice varieties × Ta strains have normal pollen and seed fertility, F_1 from reciprocal crosses exhibits a wide range of seed sterility despite high pollen and ovule fertility. The low seed fertility is due to limited anther dehiscence that leads to reduced pollen release. The inhibited anther dehiscence is caused by Tadukan cytoplasm, whose effect is retained in many BC generations produced after crossing the mst plants by pollen of Japanese rice varieties. Similar sterility caused due to indehiscent anthers is also known in two substitution lines having *O. sativa indica* Pai-co cytoplasm and genes of the Californian rice varieties Carlrose, Caloro and Caluso (Erickson 1969). About 85% of the pollen produced by these lines is fertile, but selfed seed fertility is 0.2% because of anther indehiscence. Both plants having nonviable pollen and indehiscent anthers and those having viable pollen grains but indehiscent anthers segregated in the progeny plants possess *O. sativa* var. Akebono cytoplasm and *O. glaberrima* strain *Wo401* genes (Yabuno 1977). The author opined that a dominant pollen fertility-restorer gene, *Rfg*, was present in the heterozygous state and led to anther indehiscence. Whereas the mst plants with nonviable pollen have S-cytoplasm and no dominant *Fr* gene, those with viable pollen grains but indehiscent anthers have a dominant *Fr* gene in the heterozygous condition. In the homozygous condition, the *Fr* gene causes complete pollen abortion accompanied by anther nondehiscence, but in the heterozygous state the expression is partial, leading to anther nondehiscence only. A similar genetic control of functional mst could be possible in the mst's found by Kitamura (1962) in the Ta strain and by Erickson (1969) in two substitution lines of rice. These fn-mst's have viable pollen but indehiscent anthers. Cheng and Huang (1980) investigated the anther anatomy of three such mst plants that have Pai-co or *O. rufipogon* cytoplasms and indehiscent anthers. In these anthers, the endothecium is strongly thickened and the anther wall exhibits a specific pattern of cell orientation different from the dehiscent anthers. The epidermal cells are equally highly thickened. This prevents separation of the inter-microsporangial stripe from the underlying parenchyma along each side of the connective tissue during anthesis. Furthermore, the outer surface of inter-microsporangial indehiscent anthers is ridged, whereas it is smooth in normal anthers. This impedes the mst anthers from forming the anterior pores and pollen shedding is thus inhibited. Therefore, under the influence of the *fr* gene, the anther wall structure is highly altered so as to inhibit its dehiscence completely.

27.12.2.8 Microsporogenesis

In the majority of gc-mst types, male meiosis proceeds normally and *fr* gene action initiates during microspore formation and pollen grain maturation stages (Tables 27.6, 27.21). This is especially valid for mst plants having the *O. sativa* genome in alien cytoplasms. In most of them, meiosis proceeds normally, but degeneration occurs after I or II pollen mitosis so that at anthesis physiologically weak and inviable pollen is produced. Thus *fr* gene action is considerably delayed in gc-mst rices, as it occurs almost at the end stage of gametogenesis. If the gene action is delayed through manipulation of biochemical or environmental factors, mft could be revoked in them. This would facilitate maintenance of the mst rices. However, this does not seem to be possible because even though viable pollen is produced in many gc-mst types, anther dehiscence does not occur and such plants become functionally mst. In five BC generations of red awned rice × Liantan Zao and two more Chinese rices, anomalies occur during microsporogenesis (Anonymous 1978a; Yang and Li 1984), and pollen development is arrested just prior to the

Table 27.6. Microsporogenesis in gene-cytoplasmic male sterile *Oryza sativa*

Genome	Cytoplasm	References	Microsporogenesis
Taichung-65	Chinsurah-Boro II	Chu et al. (1972)	Meiosis and tapetum normal, 25% microspores degenerate gradually after liberation, rest exhibit slow growth, reduced size and meagre starch formation
Fujisaka-5	var. *spontanea*	Katsuo (1959)	Meiosis normal, microspore abortion gradual, mostly after first-second mitotic division, abortive pollen formed at maturity.
One variety	Burmese lead rice	Watanabe et al. (1968)	Pollen stains but does not function, degeneration after second pollen mitosis
Norin-90	Bhutimura-36	Tamaru et al. (1980)	Meiosis normal, no anomaly until microspore maturation stage, gradual degeneration after trinucleate stage during anthesis
Taichung-65	*O.rufipogon*	Cheng and Huang (1978)	Meiosis normal, microspore degeneration rapid after liberation from quarters

binucleate stage. Thereafter pollen grains abort immediately, mostly at the late uninucleate stage. In many alloplasmic *O. sativa* gc-mst types, microspores abort gradually after maturation and many of them even undergo both mitotic divisions before abortion (Table 27.6). In mst plants having genome of *O. sativa* Taichung 65 or Reimei in *O. rufipogon* cytoplasm, rapid degeneration occurs immediately after microspores are liberated from the quartets (Cheng unpublished). While some microspores tend to develop up to the late uninucleate stage, microspore mitosis never occurs in them and their nuclei subsequently lose their spherical shape and cytoplasmic contents. Their walls fold inwards, giving them a deformed, sickle-shaped or crumpled appearance. Based on the S-cytoplasm source and its relation to fertility-restoring and maintainer lines, Chinese mst rices can be differentiated into three types viz. wild aborted type, Honglian type and Yunnan No. 1 type (Shuhua 1982). In them *fr* gene action occurs at different stages of pollen development. Whereas microspore abortion in the wild-aborted type is at the uninucleate microspore stage, in the Honglian type it occurs at the two-celled pollen stage and in Yunnan No. 1, it initiates just before mitosis of the reproductive nucleus.

Chinese scientists classify mst in rice into three types, viz. (i) pollen-free types – lacking pollen grains as no microspores are formed, (ii) uninucleate aborted type – uninucleate microspores abort, (iii) bi-tri nucleate aborted type – immature to mature pollen abort. Many transition-type pollen abortions that do not fit into this classification occur in rices. Moreover, uni-to bi-nucleate microspores with a varying degree of abortion occur in the same anther. This renders the classification difficult. Chaudhury et al. (1981) proposed a classification of pollen based on shape and stainability, viz. unstrained withered sterile, unstained spherical sterile, stained round sterile and stained round fertile. In some Chinese rices, pollen abortion in the offspring of distantly related parents occurs earlier than in those of the closely related parents. A nearly similar trend occurs in the gc-mst materials investigated by Chaudhury et al. (1981).

27.12.2.9 Tapetal Anomalies

These are not fully investigated in all the gc-mst cultivated Asiatic rices. Whereas some Chinese mst rices do not exhibit any tapetal anomaly at 30°/25°C and 25°/20°C (Cheng and Huang 1978), at low temperature (20°/15°C) tapetal hypertrophy, plasmodial development and tapetal disintegrations occur. These anomalies are not caused by mst genes, but by the low temperature treatments that lead also to total microspore abortion even in some mft rices. In the alloplasmic gc-mst Reimei and Taichung-65 *O. sativa* rices having *O. rufipogon* cytoplasm, the tapetum exhibits a high degree of anomaly (Cheng et al. 1981a,b 1982). During microspore formation, part of the tapetum enlarges erratically and some of its cells burst into the anther loculus, discharging their contents. In the disorganised tapetal cells, mitochondria, golgi bodies and endoplasmic reticulae are deformed. However, some of the tapetal cells persist, their cytoplasm is dense and mitochondria lose cristae and swell. This is followed by nuclear degeneration and cellular disintegration. Mostly the tapetal anomalies are confined to the cytoplasmic organelle such as endoplasmic reticulum, golgi bodies, mitochondria, Ubisch body layer and chromatin material. However, unlike mft anthers, the mst anthers have rich ^{14}C assimilated in tapetal cells. Such assimilates are extremely low in mst microspores. Although this indicates that transfer of nutrients from the tapetum to developing microspores is inhibited in the mst anthers, the inhibition could be one but not the main cause of pollen sterility, as the plantlets formed from the callus developed from sterile microspores are also mst. Thus the sterility is basically genic. In the mst line Dianrui 409, mst gene action occurs during early microsporogenesis. Meiosis is asychronous, but tapetal degeneration is normal and slightly delayed (Pan et al. 1982). Whereas 21% of the pollen appeared morphologically normal, it was unable to fertilise the ovary. In an alloplasmic Reimei rice, locular fusion occurs during the premeiotic stage (Cheng et al. 1981a,b) and di-tri-locular anthers are formed. The tapetum is abnormal and pollen sterility is complete.

27.12.2.10 Callus Formation

Microspores of the alloplasmic mst lines having a Reimei or Taichung-65 genome in *O. rufipogon* cytoplasm and those of their respective mft lines produce nearly equal proportions of calli in culture (Cheng unpublished). The reaction to culture media is similar in mst and mft genotypes. Thus *O. rufipogon* cytoplasm has no influence on calli production. From the calli of sterile microspores many plantlets develop. Of these, some are diploid and others aneuploid. Whereas plants developed from aneuploid plantlets are male-female sterile, those from diploids are mst.

27.12.3 Biochemical Differences

Unlike in other major cereals, systematic investigations on biochemical differences between mft and mst anthers have not been carried out. In the anthers of some mst rices, the respiration rate is low throughout, until the heading period, after which it enhances greatly. At the uninucleate pollen stage, the peroxidase activity in mst pollen is high but at the binucleate stage, its activity, along with cytochrome oxidase and polyphenol oxidase, is highly reduced (Anonymous 1977). In both g-mst and gc-mst anthers, decrease in sugars, starch, protein and inorganic phosphate occurs (this Vol.). Compared to mft anthers, mst anthers have lower proline and higher alanine and asparagine contents.

27.12.4 Temperature-Induced Male Sterility

i. *Low Temperature*. Constant low temperatures (10°–15°C) during reproductive growth induce sterility and delay in growth and maturity of several Japanese rice varieties (Table 27.7), but this has no or little harmful effect on pistil development (Hayase et al. 1969). The cold injury-induced mst is due to anther deformation, cessation of anther development, inhibited growth of anther loculi, pollen unripeness, partial or no anther dehiscence, meiotic breakdown, microspore degeneration, pollen remaining clumped in anther loculi and failure of pollen to

Table 27.7. Induced nongenetic male sterility (high to complete) in rice

Inducer	References	Female fertility reduction
1. *Temperature*		
Low (10°–15°C)	Hayase et al. (1969) Suzuki (1978)	Low
High (35°–41°C)	Satake and Yoshida (1978)	High
2. *Photoperiod* (14 h)	This Vol.	Low-high
3. *Chemicals*[a]		
Ethrel (4,000 to 10,000 ppm)	Perez at al. (1973), Cheng and Huang (1978), Parmar et al. (1979) Chan and Cheah (1981)	High
MH (250 ppm)	This Vol.	Very High
RH-531 (100 ppm once or after 2 or 4 days)	Perez et al. (1973)	Complete
3-(p-chlorophenyl) 6-menthoxy-s-triazine -2,4-(1H, 3H) Dione	Zhangxing and Chunnog (1980) This Vol.	Very high

[a]Phytotoxicity induced is very high.

germinate on stigmata (Ito 1980). In two rices, low temperature induces fn-mst (Sawada and Kinoshita 1977). The degree of cold injury and its manifestable effects vary significantly among rice cultivars (Suzuki 1978). Induced sterility during the booting stage is higher in early-growing spikelets than in the late-growing ones. Upper and primary branches and top spikelets of a branch exhibit higher cold-induced mst than lower and secondary branches (Nishiyama and Satake 1979). Although male meiosis is highly disturbed by low temperatures, the young microspore is the stage most sensitive to low temperature treatment (Ito 1980). In most of the mst rices, meiotic breakdown also occurs during this stage (Table 27.21). Hence the stages between microspore formation and pollen maturation are the most sensitive in rice. During cold temperature treatment, increase in reducing (Murakami et al. 1958) or nonreducing (Ito 1980) sugars, decrease in total inorganic phosphate (Nishiyama 1978), dry weight, starch, protein and amino acid contents etc. (Ito 1978) occurs. Since similarities occur in mst induced by genes or by physiological stress, the genes causing mst may disturb the physiology of the rice microspore in the same way as does the cold temperature injury.

ii. *High Temperature.* Like low temperature, high temperature (35°–41°C) applied during anthesis induces pollen sterility in rice (Satake and Yoshida 1977). The sterility induced is maximum when treatment is given during anthesis and least when given after anthesis. Thus anthesis stage is most sensitive to the temperature treatment. The treatment decreases pollen viability and shedding. This reduces the number of germinating pollen grains on stigma (Satake and Yoshida 1978) and explains why high temperature is a crucial physical constraint in rice production in tropics.

27.12.5 Chemical Male Sterility

i. *Ethrel.* Foliar sprays of 5000–8000 ppm of ethrel solution on three rice varieties, IR-8, IR-24 and IR-262–43–8, at four growth stages viz. pre, early, mid and late booting stages, were tried by Perez et al. (1973). Of these 4000 ppm treatment at the early boot stage induced maximum pollen (64%) and spike (94%) sterility. Whereas multiple chemical applications at early booting stages led to higher pollen and spike sterility and reduction in flag leaf length, ovary fertility was also enhanced. With increasing concentrations, sterility increased linearly, but ovular sterility was equally higher. Cheng and Huang (1978) also induced pollen sterility by spraying 3000 to 4000 ppm ethrel at the booting to late booting stage. The p-mst was induced by using 300 ppm ethrel solution in the upland rice variety Bisbang-A (Chan and Cheah 1981). In improved Sabarmati, Parmar et al. (1979) induced complete pollen sterility by three recurrent sprays of 8000 to 10,000 ppm ethrel at the mid primordia and boot leaf stages. The anthers were empty, shrivelled and reduced in size. Spikelet sterility of the treated plants was also complete in self- and open pollinations. A single spraying of 6000 to 8000 ppm of ethrel given a week before the boot leaf stage or two sprayings of 4000 to 6000 ppm, the first given a week before the boot leaf stage and the second at the boot leaf stage, induced very high but not complete pollen sterility in the dwarf, high-yielding, late-maturing IR-8 rice and short duration variety Pusa-2–21 (Parmar et al. 1979). In addition, the ovular sterility induced is very high. This leads to a significant reduction in grain yield. The chemical treatment causes stunted growth, severe leaf burning, tearing of leaf blades, suppression of panicle development and its emergence, defoliation, mortality of tender tillers and yellowing of the whole plant, leading to its death. Furthermore, both single ethrel treatments at high doses as well as recurrent treatments at low doses induced high sterility in rice (Perez et al. 1973; Parmar et al. 1979). This indicates a positive correlation between gametocidal concentration and longevity. The serious phytotoxic effects associated with high ethrel concentration limit its use in commercial hybrid production. Wang and Que (1981) treated young rice panicles with 300 ppm ethylene during male meiosis. This treatment caused high male as well as female sterility. Whereas 100 ppm caused partial male-female sterility, 500 ppm induced complete sterility. Similar results are obtained after ethrel treatments.

ii. *RH-531.* Its 100 ppm application once or after every 2 or 4 days at the pre-booting stage in IR-8 rice induces almost complete male as well as female sterility (Perez et al. 1973). Applied once during the early booting stage, 10,000 ppm causes similar effects; concentrations lower than this induce low pollen sterility.In the treated plants, anther dehiscence is highly erratic and abnormal. The treatments cause the early death of lower leaves, yellowing of upper leaves and reduction of panicles, flag leaf, anther and spikelet size. In some treated plants anthers are highly reduced and pollen formation is completely suppressed.

iii. *Other Chemicals.* Spraying 4000–8000 ppm solutions of 3-(p-chlorophenyl) 6-methoxy-s-triazine-2,4-(IH, 3H) dione triethanolamine salt at the boot stage induces 89–100% pollen sterility in two subspecies, Keng and Shien, of Chinese cultivated rice *O. sativa* (Zhangxing and Chunnong 1980). The chemical causes toxic and dwarfing effect on treated plants, but does not cause significant reductions in female fertility in these Chinese rices. In contrast, a 4000- to 6000-ppm spray on two Indian cultivated rice varieties, Basmati-370 and Jhona-349, causes ovular damage, phytotoxicity, leaf yellowing, dwarfing and stunted growth of the treated plants. Whereas 8000-ppm sprays are highly toxic and cause plant death, a 6000-ppm spray at the boot leaf stage induces 70–80% pollen sterility and 30–50% ovular damage (this Vol.). Pollen sterility occurs to 95–100% in the panicles when the leaves are sprayed at uninucleate pollen stage with male gametocide No.1 (Anonymous 1978b). The gametocide contains zinc methyl arsenate (CH_3AsO_3 $Zn.H_2O$) as the principal component. The chemical is absorbed and rapidly translocated to the panicle and inhibits the respiratory metabolism of the anther completely, leading to almost complete pollen abortion. The effect of the chemical on ovular fertility is not known. Since the gametocide is a cell poison, the probability of ovular damage is very high. Foliar application of DPX-3778 prevents anther dehiscence in rice (Long et al. 1973).

27.12.6 Breeding Value

Rice is basically a selfer whose flower is well suited for self-pollination. NCP is rare and varies with variety, season and environment. On an average it does not exceed 1% (Beachell et al. 1938; Virmani and Athwal 1973) but varies from 0.7 to 6%.

Genes causing hybrid sterility also exist in the rice genome and limit its crossability. Therefore both the floral structure and genetic architecture of rice have to be modified in order to increase its outcrosssing potential. A major way to enhance cross-pollination would be to develop more open florets with elongated and exposed stigma and exserted anthers. The enhanced cross-pollination would be used to exploit the positive heterosis that exists in grain yield and other quantitative traits; the heterosis ranges from 16 to 63%, with an average of 41% (Rutger and Shinjyo 1980). Even though many g-mst and gc-mst rices have been developed and in some of them satisfactory fertility restorers and sterility maintainers are available, the major difficulty in obtaining sufficient seed set on mst plants is their low NCP. Using gc-mst lines, 13% seed set was obtained at IRRI, the Philippines, 16–24% in California (Poehlman 1979), 22–26% in India (Sarma et al. 1982) and 15–45% in China (Virmani and Tan 1982). In two short-statured mst rice lines, seed set due to NCP is extremely poor (Azzini and Rutger 1982). In the third one, it averages 5.4 and 32.5% in 2 years, the increase in the latter year being due to better synchronisation of flowering of mst and mft plants. Seed set due to NCP on many mst plants is low even when they are surrounded by good pollinators (Stansel and Craigmiles 1966; Athwal and Virmani 1972; Carnahan et al. 1972). In some ethylene-imine-induced g-mst mutants, the floral morphology is altered to favour outcrossing. Thus 10–30% seed set was obtained by Fujimaki et al. (1977) on such induced mst plants. Singh and Ikehashi (1981) recorded 14.8 to 33.6% seed set on four mst plants closely surrounded by mft plants. This rate of outcrossing may be sufficient to permit intermating in composite populations. Using two gc-mst lines, Zhen Shan 97A and V20 A, obtained from China, 35–45% seed set due to NCP occurred in some plants (Virmani and Khush 1980). Nearly similar seed set on gc-mst rice lines has been reported, also from China. Seed set on open pollinated panicles varies with genotype, direction of planting and wind and distance in relation to the pollinator. Those plants with better panicle exsertion and synchronised flowering in relation to pollen source exhibit higher seed set. Because of high natural outcrossed seed set on some gc-mst lines, Virmani and Khush (1980) opine that a satisfactory outcrossed seed set on mst lines in hybrid seed production plots could be obtained by growing five rows of a mst lines (1 m-wide strip) alternatively with one row of pollen source. Increase in natural outcrossing on mst rices may be possible through breeding and selection of the lines having a short flag leaf, with exserted panicles and spikelets with exserted stigma. These traits are controlled by many genes and the environmental influence on the gene action is high. Therefore, selection for all these traits may not be simple.

In hybrid seed production gc-mst systems can be utilized only when NCP is supplemented by hand pollination. The natural crossing is enhanced by cutting the flag leaves and tearing off the leaf sheath that encloses the panicles. This process enhances NCP to 9–12%. In Indian mst rices, one third of the basal region florets do not set seeds, due to nonexsertion of the florets and the hindrance caused by flag leaf. This considerably reduces seed production due to low NCP over mst lines. The NCP in mst lines used in China is 15% (Anonymous 1980) and has led to a mass-scale cultivation of hybrid rice. Two labour-intensive practices are used there to raise hybrid seed yield (Lin and Yuan 1980) viz. (a) cutting back the flag leaf on male and female parents in order to eliminate these obstacles to pollen dispersal, the flag leaf of the mst line being completely removed but only two thirds is cut back in the mft-line, (b) supplementary pollination which is achieved by pulling a rope over the panicles of the male rows at 20–30 intervals, three to five times daily on calm days. Since F_1 seed yield is low and the operation requires labour, its feasibility on a commercial scale is questionable, especially in view of increasing labour charges.

27.12.6.1 ms Genes in Breeding

For commercial hybrid seed production, mst lines cannot be used unless the stability level of mst and outcrossing rate are highly enhanced. However, for limited seed production, the mst lines may be used in several ways. Fujimaki (1978, 1980), Ikehashi and Fujimaki (1980) and Ikehashi personal communications devised many methods (Figs. 27.5 to 27.8) in which *ms* genes facilitate rice breeding. Some of these are briefly described below:

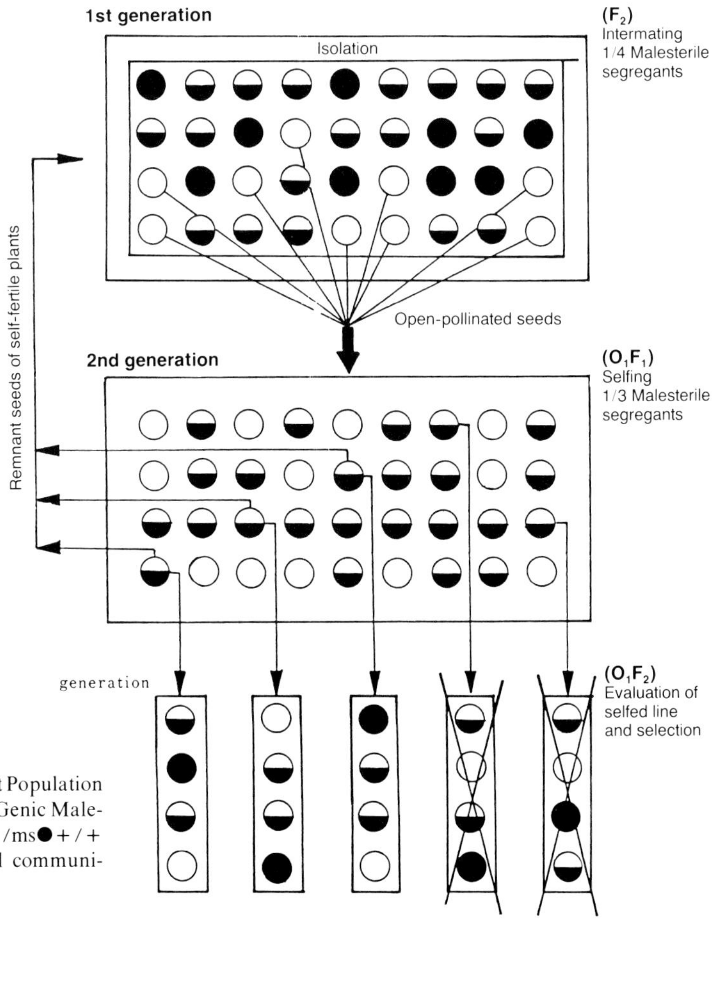

Fig. 27.5. Recurrent Population Improvement with Genic Male-sterility ○ ms/ms ◒ +/ms ● +/+ (Fujimaki, personal communication)

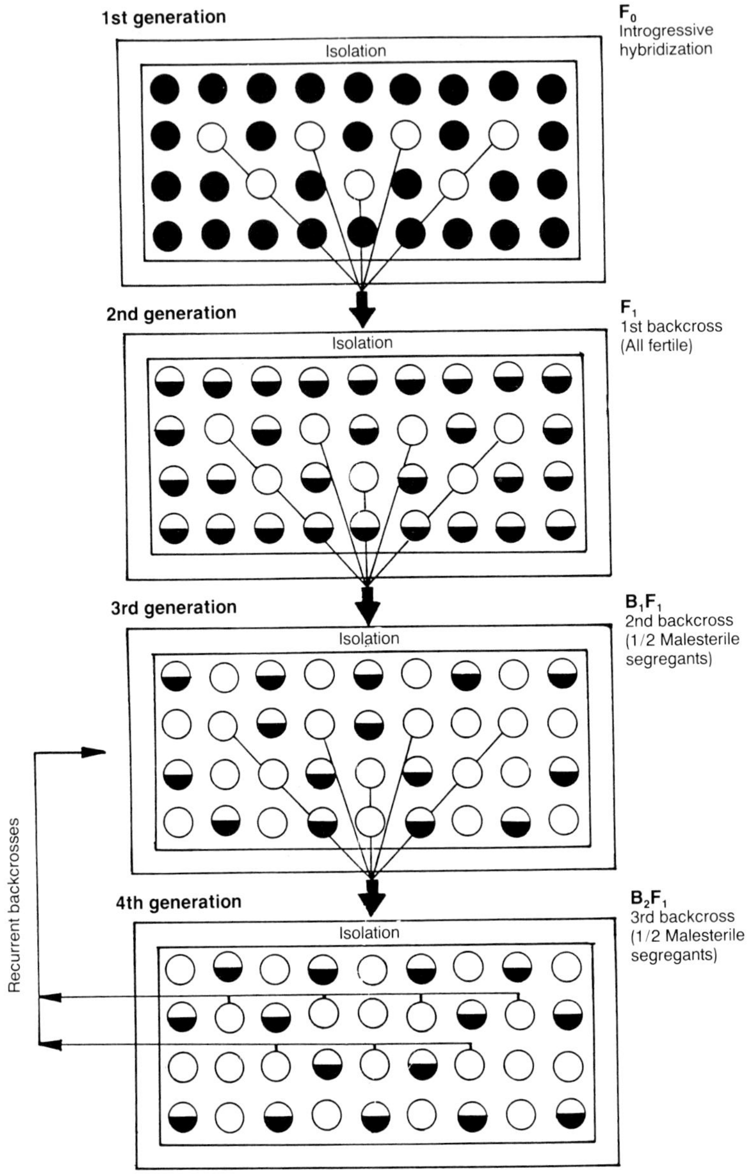

Fig. 27.6. Introgressive Improvement with Genic Malesterility. (Fujimaki, personal communication)

i. *Recurrent Selection.* Fujimaki (1978) and Ikehashi and Fujimaki (1980) presented a model (Fig. 27.7) for recurrent selection by using recessive *ms* genes. In this scheme, a foundation population is developed by a biparental, double or polycrossing or diallel crossing to which selection is applied later. This population is composed so as to be segregating for mst and is grown in an isolated environment to advance outcrossing between mst and mft plants. Seeds obtained over mst plants

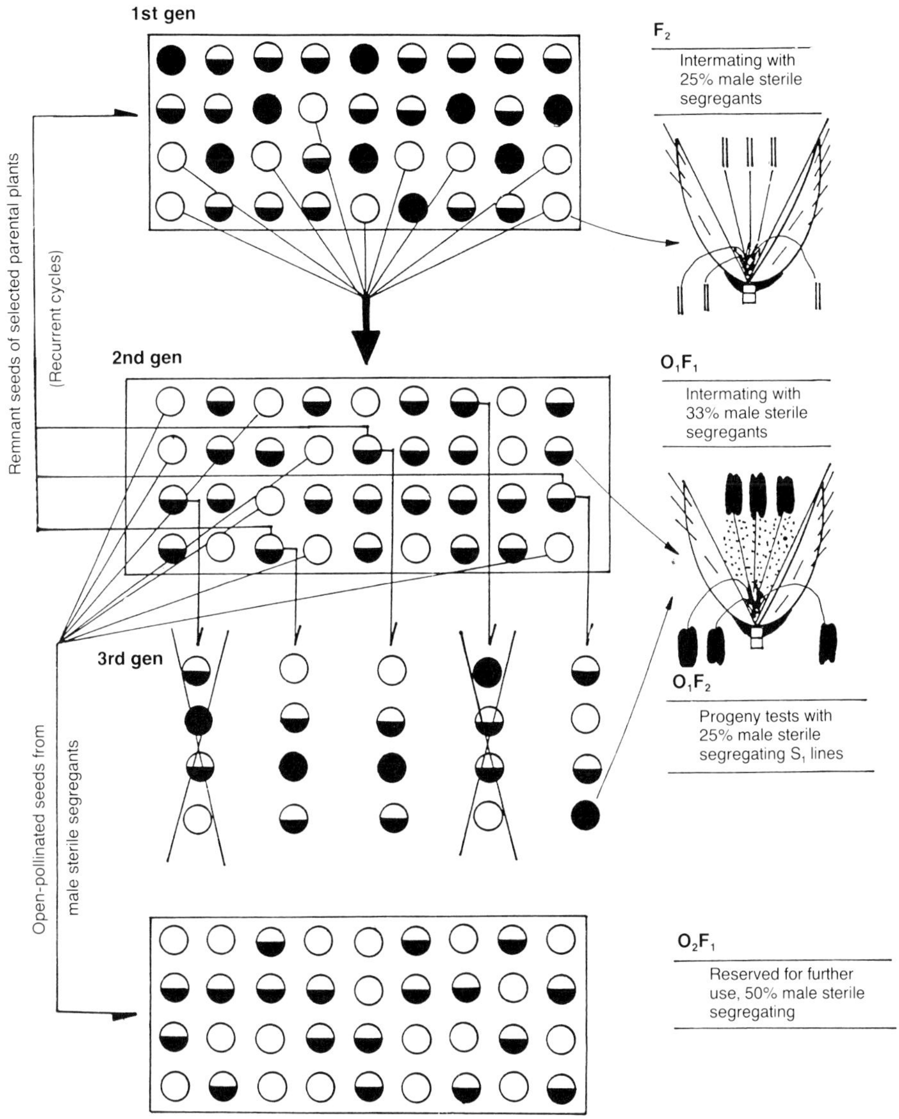

Fig. 27.7. Male sterile facilitated recurrent selection for rice breeding. (Ikehashi and Fujimaki 1980)

will yield random crossed populations. Two to three cycles of random crossing will be required for advancing recombination within foundation population before selection. In the selection phase, selfed seeds are harvested from fertile plants to develop S_1 progenies. Since every S_1 line will segregate 25% mst plants, the seed yield of these lines cannot be used for seed production evaluations. However, plant types and other agronomic traits like culm strength, leaf senescence, grain yield, quality and fineness, resistance to disease and pests, cold, drought etc. can be evaluated in these lines on a phenotypic basis (Fujimaki 1979).

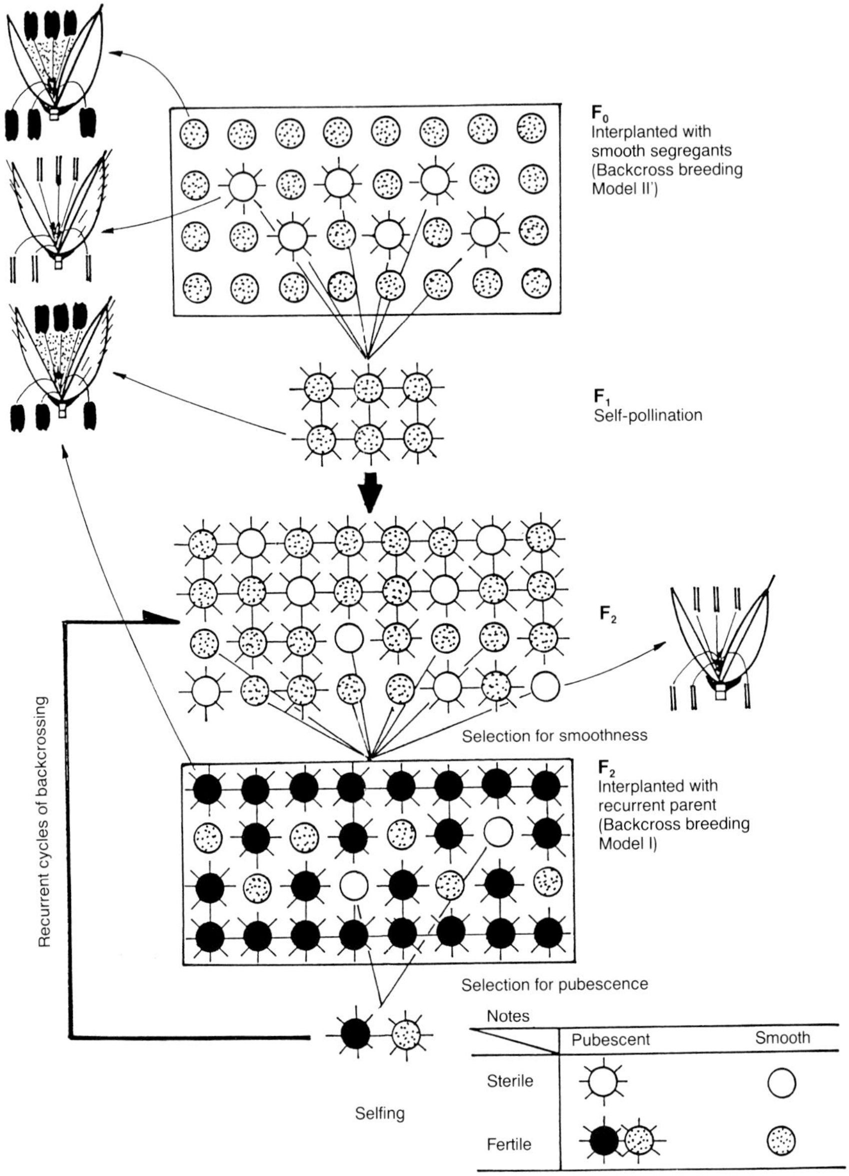

Fig. 27.8. Transfer of smoothness using male sterile facilitated backcrossing (MSBC) (Ikehashi and Fujimaki 1980)

ii. *Backcrossing*. Using the recessive *ms* genes, Fujimaki (1978) devised three models of BC breeding technique involving the recurrent selection procedure. In these techniques, *ms* gene can be used in the usual way of the BC method of breeding. Using the BC breeding model II, the trait grain smoothness that reduces dust generated during harvesting, threshing and hulling operations could be easily transferred (Fig. 27.8). This is a modified form of mst-facilitated recurrent selection and eliminates artificial BC in all the recurrent cycles. When one *ms* gene is used, two generations are needed for a BC to recover homozygous *msms* plants in a hybrid population. Since it is possible to complete one BC cycle in a year by raising a selfed generation in the greenhouse during winter, the best use of recurrent selection or BC using the mst allele can be made by linking either of them with the rapid generation advance technique given briefly below.

iii. *Rapid Generation Advance.* Although the mst recurrent selection system using the *ms* gene is a useful breeding technique, the breeding cycle is prolonged, as an intermating phase is involved during which no gene fixation is advanced. This insertion does not occur in conventional breeding. In order to bring the recurrent selection at par with the conventional breeding level, Ikehashi and Fujimaki (1980) suggested that after several intermatings or recurrent selections, rapid generation advance should be applied to the recombinant-rich hybrid populations. This will accelerate gene fixation and population development.

27.13 *Pennisetum typhoides* (Burm. f) stapf and Hubbard [*P. americanum* (L.) leek, *P. glaucum*, pearl millet]

Controversy exists presently over the nomenclature and status of this species (Terrel 1976). The traditionally established name *P. typhoides* has been maintained by Terrel (1976), who revalidated the name *P. americanum*, also agrees to retain the traditional name *P. typhoides* for this species (Terrel 1980 personal communication). Both genetic (Tables 27.21, 27.22) and chemical mst occurs in it (Table 27.23).

27.13.1 Genic male sterility

Using X-irradiations, a mst mutant having shrivelled anthers and scanty nonviable pollen was isolated by Krishnaswamy and Ayyangar (1942). While female fertility of this mutant was normal, mst was complete and was conditioned by a recessive *ms* gene. Kadam et al. (1940) found three types of genic sterility in pearl millet, viz: complete steriles, partial steriles and headless plant types having no florets at all. Of these, in some of the plants exhibiting partial sterility, the stigma appeared normal and exserted out of the floret. Their anthers were shrunken and deformed and no seeds were set on these plants. But whether female fertility was normal or impaired was not ascertained. Two cases of undetermined mst have been reported by Kajjari and Patil (1956) and Patil and Joshi (1962). While in the first (Kajjari and Patil 1956) the mst is genic, in the second (Patil and Joshi 1962) both male and female sterility is nearly complete, as only a few seeds were obtained from these after cross pollinations. The progeny raised from these seeds was mst. Since female fertility is extremely low in these plants, to consider them as true mst is questionable. From the progeny of mutagen-treated pearl millet populations three complete mst female fertile mutants were obtained by Powell and Burton (1969). Meiosis in these mutants proceeds normally but microspores degenerate after liberation. In each of them, sterility is conditioned by a recessive *ms* gene. The allelic nature of these three *ms* genes is not known. In three inbred lines, three recessive complementary *ms* genes control mst (Gill et al. 1973). Depending upon the presence of one or two homozygous recessive *ms* genes in the inbred lines, mono- or digenic segregations for mst are obtained. Rao and Koduru (1978a,b) detected mst plants in inbred lines 1P482 and Vg272. Sterility in each of them is conditioned by a single recessive gene, ms_1 and ms_2, respectively. Later, Rao and Devi (1983) found these two mst genes to be allelic. Actually the time of action of the *ms* gene was influenced by the cytoplasm type. In C-1 cytoplasm, *ms* gene acts very early, leading to degeneration before archespore differentiation. In C-2 cytoplasm, the *ms* gene action depends on two nuclear R_1 and R_2. Dominance for either or both of these genes allows the *ms* gene to act during meiosis, leading to syncyte formations. Double recessiveness for *R*-genes delays *ms* gene action up to microspore formation, leading to microspore degeneration before pollen mitosis. The action of R_1 and R_2 genes is specific to the C-2 cytoplasm and *ms* gene. These two genes are ineffective in C-1 cytoplasm or in association with the *Ms* gene.

27.13.1.1 Microsporogenesis

In mst inbred line 1P842, delayed and asynchronous meiotic development, desynapsis and a blockage of meiosis are caused by ms_2 gene (Rao and Koduru 1978a). The expression and frequency of these meiotic anomalies varies from flower to flower. In fact even within a flower, anthers exhibit nonsynchronous morphological development and delayed and highly asynchronous meiosis. Thus the *ms* gene acts during anther development and impairs ontogenic processes so that subsequent development and differentiation of archesporial tissue are highly erratic. This leads to anomalous meiosis and consequently to inviable pollen. In these mst's, in young anthers archesporial cells and PMC's at preleptotene are uninucleate, whereas in older anthers a mixture of uninucleate cells and plasmodia appears. The plasmodia have 2–12 nuclei, formed due to the fusion of uninucleate cells starting from early pachytene. In the uninucleate cells, occasionally two to four univalents and rarely 14 univalents are present. With the increase of univalency, MI orientations and AI segregations are highly disturbed but even then meiosis proceeds in most of them and microspores are formed. In the majority of the syncytes, a blockage of meiosis either during or after diakinesis occurs. Also in many uninucleate and plasmodial sporocytes, meiosis is completely suppressed and a wall is formed around these cells, leading to the formation of giant pollen grains. Since the pollen grains are sterile, selfing yields only a few seeds, produced due to apomixis. Although apomixis is known in some *Pennisetum* species (Hanna and Powell 1973, Hanna *et al.* 1976), hand pollination of these mst's with fertile pollen yields normal seed set, indicating female fertility and lack of apomixis.

In another mst material that arose spontaneously in an inbred line, Vg 272, the anthers are phenotypically normal but nondehiscent (Rao and Koduru 1978b). Except for slightly clumped bivalency, male meiosis is normal up to microspore formation. Microspores degenerate at the uninucleate stage just before the onset of the first mitotic division. Up to this stage, tapetal development is also normal. Pollen degeneration and tapetal breakdown occur simultaneously. At anthesis, anthers have empty pollen grains and an acellular cytoplasm without nuclei. Stigma receptivity and female fertility are normal. The sterility is controlled by a recessive gene and is not associated with any phenotypic trait. The *ms* gene, when transferred to various cytoplasmic backgrounds, exhibits full expressivity and penetrance. Since anther development, archesporial tissue differentiation and meiosis are normal in this mst line and the sterility is not associated with a hairy plant texture, this *ms* gene initially appeared to be different from the ms_2 gene, but the two genes were later found to be the same by Rao and Devi (1983).

In the inbred line Vg272, the *ms* gene causes microspore degeneration before first mitosis in the earlier generations. In later-segregating generations, the action of this same *ms* gene is preponed and the gene inhibits archespore differentiation and function and no PMC's are formed. This change in the time of *ms* gene action is assigned by the authors to cytoplasmic mutation from C-2 to C-1, as in later generations a back mutation occurred. Thus variation in the mode of *ms* gene action is due to cytoplasm and not to gene mutation. The *ms* gene action is highly influenced by modifier genes and environmental factors.

27.13.2 Gene Cytoplasmic Male Sterility

While inbreeding local millet from Nasik (Bombay), Kadam *et al.* (1940) observed various forms of mst. Kajjari and Patil (1956) postulated the existence of gc-mst in pearl millet on the basis that selfed mst's produced no seeds and crossed ones gave high seed set. Later, in an open pollinated pearl millet population, Menon (1958, 1959) detected a high frequency of mst's at Coimbatore (India) in the strain P.I. 819 which had underdeveloped, nondehiscing anthers and empty pollen. While backcrossing certain normal varieties on these mst's for five generations that yielded only mst populations, one variety produced equal proportion of sterile and normal plants when crossed with these sterile plants. Thus this variety appears to be heterozygous for a *Fr* gene, the mst being due to interaction of the S-cytoplasm and the two homozygous recessive *fr* genes, the dominant allele *Fr* restoring fertility to the S-cytoplasm. This source differs from many mst sources and their maintainers on which it was tested later (Appadurai et al. 1982). In the winter of 1955–56, at Tifton, Georgia, an inbred line 556 developed by Burton (1958) produced little pollen and few seeds upon selfing. Pollinating it with Tift-23 line led to mst segregation in F_2. Upon backcrossing to Tift-23, these yielded mst populations. This established the existence of gc-mst in pearl millet. When these mst plants were pollinated by the pollen of Tift-23, seed set occurred and the resultant progeny segregated mst's. Hence Tift-23 worked as an excellent sterility maintainer for this S-cytoplasm, and line 556 carried the Tift-23A type of cytoplasm. In addition to the Tift-23 (A_1), two more sources of mst (A_2 and A_3) were found from the pearl millet lines L66A and L67A from Ludhiana (India) by Athwal (1965). Crossing the same restorers and maintainers on these two and on the line 556 led Burton and Athwal (1967) to assume that the three sources of cytoplasmic sterility differed from each other, the sterility in them being conditioned by three nonallelic recessive *fr* genes interacting with a specific type of S-cytoplasm. While type 556 lodges S_1 cytoplasm and genes fr_1fr_1, Fr_2Fr_2, Fr_3Fr_3, type 66A has S_2 cytoplasm and genes fr_1fr_1, fr_2fr_2, Fr_3Fr_3 and type 67A the S_3 cytoplasm and fr_1fr_1, fr_2fr_2, fr_3fr_3 genes. Yadav (1974) found that at least two pairs of nonallelic recessive genes fr_1fr_1, fr_2fr_2 control mst in pearl millet and the genotype of a particular variety behaved differently from the S-cytoplasms of American and Indian mst pearl millet. In a composite mst stock, Prakash (1967) found polygenic control of gc-mst. The influence of the environment over mst expression is high. This distorted the existence of discrete segregations in this mst material.

Another source of gc-mst was found at Ibadan (Nigeria) in a plant from a population of ex-Bornu, an improved northern Nigerian variety of gero, a day-neutral pearl millet (Akenova and Chheda 1981). Two plants gero-1 and gero-2, that were maintainers of this mst when crossed to the known three gc-mst sources Tift-23A, 239A_2 and 426A_3, acted as maintainers for 239A_2 and 426A_3 sources, but none was a good maintainer for all the three sources. Thus gero-1 and gero-2 appear to carry all the three *fr* genes in either the homozygous or heterozygous state, as neither of the two proved a good maintainer or fertility restorer for all the three known mst sources. Hence the ex-Bornu source may represent an additional gc-mst source. To confirm this, determination of the ability of the maintainer lines for the three known sources (A_1, A_2, A_3) to maintain sterility in ex-Bornu would be useful. The crosses made with PT 732A and B (Appadurai et al. 1982) on other A and B lines indicate that PTB2A may be another new gc-mst source. From the available sterility sources, a large number of agronomically useful disease-resistant mst lines were developed in India by Vohra (1969), Utkhede (1972), Athwal et al. (1976), Pokhriyal et al. (1976), Gill et al. (1979, 1981), Andrews and Anand Kumar (1982), Appadurai et al. (1982). Details are given by Anand Kumar and Andrews (1984). In the USA, Burton (1981) developed Tift-23 DAE and Tift-23 DBE, which are early and photoperiod – insensitive versions of Tift-23DA and Tift-23 DB.

27.13.2.1 Sterility Transfer and Fertility Restoration

After detection of the S-cytoplasm in Tift-556 inbred line, Burton (1958) found Tift 23 to be the best maintainer of this sterility. Tift-23 possesses excellent fodder qualities and a high general combining ability. Therefore it was sterilized by several BC's [(Tift-556 × Tift-23) × Tift-23]. Subsequently, many other pearl millet lines were sterilized at Georgia, including Tift-18, a contrasting type of Tift-23, for Tift-18 is tall, nontillering, large cylinder-headed pearl millet, whereas Tift-23 is leafy, broad-leaved, short-internoded, small-headed and profuse-tillering. Crossing Tift-23A with 41 inbred lines indicates that nearly two-thirds of the elite Georgia lines are fertility restorers or C-lines. However, the frequency of fertility restorers in Indian lines is low (10%). L66A and L67A differ in their fertility restoration from Tift 23A (Burton and Athwal 1967). Maintainers for these two sources also maintain sterility in Tift-23A. Whereas Tift-23B, the maintainer for Tift-23A, restores fertility in L66A and L67A, the promising fertility restorers for Tift-23A, such as L4 and L239, act as excellent sterility maintainers for L66A and L67A. On the other hand, L66B, the maintainer for L66A, restores fertility in L67A, while L67B is only a partial fertility restorer for L66A. Burton and Athwal (1967,

1969) proposed the designations A_1 and A_2 as suffixes for the two different sterile cytoplasms in Tift-23A and Tift-239A, respectively. The maintainer Tift-23DB$_1$ maintains sterility for the Georgia line Tift-23DA$_1$ and restores fertility in the Indian gc-mst source incorporated into Tift-239DA$_2$ and having the maintainer Tift-239DB$_2$ (Rachie and Majmudar 1980).

Whereas Tift-23 proved to be an excellent maintainer for the A_1 source (Burton 1965), Tift-239 proved to be the most promising maintainer for the A_2 source of mst (Burton and Athwal 1967, 1969). Burton and Athwal (1967) suggested a model (Table 27.8) that gives the relationship between cytoplasm sources, (A-lines) sterility maintainers (B-lines) and fertility restorers (C-lines) in the three S-cytoplasm types of pearl millet. In all BC's involving Tift-23, the F_1's failed to shed pollen, and further backcrossing established the isogenic lines Tift-23A and Tift-23B. Using the four fertility-restorer lines Tift-238, 239, 240 and 241 in crossings with Tift-23A, one major gene was found to be responsible for fertility restoration in each line. Either these four genes differ from each other or they carry different modifiers. Burton and Athwal (1967) suggested three genes and modifiers responsible for fertility restoration. Even control by a single *Fr* gene in association with modifiers cannot be ruled out. Both modifiers and environment influence fertility restoration considerably. Furthermore, each pearl millet genotype behaves differently with the cytoplasm types of American and Indian sources.

Table 27.8. Genetic models of fertility restoration in pearl millet. (Modified after Rachie and Majmudas 1980)

Line	Cytoplasm type	Fertility-restorer genes		
Tift-23A	S_1	$fr_1 fr_1$	$Fr_2 Fr_2$	$Fr_3 Fr_3$
Tift-23B	N			
L66A	S_2	$fr_1 fr_1$	$fr_2 fr_2$	$Fr_3 Fr_3$
L66B	N			
L67A	S_3	$fr_1 fr_1$	$fr_2 fr_2$	$fr_3 fr_3$
L67B	N			
L103A	S_2	$Fr_1 Fr_1$	$fr_2 fr_2$	$Fr_3 Fr_3$
L103B	N			
Tift-239R	N	$Fr_1 Fr_1$	$fr_2 fr_2$	$fr_3 fr_3$
Tift-13R	N			
Bil-4R	N	$Fr_1 Fr_1$	$fr_2 fr_2$	$Fr_3 Fr_3$
Bil-GR	N			

27.13.2.2 Expressional Breakdown and Fertility Reversions

In mst pearl millet temperature and humidity influence the frequency of the plants' shedding pollen. This is evidenced from the comparisons of mst expression in different seasons and years made by Vittal Rao (1969), Reddy and Reddi (1970), Thakare (1977) and Saxena and Chaudhary (1977). When Tift-23A was grown at Hyderabad (India) the main primary tillers had only sterile ears, but the nodal tillers developed fertile ears (Vittal Rao 1969). This type of fertility restoration occurred regularly in secondary and tertiary tillers of first and second nodal branches. Low relative humidity and high temperature appear congenial for throwing pollen shedders in Tift-23A (Reddy and Reddi 1970), but under such climatic conditions,

Tift-23D_2A and Tift-628A are mostly stable. Thakare (1977) detected no differences due to seasonal effects in which modifier genes can restore partial pollen fertility in an otherwise purely mst line. Not only genetical factors and environmental conditions influence mst fertility reversions, certain physiological factors appear equally important. For instance, increased soil nitrogen application increases endothecial thickness and pollen sterility in maintainer lines of Tift-23A and Tift-628A (Reddy and Reddi 1974). This adds one more dimension to the arena of mst control in this crop.

Clements (1975) detected several sectors of fertile anthers in four different gc-mst lines. Hybrids obtained by crossing normal mst's × pollen from fertile sectors yielded fertile progeny, indicating that the sectors arose due to cytoplasmic mutations. Burton (1977a) evidenced the cytoplasmic nature of the fertile sector mutants, as he found 103 mft mutants that arose due to S-cytoplasm mutation to N-type and utilized these as maintainers. In appearance and performance, they were stable and identical to normal lines. Circumstantial evidence has indicated that mft or mst's with fertile sectors in pearl millet originate due to nuclear or cytoplasmic mutations. Environmental conditions trigger these mutants. This necessitates limited generation increase from breeders' seed and constant roguing to eliminate cytoplasmic and nuclear-fertile mutants arising in hybrid seed production plots.

27.13.2.3 S-Cytoplasm Disease Susceptibility

In early 1970, a downy mildew epidemic caused by *Sclerospora graminicola* reduced total hybrid grain pearl millet production from 8 to 3.5 million tons in India. The epidemic was due to the susceptibility of the S-cytoplasm of Tift-23A that was used solely in the production of all the hybrids popular at that time. Following γ-irradiations, three mutants resistant to downy mildew and stable for mst were isolated in Tift-23A (Murty 1973). Using these mutants, new hybrids, NHB-3, NHB-4, NHB-5, were reconstituted; but resistance to downy mildew broke down again in the hybrids and many new resistant mst lines were developed by Pokhriyal *et al.* (1976), Gill *et al.* (1977, 1981) and Andrews and Anand Kumar (1982). In order to breed diversified hybrids to minimise future disease epidemics, new resistant mst lines are to be constantly developed by using different S-cytoplasm sources and broad germ plasm. This is being actively done at ICRISAT, India (Anand Kumar and Andrews 1984). In addition, incorporation of resistance to ergot (*Claviceps fusiformis*) and smut (*Tolyposporium penicillariae*) in mst lines using genes from the pearl millet world germ plasm collection is also being tried at ICRISAT.

27.13.2.4 Microsporogenesis

This has not been investigated in all the mst materials. In one, the *fr* gene causes microsporogenesis breakdown mostly during tetrad formation (Burton 1958). In another material, meiosis proceeds normally but microspores degenerate immediately after their release (Singh and Sharma 1963). Although in the two materials the *fr* genes act during different stages of microsporogenesis, the tapetum in both of them is persistent.

27.13.3 Structural Male Sterility

This is known in two pearl millet mutants, in which transformation of anthers into deformed functionless pistil-like structures occurred (Goyal 1962, Manga 1977). The degree of transformation is highly variable and very few functional pollen grains are produced in a few anthers. Significant variability in the expression of anomaly exists between spikelets, flowers and plants. The tendency towards normal anther development is gradational from apex to spike base. The anomaly in each case is conditioned by one recessive gene. Female fertility of these mutants is normal.

27.13.4 Induce Genetic Male Sterility

i. *Ethidium Bromide (EB)*. Burton and Hanna (1976) obtained a low frequency of mst mutants in MI after soaking the seeds of Tift-23DB$_1$ (a maintainer for gc-mst Tift-23DA$_1$) in deionized water solution having 0, 250 and 1000 ppm ethidium bromide at 5°C for 40 h. The mst mutants (whole spikes, sectored spikes, or spikes with sterile and fertile florets interspersed) occurred in frequencies of 1 in 30,100, 1 in 515 and 1 in 274 plants treated with 0, 250 and 1000 ppm EB, respectively. The sectored heads proved mutation to have occurred. Since the initial line was homozygous and mutants appeared in the M$_1$ generation, the mutation is cytoplasmic, as the seeds obtained over mst mutants after open pollinations produced mst progeny in the greenhouse. The greenhouse-grown progenies from complete and sectored mst heads were variable and had mst and some p-mst plants. Also a few plants from mst heads were as fertile as the untreated initial line. Although 250 ppm and 1000 ppm EB seed treatments increased the frequency of gc-mst mutants by 5.8 and 10.9 times, respectively, all the induced mst plants, when pollinated with Tift-23DB$_1$ pollen, produced progenies with many pollen-shedding plants. The variation in sterility suggests that it may be possible to select stable gc-mst mutants, but two cycles of selection made by Burton and Hanna (1976) could not yield such mst mutants.

ii. *Mitomycin and Streptomycin*. Stable gc-mst mutants in a very low frequency were induced by treating seeds of Tift-23DB$_1$ with 0, 200, 500 ppm of streptomycin and 50 ppm of mitomycin (Burton and Hanna 1982). The mutation frequency increased 2.9, 3.6, and 6.2 times by soaking the seeds in 200 and 500 ppm of streptomycin and 50 ppm of mitomycin, respectively. Thus these two chemicals may be useful in producing stable mst mutants in other crops. This needs to be investigated. The sterility fertility segregations of these mutants crossed with two sterility maintainers and six fertility-restorer inbreds of pearl millet suggest that these induced mutants have sterility-maintainer and fertility-restorer requirements akin to gc-mst Tift-23 DA$_1$ and are thus similar to gc-mst mutants.

iii. *Thermal Neutrons.* This treatment produced female sterile (Hanna and Powell 1974) and not a mst mutant in Tift-23B. However, this method could be used to produce an obligate apomictic. The authors suggested a scheme for the production of hybrid seeds which presupposes a chromosomal rearrangement in which female sterile segregates to yield 3:1 female sterile: female fertile plants.

27.13.5 Chemical Male Sterility

Attempts to induce mst through use of X-rays, 2, 4-D, mendok and DPX-3778 were tried (Kumar and Joshi 1939, Krishnaswamy and Ayyangar 1941, 1942, Sharma 1961, Hanna 1977, this Vol.). In all cases, both pollen and ovular sterility were induced and the treatments proved ineffective. Moreover, some of these chemicals caused plant damage.

i. *Dalapon.* This produces p-mst, high ovular sterility and phytotoxicity (Miller 1961).

ii. *Dimethyl Arsenic Acid.* This treatment reduced pollen viability considerably (Powell and Taylorson 1967), but caused high plant and ovule damage. Since the chemical compound is a cell poison, tissue damage induced by it is serious.

iii. *DPX-3778.* This prevents anther exsertion and dehiscence but pollen remains stainable and plump. Application of 7.43 kg ha^{-1} at mid to late boot stage on inbreds Tift-23DB, Tift-23DB_2, Tift-13 and Tift-18DB followed by a second application at the same rate 5 to 6 days later reduced selfed seed set to 2, 13, 5 and 30%, respectively (Hanna 1977). No reduction in seed test weight or female fertility occurs. Response to the chemical varies with genotype, environment and growth stage.

iv. *Mendok.* Although various mendok concentrations induced both pollen and ovular abortion, Sharma (1979) was able to induce anther indehiscence by 0.5% mendok foliar spray. In the treated plants, anther emergence was delayed by a few days. Anthers clumped around the panicle and did not dehisce. Both pollen and ovular fertility were not damaged, and seeds were set after hand pollination. Treatments above 0.5% are toxic and impair gametic fertility. Repeating this treatment in a local pearl millet variety induced pollen and ovular sterility and 1 week delay in anther dehiscence (this Vol.). Mendok treatments of 1.0 and 1.5% induce plant, pollen and ovular damage.

27.13.6 Breeding Use

In importance pearl millet ranks sixth among cereal crops. In Africa and India, it is the principal grain crop in areas that are too hot and dry for other cereal crop growth. In the USA it is a high-yielding summer forage of excellent quality. The g-mst has not been utilized in breeding because of its obvious limitations. However, with the discovery of gc-mst by Burton (1958), and its excellent maintainers later (Burton 1965, Burton and Athwal 1967, 1969), its use in hybrid production became feasible. By introducing dwarf gene (d_2) into these genotypes, shoot height became reduced to one half (Burton and Fortson 1963, Burton 1969) and grain yield increased considerably. The first commercial hybrid, HB1, utilising the gc-mst system and Tift-23A as the female seed parent and Bil-3B as the restorer parent produced by Athwal was released in India in 1965 (Burton and Powell 1968). This hybrid not only outyielded many other experimental hybrids, but produced 88% more grain than the best open-pollinated checks in numerous yield trials throughout the Indian millet belt, that ranges from 11° to 31°N lat. The hybrid exhibited excellent performance under favourable growing conditions as well as under the extreme stress conditions to which this crop is usually subjected both in India and in Africa. In fact, cultivation of this hybrid led to a dramatic increase of grain yield from 3.5 million metric tons in 1965 to 8.0 million metric tons in 1970. Encouraged by such results, more hybrids using mst have been developed in India and are currently under extensive cultivation. In practice, commercial F_1 hybrid seed is produced by planting 6–12 row strips of mst genotype between two row strips of male parents. Thereafter only mst rows are harvested for seed, the resultant seed being hybrid.

At the International Crop Research Institute for the Semi-Arid Tropics (ICRISAT), Hyderabad, India, three systems of gc-mst are being used to produce pollen and seed parent material for high-yielding disease-resistant hybrids (Jain *et al.* 1977). The potential hybrid parents that are developed are tested through three stages: (a) in single plots, the test crosess are evaluated and those exhibiting incomplete pollen shedding are discarded, (b) agronomically promising and fully fertile lines are subjected to highly replicated trials including the disease nursery, (c) the most promising of these are subjected to national and international testing. Many hybrids exhibiting good agronomic performance, high yield and disease resistance to diverse environments have been developed through this three-stage testing viz. ICH-117, 118, 126, 128, 129, 143 etc; of these ICH-128, 129 had African restorers (Rachie and Majmudar 1980). In all these cases, using the mst lines 67A, 111A, and 5054A as seed parents, high-yielding hybrids have been developed. However, since a great residual genetic variability exists in the commercially used mst for morphological characters, downy mildew resistance and the frequency of pollen shedders in the A-line, the character variability is being minimized for five mst's (66A, 67A, 111A, 5054A, 5141A) by making a large number of plant-to-plant crosses and testing the progenies in the diseased nursery (Rachie and Majmudar 1980).

In the USA, pearl millet being a crop of forage value, several hybrids having enhanced forage yield have been released. Most of these have Tift-23A or Tift-23DA as the female parent. These hybrids continue to remain vegetative for a long time and do not become weeds for the subsequent crops. Fortunately, the d_2 gene not only reduces height by 50% and enhances seed production but also increases leafiness, forage quality and animal performance (Johonson et al. 1976, Burton et al.1969). Gahi 3 is an example that was released by Burton (1977b) and is a first-generation hybrid between Tift-23A and Tift 186. It is an extremely uniform, high forage-yielding, medium-stemmed, leafy hybrid with smooth nonhairy stems and leaves. It is well suited for pasture, green chop, dehydration, pelleting or silage production (Burton 1977b). The frequency of Tift-23A fertility restoration gene in Indian breeding stock is low. Thus the search for heterosis for grain production in

Indian breeding programs was initially confined to a relatively small proportion of the best-adapted breeding stocks. Therefore, new sterile mechanisms, particularly if introduced into isogenic lines, would greatly increase the proportion of high-combining, fertility-restored hybrids in breeding programs. It appears reasonably possible to have gc-mst A-lines which have restored fertility in combination with almost any pollinator. Theoretically, it is possible to develop isocytic stocks from both elite, high combining lines such as Tift-23, Tift-18, Bil-3B, J-104 and D174 with different kinds of cytosterile mechanisms which offer fertility restoration in combination with any possible pollinator (Rachie and Majmudar 1980). Using two different cytoplasms, two breeding populations may be established to facilitate breeding systems and recurrent selection techniques. Thus while one population would contain plants with A_1-cytoplasm and their maintainers, the second population would have A_2-cytoplasm with respective maintainers. These could greatly facilitate the development of superior F_1 hybrids in a species where a large amount of genetic variance for yield is nonadditive.

Powell and Burton (1966) suggested that pearl millet (*P. typhoides*) × napier grass (*P. purpureum*) hybrids could be propagated from the seed produced by planting gc-mst pearl millet between rows of napier grass at a time that would bring both into flowering simultaneously. Compared to pearl millet hybrids, these hybrid plants are slow-growing but the overall forage production in both of them is nearly equal, as the latter type hybrid plants exhibit luxuriant vegetative growth in autumn. However, development of a simple and cheap commercial seed production method is needed for this purpose. Amphidiploid hybrids produced by Burton and Powell (1968) are less fertile and more unstable than their parents. The variability for yield, grain weight and days to ear emergence is, however, higher in mst lines than in mft lines (Tyagi et al. 1982), and this variability could be used profitably in commercial breeding.

27.13.6.1 Developing Elite Inbreds

Elite inbreds with high breeding potential and genetic value have been developed in mst pearl millet. A detailed and systematic account has been given by Rachie and Majmudar (1980) from which the following is derived. Most useful elite lines are Tift-23A and B, Tift-23DA and B (dwarf), Tift-18A and B from Georgia (USA) and, L101A and BL111A and B, L66A and BL67A and B from Ludhiana (India); the latter two have a cytogenetic mst system which differs from the Georgia type. High combining Indian restorer lines (on the Georgia gc-mst source) are Bil-1 and Bil-3 from Ludhiana, D174 and D1162, both dwarfs from Delhi; J115 and J104 from Jamnagar; and K560 and K230 from Kanpur. The Georgia dwarfs D_1, D_2, D_3 and D_4 and the tall Tift-13 are likewise restorers for Tift-23A. Dwarf inbred D174 is highly vigorous, high-yielding and adapted widely enough for direct commercial use in the northern and southern parts of India (Shanmugasundaram et al. 1968). Recently a large number of novel and superior inbreds have been developed in India, Africa and the USA, which are being recombined to obtain desirable lines. The five dwarf series (D_1–D_5) developed by Burton (1967) have been highly useful in breeding in India (Bakshi et al. 1966). These genetic stocks have not only made shoot height reduction possible, but have also helped to provide the fertility restoration so urgently needed in Indian pollinators. From the recombinants of these dwarfs, a new array of highly vigorous, widely adapted, uniformly tillering, high-yielding inbreds have emerged. Whereas some of these may now prove useful directly as varieties for some Indian and African sorghum-growing regions, most of them have served as the parental stocks for hybrids (both dwarf seed-productive, and tall forage-yielding) as well as basic material for further recombinations to obtain the most promising combinations. In India, the most useful and high combining dwarfs are D_2 and D_4. After six BC's, Akenova (1982) transferred mst from Tift-23A into a short-day local variety Maiwa, whose maintainers have also been identified. Likewise, Chheda et al. (1974) tried transfer of mst from Tift-23A into a S.W. Nigerian Maiwa population for use in the commercial production of Maiwa pearl millet × elephant grass F_1 hybrids. Using gc-mst, Singh and Sharma (1964) developed a high-yielding pearly seeded strain with prolific vegetative growth and good tillering. Another similar line developed by these authors has nearly 20% seed protein content. This material is being used in recombination with other genetic stocks for simultaneous grain yield and seed protein improvement. Other inbreds with pearly amber yellow endosperm, ashy gray or steel blue seeds have 15–17% protein, whereas HB_1, has 12–14% and the local strains have only 9–12% protein content.

27.13.6.2 Hybrid Production

Development of gc-mst pearl millet lines offered commercially viable economical method of producing high grain-yielding hybrids. Five grain hybrids using Tift-23A yielded 40–113% over the check and control varieties in 1963 in India. In 1964, hybrid bajra gave 175–200% yield grains and using mst line Tift-23A, 5 hybrids (HB_1–HB_5) were subsequently released; HB-3 was popular due to early maturity, bold grains and promising performance under moisture-stress conditions. With hybrid release, yield increased from 288 to 622 kg ha^{-1} (Safeeulla, 1977), the area of hybrid plantation increased from 3.3% in 1967–1968 to about 16% in 1971–1972 and production increased from 2.5 million metric tons in 1950 to about 8 million metric tons in 1970. However, in 1970, downy mildew epidemics reduced hybrid seed production in India to 3.6 million metric tons (Jain and Pokhriyal 1975, Murty 1980). This epidemic was due to the susceptibility of the common S-cytoplasm source (Tift-23A) used in the hybrid production. Using γ-rays, 3A line downy mildew-resistant mutants were isolated from Tift-23A (Murty 1973, 1974). One of these, 5071A, exhibited high field resistance to downy mildew (Raut et al. 1973). Using this line, HB-3, HB-4 and HB-5 were reconstituted and named NHB-3, NHB-4, NHB-5, respectively. Their seeds were quickly multiplied and supplied to farmers, so that in 1974, multiplication of 600 metric tons of hybrid seed was performed to replace some of the then existing downy mildew-susceptible hybrids (Jain and Pokriyal 1975). Concurrently, another downy mildew-resistant mst line, 111A, was found and two resistant hybrids PHB-10 and PHB-14 were released in 1975 (Gill et al. 1975). However, after 2–3 years, the reconstituted hybrids having 5071A cytoplasm became partially susceptible to downy mildew (Sain and

Kanwar 1977). Two more downy mildew-resistant mst lines 5141A and 5054A were isolated, and BJ-104 and BK-560 hybrids were developed. Although of these BJ-104 is early, high tillering downy mildew-resistant with yield stability and promising performance, under moisture stress conditions in India (Pokhriyal 1977) both the hybrids grow well. Currently using spontaneous and induced mutations, the selection of downy-mildew-resistant mst lines is being continuously made (Gill et al. 1981, Andrews and Anand Kumar 1982). The resistant mst lines are tested for stability and used for hybrid seed production. In addition, the incorporation of ergot and smut resistance in mst lines is also being tried at ICRISAT, India. Details are given by Anand Kumar and Andrews (1984).

27.14 *Phleum* L. (cat's tail)

Many plants of *P. nodosum* and a few plants of *P. pratense* (Timothy grass) with shrivelled nondehiscing anthers and degenerated pollen grains were detected by Nordenskiold (1945–46). In four clones, 1–2 *ms* genes cause mst (Townley-Smith 1969). In a clone of *P. nodosum*, mst is conditioned by a recessive *ms* gene. In some of the plants female fertility is also impaired. Thus they may be synaptic mutants rather than mst's. Many mst clones having sterile pollen were isolated by May and Kasha (1971). In one such clone desynapsis leads to meiotic irregularities and pollen sterility. A maximum of 40% pollen stained, but its functionability was not determined. While desynapsis is controlled by one recessive gene, the genetic control of mst is not known. In this diploid Timothy clone, desynapsis is the cause and not the effect of mst (May and Kasha 1980). Delayed tapetal degeneration occurs in the mst clone and leads to microspore abortion.

27.15 *Saccharum* L. (sugarcane)

This genus possesses ten species, all of which are robust perennials with tall stout stems and long, usually flat leaf blades. Established cases of g-mst are not known in it. Spontaneous occurrence of mst in Indian sugarcane varieties is documented by Dutt *et al.* (1954). The mst's possess either nondehiscent or dehiscent anthers with aborted or agglutinated pollen. Genetic control of the sterility is not known. A parthenogenetic derivative of the cultivated sugarcane, *S. officinarum L.* with 110 chromosomes (n = 10, 12) obtained after pollination by *Imperata cylindrica* was completely pollen sterile (Janaki-Ammal 1941). Meiosis was characterised by the presence of a variable number of uni- to multi-valents and proceeded only up to AI. The second division was mostly deleted and the dyads either moved down the side of the spindle or passed through it, forming a restitution nucleus. Immediately thereafter, degeneration sets in. Rarely, in a few PMC's, meiosis proceeded up to the quartet stage, but disintegration followed. The *ms* gene appears to inhibit the spindle function completely, rendering it a dead spindle. The sterility appears to be genic, but the precise genetic control is not known in this highly complex polyploid. Also the status of female fertility is unknown. In some interspecific *Saccharum* hybrids, genetic male sterility is known (Purseglove 1972, pp. 227–228). Such mst plants possess aborted or no pollen in their anthers. In them, pollen abortion occurs during the microspore liberation stages. Whether the sterility is g-mst or gc-mst is not known. Some clones of noble canes *S. officinarum* also exhibit g-mst, conditioned probably by one recessive gene. The Indian sugarcane variety Co421 develops stainable pollen but nondehiscent anthers. Dusting pollen of *Coix lachryma jabi, Euchlaena mexicana, Narenga porphyrocoma* and *Sclerostachya fusca* induces anther dehiscence in this variety (Babu 1961). In 17 Indian sugarcane genotypes, mst's having 0–18% pollen fertility were identified by Subramanyam and Andal (1984). Of these, 12 genotypes have complete pollen sterility, which was stable for 3 years. The mst's have nonemergent and nondehiscent anthers. The mst does not follow any pedigree pattern and arose in a F_1 cross of mft parents or from crosses in which one parent was mst. Genetic control of the sterility is not known.

Pollinating *S. spontaneum* by the pollen of *S. barberi, S. erianthus, S. officinarum, S. narenga* or *S. sclerostachya* led to mst F_1 plants (Raghavan 1951, 1964). The progeny of the reciprocal cross was normal. F_1 plants obtained after crossing *S. spontaneum* with *S. sclerostachya* pollen when backcrossed by *S. sclerostachya* pollen produced progeny segregating 1:1 mst:mft plants. Probably the heterozygous *fr* genes interact with *S. spontaneum* cytoplasm to produce the mst progeny. A similar type of mst was obtained by Janaki-Ammal (1942) by crossing *S. officinarum* to *Narenga porphyrocoma*. Crossing a strain of *S. spontaneum* (2n = 48) to *Sclerostachya fusca* (2n = 48) produced normal F_1 plants and about 17% mst plants in F_2 (this Vol.). Both F_1 and F_2 progenies of the reciprocal cross were normal. This indicates a gc-mst control of the sterility; the sterility being conditioned by the interaction of S-cytoplasm of *spontaneum* and a *fr* gene present in the *Sclerostachya fusca* nuclear genome. In mst plants, male meiosis is highly abnormal and is arrested after TII. In a few, uneven-sized microspores are formed; they degenerate rapidly after liberation.

27.16 *Secale cereale* L. (rye)

This is a natural cross-breeding species. Inbreeding in it leads to many morphological and meiotic anomalies, including reduced homologous chromosome pairing culminating in desynapsis (Gottschalk and Kaul 1980a,b). The homozygosity of recessive genes brought about by enforced selfing has also led to the isolation of many mst's in this crop (Tables 27.21–27.22). In all these cases, the sterility is conditioned by one to two recessive *ms* genes.

27.16.1 Genic Male Sterility

Sporadic occurrence of rye plants with abortive, nondehiscent shrivelled anthers or normal anthers with aborted pollen was reported by Frimmel and Baranek (1935) and Müntzing (1946). Müntzing (1946) found that over 50% of the 610 rye plants having normal-looking anthers exhibit pollen sterility ranging from 0 to 90%. The correlation of this sterility between mother plants and offspring was positive, indicating its heritable nature. The sterility was considerably influenced by the environment. Interplant pollen sterility variation exceeded the intraplant variation, indicating genetic significance of the sterility. Unfortunately, genetic control of this highly variable pollen sterility has not been determined. Furthermore, 11 rye inbred lines selfed between 3 and 13 generations exhibited highly divergent pollen abortion and seed set (Putt 1954). Although many cytological aberrations occur, meiosis continues until the end, but the pollen abortion was abrupt. The sterility is genetic but its inheritance is very complex. However, in one family, cytoplasmic influence appears, but is not confirmed.

Three mst rye lines were detected in the S_5 generation of Wierzbinskie-794 (ms-W) and Smolickie Nowe (ms-SN), and from the S_8 generation of the hybrid between Horta and Zeelandzkie G3d (ms-HZ) (Cebrat and Zadecka 1978). While the sterility of the ms-W line is controlled by a single pair of recessive genes (Cebrat and Zadecka 1978, Kaczmark unpublished), genetic control in the other two lines is not known. In the *ms*-Hz line, instead of the usual three anthers, flowers with four anthers having a number of locules varying from two to ten occur. In this line, initial anther development is similar to that of the fertile plants, but in most of them the sporogenous cells and wall layers differentiate abnormally (Fig. 27.9 a-f). In the remaining ones, either only one differentiates or a small group develops abnormally. The anther loci with disturbed PMC differentiation have either reduced or no tapetal formation and possess abnormal nondirected meiotic process. Such anthers mostly degenerate during very early meiosis (prophase-I) and sometimes even before meiotic initiation. From the uninucleate stage, the tapetum never becomes binucleate, unlike in mft anthers. This probably leads to meiocyte degeneration and then to anther degeneration. Like *ms*-HZ plants, the plants of *MS*-W lines also have a varying number of stamens, lodging a variable number of anther cells. While the initial anther development in these plants is normal, the intermediate tapetal layers are coarse, strong and exhibit a yellow flourescence characteristic of suberized cells. This is not found in the fertile anthers. Also mst anthers develop endoderm with Casparian bands around the vascular bundle in the connective, but fertile anthers never have such an endoderm. Meiocytes of the sterile anthers exhibit a disturbed meiosis, loose chromosome scattering at AI and distracted cytokinesis. Anther degeneration usually occurs immediately after meiosis, but in a few it occurs either during or at the beginning of meiosis. Suberisation of cell walls of the intermediate tapetal layers blocks nutrient supply from the vascular bundle of the connective to the tapetum and PMC's. This causes microspore abortion. Thus the disturbed meiosis and plasmodial formation could be due to undernutrition. If this is the case, then the *ms* gene action over meiocytes is indirect, i.e. via the tapetum by causing its malfunctioning.

The plants of MS-SN genotype have anthers which lodge coarse-walled degenerating pollen grains without protoplast, indicating that meiosis probably occurs in the meiocytes but the microspores degenerate immediately after their formation. Rarely, uninucleate pollen grains develop but they are nonviable. Cells surrounding the vascular bundle in the connective are highly thickened, unlike those in the fertiles. This leads to a reduction of the nutrient supply in the steriles and to consequent pollen degeneration. In this case also, *ms* gene action over microsporogenesis is indirect via blockage of nutrient supply through connectives.

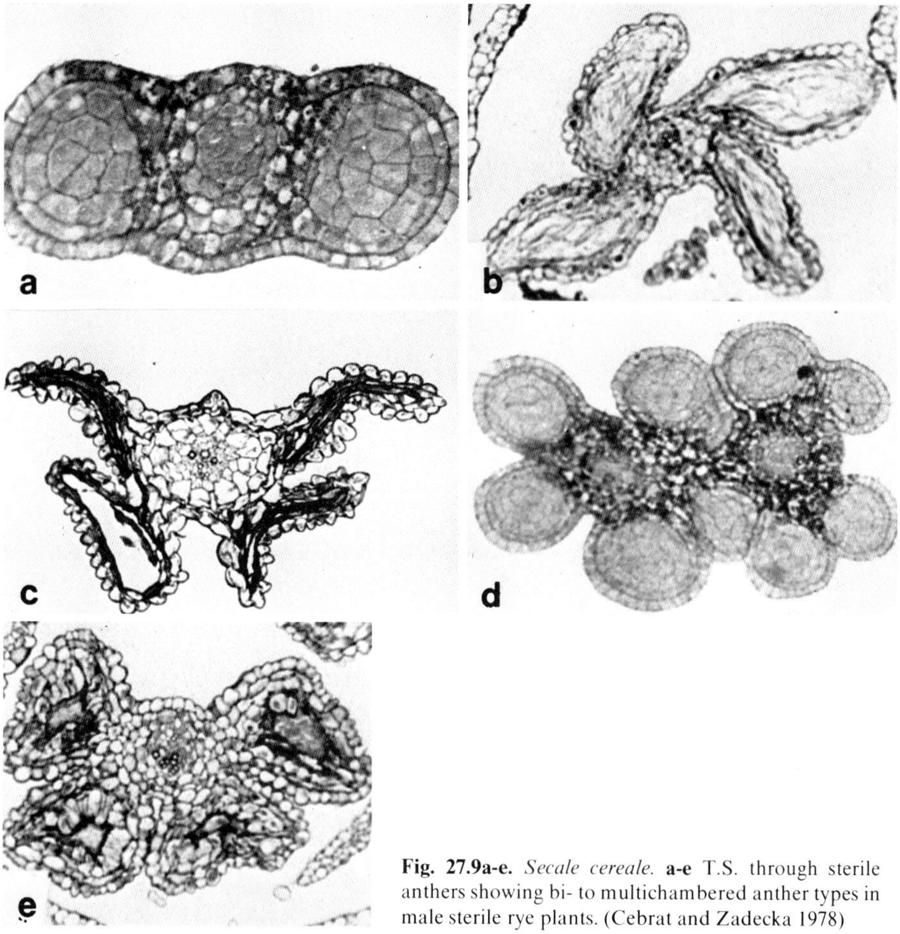

Fig. 27.9a-e. *Secale cereale.* **a-e** T.S. through sterile anthers showing bi- to multichambered anther types in male sterile rye plants. (Cebrat and Zadecka 1978)

In a mst analogue of the winter rye variety Horton, mst is controlled by a recessive gene f_2 ($\equiv ms_2$) and transmitted only through pollen (Kobylyansky and Katerova 1973a). In other mst rye plants, mst is controlled by one recessive gene. Fertility restoration is controlled by a dominant gene. Precise time and mode of action of the two *ms* genes is not known, but microspores degenerate after liberation. Since the sterility is noncytoplasmic and purely genic in nature, the conventional symbol *ms* should be used instead of *rf*.

In most of the above-mentioned cases, microspore degeneration occurs at or prior to the uninucleate pollen stage. However, in the three mst lines of rye, investigated by Adamčhuk (1971, 1972), degeneration after a normal course of male meiosis occurs at the mononuclear pollen stage in one line and at bi-, tri-nuclear stage in the other two lines (Adamčhuk 1971, 1972). Whereas in Adamčhuk's material the *ms* gene acts almost at the end of microsporogenesis, in the tetraploid rye material investigated by Kolobaeva (1974), the gene acts almost at the beginning of meiosis and inhibits the homologous chromosomal synapsis completely. Meiosis following asynapsis is highly anomalous, and nonviable, irregular, crumpled and deformed pollen are developed. Kolobaeva (1974) considered this material asynaptic, but since the female fertility of these steriles is normal as seeds are set after cross pollinations, it is a mst mutant in which the *ms* gene acts during early meiosis, causing asynapsis. However, it is desirable to check female meiosis and seed development, as apomixis may also be occurring in this material. Of the three mst sources investigated by Madej (1975, 1976) and Garlicka and Madej (1975), one (Wms) is g-mst and the other two gc-mst types. Male sterility in the former is conditioned by a single recessive *ms* gene.

27.16.2 Structural Male Sterility

In the S_2 inbred generation originating from the rye cultivar Smolickie, a line having stamenless but multi-pistillate flowers segregated (Kubicki et al. 1981). These flowers developed four pistils, three of which replace stamens, occupying the lateral position and surrounding the one centrally located pistil. The sterility is conditioned by a single recessive *ms* gene designated by the authors as *mp*. The gene pleiotropically affects growth rate, tillering and spike size. The female fertility of the mutant is normal. No seeds are developed in these extra pistils.

27.16.3 Gene-Cytoplasmic Male Sterility

27.16.3.1 Intraspecific Crosses

The first report of the occurrence of gc-mst in rye is by Putt (1954) from the USA. This was followed by many others from W. Germany, Poland, the USSR etc. Dmitrieva and Zdril'ko 1967, Zdril'ko 1969, Kobyliansky 1969, Geiger and Schnell 1970, Geiger 1971, Chekhovskaya 1971, Klyuchko and Belousov 1972, Lapinski 1972, Scoles and Evans 1979a,b, Ahokas 1980c). Geiger and Schnell (1970) found mst plants in the progeny of an inbred line crossed as male with an Argentinian cultivar, Pampa. Later their maintainer and restorer lines were developed. Scoles and Evans (1979a) transferred this mst to many spring rye lines in Canada. Three sources of mst of different origin were studied by Madej (1975, 1976) and Garlicka and Madej (1975). These are (a) 128/3 – a spontaneous mst found in the Smolickie cultivar, (b) Wms – a spontaneous mst found in the Wierzbinskie cultivar, (c) Pampa – derived from S1 lines of crosses between the Argentinian cultivar Pampa and the inbred line S17 Petkuser Normal (Geiger and Schnell 1970, 1974). These three steriles resemble each other in anther and pollen grain features but differ in inheritance pattern. While mst of 128/3 is of the gc-mst type and in the sterile (S) cytoplasm it is caused by a single recessive gene, in Wms it is of g-mst type. In Pampa, fertility restoration by line L18F is explainable on the basis of the model requiring two complementary major genes. However, Muller et al. (1978) suggested the existence of two types of cytoplasm designated type V and P, for the spontaneous and Pampa sources, respectively, of winter rye. In each case he considered mst to be conditioned by two pairs of recessive genes. The influence of environment over sterility expression is profound. From Finland, Ahokas (1980c) detected 28 mst plants in two Finnish rye cultivars, viz. Ensi (27) and Voima (1). All the mst plants of cv. Ensi except one are phenotypically similar and exhibit high female fertility. In this exceptional mst plant, female fertility is also highly reduced. This probably may be a meiotic mutant and not a mst plant. Both maintainer and restorer genes were also detected in the original populations of both the cultivars.

27.16.3.2 Interspecific Crosses

Development of mst in wild mountain rye *S. montanum* by substituting its cytoplasm by *S. cereale* was done by Lapinski (1972). Using the wild mountain rye as male parent in crosses to *S. cereale* as female, mst plants segregated in F_1 and F_2. The sterility appears to be conditioned by the interaction of the S-cytoplasm present in *S. cereale* with a recessive gene pair (*frfr*) present in *S. montanum*. Similar mst plants segregated in F_2 progeny of the crosses *S. cereale* × *S. kuprijanovi*. In both the rye species, fertility restoration is conditioned by one dominant *Fr* gene, whose action is altered by modifiers, as none of the segregating progenies is completely free of nearly sterile and fertile segregates. This indicates more than two alleles at the Fr locus.

27.16.3.3 Intergeneric Crosses

Plants of 6th and 7th substitution BC generations of alloplasmic *S. cereale* with *Triticum durum* cytoplasm exhibit pollen abortion averaging 29.3% (Vega and Lacadena 1979). Since anthers of all these plants, irrespective of the presence of fertile pollen, are indehiscent, the resultant mst is complete and is of functional type. The basic difference between the tapetal behaviour of mft and mst anthers is the lack of tapetal degradation prior to anther dehiscence in the mst plants. Tapetal degeneration is partial in some, and fusion of the neighbouring loculi occurs in them. A similar kind of fn-mst occurs in the alloplasmic rye with *T. aestivum* cytoplasm (Lein 1948). In it, both fertile and sterile pollen grains are locked up in anther sacs which never dehisce. A varying degree of pistiloidy, malformed anthers, locular fusion and plasmodial PMC formation, occurs in it (Vega and Lacadena 1979).

27.16.3.4 Fertility Restoration

Development of gc-mst rye having the S-cytoplasm of the Argentinian cultivar Pampa was reported by Geiger and Schnell (1970). Later, it was found that not only many other inbred lines of rye but the main S-cytoplasmic source, Pampa, also have restorer genes for this sterility (Geiger 1972, Geiger and Morgenstern 1975, Scoles and Evans 1979a,b). Genetic control of fertility restoration in rye is not fully investigated as the findings are variable. For instance, in some lines, restoration is reported to be conditioned by one dominant gene (Kobylyansky 1969, Klyuchko and Belousov 1972, Artemova 1980, Kobylyansky et al. 1980), in others by more than one dominant gene (Kobylyansky 1971, Katierova 1975, Kobylyansky et al. 1980, Artemova 1980). Polygenic control is documented by Katierova (1975), Zdril'ko and Adamčhuk (1975), and Los and Ruebenbauer (1982), Scoles and Evans (1979b) evidenced the presence of three dominant restorer genes *Rf*1, *Rf*2, *Rf*3 (≡Fr_1, Fr_2, Fr_3) present in each of the three inbred lines they studied. The effectiveness of restorer genes in rye depends upon parental genotype and the environment in which the hybrids are grown, temperature being the most effective (Scoles and Evans 1979b). Three unrelated rye inbred lines considered to be homozygous for dominant restorer genes were crossed by Scoles and Evans (1979b) with either spring or winter forms of gc-mst plants presumed to be homozygous recessive for restorer genes. In all these crosses, restoration was full in F_1 but in F_2 and BC progenies, mft, p-mft and mst plants segregated, indicating the existence of at least three dominant restorer genes, Fr_1, Fr_2 and Fr_3, in these lines. Partial fertility resulted when two of these three genes had at least one dominant allele. Expression of partial fertility was dependent upon the gene loci involved and the environment. When Fr_1 and Fr_2 were present as dominants, plants were fully fertile in growth chambers but partially fertile in field and greenhouse. When Fr_1 and Fr_3 were present as dominants, plants were partially fertile in the growth room but sterile in field and greenhouse. Thus according to their expressivity, these genes can relatively be ranked as $Fr_1 > Fr_2 > Fr_3$. Growth chambers are more favourable for restoration than greenhouse or field. Allied differences in the expressivity of restorer genes and environmental effects on expressivity and penetrance of genes are known in wheat (Talaat et al. 1973, Bahl and Maan 1973) sorghum and maize (Edwardson 1970). Certain deviations in segregation ratios obtained by Scoles and Evans (1979) could be due to the existence of modifier genes, interallelic interactions or segregation of some additional restorer genes. Heterosis and interallelic interactions at the restorer gene loci are known in gc-mst *Sorghum* (Miller and Picket 1964). Since rye has preserved heterozygosity (Müntzing 1963), the restorer lines used by Scoles and Evans (1979b), although highly inbred, may not have been fully homozygous for all the restorer genes. This could account for the deviant segregations.

27.16.3.5 Temperature Influence

Under diverse temperature regimes tested by Scoles and Evans (1979), both penetrance and expressivity of *fr* genes in the homozygous state are complete in the S-cytoplasm rye plants. Anthers of such plants are indehiscent and devoid of pollen. Likewise, under such temperature regimes, mft is complete and perfect in the plants having *Fr* genes in the homozygous state. Accordingly, anthers of these plants dehisce normally, releasing viable, fully functional pollen grains. However, in the sterile restorer hybrid plants, partially and fully nondehiscent anthers develop. These are confined to the upper and lower parts of the spikes. Thus penetrance and expressivity of the *Fr* gene in the presence of the *fr* gene is partial in hybrids. Moreover, the effectiveness of *Fr* genes not only depends upon the genetic background, but also upon the temperature regime because restoration is higher at 20°/15° or 25°/20° than at 15°/10°C. Thus, it is equally likely that while at lower temperature regimes fertility restoration is completely inhibited, at high temperature regimes fertility restoration could be complete. If this sterility manipulation is possible it could be of direct use in rye breeding.

27.16.3.6 Microsporogenesis

Meiosis and pollen development of the gc-mst forms 128/3 and Pampa reveal regular formation of seven bivalents at diakinesis and MI (Garlicka and Madej 1975). Sometimes univalents and nonorientated bivalents do occur. These lead to delayed disjunction, acentric fragment formation and AI single and double bridges. Fewer AII abnormalities indicate a partial regulation at AII. Usually second division is asynchronous, but tetrad formation is nearly normal and only in a few are micronuclei formed. Shortly after separation of the microspores from the tetrads, degeneration starts with typical vacuolizations, separation of cytoplasm from the cell wall followed by its gradual contraction and complete deterioration. This produces irregularly shaped shrunken and hollow pollen grains. Anther development is normal up to tetrad formation, but at the mono-nuclear stage, microspores lose contact with the tapetum, fill the interior of the anther chamber and degenerate quickly. Tapetal cells remain intact, unlike those in the mft's, and degenerate late. This tapetal degeneration occurs earlier in Pampa than in 128/3 mst's. For the above-mentioned meiotic studies, Garlicka and Madej (1975) used Carnoy's fixative, which induces artefacts (Laser and Lersten 1972). Scoles and Evans (1979c) used certain histological techniques and a Pampa S-cytoplasm source with another genetic background. In this, in the mft line, a high degree of organisation and orderly development of microspores and pollen occurs within the locule (Fig. 27.10a,b). In the mst lines anther development up to the tetrad stage is nearly normal. Thereafter, tapetum becomes vacuolate, and invades the locule, pushing microspores to periphery. Both the microspores and tapetum are transformed into an unorganized mass within the locule and at dehiscence, the locular contents remain as a compressed layer over endothecium(Fig. 27.10c). In another material, tapetal degeneration occurs after tetrad formation (Artemova 1980). Male meiosis proceeds normally until microspore liberation, after which developing microspores degenerate quickly.

27.16.3.7 Tapetal Degeneration

This is considerably delayed in mst anthers (Garlicka and Madej 1975), which causes nutrient deficiency in the developing microspores which finally abort. Thus the *fr* gene inhibits normal tapetal breakdown that starves microspores to death. Simultaneous microspore and tapetal degeneration occurs in mst spring rye inbred lines (Scoles and Evans 1979c). In this material *fr* genes act both over microspores and tapetal cells almost simultaneously.

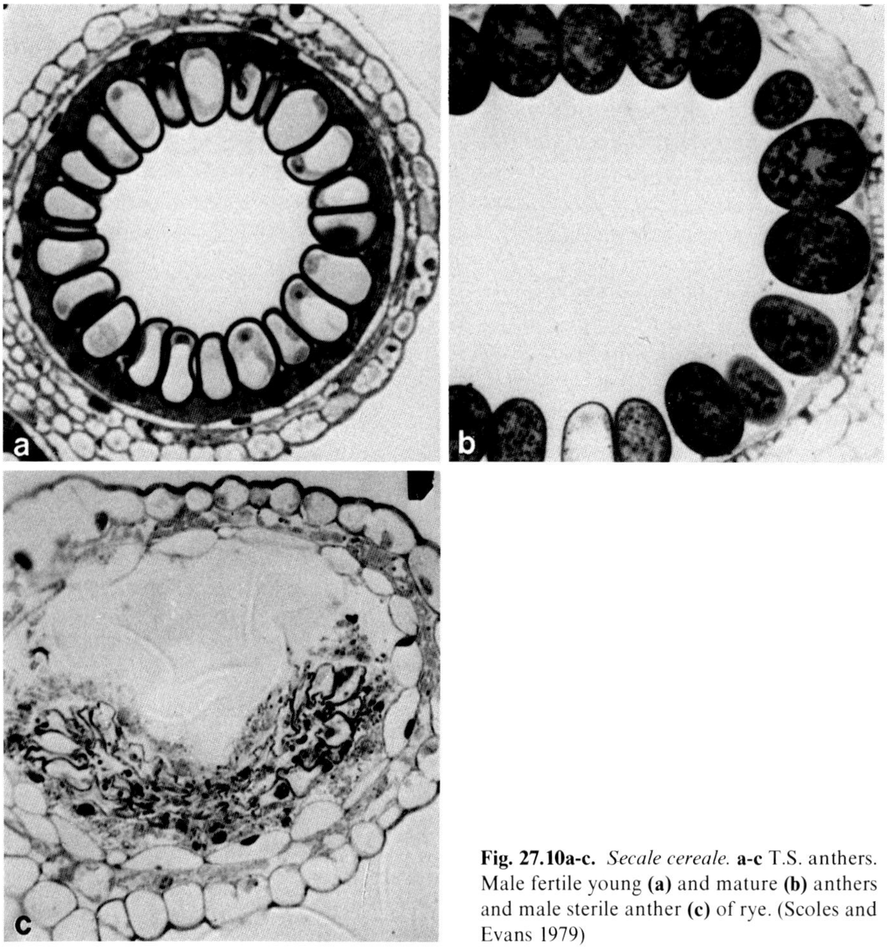

Fig. 27.10a-c. *Secale cereale.* **a-c** T.S. anthers. Male fertile young **(a)** and mature **(b)** anthers and male sterile anther **(c)** of rye. (Scoles and Evans 1979)

27.16.4 Breeding Value

Rye differs from other small grains in that it is cross-pollinated. Although pollen can and does fall into flowers of the same plant, little self-pollination takes place even when the spikes of individual plants are bagged. However, some inbreds selfed continuously for 20–25 years have improved self-fertility to about 40%. Rye breeders in Germany, Poland and the USSR have developed possible ways to utilize inbred lines or early generation material in the production of hybrid rye. Systems for controlled pollination based on gc-mst are available (Geiger and Schnell 1970, Gulyaeva 1971, Klyuchko and Belousov 1972, Kobylyansky and Katierova 1973, Madej 1976). Crossing schemes for hybrid rye are nearly similar to those used for hybrid maize and sorghum seed production. Therefore to avoid repetition these are not given here. Hybrid plants exhibit considerable heterosis, especially in economic traits (Pfahler 1968a,b; Sechler and Chapman 1967). The ergot yields obtained on mst plants were several times higher than on fertile lines in 2 years of testing (Geiger and Bansback 1979). While ergot strains used for infection differ strikingly in their virulence against mft host plants, such differences are narrow or negligible with mst forms. Therefore mst plants might open new possibilities for exploiting ergot strains with high alkaloid content and/or with favourable alkaloid content and low virulence.

27.17 *Sorghum vulgare* pers. (Syn. *S. bicolor* L. Moench)

The grain sorghum, broom corn, sorgos, sudan grass, all of which are annuals with 4x = 20, are commonly grouped under this species but are referred to as sorghum by breeders. Of the 40 known species in this genus, mst is documented only in *S. vulgare*, in which the sterility assumes diverse forms. The plants possess either hermaphrodite pollenless flowers or antherless flowers or anthers which are partially, fully or functionally mst. In addition to mst (Tables 27.21–27.23) the existence of female sterile flowers and apomixis is also known.

27.17.1 Genic Male Sterility

27.17.1.1 Spontaneous

Earliest reports about spontaneously arisen mst in sorghum are from Ayyangar and Ponnaiya (1937) and from Stephens (1937), who found it in an Indian and an American variety (Texas Blackhill Kafir), respectively. The mst in each is conditioned by a recessive gene designated ms_1 and ms_2. While ms_1 mst's have well-developed anthers, the ms_2 sterile anthers are half the normal size or less. Both these steriles are devoid of pollen. Since no PMC's are found, premeiotic degeneration of archesporial tissue is inferred in both of them. The ms_1 plants are fully female fertile, but ms_2 plants are partially fertile and produce seeds in 10–15% of the florets. Some lines of ms_2 plants having high female fertility were recovered and their use in hybrid seed production was proposed by Stephens (1937). This ms_2 gene is linked with awned (*A*) and virescent genes (*V10*) (Stephens and Quinby 1945). The gene ms_2 exhibits some variability in its expressivity as in a few scattered populations, some plants develop apparently fertile anthers. These plants can be identified by their highly pointed panicles and increased lateral branching. From two divergent sources, Karper et al. (1938) found two more *ms* genes, designated ms_3 and ms_4. Their time and mode of action is not known. One mst plant arose spontaneously in the Day variety and was designated Day male sterile by Stephens et al. (1952). Its progeny segregated normal, partial and complete mst plants. In subsequent crossings progeny comprised partial or complete mst's or only mft's. The sterility is conditioned by more than 2 *ms* genes; the influence of cytoplasm on sterility expression cannot be discounted. Another recessive gene, *an*, causes antherlessness and is inherited monogenically.

27.17.1.2 Induced

Barabâs (1962) obtained three phenotypically different mst mutants (Fig. 27.11) after treating var. Rancher with X-rays and colchicine, singly and in combinations. From X-irradiated progeny, the mst mutant isolated was short-statured with shrunken nondehiscent anthers (Fig. 27.11b) and aborted microspores. The mst mutant obtained after colchicine treatment had luxuriant vegetative growth and diminutive anthers (Fig. 27.11c) lacking pollen. Another mutant was polygynaeceous, having 10–12 pistils (Fig. 27.11d). Some of its progeny plants developed antherlike organs. The anomalies are controlled by three nonallelic recessive *ms* genes. In the colchicine-induced mutant, the mst is expressed under short-day conditions. Under long-day conditions, stainable pollen is produced by the plants. Another induced mst mutant was found by Andrews and Webster (1971) in the M3 generation obtained after irradiations of seeds. The mst's are phenotypically distinguishable because of widely opened flowers and diminitive whitish anthers. No pollen is produced in their anthers and mst is complete. Female fertility is normal. Allelism tests indicate that the gene is nonallelic to ms_1 and ms_2. The gene was designated ms_7 by the authors. However, the time and mode of action of this gene over microsporogenesis is not known. The gc-mst in Sudan grass was obtained by transferring the genome of *S. arundinaceum* into combine Kafir 60 cytoplasm (Craigmiles *et al.* 1958). The msts have diminutive, nonexserted, shrunken anthers (Fig. 27.12b) and have been used for hybrid production.

27.17.1.3 Structural Male Sterility

Antherlessness is caused by a single recessive gene, in *S. vulgare* (Ayyangar 1942) and *S. sudanensis* (Karper and Stephens 1936). The genes do not influence female fertility. Certain single (Ayyangar and Ponnaiya 1939, Townsend 1960a,b) and double (Jaisani and Drolsom 1971, 1972) recessive *ms* genes induce a random or nonrandom barren type of sterility in *S. vulgare*. Progenies of the random barren type segregate nonrandom barren types, but with advancing generations (S_1–S_3) the frequency of the nonrandom barren type of sterile plants decreases among the progenies of random barren-type sterile plants. The expression of these genes conditioning barrenness is highly influenced by modifier genes and environment.

27.17.2 Gene-Cytoplasmic Male Sterility

Enhanced vigour, increase in height, tillering, forage production and grain yield in sorghum hybrids was known to early plant breeders, who were constantly seeking a convenient hybridisation method. Although g-mst could be used to facilitate cross pollination in population improvement schemes involving recurrent selection, gc-mst was convenient for use in hybrid seed production. This was first detected by Stephens and Holland (1954), who found that by pollinating Texas Blackhill Kafir with Double Dwarf Yellow Sooner Mili or Day (a derivative of milo), F_1 progeny was fertile but some p-mst's appeared in F_2 of the cross in which milo was the female parent. In BC2 of milo × Kafir pollinated by Kafir, mst was

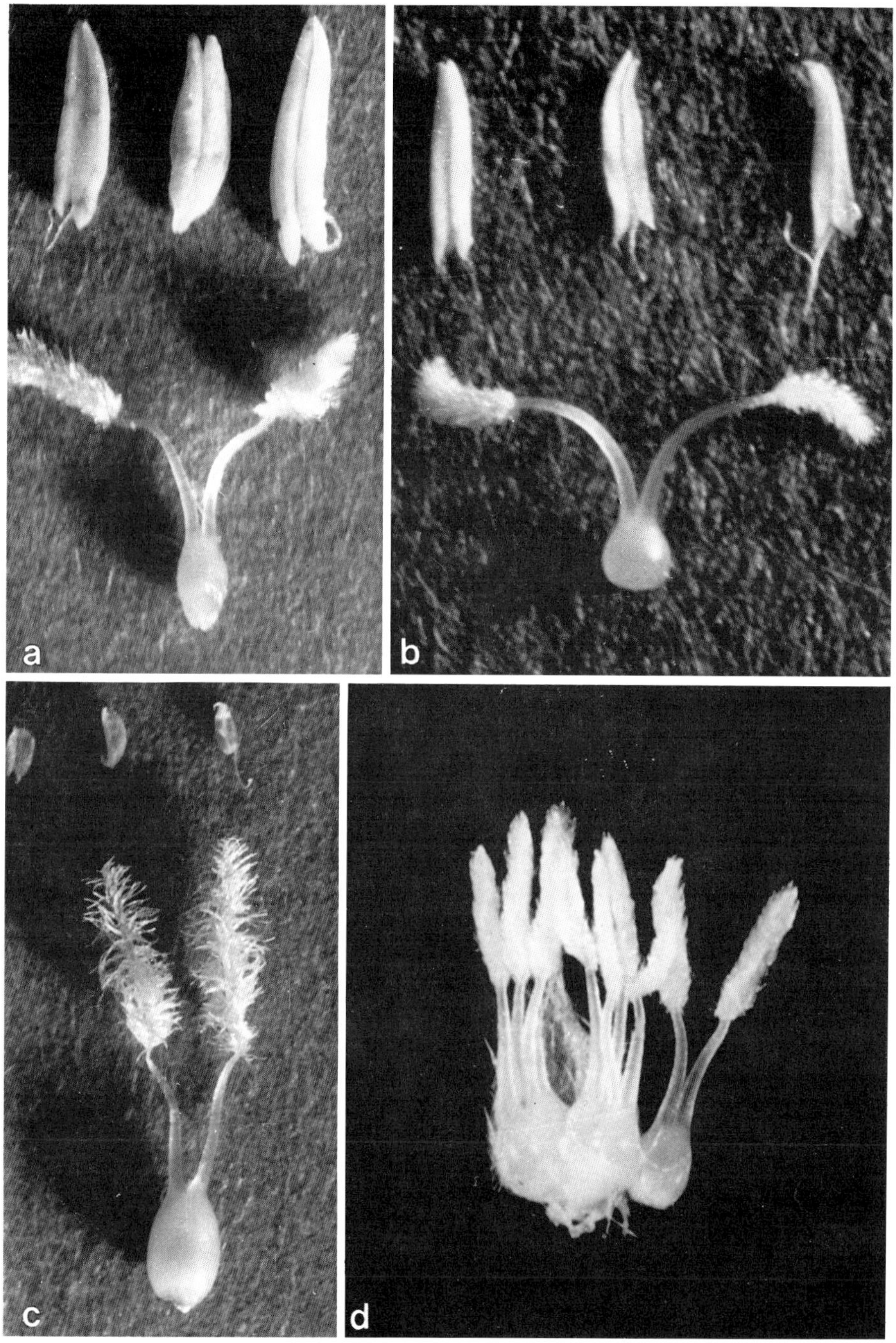

Fig. 27.11a-d. *Sorghum vulgare.* **a-d** Male and female organs of male fertile **(a)** and three male sterile mutants **(b-d)**, viz. abortive pollen **(b)**, microanther **(c)** and polygynacious **(d)** mutants of sorghum, (Barábaś 1962)

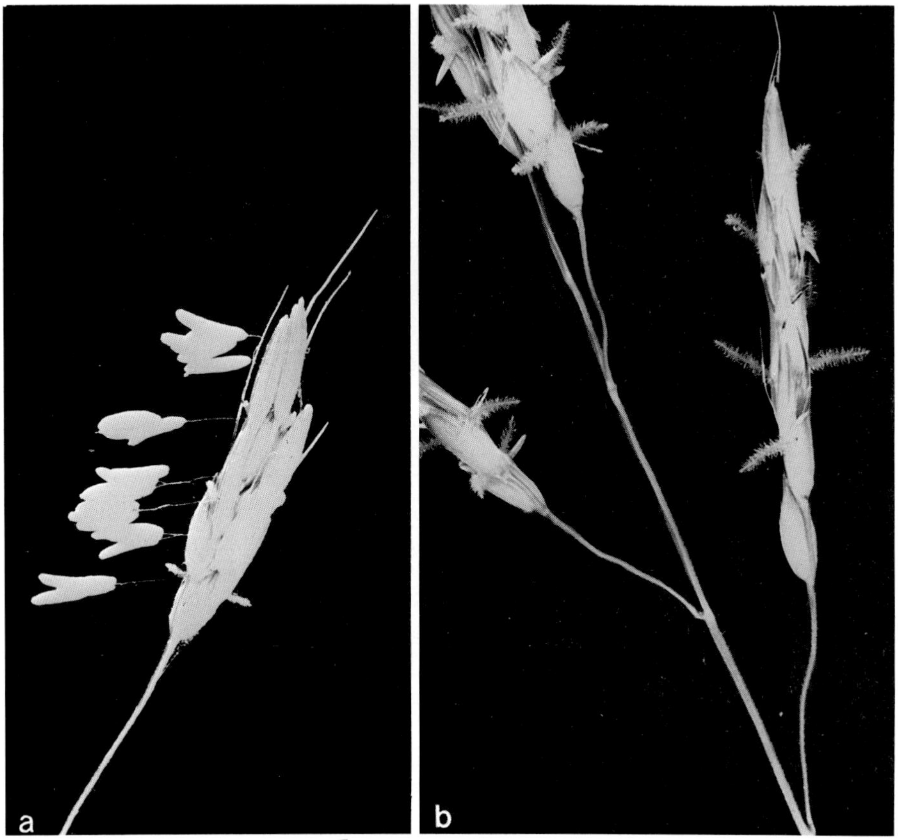

Fig. 27.12a,b. *Sorghum vulgare.* **a-b** Mature spikes of male fertile **(a)** and male sterile **(b)** sorghum plants. (Craigmiles JP, personal communication)

complete but when milo was used to pollinate F_1 or F_2 plants of the cross milo × Kafir, mft was restored in the progeny plants. Stephens and Holland (1954) suggested mst as due to interaction between milo cytoplasm and Kafir nuclear genes. The authors visualised involvement of more than two pairs of *fr* genes in association with the S-cytoplasm of milo in the induction of mst. However, genetic control of mst was not given by these authors because identification of mst's in segregating populations was distorted by drought and high temperatures. High temperature reduces mst expression, increasing pollen fertility of mst's and restored hybrids. One recessive gene, ms_2, from Texas Blackhill Kafir sorghum was found by Stephens (1937). However, it seems that actually two ms gene loci exist (Erichsen and Ross 1963a, Appadurai and Ponnaiya 1967). Quinby (1980) symbolised mst Texas Blackhill Kafir genetically as msc_1, msc_2 ($\equiv fr_1\ fr_2$), and opined that whereas for inducing mst in Kafir cytoplasm, genes at both these loci are required, only one of the recessives is needed for mst induction in milo cytoplasm. For this reason, Texas Blackhill Kafir that was recessive at either msc_1 or msc_2 was mft in its own cytoplasm but mst in milo cytoplasm (Stephens and Holland 1954). After the first authentic report of the existence of gc-mst in sorghum, many such claims followed rapidly in succession. Mittal et al. (1958) found a single mst sorghum plant in which sterility is due to g-c interaction. The plant does not have milo cytoplasm or Kafir nuclear genes. The same is true for the mst plant found by Sieglinger (Quinby and Martin 1954). Subsequent studies reveal a number of cases in which mst is conditioned by g-c interaction. In them S-cytoplasm differs from milo S-cytoplasm. The mst expression is highly influenced by inhibitors, modifiers, minor genes and environment. The sterility appears to be conditioned by different genes, as is evident from the following account.

27.17.2.1 Gene Control

i. The mst is conditioned by a pair of recessive fr_1 genes interacting with milo type(s) cytoplasm (Maunder and Pickett 1959). Fertility is restored by one dominant Fr_1 gene. Modifier or weak restorer genes reduce the expression of normal restorers and produce plants having various degrees of pollen sterility (Maunder and Pickett 1959, Hadley and Singh 1961).
ii. The mst is controlled by two pairs of recessive genes, fertility being restored by one or both dominant genes in hetero- or homozygous condition (Pi and Wu 1963).
iii. More than two pairs of recessive genes in association with S-cytoplasm condition mst (Stephens and Holland 1954).
iv. Three pairs of recessive genes cause mst, partial sterility is conditioned by one or more major genes plus modifiers, complete fertility restoration is under a multigenic control, especially in Sudan grass (Craigmiles 1961, 1962). In Sudan grass three pairs of recessive genes condition mst, fertility restoration depends upon the presence of at least a single dominant fertility restorer gene (Alam and Sandal 1967a,b). Thus fertility restoration occurs when one, two or three genes are present.
v. A single major gene and one or more modifiers with additive effect coordinate fertility restoration. Three dominant modifier genes can induce fertility in the absence of major restorer genes (Kidd 1961).
vi. Two dominant genes in conjunction with modifier genes restore fertility (Erichsen and Ross 1963a).
vii. Two independently acting major dominant genes are necessary for full restoration of fertility; the presence of either of the two genes restores partial fertility (Miller and Pickett 1964). The restoration is highly influenced by intra- and interallelic interactions.
viii. Using a derived female with milo cytoplasm and tropical sorghum genotypes, a four-gene model for fertility restoration was suggested by Tripathi et al. (1980). In one of their crosses one more gene is involved.

The above-mentioned divergent bases and interpretations for the genetic control of mst-fertility restoration in sorghums have been reconciled by Duvick (1966), who assumed that one dominant gene, $Mscl(Fr_1)$, restores fertility, but that in some environments modifiers are required to induce complete fertility. In certain genetic backgrounds and environments, one modifier has an equal rank with the restorer gene. Thus, two dominant genes are involved in restoration in such cases.

Some other genetic background and environments induce the homozygous msc_1 ($fr_1\, fr_1$) plants to develop varying amounts of viable pollen. Duplicate dominant restorer genes modify fr_1 gene expression, but it is not known whether they are the same genes that modify *Fr* gene expression also. Duvick's assumptions are compatible with Alam and Sandal's (1967a) views that three pairs of recessive fr_1 genes condition mst in Sudan grass crosses so that mst occurs in plants homozygous for three pairs of recessive genes fr_1, fr_2, fr_3, as well as in plants heterozygous for one gene locus, i.e. $Fr_1\, fr_1, fr_2\, fr_2, fr_3\, fr_3$, or $Fr_1\; Fr_1, fr_2\, fr_2, fr_3\, fr_3$. Steriles of this genotype can be restored by lines carrying dominant genes at Fr_1 and/or at the Fr_2 locus. Brooks and Brooks (1967) opined that mst in sorghum is controlled differently in different sorghum genotypes. Shambhulingappa and Magoon (1963) and Appadurai and Ponnaiya (1968) suggest high plasmon diversity in eusorghums and many *fr* genes inducing mst. Intra- and interallelic interaction and complementations influence fertility restoration considerably.

From the above it is apparent that the genetic control of gc-mst is diverse and complex in sorghum. In some, single pairs of recessive genes cause mst, in others two, three or more *fr* genes are necessary for mst induction. Like mst, genetic control of fertility restoration is equally diverse and is conditioned by one, two or more dominant *Fr* genes acting singly or in combination. *Sorghum* is a cross-pollinated plant. High heterozygosity of parental genotypes could also account for such a divergent genetic control of sterility-fertility restorations. Moreover, since cytoplasmic diversity exists in sorghum, this broadens the genetic control. Also, the influence of the environment on sterility-fertility expressions is very high. This eludes the precise determination of genetic control of mst and its restoration in this crop.

27.17.2.2 Gametophytic Control

In almost all the cases of gc-mst, control of sterility is sporophytic. However, in one case, the pollen viability estimates (57% fertile, 43% sterile) obtained from the pollen examination of a heterozygous restored line containing sterile cytoplasm led Brooks et al. (1966) to suggest gametophytic control of mst. However, sporophytic control of the sterility was demonstrated by the pollen viability tests made on other F_1's of sterile × restorer crosses, and on BC's of such F_1's on sterile plants (Brooks and Brooks 1967). The sporophytic control is clearly demonstrated by the 1:1 ratio of steriles to fertiles obtained by these authors in BC populations of these crosses. Since diversity exists in the genetic control of sterility and its restoration and also in S-cytoplasm types, the existence of gametophytic control cannot be completely ruled out in this crop. In maize such a system does exist.

27.17.2.3 Functional Male Sterility

F_1 progeny of the cross of two sorghum lines 9E × Leoti Sorgo develops normal but indehiscent anthers (Webster and Singh 1964). These anthers have a variable number of viable pollen grains. Crossing these and backcrossing for six generations with the varieties Kafir-60, Redbine-60 and Martin yielded only nondehiscent anther-type progenies. While some F_1's derived after crossing some of these BC_6 steriles with other sorghum varieties had only nondehiscent anthers, others contained plants having a high number of dehiscing anthers. Although the breeding behaviour of nondehiscent character is nearly similar to that of milo-type mst, the sterility is not as stable and reliable as the milo type. Therefore, its use in hybrid seed production, as was suggested by Webster and Singh (1964), is not feasible.

27.17.2.4 Cytoplasm Diversity

After the discovery of gc-mst by Stephens and Holland (1954) in plants having milo cytoplasm and Kafir nuclear genes, similar sterility caused by other sources of S-cytoplasm was reported. Mittal et al. (1958) and Sieglinger (Quinby and Martin 1954) isolated mst's that did not have milo cytoplasm. The S-cytoplasm of 9E, a sorghum cultivar from Ghana, causes a nondehiscent anther mst (Webster and Singh 1964) and milo lines do not restore fertility to the F_1 hybrids of the steriles. Therefore, cytoplasm of 9E differs from that of milo. This is confirmed by mt-DNA and cp-DNA studies. Rao (1962) detected another S-cytoplasm source from India and converted M35-1 (Indian winter sorghum) and IS 3691 (Yellow hegari) into mst's. Additional S-cytoplasm sources were also found in *S. durra* (G2, VZM1, VZM2) by Hussaini and Rao (1964). At Raichur (India) another indigenous mst M31-A is known to owe its origin to induced mutations. Nagur and Menon (1974) also found S-cytoplasm sources in Indian sorghums and distinguished them into four types, S_1 to S_4, based on sterility response. Compared to milo-Kafir systems, fertility restoration in Indian mst sorghums is arduous and Indian S-cytoplasm differs from milo S-cytoplasm. This is confirmed by differences in seed-soluble protein patterns, isoenzymes of alcohol dehydrogenase, glutamate dehydrogenase and esterases of seeds of the two S-cytoplasm (Tripathi *et al.* 1981, 1982a). Moreover S-cytoplasm differences have not only been found in grain sorghums but also in grass sorghums like *S. arundinaceum*, *S. verticilliflorum*, *S. niloticum* and *S. sudanense* (Ross 1965, Appadurai and Ponnaiya 1967, Alam and Sandal 1967b). The presence of S-cytoplasms in so many different sorghum species indicates high cytoplasm diversity and differentiation in this genus.

From BC and intercross progenies of introduced sorghum lines of diverse types obtained from widespread geographic locations, three mst lines whose S-cytoplasms appeared to differ from milo cytoplasm were detected by Schertz and Ritchey (1978). These lines, being in early generation, differ in nuclear genes. Therefore, the development of isocytoplasmic lines by incorporating a common genome into each of these three cytoplasmic sources would permit definitive conclusions. Quinby (1980) reported the isolation of new cytoplasms and presented a hypothesis based on genetic and hormonal control of mst. The symbol A was assigned to milo cytoplasm found by Stephens and Holland (1954), A2 to the cytoplasm obtained from Ethiopia and recognised by Schertz (1977) and A3 to the one obtained from India (Quinby 1980). Thus at least three known difinite S-cytoplasm types, A, A2, A3, exist in sorghum. The probability of having many more types is high, as a perusal of Table 27.9 indicates the presence of different S-cytoplasm types in a divergent array of sorghum genotypes. Based on the differences in progenies from reciprocal crosses, Rao (1961), Hussain and Rao

Table 27.9. Sorghum genotypes having S-cytoplasm. (Adapted after Schertz and Pring, personal communication)

Origin	IS no.	Sc. no.	Group	Race
Ethiopia	125 39C	19	Dochna-Amber	Bicolor
	126 45C	154	Durra-Bicolor	Durra-Bicolor
	126 62C	171	Candatum-Nigricans	Caudatum
India	1112C	193	Durra-Subglabrescens	Durra (Durra-Bicolor)
	1116C	195	Roxburghii	Guinea
	2177C	215	Cernuum	Guinea-Bicolor
	12680C	217	Dochna-Leoti	Bicolor
	5322C	250	Roxburghii	Guinea
	6271C	317	Dochna-Roxburghii	Guinea-Caudatum
	8232C	642	(Caffrorum-Durra)	(Kafir Caudatum)
			(Caudatum-Nigricans)	Caudatum
Nigeria	7920C	299	Conspicuum	Guinea
	3620C	303	Margaritiferum	Guinea
	7498C	353	Caudatum-Conspicuum	Caudatum
	7506C	407	Caudatum-Guineense	Bicolor
	6964C	411	Caudatum-Dochna	Bicolor
Zimbabwe (North)	12600C	91	*S.nitidum*	–
Zimbabwe (South)	2816C	120	Zerazera	Caudatum
Sudan	12565C	50	Caudatum-Conspicuum	Caudatum
	2573C	64	Caudatum-Kafir	Caudatum
	2266C	93	Caudatum	Durra
	7007C	268	Conspicuum-Nigricans	Guinea
Uganda	2757C	319	Dochna-Nigricans	Caudatum

(1964), Webster and Singh (1964), Ross and Hackerott (1972), Nagur and Menon (1974), Schertz (1977), Schertz and Ritchey (1978), Worstell et al. (1984) reported the existence of many S-cytoplasm sources other than milo S-cytoplasm in sorghum. Definitive genetic analysis has not been completed in all these newly reported cytoplasms but some certainly differ from milo S-cytoplasm and thus represent additional S-cytoplasm sources in sorghum.

27.17.2.5 Cytoplasmic Mutations

These are the major cause of cytoplasmic diversity, as many cytoplasmic mutations are known in sorghum. From the progeny of a mutant plant obtained after seedling treatment of the sorghum line Experimental 3 with colchicine, four mst's were found by Erichsen and Ross (1963a). Upon crossing these with sorghum varieties that maintained the milo type of mst, the progeny produced was mst. Thus the milo type maintainers also maintain this mst. Crossing the mst with sorghum varieties that restore fertility to milo type steriles produced mft progeny. Thus the restorers of milo type sterility restore fertility in the mst's. Such mst's were also produced after backcrossing Experimental 3 to F_1 of the cross mst mutant × Experimental 3. Hence a cytoplasmic mutation appears to have occurred in these mutants due to colchicine treatment. The mst M-31A found at Raichur (India) owes its origin to induced mutations and also appears to be a cytoplasmic mutant.

27.17.2.6 Microsporogenesis and Pollen Phenotype

In mst sorghum material studied by Singh and Hadley (1961), male meiosis proceeds normally, but the pollen development ceases prior to first pollen mitosis. Most of the microspores develop a fuzzy, weakly reticulate exinelike wall, but fertile pollen has a thick smooth exine. Narkhede et al. (1968) also found pollen abortion to occur either before or after pollen mitosis in two gc-mst lines. Whereas in these mst lines meiosis was normal, in some meiotic abnormalities occur frequently. Similar meiotic anomalies leading to multiploid sporocyte formation occurs in the mst Martin wheatland (Damon 1961) and Reliance sorghum varieties (Erichsen and Ross 1963b). It is not known if any of the multiploid pollen in normal cytoplasm is functional or not, but the zygotes developed after fertilization with such multiploid pollen produce abnormal progeny composed of aneuploids. Breakdown and regeneration of the microsporocyte callose wall leads to aneuploid and polyploid microspores in 5% PMC's of mft lines of the wheatland variety of *S. vulgare* (Damon 1961). The frequency of such anomalies is high in three mst lines closely related to mft wheatland, but no such anomalies occur in Ethiopian gc-mst lines of *S. subglabrescens* var. *leiochladum*, Snowden. Thus cytological anomalies occur in both mft and mst lines of wheatland sorghum but not in the Ethiopian sorghum lines. Hence the anomalies are genome-cytoplasm-specific and occur in mst *S. vulgare* but not in *S. subglabrescens*. In milo and colchicine-derived mst sorghum lines, aneuploid and polyploid metaphases occur (Erichsen and Ross 1963b). These anomalies are either due to failure to form or breakdown of the microsporocyte cell wall during meiotic prophase followed by cytoplasmic fusions. The varying proportion of sterile pollen in maintainer, restorer and mst × restorer hybrids (Brooks and Brooks 1969) is due to the low frequency of multiploid sporocytes in mst and maintainer lines of wheatland sorghum varieties. Since meiosis in sorghum is influenced by low temperature during pre- and mid-flowering season, anomalies could be due to low temperature. Scanning photomicrographs reveal distinct differences in mst and mft pollen in grain shape and size, exine sculpture and internal structure (Tripathi et al. 1982b). The mst pollen is almost devoid of protein and starch particles.

27.17.2.7 Tapetal Degeneration

As in other cereals, abnormal tapetal form and/or function is assigned as the cause of mst in sorghum. Singh and Hadley (1961) associated microspore degeneration with the nutritive role of the tapetum. While in mft plants, the tapetum disintegrates immediately after microspore formation and provides nutrition to the developing pollen, in mst plants the tapetum is persistent and stops food material required for pollen development. A similar nondisintegration of the tapetum occurs in two mst lines (Narkhede et al. 1968). In these, pollen degenerates either before or after pollen mitosis. Similarly, in Sudan grass, while in mft's tapetal cells degenerate during microspore maturation, in mst's they persist and the microspores abort gradually due to starvation (Alam and Sandal 1967b). Compared to mft lines, anthers of mst lines are significantly shorter and have lower dry and fresh weights at the pre-pollen stage (Brooks 1962), which represents the terminal stage in these mst lines. Furthermore, 68% of the tapetum of mst sorghum lines is narrower than the mft tapetum at the pre-pollen stage (Brooks et al. 1966). In all the five mst lines which these authors studied, a great interline variability occurs in the width and morphology of the tapetum. The inter-genotypic variability may be a differential expression of genome-cytoplasm interaction.

27.17.2.8 Environmental Influence

As indicated earlier, gc-mst expression in sorghum is highly environment-sensitive. In field-planted mst sorghums, Kidd (1961) found that while higher atmospheric temperature reduces mst frequency in milo cytoplasm plants, low temperature, especially during flowering, considerably enhances their frequency. Kongtian and Hongyi (1981) and Zhang and Fu (1982) restored mft in mst plants by subjecting them to 39°–40°C at the premeiotic stage. Isozyme electrophoresis reveals the disappearance of some bands in mst anthers at the trinucleate stage. In normal populations of sorghum that do not have *fr* genes or S-cytoplasm, low night temperatures during flowering cause mst. Downes and Marshall (1971) demonstrated in glasshouse experiments that night temperatures of 13°C or less during meiosis induce mst in sorghum. This temperature sensitivity appears to be restricted to the pre-zygotene stage (Brooking 1976).

27.17.3 Biochemical Differences

Significant differences between mst and mft anthers occur in starch, aminogram and cytochrome oxidase. The mst anthers have reduced starch and carbohydrate contents, the reduction is more pronounced after the microspores are liberated and are about to abort.

i. *Amino Acids.* Until the tetrad stage, the total amino acid content of mst and mft anthers is similar (Kern and Atkins 1972) but it is considerably reduced later in mst anthers. Unlike in mst's, in mft anthers, proline and tyrosine contents are enhanced with microsporogenesis progression. In contrast to the field-grown mst's, in greenhouse-grown plants with advancing microsporogenesis, the asparagine content increases considerably. At the pre-pollen stage, mst anthers have a higher glycine content than the mft anthers. Reduced amounts of histidine, glutamic acid, glycine, leucine, phenylalanine and threonine and enhanced quantity of alanine, proline, serine and tyrosine occur in mst anthers compared to mft anthers (Tripathi *et al.* 1981). In anthers of apomictic plants, the contents of aspartic acid, glutamic acid, isoleucine, leucine and lysine are higher than those of mst and mft anthers (Tripathi *et al.* 1981). This suggests differences in protein quality and quantity and in enzyme contents between these three genotypes.

ii. *Cytochrome Oxidase.* The mst anthers have a deficient or abnormal cytochrome oxidase system (Alam and Sandal 1969). This might contribute to pollen abortion by causing defective and inefficient oxidative phosphorylation. The mst anthers lack free histones which may be complexed with DNA and repress regulator genes that support DNA-dependent RNA synthesis. This repression would block the synthesis of certain proteins that could lead to pollen abortion. Blockage in synthesis of some proteins does occur in mst anthers of sorghum (this Vol.). Thus instead of the normal 20 protein bands present in the mft anthers, only 9 were detected in mst anthers by Alam and Sandal (1969) and 13 by Kaul (this Vol.).

iii. *mt- and cp-DNA Analysis.* Restriction endonuclease fragment analyses of mt- and cp-DNA distinguish milo from kafir cytoplasm plants (Pring *et al.* 1979b), the differences between mt-DNA's of these two sorghum lines being greater than those observed for cp-DNA's. Eco R1 digests of kafir and milo cp-DNA resemble Eco R1 pattern of cp-DNA of *Zea mays*, but BamH1 digests of the sorghum and maize cp-DNA's do not exhibit this resemblance. This suggests conservation of Eco R1 sites between related species (Pring *et al.* 1980). Restriction endonuclease analysis of cp- and mt-DNA's from many alternative S-cytoplasm sources reveal mt-DNA's as more variable than cp-DNA's. The mt-DNA analysis identified eight groups but cp-DNA analyses yielded only three groups (Pring *et al.* 1982a,b). In six cases, mt-DNA restriction patterns are correlated with the observed differences in fertility restoration (Conde *et al.* 1982). This provides circumstantial evidence that sorghum mt-DNA rather than cp-DNA is the carrier of the *c*-gene. This is further confirmed by the analysis of polypeptides synthesized by isolated mitochondria that enabled Dixon and Leaver (1982) to distinguish milo S-cytoplasm and three alternative sources of S-cytoplasm from each other, as well as milo from Kafir cytoplasm.

Analysis of mitochondrial translation products differentiated N-cytoplasm from four S-cytoplasm sorghum lines (Leaver et al. 1982). The mitochrondria from milo cytoplasm in a kafir nuclear background synthesised from two gc-mst's, sources M35-1 and IS1112, do not synthesise this 55kDa polypeptide but synthesise an additional high molecular weight polypeptide (45–82kDa), the major one being 82kDa polypeptide. Those from IS1112 cytoplasm also synthesise an additional 12kDa polypeptide not synthesised by the mitochondria from any other cytoplasm studied by Leaver et al. (1982). An analysis of mt-DNA reveals the presence of equimolar amounts of two plasmid-like DNA's of 5.3 kb and 5.7 kb in IS1112 and M35-1 S-cytoplasms. These linear molecules are designated N-1 and N-2 by Pring et al. (1982b). Their nick translation, followed by membrane hybridization, reveals substantial homology between the two DNA's and some homology to S-1 and S-2 maize plasmid-like DNA's. N-2 apparently shares a greater degree of sequence homology with S-2 than with S-1, suggesting that N-2/S-2 homologous sequences lie in the unique sequence region of S-2. In addition, four more linear DNA species ranging from nearly 1000–4000 base pairs occur in mt-DNA. The presence of plasmid-like DNA's in sorghum resembling the S-cytoplasm plasmid mt-DNA's of maize suggests an etiological relationship of the molecules to the inheritance and expression of gc-mst in sorghum. No attempt has been made to determine the relationship between the sorghum 2.2 kbp molecule and N-1, N-2 or S-1, S-2 molecules. Unlike maize, in which *fr* genes affect the relative amount of S-1 and S-2 mt-DNA's, no similar effect was found after examining 43 different sorghum lines, each carrying IS1112 C-cytoplasm with N-1 and N-2 mt-DNA molecules (Hanson and Conde 1985). In addition, N-1 and N-2 do not appear to share homology with any portion of the remaining mt-DNA of IS1112C. No genetic evidence exists that could implicate these molecules in mst expression in sorghum in a manner analogous to the S-1 and S-2 molecules of maize.

Three S-cytoplasms produce specific polypeptides which differ from those of milo and kafir (Hanson and Conde 1985). IS1112C, which has the linear N-1 and N-2 molecules, lacks the milo-specific 65 kDa polypeptide, but has several high molecular weight polypeptides (54–82 kDa) and the 12 kDa polypeptide. M35–1 mitochondria produce patterns similar to that of IS1112C mitochondria, but lacks the 12 kDa polypeptide. Mitochondria from 9E sorghum lack the 38 kDa polypeptide and produce a new 42 kDa polypeptide. The 9E specific 42 kDa polypeptide and 38 kDa polypeptide are related, as is indicated by immunoprecipitation by an antiserum against yeast cytochrome oxidase subunit I. Examining genomic sequences for cytochrome oxidase subunit I may reveal why different products are made in vitro (Hanson and Conde 1985). Correlating such differences with the gc-mst phenotype will require revertant or recombinant somatic hybrid genomes.

27.17.4 Chemical Male Sterility

No chemical appears useful in mst induction in sorghum. Sayers (1959) and Miller (1961) found severe plant damage associated with only a trace of mst induced by 0.1, 0.5, 1.0, 3.0 and 5.0% mendok treatments. No seed set occurred after hand pollination in the treated plants, indicating induction of complete fst. Similar effects occurred following dalapon treatments (Miller 1961). Nearly complete mst was induced by 60–2000 ppm MH (Mcllrath 1957) but plant growth was highly suppressed and ovular abortion was considerable. Moreover, the chemical effect is highly variable and is influenced by genotype, plant developmental stage and environment. The mst induced by colchicine resembles gc-mst, but is associated with high female sterility.

27.17.5 Hybrid Seed Production

Certain sorghum variety crosses produce extremely vigorous and high seed-yielding hybrids. *Sorghum*, unlike maize, being a selfer, does not suffer from inbreeding depression. Also high positive heterosis in economically useful traits is obtained in F_1 of certain inbred crosses. This prompted the use of mst in hybrid seed production, using both g-mst and gc-mst.

i. *Genic Male Sterility*. Stephens (1937) proposed a scheme using g-mst for commercial hybrid seed production. In this process, however, since one half of the female flowering heads were fertile, they had to be removed before pollination. This manual removal increased the production cost and rendered the method uneconomical. Stephens *et al.* (1952) suggested a three-way cross for producing hybrid seeds using Day mst plants. In this procedure, three seed stocks and two isolated crossing blocks are used in the following manner:

Block 1. Strain A, that segregates for mst and mft in a 1:1 ratio, is maintained in isolated blocks by harvesting seed from mst plants only.

Block 2. Strain B is mft but nonpollen-restoring. It can be maintained either in isolation or by bagging the heads. This strain is used to plant pollen-producing rows in the crossing block 1.

Block 3. Single-cross A × B, which is mst, is harvested from seed rows in crossing block 1 and planted as the seed rows in crossing block 2.

Block 4. Strain C, which restores pollen fertility, is planted as the pollen producing rows in crossing block 2.

Block 5. The seed produced as a result of the three-way cross in crossing block 2 is harvested and used by growers for commercial production.

ii. *Gene-Cytoplasmic Male Sterility.* Due to the inherent difficulties in the g-mst system and the roguing of mft's, g-mst has not been used in commercial seed production. Instead, gc-mst is being extensively used. Its use started when commercial sorghum hybrids using Day g-mst's were just getting into production. Stephens *et al.* (1952) utilized it immediately for hybrid seed production in the following manner.

Maintenance of mst line. For maintenance and increase, the mst line A is grown in an isolated field with the maintainer line B. Line A is pollinated by B and seed is harvested from A-line rows. Since cytoplasm is derived from the female parent, each generation of A-line seed is mst, like the A-line maternal parent.

Procuring hybrid seed. The mst line A is grown in a second isolated field and pollinated by C line (restorer), which is also grown with it. The hybrid seeds obtained thereof on line A are sold to farmers for commercial production.

Utilising gc-mst for hybrid seed production by the above procedure is more advantageous than using g-mst, because in the former the parental lines are easier to maintain and only two instead of three isolated blocks are required to produce the hybrid. In commercial hybrid seed production, six rows of mst lines are planted to two rows of pollinator lines (6:2). The 12:4 ratio suggested by Martin (1959) is also effective. Ross (1965) reported that the seed yield of mst lines in plantings having one quarter and one half sterile plants yielded nearly equal (95%) mft plants and that the proximity of pollinators played a major role in seed production. Having seed parents and pollinators to bloom at the same time is equally important (Ross 1957), as mst panicles tend to dry up when not pollinated at favourable receptive stages. Poehlman (1959) suggested planting alternate pollinator rows on different dates in order to ensure pollen supply over a longer period. Pollination success on mst depends on the availability of fertile pollen and the duration of stigma receptivity; the longer the duration the greater are the chances of pollination. This accounts for low seed set on mst Martin variety compared to combine-kafir-60. The stigmas of Martin lose receptivity much earlier than those of combine-kafir-60 (Ross and Webster 1959).

27.17.6 Breeding Value

After the discovery of gc-mst, many commercial hybrids were developed in *S. vulgare*. Almost all these are based on milo cytoplasm. In grain production, all hybrids averaged 770 million bushels/year during 1972–76 in the USA (Koehn 1977, Harvey 1977). Now over 97% of sorghum hybrid seed production is from the gc-mst source with kafir genes and milo cytoplasm combination (Harvey 1977). This sterility has been transferred to other sorghum species in order to exploit interspecific hybrid seed production (Craigmiles et al. 1958, Craigmiles 1961, 1962). Craigmiles and Stacy (1961) crossed combine-kafir 60 that had S-cytoplasm of Day-milo to *S. arundinaceum* procured from S. Rhodesia. This is referred to hereafter as Rhodesian grass. A gc-mst Rhodesian Sudan grass was developed by Craigmiles et al. (1958) and Craigmiles (1961, 1962), after placing the nucleus of this grass into the S-cytoplasm of mst combine-kafir through a series of BC's using Rhodesian grass as the recurrent male parent. From this, F_1 sudan grass hybrid was produced by allowing pollination of mst Rhodesian grass with Tift Sudan grass in an isolated field. The hybrid, named Suhi-1, is tall, vigorous, late-maturing, produces more forage throughout the season for a longer period of time, possesses wide broad leaves, semi-dry stems and is resistant to drought and diseases. Seed production of this cytosterile hybrid is economically unattractive because its extreme height does not lend itself to mechanical seed-harvesting equipment. To rectify this, recessive triple dwarf genes were transferred to the cytosterile form through backcrossing and the height was reduced from 3.6 m to 1.2 m (Craigmiles 1966). This line has worldwide implication, since locally adapted F_1 hybrids can be obtained, substituting high yielding indigenous Sudan grasses as pollinators for the sterile line. A few mst plants occurred in populations derived by pollinating *S. vulgare* pollen on *S. arundinaceum, S. verticilliflorum, S. virgatum* and *S. niloticum* strains (Ross 1961). Many gc-mst lines of sorghum have been utilised to produce interspecific hybrid seeds after pollinating these with *S. halepense* pollen (Wilhite 1960a,b, McClure 1962) and with *S. almum* (McClure 1962).

A method of producing seedlesss broomcorn *S. vulgare* using gc-mst was proposed by Hadley (1960). In this the genome of *S. vulgare* var. *technicum* is transferred into S- and N-cytoplasm *S. vulgare* plants. Genes controlling seed and glume shedding could be obtained from Tunis grass, *S. virgatum*. Since two dominant epistatic genes in either the homozygous or heterozygous state are required to induce seed and glume shedding (Karper and Quinby 1947), the crossing of nonshedding sterile lines by nonshedding fertile lines would yield sterile hybrids which shed glumes. Of the three mst lines isolated whose cytoplasm appeared to differ from the milo type (Schertz and Ritchey 1978), two did not find an immediate use in sorghum breeding. This is because while one line exhibited partial fertility, the second had plants with indehiscent anthers. This rendered the differentiation of fn-mst's from the mft's difficult in the field. In the third line, mst was perfect and stable. This line is suitable for use in breeding programs and has been released as A2TX 2753 by Schertz and Ritchey (1978).

Using the gc-mst system, mst combine-kafir 60, silver-kafir 195 and locally developed steriles have been used all through Asia to develop new and promising hybrid sorghum lines (Wall and Ross 1970) These hybrids are being grown, maintained and multiplied over almost all sorghum-growing areas. In India, hybrid sorghum development has been extremely useful in solving the food problem for people and cattle. Many commercial hybrids (CSH) have been developed. These exhibit dramatic increases (nearly 70%) in grain and fodder yield. The adoption of high-yielding, dwarf grain hybrids suitable for mechanisation and advanced management practices is now nearly universal through almost all Asia and Oceania. The majority of mst sorghums have been essentially kafir or kafir-milo derivatives. Since the chalky white grain of kafir limited the development of a better grain quality, new mst's with corneous endosperm through substitution BC in milo cytoplasm were developed at Coimbatore (India). These are 2219A (kafir-shallu), 2077A, 3675A, 3677A (kafir types). CSH-3 is based on 2219A cytoplasm. Another approach to improve grain quality is by crossing the tall Indian varieties (which are normally fertility restorers on the milo-kafir sterility system) to kafir B, identifying the sterility maintainers in F_2 by test crosses and carrying them through several BC generations until stable mst A-lines are developed. These steriles are close to kafir in several respects but have improved grain quality. The mst line 1036A (based on PJ 7R) was developed in this way and was the female parent for CSH-4 (1036A × Swarna). Several such mst lines were developed and used in hybrid production. Whereas the seed yield over mst's declines linearly as the distance from the pollen parent increases, it does not alter significantly when the usual mst:mft row ratio of 3:1 is changed to 7:1 (Robinson 1984).

27.17.6.1 Constraints and Remedies

Low temperature causes mst in many sorghum genotypes. These temperature-sensitive lines can be used as female parents in crosses with nonsensitive sorghum lines, provided female fertility and seed production are not reduced. Maintaining local sorghum lines at 14°C during preflowering and flowering periods induces pollen sterility (this Vol.), but seed set obtained after hand pollination is reduced by 16–21%. Hence low temperature treatments impair female fertility also. This, however, needs more testing using many sorghum genotypes, as genotypic differences to environmental sensitivity, especially to low temperature, are known in sorghum (Brooking 1979). Continued breeding and selection for low temperature tolerance in grain sorghum would be promising in order to increase the range of environments where grain sorghum can be usefully grown. This would include high altitude tropical regions and areas with marginal summer temperatures. Using temperature-sensitive genotypes as female in crosses to nonsensitive varieties, hybrid seed production is possible. The hybrids are intermediate to low temperature-tolerant. Thus suitable parents will have to be selected to obtain fully tolerant hybrids. In India, the first sorghum hybrid that was put into production is adapted to the areas where this crop is planted in spring or summer when the atmospheric temperature is fairly high but not to areas where the crop is cultivated in winter or the rabi season. Therefore new mst genotypes and their maintainers need to be developed that are cold-tolerant and also such as can be grown in early autumn in the warmest part of the temperate zone. No A line has so far been found that is always cent per cent mst in a seed field. However, the majority of the A lines are highly mst with only a few plants producing some pollen-shedding anthers. The isolation of complete mst's may be possible through careful selection. Whereas no difficulties exist in maintaining seeds of male parents of hybrids because the seed for the next crop can be harvested from male rows in a seed field, the maintenance of the seeds of the female parent is slightly problematic. The method of maintaining and producing seed from gc-mst line A indicated earlier is practicable, but a careful search for off-type plants and their roguing is necessary. This can be avoided if the female rows in hybrid seed fields and fields for the production of females or A-line seeds are well isolated.

Although cytoplasm diversity is known in sorghum, for hybrid seed production only two S-cytoplasms (milo and Indian) have mainly been used. Very little has been used from Ethiopian sterile source. Such a narrow cytoplasm base is hazardous and should be broadened through search for new S-cytoplasm sources from the existing sorghum gene pool and by inducing cytoplasm mutations. Schertz and Ritchey (1978) have reported three such S-cytoplasm sources. These are to be tested after the development of isocytoplasmic lines. Random mating grain sorghum populations generated and maintained through the use of *mst* provide the breeder with many opportunities. Many genetic recombinations can be produced with minimum efforts. The recurrent selection system currently used successfully in corn can be applied to such populations. A broad genetic base can be maintained to minimise genetic vulnerability and promising new breeding material from exotic populations can be introduced into the random mating populations at any time. *Sorghum* ranks fifth in world crop acreage, being exceeded by wheat, rice, maize and barley, and is second to rice among hot-weather (summer) cereals in south and south-east Asia. Keeping in view the growing world population and dwindling food resources, the gc-mst system has to be further developed and used in bridging crosses, breaking barriers to crossability and combining the most desirable gene combinations in this crop so useful for the developing world. Limited success has been achieved in interspecific crosses. Certain intergeneric crosses could be tried similarly, using the gc-mst system. Furthermore, the gc-mst system has been properly developed in *S. vulgare* and *S. arundinaceum* var. *sudanense* only. Other sorghum species also need to be investigated for the presence of *fr* genes in their genomes.

27.18 *Tragopogon* L. (goat's beard)

The exact number of species in this genus is unknown and ranges between 35 and 45. This inexactness is because the species of this genus undergo extensive intra- and interspecific hybridizations and many amphidiploids have become naturalized. Some of these exhibit mst (Ownbey and McCollum 1953), for instance, sterile plants segregated in F_2 progeny of *T. pratensis* × *T. porrifolius*. The progeny of the reciprocal cross was normal. The reciprocal differences are due to g-c interactions. Female fertility of the steriles is not precisely known.

27.19 *Tripsacum laxum*, Nash.

This genus comprises four perennial grasses with stout culms and monoecious inflorescence, the staminate flowers lie above the pistillate flowers. In *T. laxum*, which has similar sexual arrangement, a tetraploid mst plant was detected by Dodds and Simmonds (1946). In this, overcrowding of 36 bivalents at MI occurs, which results in an irregular orientation to the MI spindle and disturbed AI disjunction, leading to multifarious anomalies in the second division and finally to pollen sterility. Details of genetic control of sterility and of female fertility of this plant are not known. Since autotetraploids are mostly male-female sterile due to disturbed meiosis (Kaul 1964, 1965), female fertility of the mst tetraploid should be checked.

27.20 *Triticum* L. (wheat)

This is as old as farming. Being a staple food to mankind, it is cultivated all through the world except in areas of climatic extremes. In a drive towards increased food production, exploitation of hybrid vigour for increased grain yield began in the early 1950's. This led to the search for mst sources and both g-mst and gc-mst were detected in *T. aestivum*, *T. durum* and *T. monococcum* (Tables 27.21, 27.22). In addition, extensive cytoplasmic differentiation was uncovered (Tsunewaki 1980b, Wilson and Driscoll 1983) and many chemicals induced nongenetic mst (Table 27.23).

27.20.1 *T. aestivum* L. (bread wheat)

27.20.1.1 Genic Male Sterility

27.20.1.1.1 Spontaneous, Monogenic Recessive

Compared to other major cereals like maize, rice, and barley, the number of spontaneously arisen mst mutants in wheat is low mainly because of its hexaploid nature. In a F_3 family from a cross Kenya Farmer × Bearded Javelin 48, Pugsley and Oram (1959) found complete mst plants. Their female fertility was normal, as NCP + hand pollinations yielded high seed set. A few selfed F_4 progenies from mft F_3 also segregated some mst plants, but their genetic situation is unclear. Pugsley and Oram (1959) theorized mst in this material as either due to homozygosity for a recessive allele at one locus (gametes with that allele being less viable than those with a dominant allele) or due to a complementary action of two recessive alleles at two loci. The somatic chromosome number of these mst's varied from 40–44 (Waninge and Zeven 1968), but there was no association of chromosome number with sterility. The mst was controlled by one recessive *ms* gene. Realizing the potential use of this sterility in mass production of wheat hybrids and for increasing

NCP within populations, Suneson (1962) incorporated the sterility into some of his breeding material. Complete mst could not, however, be achieved, as the environment considerably influenced the sterility expression. Recovery of mst's from various heterozygotes ranged between 0 and 13%. From this material, Briggle (1970) transferred the recessive *ms* gene to the wheat variety Chancellor. After the initial cross and four BC's to Chancellor, accompanied by rigid selection for single gene segregations for mst, some plants having complete mst developed. The F_2 populations offered progressively more evidence that a single recessive gene was involved in mst expression and as additional BC's to Chancellor were made, transmission of the recessive gene by the male gametes occurred at the same frequency as by the female gametes. Thus Chancellor wheat variety appears to have a compatible genetic background for mst expression governed by the gene. This *ms* gene does not seem to be as effective in other wheat cultivars as in Chancellor, probably because of the different genetic backgrounds that other cultivars have. It could also equally be possible that more than one recessive gene, as well as modifier genes enhancing mst expression, were present in Pugsley's mst stock. By continually selecting for the gene causing mst and as progressive BC's were made to Chancellor, perhaps the additional gene(s) for mst and/or modifiers became eliminated.

From the multiplication stocks of the wheat line TJB 31187, Lupton and Bingham (1966–1967) in England detected four mst plants in a single row. These were probably derived from a single recessive mutation. Meiosis proceeded normally in them but pollen development failed soon after tetrad formation. Their female fertility was normal. In the heterozygote, the *ms* gene was transmitted normally through the ovule but poorly through pollen, so that the reciprocal BC's to normal parents gave different segregation ratios. The sterility expression remained uninfluenced by environment and alien cytoplasm and could be transferred successfully to other winter and spring wheat varieties. In 1968, Bingham found some more mst plants in a F_4 family of a triple cross and a single mst plant in a lot of a wheat variety Maris-Widgeon. F_1 plants involving crosses onto these plants were fertile as was the case when the above four original mst plants were crossed. In the hexaploid spring wheat Russian variety Saratovskaya-29, Krupnov (1968) found a mst mutant whose female fertility was normal but mst was complete. The sterility is conditioned by one recessive *ms* gene which causes microspore abortion. The occurrence of similar mst plants in low frequency is known in some more Russian wheat varieties like Belaya-Tserkov 1980, Bezostaya and Mironovskaya-808 (Lemekh et al. 1971). In all these, male meiosis is abnormal from prophase-I, but it continues until microspores are liberated. Thereafter microspores degenerate gradually.

27.20.1.1.2 Spontaneous, Multigenic Recessive

From India, a specific type of mst caused by the interaction of three recessive genes having an additive effect occurred in an F_2 population of a composite wheat cross (Athwal et al. 1967). The mst degree depends on *ms* gene number; the lesser the number, the lesser is the mst expression. Thus the genotypes having 1, 2 and 3 *ms* genes exhibit 36–49%, 47–68% and 73–97% mst, respectively. This p-mst can be developed by introducing 1–2 *ms* genes in this wheat genome. This expression was further influenced by environment and by modifying genes. According to Borlaug (Athwal et al. 1967), two of the three *ms* genes might have been inherited from Norin-10, which entered the pedigree as well as the parentage of the above material. In PMC's of the mst's, meiosis proceeds normally, but microspores degenerate after release. Female sex is fully functional. Selfing yielded 1–8% and open pollinations 45–71% seed set. All the plants of one F_3 line (UC 9109–9) of a cross between two mft American wheat varieties Ramona 50 and D 6301, were mst and had abnormal development and differentiation of stamens (Jan 1974). The stamens exhibit pistiloidy of varying degree (Fig. 27.13). These steriles are composed of disomic, monosomic and nullisomic plants. The disomics are less stable than others. Three independent recessive *ms* genes control mst in them. The absence of either of the *ms* genes leads to mft, unlike in the steriles studied by Athwal et al. (1967), in which each gene contributed to the sterility expression, the absence of one or two of them decreasing the sterility expression gradually. The early tillers of mst exhibit a greater degree of anther transformation than the late-formed ones. Jan and Qualset (1977) proposed that a specific inhibitory gene *i* interacts with the genes governing normal anther differentiation and modifies anthers into ovarylike structures. The sterility genes appear to block the earliest stages of anther differentiation, alter sporogenous tissue formation and induce staminal primordia to develop into pistil-like structures. This *ms* gene action is strongly manifested at the beginning of spike differentiation and less at later stages. The influence of environmental factors over *ms* gene action is high.

27.20.1.1.3 Induced, Monogenic Recessive

Alloplasmic bread wheat var. Chris (*T. tauschii*), *T. aestivum* is completely mst. Franckowiak et al. (1976) isolated 12 EMS-induced mst mutants in this variety. Four of these have empty pollen grains and the remaining eight produce a low amount of fertile-looking pollen. The sterility in one of these is conditioned by a dominant gene and in the remaining 11 by single recessive genes, three of the genes are nonallelic (Sasakuma et al. 1978). The dominant *Ms* gene is highly stable under different genetic backgrounds and does not interact with other genes of the genome or those controlling female fertility. This gene and eight recessive *ms* genes induce mst in eu- as well as alloplasmic Chris bread wheat plants. Thus these genes are cytoplasm-nonspecific and of genic type. After X-irradiation of seeds of winter wheat variety Probus, Fossati and Ingold (1970) obtained a mst mutant whose sterility is conditioned by one recessive gene, msa_1. The sterility is complete, as its anthers have empty pollen grains and its florets remain open at flowering. This favours some cross pollination in it and consequently improves its value in hybrid breeding. After treating seeds of the winter wheat varieties Mironovka-808 and RPG-186 with N-nitroso-N-ethyl urea, Latypov (1974) obtained many mst mutants in each of these varieties. In each case, one recessive gene conditioned mst. Irradiation (250 or 500 kR of γ- or X-rays) of euploid pollen at about the time of

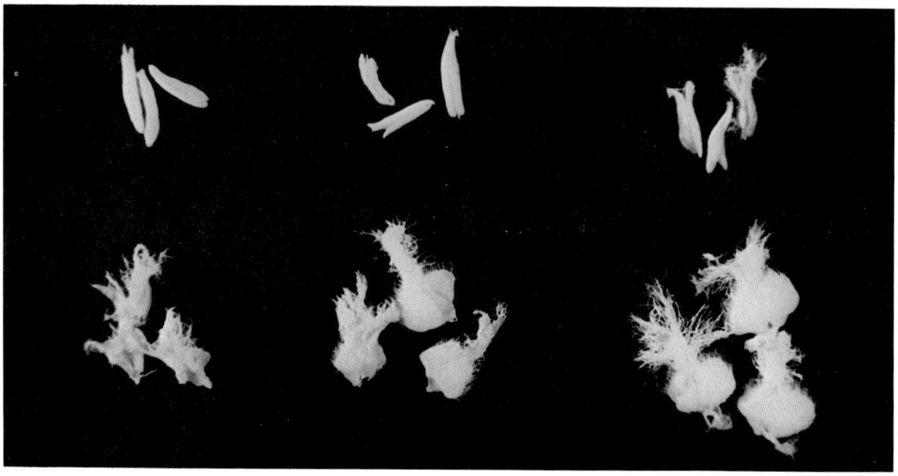

Fig. 27.13. *Triticum aestivum*. Increasing degree of anther pistilloidy in a genic male sterile wheat. (Jan and Qualset 1977)

pollen shedding and its application to emasculated monosomics 4A, 4B, 5A, 5B and 5D plants produced 795 F_1 plants; of these, 62 involved monosomic 4A and produced one mst mutant, the Cornerstone mutant (Driscoll 1976, Driscoll and Barlow 1976). The *ms* gene is located on 4A and is allelic to Pugsley's and Probus mutants. Another *ms* gene in an X-ray-induced mutant is also located on chromosome 4A (Kleijer and Fossati 1976). The *ms* gene of Cornerstone bread wheat has been incorporated into 4x *T. durum* wheat and 6x and 8x triticales by Hossain and Driscoll (1981). Diploid rye restores fertility at both the ploidy levels.

27.20.1.1.4 Monogenic Dominant

In a Chinese bread wheat, a dominant *Ms* gene induces mst (Xiangmin and Herong 1981). These plants segregated from the progeny derived from a composite hybridization with eight materials of *T. aestivum* from Taigu county in the Shagai province of China. Male meiosis either stopped completely or proceeded abnormally in mst plants and microspore abortion was complete. In another similarly inherited mst material (Deng and Gao 1982), male meiosis proceeds normally until tetrad formation. The tetrads degenerate prior to liberation. The allelic nature of these two dominant genes is not known, but since *ms* gene action differs and the two materials originated from different sources, the two *ms* genes may be nonallelic. Franckowiak et al. (1976) obtained 12 mst mutants after EMS treatment to the alloplasmic Chris bread wheat *T. aestivum* having *T. tauschii* cytoplasm. In one of these mutants, a dominant gene conditions mst (Sasakuma *et al.* 1978). The gene is highly stable and its penetrance and expressivity in *T. tauschii* cytoplasm is complete. Transfer of this sterility into *T. aestivum* var. Len was done by Busch (Altman *et al.* 1984). The transfer was confirmed by comparing electrophoretic gel patterns of mt-DNA of the parental and derived wheat lines. All restriction patterns of the putative euplasmic line verified that the cytoplasm was from *T. aestivum* (Altman et al. 1984).

27.20.1.1.5 Allelic Nature and ms Gene Location

Driscoll (1975) studied the genetic relationship and chromosomal location of the *ms* genes of Pugsley and Probus mst mutants. The mutant genes failed to complement each other and partial monosomic analysis reveals both the mutant genes as alleles, and located on chromosome 4A. A strong homo-allelic echo was received from chromosome 4B with the Probus mutant but not with the Pugsley mutant. The *ms* gene of the Cornerstone mutant is located on the 4A chromosome and is allelic to Pugsley and Probus mutants (Driscoll 1977). Some plants were of intermediate phenotype. In them many anthers failed to dehisce but some liberated functional pollen. Such intermediate types occurred in F_1 of monosomic 4B × Probus heterozygote. This seems to have involved monosomy for 4B and heterozygosity for Probus mst on chromosome 4A. By contrast, all 12 individuals of the corresponding cross involving Pugsley's mst were fully fertile. It is inferred that the Probus mutant involves the loss of two loci concerning mft, the Pugsley mutant involves a change in one of these loci and chromosome 4B possesses only one locus, the one unchanged in the Pugsley mutant (Wilson and Driscoll 1983). Thus the Probus mutant appears to have lost more than one locus on 4A that is involved in fertility. On this chromosome 4A, the *ms* gene of the Cornerstone mutant is located. In the Probus mutant, the mst is due to terminal deletion of the short arm of chromosome 4A (Kleijer 1984). Apart from 4A, the most likely chromosomes in which mst mutants could be induced are 4B, 5A, 5B and 5D. Nullisomics for these chromosomes are mst, one ditelocentric is mst and the other is mft (Wilson and Driscoll 1983). Hence deletion of a segment of the critical arm should lead to mst. The *ms* genes in Cornerstone and Probus mutants are located in the L arm of chromosome 4A and are at least 50 map units from the centromere (Barlow and Driscoll 1981). This distal region 4AL lodges simply inherited *ms* gene(s) of high mutability. This chromosome is distinct in many respects (Sears 1972). Whereas all the three nullisomics of homoeologous group 3 are mst and vigorous, distinct differences exist between nullisomic-tetrasomic lines (Sears 1966). Although vigorous nullisomic 4B (or 4D)-tetrasomic 4A is mft, nullisomic 4A-tetrasomic 4A-tetrasomic 4B (or 4D) is mst. Hence 4A has the gene(s) for mft that neither 4B nor 4D has. Pairing relationships reveal the origin of 4A from elsewhere than diploid wheat. The chromosome occurs in 4x and 6x wheat and the mutant (mst) trait expresses at both these ploidy levels; there is, however, greater selection against gametes carrying the mutant at 4x level than at the 6x level.

27.20.1.1.6 Microsporogenesis

In the majority of the genic mst wheats, mst is conditioned by one to three recessive genes. In most of these, the genes do not affect male meiosis but inhibit microspore development and pollen formation. The microspores mostly abort after liberation

(Lupton and Bingham 1966–67, Athwal et al. 1967, Bozzini and Scarascia-Mugnozza 1968, Krupnov 1968, Fossati and Ingold 1970). However, in some Russian mst wheats (Lemekh et al. 1971), the *ms* gene may act anywhere during microsporogenesis. The dominant *Ms* genes may either stop (Xiangmin and Herong 1981) meiosis completely or permit its course abnormally, but no viable pollen is ever produced.

27.20.1.1.7 Fertility Restoration

The *ms* gene in Probus, Pugsley and Cornerstone mst wheats is located in the L arm of chromosome 4A so that the hybrid crop has a 4A chromosome with a terminal segment of the L arm deleted. Thus fertility of the hybrid crop can be restored by introducing normal chromosome 4A into its genome. Since *ms* allele is recessive, the effect of this gene in hybrids is nil. The introduction of an alien chromosome is necessary, as a homogeneous population of mst plants constituting the female parent can be obtained. The ideal characteristics of such an alien chromosome, as indicated by Wilson and Driscoll (1983), are that it (i) carries a gene(s) capable of replacing the deleted gene(s) for mft, (ii) refrains from pairing with wheat chromosomes this prevents the alien fertility gene(s) from incorporating into wheat complement, (iii) does not adversely affect the plant phenotype, (iv) has a low male transmission and mis-division rate in Y line, and (v) has a low rate of addition decay in the X line. Of these, the first two are most essential. Finding a satisfactory alien chromosome is arduous. For instance, substitution of chromosome 4E of *Agropyron elongatum* for 4A produces vigorous plants, but they are mst (Dvořak 1980). The alien chromosome $4S^1$ of *Aegilops longissima* substitutes successfully for 4A and the line is mft. However, when present singly, male and female gametes not bearing this *Ae. longissima* chromosome are precluded from functioning (Maan 1975b). Thus this chromosome is not useful. Driscoll (1981a,b) tried Transec 4A chromosome, chromosome 4 of 2x wheat, modified chromosome 2 of rye and modified chromosome 4 of barley. The last two sources of alien chromosome are promising for hybrid wheat production. A latent gene(s) for mft has been detected in Cornerstone chromosome 4A (Islam and Driscoll 1984). Although this gene is unable to effect fertility in Cornerstone, it is able to do so in the presence of a structurally modified barley chromosome 4 HM disomically substituted in mst for 4A of wheat. Thus the two lines, Hmy (Islam and Driscoll 1984) and 2 Rsy (Hossain and Driscoll 1983), represent the fertile lines isolated for the xyz system for producing hybrid wheat.

27.20.1.2 Gene Cytoplasmic Male Sterility

Outcrossing in wheat produces complete to partial sterile plants. Realising the positive heterosis in F_1 wheat, intensive search began to obtain mst through interspecific and intergeneric-hybridisations in the early 1950's. While studying the effects of nuclear substitutions into cytoplasm of other grass species and genera, Kihara (1951) obtained mst plants after the *T. aestivum* var. *erythrospermum* genome was transferred to *Aegilops caudata* cytoplasm. The near-genome transfer occurred in BC_4. This alloplasmic combination was depicted by Kihara (1967) by writing the cytoplasm donor in parenthesis followed by the genome donor viz. (*Ae. ovata*) *T. aestivum*. This represents the first case of mst in wheat resulting from interaction between the cytoplasm of one species and the nucleus of another. Later, for over 30 generations, no segregation of the *c*-gene has been detected in this alloplasmic combination. In other words, the wheat genome + *Aegilops* cytoplasm interaction is stable. Returning the *Ae. caudata* nucleus to its own cytoplasm restored fertility (Kihara 1951). Later, Fukasawa (1953) transferred *T. durum* nucleus to *Ae. ovata* cytoplasm, the resulting alloplasmic combination being mst. By placing *T. aestivum* var. *Bison* genome in *T. timopheevi* cytoplasm, Wilson and Ross (1962) obtained complete mst plants. Following this, genome transfer of several wheat species into many alien cytoplasms was done. In some combinations, mst resulted (Table 27.10) with or without side-effects. Whereas in two cytoplasms (*Ae. caudata* and *Ae. ovata*), anther malformation is absent and no delay in heading date or reduction in plant vigour occurs, in the remaining cytoplasms, anthers are malformed, heading date is delayed and plant vigour is considerably reduced in all the eight hexaploid wheat genotypes. In addition, such substitutions reveal interesting cytoplasm diversity, cytoplasm-gene interactions and its effects in various *Triticum* and *Aegilops* species. These are reviewed by Tsunewaki (1980b) and Wilson and Driscoll (1983), and have been collaboratively treated by three research teams under Drs. SS Maan and KA Lucken, North Dakota State University, Fargo, USA, I. Panayotov and K. Gotsov, Institute of wheat and sunflower, Tolbuhin, Bulgaria and K. Tsunewaki, Y. Mukai and R.T. Endo, Kyoto University, Kyoto, Japan (see, Tsunewaki 1980b, Panayotov 1980, Maan and Endo 1981, Mukai, 1982, 1983). Sterility-fertility relations between *T. aestivum* and *T. spelta* genomes and *Triticum* and *Aegilops* cytoplasms as given in Table 27.11 reveal that whereas the two bread wheat varieties Chinese spring and Salamon do not differ from each other in their interaction to *Triticum* cytoplasms, they differ in

Table 27.10. Male sterility in alloplasmic wheat

Genome	Cytoplasm
T.aestivum	*Aegilops aucheri, Ae.bicornis, Ae.biuncialis, Ae.caudata, Ae.columnaris, Ae.comosa, Ae.dicoccoides, Ae.heldreichii, Ae.kotschyi, Ae.macrochaeta, Ae.mutica, Ae.ovata, Ae.recta, Ae.sharonensis, Ae.speltoides, Ae.triaristata, Ae.triuncialis, Ae.umbellulata, Ae.uniaristata, Ae.variabilis, Ae.ventricosa, Secale cereale, T.ararticum, T.boeoticum, T.compactum, T.dicoccoides, T.monococcum, T.timopheevi, T.zhukovskyii*
T.durum	*Ae.bicornis, Ae.caudata, Ae.ovata, Ae.squarrosa, Ae.uniaristata, Secale cereale*
T.macha	*Ae.columnaris, Ae.kotschyi, Ae.ovata, Ae.triaristata, Ae.triuncialis, Ae.variabilis*
T.spelta	*Ae.biuncialis, Ae.boeoticum, Ae.caudata, Ae.columnaris, Ae.heldreichi, Ae.kotschyi, Ae.mutica, Ae.ovata, Ae.sharonensis, Ae.squarrosa, Ae.triaristata, Ae.triuncialis, Ae. umbellulata,Ae.uniaristata, Ae.variabilis*
T.sphaerococcum	*Ae.biuncialis, Ae.columnaris, Ae.ovata, Ae.triaristata.*
T.timopheevi	*Ae.kotschyi, Ae.variabilis, Ae.ventricosa*
T.turgidum	*Agropyron intermedium*
T.vulgare	*Ae.caudata, Ae.ovata*

Table 27.11. Male sterility(S)-fertility(F) relationships in alloplasmic wheat. (After Mukai, personal communication)

Cytoplasm	genome (ABD)		
	T.aestivum		*T.spelta*
a) Wheat source	(Cs)	(S)	(D)
T.aestivum (6x = 42)	F	F	F
T.ararticum (4x = 28)	S	S	F
T.boeoticum (2x = 14)	S	S	S
T.dicoccoides (4x = 28)	F	F	F
T.dicoccum (4x = 28)	F	F	F
T.durum (4x = 28)	F	F	F
T.macha (6x = 42)	F	F	F
T.monococcum (2x = 14)	S	S	S
T.timopheevi (4x = 28)	S	S	F
T.turgidum (4x = 28)	F	F	F
T.zhukovsky (6x = 42)	S	S	F
b) Aegilops source			
Ae.aucheri (2x = 14)	S	S	F
Ae.biuncialis (4x = 28)	F	S	S
Ae.caudata (2x = 14)	S	S	S
Ae.columnaris (4x = 28)	F	S	S
Ae.comosa (2x = 14)	S	S	S
Ae. cornis (2x = 14)	F	F	F
Ae.crassa (4x = 28)	F	F	F
Ae.crassa (6x = 42)	F	F	F
Ae.cylindrica (4x = 28)	F	F	F
Ae.heldreichii (2x = 14)	S	S	S
Ae.juvenalis (6x = 42)	F	F	F
Ae.kotschyi (4x = 28)	F	S	S
Ae.longissima (2x = 14)	F	F	F
Ae.mutica (2x = 14)	F	S	S
Ae.ovata (4x = 28)	F	S	S
Ae.sharonensis (2x = 14)	F	S	S
Ae.speltoides (2x = 14) s_1	F	F	F
s_2	S	S	F
Ae.squarrosa (2x = 14)	F	F	F
Ae.triaristata (4x = 28)	F	S	S
Ae.triaristata (6x = 42)	F	S	S
Ae.triuncialis (4x = 28)	F	S	S
Ae.umbellulata (2x = 14)	F	S	S
Ae.uniaristata (2x = 14)	F	S	S
Ae.variabilis (4x = 28)	F	S	S
Ae.vavilovii (6x = 42)	F	F	F
Ae.ventricosa (4x = 28)	F	F	F

Cs = chinese spring, S = salmon, D = duhamelianum, s_1,s_2 = two strains, F = fertile, S = sterile.

this respect in 12 *Aegilops* cytoplasms. In contrast, Salmon and *T. spelta* exhibit similarities in 25 out of 27 cytoplasms tested. Within wheats, Salmon wheat exhibits mst in maximum *Aegilops* cytoplasm (17) followed by Duhamelianum wheat (15). Only five such cytoplasms induce mst in Chinese spring wheat. This is nearly reversed in wheat cytoplasms. In wheat Salmon exhibits the highest (5), followed closely by Chinese spring (4) and is least in Duhamelianum (2) wheat. The differences are due to differential genome-cytoplasm interactions.

Aegilotricum #2 (2n = 56) is an amphidiploid derived from the hybrid *Aegilops ovata* (2n = 28) × *Triticum durum* var. *hilderbranti* (2n = 28) (Tschermak and Bleier 1926). Pollen sterility occurred in some BC populations (*Aegilotricum* #2 × *T. durum* var *reichenbachi*) x *T. durum*) (Fukasawa 1953). When *Ae. ovata* was utilized as the male parent, an initial increase in degree of pollen sterility was followed by its decrease in BC generations. During this, continuous loss of *T. durum* chromosomes occurred. In BC_3, plants having 14 pairs of *Ae. ovata* chromosomes were obtained. In most of them pollen fertility was regained. On the other hand, when *T. durum* was used as the recurrent male parent, the degree of pollen sterility was enhanced continuously with successive BC's. In the third backcross generation, plants with 14 pairs of *T. durum* chromosomes were found and were fully mst. In them male meiosis proceeded normally, but young pollen grains degenerated. Sibs of the 28-chromosome plants that had one to three *Ae. ovata* chromosomes in addition to the *T. durum* genome had 88–92% fertile pollen. Thus mst of 28 chromosome plants was due to the interaction of the *Ae. ovata* cytoplasm and the *T. durum* genome. When *T. dicoccum* was used as the male parent in crosses to *Aegilotricum*, the progeny was mst (Fukasawa 1953). Via substitution backcrosses (SBC), Schulz-Schaeffer (1972, 1978) introduced gc-mst into alloplasmic *Agropyron intermedium* having *T. turgidum* var. *durum* cytoplasm. In some plants, either young or mature microspores aborted, in others pollen with or without exine degenerated (Young *et al.* 1979). In almost all, tapetal anomalies were rampant, in many of them orbicular malformations occurred.

27.20.1.2.1 Phenotypic and Physiological Effects

These, detailed out by Sage (1976) and Tsunewaki (1980b), are briefly considered here. In addition to mst, g-c interactions cause plant vigour and stunted growth, abnormal chlorophyll (Fukasawa 1956a, Gilmore 1970, Sanchez Monga 1970, Mukai 1981), and stomatal development (Fukasawa 1958b), pistiloidy (Kihara 1962, Maan and Lucken 1968a, Tsunewaki et al. 1980a,b), anther malformation and transformations (Fig. 27.14), microspore and pollen abortion (Fig. 27.15), impaired female fertility, kernel shrivelling on crossing (Schmidt et al. 1970, Tsunewaki et al. 1980a,c), reduced seed germination (Kihara and Tsunewaki 1961), increased frequency of germless or nongerminating pollen (Kihara and Tsunewaki 1964, Maan 1973, Tsunewaki et al. 1980a,b) parthenogenetic development and haploids, aneuploids and twin seedling production (Kihara and Tsunewaki 1962b, Tsunewaki et al. 1968, Lacadena 1968, Schmidt et al. 1970, Tsunewaki et al. 1980a,c, Kobayashi and Tsunewaki 1980, Mukai et al. 1978, Fig. 47.16). Considerable reduction in plant vigour and delay in heading occurs in eight wheats when their genome is placed in *Ae. biuncialis*, *Ae. columnaris* and *Ae. triaristata* cytoplasms (Mukai 1981). Some of the major manifestations are given below:

i. *Anther Malformation.* Stable and complete mst results when wheat genome is placed in *T. timopheevi* cytoplasm. The steriles have short, elongated, small, often arrow-shaped anthers having aborted or shrunken nonviable pollen. Anther size and shape varies, depending upon the genome-cytoplasm combination (Joppa *et*

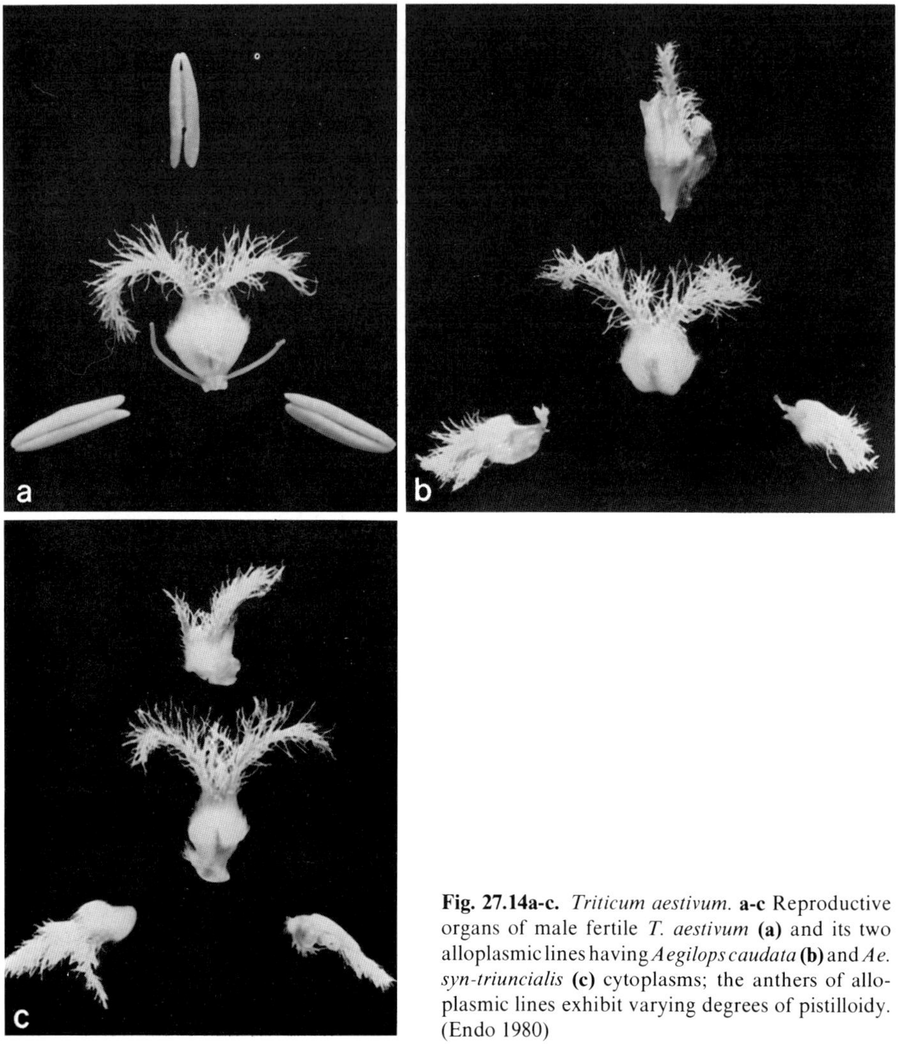

Fig. 27.14a-c. *Triticum aestivum*. **a-c** Reproductive organs of male fertile *T. aestivum* (**a**) and its two alloplasmic lines having *Aegilops caudata* (**b**) and *Ae. syn-triuncialis* (**c**) cytoplasms; the anthers of alloplasmic lines exhibit varying degrees of pistilloidy. (Endo 1980)

al. 1966). In the mst's obtained after placing wheat genomes in *Aegilops* cytoplasms, the degree of anther malformation and pollen sterility considerably varies from one g-c combination to other (Table 27.12) so that anthers are nearly normal to totally malformed. This was intensively studied by Endo (1980) and Mukai (1982) from whose data Table 27.12 is prepared. In general, anther malformation and pollen abortion are complete in wheats having *Ae. caudata* and *Ae. syntriuncialis* cytoplasms (Fig. 27.14). Wheats with *Ae. biuncialis, Ae. triuncialis* and *Ae. umbellulata* cytoplasms have normal anthers, but their pollen sterility is complete. Thus these three cytoplasms induce mst, but do not influence anther phenotype in many wheat genotypes. That the two developmental anomalies, anther malformation and pollen abortion, are conditioned by two different genes or sets of genes

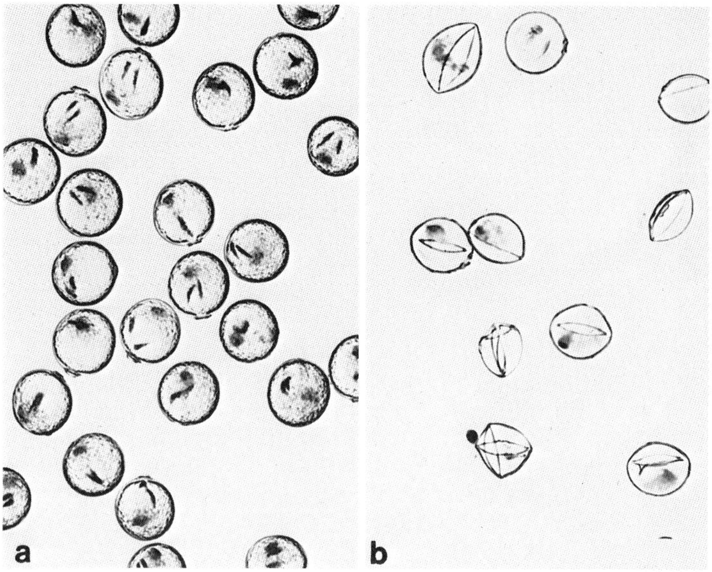

Fig. 27.15a,b. *Triticum aestivum*. **a,b** Pollen from male fertile **(a)** and male sterile **(b)** wheat. (Endo 1980)

is evidenced by the total pollen abortion in wheats having either completely or partially malformed or normal anthers (Table 27.12), or fully malformed anthers but low pollen abortion as in *T. aestivum* S-615 having *Ae. caudata* or *Ae. syntriuncialis* cytoplasm (Endo 1980).

ii. *Female Sterility*. Reduced female fertility accompanies mst in many alloplasmic wheats (Tsunewaki et al. 1980c), especially in those having rye cytoplasm in which plant vigour is also reduced. The addition of one rye chromosome to the steriles improves mft and vigour, but not female fertility. Some plants with one or more rye chromosomes possess normal-looking anthers and abundant functional pollen, but female fertility is not improved in them. Thus fst is more stable than mst in this alloplasmic combination. It is equally probable that the gene(s) conditioning female fertility is partially effective in rye cytoplasm because of incomplete penetrance.

iii. *Haploid and Twin Seedling Production*. Certain g-c interactions influence embryogenesis, leading to haploids and twin seedlings production (Fig. 27.16). Kihara and Tsunewaki (1962b) first documented that progeny of the alloplasmic bread wheat Salmon having *Ae. caudata* cytoplasm produced about 30% haploids. Later six more *Aegilops* cytoplasms were found to induce haploids and twin seedlings in this wheat (Tsunewaki et al. 1976b). Even though many mst alloplasmic wheats have these characters (Kobayashi and Tsunewaki 1980, Mukai 1981), it is uniquely high in Salmon, as it occurs in more cytoplasms than in any other wheat variety (Fig. 27.17). Except in *Ae. caudata*, the proportion and frequency of haploid production differs significantly between alloplasmic *T. aestivum* and *T. macha* in the same cytoplasm. This indicates differential g-c

Table 27.12. Anther phenotype in male sterile alloplasmic wheat having *Aegilops* cytoplasm

Phenotype	Genome	Cytoplasms
Anthers normal, pollen sterility complete	*T.aestivum* var.*erythrospermum*	*Ae.biuncialis*
		Ae.caudata
		Ae.syntriuncialis
		Ae.triuncialis
		Ae.umbellulata
	Chinese spring	*Ae.triaristata*
	Salmon	*Ae.biuncialis*
		Ae.triaristata
		Ae.triuncialis
		Ae.umbellulata
	S-615	*Ae.ovata*
	Selkirk	*Ae.biuncialis*
	T.macha	*Ae.biuncialis*
		Ae.columnaris
		Ae.triuncialis
	T.sphaerococcum var.*rotundatum*	*Ae.biuncialis*
		Ae.triuncualis
		Ae.umbellulata
	T.spelta	*Ae.kotschyi*
		Ae.triuncialis
		Ae.umbellulata
		Ae.variabilis
Anthers normal, pollen sterility high	*T.aestivum* cv.Selkirk	*Ae.umbellulata*
	T.compactum	*Ae.umbellulata*
	T.macha	*Ae.umbellulata*
	T.spelta	*Ae.biuncialis*
Complete anther malformation and total pollen abortion	*T.aestivum* var.Chinese Spring, Norin 26, Salmon, Jones fife, Selkirk, *T.macha*, *T.sphaerococcum*, *T.spelta* var.*duhamelianum*	*Ae.caudata* or *Ae.syntriuncialis*
High anther malformation and total pollen abortion	*T.aestivum*	*Ae.columnaris*
	Norin 26	*Ae.ovata*
	T.compactum	*Ae.ovata*
50% Anther malformation complete pollen sterility	*T.aestivum* var erythrospermum, Salmon, Jones, fife, Selkirk	*Ae.columnaris*, *Ae.ovata*
	T.sphaerococcum, *T.spelta*	*Ae.columnaris*

Genotypes utilized; *T.compactum* cv. No.44, *T.macha* var.*subletschumicum*, *T.spelta* var.*duhamelianum*, *T.sphaerococcum* var.*rotundatum*

interactions and contribution of nuclear genes to haploid production, as cytoplasm is common in these alloplasmic wheats. In Salmon, genes controlling mst and haploid production may be closely placed, as the two phenomena are distinctly associated. However, haploid and twin seedling production, a result of g-c interaction, is not always associated with mst. Thus it is absent in many alloplasmic mst wheats (Kobayashi and Tsunewaki 1980). The twin seedlings in *Ae. caudata* salmon arise due to parthenogenetic development of egg cells followed by fer-

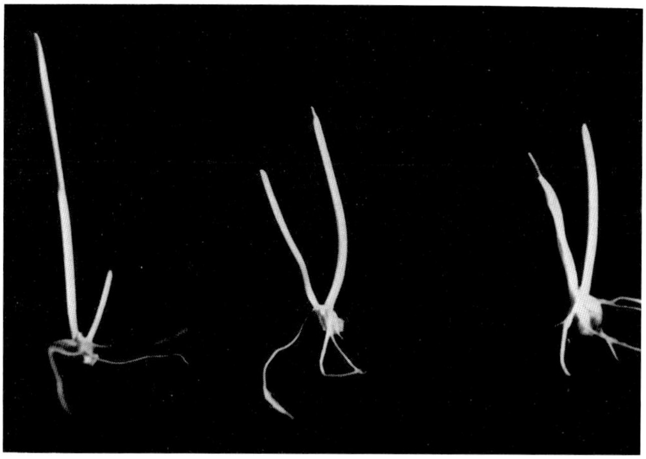

Fig. 27.16. *Triticum aestivum.* Twin seedlings from an alloplasmic 1B/1R translocation male sterile wheat line. (Mukai 1981)

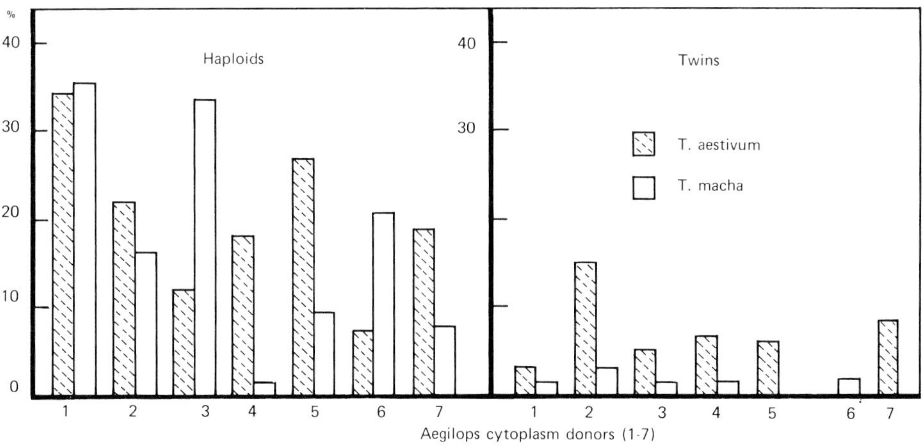

Fig. 27.17. Haploid and twin seedling production of gc-mst wheat having *Aegilops* cytoplasm. Nuclear genome of *T. aestivum* strain Salmon and *T. macha* var. subletschumicum. *Aegilops* cytoplasm donors *1–7* are: *Ae. biuncialis, Ae. caudata, Ae. columnaris, Ae. Kotschyi, Ae. triuncialis, Ae. umbellulata* and *Ae. variabilis*, respectively; + triplet seedlings also produced. (After Kobayashi and Tsunewaki 1980)

tilisation of one of the synergids (Tsunewaki et al. 1968). They likewise arise in four 1B/1R translocation wheat lines having *Ae. kotschyi* cytoplasm. Since all the diploid partners of n-2n twins have 1B/1R chromosome, egg cells with 1B/1R chromosome develop into haploid partners of these twins. This is confirmed by Tsumura and Mukai (unpublished).

iv. *Haploid Production and Translocation.* Bread wheat has two pairs of sat-chromosomes, 1B and 6B, but *T. aestivum* var. *salmon*, because of translocation, lacks the satellite of the 1B chromosome and possesses a rye-chromosome part in its place. One satellite chromosome is always present in nearly 90% of the haploids derived from *Ae. kotschyi* salmon progeny (Kobayashi and Tsunewaki 1980). Therefore, haploids appear to be derived only from those egg cells that have the 1B/1R chromosome (Fig. 27.16) and not from those with the normal 1B chromosome. This indicates that the presence of a translocated chromosome like 1B/1R may be a basic requisite for haploid production in certain alloplasmic wheats. If this is correct, some gene(s) on 1B/1R chromosome may be controlling haploid induction gametophytically but not sporophytically. However, the presence of the 1B/1R chromosome is a necessary but not sufficient condition for the egg cell to develop into parthenogenetic haploid embryos (Kobayashi and Tsunewaki 1980). For haploid and parthenogenesis frequency estimation, Mukai (1981) utilised five gc-mst bread wheat varieties having kotschyi cytoplasm, in each of which a part or a complete short arm of 1B chromosome was replaced by a segment of the 1R rye chromosome. These 1B/1R translocation lines were deficient in the NOR region of the 1B chromosome. Some of them also lacked the *Rfvi* locus responsible for mft restoration in mst wheats having *Ae. kotschyi* or *Ae. variabilis* cytoplasm. Of these, four 1B/1R alloplasmic wheat varieties produce 3–12% haploids. Since in all these translocation lines, haploid production is high, unlike in the ones lacking such translocations in which it is low or absent, the haploid production is associated with the translocation. The wheat *Ae. kotschyi* salmon remains heterozygous for 1B/1R chromosome due to preferential transmission of normal 1B chromosome. In it haploid parthenogenesis is associated with the presence of 1B/1R translocation chromosome in the egg cell, and it is induced gametophytically (Kobayashi and Tsunewaki 1980). Both heterozygotes and homozygotes of the 1B/1R chromosome develop parthenogenetic embryo in 47% and 87.5% of the ovules, respectively (Tsumura and Mukai unpublished). This indicates that haploid production is correlated with the translocation 1B/1R. Not a single haploid arose in 1150 offspring of *Rfvi Rfvi* homozygotes, but mst's having the *rfvi* gene produced exclusively haploids (Kobayashi and Tsunewaki 1980). This confirms the 1B/1R chromosome as the carrier of the *rfvi* gene essential for haploid production in the female and explains the high haploid frequency in *Ae. kotschyi* salmon plants in which the presence of the 1B/1R chromosome is the main genetic factor responsible for haploid production. However, this does not imply that the *rfvi* gene is directly responsible for haploid production, because *T. spelta* var. *duhamelianum*, which is known to have the *rfvi* gene (Mukai and Tsunewaki 1979), does not produce haploids in *Ae. kotschyi* cytoplasm. Therefore mst is not a necessary condition for haploid production in wheat.

27.20.1.2.2 Microsporogenesis and Tapetal Behaviour

Despite the occurrence of mst in many alloplasmic wheats, microsporogenesis and tapetal behaviour have been studied in only a few cases. Some of these are described below:

(Ae. ovata) T. aestivum. Chauhan and Singh (1966) found three types of abnormalities in these mst's viz. (a) in premeiotic stages, tapetal cells form periplasmodia which intermingle with PMC's whose walls are disorganised, (b) in central florets of spikelets, tapetal cells degenerate abruptly when PMC's are just in the premeiotic stage, PMC's degenerate thereafter, (c) in lateral florets, male meiosis proceeds normally, the tapetum degenerates prematurely, mature microspores lack cytoplasm and germ pore and finally degenerate. In two lines, male meiosis is normal. Microspores degenerate after liberation along with the tapetum in one and before tapetal breakdown in another line (this Vol.).

(Ae. caudata) T. durum. Male meiosis exhibits a few anomalies, but microspores degenerate abruptly after release (Herva's 1974). The mst is due to microspore abortion rather than to meiotic anomalies, as such anomalies occur in homoplasmic *T. durum* also.

(Ae. ovata) T. durum. Microsporogenesis proceeds normally until microspore formation, after which degeneration occurs after first pollen mitosis at one, two or three nuclear stage (Fig. 27.18, Fukasawa 1956b, 1959).

(T. compactum) T. aestivum. Highly abnormal microsporogenesis accompanied by microspore degeneration after liberation from tetrads occurs in this line (Xiangmin and Herong, 1981). In one such line, microspore degeneration is delayed until mitosis was completed.

(T. timopheevi) T. aestivum. In this male meiosis is normal but the microspores abort. Tapetal behaviour is variable, depending on the genotype of the nuclear donor. For instance, in the line having *T. aestivum* var. Nadadores genome, the

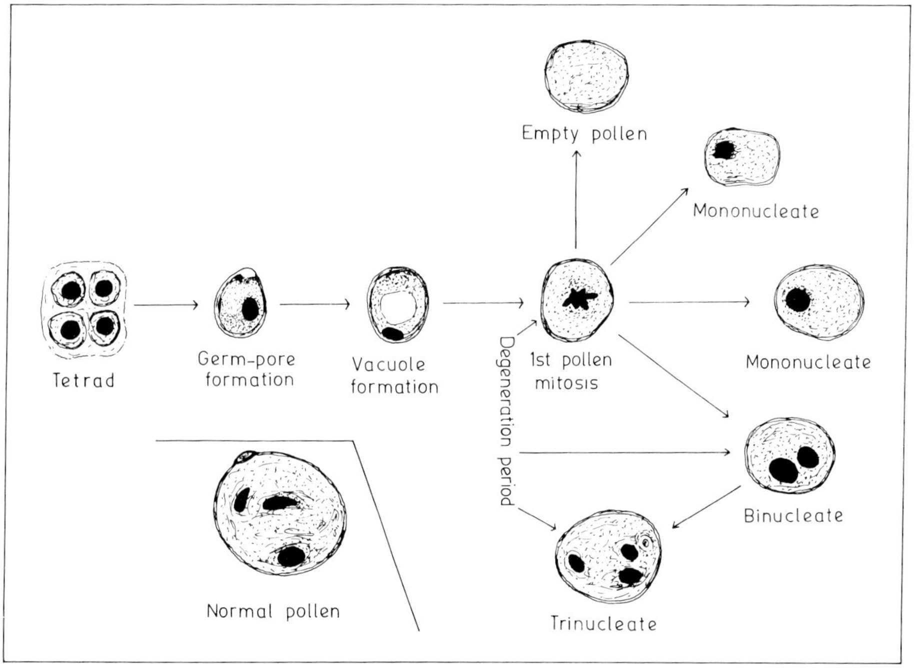

Fig. 27.18. *Triticum aestivum.* Stages of microspore and pollen abortion in alloplasmic male sterile wheat. (Fukasawa 1959)

tapetum is persistent (Jain *et al.* 1981), but it is plasmodial in those having the K_{7435} or WL_{711} genome. In the lines with the nuclei of HD_{200}, HD_{1593}, K_{7435}, WH_{157}, WL_{711} genomes, tapetal degeneration is followed or preceded by sporogenous tissue degeneration (Jain *et al.* 1981). A similar differential mode of degeneration occurs when the *T. aestivum* nucleus is placed in different cytoplasm donors of *T. timopheevi* (Stehno and Apltauerová 1980). In a Chinese alloplasmic wheat in which mst is controlled by a dominant *Fr* gene, male meiosis is normal but microspores degenerate immediately after liberation from the tetrads (Xiangmin and Herong 1981). The tapetum is persistent.

27.20.1.2.3 Grafting

By grafting the cut end of normal wheat with the cut end of mst wheat having *Ae. caudata* cytoplasm and of that having *T. timopheevi* cytoplasm, Zeven (1967) attempted to transmit mst to normal wheat. The sterility was not transferred across the grafts, as was confirmed by Lacadena (1968). Heat treatments for the inactivation of viruses failed to "cure" mst plants (Zeven 1967). To explore the possibility of asexual transmission of *c*-genes, Lacadena (1968) dissected embryos and endosperms of normal and mst wheat (*T. timopheevi*) *T. aestivum* and these were interchanged, but no influence of mst endosperm on normal embryos or of normal endosperm on mst embryos occurred.

27.20.1.2.4 Environmental Influence

Like other alloplasmic mst's, some mst wheats are highly stable and others sensitive to environment. Over a period of 30 years, mst is stable and is maintained in (*Ae. caudata*) *T. aestivum* plants. Likewise it is stable in (*T. timopheevi*) *T. aestivum*. This was tested by Baier et al. (1978) by growing 12 such lines at 15 locations for two generations in Germany. The mst was complete and stable in all these. Placing *T. aestivum* or *T. durum* genomes in *T. boeoticum* cytoplasm produces stable mst alloplasmic plants, but in *Ae. umbellulata* or *Secale cereale* cytoplasms, the plants are mst in autumn but p-mst in spring when grown in the glasshouse (Maan 1973a). *T. aestivum* var. Selkirk or Chinese spring in *T. timopheevi* are completely mst in the glasshouse in autumn but mft in growth chambers under a 12-h photoperiod. Both high temperature and long photoperiods, given especially during early meiosis, induce pollen abortion in wheat. This influence is more profound in p-mst's in which the environmental manipulations lead to the production of one or two grains per ear to seed set in over 50% of the ears.

Like mst, *Fr* genes are sensitive to some environmental factors. Thus fertility restoration is influenced by location (Monteagudo et al. 1967, Lucken and Maan 1967, Wilson 1968b, Schmidt et al. 1970), photoperiod (Welsch and Klatt 1971), day temperature (Johnson and Patterson 1973) and planting date in which photoperiod and temperature are effective. Long photoperiods or high temperatures enhance pollen sterility in many restored lines. Thus with increase in the latitude towards the north, the photoperiod increases which induces mst in the restored lines. Wilson (1968b) introduced the terms "deep-sterile", "sterile" and "shallow-sterile" to describe environments in which progressively greater degrees of restoration were expressed, and suggested that in the "deep-sterile" environment even normal wheat varieties have a tendency to ear-tip sterility. This environment is associated with long photoperiod and short growing season and corresponds areas of high latitude north.

27.20.1.2.5 Biochemical Differences

The mst anthers have deficient sugars, especially sucrose (Fukasawa et al. 1957, Erickson 1967, Savchenko et al. 1968), reduced succinic dehydrogenase activity (Fukasawa 1957b, 1959, 1961) and low proline but high asparagine, aspartic acid, glutamic acid and alanine contents (Fukasawa 1954, Kihara 1966a,b, Erickson 1967, Savchenko et al. 1968, Rai and Stoskopf 1974, Zhu and Tsao 1984). In addition, mft and mst anthers exhibit significant differences in isoenzyme patterns of peroxidase, esterase and leucine aminopeptidase (Hohler and Borner 1980). In mst anthers, conversion of proline into glutamic and aspartic acid occurs, leading to proline deficiency. However, the addition of proline before pollen mitosis does not prevent pollen abortion. Fukasawa et al. (1957) induced microspore abortion in cut culms of *T. durum* kept in darkness before onset of meiosis. Addition of 5% sucrose to the culture medium prevents this abortion. Accordingly, when alloplasmic (*Ae. ovata*) *T. durum* culms were grown in a culture medium to which 5% sucrose was added, microsporogenesis improved, but young pollen aborted gradually and mst was complete. Hence proline deficiency or sugar starvation is not the cause but the consequence of mst in wheat. Furthermore, a significant increase in kernel shrivelling, alpha-amylase activity, sprouting and protein content occurs in mst wheats compared to their fertile counterparts (Doig et al. 1975, Miller et al. 1975, Clayton and Ellis 1976, Jönsson 1976). All these alterations are caused by differential g-c interactions, as ordinarily no two cytoplasms would interact with a given genome in exactly the same way for all characters except sterility and fertility. This is substantiated by esterase isozyme variations in anthers and seeds of different mst bread wheat varieties having *Ae. comosa* cytoplasm (Karim et al. 1984).

27.20.1.2.6 Fertility Restoration

i. *Wheat Genome.* For the economic production of a hybrid variety, the two basic essential pre-requisites are gc-mst and fertility restoration. Immediately after development of alloplasmic mst bread wheat (*T. timopheevi*) *T. aestivum*, its fertility restoration was achieved by Schmidt *et al.* (1962) from a derivative of *T. timopheevi*. Later, many such lines, the majority of which are derivatives of *T. timopheevi*, were found to restore fertility to mst wheats. Since *T. timopheevi* was an S-cytoplasm contributor and this species and its derivatives had *Fr* genes, it was thought that *Fr* genes were present exclusively in this cytoplasm donor species. It was, however, soon realised that the genome of the cytoplasm donor species did not represent the only *Fr* gene source, as species other than cytoplasm donors could also act as restorers. Thus, besides *T. timopheevi*, the cytoplasm donor species considered as the best source of S-cytoplasm and *Fr* genes, Joppa and McNeal (1969), Tahir (1969) and Miri et al. (1970b) found that the nucleus of *T. aestivum* genome also has *Fr* genes. Of the restoration sources found among 6x wheats, Primepi (Oehler and Ingold 1966a,b) and *T. spelta* var. *duhamelianum* (Kihara and Tsunewaki 1964) are the best sources. Although both these sources give complete restoration, in some, F_1 fertility is influenced by genotype and environment. Common wheat fertility restorers for *T. timopheevi* cytoplasm are listed by Wilson and Driscoll (1983). The effectiveness of these sources varies considerably, although direct comparisons are not available. Most of these are inferior to Primepi and none gives complete restoration in a wide array of genotypes and environments. Investigations on fertility restoration have revealed that while cytoplasm donor species are the best source of *Fr* genes, these genes are also present in species contributing cytoplasm or genome to the alloplasmic mst wheat. Furthermore, different *Fr* gene sources restore fertility to the same g-c combination and some of them almost to the same extent. In addition, the same *Fr* gene source restores fertility to different g-c combinations, especially to those differing in cytoplasms. Thus *T. aestivum* var. *Primepi* restores fertility to gc-mst plants having *T. timonovum*, *T. timopheevi* or *Ae. ventricosa* cytoplasm. Likewise, *T. compactum* restores fertility to gc-mst plants having *Ae. caudata* and *Ae. umbellulata* cytoplasm (Table 27.13). However, whereas some sources restore fertility to all gc-mst combinations, no one alone restores fertility in all the combinations. In other words, there is no single universal restorer line for all the gc-mst wheats.

In mst (*Ae. caudata*) *T. aestivum* plants, fertility restoration by a gene for black-ear from the *Ae. caudata* genome is partial (Kihara 1951). It is about 85% by pollinating (*Ae. ovata*) *T. durum* wheat by *T. dicoccoides* var. *kotschyanum* pollen (Fukasawa 1958a). Likewise fertility restoraton in many other cases is incomplete or partial (Table 27.13). Moreover, in some it is restored in basal florets but apical florets remain mst. Even within a floret, fertile and sterile anthers are sometimes developed and within an anther a varying proportion of viable and nonviable pollen may be developed. Talaat (1969) suggested variable penetrance and expressivity of *Fr* genes in mst wheat. The control of fertility restoration by one major gene (Mukai and Tsunewaki 1979) or two major genes with the influence of modifiers (Livers 1964a, Wilson 1968b, Skurygina 1970, Miri et al. 1970b, Odenbach 1970, Schmidt et al. 1970) or by three major genes (Bajwa and Lucken 1968, McCuiston 1968) is well known in wheat. In all cases, the influence of environment on the degree of restoration is high. In mst Redcoat wheat × R5 crosses, where three *Fr* genes are required for fertility restoration in the greenhouse, only two are needed in the field (Wickersham and Patterson 1980). Two such genes, Rfc_2 and Rfc_3, act as duplicate dominant, two, Rfu_1, Rfu_2, complementary (Tsunewaki 1974) and one, Rfc_2, is almost recessive (Mukai and Tsunewaki 1980a).

ii. *Rye Genome.* Porter and Tuleen (1972a,b) evidenced the existence of *fr* genes in rye that restore mft in mst (*T. aestivum*) *T. timopheevi* wheat plants as two spontaneously arisen partially mft 56 chromosome amphidiploids were produced on mst wheat/rye F_1's after pollination by rye. Additional evidence of restoration from rye came from a relatively high frequency of partially fertile F_1's obtained

Table 27.13. Fertility restoration in male sterile wheat having *T. aestivum* or *T. durum* genome[a]

Cytoplasm source	Fertility restoration Gene source	Partial	Complete	References
Triticum				
T.araraticum	*T.aestivum*		+	Maan and Lucken (1972)
	R1, R2, R3, R4			
T.boeoticum	*T.aestivum* R5, R6		+	Maan and Lucken (1972)
T.dicoccoides var.	*T.aestivum*			
nudiglumis	R1, R2, R3, R4		+	Maan and Lucken (1972)
T.monococcum	*T.durum*, *T.aestivum*		+	Maan (1973d)
T.timonovum	*T.aestivum* Primepi		+	Drogacheva (1972)
T.timopheevi	*T.spelta*			Oehler and Ingold (1966b)
	T.aestivum Primepi		+	Maan and Lucken (1972)
	T.aestivum			
	R1, R2, R3, R4, R5, R6		+	Maan (1973d)
T.zhukovskyi	*T.aestivum*			
	R1, R2, R3, R4		+	Maan and Lucken (1972)
Aegilops spp.	*T.aestivum* R5, R6		+	Maan (1973d)
Ae.caudata	R5, R6	+		Kihara (1968b)
	T.compactum-44	+	+	Kihara (1968b)
	T.aestivum P168	+		Kihara (1968b)
Ae.caudata (F1)	ABD-13	+		Kihara and Tsunewaki (1964)
Ae.longissima	ABD-13	+		Tahir (1969)
Ae.ovata	ABD-13	+		Kihara and Tsunewaki (1964)
	P-168	+		Kihara (1968b)
	R1, R2, R3, R4	+		Kihara (1968b)
Ae.sharonensis	*T.aestivum* R5, R6		+	Maan and Lucken (1972)
Ae.speltoides	R1, R2, R3, R4		+	Maan and Lucken (1972)
Ae.umbellulata	R5, R6	+		Maan (1973d)
	T.compactum 44		+	Kihara (1968b)
	T.aestivum	+		Kihara 1968b)
	P-168	+		Kihara 1968b)
	T.aestivum			
	erythrospermum	+		Kihara (1968b)
Ae.ventricosa	*T.aestivum* Primepi		+	Oehler and Ingold (1966b)
	T.aestivum		+	Maan and Lucken (1971b)
	R4		+	Maan (1972)

[a]After Sage (1976); for other *fr* gene sources refer Table 1 of Wilson and Driscoll (1983).

from crosses of mst wheat-rye F_1's × 6x Triticale. Of 121 F_1s, 70 plants were p-mft. These investigations reveal that one or more chromosomes of rye bear *fr* genes, which either alone or in combination with the *T. aestivum* genome and *T. timopheevi* cytoplasm restore mft. The exact gene number and mode of action are not known.

27.20.1.2.7 Chromosomal Control

By backcrossing amphidiploid *Ae. ovata* × *T. durum* hybrid to *Ae. ovata* and *T. durum*, Fukasawa (1953, 1955) developed the combinations (*Ae. ovata*) *Ae. ovata* and (*T. durum*) *Ae. ovata*. While the former combination plants resembled the natural *Ae. ovata* and were mft, the latter were mostly mst and morphologically similar to *T. durum* because of its genome. During such substitution BC's, some alloplasmic plants having *Ae. ovata* cytoplasm + *T. durum* genome (28 chromosomes) and in addition one specific chromosome from *Ae. ovata* were mft. Their pollen had either 14 or 15 chromosomes, but both were functional. Thus mst was sporophytic and mft in such plants was revoked by one *Ae. ovata* chromosome. This chromosome was specific, as plants having several other *Ae. ovata* chromosomes were mst. Therefore, *Fr* gene(s) was inferred to lie on this specific chromosome. Likewise, alloplasmic *T. aestivum* and *T. durum* lines with *Ae. sharonensis* cytoplasm were mst in the absence of a specific *Ae. sharonensis* chromosome (Maan 1975b) and mft in its presence. Therefore, this chromosome has *Fr* gene(s). This specificity led to chromosome identification and *Fr* gene location over such chromosomes. So far 13 *Fr* genes have been located, six lie on chromosomes of homoeologous group 1, three on group 6, two on group 7, one each on chromosome 5 and 2 (Table 27.14). All the restorer lines derived from *T. timopheevi* and its close relatives carry one of their *Fr* genes on chromosome 1A. Other effective *Fr* genes for *T. timopheevi* cytoplasm lie on chromosome 1B of Primepi and *T. spelta*. In some g-c combinations, *Fr* genes on chromosomes 1B, 6D and 7D are inhibitory, while in others they promote fertility. Genes Rfv_1 and Rf_3 located on the short arm of chromosome 1B (Tahir and Tsunewaki 1969) act in the opposite manner in the alloplasmic combinations (*T. timopheevi*) *T. aestivum*, and (*Ae. kotschyi*) *T. aestivum* as regards fertility restoration. Among 21 chromosomal pairs in common wheat, four pairs are satellited, of which two are prominent (1B and 6B) and two faint (1A and 5D). Eight of the 13 *Fr* genes are located on NOR chromosomes (Table 27.14). This indicates that satellite genes, usually known to control metabolic activity of the nucleolar organiser, may also be involved in mst control. The same is valid for alloplasmic mst tobacco, in which the NOR region is implicated in mst.

27.20.1.2.8 Gene Control

In (*T. timopheevi*) *T. aestivum* mst wheat, fertility restoration by the hexaploid wheat *T. aestivum* Primepi and *T. spelta* var. *duhamelianum* was considered by Goujon and Ingold (1967b) and Tahir and Tsunewaki (1967), to be caused by a single dominant *Fr* gene. Later, it was found by Miller and Schmidt (1970) and Bahl

Table 27.14. Chromosomal locations of *fr* genes in male sterile wheat

No.	Gene designation	Chromosome	Carrier genotypes[a]	Cytoplasms	References
1.	Rf1	1A[b]	R1, R2, R3, R4, R5	*T.timopheevi*	Robertson and Curtis (1967) Talat (1969), Bahl and Maan (1973)
	Rf1	1A[b]	R-D, R-K	*T.timopheevi*	Yen et al. (1969)
2.	Rfu1	1B[b]	CS	*Ae.umbellulata*	Tsunewaki (1974)
3.	Rfc1	1C[b]	P168	*Ae.biuncialis*	Tsunewaki et al. (1978)
	Rfc1	1C[b]	P168	*Ae.caudata*	Tahir and Tsunewaki (1971a)
	Rfc1	1C[b]	P168	*Ae.columnaris*	Tsunewaki et al. (1978)
	Rfc1	1C[b]	P168	*Ae.ovata*	Tahir and Tsunewaki (1971a)
	Rfc1	1C[b]	P168	*Ae.triaristata*	Tsunewaki et al. (1978)
	Rfc1	1C[b]	P168	*Ae.triuncialis*	Tsunewaki et al. (1978)
	Rfc1	1C[b]	P168	*Ae.umbellulata*	Tsunewaki et al. (1978)
4.	Rfv1	1B[b]	CS	*Ae.kotschyi*	Tsunewaki (1980a)
	Rfv1	1B[b]	CS	*Ae.variabilis*	Tsunewaki (1980a)
5.	Rf2	7D	R1, R2, R3, R4, R5	*T.timopheevi*	Bahl and Maan (1973)
	Rf2	7D	R-D	*T.timopheevi*	Yen et al. (1969)
6.	Rfc2	6B[b]	Cmp	*Ae.caudata*	Tsunewaki (1974)
7.	Rfu2	2B	CS	*Ae.umbellulata*	Tsunewaki (1974)
8.	Rf3	1B[b]	Splt	*T.timopheevi*	Tahir and Tsunewaki (1969)
	Rf3	1B[b]	Primepi	*T.timopheevi*	Bahl and Maan (1973)
9.	Rfc3	1D	Cmp	*Ae.caudata*	Tsunewaki (1974)
10.	Rf4	6B[b]	R-C, R-K	*T.timopheevi*	Yen et al. (1969)
	Rf4	6B[b]	RZ	*T.timopheevi*	Bahl and Maan (1973)
11.	Rf5	6D	R-C	*T.timopheevi*	Yen et al. (1969)
12.	Rf6	5D[b]	Primepi	*T.timopheevi*	Bahl and Maan (1973)
13.	Rf7	7B	R4	*T.timopheevi*	Bahl and Maan (1973)

Abbreviations used: P168 = *T. aestivum* strain, CS = *T. aestivum*, Chinese spring, Cmp = *T. compactum*, Splt = *T. specta* var. duhamelianum, Primepi = *T. aestivum* var.

[a]After Tsunewaki (1980a).

[b]Satellited chromosomes.

and Maan (1973) that Primepi has two *Fr* genes. Each of these two genes is supposed to be incompletely dominant, one with major and the other with minor effect (Miller and Schmidt 1970). The existence of two complementary genes Rf_1 and Rf_2 (Fr_1 and Fr_2) derived from *T. timopheevi* genome conditioning restoration in hexaploid wheat having *T. timopheevi* cytoplasm was documented by Livers (1964a). Restoration is partial when one of the genes is present. Double recessiveness caused completed mst. Restoration controlled by two dominant *Fr* genes in conjunction with one recessive epistatic gene was found by Nettevich and Naumov (1970) in some mst wheats. Shebeski (1971) found restoration conditioned primarily by one major and one minor dominant gene and influenced by one inhibitor and one modifier gene. In (*Ae. ovata*) *T. durum* mst wheat, restoration by *T. spelta* var. *kotschyanum* is controlled by two major genes (Fukasawa 1958a). One more major gene to this system was added by Gilmore (1970). With *T. dicoccoides*, the existence of two *Fr* genes and one modifier is evident from F_3 and F_4 segregations. Likewise, restoration in (*Ae. ventricosa*) *T. aestivum* plants brought about by Primepi and Palmaress wheats is not conditioned by one (Ingold 1968) but by two *Fr* genes, whose action is influenced by one modifier gene and temperature during flowering.

The restorer genes may be dominant (with complete or partial penetrance) recessive, cumulative, inhibitory, epistatic or complementary in their action. Their expression is influenced by female parent, modifying genes, environment and genotype-environment interaction. Due to the tremendous environmental influence over *Fr* gene action, they appear polygenic in some, especially in those where increase in number of BC's to restorer parents enhances mft expression. Some *Fr* genes are more polyvalent than others, some are versatile, others conservative and effective only in limited gc-mst genotypes. The *Fr* genes exhibit high genome-cytoplasm-environment interaction. This makes fertility restoration highly sensitive, so that fertility-favouring environments enhance *Fr* gene penetrance, while others reduce its effectiveness. In such cases, another *Fr* gene may be required to compensate the environmental effect and restore full fertility. Therefore, restoration genetics and *Fr* gene action in wheat is relative to environmental and genetical background. This is also vivid from the graphical illustration given by Jost (1979, 1980, 1981).

27.20.1.2.9 Hormonal Control

Sage (1972) proposed that a gene-mediated hormonal threshold concentration is required for fertility restoration. This could explain genotypic differences, gene dosage effects, different effectiveness of *Fr* genes and the influence of photoperiod, temperature and water stress over fertility restoration, as these factors can more easily affect the functioning of a hormonal restoration substance than the expression of genes themselves. This hypothesis also explains the basal fertility and apical sterility of partially restored wheat lines by assuming that the chemical substance is translocated acropetally to the ears and the threshold or critical concentration of this chemical does not reach the top and the florets remain mst. This hormone has, however, not been isolated so far.

27.20.1.2.10 *Cytoplasm Differentiation*

Stable mst unaccompanied by any major deleterious phenotypic effect results when *T. aestivum* genome is placed in *Ae. speltoides* or in *T. timopheevi* cytoplasms. Thus *T. aestivum* does not differentiate between the cytoplasm of these two genera, but the two cytoplasms can be differentiated when the *T. timopheevi* genome is placed in its own and in the *Ae. speltoides* cytoplasm. The former combination plants are normal and mft, the latter are weak and mst. This differential behaviour is due to nuclear gene differences existing between genomes of the two wheat species, *T. aestivum* and *T. timopheevi*. The *T. timopheevi* nuclear genes differentiate their own and *Ae. speltoides* cytoplasms and are considered by Maan and Lucken (1972) to have "cytoplasm-differentiating genes". Most of these are identified and investigated for their role in fertility restoration and at least 13 such genes are known (Table 27.14). On the other hand, placing one genome in many cytoplasms and observing phenotypic differences in the resulting g-c combinations differentiates "genome-differentiating cytoplasm-genes" and cytoplasm types. This is well known in wheat *T. aestivum* and *T. durum*, in which cytoplasms like *Ae. bicornis*, *Ae. variabilis*, *Ae. umbellulata* produce mst accompanied by loss of vigour. Both these features occur in alloplasmic *T. durum* having *Ae. longissima*, *Ae. sharonensis*, *Ae. cylindrica*, *Ae. heldreichii* or *T. monococcum*, but mst coupled with normal vigour occurs when the *T. aestivum* genome is placed in these cytoplasms. Based on phenotypic effects like growth inhibitions, variegation, anther malformations, pistilloidy and mst, 53 cytoplasms of 34 *Aegilops* and *Triticum* species were differentiated into 12 major groups by Tsunewaki et al. (1980a,b,c), using genetic distance, cluster and principal component analysis. The species within each group show the same reaction to "cytoplasm-differentiating genes" and consequently their cytoplasms are considered as similar. The following four cytoplasm groups are suitable for hybrid wheat development:

1. *T. dicoccoides* var. nudiglumis and var. spontanesvillosum *T. timopheevi*, *T. zhukovskyi*.
2. *Ae. aucheri*, *Ae. speltoides*.
3. *T. araraticum*.
4. *Ae. caudata* var. typica, *Ae. triuncialis*.

The cytoplasm differentiation is maximum at diploid level and decreases progressively and rapidly with ploidy increase. In this respect, nuclear-gene and cytoplasmic-gene diversities resemble each other fully. Like nuclear gene differentiation, in facultative outcrossing species intraspecific cytoplasm differentiation occurs at the diploid level. Based on the genetic similarity of cytoplasms of genomically related diploid and polyploid *Aegilops* and *Triticum* species (Kihara and Tanaka 1970), Tsunewaki et al. (1978, 1980c) traced cytoplasmic descent from their putative ancestral diploid species (Fig.27.19). On this basis wheat owes its cytoplasm to *Ae. speltoides* and the D-genome to *Ae. squarrosa*. However, mt-DNA of wheat and *Ae. squarrosa* differ from each other, unlike cp-DNA (Vedal *et al.* 1978). This difference could be due to mt-DNA mutation and/or recombinations, such as are known in yeast and tobacco in *Nicotiana* (Fonty *et al.* 1978, Belliard *et al.* 1979).

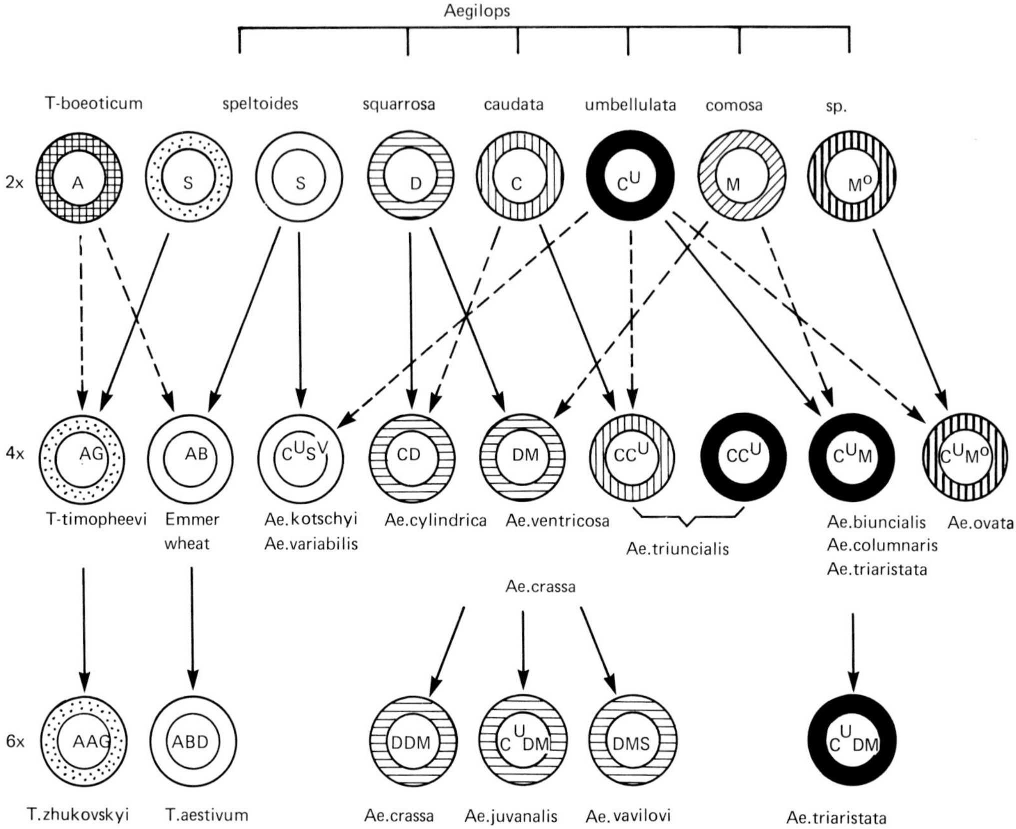

Fig. 27.19. Genome cytoplasm descent in some *Aegilops* and *Triticum* species Genome and cytoplasm donor ⟶ Genome donor only - - -› For 6x only ♀ parent indicated. (After Mukai et al. 1978)

27.20.1.3 Nongenetic Male Sterility

Certain chemical and temperature treatments or growing plants on copper-deficient soils induce mst in wheat, but most of these impair female fertility, reduce seed yield and affect plant morphology as well. This is evident from the following:

i. *Copper Deficiency.* Plants grown on Cu-deficient soils develop mini anthers with abortive, starchless pollen (Graham 1975). With higher deficiency either no flowers or staminodes are developed (Dell 1981). In mild Cu-deficient plants, endothecium development is impeded or inhibited. This prevents anther dehiscence and leads to functional mst. Grain development is highly impaired and about 66% grain yield reduction occurs in Cu-deficient plants. Therefore this type of induced sterility has no commercial breeding value.

ii. *Temperature Treatments.* In this cool season crop, high temperature given during sporogenesis diminish gametic fertility. For instance, subjecting *T. aestivum* cv. Gabo to 30°C for 3 days at the time of meiosis led to anther indehiscence and from a species that is unselected, unadapted, not commercially used in advanced agriculture and has chromosome number and homology different to wheat (Wilson and Driscoll 1983).

iii. *Ae. speltoides*. Ghiasi and Lucken (1982a) compared fertility restoration and agronomic performance of segregating gc-mst bread wheat populations having *T. timopheevi* and *Ae. speltoides* cytoplasms. No significant differences in F_1 seed set and F_3 seed yield occurred between these two cytoplasm plants. However, bread wheat populations with *T. timopheevi* cytoplasm have 10% higher mean anther extrusion than those with *Ae. speltoides* cytoplasm. *T. timopheevi*-derived restorer genes confer similar levels of mft to common wheats with the cytoplasms of either *T. timopheevi* or *Ae. speltoides*. Both the cytoplasms have similar influence on heading date, plant height, grain yield, test weight, kernel weight and grain protein percentage. The results with *Ae. speltoides* cytoplasm apply to its use in R-line development and not to its use in A-lines. Since both the cytoplasms responded similarly to many *fr* genes commonly used in hybrid wheat breeding, *Ae. speltoides* cytoplasm may interchangeably be used with *T. timopheevi* cytoplasm in restorer breeding while broadening the cytoplasm base of breeding lines of bread wheat.

iv. *Ae. kotschyi* and *Ae. variabilis*. Mukai and Tsunewaki (1979, 1980a) obtained mst by placing genomes of *T. aestivum* strain Salmon, *T. spelta* var. *dehamelianum* or *T. macha* var. *subletschchumicum*, fertility restoration being conditioned by a dominant gene Rfv_1 in *Ae. kotschyi* or *Ae. variabilis* cytoplasms. These alloplasmic lines have no deleterious effects on major agronomic traits of wheat and resemble *T. timopheevi* cytoplasm in this respect. Selfed seed fertility of F_1 hybrids equals those lines with *T. timopheevi* cytoplasm. The seed fertility of F_1 hybrids having the Rfv_1 gene in the heterozygous condition nearly equals that of the restorer lines homozygous for *Fr* genes against *T. timopheevi* cytoplasm. Thus the restoration capacity of a single Rfv_1 gene against *Ae. kotschyi* or *Ae. variabilis* cytoplasm is higher than that of two Rf_3 gene doses against *T. timopheevi* cytoplasm (even with two such gene doses restoration is incomplete) and nearly equals three *Fr* genes in homozygous condition in *T. timopheevi* cytoplasm. Since selection of plants having multiple-fertility-restoring genes is difficult, the RfV_1 gene system, utilizing the above-mentioned two cytoplasms, has an added advantage in hybrid wheat breeding over that of the (*T. timopheevi*) *T. aestivum* wheat lines, as most of the common wheat cultivars serve as restorers for *Ae. kotschyi* and *Ae. variabilis* mst and maintainer lines are to be bred instead of the mst and restorer lines required in (*T. timopheevi*) *T. aestivum* combinations. Mukai and Tsunewaki (1979) proposed a method of maintaining and using the mst lines in wheat breeding on a system almost similar to one proposed by Franckowiak et al. (1976), who tried to utilise *Ae. squarrosa* cytoplasm for hybrid wheat breeding.

v. *Other Cytoplasms*. In search of new sources from *Ae. aucheri*, *Ae. triaristata* (Panaytov and Gostov 1973) and *Ae. biuncialis*, *Ae. columnaris* (Popov et al. 1974) have male sterilizing ability but their use in hybrid wheat production is limited due to some undesirable features associated with mst.

27.20.1.5.2 Pollination Limitations

Although the floral structure of wheat may be oriented towards cross pollination (Wilson 1968a), in its present form it is more suited for self-pollination. Variability in floral characteristics favouring some wind-outcrossing exists in wheat (De Vries 1971), but selection of wheat lines with widely opened lemma and palae and with large-sized fully exposed and long-duration receptive stigma suited for cross-pollinating ability is tedious and time-consuming. Therefore, improvement, although necessary, will be slow. In the male parents, lines with large-sized and highly extruded anthers having pollen more viable and with increased longevity are to be bred. Many of these traits are difficult to measure. Consequently, anther extrusion, which is easily observable, has received the most attention. NCP on mst wheats is influenced by environment, male and female genotype and mechanics of seed production. Moderate temperature, high relative humidity, increased soil moisture and adequate air movement enhance NCP and hybrid seed yield on mst wheats. Most of these factors improve anther extrusion, stigma receptivity and pollen longevity (Imrie 1966, D'Souza 1970). Rainfall during flowering reduces pollen distribution. Sixfold difference in grain yield occurred on mst lines grown for 3 years at five locations in N. Dakota (Miller and Lucken 1976). The yield differences are due to variable NCP over mst's in these locations. Devices to enhance NCP over mst's have been suggested. Due to the size and weight of wheat pollen, better NCP occurs with tall males and short females. Therefore Lelley (1966) suggested the use of tall pollinator (C-lines) for mst (A-lines) which preferably may be awnless, and/or open lodiculed (Watson and McWhirter 1968). Wilson et al. (1980) suggested that full-dwarf × semi-dwarf hybrids would have ideal height and yield potential. These, and full-dwarf × tall wheat hybrids, may have a seed production advantage over semi-dwarf × semi-dwarf hybrids. Maintaining a 1:1 A- and C-line ratio increased NCP to 72.6% (Johnson and Patterson 1980). Certain planting arrangements in Kyoto, Japan enhanced NCP by 50% over mst wheats (Mukai and Tsunewaki 1980a). Thus the mechanics of seed production, the ideal ratios of mst to pollinator, the optimum width of mst strips, the best seed rates and planting times will vary according to the type of cross (A × B) or A × C), the characteristics of the parents and the location (Wilson and Driscoll 1983).

27.20.1.5.3 Restoration Difficulty

Coupled with some side effects, a few of which are deleterious, fertility restoration is genetically highly complex, controlled by many genes and influenced by genomic background of male-female parents, modifying and inhibitory genes and by environment. This renders the development of lines with efficient and complete restoration difficult. In addition to a long period involved in sterilising a line, the rareness of satisfactory *Fr* genes limits hybrid wheat development considerably. By transferring *T. timopheevi* cytoplasm into 25 Japanese wheat lines, mst lines were developed by Mukai and Tsunewaki (1980a). Fertility restoration in all of them is incomplete, even though Rf_3 gene was maintained in double doses. In many wheats, floret fertility and grain set in lower spikelets of F_1 hybrids (A × C) is good, but poor in upper spikelets, which are effectively mst. Thus grain yield of many F_1

wheat hybrids is often limited by reduced mft (Done and Whittington 1978). Hence, improvement of C- and A-lines so that they possess good restoration potential is highly vital. Since fertility restoration is not uniformly similar, incomplete fertility restoration in hybrid is a more common problem in soft red winter and hard red spring wheats than in hard red winter wheats. A common set of hybrids grown in several countries exhibits a high degree of instability and environmental sensitivity associated with fertility restoration (Jost 1979, 1980, 1981), so that seed set varies from country to country. It is, however, not clear which environmental component causes the variation.

27.20.1.5.4 Seed Set Variability

Wheat, a self-pollinator, depends on wind-blown pollen for outcrossing. The NCP is limited by factors described earlier that directly determine seed set over mst's. In addition to these, the seed set is influenced by the genotype of male and female parents. This is evidenced from the results of Kherde et al. (1964), who determined seed set over mst wheats having *Ae. caudata*, *Ae. ovata* and *T. timopheevi* cytoplasms planted between rows of 45 different pollinators. Whereas 20–30% seed set occurred on mst's having *Ae. caudata* and *T. timopheevi* cytoplasms, only 1–5% seeds were set on those with *Ae. ovata* cytoplasm. Under unlimited pollen supply, seed set over *Ae. ovata* and *T. timopheevi* cytoplasm mst wheats averaged 29 and 35%, respectively (Livers 1964b). Significant seed setting differences occurred between various mst lines having *T. timopheevi* cytoplasm (Mukai and Tsunewaki 1980b). Likewise, these differences are high between mst plants pollinated by different pollinators. Therefore, for optimum seed set, proper selection of male and female lines is essential. This involves the characteristics of parents mentioned earlier.

27.20.1.5.5 Sprouting

This involves coleoptile and radicle growth or endosperm autolysis, and occurs in some hybrid wheats developed after using mst lines. This is absent in the analogous B lines (Done and Whittington 1978) or nonhybrid wheats. Therefore, selection of nonsprouting female lines is vital before these are used in commercial hybrid seed production.

27.20.1.5.6 Genic Male Sterility

The long period involved in sterilizing a line and the rareness of satisfactory restorer genes have prompted search for g-mst in wheat and its possible role in hybrid wheat development. However, maintaining gc-mst stocks and the need of roguing in seed production are two major problems limiting the use of gc-mst in hybrid wheat production. Because of simple genetic control and complete mst expression, single-gene English and Russian g-mst wheat mutants are being incorporated in wheat breeding programs, especially for heterosis breeding or for selection of crosses in traditional line breeding. Since specific *Fr* genes are not required, it is possible to employ a partial diallel technique effectively by using the *ms* genes. If a *ms* gene could cause sterility in a specific genome cytoplasm combination, as in two barley varieties, Betzes and Penrad, in which the gene causes mst when incorporated in the Betzes genome but not in the Penrad genome (Pfeiffer 1972), the necessity of introducing the restorer gene in the male parents of potential hybrids is obviated. Sage (1976) suggested that a similar system in wheat would cheapen the cost of the parental line and hybrid wheat development. Such a satisfactory system has not been found so far in wheat and will involve efforts especially as only a few monogenic recessive mst mutants are available in wheat.

27.20.1.5.7 Chromosomal Male Sterility

Driscoll (1972) proposed this system which requires three lines, X, Y and Z, all of which are homozygous for *ms* allele. X- and Y-lines carry an extra chromosome in 2 and 1 dose, respectively. The extra chromosome bears a replacement gene(s) for mft (*Ms*) which does not cross over to the normal wheat chromosome. An introduced alien rye chromosome is involved in producing the female parent of the hybrid, but is not involved in the female parent or in the hybrid crop. Hence its influence on the agronomic aspects of a hybrid crop is not a consideration (Wilson and Driscoll 1983). Fertility restoration is done by normal wheat chromosomes. In addition to 21 pairs of wheat chromosomes, the X-line has a pair of alien chromosomes; it does not have any chromosomal anomalies and breeds true. On the other hand, the Y-line with one alien chromosome produces two types of eggs and pollen, one with and the other without the alien chromosome. Whereas both types of egg function, only pollen without alien chromosome functions because of certation. Thus two types of selfed offspring, three fertile Y plants and one sterile Z plant are produced. In management, as proposed by Wilson and Driscoll (1983), this system is first primed by using selfed progeny of Y plants and allowing the Z segregants to be pollinated by Y segregants. Seeds from these will yield Z plants only. These are planted in alternating strips, (A) with X seed, the resulting seed from Z will be Y, and (B) with Y seed, the seed from Z will be Z seed (Fig. 27.20). Hence theoretically it is possible to produce an unlimited amount of homogeneous mst plants after a small initial amount of Z seed is obtained. Those which contain no alien chromosomes represent female parents and are planted in alternating strips with normal wheat, which is the male parent (step C, Fig. 27.20). The chromosomal system, although it involves many steps before hybrid seed production, has certain advantages over the gc-mst system (Driscoll 1973). In the gc-mst system, the deficiency of fertility restoration affects the commercial crop, but in the chromosomal system it affects the hybrid seed production stage, which is less serious. Unlike in the gc-mst system, in which the male parent used for hybrid seed production has *Fr* genes, in the chromosomal system these are not especially needed. Hence in the chromosomal system male parents can be selected on the basis of the amount of pollen shed and the combining ability. However, the possible genetic leakages in X and Z lines warrant caution in using the chromosomal system. The leakage involving male transmission of the alien chromosome from Y plants may result in few mft plants in the female parent and some inbred and some mst

Step	Female parent		Male parent		Seed
(A)	Z	x	X	→	Y
(B)	Z	x	Y	→	Z[a]
(C)	Z	x	Normal variety	→	Hybrid seed

[a] Can be recycled through steps A and B

Fig. 27.20. Steps in hybrid wheat production. (After Driscoll 1974b)

plants in the hybrid crop. Due to their limited number, hybrid performance should not be affected.

The Cornerstone mst (Driscoll 1977) is a satisfactory mutant for producing hybrid wheat when using the XYZ nuclear mst system initially proposed by Driscoll (1972). Until recently, the second component of this system, an alien chromosome that compensates for Cornerstone mst has been difficult to find. However, with the recent discovery of a latent mft restoration gene(s) located on Cornerstone chromosome 4-A (Islam and Driscoll 1984), which is unable to restore fertility to its parental genome but restores mft in mst wheats having one modified barley chromosome, 4 Hm, production of hybrid wheat should be possible. Thus the HmY line (Islam and Driscoll 1984) and a 2-RSY line (Hossain and Driscoll 1983) are the two fertile Y lines useful in producing hybrid wheat.

27.20.1.5.8 Hybrid Wheat Performance

A survey by Briggle (1963) indicated positive heterosis for grain yield ranging over 50 to 100% in wheat. The F_1 seed was produced by hand pollination of emasculated ears grown in the glasshouse or well-tended fields with few replications. Therefore the assessment of the yield potential of commercial F_1 hybrids becomes doubtful (Rajki and Rajki 1966, Johnson and Schmidt 1968); but the promise of high yield levels inspired breeders to use mst for hybrid breeding. With the discovery of the *T. timopheevi* mst-mft system, commercial hybrid wheat breeding was visualised. Initially, experiments were conducted in small plots at suboptimal planting rates with limited and arbitrary parental selection. Since the number of tests was small, the assessment of genotype-environment interactions was not possible. The results were variable but the heterotic effect was discernible (Briggle 1963, Wienhues 1968, Livers and Heyne 1968, Wells and Lay 1970, Bhatt 1971, Bitzer et al. 1971, Bitzer and Fu 1972, Zeven 1972b, Cregan and Busch 1978). Four-year results of Livers mst source was used by Jones and Everett (1949), Rogers (1950a,b), Jones and Mangelsdorf (1951) and Rogers and Edwardson (1952) for studies and in hybrid seed production. Proof that the sterility was of gc-mst type was provided initially by Josephson and Jenkins (1948), Jones and Everett (1949) and Rogers (1950a), and was confirmed by many others later (Duvick and Noble 1978).

ii. *S-Cytoplasm*. Also known as USDA sterile cytoplasm, this was discovered by MT Jenkins of USDA from a strain of Teopod maize and developed for commercial use by Jones et al. (1957a,b). This source was distinguished from T-cytoplasm by its fertility restoration property, as T- and S-cytoplasms were restored by different *Fr* genes.

iii. *C-Cytoplasm*. After discovering S- and T-cytoplasms, breeders tested other maize cytoplasms using the restoration techniques. Whereas some maize lines resembled S- or T-cytoplasm in restoration behaviour, a few resembled neither of these two cytoplasms and belonged to another cytoplasm detected by Beckett (1971) from Brazilian maize, Charrua and RB sources. This was designated C-cytoplasm. Thus, maize possesses three major types of sterile cytoplasm, T, C and S, in addition to those which do not fall under any of these classifications. These three cytoplasms are distinct from each other, and each exhibits certain unique features at phenotypic, genotypic and molecular levels. These features detailed below and summarised in Table 27.18, have permitted their differentiation and grouping.

Table 27.18. Chromosomal location of male sterile genes in maize

Chromosome	Male sterile genes
1	ms_9, ms_{12}, ms_{14}, ms_{17}, *ad1*, *an1*, *an2*, *pa*, *ts2*, ts_3, *vg*,
2	*ts*
3	ms_3
4	*lp1*, *Sp*, *whp 1*
5	ms_5, ms_{13}
6	ms_1, ms_6, *po1*, *si*, *Si2*, Si_3, *ms*
7	ms_7, *tp*, *va1*
8	ms_8
9	ms_2
10	ms_{10}, ms_{11}, *Sp2*, *ms*

27.21.2.1.2 Differentiation Methods

N-,T-, C- and S-cytoplasms are distinguished at the phenotypic, genotypic and biochemical level, and each cytoplasm type exhibits a high degree of constancy and specificity in diverse genomic backgrounds (Table 27.19). This is clear from the following account.

Table 27.19. Distinguishing features of three sterile cytoplasm types in maize

Trait	Sterile cytoplasm types T	S	C
Anther exsertion	Nil	Sometimes exserted	Nil to little
Anther shape, dehiscence	Small, shrunken indehiscent	Normal to small, indehiscent	Shrunken, indehiscent
Microsporogenesis	Normal	Normal	Disturbed
Pollen	Transparent, degenerated	Yellow, shrivelled, adhered to walls	Aborted
Tapetal form and function	Abnormal	Normal	Abnormal
fr gene action	During early prophase-I, TI, TII, or microspore formation	During pollen maturation	After microspore development
Tapetal mitochondrial degeneration during *fr* gene action	Yes	No	No
Fertility restoration	Sporophytic, Fr_1, Fr_2	Gametophytic	Sporophytic Fr_4, Fr_5, Fr_6
Fr gene number, chromosomal location	2 (Fr_1, Fr_2), 2, 9	1 (Fr_3), 2	2–3[Fr_4, Fr_5, Fr_6 (?)] 1 (?)
HmT susceptibility	Yes	No	No
Methomyl sensitivity	Yes	No	No
Sterility	Stable	Unstable	Stable
Spontaneous reversion	Low	High	Low
Mitochondrial DNA differences	One linear episome 2 kb	Two linear episomes 5.2 and 6.2 kb	Two circular DNA's 1.57 and 1.42 kb
Translation products (protein) specific polypeptides of mol. wt:			
a) 13,000[a]	Present	Absent	Absent
b) 17,500[b]	–	Absent	Present
c) 21,000	Absent	Present	Present
d) 58,000–88,000[b]	Absent	Present	Absent

Synthesis suppressed
[a]or not suppressed
[b]by the corresponding *Fr* genes.

27.21.2.1.2.1 Fertility Restoration

The characterization of a cytoplasm is done by scoring the pollen fertility in the progeny from crosses of cytoplasmic sources and tester stocks containing *Fr* genes. This classical field method involves crossing of each cytoplasmic source with pollen from many inbred lines and backcrossing repeatedly. These inbreds restore fertility with some cytoplasmic sources but not with others. As each cytoplasm type is distinguishable on the basis of specific *Fr* genes, sets of inbred lines which contain various combinations of *Fr* genes are commonly utilised for identification of gc-mst type. On the basis of fertility restoration, Duvick (1965) found two sterile cyto-

T-sterility. The restorer line for this is rare and fertility is controlled by $fr_1\,fr_1$, $Fr_2\,Fr_2$, $Fr_3\,Fr_3$.

Heavy nitrogen fertilization enhances pollen fertility (Duvick 1959). While some partial restorers are at such a low level of restoration that they are sterile under all except the most propitious environments (cool, humid), others are fully as fertile as their genetic equivalents in normal cytoplasm in all except the most severe environments (dry, hot). Therefore, in the absence of defined environments, classification of a genotype as restorer, partial restorer or nonrestorer (sterile) for the T-cytoplasm becomes ambigous. Furthermore, the Fr_1 locus appears to possess over eight alleles having different restoring ability grades, and thus the existence of a multiple allelic series at the *fr* locus seems to be certain. However, while Blickenstaff *et al.* (1958) considered the involvement of several genes in partial fertility restoration, Stead (1960) regards restoration as due to two complementary genes different from Fr_1. Since restoration is considerably influenced by microclimatic conditions, its genetics has become difficult to understand. Thus due to high environmental interactions of partial restorer genotypes, a sharp threshold exists which, if passed, renders partial restorer nearly pollen fertile and, if missed, makes the genotype sterile. Usually, T-cytoplasm plants with heterozygous *Fr* genes are mst, but those having $Fr_1\,fr$ genes with proper modifiers and environmental conditions can be rendered fully fertile; their anthers are completely turgid and exserted like those of normal fertile plants.

S-Cytoplasm. Whereas the complementary action of two dominant genes, Fr_1 and Fr_2, restores fertility to T-cytoplasm plants, only one gene, Fr_3 (Buchert 1961, Duvick 1965), mapped on chromosome 2 (Laughnan and Gabay 1975b) restores fertility in S-cytoplasm plants. Unlike T-cytoplasm plants, in which fertility restoration is sporophytic, in the S-cytoplasm it is gametophytic. The phenotype is determined by the genotype of the pollen grain and not by the sporophyte. These S-cytoplasm plants with genotype Fr_3Fr_3 are mft, those with genotype fr_3fr_3 are mst and plants that are heterozygous fr_3fr_3 produce one-half fertile and one-half aborted pollen; fertile pollen has gene Fr_3 and aborted grains the fr_3 gene. Some inbred lines restore only certain sources of S-cytoplasm but not others (Gracen and Grogen 1974). Cytoplasm B, D, ME and MY often give fertility restoration responses different from the typical S-restoration pattern. Also BC's involving W, J, R, M, VG, and CA cytoplasms with certain inbred lines exhibit segregations (Josephson et al. 1978), suggesting the involvement of more than one *Fr* gene for the S-cytoplasm. Therefore, it appears that for some S-cytoplasm sources, either genes other than Fr_3 are necessary for restoration or multiple alleles that have slightly different fertility restoration effects occur at the Fr_3 locus. Sisco (1982) found an inbred line that restored fertility to all the unknown gc-mst maize plants with S-cytoplasm, and suggested that it has duplicate dominant restorer genes for the cytoplasm. Differences in fertility restoration and mitochondrial analysis permitted division of 25 S-cytoplasm maize accessions into five subgroups (Sisco et al. 1985). The fertility of each subgroup was quantified by pollen morphology analysis of the S-cytoplasm plants heterozygous for the *Fr* gene. LBN S-cytoplasm plants possess two double-standard RNA's of apparent molecular weight 1.9 and 0.5×10^6 (Sisco *et al.* 1984). These RNA's are associated with mitochondria and are

Table 27.19. Distinguishing features of three sterile cytoplasm types in maize

Trait	Sterile cytoplasm types T	S	C
Anther exsertion	Nil	Sometimes exserted	Nil to little
Anther shape, dehiscence	Small, shrunken indehiscent	Normal to small, indehiscent	Shrunken, indehiscent
Microsporogenesis	Normal	Normal	Disturbed
Pollen	Transparent, degenerated	Yellow, shrivelled, adhered to walls	Aborted
Tapetal form and function	Abnormal	Normal	Abnormal
fr gene action	During early prophase-I, TI, TII, or microspore formation	During pollen maturation	After microspore development
Tapetal mitochondrial degeneration during *fr* gene action	Yes	No	No
Fertility restoration	Sporophytic, Fr_1, Fr_2	Gametophytic	Sporophytic Fr_4, Fr_5, Fr_6
Fr gene number, chromosomal location	2 (Fr_1, Fr_2), 2, 9	1 (Fr_3), 2	2–3[Fr_4, Fr_5, Fr_6 (?)] 1 (?)
HmT susceptibility	Yes	No	No
Methomyl sensitivity	Yes	No	No
Sterility	Stable	Unstable	Stable
Spontaneous reversion	Low	High	Low
Mitochondrial DNA differences	One linear episome 2 kb	Two linear episomes 5.2 and 6.2 kb	Two circular DNA's 1.57 and 1.42 kb
Translation products (protein) specific polypeptides of mol. wt:			
a) 13,000[a]	Present	Absent	Absent
b) 17,500[b]	–	Absent	Present
c) 21,000	Absent	Present	Present
d) 58,000–88,000[b]	Absent	Present	Absent

Synthesis suppressed
[a]or not suppressed
[b]by the corresponding *Fr* genes.

27.21.2.1.2.1 Fertility Restoration

The characterization of a cytoplasm is done by scoring the pollen fertility in the progeny from crosses of cytoplasmic sources and tester stocks containing *Fr* genes. This classical field method involves crossing of each cytoplasmic source with pollen from many inbred lines and backcrossing repeatedly. These inbreds restore fertility with some cytoplasmic sources but not with others. As each cytoplasm type is distinguishable on the basis of specific *Fr* genes, sets of inbred lines which contain various combinations of *Fr* genes are commonly utilised for identification of gc-mst type. On the basis of fertility restoration, Duvick (1965) found two sterile cytoplasms, the Texas(T) and USDA (S) cytoplasm. Beckett (1971) and Gracen and Grogan (1974) surveyed over 30 mst sources and added a third type of sterile cytoplasm, the Charrua or C-cytoplasm. In the following, brief description of each of sterile cytoplasm type identified basically on fertility restoration is given.

i. *T-Cytoplasm.* The T-cytoplasm first discovered by Rogers and Edwardson (1952) represents a relatively stable mst source and includes P, Q, HA, RS and SC sources of cytoplasms (Gracen 1982). These mst strains were detected from independent sources. All these are relatively stable but are susceptible to race T- of *Helminthosporium maydis.* Fertility restoration is conditioned by two genes *Rf* (≡Fr_1) and Rf_2 (≡Fr_2), both of which must be present for pollen fertility restoration. However, to ensure complete fertility, the presence of one or more modifying genes is also necessary.

ii. *C-Cytoplasm.* Originally this contained Charrua and RB sources (Gracen and Grogan 1974) isolated from a Brazilian maize (Beckett 1971). In recent years, it has been expanded to include El Salvador (ES) source, Bb and PR (Parana) sources from Brazil, and E source (Gracen 1982). Others, like Cuarentenos, Italy IB, IR-1, IR-2 and P1 also appear to posses C-type cytoplasms. Fertility restoration is sporophytic, conditioned by a single gene Fr_4 (≡Rf_4 Kheyr-Pour *et al.* 1981). The occurrence of three or more genes Rf_4,Rf_5, Rf_6 (≡Fr_4, Fr_5, Fr_6), is postulated by Josephson et al. (1978) and Gracen (1982). Laughnan et al. (unpublished) have shown the existence of two duplicate independent *Fr* genes.

iii. *S-Cytoplasm.* This is the most common source of sterile cytoplasm and sources like Ca, F, G, H, I, IA, J, K, L, LF, M, ME, ML, MY, PS, R, SD, TA, VG and W have been included in this group (Gracen et al. 1979). Whereas L, ME, M and MY differ from S to some extent, cytoplasms ME and MY often give fertility restoration responses different from the typical S-restoration pattern. This led Kálmán and Dévényi (1982) to subgroup S-cytoplasms, using principle component analysis. In terms of stability of mst over various environments and towards the nuclear fertility restorer genes, of all the cytoplasmic sources, S-cytoplasm is the most diverse and versatile. It is unique in fertility restoration which is conditioned by one gene, Rf_3 (Fr_3), and is gametophytic, determined by the pollen genotype and not by the sporophyte. A gc-mst maize line AW 182BN (L) having S-cytoplasm and designated LBN, possesses two unique RNA bands of 2900 and 840 base pairs (Sisco et al. 1982, Sisco 1982). These double-stranded RNA's are seed-transmitted and correlated with an extra degree of sterility. This maize line has no exserted anthers and is fully mst, unlike others that exsert anthers and produce little viable pollen. Thus this represents a special S-cytoplasm maize line that has two unique RNA bands and mst of extreme order. Instability of the S-cytoplasm is known in a number of instances (Singh and Laughnan 1972). These have their origin through mutations at either the cytoplasm or nuclear gene level (Singh and Laughnan 1972, Laughnan and Gabay 1973, 1978). They represent first-documented mutations in nuclear *fr* genes in maize. These apparent S-cytoplasm *fr* mutant genes can be mapped to several different chromosomes (Laughnan and Gabay 1973). Except one, all of these mutant *fr* genes are associated with deleterious phenotypic side effects with reduced survival, selection and fitness values.

iv. *Unclassified Cytoplasms.* Cytoplasms B, CH, D, EP, LP, NT, OY, SG, TC and V have not been definitively classified by the fertility restoration technique (Gracen et al. 1979). Of these NY, OY and SG are fertile in most inbred backgrounds and their classification in T, C- and S-cytoplasms is not possible. A cytoplasm that resembles both T and C types is designated the TC source.

v. *Iojap Strain.* The nuclear gene ij in a homozygous state induces plastid alterations (Jenkins 1924) that cause many variable white stripes on leaves. These alterations are inherited in a uniparental mode and are independent of the genetic constitution of the nucleus (Rhoades 1943). The gene *ij* conditions loss of chlorophyll in sectors in maize (Rhoades 1946). Although some striping tendency continues to be inherited maternally, the majority of the striped individuals derived by the *ij*-effect transmit an entirely green or entirely white character to the progeny. The induction of mst by the mutation of this gene was reported by Rhoades (1950), but Duvick (1965) deduced convincing evidence which indicates that the sterility might have been derived from a pre-existing *fr* genes + S-cytoplasm (s) combination in the iojap strain.

Mutation of N- to S-cytoplasm occurred in F_2 of a cross between an inbred R 181 and a gene stock carrying the recessive allele *iojap* (Lemke *et al.* 1986). The mt-DNA mst's have S-1 and S-2 episomes and the mutated S-cytoplasm fits in the CA subgroup of S-type maize cytoplasm.

27.21.2.1.2.2 Genetic Control

T-Cytoplasm. Two dominant genes *Rf* ($\equiv Fr_1$) and Rf_2 ($\equiv Fr_2$), located on chromosome 3 and 9, respectively, restore the pollen fertility of the T-cytoplasm (Duvick et al. 1961, Snyder and Duvick 1969). Being complementary in action, both of them are required for fertility restoration. Thus T-cytoplasm plants with Fr_1 and Fr_2 are restored male fertiles, but those Fr_2 Fr_2 or Fr_1 Fr_1 or fr_2fr_2 or fr_1fr_1 are nonrestored mst. For fertility restoration, both the dominant restorer genes, either in homozygous or in heterozygous form, must be present at both loci. The restoration is therefore sporophytic, as all the pollen of S-cytoplasm plants with genes Fr_1fr_1, Fr_2fr_2 is fertile, although only one quarter of the pollen grains from such a plant contain both the restoring alleles Fr_1 and Fr_2. The restorer genes appear to have other functions as well. This is evidenced from the observations in which T-cytoplasm plants that had been fully restored to fertility exhibited a susceptibility to pathotoxin that was intermediate between normal and nonrestored cms-T plants (Watrud et al. 1975, Barratt and Flavell 1975). Many dominant genes ranging in restoration capacity from weak to strong exist at the *Fr* locus (Duvick 1965). For complete fertility restoration in all the environments, a strong allele at the *Fr* locus must be accompanied by an unknown number of modifier genes. Segregation data obtained from crosses between 28 T-mst lines and 66 N-mft lines led Chen (1981) to postulate mst to be of three types. Type I is characterized by an easy fertility restoration, fertility being controlled by the genes Fr_1 Fr_1. Type II has a stable sterility which is not easily restored, fertility is conditioned by two gene pairs Fr_1Fr_1 Fr_2 Fr_2. Type III is the stable typical

T-sterility. The restorer line for this is rare and fertility is controlled by $fr_1\,fr_1$, $Fr_2\,Fr_2$, $Fr_3\,Fr_3$.

Heavy nitrogen fertilization enhances pollen fertility (Duvick 1959). While some partial restorers are at such a low level of restoration that they are sterile under all except the most propitious environments (cool, humid), others are fully as fertile as their genetic equivalents in normal cytoplasm in all except the most severe environments (dry, hot). Therefore, in the absence of defined environments, classification of a genotype as restorer, partial restorer or nonrestorer (sterile) for the T-cytoplasm becomes ambigous. Furthermore, the Fr_1 locus appears to possess over eight alleles having different restoring ability grades, and thus the existence of a multiple allelic series at the *fr* locus seems to be certain. However, while Blickenstaff *et al.* (1958) considered the involvement of several genes in partial fertility restoration, Stead (1960) regards restoration as due to two complementary genes different from Fr_1. Since restoration is considerably influenced by microclimatic conditions, its genetics has become difficult to understand. Thus due to high environmental interactions of partial restorer genotypes, a sharp threshold exists which, if passed, renders partial restorer nearly pollen fertile and, if missed, makes the genotype sterile. Usually, T-cytoplasm plants with heterozygous *Fr* genes are mst, but those having $Fr_1\,fr$ genes with proper modifiers and environmental conditions can be rendered fully fertile; their anthers are completely turgid and exserted like those of normal fertile plants.

S-Cytoplasm. Whereas the complementary action of two dominant genes, Fr_1 and Fr_2, restores fertility to T-cytoplasm plants, only one gene, Fr_3 (Buchert 1961, Duvick 1965), mapped on chromosome 2 (Laughnan and Gabay 1975b) restores fertility in S-cytoplasm plants. Unlike T-cytoplasm plants, in which fertility restoration is sporophytic, in the S-cytoplasm it is gametophytic. The phenotype is determined by the genotype of the pollen grain and not by the sporophyte. These S-cytoplasm plants with genotype Fr_3Fr_3 are mft, those with genotype fr_3fr_3 are mst and plants that are heterozygous fr_3fr_3 produce one-half fertile and one-half aborted pollen; fertile pollen has gene Fr_3 and aborted grains the fr_3 gene. Some inbred lines restore only certain sources of S-cytoplasm but not others (Gracen and Grogen 1974). Cytoplasm B, D, ME and MY often give fertility restoration responses different from the typical S-restoration pattern. Also BC's involving W, J, R, M, VG, and CA cytoplasms with certain inbred lines exhibit segregations (Josephson et al. 1978), suggesting the involvement of more than one *Fr* gene for the S-cytoplasm. Therefore, it appears that for some S-cytoplasm sources, either genes other than Fr_3 are necessary for restoration or multiple alleles that have slightly different fertility restoration effects occur at the Fr_3 locus. Sisco (1982) found an inbred line that restored fertility to all the unknown gc-mst maize plants with S-cytoplasm, and suggested that it has duplicate dominant restorer genes for the cytoplasm. Differences in fertility restoration and mitochondrial analysis permitted division of 25 S-cytoplasm maize accessions into five subgroups (Sisco et al. 1985). The fertility of each subgroup was quantified by pollen morphology analysis of the S-cytoplasm plants heterozygous for the *Fr* gene. LBN S-cytoplasm plants possess two double-standard RNA's of apparent molecular weight 1.9 and 0.5×10^6 (Sisco *et al.* 1984). These RNA's are associated with mitochondria and are maternally transmitted. The nuclear genotype influences the amount of recoverable ds-RNA's.

C-Cytoplasm. Since the existence of C-cytoplasm has recently been fully established, the gene(s) necessary for its fertility restoration different from T and S cytoplasms have not been clearly identified so far. Like T-, and unlike S-cytoplasms, restoration of the C-cytoplasm is sporophytic. Many established maize inbred lines restore the C-cytoplasm, but chromosome location of the *Fr* gene(s) is not documented. Segregations in certain F_2 and BC progenies led Kheyr-Pour et al. (1981) to suggest one gene, Rf_4 ($\equiv Fr_4$) as involved in fertility restoration of C, RB, PR, ES and Bb cytoplasms. However, an allied study by Josephson et al. (1978) indicates involvement of three or more genes, Fr_4, Fr_5 and Fr_6. Studies of Laughnan, Gabay-Laughnan and Patterson (unpublished) involving C-mst restorer strains and series of translocation testers indicate *Fr* gene linkage with both T 1–9 and T 2–9. This suggests the existence of two *Fr* genes, one on chromosome 1 and the other on chromosome 2 of maize. These two genes exhibit duplicate but not complementary (unlike Fr_1 and Fr_2) gene action. Many mst lines with C-cytoplasm are relatively stable in some inbred backgrounds, but some combinations exhibit mst breakdown after 7–10 days of silking. This late breaking characteristic is influenced by some nuclear genes other than the Fr_4 allele (Gracen 1982). Evidence favours the involvement of a transposable element in S-cytoplasm fertility restoration. This restorer element appears to occupy other sites, in addition to the standard one on chromosome 2.

27.21.2.1.2.3 Anther and Pollen Phenotype

T- and C-cytoplasm plant anthers are mostly shrunken and indehiscent, but those of S-cytoplasm are nearly normal and indehiscent (Fig. 27.23). T-cytoplasm plants do not exsert their anthers at all, but those with S-cytoplasm do so in some cases. However, like typical mst maize anthers, such anthers have a characteristic slender, stiff, needlelike appearance and contain only aborted pollen grains. Partially fertile plants with T-cytoplasm exsert their anthers but have a characteristic twisted, misshapen appearance, due to the presence of fertile pollen in some anther regions and sterile grains in others. When these anthers are arranged in order of decreasing pollen fertility, they became progressively shorter and slimmer and attain finally a needlelike appearance, lodging mostly aborted pollen (Duvick 1965). Partially fertile plants with S-cytoplasm also exsert their anthers, but these are uniformly smooth and are not twisted or misshaped like those of T-cytoplasm plants. Whereas mature T-anthers have few degenerated pollen grains, no pollen is present in C-anthers. S-anthers have aborted pollen grains which are shrivelled, yellow and frequently adhere to the walls of normal pollen grains that may be present in the same anther. The pollen has a low level of starch grain formation (Koncz et al. 1980). Because of the gametophytic nature of the sterility, S-cytoplasm plants with $Fr_3 fr_3$ genes have 50% aborted pollen grains. Their anthers are exserted but are slimmer, reduced in size and have a pebbly appearance due to the random distribution of aborted, shrivelled and plump-fertile pollen grains within the

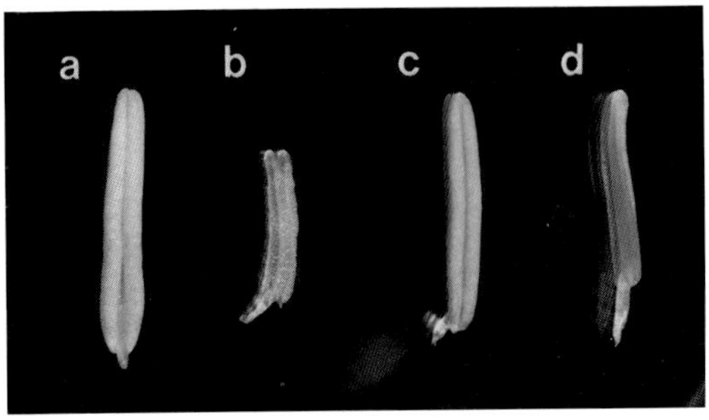

Fig. 27.23a-d. *Zea mays.* **a-d** Mature anthers of male fertile N- **(a)**, and male sterile C- **(b)**, S- **(c)** and T- **(d)** cytoplasm type maize plants. (Lee et al. 1979)

anther. Thus it is easy to differentiate Fr_3 Fr_3 from $Fr_3 fr_3$ plants by merely looking for the absence or presence of a pebbly surface on the anthers. The pebbly appearance is identical to that in anthers of plants that are heterozygous for reciprocal translocations and possess 50% sterile pollen grains. However, their pollen, unlike in S-anthers, does not adhere to the walls.

27.21.2.1.2.4 Helminthosporium Susceptibility

The causal organism of southern corn leaf blight disease is *Helminthosporium maydis* race T (HmT), which causes the disease by producing a specific HmT toxin. Both HmT and its toxin affect T-cytoplasm maize plants only and not those with N-, C- or S-cytoplasms (Hooker et al. 1970, Smith et al. 1970). The fungus causes lesions, chlorosis and necrosis of leaves and chlorotic streaks in adult plants and leads to numerous other morphological, ultrastructural, physiological and biochemical alterations, as reviewed by Gregory et al. (1977). Unlike in N-mitochondria, in T-mitochondria the HmT pathotoxin causes the uncoupling of oxidative phosphorylation, inhibition of oxogluterate oxidation and irreversible swelling in KCl medium, but has no effect on N-mitochondria (Miller and Koeppe 1971). The toxin induces activation of cytochrome oxidase and succinate cytochrome C-reductase and suppresses malate and succinate inhibition of ATP-ase in T- and not in N-mitochondria isolated from etiolated maize seedlings (Peterson et al. 1975). In isolated T-mitochondria, the toxin stimulates succinate and NADH respiration, increases mitochondrial swelling and causes loss or matrix density and unfolding of the inner membrance (Gregory et al. 1980). Since all these pathotoxin-induced multiple mitochondrial anomalies are cytoplasm-specific and occur consistently in T- and not in N-cytoplasm plants, they could form the basis for T-cytoplasm identification. These assays are sensitive, require little toxin and mitochondrial substrate, and do not necessitate waiting until the pollen formation stage (Peterson et al. 1974). They do, however, require technical skill and laboratory facilities, and therefore cannot be used for the isolation of T-cytoplasm plants in extensive worldwide maize breeding programs. Moreover, a distinction between other maize cytoplasm types (C, S, N) cannot be made by this method.

27.21.2.1.2.5 Methomyl Sensitivity

When maize seedlings with N-, C- S- or T-cytoplasm and their maintainers are treated with 300 ppm methomyl, only the inbreds and the maintainers with T-cytoplasm were severely affected (Ohmasa 1982). Seedling growth of T-cytoplasm WF9 maize inbred is reduced by 50% after 500 ppm chemical treatment, the reduction in N-cytoplasm is about 3%. Thus methomyl sensitivity may differentiate T-cytoplasm maize lines from others, even though variations in sensitivity degree exist among T-cytoplasm lines.

27.21.2.1.2.6 Molecular Basis

The classification of maize cytoplasms into N, T, C and S types on a molecular basis rests on mitochondrial DNA (mt-DNA) differences detected by agarose gel electrophoresis (Kemble et al. 1980), restriction endonuclease banding patterns (Levings and Pring 1976, Pring and Levings 1978) and mitochondrial translation products (Forde et al. 1978, 1980). The presence of low molecular weight DNA species (S1 and S2 molecules) characterises the S-cytoplasm (Pring et al. 1977).

i. *Restriction Endonuclease mt-DNA Fragment Analysis*. Following endonuclease digestion and agarose gel electrophoresis, Levings and Pring (1976) found that mt-DNA from N- and T-cytoplasm maize plants exhibit different banding patterns. Later, utilising the same procedure and using four different restriction enzymes, Hind III, EcoRI, BamI and Sal-I, that cleave double-stranded DNA at sequence-specific sites, distinct fragment patterns from the four cytoplasms N, T, C and S were obtained by Levings and Pring (1976) and Pring and Levings (1978). These authors studied mt-DNA in several different nuclear backgrounds. Restriction pattern differences were consistent, irrespective of the genetic backgrounds. Investigations of the mt-DNA fragment pattern of over 25 mst maize lines whose cytoplasm type had already been determined through the fertility restoration procedure led the authors to conclude that each cytoplasm type was associated with a unique mt-DNA and the classification of maize cytoplasm by the restriction endonuclease fragment technique matched that obtained by fertility restoration method. Using this technique Kálman and Palagyi (1980) and Koncz et al. (1980) also classified maize cytoplasms, the classification likewise matching that of fertility restoration. Each cytoplasm type gives a specific fragment pattern, readily distinguishable irrespective of the endonuclease employed (Pring and Levings 1978). The specificity is because each of these four cytoplasms possesses a unique mt-DNA complement. The restriction patterns produced are complex, but some of these are specific for each cytoplasm type. The analysis of 50 cytoplasm lines revealed mt-DNA heterogeneity in them, but Pring et al. (1980) differentiated the lines into three subgroups on the basis of the fragment pattern. Koncz et al. (1980) isolated mt-DNA's of mst maize inbred lines having T-, C- or S- cytoplasms and of the N-cytoplasm lines, and digested them with restriction endonucleases BamHI, SalI, Hind III, Pst I, EcoRI, XhoI and Bg III. The mt-DNA of each cytoplasm type exhibited significant heterogeneity. The mt-DNA restriction endonuclease digestion produced different, complex, characteristic and reproducible fragment patterns in agarose gel electrophoresis. The average molecular weight is between 140-160 MDa. The homologous fragments represent about 65-80% of the molecular weight of the mt-DNA's and the number of common bands between the mt-DNA's of different cytoplasms varies with each enzyme used. Of the various enzymes, restriction endonuclease XhoI is useful in comparing mt-DNA fragment patterns of different cytoplasm lines of maize.

ii. *Mitochondrial Translation Products.* This technique, involving the use of mitochondria isolated from 20–30 g etiolated maize shoots, incubated in vitro with 35S-methionine and of sodium dodecyl sulphate polyacrylamide gel electrophoresis, was employed by Forde et al. (1978), Forde and Leaver (1980). The mitochondria produced radioactively labelled translation products, the polypeptides of which were characteristic for each maize cytoplasm type (Forde et al. 1980). From the electrophoretic analysis of the translation products of mitochondria, the authors classified 32 cytoplasmic sources into N, T, C and S groups (Fig. 27.24). This grouping is in accordance with the classification made by Beckett (1971) and Gracen and Grogan (1974) for these cytoplasms, using fertility restoration methods, and with the conclusions drawn by Kemble et al. (1980) However, unlike the observations of Levings and Pring (1977) and Pring et al. (1979a,b), no within-cytoplasm heterogeneity in mitochondrial translation products was detected by Forde et al. (1980). From analysis of isolated mitochondria of T-

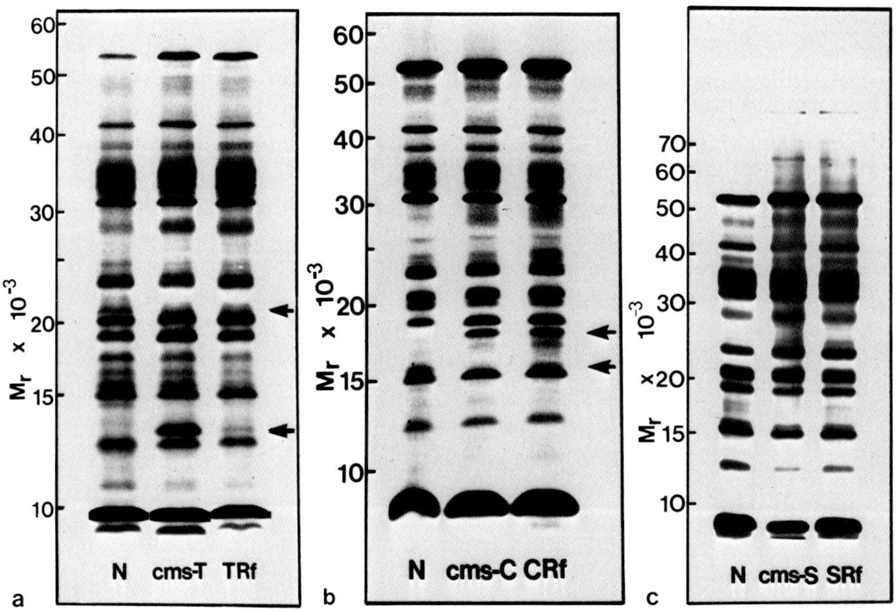

Fig. 27.24a-c. *Zea mays.* **a-c** Polyacrylamide gels of mitochondrial polypeptides from normal (*N*) and male sterile T-, C-, and S-cytoplasms and their respective restored (*Rf*) lines of maize. *Arrows* indicate polypeptides not common to the sterile cytoplasm lines. (Forde and Leaver 1980)

cytoplasm lines differing in the presence or absence of *Fr* genes that suppress the mst phenotype, Forde and Leaver (1980) found that *Fr* genes specifially suppress 13,000 M_r variant polypeptide synthesis. A 21,000 M_r polypeptide synthesized by N-mitochondria was absent among the translation products of T mitochondria from either restored or nonrestored lines. The mitochondria from S-cytoplasm plants synthesise a group of minor polypeptides ranging in molecular weight from 58,000 to 88,000. Such polypeptides are absent in N, T or C. Thus the four cytoplasm lines differ from each other in mitochondrial translation products. In addition, the frequency classes and contour lengths of mt-DNA of S- and T-cytoplasm plants differ from those of N-cytoplasm plants (Tables 27.19 to 27.23).

iii. *Agarose Gel Electrophoresis.* Mitochondrial DNA preparations obtained from 5-g etiolated seedling shoots of 31 maize lines were fractioned on agarose gels by Kemble et al. (1980). Although a number of discrete low molecular weight bands were present in all these lines, only four different DNA banding patterns were found by the authors. These correlated with N-, T-, C- and S-cytoplasms. This cytoplasmic differentiation is in complete agreement with the classification based on fertility restoration of Beckett (1971), Gracen and Grogan (1974) and the variation found in mitochondrial translation products by Forde et al. (1980). Kemble et al. (1980) verified the T-cytoplasm group by HmT pathotoxin test (Peterson et al. 1974, Flavell 1975). Unlike the heterogeneity found within N and C groups by the restriction endonuclease digestion of mt-DNA (Levings and Pring 1977, Pring et al. 1979a,b), no heterogeneity occurred within any of the four cytoplasmic groups classified on the basis of agarose gel electrophoresis. A similar nonheterogeneity was found by Forde et al. (1980), who used mt-translation product assay. Analysis of low molecular weight mt-DNA revealed significant and consistent differences in the banding pattern of the four maize cytoplasms. Compared to N-, T-cytoplasm lacks one mt-DNA band and S- and C-cytoplasms carry additional mt-DNA species which differ from each other and are not found in N- or T types (Kemble and Bedbrook 1980). Pring et al. (1977) demonstrated that the mitochondrial preparations (not subjected to restriction endonuclease digestion) from maize lines carrying S-cytoplasm possessed two low molecular weight DNA species which were absent in the mitochondria from N, T and C lines of maize. Based on these findings, Kemble and Bedbrook (1979) developed a rapid

Table 27.20. Mean frequency and length of two major classes of mt-DNA molecules in maize cytoplasms. (After Pring et al. 1979b)

Cytoplasm type	Frequency class		Contour length (μm)
N	a	48	21
	b	20	15
S	a	40	17
	b	32	36
T	a	47	25
	b	37	12

a) most frequent b) second most frequent.

Table 27.21. Genic male sterility in Gramineae

Species	References	Gene control	Gene action	Remarks
(1)	(2)	(3)	(4)	(5)
Alopecurus myosuroides	Johnsson (1944)	1-many	During MI	
A.pratensis	This Vol.	1r	Prophase I	Tapetum degenerates after PMC's disintegrate
		nk	Prophase I	
Avena byzantina	Ramage and Suneson (1958)	nk	nk	
	Garber (1922)	2r	nk	
A. fatua	This Vol.	nk	After microspore release	
A.strigosa	Sadanaga (1965)	1r	After microspore liberation	
Chionachne koenigii	Rao (1975)	1r	nk	
		1r	During staminal initiation	Pleiotropy and epistasis
Dactylis glomerata	Filion and Christie (1966)	nk	After microspore release	Abnormal tapetum
	This Vol.	po	Before pollen mitosis	
	Kremnina (1973)	nk	nk	Mutagen-induced
Echinocloa colone	Ayyangar and Srinivasa Rao (1941)	1r		
Festuca arundinacea	Goral et al. (1981)	nk	nk	One stock temperature-sensitive, other insensitive
Hordeum vulgare	This Vol.	1r	Premeiotic	Tapetum normal
	Kaul and Singh (1966)	1r	During prophase I	In basal region *ms* gene action is delayed
	Deshpande et al. (1981)	1D	During prophase I	
	Sharma and Reinbergs (1972)	1r	During cell plate formations following I and II meiotic division	
	Herd and Steer (1984)	1r	During cytokinesis after meiosis I	
	Roath and Hockett (1970)	1r	During microspore formation, prior to release	Tapetum degenerates along with meiocytes
	Singh and Kaul (1966)	1r		
	Singh and Kaul (1966)	1r	After microspore release	Tapetum persistent
	Mian et al. (1974)			

	Herd and Steer (1984)	1r	During pollen development	
	Roath and Hockett (1970),	1r	During pollen maturation	
	Ahokas (1975a), This Vol.	1r		
Lolium multiflorum	Nitzsche (1971)	1r	After pollen formation	
	Fejer (1966)	1r	During anther differentiation	
L.perenne	Hayward and Manthiratna (1972)	nk	During microspores maturation	Tapetum persistent
Oryza sativa	Lu and Rutger (1984a,b)	1r	During premeiosis, meiosis, post-meiosis and pollen maturation	Pollen phenotype based on *ms* gene action time
	This Vol.	1r	After microspore liberation	
Pennisetum typhoides	Krishnaswamy and Ayyangar (1942)	1r	nk	
	Gill et al. (1973)	3r, com.	nk	
	Rao and Koduru (1978a,b), Rao and Devi (1983)	1r	Before archespore differentiation or during microspore formation	Tapetum and sporogenous tissue degenerates simultaneously
	Powell and Burton (1969)	1r	After microspore release	X-Ray-induced
Phleum nodosum	May and Kasha (1971, 1980)	1r	During pachytene	Delayed tapetal degeneration
Saccharum officinarum	Janaki-Ammal (1941), Dutt et al. (1954)	nk	During MI spindle formation	
Secale cereale	Cêbrat and Zadecka (1978)	1-many	During premeiosis	Tapetum develops abnormally
	Kolobaeva (1974)	1r	During early prophase	Tapetum develops abnormally, rarely normally
	Putt (1954)	Complex	During early prophase	
	Adamĉhuk (1971, 1972)	1r	During microspore mitosis	
	Koblyansky and Katerova (1973a) } Garlicka and Madej (1975) }	1r	nk	
	Kubicki et al. (1981)	1r	nk	Structural mst

Table 27.21 *(continued)*

Species	References	Gene control	Gene action	Remarks
(1)	(2)	(3)	(4)	(5)
Sorghum vulgare	Ayyangar and Ponnaiya (1937)	1r	During premeiosis	
	Stephens (1937)	1r	During premeiosis	Linked with awned and virescent genes
	Karper et al. (1938)	1r	nk	photoperiod influenced
	Barabas (1962)	1r	nk	X-ray- and colchicine-induced, photoperiod-sensitive
	Stephens et al. (1952)	Many r	nk	
	Andrews and Webster (1971)	1r	nk	γ-ray-induced
Tripsacum laxum	Dodds and Simmonds (1946)	nk	In prophase	
Triticum aestivum	Pugsley and Oram (1959)	1r	nk	
(Spontaneous mutants)	Lupton and Bingham (1966, 1967)	1r	Before microspore release	
	Krupnov (1968)	1r	After microspore release	
	Lemekh et al. (1971)	1r	During prophase I to microspore formation	
	Athwal et al. (1967)	3r	After microspore liberation	Gene action additive
	Jan (1974)	3r,-com.	Before meiosis	Anther transformed into ovarylike structure
	Xiangmin and Herong (1981)	1D	During prophase I	
	Deng and Gao (1982)	1D	After tetrad formation, before microspore liberation	
(Induced mutants)	Fossati and Ingold (1970)	1r	After microspore release	X-ray-induced
	Latypov (1974)	1r	nk	NEU-induced
	Driscoll and Barlow (1976)	1r	nk	γ-ray-induced
	Kleijer and Fossati (1976)	1r	nk	X-ray-induced
	Franckowiak et al. (1976)	1r	nk	EMS-induced
	Sasakuma et al. (1978)	1r	nk	
Zea mays	Refer Tables 27.15 to 27.17	1D	nk	EMS-induced

Table 27.22. Gene cytoplasmic male sterility in Gramineae

Species/cross	References	Gene control	Gene action	Remarks
(1)	(2)	(3)	(4)	(5)
Aegilops caudata × *Ae.umbellulata* × *Ae.persica* × *Ae.typica*	Kihara and Kondo (1943)	nk	After microspore liberation	
Ae.longissima × *Ae.aucheri*	Kihara (1951)	nk	nk	
Ae.caudata × *Triticum durum*	Hervas (1974)	nk	After microspore liberation	mst Expression stable
Ae.ovata × *T.aestivum*	Chauhan and Singh (1966)	nk	During premeiosis or tetrad formation	Tapetum abnormal
Ae.ovata × *T.durum*	Fukasawa (1953, 1956b)	nk	After tetrad formation	Tapetum normal
Dactylis glomerata	Myers (1946)	1D	nk	nk
Euchlaena mexicana × *Zea mays*	Mazoti and Velasquez (1962)	nk	At prophase-I	
Festuca pratensis × *Lolium perenne*	Jenkins (1933)	nk	After microspore release	
(*Hordeum jubatum* × *H.compressum*)	Connoly and Wright-Turner (1984)	1–22	nk	
(*H.bulbosum* × *H.vulgare*)	Schooler (1967)		After microspore liberation	Tapetum persistent
Hordeum vulgare	Favret and Ryan (1966)		nk	p-mst
H.vulgare sp.*spontaneum*	Ahokas (1979a, 1982b)	1r	After microspore release	Tapetum persistent
L.multiflorum	Nitzsche (1971)		During tetrad to microspore formation	Tapetum persistent
Lolium multiflorum × *Festuca arundinacea* *L.perenne* × *F.arundinacea*	Anderson (1966)	nk	During anther development	Reciprocal crosses unsuccessful

Table 27.22 *(continued)*

Species/cross	References	Gene control	Gene action	Remarks
(1)	(2)	(3)	(4)	(5)
(*L.perenne* × *L.multiflorum*)	Wit (1974)	1–2r	nk	
× *F.pratensis* (4x)	This Vol.	3r	After microspore release	
L.perenne × *L.multiflorum*	Ahloowalia (1981)	nk	During pachytene	*ds* and *ms* genes linked
Oryza glaberrima × *O.sativa* indica	Carnahan et al. (1972)	nk	After microspore release	
O.glaberrima × *O.sativa* japonica	Yabuno (1977)	nk	After microspore or pollen development	
O.perennis × *O.sativa*	Katsuo and Mizushima (1958)	nk		
O.rufipogon × *O.sativa* indica	Chang et al. (1982)		After microspore release	Abnormal tapetum
O.rufipogon × *O.sativa* japonica	Chang and Huang (1978, 1979)		After microspore release	Temperature influenced
O.sativa	Tamaru et al. (1980)	nk	After pollen mitosis	
	Anonymous (1978a)	nk	After binucleate pollen stage	
	Shuhua (1982)	nk	After microspore, liberation, or Ist or IInd pollen mitosis	
	Pan et al. (1982)	nk	During prophase I	Tapetum normal
O.sativa indica × *O.sativa* japonica	Watanabe et al. (1968)		After pollen formation	
	Shinjyo and Omura (1966)	2r		
	Kinoshita et al. (1980)	2r	After pollen mitosis	
O.sativa japonica × *O.glaberrima*	Yabuno (1977)	1r	After microspore release	Anthers indehiscent

O.sativa spontanea × *O.sativa* japonica	Katsuo (1954), Katsuo and Mizushima (1958), Kitamura (1962)	1r	Before or after pollen mitosis	fn-mst
	Chu et al. (1972)		After release microspores degenerate	
Pennisetum typhoides	Burton (1958), Athwal (1965)	1r	During tetrad formation	
	Menon (1958, 1959), Yadav (1974)	2r		
	Akeńova and Chheda (1981)	3r		
	Prakash (1967)	po	After microspore release	
	Singh and Sharma (1968)	nk	After microspore release	Tapetum persistent
Saccharum spontaneum × *S.barberi* × *S.narenga* × *S.officinarum* × *S.sclerostachya*	Raghavan (1951, 1953, 1964)			
S.spontaneum × *Sclerostachya fusca*	This Vol.	2r	During prophase I	Tapetum normal
Secale cereale	Madej (1975)	1r	After microspore liberation	Tapetum persistent
	Garlicka and Madej (1975)	1–2r	nk	
	Geiger and Schnell (1970), Garlicka and Madej (1975)	2r	After microspore liberation	Tapetum persistent
	Scoles and Evans (1979c)	2r	After microspore liberation	Tapetum abnormal
	Artemova (1980)	nk	During microspore development	
S.cereale × *S.kuprijanovii*	Lapinski (1972)	1r		*Fr* gene action influenced by environment
× *S.montanum*	Lapinski (1972)	1r		

Table 27.22 *(continued)*

Species/cross	References	Gene control	Gene action	Remarks
(1)	(2)	(3)	(4)	(5)
Sorghum vulgare	Maunder and Pickett (1959)	1r		
	Kidd (1961)	1r		
	Duvick (1966)	1r		
	Singh and Hadley (1961)	1r	Before pollen mitosis	Tapetum persistent
	Dámon (1961)			
	Pi and Wu (1963), Erichsen and Ross (1963a), Miller and Pickett (1964), Quinby (1980)	1–2r	Before or during pollen mitosis	–
	Stephens and Holland (1954)	More than 2r		Temperature influence high
	Craigmiles (1961, 1962), Alam and Sandal (1967)	3r	During pollen mitosis	
	Tripathi et al. (1980)	4r–5r		
	Webster and Singh (1964)	nk		functional mst
	Narkhede et al. (1968)	nk	Before or after pollen mitosis	Delayed tapetal degeneration
	Raj (1968)	nk	During meiosis	
S. vulgare × *S. almum*	McClure (1962)	nk	After microspore release	–
S. vulgare × *S. halepense*	Wilhite (1960a), McClure (1962)	nk	After microspore release	–
Tragopogon pratensis × *T. porrifolius*	Ownbey and McCollum (1953)	nk		–
Triticum aestivum	DeVries and Ie (1970)		During pollen mitosis	Tapetum normal
Triticum durum × *Secale cereale*	Vega and Lacadena (1979)		During pollen maturity	Tapetum persistent and degenerates later
T. aestivum × *S. cereale*	Lein (1948)			Functional, structural sporogenous mst
T. compactum × *T. aestivum*	Xiangmin and Herong (1981)		During prophase I	

T.timopheevi × *T.aestivum*	Fedorova and Nettevich (1969)	Before microspore and pollen mitosis	
	Jain et al. (1981)	During microspores development	Tapetum abnormal
	Joppa et al. (1966)	Between microspore mitosis	Tapetum persistent
	Xiangmin and Herong (1981)	After microspore release	mst Expression highly stable
T.timonovum × *T.aestivum*; *T.zhukovskyi* × *T.aestivum*	Fedorova and Nettevich (1969)	Between microspore and pollen mitosis	
Zea mays (T-cytoplasm)	Gabelman (1949), Chang (1954)	After microspore liberation	Tapetum abnormal
	Jones et al. (1957)		Tapetum persistent
	Rhoades (1933)		
	Herich (1965), Spasojevič (1966), Orel (1967)	Between tetrad formation to microspore liberation	
	Rogers and Edwardson (1952)	During microspore to pollen mitosis	
	Diaconu (1965)		
	Khoo and Stinson (1957)	During pollen maturation	
	Lee (1976), Warmke and Lee (1977)	During microspore development	Tapetum abnormal, persistent
	Colhoun and Steer (1981)	Around diad-tetrad stage	
Zea mays (S-cytoplasm)	Lee et al. (1981)	During pollen maturity, after mitosis	Tapetum normal Tapetum normal
	Colhoun and Steer (1981)	At or just after vacuolate microspore stage	Tapetum abnormal, persistent
Zea mays (C-cytoplasm)	Lee et al. (1979)	Before or after exine formation	

Table 27.23. Chemical male sterility in Gramineae

Species/Chemicals	References	Male sterility induction	Female fertility reduction	Side effects
(1)	(2)	(3)	(4)	(5)
Avena sativa				
DPX-3778	Johnson and Brown (1976)	Partial	Partial	Reduced shoot height, spikelet number and grain yield
GA	Frohberg and Frey (1970)	High	To half	Excessive internode elongation and reduced spikelet size
MH	Frohberg and Frey (1970)	Complete	High	Toxic
Hordeum vulgare				
CCDP	Yih et al. (1971)	Nearly complete	nk	nk
Ethrel	Stoskopf and Law (1972), Law and Stoskopf (1973)	Low	High	Many morphological ill effects
	Kumar et al. (1976), Verma and Kumar (1978)	Nearly complete	Low	nil
GA	James and Lund (1960)	Complete	nk	nk
	Paleg and Aspinall (1958)	Only in tips	nk	nk
RH-531, 532, 2956	Wang and Lund (1975)	Nearly complete	Low-medium	nk
	Hockett and Feltner (1978)	Partial	Partial	Reductions in height, flowers and grain yield
Oryza sativa				
DPX-3778	Long et al. (1973)	Partial	nk	nk
Ethrel	Perez et al. (1973), Cheng and Huang (1978), Parmar et al. (1979), Chan and Cheah (1981)	Partial-high	High	Severe phytotoxicity and plant death with high dosage
Ethylene	Wang and Que (1981)	High	High	nk
3-(p-Chlorophenyl) 6-methoxy-S-triazine -2, 4-(1H, 3H) dione triethanol amine salt	Zhangxing and Channong (1980), This Vol.	Nearly complete	Very high	Highly toxic, leaf burning and plant death
RH 531, 532, 2956	Perez et al. (1973)	High	High	High phytotoxic effects

Pennisetum typhoides				
Dalapon	Miller (1961)	Partial	Very high	Highly toxic
DPX-3778	Hanna (1977)	Partial to complete	nk	nk
Dimethyl arsenic acid	Powell and Taylorson (1967)	Very low	High	Highly phytotoxic
Ethidium bromide	Burton and Hanna (1976)	Partial to complete	nk	nk
Mendok	Sharma (1979)	High	nil	nk
Sorghum vulgare				
Colchicine	Erichsen and Ross (1963a,b)	High	High	nk
Dalapon	Miller (1961)	Partial	High	nk
MH	McIlrath (1957)	Nearly complete	High	Reduced plant growth
Mendok	Sayers (1959), Miller (1961), this Vol.	Partial	Complete	nk
Triticum aestivum				
Dalapon	Kaul and Singh (1967)	Complete	High	Reduced vegetative and
	Porter and Wiese (1961)	Partial	High	reproductive growth, highly phytotoxic
DPX-3778	Johnson and Brown (1976, 1978)	Nearly complete	Low	Genotype-specific sensitivity, delays anthesis, reduces plant height and is toxic
Estrone	Fendal (1967)	Low	nk	Anomalous ovule development and reduced seed yield
Ethrel	Rowell and Miller (1971)	Complete	Low	Dwarfing, poor spike emergence
	Bennett and Hughes (1972)	Complete	Low	nk
	Stoskopf and Law (1972), Bonali et al. (1972), Vandam (1974), Fairey and Stoskopf (1975)	High	High	Dwarfing, increased tillering and high floret sterility
	Jan and Rowell (1981)	Low	nk	nk
	Dotlaĉil and Aṕltauerova (1978)	Complete	Low	Delayed vegetative growth, height reduction, imperfect heading
	Hoogland et al. (1953)	Complete	High	Reduced height and top growth
	Chopra et al. (1960)	Complete	Low	nk

Table 27.23 *(continued)*

Species/Chemicals	References	Male sterility induction	Female fertility reduction	Side effects
	Porter and Weise (1961)	Partial	High	Severe plant damage, delayed maturity, poor grain yield
	Kaul and Singh (1967)	High	Medium	nk
Mendok RH 531, 532, 2956	Jan et al. (1974, 1976), Jan and Rowell (1981), Rajendra and Bates (1981)	High	Low	Spikes compact, poor flower opening, many genetic effects
Triticale				
DPX-3778	Johnson and Brown (1976)	Nearly complete	High	nk
Zea mays				
DPX-3778	Long et al. (1973), Laible (1974)	Complete to partial	Nil	Genotype-specific sensitivity, delayed anthesis
GA3	Nickerson (1959, 1960), Hansen et al. (1976)	Complete	Nil	Increased shoot height, thin and weak stems
	Wittwer and Bukovac (1957), Nelson and Rossman (1958), Cherry et al. (1960)	Partial	Nil	
MH	Denisen and Haber (1950)	Complete	Nil	
	Eskew and Willard (1950)	Partial	nk	
	Moore (1950), Josephson (1951)	Complete	High	Delayed flowering
	Naylor and Davis (1950)	Complete	Nil	
Mendok	Cameron and Eaton (1959)	Low-high	High	

assay for detecting the presence of S-cytoplasm in maize using five etiolated seedling shoots per sample. Since only a few seedlings are required and destroyed, this method could be useful for assaying seed bulks. Later Kemble (1980) developed a nondestructive, simple and rapid assay for an unambiguous identification of the S-cytoplasm. This method utilizes a 0.5-g leaf segment per sample and thus is a single plant assay in which the test plant is not destroyed.

27.21.2.1.2.7 Relative Advantages

Methods used for the identification of maize cytoplasms require technical skill, costly equipment and expensive chemicals; some of the procedures are lengthy. The biochemical methods are relatively refined, rapid and lack ambiguity compared to the traditional fertility restoration technique. The classical field method is time-consuming, as it involves the crossing of each cytoplasmic source by pollen from many inbred lines and backcrossing repeatedly. However, this is the traditional method and it has been used for confirmation of cytoplasm types by various biochemical means. The mitochondrial translation product method devised by Forde et al. (1980) identifies cytoplasm types in all the nuclear backgrounds. Furthermore, no within cytoplasm variation occurs in mitochondrial translation products (Forde *et al.* 1980), whereas differences in fertility restoration patterns do occur within the same cytoplasm type (Beckett 1971, Gracen and Grogan 1974). These features render the mitochondrial translation product method advantageous over fertility restoration techniques for precise cytoplasm type classification. The restriction enzyme analysis method devised by Levings and Pring (1976 and Pring and Levings (1978) identifies each maize cytoplasm specifically, but exposes mt-DNA heterogeneity in N- and among C-cytoplasms. No heterogeneity in the mt-DNA banding pattern within any of the four cytoplasmic groups occurs after mitochondrial translation product analysis (Forde et al. 1980) and mt-DNA agarose gel electrophoresis (Kemble et al. 1980). The restriction enzyme analysis method is sensitive and efficient in detecting genetic differences between cytoplasmic DNA's, but it is not known if these differences are of any practical breeding value. Whereas large quantities of etiolated maize shoots are required for restriction endonuclease digestion of mt-DNA, only a small amount (5 g) of the material is needed in the agarose gel electrophoresis method of Kemble et al. (1980). Furthermore, the latter method provides a characteristic banding pattern of mt-DNA regardless of the nuclear background (Kemble and Bedbrook 1980). By this method maize cytoplasm can be classified, even if a line having sterile cytoplasm has been restored to fertility by *Fr* genes. Furthermore, this method can identify a cytoplasm type that cannot be distinguished by fertility restoration. Thus this method shares many advantages of the mitochondrial translation product. Furthermore, the agarose gel electrophoresis of Kemble et al. (1980) does not require 35S methionine, which is expensive. Also it is rapid, requires 20–30 seeds, and classifies maize cytoplasms precisely. Since it is least expensive, simple and rapid, its success has been recently demonstrated.

For a quick T-cytoplasm identification, the rapid electron transport assay using Hmt toxin and an electron acceptor dye (dichlorophenolindophenyl) (Peterson et al. 1974, Flavell 1975) may be the simplest assay method. For

S-cytoplasm identification, the rapid DNA assay that needs five or less etiolated maize shoots or 0.5 g leaf material (Kemble and Bedbrook 1979, Kemble 1980) is the simplest available method. The leaf assay method is technically simple and rapid, and permits individual plants to be assayed at any developmental stage without destroying the plants under test. Since a simultaneous assay of many samples is possible, this method may be used as a routine screening test for S-cytoplasm types of maize. Like the etiolated shoot method (Kemble and Bedbrook 1979), the leaf assay method (Kemble 1980) detects the presence of S-cytoplasm in maize plants, irrespective of their nuclear genetic background or their having been restored to full mft by nuclear fertility-restoring genes, or if they possess cytoplasms related to S-type cytoplasm but displaying different degrees of mft in the same nuclear background. Thus individually tested plants can be used in plant breeding programs with certainty as to their cytoplasmic background.

27.21.2.2 Microsporogenesis

i. *T. Anthers.* In most of the T-anthers, male meiosis proceeds normally until tetrad formation. Abortion occurs either during microspore development while the microspores are enclosed within the PMC's or during or after first vegetative division after the microspores are released from the PMC's (Lee 1976, Warmke and Lee 1977). Colhoun and Steer (1981) found that in some anthers, the meiocytes abort during early prophase I and in many others the abortion occurs at or before the diad-tetrad stage of meiosis. This variability in the stage of microsporogenesis breakdown is due to differential gene action caused by different nuclear genetic backgrounds of T-cytoplasm plants. Pollen of the fertile line has a single nucleolus, but two to six nucleoli of varying sizes occur in the nuclei of mst T-cytoplasm line pollen. Since in most of the T-cytoplasm maize mst lines, pollen mitosis is suppressed, Herich (1965) regarded the nucleolar abnormality as the cause of this mitotic suppression.

ii. *S-Anthers.* In comparison to C- and T-anthers, *fr* gene action is considerably delayed in S-anthers. In them, the pollen undergoes mitosis and becomes binucleate (Lee et al. 1980). Pollen abortion commences during the final stages of pollen maturity and after succesful completion of all the tapetal cell stages. Thus at preanthesis, the pollen grains appear nearly normal with regular starch grains. Colhoun and Steer (1981) found "exine ghosts" in locules with intact tapetal cells, suggesting abortion at an earlier stage, probably at or just after the vacuolate miscrospore stage. In general, it appears that the pollen degeneration occurs either before (Colhoun and Steer 1981) or after (Lee et al. 1980) grain filling and intine formation.

iii. *C-Anthers.* Whereas meiosis proceeds normally in these anthers, the post-meiotic development during and after the tetrad stage is variable, the sequence of deviation depending on the time of *fr* gene action. The action appears to be related to the developmental pattern of the tapetum (Lee et al. 1979). After release, the microspores destined to develop into pollen grains abort either before or after exine formation (Fig. 27.25b) or even during or just after tetrad formation but prior to release (Fig. 27.25a).

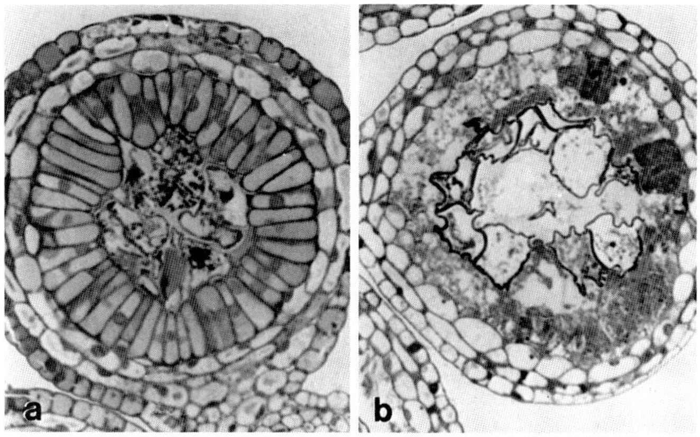

Fig. 27.25a,b. *Zea mays.* **a,b** C-cytoplasm male sterile maize anthers (T.S.) having aborted tetrads **(a)** and microspores **(b)** with intact **(a)** and degenerated **(b)** tapetum. (Lee et al. 1979)

27.21.2.3 Tapetal Changes

i. *T-Anthers.* The breakdown of tapetal mitochondria and mid-anther layers are the first visible signs of anomalies that occur shortly after meiosis (Warmke and Lee 1977). During this process, swelling, loss of cristae and matrix density of mitochondria occurs. These alterations resemble the mitochondrial response to HmT toxin. In a few T-anthers, mitochondria may remain intact during pollen abortion (Colhoun and Steer 1981). Thus the mitochondrial degeneration is not necessarily the earliest detectable event in T-anther abortion. Reduction or failure of orbicule formation on inner tapetal surface represents a significant alteration in T-anthers during pollen abortion (Colhoun and Steer 1981).

ii. *S-Anthers.* Both anther tissue and cell organelle (mitochondria and chloroplast) remain normal in these anthers. Certain membranous structures appearing as whorls or cisternae or as aggregates of concentric membranes appear at a much higher frequency in S-anthers than in restored S- or N-anthers (Lee et al. 1980). Although tapetal cell development is normal, at late stages of pollen abortion an unusual loss of rough endoplasmic reticulum from the cells (apart from cisternae associated with the organelle) occurs (Colhoun and Steer 1981).

iii. *C-Anthers.* Two different modes of microspore development occur in C-anthers (Lee et al. 1979), so that the microspores abort either before or after exine development. The difference between the two depends on whether the changes in tapetum are initiated early or late during pollen development. The early action is at the mononucleate stage. In this case, the tapetal cell wall persists until flowering, the cells vacuolate highly and expand radially into anther locules almost occluding it. Microspores abort before exine formation. The late-appearing tapetal changes occur at the binucleate tapetal stage. The tapetal cells vacuolate considerably and expand just before disintegration. In this case, the microspores abort after exine development, leaving "exine ghosts" in the locules. The two tapetal anomaly types occur in different anthers of the same plant grown in field or glasshouse. However, in one plant a mixture of both these types was found in different locules of the same anther. In addition to the two pathways of *fr* gene action, Colhoun and Steer (1981) found an intermediate pathway in which anthers have binucleate tapetum but microspores are already aborted prior to exine formation. Similarly, Ohmasa et al. (1976) found microspore abortion at the mononucleate stage but the tapetum persisted until flowering. Unlike in T-anthers, no organellar abnormalities, such as mitochondrial or plastid degenerations, occurred in C-anthers. However, certain deviations in the fine structures of C-anthers do occur. For instance, the orbicule development is either highly impaired or completely inhibited even in the late degenerating tissues (Lee et al. 1980, Colhoun and Steer 1981). Endoplasmic reticulum is present in tapetal cells, but protrusions into vacuoles or formation of multivesicular bodies is either inhibited (Colhoun and Steer 1981) or not inhibited (Lee et al. 1980).

27.21.2.4 Gene Action in Sterile Anthers

In the majority of the mst anthers, meiosis proceeds normally but *fr* gene acts after meiotic completion. In T-anthers, pollen abortion occurs between the tetrad and vacuolate pollen stages. In a few, *fr* gene action occurs earlier and meiosis is impeded during the diad-tetrad stages. Whereas in most of the anthers mitochondria degenerate along with spores, in the remaining few they remain intact. Thus in T-anthers mitochondrial degeneration is not a necessary event of microspore abortion. In C-anthers, timing of *fr* gene action alters the tapetal development and microspore abortion. This timing is influenced by genotypic background and environmental conditions. However, in all cases, microspore abortion occurs prior to or after exine formation. In S-anthers, *fr* gene action is considerably delayed and occurs either before or after intine formation. Unlike in T-anthers, no mitochondrial or plastid anomalies occur in C- and S-anthers. Pollen abortion in T- and C-cytoplasm anthers is associated with tapetal abnormalities during the early tetrad stage. In S-anthers no such anomalies occur. This, coupled with the gametophytic nature of S-pollen restoration, suggests that *fr* genes within the pollen grains are responsible for the S-pollen abortion. Hence S- is distinct from C- and T-cytoplasm, which follow a sporophytic restoration pattern and exhibit cytological evidence of a relationship between the sporophytic tissue, tapetum and gametophytic tissue, the pollen. Since pollen development proceeds to a very late stage in S lines, pollen abortion could be averted through physiological or environmental manipulations.

27.21.2.5 Pollen Inclusions

Khoo and Stinson (1957) found inhibition of microspore wall and germ pore development in T-cytoplasm mst maize. The pollen grains had some inclusions not known in C- or S-cytoplasm pollen. Earlier, Rhoades (1933) detected some inclusions in maize microspores which were rod-shaped in the sterile Peruvian source and rounded in the fertile line. He regarded these differences as a consequence rather than a cause of mst in maize; but no such inclusions in C-, S- or T-cytoplasm maize pollen were found in later investigations.

27.21.2.6 Biochemical Differences

i. *Amino Acids*. Fukasawa (1954) and Khoo and Stinson (1957) found proline in lower and alanine and asparagine in higher proportions in T-compared to N-cytoplasm anthers. This indicates that the abortive microspores are unable to utilise alanine and asparagine, which consequently accumulate in sterile anthers. Both these amino acids are active in transamination reactions in plant tissues. A relatively low amount of proline in T-anthers could be a cause of pollen abortion if it were an essential component of one or more storage proteins or enzymes of developing pollen grains. Proline is thought to be uniquely important in the determination of the overall shape of protein molecules (Fowden 1963), and as such its absence or scarcity may be crucial. But the injection of proline into young T-anthers does not restore fertility to them. On the other hand, gene *Fr* restores both pollen fertility and normal free amino acid balance in T-anthers. This indicates that the block in free amino acid utilization by T-cytoplasm microspores is removed by *Fr* gene introduction. However, since pollen of such plants, although fertile, is smaller than those with N-cytoplasm, it seems that *Fr* gene merely provides an alternative pathway to pollen fertility and does not act on the causes of the basic differences between T- and N-cytoplasms, or the gene *Fr* acts on one of the several differences existing between these two cytoplasms (Duvick 1965).

ii. *Enzyme Activity*. Compared to fertile lines, reduced amylase activity and increased polyphenoloxidase and peroxidase activities occur in S- and T-anthers (Dmitrieva and Kharazhinskaia 1962). The activity of glucose 6-phosphate dehydrogenase (GPDH) and 6-phosphogluconate dehydrogenase (PGDH) is low in mst anthers (Markova 1981). Throughout the period of microspore development, cytochrome oxidase and malate dehydrogenase activity is reduced in mst anthers having T-, C- or S-cytoplasm (Ohmasa et al. 1976.) With the progression of microspore development, enzyme activity decreases rapidly in mst anthers. In the mft and mst version of maize line VIR-29, the activity of the enzymes GPDH, PGDH, hexokinase, α-amylase and β-amylase during microsporogenesis is nearly the same until the tetrad stage (Danilenko 1981). Thereafter, enzyme activity decreases with the progression of microsporogenesis in mst anthers but increases in mft anthers. In mst maize anthers, irrespective of the sterile cytoplasm type, cytochrome oxidase activity is considerably lower than that of the corresponding mft anthers. This enzyme is mainly located in the inner mitochondrial matrix. However, whereas mitochondrial structural alterations occur in T-anthers, no such phenotypically detectable alterations are known in S- and C-anthers. The depressed enzyme activity in all mst-type maize anthers may actually be due to insufficient translocation of photosynthates from rachis to anthers, as reported by Nakashima and Hosokawa (1974b). An indication of this disturbance in carbohydrate and phosphate metabolism in mst anthers is provided by Markova (1981a,b), who found reduced activity of the enzymes glucose-6-phosphate dehydrogenase, 6-phosphogluconate dehydrogenase, and NAD^+-dependent glutamate dehydrogenase (GDH) in the mst anthers compared to the corresponding mft anthers. Such a reduction in first two enzymes is significant in WF9 maize with N-cytoplasm. The lower GPDH activity in mst anthers indicates

Walbot 1982) and lead to cytoplasm diversity and within-group heterogeneity. Using recombinant DNA technique, Spruill (1982) isolated Bam HI restriction fragments from N-mt DNA and hybridised them to Southern blots of mt-DNA obtained from N- and T-cytoplasms. Whereas the majority of cloned fragments hybridised only to single bands in mt-DNA, in seven cases, labelling of additional bands indicated sequence repetitions. Although 15 cloned fragments hybridized to co-migrating bands in N and T digestions, substantial differences in hybridization pattern occurred in the remaining 20 fragments. Autoradiographs produced by these 20 clones indicated cleavage site changes and/or sequence rearrangements of the mt-DNA.

27.21.2.9.2 Mitochondrial DNA Differentiation

In maize, mitochondria are maternally inherited (Pring and Levings 1978, Conde et al. 1979). The mitochondrial genome possesses a heterogenous array of DNA molecules existing as covalently closed circular molecules ranging from 60–90×10^6 daltons (Levings et al. 1979a,b). Among four cytoplasm types, N, C, S and T, frequency and sequence arrangement of these variously sized DNA molecules varies and each cytoplasm type exhibits specificity in DNA banding pattern (Levings and Pring 1976, Pring and Levings 1978), mitochondrial translation products (Forde et al. 1978, 1980) and agarose gel banding patterns (Kembel et al. 1980). Mitochondria of maize cytoplasm N, C, S, T types contain certain low molecular weight supercoiled circular DNA molecules of approximately 1940 base pairs (Kemble and Bedbrook 1980). This small supercoiled circular DNA is suggestive of its being an autonomously replicated plasmid. Also N-, C- and S-cytoplasms contain a DNA species of approximately 2350 bp, which is absent in the T-cytoplasm. The C-cytoplasm contains two additional circular DNA species of 1570bp and 1520 bp and the S-cytoplasm possesses two linear mt-DNA species not found in the other cytoplasms. Three DNA bands viz. supercoil (SC), open circle (OC) and linear (L) conformation of the same sequence are common to all the cytoplasms. Due to insufficient knowledge of the functional identification of proteins which are coded by higher plant mt DNA, it is not possible to pinpoint exactly the mitochondrial gene products that are associated with T-cytoplasm and HmT disease susceptibility. Fungal and mammalian system studies indicate the constituent proteins of cytochrome oxidase, cytochrome b and the oligomycin-sensitive ATPase (Schatz and Mason 1974) as the probable gene products. Evidence suggests the influence of T-cytoplasm on cytochrome oxidase and cytochrome b (Pring 1975, Watrud et al. 1976). Examination of N-, C-, S- and T-cytoplasms in different nuclear backgrounds reveals cytoplasm-specific mitochondrial proteins, but the site and function of these proteins is unclear.

Ohmasa et al. (1976) obtained correlation between gc-mst and mitochondrial enzyme activity in maize anthers. Cytochrome oxidase and malate dehydrogenase activity of N-cytoplasm anthers increased rapidly as the pollen matured reaching maximum before anthesis while enzyme activity of T-, S- and C-cytoplasm anthers did not show such a rapid increase. Since restored mft hybrid lines with T-, C-, or S-cytoplasm exhibit a mitochondrial enzyme activity as high as the lines with

N-cytoplasm, some components, probably proteins of mst mitochondria, change following fertility restoration. Cytological (Lee 1976, Aldrich et al. 1977, Warmke and Lee 1977), physiological and biochemical (Miller and Koeppe 1971, Barrett and Flavell 1975, Levings and Pring 1976, Pring and Levings 1978) studies indicate mitochondria as the carriers of determinants of mst in maize. The gc-mst system is thus a unique trait of maize mitochondria. This is further evidenced from the following account.

27.21.2.9.3 Instability and Mutability

A high frequency of spontaneous reversions from fertile to mst cytoplasms have occurred in maize. This has led to the repeated discoveries of naturally occurring strains of maize that carry T-, C- or S-cytoplasms, but reverse mutations are rare in nature under field conditions in T- and C-cytoplasm plants. Mutagens have neither produced new stable mst types in strains with normal maize cytoplasm nor has the long association of T- or C-cytoplasm with *Fr* genes been able to change the sterile cytoplasm to a normal type. This led Duvick (1965) to point out the mst cytoplasms as "notoriously stable" in maize. But this stability is not absolute and without exception, for S-cytoplasm is well known for its instability. Jones (1956) reported inbred lines WF9 having T-cytoplasm, which was interpreted as a cytoplasm mutation. Similar reversions are occasionally observed but rigorous proof for cytoplasmic mutation does not exist. Instances from the USSR indicate the origin of sterile cytoplasm following treatments with γ- rays, ultrasonic waves or neutrons (Lysikov et al. 1969, Stoilov 1970). A US patent No. 3,594,152 was issued to a Russian researcher for his technique of producing gc-mst maize by streptomycin treatment (Duvick and Noble 1978). Attempts to duplicate this are unsuccessful. However, spontaneous reversion of S-cytoplasm and induced mutations in T-cytoplasm are well known. Reversion of T-cytoplasm has been achieved by selecting the T-cytoplasm revertant after subjecting seeds to mutagenic treatments or by culturing immature young embryos in the absence or presence of HmT toxin in culture medium. Cassini et al. (1977) subjected dry seeds of T-cytoplasm maize inbred line to EMS or X-ray treatments. The M_1 plants were crossed by a restorer line to produce M_2 progeny. In the resulting selfed M_3 progeny, several HmT toxin-resistant families were isolated by subjecting the plants to the toxin at seedling and mature plant stages. After several screenings in subsequent generations, seven families were mft and fully HmT-resistant (Cassini et al. 1977, Cornu et al. 1980). They bred true to these traits and thus represent mutant derivatives.

27.21.2.9.4 Reversions

Instability of the S-cytoplasm and its spontaneous reversion to the mft state were first documented by Singh and Laughnan (1972). Such reversions to mft, initially recognized as mft tassel or as fertile tassel sections on plants in mst strains, are based on genetic changes at either the cytoplasmic or nuclear level, following which fertility is inherited in Mendelian fashion (Laughnan and Gabay 1973). The cytoplasmic and nuclear revertants are easily distinguished by conventional crossing procedures and the newly arisen mft phenotype is inherited in ensuing generations and remains stable. Some genetic backgrounds are highly favourable to these reversions and others unfavourable. Moreover, from over 300 sterility reversion cases it appears that the genetic background influences a proportion of genetic and cytoplasmic mutations. The majority of the reversions have occurred in the strain M825, which appears especially prone to such an event. Near-replacement of the nuclear genome of 14 inbred lines by the M825 genome led to an increase in spontaneous reversion rate from 1% to 8.3% (Laughnan and Gabay-Laughnan 1982). Among the fertility-reverting strains, M825 exhibits a high frequency of cytoplasmic and a low frequency of nuclear reversions; The reverse occurs in strain W84 (Laughnan and Gabay-Laughnan 1982). Strain M825, Ohio 7 exhibit an equal frequency of both types of reversions. The reversion of sterility to fertility due to nuclear genes indicates *fr* gene mutation to *Fr*. These genes appear to be different from the standard Fr_3 locus. Ten such genes are known and numbered from I to X. Gene Fr_3 has been mapped on chromosome 2 possibly in the long arm, restorers I and VIII on chromosome 8, IV and VII on chromosome 3 and IX and X on chromosome 1. It appears that each new restorer is located at a specific chromosomal site other than that of the Fr_3 locus. Thus the *Fr* genes in these ten independently occurring nuclear revertants are located in several different sites in different chromosomes. This led Laughnan and Gabay (1973, 1978) to propose that S-mst plants carry an episomal fertility element (F) that is capable of being fixed either in the cytoplasm or in the nucleus, leading to cytoplasmic or nuclear revertants, respectively. Laughnan and Gabay-Laughnan (1982) conclude that the nuclear genome has a predominant influence not only over reversion frequency but also over reversion site. Evidence indicates that reversion events involve transposition of S-1 and S-2 plasmids whose behaviour is controlled by genes.

27.21.2.9.5 Fungal Susceptibility

In the early part of the 1970 growing season, Southern corn leaf blight caused by the fungus *Helminthosporium maydis* (syn. *Bipolaris maydis*, *Drechsclera maydis*) invaded the US maize crop. By mid 1970, the yield loss was nearly 15% (Tatum 1971). The fungus was extremely virulent on T-cytoplasm maize plants but little on N-, C- or S-cytoplasm plants. This epidemic was due to a newly arisen fungal race designated *H. Maydis* race T(HmT) (Hooker et al. 1970, Scheifele et al. 1970, Smith et al. 1970) which arose due to spontaneous mutation from the older race that was not virulent on T-cytoplasm maize plants. This T-race appeared around 1970, as in the 20 previous years during which T-cytoplasm maize plants had been extensively grown, the fungal infection was too mild. The T-toxin secreted by HmT was highly specific to T-cytoplasm maize plants and not to the other cytoplasm types. Thus plants having N-, C- or S-type cytoplasms are resistant to fungus and insensitive to HmT toxin (Hooker et al. 1970). Conditions which restore mft, modify disease reaction and toxin response of T-cytoplasm maize lines do not cause resistance to the fungus or its toxins. The degree of modification varies in various inbred lines, some exhibiting a higher and others a lower level of resistance to the fungal infection (Caunter 1977) and to its toxin. The toxin produces many morphological, physiological, ultrastructural and biochemical changes as reviewed by Gregory et al. (1977). For toxin production, HmT race isolates possess one to several genes. These are in addition to some or all of the genes for pathogenicity found in HmO isolates (Yoder and Gracen 1975). Depending upon the number of polygenes in an isolate, the virulence is not host-genome-specific, but cytoplasm-type-specific. Therefore, the cytoplasmic component for resistance is more important than the nuclear component. Since within the same isolate the degree of virulence varies in various inbred lines with T-cytoplasm, virulence appears to depend on genome-T-cytoplasm interaction. The susceptibility to the HmT pathotoxin is cytoplasm-dependent. On T-cytoplasm, *Phyllosticta maydis* exhibits increased virulence compared to other cytoplasm lines (Scheifele and Nelson 1969, Scheifele et al. 1969).

27.21.2.10 T-Mitochondrial Alterations

27.21.2.10.1 In Vivo

Mitochondria possess their own genetic system. Accumulative evidence implicates them as the main carriers of the genes conditioning mst in maize (Miller and Koeppe 1971, Levings and Pring 1976, Warmke and Lee 1977, Pring and Levings 1978, Forde et al. 1978, Forde and Leaver 1980). The occurrence of certain changes in mitochondria, such as size enlargements, loss of cristae and matrix density at the tetrad stage prior to pollen abortion in T-anthers, was documented by Warmke and Lee (1977). Similar mitochondrial changes occur after protoplasts and root tips of T-cytoplasm maize are treated with HmT toxin (Aldrich et al. 1977). Prior to anther degeneration, size and number of mitochondria or plastids do not differ between N- and T-cytoplasms. A rapid increase (20–40-fold) in mitochondrial number per cell occurs in the tapetum and sporogenous cells during meiosis of both N- and T-lines. This dramatic increase, culminating at about the time when mitochondrial degeneration in T-lines occurs, led Warmke and Lee (1978) to propose that T-mt were unable to function completely under these stress conditions in the T version of the W-182 BN. Both *H. maydis* (race T) and *Phyllostica maydis* produce host-specific toxins (Hooker et al. 1970, Comstock *et al.* 1973, Yoder 1973), which preferentially affect the mitochondria of the maize lines carrying T but not N (Miller and Koeppe 1971, Comstock *et al.* 1973), C- or S-cytoplasms. This suggests an association between mitochondria and disease susceptibility. Nuclear genes that restore fertility to the T-cytoplasm modify the response of mitochondria to HmT toxins as well (Watrud et al. 1975a,b, Barrett and Flavell 1975). However, such an effect of the nuclear gene product on mitochondria is not consistent with what is currently known about fungal mitochondrial genetics (Schatz and Mason 1974).

Gregory et al. (1977) suggest that a major and perhaps primary component of southern corn leaf blight disease is HmT toxin-induced damage of mitochondria of T-cytoplasm maize plants. This is based on the observations that isolated T-mitochondria are severely damaged by HmT toxin. Ultrastructural investigations reveal that HmT toxin can damage T-mitochondria in situ, which is a result similar to that observed in toxin-treated isolated T-mitochondria. The cytoplasmic inheritance of toxin sensitivity and the marked effects of HmT toxin on nongreen tissue like seedling roots indicate the mitochondria and not the chloroplast as a possible site for toxin sensitivity. A low level of toxin causes rapid uncoupling of oxidative phosphorylation and inhibition or stimulation of respiration (depending on substrate supplied) in mitochondria isolated from the T-cytoplasm maize plant (Miller and Koeppe 1971). Spectrophotometric and ultrastructural studies of such isolated mitochondria reveal toxin-induced swelling and disruption of cristae of mitochondria (Gengenbach et al. 1973, York et al. 1980). Mitochondria isolated from N-, C- or S-cytoplasms do not exhibit such anomalies. Moreover, the insecticide Lannate 90 (methomyl) exhibits specificity to T-mitochondria similar to that for the HmT pathotoxin (Humaydan and Scott 1977) and produces physiological effects similar to that of toxin in isolated T-mitochondria (Koeppe et al. 1978). The causes of such physiological responses are not known. They are genetic, since T-mt-DNA differs from mt DNA of other cytoplasm types of maize.

27.21.2.10.2 In Vitro

The involvement of mitochondria in toxin action is further supported by studies of toxin-treated leaf mesophyll protoplast (Earle 1982). Within 12–24 h of the addition of HmT toxins to T-leaf protoplast culture, protoplast stops growing and distorts (Pelcher et al. 1975, Earle et al. 1978). Within 1–3 days, depending on toxin concentration, all toxin-treated T-protoplasts collapse, leaving a cluster of damaged dark-green chloroplasts. In these, the first ultrastructural changes are mitochondrial swellings, cristae disruption and matrix leaching. These changes resemble those occurring in HmT toxin-treated isolated T-mitochondria (York et al. 1980). The mitochondrial damage occurs in both fresh and cultured, green and nongreen T-protoplasts (Earle 1982). Nearly similar mitochondrial changes occur during early pollen abortion in T-anthers (Warmke and Lee 1977). C- and S-anthers do not exhibit these mitochondrial changes during early pollen abortion (Lee et al. 1979, 1980). Biochemical investigations of ATP level in T-protoplast have also evidenced the primary effect of HmT toxin on mitochondria. Luciferin-luciferase assays reveal that within 1–2 min after toxin treatment, steady-state ATP levels of dark-grown T- but not N-protoplasts drop sharply (Watson et al. 1979). Thus it appears that the oxidative phosphorylation of T-mitochondria is inhibited in situ as well as in isolated mitochondria. However, it is not clear how HmT toxin exactly affects the oxidative phosphorylation of T-mitochondria and how toxin sensitivity of somatic cells relates to pollen abortion in T anthers. Isolated T-mitochondria tend to lose their sensitivity to HmT toxin when held on ice for several hours (Pham and Gregory 1980). This age-associated phenomenon can be prevented by the addition of DDT, dithiothreotol or nitrogen. This indicates the occurrence of oxidation of certain mitochondrial components during ageing. Thus certain specific and subtle mitochondrial alterations are associated with HmT sensitivity. Other evidence associating mitochondria with gc-mst and disease susceptibility is from Gengenbach and Green (1975), who utilized HmT toxins to select resistant cells from T-cytoplasm callus. Mature plants differentiated from this callus were resistant to the fungus and to its toxin; mitochondria were similarly unaffected by the toxin and the plants were fertile (Gengenbach et al. 1977). Genetic studies indicate that these alterations were cytoplasmic and not nuclear. Crossing the plants as female to a nonrestoring nuclear genotype resulted in fertile progeny. HmT toxins cause irreversible swelling, uncoupling of oxidative phosphorylation, inhibition of malate/pyruvate or oxogluterate oxidation and increase the oxidation of NADH (Miller and Koeppe 1971, Peterson et al. 1974, 1975). These responses are initiated within seconds after addition of the toxin to mitochondrial preparations. Evidence from in vitro reactions of T-mitochondria to the toxin has been substantiated by in vivo physiological reactions, namely increase in respiration rate of detached leaf tissues (Bednarski et al. 1977). Rapid cytological effects of toxins on mitochondria of root cap cells and swelling and internal disorganization occur in T- but not in N-mitochondria (Aldrich et al. 1977). Electron microscope studies of maize root tip cells exposed to HmT toxins reveal that mitochondria are affected very early after such treatments (Aldrich et al. 1977). The toxins interact with the first ATP-coupled site of the electron transport chain, an inner membrane location (Peterson 1975). The inner membrane proteins are thus the obvious candidates as mitochondrial gene products (Schatz and Mason 1974).

27.21.2.10.3 In Cultures

Selection of toxin-resistant cell lines from T-cytoplasm HmT toxin-susceptible cell cultures was elegantly performed by Gengenbach and his co-workers. Immature embryos of the N-cytoplasm inbred line A188 and T-cytoplasm BC-derived A188 line were cultured and HmT toxin was incorporated in the culture media by Gengenbach and Green (1975) and Gengenbach et al. (1977). Whereas the toxin had no effect on N-cytoplasm cultures, a concentration-dependent inhibition occurred in T-cytoplasm cultures. These investigators selected toxin-resistant cultures from T-cytoplasm cultures through a series of 4–5-week interval transfers to media containing successively higher toxin concentrations. During such initial selection cycles, growth of the cultures was slightly inhibited and the fastest-growing tissues that also appeared to retain plant regeneration capacity were selected at each transfer. During the early stages of selection, toxin-resistant as well as toxin-susceptible mst plants regenerated from these cultures. After successive selection cycles, stable resistant cultures could be isolated and finally during the fifth selection cycle, two fast-growing cultures were obtained on the media having toxin sufficient to kill nearly all other cultures. From these two cell lines and from 15 other resistant lines obtained subsequently, 167 plants were regenerated by Gengenbach and his co-workers over a period of 18 months. In addition to HmT toxin resistance, these plants were resistant to *H. maydis* fungus as well. Of these 167 resistant plants, 97 were mft, as they shed fertile pollen from the exserted anthers and the remaining 70 were mst, exhibiting anomalies in tassel, anther or pollen development (Gengenbach et al. 1981). No progeny could be raised from ear pollination of regenerated resistant 70 mst plants. However, several tassel seeds were produced by one such plant that had a pistilled tassel. The progeny was disease-resistant and mft. These regenerated plants were pollinated or crossed by the N-cytoplasm inbred line A188 and many such mft plants were crossed on to T-cytoplasm plants (Gengenbach and Pring 1982). Genetic tests indicate mft and toxin resistance as stable, cytoplasmically inherited and nonseparable either by selection for resistance in T-cytoplasm cultures or by subsequent sexual crosses (Gengenbach et al. 1981). These investigations reveal that certain T-cytoplasm maize strains, on growing in tissue culture media containing HmT toxin, produced callus growth that is toxin-resistant and that plants regenerated from such T-cytoplasm callus cultures are not only resistant to *H. maydis* fungus but to its toxin as well and are mft in addition. These results were confirmed by Brettell et al. (1980), who used similar maize strains and corresponding culture conditions. They, however, added an interesting finding. Of 60 plants regenerated from unselected control cultures that were not inoculated with T-toxin, 35 were mft and toxin-resistant. In addition, one regenerant from unselected culture was mft but toxin-sensitive. Later experiments involving this exceptional plant (Dixon et al. 1982) indicate that it may have been a chimera, since only sterile sensitive and fertile resistant plants segregated in F_2. Barring this exception, mft and toxin resistance are maternally transmitted. This is further evidenced by the Umbeck and Gengenbach's, experiment (unpublished data) in which two mft toxin-susceptible plants from unselected growing cultures of T-cytoplasm A188 plants developed. Out of 224 plants, 221 are mft and toxin-resistant and only three plants are mst and toxin-susceptible. The results of induced mutagenesis and culture experiments reveal that mft and HmT resistance are transmitted as extranuclear traits, and that the mitochondria isolated from resistant mutant plants are resistant to HmT toxin also. Furthermore, the mutants are also resistant to insecticide methomyl (Cornu et al. 1981, Gengenbach et al. 1981). Thus, as per host-pathogen relations, these new types are completely identical to N-cytoplasm plants. However, in their progenies some abnormal mst plants appear, but these could not be maintained by maintainer lines (Brettell et al. 1980, Cornu et al. 1981, Gengenbach et al. 1981).

27.21.2.10.4 In DNA Structure

Restriction endonuclease analysis of mt-DNA isolated from mft and HmT toxin-resistant revertants has given inconclusive results. In one case three of the four resistant lines analysed by Gengenbach et al. (1981) exhibited slight differences in band patterns, but all were similar to T-cytoplasm rather than to N-cytoplasm. Two other revertant strains exhibited a banding pattern indistinguishable from T-cytoplasm plants (Brettell et al. 1980). Following the Kemble and Bedbrook (1980) technique, the electrophoretic analysis of undigested mt-DNA of mft toxin-resistant regenerant obtained by Brettell et al. (1980) from T-cytoplasm culture (not inoculated with HmT toxin) indicated its banding pattern similar to that of T-cytoplasm plants. The banding patterns of the toxin-resistant plants selected from T-cytoplasm cultures of Gengenbach et al. (1977) and Brettell et al. (1980) were similar to those in T-cytoplasm plants (Kemble and Bedbrook 1980). On the other hand, the mutagen-induced mft toxin-resistant lines seem to possess mt-DNA similar to N-cytoplasm plants (Paillard and Berville, unpublished). Thus they are truly mft toxin-resistant N-cytoplasm mutant plants. Following the technique of Forde et al. (1980), it was found that these plants exhibit the same mt-DNA translation products as N-mitochondria. They breed true and cross freely with normal mft maize plants. Thus they are stable induced cytoplasm mutants of maize whose reversion is not only genetic but genotypic. It is difficult to offer a general explanation for T-cytoplasm revertants obtained after tissue culture and mutagen treatments that resemble each other phenotypically, but whose mt-DNA differs from each other. Moreover, mt-DNA changes in revertants from different tissue culture sources are also divergent. These alterations could be due to mitochondrial recombinations.

27.21.2.10.5 Nature of Reversion

Plants raised from the cultures that were regenerated from the selected T-toxin-resistant cultures after five or more periods of serial subcultures of tissue cultures on medium containing T-toxin were mostly fully toxin-resistant and mft (Gengenbach et al. 1977). When these mft plants were used to pollinate toxin-susceptible T-cytoplasm plants, their progeny was toxin-susceptible and mst. For these results, the authors forwarded two explanations: (1) The T-mitochondrial genome is heterogenous where selection against susceptible mitochondrial genome segregated few plants with resistant mitochondrial genome, or (2) that one locus of the mitochondrial genome conditions both T-sterility and T-toxin sensitivity. Mutation for T-toxin resistance accompanies fertility restoration. The high frequency of resistant cultures obtained by Brettell et al. (1979) suggests the obtained resistance as not being due to mutations but to selection of resistant mitochondria from a genetically heterogenous mitochondrial population already present in the T-cytoplasm. It is not known, however, if such a heterogenous mt-DNA population exists in T-cytoplasm plants. Wheat mt-DNA is known to possess a heterogenous population of DNA molecules (Quetier and Vedal 1977), but these molecules do not characterise S- or N-cytoplasms. The presence of T-toxin in medium may favour multiplication of a small population of normal T-toxin-resistant mitochondrial genomes already present in the T-cytoplasm (Gengenbach et al. 1977, Brettell et al. 1979). In cultures, N- type mitochondria or mitochondrial genomes were preferentially selected and amplified later. This interpretation is not valid, as mft-resistant revertants isolated by Gengenbach et al. (1977) have mt-DNA characteristics of the T-cytoplasm (Gengenbach et al. 1981). Furthermore, all the regenerated plants analysed have characteristics of the T-cytoplasm mt-DNA irrespective of their sterility/fertility and sensitivity/resistance traits. The traits mft and resistance to HmT and to its toxin are co-transmitted. No clearcut breakage of association between mst and toxin susceptibility occurs under cultural conditions. Gengenbach et al. (1981) believe that the toxin resistance and mft are non-separable and cytoplasmically inherited. The mst in toxin-resistant plants obtained by Gengenbach et al. (1977) is due to anomalies in tassels, anther and pollen development. Progeny of these mst plants was mft- and disease-resistant. Thus their mst was not genetic but due to developmental anomalies. The mt-DNA gel electrophoresis pattern and translation product of the toxin-resistant revertants resembled that of T-cytoplasm toxin-susceptible plants. Thus it appears that the toxin resistance induced in genetically toxin-susceptible mst plants was due to the altered physiological state of the mitochondria or its fragmentation pattern. The alteration was enhanced by pathotoxin additions in the culture media. Such alterations in mt-DNA fragment pattern do occur under cultural conditions in T-cytoplasms of maize (Kemble et al. 1982). This alteration due to culturing and regeneration may have led to changes in pathotoxin susceptibility. Such changes are rare in conventional plant propagation and breeding. The toxin-resistant mft plants obtained after mutagenic treatments of seeds of T-cytoplasm maize by Cassini et al. (1977) represent true induced stable mutants, whose mt-DNA pattern and product differ from that of T-cytoplasm plants and resemble those of N-cytoplasm plants. Thus the gene(s) for mst and toxin susceptibility lies in mt-DNA. Loss of the 6.5–7.0 kb fragment was detected by Gengenbach and Pring (1982) in the DNA of mft-resistant lines that arose from T-cytoplasm cultures. This fragment may thus be associated with mst and toxin susceptibility expression. The authors believe that reversion to mft and toxin resistance may be conditioned by nucleotide base change or by rearrangements within a portion of 6.5–7.0 kb fragments. These alterations could have been brought about by mutagens used by Cassini et al. (1977), and have led to the production of mft toxin-resistant plants from mst toxin-susceptible cytoplasm plants. The above-results indicate that both T-mst and toxin sensitivity are commonly due to *fr-c* gene interaction. This is supported by γ-ray and EMS treatments of T-mst's (Cassini et al. 1977, Cornu et al. 1981), the derivatives being both mft- and toxin-resistant.

27.21.2.11 S-Mitochondria

27.21.2.11.1 Specific DNA Molecules

The advent of the restriction endonuclease enzyme digestion technique followed by aragose gel electrophoresis provides an opportunity to characterise mt-DNA of various maize cytoplasms, each of which exhibits characteristic banding pattern. The technique has led to the detection of two unique small molecular weight plasmid-like double-stranded DNA molecules (S-1, S-2) in S-cytoplasm maize plants (Pring et al. 1977). These molecules are present in every S-cytoplasm, irrespective of its source and genetic background, but are absent in N-, T- or C-cytoplasm. They are linear, comprise 6180 bp and 5175 bp molecules and their molecular weights are 4.15×10^6 and 3.45×10^6 Da, respectively. Each molecule has a similar reverse repeated sequence of approximately 200 bp at the terminus of the duplex molecules. Transmission of these DNA's is exclusively maternal. When plants having S-cytoplasm and Fr_3 gene are used as a pollen parent in cross with N-cytoplasm plants, the unique DNA is absent in progeny plants; but when S-cytoplasm mst plants are used as a female parent, the unique DNA's are transmitted to the offspring regardless of the male parentage. These molecules are absent in S-cp-DNA, but are consistently present in S-mt-DNA. They share limited sequence homology in addition to their terminal inverted repeat sequences , as well as homology to main band mt-DNA (Kim et al. 1979). Using eight restriction endonucleases, Kim et al. (1982) constructed physical maps of these molecules and found each DNA molecule to be a homogenous population. Heteroduplex analysis showed that almost one-third of the S2 sequences are homogeneous with S1 DNA, with a 1350-bp segment at the left terminus and a 200-bp segment at the other. Electron microscope studies reveal an unusual base sequence arrangement of these unique DNA's (Levings et al. 1978). When these DNA's are denatured and briefly self-annealed, stem loop configurations consisting of a short double-stranded stem and a large single-stranded loop appear. The occurrence of stem loop configurations due to the presence of terminal inverted repeats comprising 200 bp in which identical DNA sequences lie in reverse polarity occur at either end of the molecules. Such inverted repeats carried by S1 and S2 mt-DNA plasmids are reminiscent of the insertion sequences (IS) which are common in the mechanism by which transposable elements are inserted into bacterial chromosomes. Thus it appears that these IS-like sequences equip S1 and S2 plasmids for a similar role and the plasmids correspond to episomal F factor proposed by Laughnan and Gabay (1975, 1978). The physical structure of S1 and S2 DNA's, their apparent interaction upon cytoplasmic reversion to fertility and their sequence homology with N-mt-DNA suggest that they arose as an IS directed deletion mutation in mst-S-mt-DNA from mft-N-mt-DNA.

The S1 and S2 plasmids in the majority of the S-cytoplasm strains are present in near equimolar amounts (Pring et al. 1977). Restriction analysis indicates that the replication of these plasmids is amplified about five fold in comparison with the main high molecular weight mitochondrial genome. However, in the M825 inbred line background, the mst-VG plants exhibit a striking reduction in the molar equivalents of the S2 compared to the S1 plasmid (Levings et al. 1980). This difference is due to the genetic background, as the rate of reversion is extremely high in this inbred line (Laughnan and Gabay 1975, 1978). Thus likewise, the differential replication of S1 and S2 plasmids in M825 nuclear backgrounds may be related to the high reversion rate in these strains. This indicates that the nuclear genotype basically controls the replicating property of S1 and S2 plasmids. This is substantiated by the findings of Carlson et al. (1982), who found that in inbred line 38–11, the molar equivalents of S1 plasmids compared to S2 plasmids are significantly reduced. The authors opine that the difference is determined by the nuclear genotype rather than the cytoplasm genome. S1 and S2 molecules possess some structural features resembling the translocable drug-resistant elements from prokaryotes (Kleckner 1977). Their size, 6200 and 5200 bp for S1 and S2 DNA's, respectively lies within the 4100 (Tn4) to 9300 bp (Tn10) size of translocable element, as is the presence of short-lived terminal repeats on each (Tn-1-Tn4). The presence of a 1150-bp homologous sequence penultimate to 200 bp inverted repeat at one terminus of each S1 and S2 DNA's further points out their structural similarity to prokaryotic translocable elements, i.e. inclusion of insertion sequence (IS) as in Tn9 and Tn10. The structural similarity between S1, S2 and prokaryotic translocable elements may also be functional (Kim et al. 1982). It is not clearly known if the plasmid-like DNA's are transcribed and translocated in host cells. An indication of their expression is revealed from Thornbury's (1979) observation, who found a protein of molecular weight 130,000 restricted to S- but not to N-, C- or T-cytoplasm plants. Its association with plasmids was deduced from the fact that this protein molecule, as well as the plasmids, was absent in S-cytoplasm fertile revertants. This indicates an association of plasmids with the protein molecule. Thus the protein may be a gene product of the plasmids and may condition mst. Several high molecular weight polypeptides are synthesized by isolated S-cytoplasm mitochondria but not by N-cytoplasm mitochondria (Forde and Leaver 1980). This suggests that the S-cytoplasm may be a mutant that is deficient in processing RNA transcribed from mt-DNA or protein translated by mitochondrial polysomes. The presence of Fr_3 nuclear restorer gene specifically suppresses S-mt, even though S1 and S2 DNA molecules are present in the mitochondria. This gene does not influence in vitro synthesis of these large polypeptides (Forde and Leaver 1980). It appears that the unique sequences in S1 and S2 control mRNA or protein synthesis and/or processing, the homologous sequence mediate nuclear control of genomic organisation and expression. Although the two unique plasmidlike mt-DNA's are present only in S-cytoplasm of maize and are absent in the cytoplasms of N, T, C maize, *Tripsacum* and teosinte (Levings and Pring 1979), of 81 Latin American maize races studied by Weissinger et al. (1982), 12 South American races possess such unique DNA in the mitochondria. These plasmidlike two DNA molecules R-1 and R-2 appear identical, double-stranded, and possess terminal inverted repeats. Both these DNA molecules share considerable homology with S1 and S2 molecules of the S-cytoplasm, but unlike them are not associated with pollen sterility.

Koncz et al. (unpublished) cloned S1 and S2 DNA species in plasmid PBR 322 by the homopolymer extension method. Probes made with these recombinant plasmids were used to establish the homology between high molecular weight mt-DNA's of mft and mst cytoplasms and small mitochondrial plasmidlike molecules. Hybridisation and mapping data show that S2 DNA copies are homologous with the sequence of the normal mitochondrial genome, and spontaneous reversion of mst to mft is accompanied by the disappearance of free S1 and S2 DNA's from mitochondria of the S-cytoplasm. The hybridisation results suggest that S1 and S2 DNA's integrate into high molecular weight mt-DNA during sterility-fertility reversion. Cloning of N-cytoplasm DNA and use of S1 and S2 labelled probes by Lonsdale et al. (1981) revealed the occurrence of S1 and S2 sequence in normal genome as single copies at unique sites and each of them to be flanked by a 26 kilobase (kb) repeated sequence. Lower mol wt plasmidlike elements are also present in some fertile and sterile maize cytoplasms. The mitochondria of N-, C-, S- and T-cytoplasm plants contain a 1.94 kb pair of supercoiled DNA species (Levings *et al.* 1980, Kemble and Bedbrook 1980). In addition, N-, C- and S-mitochondria contain a DNA species of about 2.35 kb pairs that is absent in T-cytoplasm. C-mitochondria possess two circular DNA's having about 1.57 and 1.42 kb pairs that are absent in other maize cytoplasms (Kemble and Bedbrook 1980). Thus six plasmid-like mt-DNA species are known from four maize cytoplasms, but except for S1 and S2, none of them is implicated in sterility reversion. However, in 12 South American mft maize strains, similar molecules do exist (Weissinger et al. 1982). One of them is identical with the S2 plasmid and the other to some extent to the S1 plasmid, but it has a higher molecular weight. The significance of this S2-line plasmid and of modified S1-like plasmid in these mft maize strains is not clear.

27.21.2.11.2 Unique Double-Stranded RNA

In one of the S-group cytoplasms, *cms*-L in the inbred line W182BN, designated LBN, two double-stranded RNA's (ds-RNA's) of apparent molecular weight 192 and 0.5×10^4 were detected by Sisco *et al.* (1984). The ds-RNA's are associated with mitochondria and are maternally transmitted. Nuclear genotype influences the recoverable ds-RNA amount. Origin, effect and precise intracellular location of these LBN ds-RNA's are not known. Existence of single-stranded RNA's (ss-RNA's) homologous to LBN ds-RNA's was documented by Schuster et al. (1983) in other S-type cytoplasms and in RU – a cytoplasm not associated with mst (Weissinger et al. 1982). The LBN ds-RNA may be a replicative form of these ss-RNA's like that found in many plant viruses, but the ds-RNA's are not viral components, as no such particles exist in electron microscope scans of LBN leaf extracts. Also attempts to transmit the ds-RNA's by aphids or mechanical inoculation or to translate them (after denaturation) in a wheat germ system in vitro were unsuccessful (Sisco et al. 1984). Such ds-RNA's in *Vicia faba* exhibit a strong correlation with mst and cytoplasmic spherical bodies (CSB's), as both ds-RNA's and CSB's disappear in the fava beans restored to fertility. However, unlike maize, restoration is irreversible in fava beans.

27.21.2.11.3 Callus Culture

From immature, embryos callus cultures of WF9-S mst were established and maintained by subculturing (Chourey and Kemble 1982). Mostly the callus is compact and "organised" but occasionally a loosely organised "friable" form also veloped. Whereas the organised callus demonstrated hybridization at the position on agarose gel electrophoretograms corresponding to linear S1 and S2 molecules, no such hybridization occurred at the corresponding positions in preparations of mt-DNA from "friable" callus. Since plants could not be regenerated from either of the callus types, it cannot be determined if apparent integration of S1 and S2 sequences into main mt-DNA is associated with a change from sterility to fertility in S-cytoplasm plants.

27.21.2.12 *c*-Gene Location

It is fully established that the gc-mst trait does not exhibit Mendelian inheritance, and is not thus directly under the control of nuclear genes only. Since the transmission is maternal, the inheritance is extranuclear and of cytoplasmic type. The involvement of DNA containing cytoplasmic organelle mitochondria or chloroplast in the control and coordination of mst is certain in many cereals. Evidence from several diverse sources suggests that *c*-genes in maize are located on mt-DNA, although it has also been conjectured that the sterility might also be controlled by cp-DNA or viruses. Constant and specific mt-DNA differences in form and structure between N-, T-, C- and S-cytoplasm are known. Such differences in cp-DNA are inconsistent and insignificant. Therefore, the mitochondrial genome is associated with mst and *c*-gene location in maize. This is evidenced from the following.

i. *mt-DNA Structure and Translation Products.* The organization of the maize mitochondria genome is revealing but complex. It has more than one class of DNA molecule and exhibits high molecular heterogeneity. Each cytoplasm possesses a characteristic distribution of contour lengths in mt-DNA molecules (Table 27.20, Fig. 27.24). Although much is not known about the exact mechanisms responsible for this molecular heterogeneity, specific structural differences in mt-DNA may be associated with each sterile cytoplasm type. After analysing fragment patterns produced by restriction enzymes, Hind III, EcoRI, Bam I, Sal I, mt-DNA of N-, T-, C- and S-cytoplasms are readily distinguishable from each other (Pring and Levings 1978), as each of the maize cytoplasms possesses unique mt-DNA. This is confirmed by agarose gel electrophoresis (Kemble et al. 1980) and mitochondrial translation products of the four cytoplasms (Forde et al. 1980). The presence of distinct mt-DNA by each cytoplasm provides circumstantial evidence that the mitochondrial genome is associated with mst. These DNA differences persist when the four cytoplasms are examined in different nuclear backgrounds. This indicates that the mitochondrial genome is conserved in diverse nuclear genotypes.

Investigations of biosynthesis of isolated mitochondria by Forde et al. (1978, 1980) and Forde and Leaver (1980) have led to the detection of discrete differences in mitochondrially synthesised polypeptides, especially in the mean frequency and lengths of two major classes of mt-DNA molecules, of many restored and non-restored versions of mst and mft strains (Table 27.20). Isolated mitochondria of each of the four cytoplasm plants (N, T, C and S) synthesise unique polypeptides characteristic to each cytoplasm type. In addition, S-mitochondria synthesise eight characteristic high molecular weight polypeptides not synthesised by mitochondria of other cytoplasm types. These are not synthesised by revertants, indicating thereby their clear association with S-cytoplasm mst.

ii. *T-Mitochondrial Specificity.* Unlike N-mitochondria, T-mitochondria are highly sensitive to HmT fungus and to its toxin, which causes swelling and disruption of cristae and matrix density. Almost similar mitochondrial changes occur in tapetum and mid-anther layers of T-cytoplasm plants prior to pollen degeneration (Warmke and Lee 1977). T-mitochondria synthesise a 13,000 M_r variant polypeptide (Forde et al. 1978). A model for possible involvement of this variant polypeptide in T-mst and HmT toxin susceptibility was proposed by Forde and Leaver (1979, Fig. 27.26). Possible roles assigned by the authors to this polypeptide include: (1) provision of a binding site for HmT toxin and "compound x", (b) altering mitochondrial membrane conformation, thereby creating or exposing a binding site on another membrane component, or (3) weakening protein-lipid or protein-protein interaction in the membrane, rendering it

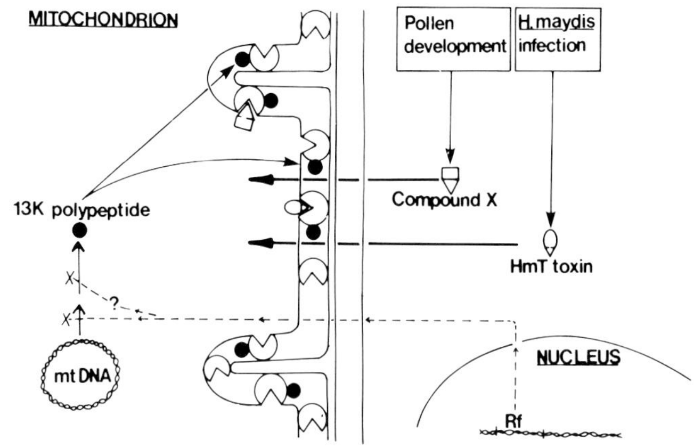

Fig. 27.26. *Zea mays.* A model for possible involvement of 13,000 M_r polypeptide in T-cytoplasm male sterility and susceptibility to HmT toxin. (Forde and Leaver 1979)

susceptible to disruption by these substances. Translation of this variant polypeptide on mitochondrial ribosomes and its cytoplasmic mode of inheritance indicate that the amino acid sequence is encoded over mt-DNA. Likewise, in yeast all proteins made on mitochondrial ribosomes are encoded by mitochondrial genes. There is no indication that m-RNA is imported from the nucleus (Borst and Grivell 1978). Since T-mitochondria fail to synthesise a 21,000 M_r polypeptide synthesised by N-mitochondria (Forde and Leaver 1980), deletion or other mutation in the structural gene for this larger peptide may have led to the synthesis of an aberrant form of protein, the 13,000 M_r polypeptide which causes mft loss in T-cytoplasm plants. Forde and Leaver (1980) evidence that gene Fr_1 represses the polypeptide synthesis. Fr_2 may regulate the synthesis of compound X during pollen development, reduce mitochondrial membrane permeability to compound X or rectify effects of 13,000 M_r polypeptide by altering properties of the inner mitochondrial membrane in some other way (Forde and Leaver 1979).

iii. *S-Mitochondrial Plasmids.* In undigested mt-DNA preparations from S-cytoplasm plants, two low molecular weight plasmidlike double-stranded DNA molecules, S1 and S2, occur (Pring et al. 1977). These plasmids have a major role in mft control of S-cytoplasm plants. Analysis of unrestricted DNA from 23 cytoplasmic revertant strains indicates that in each case the reversion event is associated with the disappearance of S1 and S2 plasmids (Levings et al. 1980). Also, restriction enzyme fragment studies (Kim et al. 1980 and Levings et al. 1980) of mt-DNA from such revertants, coupled with hybridization analyses using ^{32}P-labelled S1 and S2 plasmids as probes, have revealed that reversion to mft is associated with the integration of plasmid sequences into the main mt-DNA genome. Such studies also indicate that the independently occurring reversions to mft are associated with integration of plasmid sequences at different sites in the high molecular weight mitochondrial genome. This close relationship between mst reversion and mt-DNA plasmid disappearance is strong evidence favouring genetic control of mst in S-cytoplasm being located in mt-DNA.

iv. *Conclusions.* The circumstantial but compelling evidence given above suggests that mt-DNA is the carrier of the *c*-gene in maize. The unusual organisation of mt-DNA, its large size and its apparent packaging among several molecules (molecular heterogeneity) lend it a realistic potential for contributing to the sterility trait in maize cytoplasm. T-mt is highly susceptible to HmT and to its toxin, S-mt-DNA possesses two low molecular weight S1 and S2 molecules, which, if lost, revoke mft. In contrast to high-heterogeneity within maize and teosinte lines (Timothy et al. 1979), the restriction pattern of cp-DNA from many ancient and present-day inbred lines of maize and races of teosinte (a close relative of maize) are almost invariant (Pring and Levings 1978, Timothy et al. 1979). Two-dimensional electrophoresis, using a gradient of denaturing agents, indicates some differences in cp-DNA's of N-, C- and T-cytoplasm maize lines (Jigeng and Yi-nong 1983). This indicates that certain changes in maize cp-DNA do occur following mst.

27.21.2.13 Plasmid as Male Sterility Gene Carrier

The existence of episome-like elements to explain the unusually high instability rate of S-cytoplasm maize plants was postulated by Laughnan and Gabay (1975). This instability was expressed as fertile revertants in nuclear backgrounds that had no dominant *Fr* genes. Genetic analyses of these fertile revertants indicated the existence of a cytoplasmic mutation to fertility (up to 10% frequency) and gene mutation of low frequency. The authors postulated the existence of episome-like elements that are in the form of two low molecular weight plasmid-like double-stranded DNA molecules, S1 and S2. When fixed in the cytoplasm, the plasmids restore mft. They are also capable of translocation to and incorporation into maize nuclear chromosomes. Laughnan and Gabay (1978) mapped loci for ten such possible integrative elements. Restriction fragment studies of mt-DNA from the revertants, coupled with hybridization analyses using ^{32}P labelled S1 and S2 plasmids as probes (Kim et al. 1980, Levings et al. 1980), reveal the integration of plasmids into the main DNA genome. Since restricted mt-DNA's from cytoplasmically reverted strains have unique bands which hybridise well with labelled S1 and S2 plasmid probes, independently occurring cytoplasmic reversions to mft are associated with the integration of the plasmid sequence at different sites in the high molecular weight mt-DNA. Thus it is evident that in S-cytoplasm maize lines, mst is regulated by the insertion of plasmid into mitochondrial genome. Mitochondria of fertile S lines possess entire episome inserted into mt-DNA, in contrast to sterile lines, in which mitochondria have free plasmid copies and only the insertion sequence of the episome in their genome. This is supported by hybridisation experiments with p CK 5D1 and C, T and N mt-DNA's (Koncz et al. 1980). If S1 and S2 plasmids represent a male fertility episome whose presence in a certain form in the high molecular weight portion of the mt-DNA genome of maize is necessary for mft expression, reversion from sterility to fertility would accompany the transpositional events involving S1 and S2 molecules. This appears to be so (see Levings et al. 1980). However, it is not known if the reversions to fertility that occur at the nuclear level involve corresponding transpositions in which plasmid sequences are integrated into the nuclear chromosomes. The above account indicates that fertility reversions in S-maize are correlated with the loss of freely replicating S1 and S2 plasmid-like mt-DNA's and rearrangements of S1 and S2 sequences in high molecular weight mt-DNA. These results were obtained with lines established from individual S-cytoplasm revertants and propagated for two or more generations. The following evidence, however, argues against the direct involvement of S1 and S2 episomes:

i. R1 and R2 plasmid-like molecules, very similar to the S1 and S2 plasmids of S-cytoplasm of maize, occur in mt-DNA's of fertile cytoplasms of some Latin American maize races (Weissinger et al. 1982).

ii. S HMW mt-DNA contains virtually the entire S2 episomal sequence but only 50% S1 episomal sequence. The new restriction bands of S-fertile revertants (Levings et al. 1980) might have originated from the S1- and S2- related sequences already present in the HMW gc-mst-S mt-DNA.

iii. Sequences to S1 and S2 are lost in fertile T-cytoplasm revertants. This indicates that these sequences are not indispensable for normal pollen development in the absence of *Fr* genes. Moreover, mst in C and T lines is not due to the loss of these sequences (Kemble et al. 1982).

From the above, it appears that a strictly causal relationship cannot be drawn between the presence of autonomous S1and S2 plasmid mt-DNA's and S-mst phenotype in maize. The vegetative tassel tissues of S-revertants still possess S1 and S2 plasmids at relatively high levels. Their continued presence in seeds from revertant sectors suggests that, concomitant with the reversion event, replication of S1 and S2 ceases, but a finite number of cell divisions is required to dilute the remaining S1 and S2 levels (Carlson et al. 1983). The stability of S1 and S2 after replication is diminished, probably due to covalently bound terminal proteins on these plasmids. In microsporogenesis, blockage of transmission of nonreplicating DNA's might explain the absence of S1 and S2 plasmids in pollen from revertant sectors.

27.21.2.14 Causes of Male Sterility

In maize, as in many other plants, both nuclear and cytoplasmic genes control and coordinate mst either independently or conjointly. The precise nature of this interaction and cooperation are unknown, although the effects are vividly manifested. Harvey et al. (1972) suggested that certain genes responsible for mft initially located on organelle DNA (mt-DNA?) have been transposed to a nuclear site through the evolutionary processes. When redundancy of genetic information exists between the organelle genome and the nuclear genome, a portion of this information at one site (organelle) could be lost without any adverse effect so long as the same information was retained by the other (nucleus). Although a reciprocal situation is also possible, the nucleus seems to have established a dominant role in the information coding. Since parallel evolution is not necessarily equal, distribution of mft genes between the organelle and nuclear genome need not be equivalent among all segments of a population. This accounts for the existence of several cytoplasm types and for the requirement of specific nuclear restorer genes for each mst type. Hybridization amongst such population members could result in such an organelle nuclear genome recombination where the coded information would be absent in the organelle and in the nucleus, and lead to mst; but the current information is too sketchy to substantiate this theory.

Restriction endonuclease digestions reveal N mt-DNA to have a higher molecular weight than T mt-DNA. This suggests that gc-mst trait may be due to deletion rather than to a simple gene mutation. Also the mitochondrial genome of mft's contains a larger complement of genes necessary for mft than their cytoplasmic mst counterparts. Thus the gc-mst condition can be countermanded by the addition of certain dominant nuclear *Fr* genes that restore fertility. However, too little is known about *Fr* genes and how they substitute for *c*-genes controlling mst. Probably they have evolved from the nucleus and have created an alternative pathway to circumvent the cytoplasmic lesions. Alternatively, they may simply be cytoplasmic genes that have been transposed to a nuclear site. This is supported by the studies of S-cytoplasm, where the episomal event has been proposed to explain the newly arising nuclear restorer genes. It may, however, be that the S-case is unique and is not applicable to the restorers of T and C. For S-cytoplasm, *Fr* genes appear to possess episomal features (Laughnan and Gabay 1975) and the specific characteristics manifested by each of these restorer genes may either be due to their differences in the integration sites or because of their basic structural differences. In a bacterial system, episomes have the capacity to be transposed from one site to another or to be lost entirely. According to Laughnan and Gabay (1975), attachment or fixation of specific fertility elements either in the cytoplasm or in the nucleus leads to fertility reversions of cytoplasmic or nuclear type, respectively. However, it is not confirmed that the *c*-genes lie in mt-DNA. S-cytoplasm exhibits unusual instability in mst traits (Pring and Levings 1978). The S-mitochondria contains plasmidlike DNA molecules (Laughnan and Gabay 1975, 1978). It is equally possible that cytoplasmic S mst is caused by these intra-mitochondrial plasmid DNA's. Since plasmids induce high variability, they might be responsible for the observed instability of S-sterility. Accordingly, a model was suggested by Laughnan and Gabay (1975) implicating an episomal element in the S-cytoplasm. In this model the episome is capable of translocation from the cytoplasm to the nuclear chromosomes, causing reversion from mst to mft.

Turbin and Palilova (1970) postulated a mitochondrial site of a factor causing gc-mst, the site lodging two mitochondrial genes, one structural (*Ps*) and one repressor ($P\overset{+}{r}$). Products of the nuclear genes (*Fr*) bind mitochondrial repressor gene and allow the mitochondrial structural gene to function in N-cytoplasm plants. In gc-mst plants, the mitochondrial repressor gene mutates to a more active state ($\mathrm{P}\overset{+}{\mathrm{r}} \rightarrow \mathrm{P}\bar{\mathrm{r}}$) and therefore more active nuclear controlling gene (*Fr*) is required to control the repressor and allow the mitochondrial gene to operate. Thus in the presence of sterile cytoplasm, the combination $P\bar{r}$, *fr* leads to mst and the combination *Pr, Fr* to fertility restoration. This theory regards "rate-limiting" rather than structural gene changes as the cause of gc-mst. Since mitochondria of the T-cytoplasm are implicated in the susceptibility of the *H. maydis* T race, Peterson *et al.* (1974, 1975) suggested that the pathotoxin of this race has a binding site on the inner membrane of T-cytoplasm mitochondria and that this site is controlled by cytoplasmic DNA. *Fr* genes tend to partially correct the typical aberrant electron-transport reactions. This suggests that *Fr* gene products are the mitochondrial inner membrane proteins and that they alter a defect in this membrane. However, since *Fr* genes do not help in overcoming race T infection or toxin susceptibility of T-cytoplasm in vivo (Hilty and Josephson 1971, Gracen and Grogan 1972), the problem appears to be not so simple. On the basis of altered mitochondrial structure (irreversible swelling and changes in respiratory and oxidative phosphorylation rate) due to the HmT pathotoxin, Flavell (1974) proposed a model in which he visualised that the anthers produce a substance capable of inhibiting altered organelles (mitochondria or plastids) in the pollen of mst plants. This inhibiting substance is ineffective in pollen of N-cytoplasm plants. *Fr* genes and/or environmental variations alter the amount or the availability of the inhibiting substance, or altenatively correct the altered organelle structure. Thus any of these changes allows the aberrant mst organelles to operate at more or less normal levels, yielding a near to normal amount of pollen fertility.

Both the above models postulate the existence of a gene product that overcomes the effect of the sterility repressor that is specific for the inhibition of sterile pollen development. While Turbin and Palilova's model regards the repressor as abnormal and synthesised by mst organelles (mitochondria), Flavell's model postulates that the repressor is normal, is always present and is synthesised by the anthers of both N and mst plants. Furthermore, Turbin and Palilova postulate normal organelle structural genes in mst pollen, but Flavell postulates abnormal structural genes. If c-genes lie in mt-DNA, it should be possible to produce mutant and new types of sterile cytoplasms, at least in maize, in which mst is intensively investigated. Such possibilities have been explored and the reports of Lysikov *et al.* (1969) from Russia indicate mutation of a fertile cytoplasm to a sterile one using γ-irradiations and ultrasonic waves. A similar conversion was obtained in Bulgaria by Stoilov (1970), who used γ-rays and/or neutron irradiations. However, attempts to achieve such conversions using these and other mutagens have failed in the author's laboratory as well as elsewhere.

27.21.2.15 Ru-Cytoplasm

Of the 81 latin American maize races whose mt-DNA's were examined by agarose gel electrophoresis, 12 south American races termed Ru (Racinio de Uva) have unusual plasmid characteristics (Weissinger *et al.* 1982), as their mt-DNA has two plasmid-like molecules, R1 and R2. Like S1 and S2, R1 and R2 are double-stranded linear DNA molecules with terminal inverted repeats that appear to be the same in all the four plasmid-like molecules. However, the R2 molecule is nearly 5.4 kb long and homologous to S2, the smaller plasmid-like molecule of the S-cytoplasm. R1 exhibits a high degree of homology with S1, but its length is about 7.4 kb; it is 1.2 kb longer and has a unique sequence of about 2.6 kb (Laughnan and Gabay Laughnan 1983). Also R2 is not under the same nuclear control as S2. Comparison of mt-DNA restriction patterns of N-, S- and Ru-cytoplasms indicates 17 band differences between Ru and S in the upper 80% of gel but only one to two band

season (Nickerson 1960). Thus GA treatment either acts as a substitute to low temperature treatment or suppresses *ms* gene action directly in this mutant. The former possibility can be excluded, as only one GA treatment restores anther dehiscence in this functional mst corn mutant.

iv. *Maleic Hydrazide*. In Wisconsin hybrid No. 515,25 ml foliar application of 0.025% MH solution prevents anther formation (Naylor 1950a). Seed set on the ears is normal. Direct MH sprays do not induce mst, but it is induced by using diethanolamine salt having 30% MH by weight (Naylor 1950a,b). In some hybrid corn varieties, sterile tassel and stubby ears with functional pistillate flowers are developed after 600 ppm MH spray (Moore 1950). Female fertility of the treated plants is normal, as seed set is high over them. Similarly induced mst following 0.025, 0.05 and 0.1% MH applications in sweet corn was obtained by Denisen and Haber (1950). Eskew and Willard (1950) found p-mst after spraying nearly 1–4 kg MH acre^{-1} in corn plants, but MH applications proved ineffective in three corn hybrids (Warren and Dimmock 1954). Moore (1950) and Josephson (1951) found that MH treatments delayed flowering and induced pollen and ovular sterility in corn. However, tassels of 600 ppm MH-treated plants of sweet corn were sterile, with stubby and functional ears.

v. *Mendok*. Variable results of pollen sterility after spraying 0.5 and 1% of sodium α-β-dichloroisobutyrate to late- and mid-season flowering inbreds and hybrids of maize were obtained by Cameron and Eaton (1959). The treatments suppressed pollen shedding and damaged ears in field-grown plants. In the greenhouse, the sprays damaged the top ears. Neither complete suppression of pollen-shedding nor normal plant and ear growth were obtained in most of the treated maize lines. In two lines, 0.5% chemical applied after 37 days of planting prevented pollen shedding in eight out of ten main tassels, but the tiller tassels shed pollen normally. Thus complete pollen shedding suppression without ear injury was not possible with the mendok treatments. Dalapon sprays reduced plant height without influencing pollen sterility significantly (Funderburk and Davis 1960).

vi. *Molybdenum Deficiency*. This causes reduction in tassel size, anthers and their viability (Agarwala et al. 1979). A decreased activity of acid phosphatase, starch phosphorylase and an increased activity of catalase and peroxidase occurred in the developing pollen grains of the deficient plants. Their female fertility is not known but is usually reduced.

27.21.4 Hybrid Seed Production

A dramatic increase in maize grain yield occurred as farmers changed from open-pollinated varieties to hybrid maize. This led to a progressive improvement in grain yield as new and more productive hybrids were developed. Initially, it was tried to develop inbreds with sufficient vigour that could be maintained and utilized for adapted hybrid combinations for commercial production. Later, the hybrids already in commercial production were gradually improved. Whereas the development of inbreds necessitated forced inbreeding, hybrid seed production entailed detasseling of the female lines. Thus in single-cross hybrid seed production, detasseling of one inbred, in three-way cross hybrid seed production, detasseling of one inbred, and of the single-cross hybrid plants and in double-cross hybrid seed production, detasseling of two inbreds and of one single-cross hybrid parent are necessary. Utilization of g-mst and gc-mst replaced detasseling and revolutionized hybrid seed production.

27.21.4.1 Genic Male Steriles

Use of mst for hybrid maize seed production interested both plant breeders and farmers because in maize the male and female reproductive organs, although on the same plant, are borne separately in tassel and ear, respectively. Initially, the commercial hybrids were produced by growing mft inbreds A and B in adjacent rows in a production field. Tassels of one of the inbred lines A, previously identified as a high-quality seed-bearing strain, were physically removed to ensure that pollen of strain B fertilized A and seeds borne on ears of strain A had hybrid embryos. These were the superior hybrid seeds. With the detection of many g-mst in maize, it was thought that the process of hand detasseling would be eliminated. Since sterility in the majority of the mst mutants is controlled by recessive genes, the mst plants were produced from crosses that are either *Ms ms* × *Ms ms* or *ms ms* × *Ms ms*. In either case, mft plants would occur along with mst siblings and the mft plants would have to be either rogued out or detasseled for use in hybrid seed production. The inability to maintain a pure stock of mst lines limited the use of *ms* genes in commercial hybrid seed production.

27.21.4.2 Production of Pure Male Sterile Stocks

Maintenance of mst lines in the pure state is the major problem for their use in hybrid seed production. For this, two approaches have been devised, (1) development of a chromosomal-genic system, and (2) use of endogenous growth regulators. Under chromosomal-genic schemes, several proposals forwarded almost simultaneously are: viable deficiency in chromosome 9, tertiary trisomics, pollen abortion genes (Burnham 1971); deficiencies of chromosome segments, duplicate-deficient chromosomes originating from heterozygous pericentric inversions, or from reciprocal translocations (Patterson 1971), haploviable deficiency duplication (Phillips et al. 1971), telocentric trisomics (Doyle 1972) and double duplications (Phillips and Springer 1972). These are based on the principle that certain kinds of altered chromosomes (for instance those with small deletions) are not normally inherited through pollen (as the pollen that carries them aborts), but are inherited through the egg. When the breaking point of such a chromosome alteration is tightly linked with a *ms* gene, maintainer lines producing 100% mst progenies can be developed when the maintainers are crossed to ms/ms plant. The maintainers are heterozygous for a recessive *ms* gene and for a chromosome alteration. Pollen of the maintainer line transmits only the recessive *ms* allele carried on a normal chromosome. It does not transmit the dominant *mft* genes (*Ms*), as it is tightly linked with the breaking point of the altered chromosome; pollen with the altered chromosome aborts. Thus the mst plants are homozygous for *ms* gene and have only normal (unaltered) chromosomes and normal cytoplasm. The chromosomal-genic maintainers resemble gc-mst maintainers as they can be crossed as male to the sterile plants to yield 100% mst progenies, but the sterile females made with these maintainers have normal cytoplasm and shed less than the normal amount of viable pollen. Hybrids produced by this technique will be mft, since their male inbreds will be *Ms/ms* and so will be natural restorers for *ms/ms* mst females. Therefore, hybrid seed production in its last step is identical with that for production of a restorer hybrid by the gc-mst method. Use of a *ms* gene in conjunction with a duplicate-deficient chromosome which is not transmitted through the pollen was initially suggested by Phillips et al. (1971). Patterson (1975) demonstrated that hybrid maize seed can be produced using ms_1, ms_2, or ms_{10} genes in conjunction with appropriate duplicate deficient chromosome. Almost similar results were obtained by Phillips (1975). Of the three mst mutants, ms_{10} is promising, as it is consistent in mst expression and does not revert to mft in response to different environments or genetic backgrounds. This system was patented, but its utility in commercial hybrid seed production needs additional evaluations and also the incorporation of chromosome modifications in various suitable mst mutant lines.

Another method for pure stock mst line production was proposed by Greyson and Walden (1976). This is based on their presumption that mst in some maize lines might be due to an imbalance in endogenous growth regulators. Thus addition of the appropriate growth regulators would revoke temporary fertility in mst plants. This would allow production of seeds homozygous for the *ms* gene. The seed could serve as a source of the female parent for use in hybrid seed production. This method has limited practicability as only a few mst lines react favourably to such growth regulator applications and in most of the treated plants, morphology and fertility are highly impaired, especially by GA3 treatments, used most frequently for fertility restoration.

27.21.4.3 Gene-Cytoplasmic Male Steriles

The presence of sterile cytoplasm in maize was first described by Rhoades (1931), but its commercial exploitation began only when it was discovered by Jones (1950) that certain maize lines carry genes suppressing mst phenotype. Until then, because of the inherent difficulties in maintaining the g-mst stocks, the hybrid seed was produced by the conventional procedure of detasseling the seed-producing rows, which would then be pollinated from the pollen-producing rows. Utilisation of gc-mst for hybrid maize seed production has not followed the pattern mostly used in other crops. Although pollen production in the commercial hybrid was essential, early applications of the gc-mst method bypassed the problem of fertility restoration by mixing seed produced by hand-detasseling and by gc-mst (Duvick 1966). The seed produced with hand detasseling provided pollen for all the plants. Since most of the inbred lines developed in the USA maize belt were nonrestorers, it was easy to convert them into mst versions. These were crossed to other nonrestorer lines to produce mst single crosses which could in turn be used as seed parents for double-cross seed production. It was immaterial if the single cross used as pollen parent was restorer or nonrestorer, as long as a safe proportion of the seed planted by the farmer was from detasseled plants with normal cytoplasm. Since use of gc-mst and its fertility-restoring gene system eliminated the necessity of detasseling, its use increased due to (a) isolation of a stable gc-mst system with reliable *fr* genes, (b) perfection of techniques for utilisation of the gc-mst systems and appropriate genes, and (c) labour shortage coupled with rising costs of detasseling.

A stable and reliable gc-mst source was obtained from American maize variety, Mexican June. It had T-cytoplasm in which fertility was restored by two genes, Fr_1 and Fr_2. Whereas Fr_2 was present in many US maize belt inbreds, the Fr_1 gene needed to be introduced into these inbreds to convert them into fertility-restorer lines. Because of the ease of fertility restoration and the absence of anther exsertion, 85% of US hybrid maize in 1970 was produced in T-cytoplasm. In addition, many hybrids were restored by the introduction of *Fr* genes through the pollen parent and thus almost 100% of their plants had T-cytoplasm. However, in the early part of the growing season of 1970, *Helminthosporium maydis* race T(HmT) invaded the T-cytoplasm maize crop, causing the southern corn leaf blight. By mid-1970, the infection had spread over the southern and central maize belt of the USA and damage was extensive, since over 90% of the hybrid maize contained T-cytoplasm. Furthermore, *Phyllosticta maydis*, causing yellow leaf blight, exhibited increased virulence on T-cytoplasm plants (Scheifele and Nelson 1969, Scheifele et al. 1969). After 1970, utilisation of gc-mst was discontinued and hybrid seed maize producers returned to the practice of detasseling the seed-producing rows. However, there was a growing interest in obtaining alternative sterile cytoplasm sources, as an increase in cytoplasmic diversity would reduce the probability of an epidemic arising from cytoplasmically inherited susceptibility disease. The search for alternative sources of sterile cytoplasms led to the isolation of two sterile cytoplasms, C and S, both of which are equally as resistant as the N-cytoplasm with regard to the HmT race and to *P. maydis*. These cytoplasms became immediate prospects for use in producing hybrid seed without detasseling. S-cytoplasm soon fell out of favour because many genotypes with this cytoplasm were imperfectly sterile and the fertility was highly affected by the environment. Thus the maintenance of sterility in all environments was uncertain with this cytoplasm. Moreover, its restorer system is less efficient than that of T2, since heterozygous-restored plants shed only about 50% as much fertile pollen as the comparable normals (Buchert 1961). This is because, unlike T- and C-cytoplasm in which fertility restoration is sporophytic, in S-cytoplasm the restoration is gametophytic. Furthermore, this cytoplasm in certain genetic backgrounds exhibits a high rate of reversion to fertility, C-cytoplasm is better than S and consequently the percentage of genotypes sterilized by C is higher than that for S. Also a relatively high percentage of inbreds appear to be full restorers of C, and so are not to be rebred for use as restorers. Restorer heterozygotes in C are mostly 100% fertile, unlike 50% in S. Keeping in view the merits of this cytoplasm, Laughnan and Gabay (1978) proposed a crossing scheme that also illustrates the mode of C cytoplasm gc-mst restoration. Since S Fr_3fr_3 are semi-fertile as they produce 50% normal (Fr_3) and 50% aborted (fr_3) pollen grains, when they are crossed to S fr_3fr_3 plants, only Fr_3 pollen functions and the progeny is semi-fertile

(SFr_3Fr_3). When such plants are selfed, because of uniparental inheritance of the cytoplasmic conditions, half of the progeny is semi-fertile, having the nuclear restorer genotype Fr_3fr_3, and the other half fully fertile, despite the sterile cytoplasm, because of the double dose of restorer genes fr_3fr_3.

In the USSR and some other countries, *H. maydis* race T and *P. maydis* are not major problems limiting the use of T-cytoplasm in hybrid seed production, but in the major maize belt of the USA, most of the production is based on T-cytoplasm or on Moldarian cytoplasm. Due to HmT susceptibility of the T-cytoplasm, in the US and Canada it was not used for seed production from 1971 to 1973. In 1974, a small amount of production and in 1975 about 13% of the production was in C- and S-cytoplasms (Zuber 1975). The proven advantage of yield of sterile plants over pollen-shedding plants under yield stress conditions, especially during drought at silking time (Vincent and Wooley 1972), means that it is advantageous for farmers to plant hybrids in which up to 50–60% of the plants are mst (Duvick and Noble 1978). There are many cytoplasms that are not, C, S or T type. These are either modified versions of these cytoplasm types, or some of them may be new types. Careful survey and precise analysis will lead to the discovery of new sterile cytoplasms and increase the sterile cytoplasm base of maize. A new gc-mst system whose restorer genes have no effect on T-, C- or S-cytoplasm may be found. Thus if a series of such new cytoplasms could be developed, more diverse sets of hybrids would be available and these in blends could safely be used for hybrid seed production. Grogan (1971) and Gracen and Grogan (1974) proposed the term "multiplasm", which is applied to the technique of producing hybrids containing a blend of several kinds of mst cytoplasms. This multiline-like concept of breeding has a promising future in gc-mst maize breeding and hybrid seed production.

27.21.4.4 Procedural Methods

The method of using sterility type depends on the cross combinations and the cytoplasm type. For fertility restoration, whereas the T-cytoplasm requires two *Fr* genes, the C- and S-cytoplasms require mostly one *Fr* gene each. Many of the maize inbreds with T-cytoplasm possess one of the *Fr* genes (Fr_2) required for fertility restoration. Therefore, restoration in these steriles is brought about by the introduction of one dominant *Fr* gene. Thus, in general, each cytoplasm requires at least one dominant *Fr* gene for restoration. Based on this supposition, the following cross combinations are used for hybrid seed production (Fig. 27.27). In each it is assumed that both parental mst and hybrid fertility restoration are complete and perfect. Both these assumptions are not always fulfilled. This weakens commercial maize hybrid seed production.

i. *Single-Cross Hybrid.* The seed-producing inbred A is mst with sterile cytoplasm and recessive *fr* genes. The male parent may have fertile or sterile cytoplasms but has dominant *Fr* genes. Single-cross hybrids will have sterile cytoplasm but heterozygous *Fr* genes and will be mft.

ii. *Triple Cross.* Inbred A is mst with sterile cytoplasm and recessive fr genes, inbred B is mft with fertile cytoplasm but recessive *fr* genes. Inbred C may have

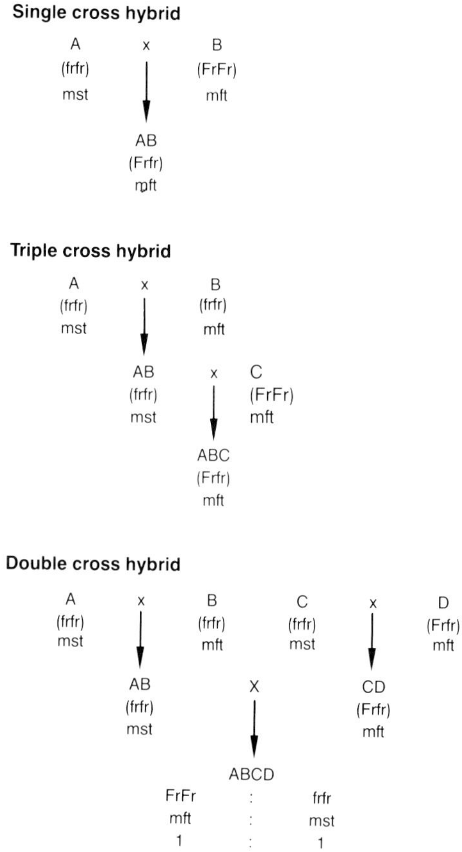

Fig. 27.27. Method of hybrid seed production by single, double and triple crosses in maize

fertile or sterile cytoplasm but should have double dominant *Fr* genes. The triple-cross hybrid will have sterile cytoplasm but will be mft.

iii. *Double Cross.* A and B inbreds are genetically similar to those in triple cross, but unlike it, inbred C should possess recessive *fr* genes. Inbred C may possess fertile or sterile cytoplasm, in the former case it is detasseled to make it a seed parent. To eliminate detasseling, it is preferred to have inbred C with sterile cytoplasm and recessive *fr* genes. Inbred D may have fertile or sterile cytoplasm, but should have double dominant *Fr* genes to be fully mft and an effective pollen parent. Since it is not always possible to have desirable genes in sterile cytoplasm, C and D parents usually possess fertile cytoplasm. In this case, inbred C is detasseled. Only 50% of the double-cross hybrid plants will be mft but these provide adequate pollen for fertilisation of the whole field having 50% mst plants. Either inbred C or D or both may possess *Fr* genes in the dominant state; if both carry them, the double cross will be 100% mft. Accordingly, either inbred C or D or both may have sterile cytoplasm, provided they possess dominant *Fr* genes.

It has now become a general practice to produce an identical hybrid by detasseling (without gc-mst) and to blend 25 to 50% of the fertile seed with 50 to 75%

of the seed produced by gc-mst (Poehlman 1979). This reduces the farmer's risk, especially if the restorer parent does not provide adequate pollen in the particular environment in which the crop is grown. It would also diminish the risk from disease and facilitate the incorporation of multilines with similar diffrent cytoplasm types.

27.21.4.5 Male Sterility Conversion

The successful conversion of a mft inbred line into a mst line by utilising paternal monoploids developed through androgenesis was reported by Goodsell (1961). Out of 150,000 F_2 progeny of a cross of purple mst × nonpurple mft inbred N6, two paternal monoploid seedlings having white roots were mst and set seeds subsequent to pollination with mft counterparts. Seed development in these monoploids indicates that a considerable diploidization occurred during the ontogeny of these plants, and thus many diploid sectors in the tassels existed. It appears that these paternal monoploids developed from a sperm nucleus which presumably failed to fuse with the egg nucleus but acquired the egg cytoplasm containing *c*-genes and then formed the initial embryonic cell which developed into a monoploid sporophyte. Since a converted mst inbred will have the same frequency of homozygous loci in the progeny from the paternal monoploid parent, and the inbred from which the monoploid plant was derived existed in the original inbred line, this method provides pure breeding mst lines in two generations only, i.e. one for making a cross of the inbred on mst marker and the second for screening monoploid plants and pollinating them for the initial generation of diploid mst seed of the pure inbred. However, a third generation will be required to increase the seed bulk. The extremely low frequency of such monoploids is, however, a serious handicap for their commercial utilization unless some genetic stock producing a high frequency of paternal monoploids or some means of artificially initiating androgenesis is found.

CHAPTER 28
Iridaceae

28.1 *Iris* L.

In this ornamental genus of over 150 species, many of which are prized by fanciers, g-mst is known only in *I. douglasiana*, *I. pallida* and *I. persica* (Table 28.1).

Table 28.1 ms Gene effects in *Iris*

Species	References	*ms* genes effect
I.douglasiana	Uno (1982)	Early flowering, reduced sepals and petal size and increased seed production
I.pallida	Maleci and Maugini (1977)	Altered karyotype
I.persica	This Vol.	Suppresses anther development and differentiation

28.1.1 *I. douglasiana* Herb (prominent iris)

This is a clonal perennial herb with creeping rhizomes and flowers having three separate units, each unit consisting of one sepal, style and stigma. Uno (1982) found that 11% of mst plants, including some which bloomed with the earliest blooming, were hermaphrodite types. The sterility is genic, as it persisted over four generations of testing, but the exact gene control is not known. Compared to hermaphrodites, the mst flowers have smaller sepals and petals and fewer pollinated stigma, leading to low seed production. Although floral phenology, nectar flow pattern, amino acid and sugar contents of nectar are similar in both the flower types, mst flowers produce much less nectar per flower. This leads to a lower rate of insect visitation over mst flowers, but no significant differences in the ovule or seed number per capsule occur between mst and hermaphrodite flowers. Seed loss through larval predation is much greater in capsules from hermaphrodite flowers than in those from mst flowers. This leads to increased seed set over mst plants. Early flowering and greater seed set in mst plants gives them a reproductive advantage over hermaphrodites which experience higher levels of larval predation in the later growing season.

28.1.2 *I. germanica* L. (bearded iris)

Genetic male sterility is not known in this species. Subjecting it to 600 and 2000 ppm MH treatments did not produce any appreciable degree of mst (Moore 1950). The chemical is toxic to the plant and damages ovules.

28.1.3 *I. pallida* Lam. (pale iris)

From Tuscany, Italy, the occurrence of sexually reproducing mst plants was documented by Simonet (1932). Although genetic control of mst is not known, the mst plants possess somatic chromosomes that could be differentiated from those of the mft plants (Maleci and Maugini 1977).

28.1.4 *I. persica* L. (Persian iris)

In an orange keeled variety, one plant with rudimentary anthers appeared (this Vol.). In it the anthers, archesporial tissue and tapetum were misdifferentiated and PMC's did not develop. The sterility is genic as it persisted in clones also. The *ms* gene suppresses anther development and differentiation.

CHAPTER 29

Labiatae

In this mint family, that comprises about 160 genera and 3000 species, mst is known only in *Coleus, Lavandula, Origanum, Salvia* and *Satureja* (Table 29.1).

29.1 *Coleus* Lour.

Esteemed highly for its beautiful coloured foliage varieties, this possesses 150 species and numerous ornamental hybrids which are characterised by a profusion of variation in colour and shape of leaves. All these are mft, but mst is known in *C. blumei* and *C. hybridus* (Table 29.1).

29.1.1 *C. blumei* Benth (French nettle)

A dominant *Ms* gene induces deep leaf lobing and a recessive gene shallow lobing. Accordingly, all the deep-lobed plants are mst (Rife 1940, 1948). Their frequency is very low. Out of numerous descendents, one plant was deep lobed and mft (Ford 1950). This indicates that the genes for leaf lobing and mst are very closely linked and no pleiotropy is involved. Precise *Ms* gene action over male meiosis is not known.

29.1.2 *C. hybridus*

This spontaneously arisen mutant plant of *C. hybridus* having sharp-toothed, broad-based, ovate, highly patterned, many-coloured leaves and blue flowers has dwarfed staminal tube and rudimentary nondehiscent anthers (this Vol.). Its PMC's degenerate during prophase-I and no pollen grains are formed. The tapetum develops normally, but degenerates after PMC degeneration. Female fertility is normal. The mst is controlled by a recessive *ms* gene. Since *C. blumei* is known to be one of the parents of *C. hybridus*, the *ms* genes of the two species are nonallelic, as the gene is dominant in *C. blumei* and recessive in *C. hybridus*.

Table 29.1. Genetic male sterility in Labiatae

Species	References	Gene control	Gene action	Remarks
Genic male sterility				
Coleus blumei	Rife (1948), Ford (1950)	ID	nk	Deep leaf lobing linked to *ms* gene
C. hybridus	This Vol.	1r	During prophase-I	Dwarfed staminal tube, rudimentary nondehiscent anthers
Lavandula angustifolia	This Vol.	2–3r	nk	
Origanum majorana	This vol.	1r	During premeiosis or tetrad formation	
O. vulgare	Lewis and Crowe (1952, 1956)	2D	nk	
Salvia nemorosa	Linnert (1955)	nk	After pachytene	Pollenless anthers, tapetum intact and normal
S. officinalis	Linnert (1955)	nk	During interphase	Aborted microspore do not degenerate
Gene-cytoplasmic male sterility				
Origanum vulgare	Kheyr-Pour (1980, 1981)	1r	nk	High degree of polymorphism in *Fr* genes and S-cytoplasm
Satureja hortensis	Correns (1904, 1908) East (1934)	1r	nk	

29.2 *Lavandula angustifolia* Miller (*L. officinalis* chaix, common lavender)

In this outbreeding highly ornamental aromatic plant, inbreeding led to segregation of mst plants whose female fertility is high (Romanenko and Vislovkhov 1981). The sterility is controlled by two to three recessive genes (this Vol.).

29.3 *Origanum* L. (marjoram)

Of the 35 known species, mst is known in only two species, *O. majorana* and *O. vulgare* (Table 29.1). Of these, the latter species is gynodioecious, occurring in Europe, especially the Mediterranean region to central Asia and has a varying proportion (1–65%) of mst and hermaphrodite plants. Others are unisexual. In contrast to the wide occurrence of mst in *O. vulgare*, it is rare in *O. majorana*.

29.3.1 *O. majorana* L. (*Majorana hortensis*, sweet garden knotted marjoram)

A few mst plants having nonprotruding stamens due to dwarfed filaments and reduced anthers segregated from a stand of this species (this Vol.). The archesporial tissue was highly reduced and only a few PMC's were developed. They underwent meiosis until tetrad formation, after which microspores aborted prior to release. This was followed by tapetal degeneration. Female fertility was reduced by 9–15%. The sterility is controlled by one recessive *ms* gene. However, in three segregating families, digenic control is evident.

29.3.2 *O. vulgare* L. (wild marjoram, pot marjoram)

This has three sexual forms: (i) mft's developing hermaphrodite flowers, each flower has four well-developed anthers, each with two visible pollen masses, (ii) mst's bearing only female flowers during the whole flowering period, the anthers being aborted on the bottom of the smaller corolla, (iii) intermediate (int) plants possessing both hermaphrodite and mst flowers on the same spicule in varying proportions. The frequency of mst plants in natural populations of this species is highly divergent. Kheyr-pour (1980) found 1–62% mst plants in about W. European populations. The plants exhibit wide g-c polymorphism for mst, and each population appears to be unique in its genome-cytoplasm combination.

29.3.2.1 Genic Male Sterility

Lewis and Crowe (1952, 1956) studied progenies of two mst plants and included only within-family crosses in their investigations. They proposed a digenic control in which two dominant genes, *F* and *H*, condition mst. Gene *F* causes anther abortion and gene *H* is epistatic to gene *F*. Thus the genotypes with *FFhh* or *Ffhh* genes are mst and those with *FFHH*, *FfHh*, *FfHH* or *ffHh* are mft. With such a genic system, 13:3 mft:mst ratios are usually expected in F_2, but distortions in these and other ratios occurred all through the segregating populations. In order to interpret their segregation ratios, the authors assumed *ffhh* genotypes to be lethal and *HH* to be three-quarters lethal and the rest self-compatible. The lethality of the double recessive was considered to maintain a balance between mst and mft plant types.

29.3.2.2 Gene-Cytoplasmic Male Sterility

Sterility with a high degree of polymorphism was first demonstrated by Kheyr-Pour (1980,1981) in a sample of four mst plants selected randomly from natural populations in the same area. In three families (P_1, V_2, V_3) within-family studies indicated mst plants as homozygous for a single recessive gene, the three *fr* genes were nonallelic. Reciprocal crosses between members of these families gave asymmetric results on the basis of the presence or absence of segregation for mst F_1 and BC_1 segregations. These results suggest that each family has a different and distinct S-cytoplasm type, which in combination with a particular *fr* gene causes mst. Furthermore, each cytoplasm appears to be restored by a specific dominant restorer nuclear gene, designated by the author *Rf*1, *Rf*2 and *Rf*3, (= Fr_1, Fr_2, and Fr_3), respectively. In the fourth family (L_1), mst was caused by interaction of a specific S-cytoplasm (L_1-type) and a few nuclear genes, polygenic in nature. Crossing L_1 mst plants to mft plants of the other three families, yielded "type-III" intermediate (*int*) phenotype plants, that possess hermaphrodite and mst flowers on the same spicule. Selfing and test crossing *int* plants produced segregations comparable to those of heterozygous mft plants. Thus these *int* plants are heterozygous for major dominant restorer genes in each S-cytoplasm type. Since segregation data of three L_1 mst plants crossed to eight mft plants belonging to the other three families indicates a high proportion of *int* plants (Kheyr-Pour 1980), the above-mentioned model cannot account for the genic control of *int* plant type phenotypes. Furthermore, the inheritance pattern of 21 additional mst plants collected from diverse populations indicated that four families exhibited monogenic inheritance of mst (Kheyr-Pour 1981) resembling the three families (P_1,V_2,V_3). Reciprocal crosses between some of these revealed S-cytoplasm differences between them. Because of monogenic inheritance, each cytoplasm has a specific major restorer gene. Therefore, a high degree of polymorphism exists in S-cytoplasm and *fr* nuclear genes in the species. This is clarified in the following account.

29.3.3 Gene-Cytoplasm Polymorphism

Both between and within the natural populations of *O. vulgare*, a wide degree of polymorphism exists for gc-mst. This is evidenced by the fact that some mst plants collected from the same populations by Kheyr-Pour (1980, 1981) exhibit not only different S-cytoplasm types but also different nonallelic *fr* genes. Also four mst plants selected randomly from natural populations in the same area differ in *fr* genes and S-cytoplasm types. One family (V_5) possessed S-cytoplasm plant types different from the two others (V_4,V_5), although all three were collected from the same population by Kheyr-Pour (1981). Thus in a sample of 24 mst plants, nearly eight S-cytoplasm types could be distinguished. Gene control of the sterility in most cases is recessive and often involves single *fr* gene. Since the progeny of all crosses between plants of diverse populations were as female fertile as within-population or within-family crosses, all the plants belong to the same Mendelian populations sharing a certain amount of common gene pool without any detectable genetic barriers. Hence the large polymorphism for gc-mst in *O. vulgare* represents the germ plasm resource of a single species that has a rich reservoir of genes producing mft, mst and *int* type plants. Its mutation rate, at both the nuclear and cytoplasmic levels, is very high. There is evidence that the *c* genes are located in the mitochondria. Thus the mitochondrial genome, like the nuclear genome, is highly polymorphic in *O. vulgare.*

29.4 *Salvia* L. (sage, clary)

Of the 700 species distributed all through temperate and tropical regions, mst of unestablished nature is known only in *S. nemorosa* and *S. officinalis* (Table 29.1).

29.4.1 *S. nemorosa* Mott.

Plants having phenotypically normal but pollenless anthers were detected by Linnert (1955). Male meiosis is normal up to the end of the pachytene but immediately thereafter sudden and drastic degeneration of the chromosomes and cell contents occurs. The plasma becomes structureless, the cell membrane shrinks and the nuclei collapse. Highly pycnotic chromosome-like pairs occur in some PMC's. No further stages of meiosis ever occur in them. The tapetum remains intact. The anomaly is genetically conditioned but the gene control is now known.

29.4.2 *S. officinalis* L. (sage)

Linnert (1955) detected mst's that had either completely or partly aborted microspores. In these, male meiosis proceeds normally but anomalies appear during interphase and are apparent during dyad formation. Each interphase nucleus possesses one to five nucleoli. It appears that during interphase each haploid set of the chromosomes becomes divided amitotically, so that many metaphase plates are formed. These anomalies increase with the further onset of meiosis and instead of tetrads, polyads are produced. Each sporad, including the micronuclei, possesses a conspicuous nucleolus. Linnert (1955) believes the anomalies to be due to nucleic disturbances, since even the smallest micronuclei possess a conspicuous nucleolus. Although mst is partial, it appears to be nearly equal in both the sexes. Consequently seed set upon crossing and selfing in these plants is low, and therefore they cannot be considered as a typical mst. Moreover, details about genetic control and environmental influence over the sterility are not known. While in *S. nemorosa* the sterility gene(s) acts just at the beginning of heterotypic division, in *S. officinalis*, the gene(s) acts after this division is completed. Whether these two genes are allelic or not is not known. To consider the sterility-inducing gene in *S. officinalis* as a *ms* gene is questionable, as it impairs ovular fertility considerably.

29.5. *Satureja hortensis* L. (summer savory)

While the offspring of mst gynomonoecious populations are always mst, those of gynodioecious plants breed true (Correns 1904, 1908). A nearly similar situation is found in *Cirsium*. Wettestein (1924) and Punnett (1927) opined that the cytoplasm of mst's inhibited the genes controlling stamen formation which are contributed by the male parent. East (1934) assumed that the true breeding hermaphrodites possess the genes *aa* and mst's *Aa*, and that in the females, the *a* gametes are either inviable or are not fertilized at all. Detailed studies indicate the mst to be conditioned by g-c interaction. The *Fr* genes are very rare and their mutation to *fr* is high.

CHAPTER 30

Leguminosae

In this major angiospermic family that has about 500 genera, mst occurs in *Arachis, Cajanus, Crotalaria, Glycine, Lathyrus, Lupinus, Medicago, Melilotus, Phaseolus, Pisum, Trifolium, Vicia* and *Vigna*. Sterility in these is mostly of g-mst type and only in a few gc-mst occurs (Tables 30.4, 30.5). In some, ch-mst has been induced (Table 30.6).

30.1 *Arachis hypogaea* L. (groundnut, peanut)

Of the 15 known species, mostly Brazilian in origin, *A. hypogaea* is the only cultivated species and ranks second to soybean in commercial importance as a source of high-quality vegetable oil. The first report on g-mst is from India by Patel et al. (1936), who detected plants with congested nodes and crumpled leaves in F_2 segregants of a cross involving corrientes-3 groundnut variety. These plants produce empty shrunken pollen and their stigmas are slightly underdeveloped. The anomaly is conditioned by duplicate genes, n_1 and n_2. The occurrence of similar steriles in west African groundnuts is documented by Hayes (1933) and in Japanese groundnuts by Katayama and Nagatomo (Ashri 1968). Some F_2 plants stunted in growth and termed brachytic dwarfs were found by Ashri (1968). These had dark green, fleshy and densely crowded leaves with thick short stems which bore five or more leaflets instead of four found in the normal plants. They developed a few reduced flowers, all of which had diminutive pollenless anthers and were completely mst. Since female gametic sterility or failure of zygote development occurred in most of them, such plants cannot be regarded as mst's. Although duplicate genes appear to control the sterility, in one case a third locus either as dependent or as modifier also occurred. While Ashri recorded digenic control, a monogenic recessive control of mst occurred in the brachytic dwarf found by Hull (1937) from the F_2 family of the groundnut variety cross Virginia Runner × Tennesee Red. One gene mechanism can also be explained by assuming that either one of the parents used was heterozygous Ms_1ms_1 Ms_2ms_2 or ms_1ms_1 Ms_2Ms_2, or, since Hull based his conclusion on advanced generations, he recovered only those lines segregating 3:1 progenies. Levi and Ashri (1975) induced mst by ethidium bromide treatments to seeds. While some may be genic, a few others appear to be gc-mst types.

30.2 *Cajanus cajan* (L) Millsp. (*C. indicus,* pigeon pea)

30.2.1 Genic Male Sterility

This is one of the most vital plant protein sources and the second most important pulse crop of India. Spontaneously arisen gametic sterility affecting both the sexes occurs often in it, but its genetic control is not known (Deshmukh 1959). In two distant pigeon pea collections only the male sex is impaired (Reddy and Thammiraju 1977; Reddy et al. 1978). The anomaly is monogenic recessively conditioned. These mst's exhibit delayed senescence and nondehiscent, flat, translucent anthers with a whitish scaly surface. Male meiosis proceeds normally up to tetrad formation, after which degeneration occurs quickly in nonseparated tetrad cells, leaving behind an intact tapetum, unlike in fertiles, in which the tapetum degenerates during tetrad formation and microspore release. Pod set on mst's by open pollination is adequate for hybrid seed production (Reddy et al. 1978). The white translucent anther colour offers an efficient and effective marker of mst's for commercial hybrid seed production of this legume, that is cultivated and consumed in nearly all the tropical countries of the world.

Wallis et al. (1980) found mst's in the advanced breeding line B15B in breeding populations of various crosses involving this line. In the mst's, the anthers are brown, shrivelled, arrowhead-shaped and devoid of pollen at maturity; their female fertility is normal. Male meiosis proceeds normally until tetrad formation. Thereafter degeneration occurs, during which nuclear membrane ruptures and cell wall collapses (Dundas et al.1981). At maturity protoplasm masses with cell wall remnants remain. The tapetum becomes highly vacuolated during prophase-I, the vacuoles occur in nearly half of each cell by MI. Whereas in mst's the tapetal walls disintegrate during tetrad degeneration, in the mft anthers, the tapetum persists until pollen grains become binucleate. Although tapetal anomalies in mst's appear in prophase-I, middle layer anomalies are delayed until prophase-II. In mft anthers these layers degenerate, but in the mst anthers the inner of the two middle layers enlarges twice by vacuolation, and after anther degeneration assumes the appearance of vacuolate tapetal cells (Fig. 30.1). An almost similar enlargement of the middle layer occurs in mst *Phaseolus atropurpureus* (Pritchard and Hutton 1972). Furthermore, in mst anthers of *C. cajan* the endothecium, unlike that in mft anthers, does not enlarge and disintegrate, instead it persists until anther maturity. Anomalies in almost all the tissues of anther are prevalent in the mst anthers. Therefore, the *ms* gene appears to act both on sporogenous as well as on nonsporogenous tissue of anther. Differences in anther phenotype and tapetal behaviour between mst materials studied by Reddy et al. (1978) and Dundas et al. (1981) suggest that two different *ms* genes condition mst in these two materials. Whereas in the mst material of the former author, the anthers are white and transluscent and the tapetum remains intact, in the latter author's material the anthers are shrivelled, and brown and their tapetum degenerates along with the tetrad cells. However, it is worthwhile to conduct allelism tests before any conclusions on genetic identity are drawn.

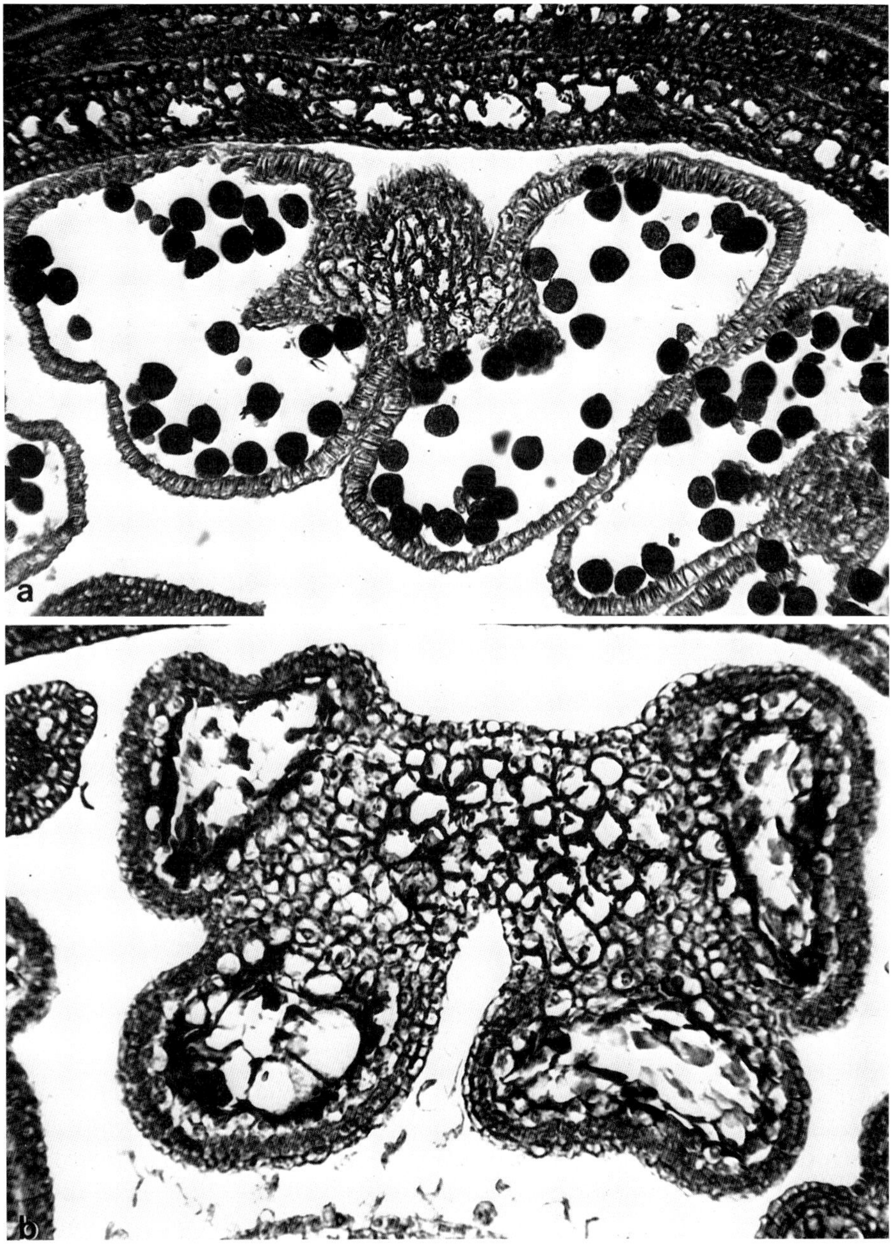

Fig. 30.1a,b. *Cajanus cajan*. **a,b** T.S. of mature anthers of male fertile **(a)** and male sterile **(b)** pigeon pea. (Dundas et al. 1981)

A third mst source was identified by Dundas et al. (1982) from an elite photoperiod-insensitive breeding line QPL-2 of Indian origin. In the mst mutant the PMC's are smaller initially but they enlarge abnormally due to the failure of adjacent PMC wall separations. Meiosis is arrested at the pachytene and the PMC cell contents disintegrated gradually. Thus whereas in the mst materials of Reddy et al. (1978) and Dundas et al. (1981) the *ms* genes act during early and late tetrad stages, in the Dundas et al. (1982) mst mutant, the *ms* gene acts during early prophase-I. In the latter material, the tapetum fails to form a complete layer of cells in the anther sac and is represented by a few small isolated cells in the spaces between PMC's and the rest of the anther wall. This mst mutant exhibits three major features not present in the other two mutants: (a) the presence of more and larger PMC's, (b) early meiotic breakdown, and (c) delayed and incomplete anther wall development. These features and other differences indicate that the *ms* gene differs from the other two *ms* genes. This needs confirmation through allelism tests.

30.2.2 Chemical Male Sterility

Mendok. Of the 0.5, 1 and 1.5% aqueous solution applications of MH, FW-450 and coumarine, either only 0.5% mendok sprayed twice or 1% sprayed once induced complete mst, lasting 15–25 days (Kaul and Singh 1967a). Significant reduction in fruit number, seeds per fruit and total yield per plant occurred in the treated plants. Reduced seed number per fruit is indicative of ovular abortion. One percent solution induced complete pollen abortion and abscission of floral buds. This abscission could be stopped by 0.001 ppm 2, 4-D sprays. However, total grain yield reduction persisted in the treated plants.

30.2.3 Breeding Value

From a germplasm collection, Reddy et al. (1977) documented five types of mst plants from 35 sources and from F_4 derivatives of *Cajanus cajan* × *Atylosia* sp. crosses. Of these, translucent mst types are promising for use in breeding. The brown shrivelled anthers of the mst plants found by Dundas et al. (1981) provide a rapid and efficient method to recognise mst plants in the field for subsequent use in breeding. The sterility of these mst plants is complete and stable and phenotypic identification by anther colour is easy. Moreover, monogenic recessive inheritance permits relatively easy maintenance. Preliminary observations have indicated that in a population of fertile and sterile sibs, seed set on sterile plants under natural conditions with insect cross pollination is 84% of the normal fertiles (Sharma and Green 1980). Thus the mst mutants have a high breeding value.

30.3 *Crotalaria* L. (rattle box)

In this cosmopolitan genus of about 325 species strongly developed in Asia, mst is known in *C. juncea* (Mitra 1976; this Vol.), *C. mucronata* (Edwardson 1967), *C. pallida* (Arora and Gupta 1984) and *C. striata* (Kempana and Shastry 1958). The species *C. mucronata, C. pallida* and *C. striata* are considered as synonyms (Terrell 1976) and due to nomenclatural confusions and subsequent revisions, it is assumed that the mst is known in *C. juncea, C. mucronata* and *C. pallida* (Tables 30.4, 30.5).

30.3.1 *C. juncea* (sunn hemp)

30.3.1.1 Genic Male Sterility

This plant produced nonviable pollen even though the male meiosis and tetrad formations are normal (Mitra 1976). Genetic control of this anomaly is unknown. Female fertility is normal. Five nonfruiting slow-growing plants that remained green past maturity appeared spontaneously in a cultivated field (this Vol.). Their stamens were reduced, indehiscent and pollenless. Their PMC's degenerated during the zygotene. The anomaly is controlled by a recessive gene ms_1. Female fertility, as evidenced by seed set obtained after hand pollination, is normal.

30.3.1.2 Chemical Male Sterility

Pollen sterility lasting for 11 days was induced in this species by Kaul and Singh (1967b) after a foliar spray of 0.5% FW-450 before and after initiation of floral primordia. The treatment depressed plant growth, diminished floral organ size and the keel petals split either partially or completely exposing the stigma. Anthers remained clumped around the style in the form of a ring and were indehiscent. Ovular fertility was reduced, as pod set was 35–50% of the control. The effects of dalapon were nearly similar to those of FW-450. Plants sprayed with 0.25% dalapon produced 65% pollen grains exhibiting in-situ exudation of contents. Pod number was also reduced highly. A concentration of 0.5% caused severe phytotoxic effects, 82% pollen sterility and nearly equal ovular damage. MH was highly toxic to both vegetative and reproductive parts and the plants were fully male-female sterile.

30.3.2 *C. mucronata* Desv. (*C. striata*)

From a breeding population, a plant with poorly developed nondehiscent anthers devoid of pollen was detected by Kempana and Shastry (1958). Except for the constant occurrence of anaphase bridges, meiosis is normal but pollen sterility is complete (Kempana 1960). The bridge formations are due to chromosome stickiness and are environmental, as they were infrequent in the material studied by Edwardson (1967). Cross and hand pollination produced seeds, and those obtained by open pollinations produced mst plants exclusively. Genetic studies indicate that pollen abortion results from the interaction of two *fr* genes (fr_1 and fr_2) with S-cytoplasm (Edwardson 1967). Fertility is restored by one or both dominant alleles Fr_1 and Fr_2 in the heterozygous or homozygous state. All attempts to transmit sterility from mst stocks and scions to normal stocks and scions failed, indicating failure of mst transmission across grafting (Edwardson 1967).

30.3.3 *C. pallida* Ait. (smooth crotalaria)

In this self-incompatible species gc-mst was detected by Gupta (1975). The pollen sterility is due to the failure of cytokinesis in PMC's, due to which no tetrads are formed (Arora and Gupta 1984). Tapetal form is normal.

30.4 *Desmodium sandwicense* Desv. (tick-clover)

30.4.1 Gene-Cytoplasmic Male Sterility

Two mst plants that probably arose from the natural hybridization of *D. intortum* × *D. sandwicense* were detected by McWhirter (1969). They had deformed abnormal-shaped pollen, indehiscent anthers and failed to form pods and seeds upon selfing. Pollinating them by *D. sandwicense* pollen and backcrossing further by this species produced mst progeny. When the mst plants were pollinated by *D. intortum* pollen, fertility was partially restored. Crossing *D. sandwicense* with pollen of the restored plants yielded normal mft plants, indicating nontransmission of mst by these restored plants. Although the exact genetic control of this mst is not clearly known, from the segregation ratios and mode of transmission of mst, the sterility appears to be due to interaction between *D. sandwicense* gene(s) and the S-cytoplasm derived from the original sterile population. Since this species has N-cytoplasm, the S-cytoplasm arose due to spontaneous mutation.

30.4.1.1 Fertility Restoration

This is controlled by complementary action of two nonallelic dominant genes Fr_1, Fr_2. They act sporophytically, but restoration is incomplete since restored F_1 hybrids produce apparently normal pollen but have reduced fertility compared to that of F_1 hybrid plants with N-cytoplasm. This incomplete fertility restoration does not appear to be due to the heterozygosity of *Fr* genes as with gametophytic expression, but is probably due to incomplete penetrance of the *Fr* gene.

30.4.1.2 Microsporogenesis

Male meiosis proceeds normally but degeneration of pollen takes place immediately after its formation in some plants and much later in others. Thus although some pollen appears phenotypically normal prior to anthesis, it degenerates immediately thereafter, so that at anthesis no viable pollen ever exists. In the restored plants with heterozygous *Fr* genes, a part of the pollen degenerates during anthesis but in the plants where anthesis is delayed, pollen sterility is severe. In such a situation the pollen fertility was complete in N-cytoplasm that were growing adjacent to the restored plants (this Vol.).

30.4.2 Grafting

The mst does not become transmitted across grafts as it did not occur in the grafts and in the selfed progeny of these grafted plants.

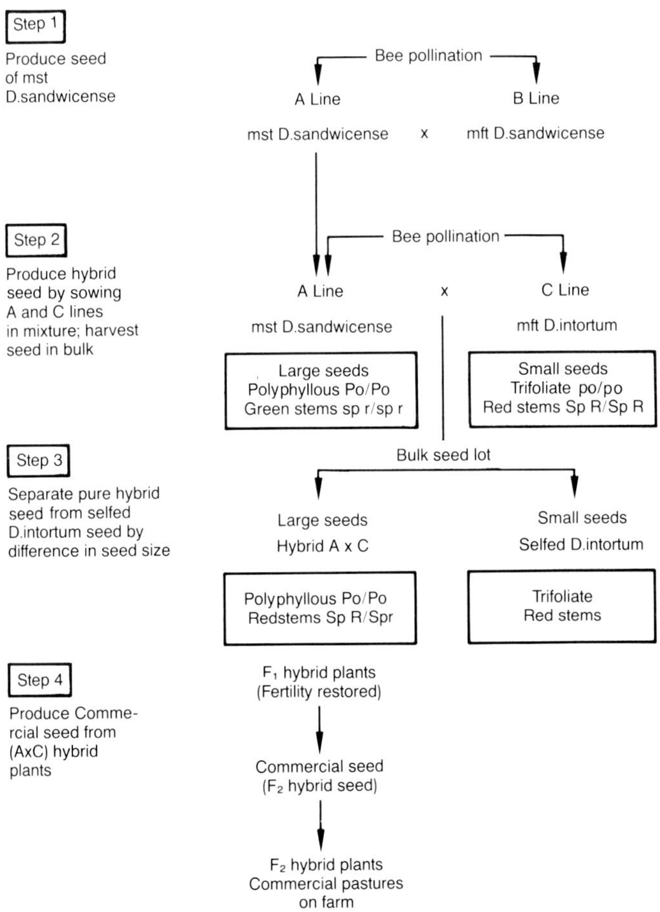

Fig. 30.2. A hybrid *Desmodium* seed production scheme. (After McWhirter 1969)

30.4.3 Breeding Value

Both *Desmodium sandwicense* and *D. intortum* from which the mst was derived exhibit adaptation to the entomophilous mode of pollination. Maintenance of mst *D. sandwicense* is possible by pollinating its normal sib through bees. F_1 seeds can also be produced by bee pollinations of mst plants with *D. intortum* pollen. Based on the bulking system proposed for wheat by Watson and McWhirter (1968), McWhirter (1969) devised a scheme for commercial interspecific *Desmodium* hybrid seed production using the gc-mst system (Fig. 30.2). This scheme has two main features: (a) it is a bulk hybrid seed production scheme in which separation of hybrid from nonhybrid seed mixture can be done by differences in seed size, as this differs considerably between parents, (b) F_2 populations produced by fertility-restored F_1 hybrids is proposed as the commercial hybrid cultivar and small quantity of F_1 seed is needed for establishing perennial fields for F_2 commercial seed production. Although this obviates the requirement of having high seed set on mst plants, it necessitates almost complete fertility restoration of F_1hybrids. However, this is not achieved in these two *Desmodium* species. The F_1 hybrids using gc-mst were found to be early-flowering and -maturing types and exhibited positive heterosis for plant growth (McWhirter 1969). Thus mst offers a useful approach for hybrid seed production in *D. sandwicense.*

30.5 *Glycine max* (L) Merr. (soybean)

Of the ten *Glycine* species, mst is documented only in *G. max*, an earlier obscure forage and now an extensively cultivated protein-rich crop used for human and animal food, as its seeds contain 40% protein and 20% oil. Of the 18 known sterility-inducing recessive mutant genes, eight cause nearly complete mst, six p-mst and the remaining four both male and female sterility (Table 30.4).

30.5.1 Genic Male Sterility

30.5.1.1 Complete

30.5.1.1.1 Spontaneous

The sterility in each of these mst mutants is conditioned by one recessive gene, five of which are nonallelic, and for the remaining allelism tests are to be done. The first report is on the ms_1 mutant that arose from the progeny of a seed collection grown at N. Carolina, USA (Brim and Young 1971). Another mutant ms_2 was detected by Bernard and Cremeens (1975) from an F_3 breeding population, the mutation occurring either in F_2 or parental material. In F_3 of the intervarietal cross Calland × Cuttler, the ms_3 mutant segregated (Palmer et al. 1980). Another mutant, ms_4, arose from the cultivar Rampage (Delannay and Palmer 1982). Linkage studies have linked ms_1 to the flower colour gene W_1 and *Wm* and st_5 (Palmer and Kaul 1983). Two additional mst mutants are designated by the cultivars from which they arose, viz. Wabash (Chaudhari and Davis 1977), and Beeson (Albertseen and Palmer unpublished). Except for Beeson, whose *ms* gene is nonallelic to ms_1 to ms_5 genes, allelism tests in others have not been done. The mutant of Rubaihayo and Gumisiriza (1978) is both male-female sterile, exhibiting strong desynapsis, and is hence a meiotic mutant. From a strain S-571, a composite of the F_4 generation derived from double-cross plants having 90–95% pollen sterility segregated (Caviness et al. 1970). Their female fertility is normal. The sterility is controlled by a single recessive gene (Fagala 1972) whose expression is influenced by temperature. A daytime temperature of 35°C leads to complete pollen sterility and of 21°–29°C to p-mst (Caviness and Fagala 1973). Pod set per plant was highest at 29°C. Although NCP on p-mst plants was about 45 times greater than that over mft plants, pod set (one to two seeded pods) was 11–30% of the normal plants. This was due to the insufficient pollen available for pollination. This mutant is a normal mst genotype and not p-mst because only 2–3% viable pollen is produced by it; the seed set on it is due to its high NCP rate.

30.5.1.1.2 Induced

From M_2 progeny obtained after 20 kR γ-irradiated Semmes variety soybean seeds, Singh et al. (1974) detected two mst mutants. Sterility is conditioned by two recessive genes, Semmes ms_1 and ms_2, whose allelic nature is not known. Of these, one mutant gene causes almost complete fst and is therefore not a *ms* gene. Another mutant ms_5 arose from the irradiated progeny of cultivar Essex (Buss 1983).

30.5.1.1.3 Microsporogenesis

Male meiosis up to TII proceeds normally in ms_1, ms_2, ms_3, ms_4, N-69-2774 and TGM 242–42. In ms_1, ms_4, N-69–2774 and TGM-242-4, failure of cytokinesis after TII leads to coenocyte formation (Palmer and Albertsen 1975; Patil and Singh 1976; Rubaihayo and Gumisiriza 1978; Albertsen and Palmer 1979; Delannay and Palmer 1982; Fig. 30.4). Of the four *ms* genes, ms_1 is fully penetrant, ms_4 exhibits variable penetrance and in others the action is partial. Thus except in ms_1, a few pollen grains are produced following cytokinesis. In Semmes in *ms-2* and ms_2, tetrads abort after cytokinesis (Graybosch and Palmer 1983) and prior to release from quartets (Patil and Singh 1976). In ms_3, microspores degenerate after wall initiation rarely before but mostly immediately after release (Palmer et al. 1980; Fig. 30.3). In Semmes *ms-1* the *ms* gene action is delayed and microspores abort after liberation during the pollen maturation stage (Patil and Singh 1976).

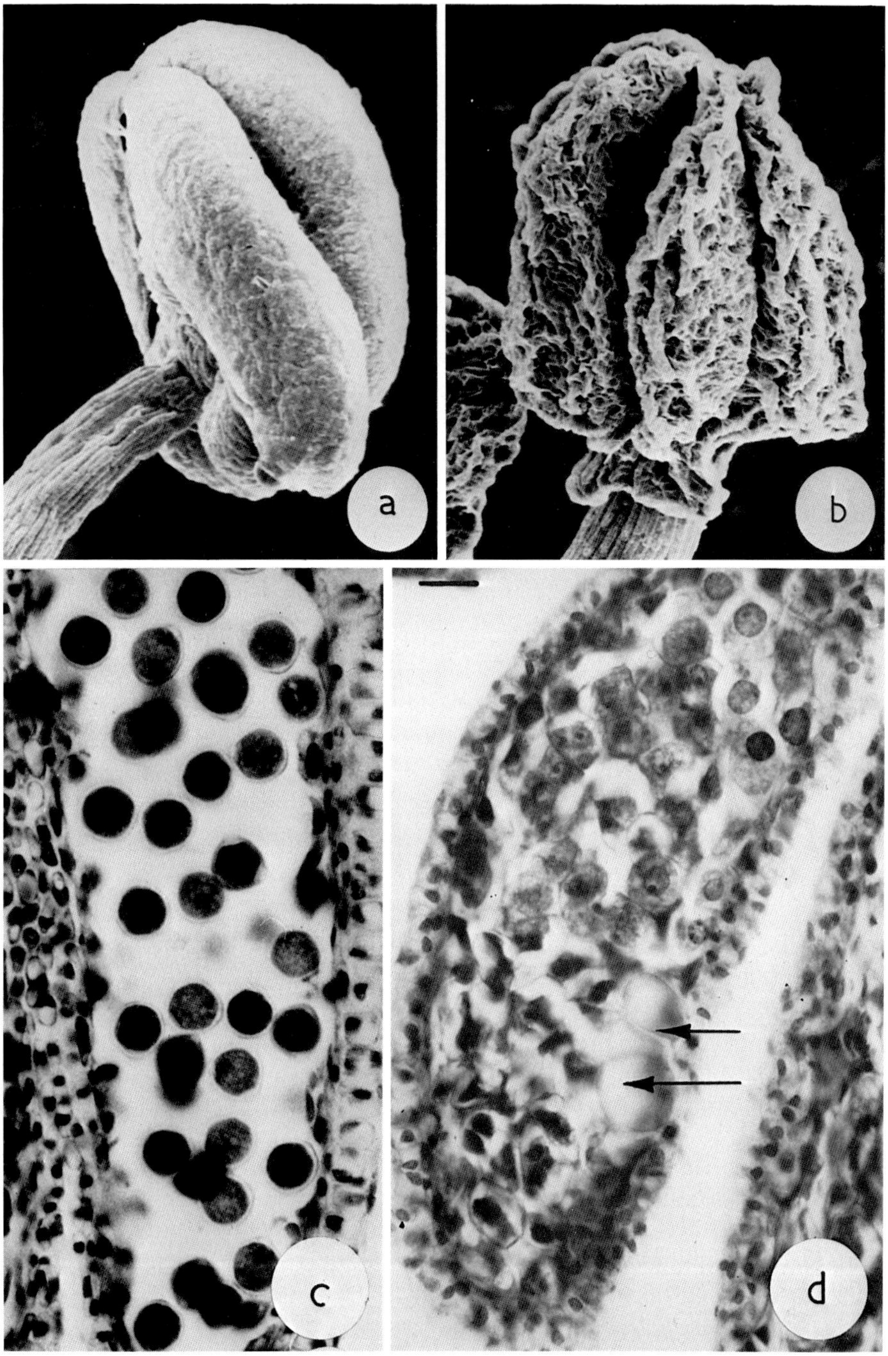

Fig. 30.3a-d. *Glycine max*. **a,b** SEM of mature fertile **(a)** and ms_2 sterile **(b)** anthers. (Graybosch et al. 1984). **c,d** T.S. of anthers of normal **(c)** and male sterile **(d)** plants showing fertile pollen **(c)** and degenerating microspores **(d)** *Arrows* indicate vacuoles adjacent to or within the tapetum. (Palmer et al. 1980)

30.5.1.1.4 Coenocyte Formation

Even though *ms* genes either disrupt or inhibit cytokinesis in ms_1 and ms_4, the tendency towards pseudo-pollen formation and pollen function exists in them, as they produce large uni- to multinucleate coenocytes (Albertsen 1976; Albertsen and Palmer 1979; Fig. 30.4). These develop a pollen wall comprising tectum, collumellae and endexine as in normal pollen but, unlike it, they possess a variable number of randomly positioned colpi and furrows (Fig. 30.4b,c). The coenocytes either degenerate or a few undergo mitosis. Some of them (3.6–4%) develop pollen tubes which apparently look normal or highly distorted (Albertsen 1976; Skorupska and Nawracali 1980). Thus coenocytic microspores act like pollen grains. Ovular fertilization by a few of them may have led to the production of aneuploid and polyloid plants as obtained by Palmer and Heer (1976). However, the existence of certain disturbances in the female gametophyte development of ms_1 plants is also known (Cutter and Bingham 1977). These would also lead to variant progeny plants and plants having over 100 chromosomes. Thus the few seeds developed by these steriles are formed by the fusion of unreduced, irregularly and partly reduced male and female gametes or due to agamospermy. Natural outcrossing in soybean is low, therefore this appears to be a remote possibility of origin for the aneuploid progeny. The average ratio between coenocytes and normal pollen grains is nearly 1:4. This indicates that each PMC of ms_1 produces one coenocyte and the penetrance of the ms_1 gene is thus complete. It is incomplete in ms_4, in which cytokinesis and pollen grain formation occurs in a few anthers. Coenocyte development is erratic and differs from anther to anther and flower to flower (Delannay and Palmer 1982). Either four nuclei of the coenocytes fuse together into one and the fused nucleus divides mitotically, or daughter cells with none or one to four nuclei are formed. The coenocytes either degenerate or persist until anthesis.

30.5.1.1.5 Tapetal Behaviour

In ms_1 and ms_4, tapetal differentiation and development are normal up to TII. These are normal later also in ms_1 (Albertsen and Palmer 1979). In ms_4, tapetal degeneration is late and starts after coenocyte degeneration is already in progress. In ms_2, the tapetum becomes vacuolated during prophase-II. The vacuole enlarges up to the tetrad stage and collapses along with the tetrads (Graybosch and Palmer 1983). In ms_3, tapetal cell mitochondria are larger and less electron-dense, with meagre cristae development. Tapetal degeneration may start either prior to (Buntman and Horner 1982) or after microspore degeneration (Palmer et al. 1980).

30.5.1.1.6 Phenotypic Alterations

Up to and during flowering most of the mst plants do not differ phenotypically from their mft counterparts, but during the later stages the mst plants continue to remain green when mft plants have already matured. The ms_1 plants develop leathery dark green leaves and a few fleshy pods, over 70% of which are single-seeded. Similar

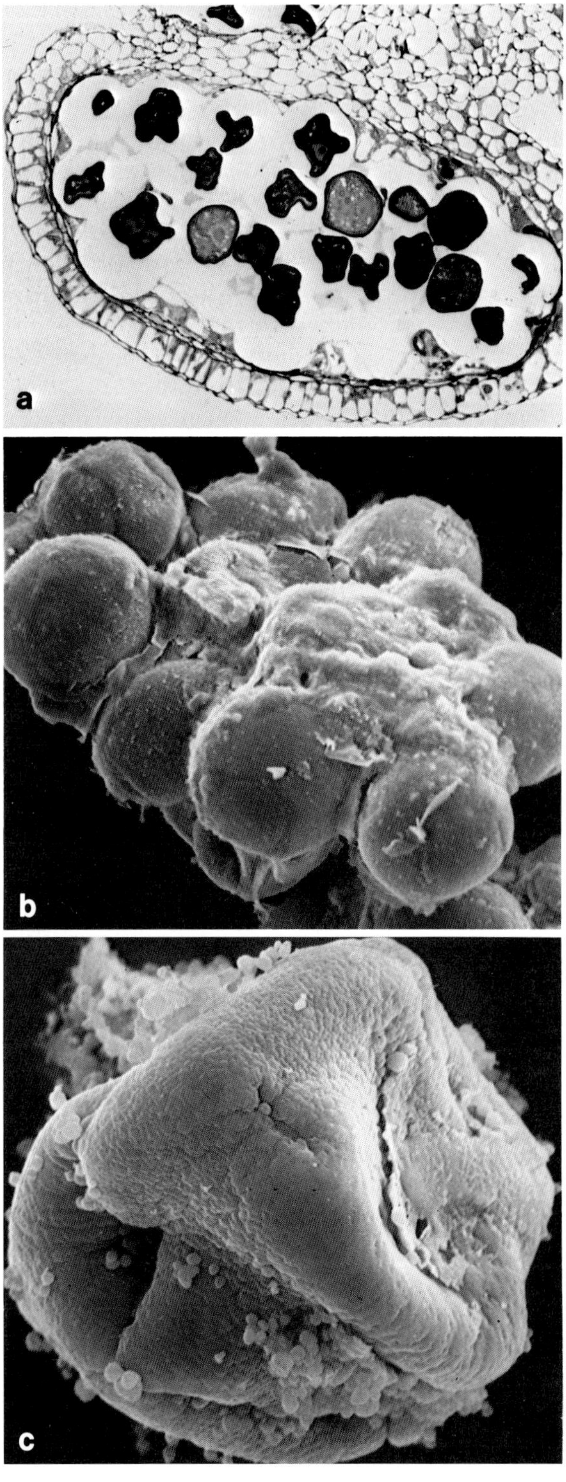

Fig. 30.4a-c. *Glycine max (ms*$_1$ male sterile mutant). **a** T.S. of anther sac with intact endothecium and degenerating coenocytes. **b,c** SEM of clumped coenocytes **(b)** and enlarged view of degenerating coenocyte **(c)**. (Albertsen and Palmer 1979)

thick and leathery leaves occur in the induced mst mutants studied by Singh et al. (1974). Except in ms_2, in other mst mutants the anthers are shrunken at anthesis. In ms_2, anthers are highly reduced, flat and white (Fig. 30.3b). By this anther phenotype, these mst's can easily be distinguished from the mft plants. In ms_3, earlier-formed flowers drop off and subsequent ones do not develop any pods.

30.5.1.1.7 Female Fertility

Of the five mst mutants (ms_1-ms_5), female fertility, except of ms_1, is nearly normal. Many supernumerary nuclei, particularly in the regions of secondary nucleus and egg apparatus, develop in 72% of ms_1 plants (Cutter and Bingham 1977); 25% of the fertilized ovules have one embryo and 60% lack it. In the remaining ovules, one to seven nuclei develop. Since 28% of the embryo sacs are normal and 25% of these develop an embryo, it appears that the majority of the seeds produced by the plant are developed from these normal embryo sacs. This leads to single-seeded pods and to the consequent seed yield reductions in ms_1 plants. Whereas the Carolina ms_1 mutant has nearly 25% female fertility (Brim and Young 1971), Urbana ms_1 has considerably higher fertility (Boerma and Cooper 1978). In the latter, 16–19% pods per plant are three-seeded, while N. Carolina ms_1 produces maximum 8% three-seeded pods.

30.5.1.1.8 Mutations and Linkage

Of the ms genes, ms_1 is genetically unstable and exhibits a high mutation rate. Thus four spontaneously arisen independent mutations occur at this locus (Palmer 1976; Palmer et al. 1978b). These were designated according to the place of their origin N. Carolina, Urbana, Tonica and Ames. They exhibit certain differences due to different genotypic backgrounds. In various genetic backgrounds and under two environments, the Urbana mst plants exhibited higher levels of female fertility than N. Carolina mst plants (Boerma and Cooper 1978). The locus ms_1 is genetically unstable and has a high spontaneous mutation rate. The gene is located on sat chromosome 8, and is linked to white flower colour *w*, and desynaptic gene st_5 (Palmer and Kaul 1983); the translocation break point lies between ms_1 and *w* (Sacks and Sadanaga 1984).

30.5.1.1.9 Polyembryony and Polyploidy

The gene ms_1 enhances polyembryony frequency up to ten times. Thus from about 0.44% spontaneous twin seedling frequency in normal soybean (Owen 1928), it is about 4% in ms_1 (Kenworthy et al. 1973). The increase depends on the genotypic background and environment, so that the frequency of multiple seedlings is 2.3% in F_5 (Beversdorf and Bingham 1977) to 28% in F_7 (Shorter and Byth 1975). The frequency of abnormal seedlings from six sources varies from 1.7 to 16.8% and that of polyploids among abnormal seedlings ranges from 6.8 to 66.7% (Chen et al. 1985). Their chromosome number varies between 20 and 200.

30.5.1.4 Biochemical Differences

Specific biochemical differences, especially in individual amino acids, between mst and mft anthers before and after sporogenous tissue degeneration have not been determined in soybean. However, rate and duration of nitrogenase activity, dinitrogen fixation, and contents of total nitrogen, lipids and carbohydrates present in the roots of mst and mft plants have been analysed. These are described in the following:

i. *Dry Matter Accumulation.* The ms_1 plants develop 87% fewer pods than the mft plants (Burton et al. 1979) and naturally the dry weight accumulated in ms_1 pods is about 87% less than those in mft pods. Until pod setting, no significant differences exist in the rate of dry matter accumulation per pod or in the total accumulation of plant dry matter in ms_1 and mft plants. In mft plants, total dry weight of shoot and leaves decreases after pod set, due to petiole and leaf abscission and due to the mobilisation of N_2 and protein from the leaves into pods. In mst plants, since leaf abscission and pod set is very low, stem and leaf weight enhances proportionately. Thus whereas in the mft plants increase in dry weight after pod set is due to increase in pod weight, in mst plants, stem petioles contribute 40%, leaf blades 35% and pods 25% of the total increase in the aboveground dry matter content between 100 and 130 after seedling emergence (Burton et al. 1979).

ii. *Carbohydrate Content.* Rate and total amount of aboveground dry matter production of ms_1 and mft plants are nearly similar (Burton et al. 1979) but a seven fold increase of carbohydrate content in the roots of ms_1 plants over mft plants was found by Wilson et al. (1978). Apparently the root starch is not metabolised by the ms_1 plants. Since ms_1 plants produce 87% fewer pods than their mft counterparts, the reduced seed production was associated by Wilson et al. (1978) with the increased carbohydrate content in the root and leaf systems of ms_1 plants. In fact, developing seeds compete so strongly for photosynthate that very low amounts are left to the roots and leaves of mft soybean plants. Prolonged retention of green leaves in ms_1 plants permits more photosynthate to become diverted to roots for utilization in metabolic processes or for conversion to metabolic reserves. Thus, both the absence of pod and seed formation and retention of green leaves for longer periods enhances photosynthates in the roots of ms_1 plants.

iii. *Lipids.* In contrast to photosynthetic products that are translocated to root tissues, root lipids are synthesised in-situ and may function as membrane components and metabolic storage products. Thus differences in lipid content and quality between ms_1 and mft roots reflect the secondary effects of *ms* gene action. Such differences between ms_1 and mft were found by Wilson (1981) between 110 and 140 days after emergence. During this period, total root lipid content increased from 1.3 to 3 times in ms_1 roots compared to mft roots. Later, a 6.2 times increase in triacylglycerol content occurred in mst rows. Although the major glycerolipid fraction in ms_1 and mft roots is phospholipid within the phospholipid fraction, phosphatidylglycerol and cardiolipin content were enhanced dramatically in the ms_1 roots compared to mft roots; the concentration ratios of palmitoleic and linolenic acids are 16:1 and 18:3, respectively. A similar concentration difference occurs in the triglycerol content. Thus the lipid biochemistry of ms_1 soybean roots is altered drastically. Whether the increased levels of storage lipid and of biosynthesis of membrane lipid components bears any relationship to mst or not is not known.

iv. *Dinitrogen Fixation.* Soybean mst plants exhibit delayed leaf senescence. Certain mft soybean lines also possess delayed leaf senescence and in others delay in senescence can be induced by premature pod removal (Nooden et al. 1978). In such mft genotypes, dinitrogenase activity increases in root nodules. Also duration of nitrogen fixation is prolonged in them (Abu-Shakra et al. 1978). Thus mst soybean plants exhibiting delayed leaf senescence should fix more nitrogen than mft plant roots. This occurs in ms_4 plant roots, in which higher and more prolonged nitrogenase enzyme activity occurred (Imsande and Ralston 1982), but in ms_1 roots no increased nitrogenase activity occurs (Wilson et al. 1978). Even though they have about 87% fewer pods/plant, insignificant differences occur in rate of dry matter accumulation per pod or in the total accumulation of plant dry matter (Wilson et al. 1978). Thus unlike ms_1, ms_4 plants exhibit a larger mass of nodules, a higher rate of total nitrogenase activity and a higher partitioning index over their mft counterparts (Imsande and Ralston 1982). This suggests that ms_4 plants direct a greater proportion of their photosynthates into dinitrogen fixation than do the mft plants. Considerable nitrogen accumulation in stem and leaves of ms_1 plants indicates that dinitrogen fixation goes on even when a certain nitrogen level is acquired by the plant. The nitrogen content, if not stored in the seeds, remains "free-floating" in aboveground plant organs, especially in shoots and leaves and/or is accumulated in the root nodules. Thus whereas under certain conditions dinitrogen fixation is switched off in mft soybean plants (Streeter 1972; Law and Brun 1974; Thibodeau and Jaworski 1975), it is not so in ms_1 plants. Whether this has any relationship with the ms_1 gene or not is not known.

30.5.1.5 Viral Infection

In a local and an American soybean cultivar, Harsoy, p-mst occurs following tobacco ring spot virus (TRSV) infection. Yang and Hamilton (1974) found viral particles in pollen intine, cytoplasm of generative cell, integument, nucellus, embryo sac wall and in megagametophyte cells. The infected anthers produce less pollen than the anthers of virus-free plants. Moreover, the stainable pollen from the infected plants has a lower germinative capacity and its germ tube elongates more slowly than those of virus-free plants. Most of the pods on infected plants are single-seeded, whereas in healthy plants they are multiseeded. Since p-mst plants produce single-seeded pods, viral infection leading to p-mst and/or to single-seededness cannot be ruled out. Being fully seed-transmitted (Athow and Bancroft 1959), maternal transmission is guaranteed. Moreover, mosaic virus is known to infect p-mst soybean plants, causing shoot stunting and leaf deformity (Caviness et al. 1970; Stelly and Palmer 1980b). However, grafting experiments indicate no association of p-mst and viral infection (Stelly and Palmer 1980b).

30.5.2 Interspecific Hybrids

The progeny of both direct and reciprocal crosses between *G. tomentella* × *G. tabacina* is mst, having poorly dehiscent anthers with sterile pollen (Palmer and Hadley 1968); but the plants are also fst and possess highly abnormal meiosis. Likewise progeny of the cross *G. gracilis* × *G. max* is both male and female sterile.

30.5.3 Induced Nongenetic Male Sterility

30.5.3.1 Mendok Treatment

In four soybean varieties, 60 to 70% pollen abortion occurs following foliar spray of 250, 500, 750 and 1000 ppm mendok 3 weeks after planting (Starnes and Hadley 1962). In addition, inhibition of plant growth, causing stunting, chlorosis and buckling of leaf mesophyll, drop-off of floral buds, reduction in flower size and tight closure of flowers, occurs in the treated plants. Because of high toxicity even at low concentrations, the use of this chemical as a male gametocide is neither desirable nor feasible.

30.5.3.2 Photoperiodic Treatment

Nielson (1942) studied its effect on flowering and microsporogenesis in var. Biloxi. Its young plants were subjected to two to ten photoinductive cycles, consisting of 8 h daylight + 16 h darkness. After induction, two sets were placed on cycles of long photoperiod, consisting of a 21-h light and 3-h dark period and one set on 16-h light and 8-h dark period. In plants receiving five or less photoinductive cycles PMC's enlarged rapidly, their cytoplasm became highly vacuolated and nuclei finally disintegrated during early prophase. In a very few, M_I-like configurations appeared, but they degenerated immediately thereafter. In the plants that received six or more photoinductive cycles of short photoperiod, PMC's degenerated during prophase in some anthers and in others tetrads degenerated before cytokinesis. In the remaining few, degeneration occurred just after cytokinesis so that microspores were never formed. In very few anthers, a few microspores were formed, but they degenerated during development. Even the plants receiving ten photoinductive cycles of short photoperiod developed a high percentage of degenerated microspores. From this it appears that the number of photoinductive cycles influences the stage and degree of microsporogenesis degeneration. Thus the plants receiving five or less photoinductive cycles exhibited the earliest and maximum microsporocyte degeneration. However, with an increase in photoinductive cycles, the stage of breakdown became delayed and the degree of breakdown reduced proportionately. A few phenotypically normal microspores developed in some plants that received eight to ten photoinductive cycles despite a high proportion of early PMC degeneration. It is, however, not known if these pollens are viable and functional. The long postinductive photoperiodic treatment could also have led to the observed mst. The influence of such a variable photoinductive and photoperiodic treatment on the female fertility of these plants is not known. Therefore it is not possible to say precisely if the photoperiodic treatments suppressed sexuality completely or unilaterally.

30.5.4 Breeding Value

Many soybean hybrids outyield parents in seed production. Therefore, the use of mst for hybrid seed production is most desirable. In soybean gc-mst is not known. Instead, monogenic recessively conditioned g-mst is fully established. The use of g-mst's in breeding is limited as only 50% plants in a female production row will be mst. To produce 100% hybrid seed, 50% plants in the rows (mft) are to be identified and removed before flowering. Since no phenotypic markers associated with mst are known, pollen of each plant is to be examined microscopically to identify mst's. This is not possible in commercial seed production plots. Soybean flowers are self-pollinated on the day when the corolla opens. The amount of natural crossing is approximately 1% for adjacent plants within a row and 0.5% between plants in adjacent rows (Webber and Hanson 1961). Effective use of g-mst in soybean requires controlled pollen movement. Using mst soybean, Boerma and Moradshahi (1975) found that pollen moved more freely within the rows than across the rows. Utilizing ms_2, Nelson and Bernard (1979) found that a 10-m buffer area of soybean should eliminate almost all pollen contamination into mst intermating blocks and a 5-m buffer area would allow less than 5% contamination. Since pollen movement from mft to mst is limited and formidable, possibilities for redesigning reproductive structures of soybean have been suggested (Bradner 1974). The use of bees and other insects for transferring pollen from the male to the female parent is also being tried. Erickson (1975a,b) identified certain soybean cultivars which are more attractive to bees. This led to greater seed yield in Corsoy and Hark soybean cultivars. However, the economic levels of seed production on mst's are not attained because of insufficient means of pollen transfer. The ms_1 seed set is nearly 20% or less, mainly because of insufficient pollen available to mst flowers. In ms_2, up to 60% outcrossing occurs in closely spaced mft and mst plants (Sadanaga and Grindeland 1981). In general, the use of g-mst is not desirable as hand pollination provides adequate crossed seeds without contamination or other extraneous efforts involved in using g-mst plants.

The available g-mst's eliminate emasculation, but hand pollination is necessary to achieve a fairly high seed yield. Thus g-mst facilitates hybridization for recurrent selection programs (Brim and Stuber 1973). Although the precise distance required for complete isolation of a crossing block is not known, outcrossing is less than 0.5% when mst plants are 12 m or more from a foreign pollen source (Boerma and Moradshahi 1975). Plants on the boundaries of the crossing block sustain maximum outcrossing with foreign pollen and can be eliminated at harvest to minimize contamination. Little is known about the optimum density for natural hybridization. Since cross pollination is more common within rows than between adjacent rows, Brim (personal communication) grew g-mst soybean populations on a square grid to create rows in all directions. He used single plant hills on 50-cm centers, subdivided the area into blocks of an equal number of hills for harvest, and bulked an equal amount of seed from mst plants in each block to enhance random pollination. As NCP evidenced by the seed set on mst plants is low and uneconomical for commercial seed production, the p-mst's offer some possibility of obtaining hybrid seeds. Caviness et al. (1970) found 45% more NCP in p-mst than in mft plants. This led to 11–30% of pod set on the field-grown p-mst plants, the rest of the seed produced being due to selfing. Thus p-mst might facilitate hybrid seed production for use in a recurrent selection program. However, the potential use of p-mst's lies in their capacity to produce seeds from self-pollination which permits sexual propagation of homozygous p-mst genotypes. In case an efficient cross pollination could be obtained in them, their use in hybrid seed production would enhance. However, p-mst's permit the synthesis of a large homogeneous population of genetically sterile plants useful in soybean breeding.

30.6 *Lathyrus* (peas, vetches)

From among 130 species of this genus, mst is known in *L. luteus*, *L. odoratus*, and *L. sativus* (Tables 30.4, 30.5).

30.6.1 *L. luteus* Moench (*L. annus, Orobus luteus*)

From selfed progeny of a seed stock of *L. luteus* obtained from France, mst plants segregated (this Vol.). Meiosis is normal up to the TII stage but microspores degenerate within PMC's. In the few late-formed flowers, microspores are liberated but they abort soon after. Female fertility is normal. The sterility is conditioned by one recessive gene, ms_1. The tapetum develops normally and disintegrates after microspore degeneration.

30.6.2 *L. odoratus* L. (sweat pea) Netchling

Bateson et al. (1903, 1908) found a recessive *ms* gene causing contabescent anther formation. Gregory (1905) detected abnormalities during male meiosis, and failure of the archesporial cells to divide (Faberge 1937) led to the formation of scanty PMC's in this plant. The bivalent congression at MI was imperfect and metaphase plate was rarely formed. Meiotic arrest occurred mostly during diakinesis to MII and, when rarely meiosis continued, it was accompanied by highly irregular cytokinesis. Female sex was normal and pollen sterility was complete (Punnett 1923, 1925). In another mst sweet pea mutant, chromosomes at MI were much shorter than those of fertiles, and chiasmata were more largely terminalised than in the mft's (Upcott 1937). Timing imbalance between chromosome cycle and anther development occurred at the pachytene and became acute by the end of the first division in the mst material. Female fertility and genetic control of sterility are not known.

30.6.3 *L. sativus* L. (grass pea, bitter vetch)

30.6.3.1 Genic Male Sterility

A single plant in the germplasm line S-372 had short filaments and indehiscent anthers (Srivastava and Somayajulu 1981). Its pollen is completely sterile, but female fertility is normal. The mst is conditioned by a recessive gene ms_1.

30.6.3.2 Gene Cytoplasmic Male Sterility

Intercrossing certain grass pea lines produced F_1 mst plants (Bardner and Vazhnitskaya 1969), no such plants appeared in the reciprocal crosses. In the Russian grass pea variety, Stepnaya 287 mst mutants were obtained by Chekalin (1972) after treating its seeds with N-nitroso ethyl urea (NEU) followed by 1, 4-bis (diazo acetyl)-butane (DAB) and γ-ray treatments. The mutants were isolated in M_2-M_4 generations and possessed white shrunken anthers, the majority of which lacked pollen completely. However, in a few anthers, a deformed mass of sporogenous cells was produced, while the remaining few had some apparently normal-looking pollen. A few meiotic anomalies occurred in the anthers of mft plants but in mst anthers the anomalies increased ten times during the uninucleate microspore stage (Chekalin et al. 1973). The abnormalities led to the production of sterile young pollen grains that lacked cytoplasm and/or nucleus and degenerated quickly. In an mst mutant, Zelenskaya and Postova (1974) found nearly normal male meiosis up to TII, but further meiotic progression stopped, mostly at the uninucleate microspore stage, after which degeneration was abrupt. Although 8–20% pollen grains stain in acetocarmine, only 0.21–0.55% germinated on a nutrient solution. They were nonfunctional on stigmas. The frequency of mst plants varied from family to family. After crossing with mft plants, in some families restoration was complete, in others partial and in a few nil. This indicates the existence of mst S-cytoplasm, *fr* and *Fr* genes. Segregations reveal mst as caused by the interaction of two *fr* genes and S-cytoplasm. The sterility is influenced by the environment. Whereas during early- and mid-flowering mst expression is complete, at the end of flowering a substantial amount of fertile pollen is produced by mst plants. This pollen is functional and seeds are borne by the mst plants (Chekalin 1975), but the resultant progeny is mst.

30.7 *Lens culinaris* Medic (lentil)

In the lentil variety L-235, Sharma and Sharma (1978) induced mst following mutagenesis with gamma rays and NMU. The mst plants had reduced internode length, profuse branching and matured later than the mft plants. They remained green past the usual maturity and had dark green thick leaves. The course of microsporogenesis in mst plants is not known but the pollen produced was fully sterile. The sterility is conditioned by a recessive gene ms_1. The lack of a distinct phenotypic marker for distinguishing mst plants before flowering limits their use in commercial heterosis breeding or hybrid seed production.

30.8 *Lotus filicaulis*

Unlike other accessions that produced fertile pollen, an accession (P.1.51, 864) obtained from Spain produced phenotypically normal but nonstainable pollen (Nettancourt and Grant 1963). F_1 plants of the cross *L. tenuis* × *L. filicaulis* are p-mst, having about 56% pollen sterility (Nettancourt and Grant 1963). Since male meiosis in 85% of meiocytes is normal and only two univalents in 13% PMC's occur at MI, high pollen sterility of F_1 plants is due to mst genes. Female fertility of the mst F_1 plants is normal. Inbreeding increased the degree of mst and frequency of sterile plants.

30.9 *Lupinus* (Tourn) L. (lupin, Texas bluebonnet)

The majority of the 200 species of this genus are economically useful as fodder, food, ornamentals or medicine. Of these, mst is known only in *L. mutabilis* and *L. regalis* (Table 30.4).

30.9.1 *L. mutabilis* sweet (sweet lupin)

Pakendorf (1970) detected some mst plants in collections from Germany and Peru. In them male meiosis proceeded normally up to tetrad formation but further development was arrested. While the tapetal layer disintegrated in the fertile plants, it persisted in the steriles. Some tapetal cells coalesced, forming a continuous strip around the periphery of the anther. The tetrads degenerated gradually without separating out from each other. This was coupled with a progressive increase of the endothecial layer. Unlike in the fertile plants, in the steriles, cells between two adjacent anther lobes did not dissolve. Eventually the tapetum also degenerated and no viable pollen was formed. Neither genetic control of the sterility nor female fertility of these steriles is known.

30.9.2 *L. regalis* Berg. (*L. hybridus* Hort.)

Plants having nondehiscent anthers and attractive bicoloured flowers of peach and apricot pink types segregated from a seed stock (this Vol.). Male meiosis proceeds normally until tetrad formation but microspores degenerate immediately after their liberation. This is followed by tapetal breakdown. Female fertility is normal.

The mst is controlled by the complementary action of two recessive genes, ms_1 and ms_2. Although in a few advanced breeding lines segregation for a single *ms* gene occurred, in most of the lines digenic control is evident.

30.10 *Medicago sativa* L. (alfalfa, lucerne)

This is a superior pasture legume for many classes of livestock because of its high quality forage, yield, and wide ecological adaptation. Both g-mst and gc-mst are known in cultivated tetraploids and wild diploids of this species complex (Table 30.4, 30.5).

30.10.1 Genic Male Sterility

The occurrence of pollen sterility in variable proportions in alfalfa plants has been recorded by Clark and Fryer (1930), Engelbert (1932), Armstrong and White (1935), Brink and Cooper (1936), Bolton and Fryer (1937), Julien (1944), Grün (1951), Petten (1951) and Sprague (1959). The sterility reduces seed and pod set (Armstrong and White 1935) and is genic, as it occurs constantly and segregates mono- and di-genically in F_2. In a mst population, Clark and Fryer (1930) found the aborted pollen percentage to be relatively constant from year to year. Brink and Cooper (1936) obtained significant correlations between the data on pollen sterility obtained in two winters for an alfalfa population. Precise gene control of the mst in these materials is not known. A complete mst plant P-1-3 was detected in Wisconsin by Childers (1952). Upon crossing to an mft plant, it produced complete and partial mst plants in the progeny. Genetic studies of the complete mst plants indicate the sterility to be controlled by two duplicate recessive genes, ms_1 and ms_2. In p-mst, the sterility appears to be conditioned by three polygenes that act additively. Precise gene control could not be determined, as the gene action was considerably influenced by sampling and environmental factors. In the variety Rhizoma, Petten (1951) detected mst plants that have yellow-green variegated flowers and sickle-like pods. The anomaly is conditioned by two duplicate recessive genes (this Vol.). In the clone 20 DRC, Childers and McLennan (1960) found mst to be conditioned by three recessive genes, ms_3, ms_4 and ms_5, inherited in a disomic manner. In some, tetrasomic inheritance of *ms* genes (Stanford 1951) or interaction of disomic and tetrasomic inheritance (Twamley 1955; Dudley and Wilsie 1957) are known. Although Stanford and Clements (1958) considered inheritance to be tetrasomic, it is really disomic. Childers and McLennan (1960) transferred the sterility from the tetraploid to the diploid level and found it to be conditioned by one recessive gene, ms_3, inherited in a disomic manner. Depending upon the ploidy level, the inheritance pattern changes, as alfalfa plants are diploids, tetraploids and octaploids, the former two types being common.

30.10.1.1 Microsporogenesis

Childers (1952) detected complete and partial mst's in alfalfa. In the former, microsporogenesis is highly irregular and meiosis breaks down during early zygotene. This is followed by meiocyte degeneration, and results in the production of empty, shrunken anthers. In p-mst's, meiosis proceeds normally until microspore formation, but 5–81% microspores abort after separation from the tetrads. They become vacuolated and their cytoplasm degenerates completely. The remaining microspores develop into normal pollen, so that in mature anthers degenerated microspores are interspersed with fertile pollen. In the mst plants of 20 DRC, male meiosis up to microspore separation occurs normally (Childers and McLennan 1960), but immediately thereafter, the microspores degenerate completely so that no trace of pollen is left in the mature anthers. The *ms* gene action is complete and highly stable both in greenhouse and in field.

30.10.1.2 Tapetal Behaviour

In complete mst plants studied by Childers (1952), male meiosis is erratic and meiocytes degenerate before the completion of heterotypic meiotic division. Even before the onset of meiosis, tapetal development is abnormal, as these cells become considerably swollen, densely cytoplasmic and degenerate gradually. In most of the anthers, tapetal disintegration occurs during the early zygotene stage. This is followed by PMC degeneration. In a few anthers, tapetal degeneration is delayed until PMC's pass the MI stage. Whereas meiotic arrest occurs early in complete mst's, in p-mst's it is delayed until microspore separation. During this stage, separation, elongation and adherence of some of the tapetal tissue to the adjacent tissue fails to occur in different degrees (Childers 1952). The tapetal cells become short, thickened and lose contact with the adjacent tissue. Concomitantly, a varying proportion of microspores become vacuolated, shrunken and their protoplasmic contents degenerate. Likewise, in the mst plants of 20 DRC, abnormal tapetal development accompanies microspore degeneration (Childers and McLennan 1960). The first indication is separation of tapetal cells from the inner locule wall and from one another. This is followed by rapid degeneration of the microspores. At this stage, the tapetum appears densely cytoplasmic with swollen, short, thick cells separated from one another. Soon the microspore wall nuclei and the whole cell organisation degenerates rapidly and the tapetal absorption by the anther locule occurs briskly, leaving an empty locule and thick-walled endothecium. Although none of these events establishes a definite cause-and-effect relationship between tapetal abnormalities and microspore degeneration, in all three materials, the *ms* gene action is almost simultaneous over tapetal and sporogenous tissue.

30.10.2 Gene-Cytoplasmic Male Sterility

The existence of gc-mst in alfalfa plants was first documented from N. America by Davis (1967). Later a search for mst plants led Davis and Greenblatt (1967) to identify one complete and nine partial mst's. The sterility is conditioned by the interaction of the S-cytoplasm with an undetermined number of recessive *fr* genes. Three genetic types of alfalfa plant were found by these authors viz. (a) mst's with S-cytoplasm and *fr* genes, (b) mft's with N-cytoplasm and *fr* genes and (c) mft's with N-cytoplasm and *Fr* genes. From Ottawa, Canada, an mst plant designated Oms-5, having 95% aborted pollen, was found by Bradner and Childers (1968) in the alfalfa variety Ontario variegated. Pollinating the mst plants by almost 100 different mft pollinators produced progeny exhibiting a wide range of pollen sterility, but the F_1 of cross Oms-5 × 5–81 had pollen sterility almost equal to the maternal parent. This led the authors to conclude that while the former pollinators were simplex, duplex, triplex and quadriplex for *Fr* genes, the latter 5–81 male parent was nulliplex and therefore a maintainer with $N_{fr\,fr}$ genotype. The findings indicate that fertility restoration of mst plants is conditioned by 1–4 *fr* genes or by polygenes. No positive proof for the existence of gc-mst in the materials investigated by Davis and Greenblatt (1967) and Bradner and Childers (1968) was given, since the authors did not make reciprocal and test crosses. However, as no loss of mst occurred in the progeny plants of two BC progenies raised by Davis and Greenblatt (1967), the sterility transmission is maternal and of the gc-mst type. The existence of gc-mst was established by Pederson and Stucker (1969), who detected mst plants in Dupuits alfalfa variety. Sterility was maintained in F_1 and BC plants, but vanished in the reciprocal crosses. Because of tetrasomic segregation, the exact gene number controlling fertility restoration could not be determined. However, the mst is conditioned by the interaction of the S-cytoplasm with two *fr* genes. In addition to the N. American source of the S-cytoplasm (AS), another sources (ES) was detected in Europe by Staszewski (1970, 1976), Sinska (1973), Staszewski and Jakubowska (1978) and Guy (1978). Based on the extent of nonviable pollen production and pollen sterility control, the two cytoplasms appear to differ from each other, but they have not been rigorously tested for cytoplasmic differences. Some evidence for their distinctions is clear from the investigations of Dattée et al. (1980).

30.10.2.1 Genetic Control

Complete gc-mst in alfalfa is conditioned by the interaction of the S-cytoplasm with the recessive *fr* genes in the homozygous state. Bradner and Childers (1968) proposed a similar control in Oms-5 mst's. Barnes et al. (1972) assumed that partial or incomplete sterility is conditioned by nuclear *Fr* genes present in simplex, duplex or possibly triplex manner. Thus duplex genotypes will have more fertile pollen than simplex genotypes. This dosage effect of *Fr* genes can account for the variation in mst found in the progeny of gc-mst plants. The proposed additive action of *Fr* genes is substantiated by the inheritance pattern of viable pollen in diploid alfalfa

(*M. fulcata*, 2n = 16) plants studied by Schonhorst (1958), who distinguished three levels of pollen fertility viz. (a) high (30–70%), (b) intermediate (29–50%) and (c) low (below 29%). Three major genes condition pollen fertility. Gene *A* is dominant, environmentally stable and causes high pollen fertility, Genes *B* and *C* are additive, together equalling gene *A* and individually producing intermediate fertility, *B* is more stable environmentally than *C* and both these are less stable than *A*. The three genes in the homozygous recessive state produce over 50% pollen sterility and the genotype exhibits extreme environmental susceptibility. Nearly similar genetic control of fertility restoration occurs in three other gc-mst strains of *M. sativa* (this Vol.).

Apart from variations in the amount of sterile pollen in the anthers, Sinska (1979) found that pollination of four mst alfalfa plants to the same male parent yielded progeny plants that exhibited different mst grades. For instance, while in the progeny of A3 mst, only sterile and semi-sterile plants segregated, in the remaining ones, fully fertile and semi-sterile plants also segregated. This variation could be because the four plants used by Sinska differed in the number of *Fr* genes having additive action. This led to different segregations when crossed to the male parent. However, segregation of Suginobu (1979), using Ontario mst lines, indicates an additive action of *Fr* genes in some and a dominant action in others. The genetic analysis is hindered by the profound environmental influence and by the instability of mst expression under different environmental conditions. The effect of modifying genes and the environmental influence enhance the difficulty of conceiving a definite inheritance pattern of mst on one hand and fertility restoration on the other. In fact, a more or less continuous range from fully sterile to fertile pollen in the progeny of some p-mst plants led Pederson and Stucker (1969) to suggest a quantitative inheritance of *Fr* genes. Interaction of the S-cytoplasm with two recessive genes, fr_1, fr_2, leads to complete mst in three Bulgarian cultivars of *M. sativa* viz. Local, Miechowska and Vernal (Staszewski 1976). Partial pollen fertility in them is restored when *fr* genes are in heterozygous state. A nonallelic dominant gene, *Id*, induces anther indehiscence in mft plants having N-cytoplasm. These plants are thus rendered fn-mst. In S-cytoplasm, the *Id* gene is not effective. Thus plants with S Fr_1fr_1 Fr_2fr_2Id are partially fertile and dehiscent.

30.10.2.2 Microsporogenesis and Tapetal Breakdown

Compared to g-mst, anther development and differentiation are poorer in gc-mst alfalfa (Fig. 30.6b,c). In the latter types, male meiosis and tapetal development proceed normally until microspores are liberated from the tetrads (Bradner and Childers 1968; Pederson and Stucker 1969). Thereafter, tapetal cells separate from each other, as also from the inner wall of the locule. They swell considerably and microspores develop a thick wall (Fig. 30.6d). Tapetal cell content degeneration and microspore abortion occur almost simultaneously, but tapetal degeneration is rapid and is completed quickly. It appears that the *fr* gene acts almost simultaneously over tapetal function and microspore development in gc-mst alfalfa. However, in the European gc-mst alfalfa, tapetal disintegration begins earlier than

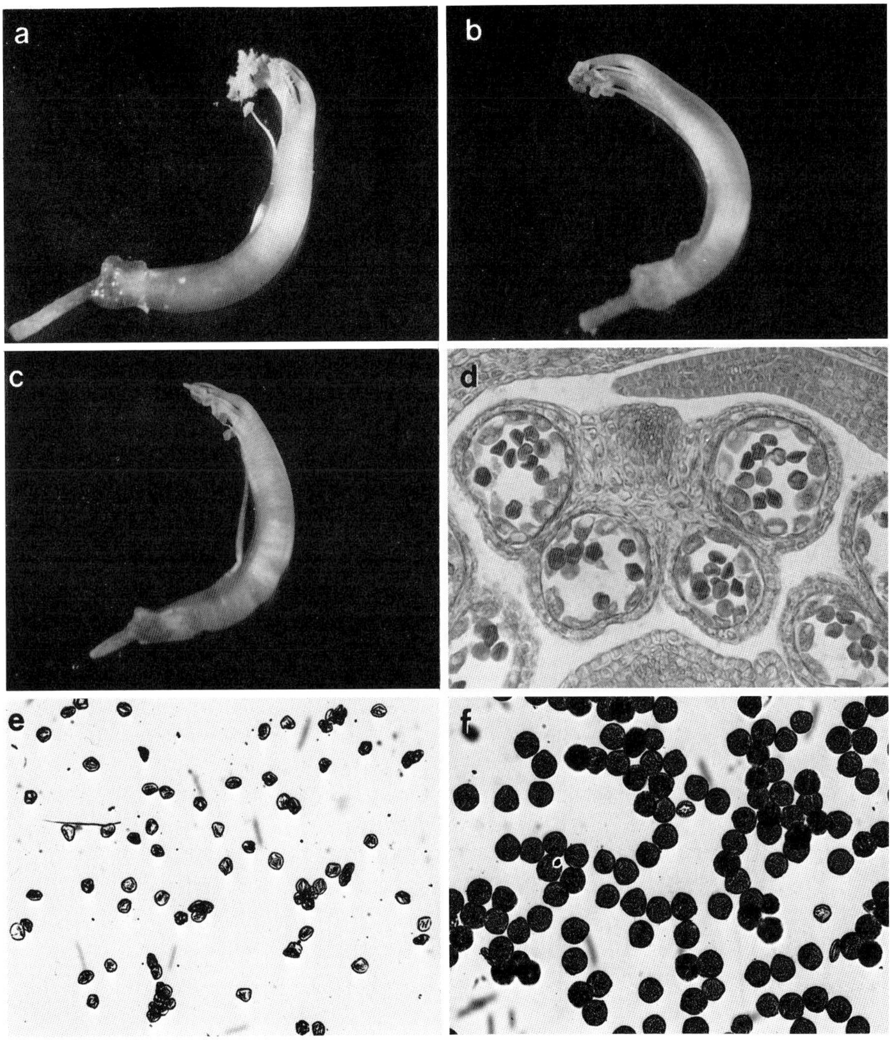

Fig. 30.6a-f. *Medicago sativa.* **a-f** Anthers of fertile **(a)** and male sterile (genic = **b**, gene-cytoplasmic = **c**), T.S. of anther and whole mount of pollen grains **(d,e)** of male sterile and pollen of male fertile **(f)** alfalfa plants. (Suginobu 1979)

microspore liberation, as Staszewski (1976) found that tapetum disintegrated when PMC's were still at the MI stage only. Meiosis in this material proceeds up to microspore formation and is arrested thereafter. At anthesis, the pollen of mst's is empty, shrivelled and deformed (Fig. 30.6e), in contrast to the large plump pollen of mft's (Fig. 30.6f).

30.10.2.3 *fr* Gene Frequency

In diverse alfalfa cultivars, the frequency of fr mutations varies between 0.01 to 0.6%. Davis and Greenblatt (1967) screened over 50,000 plants in the USA and found only one complete and one other p-mst plant (0.27%). After screening 5000 plants in Europe, Staszewski (1970) detected eight mst plants (0.06%). Sinska (1973) found six mst plants, but Lubenec and Nagovicina (1968) obtained 30 mst plants (0.6%). Thus compared to American varieties, European populations appear to possess a higher frequency of mst plants, indicating a higher rate of sterility mutations in European alfalfa. This differential mutation rate may be a characteristic feature of the American and European cytoplasmic types, as these two cytoplasms differ from each other genetically.

30.10.3 Comparison of Male Sterility Types

Gene control of complete and partial g-mst differ significantly from each other. Whereas complete g-mst is conditioned by two or three recessive *ms* genes, partial g-mst is controlled by three polygenes (Table 30.4). In contrast to this, both complete and partial gc-mst is controlled by many recessive genes (Table 30.5), their number ranging from three to five (this Vol.). Whereas in the complete mst studied by Childers (1952), the *ms* gene acts during early zygotene, in those investigated by Childers and McLennan (1960), the gene acts during microspore release (Table 30.1). On the other hand, in partial g-mst's, the *ms* gene, as in complete g-mst's, acts during early prophase-I. Thus there is no relation between mst degree and time of *ms* gene action in g-mst alfalfa. The same is valid for the complete and partial gc-mst's, as the *fr* genes in both of them act either immediately after liberation or later during microspore maturation (Table 30.1). In all the mst's, the tapetal development is abnormal and its degeneration is either before or simultaneous with *ms/fr* gene action. The influence of environment on gene action is significant in all mst's, but is profound in p-mst's.

Table 30.1. Male sterility in relation to time of *ms/fr* gene action in alfalfa

Male sterility degree	References	Gene control	Microsporogenesis breakdown stage
Genic			
Complete	Childers (1952)	2r	In zygotene
	Childers and McLennan (1960)	3r	During microspore release
Partial	Childers (1952)	3po	In early prophase-I
Gene cytoplasmic			
Complete and Partial	Davis (1967), Davis and Greenblatt (1967), Bradner and Childers (1968)	Many r	Either immediately after microspore liberation or during microspore maturation
	Staszewski (1976)	2r	Either immediately after microspore liberation or during microspore maturation

30.10.4 Pollen Production

Considerable variation in pollen production exists between various p-mst alfalfa cultivars. Barnes et al. (1972) found that all flowers of p-mst's did not release the same amount of pollen when tripped. Nearly similar observations are recorded by Davis and Greenblatt (1967) and Pederson and Stucker (1969). A random intra-plant variation in pollen production in p-mst clones of alfalfa was found by Barnes and Garboucheva (1973). Both the total amount as well as the percent of sterile pollen grains per anther vary within and between flowers and plants. In addition, anther dehiscence exhibits high inter-progeny variability even among sibs. It persists in progenies derived from cytoplasm sources of wide geographic origin (Dattée et al. 1980). The pollen sterility percent of the same clone differs from harvest to harvest within and between years (Lubenec and Nagovicina 1968). These differences are attributed to environmental fluctuations (Vengerovsky and Dvzelali 1969). Growth and temperature influence pollen production and viability of p-mst's considerably. In most of them, pollen sterility is higher in the field than in the greenhouse (Blondon et al. 1979). Sensitivity to this environmental change from lower to higher temperature is highly variable and genotype-specific. Thus 11 plants are stable, three sensitive and six intermediate to this environmental change from greenhouse to field or vice versa (Suginobu 1979). In addition to temperature, pollen sterility is also influenced by light intensity. Under low light intensity, especially during premeiotic stages, pollen sterility is higher in p-mst's. Such an influence of temperature or light intensity is not known in complete mst's of *M. sativa*. However, a few *M. falcata* plants that are complete mst's in the greenhouse produce some stainable pollen in open fields (Schonhorst 1958).

30.10.5 Grafting

Graft-induced gc-mst transmission in *M. sativa* was demonstrated by Thompson and Axtell (1978), who utilised an easy and fast method of grafting by using a self-adhesive latex bandage material to bind graft components and not the cleft-grafting technique of Bernardo et al. (1970). Whereas the autonomy of the scion flowers remained unaltered, gc-mst occurred in F_1 scion progeny of three graft combinations. This graft transmission is analogous to that known in *Petunia hybrida* (Frankel 1956), wherein the sterility transfer is first detected in scion progeny plants.

30.10.6 Chemical Male Sterility

Mendok. Its use as a male gametocide was tried by Pederson (1959), Srivastava (1961), Miller and Hittle (1960, 1963), but none of them was successful in inducing mst without impairing female fertility. Abnormal pollen production coupled with leaf burning occurred after 0.5 or 1% aqueous solution treatments (Pederson 1959). The treated plants were also female sterile. Miller and Hittle (1963) sprayed six *M. sativa* clones with 500, 1000, 5000, 10,000 ppm mendok at early and late anthesis in addition to biweekly application beginning at early bud stage. The sprays were made to run off using water as carrier. After 12–14 h of spray, the extent of plant damage as expressed by leaf mottling, crinkling and burning was proportional to the chemical concentration. Like plant damage, the pollen sterility enhanced with increasing chemical concentrations. With 10,000 ppm, 42.5% pollen became sterile but toxicity was high and no pods or seeds developed on the plants. With 5000 ppm, pollen and pod sterility was nearly 34% and seed sterility 40%. With this chemical, Srivastava (1961) obtained 60–70% pollen sterility but seed sterility was still higher. Thus in addition to pollen sterility induced by mendok, plant and ovular damage is considerable. This renders the chemical unsuitable as a male gametocide. Almost similar effects occur following GA applications (Srivastava 1961).

30.10.7 Breeding Value

Most modern commercial alfalfa varieties are synthetics. A decline in yield with generations of synthetics (Theurer and Elling 1964) is an undesirable feature of alfalfa synthetics. The decline in forage yield is due to inbreeding, which results from parents per se (Carnahan and Peaden 1967) and relatives originating from a finite number of parents (Busbice 1969). Basically, alfalfa is cross pollinated and most clones exhibit self-sterility or gametophytic incompatibility. Whereas outcrossing infuses hybrid vigour and leads to increased forage yield, selfing leads to a considerable inbreeding depression. This depression is greater than expected on the basis of the inbreeding co-efficient (Busbice and Wilsie 1966). Utilisation of F_1 plants obtained by pollinating mst plants by mft pollen is one of the best ways to avoid inbreeding depression and exploit hybrid vigour. Since alfalfa depends on insects for pollination and varies in attractiveness to pollinators (Boren et al. 1962), cage-planting designs are important in hybrid seed production. The seeds are usually produced by enclosing the plants in a cage with honey bees, leaf cutter bees or bumble bees. A high proportion of selfing may occur in field cages because individual bees prefer a single clone instead of randomly visiting all clones (Hanson et al. 1964). However, in certain situations, the crossing percentage is significantly higher with an alternate-plant design than with an alternate-row design (Kehr 1973). The alternate plants reduce the possibility of pollinators being oriented to large blocks of similar plants. Thus no significant differences in mean frequency of bees' visitation of mst and mft plants in the field was observed by Beard and Meserve (1964) and Suginobu (1979). This is because the attractiveness for bees lies in the nectar rather than the pollen. However, despite normal bee visitations, seed yield over mst plants is lower than that over mft plants. This has led to the production of three-way hybrid cultivars utilising gc-mst (Barnes et al. 1972). Even though this production system is workable and economically justifiable, the seed production costs for F_1 hybrids are higher than those for synthetic cultivars. Therefore, seed production on mst plants has to be improved by at least 20% before F_1 hybrid production is used commercially. Of 108 hybrids produced from mst plants, 21.4% outyielded the commercial DuPuits by 20% or more and 8.4% by 30% or more (Suginobu 1979). These results raise the hopes for a profitable hybrid seed production, provided desirable parental genotypes are available and are mated efficiently. Selection will have to be made dependent on general or specific combining ability. After growing plants of three mst genotypes along with a fst genotype in either mixed or alternate rows, the seed set on mst's was significantly higher in the mixed than in the alternate rows (Brown and Bingham 1984).

30.11 *Melilotus alba* Desv. (sweet white clover)

In some plants of annual and biennial varieties of this species, male meiosis proceeded normally, but no cell wall in meiocytes was formed after TII (Castetter 1925). A callose wall developed around each pollen grain, that assumed a giant appearance. Such coenocytic pollen grains degenerate at anthesis. Neither female fertility of these plants nor genetic control of this anomaly are known. Seven mst plants segregated from the seed progeny of an annual variety of the clover (this Vol.). Their flowers are long-pedicelled and the anthers dwarf, shrunken and pollenless. PMC's degenerate during the zygotene, which is followed by tapetal disintegration. The anomaly is controlled by a recessive gene ms_1. In many segregating lines, a deficit of mst's occurred. Female fertility is normal. The long-pedicelled flower trait appears to be linked to the ms_1 gene.

30.12 *Phaseolus*

This is a conglomerate of closely related, economically useful species and possesses over 200 species. Of these, mst occurs in *P. atropurpureus*, *P. lunatus* and *P. vulgaris* (Table 30.4).

30.12.1 *P. atropurpureus* Moc. and Sesse ex DC.

In the progeny of a cross between a plant with cerise-coloured flowers and a promising bred derivative of the cultivar Siratvo, Pritchard and Hutton (1972) found two mst plants whose anthers were shrivelled and indehiscent, but ovules and female fertility were normal. In them, male meiosis proceeded normally until microspore formation. Thereafter, the microspores degenerated along with the tapetum. In their anthers, endothecium did not develop thickened walls. At maturity, irregularly shaped, thin-walled, degenerated microspores and disintegrated tapetum with protoplasm coalesced into a structureless mass. Whereas no dissolution of the dividing wall between loculi took place, the middle layer persisted until floral anthesis. Thus *ms* gene(s) affected the development of the whole anther tissue and not only of tapetal or sporogeneous cells. The sterility is maternally transmitted but genetic control is not known. In a spontaneously arisen mst mutant, one recessive *ms* gene controls mst (this Vol.). The microspores abort along with the tapetum after normal meiosis.

30.12.2 *P. lunatus* L. (lima bean)

One plant per 2000 is sterile and one per 20,000 is mst (Allard 1953). The gene ms_1 in the Henderson variety induces mst and pleiotropically affects seedling and plant morphology, so that the primary axis of the seedling fails to develop beyond the production of leaves at the second or third nodes. The laminae are very thick, crinkled and darker green than mft seedlings. The petioles are short, and plant growth is axillary. The anthers are shrunken, indehiscent and have empty pollen grains. Sterility is complete and stable under many environmental conditions.

30.12.3 *P. vulgaris* L. (kidney bean)

This is an unsurpassed source of vegetable protein for man and animals and a source of essential amino acids in complementing high carbohydrate composition. Both g-mst and gc-mst are known in it (Tables 30.4, 30.5). Sterility is either complete or partial (Ibrahim et al. 1974; Agbo and Wood 1977; Van Rheenen et al. 1979; Mutschler and Bliss 1980; Singh et al. 1980; Bassett and Hung 1982; Bassett and Shuh 1982; Wyatt 1984a,b). The mst's detected by Agbo and Wood (1977) and Van Rheenen et al. (1979) have deep green foliage and profuse flowers, but poor pod and seed set. The anomaly is controlled by one recessive gene which also impairs female fertility. In the spontaneously arisen mst mutant detected by Ibrahim et al. (1974), sterility is attributed to indehiscent anthers caused by a recessive gene, *ia*. The gene alters the stigma shape from introse to extrose, and induces almost complete pollen sterility, due to which anthers remain indehiscent. However, 7–9% pollen is stainable in some plants. Another indehiscent p-mst mutant was detected by Wyatt (1984a) among greenhouse-grown snap bean breeding lines. Unlike the mst mutant of Ibrahim et al. (1974), which is a sporogenous mst mutant, this mutant is a functional mst. In this latter mutant also the anomaly is controlled by a recessive gene, *ia* (Wyatt 1984b). The gene does not affect stigma or floral morphology.

From the F_2 progeny of the cross WI 74–2047 × Swedish Brown, an mst plant segregated (Mutschler and Bliss 1980). Its pollen is shrivelled and lacks cytoplasm. A few anthers have some stainable pollen, but it is nonfunctional. Male meiosis proceeds normally until microspore liberation, immediately after which microspore contents degenerate. Female fertility of the mst plants, as revealed from hand pollination, is normal. The mst is conditioned by a dominant gene, MS_1, whose expression is inhibited by another unlinked dominant gene, $In\text{-}ms_1$. Thus mst, although basically caused by a single dominant gene, involves two unlinked genes, Ms_1 and $In\text{-}ms_1$, which interact to yield 12:3:1 F_2 ratio of mft:mst:lethal plants. The mst genotypes have Ms_1, $in\text{-}ms_1$ and $in\text{-}ms_1$ genes. This type of mst gene control is nearly akin to that of *Centranthus ruber, Lactuca sativa* and *Origanum vulgare* but unlike them, the gene $in\text{-}ms_1$ is loosely linked to one of the translocation breakpoints. This produces populations with sterile and semi-sterile plants. Seed set on mst's indicated 68% average outcrossing (Estrada et al. 1984). These outcrossing estimates are not reliable, as some degree of selfing occurs in mst's due to spontaneous mft reversions caused by temperature fluctuations. For instance, temperature regimes of 30°C day/18°C night for 2 weeks after flowering alter the expression of p-mst's to complete mst's (Estrada et al. 1984). A reverse shift occurs in the 18°C day/7°C night temperature regime. The mst is restored by returning plants to higher temperatures. F_1 progeny of mst × Swedish Brown-10 are unstable mst's. Estrada et al. (1984) obtained temperature-stable mst's by selfing cold-treated unstable mst's or by using these plants to pollinate stable steriles. Three stable mst mutants having degenerated pollen were induced after EMS and gamma rays (Hussein and Disouki 1974; this Vol.). The sterility in each is conditioned by one recessive gene.

30.12.3.1 Gene Substitution

When the gene *Hyp* (hypogeal cotyledons) and *Ext* (exterior stigma) of *Phaseolus multifluorous* were transferred to the genome of *P. vulgaris*, mst resulted (Lamprecht 1945). Transferring other genes from *P. multifluorous* into *P. vulgaris* genome or vice versa through interspecific hybridization and backcross accompanied by selection did not cause mst in either of the species. It appears that gene *Hyp* and *Ext* interact with *P. vulgaris* cytoplasm and cause mst. Thus the two genes are highly plasma-specific. Since the linkage was loose, crossing over was sufficient to produce populations segregating for sterility but without semi-sterile plants. The lines segregating for sterility but not for semi-sterility are useful in hybrid bean production. The mst flowers, when open, can be directly pollinated and from 130 such crosses 96% seed set was obtained by Mutschler and Bliss (1980). Crossing after conventional emasculation yielded only 80% seed set. This indicates the utility of mst plants in breeding and hybrid seed production. The NCP of the mst plants is high.

30.12.3.2 Gene-Cytoplasmic Male Sterility

From CIAT, Cali (Columbia), Bassett and Shuh (1982) detected a mst line with average 91% pollen abortion. The majority of the microspores abort during maturation after release and about 3–7% are stainable and 1–3% viable. The Sprite variety acts as the maintainer and no satisfactory restorer was found.

30.13 *Pisum* L.

About seven species constitute this genus, which is native to European and West Asian regions. The mst is known in *P. arvense* and *P. sativum*. In both these the sterility is genic and conditioned by a single recessive genes (Table 30.4). Gc-mst is not known in any of them.

30.13.1 *P. arvense* L. (field pea)

Two mst mutants were obtained after 3 kR gamma-ray and 0.2% EMS treatments (this Vol.). In the EMS-induced mutant, male meiosis proceeds normally up to TII, but microspores degenerate gradually after liberation from tetrads. This is followed by tapetal breakdown and degeneration. In the gamma ray-induced mutant, MI spindle development is defective and bivalent congression imperfect. Subsequent meiotic stages are irregular and variable-sized nonviable microspores develop. Tapetal development is normal. Female fertility, as evidenced by seed set after hand pollination, was reduced by 6–9% in these two mutants. In each of them the sterility is conditioned by one recessive *ms* gene. The genes are nonallelic and designated ms_1 (EMS-induced) and ms_2 (γ-ray induced). A similar monogenic recessive gene-controlled g-mst is documented by Gritton (1980), but the *ms* gene action and female fertility of this mutant are not known. The g-mst is useful for hybrid seed production, as pollen transfer is done by insects in this species.

30.13.2 *P. sativum* L. (garden pea)

In this species, 38 mst mutants are known. In all of them, the mst is conditioned by single recessive genes, but the allelic nature of most of them is not known. Whereas msk mutants were induced by the present author after 2–5 kR γ-irradiations, other mst mutants were isolated by Gottschalk and his students after X- or neutron-irradiation. Myers (1984) identified 14 mst mutants (ms_1 to ms_{14}) with almost normal ovular fertility.

30.13.2.1 Structural Male Sterility

In a 2-kR gamma ray-induced mutant, msk_9, the anther sacs are rudimentary and no archesporial tissue is differentiated (this Vol.). The female fertility of this mutant is normal. The anomaly is conditioned by a recessive gene, msk_9. In M_2 progeny of 5-kR gamma ray-treated seeds, a nearly similar mst mutant segregated that has minute dwarf anthers, differentiated sporogenous tissue but produced malformed PMC's in which no meiosis occurred. Female fertility is normal. The sterility is conditioned by one recessive gene, msk_{10}. Whereas in the above-mentioned mutants the anthers are differentiated, in the monogenically conditioned recessive mutant 192 (Gottschalk 1961), transformation of the stamens into carpel-like organs occurs. In mutants 178 and 2254A (Gottschalk 1976), no normal anther differentiation occurs. The thecae are filled with a homogenous nonspecific cellular mass that lacks differentiation into tapetum, endothecium and sporogenous tissue. The anomaly in each mutant is conditioned by a single recessive gene.

30.13.2.2 Sporogenous Male Sterility

Sixteen single gene recessive mst meiotic mutants arising after X- or γ-irradiations were found in garden pea. In all these, anther and archesporial development and differentiation is normal. The *ms* gene affects male meiosis usually at a specific stage (Klein 1969; Gottschalk and Baquar 1971; Gottschalk and Kaul 1974; Gottschalk 1976; Gottschalk and Klein 1976; this Vol.) and the mutants are categorised in order of *ms* gene action over microsporogenesis stage (Table 30.2). .The earliest action of *ms* genes is known in a neutron-induced mutant in which staminal differentiation and development is normal but no archesporial tissue is developed in anther sacs (Klein and Milutinovic 1971). Likewise in msk_1, the anthers develop normally but sporogenous tissue is malformed and no meiocytes develop. In msk_2, PMC's are formed, but leptotene threads coalesce to form knotted structures. Thereafter meiosis is arrested and degeneration ensues. Until the pachytene, meiosis proceeds normally in mutants 38B and 69, but either during the late zygotene or early pachytene, bivalents cluster forming a structureless amorphous body (Fig. 30.7). Chromatin and cytoplasm gradually disappear and cell contents are absorbed by the adjacent anther tissues. Thus pachytene is the final stage of microsporogenesis in these two mutants. In mutants *195B*, msk_3 and msk_4 bivalent adhesion and agglutinization occurs during late prophase I and MI. No metaphase bivalent congression, orientation and distribution occurs in these mutants. In msk_3, meiosis is arrested at MI, but in msk_4 erratic AI distribution and TI organisation follow. Their PMC's degenerate at MI and TI respectively. In contrast, in 195B, meiosis progresses erratically until microspore-like bodies are formed (Gottschalk and Kaul 1974). They degenerate immediately thereafter and no pollen is ever produced.

In mutant 71A, the earliest meiotic anomalies occur during AII (Gottschalk and Klein 1976). Irregular chromatid distributions and four to six nuclei per PMC usually occur in this mutant. The PMC's develop thick walls and assume giant pollen grainlike structure having tetraploid genome (coenocyte). Some of the PMC's undergo irregular mitotic division, but the meiocytes degenerate without any viable pollen production. In mutant 71A, *ms* gene acts at AII initiation, in 98A the gene acts after AII separation. A supernumerary division of AI chromosomes

Table 30.2. *ms* Gene action in male sterile pea

Mutants	*ms* gene action
ms	Archesporial tissue formation inhibited
msk_1	During sporogenous tissue formation
msk_2	During leptotene
38B, 69	During pachytene
msk_3, msk_4, 195B	During late prophase-I
71A	During AII
98A	After AII
67, 78, 395	During tetrad formation
33C, msk_5, msk_6, ms_5, ms_7, ms_8, ms_9, ms_{11}	After microspore liberation
503	After exine formation

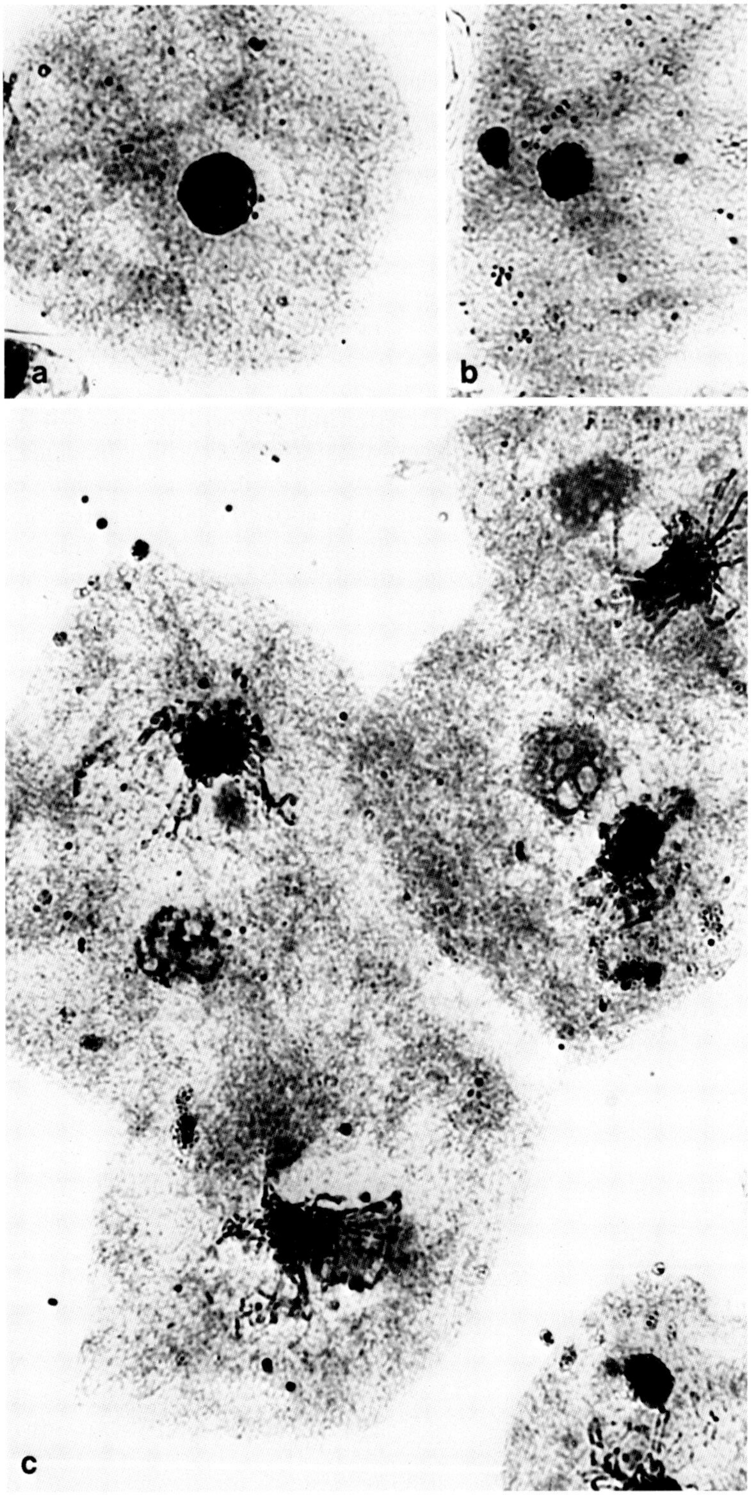
a
b
c

is initiated (Klein 1969) and the chromosomes assume a mitotic form. Some chromosomes move to either poles and others lag behind. Seven to 19 telo- and inter-phase mini-nuclei are formed within meiocytes that develop a thick wall and finally perish. In mutant 67, tetrad nuclei in each PMC are formed normally (Klein 1969), but thereafter nuclei and cytoplasm degenerate gradually. PMC's secrete a thick wall, degenerate, and only the cell walls bordering pycnotic and clumped protoplasmic contents remain. In mutant 395, meiosis proceeds up to the TII stage, quardipartition is initiated but is not completed (Klein 1969). Microspore tetrads separate partially and are connected by broad cytoplasmic strands. Degeneration follows quickly and no free microspores are ever produced. Similarly fused microspores are produced by mutants 33C, msk_5 and msk_6 after a normal course of meiosis. The microspores degenerate during development and no viable pollen are produced. Much delayed *ms* gene action occurs in the mutant 503, in which young pollen grains are formed but exine and apertures are not differentiated (Klein 1969). Their cytoplasm atrophies and cell contents coalesce to form darkly staining amorphous mass of tissue. No viable pollen ever appeared in this mutant.

In the majority of the above-mentioned mutants, *ms* gene action is distinct and almost limited to a specific meiotic stage. However, in the mutants 71A, 78 and 98A, the *ms* gene influences the whole course of meiosis in varied ways. These mutants exhibit reductions in PMC number per anther, premeiotic anomalies leading to the production of hypodiploid PMC's, poor chromosomal and chromatin stainability, irregular anaphasic distribution and incipient third reduction division at the end of microsporogenesis. PMC degeneration occurs mostly at three stages; viz. during: (a) interphase II, (b) microspore formation and (c) microspore maturation. A few pollen grains near to the tapetum develop normally, but they are nonfunctional, as evidenced by pollen germinability and TTZ viability tests. In ms_2, ms_3, ms_6 and ms_{10}, male meiosis is abnormal (Myers 1984).

30.13.2.3 Functional Male Sterility

In the mutant msk_8, staminal differentiation and development is normal but style elongates extraordinarily and protrudes so much out of floral bud that no selfing occurs (this Vol.). Thus the mutant is functionally mst. Seed set by natural pollination is low (3–5%), but hand pollination yielded normal seed set. Thus this mutant could be useful for hybrid seed production.

30.13.2.4 Gene Location and Identity

Of the 14 recessive monogenically conditioned mst's (ms_1 to ms_{14}), while ms_2 is identical to ms_4, three allelic mutations at ms_5 and 2 each at ms_6 and ms_{10} were identified by Myers (1984). Of these, the chromosome 1 carries ms_7 and ms_{11},

Fig. 30.7a-c. *Pisum sativum.* **a-c** PMC degeneration before **(a,b)** or during pachytene **(c)** in a male sterile pea. (Gottschalk and Kaul 1974)

chromosome 3, ms_6, chromosome 4, ms_9, chromosome 6 ms_2 and chromosome 7 ms_5. The genes msk_1 and msk_2 are located on chromosome 1 and msk_3 and msk_5 on chromosome 4 (this Vol.).

30.13.2.5 Chemical Male Sterility

Using Embutox-4, Fertig et al. (1964) induced pollen sterility accompanied by high female sterility. Similar results are obtained with 0.5 and 1% mendok and dalapon sprays, which in addition cause phytotoxicity (this Vol.).

30.14 *Trifolium* L. (clover)

Of about 300 species known, sporogenous, functional and chemical mst occurs in *T. pratense* and *T. repens* (Tables 30.4 to 30.6).

30.14.1 *T. pratense* L. (red clover)

Amongst true clovers, this is the most widely grown and rates second to alfalfa in importance. Both g-mst and gc-mst are known in it.

30.14.1.1 Genic Male Sterility

i. *Structural Male Sterility*. Plants with stamens transformed into pistil-like structures were found by Kazimerski and Tsikov (1981). This anomaly is conditioned by two recessive *ms* genes. The female fertility of the mst plants is normal.

ii. *Sporogenous Male Sterility*. A dominant gene *P* induces mst (Nijdam 1932). The plants had shrunken and deformed pollen but the material is lost and so far no other dominant *ms* gene has been found in the red clover genome. Smith (1971) detected a recessive ms_1 gene that causes complete pollen abortion. Whereas mft's have normal plump yellow-coloured thecae, mst's have shrunken, dark orange-coloured thecae with shrivelled pollen grains. The microsporogenesis breakdown stage is not known in the mst's. A similar mst having wrinkled petals and indehiscent anthers was found by Taylor et al. (1978); the sterility is conditioned by a recessive gene ms_2, the gene is nonallelic to ms_1.

30.14.1.2 Gene-Cytoplasmic Male Sterility

In Bulgaria, Macewicz (1976a) detected some completely and some partially stamenless plants in addition to plants having indehiscent anthers. In the stamenless plants, sterility is governed by two complementary genes, one recessive

fr_1 and the other dominant *Fr_2*, which interact with the S-cytoplasm and lead to stamenlessness. In partially stamenless plants, androecia develop abnormally but their cytoplasm degenerates even before the archesporial tissue is formed (Macewicz 1976b). Therefore, no male meiosis occurs in them. The sterility in both these types of plant and in indehiscent anther plants is complete. While the stamenlessness is due to g-c interaction, the genetic control of partial stamenlessness and indehiscent anthers is not known. Since all these plants were detected from a single composite material, the mst is of gc-mst type. Development of indehiscent anther F_1 plants in one intravarietal cross and normal F_1 plants in the reciprocal cross favour gc-mst control. Backcrossing F_1's produced partially stamenless plants whose proportion enhanced with each BC generation so that BC_4 generation is stamenless. Precise genetic control of the sterility is not known but at least three *fr* polygenes condition stamenlessness in S-cytoplasm plants (this Vol.).

30.14.1.3 Chemical Male Sterility

Mendok. 0.2% foliar spray 4 to 5 weeks before flowering led to the development of 75% flowers with indehiscent anthers in one replication and 26% in another replication (Wit 1960). These significant replication differences within each treatment are due to sampling error, variable chemical effectiveness, or population heterogeneity. The female fertility of the treated plants is not known, but since all other mendok concentrations used induced female sterility, it is certain that the high degree of mst is accompanied by nearly equal or higher female sterility. This is confirmed by Summing (1959), who found mendok-induced mst always associated with reduction in plant growth and seed production. Therefore the chemical cannot be used as a male gametocide because of its accompanying side effects and instability of action.

30.14.1.4 Breeding Value

Red clover is a self-pollinated plant. Therefore, g-mst is not of much value for a large-scale breeding program, since bee visitation is minimum and so is NCP. Smith (1971) proposed the use of the *ms* gene in conjunction with the self-fertility *sf* gene of the incompatibility system to develop inbred lines. In this case, mft segregates would have to be identified and rogued prior to pollination. Theoretically, these inbreds can be utilized only for one generation of crossing or seed increase (single crosses) because the *sf* allele would still be present in mst's. On the other hand, if restorers for gc-mst are found, commercial hybrid production could be tried, provided sufficient cross pollination is ensured.

30.14.2 *T. repens* L. (white clover)

This species contains genotypes that exhibit self-incompatibility, pseudoself-compatibility and self-compatibility. Different degrees of self-incompatibility were found among 110 first-generation inbred plants obtained from a highly self-compatible parent by Atwood (1942). Some of the inbred plants exhibited functional mst, as their anthers were extremely hard, rendering natural pollen liberation impossible. The pollen could not be artificially removed and a plant homozygous for oppositional alleles, one of the F_1 plants, was mst and female fertile (Atwood 1944). Athough phenotypically normal pollen grains were produced and they stained well with carmine and iodine solutions, the pollen grains were ineffective in fertilizing the ovary, and consequently there was poor seed set (4–7%). Details of male meiosis and genetic control of the sterility are not known. It seems that this functional mst is caused by recessive gene(s), as it segregated repeatedly in inbred lines. However, the exact number of the *ms* genes conditioning the sterility is not known.

30.14.2.1 Chemical Male Sterility

Mendok. A foliar spray of 0.2% mendok 4 to 5 weeks before flowering, led to development of nondehiscent anthers in 85% of flowers (Wit 1960). The female fertility of these mst plants is not known, but since other concentrations used induced female sterility, it is certain that the high degree of mst induced by 0.2% chemical is accompanied by nearly equal or higher fst.

30.15 *Vicia faba* L. (broad, faba, fava or horse bean)

Known to have been cultivated from Neolithic times (Schultze-Motel 1972), this is a protein-rich crop used as animal and human food in fresh, conserved or dried form. Of the three mst types (Tables 30.4 to 30.6), gc-mst is extensively investigated in it.

30.15.1 Genic Male Sterility

In an English stock of winter field beans, mst's having abortive pollen were detected by Bond et al. (1964a,b). Rarely, pollen was released but it was nonfunctional. The sterility is controlled by one *ms* gene, whose expression is complete in different environments in cytoplasms of 11 inbred lines of English winter beans, eight varieties of spring beans and in three varieties of broad beans (*V. faba* major). A p-mst from India was detected by Kaul and Singh (1967b). The plant is male-female partial sterile, exhibiting meiotic anomalies, and is thus a meiotic mutant. After treating the French commercial bean varieties Ascott and Diana, with EMS and EMS + EB, Picard et al. (1981) and Duc (1982) isolated plants in which mst is conditioned by one dominant *Ms* gene.

30.15.2 Gene-Cytoplasmic Male Sterility

In progeny of an open-pollinated English winter bean, mst plants having reduced nondehiscent anthers (Fig. 30.8b) appeared (Bond et al. 1966a). The sterility source was designated S-447. Some of them developed a little viable pollen more or less all over the flowers of the plant, whereas others produced a few fertile along with the sterile tillers. The authors name this first type semi-sterile and the second type tiller-sterile plants. These steriles were crossed with other commercial lines of spring and winter beans. Five types of F_1 populations developed: (1) all fertile, (2) all sterile, (3) both sterile and fertile, (4) sterile plants intermixed with plants having fertile and sterile tillers, (5) sterile, fertile and plants having both sterile and fertile tillers. Crossing restored plants with nonrestorers or selfing restored plants yielded a few sterile plants, the change from sterility to fertility was irreversible. Another mst source, S-350, was found by Berthelem (1970) from France. The S-447 (Bond 1966) and S-350 (Berthelem 1970) sources, discovered independently from different populations, represent two S-cytoplasm sources and different genes restore fertility in them. However, one source restores fertility in both of them. S-447 has the following five distinct features:

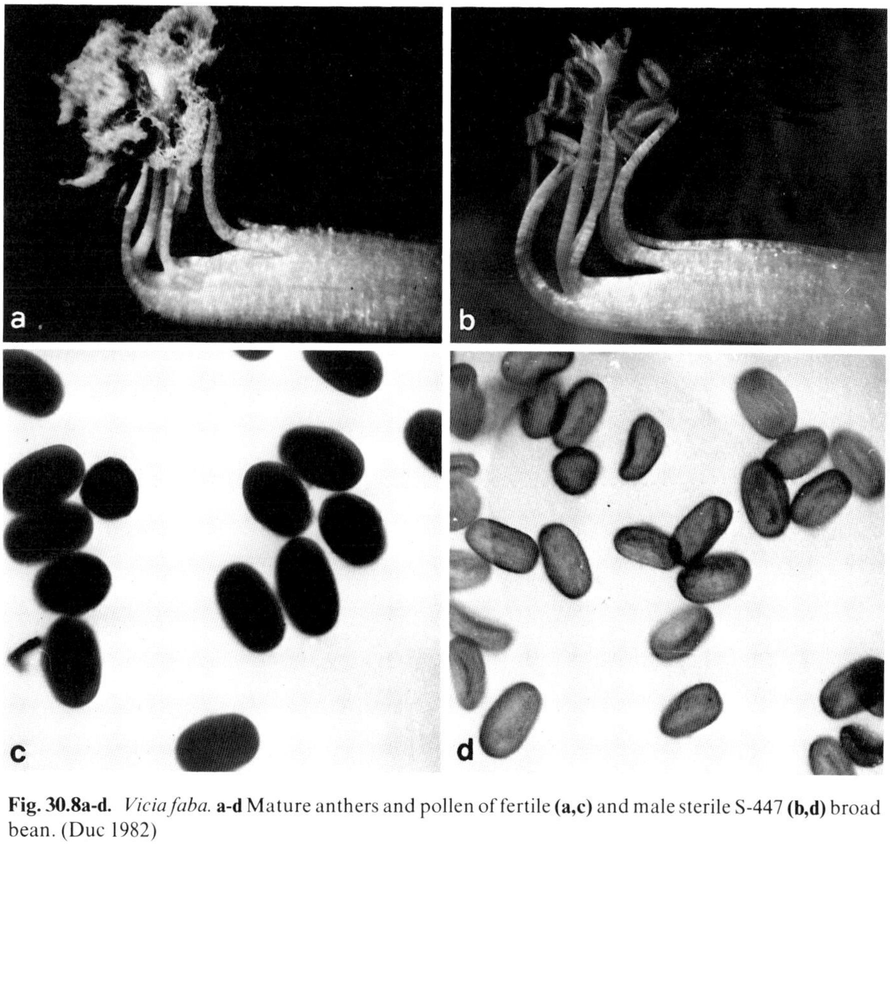

Fig. 30.8a-d. *Vicia faba.* **a-d** Mature anthers and pollen of fertile **(a,c)** and male sterile S-447 **(b,d)** broad bean. (Duc 1982)

30.15.2.1 Instability

The mst plants (S_{frfr}), when pollinated by a maintainer (N_{frfr}), rarely produce a complete mst progeny. Whereas some plants exhibit an intermediate mst level (p-mst), in others the reversion to fertility is variable. The proportion and frequency of reversion depends upon the genetic background of the maintainer and the environmental conditions, mainly temperature and light (Duc 1978). These reverted plants act fully as maintainers of sterility and, on selfing, produce mft progeny. This indicates that either the S-cytoplasm has mutated to the N-form or its action is inhibited or suppressed by some other cytoplasmic gene, or the *c*-gene is lost from the S-cytoplasm during each reversions.

30.15.2.2 Permanent Restoration

The mst plants (S_{frfr}), when pollinated by a restorer (*Fr Fr*), produce mft F_1 progeny supposed to be $S_{Fr\,Fr}$. On selfing F_1, instead of obtaining 25% mst plants, the whole progeny is uniformly mft. The 25% $S_{fr\,fr}$ plants either lose their *c*-gene or its action is inhibited by some other cytoplasmic gene or is mutated to a fertile gene. Thus once the mst plant is restored either through the *Fr* gene or through any kind of reversion, mst never reappears in the following selfed progenies. In other words, fertility restoration is permanent in S-447. This is not known in S-350 (Duc 1981).

30.15.2.3 Unidirectional Reversion

In mst broad beans, reversion from sterility to fertility is frequent (Table 30.3) but the reverse is so rare as to be almost absent. This leads to the presence of some plants with stainable pollen to fully pollen fertile plants among mst populations (Bond et al. 1966a; Berthelem and Le Guen 1975; Benevent and Le Guen 1976; Thiellement 1977a; Duc 1978). Even within tiller-sterile plants, some flowers or some stamens of a flower are partially or completely fertile. In some plants first-formed flowers are completely mst, the later-formed ones partially mst, and the last-formed one or two fully mft. The existence of intraplant, interplant and intra- and interline

Table 30.3. Phenotype alterations in *V. faba* having S-cytoplasm

No.	Genotype	Initial phenotype	Final phenotype
1	Fr_1Fr_1vv	mft	mft
2	Fr_1fr_1vu	mft	mft
[a]3	Fr_1fr_1vv	mst	mft
4	Fr_1Fr_1VV	mst	mst
5	Fr_1fr_1VV	mst	mst
6	Fr_1Fr_1Vv	p-mst	mst
7	fr_1fr_1vv	mst	mst
[a]8	fr_1fr_1Vv	mst	mft

[a]Unstable phenotypes.

variations in mst expression was found by Berthelem et al. (1978). In all cases, the phenomenon of reversion is always unidirectional, i.e. only from sterility to fertility. In S-447 faba beans, a high frequency reversion to fertility occurs, the reversion may cover the whole or a part of the inflorescence, and in over 90% of the cases, the reversion is associated with a change in cytoplasm, but not at the nuclear-genome level. The cytoplasmic change from sterility to fertility is permanent, as no mst plant is found in the selfed progenies of restored plants. Furthermore, the frequency of reversion is nuclear genotype-dependent, as a few stable mst lines exist. In this unidirectional reversion the S-447 cytoplasm resembles the S-cytoplasm of maize.

30.15.2.4 Population Heterogeneity

As indicated above, reversion from sterility to fertility is unidirectional and gradual. This leads to plants having different intermediate levels between mft and mst (Thiellement 1977b, 1979b). In intermediate (I) plants, the range of pollen production varies from nil to a few stainable grains. Some of these plants have few tillers that produce no pollen or little pollen or sufficient viable pollen. In some, earliest-formed tillers are completely mst, later-developed ones are intermediate and the last-produced ones were mft. This produces considerable phenotypic heterogeneity in the progeny of mst plants, and adds to population heterogeneity. The degree of heterogeneity narrows with generation advances, and the plants reverting to mft increase with each generation. A high degree of heterogeneity occurred in BC_9 and F_{10-12} generations of S-447 plants (Thiellement 1977b, 1979b, 1980). In BC_{10}, mst plants occurred only in the progeny of mst mother plants, 63% of these were completely mst, the remaining are heterogenous with a tendency to a high degree of mft. In BC_{11} no complete mst plants appear (Thiellement 1979b). The progeny was intermediate and in BC_{12}, although sexual variability between and within plants was considerable, the influence of the grandmother parent, termed "grandmother effect" by Thiellement (1979b), was significantly high (Thiellement 1980).

30.15.2.5 Directional Segregation

Population heterogeneity in mst progenies of field beans has been regarded by Thiellement (1980) as largely due to somatic reversions, which led to 46% intermediate type plants in BC_9 generation (Thiellement 1977b). Although each plant was affected independently, all the tillers of a plant were affected simultaneously. Thiellement (1977c, 1980) considered these modifications as evidence of hereditary changes, and assumed somatic segregation to be frequent in female gametogenesis. This leads to different phenotypes in the next progeny. The segregation was assumed to be random, but is directional because the tendency towards mft enhances with age and generation. Thus the earlier-formed tillers are mst, the later ones intermediate and the last ones mft. Similarly, as generations increase, fertility increases in intensity and frequency. For instance, in BC_{11}, no complete mst plant ever occurred even from the progeny of a completely mst mother plant (Thiellement 1979b).

30.15.2.6 Fertility Restoration

Like mst expression, restoration in *V. faba* is highly variable. Bond (1966) classified the genotypes into three types: nonrestorers, partial restorers and high level restorers. Among the nonrestorers, a high variation from poor to good maintainers was found, which led Berthelem (1970) to hypothesise the existence of recessive minor genes conditioning fertility restoration. Presently no reliable and complete maintainer useful in breeding is known. The main reason for the lack of maintainers could be the existence of a polygenic group of genes (Huglo and Duc 1980) in addition to the known major gene pair *Fr fr* with dominance and epistatic interaction. Apart from this, some fertility-restorer enhancing genes are recessive in some genotypes but dominant in others. If the fertility restoration were only polygenically conditioned, selection and breeding techniques would help in building up of genotypes with better mst-maintaining ability. On analysing large segregating populations obtained after crossing S-447 gc-mst and mft broad bean plants, fertility restoration appears to be conditioned by a single dominant gene Fr_1. Another dominant gene, *V*, is epistatic to Fr_1 and suppresses its action completely if present in homozygous condition and partially when in heterozygous condition (Table 30.3). Genotypes $fr_1 fr_1 v\ v$ are mst. Genotype $fr_1 fr_1$ Vv develops as mst but becomes mft later. Thus in the presence of Fr_1 gene, the gene *V* inhibits fertility restoration whereas in its absence, mft is gradually revoked by this dominant gene. The fertility restoration is influenced by modifier genes and environmental factors.

30.15.2.7 Environmental Influence

Temperature and light intensity influence mst expression (Nouy and Le Guen 1977; Duc 1978). The most sensitive period is during male meiosis. Duc (1980) found that S-447 plants are mst at a constant temperature regime of 10° and 20°C, but the temperature treatments of 10°–23°C, and 20°–32°C at the anthesis stage of the first floral node induce fertile pollen production in mst plants. At this treatment stage the plants were undergoing male meiosis. Pollen fertility is maximum in the 17°–27°C temperature range (Berthelem and Le Guen 1975). This indicates a high temperature influence on the expression of mst in *ms* S-447 *V. faba* plants. On the other hand, S-350 mst plants are more sensitive to light intensity than to temperature. By transferring them from 8000 lx (9 h) to 25,000 lx (16 h), pollen fertility up to 80%, in 66% plants occurred in the flowers which were to undergo male meiosis when the illumination treatment was given (Duc 1980) (others were not affected); the fertile pollen was fully functional.

30.15.2.8 Pollen Phenotype

The pollens of three mft lines, maintainer, restorer and restored hybrids, resemble each other in development and formation, whereas mst pollen differs considerably from them. Its size is small, the intine is absent (Fig. 30.8d), the ectexine sculpture is much reduced, the tectum is overlaid by a membraneous system, the infratectal structure is highly compact and a sporopollenin granular mass obstructs the central region of the aperture (Audran and Bouillot 1981; Cerceau-Larrival et al. 1982), thus modifying the aperture configuration. The aperture is nonfuctional and prevents pollen germination. Also biosynthesis of chemical constitution of sporopollenin is modified. The wall development and its autoflourescence reveal the presence of spectral maxima during the tetrad stage, a thick endexine and the absence of intine in the mst pollen (Audran and Willemse 1982).

30.15.2.9 Cytoplasm Heterogeneity

The occurrence of many intermediate sexual phenotypes between completely mst and mft plants and the phenotypic alterations occurring simultaneously on different tillers of a plant led Thiellement (1977a) to hypothesise cytoplasmic heterogeneity as a basic cause of phenotypic variability. The reversion from mst to fertility is not abrupt but gradual. Therefore, it is not due to mutations but quantitative in nature. Thiellement (1982) postulated that cytoplasmic states leading to pollen phenotypes are due to relative quantities of cytoplasmic determinants, the plasmons, among which a selective pressure is acting towards mft. The author based his results on S:N-cytoplasm type ratio that gradually reduces due to reversion.

30.15.2.10 Cytoplasmic Spherical Bodies

Edwardson et al. (1976) detected certain cytoplasmic spherical bodies (CSB) nearly 70 nm in diameter in tissues of all mst S-447 plants (Fig. 30.9). These bodies are absent in the tissues of mst S-350 plants and also in the tissues of maintainers, restorers, revertants, or restored sterile S-447 cytoplasm plants (Edwardson et al. 1976; Scalla et al. 1981). They are present neither in the reverted fertile branch of a tiller-sterile plant, nor in five out of six reverted fertile plants. However, one reverted plant contained CSB in ovules. TEM observations of Moussel et al. (1982) indicate the constant presence of CSB's in all the anther tissues of S-447 plants. Initially during early meiotic stages, their number is low, but it gradually increases and reaches maximum at the microspore formation stage. When microspores aborted, CSB's are destroyed by an autophagic process. Scalla et al. (1981) compared anther and pistil cells of four mst S-447 cytoplasm lines and their isogenic maintainer counterparts. Using the sucrose gradient centrifugation technique, that provides a reliable and fast method of detecting CSB, bodies of 62–73 nm diameter were consistently found in sterile plant tissues. They are made up of a circular unit membrane enclosing an electron-dense central zone and lodge

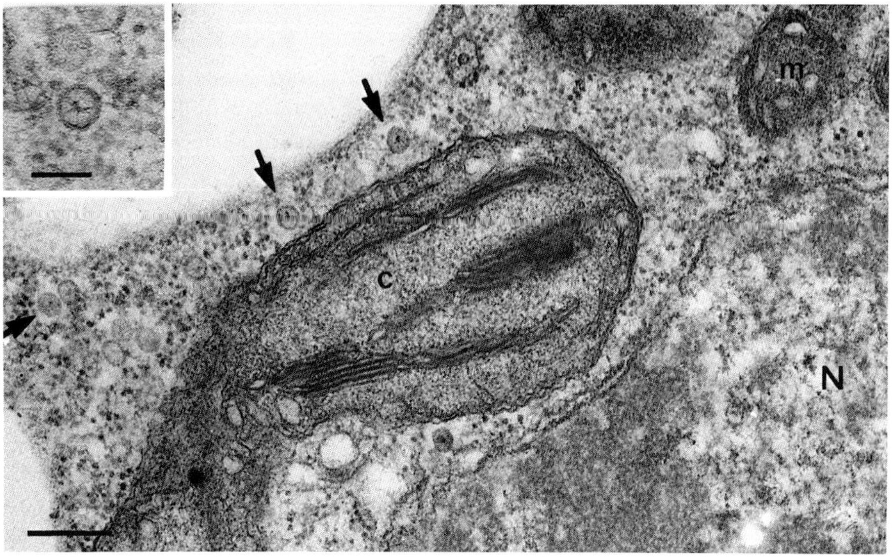

Fig. 30.9. *Vicia faba.* E.M. section through a synergid in the embryo sac of male sterile broad bean showing cytoplasmic spherical bodies (CSB) in cytoplasm (shown by *arrows*) but not in nucleus; *insert* CSB magnified; *c* chloroplast; *m* mitochondria; *N* nucleus. (Scalla et al. 1981)

a double-stranded (*ds*) RNA of molecular weight higher than 2×10^6. (Grill 1980; Grill and Garger 1981; Fig. 30.10a). The mst strain is invariably associated with CSB's having ds-RNA in young plants, anthers and pistils. When mst is lost through reversion or restoration, CSB's also disappear. The presence of seed-transmitted CSB's in all mst *V. faba* tissues and their absence in all mft tissues (except in one reverted plant) led Edwardson et al. (1976) to suggest these bodies as the sites of or possibly products of *c*-gene action in *V. faba*. It is not known if CSB's are the cause or the consequence of mst in *V. faba*, but if it is assumed that they are the causal factors, a high intracellular number of these bodies could be a necessary condition for a stable distribution in daughter cells through successive mitosis, and consequently for a better stability of mst expression (Scalla et al. 1981).

30.15.2.11 Grafting

More conclusive evidence of RNA association with mst in *V. faba* (S-447) is provided by Grill and Garger (1981), who transmitted mst (through RNA transfer) into a different fertile line of *V. faba*. These authors cultured dodder *Cuscuta subinclusa* on S-447 plants and a few strands of the established dodder were brought into contact with normal *V. faba* (Burpee, fava long-podded bean). After 2–4 weeks, the dodder was removed from these fertile plants. Of the five plants, three were mst and possessed high molecular weight ds-RNA present in S-447 plants. Whereas these plants produced sterile pollen, dodder union among mft *V. faba* did not affect their pollen viability. Thus Grill and Garger (1981) were able to

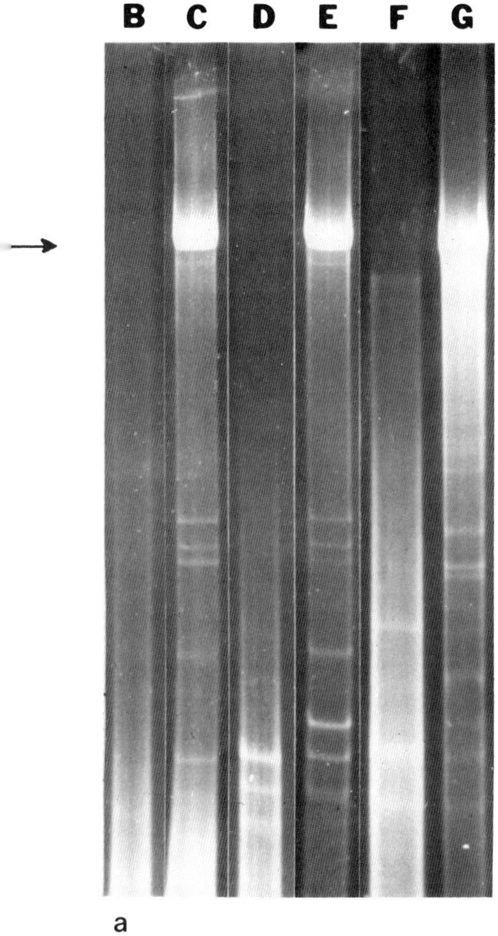

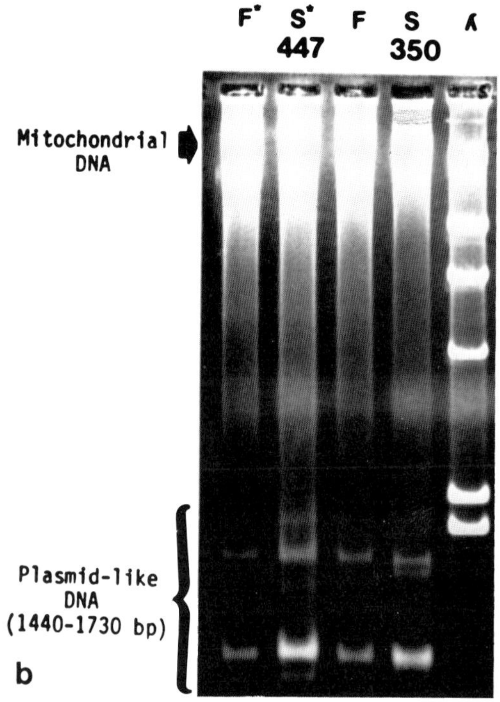

Fig. 30.10a,b. *Vicia faba.* **a** Agarose gel electrophoretogram of nucleic acid extracted from male sterile (*C,E,G*) and their corresponding fertile maintainers (*B,D,F*). The largest double-stranded RNA (*arrow*) is present only in male sterile lines (*C,E,G*). (Grill and Garger 1981). **b** Agarose gel electrophoretogram of mitochondrial DNA extracted from two sterile S-cytoplasm types (*S*) and from their corresponding fertile (*F*) lines. (*Arrow* indicates additional band in S-350 cytoplasm). (Boutry and Briquet 1982)

transmit the mst, but the transmission did not occur through direct grafting (Bond et al. 1966a; Duc 1978; Grill et al. 1983). Furthermore, the mst induced through dodder grafts is not stable, even though flowers which develop after grafting are mst.

30.15.2.12 Viral Nature

Bond et al. (1966a) hypothesised a seed-transmitted virus or defective virus (cytoplasmic agent M) responsible for mst. Edwardson et al. (1976) suggested that the virus-like nature of mst in S-447 *V. faba* ds-RNA enclosed in CSB's is associated with mst and most of the plant viruses are RNA's. The RNA was isolated by Grill and Garger (1981) from a 260-nm absorbing band in a CsCl gradient of a crude cytosol preparation. This suggests a virus-like association of nucleic acids and proteins and the transmission of mst through dodder further asserts a virus-like nature of the mst-inducing gene. In fact, dodder is known to transmit viruses and viroids (Weathers 1965). The CSB's are reminiscent of viruses like carrot mottle or tomato spotted wild virus, but the characteristic internal nucleocapsid of helicoidal or cubic symmetry is not found in CSB's. All these features suggest that the mst may be caused by a virus, but do not rule out other possibilities, as attempts to graft transmit 447-mst have not been successful (Bond et al. 1966a; Duc 1978; Scalla et al. 1982). High molecular weight ds-RNA occurs in mst sugarbeet and sunflower plants, but certain fundamental differences make ds-RNA of S-447 broad bean unique in nature. In restored sunflower plants, ds-RNA is present. Fertile maintainer sugarbeet plants have ds-RNA, but in a much lower concentration compared to the mst plants. Such RNA is present in S-447 mst broad bean plants but absent in maintainers, fertile revertants or restorers. Once ds-RNA containing membrane-bound spherical bodies (CSBs) are lost, they are not easily regained.

30.15.2.13 Mitochondrial DNA

Either mitochondria or chloroplast or both are implicated as the carriers of *c*-genes. Boutry and Briquet (1981) found the mitochondrial mt-DNA restriction patterns of S-350 and S-447 identical to each other, but different from those of normal mft cytoplasm. Both S-cytoplasms are distinguishable by restriction endonucleases Eco R_1 and Bam H_1 (Fig. 30.10b; Boutry and Briquet 1982). Out of over 30 bands, each enzyme produces two to four unique bands. These results are not in agreement with those of Grill and Garger (1981), who documented the absence of significant differences in mt-DNA of the two S-cytoplasms of faba beans. In S-350 DNA species, lower electrophoretic mobility is absent; instead a new band of higher mobility exists in it. This additional low molecular weight DNA is specific to S-350 cytoplasm and absent in S-447. Using intact mitochondria for in situ protein synthesis in the presence of (35_S) methionine, Boutry and Briquet (1982) detected an additional polypeptide of 25,000 kDa in the two types of mst plant after polyacrylamide slab gel electrophoresis and autoradiography techniques. Although these investigations demonstrate mt-DNA dissimilarities between mft and mst plants, it cannot be firmly concluded that mst is directly associated with the altered mt-DNA pattern and with the consequently changed mitochondrial protein synthesis. The S-350 exhibits about 30% decrease in respiratory state-3 of oxygen uptake during oxidation of NADH or malate + pyruvate (Boutry and Briquet 1982). This may indicate a lower capacity of the flow in the respiratory chain rather than a difference in the phosphorylation process. Furthermore S-350 plants do not have CSB's. Their cytosol has not been examined for ds-RNA's. The extra polypeptide in the protein synthesised isolated mitochondria could be a product specified by ds-RNA lodged within mitochondria or may represent a fundamental difference between the two S-cytoplasms. Whether mitochondria, as well as the cytosol, have ds-RNA can be determined when cloned probes to the ds-RNA are generated. Since synthesis of a 24,000-variant mitochondrial polypeptide is highly reduced in S-cytoplasms (Briquet et al. 1985), the involvement of this variant polypeptide in mst expression is suggested. In addition to the major mt-DNA molecule found in fertile and in two sterile cytoplasms of *V. faba* (Boutry and Briquet 1982), several small plasmid-like DNA molecules of 1 to 2 kb occur in its mt-DNA genome. The two bands found by Negruk et al. (1982) in mt-DNA of mft lines belong to only one class of 1.6 kb circular molecules of DNA, as revealed by electron microscopy. Two small supercoiled DNA molecules of mean length 1700 and 1420 bp were isolated and characterised by electron microscopy from S-350 and S-447 cytoplasms by Goblet et al. (1983) (Fig. 30.11). By agarose gel electrophoresis, S-350 cytoplasm is distinguished from the S-447 and N-cytoplasm by the presence of an additional supercoiled DNA molecule of approximately 1540 bp. The absence of this molecule indicates that no direct relation exists between this plasmid and sterile cytoplasm in fava beans.

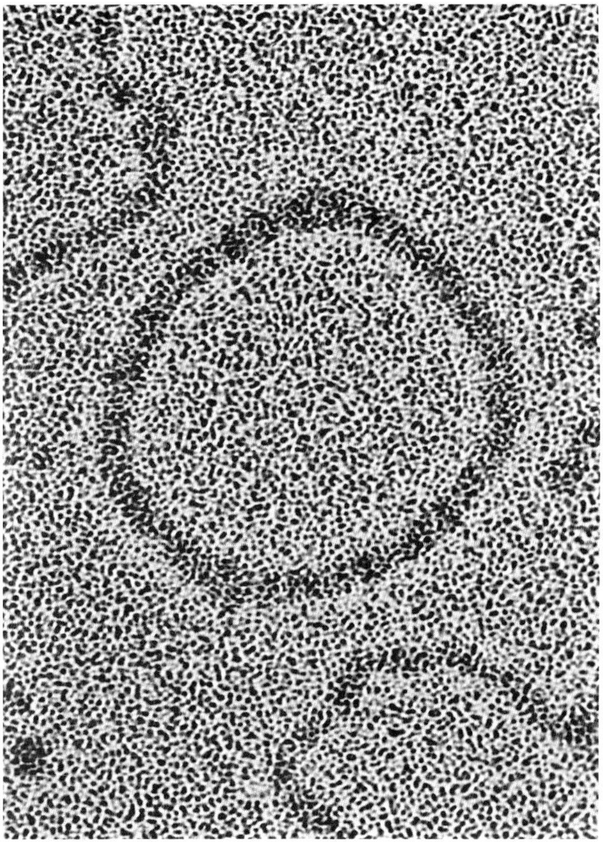

Fig. 30.11. *Vicia faba*. E.M. of circular form of 1700 bp plasmid, a low molecular weight circular DNA from S-350 cytoplasm mitochondria. (Goblet et al. 1983)

original mst plant had S-cytoplasm and *frfr* genes. Further crosses confirmed that S_{frfr} plants were mst and others, like S_{FrFr}, S_{Frfr}, N_{FrFr} and N_{frfr}, were mft. Thus only plants with homozygous for *fr* genes in combination with S-cytoplasm are mst and those with N-cytoplasm are mft, irrespective of their genetic constitution. This was confirmed by Foskett (1954), Peterson and Foskett (1953) Banga and Petiet (1958), Yen (1959), Banga (1961). From a breeding population, a mst line exhibiting swelling and chloranthy of anther filaments and strong perianth reduction segregated (Meer 1978). The mst plants were S_{frfr} type but exhibited diverse floral anomalies in addition to microspore abortion. In the cultivars Wolska and Strigunowskij, which have T- instead of S-cytoplasm, three recessive genes, *a, b* and *c*, condition mst (Schweisguth 1973). The gene *a* is independent of *b* and *c*. Plants with *aa* gene and T-cytoplasm are mst. Genes *b* and *c* are complementary in their action and all *bbcc* plants with T-cytoplasm are mst. Plants lacking either *b* or *c* gene are mft even though they have T-cytoplasm.

31.1.2.1 Cytoplasm Differentiation

Jones and Emsweller (1937) detected sterile cytoplasm(S) in the Italian Red onion variety. Another sterile cytoplasm type was detected by Berninger (1965) in the French onion cultivar Jeanne paille des vertus. Schweisguth (1973) designated it as T-type, which is also found in the cultivars Wolska and Strigunowskij.

31.1.2.2 Gene Frequency

From a random sample of 1000 onion plants of Italian Red variety, Monosmith (1926) found 56 mst plants. After *fr* gene detection in this onion, the genes were found in onions collected from England, France, Germany, India, Japan, Korea, New Zealand, the Netherlands, Russia, Syria, Turkey etc. (Davis 1957). In most of these, the *fr* genes were in a heterozygous state. Little et al. (1944) tested 29 commercial onion varieties and found 25 to have such *fr* genes. Of the eight German and American onion varieties tested by Lichter and Mündler (1961), seven possessed the *fr* genes. The frequency of *fr* genes in the American variety Scott County Globe is 0.84 to 0.96% (Peterson and Foskett 1953). It is 1–2.9% in the German variety Zittauer Gelbe (Kobabe 1958). In the Dutch variety, Rijnsburger, the *fr* gene frequency is 0.95% while that of S-cytoplasm is below 0.01% (Meer and Bennekom 1971). Whereas only one mst plant in several large fields of Stuttgarter Riesen was found by Kobabe (1958), 92 mst plants in commercial seed fields of two Dutch onion cultivars, Primeur and Wijbo were detected by Banga and Petiet (1958) in the Netherlands. In Russia, Makarov (1960) found mst plants in the cultivar Arzamas. Yen (1959) demonstrated the presence of *S* and *fr* genes in the cultivar Pukekohe Longkeeper of New Zealand. In most all these populations mst control is nearly similar to that found by Jones and Clarke (1943), but it is not known whether the same *fr* or nonallelic genes are present in so many distantly located populations. In a French onion variety, either one recessive gene or complementary action of two other *fr* genes is known to cause mst in T-cytoplasm plants. Widespread occurrence of *fr* and *Fr* genes in *A. cepa* cultivars and varieties indicates that mutation from the *Fr* to the *fr* locus occurred long back and/or has taken place more than once, probably at different loci. These *fr* genes are perpetuated in N-cytoplasm plants. Due to high cross pollination, most of them are heterozygous. In contrast to high *fr* gene frequency, S-cytoplasm is confined to the Italian Red 13–53 onion variety. All other inbreds utilised for producing hybrid onions derived their S-cytoplasm from this source. Another sterile cytoplasm source, the T-cytoplasm, is known in a French variety (Berninger 1965). Unlike high *fr* gene diversity, cytoplasm diversity in onion is low.

31.1.2.3 Anther Phenotype

In the majority of mst onions having S-cytoplasm, anther morphology is normal but at anthesis the thecae are green, small and indehiscent. On the other hand, in those having T-cytoplasm, anther form and morphology are disturbed. For instance, in mst onion varieties Rawska, and Wolska which have T-cytoplasm, three types of anthers occur (Dyki 1973a), (a) glassy-green anthers containing an aqueous fluid, (b) stamens completely dried up and much elongated pistils (c) brown and hard anthers with normally developed filaments. Whereas in type a plants, variable sized free degenerated microspores occur, in the remaining two types abnormal meiocytes and tetrads clump and remain cemented in the disintegrated tapetal tissue.

31.1.2.4 Microsporogenesis

In gc-mst Italian Red 13–53, Monosmith (1926) and Jones and Emsweller (1937) found normal meiosis, but microspores degenerated after liberation from the quartets. In almost all other cases of S-cytoplasm, microspores degenerate either immediately after release or during pollen mitosis (Table 31.1) A slight degree of

Table 31.1. Gene action in gene-cytoplasmic male sterile *Allium cepa*

Cultivar	References	Gene action
S-cytoplasm (Normal meiosis)		
Italian Red 13–53	Monosmith (1926), Jones and Emsweller (1937)	After microspore release
Globe Denvers	Tatebe (1952)	During pollen mitosis
Zittauer Gelbe	Kobabe (1958), Virnich (1967)	After microspore release
New Zealand lines	Yen (1959)	During pollen mitosis
Niphad	Patil et al. (1973)	Before microspore release or during tetrad formation
T-cytoplasm (Abnormal meiosis)		
Wolska, Rawska	Dyki (1973a)	Before microspore release or during prophase I
French stock	This Vol.	During prophase I

variability exhibited in the degeneration stage is due to environmental or physiological factors. However, in contrast to normal initial meiosis in S-cytoplasm plants, it is abnormal in the T-cytoplasm. This indicates a differential *fr-c* gene interaction in S- and T-cytoplasm plants of onion.

31.1.2.5 Tapetal Behaviour

In Italian Red 13–53 mst onion, Monosmith (1926) found hypertrophy of some centrally located tapetal cells. These swell considerably after microspore release, almost filling the anther locule. Thereafter, the cells degenerate abnormally and the microspores die within the anther locule. A nearly similar tapetal behaviour was observed by Foskett (1954) in mst Scott County Globe onion. In mst Japanese onions, the tapetal cell wall breaks down abruptly and the cell contents fuse to form a plasmodium that degenerates quickly (Tatebe 1952). During tapetal degeneration, microspores also degenerate. In the mst German variety Zittauer Gelbe, microsporogenesis proceeds normally up to the tetrad and early pollen formation (Kobabe 1958), but immediately after release from PMC's, the young pollen grains shrivel and degenerate gradually. Under certain environmental conditions, a small proportion of viable pollen grains are produced, indicating environmental influence over sterility. Comparative investigations reveal distinct differences in tapetal behaviour of sterile and fertile anthers. For instance, unlike in the mft anthers, the tapetum in mst's remains persistent for a long period of time. Its degeneration begins only when pollen grains shrivel and their cytoplasmic contents are already dead. Furthermore, the endothecium is abnormal and delays anther dehiscence. Kobabe (1958) regards sterility in this material as not due to tapetal missecretion but to stoppage of nutrients to the developing microspores. On the other hand, in mst 2x and 4x onions of this variety, the tapetal degeneration occurs only after pollen contents are resorbed, indicating pollen degeneration as due to the inability of the microspores to absorb nutrients from the tapetum (Virnich 1967). In the T-cytoplasm mst plants Wolska and Rawska, meiosis is highly abnormal and leads to the formation of diads and triads. In them, intensive vacuolization of the tapetum occurs (Dyki 1973b). This is followed by microspore and tapetal degeneration. Whereas the abnormal meiotic course in T-cytoplasm mst plants is the cause of pollen sterility, abnormal tapetal behaviour causes pollen sterility in S-cytoplasm plants. However, tapetal behaviour is variable in different S-cytoplasm mst onions. For instance, whereas in American mst onions tapetal cells enlarge to fill the anther cavity, in Japanese onions, the tapetum breaks down abruptly forming a plasmodial mass. In German mst onions, the tapetum remains intact and causes microspore starvation, but in other mst onions, the tapetum breaks down along with the microspores and hence microspore starvation cannot be the cause of pollen abortion. In all S-cytoplasm mst onions, blockage or nontransfer of nutrients across the tapetum appears to be one of the causes of mst. This is supported by the biochemical results of Saini and Davis (1969), who found much reduced nucleic acid synthesis in mst anthers. Feeding young mst onion umbels with uridine led to the development of some fertile pollen in mst plants obtained after crossing *A. cepa* Italian Red 13–53 line with *A. galanthum* and

A. drobovii. In all these mst plants, tapetal degeneration is normal. It is abnormal in *A. cepa* mother plants and F_1's of *A. cepa* × *A. pskemense*. In both these, uridine applications do not restore fertility. It appears that in those mst plants in which the tapetum was normal, uridine was transferred across the tapetum to the developing microspores and led to some pollen fertility; but in others, in which the tapetum was abnormal, such a transfer did not occur and therefore no fertile pollen developed. Hence, it is evident that certain g-c combinations affect tapetal function and behaviour, culminating in pollen abortion.

31.1.2.6 Expressional Instability

It is not known whether the same or different *fr* genes in association with S-cytoplasm cause mst in various onion varieties and cultivars; but in different genetic backgrounds, mst expression is not always complete. For instance, small amounts of fertile pollen are produced in certain onion varieties converted to gc-mst (Clarke and Pollard 1949). S_{frfr} plants of some onion varieties produce fertile pollen, unlike similar plants of other varieties (Barham and Munger 1950). This indicates incomplete penetrance of *fr* genes in certain genetic backgrounds. In some onion lines, S_{Frfr} plants are p-mst (Jiȓik and Novâk 1969). In these, stainable pollen grains, along with the residue of aborted grains, occur together in the anther sacs, which appear pebbly. In other onion cultivars, such plants are completely mft (Virnich 1967) and S_{frfr} plants are completely mst (Jiȓik 1973). This expressional instability is caused by genotypic background and environmental factors and their interaction. Certain S_{frfr} mst onions have shrivelled and nonviable pollen and a small amount of normal-looking pollen, staining feebly. Lichter and Mündler (1961) regard the stainable pollen formation as due to delay in the time of degeneration. It is not known if the stainable pollen produced is viable, as the aceto-carmine stainability test for onion pollen viability has not been found reliable by Ockendon and Gates (1976), who found onion pollen having normal staining cytoplasm as fully nonviable after germination tests on artifical media. They recommended the use of a fluorescein dye as rapid and reliable for pollen viability determination in onion.

31.1.2.7 Temperature Influence

Both mst expression and its manifestations are influenced profoundly by temperature. This is amply evidenced by Barham and Munger (1950), who cultivated mst onions at temperature ranges of 10°–15.5°, 15.5°–21.1° and 21.1°–26.6°C and maintained these under 9.2 to 14.5 and 14.5 to 17 h photoperiod at each temperature regime. They found that while photoperiod had no effect on mst expression, the temperature to which the flower buds were subjected during the time from tetrad formation until completion of microspore mitosis is crucial in determining viable pollen production. Below 21 °C, no viable pollen was produced. Between 21° and 25.5°C, some germinable pollen developed but no seeds were produced on selfing. This means that the pollen was either physiologically weak or inviable.

Stainable and phenotypically normal onion pollen may be inviable (Ockendon and Gates 1976). Therefore, it appears that most of the pollen produced may not be fully functional. Lichter and Mündler (1961) found that the daily mean temperature 12 days before flowering markedly influences the amount of pollen formed in some restored onion lines and concluded that whereas temperatures between 11.2° and 24.5°C increased fertile pollen production, lower temperatures reduced it considerably. However, their conclusion cannot be generalised because Meer and Bennekom (1969, 1978) found that while some onion populations were mst at 14°C, much less at 20°–23°C, and not at all at higher temperatures, other mst's did not react in the same way these temperature treatments. Thus, many populations and hybrids of the Dutch variety Rijnsburg and the Polish variety Wolska exhibited a high proportion of mst plants at 14° and 17°C (Meer and Bennekom 1978); others responded differently. In their reaction to temperature, genetic differences exist between different mst genotypes of onion. Since the S-cytoplasm from a common source was present in all of them, the differential temperature response is due to their different genetic backgrounds. Moreover, pollen fertility determined by carmine staining is no criterion for viability and functionability, as fertile-looking pollen produced by mst onions maintained at higher temperatures was nonfunctional (Barham and Munger 1950). Ockendon and Gates (1976) found fertile-looking onion pollen with normal cytoplasm to be nonviable.

31.1.2.8 Grafting

The mst in onion is not transmitted across grafts as after grafting mst flower stalks onto mft plants and vice versa, no changes occur either in stocks or scions (Meer and Bennekom 1970).

31.1.2.9 Biochemical Differences

Low amounts of DNA and RNA occur in the sterile anthers (Saini and Davis 1969). Due to these insufficient nucleic acids and other nutrients, the microspores are unable to develop into viable pollen grains in Italian Red mst onion. This reduced level of the biochemicals could be due to the absence of enzyme systems, precursors or nucleotides, the absence being a consequence of g-c interaction. Saini and Davis (1969) found that by feeding young mst onion umbels with nucleotides in various combinations with sugars, 23% fertility was restored in some mst F_1's, viz. *A. cepa* × *A. galanthum*, *A. cepa* × *A. drobovii*, *A. roylei* × *A. drobovii*, but not in others like mst *A.cepa* (IR 13–53), *A. cepa* × *A. pskemense*. In those mst hybrids that responded to uridine by producing some viable pollen, the tapetal degeneration was normal. Therefore, in them some enzymes associated with the synthesis of uridine are not synthesised and consequently the DNA and RNA synthesis is hampered. Compared to anthers of mft plants, those of mst plants possess higher dry matter, ash, copper and iron and lower potassium and zinc contents (Pospisilova and Jirik 1979). In the mst anthers, most of the spore tetrads lack callose wall (Nanda and Gupta 1974). This is hydrolysed by the enzyme callose, whose activity is considerably higher than that in mft anthers.

31.1.2.10 Chemical Male Sterility

i. *GA_3*. Spraying 2% GA_3 in the beginning of bolting induced mst in Rijnsburger onion (Meer and Bennekom 1973). Likewise, pollen abortion occurred following 0.05% $GA_{4/7}$ spray applied frequently between the initiation and the end of bolting (Meer 1978). Ovular fertility and seed production was reduced by almost 75% or more.

ii. *MH*. Complete mst occurred following 10 ppm spray (Chopra et al. 1960). The treatment induced meiotic and some carpel anomalies. Due to crop damage by hail storm, seed production was not determined. Sprays of 500 ppm induced complete sterility and plants had poorly formed flowers (Meer and Bennekom 1973). Injecting 0.01 and 0.025% MH solutions at the base of the stalk, when inflorescences appeared as minute protuberances at the top, induced mst (Kaul and Singh 1967c). Through controlled pollinations, 60–70% seed was set on the treated plants. This indicates low ovular sterility induced by the chemical.

iii. *Phosphon-5*. Spraying 750 and 1000 ppm phosphon-5 over an inbred onion line Iowa-90, when its florets were nearly one and one-half weeks from anthesis, led to near complete mst which lasted for 2 weeks under greenhouse conditions (Cohan and Weigle 1966). Under field conditions, sterility was partial. In them, first- and last-formed florets of umbel had normal anthers with viable pollen whereas the controls had aborted pollen. A nearly similar situation occurred in 0.45% Mendok-treated snapdragon plants. This indicates the effectiveness of the chemical at only certain specific growth stages. Furthermore, many young florets abort prematurely and total seed production is reduced considerably. Hence this chemical treatment is not economically feasible for commercial onion hybrid seed production.

iv. *Other Chemicals*. Injections of 3–4% aqueous sodium nucleate prior to meiosis induced meiotic anomalies and pollen sterility (Chopra et al. 1960) TIBA and mendok sprayed or injected did not cause pollen sterility (Meer and Bennekom 1973).

Table 31.2. Chemical male sterility in *Allium cepa*

Chemicals	References	Male sterility induction	Female fertility reduction
GA_3	Meer and Bennekom (1973)	High	High
$GA_{4/7}$	Meer (1978)	high	High
MH	Chopra et al. (1960)	Complete	nk
	Kaul and Singh (1967c)	Complete	High
	Meer and Bennekom (1973)	Complete	Complete
Phosphon-5	Cohan and Weigle (1966)	Partial to complete	High
Sodium nucleate	Chopra et al. (1960)	Complete	nk

31.1.2.11 Breeding Technique

Onions exhibit high inbreeding depression and therefore the pedigree breeding method is not effective with this crop. Techniques involving alternate inbreeding and outbreeding generations are preferable, and currently F_1 hybrids are utilized to produce many modern cultivars which are not only more uniform but often mature earlier than the conventional types. Gc-mst used to produce F_1 onion cultivars is the classic example of this technique. The first F_1 onions were produced in the USA by maintaining the mst line vegetatively from small viviparous bulbils. This system was abandoned due to the difficulty of keeping the stock free from yellow dwarf virus. For production of F_1 cultivars using gc-mst, three lines, A, B and C, are required. These are: the female A line (S_{frfr}), which is mst and produces the hybrid seed, the maintainer B line (N_{frfr}), which is genetically similar to A but produces pollen, the male restorer line C (N_{FrFr}), which is genetically dissimilar to the other two lines and is chosen for its capacity to produce a good hybrid with the A line. B and C lines are produced by growing batches for seed in insect-proof cages or in well-isolated plots in the open. Line A can only be produced by crossing with B. Blow flies are used for pollination of small batches of the parents, but natural insect fauna is relied upon for outdoor seed crops.

Starting a search for mst from the basic material would be a lengthy process but the newcomer to the breeding of this crop can obtain the S-cytoplasm and recessive *fr* genes from currently available F_1 cultivars that are fertile (S_{Frfr}) and can be selfed to give 1/4 sterile S_{frfr} types. These cannot be used directly, as they are highly heterozygous for other horticultural characteristics and several generations of breeding are required to improved homozygosity. Since the *fr* gene is widely distributed among onion varieties in many parts of world, it should not be hard to find S-cytoplasm genotypes in these countries. Of course, this needs a careful search.

31.1.2.12 Breeding Value

Since onion is cross pollinated the breeding value of mst Italian Red 13–53 (IR) was realised by Jones and Clarke (1943), as they obtained a potential means for large-scale hybridization of almost any two onion lines. Only N_{frfr} plants were needed to be found in other desirable onion varieties as they could be sterilised by repeated BC's to IR onion. The authors accordingly proposed a scheme for mst utilisation to genetically emasculate hermaphrodite plants for use in mass and commercial hybridisations. In 1953, hybrid onions were produced commercially and since then they have rapidly increased in number and acreage. Above 50% of the onions cultivated currently are of hybrid origin and the supply of hybrid seed has never been sufficient to meet the demand. The main reasons for the heavy demand are high yield, crop uniformity in agronomic traits and superiority of F_1 hybrids over heterozygous varieties in uniformity of size, shape and maturity. In Texas, the hybrid Granex has exhibited a 94% increase in yield over the standard variety Texas Early Grano (Anonymous 1956). Susceptibility to disease is a major problem with onions. Disease resistance can be more conveniently incorporated into such hybrids.

Wide usage of mst onion is not only because it is cross pollinated but also because many N_{frfr} genotypes were immediately found in Europe and Russia. Therefore it became easier to find suitable sterilisable genotypes. In England, Germany and the Netherlands, breeding at first began with IR 13–53 source, but later mst genotypes in locally adapted standard varieties were isolated and used in onion breeding. The majority of these have S-cytoplasm and relatively few have T-cytoplasm. The main advantage of the S_{frfr} type is its simple inheritance, high stability and little temperature influence. Barham and Munger (1950) found that moderately higher temperatures allowed microspores in sterile plants to develop more than usual but these microspores never or rarely developed into functional pollen even under such conditions. Clark and Pollard (1949), however, detected a little viable pollen in certain onion lines converted to mst. The average amount of selfing is four such hybrids was 4%, which is not enough to present serious problems in hybrid production. Using mst lines, Banga and Petiet (1958) suggested a breeding program involving recurrent selection for hybrid seed production. They found that the quickest results could be obtained when different varieties or selections adapted to the same or similar ecological conditions were used conjointly in the program.

Currently, all over world hybrid onions are produced by using mst lines. Field hybridisation is effected by planting alternate blocks of mst and mft rows. Usually a ratio of one male row to four female rows is used. Most of the pollination is effected by bees placed specially in crossing fields, usually four to five lines/acre. For commercial trade, only F_1 hybrids are grown. Since double-cross hybrids exhibit heterogeneity, single-cross hybrids are preferred due to their bulb uniformity and higher productivity, keeping quality and increased dry matter content. However, lack of consistent *fr* gene expression of some A lines due to different genetic backgrounds and heterogeneity, temperature and/or photoperiod influence over mst expression, have limited the use of mst lines in hybrid breeding. None of the chemicals used as male gametocides for chemical emasculation of onions seems to be useful in hybrid production. However, certain MH and GA concentrations, if standardized, may be used to chemically emasculate the lines for use in intervarietal or interspecific and recurrent BC's, because in these only little amounts of seeds are required. Meer and Bennekom (1973) suggested a method of using chemical emasculation in onion breeding in the Netherlands.

31.1.2.13 Seed Colour as Hybrid Marker

In order to easily identify F_1 seeds and reduce roguing costs, Davis (1966) suggested the use of seed coat colour genes as the generic markers for hybrid seed production. Black seed coat colour controlled by one gene is dominant to brown colour. Brown seed coat is associated with smaller, rounder and smoother seeds than black. The seed colour would differ depending on the way the brown seed parent enters the cross. If the mst line has brown seed and the pollen parent black, the F_1 embryo would be enclosed within a brown coat and hybrid seed would be brown. If the mst parent has a black seed coat and the pollen parent brown, the hybrid seed would be black. The separation of different-coloured seeds could be facilitated by using modern computerized electronic sorting machines.

31.1.3 *A. fistulosum* L. (Welsh onion)

In a strain of this nonbulbing onion, meiosis was normal but tapetal cells enlarged considerably immediately after tetrad formation (Nishi and Hiraoka 1958). This enlargement apressed the tetrads, which degenerated subsequently. The enlarged tapetal cells exhibited high vacuolar activity and in some periplasmodia were formed. The anomaly is conditioned by one recessive gene. In another similar mst strain, male meiosis proceeds normally but microspores abort immediately after release from the quartets (Virnich 1967). The tapetum is persistent.

31.1.4 *A. longicuspis* Regel.

Because of pollen sterility, this diploid onion does not produce seeds but propagates by bulbils (Novăk 1972). It exhibits normal male meiosis, but the microspores abort soon after liberation during which tapetal hypertrophy occurs and the tapetal tissue grows inside the anther cavity, crushing the microspores. Tapetal malfunction is assigned as the cause of pollen abortion by Novăk (1972). The sterility is genic but the precise gene control is not known.

31.1.5 *A. porrum* L. (*A. ampeloprasum* L.)

In this cultivated garden onion, Schweisguth (1970) found mst plants in a frequency of 10^{-4}. While in some, the single gene appears to be associated with the sterility, in others two to three or more unlinked genes appear to control the sterility. The complex gene control of the sterility may be due to the tetraploid nature of the species. In certain inbred lines of the species, mst is coupled with a high degree of fst.

31.1.6 *A. sativum* L. (garlic)

Multiplication of garlic is vegetative and takes place by means of collateral buds and bulbils. After an early bulb extirpation, sexual flowering organs develop but remain completely pollen sterile. The sterility is due to postmeiotic disturbances especially tapetal hypertrophy and first pollen mitosis blockade. Coupled with microspore abortion, tapetal hypertrophy occurs in many garlic strains (Novâk 1972). However, in one clone each, Koul and Gohil (1970) and Goriand Ferri (1982) found regular meiosis leading to normal microspore formation. The developing microspores become vacuolated, shrivel and degenerate either soon after liberation from the tetrads or just prior to pollen mitosis (this Vol.). Following microspore degeneration, young bulblets appear on the inflorescence. In some garlic clones, after 4-week antibiotic sprays comprising a mixture of tetracycline hybrochloride-tylocine (1:1, 0.001%) pollen fertility was restored (Konviĉka et al. 1978). While rickettsia-like organisms in different anther tissues developed in sterile plants, they were absent in the restored plants. The antibiotic treatment appears to suppress some gene products that cause sterility in garlics.

31.1.7 *A. schoenoprasum* L. (chive)

After an initial report on the existence of mst in different chive varieties by Kobabe (1965), eight mst plants were detected by Singh and Kobabe (1969) from farmers' fields. Their flowers, anthers and filaments were smaller than those of the fertiles. Anthers of the steriles were shrunken, indehiscent and dried up at anthesis. Initial meiosis proceeded normally but microspores degenerated immediately after their formation, before liberation from PMC's. On separation, the microspores possess a degenerated cytoplasm (Singh and Kobabe 1969). In a few, the *ms* gene action occurs before meiosis and an undifferentiated mass of sporogenous tissue develops. In all the mst's, however, tapetal growth and degeneration are slow and delayed so that at anthesis the tapetum of mst flowers appears persistent. In contrast, in mft's the tapetum degenerates soon after microspores are formed. Whereas in one variety mst is controlled by a recessive *ms* gene (this Vol.), in the remaining, the sterility is conditioned by interaction of one *fr* gene with the S-cytoplasm (Tatliogu 1982). Under field conditions, some gc-mst's revert to mft while others are stable. In the sensitive ones, temperature is the potent factor, influencing mst expression as they are mst at 17°/17°C (day/night) and 20°/17°C, but mft at 24°C constant temperature. Due to its perennial nature, the same plant was tested by the author during the second flowering season and it was demonstrated that the temperature-induced fertility is reversible.

Table 31.3. Male sterile F_1 onion hybrids

Cross combination
A.cepa × *A.ascalonicum*
A.cepa × *A.galanthum*
A.cepa × *A.drobovii*
A.cepa × *A.pskemense*
A.roylei × *A.drobovii*

31.1.8 Onion Hybrids

Saini and Davis (1969) obtained mst F_1 hybrids in crosses involving S-cytoplasm mst *A. cepa* (IR 13–53) as female (Table 31.3). Male meiosis in the steriles proceeded normally up to the quartet stage but microspores degenerated after release, leading to the production of abortive pollen grains in mature anthers. While in the mst F_1 plants of *A. cepa* × *A. pskemense*, tapetal hypertrophy and microspore degeneration occurred abnormally, in other mst hybrids, tapetal behaviour was normal

(Table 31.5). Whereas the mst hybrids exhibiting normal tapetal behaviour develop some fertile pollen after uridine feeding, those with abnormal tapetal behaviour do not produce any fertile pollen even after continued uridine applications (Saini and Davis 1969). Thus while in the former plant types, either uridine and/or other nutrients were transferred by the tapetum onto the developing microspores, in the latter this transfer does not occur and the microspores abort consequently. Hence the site of *fr* gene action is different in the two mst type plants.

Table 31.4. Genic male sterility in Liliaceae

Species	References	Gene action	Remarks
Allium ascalonicum[a]	Darlington and Hacque (1955)	After first pollen mitosis	–
	This Vol.	After microspore release	Tapetum normal, collapses after microspore degeneration
A. fistulosum[a]	Nishi and Hiraoka (1958)	Before release from tetrads	Tapetum enlarges and crushes tetrads
	Virnich (1967)	After microspore release	Tapetum persistent
A. longiscuspis	Novâk (1972)	After release from tetrads	Tapetal malfunction
A. porrum[b]	Schweisguth (1970)	After release from tetrads	–
A. sativum	Novâk (1972), Koul and Gohil (1970), Konviĉka et al. (1978), this Vol., Gori and Ferri (1982)	During microspore development to pollen mitosis	Tapetal malfunction
A. schoenoprasum[a]	Singh and Kobabe (1969), This Vol.	After microspore formation, before release	Tapetum persistent or degenerates slowly
Kniphofia sp.	Moffet (1932)	During MI	–

Gene control by 1r[a] or 1–3r[b], in others not known.

31.2 *Kniphofia* Moench. (red hot poker, torch lily)

Herr Dr. J.H. Kniphof, a German professor of medicine deeply interested in botany, is commemorated by the name of this genus that has about 70 species, most of which are handsome ornamentals. Moffet (1932) found one mst plant in an unknown species of this genus. In it the failure of spindle development and function at MI led to numerous subsequent meiotic anomalies. However, male meiosis continued but pollen development was arrested uniformly after tetrad formation, so that no cytokinesis followed. The resulting coenocyte had four colpi instead of one present in normal pollen grains. No viable pollen grains developed. The status of female fertility and genetic control of the anomaly are not known. Since only one plant in the stand was mst, it appears to be a spontaneously arisen mutant.

Table 31.5. Gene cytoplasmic male sterility in Liliaceae[a]

Species	References	Gene action	Remarks
Allium cepa	Dyki (1973)	During prophase I	Anther development and differentiation variable, tapetum abnormal
	Patil et al. (1973)	Before microspore release or during tetrad formation	
	Kobabe (1958)	After microspore release	Tapetal degeneration delayed
	Virnich (1967)	After microspore release	Tapetal malfunction
	Tatebe (1952), Yen (1959)	At pollen mitosis	Tapetal degeneration abnormal
	Monosmith (1926), Jones and Emsweller (1937)	After microspore release	Tapetal degeneration abnormal
	This Vol.	During prophase I	
A.cepa × *A.ascalonicum*	Little et al. (1944)	After microspore liberation	Tapetum normal
× *A.drobovii*	Saini and Davis (1969)	After microspore liberation	Tapetum normal
× *A.galanthum*	Saini and Davis (1969)	After microspore liberation	Tapetum normal
× *A.pskemense*	Saini and Davis (1969)	After microspore liberation	Tapetum abnormal
A.roylei × *A.drobovii*	Saini and Davis (1969)	After microspore liberation	Tapetum normal

[a]Gene control by lr gene.

CHAPTER 32
Limanthaceae

32.1 *Limnanthes douglasii* R. Br. (meadow foam)

In the variety *nivea* of this spreading branched yellow green herb, mst plants were detected from a natural population by Jain et al. (1978). They were characterised by the presence of white anthers and much shorter petals than mft plants and were thus easily distinguishable in the field. They were of two types: those with empty, small white anthers lacking PMC's due to premeiotic sporogenous degeneration and those with yellow anthers possessing normal microspore tetrad development but exhibiting postmeiotic microspore degeneration leading to nonviable pollen production. The sterility in both types is conditioned by a single dominant gene. The role of the cytoplasm cannot be completely eliminated, keeping in view the segregations obtained and assuming a restorer gene segregating among the male parents of these families. Meadow foam represents a potential new crop (Jain et al. 1977) and outbreeding is promoted in many *Limnanthes* spp. by protandry. Due to the availability of the mst gene linked closely to or causing pleiotropically the formation of white anthers and short petals, the mst is useful in hybrid breeding of this genus, in which the recent discovery of unique biochemical properties of its seed oil, a potential substitute for sperm whale oil, has generated considerable interest in its domestication and improvement.

Table 32.1. ms gene action in *Limnanthes*

Species	References	Gene control	Gene action	Remarks
Genic male sterility				
Limnanthes douglasii	Jain et al. (1978)	1D	Pre- to post-meiotic	Possesses white anthers and short petals

CHAPTER 33

Linaceae

33.1 *Linum usitatissimum* L. (flax)

This economically most valuable fibre and seed oil-yielding plant represents one of the first examples of the plant kingdom in which g-c interaction was found to cause mst exclusively. This was reported from England, when Bateson and Gairdner (1921) obtained nearly 25% mst plants in F_2 progeny of the cross of a procumbent form as female and normal *L. usitatissimum* as male parent. No mst plants segregated in the progeny of the reciprocal cross. From many reciprocal cross differences and the results of BC and selfings, Chittenden (1927), Chittenden and Pellew (1927) and Gairdner (1929) interpreted the mst as due to interaction between the cytoplasm of the procumbent form and a recessive gene from the normal (tall) form of *L. usitatissimum*. Gajeski (1937) reported that the procumbent form used by Bateson and Gairdner (1929) was not *L. usitatissimum* but *L. floccosum*. The procumbent from crossed to *L. floccosum* in both directions produced F_1 and F_2 plants, all of which were hermaphrodite and perfectly identical with the parents. This indicates complete genome-cytoplasm homology. The Index Kewensis does not list *L. floccosum* and so far no current viable seed source has been located by the present author. Using *L. floccosum* (Bucarest), obtained from Vilmorin-Andriex Seed Co. as the female parent in crosses to four other varieties of *L. usitatissimum*, Gajeski (1937) obtained nearly 25% mst plants in F_2, its F_1 was normal and had no mst plants. Both F_1 and F_2 progenies of the reciprocal crosses were mft. Similarly only normal mft plants occurred in F_1 and F_2 generations of the reciprocally crossed *L. floccosum* and *L. angustifolium*. When the mst plants obtained in F_2 of *L. floccosum* × *L. usitatissimum* were crossed to *L. angustifolium* by Gajeski (1937), the F_1 plants were normal mft but in F_2, the ratio was 3 mft: 1 mst. It is not fully resolved if the procumbent form used by Bateson and Gairdner (1921) and *L. floccosum* utilized by Gajeski (1937) represent one and the same material or two ecotypes, biotypes or genotypes, varieties or species. Since neither seed source of *L. floccosum* is available nor is the species listed in the Index Kewensis, it appears that the two may be the biotypes of the procumbent form, a mutant of *L. usitatissimum* having minute subtle genetic deviations and a major cytoplasmic difference. This contention is substantiated by the detection of mst plants by Burnham et al. (1981) in F_2 of the cross involving a large seeded flax from the Crete source as female parent and some *L. usitatissimum* source as the male parent. Likewise, no mst's segregated in the reciprocal cross progeny.

33.1.1 Genetic Control

In the crosses involving procumbent Crete and *L. floccosum* plants as female parents to *L. usitatissimum* as male parents, F_1 was normal but F_2 segregated nearly 25% mst plants. No mst plants occurred in the progenies of the reciprocal crosses. This indicates that mst is due to the interaction of two recessive *fr* genes present in the homozygous state in the male parent with the S-cytoplasm of the female parent (Table 33.1) However, this gene control does not seem to be so simple. This is evidenced by the observations of Wettstein (1946), who obtained mostly p-mst plants, a few complete mst plants and some mft plants after selfing some mft F_2 plants. Backcrosses by the normal tall plants yielded predominantly mst and never mft progeny, but BC's by the procumbent plants produced either fertile or fertile and p-mst progeny. The author interpreted these results as due to segregation of residual genes present in the procumbent parent. The genes probably controlled fertility in the procumbent cytoplasm by modifying the expression of *fr* genes in the S-cytoplasm of the procumbent parent. While using Crete mst plants as the female parents in crosses to other flax plants, Burnham et al. (1981) obtained two types of F_2 segregation ratios. Whereas one type indicated single gene differences between mst and mft plants, in the other, one to two gene differences were vivid. The two genes appeared to be complementary either for restoration of pollen fertility in F_1 or for incomplete restoration. Burnham et al. (1981) proposed two hypotheses, both of which explained equally well the differential segregation ratios. One is that the plants like *L. floccosum* (?) procumbent and Crete possess a dominant nuclear fertility allele *Ms* and a nonrestorer (NR) cytoplasm (S). Most other stocks have ms_1 ms_2 genes but a cytoplasm (CR) that restores fertility (N). Crossing mst (S_{frfr} = $NR_{ms_1ms_2}$) with the proposed maintainer (N_{frfr} = $CR_{ms_1ms_2}$) would yield mst F_1 plants. The second hypothesis also explains the reciprocal differences obtained by the authors. In this, the mst plants were assumed to have S-cytoplasm and a nuclear pollen fertility restoration gene (*Fr*). The male parents had N-cytoplasm and the nonrestorer genes fr_1, fr_2. Thus the mst plants would genetically be $S_{fr_1fr_2}$ and mft plants $Nfr_1\,fr_2$. To sum up, it appears that whereas the mst plants were produced due to interaction of the S-cytoplasm and two *fr* genes, the mft plants had either N- or S-cytoplasm and a dominant *Fr* gene in the homo- or heterozygous state. The plants like *L. floccosum* (?), procumbent and Crete had S-cytoplasm and at least one

Table 33.1. Gene-cytoplasmic male sterility in *Linum*[a]

Species/cross	References	Gene action	Remarks
L. usitatissimum	Bateson and Gairdner (1921) Burnham et al. (1981)	nk	
L. floccosum × *L. usitatissimum*	Gajewski (1937)	nk	Tapetum persistent and enlarges
	Dubey and Singh (1965)	Before or after microspore release	Tapetum persistent, no transmission across grafts

[a]Gene control 1–2r.

Fr gene, whose action is influenced by modifier genes; this influence may sometimes override the *fr* gene action. This interaction led Gagnieu (1954) to conclude that gc-mst in flax does not involve major genes. Comstock (cited by Duvick 1966) detected the presence of a gene more potent than the *Fr* gene in an introduced procumbent mst plant with open flowers.

33.1.2 Fertility Restoration

Initial results revealed fertility restoration of mst plants to be conditioned by a single dominant *Fr* gene. However, the differential segregation ratios obtained by Wettstein (1946), Miller and Thompson (1978) and Burnham et al. (1981) in different crossed led to the proposition of one, two or multiple *Fr* gene hypotheses. To explain the monogenic, digenic and reciprocal differences in the cross using Crete as female, Burnham et al. (1981) explained that gc-mst in flax can be explained equally well by either of the two proposed hypotheses, viz. (a) S-cytoplasm interacts with nuclear fertility restorer vs. nonfertility-restorer genes, (b) nuclear gene for mst is restored to fertility by a particular cytoplasm type. Based on the observations on F_1's between mst plants and stocks from the world flax collection, different nuclear genes for fertility restoration and for determining the differences in the degree of restoration appear to exist (see Comstock 1963, 1964). Kumar (1971) and Kumar and Singh (1972) found a multiple allelic series of four genes, two of which led to an intermediate degree of mft in flax. The author stressed the role of minor genes influencing *Fr* gene action. Some of their segregation data indicate the influence of inhibitory genes and of the environment over fertility restoration and sterility expression. The effectiveness of *Fr* genes is conditioned by the number and types of inhibitory or other modifier genes present in the genome. Thus whereas only 8.5% of the world collection lines had highly effective restorer genes, another 8.5% lines had restorer genes of intermediate expression (Thompson and Miller 1978). It is equally possible that many of these lines possess the same *Fr* genes. In 15 crosses in which highly effective *Fr* genes were involved, a monogenic segregation evidencing the control of fertility restoration by a dominant gene was obtained by Miller and Thompson (1978). In many other crosses involving intermediate effective *Fr* genes, fertility restoration is conditioned by more than three genes. Based on the observations over F_1's between mst plants and stocks from the world collection, there may be different nuclear genes for fertility restoration, and for determining the differences in the degree of restoration (Comstock 1963, 1964).

33.1.3 Microsporogenesis

Up to tetrad formation, male meiosis proceeds normally in mst and mft plants having a procumbent S-cytoplasm (Dubey and Singh 1965). Whereas in mft's the tapetum disintegrates after tetrad liberation, in msts's, the tapetum persists and enlarges (Table 33.1). In a few, the free cytoplasmic mass of tapetal cell prevails throughout anther development. The exact stage of microspore degeneration varies considerably among various mst lines. In some, microspores degenerate while still bound in tetrads, in others they disintegrate after separation and enlargement. In some other lines, the developing or developed microspores abort (Table 33.1).

33.1.4 Grafting

Although interspecific grafts between *L. usitatissimum* and *L. grandiflorum* are successful, no graft-induced changes in mst occur (Thompson personal communication) because the fertile scions produce mft open flowers and sterile scions develop mst flowers with closed corollas. Therefore, graft transfer of gc-mst does not occur in flax.

33.1.5 Chemical Male Sterility

By applying four sprays of 500 ppm mendok or dalapon and three to four sprays of 1000 ppm of these gametocides, Rustagi and Mohan Ram (1971) obtained nearly complete mst at the mid-flowering season in *L. usitatissimum*, but ovular fertility was damaged considerably (Table 33.2). The mst plants are characterised by completely or partially barren anthers with rhythms of pollen invariably interspersed with periods of fertility restoration or by agglutination of pollen, non-dehiscent anthers and lack of anthesis. The sprays cause toxicity, leading to hyponasty, chlorosis, burning of young leaves and flower buds. The flowers become progressively smaller and close gradually. Seed production and germination are very low. These negative features make the gametocide applications useless for flax hybrid breeding.

Table 33.2. Chemical male sterility (complete) in *Linum usitatissimum*[a]

Chemicals	References	Side effects
Dalapon	Rustagi and Mohan Ram (1971)	High toxicity, reduced vegetative and reproductive growth
Mendok	Dubey and Singh (1963) Rustagi and Mohan Ram (1971)	Highly toxic, chlorosis, burning of foliar and floral buds, flower closure obstructing cross pollination

[a]Female fertility reduction high.

33.1.6 Breeding Value

L. usitatissimum is self-pollinated and grown mainly to produce oil and fibre. Traditionally, cultivars have been developed using a pure line breeding system which has led to a slow improvement in yield. Although the first proven gc-mst was detected in 1921 by Bateson and Gairdner, it has not been used for commercial hybrid seed production. Since 25–40% increased yields from F_1 are known in flax

(Carnahan 1949; Kashim 1964; Comstock 1965; Shehala and Comstock 1971), commercial hybrid seed production would be a useful development (Hoffmann 1961; Kenaschuk 1975). The major two barriers to hybrid seed production are heavy sticky pollen that is probably transferred only by insects and closed or tubular flowers associated with mst, inhibiting cross pollination completely. Duvick (1959) suggested that selection as in tabacco might change the mst flowers from closed or tubular to funnel form. The occurrence of complete mst flax with open corollas was first reported by Dubey and Singh (1965, 1966) from India. Seed set associated with the degree of corolla opening was assumed to result from cross pollination. However, later reports by Kumar (1971) and Kumar and Singh (1970, 1972) revealed that the open corolla plants reported earlier to be mst were actually mft, as they had minor *Fr* genes and produced selfed seeds. Thus their open-flowered material was not complete mst. While examining the possibility of utilizing mst flax in commercial hybrid seed production, the extent of natural crossing (a basic requisite for hybrid seed production) was found by Dubey and Singh (1966) and Kumar and Singh (1970) to depend upon the degree of floral opening and petal size. Introductions from Greece have genes that cause mst flowers to open (Comstock 1965) and to become cross pollinated. Thompson (1977) obtained mst F_1 plants by crossing mst *L. usitatissimum* to *L. bienne* that had open corollas. Although the open corrolla trait appeared to be conditioned by many genes, progress in selecting open-flowered mst *L. usitatissimum* lines was made. This indicates polygene control with additive gene action. Since mst flowers with open corollas have invariably well-developed anthers, swelling of the filament bases causes the corollas to open (Thompson 1977). From the USDA world flax collection, Thompson amd Miller (1978) identified 30 accession sources for open-corolla genes. These lines exhibited significant variability, which gives a possibility of selecting open-flowered mst plants for use in hybrid production. By using the recurrent breeding technique to accumulate favourable genes conditioning the degree of corolla opening, lines with the most desirable open-corolla type can be developed. Thompson and Miller (1978) developed several mst flax lines that had perfectly open corollas during the early flowering period. The degree of corolla opening decreased, however, with age, decreasing in 10 weeks from 13.97 to 8.8 mm. Since with decrease of corolla diameter, cross pollination also decreases, lines exhibiting minimum decrease of corolla size have to be selected for use in hybrid breeding. Flax is usually grown as a short-season crop and most seed is set shortly after flowering begins. This means that the first measurements of corolla diameter should be important in selecting lines for a breeding program. This nearly coincides with the plant age at which most crossing would occur under field conditions.

Of the world flax collection, 79% does not have restorer genes (Thompson and Miller 1978) and can act as maintainer for mst lines. The absence of *Fr* genes make their conversion to A lines (mst) simpler, and adds to their breeding value. Thus the mst lines with open corolla can be effectively used in flax breeding provided they are free from minor restorer genes or such genes are eliminated in a breeding program. Thompson and Miller (1978) opined that many world collection lines can be directly used in a random mating flax breeding program to mix germplasm. In such a system, seeds of different maintainer lines would be bulked or lines could be used individually and planted in rows adjacent to a mst line. Seed set over mst lines could be used to produce a subsequent mst population for crossing each year. Thus gc-mst has tremendous potential in commercial hybrid flax production. However, the main obstacle is to have an adequate pollen transfer on open corolla mst lines. Whether NCP carried mainly by insects is sufficient to allow this system to produce usable quantities of hybrid seed needs to be determined. Equally important is the standardization of field layouts, isolation requirements and the other usual procedures required for profitable commercial hybrid production.

CHAPTER 34

Malvaceae

Of the 50 genera belonging to this family, mst is known in *Abelmoschus Gossypium* and *Thespesia* (Tables 34.1 to 34.3).

Table 34.1. Genic male sterility in Malvaceae

Species	References	Gene control	Gene action	Remarks
Genic				
Abelmoschus esculentus	Dutta and Nath (1974), This Vol.	1r	During anther differentiation	
Gossypium barbadense	Tang et al. (1981)	1r	After liberation at uninucleate stage	Tapetal degeneration partial
	Yang (1982)	1r, ID	nk	nk
G. hirsutum	Allison and Fisher (1964)	ID	Before PMC differentiation	Tapetum defective
	Weaver and Ashley (1971), Murthy and Weaver (1974) Weaver (1974) Bowman et al. (1978)	ID	During pollen wall formation	Tapetum normal
	Bowman and Weaver (1982)	ID	After pollen maturation	Tapetum normal
	Justus and Leinweber (1960)	1r	nk	
	Richmond and Kohel (1961)	1r	After release, during microspore development	Leaf abnormalities common
	Weaver (1968)	2r	After microspore release	Stable to environment
	Rhyne and Rhyne (1971), Kohel (1973), Murthy and Weaver (1974)	2r	During microspore maturation	mst linked to nectarlessness
	Seth et al. (1981)	1r	nk	Pollen grains shrivelled, anthers indehiscent
Thespesia populnea	This Vol.	–	During early MI	Abnormal meiosis, confluent stigmas

Table 34.2. Gene cytoplasmic male sterility in Malvaceae

Species	References	Gene action	Remarks
Gossypium hirsutum[a]	Thombre and Mehetre (1979)	During pre-meiosis	Autoplasmic mutation
G.anomalum × *G.arboreum*	Rhyne (1965), Meyer (1968), Chen and Meyer (1979)	Anthers transformed into petals	Environmental influence high
G.anomalum × *G.hirsutum*[b]	Meyer and Meyer (1965)	nk	Temperature-sensitive
G.harknessii × *G.hirsutum*	Murthi and Weaver (1974), Meyer (1975)	After PMC development, before meiosis	
G.hirsutum × *G.barbadense*	Zainyshev (1974)	nk	

Gene control by 2r[a] or 1-many r[b], in others unknown.

Table 34.3. Chemical male sterility in Malvaceae

Species/Chemical	References	Male sterility induction	Female fertility reduction
Abelmoschus esculentus			
Dalapon[a], Embutox[a], Mendok[a]	Dubey and Singh (1968)	b[b]	High
Gossypium hirsutum			
Coumarin	Dubey and Singh (1968)	b	Very high
Dalapon	Scott (1961), Meyer et al. (1958)	a	nk
	Dubey and Singh (1968)	b	Very high
MH[a]	Singh (1964), this Vol.	a,c	Reduced
	Meyer et al. (1958)	b	Reduced
Mendok	Eaton (1957)	a	Nil
	Roux and Chirinian (1959), Scott (1961)	a	nk
	Meyer et al. (1957), Pate and Duncan (1960), Richmond (1962), Semenova (1963), Bouharmounti and Pocket (1965), Raja (1967), Dubey and Singh (1968)	b	Very high to complete
MH + FW 450[a]	Meyer and Thomas (1958)	a	High

[a]Chemical causes phytotoxicity. [b] a = high, b = complete, c = partial.

34.1 *Abelmoschus esculentus* (L). Moench. (*Hibiscus esculentus*, lady's finger, okra)

γ-ray irradiation of 50 kR to 1 HR-21 cultivar produced a mst mutant in M_2 (Dutta and Nath 1974). The sterility is controlled by one recessive gene (Table 34.1). Kaul (this Vol.) detected a spontaneously arisen structural mst mutant in which stamens were transformed into staminodes so that no anthers or PMC's developed. Female fertility, as evidenced by seed set after hand pollination, is normal. Complete mst was obtained by Dubey and Singh (1968) after using two sprays of 0.5% MH, one to two sprays of 0.1% mendok, three sprays of embutox-4 or 1–2 sprays of dalapon (Table 34.3). The treatments caused 10–15% ovular sterility in addition to phytotoxicity.

34.2 *Gossypium*

This comprises 20 species native to warm regions of the new and old world. Partial to complete mst is known in commercially useful species like *G. barbadense*, *G. herbaceum* and *G. hirsutum*, caused by dominant or recessive genes, or by g-c or genome-cytoplasm interactions, environmental stress or chemical treatments (Tables 34.1 to 34.3).

34.2.1 Genic Male Sterility

The first report on the existence of mst in *G. herbaceum* is from India (Kumar 1937). The plant exhibited uni- to multi-valency, abnormal meiosis and produced inviable pollen. Neither the genetic control of sterility nor the status of female fertility of this material is known.

34.2.1.1. Recessive Gene Control

i. *Complete Type*. The first report on genetically conditioned complete mst in cotton is from Richmond and Kohel (1961), who detected the mst's in F_2 generation of a cross Texas 86 (a short-day stock of *G. hirsutum* race latifolium) and Dand PL-14 (a day-neutral selfed line of *G. hirsutum*). The sterility is conditioned by a recessive gene, ms_2, and the plants have shortened filaments and indehiscent anthers. In them the pollen grains are formed but they lack the spiny exine normally present in fertile grains and have vacuolelike inclusions that are absent in fertile grains. A second report about mst in *G. hirsutum* is from Weaver (1968), who obtained complete mst from F_2 population of the cotton varietal cross Lankart-57 × Greff. The sterility is conditioned by two recessive genes ms_5 and ms_6 (Weaver 1968), which are nonallelic to the ms_2 gene and stable in many environments. Meiosis proceeds normally but young developing microspores abort after becoming highly vacuolated. The mst's are fairly productive, storm-proof, nectarless, and produce glandless leaves. A plant with crooked stigma and highly deformed and nondehiscent anthers was found by Rhyne and Rhyne (1971). Meiosis proceeds normally but its mature microspores abort rapidly. This abortion is assigned by Murthi and Weaver (1974) to the premature tapetal degeneration. The sterility is conditioned by two recessive genes, ms_8 and ms_9 (Kohel 1973). One of them appears to be loosely linked with the nectarless gene (Ne_2). Since no complete linkage occurs between nectarlessness and mst, the ms_8 and ms_9 genes differ from the ms_5 ms_6 genes. Furthermore, while the mature microspores abort in ms_8ms_9 mst's, in ms_5 ms_6 mst's, young developing microspores become highly vacuolated and degenerate rapidly. Although ms_8ms_9 plants are completely mst, their sterility is considerably influenced by environment, as under certain environmental conditions they produce viable pollen (Weaver 1973) and yield unusual segregation ratios which indicate the involvement of the S-cytoplasm in the mst control.

From the nucleus seed multiplication cotton plot, Seth et al. (1981) obtained a mst mutant whose sterility is monogenically conditioned by a recessive *ms* gene. It has shrivelled pollen grains, indehiscent anthers and normal female fertility. It is not known if the *ms* gene in this mutant is allelic to the other known *ms* genes of cotton. Therefore, no gene symbol can be assigned to it. In the mst Chinese cotton *G. hirsutum* Dongting-A, sterility is controlled by one recessive *ms* gene (Tang et al. 1981). Male meiosis proceeds but pollen aborts at the uninucleate stage. The tapetum degeneration is slow and partial. The mst's in two other Chinese *G. hirsutum* varieties, DON-A and Che-A are likewise recessive monogenically conditioned (Yang 1982); the two *ms* genes are nonallelic.

ii. *Partial Type.* Complete to p-mst plants of *G. hirsutum* were found by Justus and Leinweber (1960). The anomaly is controlled by a recessive gene ms_1. While the original plant was a monosomic, its ms_1ms_1 derivative had a normal chromosome number. Probably the monosomic condition permitted the expression and subsequent detection of this recessive *ms* gene. Some of its leaves tended to be more deeply incised than the normal leaves. Neither *ms* gene action over microsporogenesis nor the influence of environment over *ms* gene expression are known in this material. In another material, a gene ms_3 induces p-mst (Justus et al. 1963). Its flowers either lack pollen or have a reduced amount of it. The extent of pollen production is considerably influenced by the environment, as the plants are more pollen-fertile in the greenhouse than in the field. Both day and night temperatures are higher and daily fluctuations are less in the greenhouse than in the field. Therefore temperature may be a major factor influencing p-mst in cotton.

34.2.1.2 Dominant Gene Control

From Acala-44 *G. hirsutum*, Allison and Fisher (1964) detected a mst plant standing completely erect, bearing small flowers with partially opened petals and scanty bolls. The sterility is controlled by a dominant gene, Ms_4. The plants either lack anthers or they develop to varying degrees (Fig. 34.1). Most of the developed anthers lack archesporial tissue. In some of its later segregants, a few anthers having

Fig. 34.1. *Gossypium hirsutum.* Three male sterile flowers exhibiting variable anther development compared to that of a fertile flower (*extreme right*) of cotton. (Allison and Fisher 1964)

archesporium and differentiating PMC's develop (Murthi and Weaver 1974). In some of them, meiosis initiates but the PMC's collapse by mid prophase-I. Tapetal cells are small, indicating a defective tapetal building in the mst anthers. Another dominant gene, Ms_7, was found in two *G. hirsutum* plants by Weaver and Ashley (1971). Phenotypically these plants resemble mst's, having a ms_2 gene, but unlike them, the MS_7 plants do not have any leaf abnormality. However, in some, Quinsenberry and Kohel (1968) found a few leaf abnormalities similar to ms_2 plants. Furthermore, Harvey's (1969) genetic data indicate ms_7 to be a misnomer for the ms_2 gene. Certain differences, apart from the leaf abnormality, are exhibited by ms_2 and MS_7 plants. For instance, while in ms_2 plants, filaments are short and anthers indehiscent, in MS_7 plants, the anthers are much reduced and have a little nonviable pollen. Furthermore, in ms_2 plants, nonfunctional pollen grains lacking a spiny exine wall are formed. In MS_7 plants, microspore abortion occurs at the onset of pollen wall formation (Murthi and Weaver 1974; Bowman et al. 1978). Another dominant mutant gene Ms_{10} (Bowman et al. 1978; Bowman and Weaver 1979) induces mature pollen abortion after normal meiosis in *G. hirsutum.* Intine is thick-walled and a dense layer of carbohydrates occurs between the exine and intine of MS_{10} pollen. The pollen grains having the MS_{10} gene appear less competitive than those with the recessive normal allele ms_{10} (Bowman and Weaver 1979). Two more dominant *Ms* genes occur in Chinese *G. barbadense* varieties *Junhai* and *Xinhai* (Yang 1982). Their allelic nature and gene action over microsporogenesis are not known. Another gene, Ms_{12}, was detected in this species by Turcotte and Feaster (1985).

34.2.2 Gene-Cytoplasmic Male Sterility

34.2.2.1 *G. arboreum* (*G anomalum* × *G. arboreum*)

Transferring the genome of *G. arboreum* into *G. anomalum* cytoplasm produces mst progeny (Rhyne 1965; Meyer 1968). In these plants, the staminal column develops petaloid instead of normal anthers. This anomaly is maintained as long as *G. arboreum* is used to pollinate the mst plants. Pollination by *G. anomalum* restores normal anther phenotype and pollen fertility. Likewise, transfer of A-47 *G. arboreum* genome into *G. anomalum* cytoplasm causes mst in progeny plants (Chen and Meyer 1979). These have double flowers produced mainly by the development of petaloid androecium. The fraction-1 protein of these petaloid plants has a large subunit polypeptide like *G. anomalum*, the cytoplasm source and small subunit polypeptides like *G. arboreum*, the recurrent *BC* parent. Pollination of the petaloid mst plants by *G. anomalum* not only restores normal stamen structure, but produces *G. anomalum* small subunit polypeptides in the progeny plants. Chen and Meyer (1979) hypothesised that interaction between heterologous chloroplast and nuclear genes could be responsible for the mst in which stamens do not differentiate into normal anthers.

34.2.2.2 *G. barbadense* (*G. hirsutum* × *G. barbadense*)

In Russia, Zainyshev (1974) obtained mst plants from the progeny of *G. hirsutum* (S_{4534}) × *G. barbadense* (6015), from F_3 of the cross S_{3506} × Acala 4-42 and from the inbred line 108-F. The mst plants have indehiscent anthers and aborted microspores. Whereas almost the complete progeny of the reciprocal cross *G. barbadense* × *G. hirsutum* was mft in Russia, in the Ivory Coast, one line segregated mst plants (Schwendiman 1974). Further segregations in this line indicate the existence of g-mst conditioned by a dominant gene (this Vol.). Following transfer of the *G. barbadense* genome into *G. harknessii* cytoplasm, mst plants having indehiscent minute pollenless anthers appeared. The sterility is highly stable. Likewise, Thombre (1983) transferred mst from *G. hirsutum* to *G. barbadense* through several BC's. In the mst's, the style length is fairly uniform. This increases the probability of natural crossing in them.

34.2.2.3 *G. hirsutum*

i. *G. hirsutum* (G. *harknessii* × *G. hirsutum*). Transfer of the *G. hirsutum* genome into *G. harknessii* cytoplasm induces complete mst in the alloplasmic plants (Meyer 1975). Their PMC's degenerate before meiotic initiation (Murthi and Weaver 1974), which is followed by tapetal disorganisation. Fertility restoration is by one incompletely dominant gene obtained from a segment of *G. harknessii* chromosome. However, this gene does not always give adequate restoration in many commercial upland cotton varieties. The restoration may be enhanced by transfer of a gene from Pima cotton, *G barbadense*.

ii. *G. hirsutum (G. arboreum* × *G. hirsutum).* A variable degree of mst occurs in BC and outcross progenies obtained after crossing amphiploids (*G. arboreum* var, *neglectum*, 2n = 26 × *G. thurberi*, 2n = 26) as female to *G. hirsutum*, 2n = 52 as male (Richmond and Kohel 1961). The mst plants had *G. arboreum* cytoplasm, a few genes of *G. thurberi* and the genome of *G. hirsutum*. Such mst's also segregated in BC's to *G. hirsutum* females. Thus the mst may not only be due to g-c interactions but is purely genic, caused by *ms* genes in some cross combinations. Initial segregations reveal the involvement of two dominant sterility genes in some of them (this Vol.). F_1 plants of *G. longicalyx*, 2n = 26 × *G. hirsutum*, 2n = 52 are triploid and have a reduced number of anthers and aborted microspores (Meyer 1972). These, as well as their colchicine-induced hexaploids, are totally sterile.

iii. *G. hirsutum (G. anomalum* × *G. hirsutum).* Crossing amphidiploid, 2n = 52 hybrids of *G. anomalum*, 2n = 26 × *G. thurberi*, 2n = 26 as females to *G. hirsutum*, 2n = 52 males and backcrossing the resultant progeny by *G. hirsutum* pollen produced mst plants in large numbers (Meyer and Meyer 1965). The degree and expression of mst enhances with the increase of the *G. hirsutum* genome into *G. anomalum* cytoplasm. Using *G. hirsutum* as female in crosses to BC_2 p-mst plants as male yielded plants having few anthers with viable pollen. The mst's bear external ovules, whose development is conditioned by a dominant gene that has reduced expressivity in *G. hirsutum* cytoplasm (Meyer and Buffet 1962).

iv. *G. hirsutum* (*G. armourianum* × *G. hirsutum*). Both direct and reciprocal crosses between these species produce mst hybrids (Roux 1961) whose female fertility is also reduced. Hence these are not typical mst's. Crossing amphidiploids, 2n = 52 (*G. armourianum*, 2n = 26 × *G. thurberi*, 2n = 26) as female to *G. hirsutum*, 2n = 52 as male and backcrossing the mst hybrid plants with *G. hirsutum* as male recurrent parent produced mst plants in BC_3 and BC_4 in which staminodes formed in place of anthers (this Vol.). In some, in addition, carpeloidy occurred.

34.2.2.4 Autoplasmic Mutation

A mst mutant that had lobed leaves arose spontaneously in a mft *G. hirsutum* stand (Thombre and Mehetre 1979). Since its fertility was restored by two duplicate dominant genes, the mst arose by mutation of N- to S-cytoplasm in a plant having $fr_1fr_1fr_2fr_2$ nuclear genes. This represents the first case of autoplasmic mutation in cotton. In the majority of mst anthers, no PMC's developed, in a few they developed but degenerated without entering into meiosis. This was followed by tapetal disorganisation (Mehetre and Thombre 1981.) The cytological anomaly is similar to that in alloplasmic mst (*G. harknessii*) *G. hirsutum* plants. Both maintainers and restorers for the mst plants have been identified. These, in addition to the known phenotypic marker (lobed leaf), add to the breeding value of mst plants.

34.2.2.5 Fertility Restoration

Both mst expression and fertility restoration are considerably influenced by a series of known and unknown environmental factors. In the steriles in which two or more dominant or recessive genes control mst, the mst expression is highly variable due to considerable environmental influence and modifier gene effects. Thus many segregations are blurred. Therefore, presice *Fr* gene control determination using such mst's is not possible. Utilising monosomic and monotelo-disomic lines of Daltapine *G. hirsutum*, restoration controlled by a dominant gene was found by Dasilva et al. (1981). In others, three genes with at least one dominant gene at two loci appear to be involved in the restoration.

34.2.2.6 Environmental Influence

The mst expression and intensity of anther transformation in cotton are highly influenced by temperature. In alloplasmic *G. hirsutum*, mst expression is high at 32°C and almost complete around 38°C (Meyer and Meyer 1965). Reciprocal hybrids having *G. hirsutum* cytoplasm are mft at such temperatures. Whereas low field temperatures induce high male-female sterility in cotton, a strong positive correlation exists between high temperature and number of sterile anthers (Meyer 1966b). In alloplasmic (*G. arboreum*) *G. hirsutum*, mst is positively correlated with wind velocity, pan evaporation and solar radiations. In lines having *G. anomalum* cytoplasm, mst is negatively correlated with temperature and solar radiations (Sarvella 1966). All these environmental factors are effective during 3 weeks prior to anthesis, as they primarily impair anther differentiation and development.

34.2.3 Biochemical Differences

Aspartic acid and arginine are present in the leaves of alloplasmic mst *G. hirsutum* plants having *G. anomalum* or *G. arboreum* cytoplasm but are scarce or completely absent in normal *G. hirsutum* leaves (Sarvella and Stojunovic 1968). The amount of acid-soluble amino acids is nearly the same in leaves of normal *G. hirsutum* plants and those containing *G. anomalum* cytoplasm, but it is lower in leaves of such plants having *G. arboreum* cytoplasm. At anthesis, the amount of acid-soluble amino acids in the sterile buds having *G. anomalum* cytoplasm is nearly the same as in the normal buds containing *G. hirsutum* cytoplasm, but is lower in the buds containing *G. arboreum* cytoplasm. The amount of acid-soluble amino acids is reduced when the *G. hirsutum* genome is placed in *G. arboreum* cytoplasm, but not when placed in *G. anomalum* cytoplasm. This indicates a differential genome-cytoplasm interaction at the biochemical level in alloplasmic mst cotton.

34.2.4 Nondehiscent Sterility

Two naturally occurring serious disorders, crazy top in short staple cotton (*G. hirsutum*) and Hollow boll in long staple Egyptian cotton, *G. barbadense* (King and Loomis 1927; Feaster and Turncotte 1959) cause anther nondehiscence, rendering the plants mst. The former disorder is also known as acromania (violent incurable insanity) and is characterized by high reduction and distortion of vegetative and reproductive parts, giving the plant tops a highly distorted or crazy appearance. Whereas in Crazy top disorder, many floral buds and bolls are shed off prematurely, in Hollow boll they are retained long. Hand pollination on the affected plants yields nearly normal seed set, indicating unimpaired female fertility. The affected plants recover sometimes under late-season conditions and set a fair crop of bolls. Aeration, poor water penetration and high nitrite concentration of soil appear to cause the disorder. Following 100 microgram urea or cholinechloride foliar applications, plants with the disorders produce dehiscent anthers (Scott 1962). The chemical effect occurs within 2 weeks of spray and the anthers remain dehiscent for about 2 months after the application.

34.2.5 Chemical Male Sterility

i. *Dalapon*. Induction of mst by this chemical was first documented by Meyer et al. (1958) in cotton. Anther indehiscence occurred following 250-mg application of dalapon or mendok to the first leaf above soil level (Scott 1961). Induction of complete mst accompanied high ovular damage (Dubey and Singh 1968).

ii. *Maleic Hydrazide*. About 82% pollen sterility occurs following 0.5% MH treatment before bud initiation (Singh 1964). In some lines, mst was nearly complete (Meyer et al. 1958). The chemical causes leaf burning, abnormal flower development and high ovular sterility at concentrations effective in causing mst. MH sprayed in combination with FW-450 reduces leaf burn and induces mst below the level obtained by treatment with FW-450 alone (Meyer and Thomas 1958).

iii. *Mendok*. Eaton (1957) reported mst induction in three cotton selections after 1.2% foliar sprays of mendok given a week before first flower appearance. This treatment did not affect female fertility, but some plants shed pollen in the afternoon and selfing occurred. The chemical effect lasted for 9 weeks in one and for about 6 weeks in other selections. Therefore repeated chemical applications throughout the growing season are needed to maintain the sterility. However, these cause ovular damage. With this chemical, complete mst is not obtained in Acala 4–42, a valuable strain for cotton breeders. Some sterility is maintained by an initial spray of 1% followed by 0.5% sprays at 3-week intervals (Roux and Chirinian 1959) and best results are obtained when the entire plant is wet with spray. However, this treatment impairs ovular fertility considerably. A high degree of mst accompanied by fst occurred in four cotton varieties after 0.25% and 0.40% mendok application (Richmond 1962). Such treatments to the cotton variety Pope caused male-female sterility in two successive years of testing (Pate and Duncan 1960) and the yield was highly reduced. However, gametocidal selectivity for male gametes occurred when 0.37%, at a rate of about 1 kg per acre of mendok was used. This was also found in some other cotton varieties (Eaton 1957; Meyer et al. 1957; Pate and Duncan 1960; Richmond 1962; Semenova 1963; Bouharmount and Pockett 1965; Raja 1967). In almost all these cases, female fertility was reduced (Table 34.3). Mendok cannot be considered as a suitable male gametocide in cotton as its response is variable and differential in various cotton varieties and cultivars. Moreover, its effective male-sterilizing doses impair ovular fertility considerably and reduce the fruit yield.

34.2.6 Hybrid Production

34.2.6.1 Genic Male Sterility

Cotton is a self-pollinated crop but about 15–30% cross pollination occurs mainly through insects and bees. Hybrid cotton is superior to selfed cotton in seedling vigour, earliness, fibre properties and fruit yield, and could combine higher yield with superior lint quality. Hence hybrids have been produced after emasculation and hand pollination. Use of g-mst and male gametocides is not succesful on a commercial basis (Weaver and Weaver 1977, 1979). Weaver (1968) suggested the use of ms_5ms_5 for hybrid cotton production. Female fertility of the mst plants is high, but being inherited as double recessive, only 1 out of 16 plants would be mst. Using the mst's, El-kadi and Weaver (1973) produced hybrid cotton plants. They exhibited high performance for lint yield and percentage, earliness, fibre strength and uniformity ratio. The F_1 seed could be grown in areas of high yield potential and advanced to F_2 generation for general planting (Weaver 1969). However, only 6.25% of F_2 plants would be mst. If mft double heterozygote, Ms_2ms_2 Ms_5ms_5 is used to pollinate a ms_2ms_2 ms_5ms_5, the offspring would segregate 3 mst: 1 mft. The mst's could be used for hybrid breeding. Using g-mst Dong A (*G. hirsutum*) in crosses to local Chinese cotton, Huang et al. (1981) obtained 12–34% increase in lint yield in F_1 plants. Barring the genes ms_2, ms_4, ms_5, ms_6 and Ms_7, which cause nearly complete mst, sterility induced by other *ms* genes is partial. In them, isolation of selfers from hybrids is difficult unless a recessive marker gene linked to the *ms* gene and identifiable preferably at the seedling stage is available. Mostly *ms* genes produce variable effects under uniform as well as under different environmental conditions. Therefore, commercial use of these genes has not been made.

34.2.6.2 Gene-Cytoplasmic Male Sterility

Use of gc-mst for hybrid seed production has an edge over the use of g-mst in cotton even though gc-mst is beset with many difficulties. The main difficulty is that the development of mst genotypes is highly sensitive, not only to genotype and cytoplasm background but also to the environmental conditions. While some *ms* genes cause p-mst, others produce variable to complete mst. Even American upland cotton varieties differ in the ease with which sterility can be maintained or induced by environmental manipulations, chemical treatments or by the introduction of genes or cytoplasm from other cotton species. Meyer (1969) developed pure-breeding mst strains having *G. arboreum* or *G. anomalum* cytoplasm and *G. hirsutum* genome. This sterility was later transferred to commercial cotton varieties through BC's. Meyer (1973) released the first acceptable gc-mst cotton in the USA. This was obtained by transferring *G. hirsutum* genome into the cytoplasm of *G. harknessii*, Brandagee, a wild lintless species. None of the strains and cultivars of *G. hirsutum* or *G. barbadense* possesses *Fr* genes. By transferring genes from *G. harknessii* into *G. hirsutum* genome, a restorer line was developed. Compared to mst's, the lint percentage was higher in the restored fertiles. Certain gc-mst cotton lines have more desirable micronaire, longer fibre length, slightly stronger and lower uniformity ratio (Weaver 1981). Hybrids obtained by using some of these mst lines have produced a consistently longer fibre length, greater fibre strength and more desirable micronaire than by using mft lines to produce interspecific hybrids. Weaver (1981) proposed a three-way hybrid system that yields high-quality fibre on mst plants. For exploitation of heterosis for ginning percentage, halo length and boll weight, the A-line was superior to many mft lines. Pollination of mst lines is a major obstacle for commercial hybrid cotton development. This could be solved either by breeding cotton attractive to bees or by breeding bees attractive to the cotton flower, or by an overall cotton crop management which includes maintaining bee colonies in the field and protecting them from insecticides. Propagation of cotton plants through root cuttings is simple and inexpensive. The mst's could be interplanted with a pollinator variety and hybrid seeds harvested. These F_1 could later also be propagated and maintained vegetatively. Stroman (1961) proposed ratooning of F_1 plants in Peru so that more than one crop of F_2 seed could be harvested. Ratooning of mst's in areas where such a practice is possible provides a useful method of producing F_1 seed (Weaver 1968).

34.2.6.3 Major Problems

i. *Isolation of Stable and Efficient C-Line*. The main obstacle in developing hybrid cotton is the lack of stable C lines that could consistently restore cent per cent F_1 fertility. Usually C lines become partial restorers after some generations, or lose the complete restoration potential during dry hot summers. Some stable C lines restore up to 90%, but never reach the 100% level. Also adequate restoration is much easier in interspecific crosses between short and long staple cotton than between interspecific crosses with *G. hirsutum*. Thus the *G. barbadense* genome has a fertility enhancement gene which is absent or silent in *G. hirsutum* genome.

ii. *Ensuring High Bee Visitation*. For cheap hybrid cotton production, reliable and adequate pollination of mst flowers is essential. The abundance and visitation of cotton pollinator wild bees are erratic and unreliable. At times they are abundant and provide excellent pollination, but at other times they are scarce and are greatly reduced by cultivation and insecticides, so that in the intensively sprayed areas often none can be found. Honeybees are likewise erratic visitors to cotton flowers, as sometimes they visit in swarms and pollinate mst flowers adequately but at other times they visit other more attractive flowers or only the extrafloral nectaries of cotton. This leads to poor seed set on mst's. Hence artificially saturating wide cotton-growing areas with honeybee colonies is desirable to ensure adequate cross pollination of mst flowers of cotton. For this at least, one colony per 2 acres of land is essential. Economically this is not feasible unless large of acreages of hybrid seed are produced in a small area. No economical or reliable method of pollinating small cotton fields is available at present for the hybrid production.

34.3 *Thespesia populnea* Soland (*Hibiscus populnea* L., tulip or portia tree)

A pollen-sterile purple-flowered plant with incipient or faulty spindle development, abnormal meiosis and distinct confluent stigmas occurred among a normal fertile stand (this Vol.). The anomaly appears genic, as only one out of 23 trees growing in a stand exhibited it.

CHAPTER 35
Moraceae

35.1 *Ficus* L.

Rudimentary anthers with or without sporogenous tissue occur in a few plants of *F. glabella, F. hispida, F. pumila* and *F. tomentosa* (this Vol.). In these, either PMC's are not developed or they degenerate before meiotic initiation. However, in a clone of *F. pumila* var. *minima*, PMC's abort after separation before the pachytene. The mst appears genetic, as other nearby-growing plants of these species are fully mft.

35.2 *Morus nigra* L. (common mulberry)

A single plant in a stand developed rudimentary single-celled anthers and the male flowers abcissed prematurely. The anomaly appears genetic, as no such premature abscission occurred in the nearby growing trees of this species (this Vol.)

CHAPTER 36

Myrsinaceae

36.1 *Rapanea thwaitesii* (Aubl.)

About 39–65% (50% average) of the pollen produced by a population of this species is abnormal and sterile (Vasanthy and Pocock 1981). Most of this pollen is large-sized, has degenerated cytoplasm and ill-defined apertures.The anomaly occurs between the free microspore and pollen grain formation stage. Since some of this pollen was fungus-infested, the infection might have induced the pollen sterility. Thus it is not certain if the pollen sterility was caused by genes or by fungus. Also female fertility of the material is impaired, as the authors found rudimentary ovules in a globose ovary. Therefore, it may be male-female sterility. Being partial, it has no immediate breeding value.

CHAPTER 37

Oleaceae

37.1 *Olea europaea* L. (olive)

In three cultivars, viz. Lucques, Oliviere and Tanche, of this thornless evergreen tree, mst occurs frequently (Villemur et al. 1984). The sterility is partial in Tanche, but complete in others. Complete mst is assigned by the authors to abnormal tapetal development. Genetic control of the sterility is not known.

37.2 *Jasminum pubescens* Willd (garden jasmine)

Pollen sterility of a high degree occurs in some plants of this fragrant-flowered jasmine. The anomaly is due to abnormal spindle development and partial failure of cytokinesis (Bhargava 1986) or due to desynapsis and consequent meiotic anomalies (this Vol.)

CHAPTER 38
Onagraceae

In this predominantly herbaceous family of 20 genera having 480 species, mst is known in *Epilobium, Godetia* and *Oenothera* (Table 38.5).

38.1 *Epilobium* L. (fire weed, willow herb)

In this ornamental genus of about 200 species, both gene and g-c interactions within and between species lead to anomalies like seedling lethality, failure of flower or sex development, and, pollen, gamete, zygote or seed abortion. Barring some inter-racial *E. hirsutum* crosses and a few interspecific crosses between *E. parviflorum* and *E. roseum*, segregation of mst plants in offspring of other intra- and interspecific *Epilobium* crosses demonstrates gene-genome-cytoplasm interactions. This is vividly demonstrated in Tables 38.1 to 38.5 and from the account to follow.

Table 38.1. Intraspecific F_1 male sterility in *E. hirsutum* races and strains

Strains/races Female parent	Male parent	Male sterility in F_1
Atika	Numerous races	Complete
Atika (4 strains)	15 races	Complete
Atika	Kew-albiflorum	Complete
Atika (4 strains)	Klausenberg	Partial
Essen	Coimbra	Partial
Fuchstrum	Jena	Partial
Kew albiflorum	Numerous races	
Klausenberg	Jena	Partial
Kolleda	Jena	Partial
Saaleufer	Jena	Partial
Xanthi	Numerous races	Partial
Xanthi	Saaleufer	Complete, or partial, or nil

Table 38.2. Intraspecific F_2 male sterility in *E. hirsutum* races and strains

Female parent (strain)	Male parent (strain/races)	F_2 male steriles
Giessen	5 races	25%
Vienna	5 races	25%
Inselcandidum	Caen	25%

Table 38.3. Interspecific male sterility in *Epilobium*

Parents	Male sterility F_1
Genic	
Epilobium hirsutum × *E.roseum*	Complete
E.roseum × *E.hirsutum*	Partial
E.parviflorum × *E.montanum*	Complete
E.montanum × *E.parviflorum*	Complete
E.parviflorum × *E.roseum*	Complete
E.roseum × *E.parviflorum*	Complete
E.roseum × *E.luteum*	Partial
E.luteum × *E.roseum*	Partial
Gene cytoplasmic	
E.hirsutum × *E.luteum*	Complete
E.luteum × *E.hirsutum*	Partial
E.hirsutum × *E.montanum*	Complete
E.montanum × *E.hirsutum*	Partial
E.hirsutum str. *Essen* × *E.parviflorum*	Complete
E.parviflorum × *E.hirsutum* str.*Essen*	Nil
E.hirsutum × *E.parviflorum*, Crailsheim	Complete
E.parviflorum, Crailsheim × *E.hirsutum*	Nil

Table 38.4. Genic male sterility in Onagraceae

Species/cross	References	Gene action	Tapetum
Epilobium hirsutum	Michaelis and Michaelis (1948)	nk	nk
E.hirsutum × *E.roseum*	Michaelis and Wertz (1935)	During microspore development	nk
E.parviflorum × *E.montanum*	Oberreuter (1925)	During premeiosis and tetrad formation	Normal
E.parviflorum × *E.roseum*			
E.montanum × *E.parviflorum*		During tetrad formation	Normal
E.roseum × *E.parviflorum* × *E.luteum*	Michaelis and Wertz (1925)	nk	
Oenothera alata	Pohl (1895), Gates (1907)	After microspore release	Abnormal
O.flava	Oehlkers (1927)	After microspore release	Abnormal
O.gigas	Gates (1911)	At premeiosis or during tetrad formation	Abnormal
O.lamarckiana-alata[a]	Oehlkers and Harte (1943), Harte and Bissinger (1952)	After microspore release	Abnormal
O.strigosa[a]	Oehlkers (1923, 1926), Harte and Bissinger (1952)	After microspore release	Abnormal
O.suaveolens × *O.strigosa*	Oehlkers (1927)	nk	Abnormal
O.rubricalyx × *O.alata*	Gates and Thomas (1914)	After microspore formation	Abnormal

[a]Gene control by 1r, in others not known.

Table 38.5. Gene cytoplasmic male sterility in Onagraceae

Species/cross	References	Remarks
Epilobium hirsutum[a]	Michaelis (1954)	Many races possess *S* and *fr* genes
E.hirsutum × *E.luteum*	Michaelis (1929, 1932, 1933, 1954)	
E.luteum × *E.hirsutum*[a]		
E.hirsutum × *E.montanum*[a]	Michaelis and Wertz (1935)	
E.montanum × *E.hirsutum*[a]		
E.hirsutum Jena × *E.parviflorum*[a]	Michaelis (1943)	
E.hirsutum Essen × *E.parviflorum*[a]	Michaelis (1949)	
E.hirsutum × *E.parviflorum Crailsheim*[a]	Michaelis (1943)	
E.montanum × *E.roseum*[a]	Oberreuter (1925)	Tapetum normal
Godetia whitneyi fulgida × str. *Bremen*	Hiorth (1948a,b)	Reduced to normally formed anthers
Oenothera biennis × *O.argillicola*	Stubbe (1964)	*c* Genes located on chloroplast (?)
O biennis × *O.biennis*	Stubbe (1964)	
O.biennis × *O.hookeri*	Stubbe (1964)	
O biennis × *O.parviflora*	Stubbe (1964)	
O.parviflora × *O.biennis*	Stubbe (1964)	
O.parviflora × *O.hookeri*	Stubbe (1964)	

[a]Gene action after microspore formation, in others not known.

38.1.1 Intraspecific Crosses

38.1.1.1 Genic Male Sterility

Progeny plants of the cross *E. hirsutum xanthi* × *E.h. Kew-albiflorum* exhibited complete pollen sterility but the reciprocal cross progeny had 2.8% fertile-looking pollen (Michaelis and Michaelis 1948). Since female fertility of the progeny plants was normal, the mst is not gc-mst as considered by the authors but is g-mst. The same is true for the cross *E.h. xanthi* × *E.h. Erlangen*, in which both direct and reciprocal crosses yield cent per cent pollen sterile plants. Genetic control of the mst is not known as no F_2-F_3 data are available.

38.1.1.2 Gene-Cytoplasmic Male Sterility

The occurrence of mst progeny in some crosses involving specific races and strains of *E. hirsutum* (Table 38.1) was detected by Michaelis and Michaelis (1948). Since progenies of all the reciprocal crosses are fully mft, mst is due to g-c interactions involving the female parent cytoplasm and the male parent genome. Interaction between the cytoplasms of the races *Fuchstrum*, *Klausenburg*, *Kolleda* and *Saaleufer* and the genes of *Jena* cause mst in these alloplasmic combinations. In all of them, pollen abortion occurs at different stages of microsporogenesis, indicating that either the S-cytoplasm offers different degrees of resistance to the action of *fr* gene(s) or that the S-cytoplasm of these races differs from each other. The cytoplasms of *Atika*, *Kew-albiflorum* and *Xanthi* produce mst progeny with a large number of *E. hirsutum* races (Table 38.1). Progeny of *Xanthi* × *Saaleufer* crosses segregates complete and partial mst's and also mft's, indicating heterozygosity of *fr* genes in the male parent and mst control by at least two nonallelic recessive genes. In three cases, mst plants with deformed anthers segregate in F_2 and BC_1 in 1:3 and 1:1 mst:mft, respectively, the reciprocals are mft (Table 38.2). This indicates monogenic control of mst in these three cases. Michaelis and Michaelis (1948) obtained normale fertile plants in F_1, F_2 and BC_1 of the cross *E.h. Vienna* × *E.h. Kew-albiflorum*. In BC_{10}, using Kew-albiflorum as the recurrent pollen parent, all the plants had deformed flowers and sterile pollen. Anther deformation degree and pollen sterility percent varied considerably between the plants.

38.1.2 Interspecific Crosses

38.1.2.1 Genic Male Sterility

i. *Complete.* In crosses *E. hirsutum* × *E. roseum*, the F_1 plants were completely mst and female fertile (Michaelis and Wertz 1935). In the reciprocals, some F_1 plants have about 20% pollen fertility, as evidenced by acetocarmine staining tests (Table 38.3), but the functionability of the pollen was not determined. Meiosis in these F_1 plants is normal but the microspores degenerate at maturity. In some of them, the *ms* gene action is delayed and in the majority of them, the mature pollen degenerates quickly. In a few, some normal-looking pollen is produced, but it is physiologically weak and nonfunctional. Therefore, this interspecific cross is included under g-mst. Genetic control of the sterility is not known. An interesting situation occurs when *E. parviflorum* is crossed as female to *E. montanum* or *E. roseum*. The progeny of both the crosses is mst (Oberreuter 1925). These plants possess very few PMC's which produce few tetrads that degenerate soon after formation. In the reciprocal crossings in which *E. parviflorum* is used as the pollen parent, a normal amount of PMC's are formed and the anther thecae are full of tetrads, but these tetrads degenerate soon after their formation. Thus while in *E. parviflorum*, cytoplasm the ms genes of *E. montanum* and *E. roseum* initially cause drastic reduction of the archesporial tissue and later complete degeneration of tetrads, in the reciprocal crosses, the *ms* genes of *E. parviflorum* in *E. montanum* or *E. roseum* cytoplasm act only when tetrads are fully formed. This indicates reciprocal differences in g-c interactions in these three *Epilobium* species and a differential timing of action of the *ms* gene(s). In both the cross combinations, mst is complete. Since the tapetum develops normally, the *ms* gene acts on the microsporogenous tissue directly.

ii. *Partial.* F_1 progenies of both direct and reciprocal crosses between *E. roseum* × *E. luteum* were p-mst, exhibiting 63 and 75% mean pollen sterility. These sterility differences have been considered by Michaelis and Wertz (1925) as due to cytoplasmic differences and the sterility as gc-mst, but the sterility is genic, as the pollen sterility differences obtained between direct and reciprocal crosses are not large enough to warrant cytoplasmic differences. In fact such differences are often obtained with g-mst pea and tomato crosses (this Vol.) Therefore, in such cases the sample size has to be increased and the means compared using the t-test for the significance of differences. However, the existence of minor cytoplasmic differences between the two *Epilobium* species cannot be ruled out, as the genus exhibits high cytoplasmic variability and diversity.

38.1.2.2 Gene-Cytoplasmic Male Sterility

F_1 plants of the cross *E. hirsutum* × *E. luteum* were mst, had anthers reduced to little green scales and their pollen degenerated completely after meiosis (Michaelis 1929, 1932, 1933, 1954). The reciprocal cross yielded F_1 plants that had 15–20% normal pollen. BC's of these partially sterile F_1 hybrid plants by the pollen of a vegetatively reproducing clone of *E. hirsutum* increased mst in BC generations and was complete at BC_5. After BC_{23}, the offspring became completely stabilized as mst plants. This long continuance of sterility clearly rules out maternal inheritance as the source of mst and is an evidence of a self-replicating cytoplasmic *c* gene(s). Reciprocal crosses made in advanced generations (BC_4) yielded about 22% normal pollen whereas progeny of the cross at similar BC_4 level [*E. hirsutum* × (*E. luteum* × *E. hirsutum*) × *E. luteum*] was entirely mst. Since these two crosses were identical as regards the nuclear gene content, as in each case the female parent was essentially *E. hirsutum* (only 1/32 of the female parent's chromosomal genes were residual ones from *E. luteum*), the nuclear genes of the progeny plants were essentially the same as those of the original *E. hirsutum* × *E. luteum* hybrid plants. However, the cytoplasms differed completely in these two classes of plant. In one, the hybrid genome was in *E. luteum* and in the other in *E. hirsutum* cytoplasm. Phenotypically they were similar but they differed in mft. Therefore, it is clear that the *c* gene(s) is located in the maternal cytoplasm of *E. hirsutum*. This gene interacts with the nuclear *fr* genes of *E. luteum* to produce mst plant progeny. The *c* gene action did not alter even though it had gone through substitution BC's and remained associated with the *E. hirsutum* genome.

A nearly similar situation was found by Michaelis and Wertz (1935) in the crosses involving *E. hirsutum* × *E. montanum Münich* strain. In this cross, F_1 progeny was completely mst. On the other hand, in the reciprocal crossings, F_1 progeny was p-mst and had 40–60% pollen fertility. When *E. montanum* was crossed by pollen of the alloplasmic (*E. luteum*) *E. hirsutum*, the progeny was likewise p-mst. BC's of these p-mst plants produced by *E. hirsutum* pollen enhanced mst degree progressively and it was cent per cent at BC_6. In these plants, the *E. hirsutum* genome was placed in *E. montanum* cytoplasm. Therefore, in both cases (*E. hirsutum* and *E. luteum*, and, *E. hirsutum* and *E. montanum*) mst results when a few *fr* genes of *E. luteum* or *E. montanum* are present in *E. hirsutum* cytoplasm. On the other hand, the nearly complete genome of *E. hirsutum* is required in the cytoplasm of *E. luteum* or *E. montanum* to produce mst. In crossings of *E. hirsutum* strain *Essen* × *E. parviflorum*, the F_1 progeny was mst. Reciprocal crosses produced mst F_1 plants (Michaelis 1949). Backcrossing the mst plants by *E. parviflorum* pollen produced only mst progenies. Therefore, mst is caused by the interaction of *E. hirsutum* cytoplasm and *E. parviflorum* gene(s). Crossing *E. hirsutum Jena* to *E. parviflorum* and its several strains led to F_1 plants that either died prematurely or never flowered (Michaelis 1943). On the other hand, progenies of the reciprocal crosses were normal and mft. However, crossing *E. hirsutum Jena* female to *E. parviflorum* strain *Crailsheim* male led to viable F_1 progeny plants but they were mst, the reciprocal progeny being fully fertile. Thus these anomalies are due to g-c interactions. These are severe in *E. hirsutum Jena* cytoplasm and *E. parviflorum* gene(s) combination, which results in high seedling lethality, hybrid inviability, floral development failure or floral anomalies. In the alloplasmic combination *E. hirsutum Jena* cytoplasm + *E. parviflorum Crailsheim* genome, the interaction permits flower development but inhibits viable pollen development.

38.1.3 Zygote Lethality

Whereas premature death of F_1 fertilized embryos occurred when the *E. hirsutum Jena* race was pollinated by Parys or Kirstenbosch races, offspring of the reciprocal crosses developed normally (Michaelis and Dellinghansen 1942). When F_1 and F_2 plants of *Parys* × *Jena* male were crossed with *Jena* as females, certain F_2 segregants failed to fertilize the *Jena* ovules because pollen tube growth was inhibited in *Jena* styles. In other such combinations ovules were fertilized but the embryos died prematurely. In addition, a few viable combinations that were formed exhibited much inhibited and distorted vegetative growth and high pollen sterility. The closer a plant reached Parys' phenotypic level, the greater was the vegetative distortion and pollen sterility. Thus it appears that most of the genes of the Parys strain are sensitive to Jena cytoplasm.

38.1.4 Fertility Reversion

The gc-mst clone termed irregular by Michaelis (1949) arose after crossing *E. hirsutum* × *E. parviflorum.* In it, early male meiosis is normal, but tetrads abort immediately after formation. Although the mst clone is stable, one of its subclones was mft (Jones and Gabelman 1965). Backcrossing mst plants to this fertile subclone, selfing the fertile isolates, propagating fertile subclone vegetatively and by seed for many generations rule out the possibility of dauermodifications and indicate spontaneous mutation from S- to N-type cytoplasm as the cause of origin of the mft subclone.

38.1.5 Gene-Cytoplasm Diversity

Whereas mst in progeny of direct and reciprocal crosses involving various races of *E.hirsutum* is genic (Tables 38.1, 38.2) and these races thus have similar cytoplasms, the mst in interracial and interstrain F_1 or F_2 hybrids is g-c in those where reciprocals are fully mft (Table 38.3). This indicates the existence of cytoplasm differentiation in these races. This differentiation between races and strains of *E. hirsutum* has not evolved to a high degree because mst is partial in many such hybrid combinations. In interracial and intervarietal crosses of *E. hirsutum*, the flower or sex development failure, pollen or gametic abortion, development of deformed flowers and/or sterile pollen are rampant. The extent of these anomalies is influenced not only by a varying number of modifying genes but also by environment. While some races appear to have S-cytoplasm and others *fr* genes, whose interaction in the hybrids leads to mst, many such races having S-cytoplasm induce stable and complete mst only when their genomes are placed in the cytoplasm of other races. A wide range of divergence and variability in mst expression and manifestation occurs in interspecific crosses of *Epilobium* (Table 38.3). Thus whereas in the offspring of the crosses *E. montanum* × *E. hirsutum* and *E. luteum* × *E. hirsutum*, mst is partial, in the reciprocals, it is complete. In both cases, one of the parents is *E. hirsutum*. When this species is used as the female parent in these crossings, mst in the offspring is complete. Thus it appears that with the same gene combinations its cytoplasm is more effective in inducing sterility than those of the other two species. This indicates cytoplasmic differentiation in *Epilobium* species and its possession of diverse and highly specialized cytoplasm types in *E. hirsutum*.

38.2 *Godetia whitneyi* Spach (Clarkia, fairy fan, red ribbon)

Although this genus has 35 species, mst is known in only one interstrain hybrid of *G. whitneyi* in which the progeny plants of the cross between the strains Bremen × Fulgida were normal but those from the reciprocal cross were mst, having highly reduced anthers and completely deformed microspores (Hiorth 1948a,b). Back-crossing the mst plants with Bremen pollen yielded populations that segregated 1:1 plants, having anthers reduced as much as the F_1 and plants with still more reductions, some being transformed into pistils. The mst is due to interaction of Bremen *fr* genes and Fulgida S-cytoplasm. While Hiorth (1948b) considered a single recessive gene pair to be responsible for the mst, segregations reveal more *fr* gene number involvement. In some, the influence of two modifier genes is evident.

38.3 *Oenothera* L. (evening primrose, sundrop)

Although this consists of nearly 80 species, some of which are excellent ornamentals and cytogenetically fascinating, mst is known in only seven species (Table 38.4), but it has not been investigated in detail, as is evident from the account to follow.

38.3.1 *O. gigas*

In certain plants, the PMC's did not round off, retained their immature polyhedral shape because the surrounding tissue failed to grow and remained in contact with the mother cells (Gates 1911). In others flowers, meiosis proceeded normally but some of the tetrads grouped centrally in PMC's. In others, a wall was formed after the second division stage or cytoplasm of the mother cells underwent a direct cleavage to form a tetrad like formation. In all cases, no viable pollen was produced and the plants were completely mst. Genetic control of the sterility and details of the female fertility of these plants are not known. Since the plant basically arose as a mutant and was propagated later, monogenic recessive control is inferred.

38.3.2 *O. lamarckiana* De Vries (*O. erythrosepala*, Borb)

In the variety alata, Pohl (1895) found a mutant in which mst was caused by the trisomy of a specific chromosome. A recessive gene, *fr*, causing mst was detected from F2 segregations of plants descended from the crosses involving the stringen complex of *O. strigosa* (Oehlkers 1923, 1926; Oehlkers and Harte 1943). Another gene, *ster*, inducing mst, was found to be located in the velans complex of *O. lamarckiana* alata (Harte and Bissinger 1952). In the mst genotypes, anthers were empty, shrunken and indehiscent. Both the mutant genes affected tapetal function and caused enzyme missecretion, inhibiting starch deposition in the tapetal cells. This causes a nutritional deficiency in the developing pollen, which consequently abort. The gene *fr* (= *ms*) has full expressivity and penetrance in different genotypic backgrounds. A similar type of mst is also known *O. alata*, *O. flava* (Pohl 1895; Gates 1907) and in some of the descendents of the cross *O. suaveolens* × *O. strigosa* (Oehlkers 1927). In a hyperdiploid plant that arose from the progeny of a cross *O. rubricalyx* × *O. alata*, the sporadic occurrence of mst is recorded by Gates and Thomas (1914). Meiosis in them proceeded normally up to microspore formation, after which degeneration occurred rapidly. The tapetal development was imperfect and deranged. Genetic control of the anomaly and the status of female fertility are not known. Interaction between *O. hookeri* and *O. biennis* genomes and *O. parviflora* cytoplasm and between *O. argillicola* genome and *O. biennis* cytoplasm has been considered by Stubbe (1964) to lead to the production of nongerminating physiologically weak pollen in the hybrid plants. Thus they are effectively mst, although phenotypically the pollen looks normal.

CHAPTER 39

Papavaraceae

39.1 *Argemone mexicana* L. (Mexican thistle)

A spontaneously arisen monogenic recessive mutant having light yellow petals and reduced green anthers appeared in a stand growing on recently deposited soil near a construction area (this Vol.). Archesporial tissue degenerated prior to PMC formation. The anomaly is controlled by a recessive *ms* gene. Female fertility is normal and mst is complete in this plant.

39.2 *Papaver somniferum* L. (opium poppy)

Spontaneous arisen as well as 10 and 20 kR γ-ray-induced mst's were detected by Singh and Khanna (1970). The mst's were characterised by staminal degeneration during various developmental stages so that either the anthers never developed or their filaments tapered to a thin narrow end. In some, rudimentary anthers lacking PMC's were formed. In a few plants, normal anthers having PMC's were formed. Male meiosis was normal but microspores degenerated immediately after liberation. In a few mst's, partial or completely mst anthers are interspersed among variously developed anthers. In some of these, the terminal bud is partially sterile while basal buds are fully mst. In a few mst's sectorial chimeras for mst differentiate so that part of the flower is mst and the rest mft. Inheritance of the sterility is complex, but it is monogenic recessive in a 10 kR γ-ray induced mst mutant in which anthers develop, male meiosis proceeds normally but the microspores degenerate after liberation from PMC's (this Vol.). In the F_2 generation of the cross *P. somniferum* × *P. setigerum*, pollen steriles segregated (Hrishi 1960). The sterility is correlated with flower colour.

CHAPTER 40

Pedaliaceae

40.1 *Sesamum indicum* L. (S. orientale, sesame)

40.1.1 Genic Male Sterility

Of the 20 known species in this genus, *S. indicum*, indigenous to old world tropics, is economically the most useful because of its oil-yielding, medicinal, horticultural and condimental properties. Divergent effects are caused by *st* and *ms* genes in this species (Table 40.1). In some, either the male-female sterility is total (Kumar and Abraham 1941; Kumar and Rao 1945) or partial (Malaguti and Mazzani 1958) or it is restricted to early stages of the reproductive process (Dabral 1968; Dabral and Mandloi 1974) or is confined to the male sex only, the latter having a use in breeding. From the progeny of five sesame lines obtained from a Venezuelan seed stock, 28 mst plants segregated (Osman and Yermanos 1982).These resembled the mft's in height, branching pattern, leaf size and maturity, but had flattened translucent greenish indehiscent anthers. The anther features distinguished them from the mft's. The mst anthers lack the endothecial layer and the pollen grains are devoid of exine wall surface structure, furrows and ridges. The anomaly is controlled by a recessive *ms* gene whose action is stable under greenhouse and field conditions. Female fertility of the mst plants as evidenced by seed set after hand pollinations is normal. However, in three USDA introduction lines, Brar (1982) detected mst plants with poor female fertility. They possess a distinctive plant type as they lack trichomes and glandular hair and develop abundant sessile tetrad glands on the stem, and diminutive flowers. Whereas plants of one mst line are early, dwarf and profusely branched, bearing one capsule per node, plants of the other two lines are late-flowering, nonbranching and have multiple capsules per node. The mst's develop stamens but the filaments are reduced and anthers shrunken. Some of their flowers have anthers with viable pollen but the anthers are fully indehiscent. The sterility in each of the lines is controlled by a recessive gene; the allelic nature of the three *ms* genes is not known. Moreover, mst and poor female fertility occurs under field conditions and not in the greenhouse. This may

Table 40.1. Genic male sterility in *Sesamum*

Species	References	Remarks
Sesamum indicum	Osman and Yermanos (1982), Brar (1982)	Tapetum defective, endothecium absent, stamens with reduced filaments and shrunken anthers

Gene action not known, gene control = 1r

be due to the temperature sensitivity of the *ms* gene. Although low temperature induces high pollen and some ovular sterility in mft sesame (Weiss 1971), higher temperatures led to increased pollen sterility in one mst and mft line and the absence of pollen in another mst line (Brar 1982). The mst mutant detected by Osman and Yermanos (1982) is true mst, as it is fully female fertile. The mutation from *Ms* to *ms* gene is not associated with any morphological effect. In contrast, the three mst lines of Brar (1982) possess considerable morphological differences and exhibit both sporogenous and functional types of mst. In addition, anthers of some flowers produce viable pollen as well. Also in almost all the mst plants, female fertility is impaired. Thus these are total male and partial female sterile mutants. They also exhibit environmental sensitivity of a varying degree.

40.1.2 Chemical Male Sterility

To induce mst, chemical sprays of 0.25, 0.5 and 1% MH, 0.1, 0.25 and 0.50% dalapon or 0.3% mendok before bud initiation, before and after bud initiation or before and after bud development, respectively, plus another spray at anthesis, were tried by Chauhan and Singh (1971) in the T4 variety. Of all the chemical treatments, mendok produced less inhibitory effects on shoot height, leaf number and fruit and seed yield; other chemicals were highly toxic (Table 40.2). Complete pollen sterility was induced by three sprays of 0.5% mendok, but it was accompanied by over 30% female fertility reduction. The treatment caused phytotoxicity and premature fruit drop. The tapetal cells either degenerated in premeiotic stages, followed by degeneration of sporogenous cells, or they persisted and became hypertrophied or pseudopodial at the vacuolate pollen stage (Chauhan and Kinoshita 1980c). The viability of the resultant seeds and the vigour of the seedlings are not known. Usually both these are reduced following such chemical treatments.

Table 40.2 Chemical male sterility in *Sesamum*

Chemicals	References	Male sterility induction	Female fertility reduction	Side effects
Dalapon, MH	Chauhan and Singh (1971)	Partial	High	Highly phytotoxic
Mendok	Kumar and Singh (1963), Chauhan and Singh (1971)	Complete	nk	nk

40.1.3 Breeding Value

Since the mst plants found by Osman and Yermanos (1982) exhibit complete female fertility and are stable under different growing conditions, their use in hybrid seed production is high. Sesame, a predominantly self-pollinated crop, exhibits a high level of F_1 heterosis in grain yield (Riceelli and Mazzani 1964; Sarathe and Dabral 1969; Murty 1973; Dixit 1976). Therefore, the use of mst in hyrid seed production will be highly beneficial but only 25% of the F_2 segregating populations can be used as female parents in crossing blocks for commercial hybrid seed production. Also 75% of the mft plants need to be rogued. These two factors limit the use of g-mst in commercial breeding. The mst found by Brar (1982) can be easily maintained by growing them in the greenhouse and used as female parents in the fields. Since the sterility is accompanied by female fertility reductions, it cannot be used for commercial seed production.

CHAPTER 41

Plantaginaceae

41.1 *Plantago* (Tourn) L. (plantain, rib-grass, sibwort)

This genus of about 250 species contains both outbreeding and inbreeding species which grow luxuriantly and exasperatingly as weeds in lawns and waste lands. Outbreeding is favoured because of protogyny, self incompatibility and mst. Mather (1940) regarded mst as a cause which has led to the outbreeding because for perpetuation of mst plants, they must be pollinated by hermaphrodites. Nearly all the species of *Plantago* segregate mst's, but it is well documented in the gynodioecious species, in which the proportion and frequency of mst's are very high in natural populations. The mst occurs in *P. coronopus* (Paliwal and Hyde 1959) *P. insularis* (Murr and Stebbins 1971), *P. lanceolata, P. major, P. maritima, P. ovata, P. lagopus* (Ross 1969, 1970a,b) and *P. serraria* (Fröst 1959, Table 41.1). Only a few rare msts occur in *P. major* and *P. lagopus* (Knuth 1909) but its frequency is high (65%) in *P. ovata* populations from India (Chandler 1954; Atal 1958; Mittal and Issar 1970). This suggests that *P. ovata* is gynodioecious.

Table 41.1. Genetic male sterility in *Plantago*

Species	References	Gene control	Remarks
[a]*Plantago coronopus*	Paliwal and Hyde (1959), Gorenflot (1964), Ross (1970a)	nk	Anthers minute with little or no pollen
P.insularis	Murr and Stebbins (1971)	1r	Linked to albino gene
P.lanceolata	Ross (1969), Van Damme (1983b)	2–3r, 2D	
P.maritima	Knuth (1909)	nk	
**P.ovata*	Mittal and Issar (1970)	lD	Inhibitory gene present

[a]Gene action after microspore liberation.

41.1.1 *P. coronopus* L. (horn plantain)

In one strain, the association of mst with B chromosomes was reported by Paliwal and Hyde (1959) and Gorenflot (1964) as all the mst plants possessed a B chromosome, while hermaphrodites lacked it. Male meiosis in mst's is regular until microspore formation, after which microspores degenerate before pollen mitosis. The majority of the steriles reproduce apomictically and a stimulus of pollination is essential for the apomictic development of the embryo. It is not known if mft's are likewise apomictic. Ross (1970a) studied *P. coronopus* material obtained from

Hyde and found two mst types in it. The first type has either minute and nonexserted anthers or small but slightly exserted indehiscent anthers, the other type has small but fully exserted anthers that produce little or no pollen. A few plants are intermediate between these two mst types and the hermaphrodites. Their anthers, although exserted, produce very little pollen compared to the normal hermaphrodites. Neither cytological nor genetic data sufficient for a satisfactory cytogenetic interpretation are available for these, but from the available crossing data, the involvement of the S-cytoplasm cannot be ruled out. Till the S-cytoplasm is detected, the sterility is considered as g-mst.

41.1.2 *P. insularis*

In the M_2 progeny obtained after gamma irradiations, albino mutant, was found by Murr and Stebbins (1971). The trait is controlled by a recessive gene, *al*. Subsequent segregations revealed considerable departures from the 3:1 ratio. This was due to complete linkage of the gene *al* with a pollen eliminator gene *pe*, due to which half of the pollen from al^+, *pe/al*, pe^+ heterozygotes abort. Microspores having gene *pe* do not develop, but degenerate gradually so that at anthesis empty microspores remain. Likewise in albino barley, selfing of a chlorophyll-deficient mutant led to the almost complete absence of the homozygous normal class (Holm 1954). The plantago heterozygotes were produced approximately in 1:1 ratio with the mutant phenotype. These heterozygotes have a high mst degree, probably due to the linkage with a gene akin to the *pe* gene of *P. insularis*.

41.1.3 *P. lanceolata* L.

The occurrence of mst's in this gynodioecious self-incompatible perennial rosette herb is documented by Correns (1906, 1908), Bartlett (1911, 1913), Stout (1919), Ross (1969, 1970a,b), Van Damme and Van Delden (1982) and Van Damme (1983a,b). While Correns (1908) could not explain the genetic control of mst's, Bartlett (1913) studied an exceptional mst type, and Stout (1919) considered sex expression and variability of mst degree as highly complex. From a natural population in Berkshire, England, Ross (1969) detected a mst plant in a hermaphrodite population. Hermaphroditism appeared dominant over mst, but the two phenotypes exhibited 1:1, 3:1, 7:1, 15:1 segregation ratios. In two families, the deficit of hermaphrodites was also evident. This was assumed to be due to the linkage of the *ms* gene with the incompatibility gene *S*. Ross (1969) proposed a digenic model in which mst's were considered as double recessives, but indicated the involvement of some additional genes, probably of cytoplasmic nature. An almost continuous variation in sex expression from mst to hermaphrodites occurs in this species. Although intermediates and p-mst's are present in most of the gynodioecious species, they are significant in this species. Based on mst phenotype, Van Damme and Van Delden (1982) distinguished two complete series, including phenotypes ranging from mst to mft. This distinction indicates the existence of two S-cytoplasm types, R and P. Accordingly, there are two mst types designated MS_1

CHAPTER 45

Ranunculaceae

Of the 35 known genera, mst is documented only in *Aquilegia* and *Ranunculus*. While in the former it is gc-mst, in the latter it is g-mst type (Table 45.1).

Table 45.1. Genetic male sterility in Ranunculaceae

Species	References	Gene action	Remarks
Genic			
Ranunculus acris	Whyte (1929)	After microspore release	Pollen completely nonviable
Gene-cytoplasmic			
Aquilegia flabellata × *A. truncata*	Skalinska (1931)	Between tetrad formation to microspore liberation	Persistent tapetum
A. vulgaris × *A. chrysantha*	Kappert (1944)	nk	Structural (stamens absent or highly reduced)
	Skalinska (1928a,b)	After pollen mitosis	

45.1 *Aquilegia* L. (columbine)

This genus comprises over 100 species which exhibit a high degree of interbreeding, so that each seed represents a specific genotype. However, the breeding is unilateral in some species because of gc-mst (Table 45.1). For instance, offspring of *A. flabellata* × *A. chrysantha* exhibit a high degree of male and female sterility and matrocliny for spur size and shape (Skalinska 1929). Reciprocal cross progeny is mft and exhibits no matrocliny. F_1's of the cross *A. flabellata* × *A. truncata* differ from the reciprocals in fertility. When *A. flabellata* was female, F_1 had 12 and 23% fertile ovules and nondehiscing anthers, respectively. The anthers had a high percent of aborted pollen. In the reciprocal cross, F_1 had high pollen fertility. Contrary to his previous assumptions, Skalinska (1931) later assumed that *A. flabellata* cytoplasm induced pollen abortion irrespective of the genetic constitution of the pollen grains. Since *A. californica* is a taxonomic synonym of *A. truncata* (Stebbins 1950), Skalinska's findings reveal that different strains of *A. truncata* react differently with *A. flabellata* cytoplasm with regards to mst induction and/or its manifestation and confirm the presence of S-cytoplasm and *fr* genes in *A. flabellata* and *A. truncata (A. californica)* plants.

The occurrence of matroclinous spur length and 50% male and 20% female sterility in reciprocal F_1 hybrids of *Aquilegia vulgaris* × *A. chrysantha* was demonstrated by Skalinska (1928a,b). Its F_2 was also matroclinous for spur length, but in reciprocals spur types did occur. Male meiosis in mst's proceeded normally, and degeneration occurred either at the tetrad stage or during generative nucleus formation. Since genetic sterility was in both the sexes, although more pronounced on the male side, it cannot be considered as a true case of mst, even though the influence of the cytoplasm upon male gamete abortion is distinct. Skalinska (1928a,b) assumed that nonviable gametes possessed female cytoplasm and an entire predominantly paternal genome. This could account for the rare occurrence of paternal spur types in F_2 populations. However, all the paternal genes did not become eliminated during meiosis, as is indicated by the production of homozygous paternal characters in BC's of *(A. vulgaris* × *A. chrysantha)* × *A. chrysantha* (Skalinska 1930).

45.2 *Ranunculus* L. (buttercup, crow's foot)

Of 400 species almost confined to cold and temperate regions of world and on mountains in the tropics, mst is known only in *R. acris*, *R. muricatus* and *R. scleratus* (Tables 45.1, 45.2).

45.2.1 *R. acris* L. (meadow butterfly)

Marsden Jones and Turill (1925) reported the existence of mst's in this species. Later, Whyte (1929) found abortive pollen in a single plant whose older stamens lodged some deformed pollen. Such a pollen type was constantly produced by all the younger stamens in one more plant. Meiosis proceeded normally, but microspores degenerated immediately after their release from the PMC's (Table 45.1). Along with this, tapetal cells disintegrated and their contents coalesced to form plasmodia masses that occupied the major part of the anther locules. The genetic control of this sterility is not known. It is not environmental because other plants growing adjacent to this plant produced functional pollen. Although viability or functionability of some little pollen produced by older flowers was not determined, the pollen appears to be nonviable, as no seeds were set on such plants. The female fertility of the plant was normal.

45.2.2 *R. muricatus* L. (spiny-fruited buttercup)

No spontaneous or induced g-mst is known, but chemical mst is known in this species. Three foliar applications of 0.75% MH and 0.2% dalapon induced complete pollen sterility in it (Chauhan et al. 1976), but the treated plants exhibited highly reduced vegetative and reproductive growth, ovular fertility and fruit yield (Table 45.2).

Table 45.2. Chemical male sterility in *Ranunculus muricatus* and *R.scleratus*

Chemical	References	Male sterility induction	Female fertility reduction	Side effects
Dalapon, MH	Chauhan et al. (1976)	Complete	Very high	Reduced vegetative and reproductive growth, toxicity high

45.2.3 *R. sceleratus* L. (celery-leaved buttercup)

In a dark green hairy plant, anthers were shrunken and pollen abortive (this Vol.). Carpels were normal and seed set was high. The genetic control and the cause of the sterility are not known. Like *R. muricatus,* complete mst induced by MH and dalapon in this species is accompanied by diminished vegetative growth and fruit yield (Chauhan et al. 1976).

CHAPTER 46

Rosaceae

Occurrence of mst has been documented in *Prunus communis* (Crane 1915) *P. persica* (Knowlton 1924; Anonymous 1973b; Vitkovskii and Ruban 1974), *Rosa wichuriana* (Nicolas 1927). *Rubus* (Lewis 1939), *Fragaria* (Nürnberg-Krüger 1958). In addition, mst flowers having rudimentary or malformed anthers, partially transformed or petaloid stamens are found in *Fragaria chiloensis*, *F. moschata*, *F. vesca*, *Malus baccata*, *M. floribunda*, *M. sylvestris*, *Potentilla aurea*, *P. cinerea*, *P. fruticosa*, *P. gracilis*, *P. recta*, *P. verna*, *Prunus amygdalus*, *P. avium*, *P. cerasifera*, *P. dasycarpa*, *P. fruticosa*, *P. maritima*, *P. mume*, *P. nigra*, *P. padus*, *P. persica*, *P. pumila*, *P. spinosa*, *P. tomentosa*, *P. virginiana*, *Pyrus communis*, *P. nivalis*, *Rosa alba*, *R. arvensis*, *R. bracteata*, *R. carolina*, *R. centifolia*, *R. chinensis*, *R. nitida*, *R. odorata*, *R. palustris*, *Rubus almus*, *R. deliciosus*, *R. hispidus*, *R. idaeus*, *R. odoratus*, *R. rosa*, *R. rosafolius*, *Sorbus americana*, *S. domestica* and *S. hybrida* (this Vol.). It is not known if the mst is genetic, physiological or purely environmental, but since the behaviour is constant, genetic control appears certain. In *Rosa* and *Rubus*, stamens are poorly developed and no archesporium is formed. In *Prunus persica*, stamens are developed, meiosis is normal but microspores develop poorly so that pollen degeneration occurs before exine formation. In *P. mume*, microspores abort after exine formation (Nakanishi 1982). The status of female fertility and environmental influence on mst expression are not known in them (Table 46.1). Due to the high economic utility of this family, it is profitable to utilize mst (if genetic) in hybrid production since the hybrids can be propagated vegetatively and utilized profitably.

Table 46.1. Genetic male sterility in Rosaceae

Species/Cross	References	Gene action
Fragaria vesca × *F. nilgerrensis*	Nürnberg Krüger (1956, 1958)	After microspore liberation
[a]*Malus sylvestris*	Gagnieu (1951)	After microspore liberation
Prunus mume	Nakanishi (1982)	After exine formation
P. persica	This Vol.	During pollen maturation

[a]Gene control by lr, in others not known.

46.1 *Fragaria* L. (strawberry)

In none of the 15 species of this genus mst is well documented. However, it occurs in the progeny of cross *F. vesca* × *F. nilgerrensis* (Nürnberg-Krüger 1958). Anthers of mst's are small and brown. Male meiosis proceeds normally up to microspore formation, but degeneration occurs after microspores separate out from each other. In mature anthers, shrivelled microspores without cell contents occur. Whereas BC's of these hybrids to *F. vesca* produced progeny plants having variable pollen fertility that was more than that of the F_1 plants, BC's with *F. nilgerrensis* yielded a segregation of 17 normal, 4 mst, 7 fst and 4 completely sterile plants. Reciprocal progeny of the above cross was completely male-female sterile. Nürnberg-Krüger (1958) utilised two other races of *F. vesca* as female in crosses of *F. nilgerrensis*. F_1 plants were not only pollen sterile but their seed fertility was low, ranging between 13–50%. Thus this cross did not yield true mst hybrids but fully male and partially fst plants. The reciprocal differences in segregation pattern are due to cytoplasmic effects.

46.2 *Malus sylvestris* Mill (*Pyrus malus* L., apple)

In some diploid apple varieties, pollen sterility is caused by certain gene combinations (Heilborn 1932). On the basis of natural frequency, pollen abortion in some apple clones was considered by Gagnieu (1951) as conditioned by single recessive genes, the genes being nonallelic and designated as *p1*, *p2*, *p3*, and *p4*. Male meiosis was normal but microspores degenerate abruptly after liberation.

CHAPTER 47

Rutaceae

47.1 *Citrus* L.

This genus includes grapefruit, citron, lemon, lime, orange and shaddock, but the precise number of its species is not reliably known, as some claim only four to five species, while others regard up to 145 species as belonging to this genus. Everett (1980–82) includes 16 species with botanical names of some of the common fruits such as *C. aurantifolia* (swingle, lime), *C. aurantium* L. (sour or Seville orange), *C. limon* Burm (Lemon), *C. limonia* (Mandarin lime), *C. medica* L. (citron), *C. reticulata* Blanco (*C. nobilis*, mandarin, satsuma orange, tangerine) and *C. sinensis* Osbeck (common sweet orange). Because of much confusion in delimiting and distinguishing citrus species from each other and from their hybrids, many researchers have not given the botanical names but common names. Therefore, presently common names have been retained to avoid the existing confusion and confounding of details. Citrus species hybridise easily with each other and many such hybrids produce apomictic seeds. The apomixis in hybrids has rendered pollen production nonuseful and extravagent. The mst has persisted through apomictic seed formation and extensive vegetative propagation. Therefore mst is most rampant in this genus. However, not much is known about the genetic control and female fertility of such mst's. Due to apomixis, in some of them seeds are frequently produced. The mst is rampant in *C. aurantifolia, C. limon, C. medica, C. sinensis, C. reticulata C. unshiu* and *C. yatsushiro*. In many of these, the microsporogenesis breakdown stage is known (Table 47.1) but not the genetic control of the sterility and status of female fertility. In a few, male meiosis is known, and is given below.

47.1.1 *C. aurantifolia* (swingle, lime)

In some clones, archesporial degeneration occurs in the anthers (Uphof 1931; Nakamura 1934; Krug and Bacchi 1943) and consequently no male meiosis ensues. In the remaining clones, meiosis is completed, but proceeds abnormally due to triploidy. Thus the *ms* gene suppresses male meiosis completely in a few clones. In others, meiosis proceeds abnormally. Female fertility of these clones is not known.

Table 47.1. Microsporogenesis breakdown in male sterile *Citrus*

Species	References	Breakdown stage
Citrus aurantifolia	Uphoff (1931), Nakamura (1934), Krug and Bacchi (1943)	Before meiotic initiation or after microspores formation
C.medica	Oppenheim and Frankel (1929)	After microspore liberation
C.reticulata	This Vol.	Early prophase I, or after microspore liberation
[a]*C.sinensis*	Osawa (1912), Nakamura (1934), Oppenheim and Frankel (1929)	Before meiotic initiation or after microspore liberation
C.unshiu	Osawa (1912)	Before meiotic initiation or microspore degeneration
	Nagai and Tanikawa (1928)	Before or after pollen mitosis
C.yatsushiro	Nakamura (1943)	Before or after pollen mitosis
Species hybrids		
[a]*Citrus limon* × *C.paradisi*	Frost (1948)	Archesporial degeneration
[a]*C.limon* × *C.sinensis*	Frost (1948)	Archesporial degeneration
[a]C.unshiu × *C.sinensis*	Iwamasa (1966)	Anthers not developed or early PMC degeneration
C.unshiu × *C.grandis*	Iwamasa (1966)	Anthers not developed or early PMC degeneration

[a]Tapetal breakdown follows sporogenous tissue degeneration.

47.1.2 *C. medica* L. (citron)

In many clones meiosis was nearly normal until microspores were liberated (Oppenheim and Frankel 1929), but the microspores degenerated immediately after release from PMC's.

47.1.3 *C. reticulata* Blanco (mandarin, satsuma orange)

In a clone, the first-formed flowers had rudimentary anthers in which no sporogenous tissue differentiated (this Vol.). In some of the later-formed flowers, stamens were developed in which PMC's degenerated in early prophase-I. In a few last-formed flowers, microsporogenesis proceeded irregularly until microspore liberation, but the free microspores degenerated rapidly. Probably the action of *ms* gene(s) is delayed by some environmental factors like temperature and/or photoperiod or by some intrinsic physiological factors. Nevertheless, in no case was viable pollen formed and the mst was complete. The genetic control of this sterility is not known.

47.1.4 *C. sinensis* Osbeck (common sweet orange)

In the Washington navel orange (Osawa 1912) and many other sweet orange varieties (Nakamura 1934), stamens are developed, but microsporogenous tissue degenerates before meiotic initiation. Consequently, no pollen is produced in them. In these either the embryo sac mother cell or embryo sac also degenerates prematurely, so this is not a case of true mst. In some clones, meiosis proceeded normally up to microspore liberation (Oppenheim and Frankel 1929), but microspores degenerated soon thereafter.

47.1.5 *C. unshiu* Marco

In a clone of satsuma mandarin, anthers were rudimentary and either the microsporangia were not formed at all or were formed in a very few anthers (Osawa 1912). In these anthers some PMC's were developed, but most of these degenerated before meiotic initiation. In the remaining few, meiosis continued normally until microspore separation. Thereafter the majority of these aborted; the remaining ones developed into pollen grainlike structures, but were functionless and did not germinate in vivo or in vitro (Osawa 1912; Nagai and Tanikawa 1928). Many of these degenerated either before or after pollen mitosis. Since Osawa (1912) found degeneration of embryo sac mother cell or of embryo sac also, this clone is not a typical mst clone.

47.1.6 *C. yatsushiro* Hort. ex, Tanaka

In some clones of yatsushiro-mikan, meiosis proceeded normally, but pollen grains aborted either before or after first pollen mitosis. Some pollen aborted just prior to anthesis and none of the pollen formed was viable or functional (Nakamura 1943).

47.1.7 Intraspecific Hybrids

The mst of a typical kind was obtained in the progeny of crosses by using the satsuma strains of oranges that included *C. unshiu* Mareovitch (satsuma mandarin) as the female partner in a diverse crossing program (Iwamasa 1966). Although mst expression in hybrid plants was highly influenced by the environment, the sterility was of two types, viz. either there was complete absence of anthers or these were underdeveloped and lodged no PMC's. On the other hand, when *C. unshiu* satsuma was crossed with sweet orange *C. sinensis* varieties Harado shaddock, Iyo and Hassaku, nearly one half of the hybrid progeny was mst, either due to complete absence of stamens or due to underdeveloped anthers, and the other half was normal mft. The female fertility of the mst's was normal. Similarly, when satsuma *C. unshiu* mandarin was crossed with trifoliate orange varieties *C. sinensis* PMC degeneration occurred in nearly half of the hybrid plants. This was followed by tapetal degeneration. The early archesporial degeneration resembles the one known in Washington navel orange *C. sinensis* and Jahita lime, *C. aurantifolia*. Since in all the crosses mst's segregated when satsuma was used as the female parent, satsuma appears to have S-cytoplasm and the male parents have the recessive *fr* genes. Thus the sterility is due to g-c interaction, but in the absence of details on substitution crosses, it cannot be confirmed. All the satsuma strains used by Iwamasa (1966) have originated through somatic mutation from the same basic stock and seem to have retained S-cytoplasm consistently. Complete archesporial degeneration occurred in the hybrids obtained by crossing lemon, *C. limon*, to *C. sinensis* (Frost 1948). While these hybrid plants were mst, hybrids of the reciprocal cross were fully mft. Genetic control of the sterility is not precisely known, although the sterility is of gc-mst type.

CHAPTER 48

Scrophulariaceae

48.1 *Antirrhinum majus*, L. (snapdragon)

48.1. Genic Male Sterility

Four mst mutants were detected by Baur (1924); mst in each of them is controlled by single recessive *ms* genes, the four genes being nonallelic. Two more such mutants, one with and the other without petals, were detected by Goldsmith (1968). Male meiosis and female fertility in these mutants is not known. In one spontaneously arisen and in one other 5-kR γ-ray-induced mst mutant, two nonallelic recessive genes, ms_1 and ms_2, condition the sterility (this Vol.). Whereas in ms_1 tetrads degenerate after formation, in ms_2 microspores are liberated but they abort gradually (Table 48.1), so that no viable pollen is produced. The tapetum in both of them develops normally, but it degenerates almost simultaneously with tetrad

Table 48.1. The ms gene action in Scrophulariaceae

Species	References	Gene action
[a]*Antirrhinum majus*	Baur (1924),	nk
	Goldsmith (1968),	nk
	This Vol.	During tetrad formation, after microspore liberation
	Schick and Stubbe (1932)	nk
Hebe parviflora	Frankel (1940)	At pachytene
H.salicifolia var. *communis*	Frankel (1940)	At tetrad stage
H.subalpina	Frankel (1940)	At pachytene
H.townsonii	Frankel (1940)	During pollen development, after mature pollen formation
H.traversii		
H.leiophylla × *H.salicifolia*	Frankel (1940)	During pollen development, after mature pollen formation
H.parviflora × *H.leiophylla*	This Vol.	During pollen development
Veronica alba	This Vol.	nk
V.alpina	This Vol.	nk
V.candida	This Vol.	nk
V.dubia	This Vol.	nk
V.elegans	This Vol.	nk
V.rosea	This Vol.	nk
V.rubra	This Vol.	nk
V.variabilis	This Vol.	nk

[a]Monogenic recessive mutants.

(ms_1) and microspore (ms_2) abortion. Thus the ms_1 and ms_2 genes act upon the tapetum and sporogenous tissue almost simultaneously. Their female fertility is normal. Schick and Stubbe (1932) obtained an induced mst mutant in which the sterility was conditioned by one recessive *ms* gene, *defgli*, the gene forming a part of multi-allelic series. The female fertility of this mutant is normal, unlike in its mother mutant parent *pet*, in which petaloidy of stamens and carpels is most frequent.

48.1.2 Chemical Male Sterility

A single spray of aqueous 0.5% mendok solution given when first flower buds appear induced mst, probably without impairing the female fertility in *A. majus nanum* (Kho and De Bruyn 1962). The sterility lasted for 1 week only. Higher chemical doses caused leaf burning and leaf-top death (Table 48.2). The closely related chemical dalapon did not induce mst at similar and other concentrations. With 0.45% mendok solution, p-mst occurred in the treated plants, but ovule abortion was equally high (Sink and Walker 1964). Thus no gametic sterility occurred by this chemical treatment. Furthermore, the effect of the chemical was localised to the central florets of the spike only. Since a lower concentration (0.45%) of mendok induced p-mst accompanied by high ovular abortion (Sink and Walker 1964) and a higher concentration (0.5%) caused complete mst (Kho and De Bruyn 1962), a higher ovular abortion is expected to accompany the latter treatment. Precise ovular or seed fertility of the treated plants is not known, although Kho and De Bruyn (1962) claimed normal female fertility after 0.5% treatment. As different genotypes react differently to gametocides, it may be that the ovules of the genotype used by Kho and de Bryun (1962) tolerated higher chemical doses than those of the one used by Sink and Walker (1964). If so, it is worthwhile to try mendok on different horticultural cultivars of snapdragons and find the specific male gametocidal dosage in each of them for subsequent use in breeding.

Table 48.2. Chemical male sterility in *Antirrhinum*

Species/Chemical	References	Male sterility induction	Female fertility reduction	Side effects
Antirrhinum majus				
Mendok	Kho and De Bruyn (1962)	Complete	Nil	Highly phytotoxic
	Sink and Walker (1964)	Partial	Very high	nk
Veronica alba				
V.candida				
V.rubra				
Mendok	This Vol.	Partial	High	Phytotoxic

48.2 *Hebe* Comm.

In this genus of nearly 300 species comprising showy, free-flowering attractive shrubs or small trees, Frankel (1940) documented mst in *H. leiophylla, H. parviflora, H. salicifolia, H. subalpina, H. townsonii* and *H. traversii.* No genetic study was done, as most of the species reach flowering are the earliest in the third year or later, but the occurrence of a small isolated mst population led Frankel (1940) to conclude that the sterility was controlled by recessive *ms* genes. The *ms* genes act at different stages in the various species (Table 48.1). For instance, in *H. parviflora* and *H. subalpina*, PMC degeneration occurs at the pachytene. The bivalents coalesce into an amorphous mass, which disintegrates actively followed by tapetal degeneration. The degeneration is extremely rapid, like that in 38B and 69 mst mutants of pea (Gottschalk and Kaul 1974). In *H. salicifolia* var. *communis,* degeneration is delayed until the end of the second meiotic division. Tedrads are formed but the PMC's collapse rapidly. Still later, degeneration occurs in a natural hybrid between *H. leiophylla* and *H. salicifolia*. In this, pollen grains are formed but they shrivel and collapse rapidly and regularly. In *H. traversii,* both the meiotic divisions proceed normally, but immediately after normal pollen grain production both tapetum and pollen grains degenerate quickly and the anthers become shrivelled. In *H. townsonii,* due to disturbance in the regular timing of meiosis, a chimerical arrangement of division stages occurs all through the anther locule. However, meiosis proceeds normally and pollen grains are formed, but the anthers do not dehisce and the pollen grains degenerate at floral anthesis. Thus the mst is of functional type. In the hybrids of the cross *H. parviflora* and *H. leiophylla,* male meiosis proceeds normally until pollen development initiates, after which pollen degeneration sets in immediately (this Vol.).

48.3 *Veronica* L.

A few plants having nonviable pollen occur in some populations of *V. alba, V. alpina, V. candida, V. dubia, V. elegans, V. rosea, V. rubra* and *V. variabilis* (this Vol.). In all these, anthers are shrunken and indehiscent. The genetic control of the sterility is not known. Mendok sprays of 0.25% induce partial pollen and ovular sterility in *V. alba, V. candida* and *V. rubra*. (Table 48.1).

CHAPTER 49

Solanaceae

In this family of about 80 genera, abounding in the tropics but well represented also in temperate regions, nearly 3000 species occur in every vegetated continent of the world (D'arcy 1972). Of these, about 60 genera belong to S. America, among which mst is known in *Capsicum, Datura, Lycopersicon, Nicotiana, Petunia* and *Solanum* (Tables 49.15 to 49.17).

49.1 *Capsicum annuum* L. (chillies, red or sweet pepper)

Of the 50 species comprising this genus, both spontaneous and induced mst conditioned by single recessive genes is known in *C. annuum*. In a few, mst occurs following interspecific hybridizations, induced mutations and chemical treatments (Tables 49.15 to 49.17). In addition to *C. annuum* var. *floralgem*, mst is documented in *C. frutescens* var. *tobacco* and *C. microcarpum* (Raghuvanshi and Singh 1984). The nature, genetic control and female fertility of these msts are not known.

49.1.1 Genic Male Sterility

49.1.1.1 Spontaneous

The first report about spontaneously arisen mst is by Martin and Crawford (1951) in *C. frutescens* var. *Clemson cayenne* No. 69 A and 4558, identified later as *C. annuum* by Heiser and Pickersgill (1969). These steriles are of three types; sterility in each is controlled by one recessive gene. The first type includes both male and female steriles. The second and third type plants are mst and female fertile, but while in one type mst is manifested only under field conditions, in the other type it occurs only under greenhouse conditions. Reciprocal transplantations revoke fertility in them. For instance, when plant types that are mst in the greenhouse are transferred to the field, they become mft, but when these plants are transferred back to the greenhouse, they become mst again. A reverse behaviour is exhibited by the other mst-type plants. Allelism tests of *ms* genes of the two types of plant have not been done, but they appear nonallelic because of their contrasting phenotypes under one environment. In both of them action of *ms* genes is influenced by temperature, as the greenhouse in which they were grown had a 10°–14°C higher temperature than the field; also the temperature difference between day and night was higher in the field than in the greenhouse. Therefore these two mst plant types may be temperature-sensitive mst mutants. This needs confirmation.

From among greenhouse plants of the Bell pepper variety All Big, one mst plant arose spontaneously (Shifriss and Frankel 1969). At maturity its anthers were shrunken and devoid of pollen grains. Male meiosis proceeded normally up to tetrad formation, after which microspores aborted. The mst expression was complete and parthenocarpic fruits were produced by the plants. The sterility is controlled by a recessive gene, *ms*-1. Two more similar stable mst mutants, one each in pepper varieties California Wonder and Gambo were detected by Shifriss and Rylski (1972) and Shifriss (1973). Sterility in each of them is controlled by one recessive gene. The gene in California Wonder is designated ms_2 by Shifriss and Rylski (1972). Another gene, also designated ms_2, has been documented by Daskaloff (1971) from his mst material. Whether the two ms_2 genes are allelic or not is not known. In the mst mutant Gambo, microspores abort immediately after liberation so that mature anthers are shrunken and deformed (Fig. 49.1). In some Japanese sweet pepper varieties, spontaneously arisen mst's were detected by Hirose (1962) and Hirose and Takashima (1963). Their anthers either do not contain any pollen or have a very small amount of abortive pollen (Hirose 1965). Cytological anomalies occur before meiotic initiation. On the other hand, in a pungent Japanese pepper variety, male meiosis proceeds normally up to tetrad formation, immediately after which degeneration occurs and no pollen grains are formed (Hirose and Fujime 1975). The sterility is controlled by a pair of recessive *ms* genes. Genetic analysis of this material indicates a monogenic recessive control of mst (this Vol.). In another g-mst line found by Hirose and Fujime (1981), male meiosis proceeded normally until tetrad formation. The tetrads enlarge abnormally and abort gradually. No microspores are released by this mutant.

From India, Meshram and Narkhede (1982) detected a spontaneously arisen mst mutant with luxuriant vegetative growth and broad dark green leaves.

Fig. 49.1a,b. *Capsicum annuum.* Flowers of male fertile **(a)** and genic male sterile **(b)** squash pepper. (Shrifriss 1973)

Microsporogenesis is normal until pollen mitosis, after which pollen aborts fully. The anomaly is conditioned by one recessive *ms* gene. The female fertility of this mutant is normal. In the mst plants detected by Murty and Lakshmi (1979), *ms* gene action is too early, so that meiosis is normal up to the pachytene, after which chromatin degenerates very rapidly and consequently no pollen is ever produced by this mutant. The plants have an altered leaf shape and sterility appears to be monogenic and recessively inherited. In these plants ovular fertility is considerably impaired. Therefore, this cannot be considered as an mst mutant. Desynapsis is the cause of sterility in this material.

49.1.1.2 Induced

The majority of such mst mutants were obtained by Daskaloff after using physical mutagens like X- and γ-rays. From the M_2 progeny obtained from X-irradiated pepper seeds, a mst mutant with anthers reduced from half to two thirds of the normal size and devoid of pollen grains was isolated by Daskaloff (1968). The sterility, stable under field and greenhouse conditions, and also after grafting them over mft stocks, is conditioned by a recessive gene, ms_3. From X- and γ-irradiated progenies, five more mst mutants were detected by Daskaloff (1971, 1971–1972, 1973a,b, 1974). Sterility in each of them is conditioned by a recessive gene, all these genes being nonallelic to each other and designated as ms_4, ms_5, ms_6, ms_7, and ms_8. While the ms_4 mutant has complete mst, as its anthers are devoid of pollen grains, in ms_5 a few fertile along with sterile pollen grains occur in plants grown in the greenhouse but not in the field. The functionability of the fertile-looking pollen is not known. Meiosis up to tetrad formation in mutants ms_6, ms_7 and ms_8 proceeds normally (Daskaloff 1974), but degeneration invariably occurs between microspore formation and maturation.

Pochard (1970a,b) treated the dormant seeds of *C. annuum* having 8% monoploid embryos with X-rays and EMS. From the progeny, three mst monoploids (mr_9, mc 509, mc 705) were isolated and treated with colchicine, that produced diploid mst's. Crossing these with diploid mft's indicates mst in each of the three lines as controlled by one recessive gene. Since each arose from a different mutagenic treatment and the mutants differed from each other in many morphological features, three different *ms* genes seem to have mutated (Pochard 1970a,b). No genetic or allelism tests were, however, conducted. Certain specific differences in the meiotic behaviour of the three mutants exist, while microsporogenesis in *ms* 509 is arrested at the tetrad stage; in mutants mr_9 and mc 705, the breakdown takes place at any stage from prophase-I to microspore formation. Therefore although the *ms* genes differ from each other in their time of action over microsporogenesis, allelism tests are necessary to establish their genetic identity. γ-Irradiation of vegetative monoploid buds and later diploidizing them by colchicine treatment produced some mst plants (Pochard 1970a). The sterility is conditioned by one recessive *ms* gene. Large-sized parthenocarpic fruits were developed on such mst plants.

49.1.2 Gene-Cytoplasmic Male Sterility

49.1.2.1 Intra-Varietal Hybrids

In a planting of red pepper, Peterson (1958) found a mst plant whose sterility was due to the interaction of a recessive *fr* gene with S-cytoplasm, so that mst peppers had the usual $S_{fr\ fr}$ combination. However, on the basis of changed segregation ratios in BC generation obtained from the cross of the pollen sterile line with the cultivar Flora Gemm, Peterson (1958) indicated the possibility of a second dominant restorer gene. In the mst's, many meiotic anomalies occur during PI to TII. However, despite these anomalies, meiosis is not arrested in between and microspores are differentiated; but thereafter degeneration sets in and the young developing microspores abort mostly before liberation. According to Hirose (1965), Pochard (1970b), Novâk et al. (1971), Horner and Rogers (1974), abnormal development in this mst material occurs in premeiotic stages but the final breakdown stage is postmeiotic. Thus *ms* gene action is premeiotic. While female fertility is not impaired by environment, the mst expression is highly influenced by modifier genes and temperature variations. The sterility is accentuated at high temperature. Seed set of 50% in mst's growing in proximity to mft's is attributed to inadequate insect pollination (Peterson 1958) rather than to female sterility. Nearly similar mst's were found by Shifriss and Frankel (1971) from progeny of some intraspecific crosses of *C. annuum*.

Digenic control of mst occurs in an American *C. annuum* (Novâk et al. 1971). The two recessive *fr* genes are complementary and induce mst in the presence of S-cytoplasm. Dominant alleles of both these loci are necessary to restore pollen fertility. Meiosis proceeds normally in most of the PMC's of mst's, but microspores abort soon after their formation. Tapetal disintegration is premature and occurs early during premeiosis (Novâk 1971). Gradually the tapetal cells separate out from the parietal layer and abort in conjunction with the microspores.

49.1.2.2 Interspecific Hybrids

Rusenova-Kondareva (1968) obtained mst hybrids after crossing *C. annuum* × *C. pendulum* (*C. baccatum* var. *pendulum*) and *C. annuum* × *C. pubescens*. The mst's are either stamenless or develop rudimentary stamens that have no pollen grains or sporogenous tissue. Their female fertility is nearly normal and reciprocal crossings produce normal mft hybrids. Thus their mst is due to the interaction of *C. pendulum* and *C. pubescens fr* genes with S-cytoplasm of *C. annuum*. A similar type of mst occurs in F_4 generation plants of the cross *C. peruvianum* × *C. annuum*. The mst plants maintain their sterility in BC's to *C. annuum* pollen. Thus the sterility is caused by the interaction of *C. annuum* genes and *C. peruvianum* cytoplasm. Likewise mst's segregate from progenies of the cross *C. baccatum* × *C. annuum*; the mst's are antherless or develop deformed minute anthers (Fig. 49.2). After three BC's by *C. annuum* pollen, the progeny is entirely antherless.

Fig. 49.2a-d. *Capsicum annuum.* **a-d** Flowers of male fertile **(a)** and of gene-cytoplasmic male sterile **(b-d)** pepper having deformed indehiscent anthers **(b,c)** or no anthers at all **(d)**. (Shifriss 1984, personal communication)

49.1.2.3 Microsporogenesis

A variable course of microsporogenesis occurs in gc-mst peppers. In Peterson's mst material, male meiosis proceeds normally up to the tetrad stage. However, the tetrads do not separate from each other, and wherever they separate they are not released from the meiocytes as their callose wall is not dissolved by the degrading enzymes (Horner and Rogers 1974). The tetrads are compressed by the enlarged tapetal cells and crushed to death. On the other hand, in the mst material studied by Novâk and Betlach (1970), some microspores are released and at anthesis about 30 deformed microspores occur in the mst anther sacs compared to about 500 pollen grains in mft anthers. In the mst plants studied by Pochard (1970b), development of the microsporocytes stops completely at the tetrad stage in some plants, in others it is interrupted at various stages during meiosis. In most of the plants, anomalies appear during premeiosis immediately after mitosis, but meiosis continues abnormally. The mst flowers exhibit marked alterations in tapetal and staminal structure. Male meiosis until tetrad formation proceeded normally in the mst material investigated by Hirose and Fujime (1975), but the meiocytes degenerated gradually and no microspores are released from the tetrads.

49.1.2.4 Tapetal Degeneration

In gc-mst materials studied by Novâk (1971) and Horner and Rogers (1974), tapetal abnormalities appear early when PMC's are about to undergo meiosis. No karyokinesis occurs in the tapetal cells, but they become vacuolated and enlarged. In Peterson's mst plants, sporogenous cells are surrounded by callose and undergo meiosis forming microspore tetrads that do not develop any further but secrete primexine, the layer where the pollen wall is initiated. These microspores are compressed to form a central mass and become crushed by the enlarging tapetum. (Horner and Rogers 1974). Thus in these two materials, even though malfunctioning of the tapetum appears prior to meiotic initiation, male meiosis proceeds normally until microspore formation. The tapetal cell wall and callose wall around the tetrads are not dissolved due to lack of the degrading enzymes.

49.1.3 Temperature Influence

The environmental influence over the degree of pollen sterility is appreciable in some mst pepper genotypes. One of the main effective factors is the temperature. This is evident in mst material studied by Martin and Crawford (1951) and Daskaloff (1972, 1973b). In the investigations by Martin and Crawford (1951), while one genotype is mft in the greenhouse and is mst in the field, the reverse is true for the other mst genotype. The mst mutant ms_5 is completely mst in the field but partially fertile in the greenhouse. Compared to the field, in the greenhouse the temperature is higher and its diurnal variation is lower. Although some variations in anther morphology, microspore abortion and tapetal anomalies occur in mst plants detected by Peterson (1958), temperature alterations induce a variable degree of mst which ranges from a complete to a partial level, so that selfing as well as crossing of the mst's is possible through temperature manipulations. Even in mft plants, temperature influence over pollen fertility is high. High temperature induces pollen abortion in normal anthers (Hirose 1965), the magnitude of abortion is correlated with the temperature prevailing during 2–3 weeks of pre-anthesis stage. Shifriss and Guri (1977) obtained 50–80% pollen abortion in the hottest month of summer. Thus stabilization of mst may be possible provided higher temperatures prevail during pre-anthesis period.

49.1.4 Graft Transmission

The mst is not transmitted across grafts (Daskaloff 1968). Ohta (1961b) found no alterations in the fertility of either stock or scion in the graft generation that involved grafting of maintainer scions on gc-mst stocks. Ohta (1961b, 1967, 1968) considers mst in peppers to be nonviral in nature but virus-influenced.

49.1.5 Biochemical Differences

Compared to mft anthers, mst anthers have a low amount of proline but a high content of asparagine (this Vol.). In ms_5, ms_8 and gc-mst line, a lower electrophoretic pattern in the spectra of the basic proteins occurs (Markova and Daskaloff 1974, 1976). This indicates disturbed protein metabolism in mst anthers. Activity of the isoenzymes of G1–6PDH, an enzyme of the pentose pathway, is highly reduced in mst anthers, indicating their reduced synthetic capacity and energy productivity. The retarded ATP synthesis could result in the hampering of some reactions in which this macroenergic compound participates, particularly in protein biosynthesis. The polyphenoloxidase spectra of the gc-mst anthers contain less components than that in fertile plants (Daskaloff 1974). This indicates that protein synthesis in mst anthers is blocked, leading to pollen degeneration.

49.1.6 Chemical Male Sterility

i. *Gibberellic Acid.* Two exogenous applications of 10 μl of 10^{-3} M GA_3 at 5 days' interval to young pepper plants before floral initiation leads to anomalous pollen development, carpeloidy and unrolling of petals (Sawhney 1981). In some, staminal development is inhibited and supernumerary carpel-like organs are produced. The carpelloid stamens have a basal ovary part and a distal anther region. The latter has a variable anther portion differentiated variously. In well-differentiated regions, microsporogenesis occurs but microspores abort after liberation from PMC's (Fig. 49.3b). In the ovary part of the carpelloid stamens, ovules differentiate either on the outside or on the inside and some develop megaspore mother cells (Fig. 49.3a). After megasporogenesis, apparently embryo sac formation occurs. Seeds are, however, never developed in such ovules. In a few treated plants, carpels develop instead of anthers. Their ovules undergo normal megasporogenesis. In some flowers, filament elongation is inhibited. In such flowers, stamens mature late but viable pollen is produced. In general, GA treatments affect the development of petals and stamens, but not of sepals and carpels. However, neither are all stamens in a flower affected similarly nor did all GA_3-treated plants exhibit similar anomalies. Since fruits formed after GA_3 treatment are highly abnormal and distorted, the chemical cannot be considered as a suitable male gametocide for chillies.

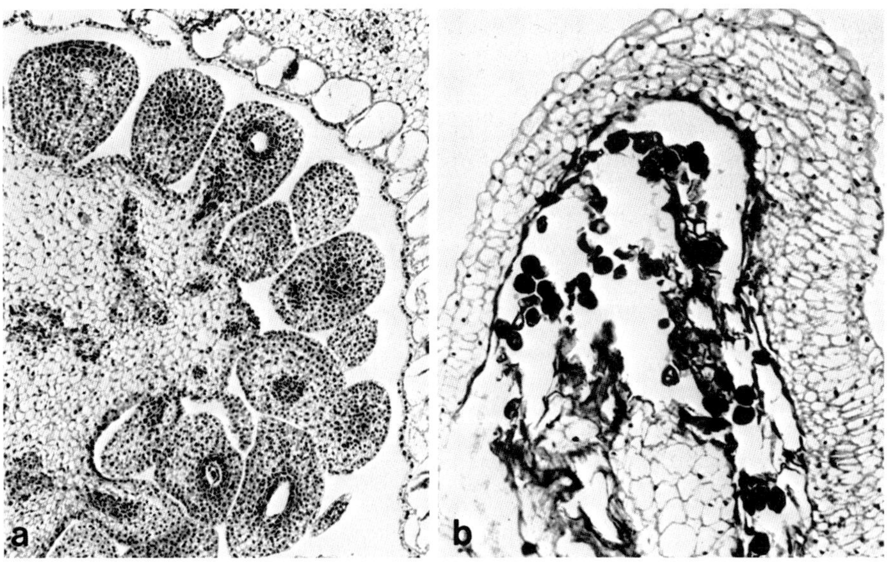

Fig. 49.3a,b. *Capsicum annuum.* **a,b** Carpelloid stamens having megaspore mother cells **(a)** and degenerated microspores **(b)**. (Sawhney 1981)

ii. *Maleic Hydrazide.* Complete mst is induced by three sprays, each of 0.4 and 0.5% MH (Chauhan and Singh 1972). The treatment causes a high degree of ovular sterility. In pollen sterile plants, tapetal cells become hypertrophied and crush the developing pollen grains (Chauhan 1976).

iii. *Other Chemicals.* Pollen sterility was induced by spraying *C. annuum* plants with sodium 2, 3-dichloro-isobutyrate (Hirose 1965; Markus 1968), and with 2, 2 dichloropropionate (Hirose 1969). These chemical treatments cause high ovular damage and decrease female fertility drastically. Similar effects are caused by dalapon (Chauhan 1980).

49.1.7 Breeding Use

In sweet peppers, Peterson (1958), Rusenova-Kondareva (1968), Daskaloff (1968, 1971, 1973), Gikalo and Studentsova (1968), Pochard (1970c), Shifriss and Frankel (1969, 1971) and Shifriss and Rylsky (1972) have recommended the use of mst for heterosis breeding. For this purpose, g-mst's have an edge over gc-mst's because of relatively better stability to various environmental conditions and a simpler inheritance pattern. During the blooming season, visitations by honeybees, bumblebees and other insects are frequent in chillies. Consequently, a high proportion of cross pollination is expected if mst and mft plants are interplanted in a field. Pollination increase of 50% over normal plants was obtained by Daskaloff (1971–1972) when mst and mft plants were interplanted in alternate rows and natural pollination by wind and insects was allowed. This gave 50% increase in fruit set per unit area of land. High positive heterosis in fruit yield was recorded by Martin and Crawford (1951) and led Duvick (1959) to emphasise the practical utility of mst plants for hybrid seed and fruit production in sweet pepper. Shifriss and Guri (1977) proposed a mixed planting design in which each mst is surrounded by fertile plants in order to obtain optimal hybrid seeds over the mst's. This they consider as superior to the planting of separate rows of sterile and fertile plants in F_1 seed production fields. Through this method these authors have claimed to have obtained enhanced fruit yield and a higher percentage of marketable fruits in Bell pepper hybrids (Shifriss and Rylski 1973).

Due to the lack of stability of gc-mst peppers, they have not been commercially utilised for hybrid fruit and seed production, unlike the genic ones. Ohta (1961a) recommended the use of Peterson's mst's for hybrid production, but since the sterility expression is considerably influenced by environmental conditions, this may not be directly used for commercial hybrid production. Although heterosis for yield has been documented in both hot and sweet pepper cultivars (Deshpande 1933; Martin 1949; Martin and Crawford 1951; Pochard 1966; Gill et al. 1973; Shifriss and Rylski 1973; Lippert 1975; Rocechetta et al. 1976; Shifriss and Sacks 1980), commercial exploitation on a large scale is much limited (Breuils and Pochard 1975) due to inefficient systems of mst-mft restorations (Peterson 1958; Daskaloff 1976; Shifriss and Guri 1979). Several gc-mst's that were partially fertile and exhibited different levels of mst have been developed by Shifriss and Guri (1979). Out of these, the one of Bikura is useful for hybrid production. Certain crosses between gc-mst sweet pepper and 14 hot pepper lines yield positive heterosis for total and red pepper yields.

49.2 *Datura stramonium* L. (thorn apple, jimson weed)

Avery et al. (1966) detected some spontaneously arisen pollen sterile plants in this weed. Kaul (this Vol.) found three mst plants in a stand. In them, male meiosis proceeded normally until microspore release, after which both the microspores and tapetum degenerated rapidly (Table 49.15). The sterility is controlled by one recessive *ms* gene. Female fertility as revealed from seed set after hand pollination, is normal.

49.3 *Lycopersicon esculentum* Mill (tomato)

This South American genus has seven established species, of which *L. esculentum*, a top ranking fruit and vegetable, cytogenetically the most explored, has over 55 *ms* genes. These genes cause sporogenous, structural and functional mst (Table 49.15).

49.3.1 Sporogenous Male Sterility

The majority of the *ms* genes act on microsporogenesis in tomato (Table 49.2, Fig. 49.6). The first report about the existence of a spontaneously arisen mst mutant is from Crane (1915). In this mutant the sterility is controlled by a recessive gene (Lewis 1942). This report was followed by the isolation of a radium-induced monogenic recessive mst mutant by Lindström (1933). Further investigations reveal the tomato genome to have a large number of *ms* genes (Rick 1944, 1945, 1948, 1966, 1980a,b; Roever 1948; Barton et al. 1955; Rick and Butler 1956; Lesley and Lesley 1958; Cleyberg et al. 1960, 1966, 1967, 1970, 1973; Rick and Boynton 1967; Yamakawa 1968; Andrasfalvy 1969, 1974; Philouze 1974a; Zamir et al. 1980, 1981; Kaul this Vol.), some of which have been located on the chromosomes (Table 49.1). All these are recessive and do not influence female fertility. The recessive nature and high rate of occurrence of mst mutants indicate a high

Table 49.1. Location of *ms* genes on tomato chromosomes[a]

Chromosome	ms Genes
1	ms_6, ms_{32}, *ms*
2	ms_2, ms_5, ms_{10}, ms_{15}, ms_{26}, ms_{35}. *ms, ps*
3	ms_9
4	*sl*
6	ms_{16}, ms_{33}, *cl-2*
8	ms_8, ms_{17}, *vms*
10	ms_{31}
11	ms_3, ms_7, ms_{12}, ms_{14}, ms_{42}, *ms, ap*

[a]Modified and extended from Rick (1974).

mutation rate of the *Ms* genes to *ms* in the tomato genome. Penetrance and expressivity of most of these *ms* genes is high and their action is not considerably influenced by the environment. However, in one mutant the temperature influence is high (Rick and Boynton 1967). In many sporogenous mst mutants, the stamens at anthesis are green, poorly developed and indehiscent (Fig. 49.4).

Fig. 49.4a-d. *Lycopersicon esculentum.* **a-d** Male fertile **(a,c)** and male sterile **(b,d)** flowers with calyx and corolla intact **(a,b)** and removed **(c,d)** showing developed **(a,c)** and underdeveloped **(b,d)** anthers. (Philouze 1969)

49.3.1.1 Microsporogenesis

The majority of *ms* genes act with a high degree of precision over a specific microsporogenesis stage and even in those genes which exhibit some degree of variability, the precision is evident (Table 49.3). Most of the *ms* genes act either during the beginning or just after the completion of meiosis (Table 49.2, Fig. 49.6). Details of the gene action are as follows.

Table 49.2. ms gene action in tomato

Degeneration stage	Investigators		
	Rick	Zamir	Kaul
Premeiotic	*ms*, ms_3, ms_{15}	ms_3, ms_{15}, ms_{29}	ms_3, ms_{15}
Meiotic prophase			*ms*
Early prophase	ms_{10}		
Late prophase	ms_5		
Diplotene	ms_{18}		ms_1
Diakinesis			ms_7, ms_{18}, ms_{33}
During tetrad formation	ms_6	ms_6, ms_{31}	ms_6, ms_{31}
Tetrad to early microspore formation	ms_2, ms_4, ms_9, ms_{11}, ms_{16}, ms_{17}	ms_{11}, ms_{34}	ms_2, ms_4, ms_9, ms_{11}
Late microspore development	ms_{13}, ms_{14}	ms_2, ms_{14}, ms_{28}	ms_{14}, ms_{24}
After microspore release	*vms*		ms_{16}, ms_{24}
During pollen development			*ms*
Prophase-I to tetrad stage	ms_8, ms_{12}	ms_5, ms_{10}, ms_{12}, ms_{33}, ms_{35}	ms_8, ms_{12}, ms_{35}
Prophase-I to microspore stage	ms_7	ms_7	—

i. *Premeiotic Breakdown.* The earliest action is exhibited by the genes ms_3, ms_{15} and ms_{29}, which act during archesporial differentiation so that mostly either only degenerated PMC's are formed or, rarely, highly erratic and irregularly shaped circular cells with disintegrated cell contents are developed (Rick 1948; Zamir et al. 1980; this Vol.). No meiosis occurs in these cells. In some of the late-formed flowers of ms_3, a few PMC's are formed. Meiosis proceeds erratically up to MI or even up to TI (Rick and Butler 1956), but these meiocytes degenerate soon thereafter. In all these three mutants, the *ms* genes predominantly act before meiosis. In another mst mutant, an erratic orientation of sporogenous tissue occurs (Rick 1948). This leads to meiotic breakdown and total collapse of the sporogenous tissue in this mutant. In five γ-ray-induced mst mutants, the *ms* genes act during anther differentiation and development, so that malformation, misdifferentiation or anomalous development of microsporophyll, microsporangium, primary sporogenous tissue and tapetum occur (Fig. 49.5). Microsporogenesis is either absent or abnormal (this Vol.).

ii. *Meiotic and Postmeiotic Breakdown.* The majority of the *ms* genes of tomato act either during prophase-I or just after meiosis is completed (Fig. 49.6). Five *ms* genes act during prophase-I and 16 during tetrad formation and pollen development.

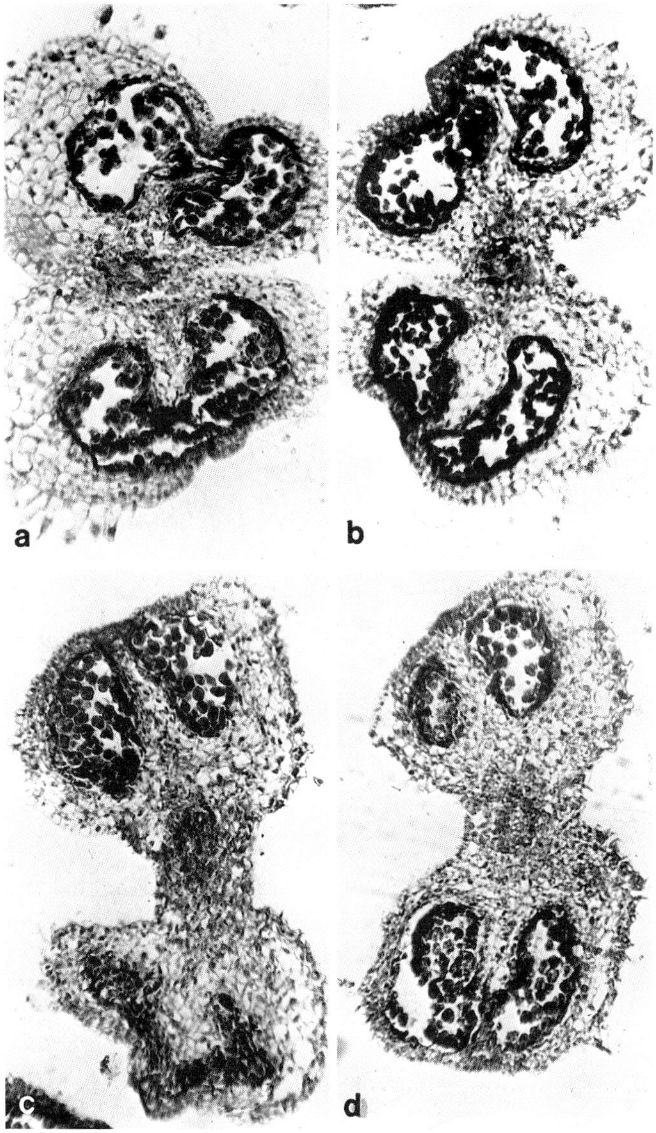

Fig. 49.5a-d. *Lycopersicon esculentum.* **a-d** Microsporangium developmental anomalies in genic male sterile tomato mutants. (This Vol.)

Whereas each of these 21 genes affects a specific stage of microsporogenesis with precision and uniformity, four other *ms* genes cause breakdown at any stage during the meiotic course. Gene ms_7 has the broadest range of action, as it acts at any time during the whole process of microsporogenesis. Thus all five genes are variable in their time of action during microsporogenesis, but all these cause pollen abortion in all the flowers of a mst genotype. The meiotic process prior to *ms* gene action is

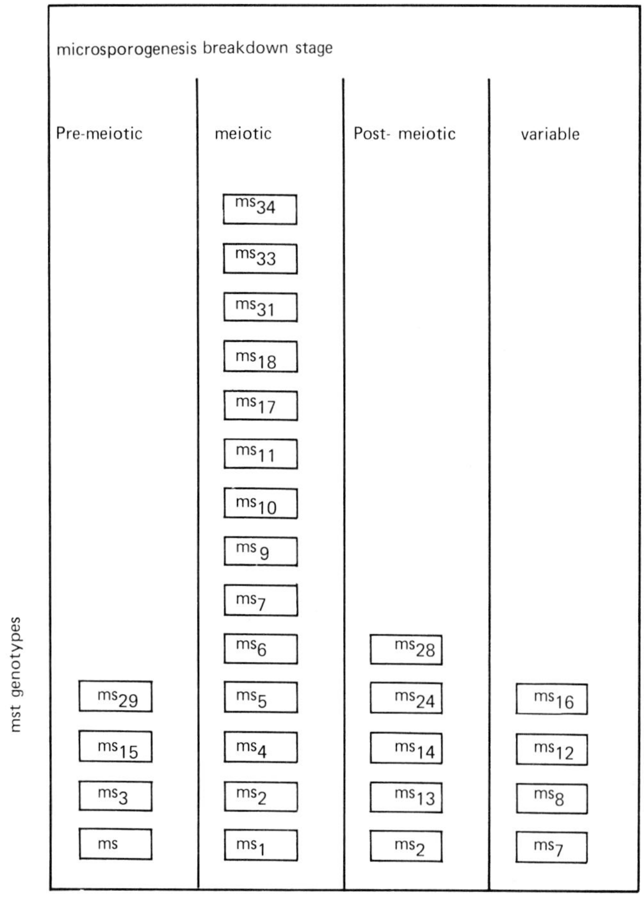

Fig. 49.6. Microsporogenesis breakdown in genic male-sterile tomato mutants

normal in all the mst's. The precision and uniformity of *ms* gene action occurs even in those genotypes where gene action is delayed until microspore release or pollen development. In such cases, aborted pollen grains appear regularly, indicating a perfect action of *ms* genes, unlike in many other plants like cotton, sorghum, sunflower, soybean and pea, in which sterility is partial. This perfection also occurs in the mst tomato mutant obtained after gamma irradiations by Tsikova and Abadzhieva (1974) in which pollen grains degenerate after formation. With regard to the development of apparently normal-looking pollen in certain microsporangia and microsporogenesis interruptions at various stages in others, the mode of action of ms_5 gene is variable and less precise than the other known tomato *ms* genes. In all the mst tomato mutants, however, microspores, wherever formed, exhibit remarkably little variation in size and form. This is in marked contrast to the extremely variably shaped and sized microspores produced by triploid, desynaptic and asynaptic tomato (Gottschalk and Kaul 1980a,b).

49.3.2 Functional Male Sterility

In the John Baer variety of tomato, a mutant having viable pollen but indehiscent anthers was detected by Larson and Paur (1948). It sets seeds only when artificially selfed or crossed, but never when bagged or left for self-pollination. Petals of this mutant are connate or pseudoconnate and lead to considerable constriction of anthers and hold them in very close contact with the pistil. These petals mostly exhibit either greater lateral growth or are sometimes gamopetalous, showing coalescence of the corolla nearly to its extremity. The greater lateral growth of the petals causes an overlapping and curling with the adjacent petals. The narrow opening between anthers and style does not prevent pollen from escaping, since by slicing away the tips of the corolla and anthers (without cutting into the microsporangium) the pollen fails to liberate. The anomaly is controlled by a recessive gene, *ps*. Another gene, *ps*-2, causes anther indehiscence (Tronichkova 1962) but the anthers cover the stigma fully so that no cross pollination occurs.

A plant resembling the John Baer mutant in having cleistogamous flowers and indehiscent anthers appeared in an Indian tomato variety Meeruti (Mittal and Thomas 1961). Its mst is also conditioned by a recessive gene, but unlike the John Baer mutant, it produces fruits without green shoulders. Fruit colour on ripening is uniform. These two traits are desirable for hybrid tomato production. Besides this, the mutant appears to be a better combiner in comparison to other cleistogamous mutants of tomato. In another cleistogamous mutant, flowers fail to reach mature size, anthers remain indehiscent and do not open at anthesis (Rick and Robinson 1951). Female fertility of the mutant was normal, as seeds were set after hand pollination. The anomaly is conditioned by a recessive gene, *cl*-2, which inhibits floral and stomial opening. In another mutant the corolla structure is highly modified, anthers are reduced and the pollen produced is scarce and nonfunctional (Rick and Robinson 1951). The anomaly is likewise controlled by a recessive gene, *ap*. This mutant cannot be considered as a functional mst mutant because over 87% of the embryo sacs abort in it. The pollen produced is scarce and of low viability (16–19%). Therefore, it is a partial sterile and not an mst mutant.

49.3.3 Structural Male Sterility

In two mutants, filaments are normal, but anther sacs are highly deformed so that no archesporial tissue or PMC's develop in them (this Vol.). Their female fertility is normal, as evidenced by the seed set after hand pollination. The anomaly in each case is controlled by a single recessive gene. The allelic nature of the two genes is not known. In the mutant ms_5, the majority of the microsporangia are abnormal (Rick 1948) from the early anther developmental stages sporogenous cells are developed in these microsporangia but the tissue extends abaxially up to the epidermis of the anther sac. In 90% of such microsporangia, PMC's degenerate during prophase I and in the remaining somewhat later. When this mutant was grown here by the present author, it developed reduced stamens having abnormal minute anthers in which no archesporial tissue differentiated. Thus this became a structural mst mutant. Rick (1948) documented its environmental susceptibility under Californian field conditions.

The mst mutant detected by Rick and Robinson (1951) either lacks stamens or they are reduced to staminodes. The anomaly is conditioned by a recessive gene, *pi*. Since the gynoecium of this mutant is also defective and a low fruit set occurs after hand pollination, the mutant is not mst but male-female sterile. On the other hand, in the mst mutant recovered in a F_2 population by Bishop (1954), stamens are vestigeal and adherent to the pistil, but the gynoecium is normal and functional. This near-antherlessness is caused by one recessive gene, *sl*, whose expressivity and penetrance are high. In another such mutant, anther development is vestigeal and no pollen is produced by the plant. The anomaly is controlled by a recessive gene, sl_2 (Hafen and Stevenson 1956).

Two more single-gene recessive mutants lacking stamens completely or having underdeveloped or feminized stamens that have no pollen sacs or have ovule-like structures at their base were reported from the USSR by Solovjeva (1970). Their female fertility is normal. The mst in each of them is controlled by a single recessive gene, but the allelic nature of these two genes is not known. The expressivity of both the *ms* genes is variable because a few flowers among the steriles lack stamens completely, others have these but they are underdeveloped and deformed. In some others, feminized stamens develop containing ovule-like organs at their base. A few flowers develop one to three normal-looking stamens that apparently have pollen grains. A nearly similar mutant having reduced staminodes was found by Simonov (1973) from Bulgaria. The sterility is likewise conditioned by a recessive gene whose expressivity is complete. Its female fertility is normal. One mutant gene, *sha*, causes extreme anther dwarfness and lack of pollen development in tomato. This gene is closely linked to the gene *pat*, which induces parthenocarpic fruit development (Bianchi and Soressi 1969), so that such plants do not develop seeds but produce parthenocarpic fruits. In a mutant studied by Sawhney and Greyson (1973a), stamens are dwarf, pale, laterally free and bear ovules at anther bases. Its female fertility is normal. After irradiating seedlings of the tomato cultivar Pelican with gamma rays, Hernandez and Wollard (1978) obtained six mutants in which the stamens are highly reduced and totally deformed. The sterility in each case is controlled by a single recessive gene whose allelic nature is not known. The female fertility of these mutants is normal.

49.3.4 Digenic Control

In all the above-mentioned mst mutants, the sterility is conditioned by single recessive genes. Lesley and Lesley (1939) found a digenic control of mst in plants obtained from an F_2 generation of a trisomic plant. Whereas in F_2, 15:1 mft:mst segregations occurred, in BC's to mst's, 7:1 segregations were found. This suggests the involvement of a third *ms* gene, and therefore the precise genetic control of mst in this mutant needs further investigation.

49.3.5 Tapetal Breakdown

In mft tomato, tapetal degeneration initiates when PMC's are in the MI-MII stages. The degeneration continues up to anthesis. Rick (1948) studied the tapetal development and degeneration of eight mst mutants (Table 49.3). Compared to mft's, sporogenous and tapetal tissue development is delayed and slower in the majority of these mutants. In many of them, tapetal degeneration is also slow, but the delayed or slow development of microsporogenous tissue does not exhibit any correlation with the breakdown stage of microsporogenous tissue (Table 49.3). Similarly, the tapetal degeneration process and pattern is not correlated with the microsporogenous breakdown stage. Delayed tapetal degeneration does not have any influence on meiotic breakdown because in mft's, tapetal degeneration initiates during the MI-MII stages. Both in ms_{10} and ms_{11}, meiotic breakdown can occur anywhere between prophase-I and the microspore stage (Table 49.2), but whereas in ms_{10}, tapetal degeneration is delayed, in ms_{11}, it is too early (Table 49.3). In ms_{13}, some tapetal cells degenerate earlier to microspore degeneration, and others degenerate later. Like tapetal degeneration, the developmental pattern of the tapetum is not correlated with the microsporogenous tissue breakdown stage. Whereas in ms_8 and ms_{10}, tapetal development is much delayed, meiotic breakdown occurs at prophase-I in ms_{10} but after microspore formation in ms_8 (Table 49.3). Tapetal development is slow in ms_6, ms_7, ms_{12} and ms_{13} but sporogenous tissue degeneration occurs during the prophase to microspore stage in ms_7 and ms_{12}, and at the tetrad to microspore stage in ms_6 and ms_{13}. It appears that the *ms* genes in almost all these mutants affect both sporogenous as well as tapetal tissue individually. In mutant ms_5, *ms* gene affects both anther development and

Table 49.3. Comparison of developmental pattern and degeneration stage of microsporogenous and tapetal tissue of some mst tomato mutants[a]

	Microsporogenous tissue		Tapetal tissue	
Mutant	Development	Breakdown stage	Development	Degeneration
ms_6	Delayed	Tetrads	Slow	Normal
ms_7	Delayed	Prophase-I microspore formation	Slow	Normal
ms_8	Moderately delayed	Microspore	Much delayed	Normal and slow
ms_9	Not delayed	Microspore	Slightly delayed	Normal and slow
ms_{10}	Moderately delayed	Prophase-I	Much delayed	Delayed, slow and incomplete
ms_{11}	Moderately delayed	Tetrad formation to microspore release	Normal or faster	Early and slow
ms_{12}	Moderately delayed	Prophase-tetrad formation	Slow	Delayed
ms_{13}	Not delayed	Microspore	Slow	Early or too late
ms[b]	Normal	Microspore	Normal	Normal
ms[b]	Delayed	Prophase-I	Delayed	Normal

[a]Adopted after modifications from Rick (1948), [b] this Vol.

differentiation. In this mutant, most of the microsporangia are abnormal, sporogenous tissue assumes a dense cytoplasmic nature and comprises a C-shaped mass of two-celled thickness (Rick 1948). An abaxial tapetal layer is not formed and the sporogenous tissue extends to the epidermal layer. However, the tapetum is formed on the adaxial side and at either end of the mass of sporogenous cells on the abaxial side. The tapetal cells develop slowly, are reduced in size and degeneration is late, but the PMC's degenerate early during prophase-I.

49.3.6 Environmental Sensitivity

The majority of the *ms* genes of tomato have complete penetrance and nearly all of them exhibit a remarkable phenotypic stability under a wide range of environmental conditions. One exception to this is ms_5 (Rick 1948), which exhibits a little variable sterility under field conditions, as 3–5% of microsporangia produce some functional pollen that was used by Rick (1948) to develop pure breeding mst populations of ms_5. Another case is a temperature-sensitive spontaneously arisen mst mutant derived from tomato variety San Marzano (Rick and Boynton 1967). In this mutant, mst is conditioned by a single recessive gene, *vms*, located on the long arm of chromosome 8. In the summer in the field, the *vms* mutant develops severely reduced deformed stamens, thicker basally and attenuated distally into a narrow threadlike element subjected to variable twisting and distortions. Prodding the theca with a blunt needle seldom dislodged any pollen. Initially, the anthers appear normal until they reach 2–3 mm length, when both the tapetum and sporogenous tissue collapse prematurely. Following meiosis, the majority of the microspores abort and most of the thecae do not contain stainable pollen. Its flowers fail to unfurl completely at anthesis, the corolla is reduced and whitish in appearance. Except for these traits, the mutant is normal morphologically and physiologically. Despite its little pollen production, it is effectively mst, as all the attempts to artificially self-pollinate it failed. Nevertheless, in seasons other than summer, normal and viable pollen is produced by the mutant both in field and greenhouse.

When the mutant plants showing mft under greenhouse conditions were transferred to the field in the following summer season, they became mst. Such a behaviour was consistently observed with rooted cuttings or seed progenies of these plants. The temperature difference accounts for the differential behaviour of this mutant. A minimal temperature of 30°C in the field and 32°C in the greenhouse is required to induce sterility in this genotype. High temperature application to a single branch of *vms* plants induces mst in the subsequently developed flowers of that branch. The sensitive stage is when flower primordia are nearly 0.075 mm in diameter, at which time the buds are highly mitotic, and precede meiosis in PMC's by nearly 10 days.

Occasionally st_2 mutant plants, which mostly possess vestigeal anthers having no pollen, produce enlarged anthers on dwarfed filaments and some viable functional pollen (this Vol.). The incidence of this instability is low (3–5%). However, when the selfed progeny of this mutant was raised after using the little viable pollen available, the progeny plants had likewise vestigeal anthers. Anther development and pollen production are highly sensitive to temperature in this mutant. At a temperature regime of 23°C day/18°C night, which the mutant receives normally, it produces abnormal stamens and nonviable pollen. At a lower temperature regime like 18°C day/15°C night, the majority of its flowers produce normal stamens and viable pollen (Sawhney 1983). However, at relatively higher temperatures like 28°C day/23°C night, carpel-like structures in place of stamens are produced by this mutant. Two tomato mutants exhibit an opposite response to environmental conditions. One is mst in the field and mft in the glasshouse, the other exhibits a reverse behaviour; the anomaly in both these is controlled by two nonallelic recessive genes (this Vol.). Not only mst mutants exhibit some degree of environmental sensitivity, but mft tomato plants also become partially to completely mst at high temperatures (Abdalla and Verkerk 1968; Levy 1972). Temperatures like 38°/27°C cause considerable reduction in pollen production, anther dehiscence and pollen germination percent (Stevens and Rudich 1978). These factors cause fruitlessness in mft tomatoes at higher temperatures.

49.3.7 Linkage

The majority of the known ms genes, (viz. *ms* 2, 3, 5, 6, 7–10, 12, 14–17, 26, 31–33, 42) do not exhibit linkage with distinct phenotypic markers, but gene ms_{35} is closely linked with gene *a*, causing the absence of anthocyanin. The two genes are 1.7 crossover units apart (Philouze 1974b). Another *ms* gene exhibits linkage with potato-leaf trait and identifies mst plants soon after seedling transplantation (Hernandez and Wollard 1978). However, this linkage occurs only when the determinate parents are crossed with mst plants. Soressi and Salamini (1975) found a close linkage between two recessive mutant genes, *sha* and *pat*, causing development of short anthers (*sha*) and parthenocarpic fruits (*pat*). Crossover between the two genes averages 0.12%. Another marker gene, *Wo* (Wooly), is closer to ms_{35}, the gene order on chromosome-2 being *aa-Wo*-ms_{35}. Durand (1981) considers the *Wo* gene a more useful marker of mst than the *aa* owing to too low recombination between the *Wo* and ms_{35} genes. A close linkage exists between ms_{10} and an enzyme marker *Pr x*-2 (Tanksley et al. 1984). This linkage provides a marker for mst identification.

49.3.8 Anther Culture

Of the available 15 mst mutants, Zamir et al. (1980) successfully cultured the anthers of four mst lines that had ms_{10}^{35} gene in homozygous state. Anthers of their fertile counterparts degenerated under such cultural conditions. Callus cells developing from mst anthers were mixoploids, ranging from 1x-12x ploidy. The cultured cells exhibited chromosomal anomalies and aneuploidy. Using mst plants heterozygous for some morphological and enzymatic markers, Zamir et al. (1981) demonstrated that the regenerants originated from the maternal diploid cells. This was supported by their isoenzyme determination using the enzyme electrophoresis technique. The frequency of callus induction is low (below 10%), for anthers cultured before or after meiotic breakdown, but it is high (37–46%) in anthers cultured shortly after the initiation of meiotic breakdown. This indicates that callus production frequency enhances during meiotic breakdown, in other words, during the period of *ms* gene action. Unlike some *ms* and *mf* genes, the ms_{10}^{35} gene favours anther callus production in tomato.

49.3.9 Graft Transmission

Andersen (1965) made reciprocal grafts between *L. esculentum* VF II, a tomato cultivar known to possess S-cytoplasm and a *Solanum pennellii* accession from Atico, Peru, known to have *fr* genes. These two species served alternatively as root stocks and scions in these grafting experiments, but the mst was not transmitted across grafts as the selfed progeny obtained through the reciprocal grafts was normal.

49.3.10 Male Sterility in Hybrids

i. *Interspecific Crosses*. Some mst plants segregated from the progeny of a *Lycopersicon* hybrid (Nishi and Hiraoka 1958). In these, the tapetum persisted until tetrad degeneration was completed. The mst was considered by the authors as gc-mst. Likewise, mst is produced from the cross *L. esculentum* × *L. hirsutum* (Siegemund 1979b). In F_1 plants meiosis proceeds normally and pairing is normal, so that the 12 bivalents have a chiasma frequency equal to their parents. Thus chromosomal homology is high between these two species. However, microspores degenerate soon after their formation (Table 49.16). Vulkova-Achkova (1980) obtained similar mst plants in progeny of the crosses involving *L. peruvianum* as female and *L. hirsutum* var. *typicum* and *S. pennellii* as male parents (Table 49.4). The mst degree enhanced with the increase of genes of male parents in the cytoplasm of the female parent. This degree was high in BC_2 generation of *L. peruvianum* × *S. pennellii* and was complete in BC_4 in both the substitution crosses.

Table 49.4. Male sterility in some intergeneric and interspecific tomato hybrids

Cross	References
Lycopersicon cheesmanii × *Solanum pennelli*	Andersen (1964)
L.esculentum × *L.hirsutum*	Siegemund (1979a,b)
L.esculentum × *L.parviflorum*	Kesicki (1980)
L.esculentum × *S.pennelli*	Andersen (1963a,b)
L.esculentum var.minor × *S.pennelli*	Andersen (1963a,b)
L.hirsutum f.*glabaratum* × *S.pennelli*	Kesicki (1980)
L.minutum × *S.pennelli*	Kesicki (1980)
(*L.minutum* × *L.hirsutum*) × *S.pennelli*	Andersen (1964)
L.peruvianum × *S.pennelli*	Vulkova-Achkova (1980)
L. peruvianum × *L.hirsutum* f.*typicum*	Vulkova-Achkova (1980)
L.peruvianum var.*glandulosum* × *L.hirsutum* f.*typicum*	Vulkova-Achkova (1980)

Flowers of *L. hirsutum* var. *typicum* × *S. pennellii* hybrids are smaller, with slightly curled petals that open partially. Their stamens become progressively smaller with each BC generation and at BC_4, 65–75% plants have abnormal flowers. Pollen viability ranges between 0.5 to 0.8% only.

In a BC derivative of the hybrid *L. esculentum* × *L. parviflorum*, a functional mst plant was detected by Kesicki (1980). The plant and its progeny exhibits an unusual phenotype. Its cotyledons are wide and have round tips. The plant has wide and obtuse leaflets and split and recurved sterile stamen tips. The traits of functional mst, cotyledon shape and leaves are inherited jointly and are controlled by three closely linked recessive genes, as indicated by F_3-F_4 segregations, although Kesicki (1980) considers it a case of pleiotropy. Due to recurved stamen tips in mst flowers, stigmas are prominently exposed and ease artificial hand pollination and NCP. At anthesis stamens are fresh green and late maturing, viable pollen is produced by the mutant but it is not naturally liberated due to anther indehiscence. Artificial self-pollinations made by puncturing and producing the stamens yield normal seed set, indicating high female fertility in those mutants.

ii. *Intergeneric Crosses*. Nearly 52% pollen abortion occurs in the offspring derived from the cross *L. esculentum* × *S. pennellii* (Andersen 1963a,b). This abortion is coupled with anther size reduction and filament elongation. In BC_1 and BC_2 generations obtained after pollinating the hybrids with *S. pennellii*, pollen abortion increased to 78% and 90%, respectively. Thus as the genes of *S. pennellii* increase, the mst degree enhances proportionately. Segregations of mst plants having different mst grades within a family, as observed by Siegemund (1979a), is due to their possession of different amounts of *S. pennellii* genes. In all mst's, anthers are long, distorted, green foliaceous and corollalike. These traits render their phenotypic identification easy. Pollen abortion is complete in the offsprings of cross *L. cheesmanii* ♀ × *S. pennellii* ♂ and very high in F_1 generation plants of *L. minutum* and *L. hirsutum* crossed by *S. pennellii* pollen (Tables 49.4, 49.16). In the latter cross with further BC's by *S. pennellii*, the pollen abortion enhances considerably, indicating g-c interaction leading to mst (Andersen 1964).

49.3.11 Chemical Control of Male Sterility

49.3.11.1 Sterility Suppression

From a F_2 segregating population, Bishop (1954) detected an mst mutant having vestigeal anthers or filaments adhered to the pistil. The sterility is controlled by a recessive gene, *sl*. Phatak et al. (1966) treated rooted stem cuttings of this mutant with 25 ppm (10^{-4}M) GA_3 solution. After 3–4 weeks of treatment, anthers with viable pollen developed in treated plants and seeds were set on them after selfing, but selfed progeny of this plant was mst and had vestigeal anthers like the original mutant plant. Thus GA_3 suppressed the action of the *ms* gene, but the suppression was temporary and restricted in time because the flower buds already visible at the beginning of the treatment did not produce anthers. Also flowers that appeared 3 weeks after the treatment were antherless, as were the next generation plants. 2-Chloroethyl-trimethyl ammonium chloride (CCA), which normally inhibits fungal gibberellin synthesis, inhibited pollen formation in the gibberellin-induced anthers of the mutant. This indicates gibberellin-induced pollen grain formation in the *sl* mutant and the suppression of *ms* gene action by the gibberellin treatment. This suppression is evidenced by the results of Sawhney and Greyson (1973b), who obtained nearly normal flowers after gibberellic acid treatments given prior to floral initiation to "stamenless mutant-2 (*sl*-2)" that has abnormal stamens which are short, pale, laterally free and bear ovule-like structures at anther bases. Using Bishop's (1954) mst mutant plants, staminal development and pollen fertility restoration was obtained by Kubicki and Potaczek (1972) after treating the seedlings at four to five-leaf stage with 62.5 to 2000 ppm GA_3. Although fertility restoration was nearly complete in the treated plants, the selfed progeny of restored plants is mst. Similar fertility restoration by gibberellin treatments was obtained by Schmidt and Schmidt (1975) in mst tomato mutants having reduced stamens.

In the stamenless mutants ms_{15} and ms_{33}, 500–1000 ppm GA treatment led to the development of normal stamens and the restoration of pollen fertility in 80% of the treated plants (Schmidt and Schmidt 1981a,b). The progeny of the selfed plants is uniformly mst (Schmidt et al. 1979). Applications of 10^{-5} to 10^{-3} M delayed flowering but led to normal stamen development in a tomato mutant characterised by short anthers and parthenocarpic fruits (Mapelli et al. 1979). However, no seeded fruits were developed by this mutant, indicating failure of self-pollination. The induction of staminal development by GA_3 applications appears to be stage-specific and development-dependent. Sawhney and Greyson (1973b) reported that GA_3 applications to buds with stamen primordia at an early stage of initiation induced the formation of normal-looking stamens, but with the progressive increase in stamen size, less and less of the normal stamen features developed following the chemical treatment and ultimately buds with stamen primordia nearly 0.8 mm long failed to respond to GA_3 and produced all the mutant characteristics, i.e. stamenlessness. Sawhney and Greyson (1979) discussed these results in terms of Waddington's (1966) concept of "determination" and "canalisation" of development; but this is restricted to the GA treatment plants only. For instance, ms_{15}, ms_{26} and ms_{33} plants produce naked ovules around the base of rudimentary sterile anthers. Schmidt and Schmidt (1981b) treated these mst plants with GA_3. The treatment restored mft and improved fruit quality, but progeny raised from the seeds of these plants was uniformly mst and had rudimentary sterile anthers. Treating these again with GA_3 (10^{-4}M) restored anther development and mft but the next selfed progeny was uniformly mst (this Vol.).

49.3.11.2 Sterility Induction

i. *2, 4-D, MH, TIBA*. Rehm (1952) was the first to induce pollen sterility that lasted for about 8 weeks after sprays of 50 ppm 2, 4-D on plants a week before first blooming. Female fertility and growth of the treated plants was considerably damaged. Likewise, 100–500 ppm MH induced complete male but very high female sterility (Jain 1956; Chopra et al. 1960). Treatments of 25 and 50 ppm of TIBA (tri-iodobenzoic acid) given a week before the appearance of first flowers induced mst that lasted for about 8 weeks (Rehm 1952). Female fertility was not appreciably impaired. Pollen of mst's exhibited various degrees of vacuolizations, plasma coagulations and many of them were shrunken and empty.

ii. *Mendok*. The complete absence of pollen grain formation for about 12 days occurred following mendok treatments (Moore 1959, 1964). About 30% of the flowers on the treated plants set fruits when pollinated with pollen from the untreated plants compared to over 50% of the flowers on the untreated controls. This indicates reduced female fertility in the treated plants. Thirty seven days after the treatment, pollen production and female fertility were normal. Complete mst occurred following applications of 0.5% Mendok (Walkof 1959). The status of female fertility of the treated plants is not known. Application of 0.075 and 0.15% of this chemical led to microspore degeneration at the uninucleate stage and to some damage to embryosac development (Aung and Byrne 1977).

iii. *Morphactin*. Jain and Mukherjee (1980) induced a high degree of mst by application of 2.5 and 5 ppm chlorflurenol 1 (chlorflurenol methyl ester 74050, CME 74050) a synthetic bioregulator with a high morphogenetic potency. Whereas male meiosis proceeded normally, microspores aborted at low chemical doses (1, 2.5, 5 ppm). At higher doses (7.5 ppm) meiosis was highly disturbed and consequent pollen sterility was high. Higher doses modified the inflorescence pattern also. The chemical effect persisted up to 2–3 weeks. Weekly sprays of 1–2.5 ppm morphactin starting from anthesis cause considerable mst induction, but female fertility and fruit form are considerably impaired. Fruits develop parthenocarpically.

49.3.12 Breeding Value

F_1 hybrid tomato cultivars have exceeded currently grown open pollinated tomatoes especially in seed germination and seedling survival per cent, seedling emergence speed, earliness and uniformity in fruit production and in the combination of ideal plant architecture, yield concentration, disease resistance, net productivity and economic yield for once-over harvest as well as in general fitness to environmental fluctuations and weather vagaries. Tomato being an inbreeder, emasculation and pollen transfer remain the two prerequisites to exploit hybrid vigour for use in commercial hybrid fruit production. Use of g-mst provides an alternative to emasculation. Although a large number of g-mst mutants in tomato are known, no useful S-cytoplasm has been detected in this species so far. Since most of these mutants are fully mst and female fertile, their value in hybrid fruit production is high. Furthermore, they constitute 50% of the BC progeny and therefore a sufficient quantity of fruits and seeds can be recovered. Penetrance of most of the *ms* genes is complete, therefore no danger of contamination by selfing exists. These advantages led Barrons and Lucas (1942) to propose the use of mst in tomato hybrid production. The following aspects, however, demand consideration.

49.3.12.1 Male Sterile Maintenance and Identification

i. *Phenotypic Markers and Linkages*. Use of g-mst tomato mutants for hybrid production requires maintenance by pollen from isogenic plants heterozygous for *ms* gene and removal of heterozygotes from the hybrid production fields. While using heterozygotes for mst maintenance of the female parent, double close spacing of plants in the production fields is advisable. Identification and removal of heterozygotes could be aided by floral morphology where the mst mutant differs in this respect from mft plants, e.g. ms_{26} and *sl*, *sl*-2. Of course, it is more efficient if heterozygotes can be identified and removed in the nursery through the use of a closely linked seedling marker gene or by pleiotropic effects on the seedlings caused by the *ms* gene. Also the presence of recessive seedling marker genes in the mst parent could be used for identification and removal of plants in the F_1 seed bed which are the result of self-pollination of the female parent.

In ms_{35} mutant, the gene *aa* causing the absence of anthocyanin is 1.7 crossover units apart from the *ms* gene (Philouze 1974a) and permits a reliable identification of selfed seeds in a cross-breeding program. Durand (1981) considers wooly gene (*Wo*) a better marker of mst than gene *aa* owing to more favourable recombination percentages. Also another *ms* gene is known to be linked with potato-type leaf trait in tomato (Hernandez and Wollard 1978). This renders the mst plants identifiable at the seedling stage approximately 2 weeks after transplantation. However, this linkage occurs only when determinate parents are crossed with mst plants. Nevertheless, this mutant was stored in the cross of PMS × Pelican and recovered in F_2. Subsequently it was found to combine favourably when bred with some determinate tomato lines.

The association or tagging of ms_{10} with the rare allele of *Prx-z*, an enzyme marker obtained from the wild tomato *L. pimpinellifolium*, provides a method for interline transfer of g-mst as homozygotes can be differentiated from heterozygotes obviating thereby the progeny testing. Also, since *Prx-z* can be assayed nondestructively from 50 mg tissue (one leaflet), selection can be made at the seedling stage (Tanksley et al. 1984), but how practicable and time-saving it is to select by electrophoresis of large populations of mst's from F_2 of BC's for field hybrid seed production is questionable. Nevertheless, the linkage permits an efficient transfer of mst into new inbred lines for hybrid seed production.

ii. *Environmental Sensitive ms Genes*. To be useful in hybrid seed production, mst mutants have to be completely sterile in the production fields. If mst mutants could be maintained by artificial restoration of pollen fertility, identification and removal of heterozygotes in the production field would be eliminated. Such mutants are not easily available in tomato. However, two mst and *vms* mutants (this Vol., Rick and Boynton 1965) which exhibit variability in mst expression under different environmental conditions can be utilized in hybrid breeding and also for obtaining 100% mst population in seed production plots without roguing.

iii. *Functional Male Sterile Mutants*. Although such mutants are available in tomato, they are not useful in hybrid seed production because they either occasionally release some pollen or pollen is released in them through insect visits or by wind joltings or by some other biotic or uncontrollable factors and thus selfing occurs. The self-pollinations make these mutants unreliable for use in hybrid production. Also artificial production of functional mst tomato flowers by treatments with growth substances (Bukovak and Honma 1965) and male gametocides have the same shortcomings as mentioned above or are also associated with reduced female fertility (Moore 1964). However, maintenance of *vms*, *sl* and *sl-2* and functional mst tomato mutants does not pose any such serious problem. For instance, growing the *sl-2* mutant at 18°C day/15°C night temperature regime revokes fertility in this mutant (Sawhney 1983).

49.3.12.2 Natural Cross Pollination (NCP)

Unfortunately NCP in most of the tomato production areas of the world is too low to provide an economical means of accomplishing the collection and transfer of pollen (Lapushner and Frankel 1967; Reeves 1973). Through detection of hybrids in the progeny of a tomato line having a recessive marker leaf gene *c* which was interplanted with a line having the dominant allele for the leaf gene, Richardson and Alvarez (1957) obtained 1.5 to 15% NCP in Mexico, and Rick (1958) 15 to 26% in southern Peru. In California, the NCP varied between 1.9 to 47% (Rick 1949) and the percentage was consistently high in the regions where natural ecology favoured the high activity of solitary bees. Apart from bee activity, NCP rate in tomato is influenced by genotype, flower size, environment, location and planting design (Richardson and Alvarez 1957; Rick 1947a,b, 1948, 1949; Soost and Rick 1957; Soressi and Lorenzoni 1970; Rick et al. 1977, 1978).

49.3.12.3 Pollen Movement

Transfer of pollen from male to female parent remains the most formidable obstacle to large-scale production of hybrid tomato (Rick 1980b). It represents the costliest operation and has not been satisfactorily solved through the usage of many mechanical devices developed from time to time. Comparative labour requirements for hybrid tomato seed production by various procedures involving normal, *ms*, GA induced exserted stigma + artificial pollination, self-incompatibility systems etc. have been given by Lapushner and Frankel (1967) and Rick (1980b). Bees, the natural tomato pollinators, have received most attention in research and it would be useful if flower colour attractive to bees is developed in mst tomatoes in order to enhance pollen transfer from male to female parent. However, species of solitary bees are the usual vectors and their pollen transfer effectiveness is governed by environmental conditions, water supply, seasonal temperature fluctuation etc. Added to this, the insecticide applications in vegetable-growing areas conflict with the use of insects as pollinators. This difficulty, coupled with the low rate of NCP has led to the production of tomato hybrid seeds by hand pollinations in many areas.

49.3.12.4 Gene-Cytoplasmic Male Sterility

This type of sterility found in tomato and its related species does not appear to be of immediate use in tomato breeding. For when *L. esculentum* is crossed as female to *S. pennellii*, and the F_1 is backcrossed as female to *S. pennellii*, mst plants are produced in increasing proportions and the mst intensity enhances as BC's continue. Thus *L. esculentum* does seem to have S-cytoplasm but the required sensitive *fr* gene(s) is in a foreign species, *S. pennellii*. In order to develop useful "sterile" and "maintainer" lines, the *L. esculentum* genome must be put into *S. pennellii* cytoplasm, but the *L. esculentum* insensitive gene(s) must be replaced with the *S. pennellii* gene(s) that are sensitive to *L. esculentum* cytoplasm and cause mst. Unilateral incompatibility may prevent this kind of gene transfer.

49.3.12.5 Self-Incompatibility and Gene Transfer

Denna (1971) proposed a scheme utilizing self-incompatibility and exploiting NCP for hybrid F_1 production in naturally self-pollinated vegetable crops like tomato. It was suggested that the strict self-incompatibility of *L. peruvianum*, which also constitutes a rich source of disease resistance, variation and adaptability, be transferred by BC's and selection to *L. esculentum*. However, in this proposition, simultaneous selection for exserted stigma and floral attractivity to bees become obligatory for achieving effective cross pollination. Such gene transfers, particularly of polygenic traits, are not so easy. Also, complex incompatibility between *L. esculentum* derivatives with differing S-alleles was encountered by Martin (1961a,b) while transferring self-incompatibility from *L. chilense* to *L. esculentum*. Therefore, for hybrid seed production the use of mst is simple and superior to the proposed self-incompatibility systems. Of course, economical and efficient pollen transfer is still required.

49.3.12.6 Achievements and Limitations

For using g-mst in tomato breeding and hybrid seed production, the mst must be complete and the steriles should carry some easily and early-identifiable genetic markers. The majority of g-mst's exhibit complete and stable mst. In mst ms_{35}, the gene causing the absence of anthocyanin is closely linked to the *ms* gene (Philouze 1974b). Another *ms* gene is closely linked to the potato-type leaf trait (Hernandez and Wollard 1978). Such mst plants were used in hybrid breeding by the two authors and much success has been achieved. In France, the ms_{35} tomato variety, Porphyre has been successfully utilised in hybrid production. These hybrids exhibit considerably higher fruit yield than their parents (Philouze 1974b). A better yield of hybrid seeds was obtained by Philouze (1969) by using $ms_{32}ms_{32}$ and $ms_{35}ms_{35}$ plants than by using their corresponding mft heterozygotes. By measuring the seed yields of mst mutants interplanted with fertile plants, Lona (1965) obtained 30% higher yield over fertile controls, whereas Rick (1950) found the yield increased by 41–43%. Using some stamenless mst lines, Paszkowska (1980) obtained many early high-yielding hybrids.

Israel was the first country that utilised g-mst tomato lines for commercial hybrid fruit and seed production (Lapushner and Frankel 1963) and later such lines were used in Bulgaria, France and the USA (Daskaloff et al. 1965; Laterrot 1976; Philouze 1976; Rick 1976). Lapushner and Frankel (1979) compared the tomato hybrid seed production methods using mst as opposed to mft seed parents. In terms of expenditure and time, use of mst lines was found to be highly profitable and economically justifiable. Further improvement and higher efficiency were visualised by the authors if homogenous stands of g-mst seed parents are available. The short anther-parthenocarpic (*sha* + *pat*) tomato mutant bears seedless fruits which are smaller with deep red colour and have higher sugar and lower acid content than the seeded fruits of the normal counterparts (Falavigna et al. 1978). Such fruits are highly valued for processing. The main drawback of this mutant, however, is the difficulty of its maintenance by seeds. Applications of 10^{-5} to 10^{-3}M GA_3 enhanced stamen development (Mapelli et al. 1979) but failed to induce seeded fruits. Fruit production of ms_{15}, ms_{26} and ms_{33}, which produce naked ovules around the base of rudimentary sterile anthers, was investigated by Schmidt et al. (1980). These mutants produced fruits of poor quality owing to the fusion of androecial tissue with the fruit. GA_3 application restored fertility and improved fruit quality. Plants raised from seed produced by mst plants treated with GA_3 are uniformly mst.

In a stamenless tomato mutant (*sl*), Phathak et al. (1966) documented anther differentiation and pollen production following GA_3 treatment. Kubicki and Potoczek (1972) utilised this gibberellin effect to self-pollinate gynoecious (*sl*) lines and suggested a method for F_1 hybrid seed production in tomatoes. The commercial applicability of such a method awaits verification and is limited because of the development of deformed and disfigured fruits developed after GA_3 applications. Furthermore, lower GA_3 concentrations induce parthenocarpic fruit development, seeds being absent in such fruits. Association of functional mst with broad rounded cotyledons and mature leaves in a BC derivative of *L. esculentum* × *L. parviflorum* does not seem to have a direct utility value in commercial hybrid seed or fruit production. Even though the mst plant exhibits high natural cross pollination and ease of hand pollination due to protruded stigma, the shape and quality of the fruits is inferior and they have low market value. Moreover, in some crosses the dominance of mft is partial and the F_1 plants are susceptible to disease and pest attacks.

49.4 *Nicotiana*

49.4.1 *Nicotiana rustica* L. (small tobacco)

49.4.1.1 Genic Male Sterility

In a stand of this strong-smelling yellow-flowered plant, a white-flowered densely branched pollen sterile plant arose (this Vol.). Male meiosis was normal but microspores aborted after liberation (Table 49.15). Female fertility as evidenced by seed set after hand pollination was normal. The anomaly is controlled by two recessive genes, ms_1, ms_2.

49.4.1.2 Chemical Male Sterility

Spraying 0.05, 0.1 and 0.5% MH and 2, 4-D at the time of main inflorescence initiation caused high toxicity, growth reduction and high male-female sterility in *N. rustica* (Jos and Singh 1976b) (Table 49.17). Application of similar concentrations plus additional 1% spray induced mst to varying degrees. When sprayed three times, 1% of the chemical induced high pollen sterility in the apical and middle regions of the inflorescence, but not in the basal regions due to the basipetal development of new buds that escaped the spray. Thus complete mst in entire inflorescence is not obtained. Additional chemical sprays reduce female fertility drastically. Also female fertility of three-times sprayed flowers is not known. Hence the use of this chemical is limited and questionable. The mature anthers of mst flowers have empty and nonstainable pollen, most of the microspores fail to separate out from the microspore mother cells and degenerate within the parental cell.

49.4.2 *N. tabacum* L. (commercial tobacco)

Both spontaneously arisen and induced g-mst and alloplasmic gc-mst are known (Tables 49.5, 49.6, 49.15, 49.16) in this tobacco-yielding species.

49.4.2.1 Genic Male Sterility

The spontaneous occurrence of a few plants with some petaloid anthers and/or petaloid-stigmatoid anthers is common in tobacco fields. Goodspeed (1930, 1954), Goodspeed and Avery (1939), Raeber and Bolton (1955), Bhat and Krishnamoorthi (1956) regard this anther transformation as genic since it occurs nearly repeatedly, although in a low frequency. The sterility is of sporogenous and structural types.

i. *Sporogenous Male Sterility*. Bhat and Krishnamoorthi (1956) detected a mst plant with diminutive vegetative and floral organs, narrow corolla tube and deformed and indehiscent anthers. Whereas in some F_2 segregations, digenic recessive and in others monogenic recessive control of this sterility is apparent, the authors concluded digenic control from F_3 results. In these steriles meiosis proceeds normally but microspores degenerate after their liberation. The diminutive vegetative and floral parts of the mst's may be due either to pleiotropy or to linkage. Since no recombination between mst and diminutive traits occurred in F_2 and F_3 segregation, pleiotropy is assumed, but this needs confirmation by analysing large segregating populations. After 2 kR γ-irradiations of tobacco seeds, some mutant plants segregated in M_2 (this Vol.). Male meiosis is abnormal, although initial chromosomal synapsis is normal. At anthesis abortive nonviable microspores are developed. The anomaly is controlled by one recessive gene. Female fertility, as evidenced by artificial pollinations, is reduced by 9–12%.

ii. *Structural Male Sterility*. Some mst's segregated from the progeny obtained from an old seed stock of American Joiner flue-cured tobacco (Raeber and Bolton

Table 49.5. Alloplasmic male sterility in *Nicotiana tabacum*

Cytoplasm donor species[a]	Anther transformation	References
N.amplexicaulis (2n = 36)	Malformed stamens and split corolla	Berbeĉ (personal communication)
N.benthamiana (2n = 38)	Feminization and petaloidy	Ramavarma et al. (1978)
N.bigelovii (2n = 48)	Feminization and petaloidy	Burk (1960), Schweppenhauser and Mann (1968)
N.clevelandii (2n = 48)	Feminization (nil to complete)	This Vol.
N.debneyi (2n = 48)	Feminization (partial to complete)	Clayton (1950), Sand (1968), Tsikov and Tsikova (1971), Hicks et al. (1977), Gerstel et al. (1980)
N.exigua (2n = 32)	Malformed stamens and split corolla	Berbeĉ (personal communication)
N.glauca (2n = 24)	Feminization	Berbeĉ (1966), this Vol.
N.goodspeedii (2n = 40)	Feminization	Tsikov et al. (1977), this Vol.
N.gossei (2n = 36)	Feminization	Gerstel (1980), Burns and Gerstel (1981)
N.knightiana (2n = 24)	Malformed stamens and split corolla	Berbeĉ (personal communication)
N.megalosiphon (2n = 40)	Feminization (partial to complete)	Clayton (1950), Chaplin (1959)
N.plumbaginifolia (2n = 20)	Reduction	Chaplin (1964), Cameron and Moav (1957)
N.raimondii (2n = 24)	Malformed stamens and split corolla	Berbeĉ (personal communication)
N.repanda (2n = 48)	Feminization (partial to complete)	Burk (1967), Gerstel et al. (1978), Burns et al. (1978)
N.rustica (2n = 48)	Nil, nonfunctional pollen	Hart (1965), this Vol.
N.suaveolens (2n = 32)	Feminization and/or stamenlessness	Izard and Hitier (1955a,b), Schweppenhauser and Mann (1968)
N.tabacum (2n = 48)	Petaloidy-stigmatoidy	Burk (1960)
N.undulata (2n = 24)	Petaloidy of variable degree	Chaplin (1964), this Vol.

[a]Genome donor *N.tabacum* (Modified and extended from Gerstel 1980).

1955). Their anthers were either rudimentary and lacked archesporial tissue or were absent. Some mst's exhibited petaloidy of stamens, reduced corolla size and colour. Although precise genetic control of this sterility was not investigated, F_2 segregations indicate monogenic recessive gene control.

49.4.2.2 Gene Cytoplasmic Male Sterility

Barring two instances in which *N. tabacum* contributes the cytoplasm (Table 47.6), in the majority of the instances mst results when *N. tabacum* genome is placed in the cytoplasm of other *Nicotiana* species (Table 49.5). Thus the sterility is allo-

Table 49.6. Male sterility in some interspecific *Nicotiana* hybrids

Species cross	References	mst in anther	Phenotype
A) *N.tomentosiformis* × *N.otophora*	Goodspeed (1954), Gerstel and Burns (1974)	F_1	Stamens abnormal, archesporial tissue degeneration, PMCs scanty, meiosis abnormal
N.langsdorffii × *N.sanderae*	East (1919,1929)	F_2	Stamens abnormal, archesporial tissue degeneration, PMCs scanty, meiosis abnormal
B) *N.tabacum* × *N.bigelovii*	Chaplin (1959), Burk (1960)	F_1	Abnormal stamens
N.tabacum × *N.glutinosa*	Burk (1960)	F_1	Petaloid anthers
N.tabacum × *N.plumbaginifolia*	Burk (1960)	BC_2	Petaloid anthers
	This Vol.	F_1	Petaloid-stigmoid anthers

plasmic. This genome transfer into alien cytoplasm is achieved by interspecific hybridisations followed by BC's using the genome donor as the recurrent parent (Fig. 3.25). Usually amphidiploid F_1's are backcrossed to the recurrent pollen parent. Depending upon the extent of meiotic stability of the amphidiploid parent, mst's may appear in BC_1 (Clayton 1950) or in BC_2 (Burk 1960). With further BC's their frequency enhances. Usually mst's tend to breed true, but mft's continue to segregate some mst's. During this transfer, diverse intermediate forms with less abnormal anthers occur in early BC generations especially with the cytoplasms of *N. debneyi, N. megalosiphon* (Clayton 1950), *N. bigelovii* and *N. repanda* (Burk 1960), *N. benthamiana* (Ramavarma et al. 1978), *N. gossei* and *N. undulata* (Burns and Gerstel 1981). However, the proportion of anther abnormality increases in successive BC's. In some, other floral anomalies also appear. In a few gc combinations, either stamens are developed but pollen is inviable or in most of them anthers are deformed, distorted or transformed into petals or into stigma-like structures (Fig. 49.7) and in the extreme cases they are completely absent (Tables 49.5, 49.6). The mst's extracted from such crosses are genetically purified through successive BC's to *N. tabacum* as the recurrent pollinator. Usually the forms with the most abnormal anthers are selected in the search for mst tobacco. In the following, various manifestations of mst are described with emphasis on the types exhibiting special and specific mst genetic features.

i. *Sporogenous Male Sterility*. In most of the alloplasmic combinations having a *N. tabacum* nuclear genome, anther transformations are common (Tables 49.5, 49.6) but in a few, anthers are untransformed and either meiocytes or microspores abort. This occurs in the alloplasmic combination of *N. tabacum* having *N. plumbaginifolia* or *N. rustica* cytoplasm (Hart 1965; Cameron and Moav 1957; Smith 1968). In them male meiosis is normal but microspores abort. In an exceptional hybrid *N. clevelandii* × *N. tabacum* sporogenous tissue is developed, but PMC's degenerate before meiotic initiation (this Vol.). Male meiosis is abnormal and pollen sterility is complete in F_1 plants produced after pollinating *N. bigelovii* and *N. megalosiphon* by *N. tabacum* pollen (Chaplin 1959). Transfer of *N. tabacum* genes into the cytoplasm of the two species enhances the degree of anther transformation, feminisation and petaloidy, so that the anther phenotype is completely lost at BC_5. Thus initially mst is sporogenous but it stabilizes at a structural level.

ii. *Structural Male Sterility*. Two characteristic features of gc-mst tobacco are alloplasmy and anther transformation, the latter leading to structural mst in almost all alloplasmic combinations in which feminisation and/or petaloidy are the common features (Table 49.5). The staminal transformation produces either malformed petal-like structures or petals that have distinct stigma-tipped extensions. The majority of (*N. bigelovii*) *N. tabacum* plants exhibit staminal feminization and petaloidy but in a few this transformation does not occur. In their anthers, sporogenous tissue develops abnormally and no pollen is produced. In (*N. repanda*) *N. tabacum*, corolla size and its pigmentation intensity are highly reduced. In (*N. debneyi*) *N. tabacum* and (*N. benthamiana*) *N. tabacum*, the corolla has deep splits or incisions. Such alterations are illustrated by Chaplin (1964) and Frankel and Galun (1977) (Fig. 49.7). These g-c interactions altering nonvegetative parts of a flower are common in tobacco and petunia. In mst tobacco, since the *N. tabacum* genome is common in the alloplasmic combinations, the manifestations may represent cytoplasmic effects. These effects are malleable by environment, specific chromosomal fragment or nuclear genes (Sect. 49.4.2.2.2). In all the alloplasmic combinations, whereas the male sex is highly disturbed and almost negated, female fertility is nearly unimpaired.

49.4.2.2.1 Hybrid Male Sterility

In alloplasmic *N. tabacum* mst is of a low order in F_1 but increases regularly with the increase of *N. tabacum* gene content in further BC's. A few F_1 plants or F_2 segregants of some *Nicotiana* crosses are mst (Table 49.6A). In the mst plants, the stamen phenotype is abnormal, sporogenous tissue is degenerated and very few PMC's are developed. In them, male meiosis is abnormal and microspore abortion is rampant. Backcrossing F_1's by the male parent enhances staminal abnormalities and in BC_3, filaments of abnormal anthers disappear and corolla morphology is also altered. Sterility in these is conditioned by the *fr* gene and S-cytoplasm. The dominant allele *Fr* at one or both loci restores fertility. From BC_2 progeny of two crosses, involving *N. tabacum* as female parent (Table 49.6B) and *N. tabacum* as recurrent pollen parent, one exceptional mst plant in each cross was detected by Burk (1960). The mst (*N. tabacum* × *N. plumbaginifolia*) × *N. tabacum*, had petaloid-stigmoid anthers. Of 123 BC_3 plants 121 were mst, having severe stamen distortions. One of the remaining two had modified anthers but viable pollen, the other had normal-looking anthers but nonviable pollen. Selfing of the plant with viable pollen or pollinating it by *N. tabacum* pollen produced mst progeny. Using this plant as pollen parent in crosses to *N. tabacum* yielded mft's. Likewise, one F_1 plant of cross *N. tabacum* × *N. glutinosa* was mst with petaloid anthers (Burk 1960). It maintained sterility until BC_7 with *N. tabacum* as the male recurrent parent.

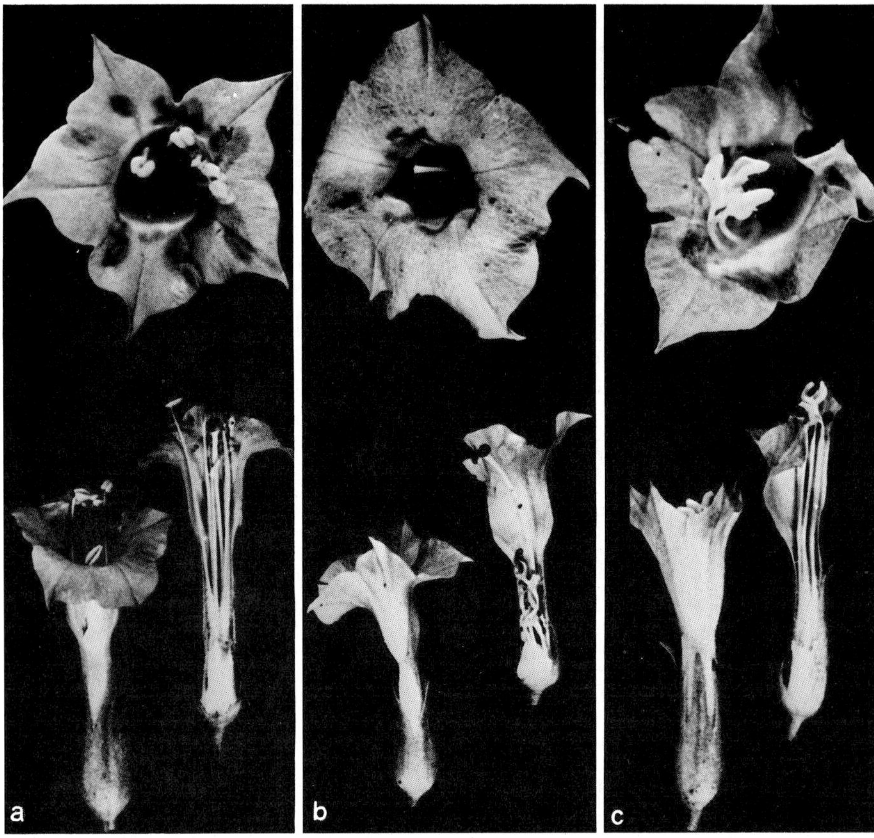

Fig. 49.7 (PART 1)

Fig. 49.7. *Nicotiana.* **a-f** Staminal form and functional alterations occurring after placing *N. tabacum* × *N. glutinosa* hybrid genome in the cytoplasms of *N. tabacum* (**a** normal anthers), *N. suaveolens* (**b** pistilloid anthers), *N. bigelovii* (**c** petalloid anthers), *N. megalosiphon* (**d** functional carpelloid anthers), *N. undulata* (**e** petalloid anthers) and *N. debneyi* (**f** stigmoid anthers and split corolla). (Frankel and Galun 1977)

These crosses returned *N. tabacum* nuclear genome in its cytoplasm but mst persisted in both the cases. One F_1 mst plant occurred in *N. tabacum* × *N. plumbaginifolia* (this Vol.). The sterility was maintained by using *N. tabacum* as male recurrent parent. The mst in these exceptional plants arose due to N-cytoplasm mutation to S-type in *N. tabacum*. This is evidenced by the maternal maintenance of sterility and fertility restoration in reciprocal crosses and by using the initial male parents viz. *N. glutinosa* and *N. plumbaginifolia* as the pollen donors to the two mst lines in subsequent BC generations. The cytoplasm mutation appears following hybridisation. Another spontaneously arisen cytoplasmic mst mutant in the field of a commercial Polish tobacco variety was detected by Berbeč (1974).

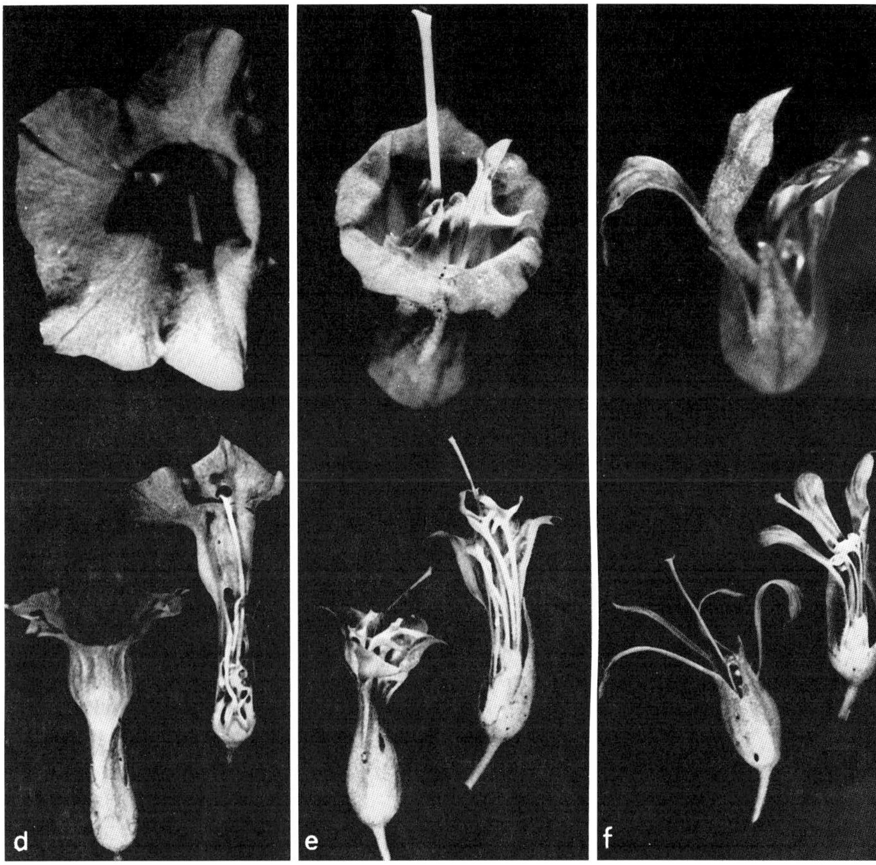

Fig. 49.7 (PART II)

49.4.2.2.2 Fertility Restoration

Placing complete or partial genome of *N. tabacum* into some *Nicotiana* cytoplasms results in mst (Table 49.5). The mst's are maintained by pollinating with genome donor pollen. Pollinating them by cytoplasm donor pollen restores mft. Thus even a part of the cytoplasm donor genome is sufficient to revoke mft in mst's. It has now been found that while substitution of alien chromosomes in a cytoplasm leads to mst in many plants, in certain alloplasmic tobacco combinations having *N. debneyi, N. repanda* and *N. undulata* cytoplasms, retention or reintroduction of a specific chromosome from the cytoplasm donor genome restores fertility.

i. ***(N. repanda) N. tabacum.*** Since *N. tabacum* and *N. repanda* do not intercross, Burk (1967) used a bridging cross method via *N. sylvestris* in order to transfer *N. tabacum* genome in *N. repanda* cytoplasm. This cross evidenced that for fertility restoration it is not essential to have a whole or part of the genome of the cytoplasmic parent into its own cytoplasm but only a part or whole of a specific chromosome. This is exemplified by three mst's. In BC_3 a heterozygous line segregated mst's and mft's. Whereas mft's continued to segregate mst's, the mst's bred true. A comparative karyotypic analysis of the male fertiles and steriles made by Gerstel et al. (1978) indicated that while the mst's had 48 chromosomes resembling those of *N. tabacum*, mft's had invariably one additional chromosome in single or double doses. This extra chromosome from *N. repanda* carries *Fr* gene(s) and has three interesting features, viz. (a) it is not an intact or complete chromosome but a part or fragment, (b) it possesses an active centromere that ensures its regular transmission in both somatic and meiotic division and (c) it has two satellites one on each end, and active nucleolar organisers at both ends. Thus it is not an ordinary fragment or normal chromosome but a highly specialized and thoroughly rearranged chromosome with coordinated and sophisticated gene action. In addition to revoking mft, this chromosome inhibits the nucleolar organiser ability of *N. tabacum* chromosomes. This inhibition occurs irrespective of whether this chromosome is present in single or in double dose. This nucleolar dominance or amphiplasty is nearly complete in PMC's but incomplete in root tip cells.

ii. ***(N. undulata) N. tabacum.*** In this alloplasmic combination complete anther-form restoration and partial mft restoration occurs after introducing a specific chromosome from cytoplasm donor parent *N. undulata* (Burns and Gerstel 1981). The restorer chromosome is satellited, associates consistently with nucleoli at meiotic prophase-I and has NOR. The NOR activity of *N. tabacum* chromosomes is weakened by amphiplasty. The *fr* gene(s) are located either at or near NOR on the short satellited arm because many restored plants have only this arm, the other long chromosome arm being missing completely or partially in them.

iii. ***(N. debneyi) N. tabacum.*** In this, a specific satellited *N. debneyi* chromosome restores the anther form, but not mft. The genetic nature of this fragment is thus different from the ones mentioned earlier and is elaborated presently. An array of diverse phenotypic forms with variously modified anthers occurs in early BC generations during the transfer of *N. tabacum* genome into alien cytoplasms. Using seed samples from advanced BC generations derived from the ancestral amphidiploid hybrid synthesised from *N. debneyi* ♀ × *N. tabacum* ♂. Sand and Christoff (1973) and Sand (1968) selected four specific flower phenotypes, designated 1A, 2C, 3E and 4H, from the hybrids between *N. debneyi* × *N. tabacum* and fixed them in a series of backcrosses to *N. tabacum*. Their morphological characteristics and inheritance pattern are investigated by Sand and Christoff (1973) and cytogenetics by Burns et al. (1978) and Gerstel et al. (1980). Extreme flower morphological types in which stigmata replaced anthers and corollas are split deep to base where designated as 1A. They breed true for mst and have *N. tabacum* chromosomes. Thus this represents a true, stable gc-mst type. In the 2C type, like the 1A type, the anthers are replaced by stigmata, but unlike 1A, their corollas are fused at the base and are not completely split down to the base. Somatic chromosomes of 2C type also resemble *N. tabacum* chromosomes but BC's of these to normal males segregate about 70% 1A phenotypes and 30% 2C types. This segregation indicates that the type 2C possesses a chromosomal segment of *N. debneyi* which is integrated in some *N. tabacum* chromosomes. Both these types do not exhibit restoration capacity and thus have either unaltered or slightly modified *N. tabacum* nuclear genome. In these two characteristics, type 3E and 4H plants differ significantly. In 3E, corollas are split but only between the lobes; anthers are deformed and often tipped by stigmas. Some PMC's develop and undergo meiosis which is partly abnormal, producing polyploids, and partly normal, permitting some analysis. The latter type cells have 49 instead of 48 chromosomes. Whereas in one line about ⅓ of PMC's have 23 II + 1 III, in another line 24 II + 1 I occur regularly. The extra chromosome is satellited resembling a satellited *N. debneyi* chromosome except that the centromere is shifted and the arm ratio altered.

The 4H-type plants are fully restored phenotypically, have normally fused petals and normal-looking anthers but their microspores abort. They have 48 somatic chromosomes of *N. tabacum* and a doubly satellited rearranged chromosome (Fig. 49.8), originating probably from *N. debneyi* genome. At MI, 24 II + 1 I occur. At prophase-I, this extra chromosome is attached to the nucleolus

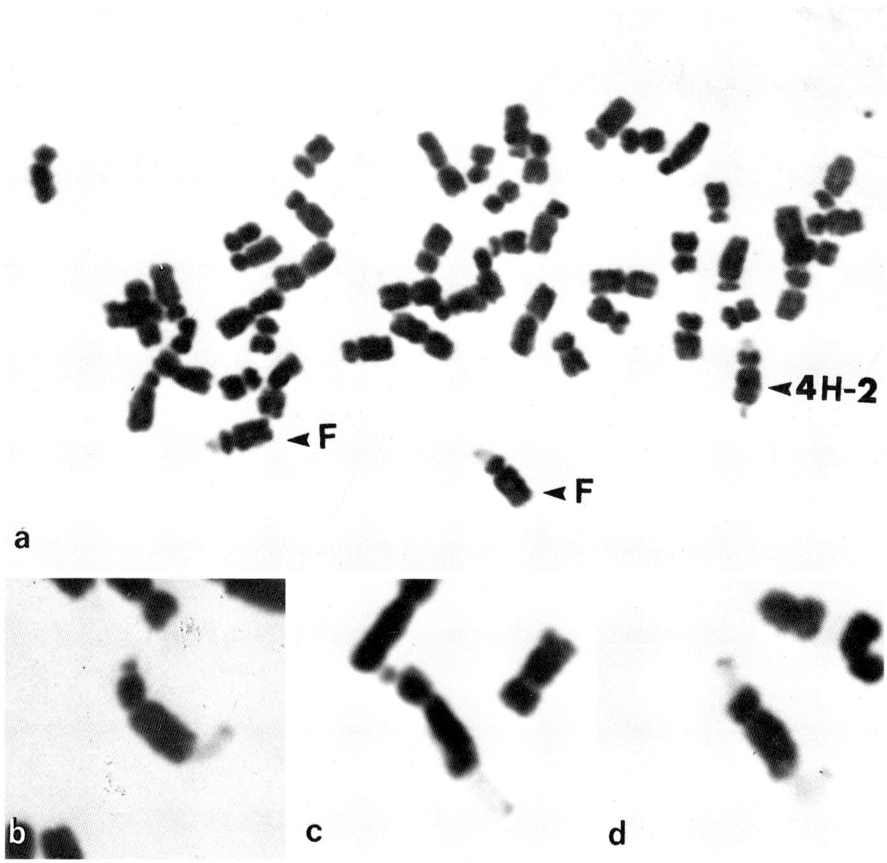

Fig. 49.8a-d. *Nicotiana tabacum.* **a-d** Somatic chromosomes of restored alloplasmic male sterile (*N. debnyi*) *N. tabacum* having two satellited *N. tabacum* chromosomes (*F*) and a doubly satellited restorer (*4H-2*) chromosome **(a)**, and enlarged double satellited restorer chromosome **(b-d)**. (Gerstel et al. 1980)

along with *N. tabacum* sat-chromosomes. However, in one line, whereas the extra chromosome was indistinguishable from one of the satellited chromosomes of *N. debneyi*, in another line, the extra chromosome was rearranged, being longer and possessing nucleolus-forming satellites at both ends like the *N. repanda* restorer chromosome. Chen and Sand (1979) have demonstrated that the extra-linkage group of a 4H type also carried the gene for a small subunit of F-I-P characteristic of *N. debneyi* genome. Thus while 1A and 2C types develop no anther tissue and have 48 chromosomes, 3E and 4H types develop anthers and possess one mft-restoring extra chromosome. It is transmitted to 20% and 7% of the progeny of 3E and 4H type plants, respectively (Gerstel et al. 1980). The chromosome has NOR, a satellite from *N. debneyi* and conditions highly restored phenotype (4H). It also slightly inhibits nucleolus organising capability of *N. tabacum* chromosomes (Burns et al. 1978; Burns and Gerstel 1981).

49.4.2.2.3 Chromosome and Gene Location

The chromosome having *Fr* gene(s) is a highly specialised, thoroughly rearranged chromosome fragment, a derivative of the genome of cytoplasm donor species. It is a specific nucleolar organiser satellited chromosome that undergoes rearrangement or loses its arm when present in the alien genome. Its gene action on restoration of anther phenotype and mft is sporophytic. Its NOR is extremely active and inhibits some of the nucleolus-forming potential of *N. tabacum* chromosomes, thus exhibiting amphiplasty. The chromosome fragment competes for a site at nucleolus and is always attached to it at diplotene-diakinesis. In root tip metaphases, secondary constriction of *N. tabacum* chromosomes is considerably shortened (Gerstel and Burns 1974). A relationship between the degree of amphiplasty and fertility restoration exists in tobacco. Amphiplasty, like fertility restoration, is very weak in (*N. debneyi*) *N. tabacum*, medium in (*N. undulata*) *N. tabacum* and strong in (*N. repanda*) *N. tabacum*. In these three alloplasmic combinations, a specific satellited chromosome fragment from the genome of the cytoplasm donor species restores anther morphology. In addition, the chromosome fragment restores complete mft in (*N. repanda*) *N. tabacum*, and partial mft in (*N. undulata*) *N. tabacum* but not in (*N. debneyi*) *N. tabacum*. Thus at least two different *Fr* genes control anther phenotype and mft restoration in these alloplasmic mst lines. The genes may exhibit additive action and may be either linked or closely placed around or adjacent to NOR. Normal anther phenotype restoration is a necessary pre-requisite for pollen fertility restoration. Therefore, a close linkage may exist between NOR and *Fr* genes. This linkage is adaptive, as it enhances population fitness and reproductive success. On the basis of transmission frequency in (*N. suaveolens*) *N. tabacum*, Schweppenhauser and Maan (1968) postulated four *Fr* genes located probably on two chromosomes (Chaplin 1959). One *Fr* gene in *N. langsdorffii* and one in *N. sanderae* can confer fertility to *N. langsdorffii* × *N. sanderae* mst hybrid plants (Smith 1962).

49.4.2.2.4 Protoplast Fusion

Using the protoplast fusion technique, restoration of mft was achieved by Aviv and Galun (1980), who utilised mft *N. tabacum* var. *xanthi*, (*N. tabacum*) (*N. tabacum*) and the alloplasmic mst line L-92 (*N. suaveolens*) *N. sylvestris*. When protoplast of L-92 was fused with X-irradiated protoplast of mft *N. tabacum*, about one quarter of the recovered calli (15 out of 63) in the second fusion experiment produced plants that had recombinant alloplasmic composition of (L-92) *N. sylvestris*, some of which were mst and others mft. Of the mft recombinants (L-92) *N. sylvestris* and (*N. tabacum*) *N. sylvestris*, one quarter were cybrids, i.e. they consisted of *N. sylvestris* nuclei with all or major components of *N. tabacum* cytoplasm. In one half of these cybrids, mft was complete and in the other half it was partial, being restored to different levels. In all fertile recombinants, the *Fr* gene(s) appears to be contributed by the cytoplasm of X-irradiated *N. tabacum* protoplast, because (*N. suaveolens*) *N. sylvestris*, *N. tabacum* and (*N. suaveolens*) *N. tabacum* combinations are mst, and X-irradiated *N. tabacum* was in the fusion mixture. In two recombinant groups (*N. tabacum*) *N. sylvestris*, mft restoration was accompanied by the transfer of *N. tabacum* chloroplasts. However, such a chloroplast transfer did not occur in mft (L-92) *N. sylvestris* plants. Therefore, fertility restoration is independent of chloroplast segregation in alloplasmic tobacco. This is also evidenced from the following: In cybrids produced by fusing the protoplast of mst *N. sylvestris* plants with X-irradiated protoplast of mft *N. tabacum*, mft was either partially or fully restored in three out of the four groups of cybrids (Aviv and Galun 1980). Whereas in two fertile cybrid groups mft restoration accompanied transfer of *N. tabacum* chloroplast, in the third group this chloroplast transfer did not occur following mft restoration. In the fourth group chloroplast transfer occurred but restoration did not occur. Therefore, mft restoration in cybrids is not associated with chloroplast segregation. Both these are independent (Aviv et al. 1980). In the cybrids, no preference for combination of one type of nucleus with either type of chloroplast was found. The independent chloroplast segregation is evident from some protoplast fusion results given in Table 49.7.

49.4.2.2.5 Somatic Hybridization

Recent progress in somatic cell engineering technology using plant protoplast fusions has led to the production of hybrids having different proportions of parental cytoplasms and one of the genomes of the parents. This has led to the synthesis of new types of hybrids which otherwise could never have been obtained through sexual crosses. Thus Gleba et al. (1974) and Gleba (1979) fused protoplasts from cytoplasmic albino *N. tabacum* and green alloplasmic mst (*N. debneyi*) *N. tabacum* (tbc). In contrast to results of other workers, all the plants obtained after regeneration were completely mst. This trait was retained by these derived plants but a shifting occurred in the chloroplast characteristics for which the two parents were contrasted viz. albino vs. green and the large subunits of F-I-P. Some of the plants were variegated with green and albino cells and contained both forms of the enzyme. Pure regenerants that had probably lost the albino components possessed

Table 49.7. Transfer of tobacco male sterility through protoplast fusion[a]

Fusion partners		Somatic hybrid phenotype	Reference
Male sterile parent	Male fertile parent		
N.tabacum[a] with cytoplasms:			
N.debneyi	*N.glutinosa*	Stigmatoid anthers (different phenotype than sterile plant)	Uchimiya (1982)
N.debneyi	*N.tabacum* var. xanthi	Variety of intermediate sterile-fertile types	Belliard and Pelletier (1978), Belliard et al. (1979)
N.debneyi	*N.tabacum* cytoplasmic albino mutant	8 variegated plants 2 male sterile 6 not analysed	Gleba (1979)
N.megalosiphon	*N.plumbaginifolia*	27 male steriles, 3 with sterile and fertile flowers	Menczel et al. (1983)
N.suaveolens	*N.sylvestris*	28 sterile hybrids	Zelcer et al. (1978)
N.suaveolens	*N.tabacum* nitrate reductase mutant	1 male sterile 22 male fertile	Glimelius et al. (1981)
N.suaveolens	*N.sylvestris*	2 different CMS types 1 self incompatible type	Aviv et al. (1980) Aviv and Galun (1980) Galun et al. (1982)

[a]Adopted after modifications from Hanson and Conde (1985).

only the *N. debneyi* type of large subunit of F-I-P and were completely mst. Gleba (1979) opined that these data may represent the first record of linkage of extrachromosomal genes in a higher plant. Zelcer et al. (1978) fused the protoplast of normal *N. sylvestris* with X-rayed protoplast (*N. suaveolens*) plants. X-raying eliminated *N. tabacum* nuclei and inhibited the unfused protoplast to divide. The X-irradiation of one protoplast parent caused a preferential transfer of cytoplasm alone into the fusion product, resulting in cybrid rather than hybrid formation. Division of the parental *N. sylvestris* protoplast was also inhibited by adding mannitol to culture medium. Following three distinct types of plants regenerated from the callus:

i. *Type A*. These plants were morphologically identical with *N. sylvestris* but, unlike it, were mst. Their stamens were stigmatoid, the anthers highly deformed and tapered. They were therefore, regarded as having genome or genomes (some were 4x) of the *N. sylvestris* and cytoplasm from the (*N. suaveolens*) *N. tabacum* (line-92) parent because mst in them resembled this alloplasmic combination of plants. This was confirmed by Aviv et al. (1980), who also found that these plants contained *N. sylvestris* nuclear genome, but contrary to normal *N. sylvestris* plants, they had chloroplasts that were sensitive to tentoxin. Fragmenting their cp-DNA by restriction endonucleases (Bgl, Eco RI) and subjecting them to gel electrophoresis separation, their pattern was found to be identical with that of the line-92 cp-DNA, the original donor of the sterile cytoplasm. Furthermore, their offspring from crosses to normal were also mst. Hence these plant types appear to have the *N. sylvestris* genome in *N. suaveolens* cytoplasm, but the possibility of the retention of a small portion of cytoplasm from the *N. sylvestris* parent by type A plants appears strong.

ii. *Type B.* These plants appeared to contain nuclear components from both parents, as they combined the morphological characteristics of *N. sylvestris* and *N. tabacum* and two of the three plants thus obtained had high chromosome numbers. These, like type A plants, had cp-DNA from the (*N. suaveolens*) *N. tabacum* parent, but their anthers were intermediate in form. Although sterile, they were distorted, in contrast to the highly deformed anthers of type A plants. Thus their cytoplasm was a mixture of *N. sylvestris* and *N. suaveolens* cytoplasms.

iii. *Type B-2.* These plants resembled *N. sylvestris* morphologically and possessed the *N. sylvestris* genome. Their anthers were phenotypically normal but they were p-mst. Their cp-DNA was of *N. sylvestris* type. Therefore, like type B1 plants, they are also cytoplasmically mixed.

Thus it is apparent that an original mixture of chloroplasts formed after protoplast fusion underwent a rapid shifting. This resulted in the plants mainly with chloroplasts of one kind of either *N. suaveolens* (in type A and B-1) or *N. sylvestris* (B-2). On the other hand, mst carriers (possibly mitochondria) probably did not segregate. This led to types B-1 and B-2, which retained components derived from both parents. As indicated earlier, mft restoration in type A plants was possible by fusing the protoplast of these steriles with X-irradiated protoplasts derived from fertile *N. tabacum* plants. The chloroplasts of the two parental donors differ in respect to tentoxin sensitivity. The chloroplasts of mst *N. sylvestris* are sensitive to tentoxin, whereas those of *N. tabacum* are insensitive. Aviv and Galun (1980) have demonstrated an independent segregation of chloroplast type and mst, as several recombinants were mst but tentoxin-insensitive and others were mft but tentoxin-sensitive.

Utilizing two types of *N. tabacum* that differed from each other in cytoplasm and a marker gene, Belliard et al. (1977, 1978) reported successful fusion of S- with N-cytoplasm. For this, the author utilized the mst Techne cultivar (*N. debneyi*) *N. tabacum*. It has split corollas, abortive anthers, and sessile leaves. These traits are controlled by recessive genes, the leaf trait served as a marker in the isolation of hybrids. The other parent used in this protoplast fusion comprised two *N. tabacum* cultivars Samson and Xanthi, which have normal *N. tabacum* cytoplasm and dominant alleles for petiolated leaves. Thus while one parent has petiolated leaves and normal fertile flowers, the other parent has sessile leaves, abnormal flowers and is mst. Since these two parent cells have distinct morphological and cytological markers, it was possible to recognise the regenerated plants with a single parental nucleus and hybrid cytoplasm produced by mixing of the two parental cytoplasms. After protoplasts of the two parents were mixed and fused, no selection was made to separate the calli developing from the hybrid protoplast of the two mft cultivars (Samson and Xanthi). Naturally the regenerated plants developed from the calli having protoplast of the parental types or of the hybrids. Of 654 such plants, 57 of the latter types had 2n = 52–102 and intermediate flowers and leaf bases. These plants are somatic hybrids with respect to nuclei and cytoplasm. Two hundred and twenty five plants possessed *N. tabacum* nucleus, but their leaves resembled one of the parents. In the corolla form and in the extent of anther abortion, their flowers were intermediate between the parental forms and could be categorised in four types, ranging stepwise from plants with normal corollas and reduced pollen fertility to mst's with deeply split corollas. The occurrence of such intermediate types suggests that the cytoplasms were mixtures rather than derived from only one parent. These were highly stable, as they exhibited an unaltered and strong maternal inheritance. Analysis using the restriction endonuclease technique indicated that only one or the other parental cp-DNA was present in the progenies of these cybrids. No mixtures or recombinant cp-DNA's occurred in them. Belliard et al. (1979) digested the mt-DNA of these intermediate types (cybrids) with Sal-I restriction endonuclease and compared them with the mt-DNA of the parents. All these cybrids possessed some DNA fragments characteristic of each parent, and barring one, they had in addition new DNA fragments which were not present in the parents. The authors opined a new type of fragments resulting from mitochondrial recombination that occurs following the simultaneous presence of *N. tabacum* and *N. debneyi* mitochondrial genome in the same cell. Moreover, it appeared that the number of mt-DNA fragments derived from the mst parent was greater in the more highly sterile intermediate plants than in feeble mst plants. These results suggest a relationship between mt-DNA and mst.

Using polyethylene glycol (PEG), Uchimiya (1982) fused protoplasts derived from suspension cultured cells of gc-mst (*N. debneyi*) *N. tabacum* and mft *N. glutinosa*. Out of 1089 colonies developed from the fused protoplasts, 29 restored whole plants. A hybrid plant exhibited leaf morphology intermediate between the parents but was mst. The narrow split-blossom of mst *N. tabacum* was absent in the mft plants. Two BC generations with pollen of *N. glutinosa* or *N. tabacum* var. Samson produced only mst progenies. This hybrid possesses *N. debneyi* cytoplasm but some genes of *N. glutinosa* incorporated in *N. tabacum* genome.

49.4.2.2.6 X-Irradiation and Cybrid Production

In protoplast fusion experiments, two types of hybrid cells are obtained viz. somatic hybrids, that combine cytoplasmic and nuclear components of both the parents, and cybrids that possess nuclear components from only one parent. In *Nicotiana*, both types of hybrid cells are obtained in protoplast fusion experiments. Cybrids are of more genetic significance and breeding value and have always been selected over hybrids. Whereas Belliard et al. (1977, 1978) obtained cybrids spontaneously in their fusion experiments involving *N. tabacum* types, Zelcer et al. (1978) utilised X-rays to inhibit the replication of nuclei of one parent. X-raying protoplast of (*N. suaveolens*) *N. tabacum* prior to fusion led to the elimination of *N. tabacum* nuclei and inhibited the unfused protoplast from dividing. However, hybrids possessing nuclear elements from both the parents also occurred in such fusion experiments. Aviv and Galun (1980) obtained mft restoration by fusing protoplasts from gc-mst *N. sylvestris* plants with X-irradiated protoplasts derived from fertile *N. tabacum* plants. Over 25% of the regenerated plants were cybrids and consisted of *N. sylvestris* nuclei combined with all or some components of *N. tabacum* cytoplasm. By X-irradiation of one of the parents used in protoplast fusion, Zelcer et al. (1978) and Aviv and Galun (1980) obtained a high proportion of cybrids, as this technique enhanced the frequency of the transfer of cytoplasm alone and suppressed hybrid formation.

49.4.2.2.7 Tentoxin Sensitivity

Tentoxin, a cyclic tetrapeptide produced by the fungus *Alternaria tenuis*, disrupts chloroplast development by binding uncompetitively to the chloroplast coupling factor-1 (CF_1) and inhibits photophosphorylation and light-driven protein and RNA synthesis in sensitive species. Nonsensitive species bind tentoxin weakly and consequently are not inhibited. The sensitivity is maternally inherited in *Nicotiana* and it is presumed that a gene for the tentoxin receptor lies within cp-DNA. The cytoplasmic determination of sensitivity is evidenced by the fact that gc-mst lines of *N. tabacum* having the nuclear genome of *N. tabacum*, a nonsensitive species and the cytoplasm of wild species like *N. glauca, N. nesophila, N. otophora* and *N.repanda*, are sensitive to the toxin. Not all *Nicotiana* species are sensitive to the toxin. Of 64 species, only 33 are sensitive to tentoxin (Durbin and Uchytil 1977). It is presumed that the gene for tentoxin receptor lies within the cp-DNA. Aviv and Galun (1980) found a positive correlation between tentoxin sensitivity and cp-DNA restriction endonuclease pattern. Aviv et al. (1980) further demonstrated this correlation with the chloroplast coded large subunit of fraction-1 protein. Thus the sensitivity test can reliably characterise a chloroplast kind and a cytoplasm type. Using this, Flick and Evans (1982) evidenced that chloroplast segregation is not random in cybrids, but in most cases the chloroplast type present in sexual hybrids was the only type found in somatic hybrids.

49.4.2.2.8 Chloroplast Segregations: Specificity and Sensitivity

In the protoplast fusion experiment, chloroplast (cp) and mitochondria (mt) are transmitted through the maternal cytoplasm. However, whereas mitochondria when present conjointly are known to undergo recombination (Fonty et al. 1978; Linnane and Nagley 1978; Mannella et al. 1979; Belliard et al. 1979), chloroplasts appear to maintain their integrity under such situations (Aviv et al. 1980; Aviv and Galun 1980). Aviv and Galun (1980) obtained somatic cybrids that are either mft or mst and have chloroplasts derived from either of the parents. Thus no relation exists between parental chloroplast types and mst-mft. Although chloroplast transmission is maternal, its segregation from the fused protoplast to individual protoplast is independent in tobacco. Aviv et al. (1980) evidenced that sterility/fertility components and parental cp-DNA's may occur in all four combinations among plants generated from one fusion experiment. This supports independent chloroplast segregation and its nonrelationship to mst in tobacco. The separation, sifting and segregation of chloroplasts is very fast as compared to mitochondria, which can coexist and recombine. Out of 16 somatic hybrids between *N. glauca* and *N. langsdorffii*, 15 contained large subunits of F-I-P (encoded in cp-DNA) of only one or the other parent; the mixture present in one plant segregated immediately in the next sexual generation (Chen et al. 1977). Chloroplast segregations were also detected by Gleba (1979). *Nicotiana* chloroplast genomes possess many distinct properties and certain unique features. The fragment patterns generated by restriction enzymes exhibit a high degree of diversity but it is species-specific (Kung et al. 1981; Rhodes et al. 1981). The specificity is maintained in protoplast fusion derivatives also and is used to characterise *Nicotiana* cytoplasms.

Zelcer et al. (1978) fused the protoplast of mft *N. sylvestris* with the X-irradiated protoplast of mst (*N. suaveolens*) *N. tabacum*. Of the callus-regenerated plants, 19, designated type A, resembled *N. sylvestris* but were mst. Because of their phenotypic and karyotype similarity to *N. sylvestris*, it was deduced that these type-A plants possess the *N. sylvestris* genome. Being mst, they were assumed to have *N. suaveolens* cytoplasm. Their chloroplasts were compared to their parents by Zelcer et al. (1978), Galun and Aviv (1979), Aviv et al. (1980) using several biochemical and genetic tests viz. gel electrophoretic pattern and restriction endonuclease fragmentation of cp-DNA and tentoxin sensitivity of chloroplasts. Both the gel migration of large subunits of F-I-P coded in cp-DNA and the cp-DNA fragmentation patterns produced by restriction endonucleases were similar in type-A plants and the (*N. suaveolens*) *N. tabacum* parent. Also like *N. suaveolens* and unlike *N. sylvestris*, type A plants were tentoxin-sensitive. This confirmed that the type A plants had cytoplasm derived from alloplasmic parent and not from *N. sylvestris*. Although the possibility that a small amount of cytoplasm from the latter parent was retained by type A plants cannot be eliminated, it is certain that they received chloroplasts from the former parent. Compared to the N-cytoplasm of *N. tabacum*, the S-cytoplasm exhibits a high frequency of variation in cp-DNA (Kung et al. 1981). The pattern of alteration is localised in one region of cp-DNA and forms a "hot spot", and ranges from having a single extra fragment to a more complex pattern, probably evolved through recombination. Some new bands not present in parents were found in gc-mst lines, a few of which are unique. These unique bands arise through cp-DNA recombination of the two species (Kung et al. 1981). This represents the first report of cp-DNA recombination in *Nicotiana*. The cp-DNA of the S-cytoplasm also exhibits deletions, insertions and mutations.

49.4.2.2.9 Mitochondrial Transmission and Recombination

Belliard et al. (1979) obtained stable cybrids or cytoplasmic hybrids by fusing *N. tabacum* var. Xanthi which has normal fertile flowers and petiolated leaves with (*N. debneyi*) *N. tabacum* var. Techne that has abnormal and mst flowers. Restriction endonuclease analysis reveals the presence of only one or the other parental cp-DNA in the progeny of these cybrids (Belliard et al. 1978). Apart from the mt-DNA characteristic of each parent, all these cybrids except one contained new mt-DNA fragments (Belliard et al. 1979) that were different from those of their parents and from the mixtures of the two. These new DNA classes seem to have arisen due to DNA recombination as is known in the yeast, *Saccharomyces cerevisiae* (Fonty et al. 1978).

49.4.2.2.10 Gene Action on Anther Development

The *fr* gene action is earliest in structural mst alloplasmic lines in which the genes act during anther ontogeny and differentiation. For instance, in type 1A (*N. debneyi*) *N. tabacum* (Sand and Christoff 1973), in which stigmata replace anthers and corolla are split deep right down to base, the *fr* gene acts on stamen rudiments of 0.5–1.0 mm in length. Differentiation into anthers and filaments is inhibited in them. During stigma differentiation, stamen initials develop stigmatoid tips subtended by a short filament bearing the initials of lateral outgrowths (Hicks et al. 1977). These tips gradually enlarge to a green bulbous recurved structure. No sporogenous tissue is ever formed in these anthers. The developmental pattern is retained by the rudiments excised at early differentiation stages and grown in cultures. Various hormonal treatments do not alter the developmental pattern (Hicks and Sand 1977). Thus early *fr* gene action is more stable than where the gene acts late during pollen maturation.

49.4.2.2.11 Anther Culture

Alloplasmic (*N. repanda*) *N. tabacum* plants are mst but a specific chromosome of *N. repanda* restores fertility to them. From these, haploid plants produced through androgenesis were obtained by Burk (Gerstel et al. 1978). The haploids produced mst or mft plants, depending on the absence or presence of the restorer chromosome. The deformed anthers of (*N. suaveolens*) *N. sylvestris* mst type A plants were cultured by Galun and Aviv (1979). The anther type of androgenic progeny resembled their diploid parents. Almost similar results were obtained by Tsikov et al. (1974), who used (*N. debneyi*) *N. tabacum* mst plants. When such androgenic haploids are diplodized by colchicine, mst is maintained and female fertility is restored in the derived diploids. In this alloplasmic combination, the percentage of embryoids recovered depends upon the developmental stage at which the pollen is placed in the nutrient medium (Tsikov et al. 1981). Whereas in pollen of mft *N. tabacum* plants, embryoids are produced by the vegetative nucleus, in the mst (*N. debneyi*) *N. tabacum*, embryoids develop exclusively from pollen grains at uninucleate stage; a stage just prior to degeneration.

49.4.2.2.12 Female Fertility

Unlike the deterioration of anther phenotype, pistils of alloplasmic mst tobacco plants are normal in form and function. Fasciation or fusion between pistils and feminized pistillate anthers occurs in some mst alloplasmic tobacco lines (Nikova

1979) but female fertility is normal in them. However, in mst (*N. plumbaginifolia*) *N. tabacum* female fertility is slightly reduced. In general, the absence of gross distortions of female organs and high seed set after hand pollination on mst tobacco lines indicate noninterference of *ms* or *fr* genes on female sex development and function in tobacco.

49.4.2.2.13 Graft Transmission

Transmission of mst across grafts does not occur in tobacco. Burk (1960) grafted mst scions on normal stocks and normal scions on mst stocks of tobacco. No alterations of scions or stocks occurred. Similar results were obtained by Sand (1960), who used mst (*N. debneyi*) *N. tabacum* plants in such grafting experiments. Sexual progenies developed from both types of scion. By crossing mst's grafted on normal stocks to normal ungrafted males, the progeny was mst. Similarly, the fertile scions on mst stocks produced fertile progeny, whether selfed or crossed with ungrafted normal plants. Thus, neither is mst transmitted across grafts nor is any modification of genetic nature induced by the scions.

49.4.2.2.14 c-Gene Location

Placement of the *N. tabacum* genome in the cytoplasm of some *Nicotiana* species having the *c*-gene produces mst progeny. This indicates that the *N. tabacum* genome possesses *fr*-nuclear genes and a N-cytoplasmic gene, rendering the plant mft. On the other hand, the species listed in Table 49.5 and many others have *c*-gene, so that mst results from the harmonious *fr-c* gene combination in the hybrids. Fertility in them is restored by the *Fr* gene contributed by the cytoplasm donor species genome. Maternal inheritance of the sterility, its unaltered persistence in many generations and its disappearance in reciprocal crosses (where possible) provide major evidence of the *c*-gene location in the cytoplasm. Chloroplast and mitochondria are the major cytoplasmic organelles known to contain DNA. Therefore one or both of them should lodge the *c*-gene. The investigations of Chen et al. (1977), Kung (1977) and Frankel et al. (1979) implicate chloroplast as the carrier of *c*-genes in tobacco. The large F-I-P subunits (encoded in chloroplast) of mst (*N. debneyi*) *N. tabacum* are identical to normal mft *N. debneyi* plants. In both, the cytoplasm is identical, being from *N. debneyi*. Utilizing the fragment patterns produced by Eco R-1 from cp-DNA of six alloplasmic lines, Frankel et al. (1979) obtained the same patterns as were found in the species which had contributed the cytoplasm in four of the six lines, viz. *N. debneyi, N. megalosiphon, N. plumbaginifolia* and *N.suaveolens* (Fig. 49.9). In (*N. bigelovii*) *N. tabacum* and (*N. glutinosa*) *N. tabacum*, cp-DNA differed from those of their cytoplasm donor parents and from that of *N. tabacum*, but resembled each other. The authors regarded this as a "directed change". The (*N. glutinosa*) *N. tabacum* represents derivative of a cross *N. tabacum* × *N. glutinosa* (Burk 1960; Burk and Durbin 1978). Its mst is not restored by *N. glutinosa* (Burk 1960; Lapushner 1978). Thus it is a unique and unusual genotype arisen due to cytoplasmic mutations.

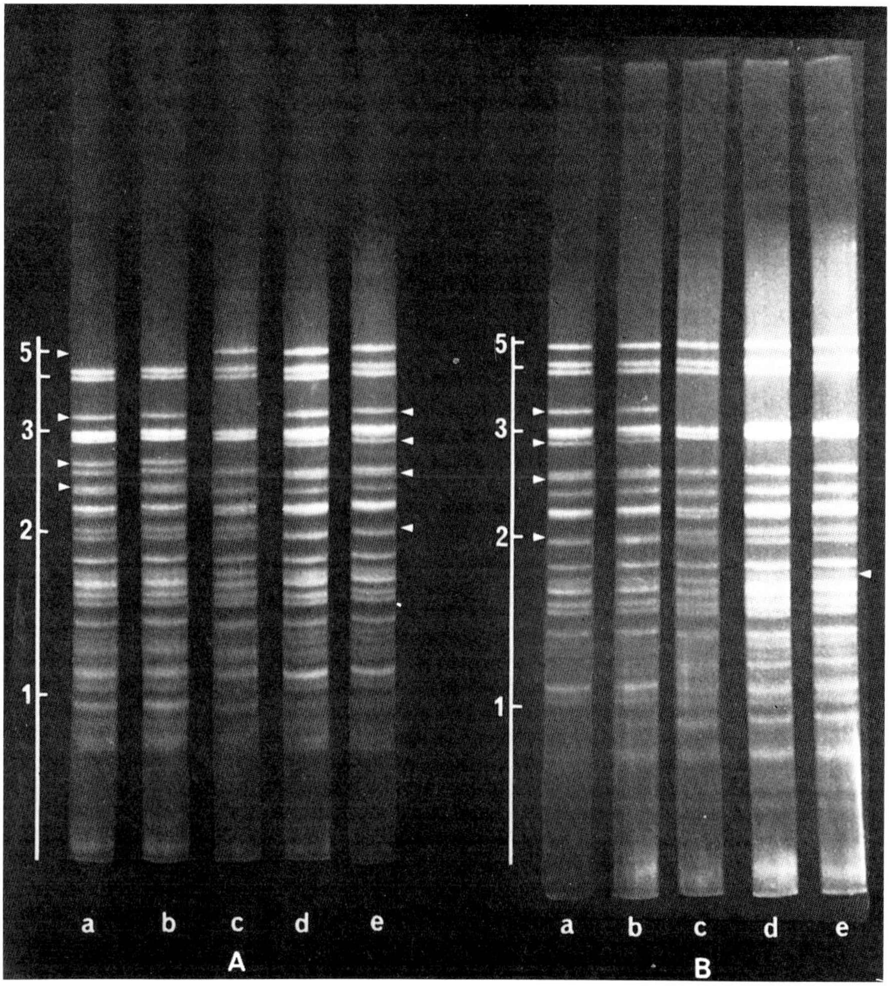

Fig. 49.9A,B. *Nicotiana.* Gel electrophoretogram of chloroplast DNA of sex isonuclear male sterile(ims) tobacco lines and their respective parental species. **A** *N. debnyi* (*a*), ims-deb (*b*), *N. tabacum* (*c*), ims-sua (*d*), *N. suaveolens* (*e*). **B** *N. megalosiphon* (*a*), ims-meg (*b*), *N. tabacum* (*c*), ims-plu (*d*), *N. plumbaginifolia* (*e*). (Fragment differences are indicated between maternal parent and *N. tabacum* by *arrows*). (Frankel et al. 1979)

Berbeĉ (1974, 1975) documented cytoplasmic mutation in *N. tabacum* conferring mst in conspecific crosses. It is the only example known in tobacco. The mst's have large subunits of F-I-P characteristic of *N. glauca*. Kung (1977) regards this cp-DNA alteration as leading to mst. Berbeĉ (1966) worked with (*N. glauca*) *N. tabacum* mst's. Therefore the absence of *N. glauca* cytoplasm in these mst plants needs to be confirmed. Thus no conclusions can be made whether a causal relationship exists between cp-DNA divergences and gc-mst, unless the specific restriction site alterations are shown to be reproducibly synthesised, following interspecific hybridization and backcrossing. Because cp-DNA in the actual original maternal plant used in the first cross was not analysed by restriction digestion by Frankel et al. (1979), the possibility remains that the maternal parent was derived from a population carrying heterogenous genomes. Intraspecific variation of cp-DNA restriction banding patterns is known in *N. debneyi* (Scowcroft and Larkin 1981). Following protoplast fusions and recovery of cybrids, Aviv et al. (1980), Aviv and Galun (1980) and Galun et al. (1982) demonstrated independent chloroplast and mst segregations in many cytoplasmic hybrids. For instance, the cybrids were either mft or mst but had chloroplasts from one of the parents. Following protoplast fusion between gc-mst (*N. suaveolens*) *N. tabacum* and mft *N. tabacum*, some mft plants with RuBPCase mobility and tentoxin sensitivity of *N. suaveolens* chloroplast were obtained, while others had phenotypic markers for *N. tabacum* chloroplast genomes (Glimelius 1981). A mst hybrid derived from fusing mst (*N. debneyi*) *N. tabacum* with mft *N. glutinosa* has the chloroplast marker RuBPCase mobility of the fertile *N. glutinosa* chloroplast (Uchimiya 1982). Some cybrids have DNA fragments characteristic of each parent and recombinant mt-DNA fragments (Belliard et al. 1979). These new mt-DNA fragments exhibit a direct relationship with mst degree, i.e. the higher their number, the greater is the mst degree. Does this indicate mitochondria as the carriers of C gene? The mt-DNA's in *Nicotiana* somatic hybrids have not been systematically studied with cloned probes such as were used to identify fragments in *Petunia* somatic hybrids. Thus identification of mt-DNA fragments in these *Nicotiana* hybrids relies on the similarity of electrophoretic mobility. Many mt-DNA bands occur in mst hybrids which are absent in the fertile parents (Galun et al. 1982), but these authors examined only two sterile and five fertile hybrids and Belliard et al. (1979) nine hybrids. An analysis of more sterile and fertile hybrids and the use of hybridisation probes to identify fragments is necessary to determine whether any specific fragment is directly associated with sterility or fertility.

49.4.2.2.15 Pollen Killer Gene

Sesquidiploid and amphidiploid hybrids of *N. plumbaginifolia* and *N. tabacum* are mst. Repeated crossing of the sesquidiploid to diploid *N. tabacum* accompanied by selection for a dominant *N. plumbaginifolia* genetic marker results in establishing types with 24 chromosomes of *N. tabacum* (24t) + 1 *N. plumbaginifolia* (1p). During studies of such lines involving gene *Bs*, that determines resistance to black shank disease *Phytophthora parastica* var. *nicotianae*, another gene *Kl* that kills pollen selectively was detected on this chromosome (Cameron and Moav 1957). This gene, *Kl*, has no deleterious effects on pollen of *N. plumbaginifolia* but in the alloplasmic combination (*N. plumbaginifolia*) *N. tabacum* having 24t + 1p, the gene causes abortion of pollen having 24 tobacco chromosomes (24t) shortly after meiosis. Thus whereas all microgametophytes with a complete complement of *N. tabacum* chromosomes (24t) abort, those having the alien *Kl* gene develop normally and are functional. This killing effect is restricted to the PMC quartets in which 1p having gene *Kl* is present. On megaspore or somatic cells, this gene has no effect.

49.4.2.2.16 Biochemical Differences

The mst in tobacco is considered to be due to the incompatibility of the *Nicotiana tabacum* genome in the cytoplasm of other *Nicotiana* species. This disharmony should be manifested at a biochemical level especially because many phenotypic alterations of male reproductive tissue and pollen grains occur due to mst. The limited available investigations prevent any conclusions. Whereas no chemical difference occurs in mst tobacco plants having *N. megalosiphon* and *N. tabacum* cytoplasms (Mann et al. 1962; Aycock et al. 1963), significant differences occur in the isoenzyme composition of peroxidase and glucose 6-phosphate dehydrogenase (G-6PDH) in anthers of mst tobacco lines having *N. debneyi*, *N. tabacum* and *N. bigelovii* cytoplasms (Markova and Tsikov 1979). Enzyme G-6PDH is an important and active enzyme of the pentose pathway. Its activity is greatly reduced in mst tobacco during staminal organogenesis. This activity could be associated with mst. In the leaves of gc-mst plants having the *N. tabacum* genome in the cytoplasm of *N. bigelovii*, *N. debneyi* and *N. megalosiphon*, reduction in glyceraldehyde-3-phosphate dehydrogenase activity occurs (Nikolov 1981). This is one of the key enzymes of the glycolytic pathway and is closely linked with photosynthesis. The decrease in enzyme activity differs in the alloplasmic gc-mst lines. Whether the difference is due to different cytoplasms or to differential gc-mst interaction is not known. Similar differences occur in multiple protein molecular forms between alloplasmic lines (Nikolov and Tsikov 1981). It is desirable to systematically investigate biochemical differences, especially amino acids, between mst and mft anthers of alloplasmic tobacco having different cytoplasms. Since the *N. tabacum* nucleus is common in them, g-c interactions at a biochemical level can easily be identified. Up to the 12-leaved stage, RNA and protein accumulation is slow in mst plants, and is followed by a decline in leaf protein content due to the translocation of proteins to the shoot apex (Oczos 1981). Moreover, ribosomal RNA synthesis is slow in mst plants until flower buds appear, but it gradually increases thereafter. These alterations are not the causes but consequences of mst.

49.4.2.3 Chemical Male Sterility

Jos and Singh (1967a) sprayed two tobacco varieties with 0.5, 1 and 1.5% of MH, 2, 4-D or mendok at the time of their inflorescence initiation. All these chemicals produced toxicity, the maximum being by 2, 4-D (Table 49.17), whose lowest concentrations stunted plant growth severely and led to high pollen and ovary abortions. Higher concentrations were completely lethal to plants. With MH treatments, growth inhibition was less drastic and even at lowest concentration (0.5%), 80% pollen abortion and still higher ovular abortion was induced by the chemical. With higher concentrations, considerable abscission of floral buds, gametic sterility and stunted plant growth occurred. Whereas 2, 4-D and MH are highly toxic to tobacco, mendok appears to be least toxic. Thus with 0.5 and 1% spray the plant phenotype was not significantly affected. Female fertility was apparently not influenced. However, hand pollinations to determine female fertility through seed set were not done. Complete mst lasted for 3–5 weeks after spraying. Before and after this period, the sterility was partial and with time the pollen sterility decreased gradually. During the period of complete mst, anthers of the treated plants were pollenless. The microsporogenous tissue degenerated immediately after its formation. Tapetal behaviour was completely normal. In a few, part of the sporogenous tissue lying close to the tapetum towards the anther apex developed a few PMC's in which meiosis proceeded abnormally and culminated in nonviable pollen formation. Of the three above-mentioned chemical gametocides, mendok appears promising for tobacco. However, determination of its effects and effectiveness on commercial tobacco varieties, as also on female fertility and hybrid production, are highly essential before its usefulness is claimed.

49.4.2.4 Breeding Value

Tobacco is self-pollinated, its NCP being low. Using mst tobacco plants, Bhat and Krishnamoorthi (1956) estimated it as about 4% on an average. Therefore, for hybrid seed production, hand pollinations are essential and these are to be done four to six times per season in order to fertilise all the flowers, since they develop sequentially. Although many mst sources are available in tobacco and most of them have been produced by transferring *N. tabacum* genome into cytoplasm of alien *Nicotiana* species (Table 49.5), nuclear genes for fertility restoration for most of the gc-mst forms have not been concomittantly transferred from cytoplasmic donor species to tobacco because the product of commercial interest is the leaf rather than the seed. Also mst is associated with many floral anomalies, reduced female fertility and slower growth in some cases. The combination (*N. megalosiphon*) *N. tabacum* has extremely reduced leaf and sucker number and much delayed flowering. Therefore, this type of mst becomes unsuitable for commercial application. In (*N. plumbaginifolia*) *N. tabacum* and (*N. bigelovii*) *N. tabacum* combinations, reduction in yield is greater than in (*N. megalosiphon*) *N. tabacum*, (*N. suaveolens*) *N. tabacum* and (*N. undulata*) *N. tabacum* (Aycock and Erdogan 1979). Usually these alloplasmic lines grow slowly during mid-season, are low-yielding, short-statured and late-flowering. The potentiality of the use of gc-mst in commercial production of flue-cured tobacco was tried by Mann et al. (1962), Aycock et al. (1963) and Chaplin et al. (1963). They used flue-cured derivatives of Chaplin's 402 mst line (*N. megalosiphon*) *N. tabacum* but did not find any distinct commercial advantage of F_1 hybrids over the available pure line varieties. Growth retardation in F_1 plants is one of the major disadvantages in the alloplasmic combination.

Using mst from several sources and transferring it into the Hicks-Broadleaf tobacco variety, NCP on mst lines is variable and very low. Therefore, it is not useful for commercial hybrid tobacco production. Furthermore, floral anomalies associated with the cytoplasms of *N. bigelovii*, *N. megalosiphon*, *N. plumbaginifolia* and *N. suaveolens* are unsuitable for hybrid seed production by hand pollination. However, in some alloplasmic combinations in which the *N. tabacum* genome is in the cytoplasms of *N. glutinosa*, *N. plumbaginifolia* or *N. undulata*, flowers having stigmas projecting well above corollas and containing petalloid anthers develop. Such flower types are useful for hybrid seed production, but even in these, natural cross pollination is very low and hand pollination is necessary for satisfactory seed set (Chaplin 1964). While *N. bigelovii, N. plumbaginifolia* and *N. megalosiphon* cytoplasms depress yield, *N. plumbaginifolia, N. megalosiphon* and *N. suaveolens* cytoplasms delay growth and maturity considerably. These features have caused a setback to hybrid production in tobacco. Likewise, many other tobacco mst strains with cytoplasms of many wild species have proved to be of little value in F_1 tobacco production due to reduced fertility, growth rate and seed set and a low quality of hybrid leaves (Chaplin and Ford 1965; Ruebenbauer and Heinrich 1968; Roman and Nalepa 1970; Povilaitis 1972). On the other hand, Tsikov and Nikova (1980) found that when gc-mst tobacco had markedly feminised stamens (pistiloidy or carpeloid stamens) or carried cytoplasm from *N. megalosiphon* and *N. goodspeedi*, their seed production was 40–50% lower than that of their normal counterparts. Whereas normal seed production occurred in tobacco plants having *N. bigelovii, N. glutinosa* cytoplasms, the production was higher with *N. debneyi* and *N. undulata* cytoplasms. Despite these demerits, low NCP, reduced seed set after hand pollinations, and slow growth of some hybrid plants, many mst tobacco lines have been released to breeders for developing hybrids. Berbeĉ (1974, 1975) found the cytoplasmic mutant plant that arose from mst *N. tabacum* to be more productive than its genetically analogous mft with respect to hybrid seed yield and other agronomic traits. Nosova (1976) obtained a variable degree of heterosis for plant height, leaf number and size, the extent of heterosis being dependent on trait, cross-combination and environment. Significant genotype-environment interaction is evident in these hybrid plants and heterosis is trait- and cross-combination-specific. Lapushner (1978) crossed mst *N. tabacum* cv. Hicks, that had cytoplasm from several different species with various intra- and interspecific pollen parents and obtained specific interactions with genotypes, cytoplasms, environments and traits. In some cases, alien cytoplasms exerted more beneficial effects than the natural cytoplasms. Therefore, for hybrid production, specific and non-random parental combinations should be used for exploiting positive heterosis.

The mst has been employed in F_1 hybrid seed production of Burley tobacco. The hybrids are grown on an estimated 60–65% of the commercial acreage (Wernsman and Matzinger 1980). For hybrid production, the fertile line is grown in another field and the mst line is hand pollinated four to six times per season to fertilise sequentially developing flowers. Usually the planting arrangements consist of four rows of female parents followed by an unplanted row used to transport pollinating personnel and equipment through the field. For the production of certified seed of pure line cultivars or hybrids, the mst line is kept at least 46 m away and isolated from its fertile counterpart contamination. However, this isolation distance may be waivered if four border rows of each cultivar are allowed to flower but are not harvested for seed.

49.5 *Petunia*, Juss

In this garden annual with showy bloom, mst is known in *P. axillaris, P. hybrida* and *P. parodii* (Tables 49.15 to 49.17).

49.5.1 *P. axillaris*, BSP (*P. nyctaginiflora*, large white petunia)

From selfed progeny of a white-flowering plant, 33% of the plants were pollen sterile (Levan 1942). In them, microspores were not liberated from PMC's and their cytoplasm degenerated. Depending upon the cytoplasmic degeneration, all to none of the microspores stained with acetocarmine. This sterility resembles the tetrad sterility known in *Solanum verrucossum* and is conditioned by one recessive *ms* gene. In a huge stand of white petunias growing near a deserted garden of Kashmir, three mst plants appeared (this Vol.). Male meiosis was normal in them but microspores degenerated after liberation from PMC's. The anomaly is controlled by one recessive *ms* gene.

49.5.2 *P. hybrida Vilm* (garden petunia)

This is a synthetic, commonly cultivated, well-established highly variable ornamental derived probably from Argentinian *P. axillaris* and *P. violacea* (*P. integrifolia*). Although a bewildering number of single and double varieties exist in this species, many more are added every year by breeders using g-mst and gc-mst, the sterility being now almost established in this species (Tables 49.15 to 49.17).

49.5.2.1 Genic Male Sterility

49.5.2.1.1 Spontaneous

Four spontaneously arisen pollen sterile plants were observed (this Vol.). In them 7–12% pollen stains with carmine but none is functional, as revealed by self- and cross-pollinations. Since the frequency of the mst plants is low, recessive gene control is assumed. A first report about the existence of recessive monogenically conditioned mst is by Frankel (1962, 1971), who detected a mst plant in S_2 progeny derived from selfing of a fertile scion grafted on gc-mst stock of Rosy Morn (RM). The author opined that g-mst had been induced in the nuclear genome of the fertile scion grafted on mst stock. The mst was complete and stable. Male meiosis proceeded normally but microspores were not released from the quartets (Izhar and Frankel 1971). Microsporogenesis breakdown followed tetrad formation and at anthesis sporogenous tissue was disorganised and dissoluted. The absence of callase enzyme activity led to the callose wall persistence around the four microspores which finally degenerated. The callase activity is differential (Fig. 49.10). Izhar (1973) detected cell budding and cleavage of young microspores in some

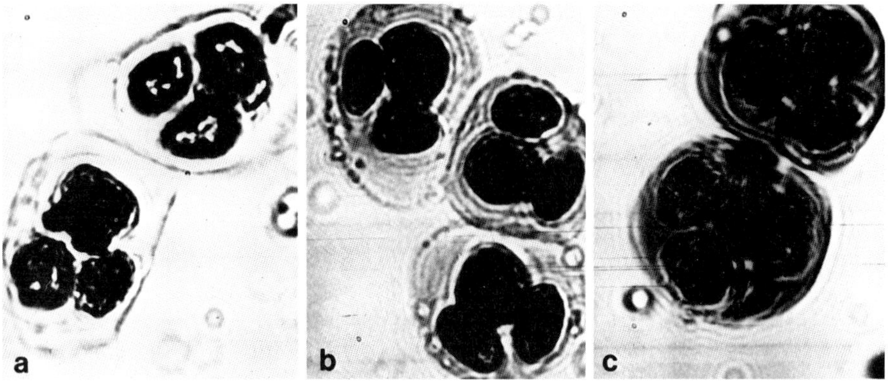

Fig. 49.10a-c. *Petunia hybrida.* **a-c** Full presence **(a)**, part presence **(b)** and complete absence **(c)** of callase enzyme activity in PMC's enclosing tetrads. (Frankel et al. 1969)

highly inbred *P. hybrida* lines. In them, the microspores swell up considerably, develop two vacuoles, bud out a small daughter cell, undergo mitosis and a daughter nucleus passes into the buds, some of which undergo fission. Some microspores cleave by furrowing. Pollen sterility is complete. Likewise microspore fission and budding occurred in the progeny plants of a strain of *P. hybrida* (this Vol.). In this strain, about 13–15% of the microspores neither cleave nor form buds. Instead they enlarge considerably and germinate in situ. The anomaly is caused by one recessive *ms* gene. In all the above-mentioned mst's, female fertility and anther morphology are normal.

From the progeny of a cultivar of *P. hybrida*, 14% of the plants develop polyspores (five to seven microspores), all of which abort after their release from the meiocytes (this Vol.). Male meiosis is abnormal. Female fertility, as revealed from seed set obtained after hand pollinations, is reduced by 7–11%. The mst is genic as the parent plant was fertile, as were the remaining 81% of the sib plants growing along with mst plants. Considerable intra- and intergenotypic variation in pollen fertility is exhibited by two diploid lines of *P. hybrida* as well as by their hybrid plants. Singh (1974) explained the variability by assuming that pollen fertility is conditioned by many genes whose distribution differed in the two lines and whose action was influenced by epistatic genes.

49.5.2.1.2 Induced

In 1948, Straub (quoted by Welzel 1954) detected a mst mutant from M_2 progeny of X-irradiated *P. hybrida* var. Himmelsröschen. The sterility is controlled by two recessive genes, ms_1 and ms_2 (Welzel 1954), both linked to the gene *parv*. The female fertility of this mutant is normal and its mst expression is not influenced by environment. Meiotic anomalies initiate during late prophase and become distinct at diakinesis. Microsporogenesis continues until the microspore formation stage, after which the young microspores abort. Cornu and Dommergues (1974) found some mst's in M_1 generation obtained after EMS treatments of F_1 hybrid zygotes produced after crossing two lines. TL-hi and Sf-la of *P. hybrida*. While the former line was obtained after doubling the chromosome complement of a haploid individual, the latter was built by a regular selfing process. Some pollen sterility also occurred in the somatic-anthocyanin F_1 mutants. Many of the pollen sterile F_1 plants exhibit abnormal meiosis due to chromosomal aberrations. High pollen sterility is likewise exhibited by the parental lines. In parent and hybrid plants intra- and intergenotypic variation in pollen fertility is significant (Singh 1974). This may be due to multigene control, gene number and distribution being different in the two lines and the phenotype being influenced by epistatic gene action. Certain segregations indicate polygene control, the gene action being environmentally influenced. Singh (1975) obtained mst mutant plants after crossing two partially pollen sterile *P. hybrida* lines and subjecting their F_1 zygotes to mutagenic treatments by exposing their ovaries 5 days after cross pollination to 0.2% EMS solution (Cornu 1970) or to 0.75 kR γ rays. Exact genetic control of the induced mutants is not known; the author opines control by recessive, dominant or partially dominant genes.

49.5.2.1.3 Tapetal Behaviour

In spontaneously arisen g-mst *P. hybrida* plants, tapetal behaviour is not known. In an induced g-mst mutant, the tapetum initially is intact, becomes periplasmodial and secretes some greasy droplets into the anther locule (Welzel 1954). Two mutant genes, ms_1 and ms_2, cause mst in it. One of these genes is assumed to impair microsporogenesis and the other tapetal function. In the mst mutant of Singh (1975), the tapetum exhibits hypertrophy and vacuolation, has abnormal inclusions and is persistent. The cells thicken and disorganise during postmeiotic stages. Due to this, the food material is retained by the tapetal cells and the young microspores are starved to death. Vacuolation and disorganisation of the tapetal cytoplasm suggest a functional breakdown before the microspores mature.

49.5.2.2 Gene-Cytoplasmic Male Sterility

Following the method of Clayton (1950), who produced mst in tobacco by means of genome transfer from the cytoplasm of one species into the cytoplasm of another, Everett (cited by Duvick 1959) produced many mst hybrids whose male and recurrent parent was *P. hybrida* or its commercial variety and the cytoplasm-contributing parent was *P. axillaris*, *P. integrifolia* and/or *P. parodii*. These mst hybrids were further backcrossed to several named varieties of petunia, using the named varieties always as pollinators. The mst plants were obtained in a few BC progenies and from these stable gc-mst's were developed through continued backcrossing to the varieties. However, the S-cytoplasm donor species is not known, since the original records are lost. Bianchi (1963) detected mst plants in the progeny of *P. axillaris* × *P. hybrida*. This indicates *P. axillaris* as the S-cytoplasm source. This species has more *Fr* genes in its genome than other *Petunia* species.

Therefore, it could be the S-cytoplasm donor of mst *P. hybrida*. The original S-cytoplasm source has become highly disseminated and both plant breeders and commercial breeders have transferred the sterility into many *Petunia* varieties. This has placed the sterility in many genetic and varietal backgrounds (Table 49.8). In Edwardson's material, two recessive nuclear genes (fr_1,fr_2) interact with the S-cytoplasm derived from the line P-266 and lead to mst (Edwardson and Warmke 1967). In certain mst's, some modifying genes and environmental factors influence sterility expression and produce pollen sterility of a variable degree. Of these, temperature is the most influential.

Table 49.8. Gene-cytoplasmic male sterility sources and varietal backgrounds in *Petunia*

Varietal background	Source
Blue Bedder	D.F. Jones, New Haven, Conn., USA
Edwardson	J.R. Edwardson, Florida Univ., Gainesville, USA
Ewart	L. Ewart, Harris Seed Co., Rochester, USA
Hamilton	B. Hamilton, Pennsylvania Univ., Philadelphia, USA
P-266	J.R. Edwardson, Florida Univ., Gainesville, USA
Rosy Morn	R. Frankel and S. Izshar, Agric. Res. Org. Bet Dagan, Israel

49.5.2.2.1 Temperature Influence

Some gc-mst lines are p-mst at 23°C day/15°C night (Izhar 1975). When flower buds of these lines, in which meiosis was about to start, were exposed to higher temperatures like 35°C/15°C night, complete mst was induced. However, when the buds in which male meiosis was progressing were subjected to such high temperature, no mst occurred. Therefore, the stage close to the onset of meiosis (premeiosis) is most susceptible to high temperatures in these gc-mst lines. This is also evidenced in a petunia line which was p-mst at 25°C day/20°C night and became completely mst when its buds at the premeiotic stage were subjected to 35°C/25°C or 15°C/10°C day/night temperature regime (this Vol.). Thus in this line, both high as well as low temperature treatments induced mst, provided the temperature treatments are given when its buds are in the premeiotic stage. Low temperature treatments reduce seed set due to impaired female fertility.

49.5.2.2.2 Fertility Restoration

This is conditioned by one dominant major gene or by a multigenic system, and is mainly of sporophytic but in some cases of gametophytic type. The existence of *Fr* genes in the petunia genome was first evidenced from mst:mft segregation data of Everett and Gabelman and of some other plant breeders working in US seed companies (Duvick 1959; Goldsmith 1968). Edwardson and Warmke (1967) detected one dominant *Fr* gene in plants grown under low temperature. Izhar (1978) confirmed this *Fr* gene control and found its action uninfluenced by temperature or the genetic background of mst line. Marrewijk (1969) detected two additional cryptomeric restorer alleles. The author postulated the involvement of a major Fr_1 gene, an accessory recessive fr_2 gene and a gene (or block of genes) Fr_3 in fertility restoration. A similar trigenic system occurs in Rosy Morn (Izhar 1977; 1978). In this multigenic system different degrees of fertility restoration occur depending on temperature. Plants having the Fr_1 gene in homo- or heterozygous state are completely fertile when grown between 12° and 15°C. In heterozygotes (Fr_1 fr_1), p-mft is restored in plants grown at 18°–21°C. In homozygotes the restoration is complete at this temperature. At 21°C, homozygotes Fr_1 Fr_1 fr_2fr_2 are fully mft. A third gene, Fr_3, is required for complete restoration in plants grown at or above 24°C. For complete and stable restoration, temperature regimes between 23° and 26°C are ideal. Even genotypes lacking the Fr_1 gene are slightly mft but completely mst below 16°C. Similar temperature sensitivity is exhibited by the two restorer lines used by Hamilton (1965). These are differentially effective as pollinators at 21° and 30°C. One of these is mft at 21°C and mst at 30°C.The sterility is due to meiotic anomalies induced at high temperature; the most sensitive stage is when floral buds are about to enter meiosis. Thus from the above account it is evident that the multigenic *Fr* system is highly temperature-sensitive as the three *Fr* genes interact with each other, with the S-cytoplasm and with temperature. In Rosy Morn, in which three *Fr* genes control restoration, the gene action is cumulative or complementary. Similar *Fr* genes, but nonallelic to Rosy Morn, occur in Red Satin and in a *P. axillaris* line (Izhar 1978). Segregations in some crosses involving multigenic restorers yield higher fertility to sterility ratios than expected. This is indicative of gametophytic restoration, which appears partial. In complete gametophytic restoration, genotype of microspore regardless of anther sporophytic tissue determines whether or not it will develop into a viable mature pollen grain. In Izhar's (1978) crosses, pollen having *Fr* genes participated in fertilisation more often than those lacking them.

49.5.2.2.3 Microsporogenesis

In the mst material studied by Edwardson and Warmke (1967), male meiosis proceeds normally up to microspore formation, after which degeneration occurs, but in the mst line 450 investigated by Frankel et al. (1969) and Izhar and Frankel (1971) microsporogenesis breakdown occurs during the early meiotic stages. Thus *ms* gene action in these two mst lines differs considerably. A high callase activity occurs in mst anthers (Frankel et al. 1969). This may degrade cell walls or prevent cell plate formations, impairing thereby the progression of the meiotic course. In the mst material studied by Hamilton (1965), male meiosis proceeds normally but microspores abort soon after their liberation from the tetrads. Bino (1985a,b) detected many irregular along with normal-looking sporogenous cells in the mst material studied by Marrewijk (1969). There is no faulty timing of callase activity, as the majority of the sporogenous cells deform before callose wall formation, and their encasing results in a distorted callose envelope. Although the first sign of breakdown appears during the leptotene, most of the meiocytes degenerate during the zygotene. A few reach MI or AI but degenerate immediately thereafter. Thus the *fr* gene(s) acts during pre- to early meiosis.

Using mst materials studied by Hamilton (1965), Edwardson and Warmke (1967), Marrewijk (1969) and three more sources, Izhar and Frankel (1976) found that the stage of microsporogenesis breakdown was highly influenced by the genotype of the recurrent restorer parent. For instance, in a BC_3 to RM mst inbred, most of the progeny plants exhibited breakdown during meiosis. With another restorer at BC_3 level, the breakdown occurred during tetrad stage. Four stable gc-mst lines and their complementary maintainer lines were synthesised by Izhar (1977). In the mst's, degeneration occurred during early meiosis or tetrad formation or after microspore liberation or around first pollen mitosis. Izhar (1978) opined that when a certain optimal number of restorer genes (*Fr*) are accumulated in the S-cytoplasm, fertility is restored, but the presence of some of these restorer genes causes microsporogenesis breakdown at different meiotic stages. It appears that with an increase in *Fr* gene number in S-cytoplasm plants, a proportionate delay in microsporogenesis breakdown occurs. When enough *Fr* alleles accumulate, delay is so much that microsporogenesis proceeds normally and mft is restored. This gene action is influenced by epistatic and inhibitory genes. Since different restorer lines may carry a different number of epistatic and inhibitory genes, the microsporogenesis breakdown stage will vary accordingly. A similar variation in breakdown stage will occur with different BC cycles to restorer lines. Both these features exist in petunia mst materials.

49.5.2.2.4 Tapetal Development

Of the various mst sources, tapetal development has been studied by Bino (1985a,b) in Blue Bedder gc-mst petunia, in which it is abnormal. The first visible sign of anomaly appears when meiocytes are in the leptotene stage during which tapetal cells remain uninucleate, shrink in size, develop large vacuole and dense cytoplasmic contents. At AI, tapetal cells abort, their cytoplasm becomes disorganised and the nucleus is deformed. Finally, the tapetum remains as a thin layer crushed in between sporogenous cells. Due to tapetal malfunction, callase enzyme is not secreted.

49.5.2.2.5 Cytoplasmic Homology

Izhar and Frankel (1976) tested mst materials of Hamilton (1965), Edwardson and Warmke (1967), Marrewijk (1969), and two more having Rosy Morn background against a uniform genetic background with restorer genes. They found no morphological differences. Even the interaction of these cytoplasms to environmental conditions was identical. Therefore the S-cytoplasm of these mst materials is similar and probably the same. Restriction digests of three mst lines, one each obtained from the USA, Israel and W. Germany give identical mt-DNA and cp-DNA restriction banding patterns. This confirms the S-cytoplasm similarity.

and mst line and obtained three mft plants with nuclear-controlled phenotypes characteristic of the sterile line. In addition, six mft somatic hybrids were also obtained. Through protoplast culture, Izhar et al. (1984) attempted reciprocal transfer of S- into N-cytoplasm using a fertile line of *P. hybrida* and a gc-mst line having *P. parodii* genome. Only 2% stable mst cybrids were recovered, indicating either nontransferrability or nonfunctioning and instability of the S-cytoplasm in the N-type. N-cytoplasm is transferrable and stable in the S-cytoplasm line. One percent of the plants were sectorial somatic hybrids and contained heteroplasm exhibiting fertility and sterility on the same plant. The investigation reveals the instability of the S-plasmon transferred to the N-cytoplasm.

Genetic and molecular analysis of the protoplast fusion products indicates that plasmon transfer is not always achieved unilaterally and that a heteroplasmic state exists in the protoplast fusion products with respect to mst and mft and to cytoplasmic organelle population (Belliard et al. 1979; Galun et al. 1982; Izhar et al. 1983; Boeshore et al. 1983; Izhar et al. 1984). However, some different results were obtained by Izhar and Tabib (1980) and Izhar et al. (1983). Whereas in the first report, mft was expressed by the heteroplasmon, in the second one, mst was exhibited by the heteroplasmic tissue of somatic hybrids. This indicates the existence of a nuclear gene (*fr*) and a cytoplasmic gene (*c*) and their fertile counterparts in the heteroplasmons of the protoplast fusion products, since different petunia lines were involved in these fusion experiments. While the S-cytoplasm transferred to the mft line exhibits instability, in reciprocal transfers mft cytoplasm exhibits a high degree of stability. Following mt-DNA analysis of the plants arising after fusion of mft *P. hybrida* protoplasts with gc-mst *P. parodii* protoplasts, Boeshore et al. (1983) concluded that parental mt-DNA molecules may remain intact or become recombined in the cybrids.

49.5.2.2.9 Graft Transmission

This has been subjected to most rigorous genetic analyses and intense investigations, beginning with the interesting experiments and significant observations (in the light of recent episomal concept of mst) of Frankel (1956), who demonstrated mst transmission through an elegantly made grafting study. The author grafted a fertile scion on a gc-mst stock. No alteration in the pollen sterility of stock or in the pollen fertility of scion occurred, but by selfing flowers of the fertile grafted scion, several mst segregants appeared in the selfed progeny. For over 40 generations now, the mst plants have exhibited stability and consistency in mst. Outcrossing mst plants by untreated maintainer pollen yielded mst progeny, and selfing the untreated maintainer line for two generations produced normal (mft) offspring only. This recovery of mst plants from the selfed progeny of a fertile scion grafted on gc-mst stock demonstrates asexual transmission of mst. Through similar grafting experiments, Edwardson and Corbett (1961) transmitted mst from a gc-mst stock to the progeny of a scion. However, whereas in Frankel's (1956) experiment, S_1 generation progeny plants of the scion were completely mst, progeny obtained by these authors was p-mst. However, the second generation of these p-mst's was completely mst. The amino acid profile of third generation grafting was identical with those of gc-mst plants. Like Edwardson and Corbett

(1961), Bianchi (1963) also obtained mst transmission across grafts. The first generation plants were p-mst. Later, in a sandwiched graft consisting of a petunia-tobacco-petunia scion, mst was transmitted across the graft successfully by Corbett and Edwardson (1964). All these authors detected no alterations in the sexuality status of stock or scion even though the combinations were maintained for about 4 years by Frankel (1962). In other words, the mst stock and mft scion remained unaltered in their sexuality types. In contrast to this stability, Hope (1979) observed mst transmission from stock to scion and obtained a graft-induced mst in the scion which was mft when grafted. This is the first such report in petunia and needs to be confirmed using stable, healthy and virus-free mft scions.

Transmission of mst across grafts appears genotype-dependent and involves a stock-scion compatibility. Thus Frankel (1962) found that not all the maintainer graft components produced mst progeny when selfed and that different maintainer pollinators produced different proportions of sterile offspring when crossed to grafted maintainers. The contention of Edwardson and Corbett (1961) that the introduced and maintainer graft components require a period of adaptation is not supported by Frankel's (1962) experiment in which he maintained a fertile scion over sterile stock for over 3 years. Grafting experiments reveal two important characteristic features viz., (i) maintenance of autonomy by fertile scion even after years of union with sterile stock, (ii) mst plants appear in the selfed progeny of grafted fertile scion and not in the progenies obtained from outcrossing with the pollen of nongrafted flowers. Whereas Frankel (1956, 1962, 1964), Edwardson and Corbett (1961) and Bianchi (1963) reported mst transmission across grafts, no such transmission was detected by Marrewijk (1968). Even mst flowers occasionally appeared on grafts not involving mst stocks. Marrewijk (1970) opined this as being due to the irregular manifestation of *fr* genes under changing environments.

The published reports about graft transmission of mst are hard to compare, as no two laboratories have used the same genotype for stock and scion. Within the same laboratory, evidence exists that the nuclear background of the transmitting or recipient genotype is important. Besides, frequency of graft transmission of mst is often low or not achieved until advanced generations. For instance, of 25 graft combinations, only four yielded F_1 mst progeny. Likewise, of 1192 F_3 plants, only two sterile plants were obtained by Edwardson and Corbett (1961). In 451 F_2 plants derived from sterile or partially sterile flowers, only 252 flowers out of 4914 examined exhibited sterility (Corbett and Edwardson 1964). Frankel (1964) pointed out that the pollen for the F_1 cross on the fertile scion must be derived from either the fertile scion itself or a flower on a scion attached to another sterile stock. While other aspects of mst graft transmission have been verified by workers using different genotypes in other laboratories, Frankel (1964) has apparently performed the only experiments testing this point, since all other reports describe the use of selfed flowers on fertile scions. Thus many factors seem to have contributed to the variability in achievements of mst transmission. In some cases, the scions might have carried *Fr* genes that masked mst expression. Where a scion lacking the *Fr* gene was used, other nuclear genes of the scion might have inhibited stable mst transmission. Alternatively, only certain stocks might have an appropriate nuclear background to transmit mst to a fertile scion. Besides, certain uncontrollable environmental factors might be needed to achieve the transmission. Thus mst transfer across grafts is not as simple as was envisaged initially.

49.5.2.2.10 Nature and Special Features

i. *Nature*. After successful transmission of mst across grafts, Frankel (1971) suggested the analogy of the *c*-gene to episomes in bacteria that are either inserted into the chromosomes forming the "integrated state" or stay in the cytoplasm in the form of an "autonomous state". In his crosses between the scions, the integrated state of episomal prevailed. In the event of its stay in cytoplasm, the dilution phenomenon of mst as observed by Edwardson and Corbett (1961) and Izhar (1984) can be explained. The episomal element concept of the *c*-gene does not explain the majority of the special features associated with gc-mst in petunia. Edwardson and Corbett (1961) and Atanasoff (1964a, 1971) suggested viral transmission of mst, but later Edwardson (1968) negated his theory, based on certain circumstantial evidence. Hess (1973) indicated an "episomal concept" of the *c*-gene, but provided no evidence to support the concept. The *c*-gene appears to be in the DNA of the cytoplasmic organelle, chloroplast or mitochondria. In maize, tobacco, sorghum, pearl millet etc., these organelles are known to carry the *c*-genes. The organelles exhibit autonomy and segregation, their formation and development is nuclear-gene-dependent but the subsequent transmission segregation and mutations are independent and nonrandom in petunia. Following mt-DNA analysis of plants arising after fusion of mft *P. hybrida* protoplasts with gc-mst *P. parodii* protoplasts, Boeshore et al. (1983) concluded that parental mt-DNA molecules may either remain intact or become recombined in the hybrids. No restriction endonuclease pattern differences occur between the cp-DNA of mst and mft petunia (Kool et al. 1983), but mt-DNA restriction patterns differ significantly between mst and mft petunia plants. These differences are reflected in their protein bands also. Therefore, the mitochondria rather than the chloroplast are the coding site of the *c*-gene conditioning gc-mst in petunia. To explain somatic and gametic segregation of mst and mft, Izhar et al. (1983) postulated sterility (*ste*) and fertility (*fte*) elements. These elements could be the cytoplasmic organelle, the mitochondria, which segregate in both somatic and gametic tissue. A *ste* element segregating in somatic hybrids may not be the same as the sterility factor transmitted through grafting. The *ste* and *fte* elements in somatic hybrid tissue may be located in mt-DNA of mst and mft parental protoplasts, while mst transmitted across grafts may be the gene product that transforms some of the fertile scion mt to mst-specifying state.

ii. *Special Features*

1. The occurrence of both spontaneous and induced g-mst as well as gc-mst, the sterility being conditioned by one or multi-gene system.
2. The co-existence of *ms* and *fr* genes in the same genotype; the genes segregate, act independently and do not interact.
3. The p-mst is influenced by temperature given during premeiosis; too low and high temperature induce complete mst.
4. Fertility restoration is conditioned either by a dominant gene or by polygenes, the latter exhibiting high temperature sensitivity.
5. The mst exhibits lack of reversion, induction and transmission through pollen.
6. A low frequency of mst transmission across grafts takes place in some compatible stock-scion combinations.

7. Transfer, maintenance and perpetuation of mst and S-cytoplasm is possible in cybrids via protoplast fusions.
8. The existence of homo- and hetero-plasmic states in somatic and gametic cells in cybrids. This is followed by the somatic segregation of *fr* and *c*-genes.

49.5.2.3 Non-interaction of *ms* and *fr* Genes

Bringing *ms* and *fr* genes together in the same genome and placing this genome in the S-cytoplasm enabled Evenor et al. (1984) and Evenor and Izhar (1984) to study the interaction of these mst-causing genes in petunia. It was found that: (i) *ms* genes can co-exist in one or two doses with *fr* genes in somatic cells or female gametes, (ii) for this co-existence *Fr* allele presence is not needed, (iii) the *Fr* gene is not epistatic to the *ms* gene. Thus both these genes are nonallelic and act independently. Microsporogenesis breakdown occurs during early meiosis in $S_{fr\,fr}$, *ms ms* genotype, which is typical of *fr* gene action and at the tetrad stage in $S_{Fr\,fr}$, *ms*, which is typical of the *ms* gene (Izhar and Frankel 1971, 1973a). Since the *fr* genes act at the beginning of meiosis and *ms* genes at the end, *fr* gene action, being early, is manifested by $S_{fr\,fr,\,msms}$ genotypes. Hence no interaction, like epistasis or inhibition, occurs in *ms* and *fr* gene action in the petunia genome. This is in contrast to garden pea, in which three recessive genes brought together in the genome interact to generate an array of genetic variability including meiotic mutations, male sterility, chromosome aberrations, breakages and irregular unions (Kaul 1978, 1983; Kaul and Murthy 1985).

49.5.2.4 Chemical Male Sterility

Mendok. Nearly seven weekly foliar sprays of 0.20, 0.25 or 0.30% of this solution reduced pollen sterility in three petunia cultivars (Sink and Gunesch 1966), but these treatments damage ovular fertility and decrease seed set drastically (Table 49.17). Splitting of the corolla tube, reduced flower opening and a rugose effect on the new foliage were caused by the treatment. The chemical resulted in stunted plant growth, shortened internodes, pistil malformation and fasciation of new shoots.

49.5.2.5 Breeding Value

Since 1955 the original mst petunia stocks have been widely disseminated among commercial petunia breeders. These have been further backcrossed to many other petunia varieties. Several attractive hybrid petunias sold are made at a considerable saving in costs of hybridisation by utilising the mst materials. Using this sterility does not, however, completely eliminate hand pollination, as cross pollinations have to be made by hand. Only hand emasculation is eliminated. It is considered to be impractical to obtain proper isolation for field hybridization of petunia. Hamilton (1965) and Izhar (1972) reported some deleterious effects associated with gc-mst. Abortion of young floral buds also known as blindness, blasting of floral buds, reduced flower size, flower dormancy and delayed flowering are common problems in hybrids produced after utilising gc-mst lines. Thus many gc-mst hybrids are inferior to their normal counterparts. GA_3 injection into the lower parts of the *P. hybrida* stalk promotes near-normal flowering without affecting mst expression (Izhar 1972). Because of the environmental sensitivity, mst breaks down sometimes. Thus many well-known petunia F_1 hybrids that are bred for mst seed parent become pollen fertile (Reimann-Philipp 1976). This prohibits mst use in commercial breeding in certain varieties. Besides, some varieties cannot be sterilised conveniently. In some, fertility restoration is controlled by a dominant gene and in others the control is multigenic. Environment modifies sterility expression in some gc-mst genotypes although most of them are quite stable.

Certain *fr-c* gene combinations suppress development of entire flowers. This is common in gc-mst Grandiflora types. The percentage of suppressed flower buds varies with variety and with environment. However, some gc-mst varieties with suppressed flower development (blindness), when crossed with some mft varieties, yield F_1 hybrids having normal flowers. On the other hand, some gc-mst lines with normal floral development produce "blind" hybrids when crossed to some mft lines. Environmental influence over the expression of "blindness" is great. Thus it may be possible to select lines with the least expression of "blindness" under a given set of environmental conditions. Nevertheless a careful testing of gc-mst hybrids is necessary prior to their release. Some breeders have used g-mst and incompatibility for hybrid production and gene transfers. The problems of maintaining such lines vegetatively have been greater than producing seed by hand emasculation and pollination. Despite these difficulties, use of g-mst and gc-mst in petunia breeding is increasing and has become widespread. One major advantage of gc-mst hybrids is that they tend to stay in bloom indefinitely and do not "seed themselves out of flower" as do some pollen fertile varieties.

49.5.3 *P. parodii*

The precise nature of mst in this species is not known, but from the descriptions of Wu and Murry (1985) the sterility is of gc-mst type and has been transferred from *P. hybrida* by Izhar and his coworkers in Israel. The sole morphological expression of mst is the development of less plump and lightly coloured anthers in the mst flowers. Although at the beginning of meiosis, mst and mft anthers have similar total protein contents, their protein patterns differ significantly (Wu and Murry 1985). Whereas mft anthers are enriched with three polypeptides of molecular weight 64,000, 63,000 and 45,000 Da, the mst anthers lack these and have instead two polypeptides of 180,000 and 66,000 Da. Both these polypeptides are present in mft anthers in very low proportions. SDS-PAGE reveals a very high amount of 64,000 and 45,000 Da peptides in mst anthers (this Vol.). Compared to mst anthers, proline and pipecolic acid contents are significantly higher in mft anthers.

49.6 *Solanum* L. (nightshade)

Due to the high degree of cross compatibility among diverse South American tuber- and nontuber-bearing *Solanum*, rampant interspecific hybridization and introgression lead to divergent mst types in this genus (Tables 49.15 to 49.17). This diversity is high in the species of tuber-bearing section *Tuberarium* (Tables 49.10, 49.14) and the g-mst is documented in *S. integrifolium, S. melongena, S. nigrum, S. tuberosum* and *S.verrucosum* (Table 49.15) and gc-mst occurs following an array of interspecific crosses (Tables 49.10, 49.16).

49.6.1 *S. integrifolium*, Poir

From a Russian collection of brinjals, two complete mst plants having normal-looking anthers but producing scanty sterile pollen occurred in the half-long purple brinjal (Habib and Rao 1973). Genetic control of the sterility is not known. The mutant arose spontaneously.

49.6.2 *S. melongena* L. (egg plant)

i. *Functional Male Sterility*. A functional mst mutant which possessed viable pollen but indehiscent anthers and an extended stigma to favour outcrossing was found by Jasmin (1954) in the variety Blackie. A limited anther dehiscence occurred in field. The trait is inherited by a recessive gene, *fs* (Nuttal 1963). The female fertility of this mutant is nearly normal, as seed set after hand crossings is high.

ii. *Chemical Male Sterility*. Jyotishi and Hussain (1968), Diki and Anikeenko (1975), Helal and Zaiki (1981) induced pollen abortion by spraying plants a week before flowering with 2, 4-D. Whereas 10 ppm sprays resulted in complete pollen sterility without significantly impairing female fertility (Jyotishi and Hussain 1968), 40–50 ppm sprays caused considerable ovular damage and consequent seed sterility (Diki and Anikeenko 1975; Helal and Zaiki 1981). Sprays of 20 ppm 2, 4-D or 400 ppm ethrel acted as selective and efficient male gametocides and did not reduce seed production appreciably (Helal and Zaiki 1981). Nearly complete pollen abortion occurred after two to three mendok sprays at early meiotic stages (Nasrallah and Hopp 1963).

49.6.3 *S. nigrum* L. (black nightshade)

i. *Genic Male Sterility*. In a diploid (2n = 24) highly foliacious nightshade plant, a mst plant having a light blue reduced cone of projecting anthers (instead of bright yellow) was found (this Vol.). Male meiosis proceeds normally until microspore formation, but the microspores abort gradually after liberation so that no viable pollen is formed. The anomaly is controlled by a recessive gene, *ms*$_1$. The tapetum is persistent and female fertility is normal.

ii. *Chemical Male Sterility. TIBA and Mendok*. Foliar sprays of 0.1 and 0.01% of tri-iodobenzoic acid led to the production of abnormal gamophyllous flowers having highly abnormal, reduced, pistilloid anthers (Rehm 1952). Although initial anther development was normal, irregularities in differentiation increased with the stages of development. The gynoecium was also abnormal and had either none or a few seeds. However, parthenocarpic fruits were produced. The application of 300 ppm mendok failed to induce complete mst, instead it altered floral morphology and depressed ovular fertility (Dulieu 1963).

49.6.4 *S. tuberosum* L. (cultivated potato)

The occurrence of pollen sterility in cultivated potato is a feature in its genetic system as it has been documented in various cultivars and clones from time to time (East 1908; Salaman 1910; Dorsey 1919; Breeze 1921; Young 1923; Stout and Clark 1924; Stow 1926; Fukuda 1927; Heyn 1930; Bleier 1931; Ellison 1936; Krantz et al. 1939; Okuno 1952a,b; Williams 1954; Schark 1957; Kishore et al. 1974). Considerable variability occurs in the amount of viable pollen production in various cultivars of cultivated potatoes. The amount of pollen sterility percent increases with higher temperatures given during microsporogenesis (Williams 1954). Of 106 *S. tuberosum* clones studied by Stout and Clark (1924), 79% were completely mst, 12% were partially sterile and only 9% had sufficient viable pollen. On the other hand, Longley and Clark (1930) found no mft potato plants in 37 clones; 81% were completely mst and the remaining p-mst.

49.6.4.1 Genetic Control

Salaman (1910) detected a single dominant gene conditioning mst in potato, but his later results (Salaman and Lesley 1922) indicate a more complex inheritance pattern. Krantz et al. (1939) found mst to be inherited by one recessive gene, whereas Okuno (1952a,b) indicated one to two recessive gene control of sterility, as in one case the control was by a sigle gene and in three others, the control was by two recessive genes. The reason for such a diverse gene control is the differential ploidy level and the presence of a variable number of *ms* genes in each clone. In fact, even intercrossing certain cultivars of 4x *S. tuberosum* ssp. *tuberosum* yields varying proportions of mst F$_1$ progeny, depending upon the cross combination used. Intercrossing certain such cultivars yielded around 30% mst F$_1$ progeny plants (Hanneman and Peloquin 1981) in certain crosses and about 70% in others. Cultivated potato *S. tuberosum* group Tuberosum, Dodds (2n = 48), is known to have evolved from the *S. tuberosum* group Andigena, Dodds (2n = 48), of S. America (Salaman 1949; Simmonds 1964; Hawks 1967). This ancestral species, *S. tuberosum* Andigena, possesses a large number of *ms* genes, as either 100% or up to 20% of F$_1$ progeny plants obtained after crossing clones of the Andean region of S. America and Mexico are mst (Grün 1973). With further evolution, the number of *ms* genes increases so that *S. tuberosum* has a much larger number of *ms* genes than its ancestral form. Determination of the precise number of *ms* genes in each potato cultivar is hampered by the complex segregation of *ms* alleles, the polyploid nature of the plant and the environmental influence over *ms* gene expression.

49.6.4.2 Microsporogenesis and Tapetum

The *ms* gene action is differential in different potato varieties and therefore the stage of microsporogenesis breakdown also differs in them (Table 49.9). The earliest *ms* gene action occurs in the material studied by Young (1923) in which anthers developed but degeneration of the anther tissue occurred immediately thereafter. Consequently no archesporial tissue was formed. In another mst material, anthers were phenotypically normal but no archesporial tissue was differentiated in them (Salaman 1910). In the material studied by Breeze (1921), PMC's were developed but they did not undergo meiosis. In many other mst materials, meiosis was abnormal and either restitution nuclei were formed at TI (Ellisson 1936) or anomalous meiosis continued to produce sterile, shrunken, deformed and inviable microspores (Williams 1954; Schark 1957). In some cultivars meiosis is normal, but the tapetum is persistent and microspores degenerate soon after their formation (Kishore et al. 1974). In Irish Cobble variety, tapetal degeneration is slow and microspores degenerate during mitosis (Cho and Kim 1984). Fukuda (1927) investigated over 50 American, British and Japanese potato cultivars. Various anomalies leading to mst occurred in these cultivars. These were (i) phenotypically normal pollen but only 10% germinable, (ii) degenerating mature pollen, (iii) anomalous meiosis, (iv) absence of second meiotic division, (v) whole PMC's becoming a giant nonviable pollen-like structure. In none of these was viable pollen developed. In 54 potato varieties of German origin, Heyn (1930) found complete PMC degeneration, abnormal meiosis or absence of second meiotic division as the main causes of mst. On the other hand, Bleier (1931) found PMC degeneration as the most common cause and anomalous meiosis as the other major cause of mst in 48 potato varieties of Dutch origin. Okuno (1952b) assumed pollen sterility in g-mst potatoes to be due to plasmolysis of the inner tapetal layers and of the PMC's. Thus whereas in the majority of cases, the *ms* gene either acts upon whole anther tissue or archesporium differentiation and development or on

Table 49.9. *ms* gene action in *Solanum tuberosum*

Gene action during	References
Anther development	Young (1923)
Archesporial differentiation	Salaman (1910), Fukuda (1927)
Premeiosis	Breeze (1921), Fukuda (1927), Heyn (1930)
Early prophase	Fukuda (1927), Bleier (1930), Heyn (1930), Ellison (1936), Williams (1954), Schark (1957)
Interphase	Fukuda (1927), Heyn (1930)
Microspore maturation	Kishore et al. (1974), Cho and Kim (1984)
Pollen development	Fukuda (1927)

male meiosis, in one instance reported by Kishore et al. (1974), the *ms* gene acts via the tapetum, inhibiting its functioning and normal disintegration. This starves the developing microspores to death. In the material studied by Schark (1957), both the tapetum and PMC's are equally affected by the *ms* gene, so that the development of both the tissues is defective. Probably a similar situation occurs in the material studied by Okuno (1952b). However, in the cv. Irish Cobble, the anther lobe is incompletely developed (Cho and Kim 1984). In the developing PMC's, parallel spindles and precocious cytokinesis leading to dyad formations are rampant.

49.6.5 S. verrucosum, Schlectd

This is a wild diploid (2n = 24) self-compatible Mexican species with many *ms* genes in its genome. These genes cause a variable type of mst that influences microspore development and form. Abdalla and Hermsen (1972) detected the following three types of microspore abortion in some mst introductions of this species.

i. *Bubble Type*. In one introduction the pollen grains are weakly stained and bubbly in appearance due to cytoplasmic breakdown (Fig. 49.11c). The sterility is polygenically conditioned.

ii. *Ordinary Type*. In some accessions, the pollen grains are shrunken and irregular in form. These either do not stain or stain feebly, or incompletely (Fig. 49.11b). The sterility is conditioned in some cases by recessive genes and in others by polygenes.

iii. *Undivided Microsporocyte Type*. In two accessions, two recessive genes, um_1 and um_2, prevent microspore formation within the meiocytes. The whole meiocyte develops a thick wall and four porelike configurations give it an appearance of pollenlike structure. In about 5% of meiocytes, over four nuclei are present, indicating the occurrence of additional nuclear divisions without wall formations in them. No viable pollen grain is produced.

The characteristic presence of the above-mentioned three mst types and differential genetic control of each type indicates the presence of many nonallelic *ms* genes in the genome of this diploid species.

49.6.6 Male Sterility in Interspecific Hybrids

In no other genus have so many different nuclear *fr* genes been identified that interact with the individual cytoplasms to produce many varied floral phenotypes and pollen sterility as in *Solanum*. This is evidenced from the recovery of mst following numerous interspecific hybridizations (Table 49.10). The account given below is based on cytoplasm donor species, as many genome donors produce mst offsprings in a *Solanum* cytoplasm.

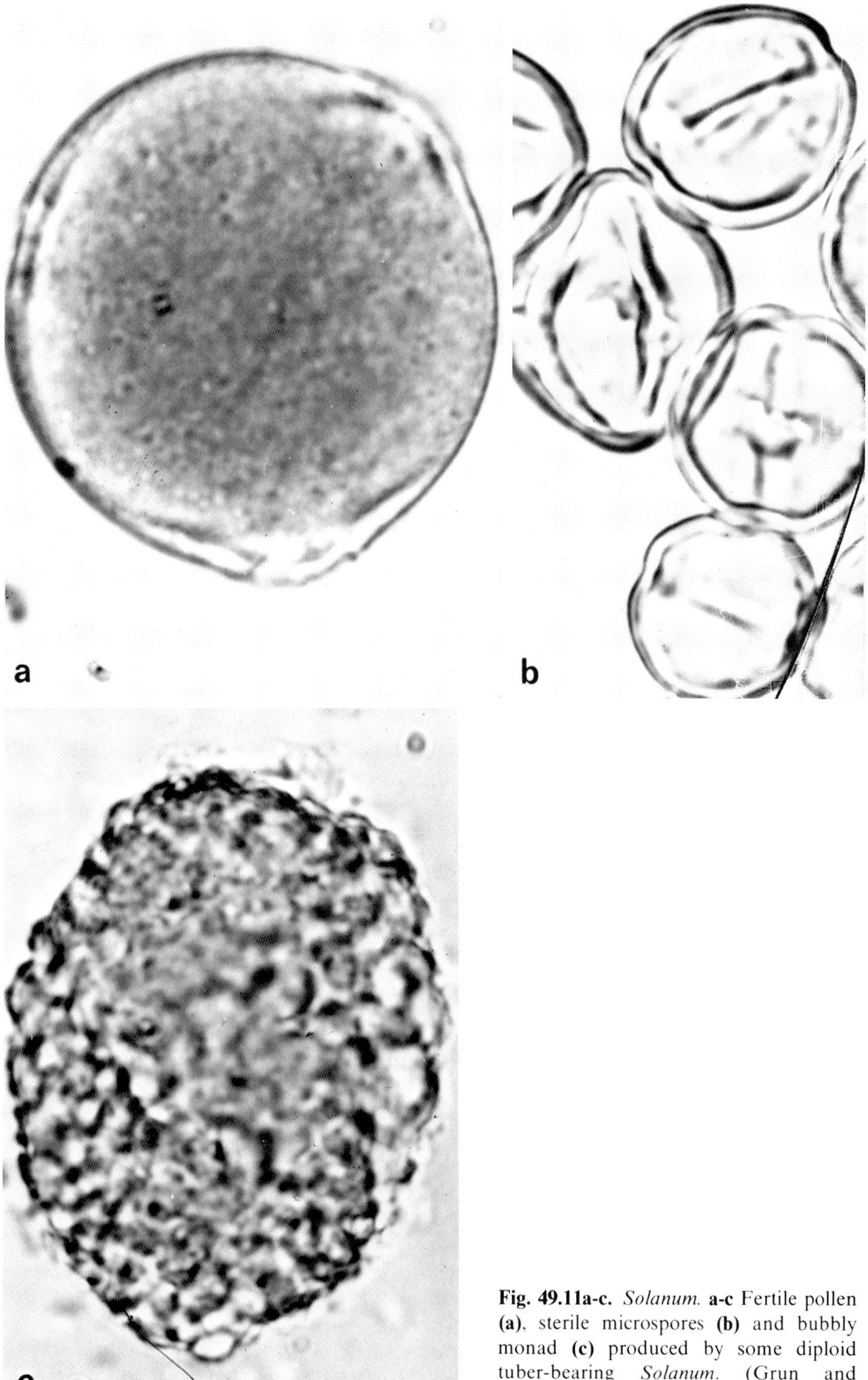

Fig. 49.11a-c. *Solanum.* **a-c** Fertile pollen **(a)**, sterile microspores **(b)** and bubbly monad **(c)** produced by some diploid tuber-bearing *Solanum.* (Grun and Aubertin 1966)

Table 49.10. Male sterility in *Solanum* hybrids

Cytoplasm	Genome	References
S.acaule (4x)	*S.chacoense* (2x)	Lamm (1953)
	S.bulbocastanum (2x)	Hermsen and Ramanna (1969)
	S.tuberosum (4x)	Lamm (1953)
S.antipoviczii (4x)	*S.chacoense* (2x)	Koopmans (1952)
	S.phureja (2x)	Koopmans (1951)
S.chacoense (2x)	*S.kurtzianum* (2x)	Grün et al. (1962)
	S.phureja (2x)	Grün et al. (1962)
	S.stenotomum (2x)	Grün et al. (1962)
	S.tuberosum (4x)	Ramanna and Hermsen (1974)
S.demissum (6x)	*S.berthaultii* (2x)	Magoon et al. (1958c)
	S.chacoense (2x)	Dionne (1961)
	S.gibberulosum (2x)	Magoon et al. (1958c)
	S.gigantophyllum (2x)	Magoon et al. (1958c)
	S.goniocalyx (2x)	Magoon et al. (1958c)
	S.phujera (2x)	Magoon et al. (1958c), Dionne (1961)
	S.saltense (2x)	Magoon et al. (1958c)
	S.simplicifolium (2x)	Magoon et al. (1958c), Dionne (1961)
	S.soukupii (2x)	Magoon et al. (1958c), Dionne (1961)
	S.tuberosum (4x)	Schnell (1948), Dionne (1961)
S.etuberosum (2x)	*S.pinnatisectum* (2x)	Hermsen and Taylor (1979), Ramanna and Hermsen (1979, 1981)
S.phureja (2x)	*S.acaule* (4x)	Koopmans (1952)
	S.chacoense (2x)	Grün et al. (1962)
	S.tuberosum (4x)	Magoon et al. (1958b)
S.polytrichon (4x)	*S.chacoense* (2x)	Toxopeus (1960, 1964)
	S.goniocalyx (2x)	Toxopeus (1960, 1964)
	S.phureja (2x)	Abdalla ans Ramanna (1971)
S.rybinni (2x)	*S.chacoense* (2x)	Koopmans (1955)
S.stenotomum (2x)	*S.chacoense* (2x)	Grün et al. (1962)
S.stoloniferum (4x)	*S.tuberosum* (4x)	Ross (1961)
S.tuberosum (4x)	*S.acaule* (4x)	Lamm (1953)
	S.andigena (4x)	De la Puente and Peloquin (1968), Grün (1973)
	S.curtilobum (5x)	Lamm (1941)
S.tuberosum (2x)	*S.ajanhuiri* (2x)	Hougas and Peloquin (1962)
	S.goniocalyx (2x)	Hougas and Peloquin (1962)
	S.phureja (2x)	Hougas and Peloquin (1962) Grün (1970)
	S.stenotomum (2x)	Hougas and Peloquin (1962)
	S.tuberosum (4x)	Howard (1970)
S.verrucosum (2x)	*S.bulbocastanum* (2x)	Hermsen and Ramanna (1976)
	S.caniarense (2x)	Abdalla (1970)
	S.chacoense (2x)	Buck (1960), Grün et al. (1962), Abdalla (1970)
	S.chomatophilum (2x)	Abdalla (1970)
	S.famatinae (2x)	Grün et al. (1962)
	S.gibberulosum (2x)	Magoon et al. (1958b)
	S.gigantophyllum (2x)	Abdalla (1970)
	S.goniocalyx (2x)	Abdalla (1970)
	S.infundibuliforme (2x)	Abdalla (1970)

Table 49.10. *(continued)*

Cytoplasm	Genome	References
	S.kurtzianum (2x)	Grün et al. (1962), Abdalla (1970)
	S.leptophyes (2x)	Abdalla (1970)
	S.macolae (2x)	Buck (1960)
	S.megistacrolobum (2x)	Abdalla (1970)
	S.multidissectum (2x)	Abdalla (1970)
	S.phujera (2x)	Buck (1960), Grün et al. (1962), Grün and Aubertin (1965), Abdalla (1970)
	S.raphanifolium (2x)	Abdalla (1970)
	S.simplicifolium (2x)	Buck (1960), Grün et al. (1962)
	S.stenotomum (2x)	Buck (1960), Grün et al. (1962) Abdalla (1970)
	S.toralapanum (2x)	Buck (1960), Grün et al. (1962)
	S.tuberosum (4x)	Magoon et al. (1958b), Abdalla (1970)
	S.vernei (2x)	Grün et al. (1962), Abdalla (1970)
	S.yabari (2x)	Magoon et al. (1958b)

(x = 12)

49.6.7 Cytoplasm Donor Species

49.6.7.1 *S. acaule*

This is a tetraploid and possesses genes for resistance to leaf-roll virus, to various races of wart and to frost. Dionne (1963) indicated its usefulness as a bridging species between *S. tuberosum* and Mexican Solanums like *S. bulbocastanum* growing mainly in semi-arid areas of Mexico and Guatemala. Some of its clones escape infection in Mexico, where most of the other resistant species become highly infected and are killed by *Phytophthora infestans*. Crossing *S. acaule* (6x) clones as female to two *S. bulbocastanum* accessions (2x) as male produced 81% triploid and 19% tetraploid F_1 plants (Hermsen 1966). The tetraploids arise from unreduced male gametes produced by the two accessions. Whereas triploids are male-female sterile, the tetraploids are mst and exhibit about 12% pollen stainability. Chromosome pairing is almost normal in the tetraploids (Hermsen and Ramanna 1969). Of this 50% is expectedly homologous and rest homoeologous. AI disjunction is also normal. Their colchicine-induced octaploids have 29% pollen stainability and both the 4x and 8x plants are mst as no seeds are set on selfing. Thus their stainable pollen is nonfunctional. Hermsen and Ramanna (1969) explain the mst as due to the carryover of cytoplasmic contents by unreduced gametes of *S. bulbocastanum* which is known to have S-cytoplasm. This needs verification.

49.6.7.2 *S. antipoviczii*

F_1 plants of *S. antipoviczii* (4x=48) × *S. chacoense* (2x=24) possess shrivelled nonviable pollen and were considered as mst by Koopmans (1952). Neither female fertility of these plants nor fertility of the plants of reciprocal crosses is known. Since the hybrid plants are triploids, meiosis is abnormal and sterility of both the sexes is expected. Hence it is desirable to reinvestigate these crosses and test them for female fertility.

49.6.7.3 *S. chacoense*

The mst occurs in F_1 plants of crosses *S. chacoense* female and *S. kurtzianum, S. phureja* and *S. stenotomum* as males. The sterility was found by Grün et al. (1962) to be due to interaction between *S. chacoense* cytoplasm and two pairs of dominant genes from the genomes of the above-mentioned male parents. In subsequent studies, Grün and Aubertin (1965) found a minimum of five independent dominant genes interacting with *S. chacoense* cytoplasm to produce mst. In the *S. chacoense* genome, two dominant fertility restorer genes were found. These genes were epistatic to the genes, causing anther indehiscence. Grün and Aubertin (1966) classified the mst plants into the following four groups on the basis of anther contents and microsporogenesis breakdown:

1. Stoppage of meiosis before or after AI leading to the formation of monads and dyads.
2. Abnormal meiosis causing aberrant disjunction resulting in the production of micronuclei and over four nonfunctional microspores per sporocyte.
3. Formation of subnormal-sized pollen grains having thickened exine and bubbly appearance (Fig. 49.11b,c).
4. Phenotypically normal pollen formed but it is neither functional nor does it shed due to the absence of terminal pores.

Grün and Aubertin (1966) consider these diverse manifestations of mst and anther types as due to modifying genes rather than to major sterility genes. These mst expressions are stable as they are not modified by environmental fluctuations and no chimeral sectors having changed anther contents or dehiscent anthers occur in the mst plants. Therefore, the role of the major genes in sterility induction is predominant.

Two F_1 plants obtained from *S. chacoense* (trisomic) × *S. tuberosum* (dihaploid) were mst (Ramanna and Hermsen 1974). In one, meiosis proceeded normally up to AI in 90% PMC's but it stopped in the remaining PMC's during early prophase-I. Subsequently, a wall was secreted around each PMC and no pollen was produced. In the second mst plant, the action of the *ms* genes was much earlier than that in the other mst plant, as meiosis stopped at early prophase in nearly 90% of PMC's; in the remaining ones, it proceeded abnormally up to the sporad stage. The meiocytes formed restitution nuclei but, unlike the other mst plant, no wall formation around the meiocytes occurred. Nuclei of these meiocytes divided rapidly so that in the older anthers more than four (12–16) nuclei, regarded as embryoids by Ramanna and Hermsen (1974), were present. The sterility is nearly akin to the undivided microsporocyte type of mst known in *S. verrucosum* and some of its interspecific hybrids. Since the action of *ms* genes differs in the two plants, these appear to differ in type and/or number of *fr* genes. Genetic control of the sterility is not known.

49.6.7.4 *S. demissum*

This medium frost-tolerant species is a herbaceous perennial allohexaploid ($6x=72$) originating from the mountains of Mexico, and has been used extensively for introducing both differential and uniform resistance to late blight. It exhibits considerable genome differentiation and relatively high crossability. After crossing the species as female to 4x *S. tuberosum* and 2x *S. chacoense, S. phureja, S. simplicifolium* and *S. soukupii*, mst's appeared in F_1 progeny (Dionne 1961) which comprised plants having various levels of polyploidy. The author selected tetraploids in which mst was nearly complete although in a few plants phenotypically normal but nonfunctional pollen was produced. The mst is due to the interaction of *S. demissum* S-cytoplasm and *fr* genes contributed by the male parents. The number and nature of the *fr* genes is not known. Male meiosis is highly abnormal in mst F_1 plants of *S. demissum* × *S. tuberosum* (Schnell 1948). The anomalies include spindle irregularities, high multivalent formations, irregular microspore number and size. These anomalies, coupled with mst, persist in BC generations also. Likewise, pollen sterility ranging between 45 and 80%, as determined after carmine staining, has been documented in F_1 plants obtained after crossing *S. demissum* as female with *S. berthaulti, S. gibberulosum, S. gigantophyllum, S. goniocalyx, S. phureja, S. saltense, S. simplicifolium, S. soukupii* (Magoon et al. 1958c). In all these tetraploid hybrids, meiosis is highly abnormal,characterised by formation of uni- tri- and quadri-valents. This produced varying-sized pollen due to the variable number of chromosomes they received. Whereas those having a higher number of chromosomes were giant in appearance, the ones with few chromosomes were miniature in size. None of the stained pollen was functional as no seed set occurred after selfing these plants. Therefore these plants were completely mst. Magoon et al. (1958c) concluded that in addition to meiotic irregularities, genetic, physiological and environmental factors are involved in pollen sterility. The female fertility of these plants is not known. Since male meiosis was abnormal due to polyploidy, it is possible that female meiosis was equally abnormal. Thus these hybrids might have been female sterile as well. This needs to be investigated.

49.6.7.5 *S. etuberosum*

Successful hybridisation between nontuberous *S. etuberosum* ($2x=24$) and a tuber-bearing species *S. pinnatisectum* ($2x=24$) was achieved by Hermsen and Taylor (1979). The hybrids are highly vigorous mst's having irregularly shaped anthers. Their pollen is shrunken, deformed and nonviable. During male meiosis, failure of chromosome pairing and a high proportion of univalents occur at MI

(Ramanna and Hermsen 1979). Further meiotic stages are also abnormal. Structural hybridity and translocation heterozygosity occurs in these hybrids (Ramanna and Hermsen 1981). From descriptions and meiotic events, the hybrid plants appear to be meiotic mutants with male-female sterility. The plants constitute a bridge between tomato and potato, the parental species possess gene pools complementary to each other and are of great value in potato breeding.

49.6.7.6 *S. phureja*

Crossing *S. phureja* (2x = 24) ♀ to *S. acaule* (4x = 28) ♂ (Koopmans 1952) and to *S. chacoense* (2x = 24) ♂ (Grün et al. 1962) leads to mst F_1 progeny. The F_1 plants of *S. phureja* × *S. acaule* possess nonviable, shrunken, deformed pollen that have degenerating cytoplasm. It is not known if the F_1 progeny of this reciprocal cross is mft or mst. Since the hybrid plants would be triploids, meiosis is expected to be abnormal and the sterility of both sexes is thus expected. On the other hand, reciprocal progeny of *S. phureja* × *S. chacoense* is mft. By using different parental strains, Grün et al. (1962) detected mst's of two different types. In one type, the flowers either lack anthers or these are present as rudiments. They were designated as "deformed flowers" although "antherless types" would have been more appropriate. The anomaly is conditioned by a pair of recessive genes. In the second type of mst's, anthers are present but they are indehiscent. Such plants were designated as "pollenless" by Grün et al. (1962), although they did have pollen. This anomaly is conditioned by a single recessive *fr* gene. The *fr* genes present in the *S. chacoense* genome interact with the *S. phureja* S-cytoplasm to produce mst progeny. Since mst expression in the two types of plants is not similar, different *fr* genes appear to condition mst in each type. This mst type is nearly similar to the one found by Koopmans (1951) in *S. rybinii* × *S. chacoense* crosses. In both these instances, the pollen parent is the same. Plants of the cross *S. phureja* × *S. tuberosum* are p-mst (Magoon et al. 1958b).

49.6.7.7 *S. polytrichon*

Although tetraploid (4x = 48), this crosses easily with *S. phureja* (2x = 24). The hybrids are triploid and mst (Ramanna and Abdalla 1970). Male meiosis is abnormal, at MI some bivalents and univalents and finally uneven-sized tetrads are formed which remain clumped. Colchicine-induced hexaploids are also mst (Abdalla and Ramanna 1971) and have a similar tetrad type of mst, but chromosome pairing is regular. Neither details about female fertility nor genetic control of mst and mft of the reciprocal crosses are known. However, since hexaploids are also mst, the sterility may be due to the interaction of *S. polytrichon* cytoplasm and *S. phureja* genome. Using the same male parent *S. phureja* to the *S. verrucosum* female, a similar type of mst occurred. This type of tetrad sterility is conditioned by a dominant gene. Since meiosis was abnormal but tetrads did not separate out, and *S. phureja* is known to have a dominant gene causing tetrad sterility, the authors assumed the sterility to be due to the interaction of genes of *S. phureja* and the cytoplasm of *S. polytrichon*. Crossing *S. polytrichon* (4x) as females

Table 49.11 Male sterility retained after colchicine diplodization *in Solanum*

S.polytrichon (4x) × *S. chacoense* (2x)	Toxopeus (1960, 1964)
S.polytrichon (4x) × *S. goniocalyx* (2x)	Toxopeus (1960, 1964)
S.verrucosum (2x) × *S.gibberulosum* (2x)	Magoon et al. (1958b)
S.verrucosum (2x) × *S.yabari* (2x)	Magoon et al. (1958b)

to *S. chacoense* (2x) and *S. goniocalyx* (2x) as males, produces mst F_1 triploid plants (Toxopeus 1960, 1964; Table 49.11). Colchicine chromosome doubling yielded female fertile amphidiploids. Although Toxopeus documented these amphidiploids as fertile while they set seeds with *S. tuberosum* pollen, they were unsuccessful as males on the commercial potato varieties. This infers the existence of mst in these amphidiploids. In all the triploid and hexaploid mst plants, pollen grains are inviable and remain clumped together at anthesis.

49.6.7.8 *S. rybinii*

F_1 of *S. rybinii* ♀ (2x = 24) × *S. chacoense* ♂ (2x = 24) are normal but in F_2 some abnormal plants segregate (Koopmans 1951). In these, anomalies are confined to anthers and ovary viz.: anthers absent or reduced to one half to one quarter or to rudiments, rudimentary anthers having stigma-like tops or anthers present but indehiscent. These anomalies do not occur in either of the parents. Backcrossing the anomalous plants with *S. chacoense* pollen increased the frequency of abnormal plants and the intensity of the abnormalities (Koopmans 1952, 1954). Crossing abnormal F_2 or BC_1 plants by *S. rybinii* led to normal progeny. The mst is caused by the interaction of *S. rybinii* cytoplasm and recessive *fr* genes of *S. chacoense* genome. Using two strains of *S. chacoense* and repeating the above-mentioned cross, Koopmans (1955) found that whereas one strain produced only normal offspring, the other produced two berries, one of which yielded a normal F_1 population and the other mst progeny. The progenies obtained after crossing F_1 normal or mst plants by F_1 normals segregated both mft and mst plants. Backcrosses of mft and mst F_1 plants by *S. rybinii* pollen also produced both mst and mft plants. The pollen percentage of mst plants in BC_1 and BC_2 was 88.6 and 99.2, respectively (Koopmans 1959). However, when the mst plants were backcrossed by *S. chacoense* pollen, the whole progeny was normal and mft. This confirms the earlier contention that mst results due to interaction of *S. chacoense* genes with *S. rybinii* cytoplasm. The number of genes conditioning mst is not known, but from these crossings it is evident that two types of cytoplasm (N and S) exist in *S. chacoense*.

49.6.7.9 *S. stenotomum*

F_1 progeny of *S. stenotomum* (2x = 24) ♀ × *S. chacoense* (2x = 24) ♂ is mst (Grün et al. 1962). In its characteristic features and inheritance pattern, the sterility resembles that found in *S. phureja* × *S. chacoense* hybrid plants.

Table 49.15 *(continued)*

Species	References	Gene control	Gene action	Remarks
Petunia axillaris	Levan (1942)	1r	Before microspore release	
	This Vol.	1r	After microspore liberation	
P. hybrida	Izhar (1973)	1r	During microspore development	
	This Vol.		After release from PMC's	
	Frankel (1962, 1971)	nk	During tetrad formation, before release	
	This Vol.	nk	During early prophase I	
	Welzel (1954)	2r	During prophase to diakinesis	X-ray-induced, persistent tapetal malfunction
	Cornu and Dommergues (1974)	Many r	Early prophase-I	EMS + γ-ray-induced
	Singh (1975)	Many r	After microspore release	Induced γ-ray mutant tapetal malfunction
Solanum integrifolium	Habib and Rao (1973)	nk	nk	
S. melongena	Jasmin (1954)	1r	nk	Functional mst
S. nigrum	This Vol.	1r	After microspore release	Tapetum persistent
S. tuberosum	Salaman (1910)	1D	During premeiosis	
	Young (1923)	1–2r	During premeiosis	
	Fukuda (1927)	Many r	During premeiosis	
	Breeze (1921)	nk	During premeiosis	
	Fukuda (1927), Heyn (1930)	nk		
	Fukuda (1927), Bleier (1930), Ellison (1936), Williams (1954), Schark (1957)	nk	During prophase-I	
	Fukuda (1927), Heyn (1930)	nk	After interphase	
	Kishore et al. (1974)	nk	After microspore release	Tapetum persistent and abnormal
	Fukuda (1927)	nk	During pollen maturation	
	Okuno (1952a,b)	1r	nk	
S. verrucosum	Abdalla and Hermsen (1972)	2r	After TII	
		Many r	After microspore release	

Table 49.16. Gene-cytoplasmic male sterility in Solanaceae

Species/cross	References	Gene control	Gene action	Remarks
Capsicum annuum	Hirose (1965), Pochard (1970b), Novâk et al. (1971), Horner and Rogers (1974)	1r	During premeiosis	Influenced by modifier genes and temperature, tapetum abnormal
	Peterson (1958)	1r		
	Hirose and Fujime (1975)	2r	After tetrad formation, before release	
	Novâk et al. (1971)	2r	After microspore liberation	Complementary gene action, premature tapetal disintegration
C.annuum × *C.pendulum*	Rusenova-Kondareva (1968)	nk	During stamen formation	
C.annuum × *C.pubescens*				
C.peruvianum × *C.annuum*	This Vol.	nk	During stamen development and differentiation	
C.baccatum × *C.annuum*	Shiffris C (personal communication), this Vol.			
Lycopersicon esculentum × *L.hirsutum*	Siegemund (1974)	1r	nk	
L.esculentum × *L. parviflorum*	Kesicki (1980)	3r	nk	*ms* gene-linked, functional mst
L.peruvianum × *L. hirsutum*	Vulkova-Achkova (1980)	nk	nk	Alloplasmic
L. cheesmanii × *Solanum pennelli*	Andersen (1964)	nk	nk	—
L.esculentum × *S.pennelli*	Andersen (1963a,b)	nk	nk	Alloplasmic
L.hirsutum × *S. pennelli*	Vulkova-Achkova (1980)	nk	nk	Alloplasmic
(*L.minutum* × *L.hirsutum*) × *S.pennelli*	Andersen (1964)	nk	nk	Alloplasmic
Nicotiana	Refer to Table 49.5			

Table 49.16 *(continued)*

Species/cross	References	Gene control	Gene action	Remarks
Petunia hybrida	Hamilton (1965)		After microspore liberation	
	Edwardson and Warmke (1967)	2r	After microspore formation	Modifier genes influence mst
	Frankel et al. (1969)		During premeiosis or early meiosis	
	Izhar (1975), this Vol.		nk	Temperature influenced ms gene expression considerably
	Izhar (1977)		During prophase-I or tetrad formation or after microspore liberation	Depending on genotypic background fr gene action is variable
			Before or after pollen mitosis	More *Fr* genes delay *fr* gene action
	Bino (1985,a,b)		Premeiosis to early meiosis	Tapetal development abnormal
P.parodii	Wu and Murry (1985)		nk	Tapetum abnormal
Solanum	(Refer to Tables 49.10 to 49.14)			

Table 49.17. Chemical male sterility in Solanaceae

Species/Chemicals	References	Male sterility induction[b]	Female fertility reduction	Side effects
Capsicum annuum				
Dalapon	Chauhan (1980)	c	Variable	nk
GA_3[a]	Sawhney (1981)	b	Very low	Fruit deformation and distortions
MH	Chauhan and Singh (1972), Chauhan (1980)	b,c	High	Fruit deformation and distortions
Mendok	Hirose (1965), Markus (1968)	a	Very high	nk
Lycopersicon esculentum				
2,4-D	Rehm (1952)	a	Very high	nk
MH	Rehm (1952), Chopra et al. (1960) Jain (1956), Chopra et al. (1960)	b	Nil	nk
Mendok	Moore (1959, 1964)	b	Medium	nk
	Walkof (1959)	b	nk	nk
	Aung and Byrne (1977)	a	Medium	nk
Morphactin	Jain and Mukherjee (1980)	a	High	Modifies flower structure and fruit shape
TIBA	Rehm (1952)	b	Low	nk
Nicotiana rustica				
2,4-D		b	Complete	Highly toxic
MH	Jos and Singh (1967b)	a	High	High toxicity
Mendok		c	nk	nk
N.tabacum				
2,4-D	Jos and Singh (1967)	a	Very high	Highly toxic
	Dubey and Singh (1968)	d	Low	nk
MH	Jos and Singh (1967a)	a	Very high	Considerable floral abscission, stunded plant growth
Mendok	Jos and Singh (1967a)	a	Low	nk
Petunia hybrida				
Mendok*	Sink and Gunesch (1966)	a	Very high	Toxic and caused anomalies
Solanum melongena				
2,4-D	Jyotishi and Hussain (1968), Helal and Zaiki (1981) Diki and	b	Low	nk
	Anikeenko (1975)	b	Very high	nk
Ethrel	Helal and Zaiki (1981)	b	Low	nk
Mendok	Nasrallah and Hopp (1963)	b	nk	nk
S.nigrum				
Mendok	Dulieu (1963)	d	High	Altered floral morphology
TIBA	Rehm (1952)	b	High to complete	nk

[a]Results highly variable.
[b]a = high, b = complete, c = variable, d = partial.

seeds can be stored under dry conditions for several years without deterioration, storage of tubers is a major problem. Furthermore, transmission of disease-causing organisms can be virtually eliminated through the use of true seed. Therefore, for commercial seed production, major reproductive alterations in *S. tuberosum* are to be made, as most of the present-day cultivars are mst and often fst also. Therefore as such they cannot be used in a breeding program. Large genetic stocks of *Solanum* species having diverse genetic variability are available in germplasm banks, especially at Lima, Peru, but their utilisation in potato breeding and improvement is limited due to sterility in the edible potato ssp. *tuberosum*. Broadening the genetic and S-cytoplasm base of *S. tuberosum* is essential and the introduction of resistant genes especially for early and late blight, mildews, viruses, scabs, ring and dry rots, internal tuber spots and many pests and worms etc. is necessary. *S.bulbocastanum, S.demissum, S.polytrichon, S.stoloniferum* and *S.verrucosum* possess resistant genes for late blight and for a few of the above-mentioned infections. Some clones of *S. bulbocastanum* are highly resistant to *Phytophthora-infestans* even in Mexico, where the most resistant species become infected or killed by the fungus. Similarly, *S. acaule* has a dominant gene for resistance to leaf-roll virus to many races of wart and to frost. *S. acaule* and *S.stoloniferum* exhibit a wide distribution and abundance, and therefore can be easily used in potato breeding. Direct transfer of the useful genes from the above-mentioned wild *Solanum* species is not possible, but useful bridging species between Mexican *Solanum* and *S. tuberosum* occur and these can be utilised using the present-day available breeding techniques and mst. Tetraploidy appears to be the optimum ploidy level for potato. Artificially induced octaploids of *S. tuberosum* are slow-growing, weak and unthrifty. Hexaploids although found in wild *Solanum* species, are not better adapted than diploids and are absent in the cultivated potatoes. Therefore, polyploidy breeding may not be successful. Instead, induction of haploidy is useful in *S. tuberosum*, since its dihaploids, compared to tetraploids, are easily crossable as female parents. As enlargement of the genetic base is a fundamental requirement for potato breeding, certain useful suggestions (i-iv) advanced by Grün and Staub (1979) are given below:

i. *Genome Transfer*. To enlarge the genetic variability in *S. tuberosum*, it is desirable to transfer its genome into the cytoplasms of wild *Solanum* species, most of which have N-cytoplasm. This necessitates a mft potato line which could be used as male parent, and four to five BC generations. Unfortunately, the majority of the best potato materials available at present have *fr* genes and S-cytoplasm. Therefore, they cannot be used as male parents in breeding procedure. Moreover, many BC generations are needed for genome transfer and the interaction of the *S. tuberosum* genome with many alien cytoplasms is unknown.

ii. *Unreduced Gamete Production*. Instead of genome transfer, transfer of genes conditioning adaptability, disease resistance, or earliness etc. from diploid *Solanum* to the tetraploid, edible potato ssp. *tuberosum* could be accomplished in crosses with potato as female in tetraploid (4x) or dihaploid (2x) form, e.g. (4x × 2x, 2x × 4x). In both these crosses, 4x offspring are produced through the functioning of unreduced gametes of 2x parents by a mechanism involving the formation of parallel spindles which lead to 1st division restitution nuclei (Quinn et al. 1974).

Table 49.17. Chemical male sterility in Solanaceae

Species/Chemicals	References	Male sterility induction[b]	Female fertility reduction	Side effects
Capsicum annuum				
Dalapon	Chauhan (1980)	c	Variable	nk
GA_3[a]	Sawhney (1981)	b	Very low	Fruit deformation and distortions
MH	Chauhan and Singh (1972), Chauhan (1980)	b,c	High	Fruit deformation and distortions
Mendok	Hirose (1965), Markus (1968)	a	Very high	nk
Lycopersicon esculentum				
2,4-D	Rehm (1952)	a	Very high	nk
MH	Rehm (1952), Chopra et al. (1960) Jain (1956), Chopra et al. (1960)	b	Nil	nk
Mendok	Moore (1959, 1964)	b	Medium	nk
	Walkof (1959)	b	nk	nk
	Aung and Byrne (1977)	a	Medium	nk
Morphactin	Jain and Mukherjee (1980)	a	High	Modifies flower structure and fruit shape
TIBA	Rehm (1952)	b	Low	nk
Nicotiana rustica				
2,4-D		b	Complete	Highly toxic
MH	Jos and Singh (1967b)	a	High	High toxicity
Mendok		c	nk	nk
N.tabacum				
2,4-D	Jos and Singh (1967)	a	Very high	Highly toxic
	Dubey and Singh (1968)	d	Low	nk
MH	Jos and Singh (1967a)	a	Very high	Considerable floral abscission, stunded plant growth
Mendok	Jos and Singh (1967a)	a	Low	nk
Petunia hybrida				
Mendok*	Sink and Gunesch (1966)	a	Very high	Toxic and caused anomalies
Solanum melongena				
2,4-D	Jyotishi and Hussain (1968), Helal and Zaiki (1981)	b	Low	nk
	Diki and Anikeenko (1975)	b	Very high	nk
Ethrel	Helal and Zaiki (1981)	b	Low	nk
Mendok	Nasrallah and Hopp (1963)	b	nk	nk
S.nigrum				
Mendok	Dulieu (1963)	d	High	Altered floral morphology
TIBA	Rehm (1952)	b	High to complete	nk

[a]Results highly variable.
[b]a = high, b = complete, c = variable, d = partial.

Of all the mst types, the tetrad type of sterility is most common in mst plants having S-cytoplasm of *S. verrucosum* (Buck 1960; Grün et al. 1962; Abdalla and Hermsen 1971), *S. polytrichon* (Toxopeus 1964; Ramanna and Abdalla 1970; Abdalla and Ramanna 1971) and haploid *S. tuberosum* (Peloquin and Hougas 1960; Abdalla 1970). The sterility is conditioned by a dominant *Fr* gene and is closely linked to another dominant gene causing pollen lobedness (Abdalla and Hermsen 1971). Because of the close linkage of these two genes, Grün et al. (1962) named the tetrad sterility "lobed sterility". This multiple genome-cytoplasmic-controlled mst phenotypes in *Solanum* make this genus an interesting target for biochemical and molecular analysis, which remains to be done.

49.6.10 Cytoplasm Differentiation and Mutations in Potato

i. *Cytoplasm Differentiation.* Cultivated potato *S. tuberosum* (4x = 48, group *Tuberosum* Dodds) ssp. *tuberosum* is considered to have evolved from *S. tuberosum* (4x = 48, group *Andigena* Dodds) ssp. *andigena*, the commonly cultivated potato of S. America (Salaman 1949; Simmonds 1964; Hawks 1967). However, the ssp. *andigena* and *tuberosum* exhibit a substantial evolutionary distance (Grün 1979). Besides, the two subspecies are morphologically distinct, occupy separate and markedly different native ranges, and when crossed as *tuberosum* ♀ × *andigena* ♂, the hybrids are usually partially to fully sterile. Furthermore, the two sub-species differ in five of the six gene-cytoplasm sensitivities listed in Table 49.14. The *fr* genes and one *fs* gene present in dominant form in *andigena* cause partial or complete male or female sterility in *tuberosum* that pre-dominantly has these in recessive form along with S-cytoplasm. Thus whereas *andigena* is a storehouse of dominant *Fr* genes but has N-cytoplasm, *tuberosum* has mostly recessive *fr* genes and S-cytoplasm. Hence the two differ genetically in *fr* nuclear and *c*-cytoplasm genes. If the N-cytoplasm of *andigena* mutates to the S-form, the plants would be totally sterile. Such forms do possibly arise in nature, but they are lost subsequently because total sterility associated with some other lethal effects. Cytoplasmic mutations are not common events and are rare in asexually propagated crops which have the least chances of reproductive progression. It appears that *andigena* and *tuberosum* may have a common parent and at the time of origin *tuberosum* had recessive *fr* genes and was fertile, having N-cytoplasm. During the course of development and stabilisation its cytoplasm mutated to the S-form and differentiated gradually. Occasional outcrossings to N. Andean *Solanum* including *andigena* introduced dominant *Fr* genes in the hybrid genome of F_1 plants which became perpetuated because of their hybrid vigour, selective advantage and vegetative propagation. The perpetuation and progression of sterility depends upon the nature of the sterility and the agronomic usefulness of the genotype. Whereas female steriles become removed from the evolutionary line of descent as their transmission through eggs is not possible, mst's would remain and perpetuate because potatoes are chosen on tuber traits and are propagated asexually. If a farmer found a mst clone having one or many agronomically useful traits, like improved tuber yield and/or quality, early maturity, or disease resistance, and propagated it, he would have a field composed exclusively of mst plants. Thus in such circumstances, mst's do not hinder cultivation or perpetuation. Since mst's do not have to expend resources for pollen, seed and fruit production, they might probably have conferred some selective advantage, viz. bigger tuber size and number. This could be one of the reasons that many of the present day cultivars are mst and are propagated and perpetuated clonally.

ii. *Cytoplasm Mutations.* These, although rare, are known in many major food crops like wheat, rice, maize, pearl millet, sorghum, onion etc. It is possible that cytoplamic diversity may occur in *S. tuberosum*, but the existence of many S-cytoplasm types and a large number of *fr* genes effective in these cytoplasms is certain. The S-cytoplasms have arisen as a result of mutations of the N-cytoplasm and have undergone differentiation in the passage of time. Thus basically only one S-cytoplasm existed in *S. tuberosum*, but it differentiated gradually. The number of *fr* genes is large and many of them are nonallelic. Genomes of many diploid and its related *Solanum* species of the North Andean region possess such nonallelic *fr* genes, but most of them have N-cytoplasm and are therefore mft. Using 21 introductions of *S. verrucosum* as females to three *S. phureja* males, Abdalla and Hermsen (1971) reported the presence of three S-cytoplasms on the basis of different *fr* gene expressions of the resultant F_1 plants. The expressions were of eclipse, striped-vacuolar and tetrad types. Since all these three mst types are conditioned by the interaction of dominant genes with the S-cytoplasm of *S. verrucosum*, their identification and detection is easier. However, seven different mst types manifested in *S. verrucosum* cytoplasm indicate the existence of many more S-cytoplasms, as well as *fr* genes, in this species. This diploid species appears to exhibit a higher cytoplasmic mutation and differentiation than cytogenetic variability, as no stable cytotypes or polyploids are known in this species.

49.6.11 Potato Breeding

Most of the present-day clones of cultivated potato *S. tuberosum* are either completely or partially mst. Many of these are female sterile (fst) as well. Compared to fst, mst is of a high order and magnitude in them. This is because of the presence of a large number of S-cytoplasm-sensitive *fr* genes, some of these being dominant, some recessive and others polygenic. Furthermore, the species possess many S-cytoplasm types sensitive to *fr* genes. Being propagated by tubers, all potato varieties are heterozygous. Homozygous plants can be obtained by several years of selfing that lead to inbreeding depression. A large number of lethal genes appear, leading to morphological anomalies and seedling death. Furthermore, potatoes require a comparatively large amount of space and labour, making it difficult to grow the large populations required for genetical investigations. Being a polyploid crop, segregations are complex and bud mutations and chimeras appear frequently. The difficulties have hampered its breeding and narrowed down its genetic base considerably. For enhancing genetic variability for subsequent use in breeding, and for commercial potato production, use of seeds in place of clones is being emphasised, as it reduces cost of transportation, tuber handling and storage, and also simplifies the virus-control problems (Sawyer 1978). Thus whereas true potato seeds can be stored under dry conditions for several years without deterioration, storage of tubers is a major problem. Furthermore, transmission of disease-causing organisms can be virtually eliminated through the use of true seed. Therefore, for commercial seed production, major reproductive alterations in *S. tuberosum* are to be made, as most of the present-day cultivars are mst and often fst also. Therefore as such they cannot be used in a breeding program. Large genetic stocks of *Solanum* species having diverse genetic variability are available in germplasm banks, especially at Lima, Peru, but their utilisation in potato breeding and improvement is limited due to sterility in the edible potato ssp. *tuberosum*. Broadening the genetic and S-cytoplasm base of *S. tuberosum* is essential and the introduction of resistant genes especially for early and late blight, mildews, viruses, scabs, ring and dry rots, internal tuber spots and many pests and worms etc. is necessary. *S.bulbocastanum, S.demissum, S.polytrichon, S.stoloniferum* and *S.verrucosum* possess resistant genes for late blight and for a few of the above-mentioned infections. Some clones of *S. bulbocastanum* are highly resistant to *Phytophthora-infestans* even in Mexico, where the most resistant species become infected or killed by the fungus. Similarly, *S. acaule* has a dominant gene for resistance to leaf-roll virus to many races of wart and to frost. *S. acaule* and *S.stoloniferum* exhibit a wide distribution and abundance, and therefore can be easily used in potato breeding. Direct transfer of the useful genes from the above-mentioned wild *Solanum* species is not possible, but useful bridging species between Mexican *Solanum* and *S. tuberosum* occur and these can be utilised using the present-day available breeding techniques and mst. Tetraploidy appears to be the optimum ploidy level for potato. Artificially induced octaploids of *S. tuberosum* are slow-growing, weak and unthrifty. Hexaploids although found in wild *Solanum* species, are not better adapted than diploids and are absent in the cultivated potatoes. Therefore, polyploidy breeding may not be successful. Instead, induction of haploidy is useful in *S. tuberosum*, since its dihaploids, compared to tetraploids, are easily crossable as female parents. As enlargement of the genetic base is a fundamental requirement for potato breeding, certain useful suggestions (i-iv) advanced by Grün and Staub (1979) are given below:

i. *Genome Transfer*. To enlarge the genetic variability in *S. tuberosum*, it is desirable to transfer its genome into the cytoplasms of wild *Solanum* species, most of which have N-cytoplasm. This necessitates a mft potato line which could be used as male parent, and four to five BC generations. Unfortunately, the majority of the best potato materials available at present have *fr* genes and S-cytoplasm. Therefore, they cannot be used as male parents in breeding procedure. Moreover, many BC generations are needed for genome transfer and the interaction of the *S. tuberosum* genome with many alien cytoplasms is unknown.

ii. *Unreduced Gamete Production*. Instead of genome transfer, transfer of genes conditioning adaptability, disease resistance, or earliness etc. from diploid *Solanum* to the tetraploid, edible potato ssp. *tuberosum* could be accomplished in crosses with potato as female in tetraploid (4x) or dihaploid (2x) form, e.g. (4x × 2x, 2x × 4x). In both these crosses, 4x offspring are produced through the functioning of unreduced gametes of 2x parents by a mechanism involving the formation of parallel spindles which lead to 1st division restitution nuclei (Quinn et al. 1974).

The 1st division restitution gametes are highly heterozygous. Therefore 4x hybrids produced by such gametes are often highly vigorous. (Kidane-Mariam and Peloquin 1974; Mendiburu and Peloquin 1977). Since cultivated potato is the female parent in both the cross combinations, the S-cytoplasm is retained in the hybrid and their breeding value is limited drastically.

iii. *Protoplast Fusion.* To avoid the limitations imposed by the S-cytoplasm and the difficulties in using the BC method of genome transfer, the protoplast fusion technique is the most promising. In this, potato protoplast could be fused with the irradiated or enucleated cytoplasm donor species. Cybrids having nuclei of potato and cytoplasm of the donor *Solanum* species would have to be selected and cultured to adult plants as has been done in mst tobacco (see Belliard et al. 1977). Fortunately, basic requirements for potato protoplast product, culture and fusion have already been worked out (see Shepard and Totten 1977; Butenko et al. 1977; Grün and Chu 1978; Melchers et al. 1978). This method may yield male and female fertile clones of potato that can be used in potato breeding and improvement.

iv. *Use of Dihaploids. S. tuberosum* is a tetraploid and its haploids are mostly described as dihaploids as they are diploids in relation to the basic haploid number of 12. Most of the dihaploids are mst even when they have been produced from pollen fertile varieties. The yield of some F_1 hybrids of dihaploids *S. tuberosum* × *S. phureja* equals those of commercial *S. tuberosum* varieties. Clones with shallow eyes, long dormancy and a relatively low number of large tubers can be obtained in BC_1, dihaploid *S. tuberosum* × (*S. phureja* × dihaploid *S. tuberosum*) (Peloquin et al. 1966). Howard (1970) suggested another way of using dihaploids in potato breeding. Duplex tetraploids can be obtained from mst dihaploids having dominant genes for resistance to diseases and pests. These, when crossed with nulliplex parents, would give a ratio of five resistant: one susceptible compared to 1:1 obtained from nulliplex × simplex crosses. Where two or three such dominant genes are involved, duplex parents are even more useful, as 70 and 57% respectively of the progeny having the desirable genes can be obtained in crosses with nulliplex with 25% and 13% from the comparable simplex parents. If two of the dominant genes are on the homologous chromosomes, the advantage is greater, as it is easier to recognise clones in dihaploids in which the two genes are in the coupling phase.

CHAPTER 50

Tiliaceae

50.1 *Corchorus capsularis* L. (jute, jews' mallow)

No spontaneously arisen mst is known in this economically most useful Indian native, but from M_2 progeny of 70 kR X-irradiated dry seed of the cultivar JRC-212, two slow-growing mst mutants having ribbon-type leaves were found by Rakshit (1967) and Singh et al. (1973). In one of them, the *ms* gene and the gene for ribbon leaf shape are closely linked with each other (Table 50.1). Mitra (1977) regards the association as due to pleiotropy, which includes many other undesirable traits like small flowers, short narrow petals having a distinct notch at the tip and diminutive anthers. Since the mutant exhibits much delayed stigma receptivity and very low seed yield (Banerjee and Rakshit 1970), it is unsuitable for hybrid breeding. Sterility in each is controlled by a recessive gene, the allelic nature of the two *ms* genes and their influence on microsporogenesis are not known.

Table 50.1. ms Gene action in ***Corchorus***

Species	References	Remarks
[a]*Corchorus capsularis*	Rakshit (1967)	Two mst mutants, one exhibits linkage with narrow leaves, others pleiotropism
	Mitra (1977)	X-Ray-induced mutant

[a]Gene control by lr.

CHAPTER 51
Urticaceae

51.1 *Pilea microphylla* Liebm (artillery plant)

In a monoecious strain of this greenhouse ornamental, staminate flowers of a plant did not open even when kept in sunny places. Unlike others in which an artillery-like pollen discharge occurred, stamens of this plant were rudimentary and lacked anther sacs. The anomaly is controlled by one *ms* gene. Female fertility of the plant is normal (this Vol.).

CHAPTER 52

Valerianaceae

52.1 *Centranthus ruber* (L.) DC (red valerian)

In this ornamental species, Sobrinho (1946) documented digenic control of mst. Basically, the mst is conditioned by a dominant *Ms* gene, but its action is inhibited by another nonallelic dominant gene *In* (Table 52.1). Thus the genotypes with *Ms Ms in in* or *Ms ms in in* are mst but those with *Ms Ms In In*, or *Ms ms In in*, or *ms ms In in* or *ms ms In In* or *ms ms in in* are mft. On the other hand, in a spontaneously arisen mst plant, highly reduced anthers lacking sporogenous tissue occurred (this Vol.). The sterility is controlled by one recessive *ms* gene. Male meiosis proceeds normally until microspores are liberated from the PMC's. Thereafter, microspores and tapetal degeneration occur almost simultaneously.

Table 52.1. Gene action in Valerianaceae

Species	References	Gene control	Gene action	Remarks
Centranthus ruber	Sobrinho (1946)	2D	nk	
	This Vol.		After microspore release	Tapetum degenerates along with microspores
Valeriana officinalis	Shugaeva (1979)	nk	During meiosis I	Tapetum degenerates along with PMC's

52.2 *Valeriana officinalis* L. (common valerian)

Most of the 270 known species of this genus are showy hardy garden plants, some of them exhibiting random pollen abortions. Of these, *V. officinalis*, prized for its pleasant spicy fragrance, exhibits nonrandom anther abortion, in 3–5% frequency, but in an experimental population of this species, 35–44% of the plants were mst (Shugaeva 1979). Their anthers are reduced and no pollen grains are formed in them. Microsporogenous and tapetal cell degenerations occur mostly either before or during first division of male meiosis (Table 52.1) and rarely during second division. Female fertility is normal as seeds are set after open pollinations. Although the precise genetic control of the sterility is not known, the segregations reveal monogenic recessive gene control.

CHAPTER 53
Violaceae

53.1 *Viola* L. (violet)

This widely distributed genus has over 300 species, most of which are ornamental perennial herbs. However, mst is known only in *V. orphanidis*, *V. tricolor* and *V. arvensis* × *V. tricolor* hybrid (Table 53.1).

Table 53.1. Genic male sterility in Violaceae[a]

Species	References	Gene action	Remarks
Viola orphanidis	Clausen (1930)	During prophase-I	Male-dominated asynapsis
V.tricolor	This Vol.	After microspore release	Tapetum degeneration simultaneous
V.arvensis × *V.tricolor*	Clausen (1926)	Before microspore release	

[a]Gene control by 1r.

53.1.1 *V. orphanidis* Boiss (common violet)

Plants indistinguishable morphologically but distinct cytologically by having 2n = 20,21,22 chromosomes were found by Clausen (1930). From thier progeny, nearly 25% mst's segregated that had either 20 or 21 or 22 chromosomes. Of these, all the 20 chromosome plants were mst. They had dry, shrunken pollenless stamens. Their PMC's were unable to complete first meiotic division. Prophase-I was delayed and synapsis inhibited. This followed entire PMC degeneration. From the degenerating PMC's, branched fibrillous attachments projected out, forming a network-like structure within the pollen sacs. Even the wall between two pollen sacs of the same stamen became dissolved. While completely unpaired chromosomes during early male meiosis were rampant, female meoisis was normal and the sex functional. The sterility is due to a *ms* gene that causes male sex-specific complete asynapsis. The absence of a specific chromosome carrying its dominant allele *Ms* results in mst in 20 chromosome plants.

53.1.2 *V. tricolor* L. (garden pansy, heart's ease)

From this ornamental and medicinal diploid (2n = 26) violet flowered population, a light blue petaled plant having indehiscent, nonhairy anthers arose spontaneously (this Vol.). Male meiosis is normal, but the maturing microspores degenerate along with the tapetum. The anomaly is controlled by one recessive *ms* gene.

53.1.3 *V. arvensis* × *V. tricolor*

Clausen (1926) documented mst in these hybrids. In them, male meiosis was highly abnormal and microspores aborted prior to anthesis. The nature and genetic control of sterility and status of female fertility are not known. Similar hybrids recovered by the present author are both male-female sterile and also exhibit highly abnormal meiosis. Therefore those obtained by Clausen (1926) may have also been total steriles.

CHAPTER 54
Vitaceae

54.1 *Vitis* L. (Grape)

54.1.1 *Vitis vinifera* L. (wine grape)

Olmo (1943) reported that flowers of the edible grape variety Almeria have completely reflexed stamens whose filaments are slender and short. Their anthers released extremely scanty pollen. In them, microsporogenesis breaks down during pollen development (Table 54.1) and the pollen does not germinate under natural or artificial conditions. The pollen lacks germ pores and exhibits poor stainability. Similar reflexed stamen-type sterility in many other vine varieties has been earlier documented by Portele (1883), Rathay (1888–1889), Olav (1930) and Kaczmarek (1938). Microsporogenesis details or genetic control of the sterility are not known. From India, Negi et al. (1974) and Singh et al. (1979) have also documented the existence of mst in grape varieties, the latter author and Neagum and Georgescu (1971) obtained reduced and variable fruit set in them. This indicates impaired female sterility.

Table 54.1. Microsporogenesis breakdown in male sterile *Vitis*[a]

Species	References	Breakdown stage	Remarks
Genic male sterile			
Vitis vinifera	Olmo (1943)	During pollen development	Pollen lacks germ pores
Gene-cytoplasmic male sterile			
Vitis vinifera × *V.riparia*	Staudt (1973)	After microspore release	Sterility is transmitted maternally

[a]Gene control not known.

54.1.2 *V. vinifera* × *V. riparia*

In the grape cultivar Aris, obtained after crossing, pollen sterility was partial and ranged between 62 and 98% (Staudt 1973). While male meiosis proceeded normally, microspores degenerated immediately after release from PMC's (Table 53.1). This degeneration was enhanced by the low temperatures prevailing during postmeiotic microsporogenesis. However, since sterility persisted in this cultivar regularly for 3 years and also occurred in its descendents, it appears genic. Detailed genetical studies to determine mst type and its control have not been done. Grape breeding is difficult because of the small flower size, poor berry and seed set and low seed germinability. This necessitates the isolation and development of stable mst genotypes for use in hybrid breeding.

54.1.3 Sex Conversion

Spraying of adenine derivatives, cytokinins and ethrel, partly suppressed male flower development in *V. vinifera* and led to a low degree of sex conversion (Table 54.2). In a grape vine clone 1000 ppm cytokinin, BAP (6-benzylamino)-9 (2-tetra-hydroxypyranyl)-9-H-purine, applied 3 weeks before anthesis, completely converted the sex from male to hermaphrodite (Negi and Olmo 1966). Development of female organs on male plants occurred following 1000 ppm BA (6-benzyladenine) and BAS (6-benzyladenosine) and BAP treatments given 18–35 days before anthesis (Iizuka and Hashizumi 1968). A similar type of sex conversion was achieved by Moor (1950) by cytokinin applications in male clones of four *Vitis* species and an interspecific hybrid. The chemical application time in relation to inflorescence development is a critical factor in the sex conversion. Kender and Remailey (1970) reported the promotion of femaleness in the staminate flowers of a grape vine clone after ethrel treatments. This led to the formation of fruits and seeds on this grape vine stock. On the other hand, Iyer and Randhawa (1965) failed to obtain sex conversions with mendok sprays. Most of these sprays in grapes are phytotoxic (this Vol.).

Table 54.2. Chemical sex conversion from maleness to hermaphroditism in *Vitis vinifera*

Chemical	Male sex suppression	References
6-Benzyladenine	Partial	Iizuka and Hashizumi (1968)
6-Benzyladenosine	Partial	Iizuka and Hashizumi (1968)
6-Benzylamino-tetrahydroxypyranyl-9H-purine	Complete	Negi and Olmo (1966)
Ethrel	Partial	Kender and Remaily (1970)
Mendok	Nil	Iyer and Randhawa (1965)

CHAPTER 55

Concepts and Conclusions

This chapter systematically synthesises and summarises the ideas conceived in this treatise and the conclusions arrived at in toto. It thus consists of concise text repetitions, followed in each case by the advances gained and discussions arising from them. Since the authors have been quoted in the preceding text and tables, these are minimised in this chapter for the sake of clarity, coherence and cogency.

55.1 Concepts

55.1.1 Classification

The inability to produce functional male gametes due to nonfunctional male sex is known as mst. It involves male sex suppression, repression or transformation and adversely affects the sex organs or pollen grain differentiation, development, form or function. Depending upon the time and target of male sterile (mst) gene action on a time-sequence basis, the action of the mst gene is earliest on differentiation and subsequent development of the initial cells destined to form the stamens. This leads either to the absence of stamens or to misformed or malformed stamens in which sporogenous tissue is absent, incipient or inefficient to produce meiocytes or PMC's. This leads to structural male sterility (st-mst). Slightly late mst gene action permits normal androecial development but inhibits differentiation, development or function of the sporogenous tissue so that microsporogenesis is either inhibited or impaired or terminated prematurely. Hence viable pollen is not formed. This results in sporogenous mst (sp-mst). Delayed mst gene action permits normal microsporogenesis and pollen grain formation but suppresses pollen functioning by delaying or inhibiting anther dehiscence or migration to the receptive stigma. This renders the plant functional male sterile (fn-mst). Examples of these sterile types are given in Tables 2.14 to 2.16, 3.9 to 3.11.

The sterility classification is complicated in some g-mst's like maize, pea, pepper, rice, sorghum and tomato and many gc-mst's of carrot, cotton, mustard, tobacco, *Solanum, Streptocarpus* etc. in which some plants of a genotype exhibit mal- or transformed stamens and a few have either rudimentary or normal stamens lacking or having incipient sporogenous tissues. In one tomato g-mst mutant, while all the early-formed flowers are antherless, some few develop rudimentary anthers and the last-developed three to four flowers have normal-looking anthers but meiosis does not occur in them. In some alloplasmic gc-mst tobacco, cotton, *Streptocarpus* and *Solanum*, many *fr*-genes are polygenic. In these steriles both frequency and extent of mst measured by degree of anther transformation increases with the addition of *fr*-genes in the compatible cytoplasm having *c*-gene(s). While in BC_1

plants, pollen is developed but anthers are indehiscent, in BC_2-BC_3 plants, microsporogenesis is abnormal and microspores abort. In BC_4-BC_5, stamens are malformed and in BC_6, staminal development is suppressed or repressed fully. In these, mst expression passes through all stages but the final anther phenotype characterises the mst-type in such cases, as it is stable and constantly maintained in subsequent generations. Thus in all of them, the sterility is of structural type.

The male sex-specific genes (mst) broadly exhibit a nuclear, cytoplasmic and nuclear-cytoplasmic inheritance pattern based on their location in nuclear or cytoplasmic genomes and the interaction of the two genomes (Fig. 1.1). The present survey of the literature clearly indicates that the nuclear mst genes are basically of two types, viz., independent (*ms*) and dependent (*fr*) on cytoplasmic (*c*) genes; the former (*ms*) genes cause genic and the latter (*fr-c*) gene-cytoplasmic (*gc*) mst. In the latter, the mst results due to *fr-c* gene interaction. The *mst* genes located in the cytoplasm and independent of nuclear gene action are unknown. Hence, the earlier instances of purely cytoplasmic inheritance of mst in *Linum* (Bateson and Gairdner 1921), *Aquilegia* (Skalinska 1928a,b), *Epilobium* (Michaelis 1932) and *Streptocarpus* (Oehlkers 1938) represent gc-mst's as *fr*-genes for them were detected later. Therefore, the term cytoplasmic male sterility (c-mst) used conventionally is invalid and is deleted in the present treatise. However, this does not invalidate the existence of other cytoplasmic genes (cytogenes) controlling vegetative and floral abnormalities (Harvey et al. 1972; Tsunewaki 1980b), female sterility (Grün 1976), disease susceptibility (Hooker 1974), herbicide tolerance (Machado et al. 1978), heterosis (Srivastava 1981), tentoxin sensitivity (Durbin and Uchytil 1977; Flick and Evans 1982) and antibiotic sensitivity (Menczel et al. 1983), Kinoshita and Mikami (1984; Medgyesy et al. 1985) do exist in plants and animals. Of all these anomalies, excluding female sterility (fst) that has negative breeding value, mst with its excellent genetic potential has provided a marvellous opportunity to dissect out independent cytoplasmic and nuclear genome functions and interactions since in all the cases, the male sex form and/or function are impaired but female sex remains uninfluenced and functional. This influences the breeding value in this sterility type through the development and use of lines indicated in Table 55.1 and described in Chapters 2.10 and 3.10.

55.1.2 Plasmon Differentiation and Gene Control

The inheritance pattern differentiates mst's into genic (g) and gene-cytoplasmic (gc) types. Over the former, the control is exclusively by *ms* genes and over the latter it is coordinated by the interaction of nuclear (*fr*) and cytoplasmic (*c*) genes. Thus the *ms* genes are cytoplasm-insensitive but *fr* genes are sensitive to a specific cytoplasm type (S-cytoplasm) having *c* genes. This differentiates cytoplasm into S- and N-types. Whereas the latter lacks *c* genes, the former has them. Therefore, plant plasmon is of normal (N) or fertile (F) type and sterile (S) type lodging *c* genes. Within the S-type, differentiation occurs in wheat, maize, rice, barley, sorghum, pearl millet, mustard, petunia, broad bean, *Solanum, Streptocarpus, Epilobium* etc.

Spontaneous mutations from S- to N-cytoplasm occur in maize, pearl millet, sunflower and broad bean and induced mutation from N to S in wheat, maize, rice, sorghum, pearl millet, barley, groundnut, sugarbeet and red clover. In N-cytoplasm, *fr* genes, if present, are ineffective either due to the presence of *Fr* and/or *C* genes, only S_{frfr} genotypes are mst, as gene *Fr* or N-cytoplasm restores fertility in these mst's (Table 55.1). Actually, the *Fr* gene suppresses mst expression, as its removal restores fertility in S_{frfr} plants. Due to exclusive nuclear gene control of sterility, the majority of the g-mst have arisen due to spontaneous mutations (Table 2.1). Only a few have been mutagen-induced (Table 2.2). Most of the g-mst's are predominantly recessive gene-controlled (Tables 2.6, 2.7). Since the appearance of gc-mst requires simultaneous or successive mutation of nuclear and cytoplasmic genes to sterility types, the probability of the occurrence of spontaneously arisen gc-mst is a quarter that of g-mst. This has been found present in the present work (Chap. 3.2.1). As gc-mst appearance requires a specific nuclear (*fr*) and cytoplasmic (*c*) gene combination, about 75% of the gc-mst's known have arisen after intra- or interspecific hybridizations (Fig 3.1). The majority of gc-mst monocots have arisen after intergeneric hybridisations (Fig. 3.2). This predominance of origin of gc-mst's after intergenotypic hybridisations favours my contention that gc-mst arises due to the interaction of nuclear *fr* and *c* cytoplasmic genes, as in the hybrid genomes the gene combination is constituted.

Table 55.1. Genetic male sterile types and breeding lines

Male sterility type (genotypic basis)	Genotype	Phenotype	Breeding line	Designation
Genic	*msms*	Male sterile	A-line	Female parent
	Msms	Male fertile	B-line	Maintainer
	MsMs	Male fertile	C-line	Restorer
Gene-cytoplasmic	S_{frfr}	Male sterile	A-line	Female parent
	N_{frfr}, S_{Frfr} or N_{frFr}	Male fertile	B-line	Maintainer
	S_{FrFr} or N_{FrFr}	Male fertile	C-line	Restorer

Because of their different origin, g-mst's and gc-mst's differ from each other in gene control. The mst is controlled by recessive dominant and polygenes in 88, 10 and 2% of g-mst's, respectively, their corresponding proportional control in gc-mst's being 65,30 and 5%, respectively. In the inter- and intraspecific gc-mst hybrids, recessive gene control is predominant, but in the intergeneric hybrids, dominant gene control is prevalent. In *Centranthus ruber, Lactuca sativa, Origanum vulgare* and *Phaseolus vulgaris*, epistatic and inhibitory genes interact with the dominant *Ms* genes. Single recessive gene control of mst occurs in 64% and dominant in 20% of angiosperms. Multiple recessive gene control known in 24% of g-mst's is unknown in gc-mst's. In 30% gc-mst's, dominant genes control mst, but such a control occurs only in 10% of g-mst's.

55.1.3 Gene Action

Irrespective of their location and genetic control, mst genes inhibit form and/or function of male sex organs as they act anywhere during differentiation and development of androecium, PMC, microspore or pollen grains (Table 55.2). These genes affect androgenesis, microsporogenesis and microgametogenesis that include all the developmental stages from undifferentiated microsporangium to the mature pollen-filled anther (Figs. 55.1, 55.2). Based on their gene action, stages sensitive to mst gene action are identified in the following. In order to construe the influence of *mst* gene action and visualise the deviations caused by it, normal anther, endothecial and tapetal development and microsporegenesis have been described preceding the deviated ones whose examples are listed in Table 55.2. This explains the *mst* gene action and identifies the gene expression over one or all of these intimately connected tissues. Compared to *fr-c* gene action, *ms* gene action over microsporogenesis has been determined in a much larger number of cases. Therefore, g-mst's provide many samples and a wide choice to classify the gene action on time-scale, developmental-phase and stage-specificity bases as given in the following.

55.1.3.1 Androecial Formation

i. *Staminal Initiation.* From a mass of ground meristematic tissue, staminal primordia divide, elongate, develop and differentiate into filaments and anther lobes. This is prevented or deviated in some mst mutants (Table 55.2a). In all these, either no stamens develop or they are transformed into nonfunctional and malformed carpel- or petal-like organs. Hence, they are classified as st-mst's in the earlier text.

ii. *Anther Sac Development and Differentiation.* This involves differentiation of homogenous meristematic basic sporogenous (archesporial) cells and nonsporogenous cells, the epidermis. Prevention of this differentiation by *ms* genes in some mutants (Table 55.2b) results in the development of a lump of an unorganised mass of cells forming small tuberlike structures at places destined to develop anthers.

iii. *Microsporangium Differentiation.* This is a highly specialised process, as it is restricted to distinct separate areas corresponding to the corner regions of anthers. At each of these regions, a discrete number of hypodermal cells differentiates to develop into an archesporial tissue. The archesporial cell divides to produce a primary parietal layer to the outside and a primary sporogenous layer to the inside. The primary parietal layer divides anticlinally and periclinally to give rise to the endothecium, the parietal layer and the tapetum. The majority of the *ms* genes included in this group (Table 55.2c) inhibit archesporial differentiation so that instead of four normal thecae, one to six abnormal thecae having defective endothecial and tapetal development develop. In many, the thecae are interconnected and produce abnormal microsporogenous cells in which cytomixis is rampant.

Table 55.2. Partial list of genic male steriles exhibiting *ms* gene action during anther. meiocyte or pollen developmental stages

ms gene action during	Species
a) Staminal initiation	*Abelmoschus esculentus, Begonia semperflorens, Capsicum annuum, Chrysanthemum balsamita, C.carinatum, C.indicum, Citrus sinensis, Daucus carota, Dianthus caesius, Hordeum vulgare, Iris persica, Lycopersicon esculentum, Nicotiana tabacum, Pilea microphylla, Pisum sativum, Silene acaulis, Tagetes erecta, Triticum aestivum, Zea mays*
b) Anther sac development and archesporial differentiation	*Abelmoschus esculentus, Arachis hypogaea, Capsicum annuum, Carica papaya, Chrysanthemum parthenium, Citrus paradisii, Cucumis sativus, Daucus carota, Hordeum vulgare, Iris persica, Lycopersicon esculentum, Nicotiana tabacum, Origanum majorana Papaver somniferum, Pisum sativum, Primula officinalis, Ricinus communis, Silene acaulis, Tagetus erecta, Triticum aestivum, Vigna sinensis, Zea mays*
a) Microsporangial differentiation	*Aquilegia chrysantha, Argemone mexicana, Capsicum annuum, Castanea crenata, C.dentata, C.mollissima, C.pumila, C.sativa C.seguinii, Cirsium arvense, C.montanum, Cucumis sativus, Daucus carota, Gossypium hirsutum, Helianthus annuus, Hordeum vulgare, Lathyrus odoratus, Lycopersicon esculentum, Nicotiana tabacum, Origanum majorana, Pisum sativum, Silene acaulis, Solanum tuberosum, Triticum aestivum, Vigna sinensis, Zea mays, Zinnia elegans*
d) Pollen mother cell formation	*Capsicum annuum, Castanea crenata, C.dentata, C.mollissima, C.pumila, C.sativa, C.seguinii, Centaurea americana, Cirsium oleraceum, Citrus sinensis, Daucus carota, Fagopyrum esculentum, Geranium argenteum, Hordeum vulgare, Lycopersicon esculentum, Manihot esculenta, Nicotiana tabacum, Origanum majorana, Pisum sativum, Silene hookeri, Solanum tuberosum, Zea mays.*
e) Premeiosis	*Capsicum annuum, Castanea carinata, C.dentata, C.mollissima, C.pumila, Citrus aurantifolia, C.sinensis, C.unshiu, Daucus carota, Epilobium montanum, E.roseum, Glycine max, Hordeum vulgare, Lathyrus sativus, Limnanthes douglasii, Lycopersicon esculentum, Nicotiana tabacum, Origanum vulgare, Petunia hybrida, Pisum sativum, Plantago coronopus, Solanum tuberosum, Triticum aestivum, Zea mays.*
f) Meiotic entry	*Capsicum annuum, Citrus aurantifolia, Epilobium montanum, E.roseum, Godetia whitneyi, Helianthus annuus, Hordeum vulgare, Lathyrus sativus, Limnanthes douglasii, Lycopersicon esculentum, Mercurialis annua, Nicotiana tabacum, Petunia hybrida, Pisum sativum, Plantago coronopus, Zea mays*
g) Meiotic initiation (leptotene to early pachytene)	*Capsicum annuum, Centaurea americana, Crotalaria juncea, Epilobium montanum, E. roseum, Hebe parviflora, H.subalpina, Helianthus annuus, Hordeum vulgare, Limnanthes douglasii, Lycopersicon esculentum, Pelargonium denticulatum, Petunia hybrida, Pisum sativum, Primula sinensis, Solanum tuberosum, Taraxacum balticum, T.silesiacum, Triticum aestivum, Zea mays*
h) Early meiosis (prophase-I)	*Alopecurus myosuroides, Cajanus cajan, Canna glauca, Cannabis sativa, Capsicum annuum, Citrus reticulata, Coleus hybridus, Epilobium montanum, E.roseum, Glycine max, Helianthus annuus, Hordeum vulgare, Limnanthes douglasii, Lycorpesicon esculentum, Mercurialis perennis, Nicotiana tabacum, Pelargonium denticulatum,*

Table 55.2. *(continued)*

ms gene action during	Species
	Petunia hybrida, Pisum sativum, Raphanus sativus, Rumex acetosa, Salvia nemorosa, Solanum tuberosum, Taraxacum balticum, Triticum aestivum, Viola orphanidis, Zea mays
i) Mid meiosis (MI to AII)	*Alopecurus myosuroides, Glycine max, Hordeum vulgare, Kniphofia sp., Pisum sativum, Raphanus sativus, Taraxacum silesiacum, Valeriana officinalis, Zea mays*
j) Tetrad Formation	*Ageratum conyzoides, A.houstonianum, Allium fistulosum, A.schoenoprasum, Antirrhinum majus, Beta vulgaris, Brassica oleracea, Cajanus cajan, Capsicum annuum, Cirsium spinosissimum, Cucumis melo, Epilobium montanum, E.roseum, Glycine max, Lathyrus luteus, Limnanthes douglasii, Lycopersicon esculentum, Manihot esculenta, Nicotiana tabacum, Parthenium hysterophorus, Pisum sativum, Solanum tuberosum, S.verrucosum, Taraxacum balticum, T.fulvum, T.nordstedtii, T.pseudofulvum, Triticum aestivum, Vaccinium angustifolium, Viola orphanidis*
k) Primexine and sporopollenin deposition	*Beta macrocarpa, Brassica campestris, Capsicum annuum, Hordeum vulgare, Lycopersicon esculentum, Petunia hybrida, Pisum sativum, Zea mays*
l) Callose breakdown and free microspore stage	*Ageratum conyzoides, Brassica campestris, Capsicum annuum Lycopersicon esculentum, Petunia hybrida, Pisum sativum*
m) Microspore development and maturation	*Allium ascalonicum, A.longicuspis, A.sativum, Antirrhinum majus, Arabidopsis thaliana, Avena strigosa, Beta vulgaris, Brassica napus, B.oleracea, Cajanus cajan, Capsicum annuum, Centaurea scabiosa, Centranthus ruber, Citrus aurantifolia, C.medica, C.reticulata, Cucumis melo, Cucurbita pepo, Dactylis glomerata, Dianthus chinensis, Epilobium roseum, Eruca sativa, Fragaria nilgerrensis, Geranium meeboldii, Glycine max, Gossypium barbadense, Hordeum vulgare, Limnanthes douglasii, Lolium perenne, Lupinus regalis, Lycopersicon esculentum, Malus sylvestris, Medicago sativa, Nicotiana tabacum, Oenothera alata, O.flava, O.lamarckiana, Papaver somniferum, Parthenium hysterophorus, Pelargonium crispum, P.denticulatum, P.radula, P.roseum, P.tomentosum, P.X.hortum, Petunia hybrida, Phaseolus atropurpureus, Pisum sativum, Plantago coronopus, Ranunculus acris, Raphanus sativus, Solanum melongena, S.tuberosum, S.verrucosum, Tagetus erecta, Taraxacum fulvum, T.nordstedtii, Triticum aestivum, Vaccinium angustifolium, Zea mays*
n) Pollen development	*Allium ascalonicum, Beta vulgaris, Capsicum annuum, Cucumis sativus, Dactylis glomerata, Glycine max, Helianthus annuus, Hordeum vulgare, Lycopersicon esculentum, Manihot esculenta, Nicotiana tabacum, Pelargonium hortotum, P.radula, Petunia hybrida, Pisum sativum, Triticum aestivum, Vaccinium angustifolium, Zea mays*
o) Pollen maturation	*Beta vulgaris, Brassica oleracea, Capsicum annuum, Cucumis sativus, Helianthus annuus, Hordeum vulgare, Lycopersicon esculentum, Nicotiana tabacum , Petunia hybrida, Pisum sativum, Prunus mume, P. persica, Solanum tuberosum, S.verrucosum, Triticum aestivum, Vitis vinifera, Zea mays*
p) Endothecium development and anther dehiscence	*Carthamus tinctorius, Chrysanthemum morifolium, Hordeum vulgare, Lycopersicon esculentum, Manihot esculenta, Pisum sativum, Sesamum orientale, Streptocarpus cooksonii, S.cyaneus, S.ketaniensis, S.primulifolius, Zea mays*

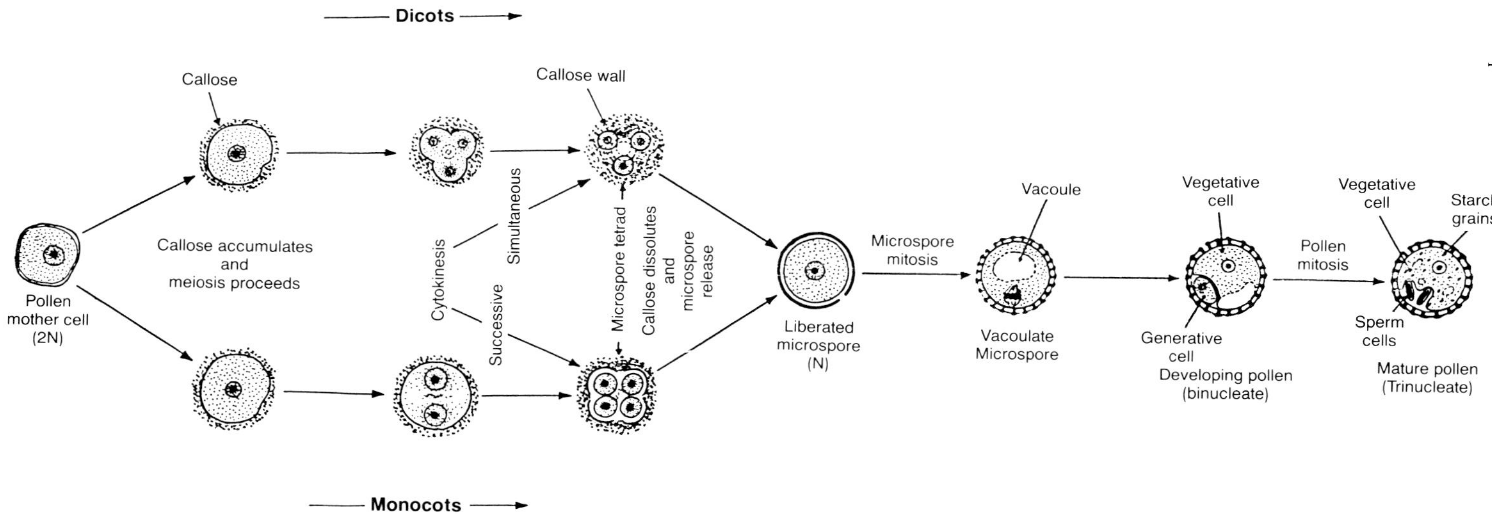

Fig. 55.1. Diagrammatic representation of normal pollen development in higher plants (After Laser and Lersten 1972)

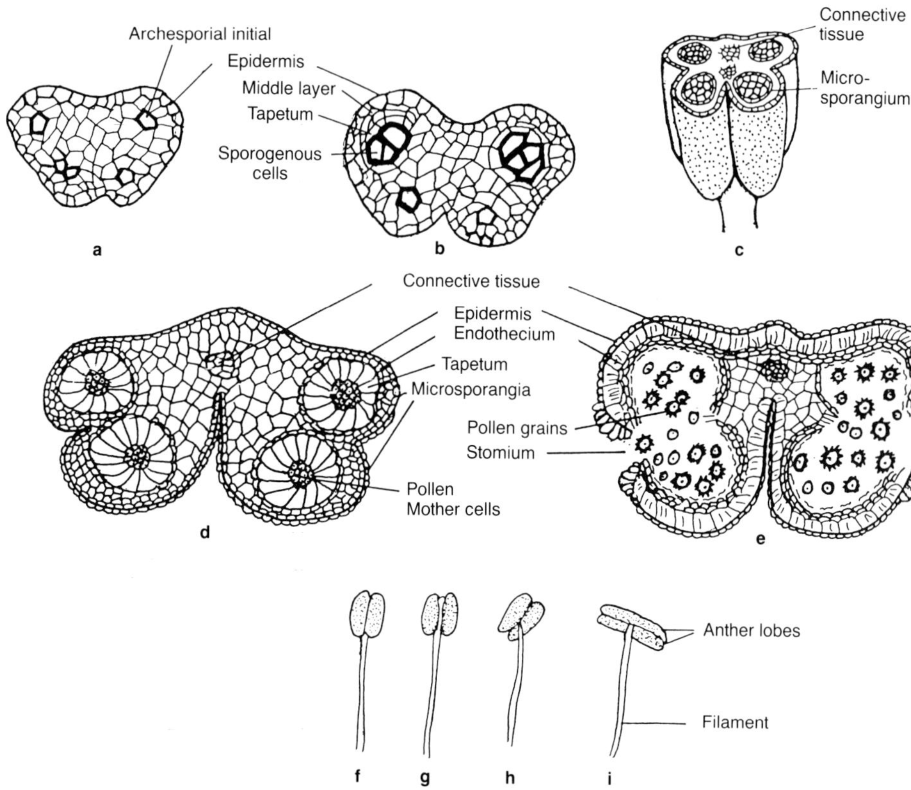

Fig. 55.2a-i. Anther development stages **(a-e)** and stamen types **(f-i)**. **a-e** Young anther showing differentiation of parietal and sporogenous **(a-c)**, tetrasporangiate anther at premeiotic **(d)** and at post-meiotic **(e)** stages. **f-i** Stamen types: basifixed **(f)**, adnate **(g)**, dorsifixed **(h)** and versatile **(i)**

iv. *Pollen Mother Cell Formation.* The primary sporogenous cell layer after a few cell divisions differentiates PMC's characterised by their large size and conspicuous nucleus and nucleoli. The *ms* genes during the stage inhibit PMC development so that the PMC's never attain an enlarged globular appearance. Their nuclear and cytoplasmic contents shrink to periphery, high vacuolation follows and content degeneration terminates their developmental activity in many mst mutants (Table 55.2d)

55.1.3.2 Premeiosis

This preparatory stage to meiosis involves DNA synthesis and replication, failure of sister chromatid and chromosome separation and ribosomal number reduction. Premeiotic DNA replication is an essential requirement for initiation and normal progression of meiosis. In some plants, the *ms* genes affect DNA replication, synapsis, chiasma formation and disjunction (Table 55.2d). In one mutant each of pea and tomato, the DNA synthesis is reduced by 45–65% and meiosis is arrested in all the meiocytes almost simultaneously in premeiosis.

i. *Meiotic Entry.* Switching-on of gene action controlling meiotic entry of PMC's is the most essential requirement for meiotic initiation. Once it is triggered, the sequence of various activities occurs in PMC's during meiotic development which is controlled by sequential release of genetic information available in the meiocyte genome. This is prevented in some mst mutants (Table 55.2e) so that no male meiosis occurs in their normally developed PMC's, which degenerate just before anthesis.

55.1.3.3 Meiosis

i. *Meiotic Initiation (Leptotene to Early Pachytene).* This involves initial recognition and synapsis of homologous DNA segments, point by point pairing, effective synapsis, chiasma formation leading to crossing over and gene recombination. In the mutants listed in Table 55.2f, asynapsis is most common and univalents degenerate along with feebly developed nucleoli.

ii. *Early Meiosis (Pachytene to Diakinesis).* This entails homologous chromosomal disjunction including release of synapsis, repulsion of homologues and chiasma terminalization. In some mst mutants, the *ms* gene prevents partial to complete terminalization (Table 55.2g). This is followed by meiotic arrest and chromosome degeneration.

iii. *Mid-Meiosis (Metaphase I to Anaphase II).* This includes congression, orientation and distribution of bivalents or individual chromosomes at the equator during MI, MII with centromeres attached to spindle fibres. This is followed by anaphase disjunction and migration of chromosomes to poles. The fundamental prerequisite for proper disjunction is complete bivalency and proper orientation to the metaphase spindle until anaphasic separation ensues. This is prevented by some *ms* mutant genes, most of which cause impaired spindle development and MI metaphase arrest (Table 55.2h).

iv. *End Meiosis (Tetrad Formation).* Culmination of meiosis results in four haploid nuclei enclosed in a single structure surrounded by callose. The four-nucleated structure undergoes cytokinesis to produce four haploid microspores. In the tetrad, the spores are isolated from each other by a thick layer of callose wall. Many *ms* genes prevent either cytokinesis or callose wall formation (Table 55.2i) so that the tetrads degenerate before microspores are released.

55.1.3.4 Postmeiosis

i. *Primexine and Sporopollenin Deposition.* While the four microspores are still enclosed in the callose sheath, individual microspores begin to secrete small areas of primexine that appear between the intact callose and microspore plasmalemma. The primexine deposition continues until the microspore is surrounded by a thin layer of primexine except in areas where colpi form. The sporopollenin deposition occurs in probacula areas which form above the primexine. This produces the basic pattern of exine sculpturing. Both primexine and sporopollenin depositions are inhibited by some *ms* genes (Table 55.2j). The microspores mostly abort before and sometimes after release from the quartets.

ii. *Callose Breakdown and Free Microspore stage.* After the basic pattern of exine sculpturing is established, callose breakdown by the enzyme callase proceeds rapidly. The enzyme frees microspores from tetrads by callose dissolution and the microspores lie free in the anther thecal fluid. Callose wall degradation and microspore release are prevented in some mutants by *ms* genes through suppression of callase enzyme synthesis (Table 55.2h). In such mutants, the microspores abort while encased in the tetrads and no viable pollen is ever produced.

iii. *Microspore Development and Maturation.* Microspore development involves completion of exine wall formation through continued sporopollenin deposition. The microspore has a centrally placed nucleus and dense cytoplasm. It enlarges 1.5- to 3-fold in volume with a vacuole occupying considerable portion. The increase in vacuolar size is due to depletion of food reserves utilised in exine formation. This stage is most sensitive to *ms* gene action. Therefore, in the majority of the g-mst mutants, *ms* genes act over this stage (Table 55.2l). The microspores abort immediately after liberation from the tetrads.

iv. *Pollen Development.* After sporopollenin deposition is complete, cellulosic intine having one or more germination apertures is laid down. The intine is first formed around pore regions and then around the entire grain surface. Microspore nucleus mitosis produces a vegetative tube nucleus occupying most of the microspore volume and a small generative nucleus. Each of the nucleis is surrounded by its own plasma membrane, giving it an independent entity. Any of these processes, especially microspore mitosis, is inhibited by *ms* genes in mutants listed in Table 55.2m. In these, the developing pollen aborts abruptly.

v. *Pollen Maturation.* For binding pollen together to attract insects and to serve as recognition substances for pollen germination on stigma, a rapid deposition of extraneous materials like lipids, carotenoid pigments, and protein components occurs over the exine surface. The vegetative cell develops organelle-rich cytoplasm and its nucleus is metabolically active. The generative cell has a small amount of cytoplasm poor in organelle and a metabolically inactive nucleus having a high DNA/RNA ratio. In this last stage of *ms* gene action over microsporogenesis, the generative cell follows vegetative cell degeneration in many mst mutants (Table 55.2m). In a few, the vegetative cell degenerates before the generative cell, e.g. *Petunia, Pisum,* and *Zea.* The end result in all of them is mature pollen abortion.

55.1.3.5 Anther Dehiscence and Endothecial Development

Maximum endothecial development occurs when the anther is ready to dehisce. The endothecium develops thickenings composed of fibrous bands of callose arising from inner tangential walls of the endothecial cells. Many *ms* genes inhibit endothecial development and/or thickening of cell walls and prevent anther dehiscence in some mutants (Table 55.2o). This leads to fn-mst.

55.1.3.6 Tapetal Behaviour

From its position, all the substances entering the sporogenous cells have to pass through the tapetum or be metabolised by it. Hence any abnormality in tapetal development and behaviour will adversely affect the development and differentiation of the developing pollen. This occurs in some mst mutants (Table 55.3), in which tapetal breakdown occurs either before or after meiosis.

i. *Premeiosis*. Barring st-mst's, in sp- and fn-mst's, premeiotic phase tapetum mostly exhibits normal behaviour, the cells are uninucleate and no nuclear aberrations are apparent. However, in some mutants, the tapetal cells divide abnormally, the intervening cell walls break down and the cell contents degenerate rapidly (Table 55.3a). The meiosis continues normally but breaks down either during late prophase or MI.

ii. *Meiosis*. The meiotic phase tapetum is most active during the course of tapetal life. Where it has to undergo nuclear division, it usually starts in early prophase and finishes by diakinesis. These nuclear divisions in some mst mutants (Table 55.3b) are abnormal, the nucleus becoming polyploid by endomitosis or by normal division followed by fusion of the daughter nuclei, dissolution of intervening cell walls, protoplasmic breakdown and degeneration of cells. Despite the tapetal anomaly, meiosis continues until tetrad or microspore formation and/or liberation.

Table 55.3. Partial list of genic male steriles exhibiting tapetal anomalies and/or malformations

Tapetal breakdown stage	Species
a) Premeiosis	*Alopecurus pratensis, Pennisetum typhoides, Pisum sativum, Salvia nemorosa, Secale cereale, Taraxacum silesiacum*
b) Meiosis	*Brassica campestris, Cajanus cajan, Cirsium spinosissimum, Cucumis melo, Fagopyrum esculentum, Helianthus annuus, Lathyrus luteus, Lupinus mutabilis, Manihot esculenta, Origanum vulgare, Primula sinensis, Raphanus sativus, Secale cereale, Vaccinium angustifolium, Viola orphanidis*
c) Postmeiosis	*Allium fistulosum, A.sativum, A.schoenoprasum, Beta macrocarpa, B.maritima, B.vulgaris, Brassica campestris, Centaurea scabiosa, Cucumis melo, Dactylis glomerata, Dianthus chinensis, Helianthus annuus, Lolium perenne, Lupinus mutabilis, Medicago sativa, Pelargonium radula, Phaseolus atropurpureus, Solanum nigrum, S.tuberosum, Vaccinium angustifolium*

iii. *Postmeiosis.* In this phase, the tapetum, after it has finished its nutritive role, degenerates with its remnants. Based on the behaviour during microsporogenesis, two major tapetal types occur in mft's, viz. (i) glandular or secretory type in which the cells remain in their original position but progressively become more disorganised and finally undergo complete autolysis, and (ii) amoeboid tapetum which is characterised by an early loss of the tapetal cell walls and intrusion of tapetal protoplasts among the developing pollen grains, followed by fusion of these protoplasts to form a tapetal periplasmodium. The common postmeiotic anomalies in the mst mutants listed in Table 55.3c are either considerable delay or premature tapetal breakdown or persistence leading to the developing microspore or pollen starvation to death.

iv. *Normal Behaviour.* In some mst mutants the tapetal development and breakdown are normal, resembling their mft counterparts, but meiosis breaks down in them either during premeiosis, meiosis or postmeiosis (Table 55.4). Since tapetal form is apparently normal in these and in others (Tables 2.21, 3.22), probably tapetal function is impaired through mis-, mal- or non-secretion of some chemical necessary for pollen development and function, but this chemical has not been precisely identified as yet. Alternatively, the *ms* gene acts directly on microsporogenous tissue and not on or via the tapetum. This appears valid, as the detectable biochemical differences between such mft and mst anthers are insignificant. Moreover, in mst's, the tapetal breakdown may be prior to, during or after the meiocyte or microspore breakdown (Tables 2.24, 3.23). Thus there is no fixed relationship between tapetal and microsporogenesis breakdown in such mst's. Instead, it is due to mst gene action, that is either on the microsporogenous tissue or on the tapetum or on both.

Table 55.4. Partial list of genic male steriles having normal tapetum but male meiotic breakdown

Breakdown stage	Species
Premeiosis	*Hordeum vulgare, Lycopersicon esculentum, Oryza sativa, Pisum sativum, Zea mays*
Meiosis	*Cajanus cajan, Centaurea imperialis, Coleus hybridus, Epilobium parviflorum, Glycine max, Hordeum vulgare, Lycopersicon esculentum, Parthenium hysterophorus, Pisum sativum, Taraxacum fulvum, T.nordstedti, T.pseudofulvum, Zea mays*
Postmeiosis	*Geranium argenteum, G.meeboldii, Glycine max, Hordeum vulgare, Lycopersicon esculentum, Parthenium hysterophorus, Pisum sativum, Taraxacum fulvum, T.nordstedti, T.pseudofulvum, Zea mays*

55.1.3.7 Concluding Remarks

Commitment to meiosis is programmed genetically and involves a series of events exhibiting a harmonious combination of universality and uniqueness and terminates with spore formation. Disturbance in these programmed events caused by *mst* genes leads to a premature end point and culminates in nondevelopment of viable and functional pollen or in anther nondehiscence. The result is mst. This is amply evidenced by the action of *ms* genes on various stages of microsporogenesis in higher plants (Table 55.2). The action of these genes at certain specific stages of microsporogenesis evidences that these stages are genetically controlled and adds to the list of meiotic mutant genes documented by Kaul and Murthy (1985). This implies the control of meiosis by a very large number of genes, the majority of which are monogenic dominant and only a few polygenic in nature.

55.1.4 Gene Locations

Plant breeders require an easy and quick method to introduce mst into a genotype and to inter-convert mst into mft and vice versa in order to control, coordinate, utilize and manipulate mst in plants for efficient use in breeding and hybrid production. This necessitates understanding the molecular basis of mst. Whereas g-mst is controlled exlusively by nuclear genes (ms), many of which have been identified and located (Table 2.31), the gc-mst results from interaction of nuclear (*fr*) and cytoplasmic *c* genes. Both the existence and location of these genes, unlike the *fr* genes, are debatable. Some circumstantial but compelling evidence implicates mitochondrial (*mt*)-DNA as the carrier of *c* genes at least in carrot, maize, wheat, tobacco, sugarbeet, sorghum, mustard, petunia and broad beans. In these, restriction endonuclease analysis has revealed marked differences between mt-DNA's from N- and S-cytoplasms, but the corresponding cp-DNA differences are either indistinguishable, inconsistent or heterogenous. The location is further substantiated by detection of intra-mitochondrial plasmids, especially in some S-cytoplasms of maize, sorghum, sugarbeet, broad bean and mustard. Further confirmation is provided by intersectional cybrids, protoplast fusion products and reciprocal transfer of S- and N-cytoplasms in tobacco, petunia and mustard. These, detailed in the chapters corresponding to these crops, are briefly described here.

On the basis of fertility restoration and several other features (Table 27.19), maize has three sterile cytoplasm types, T, C and S, in addition to the normal N-cytoplasm. Electron microscope studies reveal mt-DNA to be well-characterised and organised in each cytoplasm type, each comprising a heterogenous mixture of circular molecules, the size and frequency of each is specific for a cytoplasm type (Table 27.20). Moreover, both in vitro and in vivo mt-DNA translation products explicitly differentiate N-, C-, S- and T-cytoplasms from one another and the results are constant and highly reproducible. Likewise in sorghum, restriction endonuclease analysis of mt-DNA's and cp-DNA's of a number of alternative S-cytoplasm sources in addition to the common milo type reveals significant variability in mt-DNA but a limited one in cp-DNA (Conde et al. 1982). The mt-DNA of these alternative sources has sequences that differ from that of milo, and all these differ from N-mt-DNA sequences. S-cytoplasm mt-DNA of sugarbeet lacks low molecular weight (1.3–1.5kb) DNA molecules. Restriction analysis of the large (370 kb) mt-DNA molecules clearly differentiates N- and S-cytoplasms (Powling 1982). Many similar differences between restriction fragment patterns occur between the mt-DNA's from mft and mst *Petunia* lines (Boeshore et al. 1983), the difference in cp-mt-DNA's (Frankel et al. 1979) does not warrant cp-DNA involvement in mst (Hanson and Conde, 1985). Broad bean has two sterile cytoplasm types, S-350 and S-447. Their mt-DNA is not differentiated by restriction fragmentation, but analysis of undigested samples reveals the presence of one low molecular weight DNA species in S-350 which is absent in S-447. Both these lines, however, differ from the N-cytoplasm in the mt-DNA banding pattern. In vitro protein synthesis reveals the presence of a 25,000 molecular weight peptide in them, that is absent in mft lines. In carrots, the mt-DNA restriction endonuclease patterns of petaloid and brown anther mst S-cytoplasms differ from each other, as well as from the N-cytoplasm plants (Debonte et al. 1982), their cp-DNA nucleotide sequence being highly conserved (Debonte et al. 1983). In tobacco, by comparing mt-DNA restriction banding patterns of hybrids to the parents, mt-DNA recombination is evidenced (Charbonnier et al. 1985). In the hybrids derived from a mst line carrying *N. debneyi* cytoplasm fused with *N. tabacum,* almost all the hybrids have banding patterns different from each other and from the parental lines, due to the mitochondrial recombinations.

55.1.4.1 Cytoplasmic Hybrids

Determination of nuclear-plasmon interactions is possible through genome substitution using BC'S (Fig. 3.25) or by exposing either mitochondria or chloroplast or both to an alien nuclear environment through somatic hybridization. While the former is time-consuming and has many constraints (Hermsen 1979), the latter is quick and furnishes the only means of unidirectional transfer in either, rather than both, organelle types. The donor-recipient protoplast fusion technique (Zelcer et al. 1978), when applied at an interspecific level, produces cybrids having nuclei of the recipient species only. These plants have various nuclear organelle combinations including hybrids with "recipient" nuclei, but either or both mitochondria and chloroplast of the donor species. The latter, cybrid types are efficient potential tools for investigating nuclear-mitochondrial and nuclear-chloroplast interactions. These have revealed the association of mt-DNA in gc-mst's. Several different protoplast fusion experiments utilizing homo- and alloplasmic mft and mst lines as fusion partners have also been done. Regeneration of functional plants from these, although difficult, is possible in *Nicotiana, Petunia, Brassica* and *Zea* (Hanson and Conde 1985, Medgyesy et al. 1985). Using leaf shape as a genome marker, and floral structure and mst vs. mft as a cytoplasm indicator, Belliard et al. (1979), after protoplast fusion between *N. tabacum* and mst *N. tabacum* from a *N. debneyi* cross, developed plants from the fusion culture that were grown to flower. In the mst cybrids, cp-DNA was either of one parent or the other, there was no mixture of chloroplast types. However, novel mt-DNA occurred in them, suggesting mt-DNA recombinations. This is confirmed by Medgyesy et al. (1985), who obtained recombinant-type mt-DNA's and chloroplast co-transfer in (*N. gossei*) *N. plumbaginifolia* mst cybrids. Menczel et al. (1983) suggested that selection for streptomycin resistance, a chloroplast trait of the donor parent, resulted in cent per cent mst transfer. It is possible that genes for streptomycin resistance are located in mt-DNA. Like tobacco, novel combinations of cytoplasmic and nuclear phenotypes in petunia were obtained by Izhar et al (1984). These reveal mt-DNA recombination in mst cybrids. The cybrids, whether mst or mft, yield both sexual type progenies, but in the graft-transmitted hybrids only fertile scions and fertile progeny plants develop. The alloplasmic *Brassica napus* mst cybrids possess different chromosome numbers and novel mt-DNA type and a number of plasmid copies. While chromosome number and plasmid number vary in the progenies, mt-DNA novelty is constant in the mst's, indicating *c* gene location in mt-DNA.

In T-cytoplasm maize, mst and susceptibility to Hmt-toxin are closely associated and cytoplasmically inherited. No stable plants resistant to the toxin and expressive mst were recovered in vivo or in vitro by Gengenbach and Pring 1982 except two in cultures (Brettel et al. 1980) in which the plants were toxin-resistant and mft, but their progeny was either toxin-resistant and fertile or toxin-susceptible and mst. Therefore, the separation of mst and toxin susceptibility was temporary, and probably either due to incomplete sorting out during tissue culture and regeneration, the sorting being completed during subsequent generation or due to dauer modifications. In a few reverted plants obtained after selection or mutagen treatment in cultures, mt-DNA exhibits a novel rearrangement of sequences, probably due to mutations or recombinations.

55.1.4.2 Intramitochondrial Plasmids

Within the mt genome of *Neurospora crassa, Podospora anserina, Claviceps purpurea, Sorghum,* maize, sugarbeet, broad bean and mustard, several plasmids have been detected (Goblet et al. 1983). These, in higher plants, have been implicated in mst although their precise function and relationship with the major mt genomes are still unclear. Comparison of isolated nonrestricted mt-DNA from S- and N-cytoplasms by electrophoretic analysis reveals S-cytoplasm mt-DNA to have two DNA molecules of 6.2 (5–1) and 5.2 (5–2) kpb. These are two linear DNA molecules with exact terminal inverted repeats of 208 bp (Levings and Sederoff 1983). DNA molecules similar and closely related to S-1 and S-2 and termed R-1 and R-2 occur in some South American fertile maize races (Weissinger et al. 1982). Thus the mere presence of such linear molecules with sequences homologous to S-1 and S-2 is not sufficient to confer mst. Likewise, S-cytoplasm sorghum source (IS 1112C) has two linear plasmidlike DNA's, N-1 and N-2, about 5.7 and 5.3 kbp, respectively. In addition, it has a 2.2-kbp molecule reminiscent of the 2.3 kbp circular molecule of N-, C- and S-maize mt-DNA and up to five lower molecular weight (1 to 1.4 kbp) DNA's. N-1 and N-2 share substantial homology with each other, but less with S-1; S-2 has a higher degree of homology with N-2 than with S-1, suggesting that the N-2/S-2 homologous sequences lie in the unique sequence region of S-2. However, no genetic evidence implicates these two molecules in mst expression control in a manner analogous to S-1 and S-2 in maize. In some *Brassica campestris* and *B. napus* lines, Palmer et al. (1983) detected a 11.3 kb linear molecule plasmid that exhibited no homology, either with the main mitochondrial genome or with 11.3 kb plasmid and S-1 and S-2 mitochondrial plasmid of maize. These plasmids are absent in 14 *Raphanus sativus, B. juneea, B. carinata, B. oleracea* and *B. nigra* accessions and the plasmids in eight different *Brassica* accessions are identical at 41 restriction sites. Alloplasmic (*R. sativus*) *B. campestris* mst has a ratio of about 10:1 of plasmid mitochondrial genome. The plasmid number is very low in the mft homoplasmic maintainer *B. campestris* line. Similarly, a homoplasmic mst *B. napus* has a high amount of plasmid DNA but the corresponding fertile maintainer line lacks it. Whether this plasmid confers mst in *Brassica* is not known. Mitochondria of mst and mft *Vicia faba* cytoplasms have two small supercoiled plasmid-like DNA molecules of 1700 and 1420 bp in addition to the main mt-DNA of high molecular weight (Goblet et al. 1983). S-350 cytoplasm has an additional supercoiled DNA molecule of 1540 bp which is absent in N and S-447 cytoplasm plants. In general, the plasmid data is intriguing (Hanson and Conde 1985). For clarity, numerous S-cytoplasms in divergent nuclear backgrounds need to be analysed. Progenies from sexual crosses between plasmid-carrying and plasmid-lacking lines should be examined to determine if a high plasmid copy number is correlated with gc-mst.

55.1.5 Gene Complementation

Inheritance of mst is either exclusively nuclear (g-mst) or due to nuclear-cytoplasm interaction (gc-mst). While the former exhibits normal Mendelian inheritance, the latter is unique, as the cytoplasmic *c*-gene is contributed by the maternal parent and both the fertility-restoring (*fr*) and normal cytoplasm gene (*c*) are dominant both over *fr* and *c* genes. Therefore only those individuals are mst that have *frfr* nuclear and *c* cytoplasmic genes. The nuclear and cytoplasmic genes can be recessive or dominant. Since the majority of the mst nuclear *fr* genes are recessive, it is presumed that the cytoplasmic (*c*) genes are also recessive. This is substantiated by dominance of fertile over sterile cytoplasm. Based on this assumption, Fig. 55.1has been constructed. Thus gc-mst results from a harmoniously coordinated and compatible interaction of specific nuclear (*fr*) and cytoplasmic (*c*) genes and is not a disruption in cytoplasmic nuclear coordination as envisaged by Hanson and Conde (1985). Whereas *fr* genes are located in the nuclear genome, the *c* genes lie in the *mt*-DNA either free or in an integrated state. Two highly specialised and thoroughly rearranged free plasmidlike DNA molecules in S-cytoplasm maize exhibit a unique behaviour that led Laughnan and Gabay (1978) to postulate the existence of a fertility element behaving as an episome. In its free state, the fertility reversion is nuclear or cytoplasmic-controlled depending upon episomal integration into nuclear or mt DNA. The alterations from fertility to sterility and vice versa are due to excision and integration of sequences related to S-1/S-2 molecules. Two similar linear plasmid-like DNA's (R-1, R-2) occur in North American (Koncz et al. 1980) and South American (Weissinger et al. 1982) maize races. Due to absence of proper *fr* genes none of these are mst. Likewise some sorghum lines having two similar plasmid molecules (N-1, N-2) are mst as they have the appropriate nuclear *fr* genes required for mst induction. Whereas the *fr* genes are static and lodged in the nucleus, the *c* genes are higly specialized, thoroughly rearranged DNA molecules that function when present in the cytoplasmic organelle, especially mitochondria, but they are mobile and can exist in free or integrated state. In maize S-cytoplasm, free S-1, S-2 molecules exist. Reversion to fertility is caused by rearrangement (excision-integration) and amplification of sequences homologous to S-1 and S-2 molecules. (Hanson and Conde 1985). The appropriate *fr-c* gene combination produces some chemical that affects male sex organ development form or function. This chemical is cytoplasmically transmitted and functions across grafts only in those genotypes that have the appropriate *fr* genes. This led to many successes and failures in graft transmission of mst, especially in *Petunia*, in which it was expressed in some and not in other graft transfers. The inexpressive ones, considered as incompatible scions by Frankel (1964), actually lacked the appropriate nuclear *fr* genes. In some of the expressive ones, the mst appeared after the second or third generation of grafting. This was because the former lacked the specific *fr* genes and the latter had them in the heterozygous state. Hence for successful graft transmission, the mst-receiving partner should have the appropriate *frfr* genes and compatible cytogenetic background, as other genes influencing *fr* gene expression will modify or inhibit mst expression. This is well known in *Petunia* and *Nicotiana*.

The nature and composition of the chemical postulated to be produced by *fr-c* gene complementation are unknown. In some gc-mst's, inclusions like M (Bond et al. 1966a) and ds-RNA (Scalla et al. 1981) or iojap (Rhoades 1950) in maize or virus-like particles known in many plants (Table 6.3) may represent not the causes but the consequences of complementary *fr-c* gene interaction. In sugarbeet, the mft plants obtained after heat treatments and subsequent meristem culture (Lichter 1978) are due to dauermodifications and not to *fr* or *c* gene mutations, as mst is reverted in the subsequent generations.

55.2 Conclusions

i. *Types.* Anomalies in form and function of male sex organs accompanied by normal female sex development and function leading to male sterility (mst) range from absence to mal-development of male sex organs (structural) to their normal development, but with either impaired microsporogenesis (sporogenous) or normal microsporogenesis accompanied by anther nondehiscence (functional). These three sterility types are either nongenetic, caused by physiological stresses, physical and chemical agents and environmental factors, or genetically controlled by nuclear genes alone or in combination with cytoplasmic genes. The sterility controlled solely by nuclear (*ms*) genes is genic (*g-mst*) and that controlled by complementary action of nuclear (*fr*) and cytoplasmic (*c*) genes is gene-cytoplasmic (*gc-mst*).

ii. *Distribution and Occurrence.* Male sterility of genic (g), gene-cytoplasmic (gc) and/or chemical (ch) type is known in 720 cases comprising 617 species and species crosses (Table 55.6). Of these, 23 species exhibit all three types of mst. Conjoint occurrence of g + gc, g + ch and gc + ch mst's is known in 30, 19 and 2 species, respectively. Whereas 216 species exhibit g-mst, it is known in only 17 species crosses. The reverse is true for gc-mst, in which mst occurs in 271 species crosses but in only 16 species. The ch-mst occurs in 19 species only. This is because it has been induced in species and not in hybrids. Of the known mst species and species crosses, the nature of mst in about 23% species is doubtful, as gene control of the anomaly is not clearly known in them. In total, mst occurs in about 55% species and 45% species crosses (Fig. 55.3A). Of these, 40% are contributed each by g-mst species

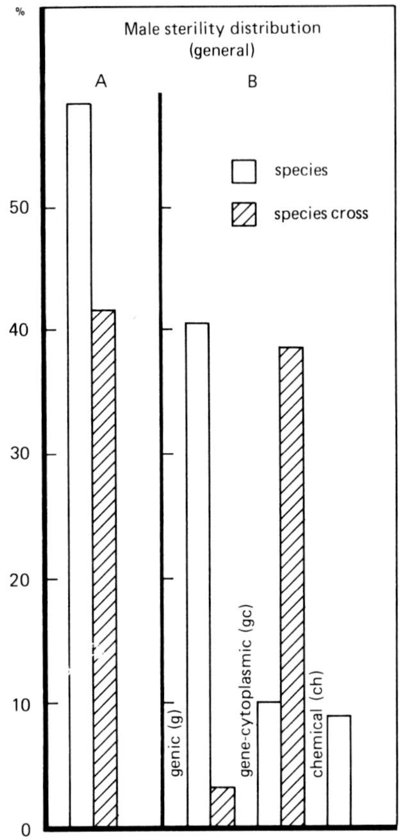

Fig. 55.3A,B. Generalised distributional pattern in total **(A)** and in three sterility types **(B)**

and gc-mst species crosses, nearly 10% each by the species exhibiting gc-mst and ch-mst (Fig. 55.4B). Although a nearly equal number of species and species crosses exhibit various mst types (Fig. 55.3A), 42% of species crosses exhibit gc-mst and only 38% species the g-mst (Fig. 55.4B). The frequency of other sterility types in species and species crosses is comparatively very low (Table 55.6).

Although genetic mst is documented in 617 species and species crosses (Tables 55.5, 55.6), its occurrence in cereals (wheat, rice, maize, barley, pearl millet, sorghum), legumes (soybean, pigeon pea, green gram, kidney bean, garden pea, broad bean, clover, alfalfa), oil yielders (mustard, rape, sunflower, groundnut, castor, cotton), vegetables (potato, tomato, onion, cucurbits, pepper, carrots, radish, turnip, pea, lettuce), ornamentals (*Ageratum, Antirrhinum, Begonia, Carthamus, Chrysanthemum, Geranium, Iris, Pelargonium, Petunia, Streptocarpus, Tagetes, Zinnia*) and in other economic plants like sugar beet, sugar cane, tobacco and citrus, is significant, as many of these have exhibited high breeding potential. Of these, the sterility is most prodigious and divergent in wheat, maize, potato, tobacco, cotton, sugarbeet, petunia and broad bean.

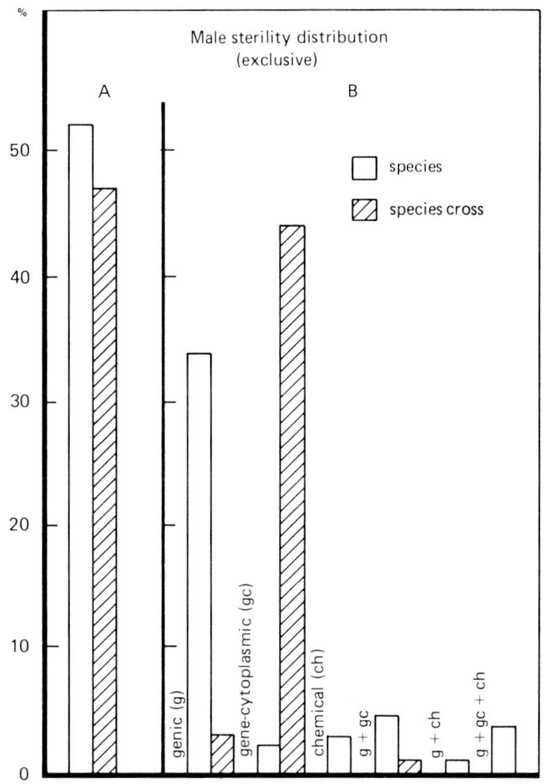

Fig. 55.4A,B. Exclusive distributional pattern in total **(A)** and in three sterility types singly and in combinations **(B)**

iii. *Origin and Gene Control.* Whereas the majority of g-mst's arise after spontaneous mutations, 64% being monogenic recessive, the gc-mst's predominantly arise after varietal or species hybridizations which offer the required *fr-c* gene combinations. Evidence indicates *c* gene location to be in the mitochondria (mt). The *c* gene is thoroughly arranged and highly specialised DNA which may be integrated in mt-DNA or remain free in the cytoplasm. In the free state it may assume episomal form and integrate with nuclear or mt-DNA. Whereas the multiple gene control known in g-mst's is unknown in gc-mst's, polygene control known in *gc-mst's* is rare in *g-mst's*. Many *ms* and *fr* genes exhibit linkage and a few pleiotropy, which facilitates their identification and early detection in the field. While *ms* genes are most prevalent and have high rates of nonallelic mutations, *fr* genes are rare and their mutation rate is low. The *mst* genes act before, during or after microsporogenesis and lead to the development of structural, sporogenous and functional mst's, but the majority of these genes act over microsporogenesis, either at the beginning or mostly at the end. Therefore, the sporogenous types are most frequent. Like other meiotic genes (Kaul and Murthy 1985), most of the mst genes are stage- and site-specific. They may act directly over the sporogenous tissue or over the tapetum or on both of them. Only 33% of mst's have normal tapetum,

Table 55.5. List of species and species crosses exhibiting male sterility[a]

Species/Cross	Male sterility type Genic	Gene Cyto-plasmic	Chemical	Page(s)
(1)	(2)	(3)	(4)	(5)
Abelmoschus esculentus (6x = 120) (Malvaceae)	+	–	+	705–707
Aegilops aucheri (2x = 14) × *Triticum aestivum* (6x = 42) (Gramineae)	–	+	–	523
Ae.bicornis (2x = 14) × *T.aestivum* (6x = 42)	–	+	–	523
Ae.bicornis (2x = 14) × *T.durum* (4x = 28)	–	+	–	523
Ae.biuncialis (4x = 28) × *T.aestivum* (6x = 42)	–	+	–	523
Ae.biuncialis (4x = 28) × *T.spelta* (6x = 42)	–	+	–	523
Ae.biuncialis (4x = 28) × *T.sphaerococcum* (6x = 42)	–	+	–	523
Ae.boeoticum × *T.spelta* (6x = 42)	–	+	–	523
Ae.caudata (2x = 14) × *Ae.persica* (4x = 28)	–	+	–	431
Ae.caudata (2x = 14) × *Ae.typica* (2x = 14)	–	+	–	431
Ae.caudata (2x = 14) × *Ae.umbellulata* (2x = 14)	–	+	–	431
Ae.caudata (2x = 14) × *Triticum aestivum* (6x = 42)	–	+	–	523
Ae.caudata (2x = 14) × *T.durum* (4x = 28)	–	+	–	523
Ae.caudata (2x = 14) × *T.spelta* (6x = 42)	–	+	–	523
Ae.caudata (2x = 14) × *T.vulgare* (6x = 42)	–	+	–	523
Ae.columnaris (4x = 28) × *T.aestivum* (6x = 42)	–	+	–	523
Ae.columnaris (4x = 28) × *T.macca* (6x = 42)	–	+	–	523
Ae.columnaris (4x = 28) × *T.spelta* (6x = 42)	–	+	–	523
Ae.columnaris (4x = 28) × *T.sphaerococcum* (6x = 42)	–	+	–	523
Ae.comosa (2x = 14) × *T.aestivum* (6x = 42)	–	+	–	523
Ae.heldreichii (2x = 14) × *T.aestivum* (6x = 42)	–	+	–	523
Ae.heldreichii (2x = 14) × *T.spelta* (6x = 42)	–	+	–	523
Ae.kotschyi (4x = 28) × *T.aestivum* (6x = 42)	–	+	–	523
Ae.kotschyi (4x = 28) × *T. macca* (6 = 42)	–	+	–	523
Ae.kotschyi (4x = 28) × *T.spelta* (6x = 42)	–	+	–	523
Ae.kotschyi (4x = 28) × *T.timopheevi* (4x = 28)	–	+	–	523
Ae.longissima (2x = 14) × *Ae.aucheri* (2x = 14)	–	+	–	431
Ae.macrochaeta (2x = 14) × *T.aestivum* (6x = 42)	–	+	–	523
Ae.mutica (2x = 14) × *T.aestivum* (6x = 42)	–	+	–	523
Ae.mutica (2x = 14) × *T.spelta* (6x = 42)	–	+	–	523
Ae.ovata (4x = 28) × *T.aestivum* (6x = 42)	–	+	–	523
Ae.ovata (4x = 28) × *T.durum* (4x = 28)	–	+	–	523
Ae.ovata (4x = 28) × *T.macca* (6x = 42)	–	+	–	523
Ae.ovata (4x = 28) × *T.spelta* (6x = 42)	–	+	–	523
Ae.ovata (4x = 28) × *T.sphaerococum* (6x = 42)	–	+	–	523
Ae.recta (6x = 42) × *T.aestivum* (6x = 42)	–	+	–	523
Ae.sharonensis (2x = 14) × *T.aestivum* (6x = 42)	–	+	–	523
A.sharonensis (2x = 14) × *T.spelta* (6x = 42)	–	+	–	523
Ae.speltoides (2x = 14) × *T.aestivum* (6x = 42)	–	+	–	523
Ae.squarrosa (2x = 14) × *T.spelta* (6x = 42)	–	+	–	523
Ae.triaristata (6x = 42) × *T.aestivum* (6x = 42)	–	+	–	523
Ae.triaristata (6x = 42) × *T.macca* (6x = 42)	–	+	–	523
Ae.triaristata (6x = 42) × *T.spelta* (6x = 42)	–	+	–	523
Ae. triaristata (6x = 42) × *T.sphaerococcum* (6x = 42)	–	+	–	523

Table 55.5 *(continued)*

(1)	(2)	(3)	(4)	(5)
Ae.triuncialis (4x = 28) × *T.aestivum* (6x = 42)	–	+	–	523
Ae.triuncialis (4x = 28) × *T.maccà* (6x = 42)	–	+	–	523
Ae.triuncialis (4x = 28) × *T.spelta* (6x = 42)	–	+	–	523
Ae.umbellulata (2x = 14) × *T.aestivum* (6x = 42)	–	+	–	523
Ae.umbellulata (2x = 14) × *T.spelta* (6x = 42)	–	+	–	523
Ae.uniaristata (2x = 14) × *T.aestivum* (6x = 42)	–	+	–	523
Ae.uniaristata (2x = 14) × *T.durum* (4x = 28)	–	+	–	523
Ae.uniaristata (2x = 14) × *T.spelta* (6x = 42)	–	+	–	523
Ae.variabilis (4x = 28) × *T.aestivum* (6x = 42)	–	+	–	523
Ae.variabilis (4x = 28) × *T.macca* (6x = 42)	–	+	–	523
Ae.variabilis (4x = 28) × *T.spelta* (6x = 42)	–	+	–	523
Ae.variabilis (4x = 28) × *T.timopheevi* (4x = 28)	–	+	–	523
Ae.ventricosa (4x = 28) × *T.aestivum* (4x = 28)	–	+	–	523
Ae.ventricosa × *T.timopheevi* (4x = 28)	–	+	–	523
Ageratum conyzoides (4x = 40) (Compositae)	+	–	–	329–330
A.houstonianum (2x = 20)	+	–	–	329–330
Agroelymus turneri (4x = 28) (Gramineae)	–	+	–	432
Agroelymus turneri (2x = 28) × *Agropyron distachymum* (4x = 28)	–	+	–	432
Agroelymus turneri (4x = 28) × *A.repens* (4x = 28)	–	+	–	432
Agroelymus turneri (4x = 28) × *Elymus canadensis* (4x = 28)	–	+	–	432
Allium ascalonicum (2x = 16) (Liliaceae)	+	+	–	685, 696
A.cepa (2x = 16)	–	+	+	685–694, 697
A.fistulosum (2x = 16)	+	–	–	694, 696, 697
A.longicuspis (2x = 16)	+	–	–	694, 696, 697
A.porrum (4x = 32)	+	–	–	694, 696, 697
A.sativum (2x = 16)	+	–	–	694–695, 696, 697
A.schoenoprasum (2x = 16)	+	+	–	695, 696, 697
A.cepa (2x = 16) × *A.ascalonicum* (2x = 16)	–	+	–	695, 697
A.cepa (2x = 16) × *A.drobovii* (4x = 32)	–	+	–	695, 697
A.cepa (2x = 16) × *A.galanthum* (2x = 16)	–	+	–	695
A.cepa (2x = 16) × *A.pskemense* (4x = 32)	–	+	–	695, 697
A.roylei (2x = 16) × *A.drobovii* (4x = 32)	–	+	–	695, 697
Alopecurus myosuroides (2x = 14) (Gramineae)	+	–	–	432–433, 570
A.pratensis (6x = 42)	+	–	–	570
Anacardium occidentale (4x = 48) (Anacardiaceae)	+ (?)	–	–	291
Antirrhinum majus (2x = 16) (Scrophulariaceae)	+	–	+	747
Aquilegia flabellata (2x = 14) × *A.truncata* (2x = 14) (Ranunculaceae)	–	+	–	738
A.vulgaris (2x = 14)	–	+	–	738
A.vulgaris (2x = 14) × *A.chrysantha* (2x = 14)	–	+ (?)	–	738, 739
Arabidopsis thaliana (2x = 10) (Cruciferae)	+	–	–	356, 371
Arachis hypogaea (4x = 40) (Leguminoseae)	+	–	–	623, 679

Table 55.5. *(continued)*

Species/Cross (1)	Male sterility type: Genic (2)	Gene Cyto-plasmic (3)	Chemical (4)	Page(s) (5)
Argemone mexicana (4x=48) (Papavaraceae)	+	–	–	728
Avena byzantina (6x=40) (Gramineae)	+	–	–	433, 570
A.fatua (6x=42)	+	–	–	433, 570
A.sativa	–	–	+	433
A.strigosa (2x=14)	+	–	–	433, 570
Begonia schmidtiana (4x=32) × *B.acuminata*	–	+	–	293, 294
Begonia schmidtiana (4x=32) × *B.hirtella* (4x=36) (Begoniaceae)	–	+	–	293, 294
Begonia schmidtiana (4x=32) × *B.subvillosa* (4x=32)	–	+	–	293, 294
B.semperflorens (2x=36)	+	+	–	293, 294
Beta macrocarpa (2x=18) (Chenopodiaceae)	+	+	–	302, 322, 324
B.maritima (2x=18)	+	+	–	302–304, 322, 324
B.patula (2x=18)	–	–	+	319
B.vulgaris (2x=18)	+	+	+	304–327
B.vulgaris (2x=18) × *B.maritima* (2x=18)	–	+	–	324
W.webbiana	–	–	+	319
Brassica campestris (2x=20) (Cruciferae)	+	+	–	356–362, 371, 373
B.campestris (2x=20) × *B.oleracea* (2x=18)	–	+	–	373
B.campestris (2x=20) × *B.napus* (4x=38)	–	+	–	357, 359
B.juncea (4x=36)	+	+	+	362–363, 371, 373, 381
B.juncea (4x=36) × *B.napus* (4x=38)	–	+	–	373
B.napus (4x=38)	+	+	+	364–370, 371, 373, 374, 381
B.napus (4x=38) × *B.carinata* (4x=34)	–	+	–	363
B.napus (4x=38) × *B.juncea* (4x=36)	–	+	–	373
B.nigra (2x=16) × *B.oleracea* (2x=18)	–	+	–	373
B.oleracea (2x=18) Var. *botrytis*	+	+	+	370, 371
Var. *capitata*	+	+	+	371, 375, 376, 377
Var. *gemmifera*	+	+	+	371, 375, 376, 377
Var. *italica*	+	–	–	372, 376
Bryonia alba (Cucurbitaceae)	–	+	–	393
B.alba × *B.dioica*	–	+	–	393, 395
Cajanus cajan (2x=22) (Leguminosae)	+	–	+	624–626, 679, 684
Callistephus sinensis (2x=18) (Compositae)	–	–	+	205
Canna glauca (2x=18) (Cannaceae)	+(?)	–	–	295
Cannabis sativa (2x=20) (Cannabinaceae)	+	–	+	296, 297

Table 55.5 *(continued)*

(1)	(2)	(3)	(4)	(5)
Capsicum annuum (2x = 24)	+	+	+	234, 239, 750–757
(Solanaceae)				829, 831, 833
C.annuum (2x = 24) × *C.pendlum* (2x = 24)	–	+	–	831
C.annuum (2x = 24) × *C.pubescens* (2x = 24)	–	+	–	831
C.baccatum (2x = 24) × *C.annuum* (2x = 24)	–	+	–	831
C.peruvianum (2x = 24) × *C.annuum* (2x = 24)	–	+	–	831
Carica papaya (2x = 18)	+	–	–	298
(Caricaceae)				
Carthamus tinctorius (2x = 24)	+	–	+	330, 332, 334–335
(Compositae)				
Castanea crenata (2x = 24)	+ (?)	–	–	418
(Fagaceae)				
C.dentata (2x = 24)	+ (?)	–	–	418
C.hybrida (2x = 24)	–	+	–	418
C.mollissima (2x = 24)	+ (?)	+	–	418
C.pumila (2x = 24)	+ (?)	–	–	418
C.sativa (2x = 24)	+ (?)	+	–	418
C.senguinii (2x = 24)	+ (?)	–	–	418
Centaurea americana	+	–	–	330, 335
(Compositae)				
C.imperialis	+	–	–	330
C.scabiosa (2x = 20)	+	+ (?)	–	335
Centranthus ruber (2x = 14)	+	–	–	840
(Valerianaceae)				
Chenopodium quinoa (4x = 32)	+	+ (?)	–	322, 328
(Chenopodiaceae)				
Cheiranthus cheiri (2x = 14)	+	–	–	372, 382–383
(Cruciferae)				
Chionachne koenigii (2x = 20)	+	–	–	435, 570
(Graminae)				
Chrysanthemum balsamita (2x = 18)	+	–	–	330, 335
(Compositae)				
C.carinatum (2x = 18)	+	–	–	330, 335
C.morifolium (6x = 54)	+	–	–	330, 335
C.parthenium (2x = 18)	+	–	–	330, 335
Cirsium acaule (2x = 34)	–	+	–	332, 336
(Compositae)				
C.arvense (2x = 34)	+	+	–	330, 336
C.eriophorum (2x = 34)	–	+	–	332, 336
C.montanum (2x = 34)	+	+	–	330, 336
C.oleraceum (2x = 34)	+	+	–	330, 332, 336
C.oleraceum (2x = 34) × *C. canum* (2x = 34)	–	+	–	336
C.oleraceum (2x = 34) × *C.palustre* (2x = 34)	–	+	–	336
C.palustre (2x = 34)	+	+	–	330, 332, 336
C.spinossimum (2x = 34)	+	+	–	330, 366
C.velenovsky	–	+	–	322
Citrus aurantifolia (2x = 18)	+ (?)	–	–	743, 744
(Rutaceae)				
C.limon (2x = 18) × *C.paradisi* (2x = 18)	–	+	–	744
C.limon (2x = 18) × *C.sinensis* (2x = 18)	–	+	–	744

Table 55.5 *(continued)*

Species/Cross (1)	Male sterility type Genic (2)	Gene Cyto-plasmic (3)	Chemical (4)	Page(s) (5)
C.medica (2x = 18)	+(?)	–	–	744
C.reticulata (2x = 18)	+(?)	–	–	744
C.sinensis (2x = 18)	+(?)	–	–	744, 745
C.unshiu (2x = 18)	+(?)	–	–	744, 745
C.yatsushiro (2x = 18)	+(?)	–	–	744, 745
C.unshiu (2x = 18) × *C.grandis* (2x = 18)	–	+	–	744
C.unshiu (2x = 18) × *C.sinensis* (2x = 18)	–	+	–	744
Citrullus vulgaris (2x = 22) (Cucurbitaceae)	+	–	+	394, 395, 396
Coleus blumei (4x = 24) (Labiatae)	+	–	–	617, 618
C.hybridus (8x = 48)	+	–	–	617, 618
Commelina virginica (2x = 12) (Commelinaceae)	–	–	+	205
Corchorus capsularis (2x = 14) (Tiliaceae)	+	–	–	838
Coriandrum sativum (2x = 22) (Umbelliferae)	–	–	+	205
Crotalaria juncea (2x = 16) (Leguminosae)	+	–	–	627, 679, 684
C.mucronata (2x = 16)	–	+	–	627, 628, 683
C.pallida (2x = 16)	–	+	–	628
Cucumis melo (2x = 24) (Cucurbitaceae)	+	–	+	394, 396, 397–399
C.sativus (2x = 14)	+	–	+	394, 396, 399–401
Cucurbita maxima (4x = 40) (Cucurbitaceae)	+	+	+	394, 396, 402, 405
C.moschata (2x = 24) × *C.foetidissima* (4x = 40)	–	+	–	395, 403
C.moschata (2x = 24) × *C.palmata* (4x = 40)	–	+	–	395, 403
C.pepo (4x = 40)	+	–	+	394, 403, 405
Cyclamen persicum (2x = 48) (Primulaceae)	–	+	–	737
Dactylis glomerata (4x = 28) (Graminae)	+	+	–	435–436, 570, 573
Datura alba (2x = 24) (Solanaceae)	–	–	+	205
D.stramonium (2x = 24)	+	–	–	758, 829
Daucus carota (2x = 18) (Cruciferae)	+	+	+	372, 373, 383–390
Desmodium sandwicense (2x = 22) × *D.intortum* (2x = 22) (Leguminosae)	–	+	–	628, 629, 683
Dianthus caesius (2x = 30) (Caryophyllaceae)	+	–	–	299
D.caryophyllus (2x = 30)	+	–	–	299
D.chinensis (2x = 30)	+	–	–	299, 300
Diplotaxis muralis (2x = 22) × *Brassica campestris* (2x = 20) (Cruciferae)	–	+	–	359, 374

Table 55.5. *(continued)*

(1)	(2)	(3)	(4)	(5)
Echinocloa colone (6x = 54) (Gramineae)	+	–	–	436, 570
Eleusine coracana	+ (?)	–	–	436
E.indica (2x = 18)	+	–	–	437
Epilobium hirsutum (2x = 36) (Onagraceae)	+	+	–	720–726
E.hirsutum (2x = 36) × *E. luteum* (2x = 36)	–	+	–	721, 722
E.hirsutum (2x = 36) × *E. montanum* (2x = 36)	–	+	–	721, 722
E.hirsutum (2x = 36) × *E.parviflorum* (2x = 36)	–	+	–	721, 722
E.hirsutum (2x = 36) × *E.roseum* (2x = 36)	–	+	–	721
E.luteum (2x = 36) × *E.hirsutum* (2x = 36)	–	+	–	721, 722
E.luteum (2x = 36) × *E.roseum* (2x = 36)	–	+	–	721
E.montanum (2x = 36) × *E.hirsutum* (2x = 36)	–	+	–	721, 722
E.montanum (2x = 36) × *E.parviflorum* (2x = 36)	+	–	–	721
E.montanum (2x = 36) × *E.roseum* (2x = 36)	–	+	–	722
E.parviflorum (2x = 36) × *E.montanum* (2x = 36)	+	–	–	721
E.parviflorum (2x = 36) × *E.roseum* (2x = 36)	+	–	–	721
E.roseum (2x = 36) × *E.hirsutum* (2x = 36)	+	–	–	721
E.roseum (2x = 36) × *E.luteum* (2x = 36)	+	–	–	721
E.roseum (2x = 36) × *E.parviflorum* (2x = 36)	+	–	–	721
Eruca sativa (2x = 22) (Cruciferae)	+	–	–	372, 390
Euchlaena mexicana (2x = 20) × *Zea mays* (2x = 20) (Graminae)	–	+	–	573
E.perennis (2x = 20) × *Z.mays* (2x = 20)	–	+	–	606
Fagopyrum esculentum (2x = 16) (Polygonaceae)	+	–	–	736
Festuca arundinacea (6x = 42) (Gramineae)	+	–	–	437, 570
F.pratensis (2x = 14) × *Lolium perenne* (2x = 14)	+ (?)	+	–	458, 573
Ficus glabella (2x = 26) (Moraceae)	+ (?)	–	–	717
F.hispida (2x = 26)	+	–	–	717
F.pumila (2x = 26)	+	–	–	717
F.tomentosa (2x = 26)	+	–	–	717
Foeniculum vulgare (2x = 22) (Umbelliferae)	–	–	+	205
Fragaria chiloensis (8x = 56) (Rosaceae)	+ (?)	–	–	741
F.moschata (2x = 14)	+ (?)	–	–	741
F.vesca (2x = 14)	+ (?)	–	–	741
F.vesca (2x = 14) × *F.nilgerrensis* (2x = 14)	–	+ (?)	–	741
Geranium albiflorum (2x = 28) × *G.pratense* (2x = 28) (Geraniaceae)	–	+	–	419
G.argenteum (2x = 28)	+	–	–	419
G.endressi (2x = 28) × *G.striatum* (2x = 28)	–	+	–	419
G.himalayense (2x = 28)	+	–	–	419
G.meeboldi (2x = 28)	+	–	–	419

Table 55.5. *(continued)*

Species/Cross (1)	Male sterility type Genic (2)	Gene Cyto-plasmic (3)	Chemical (4)	Page(s) (5)
L.esculentum (2x = 24)	+	+	+	758–775, 829, 831, 833
L.esculentum (2x = 24) × *L.hirsutum* (2x = 24)	–	+	–	768, 831
L.esculentum (2x = 24) × *L.parviflorum* (2x = 24)	–	+	–	768, 769, 831
L.esculentum (2x = 24) × *Solanum pennelli* (2x = 24)	–	+	–	768, 769, 831
L.hirsutum (2x = 24) × *S.pennelli* (2x = 24)	–	+	–	768, 769, 831
L.minutum (2x = 24) × *S.pennelli* (2x = 24)	–	+	–	768
(*L.minitum* 2x = 24 × *L.hirsutum* 2x = 24) × *S.pennelli* (2x = 24)	–	+	–	768, 769
L.peruvianum (2x = 24) × *L.hirsutum* (2x = 24)	–	+	–	768
L.peruvianum (2x = 24) × *S.pennelli* (2x = 24)	–	+	–	768
Malus baccata (2x = 34) (Rosaceae)	+(?)	–	–	741
M.floribunda (2x = 34)	+(?)	–	–	741
M.sylvestris (2x = 34)	+	–	–	741, 742
Mangifera indica (4x = 40) (Anacardiaceae)	+(?)	–	–	291
Manihot esculenta (4x = 36) (Euphorbiaceae)	+(?)	–	–	412, 413
Matthiola incana (2x = 14) (Cruciferae)	+	–	–	24
Medicago sativa (4x = 32) (Leguminoseae)	+	+	+	647–655, 680, 683, 684
Melilotus alba (2x = 16)	+	–	–	680, 683
Mercurialis annua (2x = 16, 32) (Euphorbiaceae)	–	+	–	412, 413–416
M.perennis (8x = 64)	+	–	–	412, 416
Mimulus luteus (8x = 64) (Leguminosae)	+	–	–	19
Momordica charantia (2x = 22) (Cucurbitaceae)	–	–	+	214
Morus nigra (Moraceae)	+(?)	–	–	717
Nicotiana alata (2x = 18) (Solanaceae)	+	–	–	19
N.amplexicaulis (2x = 36) × *N.tabacum* (4x = 48)	–	+	–	777
N.benthamiana (2x = 36) × *N.tabacum* (4x = 48)	–	+	–	777
N.bigelovii (4x = 48) × *N.tabacum* (4x = 48)	–	+	–	777
N.clevelandii (4x = 48) × *N.tabacum* (4x = 48)	–	+	–	777
N.debneyi (4x = 48) × *N.tabacum* (4x = 48)	–	+	–	782
N.exigua (2x = 32) × *N.tabacum* (4x = 48)	–	+	–	777
N.glauca (2x = 24) × *N.tabacum* (4x = 48)	–	+	–	777
N.goodspeedi (4x = 40) × *N.tabacum* (4x = 48)	–	+	–	777
N.gossei (4x = 36) × *N.tabacum* (4x = 48)	–	+	–	777
N.knightiana (2x = 24) × *N.tabacum* (4x = 48)	–	+	–	777
N.langsdorffii (2x = 18) × *N.sanderae* (4x = 18)	–	+	–	778
N.megalosiphon (4x = 40) × *N.tabacum* (4x = 48)	–	+	–	777

Table 55.5 *(continued)*

(1)	(2)	(3)	(4)	(5)
N.plumbaginifolia (2x = 20) × *N.tabacum* (4x = 48)	–	+	–	777
N.raimondii (2x = 24) × *N.tabacum* (4x = 48)	–	+	–	777
N.repanda (4x = 48) × *N.tabacum* (4x = 48)	–	+	–	777
N.rustica (4x = 48)	+	–	+	775, 829, 833
N.rustica (4x = 48) × *N.tabacum* (4x = 48)	–	+	–	777
N.suaveolens (4x = 32) × *N.tabacum* (4x = 48)	–	+	–	777
N.tabacum (4x = 48)	+	+	+	777, 796
N.tabacum (4x = 48) × *N.bigelovii* (4x = 48)	–	+	–	778
N.tabacum (4x = 48) × *N.glutinosa* (2x = 24)	–	+	–	778
N.tabacum (4x = 48) × *N.plumbaginifolia* (2x = 20)	–	+	–	778
N.tomentosiformis (2x = 24) × *N.otophora* (2x = 24)	–	+	–	778
N.undulata (2x = 24) × *N.tabacum* (4x = 48)	–	+	–	777
Nigella damasceana (2x = 12) (Ranunculaceae)	–	–	+	211, 214
Oenothera alata (2x = 14) (Onagraceae)	+	–	–	721, 727
O.biennis (2x = 14) × *O.argillicola* (2x = 14)	–	+	–	722
O.biennis (2x = 14) × *O.hookeri* (2x = 14)	–	+	–	722
O.flava (2x = 14)	+	–	–	721, 727
O.gigas (4x = 28)	+	–	–	721, 726
O.lamarckiana (2x = 14)	+	–	–	721, 727
O.parviflora (2x = 14) × *O.biennis* (2x = 14)	–	+	–	722
O.parviflora (2x = 14) × *O.hookeri* (2x = 14)	–	+	–	722
O.rubricalyx (2x = 14) × *O.alata* (2x = 14)	+	–	–	727
O.strigosa (2x = 14)	+	–	–	721, 727
O.suaveolens (2x = 14) × *O.strigosa* (2x = 14)	+	–	–	727
Olea europea (2x = 46) (Oleaceae)	+ (?)	–	–	719
Origanum majorana (4x = 32) (Labiatae)	+	–	–	618, 619
O.vulgare (4x = 32)	+	+	–	618, 619–621
Oryza glaberrima (2x = 24) × *O.sativa* (2x = 24) (Gramineae)	–	+	–	466, 574
O.perennis (2x = 24) × *O.sativa* (2x = 24)	–	+	–	574
O.rufipogon (2x = 24) × *O.sativa* (2x = 24)	–	+	–	467, 574
O.sativa (2x = 24)	+	+	+	459–481, 571, 574, 578
O.sativa (2x = 24) × *O.glaberrima* (2x = 24)	–	+	–	466, 574
Papaver somniferum (2x = 22) (Papavaraceae)	+	–	–	728
P.somniferum (2x = 22) × *P.setigerum* (4x = 44)	+	–	–	728
Parthenium argentatum (2x = 18) (Compositae)	+	–	–	351, 352
P.hysterophorus (2x = 18)	+	–	–	331, 353
Paspalum urvillei (4x = 40) × *P.malacophyllum* (4x = 40) (Graminae)	–	+	–	102
Pelargonium crispum (2x = 22) (Greniaceae)	+	+ (?)	–	419, 420–422
P.denticulatum (10x = 90)	+	–	–	419, 422
P.denticulatum (10x = 90) × *P.roseum* (8x = 72)	–	+	–	419, 424
P.radula (9x = 81)	+	–	–	419, 422

Table 55.6. Distribution of male sterility types in species and species crosses (data based on Table 55.5)

Male sterility type	General[a] Species	Species crosses	Total	Exclusive[b] Species	Species crosses	Total	Doubtful[c] Species	Species crosses	Total
Genic(g)	294	21	315	216	17	233	90	1	91
Gene cyto-plasmic (gc)	71	271	342	16	271	287	0	2	02
Chemical (ch)	63	0	63	19	0	19	–	–	–
g + gc	–	–	–	30	04	34	12	–	12
g + ch	–	–	–	19	0	19	–	–	–
gc + ch	–	–	–	02	0	02	–	–	–
g + gc + ch	–	–	–	23	0	23	–	–	–
Total	428	292	720	325	292	617	102	03	105

[a]Includes species having genic or gene-cytoplasmic and/or chemical male sterility.
[b]Excludes male sterility types other than those indicated.
[c]Precise genetic control unknown.

the remaining have either abnormal or persistent tapetum. Since in 75% of mst's, irrespective of tapetal type, microsporogenesis breakdown occurs during microspore and pollen development stages, *ms* and *fr* gene action over microsporogenesis is independent of tapetal behaviour.

iv. *Biochemical Characteristics.* The mst anthers exhibit reduced metabolic activity, carbohydrate, protein and proline contents, but their asparagine content is increased. In many, callase activity is either impaired or inhibited. In some, altered aminogram pattern and enzyme activity, micronutrient deficiency and nucleic acid starvation occurs in mst anthers. In many mst's, the leaf chlorophyll content is increased. All these deviations are the consequences and not the causes of *mst* gene action.

v. *Graft Transfer and Viral Basis.* Whereas some mst's exhibit graft transfer ability, others do not. In *Beta, Capsicum, Petunia,* and *Vicia,* while some reported success, others failed to achieve graft transfers under similar conditions. Graft transfer of *gc-mst* depends on the compatible stock-scion combination, the recipient should have *fr* genes to permit *c* gene products to function. The graft transmission ability of mst was associated to its viral nature, but the virus-like particles present in the S-447 cytoplasm of broad bean are products of specific *fr-c* gene complementation. These are absent in S-350 cytoplasm. Viruses, pests and fungi cause both pollen and ovular sterility.

vi. *Environmental Influence.* In 70%, mst gene expression is highly sensitive to environment. Of these, 50% are influenced by temperature and 13% by photoperiod, in the remaining the precise environmental component is unknown. Further instability in phenotypic expression, especially in alloplasmic mst's, is caused by the genes and the physiological state of the plant. In some, these even cause reversions.

vii. *Chemical mst.* None of the 30 chemicals tested as male gametocides is efficient in inducing mst, as most of these cause ovular damage, phytotoxicity and growth anomalies. Their use is further limited as they exhibit specificities of genotype, dose, duration, tissue developmental stage and physiological state. Male gametocides are therefore useful only where mst is unknown and needed for experimental hybrid production.

viii. *Special Features.* The g-mst's and gc-mst's exhibit gross differences in origin, gene control, gene expression and environmental susceptibility. Hence the two mst types are fundamentally different and should be maintained separately. These differences are basically caused by their differential gene control. While in the g-mst's the control is exclusively nuclear, in gc-mst's it is nuclear-cytoplasmic gene complementation. Despite the differences, the mst's exhibit certain features which warrant their placement as peerless genetic entities in the plant kingdom. These are male sex specificity accompanied by altered anther and/or pollen phenotype and tapetal form and/or function, unimpaired female fertility, stage-specific *mst* gene action, nuclear and cytoplasmic gene diversity, linkage and pleiotropy of *mst* genes, specific biochemical differences, genotypic and phenotypic diversity in gene action and expression and a divergent gene control but specific gene-cytoplasm complementation.

ix. *Gynodioecy.* The mst's in gynodioecious populations represent a distinct class of individuals having specialised genetic mechanism and diverse cytoplasm types in addition to enhanced reproductive fitness offered by heterozygote advantage and resource reallocations.

x. *Uses.* The mst's find considerable use in hybrid production, recurrent selection, composite and polyploidy breeding and in testing and estimation of combining ability, pollen movement, pollination percent, planting and field design for optimal productivity, apomixis, parthenocarpy and genetic control of microsporogenesis. The uses are limited by their rarity in desirable genotypes, arduous phenotypic identification, maintenance, perpetuation and transfer, expressional instability, reduced female fertility and insufficient pollination.

55.3 Recommendations

For investigations over mst, the following is recommended:

i. *Ascertaining Male Sterility.* This necessitates testing of pollen fertility, viability and functionability by staining, germination and pollination capability. Since pollen stainability or germinability do not always reliably indicate pollen fertility, selfing ability and crossability through seed set determination is essential. For this, some of the young flowers of suspected mst plants should be selfed and others crossed by pollen from mft plants. If the suspected plant is mst, its crossed flowers, in contrast to the selfed ones, would bear seeds. Also the emasculated mft flowers should be pollinated by the pollen of the suspected mst plants. No seed set would occur if the pollen is fully sterile.

ii. *Genetic Control and Type Determination.* After mst is ascertained, it is necessary to find if the sterility is genetic or nongenetic. When genetic, studies of F_1 progenies obtained after direct and reciprocal crossings and their F_2 to F_4 generations will indicate the type of mst involved and its gene control; the genetic includes genic (g) and gene-cytoplasmic (gc) mst types. Whereas gc-mst will follow Mendelian inheritance, gc-mst will follow non-Mendelian transmission, depending upon the nuclear and cytoplasmic genes of the male parent used (Fig. 55.5). Both sterility types have specific maintainers and restorers.

iii. *Gene Action Time.* Investigations over microsporogenesis will indicate the time of *ms* or *fr-c* gene action and will decide the phenotypic type of mst, viz. structural, sporogenous or functional. This can be supplemented by the visual appearance of

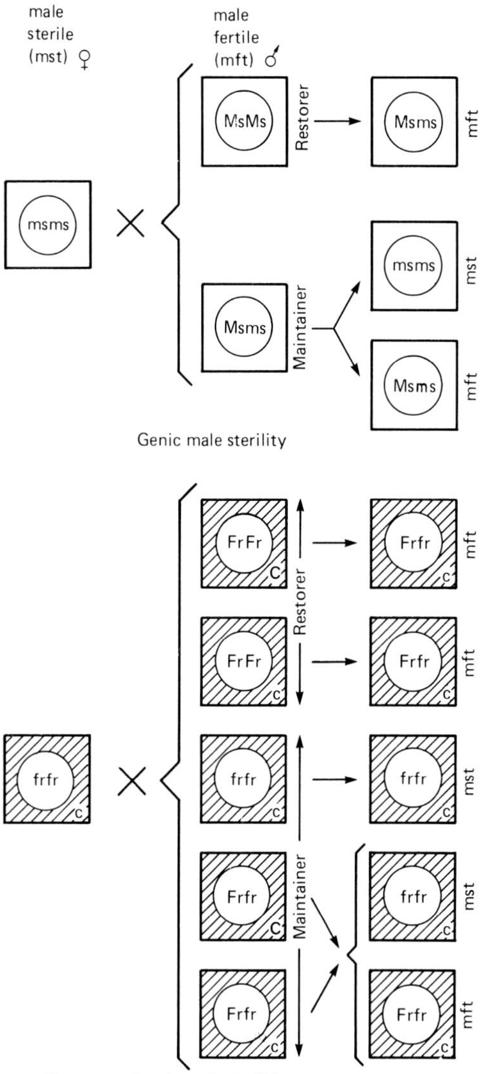

Fig. 55.5. Inheritance pattern of genetic male sterility *Squares* and *circles* represent cells and nuclei with cytoplasmic and nuclear male sterility genes, respectively

androecial form and examination of its developmental pattern and microsporogenesis breakdown stage (Table 1.2).

iv. *Gene Symbol Assignment.* Once the type and genetic control of mst are determined, the appropriate gene symbol should be assigned. If the sterility is genic controlled by one recessive gene and the report is first for the species, symbol ms_1 may be assigned. The subsequent ones detected may be symbolised as ms_2, or Ms_2, provided the gene involved is nonallelic and recessive or dominant, respectively. Likewise, each sterile cytoplasm receives a number, the first known for a species S-1, the subsequent ones S-2, S-3 and so on. For mst expression, each S-cytoplasm has *c* cytoplasmic and *fr* nuclear genes. Thus S-1 has c_1-fr_1 genes. S-2 has c_2-fr_2 genes and so on; their corresponding fertility restorers being Fr_1, Fr_2. Since these symbols indicate nature, type and gene control of mst, their adoption should be easy. In the allogamous mst's, in which a nuclear genome is placed into an alien cytoplasm, the genome donor species should be preceded by the cytoplasm donor species placed in parenthesis, because the resultant phenotype is mainly of the genome donor species. The genome-cytoplasm interaction produces mst.

v. *Environmental Influence.* Due to high genotype-environment interactions, determinations of environmental influence on mst expression and on the time and type of gene action are essential. The response and sensitivity of mst genotypes to various environments determine their breeding potential.

vi. *Outcrossing and Female Fertility Determinations.* Estimations of outcrossing percent and female fertility of mst's are essential for breeding value evaluations, as mst genotypes with high outcrossing and female fertility are useful, especially in heterosis breeding.

Ahokas H (1979b) Cytoplasmic male sterility in barley IV. Effect of msm1 cytoplasmic and partial fertility on kernel protein and lysine. Theor Appl Genet 55:269–272

Ahokas H (1979c) Cytoplasmic male sterility in barley. Acta Agric Scand 29:219–224

Ahokas H (1980a) Cytoplasmic male sterility in barley V. Physiological characterization of the msm1-Rfm1a system. Physiol Plant 48:232–238

Ahokas H (1980b) Cytoplasmic male sterility in barley VII. Nuclear genes for restoration. Theor Appl Genet 57:193–202

Ahokas H (1980c) Evidence for cytoplasmic male sterility in two Finnish rye cultivars. Hereditas 92:373–380

Ahokas H (1981) Cytoplasmic male sterility in barley X. Distribution of msm1 fertility restoration ability in the wild progenitor of barley in Israel. Ann Bot Fenn 18:313–320

Ahokas H (1982a) Cytoplasmic male sterility in barley VIII. Lipoxygenase activity and anther amino nitrogen in the msm1-Rfm1a system. Plant Physiol (Bethesda) 69:268–272

Ahokas H (1982b) Cytoplasmic male sterility in barley XI. The msm2 cytoplasm. Genetics 102: 285–295

Ahokas H (1982c) Genetics and physiology of maternal male sterility and its restoration in barley. Publ Septt Genet Univ Helsinki Finland, pp 1–10

Ahokas H (1982d) Cytoplasmic male sterility in barley: evidence for the involvement of cytokinins in fertility restoration. Proc Natl Acad Sciusk 79:7605–7608

Ahokas H (1983) Cytoplasmic male sterility in barley XV. PI296897 as a restorer of fertility in msm1 and msm2'cytoplasms. Hereditas 99:157–159

Ahokas H, Hockett EA (1977) Male sterile mutants of barley IV. Different fertility levels of msg9ci (cv Vantage) an ecoclinal response. Barley Genet Newslett 7:10–11

Ahokas H, Hockett EA (1981) Performance tests of cytoplasmic male sterile barley at two different latitudes. Crop Sci 21:607–611

Aken' Ova ME (1982) Male sterility in Nigerian bulrush millets *Pennisetum americanum* (L.) K. Euphytica 31:161–165

Aken' Ova ME, Chheda HR (1981) A new source of cytoplasmic-genic male sterility in pearlmillet. Crop Sci 21:984–985

Alam S (1967) Relationships among amino acids, proteins and RNA in male fertile and male sterile sudangrass. Ph D Thesis N. Dakota State Univ Fargo, 170pp

Alam S, Sandal PC (1967a) Cytohistological investigations of pollen abortion in male sterile sudangrass. Crop Sci 7:587–589

Alam S, Sandal PC (1967b) Inheritance of cytoplasmic male sterility in sudangrass. Crop Sci 7: 668–669

Alam S, Sandal PC (1969) Electrophoretic analysis of anther proteins from male fertile and male sterile sudangrass *Sorghum vulgare* var *sudanensis* (Piper). Crop Sci 9:157–159

Albertsen MC (1976) Developmental cytology of microsporogenesis of the *ms*1 mutant in soybean, *Glycine max* (L.) Merrill. MS Thesis Iowa State Univ Ames Iowa, pp 70

Albertsen MC, Palmer RG (1976) Environmental influence on total pollen grains per flower from normal and sterile plants of the Urbana male sterile. Soybean Genet Newslett 3:39–41

Albertsen MC, Palmer RG (1979) A comparative light and electron microscopic study of microsporogenesis in male sterile (msl) and male fertile soybeans *Glycine max* (L.) Merr. Am J Bot 66:253–265

Albertsen MC, Phillips RL (1981) Developmental cytology of 13 genetic male sterile loci in maize. Can J Genet Cytol 23:195–208

Aldrich HC, Gracen VE, York D, Earle ED, Yoder OC (1977) Ultrastructural effects of *Helminthosporium maydis* race T toxin on mitochondria of corn roots and protoplasts. Tissue Cell 9:167–177

Alexander DE, Beckett JB (1962) Search for a conversion-type mutator system at the Rf1 locus in maize. Crop Sci 2:139–140

Allard RW (1953) A gene in lima beans, *Phaseolus lunatus* pleiotropically affecting male sterility and seedling abnormality. Proc Amer Soc Hort Sci 69:467–471

Allen CE (1940) The genotypic basis of sex expression in angiosperms. Bot Rev 6:277–300

Allen ON, Allen EK (1981) The Leguminosae. A source book of characteristics, uses and nodulation. Univ Wisconsin Press, USA, pp 812

Allison DC, Fisher WD (1964) A dominant gene for male sterility in upland cotton. Crop Sci 4: 548–549

Altman DW, Storey KK, Gengenbach BG (1984) Confirmation by mitochondrial DNA analysis of a euplasmic spring wheat expressing a dominant male sterility gene. Crop Sci 24:598–601

Anand Kumar K, Andrews DJ (1984) Cytoplasmic male sterility in pearlmillet (*Pennisetum americanum* (L.) Leeke), a review. Adv Appl Biol 10:113–143

Anaschenko AV (1972a) Methods for developing hybrid sunflower involving chemical castration. In: Proc 5th Int Sunflower Conf, Clermont-Ferrand, France, pp 229–230

Anaschenko AV (1972b) A study of male sterility in the sunflower. Tr Prikl Bot Genet Sel 46: 120–131

Anaschenko AV (1981) Gene pool of sunflower and its utilization in breeding. Proc Eucarpia Symp Sunflower Breeding Prague, Oct 26–30, 1981, pp 122–137

Andersen WR (1963a) Cytoplasmic sterility in hybrids between *Lycopersicon esculentum* and *Solanum pennellii.* Rep Tomato Genet Coop 13:7–8

Andersen WR (1963b) Cytoplasmic male sterility in hybrids between *Solanum pennellii* Corr and various *Lycopersicon* species. Ph D Thesis Univ California, Davis USA, 185 pp

Andersen RG (1964) The mechanism of sterility and fertility restoration in wheat. In: Hybrid wheat seminar, crop quality council Minneapolis, Minnesota, USA, pp 23–46

Andersen WR (1964) Evidence of plasmon differentiation in *Lycopersicon* Tomato. Genet Coop Rep 14:4–6

Andersen WR (1965) Cytoplasmic male sterility and intergrafts between *Lycopersicon esculentum* and *Solanum pennellii.* Corr. Min Acad. Sci J 32:93–94

Anderson LB (1966) A male sterile syndrome in *Lolium* spp x *Festuca arundinacea.* New Zealand J Bot 4:398–405

Andrasfalvy A (1969) Tomato male sterile and functional sterile mutants for breeding purposes. Különlenyomat a Kisérlutügyi Közlemeńyek 62(c) Kertészei 1(3):3–12

Andrasfalvy A (1974) Genes, mutations and cytoplasms of special interest in modern vegetable breeding. Proc 15th Int Hort Congr 3:457–467

Andrews DJ, Anand Kumar K (1982) Induction of downy mildew resistance in pearlmillet male sterile Tift 23d2. Mut Breed Newslett 20:1–3

Andrews DJ, Webster OJ (1971) A new factor for genetic male sterility in *Sorghum bicolor* (L) Moench. Crop Sci 11:308–309

Anonymous (1913) *Streptocarpus cyaneus.* Curtis Bot Mag 139, Tab 8521

Anonymous (1956) Hybrid onion. Agric Res Serv Sp Rep, USDA, May 1956, pp 22–24

Anonymous (1964) Report from Station d-amelioration des Plantes de Clermont-Ferrand. Rapport dåctivite pour la periode 1960–1963. Inst Nat Rech Agronom, France

Anonymous (1966) *Pennisetum typhoides* (Burm. f) In: Stapf and Hubbard (eds) The Wealth of India: raw materials VII, CSIR, N. Delhi, India, pp 296–307

Anonymous (1970) Key to hybrid sunflowers found by USDA scientist. Crop Soils Mag 3:21

Anonymous (1972) Genetic vulnerability in major crops. Natl Acad Sci Washington DC Rept, pp 307

Anonymous (1973a) Annu Rep Int Inst Trop Agric, Grain legumes, improved porgramme. Ibadan, Nigeria 78 pp

Anonymous (1973b) Annu Rep Lab fruit tree culture. Tunisia Inst Nat Rech Agranom de Tunisia, pp 77

Anonymous (1977) The relationship between male sterility of rice and activities of some enzymes and rate of respiration in pollen. Acta Bot Sin 19:103–106

Anonymous (1978a) Cytological observations of pollen abortion of male sterile line red awned wild rice x Lian-Tang-Zao. Acta Bot Sin 20:150–153

Anonymous (1978b) Studies on male sterility of rice induced by male gametocide, No 1. Acta Bot Sin 20:303–313

Anonymous (1980) Hybrid rice development and seed production in China, SW 80-10022. Nat Foreign Assessment, CIA, Washington, DC, p 7

Anstey TH, Moore JF (1954) Inheritance of glossy foliage and cream petals in green sprouting broccoli. J Hered 45:39–41

Apltauerova M, Stehno Z, Dotlačil L (1981) Cytoplasmic male sterility in wheat. Vyzkumného Ustavu Rostlinné Vyroby Praha-Ruzyné 21:29–52

Appadurai R, Ponnaiya BWX (1967) Inheritance of fertility restoration in *Sorghum bicolor* (L) Moench. Madras Agric J 54:394–404

Appadurai R, Poonaiya BWX (1968) Plasma diversity in *Eu-Sorghum.* A barrier to free interbreeding. Madras Agric J 55:3–14

Appadurai R, Raveendran TS, Nagarajan C (1982) A new male sterility system in pearl millet. Indian J Genet Plant Breed 40:505–511

Appl J (1933) Artkreuzungen, Geschlechtsvererbung und Nondisjunction bei Lippenblütlern aus der Thymiangruppe. Blätter Pflanzenzücht 11:51–56

Armstrong JM, White WJ (1935) Factors influencing seed setting in alfalfa. J Agric Sci 25:161–179

Arnason TJ (1941) Sterility in potatoes. Can J Res (C) 19:145–155

Arora DK, Gupta PK (1948) Microsporogenesis in male sterile and male fertile *Crotalaria pallida* Ait. Cytologia (Tokyo) 49:685–690

Arroyo MTK, Raven PH (1975) The evolution subdioecy in morphologically gynodioecious species of *Fuchsia* sect Encliandra (Onagraceae). Evolution 29:500–511

* Artemova GV (1980) Cytological features of winter rye with cytoplasmic male sterility (CMS). Byulleten Vsesoyuznaog ordena Lenina Ordena Druzhby Naradov Inst Rasteneievodstva Imeni NI Vavilova, USSR 104:19–22

Artschwager E (1947) Pollen degeneration in male sterile sugarbeets with special reference to the tapetal plasmodium. J Agric Res India 75: 191–197

Ashri A (1968) Morphology and inheritance of sterile brachytic dwarfs in peanuts, *Arachis hypogaea.* Crop Sci 8:413–415

Assouad MW, Dommée B, Lumaret R, Valdeyron G (1978) Reproductive capacities in the sexual forms of the gynodioecious species *Thymus vulgaris* L. Bot J Linn Soc 77:29–39

Atal CK (1958) Cytoplasmic male sterility in *Psyllium (Plantago ovata* Forsk*)* Curr Sci (Bangalore) 27:268

Atanasoff D (1925) New studies on stripple-streak disease of potato. Phytopathology 15:170–177

Atanasoff D (1964a) Viruses and cytoplasmic heredity. Z Pflanzenzücht 51:197–214

Atanasoff D (1964b) Universality of interferon formation. Phytopathol Z 50:336–358

Atanasoff D (1971) The viral nature of cytoplasmic male sterility in plants. Phytopathol Z 70: 306–322

Athow KL, Bancroft JB (1959) Development and transmission of tobacco ringspot virus in soybean. Phytopathology 49:697–701

Athwal DS (1965a) Current plant breeding researches with special reference to *Pennisetum.* Indian J Genet 26A 73–85

Athwal DS (1965b) Hybrid bajra-1 marks a new era. Indian Farm Sci 15(5):6–7

Athwal DS, Virmani SS (1972) Cytoplasmic male sterility and hybrid breeding in rice. In: IRRI Annu Rep 1971 Philippines, pp 615–620

Athwal DS, Phul PS, Minocha JL (1967) Genetic male sterility in wheat. Euphytica 16:354–360

Athwal DS, Gill BS, Minocha JL (1976) Diversification of the sources of male sterility in pearl millet. Crop Improv 3:96–100

Atwood SS (1942) Genetics of self compatibility in *Trifolium repens.* J Am Soc Agron 34:353–364

Atwood SS (1944) Oppositional alleles in natural populations of *Trifolium repens.* Genetics 29:428–435

Audran HC, Bouillot J (1981) Sporoderm infrastructural and cytochemical modifications in cytoplasmic male sterile broadbean (*Vicia faba* L.) Acta Soc Bot Pol 50:223–226

Audran JC, Willemse MTM (1982) Wall development and its autofluorescence of sterile and fertile *Vicia faba* L. pollen. Protoplasma 110:106–111

Augustine JJ, Baker LR, Sell HM (1973) Female flower induction on androecious cucumber *Cucumis sativus* L. J Amer Soc Hort Sci 98:197–199

Aung LH, Byrne JM (1977) The effect of sodium 2,3-dichloroisobutyrate on microsporogenesis and fruit set in *Lycopersicon esculentum* Mill. J Exp Bot 28:1117–1124
Austenseon HM (1948) Linkage relations of the male sterile gene ms2 in barley. MS Thesis Univ Saskatchewan Canada, pp 41
Avery AG, Satina S, Rietsema J (1966) The genus *Datura*. Ronald, New York, USA
Aviv D, Galun E (1980) Restoration of fertility in cytoplasmic male sterile (CMS) *Nicotiana sylvestris* by fusion with X-irradiated *N. tabacum* Protoplasts. Theor Appl Genet 58:121–127
Aviv D, Fluhr R, Edelman M, Galun E (1980) Progeny analysis of the interspecific somatic hybrids: *Nicotiana tabacum* (CMS) + *Nicotiana sylvestris* with respect of nuclear and chloroplast markers. Theor Appl Genet 56:145–150
Aviv D, Bleichman S, Arzee-Gonen P, Galun E (1984) Intersectional cytoplasmic hybrids in *Nicotiana*. Identification of plastomes and chondriomes in *N. sylvestris* + *N. rustica* cybrid having *N. sylvestris* nuclear genomes. Theor Appl Genet 67:499–504
Awasthi NNC, Dubey DK (1983) Chemical induction of male sterility in *Vicia faba* L. (abstr). Proc 70th Indian Sci Cong Bot vol 3, 97
Aycock MK Jr, Erdogan HR (1979) Influence of five male sterile cytoplasm on the performance of maryland tobacco cultivars and hybrids. Rep Maryland Agri Exp Stat MP 940:1–36
Aycock MK Jr, Mann TJ, Matzinger DF (1963) Investigations with a form of cytoplasmic male sterility in flue cured tobacco. Tobacco Sci 7:130–135
Ayyangar GNR (1931) *Eleusine coracana* III. Sterility. Indian J Agric Sci 1:554–562
Ayyangar GNR (1942) The description of crop plant characters and their range of variation IV. The variability of Indian sorghum (Jowar). Indian J Agric Sci 12:529–540
Ayyangar GNR, Ponnaiya BWX (1937) The occurrrence and inheritance of earheads with empty anther sacs in sorghum. Curr Sci (Bangalore) 5:390
Ayyangar GNR, Ponnaiya BWX (1939) The occurrence and inheritance of tip sterility in sorghum. Curr Sci (Bangalore) 8:116–117
Ayyangar GNR, Srinivasa Rao VL (1941) Studies in Barnyard millet (*Echinocloa colone*). Madras Agric J India 29:3–12
Azzini LE, Rutger JN (1982) Amount of outcrossing on different male steriles of rice. Crop Sci 22:905–907
Babu CN (1961) Inducement of anther dehiscence in CO421, a male sterile sugarcane variety by pollen of other genera. Curr Sci (Bangalore) 30:113–114
Bahl PN, Maan SS (1973) Chromosomal location of male fertility restoring genes in six lines of common wheat. Crop Sci 13:317–320
Baier WH, Keydel KF, Odenbach N (1978) Pollensterile Winterweizenlinien unter verschiedenen Umweltbedingungen. Z Pflanzenzücht 80:131–141
Bajwa MA, Lucken KA (1968) Allelic analysis of male fertility restorer genes in six hexaploid wheats. Agron Abstr, p 2
Baker HG (1959) Reproductive methods as factors in speciation in flowering plants. Cold Spring Harbour Symp Quant Biol 24:177–191
Baker HG (1963) Evolutionary mechanisms in pollination biology. Science 139:877–883
Baker HG (1966) The evolution of floral heteromorphism and gynodioecism in *Silene maritima*. Heredity 21:689–692
Bakshi JS, Rachie KO, Singh A (1966) Development of dwarf strains of pearlmillet and an assessment of their yield potential. Curr Sci (Bangalore) 35:355–356
Bandlow G (1958) Die Pollensterilität der Beta und ihre Bedeutung für die Polyploidiezüchtung. Wiss Abh Dtsch Akad Landwirtschaftswiss Berlin 34:52–68
Bandlow G (1964a) Die Pollensterilität von *Beta vulgaris* und deren Artbastarden sowie ihre Verwertung in der Zuckerrübenzüchtung I. Z. Pflanzenzücht 52:160–208
Bandlow G (1964b) Die Pollensterilität von *Beta vulgaris* und deren Artbastarden sowie ihre Verwertung in der Zuckerrübenzüchtung II. Z Pflanzenzücht 52:209–240
Banerjee SC, Rakshit SC (1970) Use of male sterility in the production of heteroitic hybrid jute (*Corchorus capsularis* L). Indian J Agric Sci 40:691–696
Banga O (1959) Möhre, *Daucus carota* L. In: Roemer T, Rudorf W (ed) Handbuch der Pflanzenzüchtung, 2nd edn, vol 6, Paul Parey, Berlin, pp 1–22

crops. Adv Hortic Sci Appl 1:288–294
Banga O, Petiet J (1958) Breeding male sterile lines of Dutch onion varieties as a preliminary to the breeding of hybrid varieties. Euphytica 7:21–30
Banga O, Petiet J, Van Bennekom JL (1964) Genetical analysis of male sterility in carrots, *Daucus carota* L. Euphytica 13:75–93
Banga SS, Labana KS (1983) Male sterility in Indian mustard (*Brassica juncea* (L) Coss. II Genetics and cytology of MS-1. In: Proc 6th Int Rape Seed Conf, Paris 1983, pp 349–353
Banga SS, Labana KS (1984) Ethrel induced male sterility in Indian mustard (*Brassica juncea* (L) Coss. Z Pflanzenzücht 92:229–233
Banga SS, Labana KS (1985) Male sterility in Indian mustard (*Brassica juncea* (L) Coss) IV. Genetics of MS-4. Can J Genet Cytol 27:487–490
Banga SS, Labana KS, Banga SK (1984) Male sterility in Indian mustard (*Brassica juncea* (L) Coss), a biochemical characterization. Theor Appl Genet 68:515–519
Bannerot H, Boulidard L, Cauderon Y, Tempe J (1974) Transfer of cytoplasmic male sterility from *Raphanus sativus* to *Brassica oleracea*. Proc Eucarpia Meet Cruciferae Crop Sect 25: 52–54
Bannerot H, Boulidard L, Chupeau Y (1977) Unexpected difficulties met with the raddisch cytoplasm in *Brassica oleracea*. Eucarpia Cruciferae Newslett 2:16
Banu M, Diaconu P (1969) A study of meiosis in some dwarf mutants induced in maize by radiations. Genet Comunicari 12:233–242
Barabâs Z (1962) Observation of sex-differentiation in sorghum by use of induced male sterile mutants. Nature 195:257–259
Barckay PC (1963) Ariki, a new selected perennial-type ryegrass variety. Proc 25th Newzealand Grassl Assoc Conf, pp 128–137
Bardner NS, Vazhnitskaya EF (1969) The sterility of the peavine caused by foreign pollen and some features of meiosis in the anthers. In: The study of the nature of pollen sterility and its use in the breeding of agricultural plants, Kartya, Moldov, Kishinev, USSR, p 97
Barham WS, Munger HM (1950) The stability of male sterility in onions. Proc Am Soc Hortic Sci 56:401–409
Barlow KK, Driscoll CJ (1981) Linkage studies involving two chromosomal male sterility mutants in hexaploid wheat. Genetics 98:791–799
Barnes BV (1978) Pollen abortion in *Betula* and *Populus*. The Michigan Bot 17:167–172
Barnes DK (1980) Alfalfa. In: Fehr WR, Hadley HH (eds) Hybridization of crop plants. Am Soc Agron, Crop Soc Am, Wisconsin, USA, pp 177–187
Barnes WC (1961) A male sterile cucumber. Proc Am Soc Hortic Sci 77:415–416
Barnes DK, Garboucheva RA (1973) Intra-plant variation for pollen production in male sterile and fertile alfalfa. Crop Sci 13:456–459
Barnes DK, Bingham ET, Axtell JD, Davis WH (1972) The flower sterility mechanisms and pollination control. In: Alfalfa Science and technology. Am Soc Agron, Madison USA, pp 123–141
Barratt DHP, Flavell RB (1975) Alterations in mitochondria associated with cytoplasmic and nuclear genes concerned with male sterility in maize. Theor Appl Genet 45:315–321
Barrons K, Lucas HF (1942) The production of first generation hybrid tomato seed for commercial planting. Proc Amer Soc Hort Sci 40:395–404
Bartkowiak-Broda, Rouselle P, Renard M (1979) Investigations of two kinds of cytoplasmic male sterility in rape seed (*Brassica napus* L.). Genet Pol 20:487–497
Bartlett HH (1911) On gynodioecism in *Plantago lanceolata*. Rhodora 13:199–206
Bartlett HH (1913) Inheritance of sex forms in *Plantago lanceolata*. Rhodora 15:173–178
Barton DW, Butler L, Jenkins JA, Rick CM, Young PA (1955) Rules for nomenclature in Tomato genetics including a list of known genes. J Hered 46:22–26
Bassett MJ, Hung L (1982) Induction of semisterility mutations in common bean, *Phaseolus vulgaris* L. J Am Soc Hortic Sci 107:871–874
Bassett MJ, Shuh DM (1982) Cytoplasmic male sterility in common bean. J Am Soc Hortic Sci 107:791–793

Batch JJ, Morgan DG (1974) Male sterility induced in barley by photoperiod. Nature 250:165–167

Bateson W, Gairdner AE (1921) Male sterility in flax subject to two types of segregations. J Genet 11:269–275

Bateson W, Saunders ER, Punnett RC (1903) On "Male sterility in *Lathyrus odoratus.*" Rep Evol Comm Roy Soc Lond 2:88–99

Bateson W, Saunders ER, Punnett RC (1908) Male sterility in *Lathyrus odoratus.* Rep Evol Comm Roy Soc Lond 4:16

Baur E (1909) Das Wesen und die Erblichkeitsverhältnisse der "varietates albomarginatae hort" von *Pelargonium zonale.* Z Induct Abstammungs- Vererbungslehre 1:330–351

Baur E (1924) Untersuchungen über das Wesen die Entstehung und die Vererbung von Rassenunterschieden bei *Antirrhinum majus.* Bibliogr Genet 4:1–170

Bawa KS (1980) Evolution of dioecy in flowering plants. Annu Rev Ecol Syst 11:15–39

Bazin M, Chabin A, Durand R (1975) Comparison between four isoaccepting transfer ribonucleic acids and corresponding synthetases in male and female flowers of the dioecious species *Mercurialis annua* L. Dev Biol 44:288–297

Beachell HM, Adair CR, Jodan NE, Davis LL, Jones JW (1938) Extent of natural crossing in rice. J Am Soc Agron 30:743–753

Beadle GW (1931) A gene in maize for supernumerary cell division following meiosis. Cornell Univ Agric Exp Stat Mem 129:1–23

Beadle GW (1932a) A gene in *Zea mays* for failure of cytokinesis during meiosis. Cytologia 3: 142–145

Beadle GW (1932b) Genes in maize for pollen sterility. Genetics 17:413–431

Beadle GE, McClintock B (1928) A genic disturbance of meiosis in *Zea mays.* Science 68:433

Beard DF, Meserve JC (1964) Honey bee visitation to male sterile clones. Rep 19th Alfalfa Improv Conf, pp 18–20

Beasley CA, Yermanos DM (1976) Effect of storage on *in vitro* germination of jojoba pollen. Pollen Spores 18:471–477

Beckett JB (1971) Classification of male-sterile cytoplasms in maize (*Zea mays*). Crop Sci 11: 724–727

Bedbrook J, Koldner R (1979) The structure of chloroplast DNA. Ann Rev Plant Physiol 30: 593–620

Bednarski MA, Scheffer RF, Izawa S (1977) Effects of toxin from *Helminthosporium maydis* T. on respiration and associated activities in maize tissue. Physiol Plant Pahtol 11:129–141

Beglaryan NP (1970) Some cytological characteristics of a sunflower mutant induced with gibberellic acid. Tsitol Genet 4:547–552

Beiyernick MW (1885) Gynodioecie bei *Daucus carota* L. Neder Kruidk Arch Ser 2, 4:345–355

Bell GDH (1968) Plant breeding for crop improvement in Britain: methods, achievements and objectives. Proc Roy Soc (B) 171:145–173

Belliard G, Pelletier G (1978) Hybridization cytoplasmique par fusion de protoplastes chez *Nicotiana tabacum.* Physiol Veget 6:441–448

Belliard G, Pellertier G, Feraut M (1977) Fusion de protoplastes de *Nicotiana tabacum* acytoplasmes differentes: etude des hybrides cytoplasmiques neoformes. CR Acad Sci Paris 284:749–752

Belliard G, Pellerties G, Vedel F, Queties F (1978) Morphological characteristics and chloroplast DNA distribution in different cytoplasmic parasexual hybrids of *Nicotiana tabacum.* Mol Gen Genet 165:231–237

Belliard G, Vedel F, Pelletier G (1979) Mitochondrial recombination in cytoplasmic hybrids of *Nicotiana tabacum* by protoplast fusion. Nature 281:401–403

Bemis WP (1970) Polyploid hybrids from the cross *Cucurbita moschata* Poir x *C. foetidissima* HBK. J Amer Soc Hort Sci 95:529–531

Bemis WP (1973) Triple haploids in cucurbita. J Hered 64:103–104

Bemis WP, Wilson GB (1953) A new hypothesis explaining the genetics of sex determination in *Spinacea oleracea* L. J Hered 44:90–95

Benevent E, Le Guen J (1976) Problems poses par I utilisation de la sterilite male cytoplasmique en production de feverole hybride. Mem ENSA de Rennes, Paris

Bennett MD, Hughes WG (1972) Additional mitosis in wheat pollen induced by Ethrel. Nature 240:566–568

Berbeć J (1966) Badaria nad krzyowka *Nicotiana glauca* x *N. tabacum* Inf PAN Poland 7:213–340

Berbeć J (1974) A cytoplasmic male sterile mutation form of *Nicotiana tabacum* L. Z Pflanzenzücht 73:204–216

Berbeć J (1975) Investigations on the utilization of the cytoplasmically inherited male sterile strain of tobacco in the F_1 hybrid seed production. Z Pflanzenzücht 74:28–38

Bergfeld R (1956) Entwicklungsgeschichtliche Untersuchungen an einigen Blütenformmutanten der Sippe 50 von *Antirrhinum majus* L. Z Vererbungsl 87:784–794

Bergounioux-Bunisset G, Perennes C (1980) Transfert de facteurs cytoplasmiques de la fertilite male entre 2 lignees de *Petunia hybrida* perfusion protoplastes. Plant Sci Lett 19:143–149

Bernard RL, Gremeens CR (1975) Inheritance of the eldorado male sterile trait. Soybean Genet Newslett 2:37–39

Bernard RL, Jaycox ER (1969) A gene for increased natural crossing in soybeans. Agron Abstr, p 3

Bernardo EN, Sorenson EL, Painter RH (1970) Cleft grafting alfalfa. Crop Sci 10:102–103

Berninger E (1965) Contribution a letude de la sterilite male de loignon (*Allium cepa*) Ann Amelior Plant (Paris) 15:183–199

Berthelem P (1970) Tapport dắctivite station d'amélioration des plantes. INRA Rennes (France) 1968–1970

Berthelem P (1976) Amelioration genetique de la feverole. Cultivar 89:48–53

Berthelem P, Le Guen J (1975) Rapport d'activite-Station d'Amelioration des plantes. INRA Rennes (France) 1971–1974

Berthelem P, Le Guen J (1976) Rapport d'Activite 1971–1977. Stat d'amelioration des plantes. Le Rheu (France) pp 119–175

Berthelem P, Le Guen J, Bourgeois F (1977) Recherches sur les methodes de se'lection en vuedei éxploitation maximale de la vigeur hybride chez la féverole. Utilisation pralique de la stérilité mâle pour la création de variétés hybrides. Contrat CEEñ408-Compterendu i semestre 1977 mim, p 16

Berthelem P, Le Guen J, Bourgeois F (1978) Recherches sur les méthodes de sélection en vue de l'exploitation maximale de la viguer hybride chez la féverole. Utilisation pratique de la stérilité male pour la création de variété hybrides. Contrat CEEñ408. Compterendu de fin de Contrat mim, p 31

Bertin RI (1982) The evolution and maintenance of andromonoecy. Evol Theory 6:25–32

Best JO, Coffey DL (1975) Effects of ethephone on sex expression and yield of summer squash. Tenn Farm Home Sci Prog Rep 95:13–15

Beversdorf WD, Bingham ET (1977) Male sterility as a source of haploids and polyploids of *Glycine max.* Can J Genet Cytol 19: 283–287

Beyer EM (1976) Silver ion: a potent antiethylene agent in cucumber and tomato. Hort Sci 11:195–196

Beyernick W (1885) Gynodioecie bei *Daucus carota* L. Ned Kruidk Arch Ser 2, 4:345–354

Bhargava YR (1986) Genic disturbance of meiosis in *Jasminum pubescens* willd. Proc 73rd Indian Sci Congr Sec Bot (abstr), p 153

Bhatt GM (1971) Heterotic performance and combining ability in a diallel cross among spring wheats (*Triticum aestivum* L). Aust J Agric Res 22:359–368

Bhat NR, Krishnamoorthi T (1956) A male sterile mutant in *Nicotiana tabacum.* Curr Sci (Bangalore) 25:297–299

Bhattacharya A, Tokumasu S (1970) Effect of gibberellin upon sex expression and internode length in gynaecious and monoecious cucumber. J Jpn Soc Hortic Sci 39:224–231

Bhattacharya A, Kato M, Jode S (1970) Use of male sterility for heterotic effect in F_1 hybridis of muskmelon (*Cucumis melo* L). Mem Coll Agric Ehime Univ Japan 15:21–29

Bianchi A (1963) Transmission of male sterility in *Petunia* by grafting. Genen Phaenen 8:36–43

Bianchi A (1968) Male sterility and plant breeding. Eucarpia 5th Cong Milano, pp 335–345

Bianchi A, Soressi GP (1969) Mutanti di pomodora artificialmente indotti suscettibili di utilizzazione nel miglioramento genetico. Sementi Elette Italy 15:2–6

Bidwell RGS (1979) Plant physiology. MacMillan New York, pp 726
Bielousow A, Kliuizko PF (1970) Mikrosporogienez i razwitijemuzskogo gamietofita u. stierilnych i firtilnych form ozimoj fzi. Cit Gen 6:519–526
Bingham J (1968) Annu Rep Plant Breed Inst Camb, England, p 64
Bino RJ (1985a) Ultrastructural analysis of cytoplasmic male sterility in *Petunia hybrida* var Blue Bedder. Protoplasma 127:230–240
Bino RJ (1985b) Histological aspects of microsporogenesis in fertile, cytoplasmic male sterile and restored fertile *Petunia hybrida.* Theor Appl Genet 69:423–428
Bishop CJ (1954) A stamenless male sterile tomato. Am J Bot 41:540–542
Bitzer MJ, Fu SH (1972) Heterosis and combining ability in southern soft red winter wheats. Crop Sci 12:35–37
Bitzer MJ, Patterson FL, Nyquist WE (1971) Hybrid vigour and gene action in a six parent diallel cross of soft winter wheat. Can J Genet Cytol 13:131–137
Blakslee AF (1921) A graft-infectious disease of *Datura* resembling a vegetative mutant. J Genet 11:17–36
Bleier H (1931) Untersuchungen über die Sterilität der Kartoffel. Arch Pflanzenb 5:545–560
Blickenstaff J, Thompson DL, Harvey PH (1958) Inheritance and linkage of pollen fertility restoration in cytoplasmic male sterile crosses of corn. Agron J 50:430–434
Bliss FA, Gabelman WH (1965) Inheritance of male sterility in beets, *Beta vulgaris* L. Crop Sci 5:403–406
Blondon F, Cambier B, Dattee Y, Guy P (1979) Influence de la température sur la fertilité mâle et femelle de Luzerne témoins, mâle-steriles et "mainteneurs". Ann Amél Plant (Paris) 29: 89–96
Bock MC (1976) Polymorphism floral et stérilite mâle chez le ***Silene** acaulis* (L) Jacq. CR Acad Sci Paris 282:1865–1868
Boerma HR, Cooper RL (1978) Increased female fertility associated with the ms_1 locus in soybeans. Crop Sci 18:344–346
Boerma HR, Moradshahi A (1975) Pollen movement within and between rows to male-sterile soybeans. Crop Sci 15:858–861
Boeshore M, Lifshirtz I, Hanson MR, Izhar S (1983) Novel composition of mitochondrial genomes in *Petunia* somatic hybrids derived from cytoplasmic male sterile and fertile plants. Mol Gen Genet 190:459–467
Bohn GW, Principe JA (1964) A second gene in the muskmelon. J Hered 55:211–215
Bohn GW, Whitaker JW (1949) A gene for male sterility in muskmelon (*Cucumis melo* L). Proc Amer Soc Hort Sci 53:309–314
Bolton JL, Fryer JR (1937) Inter-plant variations in certain seed-setting processes in alfalfa. Sci Agric 18:148–160
Bolz G (1961) Genetisch-züchterische Untersuchungen bei *Tagetes II.* Herstellung, Genetik und Verwendung röntgeninduzierter Mutanten in der Züchtung. Z Pflanzenzücht 45:121–142
Bolz G (1968) Monohybride Vererbung der plasmatisch-genisch bedingten Pollensterilität bei *Beta vulgaris* L. Z Pflanzenzücht 60:219–234
Bonali F, Boggini G, Scalfati MC, Borghi B (1972) Primi tentalivi di indurre la sterilita mashile nel frumento tenero mediante l'impiego dell'acido2-chloroetilfosfonico "Ethrel" in vista della produzione di seme ibrido F_1. Ann Inst Sper Cereealicoltura 3:97–108
Bond DA (1966) Yields and components of yield in diallel crosses between inbred lines of winter bean (*V. faba* L.). J Agric Sci 67:325–326
Bond DA, Hawkins RP (1967) Behaviour of bees visiting male sterile field beans (*Vicia faba* L.). J Agric Sci 68:243–247
Bond DA, Fyfe JL, Toynbee-Clarke G (1964a) Male sterility in field beans (*Vicia faba* L.) I. A male-sterile bean inherited as a Mendelian recessive. J Agric Sci 63:229–234
Bond DA, Fyfe JL, Toynbee-Clarke G (1964b) Male sterility in field beans (*Vicia faba* L.) II. Yield trials of F_1 hybrid winter beans produced with the aid of male sterility. J Agric Sci 63:235–243
Bond DA, Fyfe JL, Toynbee-Clarke G (1966a) Male sterility in field beans (*Vicia faba* L.) III. Male sterility with cytoplasmic type of inheritance. J Agric Sci 66:359–367

Bond DA, Fyfe JL, Toynbee-Clarke G (1966b) Male sterility in field beans IV. Use of cytoplasmic male sterility in the production of F_1 hybrids, and their performance in trials. J Agr Sci 66:369–377

Bond DA, Lawes DA, Poulsen MH (1980) Broadbean (Faba bean) In: Fehr WR, Hadley HH (eds) Hybridization of crop plants. Am Soc Agron, Wisconsin USA 1980, pp 203–213

Bonnemaison F, Dommee B, Jacquard P (1979) Etude experimentale de la concurrence entre formes sexuelles chez le Thym, *Thymus vulgaris* L. Oecol Plant 14:85–101

Bonnet A (1975) Introduction et utilisation d'une sterilite male cytoplasmique dans des varietes precoces europeennes de Radis, *Raphanus sativus* L. Ann Amelior Plant (Paris) 25: 381–397

Bonnet A (1977) Breeding in France of a radish F_1 hybrid obtained by use of cytoplasmic male sterility. Eucarpia Cruciferae Newslett 2:5

Borchers EA (1966) Characteristics of a male sterile mutant in purple cauliflower (*Brassica oleracea* L.) Proc Amer Soc Hort Sci 88:406–410

Borchers EA (1971) Hybrid broccoli seed production utilizing the ms6 gene for male sterility. J Amer Soc Hort Sci 96:542–543

Borck KS, Walbot V (1982) Comparison of the restriction endonuclease digestion patterns of mitochondrial DNA from normal and male sterile cytoplasms of *Zea mays* L. Genetics 102: 109–128

Boren RB, Parker RL, Sorensen EL (1962) Foraging behaviour of honey bees on selected alfalfa clones. Crop Sci 2:185–188

Borghi B, Pironi W (1976) Evaluation of heterosis in *Cucurbita pepo* L. Proc 7th Eucarpia Cong Budapest. Elsevier, Amsterdam, pp 219–226

Borghi B, Bonali F, Boggini G (1973) Induction of male sterility in wheat with ethephon for hybrid seed production. In: Proc 4th Int Wheat Genet Symp, Missouri Agric Exp Stat Columbia, pp 337–347

Borowsky MI (1960) The transfer of character of male sterility in maize by vegetative hybridisation. Proc Darwin Jubilee Conf Kisinev USSR, 1960, pp 209–212

Borst P, Grivell LA (1978) The mitochondrial genome of yeast. Cell 15:705–723

Bose TK, Nitsch JP (1970) Chemical alteration of sex expression in *Luffa acutangula.* Physiol Plant 23:1206–1211

Bouharmount P, Pockett P (1965) Contribution a l'etude des effects du gametocide FW-450 sur la fertilite male et female de *Gossypium hirsutum* L. Pubs INEAC Ser Sci, p 107

Boulidard L (1975) Transfert de la sterilite male cytoplasmique due genre Raphanus au genere Brassica. Rapport d'activite 1972–1974 de genetique et d'amelioration des plantes. Versailles Inst Natl Rech Agron France, pp 1–5

Boutry M, Briquet M (1981) Modification of the mitochondrial DNA restriction pattern and of the mitochondrial protein synthesis products in two cytoplasmic male sterile lines of *Vicia faba.* Réunion du Cercla Francais de biologie cellulaire sur les éléments transposables et instabilites génétiques, Sept 28–30, 1981

Boutry M, Briquet M (1982) Mitochondrial modifications associated with the cytoplasmic male sterility in faba beans. Eur J Biochem 127:129–135

Bouverat-Bernier JP, Le Guen J (1980) Contributian étude de l'évolution de la sterilité mâle cytoplasmique chez la Féverole Mémoire. ENSA Rennes, France (see Picard et al. 1981)

Bowman DT, Weaver JB (1979) Analysis of a dominant male sterile character in upland cotton II. Genetic studies. Crop Sci 19:628–630

Bowman DT, Weaver JB, Walker DB (1978) Analysis of a dominant male sterile character in upland cotton I. Cytological studies. Crop Sci 18:730–738

Bozzini A (1974) Radiation-induced male sterility in durum wheat. Proc Symp Polyploid and Induced Mutations in Plant Breeding. IAEA, Vienna, Austria, pp 23–25

Bozzini A, Scarascia-Mugnozza GT (1968) A factor for male sterility inherited as a Mendelian recessive. Euphytica Suppl 1:83–86

Braak JP, Kho YO (1958) Some observations on the floral biology of the carrot (*Daucus carota* L.). Euphytica 7:131–139

Brabec F (1962) Über triploide fertile *Bryonia*-Bastarde. Biol Zentralbl 81:291–302

Bradner NR (1974) Hybrid soybean production. Patent No 3, 903, 645, USA

Bradner NR, Childers WR (1968) Cytplasmic male sterility in alfalfa. Can J Plant Sci 48:111–112

Brar GS (1982) Male sterility in sesame. Indian J Genet Plant Breed 42:23–27

Brar GS, Singh S, Labana KS, Chopra S (1980) Identification of male sterile lines in Indian mustard (*Brassica juncea*). Agron Abstr, p 50

Breeze MSG (1921) Degeneration in anthers of potato. Gardner's Chron USA 70:274–275

Breslavetz LP (1935) Abnormal development of pollen in different races and grafts of hemp. Genetica (The Hague) 17:154–169

Brettell RIS, Goddard BVD, Ingram DS (1979) Selection of Tms cytpolasm maize tissue cultures resistant to *Drechslera maydis* T-toxin. Maydica 24:203–213

Brettell RIS, Thomas E, Ingram DS (1980) Reversion of Texas male sterile cytoplasm maize in culture to give fertile T-toxin resistant plants. Theor Appl Genet 58:55–58

Breuils G, Pochard E (1975) Essaide fabrication de l'hybride de piment "Lamuro-Inra" avec utilization dúne stérilité mâle génétique (ms 509). Ann Amêlion Plant (Paris) 25:399–409

Brewbaker JL (1964) Agricultural Genetics Prentice-Hall, New York, pp 42–45

Brezhnev DD (1976) New aspects of heterosis in tomato selection. In: Heterosis in plant breeding. Eucarpia 7:59–62

Briggle LW (1956) Interaction of cytoplasm and genes in male sterile corn crosses involving two inbred lines. Agron J 48:569–573

Briggle LW (1957) Interaction of cytoplasm and genes in a group of male sterile corn types. Agron J 49:543–547

Briggle LW (1963) Heterosis in wheat - a review. Crop Sci 3:407–412

Briggle LW (1970) A recessive gene for male sterility in hexaploid wheat.Crop Sci 10:693–696

Briggle LW, Daum RJ, Stevens H (1964) Expression of heterosis in two wheat crosses. Crop Sci 4:220–223

Briggs FN, Knowles PF (1967) Introduction to plant breeding. Reinhold, New York

Brim CA, Stuber CW (1973) Application of genetic male sterility to recurrent selection schemes in soybeans. Crop Sci 13:528–530

Brim CA, Young MF (1971) Inheritance of a male sterile character in soybeans. Crop Sci 11: 564–566

Brink RA, Cooper DC (1936) The mechanism of pollination in alfalfa (*Medicago sativa*). Am J Bot 23:678–683

Briquet M, Faber AM, Berthelem P, Guen Le J, Duc G (1985) Microanalysis of variant mitochondrial polypeptides associated with the cytoplasmic male sterility trait in faba bean. In: Magnien E, Nettancourt de D (eds) Genetic engineering of plants and microorganisms important for agriculture. Nijhoff-Junk, the Netherlands, pp 146–147

Britikov EA, Musatava NA, Vladimirtseva SV, Protsenko MA (1964) Transformation of proline in germinating pollen. In: Linskens HF (ed) Pollen physiology and fertilization, North Holland, Amsterdam, pp 77–85

Britikov EA, Schrauwen J, Linskens HF (1970) Proline as a source of nitrogen in plant metabolism. Acta Bot Neerl 19:515–520

Brockmann I, Bocquet G (1978) Okologische Einflüsse auf die Geschlechtsverteilung bei *Silene vulgaris (*Moench) Garcke (Caryophyllaceae). Ber Dtsch Bot Ges 91:217–230

Broertjes C, Jong J (1984) Radiation-induced male-sterility in daisy-types of *Chrysanthemum morifolium* Ram. Euphytica 33:433–434

Bronckers F, Amand H (1966) Ètude des effects due gamêtocide FW450 (2,3-dichlorisobutyrate de sodium) sur *Beta vulgaris* L. Publ Techn Inst Belge Amélior Lett 111:105–108

Brooking IR (1976) Male sterility in *Sorghum bicolor* (L.) Moench induced by low temperature I. Timing of the stage of sensitivity. Aust J Plant Physiol 3:589–596

Brooking R (1979) Male sterility in *Sorghum bicolor* (L.) Moench induced by low night temperature II. Genotype differences in sensitivity. Aust J Plant Physiol 6:143–147

Brooks JS, Brooks MH (1967) Pollen abortion in relation to cytoplasmic genetic male sterility in sorghum. Crop Sci 7:47–50

Brooks MH (1962) Comparative analyses of some of the free amino acids in anthers of fertile and genetic-cytoplasmic male sterile sorghum. Genetics 47:1629–1638

Brooks MH (1965) Nucleic acid studies of fertile cytoplasmic male sterile and fertility restorer lines of *Sorghum vulgare*. Sorghum Newslett 8:54–55

Brooks MH, Brooks JS (1965) Anther tapetal width and cytoplasmic-genetic male sterility in *Sorghum vulgare*. Sorghum Newslett 8:55–56

Brooks MH, Brooks JS (1969) Pollen sterility in A, B and R lines and their hybrids. Sorghum Newslett 12:78

Brooks MM, Brooks JS, Chien L (1966) The anther tapetum cytoplasmic-genetic male sterile sorghum. Am J Bot 53:902–908

Brown CM, Earle EB (1973) Response of one winter wheat and two spring oat varieties to foliar applications of 2-chloroethyl phosphonic acid (ethrel). Agron J 65:829–832

Brown DE, Bingham ET (1984) Hybrid alfalfa seed production using a female sterile pollenizer. Crop Sci 24:1207–1208

Brozek A (1926) A genetic study of crosses between *Mimulus quinquevulnerus, M. tigrinus* and *M. tigrinoides*. J Hered 17:113–129

Brücher H (1964) El origen de la papa (*Solanum tuberosum*). Physis 24:439–452

Buchert JG (1961) The stage of genome-plasmon interaction in the restoration of fertility to cytoplasmically pollen-sterile maize. Proc Natl Acad Sci USA 47:1436–1440

Buck RW (1960) Male sterility in interspecific hybrids of *Solanum*. J Hered 51:13–14

Buckner RC, Hill HD, Burrus PB (1961) Some characteristics of perennial and annual ryegrass x tall fescue hybrids and of the amphidiploid progenies of annual ryegrass and tall fescue. Crop Sci 1:75–80

Bukovak MJ, Honma S (1965) Gibberellin-induced heterostyly in the tomato (abstr). 62nd Annu Meet Am Soc Hortic Sci, No 164, p 47

Bunn CL, Wallace DC, Eisenstadt JM (1974) Cytoplasmic inheritance of chloramphenicol resistance in mouse tissue culture cells. Proc Natl Acad Sci USA 71:1681–1685

Buntman DJ, Horner HT (1982) Microsporogenesis in normal and ms_3 mutant soybean (*Glycine max*) (abstr). Proc 94th Session Iowa Acad Sci USA, p 12

Burk LG (1960) Male sterile flower anomalies in interspecific tobacco hybrids. J Hered 51:27–31

Burk LG (1967) An interspecific bridge cross *Nicotiana repanda* through *N. sylvestris* to *N. tabacum*. J Hered 58:215–218

Burk LG, Durbin RD (1978) The reaction of *Nicotiana* species to tentoxin. J Hered 69:117–120

Burlov VV (1974) Utilization of male sterility in sunflower breeding for heterosis. Proc 6th Int Sunflower Conf, Bucharest, pp 353–360

Burnham CR (1941) Cytogenetic studies of a case of pollen abortion in maize. Genetics 26:460–468

Burnham CR (1971) Pollen abortion and related mutants as possible schemes for producing "all male-sterile progeny". Maize Genet Coop Newslett 45:133

Burnham CR (1976) Progress in establishing a true breeding line that will produce "all male sterile" progeny when crossed as genetic male sterile plants. Maize Genet Coop Newslett 50:82–83

Burnham CR, Phillips RL, Albertsen MC (1981) Inheritance of male sterility in flax involving nuclear-cytoplasmic interaction including methods of testing for cytoplasmic restoration. Crop Sci 21:659–663

Burns JA, Gerstel DU (1981) Role of nucleolar chromosomes in anther restoration of male-sterile tobacco. J Hered 72:413–418

Burns JA, Gerstel DU, Sand SA (1978) Cytoplasmic male sterility in *Nicotiana*, restoration of fertility and the nucleolus II. *N. debneyi* cytoplasm. Genetics 90:151–159

Burrows CJ (1960) Studies in *Pimelea* I. The breeding system. Trans R Soc New Zealand 88:29–45

Burton GW (1943) Interspecific hybrids in the genus *Paspalum* (Dallis grass). J Hered 34:15–23

Burton GW (1948) The method of reproduction in common Bahia grass, *Paspalum notatum*. J Am Soc Agron 40:443–452

Burton GW (1958) Cytoplasmic male sterility in pearl millet (*Pennisetum glaucum* (L.)). R, Br. Agron J 50:230–231

Burton GW (1965) Pearlmillet Tift 23A released. Crops Soils Mag 17:19

Burton GW (1969) Registration of pearlmillet inbreds Tift 23B, Tift 23A, Tift 23DB and Tift 23DA. Crop Sci 9:397

Burton GW (1972) Natural sterility maintainers and fertility restorer mutants in Tift 23A cytoplasmic male-sterile pearl millet *Pennisetum typhoides* (Burm) Stapf and Hubb. Crop Sci 12: 280–282

Burton GW (1977a) Fertile sterility maintainer mutants in cytoplasmic male sterile pearl millet. Crop Sci 17:635–637

Burton GW (1977b) Registration of Gahi 3 pearl millet. Crop Sci 17:345–346

Burton GW (1981) A gene for early maturity and photoperiod insensitivity in pearl millet. Crop Sci 21:317–318

Burton GW, Athwal DS (1967) Two additional sources of cytoplasmic male sterility in pearl millet and their relationship to Tift 23A. Crop Sci 7:209–211

Burton GW, Athwal DS (1969) Registration of pearl millet inbreds Tift 239 DB_2 and Tift $239DA_2$. Crop Sci 9:398

Burton GW, Fortson JC (1963) Inheritance and utilization of five dwarfs in pearl millet (*Pennisetum glaucum*) breeding. Proc 11th Int Genet Cong Genetics Today, vol 1, p 230

Burton GW, Hanna WW (1976) Ethidium bromide induced cytoplasmic male sterility in pearl millet. Crop Sci 16:731–732

Burton GW, Hanna WW (1982) Stable cytoplasmic male sterile mutants induced in Tift 23 DB_1 pearl millet with mitomycin and streptomycin. Crop Sci 27:651–652

Burton GW, Powell JB (1968) Pearl millet breeding and cytogenetics. Adv Agron 20:48–89

Burton GW, Monson WG, Johnson JC, Lowrey RS, Chapman HD, Marchant WH (1969) Effect of the d_2 gene on the forage yield and quality of pearl millet. Agron J 61:607–612

Burton GW, Hanna WW, Powell JB (1980) Hybrid vigour in forage yields of crosses between pearl millet inbreds and their mutants. Crop Sci 20:744–747

Burton JW (1977) Induction of sterility with ethidium bromide. Soybean Genet Newslett 4: 76–79

Burton JW, Brim CA (1981) Recurrent selection in soybeans III. Selection for increased percent oil in seeds. Crop Sci 21:31–34

Burton JW, Liles RH (1979) Selection of a maternally inherited male sterile trait in soybean. Soybean Genet Newslett 6:105–106

Burton JW, Wilson RF, Brim CA (1979) Dry matter and nitrogen accumulation in male-sterile and male-fertile soybeans. Agron J 71:548–552

Busbice TH (1969) Inbreeding in synthetic varieties. Crop Sci 9:601–604

Busbice TH, Wilsie CP (1966) Inbreeding depression and heterosis in autotetraploids with application to *Medicago sativa* L. Euphytica 15:52–67

Buschmann C, Sironval C (1978) Influence of kinetin on protochlorophyll (ide) accumulation and on the shibata shift in *Raphanus* seedlings. Planta (Berl) 139:127–132

Buss GR (1983) Inheritance of a male sterile mutant from irradiated Essex soybeans. Soybean Genet Newslett 10:104–108

Buss GR, Autio WR (1980) Observations of polyembryony and polyploidy in ms_1 and ms_2 male-sterile soybean populations. Soybean Genet Newslett 7:94–97

Butenko RG, Kuchko AA, Vitenko VA, Avetisoo VA (1977) Production and cultivation of isolated protoplasts from the mesophyll of leaves of *Solanum tuberosum* L. and *Solanum chacoense*. Sov Plant Physiol 24:660–665

Butterfass T (1960) Künstlich induzierte Pollensterilität bei Zuckerrüben. Zucker 13:164–165

Byers RE, Baker LR, Sell HM, Herner RC, Dilley DR (1972) Ethylene: a natural regulator of sex expression of *Cucumis melo* L. Proc Natl Acad Sci USA 69:717–720

Cachon H (1980) La sterilité mâle: sa recherche et son utilization en vue de la production de fleurs coupée chez *Antirrhinum, Delphinum Helianthus, Zinnia.* Proc Eucarpia Sect Ornamentals. Meeting on flower plants with sexual reproduction, Versailles, France, p 61

Camerarius RJ (1694) De sexu Plantarum epistola. In: Möbius M (ed) Oswalds Klassiker der exakten Wiss Nr 105. Engelmann, Leipzig, DDR, 1899

Cameron DR (1965) Cytoplasmic effects in *Nicotiana.* Proc 11th Int Cong Genet 1:202–204

Chung SR, Durand R, Durand B (1979) Differential cytokinin binding to dioecious plant ribosomes. FEBS Lett 102:211–215

Chun-nong H, Min-yuan X, Yuan-Zhong Z (1981) Electron microscope observations on the pollen development of male sterility in wheat induced by ethrel. Acta Bot Sin 23:12–16

Claassen CE (1948) Flowering habits and inheritance of sterility genes in cultivated safflower *(Carthamus tinctorius* L.). Ph D Thesis Nebraska Univ, Lincoln, USA, abstr, pp 1–6

Claassen CE (1952) Inheritance of sterility, flower color, spinelessness, attached pappus and rust resistance in safflower *Carthamus tinctorius*. Nebraska Agric Exp Stat Bull, p 171

Claassen CE, Hoffman A (1950) The inheritance of the pistillate character in castors and its possible utilization in the production of commercial hybrid seed. Agron J 42:79–82

Clark FJ (1940) Cytogenetic studies of divergent meiotic spindle formation in *Zea mays*. Am J Bot 27:547–559

Clark RK, Kenney DS (1969) Comparison of staminate flower production on gynoecious strains of cucumber (*Cucumis sativus*) by pure gibberellins A_3, A_4, A_7 and A_{13} and mixtures. J Amer Soc Hort Sci 94:131–132

Clarke AE, Fryer JR (1930) Seed setting in alfalfa. Sci Agric 11:138–143

Clarke AE, Pollard LH (1949) The amount of self-pollination in male sterile onion lines. Proc Amer Soc Hort Sci 53:299–301

Clausen J (1926) Genetical and cytological investigations on *Viola tricolor* L. and *V. arvensis* Murr. Hereditas 8:1–156

Clausen J (1930) Male sterility in *Viola orphanidis*. Hereditas 14:54–72

Clausen RE, Goodspeed TH (1925) Interspecific hybridization in *Nicotiana* II. A tetraploid *glutinosa-tabacum* hybrid, an experimental verification of O Winge's hypothesis. Genetics 10:279–284

Clayberg CD (1966) Further data on aw-ms_{15} linkage. Tomato Genet Coop Rep USA 16:7

Clayberg CD (1971) Screening tomatoes for ozone resistance. Hort Science 6:396–397

Clayberg CD (1974) The inheritance of cytoplasmic male sterility in *Kohleria* (Gesneriaceae). J Hered 65:232–234

Clayberg CD, Butler L, Rick CM, Young PA (1960) Second list of known genes in the tomato including supplementary rules for nomenclature. J Hered 51:167–174

Clayberg CD, Butler L, Kerr EA, Rick CM, Robinson RW (1966) Third list of known genes in the tomato with revised linkage map and additional rules. J Hered 57:189–196

Clayberg CD, Butler L, Kerr EA, Rick CM, Robinson RW (1967) Supplementary list of tomato genes as of January 1967. Tomato Genet Coop Rep USA 17:2–11

Clayberg CD, Butler L, Kerr EA, Rick CM, Robinson RW (1970) Additions to the list of genes, January 1967 to January 1970. Tomato Genet Coop Rep USA 20:6–11

Clayberg CD, Butler L, Kerr EA, Rick CM, Robinson RW (1971) Report of the gene list committee. Tomato Genet Coop Rep USA 21:2-10

Clayberg CD, Butler L, Kerr EA, Rick CM, Robinson RW (1973) Additions to the list of genes. Tomato Genet Coop Rep USA 23:3–7

Clayton EE (1950) Male sterile tobacco. J Hered 41:171–175

Clayton PB, Ellis JRS (1976) The genetic influence on preharvest sprouting in male sterile lines of winter wheat. Cereal Res Commun 4:121–125

Cleij G (1967) Influence of the cytoplasmic male sterility and fertility in beets. Euphytica 16: 23–28

Clements WM Jr (1975) Plasmon mutations in cytoplasmic male sterile pearl millet (*Pennisetum typhoides*). Genetics 79:583–588

Coe EH (1983) Mixed pollinations with white pollen. Maize Genet Coop Newslett 57:37

Coe EH, Neuffer MG (1977) The genetics of corn. In: Sprague GF (ed) Corn and corn improvement. Am Soc Agron USA, pp 111-223

Coe EH, Mc Cormick SM, Modena SA (1981) White pollen in maize. J Hered 72:318–320

Coe GE, Stewart D (1977) Cytoplasmic male sterility, self fertility and monogermness in *Beta maritima* L. J Am Soc Sugarbeet Techn 19:257–261

Cohan SM, Weigle JL (1966) Chemically induced male sterility in onion (*Allium cepa*). Hortscience 1:6

Cole K (1957) A genetical investigation of male sterility in green sprouting broccoli. Proc Genet Soc Can 2:44

Cole K (1959) Inheritance of male sterility in green sprouting broccoli. Can J Genet Cytol 1: 203–207

Colhoun CW, Steer MW (1981) Microsporogenesis and the mechanism of cytoplasmic male sterility in maize. Ann Bot 48:417–424

Comandon J, Fonbrune P de (1939) Greffe nucleaire totale, simple ou multiple, chez une amibe. Cr Séano Soc Biol 130:744–748

Cromstock JC, Martinson CA, Gengenbach BG (1973) Host specificity of a toxin from *Phyllosticta maydis* for texas cytoplasmically male sterile maize. Phytopathology 63:1357–1361

Comstock VE (1963) Studies of cytoplasmic male sterility. Annu Rep Coop seed Flax Invest Minn USDA, Minn Agric Exp Stat USA, pp 79–81

Comstock VE (1964) Development of male sterile breeding lines. Annu Rep Coop Seed Flax Invest Minn USDA, Minn Agric Exp Stat USA, pp 34–36

Comstock VE (1965) Possibilities of hybrid flax production. Proc 35th Annu Flax Inst Conf, Minneapolis, USA, pp 24–25

Conagin CHTM (1964) Cytological observations on *Pelargonium*. Bragantia *23*:23–37

Conde MF, Pring DR, Levings CS III (1979) Maternal inheritance of organelle DNAs in *Zea mays-Zea perennis* reciprocal crosses. J Hered 70:2–4

Conde MF, Pring DR, Schertz KF, Ross WM (1982) Correlation of mitochondrial DNA restriction endonuclease patterns with sterility expression in six male sterile sorghum cytoplams. Crop Sci 22:536–539

Connolly V, Wright-Turner R (1984) Induction of cytoplasmic male sterility into ryegrass (*Lolium perenne*). Theor Appl Genet 68:449–453

Connor HE (1965a) Breeding system in New Zealand grasses V. NZ J Bot 3:17–23

Connor HE (1965b) Breeding systems in New Zealand grasses VI. NZ J Bot 3:233–242

Connor HE (1971) *Cortaderia splendens* Connor sp novo (Gramineae). NZ J Bot 9:519–525

Connor HE (1973) Breeding systems in *Cortaderia* (Gramineae). Evolution 27:663–678

Connor HE, Penny ED (1960) Breeding systems in New Zealand Grasses II. NZ J Agric Res 3: 725–727

Cooper DC (1952) The transfer of DNA from the tapetum to the microsporocytes at the onset of meiosis. Am Nat 86:219–229

Cooper RL, Boerma HR (1975) Evidence of a multiple allele for male sterility. Soybean Genet Newslett 2:49–51

Corbett MK, Edwardson JR (1964) Inter-graft transmission of cytoplasmic male sterility. Nature 201:847–848

Corkill L (1964) Maori names given to pasture plants bred by grasslands division. NZ J Agric 109:138

Cornu A (1970) Recherches sur l'induction et utilisation de mutations somatiques chez le *Petunia hybrida* Hort. Thèse CNRS France No 4667

Cornu A, Dommerguess P (1974) Analyse génétique des modifications induites aux loci lies A-R du *Pétunia*. In: Proc Symp Polyploidy and induced mutations in plant breeding, IAEA, Vienna, pp 63–73

Cornu A, Gabay-Laughnan S (1982) Reversion of T male sterile cytoplasm to male fertility. In: Sheridan WF (ed) Maize for biological research. N Dakota State Univ Press, USA, pp 247–250

Cornu A, Cassini R, Berville A (1980) Induction of resistance to *Helminthosporium maydis* race T in Texas maize. Maize Genet Coop Newslett 54:42–44

Cornu A, Vuillaume E, Bodergat R (1981) Interet des variants cytoplasmiques en amelioration des plantes. In: Proc Int Symp Induced Mutations as a tool for crop improvement, IAEA, Vienna, pp 267–275

Correns C (1904) Experimentelle Untersuchungen über die Gynodioecie. Ber D Bot Ges 22: 506–517

Correns C (1906) Die Vererbung der Geschlechtsformen bei den gynodiöcischen Pflanzen. Ber Dtsch Bot Ges 24:459–474

Correns C (1908) Die Rolle der männlichen Keimzellen bei der Geschlechtsbestimmung der gynodioecischen Pflanzen. Ber Dtsch Bot Ges 26 A:686–701

Correns C (1909) Vererbungsversuche mit blass(gelb)-grünen und buntblättrigen Sippen bei *Mirabilis jalapa, Urtica pilulifera* und *Lunaria annua*. Z Induct Abstammungs Vererbungsl 1:291–329

Correns C (1916) Untersuchungen über Geschlechtsbestimmung bei Distelarten. Akad Wiss 20: 448–477

Correns C (1926) Über Fragen der Geschlechtsbestimmung bei höheren Pflanzen. Z Induct Abstammungs Vererbungsl 41:5–31

Correns C (1928) Bestimmung, Vererbung und Verteilung des Geschlechts bei den höheren Pflanzen. Handb Vererbungswiss 2:1–138

Correns C (1937) Nicht-mendelnde Vererbung. Handb Vererbungswiss 11:1–159

Cortessi HA (1967) Investigations made into male sterile beets. Euphytica 16:425–432

Craigmiles JP (1961) The development, maintenance and utilization of cytoplasmic male-sterility for hybrid sudan grass seed production. Crop Sci 1:150–152

Craigmiles JP (1962) Genetic inheritance of cytoplasmic male sterility in sudangrass. Crop Sci 2:203–205

Craigmiles JP (1966) Utilization of heterosis in sudan grass breeding. Proc 10th Int Grassland Cong Sect 3:801–803

Craigmiles JP, Stacy SV (1961) Suhi-1 Hybrid sudan grass. Leaflet NS 26 Georgia Exp Stat, USA, pp 1–8

Craigmiles JP, Harris HB, Newton JP, Dobson JW (1958) Heterosis in F_1 hybrids of *Sorghum vulgare* x *S. sudanense* and *S. vulgare* x *S. arundinaceum*. Agron J 50:714–714

Craigmiles JP, Stansel JW, Flinchum WT (1968) Feasibility of hybrid rice. Crop Sci 8:720–722

Crane MB (1915) Heredity of types of inflorescence and fruits in tomato. J Genet 5:1–12

Crane MB, Lawrence WJC (1931) Inheritance of sex, color and hairiness in *Rubus idaeus* L. J Genet 24:243–255

Crane MB, Thomas PT (1949) Reproductive versatility in *Rubus* III. Raspberry-black berry hybrids. Heredity 3:99–107

Cregan PB, Busch RH (1978) Heterosis, inbreeding and line performance in crosses of adapted spring wheats. Crop Sci 18:247–251

Crowe LK (1964) The evolution of outbreeding in plants. Heredity 19:435–457

Crowley NC (1957) Studies on the seed transmission of plant virus diseases. Aust J Biol Sci 10: 449–464

Cuelemans E (1973) Etude de l'origine et de la nature de la stérilité mâle cytoplasmique chez la betterave sucriere. These Doctorat Louvain Univ, pp 159

Currence TM (1944) A combination of semi-sterility with two simply inherited characters that can be used to reduce the cost of hybrid tomato seed. Proc Amer Soc Hort Sci 44:403–406

Curtis GJ (1967) Graft-transmission of male sterility in sugar beet (*Beta vulgaris* L.) Euphytica 16:419–424

Curtis LG, Scarčhuk J (1948) A bisex sterile variant in *Cucurbita pepo* L. J Hered 39:32

Cutter GL, Bingham ET (1975) Breeding behaviour and embryosac histology of male sterile soybeans (abstr). Am Soc Agron Annu Meet Tennessee USA, p 50

Cutter GL, Bingham ET (1977) Effect of soybean male sterile gene ms1 on organisation and function of the female gametophyte. Crop Sci 17:760–764

Dabral KC (1968) Sterility in *Sesamum indicum* L. JNKVV Res J, Jabalpur India 2:73

Dabral KC, Mandloi KC (1974) A peculiar sterility in sesame (*Sesamum indicum* L.). JNKVV Res J, Jabalpur India 8:54–56

Dahlgren KVO (1932) Über eine Form von *Primula officinalis* mit pistilloiden Staubgefässen und ihre Vererbung. Hereditas 17:115–130

Dale RMK (1981) Sequence homology among different size classes of plant mt DNAs. Proc Natl Acad Sci USA 78:4453–4457

Dalke L, Szota M, Debowski W (1983) Attempt to utilize cytoplasmic male sterility of *Beta maritima* in sugarbeet breeding programme (abstr). Proc Eucarpia Cong, Wageningen, p 51

Damon EG (1961) Studies of the occurrence of multiploid sporocytes in three varieties of cytoplasmic male sterile and normal fertile variety resistant wheatland *Sorghum*. Phyton (Horn) 17:193–203

Danielli JF (1963) The theory of cells in relation to the study of cytoplasmic inheritance in amoebae by nuclear transfer. Harvey Lect Ser 58:217–232

Danilenko NG (1981) Changes in the activity of genes coding some enzymes of carboyhdrate metabolism in plants with cytoplasmic sterility. Tez Dokl Konf Molod Uchen Biologov Riga Lativian USSR, pp 30–31

D'arcy WG (1972) Solanaceae studies II. Typification of subdivisions of *Solanum*. Ann Mo Bot Gard 59:262

Darlington CD (1958) The evolution of genetic systems. Oliver and Boyd, London

Darlington CD, Haque A (1955) The timing of mitosis and meiosis in *Allium ascalonicum*. A problem of differentiation. Heredity 9:117–127

Darvey NL (1979) Rapid method for the production of triticale and for alien incorporation into wheat. Proc 5th Int Wheat Genet Symp (1978), pp 1144–1153

Darwin C (1876) The effects of cross and self fertilization in the vegetable kingdom. Murray, London

Darwin C (1877) Different forms of flowers on plants of the same species. Murray, London

Darwin C (1889) The different forms of flowers on plants of the same species, 2nd edn Appleton, New York

Darwin C (1890) The variation of animals and plants under domestication. Murray, London

Darwin C (1893) The variation of animals and plants under domestication, 2nd edn, vol 1,2. Murray, London

Das K, Pandey BD (1961) Male sterility in brown sarson. Indian J Genet Plant Breed 21:185–190

Dashek WV, Erickson SS (1981) Isolation, assay, biosynthesis, metabolism, uptake, translocation and function of proline in plant cells and tissues. Bot Rev 47:349–385

Dasilva FP, Endrizzi JE, Stith LS (1981) Genetic study of restoration of pollen fertility of cytoplasmic male-sterile cotton. Rev Bras Genet 4:411–426

Daskaloff C, Jordanov N, Georgiev C, Daskaloff S (1965) Use of male sterility and other means to facilitate the production of hybrid tomato seed. In: Proc Symp use of male sterility in hybrid seed production. Bulg Acad Sci, Sofia, pp 63–75

Daskaloff S (1968) A male sterile pepper (*C. annuum* L.) mutant. Theor Appl Genet 38:370–372

Daskaloff S (1971) Two new male sterile pepper (*Capsicum annuum* L.) mutants. CR Acad Sci Agric Bulgaria 4:291–294

Daskaloff S (1971–72) Male sterile pepper (*C. annuum*) mutants and their utilization in heterosis breeding. Proc Eucarpia Meet *Capsicum* 7:205–210

Daskaloff S (1973a) Three new male sterile mutants in pepper (*Capsicum annuum* L.). CR Acad Sci Agric Bulgaria 6:39–41

Daskaloff S (1973b) Investigations of induced mutants in *Capsicum annuum* L. III. Mutants in the variety Zlaten Medal. Genet Plant Breed, Sofia 6:419–429

Daskaloff S (1974) Investigations on induced mutants in sweet pepper (*Capsicum annuum* L.). Proc Eucarpia Cong Budapest, pp 81–90

Daskaloff S (1976) Seed setting of male sterile mutants in connection with heterosis breeding in pepper *Capsicum annuum* L. Genet Agrar 30:407–417

Dattée Y, Annou G, Lefort G, Guy P, Genier G (1980) Sterilite male cytoplasmique chez la Luzerne: etude de la variabilite interet interadescendances de croisements entremales steriles et mainteneurs de diverses origines. Ann Amelior Plant (Paris) 30:1–16

Dauphin B, Teller G, Durand B (1979) Identification and quantitative analysis of cytokinins from shoot apices of *Mercurialis ambigua* by gas chromatography mass spectrometry computer system. Planta (Berl) 144:113–119

Dauphin-Guerin B, Teller G, Durand B (1980) Different endogenous cytokinins between male and female *Mercurialis annua* L. Planta (Berl) 148:124–129

David GW (1950) Blossoming of fruits delayed by maleic hydrazide. Science 111:303

Davies DR, Powling A (1985) The effect of heat and chemical treatments on meristem and shoot tip culture of cms and fertile sugar beet. In: Magnien E, Nettancourt de D (eds) Genetic engineering of plants and microorganisms important for agriculture. Nijhoff-Junk, the Netherlands, pp 144–145

Davies PJ (1973) Current theories on the mode of action of auxins. Bot Rev 39:139–171

Davis EW (1957) The distribution of the male sterility gene in onion. Proc Amer Soc Hort Sci 70:316–318

Davis EW (1966) An improved method for producing hybrid-onion seed. J Hered 2:55–57

Davis RL, Greenblatt IM (1967) Cytoplasmic male sterility in alfalfa. J Hered 58:301–305

Davis WH (1967) Alfalfa hybrids. Minutes 10th central alfalfa improv conf USA, pp 29–33

Davydenko OG (1981) Expression of genes for acridine induced male sterility in wheat in relation to the cytoplasm. Tsitol Genet USSR 15:48–53

Dean RT (1977) Lysosomes. Arnold, London, pp 52

De Bonte LR, Mathews BF, Wilson KG (1982) Restriction enzyme analysis of plastid and mitochondrial DNA in the genus *Daucus*. Plant Physiol (Bethesda) 69:140

De Bonte LR, Wilson KG, Mathews BF (1983) Variation of organelle DNA in the genus *Daucus*. In: Ahmad F, Downey K, Schultz J, Voellmy RW (eds) Advances in gene technology: molecular genetics of plants and animals. Academic Press, New York

De Bonte LR, Mathews BF, Wilson KG (1984) Variation of plastid and mitochondrial DNAs in the genus *Daucus*. Am J Bot 71:932–940

Dedio W (1979) Hybrid sunflowers are here to stay. Canada Agric 24:7–8

Dedio W, Putt ED (1980) Sunflower. In: Fehr W, Hadley HH (eds) Hybridization of crop plants. Am Soc Agron, Wisconsin, pp 631–643

De Fossard RA (1969) Development and histo-chemistry of the endothecium in the anthers of *in-vitro* grown *Chenopodium rubrum* L. Bot Gaz 134:125–129

Delannay X (1977) Cytological study of dioecy in *Cirsium arvense*. Phytomorphology 27:419–425

Delannay X (1978a) La gynodioécie dans le genre *Cirsium* Miller. Bull Soc Bot Belg 111:10–18

Delannay X (1978b) La gynodioécie chez les Angiosperms. Nat Belg 59:223–237

Delannay X (1979) Evolution of male sterility mechanisms in gynodioecious and dioecious species of *Cirsium* (Cynareae, Compositae). Plant Syst Evol 132:327–332

Delannay X, Palmer RG (1982) Genetics and cytology of the ms_4 male sterile soybean. J Hered 73:219–223

Delannay X, Goyon PH, Valdeyron G (1981) Mathematical study of the evolution of gynodioecy with cytoplasmic inheritance under the effect of nuclear restorer gene. Genetics 99:169–181

De la Puente CF (1966) Fertility of selected 24-chromosome *S. tuberosum* hybrids. MS Thesis, Wisconsin Univ, USA, pp 38

De la Puente CF, Peloquin SJ (1968) Male fertility of selected 24 chromosome *S. tuberosum* hybrids. Am Potato J 45:436–437

Delestaing N (1954) Contribution to the cytology of the genus *Salvia*. Rev Cytol Biol Veg 15: 195–236

Dell B (1981) Male sterility and anther wall structure in copper deficient plants. Ann Bot 48: 599–608

Deng JY, Gao ZL (1982) Discovery and determination of a dominant male-sterile gene and its importance in genetics and wheat breeding. Shanxi Nongye Kexue PR China 1:6–12

Denisen EL, Haber ES (1950) Maleic hydrazide on sweet corn. N Central Weed Control Conf Res Rept USA, pp 147

Denna DW (1971) The potential use of self incompatibility for breeding F_1 hybrids of naturally self pollinated vegetable crops. Euphytica 20:542–548

Deshmukh NY (1959) Sterile mutants in Tur (*Cajanus cajan*). Nagpur Agric Coll Mag 33:20–21

Deshpande PK, Chandra A, Makde KH (1981) Induced male sterile mutants in barley var clipper. Indian J Exp Biol 19:277–279

Deshpande RB (1933) Studies in Indian Chillies III. The inheritance of some characters in *Capsicum annuum* L. Indian J Agric Sci 3:219–300

Desjardins PR, Latterell PL, Mitchell JE (1954) Seed transmission of tobacco ring spot virus in Lincoln variety of soybean. Phytopathology 44:86

De Vries AP (1971) Flowering biology of wheat particularly in view of hybrid seed production, a review. Euphytica 20:152–170

De Vries AP, Ie TS (1970) Electron-microscopy on anther tissue and pollen of male sterile and fertile wheat *Triticum aestivum* L. Euphytica 19:103–120

Diaconu P (1965) Cytological investigations on corn cytoplasmic male sterility. Amelior Genet Technol Agric USSR 33:227–228

Dickinson HG (1973) The role of plastids in the formation of pollen grain coatings. Cytobios 8: 25–40

Dickson MH (1970) A temperature sensitive male sterile gene in broccoli, *Brassica oleracea* L. var *Italica.* J Amer Soc Hort Sci 95:13–14

Dickson MH (1975) G1117A, G1102A and G1106A cytosterile broccoli inbreds. HortScience 10:535

Diki SP, Anikeenko VS (1975) The induction of male sterility in egg plant and pepper. Referatirny Zh 3:55:202

Dionne LA (1961) Cytoplasmic sterility in derivatives of *Solanum demissum.* Am Potato J 38: 117–120

Dionne LA (1963) Studies on the use of *Solanum acaule* as a bridge between *Solanum tuberosum* and species in the series Bulbocatana, Cardiophylla and Pinnatisecta. Euphytica 12:263–269

Dixit RK (1976) Heterosis and inbreeding depression in sesame. Ind J Agric Sci 46:514–517

Dixon LK, Leaver CJ (1982) Mitochondrial gene expression and cytoplasmic male sterility in *Sorghum.* Plant Mol Biol 1:89–102

Dixon LK, Leaver CJ, Brettell RIS, Gengenbach BG (1982) Mitochondrial sensitivity to *Drechslera maydis* T-toxin and the synthesis of a variant mitochondrial polypeptide in plants derived from maize tissue cultures with Texas male sterile cytoplasm. Theor Appl Genet 63:75–80

Dmitrieva AN, Khavzhinskaia OE (1962) Physiological-biochemical differences between fertile forms of corn and their analogues with male cytoplasmic sterility. Genetika 7:147–153

Dmitrieva AN, Zdril'ko AF (1967) Cytoplasmic male sterility in rye and its physiological and biochemical characteristics. Selektsiya Semenovodstov 7:121–124

Dodds KS, Simmonds NW (1946) A cytological basis of sterility in *Tripsacum laxum.* Ann Bot 10:109–116

Doggett H (1968) Mass selection system for Sorghum, *Sorghum bicolor* (L) Moench. Corp Sci 8:391–392

Doggett H (1972) Recurrent selection in sorghum population. Heredity 28:9–29

Doggett H, Eberhart SA (1967) Recurrent selection in *Sorghum.* Crop Sci 8:119–120

Doig RI, Done AA, Rogers DF (1975) Pre-harvest sprouting in bread wheat *(Triticum aestivum)* as influenced by cytoplasmic male sterility derived from *T. timopheevi.* Euphytica 24:229–232

Dominguez-Gimenez J (1974) Fertility restoration and other characteristics in wild sunflowers. MS Thesis N Dakota Univ Fargo, USA

Dominguez-Gimenez J, Fick GN (1975) Fertility restoration of male sterile cytoplasm in wild sunflowers. Crop Sci 15:724–726

Dommée B (1981) Roles of environmental conditions and genetic factors in the breeding system of *Thymus vulgaris* L. Oecol Plant 2:137–147

Dommée B, Assouad MW, Valdeyron G (1978) Natural selection and gynodioecy in *Thymus vulgaris* L. Bot J Linn Soc 77:17–28

Done AA, Whittington WJ (1978) The restoration of male fertility in F_1 wheat hybrids. Ann Appl Biol 89:489–493

Donne J, Maitre JP, Berthelem P (1966) Contribution ai'etude des possibilites a'utiliser les insects pollinisateurs dans i'amelioration de la feverole. Memoire ENSA Rennes, France

Dorossiev L (1962) A new technique for pollination of the flowers of tomatoes, egg plants and other crops in hybrid seed production. Izv Nauk-izsled Inst Rasteniev Sofia 14:203–212

Dorossiev L (1970) Possible use of stamenless type male sterile tomato mutants in heterotic seed production. Genet Plant Breed USSR 3:171–179

Dorsey MJ (1914) Pollen development in the grape with special reference to sterility. Minn Agric Exp Stat Bull 144, pp 60

Dorsey MJ (1919) A note on the dropping of flowers in the potato. J Hered 10:226–228

Dotlacil L, Apltauerova M (1978) Pollen sterility induced by ethrel and its utilization in hybridization of wheat. Euphytica 27:353–360

Douillard R (1980) Characterization and meaning of chloroplast lipoxygenase activity. In: Mazliak P, Benveniste P, Costes C, Douce R (eds) Biogenesis and function of plant lipids. Elsevier North Holland, pp 121–126

Douillard R, Bergeron E (1979a) Quelques caractéristiques functionelles des activités lipoxygenasiques foliarires soluble on lamellaire. Consequences, structurales et functionelles. Physiol Veg 17:749–768

Douillard R, Bergeron E (1979b) Lipoxygenase activity distribution in young wheat chloroplast lamallae. In: Appelqvist LA, Ligenberg C (eds) Advances in biochemistry and physiology of plant lipids. Elsevier North Holland, pp 159–164

Downes RW, Marshall DR (1971) Low temperature induced male sterility in *Sorghum bicolor.* Aust J Exp Agric Anim Husb 11:352–356

Doyle GG (1972) A telocentric trisome and its potential use in the production of commercial hybrid corn using genetic male sterility. Maize Genet Coop Newslett 46:142–146

Driscoll CJ (1972) XYZ system of producting hybrid wheat. Crop Sci 12:516–517

Driscoll CJ (1973) A chromosomal male sterility system for producing hybrid wheat. Proc 4th Int Wheat Genet Symp, pp 669–674

Driscoll CJ (1974a) A chromosomal male sterility system of producing hybrid wheat. Wheat Inf Serv Jpn 38:28–32

Driscoll CJ (1974b) Induction and screening of chromosomal male sterile mutants for use in the production of hybrid wheat. In: Proc Symp Polyploidy and induced mutations in plant breeding. IAEA, Vienna, pp 139–142

Driscoll CJ (1975) Cytogenetic analysis of two chromosomal male sterility mutants in hexaploid wheat. Aust J Biol Sci 28:413–416

Driscoll CJ (1976) Induction and use of chromosomal male sterility in common wheat. In: Filev K (ed) Experimental mutagenesis in plants. Bulg Acad Sci Varna, pp 186–188

Driscoll CJ (1977) Registration of cornerstone male sterile wheat germplasm. Crop Sci 17:190

Driscoll CJ (1978) Induction and use of cornerstone male sterility mutant in wheat. In: Ramanujam S (ed) Proc 5th Int Wheat Genet Symp New Delhi, India, pp 499–502

Driscoll CJ (1981a) Perspectives in chromosome manipulation. The manipulation of genetic systems in plant breeding. Philos Trans R Soc Lond Ser B Biol Sci 292:535–546

Driscoll CJ (1981b) New approaches to wheat breeding. In: Evans LT, Peacock WJ (eds) Wheat science, today and tomorrow. Cambridge Univ Press, UK, pp 97–106

Driscoll CJ, Barlow KK (1976) Male sterility in plants. Induction, isolation and utilization. In: Proc Symp Induced mutations in cross breeding, IAEA, Vienna, pp 123–131

Drogacheva ZM (1972a) The efficacy of different fertility restorers under the conditions of Dagestan (from Plant Breeding Abstr 43, No 3458)

D'Souza H (1970) Untersuchungen über Eignung des Weizens als Pollenspender bei der Fremdbefruchtung, verglichen mit Roggen, *Triticale* and *Secale triticum.* Z Pflanzenzücht 63: 246–269

Dubey DK, Singh SP (1963) Effects of selective gametocides on linseed (*Linum usitatissimum* L.) Proc 50th Ind Sci Congr III, India, pp 611–612

Dubey DK, Singh SP (1965) Mechanism of pollen abortion in three male sterile lines of flax *(Linum usitatissimum).* Crop Sci 5:121–124

Dubey DK, Singh SP (1966) Use of cytoplasmic male sterility for the production of hybrid seeds in flax (*Linum usitatissimum* L.). Crop Sci 6:125–126

Dubey RS (1970) Pollen abortion in crape jasmine. Indian J Hortic 27:54–56
Dubey RS, Singh SP (1968) Gametocidal properties of certain plant growth regulators. Indian J Agric Sci 38:208–215
Duc G (1978) Modalités déxpression et hypothèses explicatives du manque de stabilité de la stérilité mâle cytoplasmique chez la féverole (*Vicia faba* L.). Thése Docteur-Ingenieur Université Paris Sud, pp 369
Duc G (1980) Effect of environment on the instability of two soruces of cytoplasmic male sterility in faba beans. FEBS Lett 2:29–30
Duc G (1981) La stérilité mâlé cytoplasmique la féverole (*Vicia faba* L.) I. Aspects génetiques de la sterilité mâle nucléo-cytoplasmique. Rev Cytol Biol Vég Botan 5:65–69
Duc G (1982) Nucleo-cytoplasmic male-sterility in Broad bean I. Genetical aspects of nucleocytoplasmic male sterility. Rev Cytol Biol Veg Botan 5(1):65–69
Dudley JW (1960) Effects of sodium 2,3-dichloroisobutyrate, a selective gametocide, on sugar beets. J Am Soc Sugar Beet Techn 11:25–36
Dudley JW, Wilsie CP (1957) Inheritance of a branched raceme and vestigeal flower in alfalfa. Agron J 49:320–323
Duffy RK (1982) Inheritance of male sterile apetalous inflorescence in *Zinnia elegans* (abstr). In: Proc 94th session Iowa Acad Sci, Fort Dodge Iowa USA, p 3
Dujardin M, Hanna W (1984) Microsporogenesis, reproductive behaviour and fertility in five *Pennisetum* species. Theor Appl Genet 67:197–201
Dulieu HL (1963) Les effects du "gametocide" 2,3-dichloroisobutyrate de sodium sur le *Solanum nigrum* L. CR Hebd Seances Acad Sci Paris 256:2679–2682
Dumas C, Dulpan JC, Säid C, Soulier JP (1983) In: Mulcahy OL, Ottaviano E (eds) Pollen: biology and implication for plant breeding. Elsevier, Amsterdam, pp 15–20
Dundas IS, Saxena KB, Byth DE (1981) Microsporogenesis and anther wall development in male sterile and fertile lines of pigeon pea (*Cajanus cajan* (L) Millsp.) Euphytica 30:431–435
Dundas IS, Saxena KB, Byth DE (1982) Pollen mother cell and anther wall development in a photoperiod insensitive male sterile mutant of pigeon pea (*Cajanus cajan* (L.) Millsp.) Euphytica 31:371–375
Dupuis F (1974) Evolution of the plastidial system during the microsporogenesis in *Impatiens balsamina* L. In: Linskens HF (ed) Fertilization in higher plants. North Holland, Amsterdam, pp 65–71
Durand B, Louis JP (1974) Male sterility and cytokinin action in *Mercurialis annua* L. (2n=16). In: Linskens HF (ed) Fertilization in higher plants. North Holland, Amsterdam, pp 327–334
Durand B, Durand R, Kahlem G, Champault A, Delaigue M (1974) Interactions moléculaires contrôlant la formation des organes sexuels chez une plante dioique *Mercurialis annua* L. Rev Cytol Biol Veget 37:763–779
Durand Y (1962) Rapport de mission en USSR sur la recherche agronomique et la culture du tournesol. Centre Technique Interporfessionnel des Oleadineaux Metropolitains Av Victor Hugo, Paris, pp 174
Durand Y (1981) Relationships between marker genes *aa* and *Wo* and male sterility gene *ms* 35. In: Genetics and Breeding of Tomato. INRA, France, pp 225–228
Durbin RD, Uchytil TF (1977) A survey of plant sensitivity to tentoxin. Phytopathology 67: 602–603
Dutt NL, Ethirajan AS, Hussainy SA (1954) Different types of male sterility in sugar cane. In: Proc 2nd Bien Conf Sugar Res Dev Workers, Jallandhar, India, pp 100–102
Dutta OP, Nath P (1974) Inheritance of male sterility in okra (*Abelmoschus esculentus*). Proc 12th Int Hortic Congr Sec 7. Vegetables, Warsaw Poland, pp 737
Duvick DN (1956) Allelism and comparative genetics of fertility restoration of cytoplasmically pollen sterile maize. Genetics 41:544–565
Duvick DN (1959) The use of cytoplasmic male sterility in habrid seed production. Econ Bot 13:167–195
Duvick DN (1965) Cytoplasmic pollen sterility in corn. Adv Genet 13:1–56
Duvick DN (1966) Influence of morphology and sterility on breeding methodology. In: Frey KJ (ed) Plant Breeding vol 1. Iowa State Univ Publ Ames USA, pp 85–138

Duvick DN (1972) Potential usefulness of new cytoplasmic male steriles and sterility systems. In: Proc 27th Ann Corn Sorghum Res Conf Am Seed Trade Assoc, Washington DC
Duvick DN, Noble SW (1978) Current and future use of cytoplasmic male sterility for hybrid seed production. In: Walden DB (ed) Maize breeding and genetics. Wiley, New York, pp 265–277
Duvick DN, Snyder RJ, Anderson EG (1961) The chromosomal location of Rf_1 a restorer gene for cytoplasmic pollen sterile maize. Genetics 46:1245–1252
Dvořak J (1980) Homoeology between *Agropyron elongatum* chromosome and *Triticum aestivum* chromosomes. Can J Genet Cytol 18:371–377
Dyki B (1973a) Cytological investigation of male sterile onion (*Allium cepa* L.) plants of the variety Wolska and Rawska. Biul Warzyw Poland 14:139–148
Dyki B (1973b) Cytological studies on microspore formation in male fertile and male sterile onions (*Allium cepa* L.) of the Wolska and Rawska varieties. Biul Warzyw Poland 15:213–221
Earle ED (1982) Effects of *Helminthosporium maydis* race T toxin on mitochondria and protoplasts from T cytoplasm maize. In: Sheridan WF (ed) Maize for biological research. N Dakota State Univ Press, pp 251–255
Earle ED, Gracen VE, Yoder Oc, Gemmill KP (1978) Cytoplasm-specific effects of *Helminthosporium maydis* race T toxin in survival of corn mesophyll protoplasts. Plant Physiol (Bethesda) 61:420–424
East EM (1908) Techniques of hybridizing the potato. Proc Am Soc Hortic Sci (1907):35–40
East EM (1919) Studies on self sterility III. The relation between self fertile and self sterile plants. Genetics 4:341–345
East EM (1929) Self sterility. Bibliogr Genet 5:331–370
East EM (1932) Studies on self sterility IX. The behaviour of crosses between self sterile and self fertile plants. Genetics 17:175–202
East EM (1934) The nucleus-plasma problem. Am Nat 68:280–303, 402–439
Eaton FN (1957) Selective gametocide opens way to hybrid cotton. Science 126:1174–1175
Ebert WW (1964) The inheritance and developmental aspects of thin-hull mutant of *Carthamus tinctorius* L. Diss Univ California, Davis, USA
Ebert WW, Knowles PF (1966) Inheritance of pericarp types, sterility and dwarfness in several safflower crosses. Crop Sci 6:579–582
Ebert WW, Knowles PF (1968) Development and anatomical characters of thin hull mutants of *Carthamus tinctorius.* Am J Bot 55:421–430
Echlin P (1971) The role of tapetum during microsporogenesis of angiosperms. In: Heslop-Harrison J (ed) Pollen development and physiology. Butterworths, London, pp 41–61
Edwards IB (1983) An assessment of the role of the cytoplasmic-genetic systems in hybrid wheat production. Plant Mol Biol Rep 1:139–144
Edwardson JR (1956) Cytoplasmic male sterility. Bot Rev 22:696–738
Edwardson JR (1962) Cytoplasmic differences in T-type cytoplasmic male sterile corn and its maintainer. Am J Bot 49:184–187
Edwardson JR (1965) Radiosensitivity in maize seed influenced by cytoplasmic factors. Radiat Bot 5:53–59
Edwardson JR (1967) Cytoplasmic male sterility and fertility restoration in *Crotalaria mucronata.* J Hered 58:266–268
Edwardson JR (1968) The nature of cytoplasmic factors inducing male sterility. Proc 12th Int Cong Genet 2:226–227
Edwardson JR, Corbett MK (1961) Asexual transmission of cytoplasmic male sterility. Proc Natl Acad Sci USA 47:390–396
Edwardson JR, Warmke HE (1967) Fertility restoration in cytoplasmic male sterile *Petunia.* J Hered 58:195–196
Edwardson JR (1970) Cytoplasmic male sterility. Bot Rev 36:341–420
Edwardson JR, Bond DA, Christies RG (1976) Cytoplasmic sterility factors in *Vicia faba* L. Genetics 82:443–449
Eenink AH (1977) Induction of male sterility in lettuce (*Lactuca sativa* L.) by application of gibberellic acid- a technical note. Euphytica 26:31–32

Eenink AH, Vereijken ALJ (1978a) Induction of male sterility in lettuce (*Lactuca sativa* L.) with GA_3, influence of temperature and GA_3-concentration. Neerl J Agric Sci 26:45–50

Eenink AH, Vereijken ALJ (1978b) Anatomical changes in flower buds of lettuce (*Lactuca sativa* L.) treated with a GA_3 solution for induction of male sterility. Acta Bot Neerl 27: 199–224

Eisa HM, Munger MH (1968) Male sterility in *Cucurbita pepo*. Proc Amer Soc Hort Sci 92:473–479

Eisa HM, Wallace TH (1969) Factors influencing petaloidy expression in the carrot *Daucus carota* L. Proc Am Soc Hortic Sci 94:647–649

Elena-Rossello JA, Kheyr-Pour A, Valdeyron G (1976) La structure génétique et la régime de la fecondation chez *Origanum vulgare* L.: rȇpartition d'un marquer enzymatique dans deux populations naturelles. CR Acad Sci Paris 283:1587–1589

El-Kadi DA, Weaver JB (1973a) Evaluation of hybrids produced from individual male sterile plants of upland cotton. Proc Beltville Cotton Prod REs Conf USA, pp 1–5

Ellerton S, Hendriksen AJT (1959) Note on the probable cause of occurrence of tetraploid plants in commercial triploid varieties of sugarbeet. Euphytica 8:99–103

Elliot C (1922) Sterility of oats. USDA Bull, p 1058

Elliot C (1925) Oat blast. Phytopathology 15:564–567

Ellison W (1936) Meiosis and fertility in certain British varieties of the cultivated potato (*Solanum tuberosum* L.) Genetica (The Hague) 18:217–254

Emerson RA (1920) Heritable characters of maize II. Pistillate flowered maize plants. J Hered 11:65–76

Emerson RA (1932) A recessive zygotic lethal resulting in a 2:1 ratios for normal vs male sterile and coloured vs colourless pericarp in the F_2 of certain maize hybrids. Science 75:566–567

Emerson RA, Beadle GW, Fraser AC (1935) A summary of linkage studies in maize. Cornell Univ Agric Exp Stat Mem USA, pp 180

Endo TR (1980) Genetic constancy of the cytoplasm. In: Tsunewaki K (ed) Genetic diversity of the cytoplasm in *Triticum* and *Aegilops*. Jpn Soc Promotion Sci Publ, pp 13–48

Endo TR, Tsunewaki K (1975a) Sterility of common wheat with *Aegilops triunialis* cytoplasm. J Hered 66:13–18

Endo T, Tsunewaki K (1975b) Genetic diversity of the cytoplasm in *Triticum* and *Aegilops* I. On the origin of the cytoplasm of *Aegilops triuncialis* L. Seiken Ziho 25–26:55–66

Engle RL (1974) Male sterility in cabbage. In: Proc crucifer improvement cooperative. New York State Agric Exp Stat Geneva USA, pp 8–9

Engelbert V (1932) A study of various factors influencing seed production in alfalfa (*Medicago sativa*). Sci Agric 12:593–603

Enns H, Dorrell DG, Hoes JA, Chubb WO (1970) Sunflower research, a progress report. In: Proc 4th Int Sunflower Conf Memphis USA, pp 162–167

Erichsen AW, Ross JG (1963a) Inheritance of colchicine induced male sterility in sorghum. Crop Sci 3:335–338

Erichsen AW, Ross JG (1963b) Irregularities at microsporogenesis in colchicine induced male sterile mutants in *Sorghum vulgare* Pers. Crop Sci 3:481–483

Erickson EH (1975a) Effect of honey bees on yield of three soybean cultivars. Crop Sci 15: 84–86

Erickson EH (1975b) Variability of floral characteristics influences honey bee visitation to soybean blossoms. Crop Sci 15:767–771

Erickson EH, Peterson CE (1979) Asynchrony of floral events and other differences in pollinator foraging stimuli between fertile and male sterile carrot inbreds. J Amer Soc Hort Sci 104: 639–643

Erickson EH, Peterson CE, Werner P (1979) Honey bee foraging and resultant seed set among male fertile and cytoplasmically male sterile carrot inbreds and hybrid seed parents. J Amer Soc Hort Sci 104:635–638 8

Erickson HT, Gabelman WH (1956) The effect of distance and direction on cross pollination in onions. Proc Amer Soc Hort Sci 68:351–357

Erickson JR (1967) Biochemical investigations of cytoplasmic male sterility in spring wheat (*Triticum aestivum* L.) (abstr). Agron Abstr, pp 8–9

Erickson JR (1969) Cytoplasmic male sterility in rice (*Oryza sativa* L.) (abstr). Agron Abstr, p 6

Eschrich W (1961) Untersuchungen über den Ab- und Aufbau der Callose. Z Bot 49:153–218

Eskew EB, Willard CJ (1950) Maleic hydrazide on corn. In: Proc N Central Weed Control Conf Res Rept USA p 187

Eslick RF (1970) Balanced male steriles and dominant preflowering selective genes for use in hybrid seed production. Barley Genet 2:292–297

Eslick RF, Hockett EA (1967) Alternative methods for the production of the female stock for hybrid barley (abstr). Crop Sci Abstr, p 6

Eslick RF, McProud WL (1974) Positioning of genetic male sterile 5 (msg5) on chromosome 3. Barley Genet Newslett 4:16–23

Eslick RF, Ramage RT, Clark DR (1974) Two genetic male sterile msg6 and ms9, bk assigned to chromosome 6. Barley Genet Newslett 4:11–15

Estrada S, Mutschler MA, Bliss FA (1984) Temperature influenced instability in a genic male sterile common bean. HortScience 19:401–402

Evenor D, Izhar S (1984) Coexistence of cytoplasmic and nuclear genes for male sterility in *Petunia*. Mol Gen Genet 194:523–527

Evenor D, Guri A, Izhar S (1984) The relationship between a nuclear restorer gene and a gene for male sterility in *Petunia*. Theor Appl Genet 67:475–478

Everett TH (1980–82) Encyclopedia of horticulture, vol 1–9. Garland, New York

Eyster LA (1921) Heritable characters of maize VII. Male sterile. J Hered 12:138–141

Eyster WH (1933) Linkage studies in maize (abstr). Am Nat 67:75

Eyster WH (1934) Genetics of *Zea mays*. Bibliogr Genet 11:187–357

Faberge AC (1937) The cytology of male sterile *Lathyrus odoratus*. Genetica (The Hague) 19: 423–430

Faegri K, Pijl Van der L (1979) The principles of pollination ecology, 3rd edn. Pergamon, Oxford

Fagala BL (1972) A genetic study of a partially male-sterile character in soybeans. MS Thesis Univ Arkansas

Fairey DT, Stoskopf NC (1975) Effects of granular ethephon on male sterility in wheat. Crop Sci 15:29–32

Falavigna A, Badino M, Soressi GP (1978) Potential of the monomendelian factor *pat* in the tomato breeding for industry. Genet Agrar 32:159–160

Falk DE, Kasha KJ (1980) A shrunken endosperm, male sterile stock to facilitate crossing in barley. Barley Genet Newslett 10:81–83

Falk DE, Kasha KJ (1982) Registration of a shrunken endosperm, male-sterile germplasm to facilitate hybridization in barley. Crop Sci 22:450

Falk DE, Swartz MJ, Kasha KJ (1980) Linkage data with genes near the centromere of barley chromosome VI. Barley Genet Newslett 10:13–16

Falk DE, Kasha KJ, Reinbergs E (1981) Presowing selection of genetic male sterile plants to facilitate hybridization in barley. Barley Genet 4:778–785

Farineau N, Roussaux J (1975) Influence de la 6 benzylaminopurine sur la differenciation plastidiale dans les cotylédons de concombre. Physiol Plant 33:194–202

Favret EA, Ryan GS (1964) Two cytoplasmic male sterile mutants induced by X-rays and EMS. Barley Genet Newslett 8:42

Favret EA, Ryan GS (1966) New useful mutants in plant breeding. In: Proc Symp Mutations Plant Breeding, IAEA, Vienna, pp 49–60

Feaster CV, Turncotte EL (1959) Hollow boll, crazy top in cotton. Prog Agric Arizona 11:6–7

Fedorova TN, Nettevich ED (1969) Study of microsporogenesis in forms of common wheat with male sterility specified by the cytoplasm of some species. Tsitologia USSR 11:1121–1128

Fehr WR, Oritz LB (1975) Recurrent selection for yield in soybeans. J Agric Univ Puerto Rico 59:222–232

Fejer SO (1958) A gene for femaleness in hermaphrodite *Lolium perenne* L. In: Proc 10th Int Cong Genet Montreal 2:81–82
Fejer SO (1966) A floral mutation in *Lolium perenne* L. Can J Genet Cytol 8:115–121
Felz H (1953) Untersuchungen an diploiden und polyploiden Zuckerrüben. Z Pflanzenzücht 32: 275–300
Fendal RK (1967) Artificial induction of male sterility in wheat. Res Prog Rep Agron Sec Washinton State Univ, p 476
Fertig SN, Loss KA, Gutenmann WH, Lisk DJ (1964) Formation of 2,4-D in 4(2,4-DB) treated timothy, birds foot, trefoil and sterile pea plants. Weeds 12:147–148
Fick GN (1978) Breeding and Genetics. In: Carter FJ (ed) Sunflower science and technology. Agron 19 Am Soc Agron Madison, pp 279–338
Fick GN, Zimmer DE (1974a) Fertility restoration in confectionary sunflowers. Crop Sci 14: 603–605
Fick GN, Zimmer DE (1974b) Monogenic resistance to *Verticillium* wilt in sunflowers. Crop Sci 14:895–896
Fick GN, Zimmer DE (1974c) Parental lines for production of confectionary sunflower hybrids. N Dakota Farm Res Bull USA 31:15–16
Fick GN, Zimmer DE (1974d) RH 271, RHA 273 and RHA 274 sunflower parental lines for producing downy mildew resistant hybrids. N Dakota Farm Res Bull USA 32:7–9
Fick GN, Zimmer DE (1976) Yield stability of sunflower hybrids and open pollinated varieties. In: Proc 7th Int Sunflower Conf Krasnoda USSR, pp 37–38
Fick GN, Zimmer DE, Dominguez-Gimenez J, Rehder DA (1974) Fertility restoration and variability for plant and seed characteristics in wild sunflowers. In: Proc 6th Int Sunflower Conf Bucharest, pp 333–338
Filion WG, Christie BR (1966) The mechanism of male sterility in a clone of orchardgrass (*Dactylis glomerata* L.). Crop Sci 6:345–347
Fisher JE (1972) The transformation of stamens to ovaries and of ovaries to inflorescences in *T. aestivum* L. under short day treatment. Bot Gaz 133:78–85
Flavell R (1974) A model for the mechanism of cytoplasmic male sterility in plants with special reference to maize. Plant Sci Lett 3:259–263
Flavell RB (1975) Inhibition of electron transport in maize mitochondria by *Helminthosporium maydis* race T pathotoxin. Physiol Plant Pathol 6:107–116
Flick CE, Evans DA (1982) Evaluation of cytoplasmic segregation in somatic hybrids of *Nicotiana* tentoxin sensitivity. J Hered 73:264–266
Fonty G, Gourst R, Wilke D, Bernardi G (1978) The mitochondrial genome of wild type yeast cells. J Mol Biol 119:213–235
Ford LE (1950) A genetic study involving male sterility in *Coleus* (abstr). Genetics 35:664
Forde BG, Leaver CJ (1979) Mitochondrial genome expression in maize: Possible involvement of variant mitochondrial polypeptides in cytoplasmic male sterility. Proc 4th John Innes Symp, pp 131–146
Forde BG, Leaver CJ (1980) Nuclear and cytoplasmic genes controlling synthesis of variant mitochondrial polypeptides in male sterile maize. Proc Natl Acad Sci USA 77:418–422
Forde BG, Oliver RJC, Leaver CJ (1978) Variation in mitochondrial translation products associated with male sterile cytoplasm in maize. Proc Natl Acad Sci USA 75:3841–3845
Forde BG, Oliver RJC, Leaver CJ, Gunn RE, Kemble RJ (1980) Classification of normal and male-sterile cytoplasms in maize I. Electrophoretic analysis of variation in mitochondrially synthesised proteins. Genetics 95:443–450
Foskett RL (1954) Nature and inheritance of male sterility in the onion variety scott country globe. Iowa State Coll J Sci Ames 28:317
Fossati A, Inglod M (1970) A male sterile mutant in *Triticum aestivum*. Wheat Inf Serv 30:8–10
Foster AE, Scholler AB (1971) Cytoplasmic male sterility in barley. In: Nilan RA (ed) Barley Genetics, vol 2. Pullman, Washington, pp 316–318
Foster CA (1972) Maintenance and multiplication of male sterile stocks without progeny testing. Barley Genet Newslett 2:114

Foster CA (1979) A balanced male sterile chloroplast mutant scheme for hybrid barley. Barley Genet Newslett 9:22–23

Foster RE (1968a) F_1 hybrid muskmelons III. Field production of hybrid seeds. Proc Amer Soc Hort Sci 92:461–464

Foster RE (1968b) F_1 hybrid muskmelons IV. Roguing-thinning to pure stands from mixed seed. J Arizona Acad Sci 5:77–79

Foster RE (1968c) F_1 hybrid muskmelon V. Monoecism and male sterility in commercial seed production. J Hered 59:205–207

Fowden L (1963) Amino acid analogues and the growth of seedlings. J Exp Bot 14:387–398

Fox TD, Leaver CJ (1981) The *Zea mays* mitochondrion gene coding cytochrome oxidase subunit II has an intervening sequence and does not contain TGA codons. Cell 26:315–323

Francis RR, Bemis WP (1970) A cytomorphological study of male sterility in a mutant of *Cucurbita maxima* Duch. Econ Bot 24:325–332

Franckowiak JD, Maan SS, Williams ND (1976) A proposal for hybrid wheat utilizing *Aegilops squarrosa* L. cytoplasm. Crop Sci 16:725–728

Frank J, Koves FS (1977) Chemical induction of male sterility in sunflower. Acta Agron Acad Sci Hung 26:318–324

Frankel OH (1940) Studies in Hebe II. The significance of male sterility in the genetic system. J Genet 40:171–184

Frankel R (1956) Graft induced transmission to progeny of cytoplasmic male sterility in *Petunia.* Science 124:684–685

Frankel R (1962) Further evidence of graft induced transmission to progeny of cytoplasmic male sterility in *Petunia.* Genetics 47:641–646

Frankel R (1964) The nature of cytoplasmic male sterility in *Petunia.* Publ Div Volcani Inst Rehovot Israel, pp 1–108

Frankel R (1971) Genetic evidence on alternative material and mendelian hereditary elements in *Petunia hybrida.* Heredity 26:107–119

Frankel R (1973a) Origin and heredity of male sterility in higher plants. In: Moav R (ed) Agricultural genetics. Wiley, New York, pp 57–69

Frankel R (1973b) Characterization of male sterility. In: Moav R (ed) Agricultural genetics. Wiley, New York, pp 70–84

Frankel R (1973c) The use of male sterility in hybrid seed production. In: Moav R (ed) Agricultural genetics. Wiley, New York, pp 95–94

Frankel R, Galun E (1977) Pollination mechanisms, reproduction and plant breeding. Springer, Berlin, Heidelberg, New York, pp 281

Frankel R, Izhar S, Nitsan J (1969) The timing of callase activity and cytoplasmic male sterility in *Petunia.* Biochem Genet 3:451–455

Frankel R, Scowcroft WR, Whitfield PR (1979) Chloroplast DNA variation in isonuclear male sterile lines of *Nicotiana.* Mol Gen Genet 169:129–135

Franken KS (1978) Chemical induction of staminate flowers on four determinant gynoecious lines of pickling cucumber. Gartenbauwissenschaften 43:280–282

Fraser AC (1933) Heritable characters of maize XLIV. Silky ears. J Hered 24:41–46

Frimmel F, Baranek J (1935) Beitrag zur Methodik der Roggenzüchtung und des Roggensaatgutbaues. Z Pflanzenzücht 20:1–22

Frohberg RC, Frey KJ (1970) An evaluation of techniques for inducing male sterility in oats. Iowa Stat J Sci Ames 44:453–459

Frost HB (1948) Cultivated varieties of citrus. In: Webber HJ, Batchelor LD (ed) The citrus industry, vol 1. Univ Calif Press, Berkley, pp 475–668

Fröst S (1959) The cytological behaviour and mode of transmission of accessory chromosomes in *Plantago serraria.* Hereditas 45:191–210

Fröst S (1963) Cytoplasmic pollen sterility in *Centaurea scabiosa.* Hereditas 49:454–456

Fu HY (1983) Variation in mitochondrial translation products and cytoplasmic male sterility in sorghum. Acta Genet Sin 10:471–476

Fujieda K (1963) Cucumber breeding III. Studies on the method of establishing the inbred gynoecious strains in cucumbers. Bull Hortic Res Stat Jpn Ser D 1:101–116

Fujimaki H (1975) Studies on the transfer of desirable genes by backcrossing II. Backcross breeding systems by the use of genetic male sterility in self pollinated crops. Jpn J Breed 25:275–282

Fujimaki H (1978) New techniques in backcross breeding for rice improvement. Trop Agric Res Serv Jpn 11:55–67

Fujimaki H (1979) Recurrent selection by using genetic male sterility for rice improvement. JARQ Jpn 13:153–156

Fujimaki H (1980) Recurrent population improvement rice breeding facilitated with male sterility. Gamma field symp No 19: induction and utilization of male sterile mutation in plant breeding. IAEA, Vienna, pp 91–102

Fujimaki H, Hirawa S, Kushibuchi K, Tanaka S (1977) Artificially induced male sterile mutants and their usage in rice breeding. Jpn J Breed 27:70–77

Fukasawa H (1953) Studies on restoration and substitution of nucleus in *Aegilotricum* I. Appearance of male sterile *durum* in substitution crosses. Cytologia 18:167–175

Fukasawa H (1954) On the free amino acids in anthers of male sterile wheat and maize. Jpn J Genet 29:135–137

Fukasawa H (1955) Studies on restoration and substitution of nucleus of *Aegilotricum* II. The interrelationships between ovata cytoplasm and fertility restoring factors. Cytologia 20:211–217

Fukasawa H (1956a) Chlorophyll variegation which appeared in male sterile Emmer wheat. Wheat Inf Serv Kyoto Jpn 3:19

Fukasawa H (1956b) Studies on restoration and substiution of nucleus (genome) in *Aegilotricum* I. Appearance of male sterile *durum* in substitution crosses. Cytologia 18:167–175

Fukasawa H (1957a) Studies on restoration and substitution of nucleus (genome) in *Aegilotricum* IV. Genome exchange between *durum* and *ovata* cytoplasm and its theoretical consideration for male sterility. Cytologia 22:30–39

Fukasawa H (1957b) Studies on restoration and substitution of nucleus (genome) in *Aegilotricum* VI. Some conspicuous characters appearing in male sterile Emmer wheats. Jpn J Genet 32:313–322

Fukasawa H (1958a) Fertility restoration of cytoplasmic male sterile Emmer wheats. Wheat Inf Serv Kyoto Jpn 7:211

Fukasawa H (1958b) Studies on restoration and substitution of nucleus (genome) in *Aegilotricum* VII. Curious abnormalities of stomata appearing in the leaves of male sterile *durum* plants. Cytologia 23:128–142

Fukasawa H (1959) Nucleus substitution and restoration by means of successive backcrosses in wheat and its related genus *Aegilops*. Jpn J Bot 17:55–91

Fukasawa H (1961) The reduction of triphenyltetrazolium chloride by dehydrogenase in cytoplasmic male sterile wheat. Nucleus Cytoplasm 2:32–35

Fukasawa H (1967) Constancy of cytoplasmic property during successive backcrosses. Am Nat 101:41–46

Fukasawa H, Mito K (1958) Paper chromatographic analysis of chlorophylls in cytoplasmic male sterile wheat. Rep Kihara Inst Biol Res Jpn 8:4–8

Fukasawa H, Mito K, Fujiwara M (1957) Preventive effect of sugar on the pollen degeneration of wheat plant. Bot Mag 70:251–257

Fukuda Y (1927) Cytological studies on the development of the pollen grain in different races of *S. tuberosum* L. with special reference to sterility. Bot Mag 41:459–474

Fukushima E, Matsuo E, Fujieda J (1968) Studies on the growth behaviour of cucumber, *Cucumis sativus* L. J Fac Agric Kyushu Univ Jpn 14:349–366

Funderburk HH, Davis DE (1960) Factors affecting the response of *Zea mays* and *Sorghum halepense* to sodium 2,3-dichloropropion. Weeds 8:6–11

Furgala B, Mussen EC, Noetzle DM, Robinson RG (1976) Observations on nectar secretion in sunflowers. In: Proc Ist Sunflower Forum Fargo, USA, pp 11–12

Furste K (1964) The induction of male sterility in *Beta* by the use of 2,3-dichloroisobutyrate sodium. Z Pflanzenzücht 51:335–346

Fyfe JL, Bailey NTJ (1951) Plant breeding studies in leguminous forage crops II. Further observations on natural cross breeding in winter beans. J Agric Sci 41:371–378

Gabelman WH (1949) Reproduction and distribution of the cytoplasmic factor for male sterility in maize. Proc Natl Acad Sci USA 35:634–640

Gabelman WH (1956) Male sterility in vegetable breeding. In: Brookhaven Symp Biol No 9. Genet Plant Breed, pp 113–122

Gabelman WH (1974) F_1 hybrids in vegetable production. In: Proc 19th Int Hortic Cong Warsaw 3:419–428

Gagnieu A (1951) Pollen production in the apple: Possibility of monofactorial genic lethality. Ann Inst Natl Rech Agronl:455–496

Gagnieu A (1954) Interaction noyan-cytoplasme dans la relisation d'un type de sterilite male chez le lin cultive. In: Proc 7th Int Bot Cong, pp 103–105

Gahan PB (1968) Lysosomes. In: Pridham JB (ed) Plant cell organelles. Academic Press, New York, pp 228–238

Gahan PB (1973) Plant lysosomes. In: Dingle JT (ed) Lysosomes in biology and pathology, vol 3, Elsevier, New York, pp 69–85

Gairdner AE (1929) Male sterility in flax II. A case of reciprocal crosses differing in F_2. J Genet 21:117–124

Gajeski WA (1937) A contribution to the knowledge of cytoplasmic influence on the effect of nuclear factors in *Linum*. Acta Soc Bot Polon 14:205–214

Galinat WC (1975) Use of male sterile 1 gene to eliminate detasseling in production of hybrid seed of bicolor sweet corn. J Hered 66:387–388

Galinat WC (1976) Genetic systems for the production of hybrid seed corn seed without detasseling. Maize Genet Coop Newslett 50:50–72

Galun E (1959) Effects of gibberellic acid and naphthalene acetic acid on sex expression and some morphological characters in the cucumber plant. Phyton (Horn) 13:1–8

Galun E (1982) Somatic cell fusion for inducing cytoplasmic exchange: a new biological system for cytoplasmic genetics in higher plants. In: Vasil IK, Scowcraft WR, Frey KJ (eds) Plant improvement and somatic cell genetics. Academic Press, New York, pp 205–209

Galun E, Aviv D (1979) Plant cell genetics in *Nicotiana* and its implications to crop plants. Monogr Genet Agrar 4:153–175

Galun E, Arzee-Gonon P, Fluhr R, Edelman M, Aviv D (1982) Cytoplasmic hybridization in *Nicotiana:* mitochondrial DNA analysis in progenies resulting from fusion between protoplast having different organelle constitutions. Mol Gen Genet 186:50–56

Ganders FR (1978) The genetics and evolution of gynodioecy in *Nemophila menziesii* (Hydrophyllaceae). Can J Bot 56:1400–1408

Ganders FR (1979) The biology of heterostyly. NZ J Bot 17:607–635

Garber RJ (1922) Inheritance and yield with particular reference to rust resistance and panicle types in oats. Minnesota Agric Exp Stat Techn Bull 7:5–62

Gardner GO (1961) An evaluation of effects of mass selection and seed irradiation with thermal neutrons on yield of corn. Crop Sci 1:241–245

Gardner CO (1972) Development of superior populations of sorghum and their role in breeding programs. In: Rao NGP, House LR (eds) Sorghum in seventies. Oxford, India, pp 180–196

Garlicka W, Madej L (1975) Microsporogenesis and anther development in male sterile and male fertile forms of rye. Aklimatyzaija 1 Nasiennctwo Poland 19:423–433

Gärtner CFV (1844) Versuche und Beobachtungen über Befruchtungsorgane. Hering, Stuttgart, Germany, pp 1–97

Gates RR (1907) Pollen development in hybrids of *Oenothera alata* x *Oe. lamarckiana* and its relations to mutation. Bot Gaz 43:81–115

Gates RR (1911) Pollen formation in *Oenothera gigas*. Ann Bot 25:909–940

Gates RR, Thomas N (1914) A cytological study of *Oenothera mut lata* and *Oe. mut semilata* in relation to mutation. Q J Micros Sci 59:523–571

Gaue I (1977) Ms-based hybrid breeding in *Lolium perenne* L. In: Proc 13th Int Grassland Congr Leipzig, DDR, pp 435–437

Gaue I (1981) Ergebnisse von Untersuchungen zur Hybridzüchtung bei *Lolium perenne*. Tagungsber Dtsch Akad Landwirtschaftswiss Berlin 191:119–126

Geiger HH (1971) Cytoplasmatisch-genetische Pollensterilität in Roggenformen iranischer Abstammung. Naturwissenschaften 58:98–99

Geiger HH, Baubach GA (1979) Untersuchungen über die Eignung pollensterilen Roggens zur parasitischen Mutterkornerzeugung. Z Pflanzenzücht 83:163–175

Geiger HH, Morgenstern K (1975) Applied genetic studies on cytoplasmic pollen sterility in winter rye. Theor Appl Genet 46:269–276

Geiger HH, Schnell FW (1970) Cytoplasmic male sterility in rye (*Secale cereale* L.). Crop Sci 10: 590–593

Geiger HH, Schnell FW (1974) Experimental basis for breeding hybrid varieties of rye. In: Proc Eucarpia meeting on breeding of rye, Poznan, Poland, pp 24–37

Gengenbach BG, Green CE (1975) Selection of T-cytoplasm maize callus cultures resistant to *Helminthosporium maydis* race T pathotoxin. Crop Sci 15:645–649

Gengenbach BG, Pring DR (1982) Isolation of revertants from CMS-T by tissue culture techniques. In: Sheridan WF (ed) Maize for biological research. N Dakota State Univ Press, pp 257–261

Gengenbach BG, Koeppe DE, Miller RJ (1973a) A comparison of mitochondria isolated from male-sterile and non-sterile cytoplasm etiolated corn seedlings. Physiol Plant 29:103–107

Gengenbach BG, Miller RJ, Koeppe DE, Arntzen CJ (1973b) The effect of toxin from *Helminthosporium maydis* (race T) on isolated corn mitochondria: swelling. Can J Bot 51:2119–2125

Gengenbach BG, Green CE, Donovan CM (1977) Inheritance of selected pathotoxin resistance in maize plants regenerated from cell cultures. Proc Natl Acad Sci USA 74:5113–5117

Gengenbach BG, Connelly JA, Pring DR, Conde MF (1981) Mitochondrial DNA variation in maize plants regenerated during tissue culture selection. Theor Appl Genet 59:425–431

Gengqi L, Konglian Z, Fanrui K (1980) Transfer of the male fertile genes of the restorer not by genetic fusion and their expression in progenies of sorghum. In: Annu Rep Inst Genet Acad Sin, pp 120–121

Georgescu CM, Römer D, Olteanu GH (1979) Peroxidase activity as a possible test for distinguishing S-plasm from N-plasm in sugar beet (*Beta vulgaris* L.). Euphytica 28:779–784

Georgieva-Todorova J (1974) A new male sterile form of sunflower (*Helianthus annuus* L.). In: Proc 6th Int Sunflower Conf Bucharest, pp 343–347

Georgieva-Todorova J (1984) Interspecific hybridization in the genus *Helianthus* L. Z Pflanzenzücht 93:265–279

Gerstel DU (1980) Cytoplasmic male sterility in *Nicotiana* (a review). N Carolina Techn Bull USA 263:1–31

Gerstel DU, Burns JA (1974) Meiosis and staminal sterility in hybrids between *Nicotiana tomentosiformis* and *N. otophora*. Tobacco Sci 18:157–159

Gerstel DU, Burns JA, Burk LG (1978) Cytoplasmic male sterility in *Nicotiana*, restoration of fertility and the nucleolus. Genetics 89:157–169

Gerstel DU, Burns JA, Sand SA (1980) Cytogenetics of flower modification of a cytoplasmic male sterile tobacco. Genetics 96:223–235

Ghiasi H, Lucken KA (1982a) Agronomic and male-fertility restoration characteristics of wheat restorer lines with *Aegilops speltoides* and *Triticum timopheevi* cytoplasms. Crop Sci 22:527–531

Ghiasi H, Lucken KA (1982b) Investigation of possible associations between partial fertility restoration and agronomic traits in *Triticum aestivum* L. Euphytica 31:253–259

Ghosh MS, Bose TK (1970) Sex modification in cucurbitaceous plants by using CCC. Phyton (Horn) 27:131–135

Gikalo G, Studentsova L (1968) Investigations of the heterosis in pepper at the experimental station of VIR. In: Heterosis in horticulture, Leningrad, USSR, pp 248–354

Gilham NW (1978) Organelle heredity. Raven, New York

Gill BS, Singh H, Singh NB (1973) Inheritance of genetic male sterility in pearl millet. J Res PAU Ludhiana 10:4–10

Gill KS, Phul PS, Jindla LN (1977) Improved bajra male sterile line Pb 111 A. Seed Techn News 7:3–6

Gill KS, Phul PS, Bhardwaj BL (1979) Induction of resistance to downy mildew by irradiation in male sterile lines of pearl millet. In: Proc symp role of induced mutations in crop improvement, Hyderabad, pp 17–22

Gill KS, Phul PS, Jindal LN, Singh NB (1981) Pb 204A a new downy mildew resistant male sterile line in pearl millet. Seeds Farms 7:19–20

Gilmore EC Jr (1964) Suggested method of using reciprocal recurrent selection in some naturally self pollinated species. Crop Sci 4:323–325

Gilmore EC Jr (1970) Fertility restoration of *Triticum timopheevi* cytoplasmic male steriles by *T. dicoccoides* var *kotschyanum* (abstr). Agron Abstr, p 10

Gleba YY (1979) Nonchromosomal inheritance in higher plants as studied by somatic cell hybridization. In: Sharp WR, Larsen PO, Paddock EF, Raghavan V (eds) Plant cell and tissue culture-principles and applications. Ohio State Univ Press, Ohio, USA, pp 774–788

Gleba YY, Shydkaya LG, Butenko RG, Sytnik KM (1974) Cultivation of isolated protoplasts. Sov Plant Physiol 21:486–492

Gleba YY, Piven NM, Komarnitskii IK, Sytnik AK (1978) Parasexual cytoplasmic *Nicotiana tabacum* and *N. debneyi* hybrids (cybrids) obtained by protoplast fusion. Dok Akad Nauka USSR 240:1223–1226

Gleeson SK (1982) Heterodichogamy in walnuts: inheritance and stable ratios. Evolution 36: 892–903

Glimelius K, Chen K, Bonnett HT (1981) Somatic hybridization in *Nicotiana:* segregation of organellar traits among hybrid and cybrid plants. Planta (Berl) 153:504–510

Goblet JP, Boutry M, Duc G, Briquet M (1983) Mitochondrial plasmid-like molecules in fertile and male sterile *Vicia faba* L. Plant Mol Biol 2:305–309

Godley EJ (1955) Monoecy and incompatibility. Nature 176:1176–1177

Godley EJ (1979) Flower biology in New Zealand. NZ J Bot 17:441–466

Goldschmidt R (1955) Theoretical genetics. Univ Calif Press, Berkley, USA

Goldsmith GA (1968) Current developments in the breeding of F_1 hybrid annuals. Hort Science 3:269–271

Golubovskaya IN (1977) In: All union society geneticists and breeders. Dokl Akad Nauka Leningrad, USSR, pp 112

Golubovskaya IN (1979) Genetic control of meiosis. Int Rev Cytol 58:247–290

Goncharova KD (1974) Sterile pollen in hemp. Len Konoplya Ukrainian USSR 12:22–23

Goodsell SF (1961) Male sterility in corn by androgenesis. Crop Sci 1:227–228

Goodspeed TH (1930) Occurrence of triploid and tetraploid individuals in X-ray progenies of *Nicotiana tabacum.* Univ Calif Press 11:299–308

Goodspeed TH (1954) The genus *Nicotiana.* Chron Bot 16:pp 536

Goodspeed TH, Avery P (1939) Trisomic and other types of *Nicotiana sylvestris.* J Genet 38: 381–458

Goral S, Pawluk A, Staszewski Z (1981) Male sterility in *Festuca arundinacea.* Hodowla Rastin Aklimatyzacja i Nasiennictovo Poland 25:121–127

Gorenflot MR (1964) Sterilite et chromosome surnomeraire chez le "*Plantago coronopus* L." CR Acad Sci Paris 258:321–324

Gori P (1983) Ultrastructural changes in the wall of the sporogenous cells in *Allium sativum,* clone Piemonte during microsporogenesis. Ann Bot 51:139–143

Gori P, Ferri S (1982) Ultrastructural study of the microspore development in *Allium sativum,* cl. Piemonte. J Ultrastruct Res 78:341–349

Gorska-Brylass A (1967) Transitory callose envelope surrounding the generative cell in pollen grains. Acta Soc Bot Pol 36:419–422

Gőshev A (1974) Male sterility in.some Bulgarian varieties and hybrid lines of *Brassica capitata* Lizq. Genet Selelk Bulgaria 7:171–174

Goss JA (1968) Development, physiology and biochemistry of corn and wheat pollen. Bot Rev 34:333–358

Gottschalk W (1960) Über strahlenbedingte erbliche Sterilitätserscheinungen bei pflanzlichen Versuchsobjekten. Atompraxis 6:341–347
Gottschalk W (1961) Über die mutative Abänderung pflanzlicher Organisationsmerkmale. Planta (Berl) 57:313–330
Gottschalk W (1968) Investigations on the genetic control of meiosis. The Nucleus India (seminar vol), pp 345–361
Gottschalk W (1971) Our present knowledge on the genetic control of microsporogenesis in *Pisum*. Pisum Newslett 3:9–10
Gottschalk W (1976) Genetically conditioned male sterility. In: Proc Induced mutations in cross breeding, IAEA, Vienna, pp 133–140
Gottschalk W, Baquar SR (1971) The effect of mutated genes on the germ cell formation in *Pisum*. Pak J Sci Ind Res Pakistan 14:223–227
Gottschalk W, Baquar SR (1972) Breakdown of meiosis in a mutant of *Pisum sativum*. Cytobiologie 5:42–50
Gottschalk W, Jahn A (1964) Cytogenetische Untersuchungen an desynaptischen und männlichsterilen Mutanten von *Pisum*. Z Vererbungsl 95:150–166
Gottschalk W, Kaul MLH (1974) The genetic control of microsporogenesis in higher plants. Nucleus (Calcutta) 17:133–166
Gottschalk W, Kaul MLH (1980a) Asynapsis and desynapsis in flowering plants I. Asynapsis. Nucleus (Calcutta) 23:1–15
Gottschalk W, Kaul MLH (1980b) Asynapsis and desynapsis in flowering plants II. Desynapsis. Nucleus (Calcutta) 23:99–120
Gottschalk W, Klein HD (1976) The influence of mutated genes on sporogenesis. A survey on the genetic control of meiosis in *Pisum sativum*. Theor Appl Genet 48:23–34
Gottschalk W, Wolff G (1983) Induced mutations in plant breeding. Springer, Berlin, Heidelberg, New York, pp 238
Goud JV, Nayar KM, Rao MG (1969) Induced male sterility in *Sorghum*. Mysore J Agric Sci 3: 1–4
Goujon C, Ingold M (1967) Restauration de ferilite de quelques lignées mâle-stériles de bles hexaploides. CR Acad Sci Paris 264:334–336
Gouyon PH, Vernet P (1982) The consequences of gynodioecy in natural populations of *Thymus vulgaris* L. Theor Appl Genet 61:315–320
Goyal RD (1962) Abnormal ovaries in male sterile *Pennisetum typhoides*. Sci Cult 28:23–24
Gracen VE (1982) Types and availability of male sterile cytoplasms. In: Sheridan WF (ed) Maize for biological research. N Dakota State Univ Press, USA
Gracen VE, Grogan CO (1972) Reactions of corn (*Zea mays*) seedlings with nonsterile, Texas male sterile, and restored Texas male sterile cytoplasms to *Helminthosporium maydis* toxin. Plant Dis Rep 56:432–433
Gracen VE, Grogan CO (1974) Diversity and suitability for hybrid production of different sources of cytoplasmic male sterility in maize. Agron J 66:654–657
Gracen VE, Grogan CO, Forster MJ (1972) Permeability changes induced by *Helminthosporium maydis* race T toxin. Can J Bot 50:2167–2170
Gracen VE, Kheyr-Pour A, Earle ED, Gregory P (1979) Cytoplasmic inheritance of male sterility and pest resistance. In: Proc 34th Annu Corn Sorghum Conf, USA, pp 77–91
Graham RD (1975) Male sterility in wheat plants deficient in copper. Nature 254:514–515
Graham RD (1976) Physiological aspects of time of application of copper to wheat plants. J Exp Bot 27:717–724
Graham RD (1978) Tolerance of triticale, wheat and rye to copper deficiency. Nature 271: 542–543
Graham RD, Nambiar EKS (1981) Advances in research on copper deficiency in cereals. Aust J Agric Res 32:1009–1031
Grant V (1951) The fertilization of flowers. Sci Am 184:52–56
Grant V (1956) The genetic structure of races and species in *Gilia*. Adv Genet 8:55–88
Grant V (1958) The regulation of recombination in plants. Cold Spring Harbor Symp Quant Biol 23:337–363

Grant V (1963) The origin of adaptations. Columbia Univ Press, New York

Graves CJ, Sutcliffe JF (1974) An effect of copper deficiency on the initiation and development of flower buds of *Chrysanthemum morifolium* grown in culture solution. Ann Bot 38:729–738

Gray JC, Kung SD, Wildman SG, Sheen SJ (1974) Origin of *Nicotiana tabacum* L. detected by polypeptide composition of fraction 1 protein. Nature 252:226–227

Graybosch RA, Palmer RG (1983) Cytological studies on a male sterile mutant (ms_2) in soybean (*Glycine max*) (abstr). Bot Soc Am, USA, p 22

Graybosch RA, Bernard RL, Cremeens CR, Palmer RG (1984) Genetic and cytological studies of a male sterile, female fertile soybean mutant. A new male sterile gene (ms_2) in *Glycine max* (L.) Merr. J Hered 75:383–388

Gregorius HR, Ross MD, Gillet E (1982) Selection in plant populations of effectively infinite size III. The maintenance of females among hermaphrodites for a biallelic model. Heredity 48:329–343

Gregory P, Early ED, Gracen VE (1977) Biochemical and ultrastructural aspects of southern corn leaf blight disease. In: Hedin PA (ed) Host plant resistance to pests. ACS Symp Ser, pp 90–114

Gregory P, Matthews DE, York DW, Earle ED, Gracen VE (1978) Southern corn leaf blight disease: studies on mitochondrial biochemistry and ultrastructure. Mycopathologia 66: 105–112

Gregory P, Earle ED, Gracen VE (1980) Effects of purified *Helminthosporium maydis* race T-toxin on the structure and function of corn mitochondria and protoplasts. Plant Physiol (Bethesda) 66:471–481

Gregory RP (1905) The abortive development of the pollen in certain sweet peas. Proc Camb Philos Soc 13:148–157

Greyson RI, Walden DB (1976) Possibilities for gibberellin male sterile relationships in corn-a proposal. Maize Genet Coop Newslett 50:116–117

Greyson RI, Walden DB, Cheng PC (1980) LM, TEM and SEM observations of anther development in the genic male sterile (ms9) mutant of corn, *Zea mays*. Can J Genet Cytol 22: 153–166

Grill LK (1980) Molecular biologists look at green plants. In: Proc 6th Mol Biol Annu Symp, Heidelberg, p 9A

Grill LK, Garger SJ (1981) Identification and characterization of double stranded RNA associated with cytoplasmic male sterility in *Vicia faba*. Proc Natl Acad Sci USA 78:7043–7046

Grill LK, Garger SJ, Turpen TH, Lomell SA, Marsden MPF, Murry LE (1983) Involvement of viruses and virus-like agents with the male sterility trait of plants. In: Liss AR (ed) Plant Molecular Biology. Academic Press, New York, pp 101–116

Gritton ET (1980) Field pea. In: Fehr W, Hadley HH (eds) Hybridization of crop plants. Am Soc Agron. Madison, USA, pp 357–366

Groff D, Odland ML (1963) Inheritance of closed flower in the cucumber. J Hered 54:191–192

Grogan CO (1971) Multiplasm, a proposed method for the utilization of cytoplasms in pest control. Plant Dis Rep 55:400–401

Grün P (1951) Variations in the meiosis of alfalfa. Am J Bot 38:475–482

Grün P (1970a) Changes of cytoplasmic factors during the evolution of the cultivated potato. Evolution 24:188–198

Grün P (1970b) Cytoplasmic sterilities that separate the cultivated potato from its putative diploid ancestors. Evolution 24:750–758

Grün P (1973) Cytoplasmic sterilites that separate the group *tuberosum* cultivated potoato from its putative tetraploid ancestor. Evolution 27:633–643

Grün P (1976) Cytoplasmic genetics and evolution. Columbia Univ Press, New York, pp 416

Grün P (1979) Evolution of the cultivated potato: a cytoplasmic analysis. In: Hawkes JG, Lester RN, Skelding AD (eds) The biology and taxonomy of the Solanaceae. Linn Soc Symp No 7. Academic Press, New York, pp 655–660

Grün P, Aubertin M (1965) Evolutionary pathways of cytoplasmic male sterility in *Solanum*. Genetics 51:399–409

Grün P, Aubertin M (1966) Cytological expressions of a cytoplasmic male sterility in *Solanum*. Am J Bot 53:295–301

Grün P, Chu LJ (1978) Development of plants from protoplasts of *Solanum* (Solanaceae). Am J Bot 65:538–543

Grün P, Staub J (1979) Evolution of tetraploid cultigens from the view of cytoplasmic inheritance. In: Proc Conf Exploration, taxonomy, maintenance of potato germplasm III. CIP, Lima, Peru, pp 141–151

Grün P, Aubertin M, Radlow A (1962) Multiple differentiation of plasmons in diploid species of *Solanum*. Genetics 47:1321–1333

Guan CY, Wang GH, Zhao JT, Li HZ (1981) Preliminary study on male sterility in swede rape induced by male-gametocide No 1. Hereditas PR China 3(5):15–17

Gulyaeva ZB (1971) A study of male sterility in rye. Genetika Beograd 8:18–24

Gundaev AI (1966) Prospects of selection in sunflower for heterosis. Maslich Kult 3:15–21 (transl Ind Natl Sci Doc Centre New Delhi, 1969)

Gundaev AI (1971) Basic principles of sunflower selection. In: Genetic principles of plant selection. Nauka Moscow, USSR (transl: State Sect Dept, Ottawa Canada 1972, pp 417–465)

Gunn CR, Skardl WH, Spencer HC (1978) Classification of *Medicago sativa* L. using legume characters and flower colours. USDA Tech Bull 9:1574

Gupta PK (1975) Cytoplasmic male sterility in self-incompatible *Crotalaria pallida* Ait. Incompatibility Newslett 5:13–14

Gururajan KN, Srinivasan K (1982) Note on the use of Okra-leaf male sterile line in the production of hybrid cotton. Indian J Agric Sci 52:20–21

Gustafsson A (1935) Studies on the mechanism of parthenogenesis. Hereditas 21:1–112

Guy P (1978) INRA station d'Amélioration Fourragère Rep Lusignan, France

Habib AF, Rao TS (1973) A case of spontaneous male sterility in scarlet egg plant *Solanum integrifolium* L. Curr Sci (Bangalore) 2:51–52

Hadley HH (1960) A genetic method for producing seedless hybrid broomcorn. Agron J 52: 610–611

Hadley HH, Singh SP (1961) Partial male fertility in sorghum hybrids with cytoplasms from male sterile combine Kafir-60. Crop Sci 1:457–458

Hadley HH, Starnes WJ (1964) Sterility in soybeans caused by asynapsis. Crop Sci 4:421–424

Hafen L, Stevenson EC (1955) New male sterile and stamenless mutants. Tomato Genet Coop Rep 5:7

Hafen L, Stevenson EC (1956) The relationship of temperature, method of pollination and male sterility to the time required for producing hybrid tomato seed. Proc Amer Soc Hort Sci 67:355–360

Hafen L, Stevenson EC (1958) Preliminary studies of five stamenless mutants. Tomato Genet Coop Rep 8:17

Haigh JC (1954) Plant breeding report. In: Natl Veg Res Stat Rep 5:13

Hakansson A (1943) Meiosis in a nullisomic and in an asyndetic *Godetia whitneyi*. Hereditas 29: 179–190

Halevy AH (1963) Interaction of growth retarding compounds and gibberellin or indoleacetic acid oxidase and peroxidase of cucumber seedlings. Plant Physiol (Bethesda) 38:731–737

Halevy AH (1967) Effect of growth retardants on drought resistance and longevity of various plants. In: Proc 17th Int Hortic Congr 3:277–283

Halevy AH, Rudich J (1967) Modification of sex expression in muskmelon by treatment with the growth retardant B-995. Physiol Plant 20:1052–1058

Halevy AH, Wittwer SH (1966) Effect of growth retardants on longevity of vegetables, mushrooms and cut flowers. Proc Amer Soc Hort Sci 88:582–594

Hall IV, Aalders LE (1961) Note on male sterility in the common lowbush *Vaccinium angustifolium*. Can J Plant Sci 41:865

Hall RH (1973) Cytokinins as a probe for developmental processes. Ann Rev Plant Physiol 24: 415–444

Hamilton BA (1965) Cytological study of temperature-cytoplasmic-genetic interaction in fertility restorer line in *Petunia.* Ph D Thesis Pennsylvania State Univ
Hamilton RI (1926) Improving sunflowers by inbreeding. Sci Agric 6:190–192
Hamilton WD (1967) Extraordinary sex ratios. Science 156:477–488
Han DW, Lee JJ (1972) Effect of ethrel, RH 531 and gibberellin on the sterile pollen induction in lettuce. Ph D Thesis Municipal Coll Agric Seoul Korea (from Plant Breed Abstr 4774, 1974)
Hanna WW (1977) Effect of DPX 3778 on anther dehiscence in pearl millet. Crop Sci 17:965–967
Hanna WW, Powell JB (1973) Stubby head, an induced facultative apomictic in pearl millet. Crop Sci 12:726–728
Hanna WW, Powell JB (1974) Radiation induced female sterile mutant in pearl millet. J Hered 65: 247–249
Hanna WW, Powell JB, Burton GW (1976) Relationship to polyembryony, frequency, morphology, reproductive behaviour and cytology of autotetraploids in *Pennisetum americanum.* Can J Genet Cytol 18:529–536
Hanneman J, Peloquin SJ (1981) Genetic-cytoplasmic male sterility in progeny of 4X-2X crosses in cultivated potatoes. Theor ApplGenet 59:53–55
Hansche PE, Gabelmann WH (1963a) Digenic control of male sterility in carrots *Daucus carota* L. Crop Sci 3:383–386
Hansche PE, Gabelman WH (1963b) Phenotypic stability of pollen sterile carrots, *Daucus carota* L. Proc Amer Soc Hort Sci 82:341–350
Hansen DJ, Bellman SK, Sacher RM (1976) Gibberellic acid controlled sex expression of corn tassels. Crop Sci 16:371–374
Hanson CH, Graumann HO, Elling LJ, Dudley JW, Carnahan WL, Kehr WR, Davis RL, Frosheiser FI, Hovia AW (1964) Performance of two-clone crosses in alfalfa and an unanticipated self pollination problem. USDA Techn Bull 1300
Hanson MR, Conde MF (1985) Functioning and variation of cytoplasmic genomes: lessons from cytoplasmic nuclear interactions affecting male fertility in plants. Int Rev Cytol 94:214–245
Hara S (1946) Linkage between factors for sterility and anthocyanin pigmentation in rice plant. Jpn J Genet 21:32
Harney PM, Kung HCC (1967) Development of non-dehiscent anthers in partially male sterile plants of *Pelargonium* x *Hortum* Bailey cv Jacqueline. Can J Genet Cytol 9:359–366
Hart GE (1965) Studies on extrachromosomal male sterility in *Nicotiana.* Ph D Thesis Univ California, Berkley USA
Harte C, Bissinger B (1952) Entwicklungsgeschichtliche Untersuchung der durch die Faktoren *fr* und *ster* bedingten Pollensterilität bei *Oenothera.* Z Vererbungsl 84:251–269
Harvey LH (1969) Linkage of male sterility in upland cotton *Gossypium hirsutum* L. Ph D Thesis Univ Georgia, Athens USA
Harvey PH (1977) *Sorghum* germplasm base in US. In: Proc 32nd Annu Corn Sorghum Res Conf. Am Soc Trade Assoc Washington DC 32:186–198
Harvey PH, Levings CS, Wernsman EA (1972) The role of extrachromosomal inheritance in plant breeding. Adv Agron 24:1–27
Hashimoto D, Yamamoto Y (1976) Studies on cool injury in bean plants VII. Sensitive stages to sterile type low temperature injury during floral development in relation to nitrogen status of soybean plants. Proc Crop Sci Soc Jpn 45:287–297
Hawkes JG (1967) The history of potato. J Roy Hortic Sci Soc 92:207–365
Hayase H, Satake T, Nishiyama I, Ito N (1969) Male sterility caused by cooling treatment at the meiotic stage in rice plants II. The most sensitive stage to cooling and the fertilizing ability of pistil. Proc Crop Sci Soc Jpn 38:706–711
Hayes RT (1933) The classification of groundnut varieties with a preliminary note on the inheritance of some characters. Trop Agric W Indies 10:318–327
Hayward MD, Manthiratna MAPP (1972) Pollen development and variation in the genus *Lolium* I. Pollen size and tapetal relationships in male-fertiles and male-steriles. Z Pflanzenzücht 67:131–144

Heaton TC, Knowles PF (1980) Registration of UC-148 and UC-149 male-sterile safflower germplasm. Crop Sci 20:554

Heaton TC, Knowles PF (1982) Inheritance of male sterility in safflower. Crop Sci 22:520–522

Hecker RJ, Smith GA (1975) Tests of granular ethephon as a male gametocide on sugarbeet. Can J Plant Sci 55:655–656

Hecker RJ, Talat B, Bhatnagar PS, Smith GA (1972) Tests for chemical induction of male sterility in sugarbeet. Can J Plant Sci 52:937–940

Heilborn O (1932) Lethal gene combinations and pollen sterility in diploid apple varieties. A critique theory. Hereditas 16:1–18

Heilbronn A (1953) Über die Rolle des Plasmas bei der Geschlechtsbestimmung der *Bryonien.* Rev Fac Sci Univ Istambul B Sci Nat 18:205–206

Heilbronn A (1954) Plasmon und Geschlecht. 8th Int Cong Bot:105–107

Heilbronn A, Basarman M (1942) Über die F_2 der *Bryonia*bastarde und ihre Bedeutung für das Problem der Geschlechtsrealisation. Rev Fac Sci Univ Istambul B Sci Nat 7:138–144

Heiser CB (1982) Registration of Indiana-I CMS sunflower germplasm. Crop Sci 22:1089

Heiser CB, Pickersgill B (1969) Names for the cultivated *Capsicum* species (Solanaceae). Taxon 18:277–283

Heiser CB Jr (1976) The sunflower. Oklahoma Univ Press, Norman USA

Helal RM, Zaiki ME (1981) Effect of 2,4-D and ethephon foliar sprays on induction of pollen sterility in eggplant. Egypt J Hortic 8:101–108

Helmerick RH, Finkner RE, Doxtator CW (1965) Combining ability in autotriploid sugarbeets *Beta vulgaris* L. J Am Soc Sugar Beet Techn 13:538–547

Henderson RG (1931) Transmission of tobacco ringspot by seed of *Petunia.* Phytopathology 21: 225–229

Hensz RA, Mohr HC (1959) Functional male sterility in watermelon induced by chemical treatment. Proc Amer Soc Hort Sci 74:552–554

Herbert W (1847) On hybridization amongst vegetables. J Hortic Soc Lond 2:1–28

Herd YR, Steer MW (1984) Microsporogenesis in genic male sterile lines of barley (*Hordeum vulgare*). Can J Bot 62:1127–1135

Herich R (1960) Gibberellin and sex differentiation of flowering plants. Nature 188:599–600

Herich R (1965) Nucleoli and cytoplasmic male sterility. Z Indukt Abstammungs Vererbungsl 96:22–27

Herrmann RB (1970) Multiple amounts of DNA related to the size of chloroplasts I. An autoradiographic study. Planta (Berl) 90:80–96

Hermsen JGT (1965) Towards a more efficient utilization of genic male sterility in breeding hybrid barley and wheat. Euphytica 14:221–224

Hermsen JGT (1966) Crossability, fertility and cytogenetic studies in *Solanum acaule* x *S. bulbocastanum.* Euphytica 15:149–155

Hermsen JGT (1968) A discussion on cytoplasmic restoration of ms-sterility. Euphytica 17 Suppl 1:63–67

Hermsen JGT (1973) Nature and inheritance of unilateral incompatibility in interspecific *Solanum* crosses. Incompatibility Newslett 2:22–23

Hermsen JGT (1979) Factors controlling interspecific crossability and their bearing on the strategy for breaking barriers to intercrossing of tuber bearing *Solanum* species. In: Proc Conf broadening genetic base of crops, Wageningen. Pudoc, the Netherlands, pp 311–318

Hermsen JGT, Ramanna MS (1969) Meiosis in different F_1 hybrids of *Solanum acaule* Bitt x *S. bulbocastanum* Dunn and its bearing on genome relationship, fertility and breeding behaviour. Euphytica 18:27–35

Hermsen JGT, Ramanna MS (1976) Barriers to hybridization of *Solanum bulbocatanum* Dunn and *S. verrucosum* schlechtd and structural hybridity in their F_1 plants. Euphytica 25:1–10

Hermsen JGT, Taylor LM (1979) Successful hybridization of non-tuberous *Solanum etuberosum* Lind and tuber bearing *S. pinnatisectum* Dunn. Euphytica 28:1–7

Hernandez T, Wollard D (1978) The use of male sterility in the production of F_1 hybrid tomatoes (abstr). 75th Hortic Sci Ann Meet, p 359

Herong C, Xiangmin P, Yun D, Yunxiang C (1981) New male sterile type derived from a cross of common and club wheat. In: 1980 Annu Rep Inst Genet Acad Sin, pp 131–132

Hervás JP (1974) Male sterile Emmer wheat: Development of meiosis with *Aegilops caudata* cytoplasm. Biol Zentralbl 93:649–654

Heslops-Harrison J (1952) A reconsideration of plant teratology. Phyton (Horn) 4:19–34

Heslop-Harrison J (1956) Auxin and sexuality in *Cannabis sativa.* Physiol Plant 9:588–597

Heslop-Harrison J (1957) The experimental modification of sex expression in flowering plants. Biol Rev Camb Philos Soc 32:38–90

Heslop-Harrison J (1959a) Photoperiod and fertility in *Rottboellia exaltata.* Ann Bot New Ser 23:345–349

Heslop-Harrison J (1959b) Growth substances and flower morphogenesis. J Linn Soc Lond 56: 269–281

Heslop-Harrison J (1961) The experimental control of sexuality and inflorescence structure in *Zea mays* L. Proc Linn Soc Lond 172:108–123

Heslop-Harrison J (1963) Sex expression in flowering plants. Brookhaven Symp Quant Biol 16: 109–122

Heslop-Harrison J (1964) Cell walls, cell membranes and protoplasmic connections during meiosis and pollen development. In: Linskens HF (ed) Pollen physiology and fertilization. North Holland, Amsterdam, pp 39–47

Heslop-Harrison J (1968) Synchronous pollen mitosis and formation of the generative cell in massulate orchids. J Cell Sci 3:457–466

Heslop-Harrison J (1971) Features of male sterility in angiosperms. Proc Ann Corn Sorghum Res Conf 26:14–21

Heslop-Harrison J (1972) Sexuality in angiosperms. In: Steward FC (ed) Plant physiology, a treatise VIC. Academic Press, New York, pp 133–289

Heslop-Harrison J, Heslop-Harrison Y (1958) Longday and auxin induced male sterility in *Silene pendula* L. Acta Biol Ser A 5:79–94

Heslop-Harrison J, Heslop-Harrison Y (1961) Studies on flowering plant growth and organogenesis IV. Effects of gibberellic acid on flowering and the secondary sexual difference in stature in *Cannabis sativa.* Proc Roy Irish Acad Sec B 61:219–231

Heslop-Harrison J, Heslop-Harrison Y (1970) Evaluation of pollen viability by enzymatically induced fluorescence; intra-cellular hydrolysis of fluorescein diacetate. Stain Techn 45: 115–120

Heslop-Harrison Y, Knox RB, Howlett B (1973) Pollen wall proteins: gametophytic and sporophytic fractions in the pollen walls of the Malvaceae. Ann Bot New Ser 37:403–412

Heslop-Harrison J, Knox RB, Heslop-Harrison Y (1974) Pollen wall proteins: exine held fractions associated with the incompatibility response in cruciferae. Theor Appl Genet 44:133–137

Hess D (1973) Transformationsversuche an höheren Pflanzen: Untersuchungen zur Realisation des Exosom-Modells der Transformation bei *Petunia hybrida.* Z Pflanzenphysiol 68:432–440

Heu MH, Chae YA (1970) Studies on the male sterility in rice (*Oryza sativa* L.). Res Rep Rural Dev (Crop) Korea 13:9–13

Heuch I (1979) Equilibrium populations of heterostylous plants. Theor Pop Biol 15:43–57

Heyn FW (1973) Beiträge zum Auftreten unreduzierter Gameten und zur Genetik einiger Merkmale bei den Brassicae. Diss Georg-August Univ Göttingen, pp 102

Heyn FW (1976) Transfer of restorer gene from *Raphanus* to cytoplasmic male sterile *Brassica napus.* Eucarpia Cruciferae Newslett 1:15–16

Heyn FW (1978) Cytoplasmic genetic male sterility in *Brassica napus.* Eucarpia Cruciferae Newslett 3:34–35

Heyn H (1930) Beitrag zur Cytologie der Kartoffel (*Solanum tuberosum* L.). Wiss Arch Landw 4:123–168

Hicks GS, Sand SA (1977) *In vitro* culture of the stamen primordia from male sterile tobacco. Plant Sci Lett 10:257–263

Hicks GS, Bell J, Sand SA (1977) A developmental study of the stamens in a male sterile tobacco hybrid. Can J Bot 55:2234–2244

Higgins TJV, Zwar JA, Jacobsen JV (1976) Gibberellic acid enhances the level of translatable mRNA for L-amylase in barley aleurone layers. Nature 260:166–169

Hill J, Robson AD, Loneragan JF (1978) The effects of copper and nitrogen supply on the retranslocation of copper in four cultivars of wheat. Austr J Agric Res 29:925–939

Hillard OM (1966) Studies in *Streptocarpus*. JISS Afr Bot 32:87–123

Hillyer IG (1956) Effects of growth substances in flowering of cucurbitaceous plants. Ph D Thesis Michigan State Univ

Hilty JW, Josephson LM (1971) Reaction of corn inbreds with different cytoplasms to *Helminthosporium maydis*. Plant Dis Rep 55:195–198

Hinata K, Konno N (1976) Cytoplasmic male sterile strain of yukina (*Brassica campestris* L.) produced by nucleus substitution. Jpn J Breed 26:127–128

Hiorth G (1948a) Über Hemmungssysteme bei *Godetia whitneyi* II. Z Indukt Abstammungs Vererbungsl 82:12–63

Hiorth G (1948b) Über Hemmungssysteme bei *Godetia whitneyi* III. Z Indukt Abstammungs Vererbungsl 82:276–330

Hirata K (1929) Cytological basis of the sex determination in *Cannabis sativa* L. Jpn J Genet 4: 198–201

Hirose T (1962) Studies on the pollination of red-pepper III. The occurrence of infertile pollen grains. Sci Rep Kyoto Agric Univ Japan 14:45–50

Hirose T (1965) Fundamental studies on the breeding of peppers. Techn Bull Fac Agric Kyoto Univ Japan 2:1–186

Hirose T (1969) Studies on chemical emasculation in pepper I. Suppression of anther dehiscence and other morphological changes caused by sodium 2,2-dichloropropionate. J Jpn Soc Hortic Sci 38:29–35

Hirose T, Fujime Y (1975) A new male sterility in pepper (abstr). Hortic Sci Kagawa Univ Japan 10:314

Hirose T, Fujime Y (1981) Studies on the hybrid seed production by using a new male sterile line in pepper (*Capsicum annuum* L.) II. On the process of pollen degeneration. Jpn Soc Hortic Sci 50:66–70

Hirose T, Takashima S (1963) The production of F_1 hybrids by means of male sterility I. Morphology of male sterile strains. Sci Rep Agric Kyoto Univ Japan 15:21–26

Ho DTH, Varner JE (1974) Hormonal control of messenger ribonucleic acid metabolism in barley aleurone layers. Proc Natl Acad Sci USA 71:4783–4786

Ho TY, Ross MD (1973) Maintenance of male sterility in plant populations II. Heterotic models. Heredity 31:282–286

Hoaglund AR, Elliot FC, Rasmusson LW (1953) Some histological and morphological effects of maleic hydrazide on a spring wheat. Agron J 45:468–472

Hockett EA (1971) Coordinators report on genetic male sterility in barley. Barley Genet Newslett 1:175–189

Hockett EA (1972a) Report of genetic male sterility in barley. Barley Genet Newslett 2:178–179

Hockett EA (1972b) Coordinators. report on genetic male sterile barley collections. Barley Genet Newslett 2:139–144

Hockett EA (1977) The genetic male sterile barley collection. Barley Genet Newslett 7:97–100

Hockett EA (1980) Fertility restoration of cytoplasmic male sterile barley. Barley Genet Newslett 10:31–33

Hockett EA, Ahokas H (1979) Male and female fertility levels of genetic male sterile barley grown at two different latitudes. Hereditas 91:65–71

Hockett EA, Eslick RF (1968) Genetic male sterility in barley I. Nonallelic genes. Crop Sci 8: 218–220

Hockett EA, Eslick RF (1969) Spontaneous frequencies of genetic and other sterilities in barley, *Hordeum vulgare* L. Crop Sci 9:23–24

Hockett EA, Eslick RF (1970a) Genetic male sterile genes useful in hybrid barley production. In: Nilan RA (ed) Barley Genetics II. Washington State Univ Press, pp 298–307

Hockett EA, Eslick RF (1970b) Natural outcrossing on genetic male sterile barley. Crop Sci 10:152–154

Hockett EA, Feltner KC (1978) Sterility induced in barley by synthetic pyridone. Crop Sci 18: 296–301

Hockett EA, Reid DA (1981) Spring and winter genetic male sterile barley stocks. Crop Sci 21: 655–659

Hockett EA, Eslick RF, Reid DA, Wiebe GA (1968) Genetic male sterility in barley II. Available spring and winter stocks. Crop Sci 8:754–755

Hockett EA, Baenzinger PS, Steffens GL (1978) A proposal for increased research on chemical induction of fertility in genetic male sterile barley. Euphytica 27:109–111

Hoff BJ, Chandrapanya D (1973) Inheritance of two male sterile characters in rice (abstr). Am Soc Agron, p 7

Hoffmann W (1961) Lien (*Linum usitatissimum*). Handb Pflanzenzücht 5:264–266

Hofmyer JDJ (1938) Genetical studies of *Carica papaya* L., vol 1,2. Dep Agric Sci Bull S Africa 187:5–64

Hogaboam GJ (1957) Factors influencing phenotypic expression of cytoplasmic male sterility in the sugar beet (*Beta vulgaris* L.). J Am Soc Sugar Beet Tech 9:457–465

Höhler B, Börner TH (1980) Vergleichende Untersuchungen von Isoenzymen des Antherengewebes fertiler und cytoplasmatisch männlich steriler Weizenpflanzen. Biochem Physiol Pflanz 175:562–569

Holm G (1954) Chlorophyll mutations in barley. Acta Agric Scand 4:457–471

Honing JA (1928) *Canna* crosses II. The chromosome numbers of *Canna glauca* x *C. indica* F_1, *C. aureo-vittata* and *C. aureo-vittata* gigas. Meded Landbouwshoogesch Wageningen 32:1–14

Hooker AL (1974) Cytoplasmic susceptibility in plant disease. Ann Rev Phytopathol 12:167–169

Hooker AL, Smith DR, Lim SM, Beckett JB (1970) Reaction of corn seedlings with male sterile cytoplasm to *Helminthosporium maydis.* Plant Dis Rep 54:708–712

Hope C (1979) see Izhar S (1984)

Hori T, Tsunewaki K (1967) Study on substitution lines of several emmer wheats having the cytoplasm of *Triticum boeticum.* Seiken Zihô 19:55–59

Hori T, Tsunewaki K (1969) Basic studies on hybrid wheat breeding II. Effects of three alien cytoplasm on main quantitative characters of hexaploid wheats. Jpn J Breed 19:425–430

Horn W (1974) F_1 hybrids in floriculture. In: Proc 19th Int Hortic Cong 4:267–277

Horner HT (1977) A comparative light- and electron- microscopic study of microsporogenesis in male-fertile and cytoplasmic male-sterile sunflower (*Helianthus annuus*). Am J Bot 64: 745–759

Horner JR, Rogers MA (1974) A comparative light and electron microscope study of microsporogenesis in male-fertile and cytoplasmic male-sterile pepper (*Capsicum annuum*). Can J Bot 52:435–441

Hornsey KL (1973) Attempted pollen transmission of cytoplasmic male sterility and the spontaneous occurrence of male sterility in O-type lines of sugar beet (*Beta vulgaris* L.). Theor Appl Genet 43:31–34

Horovitz A, Beiles A (1980) Gynodioecy as a possible populational strategy for increasing reproductive output. Theor Appl Genet 57:11–15

Horovitz A, Dulberger R (1983) The genetic basis of gender in *Silene vulgaris.* Heredity 51: 371–376

Horovitz A, Galil J (1972) Gynodioecism in east mediterranean *Hirschfeldia incana,* Cruciferae. Bot Gaz 133:127–131

Hosokawa S, Takeda T, Otani Y, Ikehata M (1954) Cytohistological studies on male sterility of sugar beets with special reference to pollen degeneration and tapetal plasmodium. Jpn J Breed 4:196–202

Hosokawa S, Tsuda C, Takeda T (1963) Histochemical studies on the cytoplasmic male sterility of sugar beets. Jpn J Breed 13:117–129

Hossain MA, Driscoll CJ (1981) Transfer of cornerstone male sterility mutant to tetraploid wheat and hexaploid and octoploid triticales. Can J Genet Cytol 23:493–496

Hossain MA, Driscoll CJ (1983) Fertility compensation of cornerstone male sterility of wheat by rye. Genetics 104:181–189

Hougas RW, Peloquin SJ (1962) The potato and its wild relatives VII. Exploitation of *Solanum* germplasm. Texas Res Found, Renner USA, pp 606

Howard HW (1970) Genetics of potato, *Solanum tuberosum*. Logos, London, pp 126

Howlett BJ, Knox RB, Heslop-Harrison J (1973) Pollen wall proteins: release of the allergen antigen E from intine and exine sites in pollen grains of ragweed and *Cosmos*. J Cell Sci 13:603–619

Hrishi NJ (1960) Cytological studies on *Papaver somniferum* and *P. setigerum*. Genetica (The Hague) 31:1–4

Hsu KJ (1945) On sterility resulting from crossing different types of rice. Indian J Genet Plant Breed 5:51–57

Huang GW, Huang YX, Shi SC, Gao DK, Zhang DM, Gou YZ, Liu ZG, Fen GB (1981) Utilization of recessive genes for male sterility in hybrid seed production of cotton. Sci Agric Sin 5:5–11

Hughes wG (1975) Use of ethephon in wheat breeding. Ann Appl Biol 81:112–114

Hughes WG (1976) Ethrel induced male sterility in wheat (*Triticum aestivum* L.) In: Proc 7th Eucarpia Cong. Elsevier, Amsterdam, pp 111–114

Hughes WG, Bodden JJ (1978) An assessment of the production and performance of F_1 hybrid wheats on *Triticum timopheevi* Cytoplasm. Theor Appl Genet 53:219–228

Hughes WG, Bennett MD, Botten JJ, Galanopoulou S (1974) Effects of time of application of etheral on male sterility and ear emergence in wheat *Triticum aestivum*. Ann Appl Biol 76: 243–253

Huglo B, Duc G (1980) Effect du génome mainteneur sur le determinisme delinstabilité de la stérilité mâle cytoplasmique (cyt 447) chez la Féverole. Mem ENSA Rennes, pp 47

Hull DL (1980) Individuality and selection. Annu Rev Ecol Syst 11:311–332

Hull FH (1937) Inheritance of rest period of seeds and certain other characters in the pea nut. Florida Univ Agric Exp Stat Techn Bull, pp 314

Humaydan HS, Scott EW (1977) Methomyl insecticide selective phytotoxicity on sweet corn hybrids and inbreds having the Texas male sterile cytoplasm. HortScience 12:312–313

Hussaini SH, Rao PV (1964) A note on the spontaneous occurrence of cytoplasmic male sterility in Indian sorghum. Sorghum Newslett 7:27–28

Hussein HAS, Disouki IAM (1974) Mutation breeding experiments in *Phaseolus vulgaris* L. I. EMS and gamma ray induced seed colour mutants. Z Pflanzenzücht 76:190–199

Hutchins AE (1944) A male and female sterile variant in squash, *Cucurbita maxima* Duchesne. Proc Amer Soc Hort Sci 44:494–496

Hyde BB (1953) Differentiated chromosomes in *Plantago ovata*. Am J Bot 40:809–815

Ibrahim AM, Coyne DP, Emery G (1974) Male sterility and stigma shape alterations in *Phaseolus vulgaris*. Annu Rep Bean Improv Coop Madison 17:48–49

Ie TS (1972) Cytoplasmic particles in *Trapaeolum majus*. Planta (Berl) 106:227–236

Iizuka M, Hashizumi T (1968) Induction of female organs in staminate grape by 6-substituted adenine derivatives. Jpn J Genet 43:393–394

Ikehashi H, Fujimaki H (1980) Modified bulk population method for rice breeding. In: Innovative approaches to rice breeding. IRRI Philippines, pp 163–182

Ilan I (1971) Evidence for hormonal regulation of the selectivity of ion uptake by plant cells. Physiol Plant 25:230–233

Iltis HH, Doebley JF (1980) Taxonomy of *Zea* (Gramineae) II. Subspecific categories in *Zea mays* complex and a generic synopsis. Am J Bot 67:994–1004

Imanishi SN (1968) Heterosis-breeding by the use of male sterility II. Sugarbeet. In: Proc 10th Symp Jpn Breed Soc Osaka, pp 7–16

Imrie BC (1966) Stigma receptivity in cytoplasmic male sterile wheat. Aust J Exp Agric Anim Husb 6:175–178

Imsande J, Ralston EJ (1982) Dinitrogen fixation in male-sterile soybeans. Plant Physiol (Bethesda) 69:745–746

Ingold M (1968) Male sterility and restorer system in wheat. Euphytica 17 Suppl 1, 17:69–74

Ishikawa J (1927) Studies on the inheritance of sterility in rice. J Coll Agric Hokkaido Univ 20:79–201

Ishikawa J (1929) On the inheritance of sterility in rice. Jpn J Genet 4:156–157

Ishimine Y, Shinjyo C (1978) Distribution of fertility-restoring genes in commercial lowland-rice varieties cultivated in Japan for two male-sterile cytoplasms of rice. Bull Coll Agric Univ Ryukyus Jpn 25:665–672

Islam AKMR, Driscoll CJ (1984) Latent male fertility in cornerstone chromosome 4A. Can J Genet Cytol 26:98–99

Itai C, Richmond A, Vaadia Y (1968) The role of root cytokinins during water and salinity stress. Isr J Bot 17:187–195

Ito N (1978) Male sterility caused by cooling treatment at the young microspore stage in rice plants XVI. Changes in carbohydrates, nitrogenous and phosphorous compounds in rice anthers after cooling treatment. Jpn J Crop Sci 47:318–323

Ito N (1980) Physiological abnormalities in anthers of sterile-type cool injury of rice plant. Res Bull Hokkaido Natl Agric Exp Stat Jpn 125:41–75

Iwahori S, Lyons JM, Smith OE (1970) Sex expression in cucumber plants as affected by 2-chloroethylphosphonic acid, ethylene and growth regulators. Plant Physiol (Bethesda) 46:412–415

Iwamasa M (1966) Studies on the sterility in genus *Citrus* with special reference to seedlessness. Bull Hortic Res Stat Jpn B6:1–76

Iwasa S (1963a) Studies on the alloplasmatic effect in the tribe *Brassiceae* I. On the *carinata* cytoplasmic *Brassica pekinensis* induced by successive backcrosses. J Fac Agric Kyushu Univ Jpn 12:201–212

Iwasa S (1963b) Studies on the alloplasmatic effect in the tribe Brassiceae II. Several conspicuous characteristics appeared in the *carinata* cytoplasmic *pekinensis* plant. J Fac Agric Kyushu Univ Jpn 12:213–228

Iwasa S (1963c) Studies on the alloplasmatic effect in the tribe Brassiceae III. On the effect on the manifestation of genome complements in some F_1 hybrids and in the strains having an extrachromosome. J Fac Agric Kyushu Univ Jpn 12:229–240

Iyer CPA, Randhawa GS (1965) Chemical induction of male sterility in grapes. Curr Sci (Bangalore) 34:411–412

Izard C, Hitier H (1955a) Observations on a complex hybrid capable of producing male-sterile plants. Ann Inst Exp Tobacco Bergerac 2:93–105

Izard C, Hitier H (1955b) Cytogenetique dun hybride complexe presentat le syndrome de sterilite male. Ist Int Sci Tobacco Cong 1:211–213

Izhar S (1972) Promotion of flowering in cytoplasmic male sterile petunia by gibberellin. Hort Science 7:555

Izhar S (1973) Cell budding and fission in microspores of *Petunia.* Nature 244:35–37

Izhar S (1975) The timing of temperature effect on microsporogenesis in cytoplasmic male sterility *Petunia.* J Hered 66:313–314

Izhar S (1977) Cytoplasmic male sterility in *Petunia.* The interaction between the plasmagene, genetic factors and temperature. J Hered 68:238–240

Izhar S (1978) Cytoplasmic male sterility in *Petunia* III. Genetic control of microsporogenesis and male fertility restoration. J Hered 69:22–26

Izhar S (1984) Male sterility in *Petunia.* In: Sink KS (ed) Petunia. Springer, Berlin, Heidelberg, New York, pp 256

Izhar S, Frankel R (1971) Mechanism of male sterility in *Petunia.* The relationship between pH, callase activity in the anthers and the breakdown of the microsporogenesis. Theor Appl Genet 44:105–108

Izhar S, Frankel R (1973a) Duration of meiosis in *Petunia* anthers *in vivo* and in floral bud culture. Acta Bot Neerl 22:14–22

Izhar S, Frankel R (1973b) Mechanism of male sterility in *Petunia* II. Free amino acids in male fertile and male sterile anthers during microsporogenesis. Theor Appl Genet 43:13–17

Izhar S, Frankel R (1976) Cytoplasmic male sterility in *Petunia* I. Comparative study of different plasmatype sources. J Hered 67:43–46

Izhar S, Power JB (1979) Somatic hybridization in *Petunia:* a male sterile cytoplasmic hybrid. Plant Sci Lett 14:49–55

Izhar S, Tabib Y (1980) Somatic hybridization in *Petunia* II. Heteroplasmic state in somatic hybrids followed by cytoplasmic segregation into male sterile and male fertile lines. Theor Appl Genet 57:241–245

Izhar S, Schlicter M, Swartzberg D (1983) Sorting out the cytoplasmic elements in somatic hybrids of *Petunia* and the prevalence of the heteroplasmon through several meiotic cycles. Mol Gen Genet 190:468–474

Izhar S, Tabib Y, Swartzberg D (1984) Reciprocal transfer of male sterile and normal plasmons in *Petunia*. Theor Appl Genet 68:455–457

Jacob F, Schaeffer P, Wollman EL (1960) Episomal elements in bacteria. In: Proc Symp Soc Gen Microbiol 10:67–91

Jacobsen HJ, Ingensiep HW, Herlt M, Kaul MLH (1980) Tissue culture studies in *Pisum sativum*. In: Sala F, Parisi B, Cella R, Ciferi O (eds) Plant cell cultures: results and perspectives. Elsevier North Holland, Amsterdam, pp 319–324

Jacobsen JV, Zwar JA (1974) Gibberellic acid causes increased synthesis of RNA which contains poly (A) in barley aleurone tissue. Proc Natl Acad Sci USA 71:3290–3293

Jain AK, Rathore RKS, Chauhan SVS, Kinoshita T (1981) Anther ontogeny in an exotic and six indigenous cytoplasmic male sterile lines of wheat (*Triticum aestivum*) possessing *Triticum timopheevi* cytoplasm. Jpn J Breed 31:251–260

Jain HK, Pokhriyal SC (1975) Improved pearlmillet hybrids. Mutat Breed Newslett 6:11–12

Jain SK (1956) Natural incidence of male sterility and its chemical induction in crop plants. Assoc Diss IARI New Delhi India, pp 267

Jain SK (1959) Male sterility in flowering plants. Bibliogr Genet 18:101–166

Jain SK (1961) On the possible adaptive significance of male sterility in predominantly inbreeding populations. Genetics 46:1237–1240

Jain SK (1968) Gynodioecy in *Origanum vulgare:* computer simulation of a model. Nature 217: 764–765

Jain SK (1976) The evolution of inbreeding in plants. Annu Rev Ecol Syst 7:469–495

Jain SK, Suneson CA (1966) Increased recombination and selection in barley populations carrying a male sterility factor I. Quantitative variability. Genetics 54:1215–1224

Jain SK, Pierce RO, Hauptli H (1977) Meadowfoam: potential new oil crop. Calif Agric 31: 18–20

Jain SK, Boussy IA, Hauptli H (1978) Male sterility in meadowfoam. J Hered 69:61–63

Jain VK, Mukherjee D (1980) Effect of chlorflurenol methylester 74050 on sex expression and parthenocarpic fruit development in tomato. Phyton (Horn) 38:89–93

Jaisani BG, Drolsom PN (1971) Random type sterility in *Sorghum*. Crop Sci 11:167–171

Jaisani BG, Drolsom PN (1972) Studies of nonrandom and barren-type sterility in *Sorghum bicolor*. Crop Sci 12:37–40

James NI, Lund S (1960) Meristem development of winter barley as affected by vernalization and potassium gibberellate. Agron J 52:508–510

Jan CC (1974) Genetic male sterility in wheat (*Triticum aestivum* L.): expression, stability, inheritance and practical use. Ph D Thesis, Univ California, Davis

Jan CC, Rowell PL (1981) Response of wheat tillers at different growing stages to gametocide treatment. Euphytica 30:501–504

Jan CC, Qualset CO (1977a) Genetic male sterility in wheat (*Triticum aestivum* L.). Reproductive characteristics and possible use in hybrid wheat breeding. Hilgardia 45:153–171

Jan CC, Qualset CO (1977b) Genetic male sterility in wheat: inheritance. Crop Sci 17:791–794

Jan CC, Qualset CO, Vogt HE (1974) Chemical induction of sterility in wheat. Euphytica 23: 78–85

Jan CC, Qualset CO, Vogt HE (1976) Chemically induced sterility in wheat for hybrid seed production. Euphytica 25:375–386

Janaki-Ammal EK (1941) The breakdown of meiosis in a male sterile *Saccharum*. Ann Bot 5: 83–87

Janaki-Ammal EK (1942) Intergeneric hybrids of *Saccharum* IV. *Saccharum-Narenga*. J Genet 44:23–32

Jarvi AJ, Eslick RF (1975) Shrunken endosperm mutants in barley. Crop Sci 15:363–366

Jasmin JJ (1954) Male sterility in *Solanum melongena* L.: Preliminary report on a functional type of male sterility in egg plants. Proc Amer Soc Hort Sci 63:443

Jaynes RA (1963) Male sterility in *Castanea* (abstr). Genetics Today 1:203

Kenaschuk EO (1975) Flax breeding and genetics. Oil seeds and pulse crops in western Canada. Western Coop Fertilizer Publ, p 7

Kender WJ, Remaily G (1970) Regulation of sex expression and seed development in grapes with 2-chloroethylphosphonic acid. HortScience 5:491–492

Kenworthy NJ, Brim CA, Wernsman EA (1973) Polyembryony in soybeans. Crop Sci 13:637–639

Kern JJ, Atkins RE (1972) Free amino acid content of the anthers of male sterile and fertile lines of grain *Sorghum bicolor* (L.) Moench. Crop Sci 12:835–838

Kesicki E (1979) New type of functional male sterility. In: Proc Symp Production of tomatoes for processing. Evora, Portugal, pp 15–26

Kesicki E (1980) A new type of functional male sterility derived from interspecific hybrids. Acta Hortic 100:365–371

Kesseli R, Jain SK (1984) An ecological genetic study of gynodioecy in *Limnanthes douglasii* (Limnanthaceae). Am J Bot 71:775–786

Kherde M, Porter KB, Merkle OG, Atkins IM (1964) Cross-fertilization of male-sterile wheat (*T. aestivum*) and forage production of sterile hybrids having different cytoplasms (abstr). Am Soc Agron, p 70

Kheyr-Pour A (1975) Déterminisme génétique et répartition écologique de la stérilité mâle cytoplasmique chez *Origanum vulgare* L. Thesis Univ Sci Techn Lanaedoc, Montpellier, France

Kheyr-Pour A (1980) Nucleo-cytoplasmic polymorphism for male sterility in *Origanum vulgare* L. J Hered 71:253–260

Kheyr-Pour A (1981) Wide nucleo-cytoplasmic polymorphism for male sterility in *Origanum vulgare* L. J Hered 72:45–51

Kheyr-Pour A, Gracen VE, Everett HL (1981) Genetics of fertility restoration in the C-group of cytoplasmic male sterility in maize. Genetics 98:379–388

Kho YO, De Bruyn JW (1962) Gametocidal action of dichloracetic acid. Euphytica 11:287–292

Khoo V, Stinson HT (1957) Free amino acid differences between cytoplasmic male sterile and normal fertile anthers. Proc Natl Acad Sci USA 43:603–607

Kibalenko AP, Sidorshina TN (1971) Viability of pollen and seed formation in sugarbeet as dependent on plant boron nutrition. Dopov Akad Nauk Ikr RSR Ser B USSR 6:558–562

Kidane-Mariam HM, Peloquin SJ (1974) The effect of direction of hybridization (4x X 2x vs 2x X 4x) on yield of cultivated potatoes. Am Potato J 51:330–336

Kidane-Mariam HM, Peloquin SJ (1975) Method of diplandroid formation and yield of progeny from reciprocal (4x-2x) crosses. J Amer Soc Hort Sci 100:602–603

Kidd HJ (1961) The inheritance of restoration of fertility in cytoplasmic male sterile sorghum - a preliminary report. Sorghum Newslett 4:47–49

Kidd HJ, Finley DE (1962) Genetical and environmental variation in cytoplasmically inherited male sterility in sorghum (abstr). Am Soc Agron, p 70

Kiermayer O (1958) Induktion von Eingeschlechtlichkeit sowie Gamophyllie an Blüten durch 2,3,5-trijodbenzoesäure. Planta (Berl) 52:393–404

Kiermayer O (1959) Induktion männlich-steriler Blüten bei *Helianthus annuus* durch 2,3,5-trijodbenzoesäure (TIBA) Naturwissenschaften 46:457

Kihara H (1951) Substitution of nucleus and its effect on genome manifestation. Cytologia 16: 177–193

Kihara H (1954) Studies in wheat. Yokendo, Tokyo, pp 753

Kihara H (1958) Fertility and morphological variation in the substitution and restoration backcrosses of the hybrids *Triticum vulgare* x *Aegilops caudata.* Proc 10th Int Cong Genet:142–171

Kihara H (1959) Fertility and morphological variation in the substitution backcrosses of the hybrid *Triticum vulgare* x *Aegilops caudata.* In: Proc 10th Int Cong Genet 1:142–171

Kihara H (1962) Cytoplasmic male sterility of common wheat. 1961 Annu Rep Natl Inst Genet Jpn 12:48–50

Kihara H (1963a) Interspecific relationships in *Triticum* and *Aegilops.* Seiken Zihô 15:1–12

Kihara H (1963b) Nucleus and chromosome substitution in wheat and *Aegilops* II. Chromosome substitution. Seiken Zihô 15:13–23

Kihara H (1966a) Nucleus and chromosome substitution in wheat and *Aegilops* I. Nucleus substitution. In: Proc 2nd wheat Genet Symp, pp 313–326
Kihara H (1966b) Factors affecting the evolution of common wheat. Indian J Genet Plant Breed 26A:14–28
Kihara H (1967) Cytoplasmic male sterility in relation of hybrid wheat breeding. Züchter 37: 86–93
Kihara H (1968) Cytoplasmic relationships in the Triticinae. In: Proc 3rd Int Wheat Genet Symp Canberra Australia, pp 125–134
Kihara H (1973) Characteristics of *Aegilops squarrosa* cytoplasm. In: Sears ER, Sears LMS (eds) Proc 4th Int Wheat Genet Symp, pp 351–353
Kihara H (1979) Nucleo-cytoplasmic hybrids and nucleo-cytoplasmic heterosis. Seiken Zihô 27–28:1–13
Kihara H, Kondo N (1943) Studies in amphidiploids of *Aegilops caudata* x *Ae. umbellulata* induced by colchicine. Seiken Zihô 2:41–42
Kihara H, Tanaka M (1970) Addendum to the classification of the genus *Aegilops* by means of genome-analysis. Wheat Inf Serv 30:1–2
Kihara H, Tsunewaki K (1961) Pistillody of *Triticum durum* induced by an alien cytoplasm. Seiken Zihô 12:1–10
Kihara H, Tsunewaki K (1962a) Pistillody of *Triticum durum* induced by an *Aegilops* cytoplasm. Annu Rep Natl Inst Genet Jpn 12:1–50
Kihara H, Tsunewaki K (1962b) Increased occurrence of haploids and twin seedlings due to an alien cytoplasm. Annu Rep Natl Inst Genet Jpn 13:45–46
Kihara H, Tsunewaki K (1964) Some fundamental problems underlying the program for hybrid wheat breeding. Seiken Zihô 16:1–14
Kim BD, Levings CS, Pring DR, Conde MF, Laughnan JR, Gabay-Laughnan SJ, Mans RJ (1979) Homologous sequences among plasmid and mitochondrial DNAs from cytoplasmically male-sterile maize (abstr). Fed Proc 38:298
Kim BD, Conde MF, Gabay-Laughnan SJ, Laughnan JR, Levings III CS, Mans RJ, Pring DR (1980) Integration of plasmid sequences into mitochondrial chromosomal DNA upon reversion of S-type cytoplasmic male sterile plants to fertility in maize. In: Scott WA, Werner R, Joseph DR, Schultz J (eds) Mobilization and reassembly of genetic information. Academic Press, New York, pp 459
Kim BD, Mans RJ, Conde MF, Pring DR, Levings CS (1982) Physical mapping of homologous segments of mitochondrial episomes from S male-sterile maize. Plasmid 7:1–14
Kimura M (1964) Gibberellin-like activities of certain auxin and diterpenes. Ph D Thesis Univ California, Los Angeles, USA
King CJ, Loomis HF (1927) Factors influencing the severity of the crazytop disorder of cotton. USDA Agric Bull, p 1484
King JR (1960) The peroxidase reaction as an indication of pollen viability. Stain Techn 35: 225–227
Kinman ML (1970) New developments in the USDA and state experiment station in sunflower breeding programs. In: Proc 4th Int Sunflower Conf Memphis, pp 181–183
Kinoshita T (1971) Genetical studies on the male sterility of sugarbeets (*Beta vulgaris* L.) and its related species. J Fac Agric Hokkaido Univ 56:435–541
Kinoshita T (1976) Genetical studies on cytoplasmic male sterility induced by gamma ray irradiation in sugar beets. Jpn J Breed 26:255–265
Kinoshita T (1977) Genetic relationship between pollen fertility restoring genes and cytoplasmic factors in the male sterile mutants of sugarbeets. Jpn J Breed 27:19–27
Kinoshita T (1980) Induction of cytoplasmic male sterility by gamma ray and chemical mutagens in sugar beets. Gamma Field Symp 19:27–48
Kinoshita T (1983) Cytoplasmic mutations induced by chemical mutagens. Wheat Inf Serv 56: 44–46
Kinoshita T, Mikami T (1984) Alloplasmic effects on callus proliferation and streptomycin resistance in common wheat. Seiken Zihô 32:31–38

Kinoshita T, Nagao S (1969) Use of male sterility in triploid sugar beets. Jpn J Genet 44:351–361

Kinoshita T, Takahashi M (1966) Inheritance of a pollen sterility induced by the irradiation. Bull Sugar Beet Res Suppl 7:40–42

Kinoshita T, Takahashi M (1972) Genetic male sterility found in *Beta maritima* L. Jpn J Breed 22:11–19

Kinoshita T, Takashi M, Childers WR (1972) Cytoplasmic male sterility in autotetraploids beets XV. Studies in polyploid varieties of sugar beets. Jpn J Breed 22:159–167

Kinoshita T, Mori K, Takashashi M (1980) Inheritance studies on cytoplasmic male sterility induced by nuclear substitution. Genetical studies on rice plant LXX. J Fac Agric Hokkaido Univ 60:23–41

Kirk JTO (1971) Chloroplast structure and biosynthesis. Annu Rev Biochem 40:161–196

Kishore H, Sardana S, Sharma AK, Misra PC (1974) Behaviour of tapetum during microsporogenesis in male sterile and normal potato varieties. Veg Sci 1:36–41

Kitada K, Kurata N, Satoh H, Omura T (1983) Genetic control of meiosis in rice *Oryza sativa* L. I. Classification of meiotic mutants induced by NMU and their cytogenetical characteristics. Jpn J Genet 58:231–240

Kitamura E (1962) Studies on cytoplasmic sterility of hybrids in distantly related varieties of rice *Oryza sativa* L. I. Fertility of F_1 hybrids between strains derived from a certain Philippine x Japanese variety crosses and Japanese varieties. Jpn J Breed 12:81–84

Kleckner N (1977) Translocatable elements in prokaryotes. Cell 11:11–12

Kleijer G (1984) Genetic and cytogenetic studies with a male sterile mutant of common wheat (*Triticum aestivum*). Euphytica 33:107–112

Kleijer G, Fossati A (1976) Chromosomal location of a gene for male sterility in wheat (*Triticum aestivum*). Wheat Inf Serv 41–42:12–13

Klein HD (1968) Die Beeinflussung der Meiosis durch mutierte Gene. PhD Diss Math Natl Fak Bonn Univ FR Germany, pp 97

Klein HD (1969) Male sterility in *Pisum*. Nucleus (Calcutta) 12:167–172

Klein HD, Milutinovič (1971) Genbedingte Störungen der Infloreszenz und Blütenbildung. Theor Appl Genet 41:255–258

Kley Van der FK (1954) Male sterility and its importance in breeding heterosis varieties. Euphytica 3:117–124

Kley Van der FK (1955) The occurrence and rates of reproduction of various male sterility genes. Genetica (The Hague) 27:453–464

Kloen D (1964) On the occurrence of male sterility in West European fodder and sugarbeets. Euphytica 13:268–272

Klyuchko PF, Belousov AA (1972) Results of a genetic study of cytoplasmic male sterility in winter rye. Genetika 8(7):9–15

Knapp E (1955) Zur plasmonisch kontrollierten Pollensterilität der Zuckerrüben. Züchter 25: 231–236

Knapp E (1956) Beta Rüben der Zuckerrüben. Handb Pflanzenzücht 3:196–284

Knapp E (1969) Zur Genetik der plasmatisch kontrollierten Pollensterilität der Zuckerrübe. Int Inst Res Bull 4:145–159

Knowles G (1953) Selective control of wild oats in cereal crops by maleic hydrazide. Can J Agric Sci 33:402

Knowlton HE (1924) Pollen abortion in the peach. Proc Amer Soc Hort Sci 21:67–70

Knox RB, Heslop-Harrison J (1966) Control of pollen fertility through the agency of the light regime in the grass *Dichanthium aristatum*. Phyton (Horn) 11:257–267

Knuth P (1898–1905) Handbuch der Blütenbiologie I–III. Engelmann, Leipzig

Knuth P (1909) Handbook of flower pollination VIII. Clarendon, London

Ko T, Yamagata H (1980) Studies on the utility of artificial mutations in plant breeding XII. Induction and gene analysis of male-sterile mutants in rice. Jpn J Breed 30:367–374

Kobabe G (1958) Entwicklungsgeschichtliche und genetische Untersuchungen an neuen männlich sterilen Mutanten der Küchenzwiebel. Z Pflanzenzücht 40:353–384

Kobabe G (1965) Möglichkeiten zur züchterischen Verbesserung von Schnittlauchsorten. Gemüse 1:171–172

Kobabe G (1967) Männliche Sterilität bei Spinat (*Spinacia oleracea* L.). Z. Pflanzenzücht 57: 114–122

Kobabe G (1972) Die Vererbung der männlichen Sterilität beim Spinat (*Spinacia oleracea* L.) und Möglichkeiten der Nutzung dieser Eigenschaft in der Hybridzüchtung. Z Pflanzenzücht 67:233–242

Kobayashi M, Tsunewaki K (1980) Utilization of the alien cytoplasms in wheat breeding sect C. Production of an effective haploid inducer utilizing the *kotschyi* and *variabilis* cytoplasms. In: Tsunewaki K (ed) Genetic diversity of the cytoplasms in *Triticum* and *Aegilops*. Jpn Soc Promotion Sci Publ Japan, pp 250–266

Kobyliansky VD (1969) A contribution to the genetics of cytoplasmic male sterility in winter rye. Genetika 5(9):43–46

* Kobyliansky VD (1971) The production of sterile analogues of winter rye sterility maintainers and fertility restorers. Tr Prikl Bot Genet Sel USSR 44:76–84

Kobyliansky VD, Katerova AG (1973a) Heredity of cytoplasmic and nuclear male sterility in diploid winter rye. Genetika 9(7):5–11

* Kobyliansky VD, Katerova AG (1973b) Mikrosporogieniez u Form rzi z citoplazmaticzieskoj i Jadiernoj muzskoj stierilnostiu. Byull Wsies Inst Rastien USSR 31:43–51

* Kobyliansky VD, Artemova GV, Katerova AG (1980) Inheritance of cytoplasmic male sterility in rye. Tr Prikl Bot Genet Set USSR 67:130–133

Koch H, Peters R (1953) Neue Gesichtspunkte der Rapszüchtung. Wiss Z Martin Luther Univ Halle 2:363–367

Koehn ML (1977) USDA agricultural statistics. US Gov Print Office Washington DC 32:186–198

Koelling PD, Kenworthy WJ, Caron DM (1981) Pollination of male sterile soybeans in caged plots. Crop Sci 21:559–561

Koeppe DE, Cox JK, Malone CP (1978) Mitochondrial heredity: a determinant in the toxic response of maize to the insecticide methomyl. Science 201:1227–1229

Kohel RJ (1973) Genetic nomenclature in cotton. J Hered 64:291–295

Kohel RJ, Richmond TR (1963) Tests for cytoplasmic-genetic interaction involving a genetic male-sterile stock of the genotype ms_2ms_2. Crop Sci 3:361–362

Kolobaeva EA (1974) Features of meiosis in an asynaptic form of tetraploid rye, *Secale cereale*. Biologiya USSR 1:135–141

Kölreuter DJG (1763) Vorläufige Nachricht von einigen das Geschlecht der Pflanzen betreffenden Versuchen und Beobachtungen. Fortsetzung 1. Ostwalds Klassiker der Exakten Wissenschaften Nr. 41, Engelmann, Leipzig

Koncz CS, Sain B (1980) Homology between rRNA of *Escherichia coli* and mitochondrial DNA of maize. FEBS Lett 109:141–144

Koncz CS, Kálmán L, Vargha J (1980) Key for classification of cytoplasmic male sterile types in maize. Plant Sci Lett 17:317–326

Koncz CS, Sümegi A, Udvardy M, Racsmány M, Dudits D (1981) Cloning of mt DNA fragments homologous to mitochondrial S2 plasmid-like DNA in maize. Mol Gen Genet 183:449–458

Kongtian Z, Hongyi F (1981) Effect of high temperature on fertility of the male sterile lines in *Sorghum*. 1980 Annu Rep Inst Genet Acad Sin, p 120

Kono Y, Takeuchi S, Kawarda A, Daly JM, Konche HW (1980) Structure of the host-specific pathotoxins produced by *Helminthosporium maydis* race T. Tetrahedron Lett 21:1537–1540

Konokov PF, Mokhov AI (1972) Comparison of the biology of flowering and fertilization in sterile and fertile forms of carrot. Nauch Dokl Vyssh Shkoly Biol 12(9):73

Konstantinova LN (1980) Cytological disturbances in anther development following gametocide-induced male sterility in sunflower (*Helianthus annuus*). Tr Prik Bot Genet i Selek USSR 67: 134–140

Konvička O, Nienhaus F, Fischbeck G (1978) Untersuchungen über die Ursachen der Pollensterilität bei *Allium sativum* L. Z Pflanzenzücht 80:265–276

Kool AJ, Bot PVM, Marrewijk van GAM (1983) Analysis of cytoplasmic variations associated with cytoplasmic male sterility in *Petunia hybrida* plants (abstr). 10th Eucarpia Cong, Wageningen, p 115

Koopmans A (1951) Cytogenetic studies on *Solanum tuberosum* L. and some of its relatives. Genetica (The Hague) 25:193–337

Koopmans A (1952) Changes in sex in the flowers of the hybrid *Solanum rybinii* x *S. chacoense.* Genetica (The Hague) 26:359–380

Koopmans A (1954) Changes in sex in the flowers of the hybrid *Solanum rybinii* x *S. chacoense.* Genetica (The Hague) 27:273–285

Koopmans A (1955) Changes in sex in the flowers of the hybrid *Solanum rybinii* x *S. chacoense* III. Data about the reciprocal cross *Solanum chacoense* x *S. rybinii.* Genetica (The Hague) 27:465–471

Koopmans A (1959) Changes in sex in the flowers of the hybrid *Solanum rybinii* x *S. chacoense* IV. Further data from the reciprocal cross *S. chacoense* x *S. rybinii.* Genetica (The Hague) 30: 384–390

Kototani M, Yamada K (1958) On the studies of male sterility in cabbages I. Morphology and histological studies on degeneration in male sterile cabbages. Jpn J Breed 8:66–72

Koul AK, Gohil RN (1970) Causes averting sexual reproduction in *Allium sativum* L. Cytologia 35:197–202

Kovačik A, Kryzanek R (1969) Biological and biochemical analysis of sterile and fertile pollen of sunflower. Genet i Šlecht USSR 5:179–184

Kovačik A, Skaloud V (1973) The pollen sterility in sunflower, *Helianthus annuus.* Genet i Šlecht USSR 9:173–180

* Kovačik A, Skaloud V (1981) Possibilities of the evolution of gene linkage and their use in the determination of the heredity of traits in sunflowers. Vedecke Prace Vyzkumneho Ustavu USSR 21:11–27

Kovalenks VI (1966) Isolation of fixators of cytoplasmic male sterility from half sterile sugar-beet varieties and lines. Genetika 9:118–123

Krantz FA, Becker CL, Finemann ZM (1939) Incidence and inheritance of pollen sterility in the potato. J Agric Res 58:593–601

Kremnina AM (1973) Male sterility in cocksfoot induced by chemical mutagens. Sb nauch rabot VNII Kormoc USSR 6:160–166

Krishnamoorthy HM (1972) Effect of GA_3, GA_{4+7}, GA_5 and GA_9 on sex expression of *Luffa acutangula* var H 2. Plant Cell Physiol 13:381–383

Krishnamoorthy KV, Bhat NR (1959) Staminal teratology in the genus *Nicotiana.* J Indian Bot Soc 38:84–92

Krishnaswamy N, Ayyangar GNR (1941) Chromosomal alteration induced by X-rays in bajra (*Pennisetum typhoides* S and H). J Indian Bot Soc 20:111–117

Krishnaswamy N, Ayyangar GNR (1942) Certain abnormalities in millets induced by X-rays. Proc Indian Acad Sci B 16:1–9

Kromer GW, Gazelle SA (1977) Situation and outlook-Sunflower seed. Econ Res Serv USDA FOS-288:22–24

Krug CA, Bacchi O (1943) Triploid varieties of *Citrus.* J Hered 34:277–283

Krupnov VA (1968) Genic male sterility in common wheat. Genetika 4(10):28–35

Kubicki B (1965) Investigations on sex determination in cucumbers (*Cucumis sativus*) I. The influence of α-napthealeneacetic acid and gibberellin on differentiation of flowers in monoecious cucumbers. Genet Pol 6:153–176

Kubicki B (1969a) Investigations on sex determination in cucumber (*Cucumis sativus* L.) III. Variability of sex expression in the monoecious and gynoecious lines. Genet Pol 10:3–22

Kubicki B (1969b) Investigations on sex determination in cucumber (*Cucumis sativus* L.) IV. Multiple alleles of locus ACR. Genet Pol 10:23–68

Kubicki B (1969c) Investigations on sex determination in cucumber (*Cucumis sativus* L.) V. Genes controlling intensity of femaleness. Genet Pol 10:69–86

Kubicki B (1969d) Investigations on sex determination in cucumber (*Cucumis sativus* L.) VI. Androecism. Genet Pol 10:98–99

Kubicki B (1969e) Investigations on sex determination in cucumber (*Cucumis sativus* L.) VII. Andromonecism and hermaphroditism. Genet Pol 10: 101–121

Kubicki B (1969f) Investigations on sex determination in cucumber (*Cucumis sativus* L.) III. Trimonoecism. Genet Pol 10:123–143

Kubicki B (1969g) Sex determination in muskmelon (*Cucumis melo* L.). Genet Pol 10:145–165

Kubicki B (1969h) Comparative studies on sex determination in cucumber (*Cucumis sativus* L.) and muskmelon (*Cucumis melo* L.). Genet Pol 10:167–184

Kubicki B, Potaczek H (1972) Gibberellin-induced perfect flowers in a tomato female form and the possibility of obtaining gynoecious lines for hybrid seed production. Genet Pol 13:67–74

Kubicki B, Kubicka H, Niemirowicz-Szczytt K (1981) Stamenless, multipistillate form of rye (*Secale cereale* L.). Genet Pol 22:411–417

Kuckuck H, Schick R (1930) Die Erbfaktoren bei *Antirrhinum majus* und ihre Bezeichnung. Z Vererbungsl 56:51–83

Kuleshov G, Kremnina A, Korovina V (1980) Particularities regarding the manifestation of cytoplasmic male sterility in *Dactylis glomerata* L. In: Proc 13th Int Grassland Cong, pp 427–430

Kumar J, Verma MM, Singh K, Gagneja MR (1976) Effectiveness of ethrel as an androcide in barley. Crop Improv India 3:39–42

Kumar LSS (1937) Sterility in cotton. J Univ Bombay 5:1–18

Kumar LSS, Abraham A (1941) A cytological study of sterility in sesame. Indian J Genet Plant Breed 1:41–60

Kumar LSS, Joshi WV (1939) Experiments on the effect of X-rays on *Pennisetum typhoideum* Rich. Indian J Agric Sci 9:675–684

Kumar LSS, Rao DS (1945) Inheritance of sterility in *Solanum indicum* L. Indian J Genet Plant Breed 5:58–59

Kumar R, Singh G (1983) Gamma-rays and EMS induced male-sterile mutants in *Hordeum vulgare*. J Hered 74:479–480

Kumar S (1971) Male sterility in flax- a multiple allelic interaction. Agra Univ J Res Sci 20:7–10

Kumar S, Singh SP (1963) Chemical induction of male sterility in sesame (abstr) In: Proc 50th Ind Sci cong III, p 12

Kumar S, Singh SP (1970) Inheritance of male-sterility in some introduced varieties of linseed (*Linum usitatissimum* L.). Indian J Agric Sci 40:184–191

Kumar S, Singh SP (1972) Inheritance of partial male fertility in linseed (*Linum usitatissimum* L.). Indian J Agric Sci 42:34–38

Kung SD (1977) Expression of chloroplast genomes in higher plants. Ann Rev Plant Physiol 28:401–437

Kung SD, Zhu YS, Chen K, Shen GF, Sisson V (1981) *Nicotiana* chloroplast genome II. Chloroplast DNA alteration. Mol Gen Genet 183:20–24

Künzel G (1983) Differences between genetic and physical centromere distances in the case of two genes for male sterility in barley. Theor Appl Genet 64:25–29

Kuptsov AI (1935) A unisexual female sunflower. Bull Appl Bot Leningrad USSR 14:149–150

Kwasinikow BW, Dolgich ST, Stozarowa IA, Tarasenkow LL (1970) Induzierung von Sterilitätsmutationen bei Gemüsepflanzen. In: Proc Eucarpia Meet Versailles, pp 115–122

Lacadena JR (1968) Hybrid wheat VII. Test on the transmission of cytoplasmic male sterility in wheat by embryo-endosperm grafting. Euphytica 17:439–444

Ladeinde TAO, Watt E, Onajole AAO (1980) Segregation pattern of three different sources of male-sterile genes in *Vigna unguiculata*. J Hered 71:431–432

Lafever HN, Patterson FL (1964) Aneuploids for commercial hybrid oats. Crop Sci 4:635–638

Laibach F, Kribben FJ (1950) Der Einfluss von Wuchstoff auf die Bildung männlicher und weiblicher Blüten bei einer monoezischen Pflanze (*Cucumus sativus* L.). Ber Dtsch Bot Ges 62: 53–55

Laibach F, Kribben FJ (1951) Der Einfluss von Wuchstoff auf das Geschlecht der Blüten bei einer monoezischen Pflanze (*Cucumis sativus* L.). Beitr Biol Pflanz 28:64–67

Laible CA (1974) Chemical methods for pollen control. In: Proc 29th Corn Sorghum Res Conf 29:174–184

Lamm R (1941) Varying cytological behaviour in reciprocal *Solanum* crosses. Hereditas 27: 202–208

Lamm R (1953) Investigations on some tuber bearing *Solanum* hybrids. Hereditas 39:97–112
Lamprecht H (1933) Ein unifoliata-Typus von *Pisum* mit gleichzeitiger Pistilloidie. Hereditas 18: 56–64
Lamprecht H (1945) Intra- and inter-specific genes. Agric Hortic Genet 3:45–60
Lamprecht H (1951) Über partielle und Semisterilität insbesondere bei *Pisum sativum*. Z Pflanzenzücht 30:422–433
Lapinski M (1972) Cytoplasmic-genic type of male sterility in *Secale montanum* Guss. Wheat Inf Serv 35:25–28
* Lapushner D (1978) Hybrid vigour in *Nicotiana* as it relates to the interaction between nuclear and cytoplasmic genetic factors. Ph D Thesis Hebrew Univ Jerusalem Israel
* Lapushner D, Frankel R (1963) The use of genic male sterility in the production of hybrid tomato seed. Hassadeh Isr 44:164–168
* Lapushner D, Frankel R (1964) Hybrid tomato breeding for fresh market and export. Isr Min Agric Ext Serv Sikumin 3:17–18
Lapushner D, Frankel R (1967) Practical aspects and the use of male sterility in the production of hybrid tomato seed. Euphytica 16:300–310
Lapushner D, Frankel R (1979) Rationale and practice of tomato F_1 hybrid breeding and seed production in Israele. Monogr Genet Agrar 4:259–273
Larsen K (1977) Self incompatibility in *Beta vulgaris* L. I. Four gametophytic, complementary S-loci in sugar beet. Hereditas 85:227–248
Larsen K (1981) Male sterility caused by beet yellow virus. Z Pflanzenkr Pflanzensch 87:111–115
Larsen RF, Paur S (1948) The description and inheritance of a functionally sterile flower mutant in tomato and its probable value in hybrid tomato seed production. Proc Amer Soc Hort Sci 52:355–364
Laser KD (1972) A light and electron microscopic study of microsporogenesis in cytoplasmic male sterile *Sorghum bicolor*. Ph D Thesis Iowa State Univ Ames
Laser KD, Lersten NR (1972) Anatomy and cytology of microsporogenesis of cytoplasmic male sterile angiosperms. Bot Rev 38:425–454
Lasure LL, Griffin HD (1975) Inheritance of sex in *Achyla bisexuallis*. Am J Bot 62:216–220
Laterrot H (1976) Utilisation conjointe de la Stérilité male et de la résistance au virus de la mosaique du tabac chez la tomate. Ann Amelior Plant (Paris) 26:517–521
Latypov AZ (1974) Digenic nature of male sterile mutants of winter wheat. Sb Nauch Tr Belovrus Acad USSR 123:3–7
Laughnan JR, Gabay SJ (1973) Mutations leading to nuclear restoration of fertility in S-male sterile cytoplasm maize. Theor Appl Genet 43:109–116
Laughnan JR, Gabay SJ (1975) An episomal basis for instability of S-male sterility in maize and some implications for plant breeding. In: Birky CW, Perlman PS, Byers TJ (eds) Genetics and biogenesis of cell organelles. Ohio State Univ Press Columbus, pp 330–349
Laughnan JR, Gabay SJ (1978) Nuclear and cytoplasmic mutations to fertility in S-male sterile maize. In: Walden DB (ed) Maize breeding and genetics. Wiley, New York, pp 427–446
Laughnan JR, Gabay-Laughnan S (1982) Nuclear control over reversions to male fertility in S male sterile maize. In: Sheridan WF (ed) Maize for biological research. N Dakota State Univ Press, Fargo USA, pp 239–242
Laughnan JR, Gabay-Laughnan S (1983) Cytoplasmic male sterility in maize. Ann Rev Genet 17:27–48
Laughnan JR, Gabay-Laughnan S, Carlson JE (1981) Characteristics of cms-S reversion to male fertility in maize. In: Proc Stadler Symp 13:93–114
Laughnan JR, Gabay-Laughnan S, Carlson JE (1982) Cytoplasmic male-sterile systems in maize and recent approaches to their molecular interpretation. In: Sheridan WF (ed) Maize for biological research. N Dakota State Univ Press Fargo, pp 225–237
Law J, Stoskopf NC (1973) Further observations on ethephon (ethrel) as a tool for developing hybrid cereals. Can J Plant Sci 53:765–766
Law RJ, Brun NA (1974) Symbiotic nitrogen fixation in soybeans I. Effect of photosynthetic source-sink manipulations. Crop Sci 14:11–16

Leaver CJ, Gray MW (1982) Mitochondrial genome organisation and expression in higher plants. Ann Rev Plant Physiol 33:373–402

Leaver CJ, Forde BG, Dixon LK, Fox TD (1982) Mitochondrial genes and cytoplasmically inherited variation in higher plants. In: Slonimski P, Borst P, Attardi G (eds) Mitochondrial genes. Cold Spring Harbour Symp Quant Biol, pp 457–470

Leback P, Vedel F (1981) Sal 1 restriction enzyme analysis of chloroplast and mitochondrial DNAs in the genus *Brassica*. Plant Sci Lett 23:1–9

Lechner L (1956) Die Pferdebohne (*Vicia faba* L.). Handb Pflanzenzücht 4:1–5

Leclercq P (1966) Une stérilité mâle utilisable pur la production d'hybrides simples de tournesol. Ann Amelior Plant (Paris) 16:135–144

Leclercq P (1969) Une stérilité cytoplasmique chez le tournesol. Ann Amelior Plant (Paris) 19: 99–106

Leclercq P (1971) La stérilité mâle cytoplasmique du tournesol I. Premiéres études sur la restauration de la fertilité. Ann Amélior Plant (Paris) 21:45–54

Lederberg J (1952) Cell genetics and heredity symbiosis. Physiol Rev 32:403–430

Lederberg J (1955) Recombination mechanisms in bacteria. J Cell Comp Physiol Suppl 2,45: 75–107

Lee CW, Janick J (1978) Muskmelon hybrid seed production facilitated by ethephon. Hort Science 13:195–196

Lee SLJ (1976) The cytology of pollen abortion in T (Texas) cytoplasmic male sterile corn (*Zea mays*). Ph D Thesis Univ Florida

Lee SLJ, Gracen VE, Earle ED (1979) The cytology of pollen abortion in C-cytoplasmic male-sterile corn anthers. Am J Bot 66:141–148

Lee SLJ, Earle ED, Gracen VE (1980) The cytology of pollen abortion in S-cytoplasmic male-sterile corn anthers. Am J Bot 67:237–245

Lein A (1948) Über alloplasmatischen Roggen (Roggen mit Weizenplasma). Züchter 19:101–108

Lelley J (1966) Befruchtungsbiologische Beobachtungen in Zusammenhang mit der Saatguterzeugung von Hybridweizen. Züchter 36:314–316

* Lemekh IM, Pavlishin MM, Savits'ka EV (1971) Some biological characters of pollen sterile plants of *Triticum aestivum* in realtion to their use for breeding. Nauk Pratsi Zemle Robstva Tvarnnitstva Zakhidn USSR 17:220–225

Lemke CA, Gracen VE, Everett HL (1985) A new source of cytoplasmic male sterility in maize induced by the nuclear gene, *iojap*. Theor Appl Genet 71:481–485

Lesins K, Gillies CB (1972) Taxonomy and cytogenetics of *Medicago*. In: Alfalfa science and technology. Am Soc Agron USA, pp 353–386

Lesley JW, Lesley MM (1939) Unfruitfulness in tomato caused by male sterility. J Agric Res 58:621–630

Lesley MM, Lesley JW (1958) Male sterile mutants in the tomato from seeds treated with the radioactive isotope ^{32}P. Proc Amer Soc Hort Sci 71:339–343

Leung H (1981) Breeding for multiple disease resistance in cytoplasmic male sterile *Brassica campestris* L. MS thesis Univ Wisconsin Madison USA

Leung H, Williams PH (1982) Cytoplasmic male sterile *Brassica campestris* breeding lines with resistance to clubroot, turnip mosaic and downy mildew. Cruciferae Newslett 7:20

Leung H, Williams PH (1983) Cytoplasmic male sterile *Brassica campestris* breeding lines with resistance to clubroot, turnip mosaic and downy mildew. Hort Science 18:774–775

Leung H, Niu X, Erickson EH, Williams PH (1983) Selection and genetics of nectary development in cytoplasmic male sterile *Brassica campestris*. J Amer Soc Hort Sci 108:702–706

Levan A (1935) Zytologische Studien an *Allium schoenoprasum*. Hereditas 22:1–126

Levan A (1942) A gene for the remaining in tetrads of ripe pollen in *Petunia*. Hereditas 28:429–435

Levi A, Ashri A (1975) Ethidium bromide- an efficient mutagen in higher plants. Mutant Res 28:397–404

Levings CS, Pring DR (1976) Restriction endonuclease analysis of mitochondrial DNA from normal and Texas cytoplasmic male sterile maize. Science 193:158–160

Levings CS, Pring DR (1977) Diversity of mitochondrial genomes among normal cytoplasms of maize. J Hered 68:350–354

Levings CS, Pring DR (1979a) Mitochondrial DNA of higher plants. In: Setlow JK, Hollander A (eds) Genetic engineering I. Plenum, New York, pp 205–222

Levings CS, Pring DR (1979b) Molecular basis of cytoplasmic male sterility in maize. In: Scandalois JG (ed) Physiological genetics. Academic Press, New York, pp 171–193

Levings CS, Sederoff RR (1983) Nucleotide sequence of the S-2 mitochondrial DNA from the S-cytoplasm of maize. Proc Natl Acad Sci USA 80:4055–4059

Levings CS, Pring DR, Conde MF, Laughnan JR, Gabay-Laughnan SJ (1979a) The basis for cytoplasmic instability in S-cytoplasm. Maize Genet Newslett 53:83–84

Levings CS, Shah DM, Hu WWL, Pring DR, Timothy DH (1979b) Molecular heterogeneity among mitochondrial DNAs from different maize cytoplasms. In: Cummings D, Borst P, David I, Weissman S, Fox CF (eds) Extrachromosomal DNA (ICN-UCLA symp). Academic Press, New York, pp 63–74

Levings CS, Hu WWL, Timothy DH, Pring DR (1978) Terminal inverted repeats in the unique DNAs associated with the S-cytoplasm. Maize Genet Coop Newslett 52:96–98

Levings CS, Kim BD, Pring DR, Conde MF, Mans RJ, Laughnan RJ, Gabay-Laughnan SJ (1980) Cytoplasmic reversion of cms-S in maize: association with a transpositional event. Science 209:1021–1023

* Levy A (1972) Morphological and physiological characters of tomato varieties (*Lycopersicon esculentum* Mill.) characterised by differential sensitivity to high temperature. MS Thesis Hebrew Univ Rehovot, Israel

Lewis D (1939) Genetical studies in cultivated raspberries I. Inheritance and linkage. J Genet 38:367–379

Lewis D (1941) Male sterility in natural populations of hermaphrodite plants. New Phytol 40: 56–63

Lewis D (1942) The evolution of sex in flowering plants. Biol Rev 17:46–67

Lewis D, Crowe LK (1952) Male sterility as an outbreeding mechanism in *Origanum vulgare* (abstr). Heredity 6:136

Lewis D, Crowe LK (1956) The genetics and evolution of gynodioecy. Evolution 10:115–125

Lewis KR, John B (1963) Chromosome marker. Churchill, London

Lhermitte A (1979) Third cycle thesis, molecular biochemistry. Orleans Univ

Liberal MT (1966) Fertilidade de populacoes em F_1 e F_2, evnolvendo haploides de *Solanum tuberosum* L. e diversas especies diploides. Pesqui Agropecu Brass Brazil 1:165–172

Lichter R (1978) The restoration of male fertility in cytoplasmic male-sterile sugar beet by heat treatment and meristem culture. Z Pflanzücht 81:159–165

Lichter R, Mündler M (1961) Untersuchungen über die Pollensterilität der Küchenzwiebel (*Allium cepa* L.) im besonderen über den Einfluss von Witterung und genetischer Konstitution. Z Pflanzenzücht 45:393–405

Lieberman M (1979) Biosynthesis and action of ethylene. Annu Rev Plant Physiol 30:533–591

Lin SC (1977) Rice breeding in China. IRRI Newslett Philippines 2:27–28

Lin SC, Yuan LP (1980) Hybrid rice breeding in China. In: Innovative approaches to rice breeding. IRRI Publ, Philippines, pp 35–51

Lin SM, Hooker AL (1972) Disease determinant of *Helminthosporium maydis* race T. Phytopathology 62:968–971

Lindqvist K (1960) Inheritance studies in lettuce. Hereditas 46:387–470

Lindström EW (1925) Heritable characters of maize XVI(XXI). A new dominant heredity character-teopod. J Hered 16:135–140

Lindström EW (1933) Heredity radium-induced variations in tomato. J Hered 24:129–137

Linnane AW, Nagley P (1978) Mitochondrial genetics in perspecitve: the derivation of a genetic and physical map of the yeast mitochondrial genome. Plasmid 1:324–345

Linnert G (1955) Cytologische Grundlagen für Sterilitätserscheinungen in der Gattung *Salvia.* Züchter 25:237–241

Linnert G (1958) Kernsteuerte Gynodiözie bei *Salvia nemorosa.* Z Indukt Abstammungs Vererbungsl 89:36–51

Linskens HF (1955) Physiologische Untersuchungen zur Pollenschlauch-Hemmung selbststeriler *Petunien.* Z Bot 43:1–44

Linskens HF (1956) Physiologie der Fortpflanzung und Sexualität. Fortschr Bot 18:329–346
Linskens HF (1964) Pollenphysiologie. Ann Rev Plant Physiol 15:255–270
Linskens HF (1966) Variation in the protein and enzyme pattern during pollen development. Planta (Berl) 69:79–91
Linskens HF (1974) The physiological basis of incompatibility in angiosperms. Biol J Linn Soc Lond Suppl 1, 16:143–152
Linskens HF, Kroh M (1967) Inkompatibilität der Phanerogamen. In: Ruhland W (ed) Encyclopedia of plant physiology 18:506–530
Linskens HF, Schrauwen JAM (1968) Quantitative nucleic acid determinations in the microspore and tapetum fractions of lily anthers. Proc K Ned Akad Wet Sect C 71:267–279
Lippert LF (1975) Heterosis and combining ability in chilli peppers by diallel analysis. Crop Sci 15:323–325
Lippert LF, Hall MC (1961) Gametocidal action of FW-450 on cantaloupe. Proc Am Soc Hortic Sci 78:319–323
Little TM, Jones HA, Clarke AE (1944) The distribution of the male sterility gene in varieties of onion. Herbertia 11:310–312
* Litvinova MK (1973) Cytoembryological investigations of male sterility in carrot. Nauch Tr Vononezh USSR 56:14–18
Litvinova MK, Klyuchnikova ZA, Ermokova NA, Kuznetsov YUP (1980) Study of sterile families of carrot used in breeding for heterosis. Genet Selekt i Semenovod Saratov USSR 89: 97–100
Livers RW (1964a) Fertility restoration and its inheritance in cytoplasmic sterile wheat. Science 144:420
Livers RW (1964b) Seed yield of field grown male-sterile wheats subjected to wind-borne pollen (abstr). Am Soc Agron Madison, USA, p 72
Livers RW (1968) Hybrid wheat release. World Farming 10:32
Livers RW, Heyne EG (1968) Hybrid vigour in hard red winter wheat. In: Finley KW, Shepherd K (eds) Proc 3rd Int Wheat Genet Symp. Aust Acad Sci Canberra, pp 431–436
Lloyd DG (1974) The genetic contribution of individual males and females in dioecious and gynodioecious angiosperms. Heredity 32:45–51
Lloyd DG (1975a) The maintenance of gynodioecy and androdioecy in angiosperms. Genetica (The Hague) 45:325–339
Lloyd DG (1975b) Breeding systems in *Cotula* L. IV. Reversion from dioecy to monoecy. New Phytol 74:125–145
Lloyd DG (1976) The transmission of genes via pollen and ovules in gynodioecious angiosperms. Theor Pop Biol 9:299–316
Lloyd DG (1982) Selection of combined versus separate sexes in seed plants. Am Nat 120:571–585
Loehwing WF (1938) Physiological aspects of sex in angiosperms. Bot Rev 4:581–625
Löhnis MP (1937) Plant development in the absence of boron. Meded Landbouwhogesch Wageningen 41:3–37
Löhnis MP (1940) Histology of boron deficiency in plants. Meded Landbouwhogesch Wageningen 44:3–36
Lona JL (1965) Tests of natural cross-pollination and hybrid seed production. Rep Tomato Genet Coop 15:42–43
Long JD, Weed MB, Fitzgerald DJ, Barrier GE (1973) Chemical control of pollination (abstr). Proc Am Soc Agron Madison USA, pp 69
Longley AE (1926) Relationship of polyploidy to pollen sterility in the genera *Rubus* and *Fragaria*. Mem Hort Soc New York 3:15–17
Longley AE, Clark CF (1930) Chromosome behaviour and pollen production in potato. J Agric Res 41:867–888
Lonsdale DM, Thompson RD, Hodge TP (1981) The integrated forms of the S_1 and S_2 DNA elements of maize male sterile mitochondrial DNA are flanked by a large repeated sequence. Nucleic Acids Res 9:3657–3669

Lonsdale DM, Fauron CMR, Hodge TP, Pring DR, Stern DB (1983) Structural alterations in the mitochondrial genome of maize associated with cytoplasmic male sterility. In: Chater KF, Cullis CA, Hopwood DA, Johnston AWB, Woolhouse HW (eds) Genetic rearrangement. 5th John Innes Symp Woolhouse HW, Croom H, London, pp 183–205

Looney NE (1968) Inhibition of apple ripening by succinic acid, 2,2-dimethyl hydrazide and its reversaal by ethylene. Plant Physiol (Bethesda) 43:1133–1137

Los T (1982) Genetic studies on the pampa-type cytoplasmic male sterility in rye. Genet Pol 23: 17–19

Loseva ZL, Palilova AN, Famichenko NS (1974) Pecularities of the oxidative phosphorylation of cytoplasmic male-sterile plants. CR Acad BSSR 18:275–277

Louis JP (1983) Sexual differentiation in *Mercurialis annua* L. (2n=16): Evidence for a control system of nucleo-cytoplasmic relationships by the study of tyrosyl-tRNAs from developmental mutants of male sterility. Mol Gen Genet 191:473–479

Louis JP, Durand B (1978) Studies with the dioecious angiosperms *Mercurialis annua* L. (2n= 16): correlation between genic and cytoplasmic male sterility, sex segregation and feminizing hormones (cytokinins). Mol Gen Genet 165:309–322

Löve A (1943) A Y-linked inheritance of asynapsis in *Rumex acetosa.* Nature 152:358–359

Lowig E (1928) Sterility in some useful species-*Secale montanum* and *Iris* spp. Flora 23:62–104

Lu YG, Rutger JN (1984a) Anther and pollen characteristics of induced genetic male sterile mutants in rice (*Oryza sativa* L.). Environ Exp Bot 24:209–218

Lu YG, Rutger JN (1984b) Cytological observations on induced genetic male sterile mutants in rice (*Oryza sativa* L.). Sci Sin 27:483–493

* Lubenee PA, Nagovicina AV (1968) Vydelenie muzskosterilnych rastenij i ispolźovanie ich v sozdanii geterozisynch gibridov lijucerny. Vestnik Sel'skochozyi Nauk USSR 2:33–38

Lucken KA, Mann SS (1967) Effects of genotypes and environment on fertility restoration of cytoplasmically male-sterile spring wheat (*T. aestivum*) (abstr). Proc Am Soc Agron Madison USA, p 14

Lupton FGH, Bingham J (1966–67) Winter wheat. In: Plant Breed Inst Cambridge Annu Rep, p 67

* Lysikov VN, Blyandur OV, Kontop AI, Byrka SG, Orlov Y (1969) A study of the possibility of obtaining maize mutants with cytoplasmic male sterility. Kishinev Moldavian Kartja Mod USSR, pp 40–47

Maan SS (1972) Cytoplasmic differentiation among species of *Triticum Aegilops* and *Secale* (abst). Agron Abstr, p 15

Maan SS (1973a) Cytoplasmic variability in *Triticinae.* In: Sears ER, Sears LMS (eds) Proc 4th Int Wheat Genet Symp Missouri Agric Exp Stat Columbia, pp 367–373

Maan SS (1973b) Cytoplasmic male sterility and male fertility restoration system in wheat. In: Proc Symp Genetics and breeding of durum wheat. Univ Bari, pp 117–137

Maan SS (1973c) Cytoplasmic and cytogenetic relationships among tetraploid *Triticum* species. Euphytica 22:287–300

Maan SS (1975a) Cytoplasmic variability and speciation in Triticinae. In: Wali MK (ed) Prairie: a multiple view. N Dakota Univ Press Grand Forks, pp 255–281

Maan SS (1975b) Exclusive preferential transmission of an alien chromosome in common wheat. Crop Sci 15:287–292

Maan SS (1976) Cytoplasmic homology between *Aegilops squarrosa* L. and *Ae cylindrica* Host. Crop Sci 16:757–761

Maan SS (1977a) Cytoplasmic homology between *Aegilops mutica* Boiss and *Ae. ovata* L. Euphytica 26:601–613

Maan SS (1977b) Cytoplasmic control of cross incompatibility and zygotic and sporophytic sterility in interspecific hybrids. In: Proc 8th Cong Eucarpia Madrid Spain, pp 201–214

Maan SS (1978a) Cytoplasmic relationships among the D- and M-genome *Aegilops* species. In: Ramanujam S (ed) Proc 5th Int Wheat Genet Symp New Delhi India, pp 231–260

Maan SS (1978b) Cytoplasmic control of cross-incompatibility and zygotic and sporophytic sterility in interspecific hybrids. In: Proc 8th Eucarpia Cong Madrid, pp 201–214

Maan SS (1980) Cytoplasmic homology between *Aegilops triuncialis* and *Ae. umbellulata*. Can J Genet Cytol 22:197–212

Maan SS, Endo TR (1981) Cytoplasmic homology between diploid *Aegilops uniaristata* and related polyploids: *Ae. crassa* and *Ae. juvenalis*. Cereal Res Commun 9:9–15

Maan SS, Lucken KA (1966) Development and use of aneuploid set of male sterile Chinese spring wheat in *Triticum timopheevii* Zhuk cytoplasm. Can J Genet Cytol 8:398–403

Maan SS, Lucken KA (1967) Additional cytoplasmic male sterility-fertility restoration sytem in *Triticum*. Wheat Inf Serv 23–24:6–9

Maan SS, Lucken KA (1968a) Cytoplasmic male sterility and fertility restoration in *Triticum* L. In: Proc 3rd Int Wheat Genet Symp, pp 135–140

Maan SS, Lucken KA (1968b) Male sterile durum: Interaction of *Triticum boeoticum* and *T. monococcum* cytoplasms and *T. durum* nucleus. Wheat Inf Serv 26:5

Maan SS, Lucken KA (1970) Interaction of *Triticum boeoticum* cytoplasm and genomes of *T. aestivum* and *T. durum*: Restoration of male fertility and plant vigour. Euphytica 19: 498–508

Maan SS, Lucken KA (1971a) Nucleo-cytoplasmic interactions involving *Aegilops* cytoplasms and *Triticum* genomes. J Hered 62:149–152

Maan SS, Lucken KA (1971b) Male sterile wheat with rye cytoplasm: restoration of male fertility and plant vigour. J Hered 62:353–355

Maan SS, Lucken KA (1972) Interacting male-sterility-male fertility restoration systems for hybrid wheat research. Crop Sci 12:360–364

Maan SS, McCracken EU (1968) Meiotic instability of common wheat strains derived from *Triticum timopheevii* zhuk crosses. Euphytica 17:445–450

Macewicz J (1976a) Cytological and genetical studies of male sterility in *Trifolium pratenese* L. I. Genetical characteristics of male sterile plants. Hod Röslin Aklimat Nasien Poland 20: 175–194

Macewicz J (1976b) Cytological and genetical studies of male sterility in *Trifolium pratense* L. II. Analysis of androecium differentiation in male sterile plants. Hod Röslin Aklimat Nasien Poland 20:195–209

Machado VS, Bandeen JD, Stephenson GR, Lavigne P (1978) Uniparental inheritance of chloroplast atrazine tolerance in *Brassica campestris*. Can J Plant Sci 58:977–981

Mac Phee J (1925) Meiotic cytokinesis of *Cannabis*. Bot Gaz 78:335–341

Madej L (1975) Research on male sterility in rye. Hod Röslin Aklimat Nasien Poland 19:421–422

Madej L (1976) The genetic characteristics of three sources of male sterility in rye (*Secale cereale* L.). Hod Röslin Aklimat Nasienn Poland 20:157–174

Madjolelo SDP, Grogan CO, Sarvella PA (1966) Morphological expression of genetic male sterility in maize (*Zea mays* L.). Crop Sci 6:379–380

Madsen GC (1947) Influence of photoperiod on microsporogenesis in *Cosmos sulphureus* Cav. var. *Klondike*. Bot Gaz 109:120–132

Maekawa M, Kinoshita T, Takahashi M (1980) New genes for pollen fertility restoration which interact with boro type male sterile cytoplasm. Genetical studies on rice plant. Mem Fac Agric Hokkaido Univ 12:89–100

Magoon ML, Cooper DC, Hougas RW (1958a) Cytogenetic studies of some diploid *Solanums*, section Tuberarium. Am J Bot 45:207–221

Magoon ML, Cooper DC, Hougas RW (1958b) Induced polyploids of *Solanums* and their crossability with *S. tuberosum*. Bot Gaz 119:224–231

Magoon ML, Hougas RW, Cooper DC (1958c) Cytogenetic studies of tetraploid hybrids in *Solanum* from hexaploid diploid matings. J Hered 49:171–178

Magoon ML, Hougas RW, Cooper DC (1958d) Cytogenetical studies of complex hybrids of *Solanum*. J Hered 49:285–293

Magoon ML, Jos JS, Vasudevan KN (1968) Male sterile cassava. Nucleus (Calcutta) 11:1-6

Mahadevappa M (1973) A cent percent male sterile hybrid in rice at RSS Mandyo offers a great scope for research in inducing desirable characters. Curr Res (Bangalore) 2:79

Mahadevappa LMM, Coffman WR (1980) Karnatka male sterile rice KMS I. IRRI Newslett Philippines 5(3):3

* Makarov AA (1960) Cytoplasmic pollen sterility in the onion (from: Plant Breed Abstr 30:825–826)

* Malaguti G, Mazzani B (1958) Cases of sterility in sesame. Agron Trop Venezuela 8:63–65

Maleci LB, Maugini E (1977) Karyotypes of male-fertile and male-sterile *Iris pallida* Lamm growing in Tuscany. Caryologia 30:237–245

Malecka J (1971) Processes of degeneration in the anthers' tapetum of two male sterile species of *Taraxacum*. Acta Biol Crac Ser Bot Poland 14:1–10

* Malinovsky BN, Zoz NN, Kitaev AA (1973) Induction of cytoplasmic male sterility (CMS) in sorghum by chemical mutagens. Genetika 9:19–27

Mallick EH (1980) Induced cytoplasmic male sterility and fertility restoration and production of hybrid rice. Genet Agrar 34:207–213

Manga V (1977) Multiple carpel mutant in pearl millet. Proc Indian Acad Sci 86:93–96

Mangelsdorf PC (1932) Mechanical separation of gametes in maize. J Hered 23:288–295

Mangelsdorf PC, MacNeish RS, Galinat WC (1964) Domestication of corn. Science 143:538–545

Mann TJ, Jones GL, Matzinger DF (1962) The use of cytoplasmic male sterility in flue cured tobacco hybrids. Crop Sci 2:407–410

Mannella CA, Pittenger TH, Lambowitz AM (1979) Transmission of mitochondrial deoxyribonucleic acid in *Neurospora crassa* sexual crosses. J Bacteriol 137:1449–1451

Mapelli S, Torti G, Badino M, Soressi GP (1979) Effect of GA_3 on flowering and fruit set in a mutant of tomato. Hort Science 14:736–737

Markova MD (1981a) Isoenzyme composition of the NAD^+ dependent glutamate dehydrogenase in cytoplasmic male sterile lines of maize. Dokl Bolg Akad Nauk 34:1177–1180

Markova M (1981b) Enzyme sterility and multiple molecular forms of some enzymes in cytoplasmic male-sterile lines of maize. Genet Plant Breed 14:187–197

Markova M, Daskaloff S (1974) Electrophoretic studies of protein in reproductive tissues of the cytoplasmic male sterile form of pepper (*Capsicum annuum* L.). In: Proc Eucarpia Meet Budapest, pp 105–114

Markova M, Daskaloff S (1976) Biochemical investigations of male sterile mutant forms of pepper *Capsicum annuum* L. Z Pflanzenzücht 77:296–303

Markova M, Tsikov D (1979) Isoenzyme composition of peroxidase and glucose 6-phosphate dehydrogenease in cytoplasmic male sterile tobacco lines. Genet Selek Sofia 12:185–193

* Márkus F (1968) Kiserletk a füszerpaprika himsterilitásanak mestérges elöidézésére vegyszeres kezekéssel. A Duna-Tisza Közi mezögazdasagi Kiserleti Intézet Bulletinjeböl 2:27–36

Marrewijk GAM van (1968) Plasmatische mannelijke steriliteit en fertilitatsh erstel bij de petunia. Meded Landouwhogesch Wageningen 68(16), pp 129

Marrewijk van GAM (1969) Cytoplasmic male sterility in *Petunia* I. Restoration of fertility with special reference to the influence of environment. Euphytica 18:1–20

Marrewijk van GAM (1970) Cytoplasmic male sterility in petunia II. A discussion on male sterility transmission by means of grafting. Euphytica 19:25–32

Marsden-Jones EM, Turrill WB (1925) Studies in *Ranunculus* III. Further experiments concerning sex in *R. acris*. J Genet 31:363–378

Marsden-Jones EM, Turill WB (1957) The bladder compions. Royal Soc Lond (Publ, pp 29)

Marshall DR, Thomson NJ, Nicholls GH, Patrick CM (1974) Effects of temperature and day length on cytoplasmic male sterility in cotton (*Gossypium*). Aust J Agric Res 25:443–447

Marston ME (1964–65) The morphology of a *Streptocarpus* hybrid and its regeneration from leaf cuttings. Sci Hortic 17:114–120

Martin FW (1961a) The inheritance of self-incompatibility in hybrids of *Lycopersicon esculentum* Mill x *L. chilense* Dun. Genetics 46:1443–1454

Martin FW (1961b) Complex unilateral hybridization in *Lycopersicon hirsutum*. Proc Natl Acad Sci USA 47:855–857

Martin JA (1949) Preliminary study on the influence of F_1 and F_2 seed on the yield of Cayenne peppers (abstr). Proc Assoc South Agric Workshop 46:111

Martin JA, Crawford JH (1951) Several types of sterility in *Capsicum frutescens*. Proc Amer Soc Hort Sci 57:335–338

Martin-Tanguy J, Perdrizet E, Prevost J, Martin C (1982) Hydroxycinnamic acid amides in fertile and cytoplasmic male sterile lines of maize. Phytochemistry 21:1939–1945

Marty DS (1975) Heterosis, combining ability and reciprocal effects for agronomic and chemical characters in sesame. Theor Appl Genet 4:294–299

Mascarenhas JP (1975) The biochemistry of angiosperm pollen development. Bot Rev 41:259–314

Mashnekov AS, Golubovskaya IN (1980) Meiotic mutations of maize induced by N-nitroso-N-alkylureas. Genetika 16:1632–1640

Mather K (1940) Outbreeding and separation of the sexes. Nature 145:484–486

Mather K (1973) Genetical structure of populations, Chapman and Hall, London

Mathias R (1985) A new dominant gene for male sterility rape seed, *Brassica napus* L. Z Pflanzenzücht 94:170–173

Matouš J (1962) Über die Möglichkeit der Ausnutzung von Heterosissorten und die Frage der Pollensterilität bei *Cyclamen persicum* Mill. Acta Pruhoniciana 4:13–51

Matouš J (1969) Die cytoplasmatische männliche Sterilität bei *Begonia semperflorens* Hort. und Möglichkeiten ihrer Ausnutzung in der Züchtung. Acta Pruhoniciana 19:1–45

Matsumura S (1953) Improvement of sugar beets by means of triploidy. Science-sha Tokyo (from: Kinoshita T 1971)

Maunder AB, Pickett RC (1959) The genetic inheritance of cytoplasmic-genetic male sterility in grain *Sorghum*. Agron J 51:47–49

May KW, Kasha KJ (1971) Desynapsis in *Phleum nodosum* (abstr). Can J Genet Cytol 13:639

May KW, Kasha KJ (1980) The cytological expression and inheritance of desynapsis in a clone of diploid Timothy (*Phleum nodosum* L.). Euphytica 29:233–240

Maynard-Smith J 1978) Evolution of sex. Cambridge Univ Press

* Mazoti LB, Velasquez RS (1962) Interacciones nuclocitoplasmaticas. Rev Fac Agron 38:1–19

* Mazzani B, Gozalez V, Martinez A (1971) Preliminary test of gametocidal effects of chemicals in sesame. Agron Trop Venezuela 21:39–47

McClintock B (1929) A cytological and genetical study of triploid maize. Genetics 14:180–222

McClure WJ (1962) Frequency of interspecific crossing between *Sorghum vulgare* Pers. and *Sorghum halepense* (L.) Pers, and between *Sorghum vulgare* Pers and *Sorghum almum* Parodi. MS Thesis Oklahoma State Univ

McCollum GD (1966) Occurrence of petaloid stamens in wild carrot (*Daucus carota*) from Sweden. Econ Bot 20:361–367

McCollum GD (1979) Sterility in successive backcrosses of *Raphanobrassica* (2n=4x=36) with recurrent *Brassica oleracea* (2n=2x=18). Can J Genet Cytol 21:479–485

McCollum GD (1981) Induction of an alloplasmic male sterile *Brassica oleracea* by substituting cytoplasm from early scarlet globe raddish (*Raphanus sativus*). Euphytica 30:856–859

McCollum GD (1982) Experimental hybrids between *Allium fistulosum* and *A. royelei*. Bot Gaz 143:238–242

McCuiston WL (1968) The inheritance for restorer factors in two lots of Nebraska 542437 winter wheat. Ph D Thesis Oklahoma State Univ, pp 80

McGregor SE (1976) Insect pollination of cultivated crop plants. USDA Handbook US Gov Print Press, Washington, pp 496

McKelvie AD (1963) Studies in the induction of mutations in *Arabidopsis thaliana* (L) Heynh. Rad Bot 3:105–124

McKinney HH, Greeley LW (1965) Biological characteristics of barley stripe mosaic virus strains and their evolution. Techn Bull USDA, p 1324

McIlrath WJ (1957) The use of maleic hydrazide for the production of male sterility in grain sorghum. In: Zukel JW (ed) A literature summary on maleic hydrazide. Naugatuck chemical division, US Rubber Conn, pp 106

McMurray AL, Miller CH (1968) Cucumber sex expression modified by 2-chloroethanephosphonic acid. Science 162:1396–1397

McMurray AL, Miller CH (1969) The effect of 2-chloroethylphosphonic acid (ethrel) on the sex expression and yields of *Cucumis sativus.* J Amer Soc Hort Sci 94:400–402

McNay JW, Pring DR, Lonsdale DM (1983) Polymorphism of mitochondrial DNA 'S' regions among normal cytoplasms of maize. Plant Mol Biol 2:177–187

McRae DH, Usdin VR (1958) Preliminary studies with C14-lebelled FW-450 on cotton. In: Proc 11th Annu Cotton Improv Conf Houston, pp 59–66

McWhirter KS (1969) Cytoplasmic male sterility in *Desmodium.* Aust J Agric Res 20:227–241

Medgyesy P, Golling R, Nagy F (1985) A light sensitive recipient for the effective transfer of chloroplast and mitochondrial traits by protoplast fusion in *Nicotiana.* Theor Appl Genet 70: 590–594

* Meer van der QP (1978) Results of recent investigations on male sterility in onion (*Allium cepa* L.). Biuldyn Warzwniczy 22:73–79

Meer van der QP, Bennekom van JL (1969) Effect of temperature on the occurrence of male sterility in onion (*Allium cepa* L.). Euphytica 18:389–394

Meer van der QP, Bennekom van JL (1970) Failure of graft transmission of male-sterilizing substance in onion (*Allium cepa* L.). Euphytica 19:430–432

Meer van der QP, Bennekom JL (1971) Frequencies of genetical factors determining male sterility in onion (*Allium cepa* L.) and their significance for the breeding of hybrids. Euphytica 20: 51–56

Meer van der QP, Bennekom van JL (1973) Gibberellic acid as a gametocide for the common onion (*Allium cepa* L.) Euphytica 22:239–243

Meer van der QP, Bennekom van JL (1978) Effect of temperature on sex expression in onion (*Allium cepa* L.). Neth J Agric Sci 26:41–44

Meer van der QP, Van-dam R (1979) Gibberellic acid as a gametocide for cole crops. Euphytica 28:717–722

Mehetre SS, Thombere MV (1981) Stages of pollen abortion in male sterile stocks of *Gossypium hirsutum* L. cotton. J Maharashtra Agric Univ 6:159–161

Melchers G, Sacristân MD, Holder AA (1978) Somatic hybrid plants of potato and tomato regenerated from fused protoplasts. Carlsberg Res Commun 43:203–218

Melville R (1964) Continuation to Burrows and Hobbs paper: gynodioecy in New Zealand *Gentiana.* Nature 203:204

Menczel L, Nagy F, Lazar G, Maliga P (1983) Transfer of cytoplasmic male sterility by selection for streptomycin resistance after protoplast fusion in *Nicotiana.* Mol Gen Genet 189:365–369

Mendiburu AO, Peloquin SJ (1977) The significance of 2n gametes in potato breeding. Theor Appl Genet 49:53–61

Menon PM (1958) Studies in cytoplasmic inheritance in *Pennisetum typhoides* Sand H. Ph D Thesis Univ Madras India

Menon PM (1959) Occurrence of cytoplasmic male sterility in pearl millet (*Pennisetum typhoides* S and H). Curr Sci (Bangalore) 28:165–167

Mepham RH, Lane GR (1970) Observations on the fine structure of developing microspores of *Tradescantia bracteata.* Protoplasma 70:1–20

Mese MD, Azzini LE, Rutger JN (1984) Isolation of genetic male-sterile mutants in a semidwarf Japonica rice cultivar. Crop Sci 24:523–525

Meshram LD, Narkhede MN (1982) Natural male sterile mutant in hot chilli (*Capsicum annuum* L.). Euphytica 31:1003–1005

Mesquida J, Renard M (1981) Pollisation due colza d'hiver male sterile (*Brassica napus* L. var oleifera Metzger) par l'abeille domestique (*Apis mellifica* L.). Effects sur la phenologie et le rendement. Apidologie 12:345–362

Meyer JR, Thomas RO (1958) Cotton gametocide experiments. In: Proc 11th Annu Cotton Improv Conf Houston, pp 84–87

Meyer JR, Meyer VG (1961) Cytoplasmic male sterility in cotton (abstr). Genetics 46:883

Meyer JR, Roux JB, Thomas RO (1957) Some effects of the gametocide FW-450 on cotton. Missouri Agric Exp Stat, pp 26

Meyer JR, Roux JB, Thomas RO (1958) A preliminary report on the induction of male sterility in cotton by maleic hydrazide. Missouri Agric Exp Stat, Inf Sheet, p 589

Meyer VG (1965) Cytoplasmic effects on anther numbers in interspecific hybrids of cotton. J Hered 66:292–294
Meyer VG (1966a) The effect of *Gossypium arboreum* cytoplasm on expression of the gene for external ovules. J Hered 57:72
Meyer VG (1966b) Environmental effects on the differentiation of abnormal cotton flowers. Am J Bot 53:976–980
Meyer VG (1966c) Flower abnormalities. Bot Rev 32:165–218
Meyer VG (1967) Multiple carpel of cotton a mutant regulator gene? J Hered 58:218–219
Meyer VG (1968) An association between male sterility and fibre strength in hybrids of upland cotton. Crop Sci 8:117–118
Meyer VG (1969) Some effects of genes, cytoplasm and environment on male sterility in cotton (*Gossypium*). Crop Sci 9:237–242
Meyer VG (1972) Cytoplasmic effects on anther numbers in interspecific hybrids of cotton. J Hered 63:33–34
Meyer VG (1973) Registration of sixteen germplasm lines of upland cotton (Reg No GP 3 to GP 18). Crop Sci 13:778–779
Meyer VG, Buffett M (1962) Cytoplasmic effects on external ovule production in cotton. J Hered 53:251–253
Meyer VG, Meyer JR (1964) Cytoplasmic effects of the differentiation of anthers and ovules of cotton. Am J Bot 51:693–696
Meyer VG, Meyer JR (1965) Cytoplasmically controlled male sterility in cotton. Crop Sci 5: 444–448
Mezushima U, Katsuo K (1958) Elimination of self-incompatibility in the common cabbage *B. oleracea* by means of substitution of nucleus. In: Proc 10th Int Cong Genet Montreal 2:191
Mian HR, Kuspira J, Walker GWR, Muntagewerff N (1974) Histological and cytochemical studies on five genetic male sterile lines of barley (*Hordeum vulgare*). Can J Genet Cytol 16: 355–379
Mian HR, Kuspira J, Walker GWR, Munthewerff N (1981) Macromolecular changes and submicroscopic structure in the differentiation of msg 16 male sterile barley anthers. Barley Genet, pp 804–813
Michaelik B (1974) Partial male sterility in carrots. In: Proc 19th Int Hortic Cong Sect 8. Vegetables, pp 721–761
Michaelik B (1978) Stability of male sterility in carrot under different growth conditions. Bull Acad Sci Ser Biol 26:827–832
Michaelis P (1929) Über den Einfluss von Kern und Plasma auf die Vererbung. Biol Zentralbl 49:302–316
Michaelis P (1932) Über die Beziehung zwischen Kern und Plasma bei den reziprok verschiedenen *Epilobium*-Bastarden. Z Indukt Abstammungs Vererbungsl 62:95–102
Michaelis P (1933) Entwicklungsgeschichtlichgenetische Untersuchungen an *Epilobium* II. Die Bedeutung des Plasmons für die Pollenfertilität des *Epilobium luteum-hirsutum*-Bastardes. Z Indukt Abstammungsl Vererb 65:1–71
Michaelis P (1940) Über reziprok verschiedene Sippenbastarde bei *Epilobium hirsutum* II. Über die Konstanz des Plasmons der Sippe Jena. Z Indukt Abstammungs Vererbungsl 78: 223–237
Michaelis P (1943) Untersuchungen an reziprok verschiedenen Artbastarden bei *Epilobium* I. Über die Bastarde verschiedener Sippen der Arten *E. hirsutum* mit *E. parviflorum*, resp. *E. montanum*. Flora 37:1–23
Michaelis P (1949) Über Abänderungen des plasmatischen Erbgutes. Z Indukt Abstammungs Vererbungsl 83:36–85
Michaelis P (1954) Cytoplasmic inheritance in *Epilobium* and its theoretical significance. Adv Genet 6:287–401
Michaelis P (1964) Virus und zytoplasmatische Vererbung. Z Pflanzenzücht 52:333–354
Michaelis P (1968) Beiträge zum Problem der Plastiden-Abänderung IV. Über das Plasma und Plastidenabänderungen auslösende, isotopen (^{32}P) induzierte Kerngen mp_1 von *Epilobium*. Mol Gen Genet 101:257–306

Michaelis P, Dellinghausen von M (1942) Über reziprok verschiedene Sippenbastarde bei *Epilobium hirsutum* IV. Weitere Untersuchungen über die genetischen Grundlagen der extrem stark gestörten Bastarde der *E. hirsutum* Sippen Jena. Z Indukt Abstammungs Vererbungsl 80:373–482

Michaelis P, Michaelis G (1948) Über die Konstanz des Zytoplasmons bei *Epilobium*. Planta (Berl) 35:467–512

Michaelis P, Wertz E (1935) Entwicklungsgeschichtlich-genetische Untersuchungen an *Epilobium* IV. Vergleichende Untersuchungen über das Plasmon von *Epilobium hirsutum, E. luteum, E. montanum* und *E. roseum*. Z Indukt Abstammungs Vererbungsl 70:138–159

Mikami T (1982) Genetical studies on the artificial mutants of the cytoplasmic male sterility in sugar beets. Mem Fac Agric Hollaid Univ 187–223

Mikami T, Kinoshita T, Takahashi M (1974) The genetic relations between pollen restoration and the monogerm character in the crossing with male sterile mutants of sugar beet. Proc Am Soc Sugarbeet Techn 16:75–82

Mikami T, Kinoshita T, Takahasi M (1978) Cytological observations on microspore degeneration in cytoplasmic male sterility induced by gamma irradiation in sugarbeet. Mem Fac Agric Hollaido Univ 11:129–137

Mikami T, Sugiura M, Kinoshita T (1984a) Molecular heterogeneity in mitochondrial and chloroplast DNAs from normal and male sterile cytoplasms in sugarbeets. Curr Genet 8:319–322

Mikami T, Shinozaki K, Sugiura M, Kinoshita T (1984b) Characterisation of chloroplast DNA from sugarbeet with normal and male sterile cytoplasms. Jpn J Genet 59:497–504

Mikami T, Kishima Y, Sugiura M, Kinoshita T (1984c) Chloroplast DNA diversity in the cytoplasms of sugarbeet and its related species. Plant Sci Lett 36:231–235

Miller DA (1961) A study of various chemicals as gametocidal agents in *Sorghum*. Sorghum Newslett 4:22–24

Miller DA, Hittle CN (1960) A study of chemical induction of gametic sterility in alfalfa (abstr). Agron Abstr Am Soc Agron, p 51

Miller DA, Hittle CN (1963) Effect of sodium 2,3-Dichloroisobutyrate on gametic production and morphological characteristics of alfalfa. Crop Sci 3:397–401

Miller DA, Pickett RC (1964) Inheritance of partial male fertility in *Sorghum vulgare* Pers. Crop Sci 4:1–4

Miller JF, Lucken KA (1976) Hybrid wheat seed production methods for North Dakota. Crop Sci 16:217–221

Miller JF, Rogers KJ, Lucken KA (1975) Effect of field production techniques on hybrid wheat seed quality. Crop Sci 15:329–332

Miller JF, Schmidt JW (1970) Inheritance of genes controlling male fertility in two different lines of wheat. Agron Abstr Am Soc Agron, p 23

Miller JF, Thompson TE (1978) Development of open-flower male-sterile flax and the inheritance of male-fertility restoration. In: Proc 47th Ann Flax Inst Conf Fargo, pp 18–21

Miller RJ, Koeppe DE (1971) Southern corn leaf blight: susceptible and resistant mitochondria. Science 173:67–69

Milohvič J (1976) Effect of cytoplasmic male sterility on the quality of seed in wheat. Poljopr Znanstvena Smotra Zagreb 38(48):27–38

* Minina EG (1938) On the phenotypical modification of sexual characters in higher plants under the influence of the condition of nutrition and other external factors. CR Acad Sci USSR 21:298–301

Minina EG, Tylikina LG (1942) Physiological study of the effect of gases upon sex differentiation in plants. Dokl Akad Novk USSR 55:165–168

Minocha JL (1976) Male sterility in crop plants. In: Malik CP, Srivastava AK, Bhattacharya NC, Singh R (eds) Physiology of sexual reproduction in flowering plants. Kalayani Publ Delhi India, pp 196–200

Miri RK, Amawate JS, Jain HK (1970a) NP 839, NP883 and NP 880 new sources of fertility restoration in male sterile wheat. Wheat Inf Serv 31:9–11

Miri RK, Amawate JS, Jain HK (1970b) *T. timopheevi* system of cytoplasmic-genetic male sterility and fertility restoration for the development of hybrid wheat. Indian J Genet Plant Breed 30:383–394

Mitra GC (1976) Male sterility in Sunnhemp (*Crotalaria juncea* L.). Curr Sci (Bangalore) 45: 873–874

Mitra GC (1977) Inheritance of a male sterile complex mutant in white jute (*Corchorus capsilaris* L.). Genetica (The Hague) 47:71–72

Mitsuishi S (1965) Equilibrium system for eu- and aneu-ploid in the tetraploid lines of sugarbeets. Jpn J Breed 15:33–39

Mittal SP, Thomas PA (1961) A functional sterile mutant in tomato variety Meeruti. Indian J Hortic Sci 18:150–151

Mittal SP, Issar SC (1970) Fertility restorer for cytoplasmic male sterility in isaphagul (*Plantago ovata* Forsk). Sci Cult 36:550–551

Mittal SP, Swarup V, Joshi AB (1958) Cytoplasmic male sterility in Jowar (*Sorghum vulgare* Pers). Curr Sci (Bangalore) 27:314

Mittal SP, Thomas TA, Singh HB (1963) Use of male sterility in the exploitation of hybrid vigour. Indian Hortic Sci 19:50–60

Miyazawa B (1932) On the two cases of semisterility in *Oryza sativa.* Bull Miyazaki Coll Agric Forestry 4:193–197

Mizushima U, Katsuo K (1958) Elimination of self incompatibility in the common cabbage *B. oleracea* by means of substitution of nucleus. In: Proc 10th Int Cong Genet Montreal 2:191

Moffett AA (1932) Studies on the formation of multinuclear giant pollen grains in *Kniphofia.* J Genet 25:315–337

Moh CC, Nilan RA (1953) Multiovary in barley: a mutant induced by atomic bomb irradiation. J Hered 44:183–184

Mohan Ram HY, Jaiswal VS (1970) Induction of female flowers on male plants of *Cannabis sativa* L. by 2-chloroethanephosphonic acid. Experientia (Basel) 26:214–216

Mohan Ram HY, Jaiswal VS (1971) Feminization of male flowers of *Cannabis sativa* by a morphactin. Naturwissenschaften 58:149–150

Monosmith HR (1926) Male sterility in *Allium cepa* L. Ph D Thesis Univ Calif

Monteagudo A, Sanchez-Monge E, Lacadena JR (1967) Pollen-restoring ability of some wheats. Wheat Newslett 13:50–51

Moore JF (1959) Male sterility induced in tomato by sodium 2,3-dichloroisobutyrate. Science 129:1738–1740

Moore JF (1964) Male sterility induced in field grown tomatoes with sodium α, β-dichloroisobutyrate. Proc Amer Soc Hort Sci 84:474–479

Moore RH (1950) Several effects of maleic hydrazide on plants. Science 112:52–53

Morelock TE (1974) Influence of cytoplasm source on expression of male sterility in carrot, *D. carota.* Ph D Thesis Univ Wisconsin

Moringa T (1934) Interspecific hybridization in *Brassica* VI. The cytology of F_1 hybrids of *B. juncea* and *B. nigra.* Cytologia 6:62–67

Moss GI, Heslop-Harrison J (1968) Photoperiod and pollen sterility in maize. Ann Bot 32:833–846

Moussel B, Moussel C, Audran JG, Bouillot J, Duc G (1982) La stérilité mâle nucléo-cytoplasmique chez la Féverole (*Vicia faba* L.) III. Répartition an sein des tissus de l'anthére et évolution quantitative au cours de l'ontogenése pollinique de particules associées o'la stérilité-mâe nucléocytoplasmique de type 447. Rev Cytol Biol Vég Bot 5:81–93

Mukai Y (1981) Genetic studies of the wheat haploids induced by alien cytoplasms. Mem Osaka Kyoiku Univ 30:31–55

Mukai Y (1982) Genetic characterization of the cytoplasms of tetraploid *Aegilops* species. Mem Osaka Kyoiku Univ 31:13–34

Mukai Y (1983) Determination of the chromosome arm carrying a male fertility restoring gene against the cytoplasm of *Aegilops uniaristata* in wheat. Mem Osaka Kyoiku Univ 32:43–53

Mukai Y, Tsunewaki K (1976) Genetic diversity of the cytoplasm in *Triticum* and *Aegilops* IV. Distribution of the cytoplasm inducing variegation in common wheat. Theor Appl Genet 48: 9–16

Mukai Y, Tsunewaki K (1979) Basic studies on hybrid wheat breeding VIII. A new male sterility fertility restoration system in common wheat utilizing the cytoplasm of *Aegilops kotschyi* and *Ae. variabilis.* Theor Appl Genet 54:153–160

* Neagum MI, Georgescu M (1971) Contribution to the determination of combining ability in the variety Coarna Rosie. In: Plant Breed Abstr 44:(1974)3469

Negi SS, Olmo HP (1966) Sex conversion in male *Vitis vinifera* L. by a kinin. Science 152:1624–1625

Negi SS, Olmo HP (1972) Certain embryological and biochemical aspects of cytokinin SD 8339 in converting sex of a male *Vitis vinifera* (sylvestris). Am J Bot 59:851–857

Negi SS, Randhawa GS, Maheshwarappa RG, Suresh ER (1974) Some studies on grape hybridization in South India. Indian J Hortic 31:1–8

Nagruk VI, Cherny DI, Nikiforova ID, Aleksandrov AA, Butenko RG (1982) Isolation and characterization of minicircular DNAs found in mitochondrial fractions of *Vicia faba*. FEBS Lett 142:115–117

Nelson OE, Clary GB (1952) Genic control of semisterility in maize. J Hered 43:205–210

Nelson PM, Rossman EC (1958) Chemical induction of male sterility in inbred maize by use of gibberellins. Science 127:1500–1501

Nelson RL, Bernard RL (1979) Pollen movement to male-sterile soybean in southern Illinois. Soybean Genet Newslett 6:100–103

Nettancourt de D (1977) Incompatibility in angiosperms. Springer, Berlin, Heidelberg, New York

Nettancourt de D, Grant WF (1963) The cytogenetics of *Lotus* (Leguminosae) II. A diploid interspecific hybrid between *L. tenuis* and *L. filicaulis*. Can J Genet Cytol 5:338–347

Nettevich ED (1968) The problem of utilizing heterosis of wheat (*Triticum aestivum*). Euphytica 17:54–62

Nettevich ED, Feodorova TN (1966) Cytoplasmic male sterility of wheat obtained by crossing *Triticum zhukovskyi* Men and *Triticum timonovum* Heslot with *Triticum aestivum* L. Genetika 5:82–84

* Nettevich ED, Naumov AA (1970) The genetic characteristics of fertility restoration in wheat forms with cytoplasmic male sterility. Nauch Tr NIIs Tsentr Nechernozemm USSR 25: 77–85

Neuffer MG, Jones L, Zuber MS (1968) The mutants of maize. Crop Sci Soc Madison

Nickerson NH (1959) Sustained treatment with gibberellic acid of five different kinds of maize. Ann Miss Bot Gard 46:19–37

Nickerson NH (1960) Sustained treatment with gibberellic acid of maize plants caryying one of the dominant genes teopod and corn grass. Am J Bot 47:809–815

Nickerson NH, Embler TN (1960) Studies involving sustained treatment of maize with gibberellic acid I. Further notes on response of races. Ann Miss Bot Gard 47:227–242

Nicolas JH (1927) Sterility encountered in rose breeding. Mem Hortic Soc New York 3:55–57

Nielsen CS (1942) Effects of photoperiod on microsporogenesis in Biloxi soyabean. Bot Gaz 104:99–106

Nielsen EL (1956) Cytology and breeding behaviour of natural *Agroelymus hybrid*. Bot Gaz 118:79–88

Nielsen EL (1961a) Cytology, seed set and breeding behaviour of the putative hybrid, *Agroelymus turneri*. Am J Bot 48:340–344

Nielsen EL (1961b) Cytology and fertility of derivatives of *Agroelymus turneri*. Cytologia 26: 137–154

Nielson K, Nemazi A (1967) Selection for the type 0 character in *Beta vulgaris*. Am Sugar Beet Techn 14:368–376

Nieuwhof M (1961) Male sterility in some cole crops. Euphytica 10:351–356

Nieuwhof M (1968a) Identity tests of Ms-genes of Brussels sprouts (*Brassica oleracea* L. var *gemmifera* DC) and cauliflower (*B. oleracea* var *botrytis* L.). Euphytica 17:202–206

Nieuwhof M (1968b) Effect of temperature on the expression of male sterility in Brussels sprouts (*Brassica oleracea* L. var *gemmifera* DC). Euphytica 17:265–273

Nijdam FE (1932) Kruisinger mit *Trifolium pratense* L. Genetica 14:272–278

Nikolov S (1981) Glyceraldehyde-3-phosphate dehydrogenase activity in male sterile tobacco forms. CR Acad Bulg Sci 34:1585–1587

Nikolov S, Tsikov D (1981) An electrophoretic investigation of soluble proteins in male sterile tobacco forms. Genet Plant Breed 14:417–429

Nikova VM (1979) Influence of different resources of cytoplasmic male sterility in tobacco on the development of the pistil. CR Acad Sci Bulgaria 32:209–212

Nishi S, Hiraoka T (1958) Studies on F_1 hybrid vegetable crops (1) Studies on the utilization of male sterility on F_1 seed production I. Histological studies on the degenerative process of male sterility in some vegetable crops. Bull Natl Inst Agric Sci Jpn Ser E 6:1–41

Nishiyama I (1970) Male sterility caused by cooling treatment at microspore stage in rice plant VI. Electron microscopical observations on normal tapetal cells at critcal stage. Proc Crop Sci Soc Jpn 39:474–479

Nishiyama I (1976a) Male sterility caused by cooling treatment at the young microspore stage in rice plants XI. Ultrastructure of tapetal hypertrophy without primary wall. Proc Crop Sci Soc Jpn 45:254–262

Nishiyama I (1976b) Male sterility caused by cooling treatment at the young microspore stage in rice plants XII. Classification of tapetal hypertrophy on the basis of ultrastructure. Proc Crop Sci Soc Jpn 45:270–278

Nishiyama I (1978) Male sterility caused by cooling treatment at the young microspore stage in rice plants XVIII. Some enzyme activities in anthers during and after the cooling. Jpn J Crop Sci 47:551–556

Nishiyama I, Satake T (1979) Male sterility caused by cooling treatment at the young microspore stage in rice plants XIX. The difference in susceptibility to coolness among spikelets on a panicle Jpn J Crop Sci 47:181–186

Nitzsche W 91971) Cytoplasmatische männliche Sterilität bei Weidelgras (*Lolium* spp). Z Pflanzenzücht 65:206–220

Noack KL (1932) Über *Hypercium*-Kreuzungen II. Beobachtungen an *Hypercium*-Artbastarden. Ber Dtsch Bot Ges 50:256–268

Nooden LD, Rupp DC, Derman BD (1978) Separation of seed development from monocarpic senescence in soybeans. Nature 271:354–356

Nordensköld H (1945–46) Cytogenetic studies in the genus *Phleum.* Acta Agric Suec 1:1–137

North C (1979) Plant breeding and Genetics in Horticulture. MacMillan, London

Nosova PP (1976) Study of heterosis effect in some inter-varietal hybrids of tobacco (abstr), p 30. In: Proc Eucarpia VII Cong Elsevier Sci Publ Co Amsterdam

* Nouy B, Le Guen J (1977) Contribution à l'étude de l'évolution de la stérilité male cytoplasmique en production de féverole hybride. Mém ENSA Rennes, pp 34

Nověak F (1971) Cytoplasmic male sterility in sweet pepper (*Capsicum annuum* L.) II. Tapetal development in male sterile anthers. Z Pflanzenzücht 65:221–232

Nověak F (1972) Tapetal development in anthers of *Allium sativum* L. and *Allium longicuspis* Regel. Experientia (Basel) 28:1380–1381

Nověak F, Betlach J (1970) Meiotic irregularities in pollen sterile sweet pepper (*Capsicum annuum).* Cytologia 35:335–343

Nověak F, Betlach J, Dubovsky J (1971) Cytoplasmic male sterility in sweet pepper (*Capsicum annuum* L.) I. Phenotype and inheritance of male sterile character. Z Pflanzenzücht 65:129–140

Nugent PE, Snyder RJ (1967) The inheritance of floret doubleness, floret center color and plant habit in *Pelargonium hortorum* Bailey. Proc Amer Soc Hort Sci 91:680–690

Nürenberg-Krüger U (1956) Die Beeinflussing des Geschlechts bei zwittrigen Fragaria-Arten durch Bastardierung. Ber Dtsch Bot Ges 68:16–31

Nürenberg-Krüger U (1958) Genetische Untersuchungen an diploiden *Fragaria*-Arten I. Die Kreuzung zwischen *F. vesca* and *F. nilgerrensis,* ihre Auswirkung auf die Morphologie der Pflanzen und die Ausbildung des Geschlechts. Z Indukt Abstammungs Vererbungsl 89: 747–773

Nuttal VM (1963) The inheritance and possible usefulness of functional male sterility in *Solanum melongena* L. Can J Genet Cytol 5:197–199

Oberreuter M (1925) Untersuchung der Pollensterilität der reziprok verschiedenen Epilobiumbastarde. Ber Dtsch Bot Ges 43:47–51

Ockendon DJ, Gates PJ (1976) Reduced pollen viability in the onion (*Allium cepa*). New Phytol 76:511–517

Oczos A (1981) Physiological causes of occurrence of cytoplasmic male sterility in tobacco. Pamietnik Pulawski 74:139–145

Odenbach W (1970) Versuche zur Wiederherstellung der Fertilität pollensteriler hexaploider Weizen I. Timopheevi-cytoplasma induzierte Pollensterilität sowie deren Überwindung unter Freilandbedingungen in Berlin-Dahlem. Einführung und erste Beobachtungen. Z Pflanzenzücht 64:73–89

Oehler E, Ingold M (1966a) New cases of male sterility and new restorer sources in *T. aestivum.* Wheat Inf Serv 22:1–3

Oehler E, Ingold M (1966b) Männliche Sterilität und Wiederherstellung der Fertilität bei den Triticinae. Arch Julius Klaus-Stift Vererbungsforsch Sozialanthropol Rssenhyg 41(2/4):26–36

Oehlkers F (1923) Vererbungsversuche an *Oenothera* II. Z Indukt Abstammungs Vererbungsl 31:201–260

Oehlkers F (1926) Erblichkeit und Zytologie einiger Kreuzungen mit *Oenothera strigosa.* Jahrb Wiss Bot 65:401–446

Oehlkers F (1927) Entwicklungsgeschichte der Pollensterilität einiger *Oenothera.* Z Indukt Abstammungs Vererbungsl 43:265–284

Oehlkers F (1938) Bastardierungsversuche in der Gattung *Streptocarpus* Lindl. I. Plasmatische Vererbung und die Geschlechtsbestimmung von Zwitterpflanzen. Z Bot 32:305–392

Oehlkers F (1940) Bastardierungsversuche in der Gattung *Streptocarpus* Lindl. III. Neue Ergebnisse über die Genetik der Wuchsgestalt und Geschlechtsbestimmung. Ber Dtsch Bot Ges 58: 76–91

Oehlkers F (1941) Bastardierungsversuche in der Gattung *Streptocarpus* Lindl. IV. Neue Untersuchungen über die Geschlechtsbestimmung bei *Streptocarpus.* Z Bot 37:158–182

Oehlkers F (1952) Neue Überlegungen zum Problem der auskaryotischen Vererbung. Z Indukt Abstammungs Vererbungsl 84:213–250

Oehlkers F (1956) Bastardierungsversuche in der Gattung *Streptocarpus* Lindl. IX. Neue Untersuchungen über die Geschlechtsbestimmung bei *Streptocarpus.* Z Indukt Abstammungs Vererbungsl 92:445–456

Oehlkers F (1964) Cytoplasmic inheritance in the genus *Streptocarpus* Lindley. Adv Genet 12: 329–370

Oehlkers F, Ebell U (1958) Bastardierungsversuche in der Gattung *Streptocarpus* Lindley. Z Indukt Abstammungs Vererbungsl 89:559–586

Oehlkers F, Harte C (1943) Über die Aufhebung des Gonen- und Zygotenausfalls bei *Oenothera.* Flora 37:106–124

Oettler G (1985) The effect of wheat cytoplasm on the synthesis of wheat x rye hybrids. Theor Appl Genet 467–471

Offerijns FJM (1938) Meiosis in the pollen mother cells of *Canna glauca* Pure yellow. Genetica (The Hague) 20:59–65

Ogura H (1968) Studies on the new male-sterility in Japanese radish with special reference to the utilization of this sterility towards the practical raising of hybrids seeds. Mem Fac Agric Kogoshima Univ 6:39–78

Ogura H, Etoh T (1975) Manifestation of male sterility in autotetraploids derived from diploid male sterile raddishes (*Raphanus sativus*). Mem Fac Agric Kogoshima Univ 11:59–60

Ohkawa Y (1980) Screening of male sterile cytoplasm in Japanese local varieties of turnip rape, *Brassica campestris* L. Annu Rep Genet Div Natl Inst Agric Sci Tsukuba, pp 28–29

Ohkawa Y, Shiga T (1978) The origin of male sterility inducing cytoplasm and fertility restoring genes in oil seed rape *B. napus* L. II. Cytoplasm of plants having A genome and AB genome. Jpn J Breed 28:174–175

Ohkawa Y, Shiga J (1981) Possibility of hybrid seed production by use of cytoplasmic male sterility in *Brassica campestris.* Chinese cabbage. In: Proc 1st Int Symp Asian Veg Res Dev Center Shanhua Taiwan, pp 301–311

Ohkawa Y, Shiga T, Ishiga T (1979) The origin of male sterility inducing cytoplasm and fertility restoring genes in rape seed *B. napus* L. III. Cytoplasm of plants having A genome introduced from Europe. Jpn J Breed 29:108–109

Ohmasa M (1982) Biochemical studies of cytoplasmic male sterility of maize: relation between cytoplasmic male sterility and mitochondrial enzyme activities. Bull Natl Inst Agric Sci 33: 201–233

Ohmasa M, Watanabe Y, Murata N (1976) A biochemical study of cytoplasmic male sterility of corn: alteration of cytochrome oxidase and malate dehydrogenase activities during pollen development. Jpn J Breed 26:40–50

Ohta T, Matsumura S (1960) Cytological and histological observations on male sterility in sugarbeet. Seiken Zihô 11:78–82

Ohta Y (1961a) The use of cytoplasmic male sterility in *Capsicum* breeding. Seiken Zihô 12:59–60

Ohta Y (1961b) Grafting and cytoplasmic male sterility in *Capsicum*. Seiken Zihô 12:35–39

Ohta Y (1967) Effect of thermoshock and virus inoculation on cytoplasmic sterility in *Capsicum* and *Petunia*. Annu Rep Natl Inst Genet Japan 18:26–32

Ohta Y (1968) Effect of virus inoculation and thermoshock treatment on cytoplasmic male sterility in *Capsicum*. Seiken Ziho 20:63–67

Ohta Y (1970) Cytoplasmic male sterility and virus infection in *Capsicum annuum* L. Jpn J Genet 45:277–283

Ohta Y (1975) Viruses, plasmids or episome, like elements and extrachromosomal inheritance in relation to plant breeding research. Seiken Zihô 25–26:41–54

Okuno S (1952a) On the heritable male sterility occurring in potato (*Solanum tuberosum* L.). Jpn J Genet 27:107–110

Okuno S (1952b) Cytological studies on potatoes with some remarks on genetical experiments II. Jpn J Genet 27:3–21

Olav E (1930) Open pollination vs hand pollination of pollen sterile grapes. New York Agric Exp Stat Techn Bull 162:1–34

Oldemeyer RK (1957) Sugarbeet male sterility. J Am Soc Sugarbeet Techn 9:381–386

Oldemeyer RK, Smith PB (1965) Importance of sugarbeet hybrid development. J Int Inst Sugarbeet Res 1:16–27

Olmo HP (1943) Pollination of the Almeria grape. Proc Amer Soc Hort Sci 42:401–406

Omura M, Akihama T (1980) Male sterility in chestnuts: a tentative plan for the seed propagation of fruit trees. Gamma Field Symp 19:77–89

Oppenheim JD, Frankel OH (1929) Investigations into the fertilization of the Jaffa Orange I. Genetica 11:369–374

Orel LI (1967) A cytological study of maize pollen with cytoplasmic male sterility. Genetika 12:3–11

Osawa I (1912) Cytological and experimental studies in *Citrus*. J Coll Agric Tokyo Imp Univ 4:83–116

Osman H Elg, Yermanos DM (1982) Genetic male sterility in sesame: reproductive characteristics and possible use in hybrid seed production. Crop Sci 22:492–498

Ostenfelds CH (1923) Genetic studies in *Polemonium coeruleum*. Hereditas 4:17–26

Overman MA, Warmke HE (1972) Cytoplasmic male sterility in *Sorghum* II. Tapetal behaviour in fertile and sterile anthers. J Hered 63:227–234

Owen FV (1928) A sterile character in soybean. Plant Physiol (Behtesda) 3:223–226

Owen FV (1942) Male sterility in sugarbeets produced by complementary effect of cytoplasmic and mendelian inheritance (abstr). Am J Bot 29:692

Owen FV (1945) Cytoplasmically inherited male sterility in sugar beets. J Agric Res 71:423–440

Owen FV (1948) Utilization of male sterility in breeding superior yielding sugarbeets. Proc Am Soc Sugarbeet Techn 5:156–161

Owen FV (1949) Interpretation of cytoplasmically inherited male sterility in sugarbeets. In: Proc 8th Int Cong Genet, pp 638–639

Owen FV (1950) The sugarbeet breeder's problem of establishing male sterile populations for hybridization purposes. Proc Am Soc Sugarbeet Techn 6:191–194

Owen FV (1952) Mendelian male sterility in sugarbeets. Proc Am Sugarbeet Techn 10:124–132

Owen FV (1954) Hybrid sugarbeets made by utilizing both cytoplasmic and Mendelian male sterility. Proc Am Soc Sugarbeet Techn 8:64–67

Owens KW, Peterson CE, Tolla GE (1980a) Induction of perfect flowers on gynoecious muskmelon by silver nitrate and aminoethoxyvinylglycine. Hort Science 15:654–655

Owens KW, Tolla GE, Peterson CE (1980b) Induction of staminate flowers on gynoecious cucumber by aminoethoxyvinylglycine. Hort Science 15:256–257

Ownbey M, McCollum GD (1953) Cytoplasmic inheritance and reciprocal amphiploidy in *Tragopogon.* Am J Bot 40:788–796

Ozaki K, Tai K (1961) Nitrogen metabolism of paddy rice. Free proline in the pollen. Soil Plant 6:184–185

Pakendorf KW (1970) Male sterility in *Lupinus mutabilis* Sweet. Z Pflanzenzücht 63:227–236

Paleg L, Aspinali D (1958) Inhibition of the development of barley spikes by gibberellic acid. Nature 181:1743–1744

Paliwal RL, Hyde BB (1959) The association of a single B chromosome with male sterility in *Plantago coronopus.* Am J Bot 46:460–466

Palmer JD, Shields CR, Cohen DB, Orton TJ (1983) An unusual mitochondrial DNA plasmid in the genus *Brassica.* Nature 391:725–728

Palmer RG (1971) Cytological studies of ameiotic and normal maize with reference to premeiotic pairing. Chromosoma 35:233–246

Palmer RG (1974) Aneuploids in the soybean *Glycine max.* Can J Genet Cytol 16:441–447

Palmer RG (176) Cytogenetics in soybean improvement. In: Proc 6th Soybean Seed Res Conf Am Seed Trade Ass 6:56–66

Palmer RG (1979a) Inheritance of male-sterile female-fertile mutant ms_3. Soybean Genet Newslett 6:63–64

Palmer RG (1979b) Inheritance of male-sterile, female-fertile mutant ms_4. Soybean Genet Newslett 6:64–66

Palmer RG, Albertsen MC (1975) The nature of sterility in ms_1 male sterile mutant. Soybean Genet Newslett 2:15–16

Palmer RG, Hadley HH (1968) Interspecific hybridization in *Glycine,* subgenus Leptocyamus. Crop Sci 8:557–563

Palmer RG, Heer HE (1976) Polyploids among the progeny from male sterile (ms_1ms_1) soybeans. Soybean Genet Newslett 3:35–36

Palmer RG, Kaul MLH (1983) Genetics, cytology and linkage studies of a desynaptic soybean mutant. J Hered 74:260–264

Palmer RG, Winger CL (1975) Three independent male sterile mutations at the ms_1 locus. Soybean Genet Newslett 2:16–18

Palmer RG, Winger CL, Albertsen MC (1976) Independent mutations for male sterility at the ms_1 locus in soybeans. Agron Abstr Am Soc Agron, p 59

Palmer RG, Albertsen MC, Heer H (1978a) Pollen production in soybeans with respect to genotype, environment and stamen position. Euphytica 27:427–433

Palmer RG, Winger CL, Albertsen MC (1978b) Four independent mutations at the ms_1 locus in soybeans. Crop Sci 18:727–729

Palmer RG, Johns CW, Muir PS (1980) Genetics and cytology of the ms_3 male-sterile soybean. J Hered 71:343–348

Palmer RG, Albertsen MC, Johns CW (1983) Pollen movement to two male sterile soybean mutants grown in two locations. J Hered 74:55–57

Pan KQ, Wang ZX, He JR (1982) Cytological observations on pollen abortion in male sterile rice line Dianrui 409. Yunan Nongye 2:28–30

Panayotov I (1980) New Cytoplasmic male sterility sources in common wheat: their genetical and breeding considerations. Theor Appl Genet 56:153–160

Panayotov I, Gostov K (1973) Interactions between nucleus of *Triticum aestivum* L. and cytoplasms of certain species of *Triticum* and *Aegilops.* In: Proc 4th Int Wheat Genet Symp, pp 381–383

Panayotov I, Gotsov K (1975) Results of nucleus substitution in *Aegilops* and *Triticum* species by means of successive backcrosses with common wheat. Wheat Inf Serv 40:20–22.

Panayotov I, Gotsov K (1976) Interaction between *Aegilops* cytoplasms and *Triticum* genomes and evolution of *Aegilops.* Cereal Res Commun 4:297–306

Panayotov I, Popov P, Gotsov K (1975) Interactions between the nucleus of *Triticum aestivum* L. and the cytoplasms of some *Aegilops* and *Triticum* species. Genet Sel 8:157–165

Panchenko AY (1966) Methods of producing planting seed and yearly varietal renovation. In: Proc 2nd Int Sunflower Conf Manitoba, pp 27–36

Parkey W (1957) Cytoplasmic influence in the production of the pistillate sex expression in castorbeans. Agron J 49:427–428

Parmar KS, Siddiq EA, Swaminathan MS (1979) Chemical induction of male sterility in rice. Indian J Genet Plant Breed 39:529–541

Parmar KS, Siddiq EA, Swaminathan MS (1981) Evaluation of known and new sources of cytoplasmic male sterility-restorer systems in cultivated rice, *Oryza sativa* L. Z Pflanzenzücht 86:1–10

Paschal EH, Wilcox JR (1975) Heterosis and combining ability in exotic soybean germplasm. Crop Sci 15:344–349

Paszkowska I (1980) Developing heterotic varieties of bush tomatoes. Akad Landwirt DDR 168:397–400

Pare JB, Duncan EN (1960) Evaluation of sodium 2,3-dichloroisobutyrate as a selective male gametocide in cotton. Agron J 52:506–508

Pate JB, Joyner JF (1958) The inheritance of a male sterility factor in kenaf *Hibiscus cannabinus* L. Agron J 50:402–403

Patel JS, John CM, Sheshadri CR (1936) The inheritance of characters in the groundnut, *Arachis hypogaea.* Proc Indian Acad Sci 3B:214–236

Patil AB, Singh BB (1976) Male sterility in soybean. Indian J Genet Plant Breed 36:238–243

Patil BD, Joshi MG (1962) Sterility in pearl millet (*Pennisetum typhoids* Stapf and Hubb). Curr Sci (Bangalore) 32:38

Patil JA, Jadhav AS, Rane MS (1973) Male sterility in Mahrashtra onion (*Allim cepa* L.). Res J Mahatma Phule Agric Univ 4:29–31

Patterson EB (1971) Proposed procedures for the use of genic male sterility in hybrid maize production. Illinois Corn Breedeers School Mimeo, pp 24

Patterson EB (1973) Genic male sterility and hybrid maize production. In: Proc 7th Meet Maize Sorghum Sect C. Eucarpia Zagreb, pp 9–14

Patterson EB (1975) An alternative approach for producing corn hybrids. Illinois Res 17(3):16–18

Patterson FL, Bitzer MJ (1966) Hybrid wheat research at Purdue university. Purdue Univ Res Prog Rep, p 222

Paun L (1974) The cytologic mechanism of male sterility in sunflower. In: Vranceanu AV (ed) Proc 6th Int Sunflower Conf Buchares, pp 17–25

Pavithran K (1981) Genetics of male sterility and nude panicle in rice (abstr). In: Proc 4th Int Sabrao Cong Kuala Lumpur, p 6

Pavithran K, Mohandas C (1976) Genetics of male sterility in rice. J Hered 67:252

Pearson OH (1972) Cytoplasmically inherited male sterility characters and flavour components from the species cross *Brassica nigra* (L) Koch x *B. oleracea* L. J Am Soc Hortic Sci 97: 398–402

Pearson OH (1977) Cytosterile cabbage and broccoli. Veg Improv Newslett 19:2–3

Pearson OH (1981) Nature and mechanisms of cytoplasmic male sterility in plants: a review. Hort Science 16:482–487

Pederson MW (1959) Effect of sodium 2,3-dichloroisobutyrate on alfalfa gametes. Agron J 51: 573–574

Pederson MW, Stucker RL (1969) Evidence of cytoplasmic male sterility in alfalfa. Crop Sci 9:767–770

Peitian W, Jiayu C, Liping X, Lianqing W (1981a) Correlation between purple characters and fertile restorer genes distribution in rice. In: Annu Rep Inst Genet Acad Sin 1980, pp 125–126

Petian W, Jiayu C, Lipng X, Lianqing W (1981b) The distribution rules of cytoplasmic factors and genes concerning pollen fertility in cladogram of rice. In: Ann Rep Inst Genet Acad Sin 1980, pp 126–128

Pelcher LE, Kao KN, Gamborg OL, Yoder OC, Gracen VE (1975) Effects of *Helminthosporium maydis* race T toxin on protoplasts of resistant and susceptible corn (*Zea mays* L.). Can J Bot 53:427–431

Pellew C (1917) Types of segregation. J Genet 6:317–339

Peloquin SJ, Hougas RW (1960) Genetic variation among haploids of the common potato. Am Potato J 37:289–297

Peloquin SJ, Iwanaga M (1981) Restorer genes for genetic-cytoplasmic male sterility in cultivated potatoes (abstr). 65th Ann Am Potato Assoc Canada, p 513

Peloquin SJ, Hougas RW, Gabert AC (1966) Haploidy as a new approach to the cytogenetics and breeding of *Solanum tuberosum.* In: Riley R, Lewis KR (eds) Chromosome manipulation and plant genetics. Oliver and Boyd, Edinburgh, pp 21–28

* Peretyatko NA (1875) Nuclear pollen sterility in sugarbeet. Sel Semenovodstva Ukrainian USSR 31:53–60

Perez AT, Chang TT, Beachell HM, Vergara BS, Marciano AP (1973) Induction of male sterility in rice with ethrel and RH-531. SABRAO Newslett 5:133–139

Perez-Ugalde G, Hougas RW, Peloquin SJ (1964) Fertility of *S. phureja*-haploid *S. tuberosum* F_2 hybrids. Am Potato J 41:256–262

Perrot V, Dommée B, Jacquard P (1982) Etude éxperimentale de la concurrence entre individus d'autofédonation et dallofécondation chez le theym (*Thymus vulgaris* L.). Oecol Plant 3:171–184

Persson A, Rapport L (1958) Gibberellin-induced systemic fruit set in a male sterile tomato. Science 127:816–817

Peterson CE, Andher LD (1960) Induction of staminate flowers on gynoecious cucumbers with gibberellin A_3. Science 131:1673–1674

Peterson CE, Foskett RL (1953) Occurrence of pollen sterility in seed fields of scott county globe onions. Proc Amer Soc Hort Sci 62:443–448

Peterson CE, Weigle JL (1958) A new method for producing hybrid cucumber seed. Michigan Agric Exp Stat Q Bull USA 40:960–965

Peterson DF (1952) Inbred seed production in progeny classification for type O plants in open-pollinated varieties. Proc Am Soc Sugarbeet Techn 7:351–353

Peterson PA (1958) Cytoplasmically inherited male sterility in *Capsicum.* Am Nat 92:111–119

Peterson PA, Flavell RB, Barratt DHP (1974) A simple biochemical assay for Texas cytoplasm in corn by use of *Helminthosporium maydis* race T pathotoxin. Plant Dis Rep 58:777–780

Peterson PA, Flavell RB, Barratt DHP (1975) Altered mitochondrial membrane activities associated with cytoplasmically inherited disease sensitivity in maize. Theor Appl Genet 45:309–314

Petrochenko UA (1964) Dynamics of the activity and quality of catalase in maize pollen. Ukr Bot Zh USSR 21:11–15

Petten FD (1951) The selection of male sterile lines in alfalfa. MSc Thesis British Columbia Univ

Pfahler PL (1968a) Heterosis and homeostasis in *Secale montanum* Guss x *S. cereale* L hybrids I. Vegetative growth and development. Crop Sci 8:663–666

Pfahler PL (1968b) Heterosis and homeostasis in *Secale montanum* Guss x *S. cereale* L. hybrids II. Reproductive characteristics. Crop Sci 8:667–670

Pfeifer RP (1972) A method to produce hybrid barley. Agron Abstr Am Soc Agron, p 17

Phatak SC, Wittwer SH, Honma S, Bukovac MJ (1966) Gibberellin induced anther and pollen development in a stamenless tomato mutant. Nature 209:635–636

Phillips RL (1975) Nuclear male-sterility system of hybrid seed corn production. Maize Genet Coop Newslett 49:118–119

Phillips RL, Springer WD (1972) A cytogenetic method for utilizing nuclear male sterility in hybrid corn production. Maize Genet Coop Newslett 46:124–125

Phillips RL, Burnham CR, Patterson EB (1971) Advantages of chromosomal interchanges that generate haplo-viable deficiency-duplications. Crop Sci 11:525–528

Philouze J (1969) Etude de deux gênes de stérilité mâle chez la tomate: ms 32 et ms 35. Ann Amelior Plant (Paris) 19:443–457

Philouze J (1973) Mapping of bs2 on chromosome 7 and of bs and ms 32 on chromosome 1. Rep Tomato Genet Coop 28:28–29

Philouze J (1974a) Génes marquers lies aux génes de stérilité mâle ms 32 et ms 35 chez la Tomate. Ann Amelior Plant (Paris) 24:77–82
Philouze J (1974b) Utilization pratique d'une lignée mâle-sterile ms 35 chez la Tomate. Ann Amel Plant (Paris) 24:129–144
Philouze J (1976) Les hybrides de la tomate: leur intérêt, les techniques d'hybridation, d'utilisation de la stérilité mâle. Pepinier Hortic Mariach France 164:13–18
Phinney BO (1961) Dwarfing genes in *Zea mays* and their relation to the gibberellins. In: Klein RM (ed) Plant growth regulations. Iowa State Univ Press Ames, pp 489–501
Phipps IF (1928) Heritable characters in maize XXXI. Tassel seed-4. J Hered 19:399–404
Pi CP, Wu KD (1961) The inheritance of cytoplasmic-genic male sterility in sorghum. Sorghum Newslett 4:9–10
Pi CP, Wu KD (1963) The inheritance of cytoplasmic genetic male sterility in sorghum. Bot Bull Acad Sin 4:15–22
Picard J (1960) Données sur l'amélioration de la féverole de printemps *Vicia faba* L. Ann Amelior Plant (Paris):121–153
Picard J, Berthelem P, Duc G, Guen J Le (1981) Male sterility and future prospects for hybrid varieties in *Vicia faba*. In: Proc Int Conf faba beans, Cairo, pp 28–36
Pike LM, Mulkey WA (1971) Use of hermaphrodite cucumber lines in development of gynoecious hybrids. Hort Science 6:339–340
Pike LM, Peterson CE (1969) Gibberellin A_4/A_7 for induction of staminate flowers on the gynoecious cucumber. Euphytica 18:106–109
Piquemal G (1970) How to produce hybrid sunflower seeds by inducing male sterility with gibberellic acid. In: Proc 4th Int Sunflower Conf Memphis Tennessee, pp 127–135
* Pirev MN (1966) A histochemical study of microsporogenesis in pollen-fertile and pollen sterile forms of sunflower. Bull Acad Sci Moldav 11:91–94
Pochard E (1966) Donnes experimentales sur la selection, du piment (*Capsicum annuum* L.). Ann Amelior Plant (Paris) 16:185–197
Pochard E (1970a) Obtention de trois nouvelles mutations de sterilité mâle chez le piment (*Capsicum annuum* L.) par traitements mutagenes appliques a un material monoploide. Eucarpia sec Horticoles Versailles, pp 93–95
Pochard E (1970b) etude comparée de la gametôgênêse chez les plantes normales, les plantes possédant la Sterilité cytoplasmique Peterson et les mutant mâle-stériles mr 9, mc 509, et mc 705 (*Capsicum annuum* L.) In: Proc Eucarpia sec Horticoles Versailles France, pp 169–173
Pochard E (1971–72) Recherche sur le "piments". Stat d'Amel Plant Maraich Montfavet Rapp Act, pp 57–68
Poehlman JM (1959) Breeding field crops. Holt, New York
Poehlman JM (1979) Breeding field crops, 2nd ednt. Avi, Connecticut
Pogorletskii BK (1973) Genetic marking of male sterility in sunflower. Genetika 9(5):23–29
Pogorletskii BK, Barlov VV (1973) Male sterility in sunflower and the possibility of using it in breeding and seed production. Sb Nauch Tr Uses Selekt USSR 10:115–121
Pohl J (1895) Über Variationsweite der *Oenothera lamarckiana*. Österr Bot Z 45:166–169, 205–212
Pokhriyal SC (1977) Revamping bajra seed. Seeds Farms 3:25–28
Pokhriyal SC, Unnikrishnan KV, Singh B, Ramadass C, Patil RR (1976) Combining ability of downy mildew resistant lines in pearlmillet. Indian J Genet Plant Breed 36:403–409
Popov P, Panayotov I, Gotsov K (1974) Fertility restorer sources of *T. timopheevi* cytoplasm and methods for their identification. C R Acad Bulg Sci 27:981–983
Portele E (1883) Mitteilung der Landwirtschaft, see Sartorius (1926)
Porter KB, Tuleen NA (1972a) Pollen fertility restoration of male sterile wheat (*T. aestivum* L.) from factors in rye (*Secale* spp). Agron Abstr Am Soc Agron, p 17
Porter KB, Tuleen NA (1972b) Cross breeding male sterile wheat (*T. aestivum* L.) and rye (*Secale* spp). Crop Sci 12:706–708
Porter KB, Tuleen NA (1976) The performance, fertility and cytology of progeny of male-sterile wheat-rye F_1/triticale hybrids. Proc Int Triticale Symp 1973, Lubbock Texas, pp 73–78
Porter KB, Weise AF (1961) Evaluation of certain chemicals as selective gametocides for wheat. Crop Sci 1:381–382

* Pospisilova J, Jirik J (1979) Differences in the contents of some elements in stable A (S_{msms}) and B (N_{msms}) breeding lines of onion. Bull Vyzkummy Slecht Zelin Olomouc 21–22:3–28
Povilaitis B (1972) Crosses between cytoplasmically inherited male sterile cultivars and short day mutants in *Nicotiana tabacum*. Lighter 42:11–14
Powell JB, Burton GW (1966) A suggested commercial method of producing an interspecific hybrid forage in *Pennisetum*. Crop Sci 6:378–379
Powell JB, Burton GW (1968) Polyembryony in pearl millet *Pennisetum typhoides*. Crop Sci 8: 771–773
Powell JB, Burton GW (1969) Cases of partial and complete male-sterility in pearl millet and their causes. Proc Assoc South Agric Workers 66:83
Powell JB, Taylorson RB (1967) Induced sterility and associated effects of dimethylarsinic acid treatment on pearl millet. Crop Sci 7:670–672
Power JB, Frearson EM, George D, Evans PK, Berry SF, Hayward C, Cocking EC (1976a) The isolation, culture and regeneration of leaf protoplasts in the genus *Petunia*. Plant Sci Lett 7:51–55
Power JB, Frearson EM, Hayward C, George D, Evans PK, Berry SF, Cocking EC (1976b) Somatic hybridization of *Petunia hybrida* and *P. parodii*. Nature 263:500–502
Powers L, Gardner EJ (1945) Frequency of aborted pollen grains and microcytes in Guayule, *Parthenium argentatum* Gray. J Am Soc Agron 37:184–193
Powers Le Roy (1936) The nature of the interaction of genes affecting four quantitative characters in a cross between *Hordeum deficiens* and *H. vulgare*. Genetics 21:398–420
Powling A (1981) Species of small DNA molecules found in mitochondria from sugarbeet with normal and male-sterile cytoplasms. Mol Gen Genet 183:82–84
Powling A (1982) Restriction endonuclease analysis of mitochondrial DNA from sugarbeet with normal and male sterile cytoplasms. Heredity 49:117–120
Powling A, Ellis THN (1983) Studies on the organelle genomes of sugarbeet with male-fertile and male-sterile cytoplasms. Theor Appl Genet 65:323–328
Prakash S (1967) Ph D Thesis Agra Univ India (see Yadav 1974)
Pratt D (1958) Linkage of ten male sterile genes. Tomato Genet Coop Rep 8:25–26
Pring DR (1975) Cytochromes of fertile and cytoplasmic male sterile lines. Plant Physiol (Bethesda) 55:202–206
Pring DR, Levings CS (1978) Heterogeneity of maize cytoplasmic genomes among male sterile cytoplasms. Genetics 89:121–136
Pring DR, Levings CS, Hu WWL, Timothy DH (1977) Unique DNA associated with mitochondria in the "S"-type cytoplasm of male sterile maize. Proc Natl Acad Sci USA 74:2904–2908
Pring DR, Conde MF, Levings CS (1979a) Heterogeneity within the C group of male sterile cytoplasms. Maize Genet Coop Newslett 53:42
Pring DR, Levings CS, Conde MF (1979b) The organelle genomes of cytoplasmic male sterile maize and sorghum. In: Proc 4th John Innes Symp, pp 111–120
Pring DR, Conde MF, Levings CS (1980) DNA heterogeneity with the C group of maize male sterile cytoplasms. Crop Sci 20:159–162
Pring DR, Schertz KF, Conde MF (1981) Plasmid-like and other cytoplasmic DNA characteristics of male-sterile sorghum. Sorghum Newslett 24:132
Pring DR, Conde MF, Schertz KF (1982b) Organelle genome diversity in sorghum: male sterile cytoplasms. Crop Sci 22:414–421
Pring DR, Conde MF, Schertz KF, Levings CS (1982a) Plasmid-like DNAs associated with mitochondria of cytoplasmic male-sterile *Sorghum*. Mol Gen Genet 186:180–184
Pritchard AJ, Hutton EM (1972) Anther and pollen development in male-sterile *Phaseolus atropurpureus*. J Hered 63:280–282
Puente de la CF (1966) Fertility of selected 24 chromosome *S. tuberosum* hybrids. MS Thesis Univ Wisconsin, p 38
Pugsley AT, Oram RN (1959) Genic male sterility in wheat. Aust Plant Breed Genet Newslett 14: 10–11
Pundir NS, Singh SP (1965) Induction of male sterility in muskmelon by the use of FW-450. Agra Univ J Res Sci 14:177–184

Stelly DM, Palmer RG (1979) A partially male sterile mutant in soybeans. Soybean Genet Newslett 6:47–49

Stelly DM, Palmer RG (1980a) A partially male-sterile mutant line of soybeans *Glycine max* (L.) Merr. Characterization of the msp phenotype variation. Euphytica 20:539–546

Stelly DM, Palmer RG (1980b) A partially male-sterile mutant line of soybeans, *Glycine max* (L) Merr. Inheritance. Euphytica 29:295–303

Stelly DM Palmer RG (1982) Variable development in anthers of partially male-sterile soybeans. J Hered 73:101–108

Stephens JC (1937) Male sterility in sorghum: its possible utilization in the production of hybrid seeds. J Am Soc Agron 29:690–696

Stephens JC, Holland RF (1954) Cytoplasmic male sterility for hybrid sorghum seed production. Agron J 46:20–23

Stephens JC, Quinby JR (1945) The Ms_2 A V10 linkage group in sorghum. J Agric Res Washington 70:209–218

Stephens JC, Kuykendall JH, George DW (1952) Experimental production of hybrid sorghum seed with a three way cross. Agron J 44:367–373

Stephens WR, Quinby JR (1933) Bulk emasculation of sorghum flowers. J Am Soc Agron 25:233–234

Stevens MA, Rudich J (1978) Genetic potential for overcoming physiological limitations on adaptibility, yield and quality in the tomato. HortScience 13:673–678

Stevenson FJ (1928) Natural crossing in barley. J Am Soc Agron 20:1193–1196

Stewart RL, Wentz JB (1926) A recessive glabrous character in soybeans. J Am Soc Agron 18: 997–1009

Stieglitz H, Stern H (1973) Regulation of B-1,3-glucanase activity in developing anthers of *Lilium*. Dev Biol 34:169–173

Stoilov M (1970) Possibility of producing male sterile forms and restoring fertility in maize by means of gamma-ray and fast neutron irradiation. In: Proc Symp Induced mutations in plants. IAEA FAO, pp 509–515

Storey WB (1953) Genetics of the papaya. J Hered 44:70–78

Stort MNS (1984) Sterility barriers of some artificial F_1 orchid hybrids: Male sterility 1. Microsporogenesis and pollen germination. Am J Bot 71:309–318

Story R (1955) *Streptocarpus kentaniensis*. In: Flowering plants of Africa, plate 30, p 1196

Stoskopf NC, Law J (1972) Some observations on ethrel as a tool for developing hybrid cereals. Can J Plant Sci 52:680–683

Stout AB (1919) Intersexes in *Plantago lanceolata*. Bot Gaz 68:109–133

Stout AB, Clark CF (1924) Sterilities of wild and cultivated potatoes with reference to breeding from seed. USDA Bull, pp 11–95

Stow I (1926) A cytological study on pollen sterility in *Solanum tuberosum* L. Jpn J Bot 3:217–239

Stoyanova Y, Ivanov P (1975) Inheritance of oil and protein content in first hybrid progeny of sunflower. Rasten Novk Bulgaria 12:30–35

Streeter JG (1972) Nitrogen nutrition of field grown soybean plants I. Seasonal variation in soil nitrogen and nitrogen composition of stem exudate. Agron J 64:311–314

Stroman GN (1961) An approach to hybrid cotton as shown by intra- and interspecific crosses. Crop Sci 1:363–366

Stubbe W (1964) The role of the plastome in evolution of the genus *Oenothera*. Genetica (The Hague) 35:28–33

Subramanyam KN, Andal R (1984) Male sterility in sugarcane. Curr Sci (Bangalore) 53:42–43

Suemoto H (1968) The origin of the cytoplasm of tetraploid wheat. Proc Int Wheat Genet Symp 3:141–152

Suginobu K (1979) Studies on the application of male sterility to heterosis breeding of alfalfa (*Medicago sativa* L.). Res Bull Hokkaido Natl Agric Exp Stat 124:1–79

Suginobu K (1980) Utilization of male sterility in forage crops. Gamma Field Symp 19:63–75

Summers DB (1973) A biosystematic study of *Solanum chacoense* Bither and *Solanum commersonni* Dunal. Ph D Thesis Pennsylvania State Univ

Summing BG (1959) The control of growth and development in red clover (*Trifolium pratense* L.). Can J Plant Sci 39:9–24

Suneson CA (1940) A male sterile character in barley. A new tool for the plant breeder. J Hered 31:213–214

Suneson CA (1945) The use of male-sterile in barley improvement. J Am Soc Agron 37:72–73

Suneson CA (1956) An evolutionary plant breeding method. Agron J 48:188–191

Suneson CA (1962) Use of Pugsleys sterile wheat in cross breeding. Crop Sci 2:534–553

Suneson CA, Houston BR (1942) Male sterile barley for study of floral infection. Phytopathology 32:431–432

Suneson CA, Wiebe GA (1962) A Paul Bunyan' plant breeding enterprise with barley. Crop Sci 2:347–348

Suneson CA, Hockett EA, Eslick RF, Stewart VR, Wiebe GA (1970) A gene pool of naked seeded barleys (C.C.XXVIII) for use in food and hybrid breeding programms. Barley Genet 2:32–34

Susa T (1927) Sterility in certain grapes. Mem Hortic Soc New York 3:223–228

Suzuki S (1978) Anther and pollen abnormalities induced by cold treatment and their varietal differences in rice plants. Jpn J Breed 28:21–32

Swaminathan MS, Siddiq EA, Sharma SD (1972) Outlook for hybrid rice in India. In: Rice Breed IRRI Philippines Publ, pp 609–613

Swenson SP (1940) Genetic and cytologic studies of a brachytic mutation in barley. J Agric Res Washington 60:687–713

Tahir CM (1969) Sources of fertility restoring genes for *Ae. ovata* and *T. timopheevi* cytoplasms. Wheat Inf Serv 29:15–17

Tahir CM (1970) Distribution of fertility-restoring genes in hexaploid wheats for *Ae. ovata* and *T. timopheevi* cytoplasm. Jpn J Genet 44:1–9

Tahir CM (1971a) Genetic influence of male-sterile cytoplasms and fertility restoring genes on performance of common wheat varieties. Jpn J Breed 21:189–194

Tahir CM (1971b) Distribution of fertility-restoring genes in hexaploid wheats for *Ae. ovata* and *T. timopheevi* cytoplasms II. Jpn J Breed 21:49–51

Tahir CM, Tsunewaki K (1969) Monosomic analysis of *Triticum spelta* var *duhamelianum*, a fertility-restorer for *T. timopheevi* cytoplasm. Jpn J Genet 44:1–9

Tahir CM, Tsunewaki K (1971) Monosomic analysis of fertility restoring genes in *Triticum aestivum* strain P 168. Can J Genet Cytol 13:14–19

Talaat EH (1969) Chromosomal location of genes controlling pollen fertility restoration in three restorer lines of wheat. Ph D Thesis N Dakota State Univ Fargo

Talaat EH, Maan SS, Lucken KA (1973) Chromosomal location of genes controlling pollen fertility restoration in three restorer lines of wheat. Egypt J Genet Cytol 2:195–205

Takagi Y (1970) Monogenic recessive male sterility in oil rape (*Brassica napus* L.) induced by gamma irradiation. Z Pflanzenzücht 64:242–247

Takahashi M, Kinoshita T (1965) Chromosome numbers in the progeny of the crosses involving diploid, triploid and tetraploid sugar beets. Bull Sugarbeet Res Suppl 5:118–121

Takeshita M, Kato M, Tokumasu S (1980) Application of ovule culture to the production of intergeneric or interspecific hybrids in *Brassica* and *Raphanus*. Jpn J Genet 55:373–387

Takezaki Y (1932) Inheritance of a few sterile plants I. Male sterile rice. Agric and Hortic Jpn 7:387–397

Tamai T, Tokumasu S, Shinohra K (1958) Studies on the breeding of *Pelargonium* species used for the essential oil production I. Artificially induced tetraploid plant in *Pelargonium roseum*. Jpn J Breed 7:131–138

Tamaru N, Kinoshita T, Takahashi M (1980) Observations on pollen development in the cytoplasmic male sterile plants induced by the nuclear substitution. Genetical studies on rice plant LXXIII. Mem Fac Agric Hokkaido Univ Jpn 12:124–128

Tang ZS, Du SG, Wang ZX, Shou Z, Xu ZY, Liu Y (1981) Cytological study of microspore abortion in male-sterile cotton. Sci Agric Sin PR China 5:47–51

Tanksley SD, Rick CM, Vallejos CE (1984) Tight linkage between a nuclear male sterile locus and an enzyme marker in tomato. Theor Appl Genet 68:109–113

Tasei JN (1976) Les insectes pollinasateurs de la feverole d'hiver (*Vicia faba* var *equina* L.) et la pollinisation des plants male steriles en production de semence hybride. Apidologie 7:1–38
Tatebe K (1952) Cytological studies on pollen degeneration in male sterile onions. J Hortic Soc Jpn 21:73–75
Tatlioglu T (1982) Cytoplasmic male sterility in chives (*Allium schoenoprasum* L.). Z Pflanzenzücht 89:251–262
Tatlioglu T (1985) Influence of temperature on the expression of cytoplasmic male sterility in chives (*Allium schoenoprasum* L.) Z Pflanzenzücht 94:156–161
Tatum LA (1971) The southern corn leaf blight epidemic. Science 171:1113–1116
Taylor NL, Kitbamroog C, Anderson MK (1978) Genetic male sterility in red clover. Crop Sci 18: 1033–1036
Terrell EE (1966) Taxonomic implications of genetics in ryegresses (*Lolium*). Bot Rev 32:138–164
Terrell EE (1976) The correct names for pearlmillet and yellow foxtail. Taxon 25:297–304
Terrell EE (1977) A checklist of names of 3,000 vascular plants of economic importance. Agric Handbook 505 USDA Washington, pp 201
Terrell EE, Winters HF (1974) Changes in scientific names for certain crop plants. Hort Science 9:324–325
Thakare RB (1977) Breakdown of male-sterility in pearl millet CMS line Tfit-23A. Crop Improv 4:117–118
Theurer JC (1968) Linkage tests of mendelian male sterility and other genetic characters in sugarbeets, *Beta vulgaris* L. Crop Sci 8:698–701
Theurer JC (1970) Variability in partial male-fertile sugar beet. J Am Soc Sugarbeet Techn 16: 253–263
Theurer JC (1979) Effect of DPX 3778 as a male gametocide in sugarbeets. Can J Plant Sci 59: 463–468
Theurer JC, Mumford C (1976) The effect of sterile cytoplasm on curly top disease resistance. J Am Soc Sugarbeet Techn. 19:45–48
Theurer JC, Ryser GK (1969) Double-cross sugarbeet hybrids utilizing pollen restorers. Crop Sci 9:610–612
Theurer JC, Hecker RJ, Ottley EH (1968) Attempted graft transmission of cytoplasmic male sterility in sugarbeets (*Beta vulgaris* L.). J Am Soc Sugarbeet Techn 14:695–703
Theurer JC, Ryser GK, Stout M (1969) A comparison of genic male-sterile four-way sugarbeet habrids. Crop Sci 9:741–743
Theurer VC, Elling LJ (1964) Comparative performance of diallel crosses and related second-generation synthetics of alfalfa, *Medicago sativa* L. III. Forage yield. Crop Sci 4:25–28
Thibodeau PS, Jaworski EG (1975) Patterns of nitrogen utilization in soybean. Planta (Berl) 127:133–147
Thiellement H (1977a) La stérilité mâle cytoplasmique chez *Vicia faba* L. I. Historique et hypotheses. Ann Amelior Plant (Paris) 27:203–209
Thiellement H (1977b) La stérilité mâle cytoplasmique chez *Vicia faba* L. II. Hétérogenéité phénotypique dune ligne mâle-stérile. Ann Amelior Plant (Paris) 27:547–553
Thiellement H (1977c) La stérilité mâle cytoplasmique chez *Vicia faba* L. III. Hérédité des phénotypes polliniques. Ann Amelior Plant (Paris) 27:555–562
Thiellement H (1977d) De la stérilité mâle cytoplasmique chez *Vicia faba* L. Thése Docteur 3 éme cycle Univ Paris-sud, p 50
Thiellement H (1979a) Sur quelques descendances de plantes présentant simultanément deux phénotypes polliniques, a déterminisme nonmendélien, chez *Vicia faba* L. CR Acad Sci Paris 288 Ser D:827–830
Thiellement H (1979b) Cytoplasmic male sterility in *Vicia faba* L. IV. A grandmother effect. Ann Amelior Plant (Paris) 29:163–171
Thiellement H (1980) Cytoplasmic male sterility in *Vicia faba* L. V. Intermediate phenotypes. Ann Amelior Plant (Paris) 30:123–132
Thiellement H (1982) Cytoplasmic male sterility in *Vicia faba* L. VI. Genetical arguments for cytoplasmic heterogeneity. Theor Appl Genet 61:47–52

Thombre MV (1983) Transfer of male sterility from *Gossypium hirsutum* to *G. barbadense.* Curr Sci (Bangalore) 52:609

Thombre MV, Mehetre SS (1979) Male sterility in American cotton (*Gossypium hirsutum* L.). Curr Sci (Bangalore) 48:172

Thompson D, Walbot V, Coe EH (1983) Plastid differentiation in iojap - and chloroplast mutator affected maize plants. Am J Bot 70:940–950

Thompson DJ (1960) Studies on the inheritance of male-sterility and other characters in the carrot *Daucus carota* L var *sativa.* Ph D Thesis Cornell Univ Ithaca New York

Thompson DJ (1961) Studies on the inheritance of male sterility in the carrot *Daucus carota* L. var *sativa.* Proc Amer Soc Hort Sci 78:332–338

Thompson KF (1972) Cytoplasmic male sterility in oilseed rape. Heredity 29:253–257

Thompson RD, Kemble RJ, Flavell RB (1980) Variation in mitochondrial DNA organisation between normal and male-sterile cytoplasms of maize. Nucleic Acids Res 8:1999–2008

Thompson TE (1975) Negative graft transfer of cytoplasmic male sterility in flax. Proc Flax Inst Conf, USA, pp 17–18

Thompson TE (1977) Cytoplasmic male-sterile flax with open corollas. J Hered 68:185–187

Thompson TE, Axtell JD (1978) Graft induced transmission of cytoplasmic male sterility in alfalfa. J Hered 69:159–164

Thompson TE, Miller JF (1978) Cytoplasmic male-sterility in flax: restorer gene frequency and degree of corolla opening. Crop Sci 18:983–986

Thornbury DW (1979) A comparison of mitochondrial proteins from diverse cytoplasms of maize by gel electrophoresis. Ph D Thesis Univ Florida Gainesville, pp 61

Timin NI, Dobrutskaya EG (1981) Cytoplasmic male sterility under different environmental conditions. Ekol Genet Moldavian USSR 147:1–17

Tomothy DH, Levings CS, Pring DR, Conde MF, Kermicle JL (1979) Systematic variation and evolution of organelle DNAs in the genus *Zea: teosinte.* Proc Natl Acad Sci USA 76:4220–4224

Tokumasu S (1951) Male sterility in Japanese radish (*Raphanus sativus* L.). Sci Bull Fac Agric Kyushu Univ 13:83–89

Tokumasu S (1957) Histological studies on pollen degeneration in male sterile Japanese radish (*Raphanus sativus* L.). Jpn J Breed 6:45–50

Tokumasu S (1970) Comparison of anther and pollen development between male sterile diploids and male fertile tetraploids in *Pelargonium roseum.* Jpn J Breed 20:211–218

Tokumasu S (1974) Expression of male sterility in *Pelargonium crispum* L'Her Ex Ait. Euphytica 23:209–217

Tokumasu S (1976) A comparative study of anther development in male fertile and male sterile plants of *Pelargonium crispum* L'Her Ex Ait. Euphytica 25:151–159

Tokumasu S, Tamai T (1964) Studies on the breeding of *Pelargonium* species used for the essential oil production III. Cytological observations of *Pelargonium denticulatum* and *P. roseum.* Jpn J Breed 14:107–111

Tolla GE, Peterson CE (1979) Comparison of gibberellin A4/A7 and silver nitrate for induction of staminate flowers in a gynoecious cucumber line. HortScience 14:542–544

Towner JW (1961) The inheritance of femina, a male sterile character in *Tagetes erecta.* Proc Am Soc Hortic Sci, AAS, Pacific Davis, p 2

Townley-Smith TF (1969) Inheritance and cytological studies of male sterility in timothy (*Phleum pratense* L.). Forage Notes, Ottawa, vol 14, p 11

Townsend CE (1960a) Sterility in sudangrass I. Barren and chimeral partial sterility. Agron J 52: 153–155

Townsend CE (1960b) Sterility in sudangrass II. Random and non-random partial sterility. Agron J 52:639–642

Toxopeus HJ (1960) Studies on the resistance of tuber bearing solanaceae from Mexico to *Phytophthora infestans.* Euphytica 9:39–56

Toxopeus HJ (1964) Treasure-digging for blight resistance in potatoes. Euphytica 13:206–222

Trees SC, Rutger JN (1978) Inheritance of four genetic male steriles. J Hered 69:270–272

Tripathi DP, Dongre AB, Mehta SL, Rao NGP (1982a) Soluble protein and esterase isoenzyme pattern on isoelectric focussing from seeds and anthers of diverse cytoplasmic genic male sterile sorghums (*Sorghum bicolor* L Moench). Z Pflanzenzücht 88:69–78

Tripathi DP, Mehta SL, Mamprakash S, Rao NGP (1982b) Characterization of diverse sorghum male sterile pollen grains by scanning electron microscopy. Indian J Genet Plant Breed 42: 75–81

Tripathi DB, Mehta SL, Rana BS, Rao NGP (1980) Characterisation of diverse cytoplasmic genetic male steriles in sorghum (*Sorghum bicolor* L. Moench). Sorghum Newslett 23:107–108

Tripathi DB, Mehta SL, Rao NGP (1981) Amino acid in anthers of Milo and in cytoplasmic genetic male sterile sorghum (*Sorghum bicolor* L. Moench) of Indian origin. Theor Appl Genet 59:113–116

Tronichkova E (1962) Ausnutzung der männlichen Sterilität in der Hybridsaatgutproduktion bei *Tomaten*. In: Daskaloff Ch (ed) Heterosis. Academic Press, Sofia, pp 37–44

Tschermak E, Bleier H (1926) Über fruchtbare *Aegilops*-Weizenbastarde (Beispiele für Entstehung neuer Arten durch Bastardierung). Ber Dtsch Bot Ges 44:100–132

Tsikov D, Nikova V (1980) Male sterility in tobacco IV. Seed production. Genet Selek 13:198–205

Tsikov D, Tsikova E (1971) Male sterility in backcrosses of reciprocal amphidiploids of *Nicotiana debneyi* Domm and *N. tabacum* L with *N. tabacum* L. CR Acad Sci Bulgaria 4:17–19

Tsikov D, Tsikova E, Nikova V (1977) Male sterility in tobacco III. On anther feminization. Genet Plant Breed 10:129–140

Tsikov DK, Zagorska NA, Tsikova E (1974) Haploids of cytoplasmic male sterile tobacco obtained from anthers *in vitro*. CR Acad Sci Bulgaria 27:1727–1730

Tsikova E, Abadzhieva M (1974) A comparative study of the development of the tapetal layer in tobacco and tomato with cytoplasmic and nuclear male sterility. Genet Selek 7:449–459

Tsikova E, Georgieva ID, Zagorska N, Yazova M (1981) Cytological and cytochemical study of androgenesis in cytoplasmically male sterile tobacco. Genet Selek 14:3–12

Tsuchiya T (1969) Characteristics and inheritance of radiation-induced mutations in barley. Some extreme mutations. In: Proc Symp Induced mutations in plants. IAEA Vienna, pp 573–590

Tsuchiya T (1972) Report from the international committee of nomenclature and symbolisation of barley genes. Barley Genet Newslett 2:16–20

Tsuda C, Nakashima H, Hosokawa S (1978) Histochemical studies on the cytoplasmic male sterility of some crops V. Changes of phosphorylase activities and inorganic phosphorus in sugar-beet anther. Mem Fac Agric Hokkaido Univ Sapporo 11:182–188

Tsunewaki K (1964) Genetic studies of a 6X-derivative from a 8X *Triticale*. Can J Genet Cytol 6:1–11

Tsunewaki K (1973) Genetic differentiation of the cytoplasms in wheat and *Aegilops*. Genetics 74: s 280–281

Tsunewaki K (1974) Monosomic analysis of two restorers to *Ae. caudata* and *Ae. umbellulata* cytoplasms. Jpn J Genet 49:425–433

Tsunewaki K (1980a) Utilization of alien cytoplasm in wheat breeding. Sect A: Use of *Timopheevi* cytoplasm for hybrid wheat breeding. In: Tsunewaki K (ed) Genetic diversity of the cytoplasm in *Triticum* and *Aegilops*. Jpn Soc for promotion of science Publ Tokyo Japan, pp 211–236

Tsunewaki K (ed) (1980b) Genetic diversity of the cytoplasm in *Triticum* and *Aegilops*. Jpn Soc Promot Sci Publ Tokyo, pp 290

Tsunewaki K, Endo T (1973) Genetic relatedness among five cytoplasms in *Triticum* and *Aegilops*. Proc 4th Int Wheat Genet Symp 1:391–397

Tsunewaki K, Noda K, Fujisawa T (1968) Haploid and twin formation in a wheat strain Salmon with alien cytoplasms. Cytologia 33:526–538

Tsunewaki K, Endo T, Mukai Y (1974) Further discovery of alien cytoplasms inducing haploids and twins in common wheat. Theor Appl Genet 45:104–109

Tsunewaki K, Mukai Y, Fujigaki J (1976a) Basic studies on hybrid wheat breeding VI. Production of male sterile lines of common wheat cultivars. Jpn J Breed 26:25–31

Tsunewaki K, Mukai Y, Endo T, Tsuji S, Murata M (1976b) Genetic diversity of the cytoplasm in *Triticum* and *Aegilops* V. Classification of 23 cytoplasms into eight plasma types. Jpn J Genet 51:175–191

Tsunewaki K, Mukay Y, Endo T, Tsuji S, Murata M (1976c) Genetic diversity of the cytoplasm in *Triticum* and *Aegilops* VI. Distribution of the haploid-inducing cytoplasms. Jpn J Genet 51:193–200

Tsunewaki K, Mukai Y, Endo TR (1978) On the descent of the cytoplasm of polyploid species in *Triticum* and *Aegilops*. In: Proc 5th Int Wheat Genet Symp 1:261–272

Tsunewaki K, Mukai Y, Endo TR (1980a) Genetic diversity of the cytoplasm Sect A. Detailed studies of the alloplasmic wheats produced in Kyoto university. In: Tsunewaki K (ed) Genetic diversity of the cytoplasm in *Triticum* and *Aegilops*. Jpn Soc Promot Sci Publ Tokyo, pp 49–100

Tsunewaki K, Mukai Y, Panaytov I (1980b) Genetic diversity of the cytoplasm Sect B. Comparison between the alloplasmic wheats produced in Kyoto university, Japan and those produced in the institute for wheat and sunflower Bulgaria. In: Tsunewaki K (ed) Genetic diversity of the cytoplasm in *Triticum* and *Aegilops*. Jpn Soc Promot Sci Tokyo Jpn Publ, pp 101–130

Tsunewaki K, Mukai Y, Endo TR (1980c) Genetic diversity of the cytoplasm. Sect D. General discussion and conclusion. In: Tsunewaki K (ed) Genetic diversity of the cytoplasms in *Triticum* and *Aegilops*. Jpn Soc Promot Sci Publ Tokyo, pp 159–210

Turbin NV, Palilova AN (1970) A theoretical model for the interaction of chromosome, genes and cytoplasm in the case of cytoplasmic male sterility in maize. Rep All Union Lenin Acad Agric Sci USSR 11:2–4

Turbin NV, Palilova AN, Smolškaya AV, Serpokrylova LS (1969) The influence of sterile cytoplasm on the course of microsporogenesis in maize lines. Dokl Akad Nauk USSR 188:108–111

Turcotte EL, Feaster CV (1985) Inheritance of male sterile mutant MS_{12} in American Pima cotton. Crop Sci 25:688–690

Twamley BE (1955) Flower colour inheritance in diploid and tetraploid alfalfa. Can J Agric Sci 35:461–476

Tyagi CS, Lal S, Karwasra RR (1982) Genetic analysis of some quantitative characters in pearl millet. Indian J Agric Sci 52:215–219

Uchida A, Suda K (1973) Ethidium bromide induced loss and retention of cytoplasmic drug resistance factors in yeast. Mutat Res 19:57–63

Uchimiya H (1982) Somatic hybridization between male sterile *Nicotiana tabacum* and *N. glutinosa* through protoplast fusion. Theor Appl Genet 61:69–72

Uno GE (1982) Comparative reproductive biology of hermaphroditic and male sterile *Iris douglasiana* Herb (Iridaceae). Am J Bot 69:818–823

Unrau J, White WJ (1944) The yield and other characters of inbred lines and single crosses of sunflowers. Sci Agric 24:516–528

Upcott M (1937) Timing unbalance at meiosis in pollen sterile *Lathyrus odoratus*. Cytologia Fujii Jubilee 1:299–310

Uphof JCT (1931) Wissenschaftliche Beobachtungen und Versuche an Agrumen I. Über die Blütenverhältnisse der Jahitilimonella. Gartenbauwissenschaften 4:513–520

Urie AL, Zimmer DE (1970) Yield reduction in safflower hybrids caused by female selfs. Crop Sci 10:419–422

Uthkede RS (1972) Some breeding procedures used for improvement of pearlmillet (*Pennisetum typhoides* (Burm. f Stapf and Hubb). Indian J Agric Sci 42:452–456

Valdeyron G, Dommée B, Valdeyron A (1973) Gynodioecy: another computer stimulation model. Am Nat 107:454–459

Valdeyron G, Dommée B, Vernet Ph (1977) Self fertilization in male fertile plants of a gynodioecious species: *Thymus vulgaris* L. Heredity 39:243–249

Valleau WW (1932) Seed transmission and sterility studies of two strains of tobacco ringspot. Kentucky Agric Exp Stat Bull 327:43–80

Valmayor RV, Mendoza EM, Millare VE (1956) A cytological study of abaca and its relatives. Philippines Agric 40:269–276

Vandam J (1974) Influence dún gamétocide chimique sur l'expression de la fertilite du foment. Bull Rech Agron Gemblouxa:343–351

Van Damme JMM (1983a) On gynodioecy in *Plantago lanceolata*. Ph D Thesis Univ Groningen, pp 132

Van Damme JMM (1983b) Gynodioecy in *Plantago lanceolata* L. II. Inheritance of three male sterility types. Heredity 50:253–273

Van Damme JMM, Van Delden W (1982) Gynodioecy in *Plantago lanceolata* L. I. Polymorphism for plasmon type. Heredity 49:303–318

Van Damme JMM, Van Delden W (1984) Gynodioecy in *Plantago lanceolata* L. IV. Fitness components of sex types in different life cycle stages. Evolution 38:1326–1336

Van Der Mark F, Kool AJ (1985) High molecular weight polypeptides of in-organelle mitochondrial protein synthesis in *Petunia hybrida*. In: Magnien E, Nettancourt de D (ed) Genetic engineering of plants and microorganisms important for agriculture. Nijhoff and Junk, the Hague, pp 140–144

Van Dijk H (1985) Alloenzyme genetics, self-incompatibility and male sterility in *Plantago lanceolata*. Heredity 54:53–63

Van Rheenen HA, Muigai SGS, Kitivo DK (1979) Male sterility in beans (*Phaseolus vulgaris* L.). Euphytica 28:761–763

* Varenica ET, Zimina TK, Sotnik VM (1975) Inducirovanijé mužskoj sterilnosti u ozimoj mjagkoj pšenicy s pomoščju ethrela. Věstnik sel-choz nauki 7:31–34

Vasanthy G, Pocock SAJ (1981) A comparative study of anomalous and normal pollen of *Rapanea*. Pollen Spores 23:349–379

Vasil IK (1967) Physiology and cytology of anther development. Biol Rev 42:327–373

* Vavilov PV, Rogozhin AN, Sosnovskii PF (1979) Use of diploid and tetraploid male sterile forms of fodder beet in breeding for heterosis. Izv Timiry Selsk Stev Akad USSR 4:63–68

Vear F (1981) Determination of sunflower lines useable as a combining ability testers according to their aptitude to be male sterilized with giberellin. Sunflower Newslett 5:5–8

Vedel F, Quetier F, Dosba F, Doussinault G (1978) Study of wheat phylogeny by EcoRI analysis of chloroplastic and mitochondrial DNAs. Plant Sci Lett 13:97–102

Vedel F, Quetier F, Cauderon Y, Dosba F, Boussinault G (1981) Studies on maternal inheritance in polyploid wheats with cytoplasmic DNAs as genetic markers. Theor Appl Genet 59:239–245

Vedel F, Mathieu C, Lebacq P, Ambard-Bretteville F, Remy R (1982) Comparative macromolecular analysis of the cytoplasms of normal and cytoplasmic male sterile *Brassica napus*. Theor Appl Genet 62:255–262

Veen JH van Der, Wirtz P (1968) EMS-induced genic male sterility in *Arabidopsis thaliana:* a model selection experiment. Euphytica 17:371–377

Vega MP de la, Lacadena JR (1979) Cytohistological studies on anther and pollen development in alloplasmic rye. Cytologia 44:295–304

Velkov V, Stoyanova Y (1974) Biological pecularities of cytoplasmic male sterility and schemes of its use. In: Proc 6th Int Sunflower Conf Buchares, pp 361–365

* Vengrenovskij SI, Dzelali NI (1969) O polucenii muzskisterilńych form dlá selekcii gibridnoj ljucerny. Nauca Tr Vses Selek Genet Ucta 16:63–66

Verdcourt B (1970) Studies in Leguminoseae, Papilionoideae for the "Flora of Tropical East Africa". Kew Bull 24:507–569

Verma MM, Kumar J (1978) Ethrel- a male gametocide that can replace the male sterility genes in barley. Euphytica 27:865–968

Verma SC (1984) Genic male sterility in *Eruca sativa* (Cruciferae). Incompatibility Newslett 16:9

Villemeur P, Musho US, Delmas JM, Maamar M, Ouksili A (1984) Contribution a l'etude de la biologie florale de lolivier (*Olea europaea* L.): sterilite male, flux pollinique et periode effective de pollinisation. Fruits 39:467–473

Villerts A (1942) Über die Verschiedenheit reziproker Art-Bastarde der Gattung *Begonia*. J Genet 43:223–236

Vincent GB, Wooley DG (1972) Effect of moisture stress at different stages of growth II. Cytplasmic male-sterile corn. Agron J 64:599–602

Virmani SS, Athwal DS (1973) Genetic variability in floral characteristics influencing outcrossing in rice *Oryza sativa* L. Crop Sci 13:66–67

Virmani SS, Khush GS (1980) Natural outcrossing on cytoplasmic male sterile lines of rice under tropical conditions. IRRI Newslett Philippines 5(5):5–6

Virmani SS, Tan ZH (1982) Stigma receptivity of some cytoplasmic male-sterile lines of rice. IRRI Newslett Philippines 7(6):3

Virnich H (1967) Untersuchungen über das Verhalten der männlichen Sterilität und anderer Eigenschaften bei polyploiden Zwiebeln (*Allium cepa* L.) als Grundlage für eine Nutzung in der Hybridzüchtung. Z Pflanzenzücht 58:205–244

Vitkovskii VL, Ruban RV (1974) Peach varieties with male sterility. Genet e' Selekt USSR 52: 83

Vittal-Rao S (1969) An unusual occurrence of breakdown of male sterility in bajra (*Pennisetum typhoides* (Burm) Stapf and Hubb). Andhra Agric J 16:1–5

Vohra RR (1969) Development of dwarf seed parent 18D2A. Proc All Indian Millets Workshop ICAR Delhi, pp 12–19

* Volf VG (1966) Ispolzovanie mujskoy sterilinosty v selektzü podsolnechyka Selek rast s isopolz. Cytopl Mujskoy Sterilnosty Kiev USSR, pp 423–433

Voronina SA, Krupnov VA (1972) Inheritance of male sterility in spring bread wheat. Tsitol Genet 6:248–252

Vranceanu V (1967) Ereditatea surselor de androsterlitate la floavea-soarelui. Probl Agric 12:4–11

Vranceanu V (1970) Advances in sunflower breeding in Romania. In: Proc 4th Int Sunflower Conf Memphis Tenn, pp 136–148

Vranceanu V, Stoenescu F (1969) Sterilitatea mascula la floareasoarelui si perspectiva utilizariiei in producerea semintelor hibride. Agric Silvice 35:559–571

Vranceanu V, Stoenescu F (1970a) Immunitate la mana florii-soareliu, Conditionate monogenic. Probl Agric 2:34–40

Vranceanu V, Stoenescu F (1970b) Obtinerea de hybrizide de florareasoarelui pe baza androsterilitate genica. Agric Silvice 36:281–290

Vranceanu V, Stoenescu F (1971a) Aspecte genetice provind rezistenta la cadere a florii-soarelui. Agric Silvice 37:183–190

Vranceanu V, Stoenescu F (1971b) Ereditatea imunitatii la mana a florii-soarelui. Agric Silvice 37:209–217

Vranceanu V, Stoenescu F (1971c) Pollen fertility restorer gene from cultivated sunflower (*Helianthus annuus* L.). Euphytica 20:536–541

Vranceanu V, Stoenescu F (1978) Genes for pollen fertility restoration in sunflowers. Euphytica 27:617–627

Vulkova-Achkova ZV (1980) Cytoplasmic male sterility in *Lycopersicon.* Agrochi Akad Nauk Moscova USSR 33:675–678

* Vulpe VV (1968) Tipuri de androsterilitate citoplasmatic la floareasoarelui. Common Bot 7:121–133

Waddington CH (1966) Principles of development and differentiation. Mac Millan, New York

Walkof C (1959) In: Progress report on FW-450 chemical gametocide. Rohm and Haas, Philadelphia, p 9

Wall JR (1967) Correlated inheritance of sex expression and fruit shape in *Cucumis.* Euphytica 16:69–76

Wall JS, Ross WM (1970) Sorghum production and utilization: Major feed and food crops in agriculture and food series. Avi, pp 702

Wallis ES, Saxena KB, Byth DE (1980) A new source of genetic male sterility in pigeon pea. Proc Int Workshop on Pigeon pea ICRISAT Hyderabad, India, p 17

Walton JD (1983) Which mitochondrial proteins are involved in susceptibility to *Helminthosporium maydis* race T? Plant Mol Biol Rep 1:101–103

Walton JD, Earle ED, Yoder OC, Spanswick RM (1979) Reduction of steady state ATP levels in susceptible maize mesophyll protoplasts by *Helminthosporium maydis* race T toxin. Plant Physiol (Bethesda) 63:806–810

Wang LQ, Cheng JY, Shi JG, Gao ZL (1980) Studies on the male sterile material "2-2-3" controlled by a dominant gene of wheat and its use in breeding. Sci Agric PR China 2:1–8

Wang RC, Lund S (1975) Studies on male sterility in barley induced by sodium 1-(p-chlorophenyl) 1,2-dihydro 4,6-dimethyl-2-oxonicotinate. Crop Sci 15:550–553

Wang X, Que RF (1981) Induction of pollen sterility by ethylene and gametocides in rice. Acta Phytophysiol Sin 7:381–383

Waninge J, Zeven AC (1968) Chromosome numbers in Pugsley's male sterile wheat. Euphytica 17: 378–380

Warmke HE, Lee SJ (1977) Mitochondrial degeneration in Texas cytoplasmic male sterile corn anthers. J Hered 68:213–222

Warmke HE, Lee SLJ (1978) Pollen abortion in T-cytoplasmic male sterile corn (*Zea mays*): a suggested mechanism. Science 200:561–563

Warmke HE, Overman MA (1972) Cytoplasmic male sterility in *Sorghum* I. Callose behaviour in fertile and sterile anthers. J Hered 63:103–108

Warnstorf C (1896) Blütenbiologische Beobachtungen aus der Ruppiner Flora im Jahre 1895. Bot Ver Germany 38:15–63

Warren FS, Dimmock F (1954) The use of chemicals and of male sterility to control pollen production in corn. Can J Agric Sci 34:48–52

Watanabe Y, Sakaguchi S, Kudo M (1968) On the male-sterile rice plant possessing the cytoplasm of Burmese variety "Lead Rice". Jpn J Breed Suppl2, 18:77–78

Waterkeyn L (1964) Callose microsporocytaire et callose pollinque. In: Linskens HF (ed) Pollen physiology and fertilization. North Holland, Amsterdam, pp 52–58

Watrud LS, Hooker AL, Koeppe DE (1975a) The effect of nuclear restorer genes of Texas male sterile cytoplasm on host response to *Helminthosporium maydis* race T. Phytopathology 65: 178–182

Watrud LS, Baldwin JR, Miller RJ, Koeppe DE (1975b) Induction of a sensitive response to *Helminthosporium maydis* race T-toxin in resistant mitochondria of corn (*Zea mays* L.) by removal of the outher mitochondrial membrane. Plant Physiol (Bethesda) 56:216–221

Watrud LS, Laughnan JR, Gabay SJ, Koeppe DE (1976) Effects of nuclear restorer genes on the cytochrome content of corn pollen mitochondria. Can J Bot 54:2718–2725

Watson CV, Nath J, Nanda D (1977) Possible mitochondrial involvement in mechanism of cytoplasmic male sterility in maize (*Zea mays*). Biochem Genet 15:1113–1124

Watson IA, McWhirter KS (1968) A bulk method of producing hybrid wheat. Crop Sci 8:124–125

Watts VM (1962) A marked male sterile mutant in water melon. Proc Amer Soc Hort Sci 81: 498–505

Weathers LG (1965) Transmission of exocortis virus of citrus by *Cuscuta subinclusa*. Plant Dis Rep 49:189–190

Weatherwax P (1925) Anomalies in maize and its relatives III. Carpelloidy in maize. Bull Torrey Bot Club 52:167–180

Weaver DB, Weaver JB (1977) Inheritance of pollen fertility restoration in cytoplasmic male-sterile upland cotton. Crop Sci 17:497–499

Weaver JB (1968) Analysis of a genetic double recessive completely male-sterile cotton. Crop Sci 60:597–600

Weaver JB (1969) The present status of male sterile cotton. Beltwide Cotton Prod Res Conf, pp 13–19

Weaver JB (1971) Methods of utilizing male sterility in cotton. Beltwide Cotton Prod Res Conf, pp 21

Weaver JB (1973) Linkage and crooked stigma observed with Rhynes' male sterile. Beltwide Cotton Prod Res Conf, pp 6

Weaver JB (1981) Effect of cms on fiber properties in cotton. Agron Abstr Am Agron Soc, p 75

Weaver JB, Ashley T (1971) Analysis of a dominant gene for male-sterility in upland cotton *Gossypium hirsutum* L. Crop Sci 11:596–598

Weaver JB, Weaver DB (1979) Cracked root mutant in cotton: Inheritance and linkage with fertility restoration. Crop Sci 19:307–309

Webb CJ (1976) Flowering periods in the gynodioecious species *Gingidia decipiens* (Umbelliferae). NZ J Bot 14:207–210

Webb CJ (1981) Gynodioecy in *Gingidia flavellata* (Umbelliferae). NZ J Bot 19:111–114
Webb CJ, Lloyd DG (1980) Sex ratios in Newzealand apioid Umbelliferae. NZ J Bot 18:121–126
Webber CR, Hanson WD (1961) Natural hybridization with and without ionizing radiation in soybeans. Crop Sci 1:389–392
Webster OJ (1965) Genetic studies in *Sorghum vulgare*. Crop Sci 5:207–210
Webster OJ, Singh SP (1964) Breeding behaviour and histological structure of a nondehiscent another character in *Sorghum vulgare* Pers. Crop Sci 4:656–658
Weijer J (1952) A catalogue of genetic maize types together with a maize bibliography. Bibliogr Genet 14:189–425
Weiss EA (1971) Castor, sesame and safflower. Barnes and Noble, New York, pp 55
Weissinger AK Timothy DH, Levings CS, Hu WWL, Goodman MM (1982) Unique plasmid like mitochondrial DNAs from indigenous maize races of Latin America. Proc Natl Acad Sci USA 79:1–5
Welch JE, Grimball EL (1947) Male sterility in the carrot. Science 106:594
Wells DG, Lay CL (1970) Hybrid vigour in hard red spring wheat crosses. Crop Sci 10:220–223
Welsh JR, Klatt AR (1971) Effect of temperature and photoperiod on spring wheat pollen viability. Crop Sci 11:864–865
Welzel G (1954) Entwicklungsgeschichtlichgenetische Untersuchungen an pollensterilen Mutanten von *Petunia*. Z Indukt Abstammungs Vererbungsl 86:35–53
Wernsman EA, Matzinger DF (1980) Tobacco. In: Fehr WR, Hadley HH (eds) Hybridization of crop plants. Am Soc Agron, Crop Sci Soc Am Publ, pp 657–680
Westergaard M (1958) The mechanism of sex determination in dioecious flowering plants. Adv Genet 9:217–281
Wettstein von F (1924) Über Fragen der Geschlechtsbestimmung bei Pflanzen. Naturwissenschaften 12:761–768
Wettstein von F (1926) Über plasmatische Vererbung sowie Plasma und Genwirkung. Nachr Ges Wiss Göttingen Math Phys 61:250–281
Wettstein von F (1928) Über plasmatische Vererbung und über das Zusammenwirken von Genen und Plasma. Ber Dtsch Bot Ges 46:32–49
Wettsein von F (1946) Untersuchungen zur plasmatischen Vererbung I. *Linum*. Biol Zentralbl 65: 149–166
Whelan EDP (1972) A cytogenetic study of a radiation-induced male sterile mutant of cucumber. J Amer Soc Hort Sci 97:506–509
Whelan EDP (1974) Linkage of male sterility, glabrate seedling and determinate plant habit in cucumber. Hort Science 9:576–577
Whelan EDP (1976) Sterility problems in interspecific hybridization of *Helianthus* species. In: Proc Inst Sunflower Forum, Fargo, pp 5
Whelan EDP (1978a) Hybridization between annual and perennial diploid species of *Helianthus*. Can J Genet Cytol 20:523–530
Whelan EDP (1979) Interspecific hybrids between *Helianthus petiolaris* Nutt and *H. annuus* L.: effect of backcrossing on meiosis. Euphytica 28:297–308
Whelan EDP (1980) A new source of cytoplasmic male sterility in sunflower. Euphytica 29:33–46
Whelan EDP (1981) Cytoplasmic male sterility in *Helianthus giganteus* L. x *H. annuus* L. Crop Sci 21:855–858
Whelan EDP, Dorrell DG (1980) Interspecific hybrids between *Helianthus maximiliani* Schrad and *H. annuus* L. Effect of backcrossing on meiosis, anther morphology and seed characteristics. Crop Sci 20:29–34
Whited DA (1967) Biochemical and histochemical properties associated with genetic male sterility at the Ms locus in barley (*Hordeum vulgare* L.). Ph D Thesis N Dakota State Univ Fargo
Whitehouse HLK (1950) Multiple allelomorph incompatibility of pollen and style in the evolution of angiosperms. Ann Bot 14:199–216
Whitwood WN (1975) A mutation and a genetic study in *Cucurbita pepo* L. Ph D Thesis Iowa State Univ Ames

Whyte RO (1929) Studies in *Ranunculus* II. The cytological basis of sex in *R. acris.* J Genet 11: 183–191

Wickersham DS, Patterson FL (1980) Male fertility restoration in crosses of R_5 with soft red winter wheat. Crop Sci 20:100–102

Wiebe GA (1960) A proposal for hybrid barley. Agron J 52:181–182

Wiebe GA (1967) Hybrid barley. In: Meet West Soc Crop Sci (USA), pp 4–7

Wiebe GA, Ramage RT (1971) Hybrid barley. Barley Genet 2, 287–291

Wienhues F (1968) Long term yield analyses of heterosis in wheat and barley variability of heterosis, fixation of heterosis. Euphytica Suppl 1, 17:49–62

Wilhite JR (1960a) Frequency of interspecific crossing between *Sorghum vulgare* and *Sorghum halepense.* Sorghum Newslett 3:45

Wilhite JR (1960b) Frequency of interspecific crossing utilizing cytoplasmic male sterile lines of *Sorghum vulgare* exposed to pollen from *Sorghum halepense.* MS Thesis Oklahoma State Univ

Williams EJ (1954) Seed failure in the potato (*Solanum tuberosum* L.) varieties *chippewa* and *sebago.* Ph D Thesis Wisconsin Univ Madison

Williams PH, Heyn FW (1981) The origin and development of cytoplasmic male sterile chinese cabbage. Proc Ist Int Symp Asian Veg Res Dev Center Shanhua Taiwan PR China, pp 293–300

Williams PH, Leung H (1981) Methods of breeding for multiple disease resistant chinese cabbage. In: Talekar NS, Griggs TD (ed) Proc Ist Int Symp Chinese Cabbage. AVRDC Taiwan, pp 393–403

Williams MW, Stahly EA (1970) N-Malonyl-D-tryptophan in apple fruits treated with succinic acid 2,2-dimethylhydrazide. Plant Physiol (Bethesda) 46:123–125

Wilson CL, Stiers DL, Smith GG (1970) Fungal lysosomes or spherosomes. Phytopathology 60: 216–227

Wilson JA (1968a) Problems in hybrid wheat breeding. Euphytica Suppl 1, 17:13–33

Wilson JA (1968b) Hybrid wheat developments with *Triticum timopheevi* Zhuk derivatives. In: Proc 3rd Int Wheat Genet Symp, pp 423–430

Wilson JA, Ross WM (1962) Crossbreeding in wheat, *Triticum aestivum* II. Hybrid seed set on a cytoplasmic male sterile winter wheat composite subjected to cross pollination. Crop Sci 2: 415–417

Wilson JA, Wilson J, Schmidt HJ (1980) Full-dwarf x semi-dwarf wheats. In: Proc 3rd Int Wheat Conf Madrid Spain, pp 210–214

Wilson P, Driscoll CJ (1983) Hybrid wheat. In: Frankel R (ed) Heterosis. Springer, Berlin, Heidelberg, New York, pp 94–123

Wilson RF (1981) Root lipid composition and metabolism in genetic male sterile soybeans. Crop Sci 21:69–73

Wilson RF, Burton JW, Buck JA, Brim CA (1978) Studies on genetic male-sterile soybeans I. Distribution of plant carbohydrates and nitrogen during development. Plant Physiol (Bethesda) 61:838–841

Wit F (1958) Tetraploid Italian rye grass (*Lolium multiflorum* Lam). Euphytica 7:47–58

Wit F (1959) Hybrids of ryegrass and meadow fescue and their value for grass breeding. Euphytica 8:1–12

Wit F (1960) Chemically induced male sterility, a new tool in plant breeding. Euphytica 9:1–9

Wit F (1974) Cytoplasmic male sterility in ryegrasses (*Lollium* spp) detected after intergeneric hybridization. Euphytica 23:31–38

Wittenbach VA (1978) Breakdown of ribulose biphosphate carboxylase and change in proteolytic activity during dark-induced senescence of wheat seedlings. Plant Physiol (Bethesda) 62:604–608

Wittwer SH, Bukovac MJ (1957) Gibberellin and higher plants X. Field observations with certain vegetable crops. Michigan Agric Exp Stat Q Bull 40:352–364

Wittwer SH, Bukova MJ (1962) Staminate flower formation on gynoecious cucumber as influenced by the various gibberellins. Naturwissenschaften 49:305–306

Wittwer SH, Hillyer IG (1954) Chemical induction of male sterility in cucurbits. Science 120: 893–894

Woodworth CM (1932) Genetics and breeding in the improvement of the soybean. Illinois Agric Exp Stat Bull 384:297–404

Woodworth CM (1933) Genetics of the soybean. J Am Soc Agron 25:36–51

Worstell JV, Kidd HJ, Schertz KF (1984) Relationships among male sterility inducing cytoplasms of sorghum. Crop Sci 24:186–189

Wu FS, Murry LE (1985) Changes in protein and aminoacid content during anther development in fertile and cytoplasmic male sterile *Petunia*. Theor Appl Genet 71:68–73

Wunderlich R (1952) Über das Antherentapetum mit besonderer Berücksichtigung seiner Kernzahl. Österr Bot Z 101:1–63

Wyatt JE (1984a) An indehiscent anther mutant in the common bean *Phaseolus vulgaris* L. J Amer Soc Hort Sci 109:484–487

Wyatt JE (1984b) Inheritance of an indehiscent anther character in the common bean. Hort Science 19:670–671

Xaingmin DYP, Herong C (1981) Cytological study on pollen abortion of the male sterile wheat. In: 1980 Ann Rep Inst Genet, PR China, pp 132–133

Yabuno T (1977) Genetic studies on the interspecific cytoplasmic substitution lines of Japonica varieties of *Oryza sativa* L. and *O. glaberrima* Steud. Euphytica 26:451–463

Yadav RP (1974) Inheritance of pollen fertility in pearlmillet. J Palynol 10:38–50

Yamaguchi T, Kanno C (1963) Studies on the differences of character manifestation in reciprocal crosses of rape (*Brassica napus* L.) VI. Cyto-histological observations on the development of floral organs in reciprocal F_1 hybrids. Bull Tokaikinki Natl Agric Exp Stat 9:162–182

Yamakawa K (1968) Induction of male-sterile tomato mutants by gamma irradiation. J Jpn Soc Hortic Sci 37:333–339

Yamashita A, Ukai Y (1979) Male sterile mutations of barley induced by ionizing radiations and chemical mutagens I. Frequency, selection and brief description of sterile mutants induced by gamma ray irradiation and ethylene-imine treatment on seeds. Jpn J Breed 29:101–114

Yampolski C, Yampolski H (1922) Ditribution of the sex forms in the phaenerogamic flora. Bibliogr Genet 3:1–62

Yang AF, Hamilton RI (1974) The mechanism of seed transmission of tobacco ringspot virus in soybean. Virology 62:26–37

Yang MX, Li KJ (1984) Cytological studies on pollen formation and development of male sterile lines of *Oryza sativa*. Acta Bot Sin PR China 26:105–108

Yang RCA, Lis J, Wu R (1979) Elution of DNA from agarose gels after electrophoresis. In: Wu R (ed) Methods in enzymology, vol 68. Academic Press, New York, pp 176–182

Yang YD (1982) Genetic analysis of some male-steriles of cotton. Hereditas PR China 4:18–20

Yen DE (1959) Pollen sterility in Pukekohe longkeeper onions. NZ J Agric Res 2:605–612

Yen FS, Evans LE, Larter EN (1969) Monosomic analysis of fertility restoration in three restorer lines of wheat. Can J Genet Cytol 11:531–546

Yermanos DM, Knowles PF (1960) Effects of gibberellic acid treatment on safflower. Agron J 52:596–598

Yih RY, McNulty PJ, Seidel MC, Viste KL (1971) Plant growth regulating properties of 3-carboxy-2-Pyridones. Hort Science 6:460–461

Yinong L, Jigeng L (1981) Studies on the ultrastructure of chloroplasts from the C-cytoplasmic male-sterile lines in maize. In: Ann Rep Inst Genet PR China, pp 105–106

Yoder OC (1973) A selective toxin produced by *Phyllosticta maydis*. Phytopathology 63:1361–1366

Yoder OC, Gracen VE (1975) Segregation of pathogenicity and host specific toxin production in progenies of crosses between races T and O of *Helminthosporium maydis (Cochliobolus heterostrophus)*. Phytopathology 65:273–276

Yoder OC, Payne GA, Gregory P, Gracen VE (1977) Bioassays for detection and quantification of *Helminthosporium maydis* race T-toxin: a comparison. Physiol Plant Pathol 10:237–245

York DW, Earle ED, Gracen VE (1980) Ultrastructural effects of *Helminthosporium maydis* race T toxin on isolated corn mitochondria and mitochondria within corn protoplasts. Can J Bot 58:1562–1570

Young BA, Schulz-Schaeffer J, Caroll TW (1979) Anther and pollen development in male sterile intermediate wheatgrass plants derived from wheat x wheat grass hybrids. Can J Bot 57:602–618

Young DA (1972) The reproductive biology of *Rhus integrifolia* and *Rhus ovata* (Anarcardiaceae). Evolution 26:406–414

Young HM (1927) Heritable characters of maize XXX. Weak tassel. J Hered 18:505–510

Young WJ (1923) The formation and degeneration of germ cells in the potato. Am J Bot 10:325–335

Yüce S (1973) Haploidie bei der Zuckerrübe. Ph D Thesis Univ Gießen, pp 73

Yun D, Xiangmin P, Herong C (1981) Cytological study on pollen abortion of the male sterile wheat. In: 1980 Ann Rep Inst Genet Sin PR China, pp 132–133

* Zainyshev A (1974) Male sterility in cotton. Genet Selekt USSR. 23:216–226

Zamir D, Jones RA, Kedar N (1980) Anther culture of male-sterile tomato (*Lycopersicon esculentum* Mill) mutants. Plant Sci Lett 17:353–361

Zamir D, Tanksley SD, Jones RA (1981) Genetic analysis of the origin of plants regenerated from anther tissue of *Lycopersicon esculentum* Mill. Plant Sci Lett 21:223–227

Zdril'ko AF (1969) A study of cytoplasmic male-sterility in rye. Seleke Semenov Mezced Nauc Sborn 14:31–33

Zdril'ko AF, Adamchuk GK (1975) Producing forms with male sterility to obtain heterotic hybrids of rye. Byull Vses Inst Rastenievod 48:54–57

Zelcer A, Aviv D, Galun E (1978) Interspecific transfer of cytoplasmic male sterility by fusion between protoplasts of normal *Nicotiana sylvestris* and X-ray irradiated protoplasts of male-sterile *N. tabacum.* Z Pflanzenphysiol 90:397–407

Zelenskaya LA, Pestova TM (1974) A study of mutant forms with male sterility in *Lathyrus sativus.* Uspekhi Khim Mutageneza V. Selektsii Moscow USSR, pp 186–189

Zenkteler M (1962) Microsporogenesis and tapetal development in normal and male sterile carrots (*Daucus carota* L.). Am J Bot 49:341–348

Zeven AC (1967) Transfer and inactivation of male sterility and sources of restorer genes in wheat. Euphytica 16:183–189

Zeven AC (1968) Cross pollination and sources of restorer genes in wheat and a semi-hybrid variety. Euphytica 17:75–81

Zeven AC (1971a) Kunstmatig productie van zaad bij *Streptocarpus,* Constant Nymph. Vakblad Bloemistery (Netherland) 26(27):9

Zeven AC (1971b) Artificial seed production of *Streptocarpus,* Constant Nymph. Gloxinian (Netherlands) 21:28

Zeven AC (1972a) Inheritance of functional male sterility in *Streptocarpus* Constant Nymph and its mutants. Euphytica 21:265–270

Zeven AC (1972b) Plant density effect on expression of heterosis for yield and its components in wheat and F_1 versus F_3 yields. Euphytica 21:468–488

Zhang KT, Fu HY (1982) Effect of high temperature treatment on male sterility in sorghum. Acta Genet PR China 9:71–77

Zhangxing Z, Chunnong H (1980) A study of male sterility induced in rice (*Oryza sativa* L.) by 3-(p-chlorophenyl)6-methoxy-s-triazine-2,4-(1H, 3H) Dione. J Hangzhou Univ Zhejiang PR China 4:81–87

Zhouchang L, Huixin L, Futai C (1981) A comparative study on activity of ribulose biphosphate carboxylase and cytoplasmic male sterility. In: Ann Rep Inst Genet PR China, 1980, p 11

Zhu GL, Tsao TH (1984) Comparative analysis of free amino acids in anthers of nuclear male sterile Taigu wheat plants and their normal fertile counterparts. Acta Phytophysiol PR China 10:185–189

Zuber MS (1975) Corn germplasm base in the US. Is it narrowing, widening or static? In: Proc 30th Annu Corn Sorghum Res Conf, pp 277–286

Zuochang L, Huixin L, Futai C (1981) A comparative study on activity of ribulose biphosphate carboxylase and cytoplasmic male sterility. In: Ann Rep Inst Genet PR China, 1980, p 111

* Zvyagina GN (1973) [Combining ability of male sterile analogues of red clover with variety Moscow 1] Nauch Tr NII Kh Tsentr Rnov Nechern 29:98–100

Subject Index

Title Index*

* Confined to part I of this volume.

Plant Content Index*

* Confined to part II of this volume and supplements Table 55.5.

[a] Refers to part I of this volume.

Philouze J (1974a) Génes marquers lies aux génes de stérilité mâle ms 32 et ms 35 chez la Tomate. Ann Amelior Plant (Paris) 24:77–82

Philouze J (1974b) Utilization pratique d'une lignée mâle-sterile ms 35 chez la Tomate. Ann Amel Plant (Paris) 24:129–144

Philouze J (1976) Les hybrides de la tomate: leur intérêt, les techniques d'hybridation, d'utilisation de la stérilité mâle. Pepinier Hortic Mariach France 164:13–18

Phinney BO (1961) Dwarfing genes in *Zea mays* and their relation to the gibberellins. In: Klein RM (ed) Plant growth regulations. Iowa State Univ Press Ames, pp 489–501

Phipps IF (1928) Heritable characters in maize XXXI. Tassel seed-4. J Hered 19:399–404

Pi CP, Wu KD (1961) The inheritance of cytoplasmic-genic male sterility in sorghum. Sorghum Newslett 4:9–10

Pi CP, Wu KD (1963) The inheritance of cytoplasmic genetic male sterility in sorghum. Bot Bull Acad Sin 4:15–22

Picard J (1960) Données sur l'amélioration de la féverole de printemps *Vicia faba* L. Ann Amelior Plant (Paris):121–153

Picard J, Berthelem P, Duc G, Guen J Le (1981) Male sterility and future prospects for hybrid varieties in *Vicia faba*. In: Proc Int Conf faba beans, Cairo, pp 28–36

Pike LM, Mulkey WA (1971) Use of hermaphrodite cucumber lines in development of gynoecious hybrids. Hort Science 6:339–340

Pike LM, Peterson CE (1969) Gibberellin A_4/A_7 for induction of staminate flowers on the gynoecious cucumber. Euphytica 18:106–109

Piquemal G (1970) How to produce hybrid sunflower seeds by inducing male sterility with gibberellic acid. In: Proc 4th Int Sunflower Conf Memphis Tennessee, pp 127–135

* Pirev MN (1966) A histochemical study of microsporogenesis in pollen-fertile and pollen sterile forms of sunflower. Bull Acad Sci Moldav 11:91–94

Pochard E (1966) Donnes experimentales sur la selection, du piment (*Capsicum annuum* L.). Ann Amelior Plant (Paris) 16:185–197

Pochard E (1970a) Obtention de trois nouvelles mutations de sterilité mâle chez le piment (*Capsicum annuum* L.) par traitements mutagenes appliques a un material monoploide. Eucarpia sec Horticoles Versailles, pp 93–95

Pochard E (1970b) etude comparée de la gametôgénèse chez les plantes normales, les plantes possédant la Sterilité cytoplasmique Peterson et les mutant mâle-stériles mr 9, mc 509, et mc 705 (*Capsicum annuum* L.) In: Proc Eucarpia sec Horticoles Versailles France, pp 169–173

Pochard E (1971–72) Recherche sur le "piments". Stat d'Amel Plant Maraich Montfavet Rapp Act, pp 57–68

Poehlman JM (1959) Breeding field crops. Holt, New York

Poehlman JM (1979) Breeding field crops, 2nd ednt. Avi, Connecticut

Pogorletskii BK (1973) Genetic marking of male sterility in sunflower. Genetika 9(5):23–29

Pogorletskii BK, Barlov VV (1973) Male sterility in sunflower and the possibility of using it in breeding and seed production. Sb Nauch Tr Uses Selekt USSR 10:115–121

Pohl J (1895) Über Variationsweite der *Oenothera lamarckiana*. Österr Bot Z 45:166–169, 205–212

Pokhriyal SC (1977) Revamping bajra seed. Seeds Farms 3:25–28

Pokhriyal SC, Unnikrishnan KV, Singh B, Ramadass C, Patil RR (1976) Combining ability of downy mildew resistant lines in pearlmillet. Indian J Genet Plant Breed 36:403–409

Popov P, Panayotov I, Gotsov K (1974) Fertility restorer sources of *T. timopheevi* cytoplasm and methods for their identification. C R Acad Bulg Sci 27:981–983

Portele E (1883) Mitteilung der Landwirtschaft, see Sartorius (1926)

Porter KB, Tuleen NA (1972a) Pollen fertility restoration of male sterile wheat (*T. aestivum* L.) from factors in rye (*Secale* spp). Agron Abstr Am Soc Agron, p 17

Porter KB, Tuleen NA (1972b) Cross breeding male sterile wheat (*T. aestivum* L.) and rye (*Secale* spp). Crop Sci 12:706–708

Porter KB, Tuleen NA (1976) The performance, fertility and cytology of progeny of male-sterile wheat-rye F_1/triticale hybrids. Proc Int Triticale Symp 1973, Lubbock Texas, pp 73–78

Porter KB, Weise AF (1961) Evaluation of certain chemicals as selective gametocides for wheat. Crop Sci 1:381–382

* Pospisilova J, Jirik J (1979) Differences in the contents of some elements in stable A (S_{msms}) and B (N_{msms}) breeding lines of onion. Bull Vyzkummy Slecht Zelin Olomouc 21–22:3–28

Povilaitis B (1972) Crosses between cytoplasmically inherited male sterile cultivars and short day mutants in *Nicotiana tabacum*. Lighter 42:11–14

Powell JB, Burton GW (1966) A suggested commercial method of producing an interspecific hybrid forage in *Pennisetum*. Crop Sci 6:378–379

Powell JB, Burton GW (1968) Polyembryony in pearl millet *Pennisetum typhoides*. Crop Sci 8: 771–773

Powell JB, Burton GW (1969) Cases of partial and complete male-sterility in pearl millet and their causes. Proc Assoc South Agric Workers 66:83

Powell JB, Taylorson RB (1967) Induced sterility and associated effects of dimethylarsinic acid treatment on pearl millet. Crop Sci 7:670–672

Power JB, Frearson EM, George D, Evans PK, Berry SF, Hayward C, Cocking EC (1976a) The isolation, culture and regeneration of leaf protoplasts in the genus *Petunia*. Plant Sci Lett 7:51–55

Power JB, Frearson EM, Hayward C, George D, Evans PK, Berry SF, Cocking EC (1976b) Somatic hybridization of *Petunia hybrida* and *P. parodii*. Nature 263:500–502

Powers L, Gardner EJ (1945) Frequency of aborted pollen grains and microcytes in Guayule, *Parthenium argentatum* Gray. J Am Soc Agron 37:184–193

Powers Le Roy (1936) The nature of the interaction of genes affecting four quantitative characters in a cross between *Hordeum deficiens* and *H. vulgare*. Genetics 21:398–420

Powling A (1981) Species of small DNA molecules found in mitochondria from sugarbeet with normal and male-sterile cytoplasms. Mol Gen Genet 183:82–84

Powling A (1982) Restriction endonuclease analysis of mitochondrial DNA from sugarbeet with normal and male sterile cytoplasms. Heredity 49:117–120

Powling A, Ellis THN (1983) Studies on the organelle genomes of sugarbeet with male-fertile and male-sterile cytoplasms. Theor Appl Genet 65:323–328

Prakash S (1967) Ph D Thesis Agra Univ India (see Yadav 1974)

Pratt D (1958) Linkage of ten male sterile genes. Tomato Genet Coop Rep 8:25–26

Pring DR (1975) Cytochromes of fertile and cytoplasmic male sterile lines. Plant Physiol (Bethesda) 55:202–206

Pring DR, Levings CS (1978) Heterogeneity of maize cytoplasmic genomes among male sterile cytoplasms. Genetics 89:121–136

Pring DR, Levings CS, Hu WWL, Timothy DH (1977) Unique DNA associated with mitochondria in the "S"-type cytoplasm of male sterile maize. Proc Natl Acad Sci USA 74:2904–2908

Pring DR, Conde MF, Levings CS (1979a) Heterogeneity within the C group of male sterile cytoplasms. Maize Genet Coop Newslett 53:42

Pring DR, Levings CS, Conde MF (1979b) The organelle genomes of cytoplasmic male sterile maize and sorghum. In: Proc 4th John Innes Symp, pp 111–120

Pring DR, Conde MF, Levings CS (1980) DNA heterogeneity with the C group of maize male sterile cytoplasms. Crop Sci 20:159–162

Pring DR, Schertz KF, Conde MF (1981) Plasmid-like and other cytoplasmic DNA characteristics of male-sterile sorghum. Sorghum Newslett 24:132

Pring DR, Conde MF, Schertz KF (1982b) Organelle genome diversity in sorghum: male sterile cytoplasms. Crop Sci 22:414–421

Pring DR, Conde MF, Schertz KF, Levings CS (1982a) Plasmid-like DNAs associated with mitochondria of cytoplasmic male-sterile *Sorghum*. Mol Gen Genet 186:180–184

Pritchard AJ, Hutton EM (1972) Anther and pollen development in male-sterile *Phaseolus atropurpureus*. J Hered 63:280–282

Puente de la CF (1966) Fertility of selected 24 chromosome *S. tuberosum* hybrids. MS Thesis Univ Wisconsin, p 38

Pugsley AT, Oram RN (1959) Genic male sterility in wheat. Aust Plant Breed Genet Newslett 14: 10–11

Pundir NS, Singh SP (1965) Induction of male sterility in muskmelon by the use of FW-450. Agra Univ J Res Sci 14:177–184

Punnett RC (1923) Linkage in the sweet pea (*Lathyrus odoratus*). J Genet 13:101–123
Punnett RC (1925) *Lathyrus odoratus*. Bibliogr Genet 1:69–82
Punnett RC (1927) Mendelism, 2nd edn. Mac Millan, Cambridge, pp 236
Purseglove JW (1968) Tropical crops. Dicotyledons I. J Wiley, New York
Purseglove JW (1972) Tropical crops II. Monocotyledons. Longman Green, London
Putt ED (1954) Cytogenetic studies of sterility in rye. Can J Agric Sci 34:81–119
Putt ED (1962) The value of hybrids and synthetics in sunflower seed production. Can J Plant Sci 42:488–500
Putt ED (1964) Recessive branching in sunflowers. Crop Sci 4:444–445
Putt ED, Heiser CB (1966) Male sterility and partial male sterility in sunflowers. Crop Sci 6: 165–168
Quétier F, Vedel F (1977) Heterogeneous population of mitochondrial DNA molecules in higher plants. Nature 268:365–368
Quinby JR (1980) Interaction of genes and cytoplasms in male sterility in sorghum. In: Proc 35th Ann Corn Sorghum Res Conf, pp 175–184
Quinby JR, Martin JH (1954) Sorghum improvement. Adv Agron 6:305–359
Quinn AA, Mok DWS, Peloquin SJ (1974) Distribution and significance of diplandroids among the diploid *Solanums*. Am Potato J 51:16–21
Quisenberry JE, Kohel RJ (1968) A leaf abnormality for the identification of a genetic male sterile in *Gossypium hirsutum* L. Crop Sci 8:369–370
Rachie KO, Majmudar JV (1980) Pearlmillet. The Pennsylvania State Univ Press, pp 307
Rachie KO, Rawal KM (1976) Integrated approaches to improving cowpeas, *Vigna unguiculata* (L.) Walp. Techn Bull 5th Int Inst Trop Agric Ibadan, pp 1–36
Rachie KO, Rawal K, Franckowiak JD (1974) A rapid method of hand crossing cowpeas. Techn Bull 2nd Int Inst Trop Agric Ibadan, pp 1–5
Rachie KO, Rawal K, Franckowiak JD, Akinpelu MA (1975) Two outcrossing mechanisms in cowpea, *Vigna unguiculata* (L.) Walp. Euphytica 24:159–163
Raeber JG, Bolton AA (1955) A new form of male sterility in *Nicotiana tabacum* L. Nature 176: 314–315
Raghavan TS (1951) Cytoplasmic inheritance in *Saccharum*. Curr Sci (Bangalore) 20:138–140
Raghavan TS (1953) Some cytogenetical features and breeding behaviour of a *Saccharum spontaneum* derivative. In: Proc 7th Int Soc Sugar Crop Techn Congr, pp 475–485
Raghavan TS (1964) Cytoplasmic inheritance of *Saccharum*. Indian J Sugarcane Res Dev 8:134–139
Raghuvanshi RK, Singh D (1984) Embryological manifestations of male sterility in *Capsicum* L. Acta Bot Indica 12:213–215
Rai RK, Stoskopf NC (1974) Amino acid comparisons in male sterile wheat derived from *Triticum timopheevi* zhuk cytoplasm and its fertile counterpart. Theor Appl Genet 44:124–127
Raj AY (1968) Histological studies in male sterile and male fertile sorghum. Indian J Genet Plant Breed 28:335–341
Raja VDG (1967) Phytogametocidal action of FW-450 (mendok) on the characters of Madras, Combodia and Uganda cotton. Madras Agric J 54:302–311
Rajendra BR, Bates LS (1981) Genetic male and chemically induced sterility in *Triticum aestivum*. A comparison of events leading to sterility via cytogenetic and SEM analysis of microsporogenesis. J Hered 72:52–56
Rajendra BR, Tomb AS, Mujeeb KA, Bates LS (1978) Pollen morphology of selected Triticeae and two intergeneric hybrids. Pollen Spores 20:145–156
Rajki E, Rajki S (1966) Research work on hybrid wheat at Martonvásár. Acta Agron Acad Sci Hung 15:199–214
Rakshit S (1967) Induced male sterility in jute (*Corchorus capsularis* L.). Jpn J Genet 42:139–143
Ramaer H (1935) Cytology of *Hevea*. Genetica (The Hague) 17:193–236
Ramage RT (1965) Balanced tertiary trisomics for use in hybrid seed production. Crop Sci 5: 177–178
Ramage RT (1975) Hybrid barley. In: Gaul H (ed) Proc 3rd Int Barley Genet Symp Garching, pp 761–770

Ramage RT (1977) Varietal improvement of wheat through male sterile facilitated recurrent selection. ASPAC Techn Bull 37:6–18
Ramage RT, Suneson CA (1958) A nullisomic oat. Agron J 50:52–53
Ramage RT, Tuleen NA (1964) Balanced tertiary trisomics in barley serve as a pollen source homogenous for a recessive lethal gene. Crop Sci 4:81–82
Raman VS (1955) Cytology of Indian Jasmines III. Meiosis. Cytologia 20:133–147
Ramanna MS, Abdalla MMF (1970) Fertility, late blight resistance and genome relationship in an interspecific hybrid, *Solanum polytrichon* Rydb x *S. phureja* Juz et Buk. Euphytica 19:317–326
Ramanna MS, Hermsen JGT (1974) Unilateral eclipse sterility in reciprocal crosses between *Solanum verrucosum* Schlechtd. and diploid *S. tuberosum* L. Euphytica 23:417–421
Ramanna MS, Hermsen JGT (1979) Unique meiotic behaviour in F_1 plants from a cross between a nontuberous and a tuberous *Solanum* species in section petota. Euphytica 28:9–15
Ramanna MS, Hermsen JGT (1981) Structural hybridity in the series etuberosa of the genus *Solanum* and its bearing on crossability. Euphytica 30:15–31
Ramanujam S (1935) Male sterility in rice. Madras Agric J 23:190
Ramanujam S, Parthasarathy N (1935) An asynaptic mutant in rice (*Oryza sativa* L.). Proc Indian Acad Sci B 2:80–87
Ramavarma KT, Apparao K, Narayanan AI (1978) Male sterile flower anomalies in an interspecific *Nicotiana* hybrid. Tobacco Res 4:29–33
Randall TE, Rick CM (1945) A cytogenetic study of polyembryony in *Asparagus officinalis* L. Am J Bot 32:560–569
Rani A (1976) Effect of a morphactin on growth and floral morphogenesis in *Nigella damascena*. Curr Sci (Bangalore) 45:427–428
Rao MK, Devi KU (1983) Variation in expression of genic male sterility in pearl millet. J Hered 74:34–38
Rao MK, Koduru PRK (1978a) Cytogenetics of a factor for syncyte formation and male sterility in *Pennisetum americanum*. Theor Appl Genet 53:1–7
Rao MK, Koduru PRK (1978b) Inheritance of genetic male sterility in *Pennisetum americanum* (L.) Leeke. Euphytica 27:777–783
Rao NGP (1961) Occurrence of cytoplasmic genetic male sterility in some Indian *sorghums*. Indian J Genet Plant Breed 22:257–259
Rao PN (1975) Male sterility *Chionachne koenigii*. Maydica 20:197–200
Rao VSP, Jain HK (1975) Developmental basis of male sterility in bread wheat. Z Pflanzenzücht 74:168–171
Rathay E (1888–1889) Die Geschlechtsverhältnisse der Reben I, II. Frick, Wien
Rathore RKS, Bishnoi SK, Nainalu VG (1986) Differential behaviour of male sterile lines in wheat (abstr). In: Proc 73rd Indian Sci Cong Sec Bot, p 173
Raut RN, Sharma B, Pokhriyal SC, Singh MP, Jain HK (1973) Induced mutations: some basic findings and applied results. Indian J Genet Plant Breed 34:311–315
Rawal KM, Rachie KO (1974) Recent progress in population improvement of cowpeas, *Vigna ungiuculata* (L.) Walp. In: Proc 2nd Annu Conf Genet Soc Ibadan, pp 57–61
Rawal KM, Bryant P, Rachie KO, Porter WM (1978) Cross pollination studies of male sterile genotypes in cowpeas. Crop Sci 18:283–285
Rawat DS, Anand IJ (1979) Male sterility in Indian mustard. Indian J Genet Plant Breed 39: 412–415
Razzaque CA (1975) Genetics of male sterility in rice. Indian J Genet Plant Breed 34:303–308
Reddy BB, Reddi MV (1970) Studies on the breakdown of male sterility and other related aspects in certain cytoplasmic male sterile lines of pearl millet (*Pennisetum typhoides* Stapf and Hubb). Andhra Agric J 17:173–180
Reddy BB, Reddi MV (1974) Cytohistological studies on certain male sterile lines of pearl millet (*Pennisetum typhoides* S and H). Cytologia 39:585–589
Reddy BVS, Reddy LJ, Murthi AN (1977) Reproductive variants in *Cajanus cajan* (L.) Millsp. Trop Grain Legume Bull 7:11
Reddy BVS, Green JM, Bisen SS (1978) Genetic male sterility in pigeon pea. Crop Sci 18:362–364

Reddy PS, Thammiraju B (1977) Inheritance of fertility restoration in sunflower (*Helianthus annuus* L.). Euphytica 26:409–412

Reed DJ, Moore TC, Anderson JD (1965) Plant growth retardant B-995: a possible mode of action. Science 148:1469–1471

Reeves AF (1973) An observation on natural outcrossing in the tomato (*Lycopersicon esculentum* Mill.) in north west Arkansas. Proc Ark Natl Acad Sci USA 27:24–25

Rehm S (1952) Male sterile plants by chemical treatments. Nature 170:38–39

Reimann-Philipp R (1976) Breeding of F_1 hybrids in flowers. In: Proc 7th Eucarpia Cong Elsevier, Amsterdam, pp 135–146

Reimann-Philipp R (1983) Heterosis in ornamentals. In: Frankel R (ed) Springer, Berlin, Heidelberg, New York, pp 234–259

Reimann-Philipp R, Fuchs G (1971) F_1 Hybriden bei *Ageratum houstonianum.* Gartenwelt 71: 443–444

Reimann-Philipp R, Seidel H (1963) Untersuchungen bei *Begonia semperflorens* Link und Otto über die Eignung des Merkmals der Anterendeformation der "cinderella"-Sorten für ihre Verwendbarkeit als männlich steriler Kreuzungspartner in der Züchtung von F_1-Hybriden. Z Pflanzenzücht 50:59–70

Rémy R, Ambard-Bretteville F (1983) Two dimensional analysis of chloroplast proteins from normal and cytoplasmic male sterile *Brassica napus.* Theor Appl Genet 64:249–253

Rheenen Van HA, Muigai SGS, Kitivo DK (1979) Male sterility in beans (*Phaseolus vulgaris* L.). Euphytica 28:761–763

Rhoades MM (1931) Cytoplasmic inheritance of male sterility in *Zea mays.* Science 73:340–341

Rhoades MM (1933) The cytoplasmic inheritance of male sterility in *Zea mays.* J Genet 27:71–93

Rhoades MM (1943) Genic induction of an inherited cytoplasmic difference. Proc Natl Acad Sci USA 29:327–329

Rhoades MM (1946) Plastid mutation. Cold Spring Harbor Symp Quant Biol 11:202–207

Rhoades MM (1950) Gene induced mutation of a heritable cytoplasmic factor producing male sterility in maize. Proc Natl Acad Sci USA 36:634–635

Rhoades MM, Rhoades VH (1939) Genetic studies with factors in the 10th chromosome in maize. Genetics 24:302–314

Rhodes AM (1959) Species hybridization and interspecific gene transfer in the genus *Cucurbita.* Proc Amer Soc Hort Sci 74:546–551

Rhodes PR, Zhu YS, Kung SD (1981) *Nicotiana* chloroplast genome I. Chloroplast DNA diversity. Mol Gen Genet 182:106–111

Rhyne CL (1965) Cytoplasmic inheritance of reduced androecium in cotton. J Hered 61:67–70

Rhyne CL, Rhyne P (1971) Another male sterile cotton? In: Proc Beltwide Cotton Prod Res Conf, pp 69–70

Riccelli M, Mazzani B (1964) Manifestation of heterosis in development, earliness and yield in diallel crosses of 32 sesame cultivars. Agron Trop Venezuela 14:101–125

Riceman DS, Anderson AJ (1943) The symptoms and effects of copper deficiency in cereals and pasture plants in south Australia. J Agric S Aust 47:64–72

Richardson RW, Alvarez LE (1957) Pollination relations among vegetable crops in Mexico I. Natural cross pollination in cultivated tomatoes. Proc Amer Soc Hort Sci 69:366–371

Richmond TR (1962) Effect of sodium 2,3-Dichloroisobutyrate on six characteristics of American upland cotton. Crop Sci 2:58-60

Richmond TR, Kohel RL (1961) Analysis of a completely male sterile character in American uplan cotton. Crop Sci 1:397–401

Rick CM (1944) A new male sterile mutant in tomato. Science 99:543

Rick CM (1945) A survey of cytogenetic causes of unfruitfulness in tomato. Genetics (The Hague) 30:347–362

Rick CM (1947a) A flower marker for plant breeding operations. Science 106:645

Rick CM (1947b) The effect of planting design upon the amount of seed produced by male-sterile tomato plant as a result of natural cross pollination. Proc Amer Soc Hort Sci 50:273–284

Rick CM (1948) Genetics and development of nine male-sterile tomato mutants. Hilgardia 18: 599–633

Rick CM (1949) Rates of natural cross-pollination of tomatoes in various localities in California as measured by the fruits and seed set on male-sterile plants. Proc Amer Soc Hort Sci 54:237–252
Rick CM (1950) Pollination relations of *Lycopersicon esculentum* in native and foreign regions. Evolution 4:110–122
Rick CM (1958) The role of natural hybridization in the derivation of cultivated tomatoes of western south America. Econ Bot 12:346–367
Rick CM (1966) Abortion of male and female gametes in the tomato determined by allelic interactions. Genetics 53:85–96
Rick CM (1974) The tomato. In: King RC (ed) Handbook of genetics. Plenum, New York, pp 247–280
Rick CM (1976) Tomato, *Lycopersicon esculentum*. In: Simmonds NW (ed) Evolution of crop plants. Longman, London, pp 268–273
Rick CM (1980a) Linkage report. Rep Tomato Genet Coop 30:2–17
Rick CM (1980b) Tomato. In: Fehr WR, Hadley HH (eds) Hybridization of crop plants. Am Soc Agron, Crop Soc Am Wisconsin, pp 669–680
Rick CM, Boynton JE (1965) A temperature sensitive male sterile mutant Vms. Tomato Genet Coop Rep 15:52–53
Rick CM, Boynton JE (1967) A temperature sensitive male sterile mutant of the tomato. Am J Bot 54:601–611
Rick CM, Butler L (1956) Cytogenetics of the tomato. Adv Genet 8:267–382
Rick CM, Robinson J (1951) Inherited defects of floral structure affecting fruitfulness in *Lycopersicon esculentum*. Am J Bot 38: 639–652
Rick CM, Fobes JF, Holle M (1977) Genetic variation in *Lycopersicon pimpinellifolium:* evidence of evolutionary change in mating systems. Plant Syst Evol 127:159–170
Rick CM, Holle M, Thorp RW (1978) Rates of cross-pollination in *Lycopersicon pimpinellifolium:* impact of genetic variation in floral characters. Plant Syst Evol 129:31–44
Rife DC (1940) Genetic studies of *Coleus* II. Leaf form and a bud mutation. J Hered 31:293–296
Rife DC (1948) Simply inherited variations in *Coleus*. J Hered 39:85–91
Roath WW, Hockett EA (1970) Pollen development in genetic male-sterile barley. In: Nilan RA (ed) Barley Genetics II. Washington State Univ Press, pp 308–315
Roath WW, Hockett EA (1971) Genetic male sterility in barley III. Pollen and anther characteristics. Crop Sci 11:200–203
Roath WW, Hockett EA (1972) Genetic male sterility in barley IV. Self-incompatibility tests. Crop Sci 12:250–251
Robertson DW, Deming GW (1931) Natural crossing in barley at Fort Collins Colorado. J Am Soc Agron 23:402–406
Robertson LD, Curtis BC (1967) Monosomic analysis of fertility restoration in common wheat (*Triticum aestivum* L.). Crop Sci 7:493–495
Robinson RG (1984) Distance from pollen source and yields of some male sterile sunflower and sorghum. Can J Plant Sci 64:857–861
Robinson RW, Shanon S, Guardia De La MD (1969) Regulation of sex expression in cucumber. BioScience 19:141–142
Robinson RW, Whitaker TW, Bohn GW (1970) Promotion of pistillate flowering in *Cucurbita* by 2-chloroethylphosphohic acid. Euphytica 19:180–183
Rocehetta G, Giorg G, Giovànnelli G (1976) Correlation analysis between morphological traits and productivity in cultivated *Capsicum* for an understanding of the heterosis phenomenon. Genet Agric 30:355–374
Rodriquez BP, Lambeth VN (1972) Synergism and antagonism of GA and growth inhibitors on growth and sex expression in cucumber. J Am Soc Hortic Sci 97:90–92
Roever WE (1948) A promising type of male sterility for use in hybrid tomato seed production. Science 107:506
Rogers JS (1950a) In: Rep 8th southern corn improvement conf, Aug 1949, Mimeo, pp 12
Rogers JS (1950b) In: Rep 9th southern corn improvement conf, Aug 1950, Mimeo, pp 23–25

Rogers JS, Edwardson JR (1952) The utilization of cytoplasmic male sterile inbreds in the production of corn hybrids. Agron J 44:8–13

Rohrbach U (1965a) Beiträge zum Problem der Pollensterilität bei *Beta vulgaris* L. II. Untersuchungen über die Wirkung der Herkunft auf dem Phänotyp des Merkmals Pollensterilität. Z Pflanzenzücht 54:23–45

Rohrbach U (1965b) Beiträge zum Problem der Pollensterilität bei *Beta vulgaris* L. I. Untersuchungen über die Ontogenese des Phänotyps. Z Pflanzenzücht 53:105–124

Rohrbach U (1965c) Beiträge zum Problem der Pollensterilität bei *Beta vulgaris* L. IV. Untersuchungen über die Auswirkungen der Pollensterilität auf Samen- und Wurzelertrag der Zuckerrüben. Z Pflanzenzücht 54:193–205

* Roman T, Nalepa S (1970) Badina cytologiczne meskosterylnych form tytoniu (*Nicotiana tabaccum* L.). Biul CLPT 1–2:23–31

* Romanenko LG, Vislovkhov NP (1981) Inbreeding in the breeding of *Lavandula angustifolia*. Genet Selekt Ukranian USSR 73:219–221

Rommel M (1965) Cytogenetics of autotetraploid sugarbeet (*Beta vulgaris* L.) I. Tetraploid varieties. Züchter 35:219–222

Rood SB, Pharis RP, Major DJ (1980) Changes of endogenous gibberellin like substances with sex reversal of the apical infloresence of corn. Plant Physiol (Bethesda 66:793–796

Roscoe MV (1927) Cytological studies in the genus *Wisteria*. Bot Gaz 84:171–186

Ross H (1966) The use of wild *Solanum* species in German potato breeding of the past and today. Am Potato J 43:63–80

Ross MD (1967) Two genetic mechanisms governing outbreeding in *Plantago lancoelata*. Genetics 56:584–585

Ross MD (1969) Digenic inheritance of male sterility in *Plantago lanceolata*. Can J Genet Cytol 11:739–744

Ross MD (1970a) Breeding systems in *Plantago*. Heredity 25:129–133

Ross MD (1970b) Evolution of dioecy from gynodioecy. Evolution 24:827–828

Ross MD (1978) The evolution of gynodioecy and subdioecy. Evolution 32:174–188

Ross MD (1980) The evolution and decay of overdominance during the evolution of gynodioecy, subdioecy and dioecy. Am Nat 116:607–620

Ross MD (1982) Five evolutionary pathways to subdioecy. Am Nat 119:297–318

Ross MD, Shaw RF (1971) Maintenance of male sterility in plant populations. Heredity 26:1–8

Ross MD, Weir BS (1975) Maintenance of male sterility in plant populations III. Mixed selfing and random mating. Heredity 35:21–29

Ross RW, Peloquin SJ, Hougas RW (1964) Fertility of hybrids from *Solanum phureja* and haploid *S. tuberosum* matings. Eur Potato J 7:81–89

Ross WM (1957) Stigma receptivity in cytoplasmic male sterile sorghum. Agron J 49:219–220

Ross WM (1961) Cytoplasm substitutions. Sorghum Newslett 4:29

Ross WM (1965) Cytoplasmic male sterility and fertility restoration of some major sorghum groups. In: Proc grain sorghum research and utilization conference Amarillo Texas, pp 57–62

Ross WM Hackerott HL (1972) Registration of seven isocytoplasmic sorghum germplasm lines (Reg Nos GP 9 to GP 15). Crop Sci 12:720–721

Ross WM, Webester OJ (1959) Additional notes on stigma receptivity in cytoplasmic sterile sorghum. Agron J 51:632

Roundy TE, Theurer JC (1974) Inheritance of a yellow-leaf mutant and a pollen fertility restorer in sugarbeet. Crop Sci 14:62–63

Rousselle P (1982) Preliminary results of a programme to introduce the Ogura male sterility into Swede rape from raddish. Agronomie 2:859–864

Rousselle P, Renard M (1982) Value of cultivar Bronowski in producing cytoplasmically male sterile plants in swede rape (*Brassica napus*), Agronomie 2:951–955

Roux JB (1961) Contribution a l'etude d'une caracterc de sterilite male chez les cotonnier. Coton Fibres Trop 16:309–311

Roux JB, Chirinian G (1959) Essis d'induction de la sterilite male sur le Cotonnier. Coton Fibres Trop 14:363–370

Rowell PL, Miller DG (1971) Induction of male sterility in wheat with 2-chloroethylphosphonic acid (etheral). Crop Sci 11:629–631

Rowley JR (1976) Dynamic changes in pollen wall morphology. In: Ferguson IK, Muller J (eds) The evolutionary significance of exine. Academic Press, London, pp 39–65

Roy AT, Jana MK (1983) Male sterility in black gram (*Vigna mungo*) (abstr). Proc 70th Indian Sci Congr Bot, vol 3, p 90

Rubaihayo PR, Gumisiriza G (1978) The causes of genetic male sterility in 3 soybean lines. Theor Appl Genet 53:257–260

Rubashevskia MK (1931) Observations on the wild carrot in cultivation and under natural conditions. Bull Appl Bot Genet Plant Breed 26:194–252

Rudich J, Halevy AH (1974) Involvement of abscissic acid in the regulation of sex expression in the cucumber. Plant Cell Physiol 15:635–642

Rudich J, Peles A (1976) Sex expression in watermelon as effected by photoperiod and temperature. Sci Hortic 5:339–344

Rudich J, Halevy AH, Kedar N (1969) Increase in femaleness of three cucurbits by treatment with ethrel, an ethylene releasing compound. Planta (Berl) 86:69–76

Rudich J, Halevy AH, Kedar N (1972) Interaction of gibberellin and SADH on growth and sex expression of muskmelon. J Amer Soc Hort Sci 97:369–372

Rudich J, Kedar N, Halevy AH (1970) Changed sex expression and possibilities for F_1-hybrid seed production in some cucurbits by application of ethrel and alar(B-995). Euphytica 19: 47–53

Rudich J, Baker LR, Scott JW, Sell HM (1976) Phenotypic stability and ethylene evolution in androecious cucumber. J Amer Soc Hort Sci 101:48–51

Rudarf W (1954) Neue Beobachtungen an Bastarden von *Phaseolus vulgaris* L x *Phaseolus multiflorus* Lam und *Ph. multiflorus* x *Ph vulgaris.* Proc 9th Int Genet Cong 2:844–845

* Ruebenbauer T, Heinrich J (1968) Badania nad uzyskaniem nasion mieszańcowych tytoniu przy Zastosowaniu roślin cytoplazmatycznie meskosterylnych. Biul CLPT USSR 1–2:3–13

* Ruebenbauer T, Schultis L (1960) Zastosovanie selektywnego gemetocydu FW-450 dla chemeicznego kastrowania kwaitow buraka. Hodwla Róslin Aklimat Nasiennicto USSR 4:199–204

Rundfeldt H (1960) Untersuchungen zur Züchtung des Kopfkohls (*B. oleracea* L. var *capitata*). Z Pflanzenzücht 44:30–62

Rusenova-Kondareva I (1968) Results of interspecific hybridization in the genus *Capsicum* L. Bulg Acad Sci Sofia, pp 230

Rustagi PN, Mohan Ram HY (1971) Evaluation of mendok and dalapon as male gametocides and their effects on growth and yield of linseed. New Phytol 70:119–133

Rutger JN, Jensen NF (1967) On the use of genetic male sterility in screening for cytoplasmic-genetic sterility in barley. Euphytica 16:350–353

Rutger JN, Shinjo C (1980) Male sterility in rice and its potential use in breeding. In: Innovative approaches to rice breeding. IRRI Philippines Publ, pp 53–66

* Ryabota AN, Dumacheva LP, Vol FVG (1975) The inheritance of male sterility in sunflower after unrelated crosses. Selekt Semenov USSR 29:100–105

Ryder EJ (1963) An epistatically controlled pollen sterile in lettuce (*Lactuca sativa* L.). Proc Amer Soc Hort Sci 83:585–589

Ryder EJ (1967) A recessive male sterility gene in lettuce (*Lactuca sativa*). Proc Amer Soc Hort Sci 91:366–368

Ryder EJ (1971) Genetic studies in lettuce (*Lactuca sativa* L.). J Am Soc Hortic Sci 96:826–828

Sacks JM, Sadanaga K (1984) Linkage between the male sterility gene (ms1) and translocation breakpoint in soybean (*Glycine max*). Can J Genet Cytol 26:401–404

Sadanaga K (1965) Male sterility in diploid oat. Crop Sci 5:279

Sadanaga K, Grindeland R (1981) Natural cross-pollination in diploid and autotetraploid soybeans. Crop Sci 21:503–506

Safeeulla KM (1977) Genetic vulnerability - the basis of recent epidemics in India. In Day PR (ed) The genetic basis of epidemics in agriculture I. Ann NY Acad Sci 287:72–85

Sage GCM (1972) The inheritance of fertility restoration in male sterile wheat carrying cytoplasm from *Triticum timopheevi.* Theor Appl Genet 45:126–133

Sage GCM (1976) Nucleo-cytoplasmic relationships in wheat. Adv Agron 28:267–300

Sain D, Kanwar ZS (1977) Screening and evaluation of pearl millet male sterile lines, pollinators and their F_1's for downy mildew resistance. Indian J Agric Sci 47:296–298

Saini HS, Aspinall D (1982) Abnormal sporogenesis in wheat (*Triticum aestivum* L.) induced by short periods of high temperature. Ann Bot 49:835–846

Saini HS, Sedgley M, Aspinall D (1984) Development anatomy in wheat of male sterility induced by heat stress, water deficit or abscissic acid. Aust J Plant Physiol 11:243–253

Saini SS, Davis GN (1969) Male sterility in *Allium cepa* and some species hybrids. Econ Bot 23: 37–44

Saito T, Ito H (1963) Factors responsible for the sex expression of Japanese cucumber XIII. Physiological factors associated with the sex expression of flowers II. Role of gibberellin. Jpn J Soc Hortic Sci 32:287–290

Salaman RN (1910) Male sterility in potatoes, a dominant mendelian character with remarks on the shape of the pollen in wild and domestic varieties. J Linn Soc 39:301–312

Salaman RN (1939) The potato virus X, its strains and reactions. Philos Trans R S Ser B Biol Sci 229:137–217

Salaman RN (1949) The history and social influence of the potato. Cambridge Univ Press

Salaman RN, Lesley JW (1922) Genetic studies in potatoes; sterility. J Agric Sci 12:31–39

Salaman RN, Pelley Le RH (1930) Para-crinkle: a potato disease of the virus group. Proc R Soc Ser B 106:140–175

Salmon N (1963) Behandlung von *Beta vulgaris* Tel Aviv, Israel. Ref Chem Zentralbl 134:156–162

Saltviet ME, Bradford KJ, Dilley DR (1978) Silver ion inhibits ethylene synthesis and action in ripening fruits. J Amer Soc Hort Sci 103:472–475

Sampath S, Mohanty HK (1954) Cytology of semi-sterile rice hybrids. Curr Sci (Bangalore) 23: 182–183

Sampson DR (1966a) Genetic analysis of *Brassica oleracea* using nine genes from sprouting broccoli. Can J Genet Cytol 8:401–403

Sampson DR (1966b) Linkage of genetic male sterility with a seedling marker and its use in producing F_1 hybrid seed of *Brassica oleracea* (cabbage, broccoli, kale etc). Can J Plant Sci 46: 703

Sampson DR (1970) Close linkage of genes for male sterility and anthocyanin synthesis in *Brassica oleracea* promising for F_1-hybrid seed production; multivalent at meiosis not involved in the linkage. Can J Genet Cytol 12:677–684

Sams CE, Krüger WA (1977) Ethephon alteration of flowering and fruit set pattern of summer squash. Hort Science 12:162–164

Sanchez-Monge E (1970) Effetos de la aloplasmia en los trigos cultivados VII. Jornadas Genetica Luso-Españolas Pamplona Spain 12:33–44

Sanchez-Monge E (1975) Hexaploid triticale with different cytoplasms. In: Proc 1973 Int Symp Lenningrad, pp 175–180

Sand SA (1960) Autonomy of cytoplasmic male sterility on grafted scions of tobacco. Science 131:665

Sand SA (1968) Genetic modifications of cytoplasmic male sterility in tobacco. J Hered 59:175–177

Sand SA (1978) Quantitative and qualitative effects on the debneyi system of cytoplasmic male sterility in *Nicotiana*. Genetics 88:86

Sand SA, Christoff GT (1973) Cytoplasmic-chromosomal interactions and altered differentiation in tobacco. J Hered 64:24–30

Sano Y (1983) A new controlling gene sterility in F_1 hybrids of two cultivated rice species. Its association with photoperiodic sensitivity. J Hered 74:435–439

Sansome FW (1936) Some experiments with *Geranium* species. J Genet 33:359–363

Sapra VT, Sharma GC, Hughes JL (1974) Chemical induction of male sterility in hexaploid triticales. Euphytica 23:685–690

Sarathe ML, Dabral KC (1969) Heterosis in *Sesamum orientale* L. Sci Cult 35:572–573

Sarma NP, Mohanty PK, Jachuk PJ (1982) Outcrossing potential of a cytosterile stock of rice. IRRI Newslett Philippines 7(6):3

Sartorius O (1926) Zur Entwicklung und Physiologie der Rebblüte. Angew Bot 8:29–36, 65–89
Sarvella P (1966) Environmental influences on sterility in cytoplasmic male sterile cottons. Crop Sci 6:361–364
Sarvella P, Stojanovic BJ (1968) Amino acids present in male-sterile cottons (*Gossypium*). Can J Genet Cytol 10:369–373
Sarvella P, Stojanovic BJ, Grogan GO (1967) Amino acids at different stages in normal male sterile and restored maize (*Zea mays* L.). Z Pflanzenzücht 57:361–370
Sasahara T, Katsuo K (1965) Studies on the cytoplasmic differences among rice varieties, *Oryza sativa* L. III. On the abortive pollen of Fujisaka No 5-type plants with the cytoplasm of chinese wild variety *Oryza sativa* L.f. *spontanea*. Jpn J Breed 15:191–196
Sasaki M (1956) Pistilloidy in a rye strain with wheat plasma. Jpn J Genet 31:310
Sasakuma T, Maan SS (1978) Male sterility-fertility restoration system in *Triticum durum*. Can J Genet Cytol 20:389–398
Sasakuma T, Maan SS, Williams ND (1978) EMS induced male sterile mutants in euplasmic and alloplasmic wheat. Crop Sci 18:850–853
Satake T (1976) Determination of most sensitive stage to sterile-type cool injury in rice plants. Res Bull Hokkaido Natl Agric Exp Stat Jpn 113:1–35
Satake T, Hayase H (1970) Male sterility caused by cooling treatment at the young microspore stage in rice plants V. Estimation of pollen developmental stage and the most sensitive stage to coolness. Proc Crop Sci Soc Japan 39:468–473
Satake T, Hayase H (1974) Male sterility caused by cooling treatment at the young microspore stage in rice plants X. A secondary sensitive stage at the begining of meiosis. Proc Crop Sci Soc Jpn 43:36–39
Satake T, Yoshida S (1977) Mechanism of sterility caused by high temperature at flowering time in indica rice. JARQ Japan 11:127–128
Satake T, Yoshida S (1978) High temperature-induced sterility in Indica rices at flowering. Jpn J Crop Sci 47:6–7
Savchenko NI, Borisyok VA, Enko II (1974) Pigments in vegetative organs of wheat plants with cytoplasmic male sterility. Tsitol Genet USSR 8:487–491
Savchenko NI, Lastovych AS (1965) Morphological and cytological peculiarities of wheat forms with cytoplasmic male sterility. Ukrajins' K Bot Zh USSR 22:35–42
Savchenko NI, Belons VE, Scraya LV (1968) Biochemical pecularities of winter lines with cytoplasmic male sterility. Tsitol Genet Kiev USSR 2:226–231
Savitsky VF (1952a) A genetic study of monogerm and multigerm characters in beets. Proc Am Soc Sugarbeet Techn 6:330–338
Savitsky VF (1952b) Methods and results of breeding work with monogerm beets. Proc Am Soc Sugarbeet Techn 6:344–350
Savitsky VF (1954) Obtaining tetraploid monogerm self fertile, self sterile and male sterile beets. Proc Am Soc Sugarbeet Techn 8:50–58
Savitsky VF (1958) Differences in interaction of cytoplasm-genes in different male sterile races in *Beta vulgaris* L. Proc 10th Int Cong Genet 2:251
Savitsky VF (1963) Tetraploid inheritance of monogerm character and male sterility in *Beta vulgaris* L. (sugarbeet). In: Proc 11th Int Cong Genet, the Hague, vol 1, p 215
Sawada S, Kinoshita T (1977) Comparison of the male sterility induced by low temperatures with the genetical male sterility in rice I. Observation on indehiscent anthers at anthesis. J Fac Agric Hokkaido Univ 58:503–509
Sawhney VK (1974) Morphogenesis of the stamenless-2 mutant in tomato III. Relative levels of gibberellins in the normal and mutant plants. J Exp Bot 25:1004–1009
Sawhney VK (1981) Abnormalities in pepper (*Capsicum annuum*) flowers induced by gibberellic acid. Can J Bot 59:8–16
Sawhney VK (1983) Temperature control of male sterility in a tomato mutant. J Hered 74:51–54
Sawhney VK, Greyson RJ (1973a) Morphogenesis of the stamenless-2 mutant in tomato I. Comparative description of the flowers and ontogeny of stamens in the normal and mutant plants. Am J Bot 60:514–523

Sawhney VK, Greyson RI (1973b) Morphogenesis of the stamenless-2 mutant in tomato II. Modifications of sex organs in the mutant and normal flowers by plant hormones. Can J Bot 51: 2473–2479

Sawhney VK, Greyson RI (1979) Interpretations of determination and canalisation of stamen development in a tomato mutant. Can J Bot 57:2471–2477

* Sawyer RL (1978) In: Ann Rep Int Potato Center Peru Chile, pp 3–4

Saxena MBL, Chaudhary BS (1977) Breakdown of male sterility in some male-sterile lines of pearl millet (*Pennisetum typhoides*) under conditions of arid zones. Ann Arid Zone 16:427–432

Sayers ER (1959) The effect of FW-450 as a gametocide on *Sorghum vulgare*. Undergrad Spec Probl Rep Univ Illinois Urbana, pp 19

Scalla R, Duc G, Rigand J, Lefebvre A, Meignoz R (1981) RNA containing intracellular particles in cytoplasmic male sterile faba beans (*Vicia faba* L.). Plant Sci Lett 22:269–277

Scalla R, Duc G, Lefebvre A, Rigaud J, Meignoz R (1982) La stérilité mâle nucléo-cytoplasmique chez la Féverole (*Vicia faba* L.) II. Etude cytologique et biochimique de particules de type viral associées àla stérilité mâle. Rev Cytol Biol Vég 5:71–79

Schark AE (1957) A comparison of microsporogenesis in fertile and sterile potato varieties. Ph D Thesis Iowa State Coll Ames, pp 62

Schatz G, Masson TL (1974) The biosynthesis of mitochondrial proteins. Annu Rev Biochem 43:51–87

* Scheglov Yu S, Zvyagina GN (1975) Vestn Selsk Vennoi Nauk USSR 6:31–36

Scheifele GL, Nelson RR (1969) The occurrence of *Phyllosticta* leaf spot of corn in Pennsylvania. Plant Dis Rep 53:186–189

Scheifele GL, Nelson RR, Koons C (1969) Male sterility cytoplasm conditioning susceptibility of resistant inbred lines of maize to yellow leaf blight caused by *Phyllosticta zeae*. Plant Dis Rep 53:656–659

Scheifele GL, Whitehead W, Rowe C (1970) Increased susceptibility to southern leaf spot (*Helminthosporium maydis*) in inbred lines and hybrids of maize with Texas male-sterile cytoplasm. Plant Dis Rep 54:501–503

Schertz KF (1977) Registration of A_2Tx 2753 and BTx 2753 sorghum germplasm. Crop Sci 17: 983

Schertz KF, Pring DR (1982) Cytoplasmic sterility systems in sorghum. In: Proc Int Symp Sorghum ICRISAT Hyderabad, pp 373–383

Schertz KF, Ritchey JM (1978) Cytoplasmic-genic male sterility systems in *Sorghum*. Crop Sci 18:890–893

Schick R, Stubbe H (1932) Die Gene von *Antirrhinum majus* II. Z Indukt Abstammungs, Vererbungsl 62:249–290

Schmidt JW, Johnson VA, Maan SS (1962) Hybrid wheat. Nebraska Exp Stn Quart Rept Fall 9: 1–12

Schmidt JW, Johnson VA, Morris MR, Mattern PJ (1970) Cytoplasmic male sterility and fertility restoration. Seiken Zihô 22:113–118

Schmidt R (1976) Filament histology and anther deshiscence. Bot J Linn Soc 73:303–315

Schmidt V, Schmidt H (1975) Anatomical and cytological investigations of male-sterile mutants of tomato with tendency to stamenless character. In: Proc 12th Int Bot Cong Leningrad 1:233

Schmidt V, Schmidt H (1981a) Untersuchungen an pollensterilen, stamenlessähnlichen Mutanten von *Lycopersicon esculentum* Mill. I. Die Anterenentwicklung in den Mutanten ms-15 und ms-33. Biol Zbl 100:679–689

Schmidt H, Schmidt V (1981b) Untersuchungen an pollensterilen, stamenlessähnlichen Mutanten von *Lycopersicon esculentum* Mill. II. Normalisierung von ms-15 und ms-33 mit Gibberellinsäure (GA_3). Biol Zentralbl 100:691–696

Schmidt V, Schmidt H, Hagemann R (1979) Eigenschaften pollensteriler Mutanten der Tomate unter dem Aspekt ihrer Eignung für Züchtungsprogramme. Tag Ber Akad Landw Wiss Berlin DDR 168:179–183

Schneider G (1964) Eine neue Gruppe von synthetischen Planzenwachstumsregulatoren. Naturwissenschaften 51:416–417

Schneider G (1970) Morphactin: physiology and performance. Ann Rev Plant Physiol 21:499–536

Schnell LO (1948) A study of meiosis in the microsporocytes of interspecific hybrids of *Solanum demissum* x *S. tuberosum* carried through from backcrosses. J Agric Res 76:185–212

Schonhorst MR (1958) The inheritance of viable pollen in alfalfa. Ph D Thesis Purdue Univ Lafayette

Schooler AB (1967) A form of male sterility in barley hybrids. J Hered 58:206–211

Schultz ES, Folsom D (1925) Infection and dissemination experiments with degeneration diseases of potatoes. J Agric Res 30:493–528

Schultze-Motel von J (1972) Die araelogischen Reste der Ackerbohne *Vicia faba* und die Genese der Art. Kulturpflanze 19:321–358

Schulz H (1908) In: Knuth's handbook of flower pollination, vol 2. Clarendon, Oxford

Schulz-Schaeffer J (1972) An approach towards the development of hybrid intermediate wheat grass *Agropyron intermedium* (Host) Beauv. Z Pflanzenzücht 67:202–220

Schulz-Schaeffer J (1978) Registration of Montana-1 male-sterile intermediate wheat grass germplasm (Reg No GP1). Crop Sci 18:920

Schuster AM, Sisco PH, Levings CS (1983) Two unique RNAs in cms-S and RU maize mitochondria. In: Goldberg R (ed) UCLA Symp Mol Cell Biol. Plant Mol Biol 12:437–443

Schuster W (1961) Untersuchungen über künstlich induzierte Pollensterilität bei Sonnenblumen (*Helianthus annuus* L.). Z Pflanzenzücht 46:389–404

Schuster W (1964) Inbreeding and heterosis in sunflower (*Helianthus annuus* L.). Schmitz, Giessen, pp 135

Schwartz D (1951) The interaction of nuclear and cytoplasmic factors in the inheritance of male sterility in maize. Genetics 36:676–696

Schweisguth B (1970) La stérilité mâle chez le poireau *Allium porrum* L. Utilization en amelioration des plantes. La Stérilité mâle chez des plants horticoles. Eucarpia Versailles, p 63–69

Schweisguth B (1973) Étude d'un nouveau type de stérilité mâle chez Lóignon *Allium cepa* L. Ann Amélior Plant (Paris) 23:221–233

Schwemmle J (1928) Genetische und cytologische Untersuchungen bei *Oenothera*. Jahrb Wiss Bot 67:849–876

Schwendiman J (1974) Hybrid lines from the cross between *G. hirsutum* L. and *G. barbadense* L. II. Factors causing fertility anomalies. Cotton Fibres Tropicales 29:284–295

Schweppenhauser MA, Mann TJ (1968) Restoration of staminal fertility in *Nicotiana* by introgression. Can J Genet Cytol 10:401–411

Scoles GJ, Evans LE (1979a) The transfer of a cytoplasmic male sterility system to spring rye. Can J Plant Sci 59:163–169

Scoles GJ, Evans LE (1979b) The genetics of fertility restoration in cytoplasmic male sterile rye. Can J Genet Cytol 21:417–422

Scoles GJ, Evans LE (1979c) Pollen development in male-fertile and cytoplasmic male-sterile rye. Can J Bot 57:2782–2790

Scoles GJ, Evans LE (1979d) Hybrid vigour obtained by crossing inbred lines of spring rye. Can J Plant Sci 58:745–751

Scott DH, Riner ME (1946) Inheritance of male sterility in winter squash. Proc Am Soc Hortic Sci 47:375–377

Scott RA (1960a) Biochemical control of anther dehiscence in cotton treated with dalapon or GW-450. In: Proc cotton defoliation and physiology conf Memphis, pp 4–6

Scott RA (1960b) Crazy top-cause and cure. Arizona Farmer USA, pp 6–7

Scott RA (1961) Mechanism and reversal of gametocide response in cotton. Plant Physiol (Bethesda) 26:529–538

Scott RA (1962) Reversal of nondehiscence of anthers in cotton by foliar application of choline chloride or urea. Agron J 54:499–505

Scowcroft WR, Larkin PJ (1981) Chloroplast DNA assorts randomly in intraspecific somatic hybrids of *Nicotiana debneyi*. Theor Appl Genet 60:179–184

Sears ER (1947) Genetics and farming. USDA Yearbook, pp 245–255

Sears ER (1966) Nullisomic-tetrasomic combinations in hexaploid wheat. In: Riley R, Lewis KR (eds) Chromosome manipulation and plant genetics. Oliver and Boyd, Edinburgh, pp 29–45

Sears ER (1972) Chromosome engineering in wheat. Stadler Symp, Univ Miss, Columbia Missouri, pp 23–48

Sechler D, Chapman WH (1967) Hybrid vigour in intervarietal rye crosses. Proc Soil Crop Sci Soc Fla 27:81–87

Seetharam A, Kumari PK (1974) GA induced male sterility in sunflower. Sci Cult 40:398–399

Seetharam A, Kumari PK (1975) Induction of male sterility by gibberellic acid in sunflower. Indian J Genet Plant Breed 35:136–138

Selim AG (1931) A cytological study of the *Oryza sativa* L. Cytologia 2:1–26

Semenova LE (1963) Use of artificial male sterility for obtaining interspecific hybrids in cotton. Tr Prikladnoi Bot Genet Selek USSR 35:33–40

Sen NK, Bhowal JG (1962) A male sterile mutant cowpea. J Hered 53:44–46

Seth S, Meharchandani N, Kairon MS (1981) Non-dehiscent anther condition in cotton. Curr Sci (Bangalore) 50:289–290

Sethi GS, Bhateria SD (1977) Morphology and cytogenetics of some new macromutants in barley. Indian J Genet Plant Breed 37:73–79

Shambulingappa KG, Magoon ML (1963) Cytomorphological studies in the male sterile *Sorghum*. Indian J Gent Plant Breed 23:275–289

Shanmugasundaram A, Rajaskaran R, Selvaraj S, Ramamurthy TG (1968) Note on a new dwarf variety of cumbu. Madras Agric J 55:145–146

Sharma AK, Sharma A (1972) Chromosome techniques-theory and practice. Butterworths, London, pp 575

Sharma D, Green JM (1980) Pigeon pea. In: Fehr WR, Hadley HH (eds) Hybridization of crop plants. Am Soc Agron, Crop Soc Am Publ Madison, pp 470–481

Sharma RK, Reinbergs E (1972) Gene controlled abnormal cytokinesis causing male sterility in barley. Indian J Genet Plant Breed 32:408–410

Sharma RK, Reinbergs E (1976) Male sterility genes in barley and their sensitivity to light and temperature intensity. Indian J Genet Plant Breed 36:59–63

Sharma SK, Sharma B (1978) Induction of male sterility in lentil (*Lens culinaris* Medic). Legume Res 2:45–48

Sharma YP (1961) Preliminary report on breeding hybrid *Pennisetum*. Balwant Vidyapeth J Agric Sci Res 3:62–68

Sharma YP (1979) Note of functional male sterility induced by growth-regulators in pearl millet. Indian J Agric Sci 49:294–295

* Shebeski LH (1971) Cereal report. In: Inf Bull Near East V Cereal Improv Prod Project 8:1–7

Shehala AH, Comstock VE (1971) Heterosis and combining ability estimates in F_2 flax populations as influenced by plant density. Crop Sci 11:534–536

Shepard JF, Totten RE (1977) Mesophyll cell protoplasts of potato. Plant Physiol (Bethesda) 60:313–316

Shibuya T (1973) Studies on the genetic mechanism and genes responsible for partial male sterility in rice and their utilization in assessing cold tolerance. Bull Yamagata Univ Agric Sci 6: 1–55

Shifriss C (1950) Spontaneous mutations in the American varieties of *Cucumis sativus* L. Proc Amer Soc Hort Sci 55:351–357

Shifriss C (1973) Additional spontaneous male sterile mutant in *Capsicum annuum* L. Euphytica 22:527–529

Shifriss C, Frankel R (1969) A new male sterility gene in *Capsicum annuum* L. J Am Soc Hortic Sci 94:385–387

Shifriss C, Frankel R (1971) New sources of cytoplasmic male sterility in cultivated peppers. J Hered 62:254–256

Shifriss C, Guri A (1977) A study of the potential of cytoplasmic genic male sterility for hybrid seed production in *Capsicum annuum* L. Eucarpia Cong *Capsicum* 77:275–284

Shifriss C, Guri A (1979) Variation in stability of cytoplasmic-genic male sterility in *Capsicum annuum* L. J Amer Soc Hort Sci 104:94–96

Shifriss C, Rylski I (1972) A male sterile (ms-2) gene in California Wonder pepper (*Capsicum annuum* L.). Hort Science 7:36

Shifriss C, Rylski I (1973) Comparative performance of F_1 hybrids and open pollinated Bell pepper varieties (*Capsicum annuum* L.) under suboptimal temperature regimes. Euphytica 22:530–534

Shifriss C, Sacks JM (1980) The effect of distance between parents on the yield of sweet pepper x hot pepper hybrids *Capsicum annuum* L. in a single harvest. Theor Appl Genet 58:253–256

Shifriss O (1945) Male sterilities and albino seedlings in cucurbits. J Hered 36:47–52

Shifriss O (1950) Spontaneous mutations in the American varieties of *Cucumis sativus* L. Proc Amer Soc Hort Sci 55:351–357

Shifriss O (1960) Conventional and unconventional systems controlling sex variations in *Ricinus*. J Genet 57:361–388

Shifriss O (1961a) Sex control in cucumbers. J Hered 52:5–12

Shifriss O (1961b) Gibberellin as a sex regulator in *Ricinus communis*. Science 133:2061–2062

Shifriss O, George WL (1964) Sensitivity of female inbreds of *Cucumis sativus* to sex reversion by gibberellin. Science 143:1452–1453

Shifriss O, Georg WL, Quinones JA (1964) Gynodioecism in cucumbers. Genetics 49:285–291

Shiga T (1976) Cytoplasmic male sterility and its utilization for heterosis breeding in rape seed, *Brassica napus* L. Jpn Agric Res 10:178–182

Shiga T (1980) Male sterility and cytoplasmic differentiation. In: Tsunoda S, Hinata K, Gomez-campo C (eds) *Brassica* crops and wild allies. Jpn Sci Soc Press, Tokyo, pp 205–221

Shiga T, Baba S (1971) Cytoplasmic male sterility in rape plant (*Brassica napus*). Jpn J Breed 21:16–17

Shiga T, Baba S (1973) Cytoplasmic male sterility in oil seed rape, *Brassica napus* L. and its utilization to breeding. Jpn J Breed 23:187–197

Shiga T, Takayanagi K, Ohkawa Y (1976) Differences in cytoplasmic male sterile lines of rape seed *B. napus* L., derived from different origins. Jpn J Breed 26:121–122

Shiga T, Ohkawa Y, Takayanagi K (1977) The origin of male sterility inducing cytoplasm and fertility restoring genes in rape seed, *B. napus* L. I. Cytoplasm of plants having A genome and artifical *B. napus*. Jpn J Breed 27:64–65

Shiga T, Ohkawa Y, Ishige T (1978) Cytoplasm and pollen fertility restoration of European oil seed rape (*Brassica napus* L.) cultivars. Jpn J Breed 28:100–101

Shinjyo C (1969) Cytoplasmic genetic male sterility in cultivated rice, *Oryza sativa* L. II. The inheritance of male sterility. Jpn J Genet 44:149–156

Shinjyo C (1972) Distributions of male sterility-inducing cytoplasms and fertility-restoring genes in rice I. Commercial lowland-rice cultivated in Japan. Jpn J Genet 47:237–243

Shinjyo C (1975) Genetical studies of cytoplasmic male sterility and fertility restoration in rice, *Oryza sativa* L. Bull Coll Agric Univ Ryukus Japan 22:1–57

Shinjyo C, Omura T (1966) Cytpolasmic-genetic male sterility in cultivated rice *Oryza sativa* L I. Fertilities of F_1, F_2 and offspring obtained from their mutual reciprocal backcrosses and segregation of completely male sterile plants. Jpn J Breed 16:179–180

Shinjyo C, Watanabe Y (1977) Allelism test of two genes showing fertility restoring effect for male sterile cytoplasm of lead rice and their inheritance. Jpn J Breed Suppl 2, 27:70–71

Shore BF (1978) Breeding systems in *Cappodetus serratus* JR et G Forst. NZ J Bot 16:179–184

Shorter R, Byth DE (1975) Multiple seedling development in soybeans. J Hered 66:323–326

Shrivastava JN, Chauhan SVS (1979) Studies in *Raphanus sativus* L plants infested by *Lipaphis erysimi* Kalt. Curr Sci (Bangalore) 48:639–641

Shugaeva EV (1979) Male sterility of *Valeriana officinalis*. Genetika 15(1):138–143

Shuhua XU (1982) Cytological observations on pollen development in the main male sterile type of rice (*Oryza sativa* L.) developed in China. Agric Sci PR China 2:9–14

Shukla RS (1986) Cytological investigation in virus diseased plants of *Solanum viarum* (Dunal) (abstr). In: Proc 73rd Ind Sci Cong Sec Bot, p 154

Shull GH (1927) Inherited pollen sterility in Shepherd's purse. Mem Hortic Soc New York 3: 353–356

Shumway LK, Bauman LF (1966) The effect of hot water treatment, X-ray irradiation and mesocotyle grafting on cytoplasmic male sterility of maize. Am J Bot 54:773–780
Siedow JN, Girvin ME (1979) The effect of propylgallate on plant mitochondrial electron transfer and lipoxygenase activity. Plant Physiol (Bethesda) 63:5
Siegemund F (1976) Männliche Sterilität und Virus. Arch Züchtungsforsch 6:39–45
Siegemund F (1979a) Männliche Sterilität bei Art- und Gattungsbastarden der Tomate I. Morphologische Merkmale und Entwicklung der Pollensterilität bei Gattungsbastarden aus *Lycopersicon esculentum* und *Solanum pennelli*. Arch Züchtungsforsch 9:133–142
Siegemund F (1979b) Männliche Sterilität bei Art- und Gattungsbastarden der Tomate II. Der Ablauf der Meiose in der Mikrosporogenese von Bastarden aus *Lycopersicon esculentum* und *L. hirsutum*. Arch Züchtungsforsch 9:245–250
Silvolap JM (1968) Amino acid compositions of reproductive organs of plants with different types of male sterility. Sel Skokhozia Biol USSR 3:434–437
Simmonds NW (1964) Studies of the tetraploid potatoes II. Factors in the evolution of the tuberosum group. J Linn Soc Bot 59:43–56
Simmonds NW (1971) The breeding system of *Chenopodium quinoa* I. Male sterility. Heredity 27:73–82
Simonet M (1932) Recherches cytologiques et génétiques chez les *Iris*. Bull Biol Fr Belg 66:257–444
* Simonov AA (1968) Inheritance of John Baer type of functional sterility on transferring it to other tomato cultivars. Trudy Prikl Bot Genet Selek USSR 40:137–143
Simonov AA (1973) Inheritance of stamenless flowers in tomatoes. Genetika 9(7):21–28
Singh A, Laughnan JR (1972) Instability of S-male sterile cytoplasm in maize. Genetics 71: 607–620
Singh B, Sachan JN, Singh D (1977) New cytoplasmic male sterile lines of sunflower. Curr Sci (Bangalore) 46:160–161
Singh BB (1972) High frequency of natural cross pollination in a mutant strain of soybean. Curr Sci (Bangalore) 41:832–833
Singh BB, Jha AN (1978) Abnormal differentiation of floral parts in a mutant strain of soybean. J Hered 69:143–144
Singh BB, Gupta SC, Singh BD (1974) Induced male-sterile mutants in soybeans. Soybean Genet Newslett 1:19–20
Singh D (1964) Induction of male sterility in cotton by application of certain growth regulators. MSc Thesis Agra Univ
Singh DP, Sharma AK, Banerjee SC (1973) X-ray induced mutations in jute (*C. capsularis* and *C. olitorius*). Genet Agrar 27:115–147
Singh HN, Gill HS, Tewari RN (1966) Induction of closed anther character in tomato cultivars, Pusa Ruby, Money Maker and Sioux. Curr Sci (Bangalore) 35:292
Singh IS (1974) Genetic analysis of pollen fertility in two lines of *Petunia*. Ann Amelior Plant (Paris) 24:159–167
Singh IS (1975) Induced pollen sterility in *Petunia*. Mode of inheritance and tapetal behaviour. Ann Amelior Plant (Paris) 25:303–319
Singh R, Randhawa GS, Agarwal PK (1979) Effect of different pollinators on seed set in certain male-sterile grape cultivars. Indian J Hortic Sci 36:379–382
Singh RJ, Ikehashi H (1979) Induction of monogenic recessive male sterility in IR_{36} by ethyleneimine treatment. IRRI Newslett Philippines 4(3):3–4
Singh RJ, Ikehashi H (1981) Monogenic male sterility in rice. Induction, identification and inheritance. Crop Sci 21:286–289
Singh SP, Chauhan BPS (1974) Male sterility in crop improvement I. Inheritance of male sterility in *Sorghum*. In: Kachroo P (ed) Advancing frontiers in cytogenetics in evolution and improvement of plants. Hindustan, Delhi, pp 54–62
Singh SP, Hadley HH (1961) Pollen abortion in cytoplasmic male sterile *Sorghum*. Crop Sci 1: 430–432
Singh SP, Kaul CL (166) Male sterile barley. Curr Sci (Bangalore) 35:289–290

Singh SP, Rhodes AM (1961) A morphological and cytological study of male sterility in *Cucurbita maxima*. Proc Am Soc Hortic Sci 78:375–378

Singh SP, Sharma YP (1963) Preliminary observations on the breeding of *Pennisetum* at BR College Bichpuri Agra. Sorghum Newslett 6:26–28

Singh SP, Sharma YP (1964) White bajra. Sci Cult 30:238–239

Singh SP, White JW, Gutierrez JA (1980) Male sterility in dry beans. Annu Rep Bean Improv Coop 23:55–57

Singh UP, Khanna KR (1970) Male sterility in opium poppy. Sci Cult 36:354

Singh VP, Kobabe G (1969) Cyto-morphological investigations on male sterility in *Allium schoenoprasum* L. Indian J Genet Plant Breed 29:241–247

Singleton WR, Jones DF (1930) Heritable characters of maize XXXV. Male sterile. J Hered 21: 266–268

Sink KC (1966) The inheritance of the apetalous mutant in the snapdragon. In: Proc 17th Int Hortic Congr 1:213

Sink KC (1973) The inheritance of apetalous flower type in *Petunia hybrida* Vilm and linkage tests with the genes for flower doubleness and grandiflora characters and use in hybrid seed production. Euphytica 22:520–526

Sink KC, Gunesch ML (1966) Gametocidal effects of sodium alpha, beta-dichloroisobutyrate (DCIB) on petunias. Proc Am Soc Hortic Sci 88:657–661

Sink KC, Walker DE (1964) Gametocide induced male sterility by FW-450 in *Antirrhinum majus*. Natl Snapdragon Soc Bull, p 17

Sinnott EW, Durham GB (1922) Inheritance in the summer squash. J Hered 13:177–186

Sinska J (1973) Cytoplasmic pollen sterility in alfalfa. Genet Slecht USSR 46:215–221

Sinska J (1976) Ergebnisse der Forschung der Pollensterilität der *Luzerne*. C R R'eűnion d'Eucarpia Piestany, p 216–231

Sinska J (1979) The phenotypic manifestation of pollen sterility in progeny of F_1 of alfalfa hybrids. Polnopodarstvo, Czechoslovakia 25:812–817

Sirks MJ (1923) Die Verschiebung genotypischer Verhältniszahlen innerhalb Populationen laut mathematischer Berechnung und experimenteller Prüfung. Meded Landbowhoogesch Wageningen 26:1–40

Sirks MJ (1924) Die gynanthera Form des Goldlacks und ihre Vererbung. Genetica (The Hague) 6:537–548

Sirks MJ (1931) Plasmatic influences upon the inheritance in *Vicia faba* I. The elimination of a whole linkage group in the plasma of *Vicia faba* minor. Proc Kon Akad Wet Amsterdam 34: 1057–1062

Sirks MJ (1932) Beiträge zu einer genotypischen Analyse der Ackerbohne, *Vicia faba* L. Genetica (The Hague) 13:209–632

Sirks MJ (1938) Plasmatic inheritance. Bot Rev 4:113–131

Sisco PH (1982) Differences in fertility restoration and nucleic acids among S-type male-sterile cytoplasms of maize. Diss Abstr Int 43(4):938

Sisco PH, Garicia-Arenal F, Zaitlin M, Earle ED, Gracen VE (1984) LBN, a male sterile cytoplasm of maize, contains two double stranded RNAs. Plant Sci Lett 34:127–134

Sisco PH, Gracen VE, Everett HL, Earle ED, Pring DR, Mcnay JW, Levings CS (1985) Fertility restoration and mitochondrial nucleic acids distinguish at least five subgroups among cms-S cytoplasms of maize (*Zea mays* L.). Theor Appl Genet 71:5–19

Sisco PH, Zaitlin M, Gracen VE (1982) "LBN" - a newly designated member of the S-group of male sterile cytoplasm. Maize Genet Coop Newslett 56:82–83

Skalinska M (1928a) Sur les cause dűne disjunction on typique hybrides du genre *Aquilegia*. Acta Soc Bot Pol 5:141–173

Skalinska M (1928b) Etudies sur la sterilite partielle des hybrides du genre *Awuilegia*. Proc 5th Int Congr Genet 2:1343–1372

Skalinska M (1929) Das Problem des Nichterscheinens des väterlichen Typus in der Spaltung der partiell sterilen *Aquilegia* Spezies-Bastarde. Acta Soc Bot Pol 6:138–164

Skalinska M (1930) On the significance of the cytoplasm in matriclinous hybrids of *Aquilegia*. Acta Biol Exp 5:1–18

Skalinska M (1931) A new case of unlike reciprocal hybrids in *Aquilegia*. Rep 5th Int Bot Cong, p 250

Skorupska H, Nawracala J (1980) Observations of pollen grains of soybean plants in the male sterile line Urbana, ms_1. Genet Pol 21:63–67

Skoyen IO (1969) Sugar beet responses to selective gametocide sodium 2,3-dichloroisobutyrate and related chemicals. J Am Sugarbeet Techn 15:425–437

Skurygina NA (1970) On the genetic control of the factor of cytoplasmic male sterility in wheat. Genetica (The Hague) 6(11):43–51

Slonimski PP, Perrodin G, Croft JH (1968) Ethidium bromide induced mutations of yeast mitochondria: complete transformation of cell into respiratory deficient non-chromosomal "Peti-ties". Biochem Biophys Res Commun 30:232–239

Sloyanova I, Velkov VN (1974) The use of cytoplasmic male sterility for the production of hybrid seed in sunflower. Genet Selekt 7:148–156

Smilde KW, Henkens CHH (1967) Sensitivity to copper defficiency of different cereals and strains of cereals. Netherlands J Agric Sci 15:249–258

Smith CO (1927) A pistillate *Prunus*. J Hered 18:537–541

Smith DR, Hooker AL, Lim SM (1970) Physiologic races of *Helminthosporium maydis*. Plant Dis Rep 54:819–822

Smith HH (1962) Studies on the origin, inheritance and mutation of genic-cytoplasmic male sterility in *Nicotiana* (abstr). Genetics 47:985–986

Smith HH (1968) Recent cytogenetic studies in the genus *Nicotiana*. Adv Genet 14:1–54

Smith HH (1979) The genus as a genetic resource. In: Durbin RD (ed) *Nicotiana:* procedures for experimental use. USDA Techn Bull 1586:1–16

Smith ID (1965) Hybrid wheat is almost in the bag. Doane's Agric Rep USA, pp 21–23

Smith L (1938a) Cytogenetic studies in *Triticum monococcum*. Genetics 23:168

Smith L (1938b) Mutants and linkage studies in *Triticum monococcum* and *T. aegilopoides*. Minnisota Agric Exp Stat Res Bull USA, p 298

Smith L (1947) Possible practical methods for producing hybrid seed of self-pollinated crops through the use of male sterility. J Am Soc Agron 39:260–261

Smith RR (1971) Inheritance of a male-sterile character in red clover (*Trifolium pratense* L.). Crop Sci 11:326–327

Sneep J (1958) The breeding of hybrid varieties and the production of hybrid seed in spinach. Euphytica 7:119–122

Snyder RJ, Duvick DN (1969) Chromosomal location of Rf_2, a restorer gene for cytoplasmic pollen sterile maize. Crop Sci 9:156–157

Sobrinho LG (1946) Sur le comportement genetique de *Centranthus ruber* DC I. Portugaliae Acta Biol Ser A 1:385–390

Soest JL (1965) *Taraxacum* sect Palustria Dahlstedt. Acta Bot Neerl 14:1–53

Solberg OT (1976) On the relative advantages of cross- and self-pollination. Ann Miss Bot Garden 63:262–276

Solovjeva N (1970) The use of tomato male sterility in selection. Eucarpia Sect Horticole Versailles 83:73–82

Soost RK, Rick CM (1957) Effect of varieties of pollen and ovule parents on natural cross pollination of tomatoes. Proc Amer Soc Hort Sci 70:357–365

Sorrells ME, Fritz SE (1982) Application of a dominant male sterile allele to the improvement of self pollinated crops. Crop Sci 22:1033–1035

Soressi GP (1968) Brown seed: a recessive genetic marker useful in tomato hybrid seed production. In: Rep Meet Eucarpia Tomato Work Group Wageningen, pp 24–26

Soressi GP, Lorenzoni C (1970) Crossing rate after natural or hand pollination with and without flower emasculation. Rep Tomato Genet Coop 20:61–63

Soressi GP, Salamini F (1975) A monomendelian gene inducing parthenocarpic fruits. Rep Tomato Genet Coop 25:22

Sovetskaya DFP (1971) Method of producing cytoplasmic male sterility in maize. US Patent 3594, 152, patented July 20, 1971

Spasojevic V (1966) Contribution to the study of cytoplasmic male sterility in maize. J Sci Agric Res 19:30–42

Spirova M (1975) New data on the male sterility in sunflower induced by gibberellic acid. Rastenievud Nauk Bulgaria 12:10–17

Spirova M (1978) Inheritance of sunflower nuclear male sterility obtained as a result of inbreeding. Genet Sel Sofia 11:391–398

Sprague EW (1959) Cytological and fertility relationships of *Medicago sativa, M. falcata* and *M. gaetula.* Agron J 51:249–252

Sprague GF (1939) Heritable characters in maize: vestigial glume. J Hered 30:143–145

Spruill WM (1982) Recombinant DNA analysis of the maize mitochondrial genome: organisation in normal and Texas male sterile cytoplasms. Diss Abstr Int 42:4302–43)3

Spruill WM, Levings CS, Sederoff RR (1980) Recombinant DNA analysis indicates that the multiple chromosomes of maize mitochondria contain different sequences. Dev Genet 1:363–388

Srivastava HK (1981) Intergenomic interaction, heterosis and improvement of crop yield. Adv Genet 34:117–195

Srivastava PS (1961) The effect of sodium 2,3-dichloroisobutyrate and potassium-giberallate on the growth of alfalfa and red clover. MS Thesis Illinois

Srivastava YC, Somayajulu PLN (1981) Male sterility in *Lathyrus.* Indian J Genet Plant Breed 41:164–166

Stace Smith R (1970) Tobacco ring spot virus. Description of plant viruses, No 17 CMI/AAB, p 57

Staes G (1889) Die Blumen von *Daucus carota.* Bot Jahrb 1:124

Stanford EH (1951) Tetrasomic inheritance in alfalfa. Agron J 51:274–277

Stanford EH, Clement WM (1958) Cytology and crossing behaviour of a haploid alfalfa plant. Agron J 50:589–592

Stanley RG, Linskens HF (1974) Pollen: biology, biochemistry and management. Springer, Berlin, Heidelberg, New York

Stansel JW, Craigmiles JP (1966) Hybrid rice-problems and potential. Rice J 69:14–15

Stapledon RG (1931) Self and cross fertility and vigour in cocksfoot grass (*Dactylis glomerata* L.). Welsh Plant Breed Stat Bull Ser H 12:161–180

Starnes WJ, Hadley HH (1962) Some effects of the gametocide alpha, beta-Dichloroisobutyrate on soybeans. Crop Sci 2:305–310

* Staszewski Z (1970) Badania nad heterozja i formami meekojalowymi lucerny *Medicago sativa* L. Hodowal Roslin Aklimat Nasien 14:483–498

* Staszewski Z (1976) Studies on male sterility inheritance and breeding of maintaining lines in *Medicago sativa* L. Hodowal Roslin Aklimat Nasien 20:105–119

* Staszewski Z, Jakubowska B (1976) Comparison of microsporogenesis and tapetum development of male sterile and male fertile forms of alfalfa. Hodowal Roslin Aklimat Nasien 20:122–130

* Staszewski Z, Jakubowska B (1978) In: Annu Rep Inst Plant Breed Radzikow, p 17–18

Staudt G (1973) Cytologische Untersuchungen zur Sterilität von Rebensorten. Z Pflanzenzücht 70:148–156

Stead B (1960) The gene action involved in the fertility restoration of Texas type cytoplasmic male-sterile maize and its performance compared with normal cytoplasm. Ph D Thesis Univ Wisconsin Madison

Stebbins GL (1950) Variation and evolution in plants. Columbia Univ Press, New York, pp 643

Stehno Z, Apltauerova M (1980) Pollen sterility of wheat (*Triticum aestivum* L.) induced by the cytoplasm of *Aegilops* sp. Sci Agric Bohemoslo Vaca 29:245–253

Stein H, Gabelman WH (1959) Pollen sterility in *Beta vulgaris* associated with red pigmentation of the anthers. J Am Soc Sugarbeet Techn 10:612–618

Stein H, Gabelman WH, Struckmeyer BE (1959) Reversion in cytoplasmic male sterile plants of *Beta vulgaris.* J Am Soc Sugabeet Techn 10:619–623

Stelly DM (1979) Investigations of a partially male sterile line of soybeans *Glycine max* (L) Merill. MS Thesis Iowa State Univ Ames

Family Index